Plant Metabolites and Regulation Under Environmental Stress

Plant Metabolites and Regulation Under Environmental Stress

Edited by

Parvaiz Ahmad

Mohammad Abass Ahanger

Vijay Pratap Singh

Durgesh Kumar Tripathi

Pravej Alam

Mohammed Nasser Alyemeni

ELSEVIER

ACADEMIC PRESS

An imprint of Elsevier

Academic Press is an imprint of Elsevier
125 London Wall, London EC2Y 5AS, United Kingdom
525 B Street, Suite 1800, San Diego, CA 92101-4495, United States
50 Hampshire Street, 5th Floor, Cambridge, MA 02139, United States
The Boulevard, Langford Lane, Kidlington, Oxford OX5 1GB, United Kingdom

Library of Congress Cataloging-in-Publication Data
A catalog record for this book is available from the Library of Congress

British Library Cataloguing-in-Publication Data
A catalogue record for this book is available from the British Library

ISBN: 978-0-12-812689-9

For information on all Academic Press publications visit our website at https://www.elsevier.com/books-and-journals

Publisher: Andre G. Wolff
Acquisition Editor: Mary Preap
Editorial Project Manager: Mary Preap
Production Project Manager: Mohanapriyan Rajendran
Designer: Matthew Limbert

Typeset by TNQ Books and Journals

Contents

List of Contributors

Mohammad A. Ahanger, PG College Rajouri, Jammu, India

Parvaiz Ahmad, SP College, Srinagar, India; King Saud University, Riyadh, Saudi Arabia

Salama Aissam, Université Cadi Ayyad, Marrakech, Morocco

Nudrat A. Akram, GC University, Faisalabad, Pakistan

Sergio Alemano, Facultad de Ciencias Exactas, Físico-Químicas y Naturales, Universidad Nacional de Río Cuarto, Río Cuarto, Argentina

Amanda C.E. Amaro, Universidade Estadual Paulista, Brazil

Andrea Andrade, Facultad de Ciencias Exactas, Físico-Químicas y Naturales, Universidad Nacional de Río Cuarto, Río Cuarto, Argentina

Muhammad Saleem Arif, Government College University Faisalabad, Faisalabad, Pakistan

Muhammad A. Ashraf, Government College University Faisalabad, Faisalabad, Pakistan

Aditya Banerjee, St. Xavier's College (Autonomous), Kolkata, India

Daniel Baron, Universidade Federal de São Carlos, Brazil

María Blanch, Institute of Food Science Technology and Nutrition (ICTAN-CSIC), Madrid, Spain

Carmen S.F. Boaro, Universidade Estadual Paulista, Brazil

Andrés A. Borges, Instituto de Productos Naturales y Agrobiología-CSIC, La Laguna, Spain

Felipe G. Campos, Universidade Estadual Paulista, Brazil

Abdelghani Chakhchar, Université Cadi Ayyad, Marrakech, Morocco

Roberto de Armas, Havana University, Havana, Cuba

Cherkaoui El Modafar, Université Cadi Ayyad, Marrakech, Morocco

Abdelhamid El Mousadik, Université Ibn Zohr, Agadir, Morocco

Abderrahim Ferradous, Centre Régional de la Recherche Forestière Marrakech, Marrakech, Morocco

Gisela Ferreira, Universidade Estadual Paulista, Brazil

Abdelkarim Filali-Maltouf, Université Mohammed V Agdal, Rabat, Morocco

Blanca Fontaniella, Complutense University of Madrid, Madrid, Spain

Francisco J. García-Machado, Instituto de Productos Naturales y Agrobiología-CSIC, La Laguna, Spain

Cristina Garrido-Orduña, Instituto de Productos Naturales y Agrobiología-CSIC, La Laguna, Spain

Francisco V. González, University of La Laguna, La Laguna, Spain

Mallu Govardhana, Jain University, Bangalore, India

Fahima Gul, SP College, Srinagar, India

Bernhard Huchzermeyer, Northeastern University, Boston, MA, United States; Leibniz Universität Hannover, Hannover, Germany

Iqbal Hussain, Government College University Faisalabad, Faisalabad, Pakistan; Graduate School of Life Sciences, Tohoku University, Sendai, Japan

Saad Ibnsouda-Koraichi, Université Sidi Mohamed Ben Abdellah, Fèz, Morocco

Muhammad Iqbal, Government College University Faisalabad, Faisalabad, Pakistan

Nellickal S. Jayamohan, Jain University, Bangalore, India

Yachana Jha, Sardar Patel University, Anand, India

David Jiménez-Árias, Instituto de Productos Naturales y Agrobiología-CSIC, La Laguna, Spain

Muhammad Kamran, Government Postgraduate College of Science, Faisalabad, Pakistan

Hans-Werner Koyro, Justus-Liebig-Unininversität, Giessen, Germany

Belur S. Kumudini, Jain University, Bangalore, India

Mouna Lamaoui, Université Cadi Ayyad, Marrakech, Morocco

María E. Legaz, Complutense University of Madrid, Madrid, Spain

Analía Llanes, Facultad de Ciencias Exactas, Físico-Químicas y Naturales, Universidad Nacional de Río Cuarto, Río Cuarto, Argentina

Juan Cristo Luis Jorge, University of La Laguna, La Laguna, Spain

Virginia Luna, Facultad de Ciencias Exactas, Físico-Químicas y Naturales, Universidad Nacional de Río Cuarto, Río Cuarto, Argentina

Saqib Mahmood, Government College University Faisalabad, Faisalabad, Pakistan

Giselle M.A. Martínez-Noël, INBIOTEC and FIBA, Mar del Plata, Argentina; Consejo Nacional de Investigaciones Científicas y Técnicas (CONICET), Buenos Aires, Argentina

Miguel A. Matilla, Estación Experimental del Zaidín, Consejo Superior de Investigaciones Científicas, Granada, Spain

Ana M. Millanes, Universidad Rey Juan Carlos, Móstoles, Spain

Anurag Mishra, NEB Division, Department of Science & Technology, New Delhi, India

Avinash Mishra, CSIR-Central Salt and Marine Chemicals Research Institute, Bhavnagar, India; Academy of Scientific and Innovative Research, Council of Scientific and Industrial Research, New Delhi, India

Saray Morales-Sierra, Instituto de Productos Naturales y Agrobiología-CSIC, La Laguna, Spain

Pinar Nartop, Namık Kemal University, Tekirdağ, Turkey

Ashok Panda, Division of Plant Omics, CSIR-Central Salt and Marine Chemicals Research Institute (CSIR-CSMCRI), Bhavnagar, India; Academy of Scientific and Innovative Research, Council of Scientific and Industrial Research (CSIR), CSIR-Central Salt and Marine Chemicals Research Institute (CSIR-CSMCRI), Bhavnagar, India

Asish K. Parida, Division of Plant Omics, CSIR-Central Salt and Marine Chemicals Research Institute (CSIR-CSMCRI), Bhavnagar, India; Academy of Scientific and Innovative Research, Council of Scientific and Industrial Research (CSIR), CSIR-Central Salt and Marine Chemicals Research Institute (CSIR-CSMCRI), Bhavnagar, India

Savita V. Patil, Jain University, Bangalore, India

Shagufta Perveen, Government College University Faisalabad, Faisalabad, Pakistan

Jaykumar Rangani, Division of Plant Omics, CSIR-Central Salt and Marine Chemicals Research Institute (CSIR-CSMCRI), Bhavnagar, India; Academy of Scientific and Innovative Research, Council of Scientific and Industrial Research (CSIR), CSIR-Central Salt and Marine Chemicals Research Institute (CSIR-CSMCRI), Bhavnagar, India

Rizwan Rasheed, Government College University Faisalabad, Faisalabad, Pakistan

Muhammad Riaz, Government College University Faisalabad, Faisalabad, Pakistan

Aryadeep Roychoudhury, St. Xavier's College (Autonomous), Kolkata, India

Maham Saddique, University of Agriculture Faisalabad, Faisalabad, Pakistan

Elena Sánchez-Elordi, Complutense University of Madrid, Madrid, Spain

Sanchita, CSIR-Central Institute of Medicinal and Aromatic Plants, Lucknow, India

Rocío Santiago, Federal University of Pernambuco, Recife, Brazil

Günther F.E. Scherer, Leibniz University of Hannover, Institute of Horticultural Production Systems, Hannover, Germany

Muhammad Shahbaz, University of Agriculture Faisalabad, Faisalabad, Pakistan

Ashok Sharma, CSIR-Central Institute of Medicinal and Aromatic Plants, Lucknow, India

R.B. Subramanian, Sardar Patel University, Anand, India

Ágnes Szepesi, University of Szeged, Szeged, Hungary

Réka Szőllősi, University of Szeged, Szeged, Hungary

Bhakti Tanna, CSIR-Central Salt and Marine Chemicals Research Institute, Bhavnagar, India; Academy of Scientific and Innovative Research, Council of Scientific and Industrial Research, New Delhi, India

Sakshi Tewari, Department of Biochemical Engineering and Biotechnology, Indian Institute of Technology, Delhi, Hauz Khas, New Delhi, India

Jorge A. Tognetti, Universidad Nacional de Mar del Plata (UNMdP), Mar del Plata, Argentina; Comisión de Investigaciones Científicas (CIC), Provincia de Buenos Aires, Argentina

Carlos Vicente, Complutense University of Madrid, Madrid, Spain

Said Wahbi, Université Cadi Ayyad, Marrakech, Morocco

Rinukshi Wimalasekera, University of Sri Jayewardenapura, Nugegoda, Sri Lanka

Abbu Zaid, Aligarh Muslim University, Aligarh, India

Chapter 1

Advancement of Metabolomics Techniques and Their Applications in Plant Science: Current Scenario and Future Prospective

Ashok Panda[1,2], **Asish K. Parida**[1,2], **Jaykumar Rangani**[1,2]

[1]*Division of Plant Omics, CSIR-Central Salt and Marine Chemicals Research Institute (CSIR-CSMCRI), Bhavnagar, India;* [2]*Academy of Scientific and Innovative Research, Council of Scientific and Industrial Research (CSIR), CSIR-Central Salt and Marine Chemicals Research Institute (CSIR-CSMCRI), Bhavnagar, India*

1. INTRODUCTION

Metabolites are the result of the interaction of an organism's genome with its environment, and metabolome is defined as the total quantitative collection of the small-molecular-weight metabolites present in organisms, which participate in the metabolic reactions required for growth, maintenance, and normal function (Miura et al., 2012). Metabolomics refers to the comprehensive and quantitative analysis of metabolites and gives knowledge about the metabolite composition of any biological system (Yi et al., 2014). Metabolomics is the profiling of the endogenous and exogenous metabolites of an organism under different environmental conditions. Therefore the metabolomics approach helps us for better understandings of various natural products, in natural drug discovery, chemotaxonomy, and more importantly, metabolite alterations in response to abiotic and biotic stresses in the plants. Metabolomics has become a powerful tool that deepens our knowledge about the plant's primary and secondary metabolism and it has significant contributions to the advancement of plant biology. It is considered as a complementary technique along with the functional genomics approaches, e.g., transcriptomics and proteomics, to study the biological status and importance of any species (Tugizimana et al., 2013). Metabolomics, on the other hand, scans the last products of gene expression and provides phenotypic valuation of the plants. In comparison to any other organism, the metabolite composition of the plant are most diverse and enormous. Plant metabolites are structurally diverse constituting of highly complex spectrum of compounds having different size, solubility, polarity, volatility, quantity, and stability (Vogt, 2010). Apart from the natural conditions, the metabolite configurations of the plants become more diverse and complex when they experience certain environmental alternations often known as the abiotic stresses. The complexity of metabolites and their pathways is because they are not directly synthesized from genes, and various enzymatic activities are needed for their biosynthesis. In addition, various metabolites are interrelated, which form more complex metabolic networks. Therefore integrated metabolomic approaches are required to gain a comprehensive knowledge of the metabolite configurations and alterations in plants. Metabolomics can identify the phenotypic effects of stresses on plants by measuring the abundance of metabolites, which provides an efficient measure of the phenotypic responses to environmental stresses (Hong et al., 2012). Metabolomic approaches are capable of measuring a broader array of small molecules in plants when they are subjected to adverse environmental conditions (Zhao et al., 2016). The metabolites include primary metabolites directly involved in growth and development of plants such as sugars, amino acids, lipids and intermediates of photosynthesis, glycolysis, and tricarboxylic acid (TCA) cycle. The alterations of primary metabolites are considered as an indication of photosynthetic dysfunction and osmotic readjustment. On the other hand, the metabolites that are not directly involved in growth and development of the plants, but respond to particular stress conditions like high salinity, drought, temperature extremes, and pathogen infections are known as secondary metabolites, e.g., antioxidants, reactive oxygen species scavengers, coenzymes, and regulatory molecules (Arbona et al., 2013). Induced biosynthesis of secondary metabolites by various stresses is an efficient mechanism for protection against abiotic and biotic threats, establishing a link between metabolites and stress responses.

Plant Metabolites and Regulation Under Environmental Stress. http://dx.doi.org/10.1016/B978-0-12-812689-9.00001-7

Current metabolomics approaches mainly concern three types of analyses, namely, targeted, semitargeted, and untargeted analyses. These three approaches are alternatively known as metabolic profiling, metabolite profiling, and metabolomics (Dunn et al., 2011). Implementation of these strategies depends on the level of quantification, methods employed for sample preparation, experimental precision, number of detected metabolites, and the objective of the analysis. Chemical identification and structural elucidation of the detected metabolites is a prerequisite for untargeted analysis when compared to both targeted and semitargeted analysis (Dunn et al., 2013). In the untargeted approach, analytical methods have been developed that are suitable for acquiring data on diverse range of metabolites. However, in targeted and semitargeted analyses, the chemical identities of the metabolites to be assayed are known before data acquisition, and analytical methods are accordingly developed to provide higher accuracy selectivity (Dunn et al., 2013). Prior information related to metabolite classes can help to choose an appropriate method for sample preparation as well as analytical platforms for the detection and identification of the metabolites of interest (Wishart, 2011). Metabolomics approaches that are being currently used in plant metabolomics include metabolic fingerprinting, metabolite profiling, and targeted analysis (Shulaev, 2006). Continued research on plant secondary metabolites reveal a vast array of and structurally diverse secondary metabolites in plants. These metabolites have been isolated by various chromatographic techniques and structurally identified and elucidated by various spectroscopy techniques such as circular dichroism; ultraviolet (UV), infrared (IR), and nuclear magnetic resonance (NMR) spectroscopies; and mass spectrometry (MS). In general two analytical techniques are used in metabolomics: MS and NMR. MS coupled with preseparation techniques such as liquid chromatography (LC-MS) and gas chromatography (GC-MS) have been conventionally used in metabolomics study (Nakabayashi and Saito, 2013). In MS-based metabolomics various ionization techniques have been used, e.g., atmospheric pressure ionization (API), which includes atmospheric pressure chemical ionization and electrospray ionization (ESI). Ionization is the process of production of gas-phase ions suitable for resolution in the mass analyzers or mass filters. Ionization techniques for MS are classified as "hard" and "soft" ionizations techniques. The primary difference between the use of "soft" and "hard" ionization depends on the analyte of interest and its property. "Hard" ionization includes electron ionization/impact (EI) and fast atom bombardment (FAB) etc. which use fast moving electrons for the ionization of analyte molecules in the gas phase. EI is often suitable for small-sized molecules with high volatilities. However, EI fails to ionize larger molecules with low volatilities. FAB is an alternative technique that easily allows the ionization of nonvolatile samples. In this technique, ionization is done by the bombardment of the beam of atoms, specially Ar or Xe, in a vacuum. As an alternative to the hard ionization techniques, various modern ionization techniques are considered as the soft ionization techniques, e.g., matrix-assisted laser disruption/ionization (MALDI), ESI, and chemical ionization (Fuchs et al., 2010). MALDI is a fast and sensitive ionization technique that provides basic mass spectra without any significant analyte fragmentation (Fuchs et al., 2010). MALDI is a laser disruption ionization (LDI) method that can ionize any biological sample. Traditional MALDI sources were equipped with UV lasers such as nitrogen laser (337 nm) or neodymium-doped yttrium aluminum garnet (Nd-YAG; 355 nm). The primary limiting factor of MALDI is the use of a matrix that can limit the spatial resolution due to the diffusion of metabolites within the tissue during matrix application (Miura et al., 2012). The majority of the commercially available MS devices use either MALDI or ESI (or both) as ion sources. With the combinations of both the ionization techniques, maximum information can be achieved. Therefore these techniques are considered as complimentary but not competitive. Various ionization techniques which are implemented for the ionization of the analytes are given in Table 1.1 with their advantages, disadvantages, and applications. In spite of the extensive research, details of ion segregation have still not yet been completely understood. However, some new matrix compounds have been introduced for higher sensitivity and reproducibility than the previously used matrices (Jaskolla et al., 2009). In addition, the association of MALDI-MS with various chromatographic techniques provide better opportunity for mixture analysis in an easy way (Fuchs et al., 2009). An advanced form of ionization technique compared with MALDI is direct ionization on silicon (DIOS), an LDI method without the use of the matrix. This method vaporizes and ionizes the analytes trapped in porous silicon by laser irradiation. DIOS does not use matrix, thereby limiting matrix interference with detection of low-molecular-weight metabolites (Miura et al., 2012). Lately, nanostructure initiator MS (NIMS) has been introduced to give an advanced form DIOS (Patti et al., 2010). In NIMS, liquid analytes are absorbed on the surface of nanostructured silicon and then are released by laser irradiation for mass analysis. NIMS can also detects several peaks from a single cell in the low-mass range. Imaging of endogenous metabolites such as sterols or carbohydrates having low ionization efficiency by other ionization techniques such as MALDI can be achieved by NIMS. The advantages, disadvantages, and applications of various metabolomic techniques are given in Table 1.2.

2. RECENT ADVANCEMENTS IN METABOLOMICS STUDY

The transcriptomics, proteomics, and metabolomics studies are being emerged and are advancing day by day. The interpretation of metabolites in metabolomics is quite difficult due to the lack of direct link between proteome and metabolome, and sometimes it is challenging to trace the metabolites in the metabolic pathway (Sumner et al., 2015). In addition to the

TABLE 1.1 Different Ionization Techniques Implemented in Mass Spectrometry for Metabolomics Analysis

Ionization Technique	Mass Range	Advantage	Disadvantage	Application	References
EI	Less than 1000 Da	Can be applied on all volatile compounds; reproducible mass spectra; structural information	Sample must be thermally volatile and stable	Elemental composition; quality assessment of natural products	Mirsaleh-Kohan et al. (2008), Castro et al. (2010), and Abate et al. (2010)
CI	Less than 1000 Da	Gives molecular weight information through molecular ions; simple mass spectra; reduced fragmentation when compared with EI	Sample must be volatile and stable; as there is less fragmentation than EI, the fragment patterns are less informative	Commonly used for the analysis of cholesterol esters of fatty acid	Munson (2000)
DCI	Less than 1500 Da	Reduced thermal decomposition; rapid analysis; simple instrumentation	Not particularly reproducible; rapid heating required for fast scanning; not suitable for large and heat-labile compounds	Characterization of large involatile molecules and nonpolar polymers	Vincenti (2001)
NCI	Less than 1000 Da	High sensitivity; effective ionization; less fragmentation than EI and CI	Poor reproducibility	Mainly implemented for analysis of atmospheric samples	Raina and Hall (2008)
ESI	Less than 200,000 Da	This method can be used for molecules having large masses; softest ionization method available; mass of the analytes can be accurately obtained by the use of quadrupole mass analyzers	Cannot analyze the mixtures very well, multiple charges that are attached to the molecular ions can make for confusing spectral data	Extensively used for production of gas phase with fragmentation for heat-labile large supramolecules	Banerjee and Mazumdar (2012)
DESI	Less than 1000 kDa	Combined feature of electrospray and desorption ionization; sample molecules can be off from the intact sample surface	Ionization is limited by the compound that is ionized by ESI	Chemical composition of atmospheric aerosols; identification of secondary metabolites	Dill et al. (2009) and Harris et al. (2011)
DCBI	Less than 600 Da	Does not require sample pretreatment; better localization of sampling area; less congestion in mass spectra	Compounds having lager molar mass cannot be analyzed; analysis depends on the volatilities of the analytes	Mass spectra for pesticide, veterinary drug, and natural drugs	Wang et al. (2010)
DEMI	Less than 1000 Da	Wide polarity range; DART; multiple operational mode (DESI-spray-only mode, MICI-only, dual-mode DEMI); combination of DESI and DART	Limited use in direct MSI due to the production of reactive stream	Multimode detection of intact drug	Nyadong et al. (2009) and Galhena et al. (2010)

Continued

TABLE 1.1 Different Ionization Techniques Implemented in Mass Spectrometry for Metabolomics Analysis—cont'd

Ionization Technique	Mass Range	Advantage	Disadvantage	Application	References
EASI	Less than 1000 Da	Simple ion source; no voltage required; low charge density of the droplets, which decreases solvent chemical noise	Disturbance in spatial resolution of analytes	Identification of inks; analysis of surface-modified by polymers; targeted analysis from complex matrix; analysis of drug tablets and vegetable oils	Haddad et al. (2006), Lalli et al. (2010), and Porcari et al. (2016)
PESI	Less than 1000 Da	Solid needle is used, therefore free from clogging problems and tolerant to high concentration of salt and urea; high ionization efficiency; small sample consumption	–	Detection of biomolecules from solutions having high concentration of salts	Mandal et al. (2010) and Harris et al. (2011)
FAB	300-6000 Da	This method is rapid and simple; suitable for high-mass and thermally stable compounds, also suitable for nanomolar samples, tolerant to variation in samples	High chemical background; not suitable for multiple charged compounds; low sensitivity	Plant hormone analysis	Ljung et al. (2010)
SIMS	300-13,000 Da	Same as FAB except for improved sensitivity	Target can get more heated due to more energetic primary beams; ion source requires more maintenance	Lipid imaging, polymer analysis, application in plants yet to be functionally elucidated	Passarelli and Winograd (2011)
APCI	Less than 2000 Da	Suitable for less polar compounds; has an excellent LC/MS interface	High cost of specialized instrumentation	Metabolite profiling, pesticide residue analysis	Portolés et al. (2010)
MALDI	Less than 500,000 Da	Rapid and convenient mass determination	Requires mass analyzers that are compatible with pulsed ionization techniques	Analysis of small molecules; plant metabolomics study in response to stress response, natural drug discovery; protein analysis	Harada et al. (2009) and Miura et al. (2012)

APCI, atmospheric pressure chemical ionization; *CI*, chemical ionization; *DART*, direct analysis in real time; *DCBI*, desorption corona beam ionization; *DCI*, desorption chemical ionization; *DEMI*, desorption electrospray/metastable-induced ionization; *DESI*, desorption electrospray ionization; *EASI*, easy ambient sonic-spray ionization; *EI*, electron ionization/impact; *ESI*, electrospray ionization; *FAB*, fast atom bombardment; *MALDI*, matrix-assisted laser disruption/ionization; *MICI*, metastable-induced chemical ionization; *MSI*, mass spectrometry imaging; *NCI*, negative chemical ionization; *PESI*, probe electrospray ionization; *SIMS*, secondary ion mass spectrometry.

identification of various small molecules, metabolomics also provides the relative quantification of metabolites and emphasizes the correlation between their cellular concentration, physiological condition of the plant, and the residing environment (Nicholson and Lindon, 2008). NMR spectroscopy and MS represent the more potent tools for the efficient separation and identification of metabolites in the biological samples taking the concept of metabolomics to a new level. In the process of NMR, the metabolites are detected using specific atomic nuclei without any prior separation step. The number of the metabolites that are detected in NMR is usually from 20 to 60. However, advancement in NMR technique increases the number of metabolite detection to more than 100 (Nagana Gowda and Raftery, 2014). Various advanced technical alternatives of

TABLE 1.2 Analytical Characteristics of Various Metabolomics Techniques, Their Advantages, Disadvantages and Applications

Sl. No.	Analytical Platforms	Advantage	Disadvantage	Application	References
1.	GC-MS	Volatile compounds are directly analyzed; databases are available; suitable for analysis of mixed samples	Not suitable for non-volatile and thermounstable compounds; requires derivatization depending on the type of sample to be analyzed; derivatization can mask the results	Low-molecular-weight hydrophobic compounds (essential oils, hydrocarbons, esters)	Hagel and Facchini (2008) and Ernst et al. (2014)
2.	LC-MS	Derivatization not required; thermounstable compounds can be analyzed	Not all compounds in a plant extract will ionize under the same conditions; the technique is therefore more suited for targeted analysis	Suitable for analysis of relatively polar compounds with low, moderate, or high molecular weights	Allwood and Goodacre (2010) and Ernst et al. (2014)
3.	DIMS				
	(1) FT-ICR-MS	Ions are trapped by strong magnetic fields; the calibration of these machines is very stable, and the detection is nondestructive; direct separation and characterization of thousands of components in complex mixtures	Long acquisition time, isomers cannot be resolved; high instrument cost; limited dynamic range	Ion chemistry, high-resolution MALDI, surface characterization by laser desorption	Ernst et al. (2014)
	(2) TOF	The analytes are separated by the time taken by them to reach the detector depending on the *m/z* ensuring better separation and resolution; high sensitivity; suitable for the pulsed ion generation of MALDI	Use of fast digitizer can reduce the dynamic range; limited precursor ion selectivity	MALDI and GC-MS system	Fuchs and Schiller (2009)
4.	NMR	Rapid analysis; simple sample preparation; ease of quantification, absolute quantification of metabolites; provides structural information with stereochemical details	Low sensitivity; signal overlap for complex spectra of plant extract	Metabolites, body fluids, microbial analysis	Kim et al. (2011)
	(1) 1D NMR	Shorter acquisition time; no prior separation step	Signal overlap	Metabolite identification, differentiation between two samples	Farag et al. (2012)
	(2) 2D NMR	Reduces signal overlap by spreading the resonances in a second dimension	Longer acquisition time	Metabolite identification, differentiation between two samples	Forseth and Schroeder (2011)
	(3) Isotope-labeled NMR	Quantification of low-concentration metabolites in biological samples; increased sensitivity and resolution of NMR	High cost in purchasing the isotope label; limited applicability for large metabolites	Interpretation of metabolic pathway by tracing the metabolites	Fan and Lane (2011) and Bothwell and Griffin (2011)

Continued

TABLE 1.2 Analytical Characteristics of Various Metabolomics Techniques, Their Advantages, Disadvantages and Applications—cont'd

Sl. No.	Analytical Platforms	Advantage	Disadvantage	Application	References
	(4) Microcoil NMR	Minute fraction of sample from LC-MS can be analyzed; simultaneous detection of metabolites by both NMR and MS from single sample	High salt and protein concentration cause hindrance in the detection	Small-molecule liquid-state NMR spectroscopy, implication in structural biology and structural genomics	Cloarec et al. (2007) and Nagana Gowda and Raftery (2014)
	(5) HR-MAS-NMR	Structural information even in four-dimensional multitudes; minimal sample preparation	Successful implementation yet to be practiced	Identification of different classes of metabolites, chemotype variation study	Bharti et al. (2011)
	(6) NMRI	Generate image of the metabolites using resonance intensity of paramagnetic nuclei	High cost, low sensitivity, longer scan duration	Distribution of lipids in seeds, C-allocation studies	Borisjuk et al. (2013), Van As and Van Duynhoven (2013), and Sumner et al. (2015)
	HSQC	Quick confirmation of whether the usual suspects are present or not in the samples; substantial reduction in data acquisition time	Longer spectrum acquisition time, spectrum in single dimension	Identification and characterization of the metabolites from different samples	Ye et al. (2010) and Tayyari et al. (2013)
	J-resolved	Separate the effects of chemical shift and J-coupling into two independent dimensions, which reduces spectral crowding and, high resolution spectra on intact biological samples	Increased file size, longer spectra acquisition time, metabolite analysis is restricted to only homogeneous system, e.g., biofluids and tissue extract	Identification of metabolites with better resolution	Ludwig and Viant (2010)
	SOFAST-^{1}H-^{15}N HMQC	Monitoring metabolic pathway in real time	Single dimension data acquisition	Pathway analysis	Motta et al. (2010)

1D, one dimensional; *2D*, two dimensional; *DIMS*, direct inject mass spectrophotometry; *FT-ICR-MS*, Fourier-transform ion cyclotron resonance mass spectrometry; *GC*, gas chromatography; *HMQC*, heteronuclear multiple-quantum correlation spectroscopy; *HR-MAS*, high-resolution magic angle spinning; *HSQC*, heteronuclear single quantum coherence spectroscopy; *LC*, liquid chromatography; *MALDI*, matrix-assisted laser disruption/ionization; *MS*, mass spectrometry; *NMR*, nuclear magnetic resonance; *NMRI*, nuclear magnetic resonance imaging; *SOFAST*, band-selective optimized flip angle short transient; *TOF*, time-of-flight.

NMR such as one and two-dimensional (2D) NMR, microcoil NMR, isotype-labeled NMR, and NMR imaging are being more valuable tools for plant metabolomics. Advanced approach like dynamic nuclear polarization (DNP) developed by combining NMR and electron paramagnetic resonance (EPR) help in the identification and characterization of metabolites by acquiring hypersensitive NMR spectra (Frydman and Blazina, 2007). The investigation of intact tissue has become possible by the execution of high-resolution solid-state magic angle spinning NMR (MAS-NMR) (Pereira et al., 2014). The implication of non-Fourier transform schemes is that multiple spectra can e acquired in a single scan, thereby accelerating the functionality of multidimensional NMR spectroscopy which is also known as ultrafast NMR spectroscopy (UF NMR) (Herrera et al., 2009). One of the most efficient NMR technique is the ultrafast technique which is used to reduce the T_{scan} time as in band-selective optimized flip angle short transient heteronuclear multiple-quantum correlation spectroscopy (SOFAST-HMQC) in which the T_{scan} time reduces down to ≤ 100 ms (Gal et al., 2007). The better identification of potential biomarkers in metabolomics study can be achieved by the implementation of the statistical total correlation spectroscopy (STOCSY) method in which a pseudo-2D NMR spectrum is generated that displays the correlation among the intensities of the various peaks across the whole sample. The introduction of nano-ion-pairing reverse-phase high-performance liquid chromatography (IP-RP-HPLC) hyphenated with nano electrospray ionization high-resolution mass spectrometry (nano-ESI-HRMS) significantly reduce the detection limit of sample, thereby providing comprehensive knowledge about the central metabolism from a minimal amount of the cell extract (Kiefer et al., 2011). Among the advancements of LC, implementation of supercritical fluid chromatography (SFC) seems to be a promising alternative to traditional LC methods. SFC method is implemented by enhancing the dynamic performance of the mobile phase by using low-viscosity supercritical fluid for fast diffusion of the compounds in the mobile phase. Along with the advancement of the quantification techniques, imaging techniques such as mass spectrometry imaging (MSI) and NMR imaging have been developed to provide spatial and temporal localization of the metabolites in particular cellular and subcellular compartments (Sumner et al., 2015). For the MSI of the metabolites in cellular levels, ionization techniques using 2D and three-dimensional (3D) coordinates of a biological tissue have been developed (Fletcher et al., 2011). NMR imaging is a powerful and chemically specific tool that measures the resonance intensity of paramagnetic nuclei of the metabolites present in biological samples and generates the images.

2.1 Advances in the Techniques of Nuclear Magnetic Resonance

Methods of NMR are being emerged as unique tools in the field of metabolomics. Even though NMR has low sensitivity, its wide range of benefits makes it a reproducible method in the study of metabolomics. In the last few years NMR has been implemented in plant metabolomics for various applications such as to monitor the growth stages of plants, for assessing the responses of the plants to various environmental stresses, for chemotaxonomic classification, for the determining the geographical origin of the plant samples, for establishing substantial equivalence of genetically modified plants, and for quality control of the nutraceuticals (Kim et al., 2011). NMR spectroscopy is an extremely productive and quantitative tool. As the sensitivity of NMR is remain same over the signals, the solitary internal standard can be used for absolute metabolite quantification, which is independent of the chemical nature and properties of the metabolites. Moreover, unambiguous structural identification of unknown metabolites can be easily done by the combinations of several techniques with the NMR. The special attention are given for maintaining appropriate pH of the extraction medium during extraction of biological samples for NMR analysis, as the chemical shift of the metabolite is pH dependent (Okazaki and Saito, 2012). Apart from this, one of the most significant advantage of NMR is the noninvasiveness, for which the samples can be recovered for further analysis like GC/MS and/or LC/MS (Putri et al., 2013). Because of relatively simple and easy sample preparation for NMR analysis, it can be successfully implemented for the study of the metabolite profile of intact biological fluids (Nakanishi et al., 2011) and semisolid and solid samples (Andronesi et al., 2008) with no need of prior separation step (Fan and Lane, 2011). Moreover, the NMR signals are directly related to the molar concentrations reflecting the concentrations of the metabolites present in a sample. However, accurate assessment of NMR signals is stalled due to the overlapping of signals from many similar molecules in the biological samples, which is overcome by the development of cryogenic probes and multidimensional NMR techniques (Kovacs et al., 2005; Chikayama et al., 2010). By the development of the cryogenic probes, noise-generating electronics in NMR are cooled down to −253°C. The new approaches like DNP developed by combining NMR and EPR acquire hypersensitive NMR spectra for the identification and characterization of metabolites (Frydman and Blazina, 2007). The metabolic flux analysis is an emerging scope of NMR that enables the detection and contribution of the particular metabolite in the metabolic pathway (Wise et al., 2008), which is useful in elucidating the stress response mechanism of the plants. By the implementation of high-resolution solid-state MAS-NMR, analysis of intact tissues becomes possible (Pereira et al., 2014). Apart from the low sensitivity, another bottleneck of NMR is the user's skill for accurate interpretation of the peak height analysis and peak integral analysis. As an alternative, Chylla et al. (2011) developed an algorithm known as fast maximum likelihood reconstruction to overcome the aforementioned constraint (Chylla et al., 2011). NMR analyses gives a global overview of all the metabolites whether primary (sugars, organic acids, and amino acids, etc.) or secondary (flavonoids, alkaloids, terpenoids, etc.) present in a biological sample, provided they are NMR-detectable.

2.1.1 One- and Two-Dimensional Nuclear Magnetic Resonance

Of the many NMR active nuclei like ^{1}H, ^{13}C, ^{31}P, and ^{15}N, the most widely used NMR analysis is ^{1}H NMR analysis because of the ubiquitous nature of ^{1}H and its high sensitivity (DeSilva et al., 2009). One of the main advantages of ^{1}H NMR is the simple and fast data acquisition process, which provides high repeatability of spectra that covers a wide range of metabolites. High reproducibility in metabolite detection can be achieved by the implementation of high-throughput Carr–Purcell–Meiboom–Gill (CPMG) pulse sequences with water signal suppression technology and one-dimensional (1D) nuclear Overhauser enhancement spectroscopy (Nagana Gowda and Raftery, 2014). In the process of CPMG sequence, to increase the detectability of small-molecular-weight metabolites, macromolecular signals are reduced (Smith et al., 2007). When many structural isomers are being analyzed in a single extract, NMR plays an essential role in the discrimination of isomer type exclusively when reference standard materials are not available (Mahrous and Farag, 2015). Using the ^{1}H NMR spectrum of a given plant extract typically 30–150 metabolites can be simultaneously detected. Multivariate data analysis, such as principal component analysis (PCA), orthogonal projections to latent structures (O-PLS), or hierarchical cluster analysis (HCA), help in the analysis of data set created by integrating the peak values observed in the spectrum (Gad et al., 2013; Mahrous and Farag, 2015). One-dimensional NMR has significant roles in metabolomics study. However, peak overlap or peak position biases is a major problem associated with 1D NMR. The peak overlap primarily happens due to the large number of ^{1}H peaks per metabolite and a large metabolite range (Van et al., 2008). Chemical attributes like ionic strength and pH can introduce changes in the peak position of certain metabolites (Alonso et al., 2014). Apart from peak overlapping another major challenge in 1D NMR is automatic metabolite detection. Various skilful automation technical approaches have been made to tackle this problem like the MetaboHunter (Tulpan et al., 2011) and the concept of the valid cluster (Jacob et al., 2013).

The limitations of 1D NMR have been resolved by the implementation of 2D NMR, which enhances the peak resolution and improves the peak overlap problem. Two strategies are employed to untangle the overlapping NMR signals. One strategy is a J-resolved (JRES) experiment (JRES NMR), which simplify the ^{1}H NMR spectrum by creating a projection of ^{1}H broad band decoupled spectrum (Ludwig and Viant, 2010). In this approach, one dimension is represented by the proton spectrum, and another dimension is represented by the J value (coupling constant) (Ludwig and Viant, 2010). JRES NMR has been adopted by many workers to differentiate between caffeoyl esters (Kim et al., 2011) and cis- and trans-sinapyl acid esters of *Brassica rapa* (Liang et al., 2006). JRES NMR has also been used in the classification of various phenolic compounds (Ali et al., 2009). The efficiency of JRES NMR in metabolite identification relies on the prior knowledge of the chemical composition of the extract, because the coupling constant obtain from the spectrum cannot be used for searching the metabolites available in any metabolomic database. Another strategy in NMR is diffusion-ordered NMR spectroscopy (DOSY), which selectively suppresses the signals from certain compounds using relaxation or diffusion filters (Novoa-Carballal et al., 2011). In the principle of DOSY, it detects the transitional diffusion coefficient of different molecules of different sizes. DOSY seems to be suited for the analysis of the plant samples as the proton spectrum of each compound of different molecular weight and composition present in the sample can be recorded separately. This technique is otherwise known as "NMR chromatography" (Novoa-Carballal et al., 2011). In 2D DOSY the proton spectrum is plotted in one dimension and the diffusion coefficient of each NMR signal is plotted in the second dimension. A major drawback associated with 2D DOSY is the ambiguity in the analysis of diffusion coefficient of overlapped signals. This limitation can be overcome by the implementation of 3D correlation spectroscopy-DOSY and 3D total correlation spectroscopy-DOSY (TOCSY-DOSY) in which overlapped peaks are resolved by spreading them in a second dimension (Mahrous and Farag, 2015). However, the use of 2D NMR is constrained by longer data procurement time, larger data sizes, and less convenient for data exploration when compared with 1D NMR. However, the data crowding can be minimized with lesser data acquisition time by using ^{1}H-^{1}H homonuclear approach. On the other hand, heteronuclear NMR methods have much longer acquisition time but resolves the overlapping signals in a second dimension that extends to a region of more than 200 ppm (Xi et al., 2008). Two-dimensional NMR along with chemometric data analysis of NMR peaks make this technique more useful in identifying the metabolites that are correlated with the physiological alterations of the plants in stress conditions (Mahrous and Farag, 2015).

2.1.2 Isotope-Labeled Nuclear Magnetic Resonance

The selectivity and resolution of NMR analysis have been improved significantly by the application of isotope-labeled NMR technique. Isotope labeling by the use of nuclei such as ^{13}C, ^{15}N, ^{2}H, and ^{31}P, either in vivo or ex vivo, has upgraded the ability to identify and quantify low-concentration metabolite with high accuracy in biological samples (Fan and Lane, 2011). In vivo isotope labeling provides scope to trace the metabolic pathway and hence offers the understanding of network of metabolic pathway involved in stress tolerance and adaptation (Bothwell and Griffin, 2011; Moseley et al., 2011). Ex vivo isotopic labeling enhances the NMR efficiency and gives additional opportunities to unravel the metabolite complexity of plants as it identifies maximum number of metabolites. Stable isotopic tagging of the particular functional group provides a new way to enhance the sensitivity and resolution of NMR data (Gowda et al., 2010). For example, tagging amine functional group with ^{13}C isotope-based acetylation or formylation and detection using ^{1}H-^{13}C heteronuclear single quantum coherence spectroscopy (HSQC)-2D experiments are useful approaches for the analysis of the amine-containing metabolites in complex mixtures (Ye et al., 2010). In the year 2013, a new technique known as smart isotope tag with ^{15}N-cholamine has been developed that maximizes the combined strengths of two powerful analytical techniques like NMR and MS, for many metabolomic applications (Tayyari et al., 2013).

2.1.3 Microcoil Nuclear Magnetic Resonance

Development of microcoil probe has enhanced the sensitivity of NMR methods especially in the mass-limited samples (Grimes and O'Connell, 2011; Bird et al., 2012). Various efficient microcoil probes with different coil sizes and with single- or dual-volume cells have been developed and successfully used for the detection of various metabolites (Henry et al., 2008;

Bergeron et al., 2008). Minute amount of samples, i.e., the metabolite fraction eluted from the LC columns, can also be analyzed by signal enhancement using microprobe (Cloarec et al., 2007). Due to minute sample volume requirement the simultaneous detection of the metabolites can be possible by both NMR and MS by splitting the chromatography fractions. This advantage of microcoil-NMR offers new opportunities to exploit the combined strengths of these two methods in detection and identification of metabolites (Lenz and Wilson, 2007). Advancement of microprobe with integration of automated accessories enables high-throughput analysis of small-volume of samples (as low as 20 μL) for metabolomic applications (Garcia et al., 2011). Although microcoil probe enhances the sensitivity of NMR methods of mass-limited samples; but, care must be taken in concentrating the samples to increase the microcoil NMR detection, as high salt and protein concentration interferes with the efficiency of the process (Nagana Gowda and Raftery, 2014).

2.1.4 *High-Speed Nuclear Magnetic Resonance Methods*

To achieve high-throughput 2D NMR experiments, new developments have been made in this field. By combining nonlinear sampling, forward maximum entropy reconstruction, and J-compensated ^{1}H-^{13}C HSQC spectra, a 22-fold reduction in time can be achieved without any loss in quantitative information (Rai and Sinha, 2012). Data acquisition in 2D NMR can be achieved in few seconds by applying SOFAST approach, which is an amalgamation of accelerated T_1 relaxation time and optimized flip angle. Monitoring the metabolic pathway in real time is achieved by SOFAST-^{1}H-^{15}N HMQC (Motta et al., 2010). The speed of data acquisition is enhanced by the implementation of covariance NMR spectroscopy, introduced by Bruschweiler and Zhang (2004). Single-scan acquisition method proposed by Frydman et al. (2003) offers opportunities for the subsecond acquisition of 2D NMR spectra. Advancement of metabolomics is possible due to the advancement of multidimensional NMR, which plays some important roles in present-day spectroscopy. The disadvantage of 1D NMR is it lacks the resolution needed to segregate multiple overlapping lines. Although 2D NMR is capable of separating the individual atomic peaks, but it is an intrinsic time-consuming process (Gal et al., 2007). These bottlenecks of traditional 2D experiments can now be overcome by accelerating multidimensional NMR spectroscopy, which includes non-Fourier transform schemes in which acquisition of multiple NMR spectra is possible in a single experiment, and "single-scan". The multidimensional NMR spectroscopy is also called as UF NMR (Herrera et al., 2009). These techniques permit the collection of complete multidimensional NMR data sets within a single continuous acquisition. The basic principle behind the ultrafast techniques is that it either decreases the T_{scan} repetition delay or reduces the overall number of scans (N_{scan}) needed to retrieve the spectra. One of the most efficient ultrafast technique, HMQC (SOFAST HMQC) reduces the T_{scan} time down to ≤ 100 ms while preserving high sensitivity (Gal et al., 2007). Frequent possible repetition in SOFAST techniques combined with single scan capabilities of ultrafast NMR, which can further increase the efficiency and sensitivity of NMR. For an efficient understanding of the biological activity of various natural products, structural analysis is a prerequisite, and in this direction, solution-state NMR gives substantial inputs. The development of the techniques like pure shift yielded by chirp excitation-TOCSY (PSYCHE-TOCSY) has been used to study the chemical shift for correlation of complex natural products. The advantage of this technique is that it resolves the overlapping of ^{1}H-^{1}H scalar coupling multiples, which renders the chemical shift analysis of complex natural products (Kakita and Hosur, 2016). PSYCHE-TOCSY, which suppresses the effects of homonuclear coupling and allows ^{1}H spectra to be produced, has chemical shift only, with no multiple structures. Better identification of potential biomarkers in the metabolomic study can be achieved by implementation of the STOCSY analysis method. STOCSY generates a pseudo-2D NMR spectrum that displays the correlation between the intensities of the various peaks across the whole sample by taking advantage of multicollinearity of the intensity variables in a set of spectra (Cloarec et al., 2007). High intermolecular correlations between two or more molecules involved in the same pathway can be analyzed by combining STOCSY with supervised pattern recognition, and particularly O-PLS-discriminant analysis (DA). This is a two-step process and in the first step, O-PLS-DA extracts the part of NMR spectra related to discrimination. This information is then cross-combined with the STOCSY results for the identifications of the molecules responsible for the metabolic variation. The implication of STOCSY has been further extended to enhance the recovery of latent biochemical information and for the improvement of the biological classification and interpretation. This approach is known as STOCSY scaling [STOCSY(S)] (Maher et al., 2012).

In the solution-state NMR techniques, due to several intrinsic impediments such as (1) bigger molecular size and (2) solubility of the compounds to be detected, solid-state NMR techniques have been developed, which provide comprehensive knowledge about the molecular structure, metabolite dynamics, and interactions between various metabolites (Schanda and Ernst, 2016). With the development of solid-state NMR, many unanswered questions like dynamic functions of membrane proteins, molecular arrangements of molecular transporters, and interactions of cell organelles and cell wall now can be addressed (Schanda and Ernst, 2016). The detailed structural analysis of larger crystalline and soluble biomolecules, which cannot be achieved by the crystallographic techniques can be possible by the application of advanced high-resolution magic angle spinning (HR-MAS) solid-state NMR spectroscopy in combination with other techniques like cryoelectron microscopy, fiber diffraction, and small-angle X-ray scattering (Goldbourt, 2013). In HR-MAS solid-state NMR, the samples are spun in speed with an HR-MAS probe at an angle of 54.7 degrees which is known as magic angle. HR-MAS-NMR is well established in acquiring the structural information of proteins in their tertiary and quaternary folds even in the four-dimensional multitudes. However, this technique is yet to be successfully used in the field of metabolomics. However, this high-throughput technology has been adopted by many researchers to identify different classes of metabolites including amino acids, organic acids, fatty acids, organosulfur compounds, and carbohydrates and to describe the changes in the metabolome of plant tissue exposed to various stresses (Pereira et al., 2014; Rosati et al., 2014).

Nuclear magnetic resonance imaging (MRI) also known as simply magnetic resonance imaging is a chemically specific imaging technology that generates images of metabolites using resonance intensity of paramagnetic nuclei (Sumner et al., 2015). Most MRI experiments use ^{1}H resonance of water due to the high abundance and sensitivity (Windt et al., 2010). Apart from using ^{1}H resonance signals in studying the movement and transportation of water, ^{1}H signals from metabolites such as sugars and amino acids can be used for imaging (Melkus et al., 2011). Resonance signals of other paramagnetic nuclei like ^{13}C, ^{15}N, ^{17}O, ^{19}F, ^{23}Na, and ^{31}P can also be used for the imaging of different metabolites (Lambert et al., 2006). One of the most advanced applications of MRI in plants is the imaging of the distributions of the lipids in seeds (Borisjuk et al., 2013). Metabolite imaging and C-allocation studies using positron emission tomography are the emerging frontier implementations of MRI in the field of plant metabolomics (Van As and Van Duynhoven, 2013). The most advanced technology which is integrated with NMR and MRI is magnetic particle imaging (MPI). After its first introduction in 2005 by Gleich and Weizenecker, it is gaining importance among the NMR and MRI researchers (Gleich and Weizenecker, 2005; Saritas et al., 2013). MPI uses superparamagnetic iron oxide nanoparticles), which provide 3D visualization of the distribution of metabolites in space with high temporal and spatial resolution (Panagiotopoulos et al., 2015).

NMR spectroscopy–based methods have been practiced in several areas such as biomedical science, toxicology and drug development, food and beverages, environmental science, physiology, and more. Interactions of living organisms with their environments have been explored using metabolomics with NMR as a potential tool. Metabolomics study focus on the alterations of metabolite status under the influence of various environmental stresses. NMR spectra may be less sensitive than the MS method, but its productivity, reproducibility, quantitative aspects, and structural information contents are superior to those of MS. Moreover, NMR spectral analysis is essential for the confirmation of chemical structures (Schripsema, 2010). By the implementation of NMR, demonstration of the existence of a particular metabolic pathway, isotope labeling experiments, metabolite configuration of tissue extracts, localization as well as tracking of the distribution pattern of metabolites in tissues, and the determination of metabolite structure can be possible (Obata and Fernie, 2012). Due to many of advantages of NMR, it is being applied for profiling of abundant metabolites while studying changes in noncomparable profiles is highly useful for metabolite fingerprinting (Dixon et al., 2006).

2.2 Advances in the Techniques of Mass Spectroscopy

It is now well accepted that a single technique cannot provide a satisfactory conception of the whole metabolome of an organism. Therefore more than one technique and the combination of them is required to gain a comprehensive knowledge about the metabolite profile. Determination of *m/z* ratio of different metabolites makes MS one of the most potent technique for the structural elucidation of various metabolites. Sensitivity in metabolite identification is one of the major advantage of MS over NMR. As compared to other chromatographic techniques, MS increases the number of detectable compounds (Okazaki and Saito, 2012). MS in combination with different ionization techniques like ESI and MALDI are used for better identification and characterization of the metabolites. Furthermore, MALDI can be coupled with time-of-flight (TOF) analyzer for better resolution. Fragmentation of metabolites by collision-induced dissociation (CID), electron transfer dissociation (ETD), or electron capture dissociation (ECD) coupled with tandem mass spectrometric analyses (MS/MS) is also widely implemented in the field of metabolomics (Rodziewicz et al., 2014). Advancement of MS, like MS/MS, helps in identification and screening of putatively separated metabolites based on the *m/z* ratio by analyzing MS/MS spectra and the retention time. MS/MS involves multiple steps of MS in which the ions are first separated by their *m/z* ratio. The ions that are separated by *m/z* ratio are subjected to fragmentation by CID, and the resulting fragmented ions are detected in the second stage.

2.2.1 Gas Chromatography

In GC-MS, the analytes are vaporized to pass through the fused silica capillary with the carrier gas. Volatile samples that are responsive to chemical derivatization are prerequisites for GC-MS analysis (Okazaki and Saito, 2012). Certain samples such as essential oils and terpenoids are appropriate for analysis by GC-MS. Metabolites with polar functional groups such as amines, amino acids, sugars, organic acids and fatty acids can also be analyzed with the help of appropriate derivatization process. It can be specified that by using GC-MS, the primary metabolites, i.e., the metabolites from primary metabolism, can be analyzed. EI is the most commonly implemented ionization method in GC-MS. EI is known for its high reproducibility and it is less affected by ion suppression. In addition, EI produces mass spectra as a result of high degree of metabolite fragmentation, which is beneficial for the identification and characterization of different metabolites. Spectral peaks generated by GC-MS can be analyzed efficiently with the help of various mass spectral libraries of GC-EI-MS. In GC-EI-MS, detection of molecular ions that have diminishing effects on the chemical structure of the identified metabolites is not possible, and this makes GC-EI-MS more suitable for the targeted analysis of primary metabolites (Kusano et al., 2007; 2011; Ralston–Hooper et al., 2008). Nowadays, GC-MS-based metabolite profiling is being commonly used in the metabolic studies of plants, animals, and microorganisms (Kusano et al., 2011).

2.2.2 *Liquid Chromatography-MS*

Speed and accuracy of MS is enhanced by the coupling of LC separation technique with MS. Among the other LC separation techniques, HPLC is a more adaptable technique that can be coupled to MS. LC-MS is one of the important technical approaches that is used in the study of the metabolome. The method of LC-MS is suitable for metabolites with high molecular weight and thermosensitive and chemically unstable functional groups as it does not need volatilization of the compound. LC-MS is also very useful in the profiling of secondary metabolites (e.g., alkaloids, saponins, and phenylpropanoids) and lipids (e.g., sphingolipids and glycerolipids) (Matsuda et al., 2010; Okazaki et al., 2011). Many secondary metabolites and complex lipids that are not responsive to volatilization by GC can be separated and characterized by LC-MS using reverse-phase column (Matsuda et al., 2010; Okazaki et al., 2011). MS/MS provides more information on mass spectra. Furthermore, validation of isolated metabolites using multiple instrumentation is a time-intensive process in the interpretation of untargeted metabolomics data. Implementation of LC-MS with the combination of solid-phase NMR narrows down the bottleneck of untargeted data interpretation (Sawada and Hirai, 2013). Integration of triple quadrupole mass spectrometer with LC is frequently done in targeted metabolomics (Albinsky et al., 2010). The nonavailability of mass spectral libraries for API-MS and commercially accessible genuine standards is the main hindrance in the efficiency of LC-MS. Since 2000, significant advancements have been made both in chromatography and MS techniques for efficient analysis of complex biological samples. Among the chromatographic techniques, HPLC underwent rapid development. Nanoflow reverse-phase HPLC coupled to nano-ESI MS has been implemented in proteomics and gives better resolution because of miniaturization and sensitivities at the low zeptomole level (Luo et al., 2007). Introduction of nano IP-RP-HPLC hyphenated with nano-ESI-HRMS significantly reduces the limit of detection, thereby providing a comprehensive knowledge about the central metabolism from a very little amount of cell extract (Kiefer et al., 2011). The sample needed for analysis can be further reduced by the integration of nano-IP-RP-HPLC which is more sensitive instrumentation like triple quadrupoles (Kiefer et al., 2011). Lately, instrumentation of LC has moved to the next level by the introduction of ultrahigh-performance liquid chromatography (UHPLC). This technique uses very low system void volumes with 2 μm particles that can withstand pressures up to 1000–1300 bar. Some alternatives of UHPLC have been suggested for improving the performance of LC. One such alternative is silica-based monoliths with several unique features in terms of permeability and efficiency was first launched in the year 2011 (Hormann et al., 2012). SFC is one of the advancement of LC, which seems to be a promising alternative to traditional LC methods. In SFC technique, the kinetic performance of the mobile phase is enhanced by using low-viscosity supercritical fluids, which result in fast diffusion of the compounds in the mobile phase. Supercritical fluids possess features such as low viscosity and high diffusivity and their temperature and pressure are above the critical values. This technique is beneficial over high-temperature LC in which there was a risk of thermal degradation (Grand-Guillaume Perrenoud et al., 2012). Upgradation of LC techniques such as hydrophilic interaction liquid chromatography aids in the separation of small polar compounds. This technique efficiently separates polar molecules on the polar stationary phase. To make LC-MS more efficient and productive, production of additional reference compounds and more comprehensive databases are a prerequisites (Nakabayashi et al., 2010; Horai et al., 2010; Okazaki and Saito, 2012).

2.2.3 *Capillary Electrophoresis-Mass Spectrometry*

Capillary electrophoresis-mass spectroscopy (CE-MS) is employed to analyze the ionic metabolites. As in LC-MS, API is the most appropriate method used for ionization in CE-MS. The constituent ions move on their electrostatic force depending on the size and charge of ions, the type of electrolyte used and the electro-osmosis from the capillary action (Okazaki and Saito, 2012). In CE-MS, samples usually get divided for cationic and anionic analyses. For better metabolite separation, cationic analysis method using formic acid as the analyte has been implemented by Soga et al. (2006). However, high-resolution analysis of anionic metabolites is quite challenging when compared with that of cationic metabolites. Later, use of platinum ESI electrospray needle has improved the efficiency of CE-MS in analyzing the anionic metabolites (Soga et al., 2009). The advantage of CE-MS is, it does not require the derivatization step as in GC-MS, however, it detects similar compounds as GC-MS (Okazaki and Saito, 2012). CE-MS analyzes the metabolites that are important from the physiological point of view and that are similar in all organisms. Hence, CE-MS has been employed in metabolite profiling of various organisms (Urano et al., 2009; Ishikawa et al., 2010). Considering plant stress response study, CE-MS is a promising technique that enriches our knowledge about the stress tolerance mechanisms in plant and also allows us to compare the different plant groups.

2.2.4 Fourier-Transform Ion Cyclotron Resonance Mass Spectrometry

Fourier-transform ion cyclotron resonance mass spectrometry (FT-ICR MS) is a kind of MS method that analyzes the mass-to-charge ratio of ions based on the cyclotron frequency of ions in a fixed magnetic field. FT-ICR MS is gaining importance in the field of metabolomics because (1) compounds with comparable molecular weight can be identified separately without any prior chromatographic separation step, which enhance the recognition of a number of metabolites and (2) chemical structure of the metabolite can be determined by analyzing the peak, which is supportive in explaining the chemical structure of unknown metabolites. Fragmented metabolite ions can be detected with high accuracy and resolution by coupling MS/MS with FT-ICR MS, which is useful in determining the structure of metabolite. Fruitful implementation of FT-ICR MS has not been achieved because of the extremely large amount of data generated with difficulties in hardware handling. Few reports have been documented that practices this technology for metabolomics analysis in plants. Hirai et al. (2005) have used FT-ICR-MS technique for nontargeted metabolic profiling in *Arabidopsis*. Mungur et al. (2005) have also used this technique for analysis of metabolite finger printing alterations in transgenic tobacco overexpressed with glutamate dehydrogenase gene of *Escherichia coli*. However, the utility of FT-ICR-MS has been enhanced by combining it with LC as reported in *Lotus japonicus* (Suzuki et al., 2008) and tomato (Iijima et al., 2008). The strength of this technique has increased by the isotopic labeling of metabolites and analysis of peaks documented in high magnetic field (Miura et al., 2010). This technique of isotope labeling enables the reduction of the background noise peaks, eliminating the indistinctness of different metabolites, and also assists to assign a correct chemical formula and mass (Giavalisco et al., 2009). As alternative to the time-consuming hyphenated MS procedures direct inject mass spectrophotometry (DIMS) procedures are introduced in which the samples are directly infused into the mass analyzers without any prior chromatographic or electrophoretic separation steps. DIMS technique provides a comprehensive knowledge of a wide range of metabolites including lipids and its derived compounds, amino acids, sugars, and organic acids (González-Domínguez et al., 2017). However, along with the aforementioned benefits, the DIMS techniques also have important drawbacks that reduces the analytical efficiency. Chemical isomers cannot be resolved without any prior separation step as in traditional MS procedures. However, using ultrahigh-resolution instruments such as FT-ICR-MS or orbitrap-MS, only the monoisotopic masses of the chemical isomers can be distinguished. FT-ICR MS is a kind of MS method that analyzes the mass-to-charge ratio of ions based on the cyclotron frequency of ions in a fixed magnetic field. FT-ICR MS is gaining importance in the field of metabolomics because (1) compounds with comparable molecular weight can be identified separately without any prior chromatographic separation step and (2) the chemical structure of the metabolite can be determined by analyzing the peak, which helps in explaining the chemical structures of unknown metabolites. Fragmented metabolite ion can be detected with high accuracy and resolution by coupling MS/MS with FT-ICR MS, which is useful in determining the structure. Limitations such as isomeric separations are resolved by implementing some of the advanced MS protocols such as ion mobility MS, which allows isobaric multidimensional separations by movement of charged molecules in the mass domain (González-Domínguez et al., 2017).

3. IMPORTANCE OF SAMPLE PREPARATION IN METABOLOMICS

Every metabolomic experiment consists of three basic experimental steps: (1) sample preparation, (2) data acquisition by analytical techniques, and (3) data filtering using appropriate chemometric methods (Kim and Verpoorte, 2010). Finally, the metabolites are identified by peak analysis through spectral searching against the metabolite database. Sample preparation in plant metabolomics is a fundamental and critical step which has significant consequences on the accuracy of the results (Choi and Verpoorte, 2014). When compared with data acquisition, preparation of the sample for the metabolomic analysis is manual and needs special attention to get a superior analytical output. In general, sample preparation for plant metabolomic study includes harvesting, drying, extraction, and finally, analysis. However, the drying and extraction steps might be omitted depending on the analytical method and properties of the sample to be analyzed (Kim and Verpoorte, 2010). The harvesting and drying steps are similar for all the analytical platforms, but this is not the case for extraction, which strictly depends on the type of analytical platform to be used for analysis. However, the requirement of the drying step may depend on the type of biological samples to be analyzed. Harvesting of plant material must be done carefully, as the metabolites are amenable to rapid changes due to enzymatic degradation during harvesting and extraction processes and may alter the outcome. Once the material is removed from the source, it is immediately frozen to avoid metabolic alteration due to enzymatic reaction triggered by wounding of the plant (Kim and Verpoorte, 2010). Apart from this, various other things such as time of sampling and the plant parts to be used for sampling such as the leaf, stem, or root must be considered at the time of sampling. There is variations in metabolite composition not only between two different organs but also within the same organ. For example, metabolite composition is significantly different between young and old leaves. In addition, within the same leaf there is difference in metabolites between the veins and the rest of the leaves. Drying of the harvested samples is essential for metabolomics study as water provides a medium for enzymatic reaction resulting in metabolite degradation. Various drying methods have been practiced such as air drying, freeze drying and hot air oven drying (Harbourne et al., 2009). Apart from this the presence of water molecules may produce signals in ^{1}H NMR analysis when samples are analyzed without any pulse sequence for the removal of the signals due to water molecules. Extraction of the metabolites is the most important step in metabolomics. It is done by considering many aspects such as the chemistry of the solvents, the ratio of solvent and sample, the temperature during extraction, and the duration of extraction. As the metabolome consists of a wide range of metabolites with different polarities, a single extraction method cannot extract the whole range of compounds. The extraction of metabolites depends greatly on the choice of solvent and the ratio of solvent and sample. Solvents are chosen depending on their characteristics such as polarity, compatibility with the analytes, and toxicity (Kim and Verpoorte, 2010). Most importantly, metabolomic studies should be designed in such a way that they can extract as many metabolites as possible. Thus, solvent should be chosen accordingly. Different analytical platforms use different extraction process and hence different solvents. For instance, in the case of GC-MS, the choice of solvents is limited to volatile compounds, and for polar metabolites derivatization is inevitable. On the other hand, in LC-MS the limitation of extraction solvent is reduced and the solvent with which samples are injected should preferably be similar with the mobile phase (De Vos et al., 2007). Compared to MS, NMR has more moderation in solvent selection, e.g., midpolar solvents like aqueous methyl alcohol are preferred. Apart from solvent selection, pH of the extraction medium also affects the metabolomic output. For instance, in MS ionization of the analytes is affected by the pH. As a single extraction solvent cannot be efficient for complete metabolome extraction, the two-phase solvent system has been introduced (Kim and Verpoorte, 2010). For the simultaneous extraction of both polar and nonpolar metabolites, the extraction system composed of chloroform, methyl alcohol, and water (2: 1: 1, v/v) has been proved to be very efficient. Along with pH and solvent selection, the choice of extraction methods is also a critical step in metabolomics. The duration of extraction along with solubility and rate of dissolution are the important determining factors in the selection of extraction methods. Increasing the duration of extraction and increasing the energy of the extraction system by raising the temperature of the system have been adapted as alternatives for the samples having a low rate of dissolution (Kim and Verpoorte, 2010). Dissolution rate is affected by the interaction of the metabolites with the biological matrix such as lignin or cellulose, which may be considered as an adsorbent. Soxhlet extraction procedure has been conventionally used in which samples are continuously heated, and the samples are extracted with the concentrated solvents. The main disadvantage of this method is continuous and prolonged heating, which may cause degradation of the metabolites. As an alternative to the Soxhlet method microwave-assisted extraction (MAE) has been introduced, which is a fast extraction method and thus advantageous as compared to the Soxhlet method. However, microwave heating can cause an immediate rise in temperature of the extraction system that may acts as a limiting factor. Beside this, some other properties of the solvents such as dielectric constant may interfere with the efficiency of MAE (Kim and Verpoorte, 2010). In addition to this, supercritical fluid extraction method has been materialized for the efficient extraction of volatile compounds (Pourmortazavi and Hajimirsadeghi, 2007). This technique is beneficial as it uses less amount of hazardous solvents and its ease of hyphenation to GC, LC, or SFC. Although various alternatives for metabolite extraction have been proposed from time to time, solvent extraction method is the most popular extraction method.

4. PROCESSING OF METABOLOMICS DATA

Definition of metabolomics indicates that it gives knowledge about molecular-level intricacy downstream to genomics, transcriptomics, and proteomics of an organism either in natural conditions or in response to various abiotic stresses. Metabolomic study generates a huge data set as the number of small molecular metabolites differ significantly depending upon the type of tissue and organ taken for the study. Metabolomics is comparatively a new discipline, therefore a complete directory of metabolites are not available for many organisms (Pearson, 2007). On the other hand, NMR and MS spectra for all known metabolites are not yet available, making it difficult for the proper identification of the metabolites

(Kind et al., 2009). Analysis of metabolomic data is additionally complicated by the metabolite composition and complexity of plant samples. The aim of metabolomics is to identify the difference in the metabolites between two samples induced by environmental alterations in spite of the significant background metabolite similarities between them (Aoki-Kinoshita, 2006). This is because few metabolites play a central role in defining the chemical characteristic of a particular species and some of the metabolites are produced by the induction of one or the another environmental condition such as age, growth phase, media, and various abiotic stresses. In addition, various instabilities and variations instrumental in sample preparation cause wavering in spectral data, such as changes in peak position or peak width. Therefore a robust methodology along with established protocol for sample preparation is required to analyze the metabolomics data to get an in-depth knowledge about the metabolomic fluctuations and coordination under various environmental variations (Worley and Powers, 2013). One type of approach is metabolite fingerprinting that identifies metabolites by differential spectral analysis acquired from normal and perturbed biological systems (Han et al., 2009). Metabolic fingerprinting thus defines the relative alterations between the metabolomes of two or more systems to assume a biological association between them as well as the effects of environmental stresses. The basis of metabolite fingerprinting is that it practices multivariate data analysis methods to classify the biologically relevant spectra for further targeted analysis.

Two chemometric approaches have been popularly implemented in metabolomics for targeted analysis: (1) PCA and (2) partial least squares projection to latent structure (PLS). PCA is the most widely instigated multivariate statistical approach used as a dimension reduction method that identifies the variance among samples (Gu et al., 2011). PCA classifies all substantial variances into different principal components (PCs) and each PC represents one significant trend of independent variance (Xu and Goodacre, 2012). PCA is advantageous because the variance of interest can be separated and summarized into multiple PCs. However, PCA is not always able to separate the variations effectively when more than one influential factor is closely associated or coexist. The results of PCA of the data sets having more closely related variations become rather difficult to interpret because the number of variables is spread over some PCs (Xu and Goodacre, 2012). In spite of multiple independent influential factors, the subspace that results from PCA can be rotated if the variations caused by these factors is similar, making the results of the PCA more difficult to follow. To depict the underlying relationships between various sets of variations, multiblock PCA has been implemented with the purpose of revealing the "common trend" between these data. These methods have been extensively exploited for the integration of data obtained from multiple analytical platforms, e.g., from the combined analysis of denaturing gradient gel electrophoresis and GC-MS (Zomer et al., 2009; Ferreira et al., 2010). analysis of variance-PCA (ANOVA-PCA) is another PCA extension that analyzes multiple influential factors in simulated data sets (Harrington et al., 2005). Similarly, consensus PCA is another PCA approach that performs PCA on multiple blocks of data measured on the same objects (Westerhuis et al., 2008). The bilinear statistical approach PLS discriminant analysis (PLS-DA) is the most popular administered data analysis method in metabolomics. In PLS-DA, the data variables are represented on X matrix, whereas the class variables are represented on Y matrix (Gu et al., 2011). In addition to the aforementioned extensions of PCA, orthogonal signal correction is another PLS-based data analysis system that removes uncorrelated information in X matrix and analyzes the variables based on the new X matrix.

Time taken for the data acquisition is no longer considered as the limiting factor because of the speed of the advanced technical platforms. However, assessment of this massive amount of data is the limiting factor for metabolomics (Ernst et al., 2014). For the evaluation of the massive data high-throughput metabolomics techniques must be self-equipped with sophisticated computational and statistical tools for the better implementation of metabolomics. These sophisticated tools have been developed in the fields of chemometrics, chemoinformatics, and bioinformatics, and are implemented in various other omics studies (Trygg et al., 2007). Continuous development in computational metabolomics, marks it as a promising factor in making metabolomics as a standard analytical approach in different areas of science (Wishart, 2007). Acquaintance with the characteristics of computational informatics is a prerequisite for carrying out any metabolomics experiments. Data preprocessing in metabolomics includes noise filtering, peak detection, normalization, alignment, deisotoping, and identification. Noise filtering is defined to reduce or separate compound signals originating from background signals and/or from chemical or instrumental interference (Ernst et al., 2014). Peak detection mainly deals with the correct identification of a compound signal and its intensity, and deisotoping the isotopic peaks that actually originate from same type of compounds. Isotopes may interfere with data interpretation because they may assumed to be different compounds. Normalization, as well as alignment, is possible with or without the addition of an internal standard as the internal standard may interfere with the retention time (Smith et al., 2006). Various data processing software have been used for the analysis of metabolomics data, e.g., XCMS for LC-MS. Downstream processing of data by this method incorporates retention time alignment (nonlinear), filtration, peak detection, and matching. A number of endogenous metabolites have been identified without using internal standards, and these metabolites can further be quantified using internal standards. Different preprocessing software such as Tagfinder (Luedemann et al., 2008), Targetsearch (Cuadros-Inostroza et al., 2009), etc., have been developed and implemented at different laboratories for more robust GC-MS experiments.

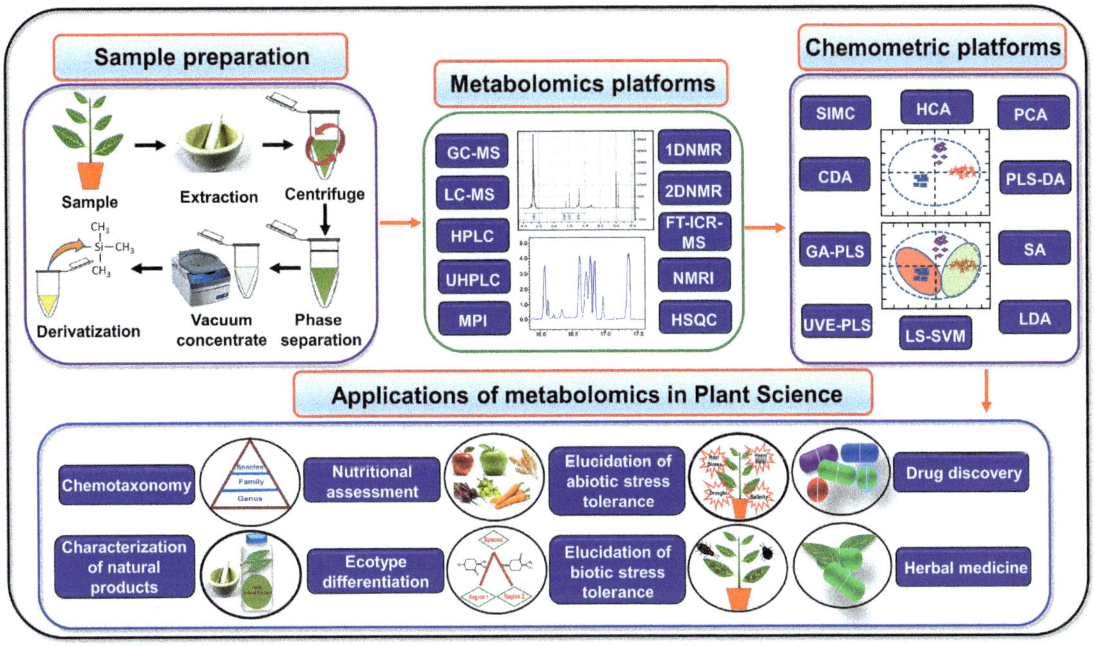

FIGURE 1.1 Schematic illustrations portraying metabolomics workflow and various applications of metabolomics in plant science.

For downstream processing of the spectral data two general approaches can be taken into consideration: (1) incorporation of all data into different recognition profiles by dividing the signals into separate bins for each sample, and (2) discarding the irrelevant data by identifying and individually quantifying significant features (Smith et al., 2006). New advanced analytical strategies are steadily being developed by the researchers worldwide that provide access to the maximum information from mass spectra. The discrimination between various metabolites can be done either by (1) differentiating the metabolites of different nominal mass, (2) differentiating the metabolites having same nominal mass, but different molecular formula and monoisotopic mass, and (3) differentiating the metabolites having same nominal and monoisotopic masses, but different chemical structures (Dunn et al., 2013). Using such strategies, software for better data exchange with various developed databases has been developed (Chambers et al., 2012). The metabolome databases have been developed with regard to higher accuracy in m/z matching and molecular formula determination, for better identification of metabolites. A number of metabolites are commercially available to be incorporated in spectral libraries, and our present knowledge about sample-specific metabolomes are considered to be the major bottlenecks in the development of efficient databases. Databases such as KEGG, MetaCyc (Caspi et al., 2016), HMDB (Wishart et al., 2009), MZedDB (Draper et al., 2009), MMD (Brown et al., 2009), and METLIN (Sana et al., 2008) have been referred for known metabolites of both species-specific or species-nonspecific origin. Among the aforementioned databases, some provide information about metabolic reaction and cellular concentration of specific metabolite, e.g., KEGG and HMDB (Dunn et al., 2013). Reference spectra are prerequisites for metabolite identification, and these spectra are provided by only limited metabolite databases. However, these spectra can also be obtained from mass spectral libraries. Various nonspecific (e.g., NIST08) and metabolite-specific libraries (e.g., GMD and FiehnLib) are available for GC-MS data analysis. At present, these are some of the most widely applied mass spectral libraries, because of their high metabolite coverage. The metabolomics workflow from sample preparation to data interpretation and pathway analysis is given schematically in Fig. 1.1.

5. APPLICATIONS OF METABOLOMICS IN PLANT SCIENCE

Modern plant metabolomics has evolved by continuous refinement and has increasing scope with the increase in number of plant metabolites identified. The advancement of plant metabolomics has provided the opportunity to have a comprehensive knowledge of plant biology and physiology (Sumner et al., 2015). Compared with the other omics techniques, metabolomics has now become a well-known technique for the study of all types of organisms and it can complement to other omics techniques (Schripsema, 2010). Due to the presence of a wide array of secondary metabolites in plants compared to animals, it is quite difficult to design a potent routine procedure for metabolomics sample preparation. Plant metabolism is highly flexible both in terms of abundance of metabolites and identities of the metabolites. Genetic, environmental, and developmental effects can be assessed by differentially quantifying metabolite data among biological samples, and it is proposed that metabolomics data will yield information concerning the genetic, environmental, spatial or developmental control of metabolism (Fiehn et al., 2008). Metabolic alterations in plants can be studied in many ways, and a variety of techniques are available to conduct plant metabolomics studies. It is now well known that not a single method can yield all the information necessary for a comprehensive metabolite analysis. Therefore an integrated approach for the metabolite study is a prerequisite for global metabolic profiling. Metabolomics has been implemented in many areas of plant science such as natural product and drug discovery from medicinal plants, pathway analysis of various important secondary metabolites, nutritional assessment of crop plants, chemotaxonomic classification of plants, and most importantly, deciphering the response of plants to various biotic and abiotic stresses. Multiple applications of different metabolomic approaches in plant sciences are depicted in Fig. 1.1.

5.1 Metabolomics in Deciphering Abiotic Stress Tolerance Mechanisms of Plants

Metabolomics has profound roles in deciphering the abiotic stress tolerance mechanisms in plants. Various advanced forms of metabolomics techniques have been implemented for the successful interpretation of the metabolic alteration in response to various abiotic stresses such as high salinity, drought, waterlogging, and heavy metal contamination. Plants respond to abiotic stresses in various manners, and alternation in metabolite profile is one way adopted by the plants for obtaining tolerance against different abiotic stresses. The study of this metabolite alternation in plants provides a way for understanding the mechanisms behind the plant responses to various prevailing abiotic stresses. Several metabolomics techniques such as GC-MS, LC-MS, NMR, and their subsequent advanced forms have been implemented for the elucidation of stress tolerance mechanisms in plants (Table 1.3). In addition to their solitary implementation, many techniques are used as an integrated approach for better output. GC-MS in combination with various ionization techniques produces highly reproducible fragmentation pattern of the analytes, thereby providing better separation and identification of differentiated metabolites. Another advantage of GC-MS is that it has long been used for metabolite profiling of plants and thus established protocols and instrumentation setup, chromatographic techniques for spectral analysis, and databases for metabolite identification are readily available. Overcoming the limitation of GC-MS due to volatilization, LC-MS provides an alternative for the identification of volatile compounds differentiated in response to various abiotic stresses. Different separation options, depending upon the choice of columns such as reverse phase, ion exchange, and hydrophobic interaction columns provide separation of metabolites based on their chemical properties. Further improvement in metabolite identification and differentiation has been made by using TOF-MS, which give faster acquisition and improved deconvolution for mixed samples (Obata and Fernie, 2012). ESI is used for connecting the LC and MS with many types of MS quadrupoles. TOF, quadrupole TOF (qTOF), triple quadrupole, ion trap, linear trap quadrupole (LTQ)-Orbitrap, and Fourier-transform ion cyclotron resonance have been implemented for the identification and characterization of stress-induced metabolites (Allwood and Goodacre, 2010). Along with the GC and LC-MS, NMR has also been implemented for the elucidation of stress tolerance mechanisms with respect to metabolite alteration. Among the abiotic stresses, drought and salinity are the major stresses predominant in the environment nowadays. Therefore considerable research efforts have been expended to understand the plant's response toward these common stresses. These studies have indicated the important roles of metabolites in regulating photosynthesis and other metabolic processes in plants under drought and high salinity conditions (Verslues and Juenger, 2011). Metabolomics study in *Arabidopsis thaliana* imposed to drought stress suggest that many important metabolites such as amino acids like proline, oligosaccharides, γ-aminobutyric acid (GABA), and TCA cycle intermediates accumulate in response to drought (Urano et al., 2009). Differential metabolite accumulation also take place depending on the type of tissue that encounter the stress, i.e., young and mature leaves. For instance, the metabolite profiles of matured leaves are widely distinct when compared with those of the young leaves, and this has been confirmed by metabolite profiling through GC-MS and LC-MS analyses (Skirycz et al., 2010). Amino acid metabolism responds differently in different tissues. Most of the amino acids are accumulated in severely desiccated leaves and are decreased in mildly desiccated leaves (Obata and Fernie, 2012). Apart from drought, salinity also poses severe threat to the agricultural yield worldwide. High soil salinity causes hyperionic and/or hyperosmotic effects that causes disordered metabolism, reduced ion uptake, and retarded growth and altered physiology. Metabolite profiling of *Arabidopsis* under high salinity and its comparison with its halophytic counterpart *Thellungiella halophila* has been done using GC-MS and various data mining approaches such as PCA and HCA (Lugan et al., 2010). This study suggests that the accumulation of proline in both the species along with inositols, hexoses, and complex sugars have some role in salt tolerance. However, the concentrations of the metabolites were found to be higher in *T. halophila* than in *Arabidopsis.* The metabolite alterations in various crop species such as tomato, grapevine, maize, and rice have also been analyzed under different abiotic stress conditions using various metabolomic tools (Cramer et al., 2007; Gagneul et al., 2007; Bowne et al., 2011). For the assessment of the metabolomic responses in the aforementioned crops various metabolomic approaches such as GC-MS, LC-MS, and NMR have been implemented along with various chemometric approaches such as PCA, HCA,

TABLE 1.3 Implementation of Various Metabolomic Approaches in Deciphering Different Abiotic Stress Responses in Plants

Abiotic Stress	Plant Species	Method of Analysis	Chemometric Methods	Key Metabolites Involved in Stress Tolerance	References
Salinity	*Nitraria tangutorum*	GC-MS	PLS-DA	Hexadecanoic acid, malic acid, succinate, asparagine	Ni et al. (2015)
	Nicotiana tabacum	^{1}H-NMR	PCA, O-PLS-DA	Fructose, glutamine, malate, nicotine	Zhang et al. (2011)
	Hordeum vulgare	GC-MS	PCA	Galacturonate, maltose, shikimate, raffinose, aspartate	Widodo et al. (2009)
		GC-MS	PCA, HCA, PLS-DA	Threonine, asparagine, mannose, glucose-6-P,	Wu et al. (2013)
	Cuminum cyminum	GC-MS, LC-MS, LC-TOF-MS	PCA	Glycine, glutamate, flavonoid, linolenic acid, palmitic acid	Pandey et al. (2015)
	Suaeda corniculata	^{1}H-NMR	PCA	Isoleucine, malate, fumarate, dimethylamine,	Pang et al. (2016)
	Zea mays	^{1}H-NMR	SIMCA, AMIX, PCA, PLS-DA	Malate, glycine betaine, glutamate, alanine, succinate	Gavaghan et al. (2011)
		GC-MS	–	ABA, auxin	Zörb et al. (2013)
	Aeluropus lagopoides	CE-MS	–	Asparagine, glutamine, glycine, phenylalanine	Sobhanian et al. (2010)
	Triticum aestivum	GC-MS	NIST, Xcalibur program	Oxalate, isocitrate, tryptophan, lysine itaconate	Bowne et al. (2012)
	Eutrema salsugineum	GC-MS	PCA	Caffeate, raffinose, galactinol, fumarate, δ-aminobutyric acid	Kazachkova et al. (2013)
	Lotus sp.	GC/EI-TOF-MS	PPCA	Proline, glycerophosphoglycerol, citric acid, malic acid, serine	Sanchez et al. (2011)
Drought	*Glycine max*	^{1}H-NMR, ^{1}H-^{1}H TOCSY	PCA	Myoinositol pinitol, allantoin, GABA, glutamine	Silvente et al. (2012)
	H. vulgare	GC-MS	PCA, NIST, GMD	Galactinol, mannitol, α-ketoglutaric acid, isocitric acid	Wenzel et al. (2015)
		GC-MS	NIST08, PCA, Xcalibur	Raffinose, mannitol, pyroglutamic acid, putrescine, myoinositol	Lanzinger et al. (2015)
	Thellungiella salsuginea	GC-MS	PCA	Galactinol, raffinose, spermidine, putrescine, guanosine, caffeoylquinate	Lugan et al. (2010)
	Oryza sativa	GC/EI-TOF-MS	TagFinder, NIST08	Polyamines	Do et al. (2013)
	Haberlea rhodopensis	GC-MS, LC-MS, LCQ Deca XP Plus ion trap	PCA	Xyloglucan, aspartate, glutamate, flavonoid, ascorbate,	Gechev et al. (2013)
	Craterostigma plantagineum	HPLC	PCA	Polyamines	Alcázar et al. (2014)
	Solanum tuberosum	HPLC-FLD, HPLS-MS, ESI-MS	Chromeleon, Xcalibur	Proline, carbohydrate, polyamines, betaines, polyols	Evers et al. (2010)
	Eucalyptus sp.	GC-MS	GMD, Agilent Fiehn, NIST	cyclohexane pentols, shikimic acid, galactinol, galactose, fructose, myoinositol, proto-quercitol	Warren et al. (2012)

Continued

TABLE 1.3 Implementation of Various Metabolomic Approaches in Deciphering Different Abiotic Stress Responses in Plants—cont'd

Abiotic Stress	Plant Species	Method of Analysis	Chemometric Methods	Key Metabolites Involved in Stress Tolerance	References
	Punica granatum	HS-SPME-GC-MS	–	Proline, 1-hexanol, cis-3-hexenol, trans-2-hexenal	Catola et al. (2016)
	Pisum sativum	^{1}H-NMR	FELIX, QSEQ, PCA, PLS-LDA	Threonine, δ-aminobutyrate, trigonelline, asparagine, GABA, malate	Charlton et al. (2008)
	Zea mays	^{1}H-NMR	PCA, SIMCA	Alanine, glutamate, malate, formate, δ-amino-butyrate, triacylglyceride	Sun et al. (2016)
		GC-TOF-MS	PCA	Vanillic acid, methionine, putrescine, proline, histidine, tryptophan	Witt et al. (2012)
	Pinus pinaster	GC/MS, LC/MS	HCA, O-PLS-DA	Fructose, glucose, rhamnose, raffinose, pyruvate, ornithine, glutamine	de Miguel et al. (2016)
	Atriplex halimus	ESI-TOF-MS	PCA	Malonate, fumarate, hydroxypyruvate, aspartate, eutypine choline sulfate	Alla et al. (2012)
	Physcomitrella patens	GC-quadrupole MS	AMDIS, GMD	Fructose, glucose, maltose, galactonic acid, galactinol, myoinositol, asparagine, threonine	Erxleben et al. (2012)
	N. tabacum	GC/MS, LC/MS	PCA	Mannitol, trehalose, galactinol, myoinositol, GABA	Rabara et al. (2015)
	Brachypodium distachyon	GC/MS	PCA, HCA	Glucose, mannobiose, maltose, sucrose, glycerol, norvaline, galactose	Shi et al. (2015)
	Lentils	GC/EI-TOF-MS	PCA, GMD	Putrescine, cadaverine, butanoic acid, erythronic acid, isocitric acid, nicotinic acid	Muscolo et al. (2015)
	Vitis vinifera	SPME-GC-MS	PCA, PLS-DA	Citric acid, 2-methyl-butanal phenyl- acetaldehyde, kaempferol-3-O-glucoside, quercetin-3-O-glucoside	Griesser et al. (2015)
	V. vinifera	GC/MS	PCA, AMDIS	Caffeic acid, proline, shikimic acid, malate, trehalose	Grimplet et al. (2009)
Heat stress	*Solanum lycopersicum*	GC/MS	PCA, mass spectral libraries	Alanine, allantoin, myoinositol, putrescine, rhamnose	Luengwilai et al. (2012)
	S. lycopersicum	HPLC	–	Polyamines	Cheng et al. (2009)
	G. max	GC/MS, LC/MS, UPLC	PCA, mass spectral libraries	Apigenin, glycitein, genistein, naringenin-7-O- glucoside, ferulate	Chebrolu et al. (2016)
	Arabidopsis thaliana	GC-MS	PCA	Asparagine, leucine, isoleucine, fumarate, malate	Kaplan et al. (2004)
Heavy metal	*S. lycopersicum*	^{1}H-NMR, HPLC-PDA	MeRy-B	Asparagine, GABA, choline, isoleucine, proline, trigonelline	Hediji et al. (2010)
	Drimia elata	HPLC, LC-MS	–	Glutamate, hydroxylysine, valine, histidine, proline, methionine	Begum et al. (2016)
	Sedum alfredii	GC-MS	PCA, O-PLS-DA, MET-IDEA, GMD	Oleic acid, 1-monohexadecanoyl-glycerol, 9-hexadecenoic acid	Luo et al. (2017)

ABA, abscisic acid; *AMDIS*, automated mass spectral deconvolution and identification system; *AMIX*, analysis of mixture software package; *CE*, capillary electrophoresis; *DA*, discriminant analysis; *EI*, electron ionization/impact; *FELIX*, free electron laser for infrared experiments; *FLD*, fluorescence detector; *GABA*, γ-aminobutyric acid; *GC*, gas chromatography; *GMD*, golm metabolome database; *HCA*, hierarchical cluster analysis; *HPLC*, high-performance liquid chromatography; *HS*, headspace; *MET-IDEA*, metabolomics ion-based data extraction algorithm; *MS*, mass spectrometry; *NIST*, National Institute of Standards and Technology; *NMR*, nuclear magnetic resonance; *O-PLS*, orthogonal projections to latent structures; *PCA*, principal component analysis; *PLS*, projection to latent structure; *QSEQ*, sequential quadrature detection; *SIMCA*, soft-independent modeling of class analogy; *SPME*, solid-phase microextraction; *TOCSY*, total correlation spectroscopy; *TOF*, time-of-flight; *UPLC*, ultra-performance liquid chromatography.

and soft-independent modeling of class (SIMC). The metabolomic alterations in responses to salinity have been compared between extremophile *Lotus creticus* and glycophyte *Lotus corniculatus* and this study suggests that some genotypic features are responsible for the extremophile nature of *L. creticus* rather than metabolic changes (Sanchez et al., 2011). The heavy metals such as cadmium (Cd), lead (Pb), zinc (Zn), nickel (Ni), and chromium (Cr) can induce enzyme inhibition, cellular perturbations, altered metabolism, physiology, and in extreme cases death of plants. Metabolite changes in response to heavy metals has been studied in *B. rapa,* which is known as a metal accumulator (Jahangir et al., 2008a). ^{1}H-NMR, 2D-J-resolved spectra analyzed with PCA and PLS-DA suggest that the higher accumulations of primary metabolites such as carbohydrates and amino acids along with secondary metabolites such as glucosinolates and hydroxycinnamic acids occur in *B. rapa* in response to heavy metal stress (Jahangir et al., 2008a). Similarly, *Arabidopsis* treated with Cd shows the higher accumulations of alanine, proline, serine, putrescine, sucrose, and compatible solutes like GABA, raffinose, and trehalose (Sun et al., 2010). Nutritional deprivation especially limitations of macronutrients such as carbon (C), nitrogen (N), phosphorus (P), and sulfur (S) have direct impact on the growth and physiology of the plants since organic molecules are composed of these elements. Nitrogen is required for the synthesis of amino acids and nucleotides, which are the building blocks of proteins and nucleic acids respectively. The metabolomics changes in response to nitrogen deficiency have been studied in tomato plant using GC-MS approach (Urbanczyk-Wochniak and Fernie, 2005). This study suggests that the levels of malate, fumarate, and most of the amino acids are significantly reduced in tomato leaves following nitrogen deficiency. The level of TCA cycle intermediate like 2-oxoglutarate which is a key regulator of nitrogen and carbon metabolism also decrease in tomato leaf under nitrogen deficiency condition (Urbanczyk-Wochniak and Fernie, 2005). The metabolomics responses under nitrogen deficiency have also been investigated in *Arabidopsis* using GC-MS approach (Tschoep et al., 2009).

5.2 Metabolomics in Deciphering Biotic Stress Tolerance Mechanisms of Plants

Metabolomics has also been implemented for the study of metabolite alterations in response to pathogen infections in resistant and susceptible crop varieties. The metabolomics studies demonstrate significant variations in hormones, as well as primary and secondary metabolites in plants in response to pathogen infections. As the plants are rich in metabolites, they provide nutrient-rich environments for the microbial pathogens, symbionts as well as herbivorous insect grazers. Therefore the plants must develop a defensive mechanism to protect themselves from infectious pathogens or develop a symbiotic relationship by sharing important metabolites. There are many ways of interactions between plants and the interacting species and depending on the way of interaction the metabolic alterations occurs in plants. Therefore plant–host interactions are extremely important in terms of both metabolite richness and alterations associated with the interaction. Previously, the metabolite alterations in response to plant–pathogen interaction mainly focus on the implementation of ^{1}H NMR as a fingerprinting technique along with some targeted metabolite identification approaches (Allwood et al., 2008). Several studies have also been carried out for elucidating the metabolic alterations in plants in response to various fungal infections. A number of secondary metabolites including hormones and polyphenolic compounds play important roles in plants in response to fungal infection. As compared to other metabolites, the phytohormones play vital role because they can easily circulate throughout the plant parts and even minute concentration of phytohormones can regulate the growth and development of plants upon pathogen infection. The plant hormones such as auxin, gibberellin, cytokinin, abscisic acid, and ethylene play important roles in defense mechanism against various biotic and abiotic stresses. Among the phytohormones salicylic acid specifically has been found to be involved in the defense response of plant against various pathogenic infections and also involved in the induction of pathogen-associated genes. In addition to plant hormones, the active roles of phenolics in pathogen defense have also been demonstrated in the literature. Either the antifungal phenolics are exist in plants before infection and distributed upon pathogen attack or their biosynthesis is induce by the interaction of pathogens. A number of metabolomics studies have been conducted in plants for the elucidation of the metabolite changes in plants upon fungal and pathogen attack. Mass spectroscopy approaches have been used to study the metabolite alterations in plants in response to *Fusarium* head blight (FHB) infection in barley genotypes (Kumaraswamy et al., 2011; Bollina et al., 2011). The metabolite alteration in wheat in response to FHB infection have also been reported (Cuperlovic-Culf et al., 2016; Gunnaiah et al., 2012). In addition to MS approaches NMR study has also been conducted for screening the FHB resistance in wheat (Browne and Brindle, 2007). The volatile organic compounds such as phenolic acids and ferulic acid are abundant in wheat bran infected with FHB. Solanaceous plants, especially *Solanum lycopersicum,* has been proved to be an excellent model for the study of plant–pathogen interaction. Although it is most widely used vegetable worldwide it often suffers pathogen attack from a number of pathogenic agents including viruses, bacteria, and fungi (Arie et al., 2007). The comparison of the metabolic profiles of the infected and control plants provides access to the metabolic pathway responsible for the plant–pathogen interaction (López-Gresa et al., 2010). Metabolite alterations in tomato in response to two different infecting agents such as *Citrus exocortis* viroid, which causes plant stunting, extreme leaf epinasty and rugosity, and *Pseudomonas syringae,* which causes rapid necrosis, have been investigated using NMR spectroscopy and LC-MS in combination with various multivariate data analysis such as PCA, PLS-DA, and soft-independent modeling of class analogy (SIMCA) (López-Gresa et al., 2010). A number of metabolites have been discriminated between control and infected plants. However, no significant differences have been observed between infected and control plants after 48 h of infections. This study suggests that the plants responds differently in the later stages of infection (López-Gresa et al., 2010). The constitutive accumulation of antimicrobe metabolites known as phytoanticipins, which includes glycosides and glucosinolates, is considered as plant's defense metabolism. One of the defense metabolism of the plants in response to pathogens is the changes in primary metabolites by converting sugar importers to exporters. Rapid defense responses to pathogen infections in terms of secondary metabolism includes induced biosynthesis of defense hormones like salicylic acid and jasmonate, changes in the antioxidant status, biosynthesis of flavonoids, phenylpropanoids, and hydroxycinnamic acids (Allwood et al., 2010). Metabolic interaction between plant and pathogen mainly involves (1) constitutive metabolism that occurs in the cells of both plant and pathogen, (2) metabolite alterations that occur either in the plant or in the pathogen, as well as the metabolite exchange between the interacting organisms, and (3) excretion of the metabolites from both the interacting partners, and the excreted metabolites are not internalized and remain in the extracellular medium (Allwood et al., 2010). Metabolomics techniques such as NMR, GC-MS/MS, and LC-MS/MS have been implemented to assess the response of Hwacheong rice cultivar in response to infection by rice blast fungal pathogen *Magnaporthe grisea. M. grisea* is one of the most destructive pathogen for cultivated rice, and it also hampers the cultivation of wheat, barley, rye, and millet (Jones et al., 2011). From this study it has been concluded that *M. grisea* induces changes in the metabolite profile of rice. NMR study has been used predominantly to prove that metabolomics can be a major approach for the assessment of the plant–insect interaction. The plant response to *Frankliniella occidentalis* commonly known as flower thrips has been studied using the NMR approach (Leiss et al., 2011). Multivariate statistical analyses such as PCA and PLS-DA have been employed to process the data obtained from 1D and 2D NMR for the metabolomic profiling of thrips-resistant and thrips-sensitive plants for the identification of key metabolites responsible for thrips resistance (Leiss et al., 2011). Thrips resistance in the crossbreeds of *Senecio jacobeae* and *Senecio aquaticus* has also been investigated using NMR, and this study confirms that the pyrrolizidine alkaloids such as jacobine and jaconine at higher concentrations were found to be the key compounds responsible for the resistance (Leiss et al., 2009). Apart from the pyrrolizidine alkaloids flavonoids like kaempferol are also known to be involved in plant resistance against herbivores such as caterpillars, aphids, and fungal pathogens (Leiss et al., 2011). Leiss et al. (2011) have shown that resistant cultivar of hybrids of *S. jacobeae* and *S. aquaticus* exhibits higher amounts of phenols including chlorogenic and feruloyl quinic acid. Chlorogenic acid is known to exert negative effects on different caterpillar species due to its antifeedant and digestibility reducer activities. The metabolite alterations of wild and cultivated varieties of tomato species with respect to thrips resistance has been compared. The cultivated tomato has been shown to be more susceptible to flower thrips when compared with the wild variety (Mirnezhad et al., 2010). In addition to the metabolomic study on thrips resistance, effects of different larval instars *Plutella xylostella* and *Spodoptera exigua* on *B. rapa* have also been studied. Different larval instars induce different metabolite alterations. Important metabolites that are induced due to insect attack are glucose, ferulic acid, and gluconapin in the second larval instar whereas alanine and sinapoyl malate are induced in the fourth larval instar. Induction of phenylpropanoid and indole alkaloid biosynthesis–related metabolites have been reported in *Catharanthus roseus* infected with phytoplasma (Choi et al., 2004). The metabolite response also varies with the type of bacterial species infecting the plants. For instance, in *B. rapa* threonine and GABA have been detected as the discriminating metabolites in the samples infected with Gram-positive bacteria, and on the other hand, induced levels of sinapoyl-malate, caffeoyl-malate, and histidine have been observed in the samples infected with Gram-negative bacteria (Jahangir et al., 2008b). The effects of the tumor-induced strain of *Agrobacterium tumefaciens* on *B. rapa* have been studied showing elevated levels of phenylpropanoid, flavonoid, sugars, and free amino/organic acid contents (Simoh et al., 2009). Metabolite alterations in *B. rapa* with respect to various fungal pathogen infections such as *Leptosphaeria maculans, Fusarium oxysporum,* or *Aspergillus niger* have been reported and this study suggest the presence of phenylpropanoids, glucosinolates, and flavonoids for antifungal defense (Abdel-Farid et al., 2009). Differentiation of pathogen susceptibility of different cultivars has also been carried out on the basis of metabolite composition, e.g., differentiation between grapevine varieties is based on quercetin-3-O-glucoside and a trans-feruloyl derivative (Ali et al., 2009).

5.3 Metabolomics in Characterization of Herbal Medicines and Natural Products

Herbal medicines have been used as an important method for disease treatment from ancient times. One important feature of herbal medicines is that, they are either extracted from a single herb or may be an amalgamation of several herbs as a combined formula. While evaluating the quality and authenticity of the herbs or herbal medicines, one or two biologically active compounds are considered for the identification and quantification in multicomponent herbal preparations.

The factors like harvest season, habitat, and drying processes may affect the chemical constituents of the herbal extract. Therefore to ensure the reliability and repeatability of the herbal products, it is necessary to check and understand the bioactivities and possible side effects of the product (Gad et al., 2013). The identification and authentication of the plant source for the preparation of herbal medicines is necessary to improve the efficacy of herbal products. Metabolomics has wide implications in biomedicine, nutritional assessment, plant research and drug discovery (Suhre et al., 2013; Gad et al., 2013). Various morphological, genetic, or metabolomics approaches have been employed in natural medicine research (Wen et al., 2010). Metabolomics is the study, mainly concerning the changes of small molecular weight metabolites in a given sample influenced by environmental or endogenous factors (Patti et al., 2012). Current advanced analytical platforms require high cost sophisticated instruments and skills for metabolomics analysis. Therefore we need a simple, fast, and low-cost analytical platform for easing the metabolomics study for the development and production of herbal medicines in developing countries like India (Hirayama et al., 2014). Along with the many preexisting metabolomics techniques, a new concept was implemented in the field of medicine (herbal medicine) lately known as smartphone metabolomics. In biomedical fields, many smartphone-based platforms have been reported for biomarker detection and infectious disease diagnosis. The smartphone-based platforms are now popular for their multifunctionality and wide distribution, user-friendly interface, and they are an economic analytical platforms (Oncescu et al., 2013; Laksanasopin et al., 2015; Kwon et al., 2016). The characterization of the medicinal plant *Schisandra sinensis* has been done by using the smartphone metabolomics approach (Kwon et al., 2016). Metabolomic techniques and various chemometric approaches that have been used in the characterization of the herbal product as well for the differentiation of herbal species are given in Table 1.4.

TABLE 1.4 Applications of Various Metabolomic Approaches in Characterization of Herbal Medicines

Plant Species	Plant Parts Used	Method of Analysis	Chemometric Methods	Key Compounds	References
Aconitum kusnezoffii	Roots	HPLC-DAD	SA, HCA	Glucose, fructose, and various amino acids	Zhao et al. (2009)
Artemisia selengensis	Whole plant	HPLC-DAD	SA, PCA, HCA	Proline, alanine, trehalose, fructose, GABA	Peng et al. (2011)
Atractylis chinensis	Rhizomes	HPLC	PCA, k-NN, LDA	Various metal ions	Ni et al. (2008)
Bupleurum ssp.	Roots	HPLC-ELSD, HPTLC	SA, SIMCA, ANNs, k-NN	Saikosaponin	Tian et al. (2009)
Camellia sinensis	Leaves	RP-HPLC sensory techniques	PLS, PCA-ANNs	Caffeine	Chen et al. (2010a)
	Leaves	FT-NIR	PLS	Caffeine	Sinija and Mishra (2009)
Cinnamomum cassia	Bark and twig	HPLC	SA, PLS-DA	Coumarin, cinnamaldehyde, cinnamic acid, eugenol, cinnamyl alcohol	Ding et al. (2011)
Coptis chinensis	Roots	UPLC-PAD	SA, HCA, PCA	Jatrorrhizine, epiberberine, coptisine, palmatine	Kong et al. (2009)
Curcuma kwangsiensis	Roots	HPLC, GC-MS	PCA, LDA, BP-ANN, LS-SVM	Caryophyllene, azulene, benzofuran, neocurdione	Ni et al. (2012)
Epimedium wushanense	Leaves	HPLC	PCA, HCA, SIMCA, BP- ANN	Sagittatoside, epimedokoreanoside, icariin, quercetin	Wang et al. (2012)
Ganoderma lucidum	Whole plant	HPLC	SA, PCA, HCA, SIMCA, PLS-DA	Saponins, triterpene	Chen et al. (2008b)

Continued

TABLE 1.4 Applications of Various Metabolomic Approaches in Characterization of Herbal Medicines—cont'd

Plant Species	Plant Parts Used	Method of Analysis	Chemometric Methods	Key Compounds	References
Labisia pumila	Leaves	UPLC-ESI-MS/MS	PCA, PCA-DA	Quercetin, catechin, epigallocatechin, kaempferol	Chua et al. (2011)
Mallotus sp.	Whole plant	HPLC-MS	PCA, PLS	Myricetin, quercetin, flavonoids, tannins	Nguyen Hoai et al. (2009)
Panax notoginseng	Roots	UHPLC-ESI-MS	PCA, PLS-DA	Dencichine, panaxatriol saponins, panaxadiol saponin	Toh et al. (2010)
Peganum harmala	Seeds	HPLC-DAD	SA, PCA, HCA, LDA	Deoxyvasicinone, deoxyvasicine, vasicinone, harmalol	Cheng et al. (2010)
Pericarpium citri reticulatae	Fruits	HPLC-DAD	PCA, PP, HCA, PLS-DA	TCA cycle intermediates, various amino acids, specially proline	Tistaert et al. (2011)
Polygala japonica	Whole plant	HPLC-DAD-ELSD	PCA	Sagittatoside, quercetin, icariin	Wang et al. (2010)
Pueraria lobata	Roots	RRLC-DAD	PCA	Puerarin, daidzin, genistein, daidzein	Zhao et al. (2011)
Rehmannia glutinosa	Roots	UHPLC-ESI-MS	PCA, O-PLS-DA	Rehmapicroside, jionoside, leucosce ptoside, forsythoside	Li et al. (2010)
Aristolochia	Roots	GC-MS	PCA	Camphene, trans-nerolido, viridiflorene, α-phellandrene	Francisco et al. (2008)
C. kwangsiensis	Rhizomes	GC-MS	PCA, PLS-DA	Terpinolene, eucalyptol, β-linalool, isobornyl acetate	Xiang et al. (2011)
Portulaca oleracea	Whole plant	GC-MS	SA, HCA	n-Hexadecanoic acid, eicosyne, cedrol, myristic acid	Zhu et al. (2010)
Teucrium flavum subsp. *glaucum*	Leaves	GC and GC-MS	PCA, CA	β-Bisabolene, spathulenol, limonene, linalool	Djabou et al. (2011)
Thymus vulgaris	Leaves	GC and GC-MS	HCA	Terpinolene, camphene, sabinene, myrcene	Chizzola et al. (2008)
Shao-Fu-Zhu-Yu (herbal formulation)	Decoction	GC-MS	PCA	Eucalyptol, bergamiol, benzenepropanal, spathulenol	Su et al. (2008)
Echinacea purpurea	Roots and aerial parts	^{1}H-NMR	PCA	Cichoric acid, echinacoside, caffeic acid	Frederich et al. (2010)
Zea mays	Seeds	^{1}H-NMR	PCA, HCA, SIMCA	Caffeic acid, fumaric acid, niacin, syringic acid	Kuhnen et al. (2010)
Astragalus membranaceus	Roots	FTIR	PLS-DA	Menthone, piperitone, carvacrol	Zhang and Nie (2010)
Eleutherococcus senticosus	Roots	NIR	PCA, DA, SIMCA, PLS-DA	Oxalate, isocitrate, tryptophan, lysine itaconate	Lucio-Gutiérrez et al. (2011)
Hoodia gordonii	Stem	FT-NIR	O-PLS, PLS	Oxalate, tryptophan, proline, fumaric acid	Vermaak et al. (2010)

TABLE 1.4 Applications of Various Metabolomic Approaches in Characterization of Herbal Medicines—cont'd

Plant Species	Plant Parts Used	Method of Analysis	Chemometric Methods	Key Compounds	References
Lavandula angustifolia × L. latifolia	Whole plant	FTIR	PCA, PLS, PLS-DA	Camphor, borneol lavandulol, terpineol, linalyl acetate	Bombarda et al. (2008)
Mentha pulegium	Whole plant	FTIR	CDA	Limonene, menthone, piperitone, carvacrol	Kanakis et al. (2012)
Rosa laevigata	Fruits	NIR	PLS, UVE-PLS, UVE-GA-PLS	Polysaccharide	Yan et al. (2011)
Rhizoma corydalis	Rhizomes	NIR	LS-SVM, RBF-ANN, PLS-DA, k-NN	Galacturonate, maltose, shikimate, raffinose, aspartate	Lai et al. (2011)
Actinidia chinensis	Fruits	UV–visible	PCA, CA, LDA	Sinapic acid, luteolin, glycosylated flavonoids, quercetin	Sârbu et al. (2012)
Coffee	Roasted beans	UV–visible	PCA, SIMCA, LDA	Caffeine	Souto et al. (2010)
Achyrocline satureioides	Aerial parts	ICP-OES	PCA, CA, LDA	Mineral elements	Cantarelli et al. (2010)
Ayurvedic medicine	Formulation	ICP-OES, GFAAS	PCA, HCA	Mineral elements	Giacomino et al. (2011)

ANN, artificial neural networks; *BP-ANN*, back-propagation artificial neural network; *CA*, cluster analysis; *DA*, discriminant analysis; *DAD*, diode array detector; *ELSD*, evaporative light scattering detector; *ESI*, electrospray ionization; *FT-NIR*, Fourier-transform near infrared; *FTIR*, Fourier-transform infrared; *GA*, genetic algorithm; *GABA*, γ-aminobutyric acid; *GC*, gas chromatography; *GFAAS*, graphite furnace atomic absorption spectrometry; *HCA*, hierarchical cluster analysis; *HPLC*, high-performance liquid chromatography; *HPTLC*, high-performance thin layer chromatography; *ICP-OES*, inductively coupled plasma optical emission spectrometry; *k-NN*, k-nearest neighbor; *LDA*, linear discriminant analysis; *LS-SVM*, least-squares support vector machines; *MS*, mass spectrometry; *MS/MS*, tandem MS; *NMR*, nuclear magnetic resonance; *PAD*, photodiode array detector; *PCA*, principal component analysis; *PLS*, projection to latent structure; *PP*, projection pursuit; *RBF-ANN*, radial basis function–artificial neural networks; *RP*, reverse phase; *RRLC*, rapid resolution liquid chromatography; *SA*, similarity analysis; *SIMCA*, soft-independent modeling of class analogy; *TCA*, tricarboxylic acid; *UPLC*, ultrahigh-performance liquid chromatography; *UV*, ultraviolet; *UVE*, uninformative variable elimination.

Various chromatographic methods such as HPLC, GC, and CE are hyphenated with various spectroscopy techniques such as MS, NMR, infrared spectroscopy, and UV spectroscopy for taxonomic classification, for determination of geographical origin and quality assessment of different medicinal plants (Gad et al., 2013). Among the chromatographic methods for the quality assessment of herbal products, HPLC is the most popularly executed analytical technique. Many closely related herbal species have been differentiated based on their metabolite profiles, and for this HPLC coupled with different analytical techniques has been implemented. The primary bioactive compounds of *Radix bupleuri* (saikosaponins) have been identified using high-performance thin layer chromatography and HPLC-evaporative light scattering detector (Tian et al., 2009). HPLC-MS technique has been employed for the quality assessment of *Mallotus* spp. and this technique also differentiates different species of *Mallotus* collected from different biological origin (Tistaert et al., 2011). HPLC fingerprinting has also been employed for the quality comparison and component analysis of cinnamon bark and cinnamon twigs (Ding et al., 2011). Phenolic and flavonoid compounds of *Labisia pumila* have been determined by the application of UHPLC-ESI-MS/MS method (Chua et al., 2011). Simultaneous identification and quantification of multiconstituent traditional Chinese medicine has been achieved by combined approach of LC and LTQ Orbitrap (LC-LTQ-Orbitrap) (Peng et al., 2011). Applications of GC-MS for quality determination of herbal medicine and herbal species has been well documented (Vagionas et al., 2007). The combined platform of chemometric analysis and various spectroscopic analysis has been implemented effectively for herbal medicines substantiation. GC-MS along with various chemometric techniques has been successfully applied for the quality assessment of some herbal medicines. Along with the instrumentation, software development has also been carried out to get a better resolution of volatile chemical ingredients of herbs by peak alignment. To get a comprehensive understanding of metabolite profile of herbal species, NMR has proved to be a powerful tool along with other chromatographic and MS methods. Differentiation of closely related species, quality assessment of herbal drugs, and discrimination based on geographic isolation have also been achieved by the implementation of NMR along with various multivariate chemometric analyses. Apart from this, various vibrational spectroscopy methods have been implemented since 2010 for better assessment of the quality of herbal drugs and various natural products. The potential implication of various vibrational spectroscopy methods such as attenuated total reflectance-Fourier-transform infrared(FTIR), mid-infrared, FT-Raman , and near-infrared (NIR) spectroscopy for the qualitation of many herbal species has been carried out and well documented (Vlachos et al., 2006; Kokalj et al., 2011). There has been proficient implementation of infrared spectroscopy in the field of natural products for differentiation of various plants species based on their chemotypes. These studies are useful for the prediction of the contents of essential metabolites with potential use in herbal medicines, especially in teas along with other herbal plant species. Quality assessment of green tea has been done by NIR combined with PLS and SIMCA to check the content of essential metabolites such as caffeine, epigallocatechin gallate, and epicatechin polyphenols for evaluation of the antioxidant potential of green tea (Sinija and Mishra, 2009). Most of the metabolomics studies on the herbal medicines are based on the absorption spectra of the metabolites having conjugated or aromatic systems such as flavonoids, lignins, and polyphenolics. UV absorption is employed for the characterization of terpenoids, because they do not have light-absorbing chromophores. However, identifications of flavonoids, lignins, and coumarins within the UV region are more advantageous than that of terpenoids, because of easy, simple and inexpensive procedures with high reproducibility (Gad et al., 2013). The determination of total polyphenol content using UV–Visible spectrophotometry with cluster analysis and PCA has been first applied in *Plantago* spp. to establish a relationship among different taxa of the genus (Grubešić et al., 2005).

The data mining approaches such as PCA and SIMCA have been implemented for the characterization of herbal medicines. The antioxidant potential of green tea has been assessed by using different metabolomics approaches combined with various statistical analyses. The chromatographic fingerprinting along with various regression models has been implemented for the determination of the antioxidant potential of green tea (Dumarey et al., 2008). A comparison of different regression models suggests that PLS and uninformative variable elimination PLS (UVE-PLS) are more suitable for the assessment of antioxidant capacity rather than the multiple linear regression (MLR) and principal component regression (PCR). UVE-PLS is considered as the most advanced extension of PCA. Similarly, discrimination between two closely related species of *Pericarpium citri* has been done by HPLC and PLS-DA, and various secondary metabolites such as hesperidin, nobiletin, and tangeretin have been identified as potential biomarkers in *Pericarpium* spp. (Yi et al., 2009). HCA analysis of various alkaloids has been done for the chemotaxonomic differentiation of *Aconitum kusnezoffii* from its adulterants (Zhao et al., 2009). The unique combined approach of HPLC-diode array detector and GC-MS has been implemented along with PCA for the analysis of complex Chinese medicine *Rhizoma Curcumae,* which is usually represented by three common types. Similarly, for the assessment of the phytochemical diversity along with variability of saponins obtained from various parts of species of ginseng UHPLC-ESI-MS has been used, and the variations have been identified using PCA plots (Dan et al., 2008). In the same way, quality assessment of *Atractylis chinensis* has been done by metabolite profiling using HPLC (Gad et al., 2013). Optimization of the extraction protocol by correlating steaming duration and production of bioactive ginsenosides in *Panax notoginseng* has been done using UHPLC-TOF-MS coupled with PCA and PLS-DA (Toh et al., 2010). Similarly, chemical biomarker–based discrimination has been carried out in *Radix Rehmanniae* (Li et al., 2010). Likewise, classification of *Artemisia selengensis* has been done using PCA and HCA, based on the determination of rutin as one of the active compound, and the results were confirmed by HCA analysis (Peng et al., 2011). Cinnamaldehyde is considered as the most abundant biomarker in cinnamon affirmed by PCA and PLS-DA analysis, which allowed to have a good discrimination between bark and twig samples (Ding et al., 2011). Qualitation and quantitation studies of *Lavandula angustifolia* for essential oils and various essential metabolites have been done by FTIR coupled with PCA, PLS, and PLS-DA (Bombarda et al., 2008). Combination of metabolomics approaches such as NIR with various chemometric approaches such as PCA, PLS-DA, and DA has been used for geographic discrimination of *Ganoderma lucidum.*

5.4 Role of Metabolomics in Chemotaxonomy and Differentiation of Ecotypes

Metaboliteo profiling is one of the emerging tool for the chemotaxonomic differentiation of various plant species. A number of reports have been documented lately describing metabolomics as a potent tool for speciation study (Bino et al., 2004; Kim et al., 2012; Saito and Matsuda, 2010). Among the techniques employed, metabolomics has proved to be a powerful and potent tool for the authentication and identification of the species, even if they belong to same species but have different geographical origin (Wen et al., 2011). For instance, NMR with various chemometric analyses have been practiced for the discrimination of *Artemisia annua* samples of different origin. The major compound that is taken for the consideration is artemisinin, which is used for geological discrimination depending upon the antiplasmodial activity. PCA has been effectively used for the analysis of HPLC finger printing data, for the taxonomic discrimination of the herbal species from other members of the same species. Such approaches have been applied to differentiate *Angelica sinensis* from other members of Apiaceae (Lu et al., 2005). GC-MS along with PCA score plot has been used for the quality assessment and discrimination of different ecotypes of *Curcuma*. The metabolites contributing to the discrimination between ecotypes are mainly curzerenone, germacrone, curdione, and epicurzerenone which are identified from PCA and PLS-DA loading plots (Xiang et al., 2011). Similarly, for the discrimination of *G. lucidum* from different cultivation sources ^{1}H-NMR spectroscopy with OPLS-DA have been implemented (Wen et al., 2010). The geographical discrimination of Greek *Mentha pulegium* have been carried out by implementing FTIR along with canonical discriminant analysis, and this differentiation depends mainly on the volatile compounds like menthone, isomenthone, pulegone, and piperitone (Kanakis et al., 2012). Apart from the aforementioned techniques, various other approaches such as least-squares support vector machines, radial basis function–artificial neural networks, PLS-DA, and k-nearest neighbor methods have been used for the classification of *Rhizoma corydalis*. However, no significant statistical variations has been found between these four methods (Lai et al., 2011). Discrimination analysis between *Eleutherococcus senticosus* and other herbal species of Araliaceae family has been carried out based on spectral fingerprinting by NIR combined with other chemometric procedures such as PCA, PLS-DA, and DA (Lucio-Gutiérrez et al., 2011). Untargeted metabolomic approaches have been implemented for the chemotaxonomic assessment of *Salvia* sp. by using α-bisabolol as a potential metabolite marker (Sandasi et al., 2012). Chemotaxonomic classification and correlation with the genotypes of *Lespedeza* sp. has also been carried out using LC-ESI-MS (Kim et al., 2012). GC-MS fingerprinting were coupled with similarity analysis (SA) and HCA of *Portulaca oleracea* has been carried out for the ecotype differentiation of this species collected from different origins (Zhu et al., 2010). Use of PCA and HCA has also been documented in *Schizonepeta tenuifolia* for the differentiation of this species collected from different origins (Chun et al., 2011). GC-MS analysis followed by SA and PCA has been undertaken for the geographical speciation of *Scutellaria barbata* collected from various areas of China for discrimination from its adulterants (Pan et al., 2011). Coupling of NIR, DA, and PLS-DA has been applied for the geological discrimination of *Scutellaria baicalensis* based on the marker metabolites such as baicalin, wogonoside, baicalein, and wogonin (Li et al., 2011).

6. CONCLUSION AND FUTURE PROSPECTS

Advancement of the metabolomics techniques and their successful implementation enrich our knowledge about the metabolite composition of any plant species including herbal species, and the metabolic alteration under various environmental perturbations. To gain enduring knowledge about the mechanisms of plant's responses to various abiotic stresses, integrated application of multiple "omics" technologies is necessary. Linking the metabolomics data with other omics such as transcriptomics and proteomics provides better visualization of the metabolic pathways and metabolic compositions of plant species. Broader applications of metabolomics in plant science can deepen our knowledge about natural products, herbal medicines, and most importantly, plant's responses to various abiotic stresses. Continuous advancement of the metabolomics approaches will provide rapid and inexpensive assays for the development of stress-tolerant crops having higher nutritional values. Additionally, metabolomic approaches can enhance plant breeding by providing value addition to certain species and/or by gaining knowledge about the metabolite composition of herbal plant species. Apart from the instrumentation, the emerging fields of systems biology also aim to identify all the existing elements in a system (plant). In addition, they also determine the relationship between these elements under various environmental stresses. The ultimate aim of the integrated study is to gather information by implementing various computational methods and predict the behavior of the system under stress conditions. For the establishment of efficient correlation between various cellular components of a system and to determine the status of the system, analysis of as many cellular and phenotypic levels as possible is required. Therefore integrated data sets along with sophisticated mathematical algorithms are needed. The gap between the number of compounds analyzed by metabolomics and the quality of annotated compounds seems to be expanding, since the developed technologies have become more and more sophisticated, and often also specific for particular questions. As a consequence, the key challenges in metabolomics, as in all "omics" technologies will be to establish and combine methods and strategies that allow us to increase the number of measured and annotated compounds—this means qualitative, quantitative, and spatial resolution without losing sample throughput and quality control. Most of the studies on plant metabolomics have been carried out in the model species. However, to get maximum benefit from the advancement of metabolomics, it has to be implemented in the field of agriculture, expanding its utility in the crops indeed. The complex genetic makeup of the crops is considered as a major drawback with the crops. In spite of this, metabolomics has been successfully implemented for enhanced breeding of important crop species. Metabolomics is yet to be used in herbal medicines and drug discovery and applied in various herbal species. Metabolomics has proved its value; however, it still faces significant challenges including metabolite identifications to a great scale. Since continuous improvements in the metabolomic techniques are going on, these techniques will be much advanced in the coming years.

ACKNOWLEDGMENTS

The grant of SERB (SB/SO/PS-14/2014), DST, Government of India, New Delhi, to AKP is duly acknowledged. JR is thankful to the Council of Scientific and Industrial Research (CSIR), New Delhi, India, for financial support in the form of Junior Research Fellowship. The manuscript has been assigned CSIR-CSMCRI registration number PRIS 024/2017.

REFERENCES

Abate, S., Ahn, Y.G., Kind, T., Cataldi, T.R.I., Fiehn, O., 2010. Determination of elemental compositions by gas chromatography/time-of-flight mass spectrometry using chemical and electron ionization. Rapid Commun. Mass Spectrom. 24, 1172–1180. https://doi.org/10.1002/rcm.4482.

Abdel-Farid, I.B., Jahangir, M., van den Hondel, C.A.M.J.J., Kim, H.K., Choi, Y.H., Verpoorte, R., 2009. Fungal infection-induced metabolites in *Brassica rapa*. Plant Sci. 176, 608–615. https://doi.org/10.1016/j.plantsci.2009.01.017.

Alcázar, R., Bitrián, M., Bartels, D., Koncz, C., Altabella, T., Tiburcio, A.F., 2014. Polyamine metabolic canalization in response to drought stress in Arabidopsis and the resurrection plant *Craterostigma plantagineum*. Plant Signal. Behav. 6, 243–250. https://doi.org/10.4161/psb.6.2.14317.

Ali, K., Federica, M., Eva, Z., Martina, R., Young, H.C., Robert, V., 2009. NMR metabolic fingerprinting based identification of grapevine metabolites associated with downy mildew resistance. J. Agric. Food Chem. 57, 9599–9606. https://doi.org/10.1021/jf902069f.

Alla, M.M.N., Khedr, A.H.A., Serag, M.M., Abu-Alnaga, A.Z., Nada, R.M., 2012. Regulation of metabolomics in *Atriplex halimus* growth under salt and drought stress. Plant Growth Regul. 67, 281–304. https://doi.org/10.1007/s10725-012-9687-1.

Allwood, J.W., Clarke, A., Goodacre, R., Mur, L.A., 2010. Dual metabolomics: a novel approach to understanding plant-pathogen interactions. Phytochemistry 71, 590–597. https://doi.org/10.1016/j.phytochem.2010.01.006.

Allwood, J.W., Goodacre, R., 2010. An introduction to liquid chromatography-mass spectrometry instrumentation applied in plant metabolomic analyses. Phytochem. Anal. 21, 33–47. https://doi.org/10.1002/pca.1187.

Allwood, J.W., Ellis, D.I., Goodacre, R., 2008. Metabolomic technologies and their application to the study of plants and plant-host interactions. Physiol. Plant 132, 117–135. https://doi.org/10.1111/j.1399-3054.2007.01001.x.

Albinsky, D., Sawada, Y., Kuwahara, A., Nagano, M., Hirai, A., Saito, K., Hirai, M.Y., 2010. Widely targeted metabolomics and coexpression analysis as tools to identify genes involved in the side-chain elongation steps of aliphatic glucosinolate biosynthesis. Amino Acids 39, 1067–1075. https://doi.org/10.1007/s00726-010-0681-5.

Alonso, A., Rodriquez, M.A., Vinaixa, M., Tortosa, R., Correig, X., Julia, A., Marsal, S., 2014. Focus: a robust workflow for one-dimensional NMR spectral analysis. Anal. Chem. 86, 1160–1169. https://doi.org/10.1021/ac403110u.

Andronesi, O.C., Blekas, K.D., Mintzopoulos, D., Astrakas, L., Black, P.M., Tzika, A.A., 2008. Molecular classification of brain tumor biopsies using solid state magic angle spinning proton magnetic resonance spectroscopy and robust classifiers. Int. J. Oncol. 33, 1017–1025. https://doi.org/10.3892/ijo_00000090.

Aoki-Kinoshita, K.F., 2006. Overview of KEGG applications to omics-related research. J. Pestic. Sci. 31, 296–299. https://doi.org/10.1584/jpestics.31.296.

Arbona, V., Manzi, M., de Ollas, C., Gómez-Cadenas, A., 2013. Metabolomics as a tool to investigate abiotic stress tolerance in plants. Int. J. Mol. Sci. 14, 4885–4911. https://doi.org/10.3390/ijms14034885.

Arie, T., Takahashi, H., Kodama, M., Teraoka, T., 2007. Tomato as a model plant for plant-pathogen interactions. Plant Biotechnol. 24, 135–147. https://doi.org/10.5511/plantbiotechnology.24.135.

Banerjee, S., Mazumdar, S., 2012. Electrospray ionization mass spectrometry: a technique to access the information beyond the molecular weight of the analyte. Int. J. Anal. Chem. 2012, 1–40. https://doi.org/10.1155/2012/282574.

Begum, M.C., Islam, M.S., Monirul Islam, R.A., Parvez, A.H., Kabir, M.S., 2016. Biochemical and molecular responses underlying differential arsenic tolerance in rice (*Oryza sativa* L.). Plant Physiol. biochem. https://doi.org/10.1016/j.toxicon.2008.05.012.

Bergeron, S.J., Henry, I.D., Santini, R.E., Aghdasi, A., Raftery, D., 2008. Saturation transfer double-difference NMR spectroscopy using a dual solenoid microcoil difference probe. Magn. Reson. Chem. 46, 925–929. https://doi.org/10.1002/mrc.2275.

Bharti, S.K., Bhatia, A., Tewari, S.K., Sidhu, O.P., Roy, R., 2011. Application of HR-MAS NMR spectroscopy for studying chemotype variations of *Withania somnifera* (L.) Dunal. Magn. Reson. Chem. 49, 659–667. https://doi.org/10.1002/mrc.2817.

Bino, R.J., Hall, R.D., Fiehn, O., Kopka, J., Saito, K., Draper, J., Nikolau, B.J., Mendes, P., Roessner-Tunali, U., Beale, M.H., Trethewey, R.N., Lange, B.M., Wurtele, E.S., Sumner, L.W., 2004. Potential of metabolomics as a functional genomics tool. Trends Plant Sci. 9, 418–425. https://doi.org/10.1016/j.tplants.2004.07.004.

Bird, S.S., Sheldon, D.P., Gathungu, R.M., Vouros, P., Kautz, R., Matson, W.R., Kristal, B.S., 2012. Structural characterization of plasma metabolites detected via LC-electrochemical coulometric array using LC-UV fractionation, MS, and NMR. Anal. Chem. 84, 9889–9898. https://doi.org/10.1021/ac302278u.

Bollina, V., Kushalappa, A.C., Choo, T.M., Dion, Y., Rioux, S., 2011. Identification of metabolites related to mechanisms of resistance in barley against *Fusarium graminearum*, based on mass spectrometry. Plant Mol. Biol. 77, 355–370. https://doi.org/10.1007/s11103-011-9815-8.

Bombarda, I., Dupuy, N., Da, J.P.L., Van Gaydou, E.M., 2008. Comparative chemometric analyses of geographic origins and compositions of *Lavandin var. Grosso* essential oils by mid infrared spectroscopy and gas chromatography. Anal. Chim. Acta 613, 31–39. https://doi.org/10.1016/j.aca.2008.02.038.

Borisjuk, L., Rolletschek, H., Neuberger, T., 2013. Nuclear magnetic resonance imaging of lipid in living plants. Prog. Lipid Res. 52, 465–487. https://doi.org/10.1016/j.plipres.2013.05.003.

Bothwell, J.H.F., Griffin, J.L., 2011. An introduction to biological nuclear magnetic resonance spectroscopy. Biol. Rev. 86, 493–510. https://doi.org/10.1111/j.1469-185X.2010.00157.x.

Bowne, J., Bacic, A., Tester, M., Roessner, U., 2011. Abiotic Stress and Metabolomics. Annu. Plant Rev. 43 Biol. Plant Metabolomics, vol. 43, pp. 61–85. https://doi.org/10.1002/9781444339956.ch3.

Bowne, J.B., Erwin, T.A., Juttner, J., Schnurbusch, T., Langridge, P., Bacic, A., Roessner, U., 2012. Drought responses of leaf tissues from wheat cultivars of differing drought tolerance at the metabolite level. Mol. Plant 5, 418–429. https://doi.org/10.1093/mp/ssr114.

Brown, M., Dunn, W.B., Dobson, P., Patel, Y., Winder, C.L., Francis-McIntyre, S., et al., 2009. Mass spectrometry tools and metabolite-specific databases for molecular identification in metabolomics. Analyst 134, 1322–1332. https://doi.org/10.1039/b901179j.

Browne, R.A., Brindle, K.M., 2007. 1H NMR-based metabolite profiling as a potential selection tool for breeding passive resistance against Fusarium head blight (FHB) in wheat. Mol. Plant Pathol. 8, 401–410. https://doi.org/10.1111/j.1364-3703.2007.00400.x.

Bruschweiler, R., Zhang, F.J., 2004. Covariance nuclear magnetic resonance spectroscopy. J. Chem. Phys. 120, 5253–5260. https://doi.org/10.1063/1.1647054.

Caspi, R., Billington, R., Ferrer, L., Foerster, H., Fulcher, C.A., Keseler, I.M., Kothari, A., Krummenacker, M., Latendresse, M., Mueller, L.A., Ong, Q., Paley, S., Subhraveti, P., Weaver, D.S., Karp, P.D., 2016. The MetaCyc database of metabolic pathways and enzymes and the BioCyc collection of pathway/genome databases. Nucleic Acids Res. 44, D471–D480. https://doi.org/10.1093/nar/gkv1164.

Castro, J., Pregibon, T., Chumanov, K., Marcus, R.K., 2010. Determination of catechins and caffeine in proposed green tea standard reference materials by liquid chromatography-particle beam/electron ionization mass spectrometry (LC-PB/EIMS). Talanta 82, 1687–1695. https://doi.org/10.1016/j.talanta.2010.07.054.

Catola, S., Marino, G., Emiliani, G., Huseynova, T., Musayev, M., Akparov, Z., Maserti, B.E., 2016. Physiological and metabolomic analysis of *Punica granatum* (L.) under drought stress. Planta 243, 441–449. https://doi.org/10.1007/s00425-015-2414-1.

Chambers, M.C., Maclean, B., Burke, R., Amodei, D., Ruderman, D.L., Neumann, S., Gatto, L., Fischer, B., Pratt, B., Egertson, J., Hoff, K., Kessner, D., Tasman, N., Shulman, N., Frewen, B., Baker, T., Brusniak, M.Y., Paulse, C., Creasy, D., Flashner, L., Kani, K., Moulding, C., Seymour, S., Nuwaysir, L.M., Lefebvre, B., Kuhlmann, F., Roark, J., Rainer, P., Detlev, S., Hemenway, T., Huhmer, A., Langridge, J., Connolly, B., Chadick, T., Holly, K., Eckels, J., Deutsch, E.W., Moritz, R.L., Katz, J.E., Agus, D.B., MacCoss, M., Tabb, D.L., Mallick, P., 2012. A cross-platform toolkit for mass spectrometry and proteomics. Nat. Biotechnol. 30, 918–920. https://doi.org/10.1038/nbt.2377.

Charlton, A.J., Donarski, J.A., Harrison, M., Jones, S.A., Godward, J., Oehlschlager, S., Arques, J.L., Ambrose, M., Chinoy, C., Mullineaux, P.M., Domoney, C., 2008. Responses of the pea (*Pisum sativum* L.) leaf metabolome to drought stress assessed by nuclear magnetic resonance spectroscopy. Metabolomics 4, 312–327. https://doi.org/10.1007/s11306-008-0128-0.

Chebrolu, K.K., Fritschi, F.B., Ye, S., Krishnan, H.B., Smith, J.R., Gillman, J.D., 2016. Impact of heat stress during seed development on soybean seed metabolome. Metabolomics 12, 1–14. https://doi.org/10.1007/s11306-015-0941-1.

Chen, Q., Zhao, J., Liu, M., Cai, J., Liu, J., 2008a. Determination of total polyphenols content in green tea using FT-NIR spectroscopy and different PLS algorithms. J. Pharm. Biomed. Anal. 46, 568–573. https://doi.org/10.1016/j.jpba.2007.10.031.

Chen, Y., Zhu, S.B., Xie, M.Y., Nie, S.P., Liu, W., Li, C., Gong, X.F., Wang, Y.X., 2008b. Quality control and original discrimination of *Ganoderma lucidum* based on high-performance liquid chromatographic fingerprints and combined chemometrics methods. Anal. Chim. Acta 623, 146–156. https://doi.org/10.1016/j.aca.2008.06.018.

Cheng, L., Zou, Y., Ding, S., Zhang, J., Yu, X., Cao, J., Lu, G., 2009. Polyamine accumulation in transgenic tomato enhances the tolerance to high temperature stress. J. Integr. Plant Biol. 51, 489–499. https://doi.org/10.1111/j.1744-7909.2009.00816.x.

Cheng, X.M., Zhao, T., Yang, T., Wang, C.H., Bligh, S.W.A., Wang, Z.T., 2010. HPLC fingerprints combined with principal component analysis, hierarchical cluster analysis and linear discriminant analysis for the classification and differentiation of *Peganum* sp. indigenous to China. Phytochem. Anal. 21, 279–289. https://doi.org/10.1002/pca.1198.

Chizzola, R., Michitsch, H., Franz, C., 2008. Antioxidative properties of *Thymus vulgaris* leaves: comparison of different extracts and essential oil chemotypes. J. Agric. Food Chem. 56, 6897–6904. https://doi.org/10.1021/jf800617g.

Choi, Y.H., Verpoorte, R., 2014. Metabolomics: what you see is what you extract. Phytochem. Anal. 25, 289–290. https://doi.org/10.1002/pca.2513.

Choi, Y.H., Tapias, E.C., Kim, H.K., Lefeber, A.W.M., Erkelens, C., Verhoeven, J.T.J., Brzin, J., 2004. Metabolic discrimination of *Catharanthus roseus* leaves infected by phytoplasma using 1H-NMR spectroscopy and multivariate data analysis. Plant Physiol. 135, 2398–2410. https://doi.org/10.1104/pp.104.041012.

Chua, L.S., Latiff, N.A., Lee, S.Y., Lee, C.T., Sarmidi, M.R., Aziz, R.A., 2011. Flavonoids and phenolic acids from *Labisia pumila (Kacip Fatimah)*. Food Chem. 127, 1186–1192. https://doi.org/10.1016/j.foodchem.2011.01.122.

Chun, M.H., Kim, E.K., Yu, S.M., Oh, M.S., Moon, K.Y., Jung, J.H., Hong, J., 2011. GC/MS combined with chemometrics methods for quality control of *Schizonepeta tenuifolia* Briq: determination of essential oils. Microchem. J. 97, 274–281. https://doi.org/10.1016/j.microc.2010.09.015.

Chikayama, E., Sekiyama, Y., Okamoto, M., Nakanishi, Y., Tsuboi, Y., Akiyama, K., Saito, K., Shinozaki, K., Kikuchi, J., 2010. Statistical indices for simultaneous large-scale metabolite detections for a single NMR spectrum. Anal. Chem. 82, 1653–1658. https://doi.org/10.1021/ac9022023.

Chylla, R.A., Hu, K., Ellinger, J.J., Markley, J.L., 2011. Deconvolution of two dimensional NMR spectra by fast maximum likelihood reconstruction: application to quantitative metabolomics. Anal. Chem. 83, 4871–4880. https://doi.org/10.1021/ac200536b.

Cloarec, O., Campbell, A., Tseng, L.-H., Braumann, U., Spraul, M., Scarfe, G., Weaver, R., Nicholson, J.K., 2007. Virtual chromatographic resolution enhancement in cryoflow LC-NMR experiments via statistical total correlation spectroscopy. Anal. Chem. 79, 3304–3311. https://doi.org/10.1021/ac061928y.

Cramer, G.R., Ergul, A., Grimplet, J., Tillett, R.L., Tattersall, E.A.R., Bohlman, M.C., Vincent, D., Sonderegger, J., Evans, J., Osborne, C., Quilici, D., Schlauch, K.A., Schooley, D.A., Cushman, J.C., 2007. Water and salinity stress in grapevines: early and late changes in transcript and metabolite profiles. Funct. Integr. Genomics 7, 111–134. https://doi.org/10.1007/s10142-006-0039-y.

Cuadros-Inostroza, Á., Caldana, C., Redestig, H., Kusano, M., Lisec, J., Peña-Cortés, H., Willmitzer, L., Hannah, M.A., 2009. TargetSearch - a Bioconductor package for the efficient preprocessing of GC-MS metabolite profiling data. BMC Bioinforma. 10, 428. https://doi.org/10.1186/1471-2105-10-428.

Cuperlovic-Culf, M., Rajagopalan, N., Tulpan, D., Loewen, M.C., 2016. Metabolomics and cheminformatics analysis of antifungal function of plant metabolites. Metabolites 6. https://doi.org/10.3390/metabo6040031.

Dan, M., Su, M., Gao, X., Zhao, T., Zhao, A., Xie, G., Qiu, Y., Zhou, M., Liu, Z., Jia, W., 2008. Metabolite profiling of *Panax notoginseng* using UPLC-ESI-MS. Phytochemistry 69, 2237–2244. https://doi.org/10.1016/j.phytochem.2008.04.015.

de Miguel, M., Guevara, M.Á., Sánchez-Gómez, D., de María, N., Díaz, L.M., Mancha, J.A., Fernández de Simón, B., Cadahía, E., Desai, N., Aranda, I., Cervera, M.T., 2016. Organ-specific metabolic responses to drought in *Pinus pinaster* Ait. Plant Physiol. Biochem. 102, 17–26. https://doi.org/10.1016/j.plaphy.2016.02.013.

De Vos, R.C., Moco, S., Lommen, A., Keurentjes, J.J., Bino, R.J., Hall, R.D., 2007. Untargeted large-scale plant metabolomics using liquid chromatography coupled to mass spectrometry. Nat. Protoc. 2, 778–791. https://doi.org/10.1038/nprot.2007.95.

DeSilva, M.A., Shanaiah, N., Nagana Gowda, G.A., Rosa-Pérez, K., Hanson, B.A., Raftery, D., 2009. Application of 31P NMR spectroscopy and chemical derivatization for metabolite profiling of lipophilic compounds in human serum. Magn. Reson. Chem. 47, 74–80. https://doi.org/10.1002/mrc.2480.

Dill, A.L., Ifa, D.R., Manicke, N.E., Ouyang, Z., Cooks, R.G., 2009. Mass spectrometric imaging of lipids using desorption electrospray ionization. J. Chromatogr. B Anal. Technol. Biomed. Life Sci. 877, 2883–2889. https://doi.org/10.1016/j.jchromb.2008.12.058.

Ding, Y., Wu, E.Q., Liang, C., Chen, J., Tran, M.N., Hong, C.H., Jang, Y., Park, K.L., Bae, K., Kim, Y.K., Kang, J.S., 2011. Discrimination of cinnamon bark and cinnamon twig samples sourced from various countries using HPLC-based fingerprint analysis. Food Chem. 127, 755–760. https://doi.org/10.1016/j.foodchem.2011.01.011.

Dixon, R.A., Gang, D.R., Charlton, A.J., Fiehn, O., Kuiper, H.A., Reynolds, T.L., Tjeerdema, R.S., Jeffery, E.H., 2006. Applications of metabolomics in agriculture. J. Agric. Food Chem. 54, 8984–8994. https://doi.org/10.1021/jf061218t.

Djabou, N., Battesti, M.J., Allali, H., Desjobert, J.M., Varesi, L., Costa, J., Muselli, A., 2011. Chemical and genetic differentiation of Corsican subspecies of *Teucrium flavum* L. Phytochemistry 72, 1390–1399. https://doi.org/10.1016/j.phytochem.2011.03.024.

Do, P.T., Degenkolbe, T., Erban, A., Heyer, A.G., Kopka, J., Köhl, K.I., Hincha, D.K., Zuther, E., 2013. Dissecting rice polyamine metabolism under controlled long-term drought stress. PLoS One 8. https://doi.org/10.1371/journal.pone.0060325.

Draper, J., Enot, D.P., Parker, D., Beckmann, M., Snowdon, S., Lin, W., Zubair, H., 2009. Metabolite signal identification in accurate mass metabolomics data with MZedDB, an interactive m/z annotation tool utilising predicted ionisation behaviour "rules.". BMC Bioinforma. 16, 1–16. https://doi.org/10.1186/1471-2105-10-227.

Dumarey, M., van Nederkassel, A.M., Deconinck, E., Vander Heyden, Y., 2008. Exploration of linear multivariate calibration techniques to predict the total antioxidant capacity of green tea from chromatographic fingerprints. J. Chromatogr. A 1192, 81–88. https://doi.org/10.1016/j.chroma.2008.03.052.

Dunn, W.B., Broadhurst, D.I., Atherton, H.J., Goodacre, R., Griffin, J.L., 2011. Systems level studies of mammalian metabolomes: the roles of mass spectrometry and nuclear magnetic resonance spectroscopy. Chem. Soc. Rev. 40, 387–426. https://doi.org/10.1039/b906712b.

Dunn, W.B., Erban, A., Weber, R.J.M., Creek, D.J., Brown, M., Breitling, R., Hankemeier, T., Goodacre, R., Neumann, S., Kopka, J., Viant, M.R., 2013. Mass appeal: metabolite identification in mass spectrometry-focused untargeted metabolomics. Metabolomics 9, 44–66. https://doi.org/10.1007/s11306-012-0434-4.

Ernst, M., Silva, D.B., Silva, R.R., Vêncio, R.Z.N., Lopes, N.P., 2014. Mass spectrometry in plant metabolomics strategies: from analytical platforms to data acquisition and processing. Nat. Prod. Rep. 31, 784–806. https://doi.org/10.1039/c3np70086k.

Erxleben, A., Gessler, A., Vervliet-Scheebaum, M., Reski, R., 2012. Metabolite profiling of the moss *Physcomitrella patens* reveals evolutionary conservation of osmoprotective substances. Plant Cell Rep. 31, 427–436. https://doi.org/10.1007/s00299-011-1177-9.

Evers, D., Lefevre, I., Legay, S., Lamoureux, D., Hausman, J.-F., Rosales, R.O.G., Marca, L.T., Hoffmann, L., Bonierbale, M., Schafleitner, R., 2010. Identification of drought-responsive compounds in potato through a combined transcriptomic and targeted metabolite approach. J. Exp. Bot. 61, 2327–2343. https://doi.org/10.1093/jxb/erq060.

Fan, T.W.M., Lane, A.N., 2011. NMR-based stable isotope resolved metabolomics in systems biochemistry. J. Biomol. NMR 49, 267–280. https://doi.org/10.1007/s10858-011-9484-6.

Farag, M.A., Porzel, A., Wessjohann, L.A., 2012. Comparative metabolite profiling and fingerprinting of medicinal licorice roots using a multiplex approach of GC-MS, LC-MS and 1D NMR techniques. Phytochemistry 76, 60–72. https://doi.org/10.1016/j.phytochem.2011.12.010.

Ferreira, D.L.S., Kittiwachana, S., Fido, L.A., Thompson, D.R., Escott, R.E.A., Brereton, R.G., 2010. Window consensus PCA for multiblock statistical process control: adaption to small and time-dependent normal operating condition regions, illustrated by online high performance liquid chromatography of a three-stage continuous process. J. Chemom. 24, 596–609. https://doi.org/10.1002/cem.1322.

Fiehn, O., Wohlgemuth, G., Scholz, M., Kind, T., Lee, D.Y., Lu, Y., Moon, S., Nikolau, B., 2008. Quality control for plant metabolomics: reporting MSI-compliant studies. Plant J. 53, 691–704. https://doi.org/10.1111/j.1365-313X.2007.03387.x.

Fletcher, J.S., Lockyer, N.P., Vickerman, J.C., 2011. Developments in molecular SIMS depth profiling and 3D imaging of biological systems using polyatomic primary ions. Mass Spectrom. Rev. 30, 142–174. https://doi.org/10.1002/mas.20275.

Forseth, R.R., Schroeder, F.C., 2011. NMR-spectroscopic analysis of mixtures: from structure to function. Curr. Opin. Chem. Biol. 15, 38–47. https://doi.org/10.1016/j.cbpa.2010.10.010.

Francisco, C.S., Messiano, G.B., Lopes, L.M.X., Tininis, A.G., de Oliveira, J.E., Capellari, L., 2008. Classification of *Aristolochia* species based on GC-MS and chemometric analyses of essential oils. Phytochemistry 69, 168–175. https://doi.org/10.1016/j.phytochem.2007.07.007.

Frederich, M., Jansen, C., De Tullio, P., Tits, M., Demoulin, V., Angenot, L., 2010. Metabolomic analysis of *Echinacea* spp. by ^{1}H nuclear magnetic resonance spectrometry and multivariate data analysis technique. Phytochem. Anal. 21, 61–65. https://doi.org/10.1002/pca.1156.

Frydman, L., Blazina, D., 2007. Ultrafast two-dimensional nuclear magnetic resonance spectroscopy of hyperpolarized solutions. Nat. Phys. 3, 415–419. https://doi.org/10.1038/nphys597.

Frydman, L., Lupulescu, A., Scherf, T.J., 2003. Principles and features of single-scan two-dimensional NMR spectroscopy. J. Am. Chem. Soc. 125, 9204–9217. https://doi.org/10.1021/ja030055b.

Fuchs, B., Süß, R., Nimptsch, A., Schiller, J., 2009. MALDI-tof-ms directly combined with TLC : a review of the current state. Chromatographia 69, 95–105. https://doi.org/10.1365/s10337-008-0661-z.

Fuchs, B., Süß, R., Schiller, J., 2010. An update of MALDI-TOF mass spectrometry in lipid research. Prog. Lipid Res. 49, 450–475. https://doi.org/10.1016/j.plipres.2010.07.001.

Fuchs, B., Schiller, J., 2009. Application of MALDI-TOF mass spectrometry in lipidomics. Eur. J. Lipid Sci. Technol. 111, 83–98.

Gad, H.A., El-Ahmady, S.H., Abou-Shoer, M.I., Al-Azizi, M.M., 2013. Application of chemometrics in authentication of herbal medicines: a review. Phytochem. Anal. 24, 1–24. https://doi.org/10.1002/pca.2378.

Gagneul, D., Aïnouche, A., Duhazé, C., Lugan, R., Larher, F.R., Bouchereau, A., 2007. A reassessment of the function of the so-called compatible solutes in the halophytic plumbaginaceae *Limonium latifolium*. Plant Physiol. 144, 1598–1611. https://doi.org/10.1104/pp.107.099820.

Gal, M., Schanda, P., Brutscher, B., Frydman, L., 2007. UltraSOFAST HMQC NMR and the repetitive acquisition of 2D protein spectra at Hz rates. J. Am. Chem. Soc. 129, 1372–1377. https://doi.org/10.1021/ja066915g.

Galhena, A.S., Harris, G.A., Nyadong, L., Murray, K.K., Fernández, F.M., 2010. Small molecule ambient mass spectrometry imaging by infrared laser ablation metastable-induced chemical ionization. Anal. Chem. 82, 2178–2181. https://doi.org/10.1021/ac902905v.

Gavaghan, C.L., Li, J.V., Hadfield, S.T., Hole, S., Nicholson, J.K., Wilson, I.D., Howe, P.W.A., Stanley, P.D., Holmes, E., 2011. Application of NMR-based metabolomics to the investigation of salt stress in maize (*Zea mays*). Phytochem. Anal. 22, 214–224. https://doi.org/10.1002/pca.1268.

Garcia, E., Andrews, C., Hua, J., Kim, H.L., Sukumaran, D.K., Szyperski, T., Odunsi, K., 2011. Diagnosis of early stage ovarian cancer by ^{1}H NMR metabonomics of serum explored by use of a microflow NMR probe. J. Proteome Res. 10, 1765–1771. https://doi.org/10.1021/pr101050d.

Gechev, T.S., Benina, M., Obata, T., Tohge, T., Sujeeth, N., Minkov, I., Hille, J., Temanni, M.R., Marriott, A.S., Bergstrom, E., homas Oates, J., Antonio, C., Mueller-Roeber, B., Fernie, A.R., Toneva, V., 2013. Molecular mechanisms of desiccation tolerance in the resurrection glacial relic *Haberlea rhodopensis*. Cell. Mol. Life Sci. 70, 689–709. https://doi.org/10.1007/s00018-012-1155-6.

Gleich, B., Weizenecker, J., 2005. Tomographic imaging using the nonlinear response of magnetic particles. Nature 435, 1214–1217.

Giacomino, A., Abollino, O., Malandrino, M., Karthik, M., Murugesan, V., 2011. Determination and assessment of the contents of essential and potentially toxic elements in Ayurvedic medicine formulations by inductively coupled plasma-optical emission spectrometry. Microchem. J. 99, 2–6. https://doi.org/10.1016/j.microc.2011.01.002.

Giavalisco, P., Kohl, K., Hummel, J., Seiwert, B., Willmitzer, L., 2009. ^{13}C isotope-labelled metabolomes allowing for improved compound annotation and relative quantification in liquid chromatography–mass spectrometry-based metabolomic research. Anal. Chem. 81, 6546–6551. https://doi.org/10.1021/ac900979e.

Grand-Guillaume Perrenoud, A., Veuthey, J.-L., Guillarme, D., 2012. Comparison of ultra-high performance supercritical fluid chromatography and ultra-high performance liquid chromatography for the analysis of pharmaceutical compounds. J. Chromatogr. A. 1266, 158–167. https://doi.org/10.1016/j.chroma.2012.10.005.

Griesser, M., Weingart, G., Schoedl-Hummel, K., Neumann, N., Becker, M., Varmuza, K., 2015. Severe drought stress is affecting selected primary metabolites, polyphenols, and volatile metabolites in grapevine leaves (*Vitis vinifera* cv. Pinot noir). Plant Physiol. biochem. 88, 17–26. https://doi.org/10.1016/j.plaphy.2015.01.004.

Grimplet, J., Wheatley, M.D., Jouira, H.B., Deluc, L.G., Cramer, G.R., Cushman, J.C., 2009. Proteomic and selected metabolite analysis of grape berry tissues under well-watered and water-deficit stress conditions. Proteomics 9, 2503–2528.

Grimes, J.H., O'Connell, T.M., 2011. The application of micro coil NMR probe technology to metabolomics of urine and serum. J. Biomol. NMR 49, 297–305. https://doi.org/10.1007/s10858-011-9488-2.

Grubešić, R.J., Vuković, J., Kremer, D., Vladimir-Knežević, S., 2005. Spectrophotometric method for polyphenols analysis: prevalidation and application on Plantago L. species. J. Pharm. Biomed. Anal. 39, 837–842. https://doi.org/10.1016/j.jpba.2005.05.004.

Goldbourt, A., 2013. Biomolecular magic-angle spinning solid-state NMR: recent methods and applications. Curr. Opin. Biotechnol. 24, 705–715. https://doi.org/10.1016/j.copbio.2013.02.010.

González-Domínguez, R., Sayago, A., Recamales, F., 2017. Direct infusion mass spectrometry for metabolomic phenotyping of diseases. Bioanalysis 9, 131–148.

Gowda, G.A.N., Tayyari, F., Ye, T., Suryani, Y., Wei, S., Shanaiah, N., Raftery, D., 2010. Quantitative analysis of blood plasma metabolites using isotope enhanced NMR methods. Anal. Chem. 82, 8983–8990. https://doi.org/10.1021/ac101938w.

Gu, H., Pan, Z., Xi, B., Asiago, V., Musselman, B., Raftery, D., 2011. Principal component directed partial least squares analysis for combining nuclear magnetic resonance and mass spectrometry data in metabolomics: application to the detection of breast cancer. Anal. Chim. Acta 686, 57–63. https://doi.org/10.1016/j.aca.2010.11.040.

Gunnaiah, R., Kushalappa, A.C., Duggavathi, R., Fox, S., Somers, D.J., 2012. Integrated metabolo-proteomic approach to decipher the mechanisms by which wheat qtl (Fhb1) contributes to resistance against Fusarium graminearum. PLoS One 7. https://doi.org/10.1371/journal.pone.0040695.

Haddad, R., Sparrapan, R., Eberlin, M.N., 2006. Desorption sonic spray ionization for (high) voltage-free ambient mass spectrometry. Rapid Commun. Mass Spectrom. 20, 2901–2905. https://doi.org/10.1002/rcm.2680.

Han, J., Datla, R., Chan, S., Borchers, C.H., 2009. Mass spectrometry-based technologies for high-throughput metabolomics. Bioanalysis 1, 1665–1684. https://doi.org/10.4155/bio.09.158.

Harada, T., Yuba-Kubo, A., Sugiura, Y., Zaima, N., Hayasaka, T., Goto-Inoue, N., Wakui, M., Suematsu, M., Takeshita, K., Ogawa, K., Yoshida, Y., Setou, M., 2009. Visualization of volatile substances in different organelles with an atmospheric-pressure mass microscope. Anal. Chem. 81, 9153–9157. https://doi.org/10.1021/ac901872n.

Harbourne, N., Marete, E., Jacquier, J.C., O'Riordan, D., 2009. Effect of drying methods on the phenolic constituents of meadowsweet (*Filipendula ulmaria*) and willow (*Salix alba*). LWT - Food Sci. Technol. 42, 1468–1473. https://doi.org/10.1016/j.lwt.2009.05.005.

Harrington, P.D.B., Vieira, N.E., Espinoza, J., Nien, J.K., Romero, R., Yergey, A.L., 2005. Analysis of variance-principal component analysis: a soft tool for proteomic discovery. Anal. Chim. Acta 544, 118–127. https://doi.org/10.1016/j.aca.2005.02.042.

Harris, G.A., Galhena, A.S., Fern, F.M., 2011. Ambient sampling/ionization mass spectrometry : applications and current trends. Anal. Chem. 83, 4508–4538. https://doi.org/10.1021/ac200918u.

Hagel, J.M., Facchini, P.J., 2008. Plant metabolomics: analytical platforms and integration with functional genomics. Phytochem. Rev. 7, 479–497. https://doi.org/10.1007/s11101-007-9086-9.

Hediji, H., Djebali, W., Cabasson, C., Maucourt, M., Baldet, P., Bertrand, A., Boulila, Z.L., Deborde, C., Moing, A., Brouquisse, R., Chaibi, W., Gallusci, P., 2010. Effects of long-term cadmium exposure on growth and metabolomic profile of tomato plants. Ecotoxicol. Environ. Saf. 73, 1965–1974. https://doi.org/10.1016/j.ecoenv.2010.08.014.

Henry, D., Park, G.H.J., Kc, R., Tobias, B., Raftery, D., 2008. Design and construction of a microcoil NMR probe for the routine analysis of 20-μL samples. Concepts. Magn. Reson. Part B 33B, 1–8. https://doi.org/10.1002/cmr.b.

Herrera, A., Fernández-Valle, E., Martínez-Álvarez, R., Molero, D., Pardo, Z.D., Sáez, E., Gal, M., 2009. Real-time monitoring of organic reactions with two-dimensional ultrafast TOCSY NMR spectroscopy. Angew. Chem. - Int. Ed. 48, 6274–6277. https://doi.org/10.1002/anie.200902387.

Hirayama, A., Wakayama, M., Soga, T., 2014. Metabolome analysis based on capillary electrophoresis-mass spectrometry. Trends Anal. Chem. 61, 215–222. https://doi.org/10.1016/j.trac.2014.05.005.

Hirai, M.Y., Klein, M., Fujikawa, Y., Yano, M., Goodenowe, D.B., Yamazaki, Y., Kanaya, S., Nakamura, Y., Kitayama, M., Suzuki, H., Sakurai, N., Shibata, D., Tokuhisa, J., Reichelt, M., Jonathan, R., 2005. Elucidation of gene-to-gene and metabolite-to gene networks in Arabidopsis by integration of metabolomics and transcriptomics. J. Biol. Chem. 280, 25590–25595. https://doi.org/10.1074/jbc.M502332200.

Hong, Y.-S., Martinez, A., Liger-Belair, G., Jeandet, P., Nuzillard, J.-M., Cilindre, C., 2012. Metabolomics reveals simultaneous influences of plant defence system and fungal growth in *Botrytis cinerea infected Vitis vinifera* cv. Chardonnay berries. J. Exp. Bot. 63, 695–709. https://doi.org/10.1093/jxb/err313.

Horai, H., Arita, M., Kanaya, S., Nihei, Y., Ikeda, T., Suwa, K., Ojima, Y., Tanaka, K., Tanaka, S., Aoshima, K., Oda, Y., Kakazu, Y., Kusano, M., Tohge, T., Matsuda, F., Sawada, Y., Yokota Hira, M., Nakanishi, H., Ikeda, K., Akimoto, N., Maoka, T., Takahashi, H., Ara, T., Sakurai, N., Suzuki, H., Shibata, D., Neumann, S., Iida, T., Tanaka, K., Funatsu, K., Matsuura, F., Soga, T., Taguchi, R., Saito, K., Nishiokaa, T., 2010. MassBank: a public repository for sharing mass spectral data for life sciences. J. Mass Spectrom. 45, 703–714. https://doi.org/10.1002/jms.1777.

Hormann, K., Müllner, T., Bruns, S., Höltzel, A., Tallarek, U., 2012. Morphology and separation efficiency of a new generation of analytical silica monoliths. J. Chromatogr. A. 1222, 46–58. https://doi.org/10.1016/j.chroma.2011.12.008.

Iijima, Y., Nakamura, Y., Ogata, Y., Tanaka, K., Sakurai, N., Suda, K., Suzuki, T., Suzuki, H., Okazaki, K., Kitayama, M., Kanaya, S., Aoki, K., Shibata, D., 2008. Metabolite annotations based on the integration of mass spectral information. Plant J. 54, 949–962. https://doi.org/10.1111/j.1365-313X.2008.03434.x.

Ishikawa, T., Takahara, K., Hirabayashi, T., Matsumura, H., Fujisawa, S., Terauchi, R., Uchimiya, H., Kawai-Yamada, M., 2010. Metabolome analysis of response to oxidative stress in rice suspension cells overexpressing cell death suppressor bax inhibitor-1. Plant Cell Physiol. 51, 9–20. https://doi.org/10.1093/pcp/pcp162.

Jacob, D., Deborde, C., Moing, A., 2013. An efficient spectra processing method for metabolite identification from ^{1}H-NMR metabolomics data. Anal. Bioanal. Chem. 405, 5049–5061. https://doi.org/10.1007/s00216-013-6852-y.

Jahangir, M., Abdel-Farid, I.B., Choi, Y.H., Verpoorte, R., 2008a. Metal ion-inducing metabolite accumulation in *Brassica rapa*. J. Plant Physiol. 165, 1429–1437. https://doi.org/10.1016/j.jplph.2008.04.011.

Jahangir, M., Kim, H.K., Choi, Y.H., Verpoorte, R., 2008b. Metabolomic response of *Brassica rapa* submitted to pre-harvest bacterial contamination. Food Chem. 107, 362–368. https://doi.org/10.1016/j.foodchem.2007.08.034.

Jaskolla, T.W., Papasotiriou, D.G., Karas, M., 2009. Comparison between the matrices a-cyano-4-hydroxycinnamic acid and 4-chloro-a-cyanocinnamic acid for trypsin, chymotrypsin, and pepsin digestion by MALDI-TOF mass spectrometry. J. Proteome 8, 3588–3597.

Jones, O.A.H., Maguire, M.L., Griffin, J.L., Jung, Y.H., Shibato, J., Rakwal, R., Agrawal, G.K., Jwa, N.S., 2011. Using metabolic profiling to assess plant-pathogen interactions: an example using rice (*Oryza sativa*) and the blast pathogen Magnaporthe grisea. Eur. J. Plant Pathol. 129, 539–554. https://doi.org/10.1007/s10658-010-9718-6.

Kakita, V.M.R., Hosur, R.V., 2016. Hadamard homonuclear broadband decoupled TOCSY NMR: improved efficacy in detecting long-range chemical shift correlations. Euro. J. Chem. Phys. Phys. Chem. https://doi.org/10.1111/j.1529-8817.2012.1208-12-055.x.

Kanakis, C.D., Petrakis, E.A., Kimbaris, A.C., Pappas, C., Tarantilis, P.A., Polissiou, M.G., 2012. Classification of Greek *Mentha pulegium* L. (Pennyroyal) samples, according to geographical location by fourier transform infrared spectroscopy. Phytochem. Anal. 23, 34–43. https://doi.org/10.1002/pca.1322.

Kaplan, F., Kopka, J., Haskell, D.W., Zhao, W., Schiller, K.C., Gatzke, N., Sung, D.Y., Guy, C.L., 2004. Exploring the temperature-stress metabolome. Plant Physiol. 136, 4159–4168. https://doi.org/10.1104/pp.104.052142.1.

Kazachkova, Y., Batushansky, A., Cisneros, A., Tel-Zur, N., Fait, A., Barak, S., 2013. Growth platform-dependent and independent phenotypic and metabolic responses of *Arabidopsis thaliana* and its halophytic relative *Eutrema salsugineum*, to salt stress. Plant Physiol. 162, 1583–1598. https://doi.org/10.1104/pp.113.217844.

Kiefer, P., Delmotte, N., Vorholt, J.A., 2011. Nanoscale ion-pair reversed-phase HPLC-MS for sensitive metabolome analysis. Anal. Chem. 83 (3), 850–855. https://doi.org/10.1021/ac102445r.

Kim, H.K., Verpoorte, R., 2010. Sample preparation for plant metabolomics. Phytochem. Anal. 21, 4–13. https://doi.org/10.1002/pca.1188.

Kim, H.K., Choi, Y.H., Verpoorte, R., 2011. NMR-based plant metabolomics: where do we stand, where do we go? Trends Biotechnol. 29, 267–275. https://doi.org/10.1016/j.tibtech.2011.02.001.

Kim, Y.M., Lee, J., Park, S.H., Lee, C., Lee, J.W., Lee, D.H., Kim, N., Lee, D., Kim, H.Y., Lee, C.H., 2012. LC-MS-based chemotaxonomic classification of wild-type Lespedeza sp. and its correlation with genotype. Plant Cell Rep. 31, 2085–2097. https://doi.org/10.1007/s00299-012-1319-8.

Kind, T., Wohlgemuth, G., Lee, D.Y., Lu, Y., Palazoglu, M., Shahbaz, S., 2009. FiehnLib: mass spectral and retention index libraries for metabolomics based on quadrupole and time-of-flight gas chromatography/mass spectrometry. Anal. Chem. 81, 10038–10048. https://doi.org/10.1021/ac9019522.

Kokalj, M., Kolar, J., Trafela, T., Krefta, S., 2011. Differences among Epilobium and Hypericum species revealed by four IR spectroscopy modes: transmission, KBr tablet, diffuse reflectance and ATR. Phytochem. Anal. 22, 541–546. https://doi.org/10.1002/pca.1315.

Kong, W.J., Zhao, Y.L., Xiao, X.H., Jin, C., Li, Z.L., 2009. Quantitative and chemical fingerprint analysis for quality control of *Rhizoma Coptidis chinensis* based on UPLC-PAD combined with chemometrics methods. Phytomedicine 16, 950–959. https://doi.org/10.1016/j.phymed.2009.03.016.

Kovacs, H., Moskau, D., Spraul, M., 2005. Cryogenically cooled probes—a leap in NMR technology. Prog. Nucl. Magn. Reson. Spectrosc. 46, 131–155. https://doi.org/10.1016/j.pnmrs.2005.03.001.

Kuhnen, S., Ogliari, J.B., Dias, P.F., Santos, M.D.S., Ferreira, A.G., Bonham, C.C., Wood, K.V., Maraschin, M., 2010. Metabolic fingerprint of Brazilian maize landraces silk (stigma/styles) using NMR spectroscopy and chemometric methods. J. Agric. Food Chem. 58, 2194–2200. https://doi.org/10.1021/jf9037776.

Kumaraswamy, K.G., Kushalappa, A.C., Choo, T.M., Dion, Y., Rioux, S., 2011. Mass spectrometry based metabolomics to identify potential biomarkers for resistance in barley against Fusarium head blight (*Fusarium graminearum*). J. Chem. Ecol. 37, 846–856. https://doi.org/10.1007/s10886-011-9989-1.

Kusano, M., Fukushima, A., Kobayashi, M., Hayashi, N., Jonsson, P., Moritz, T., Ebana, K., Saito, K., 2007. Application of a metabolomic method combining one-dimensional and two-dimensional gas chromatography-time-of-flight/mass spectrometry to metabolic phenotyping of natural variants in rice. J. Chromatogr. B. 855, 71–79. https://doi.org/10.1016/j.jchromb.2007.05.002.

Kusano, M., Redestig, H., Hirai, T., Oikawa, A., Matsuda, F., Fukushima, A., Arita, M., Watanabe, S., Yano, M., Hiwasa Tanase, K., Ezura, H., Saito, K., 2011. Covering chemical diversity of genetically-modified tomatoes using metabolomics for objective substantial equivalence assessment. PloS One 6, e16989. https://doi.org/10.1371/journal.pone.0016989.

Kwon, H.N., Phan, H., Xu, J., Ko, Y., 2016. Application of a smartphone metabolomics platform to the authentication of *Schisandra sinensis*. Phytochem. Anal. 27, 199–205. https://doi.org/10.1002/pca.2617.

Lai, Y., Ni, Y., Kokot, S., 2011. Discrimination of *Rhizoma corydalis* from two sources by near-infrared spectroscopy supported by the wavelet transform and least-squares support vector machine methods. Vib. Spectrosc. 56, 154–160. https://doi.org/10.1016/j.vibspec.2011.01.007.

Laksanasopin, T., Guo, T.W., Nayak, S., Sridhara, A.A., Xie, S., Olowookere, O.O., Cadinu, P., Meng, F., Chee, N.H., Kim, J., Chin, C.D., Munyazesa, E., Mugwaneza, P., Rai, A.J., Mugisha, V., Castro, A.R., Steinmiller, D., Linder, V., Justman, J.E., Nsanzimana, S., Sia, S.K., 2015. Supplementary Information: a smartphone dongle for diagnosis of infectious diseases at the point of care. Sci. Transl. Med. 7, 1–10. https://doi.org/10.1126/scitranslmed.aaa0056.

Lalli, P.M., Sanvido, G.B., Garcia, J.S., Haddad, R., Cosso, R.G., Maia, D.R.J., Zacca, J.J., Maldanerc, A.O., Eberlin, M.N., 2010. Fingerprinting and aging of ink by easy ambient sonic-spray ionization mass spectrometry. Analyst 135, 745–750. https://doi.org/10.1039/b923398a.

Lambert, J., Lampen, P., von Bohlen, A., Hergenröder, R., 2006. Two- and three-dimensional mapping of the iron distribution in the apoplastic fluid of plant leaf tissue by means of magnetic resonance imaging. Anal. Bioanal. Chem. 384, 231–236. https://doi.org/10.1007/s00216-005-0119-1.

Lanzinger, A., Frank, T., Reichenberger, G., Herz, M., Engel, K.H., 2015. Metabolite profiling of barley grain subjected to induced drought stress: responses of free amino acids in differently adapted cultivars. J. Agric. Food Chem. 63, 4252–4261. https://doi.org/10.1021/acs.jafc.5b01114.

Leiss, K.A., Choi, Y.H., Abdel-Farid, I.B., Verpoorte, R., Klinkhamer, P.G.L., 2009. NMR metabolomics of thrips (*Frankliniella occidentalis*) resistance in *Senecio* hybrids. J. Chem. Ecol. 35, 219–229. https://doi.org/10.1007/s10886-008-9586-0.

Leiss, K.A., Choi, Y.H., Verpoorte, R., Klinkhamer, P.G.L., 2011. An overview of NMR-based metabolomics to identify secondary plant compounds involved in host plant resistance. Phytochem. Rev. 10, 205–216. https://doi.org/10.1007/s11101-010-9175-z.

Lenz, E.M., Wilson, I.D., 2007. Analytical strategies in metabonomics. J. Proteome Res. 6, 443–458. https://doi.org/10.1021/pr0605217.

Li, S.L., Song, J.Z., Qiao, C.F., Zhou, Y., Qian, K., Lee, K.H., Xu, H.X., 2010. A novel strategy to rapidly explore potential chemical markers for the discrimination between raw and processed *Radix Rehmanniae* by UHPLC-TOFMS with multivariate statistical analysis. J. Pharm. Biomed. Anal. 51, 812–823. https://doi.org/10.1016/j.jpba.2009.10.002.

Li, W., Xing, L., Cai, Y., Qu, H., 2011. Classification and quantification analysis of *Radix scutellariae* from different origins with near infrared diffuse reflection spectroscopy. Vib. Spectrosc. 55, 58–64. https://doi.org/10.1016/j.vibspec.2010.07.004.

Liang, Y.S., Choi, Y.H., Kim, H.K., Linthorst, H.J.M., Verpoorte, R., 2006. Metabolomic analysis of methyl jasmonate treated *Brassica rapa* leaves by 2-dimensional NMR spectroscopy. Phytochemistry 67, 2503–2511. https://doi.org/10.1016/j.phytochem.2006.08.018.

Ljung, K., Sandberg, G., Moritz, T., 2010. Methods of plant hormone analysis. In: Plant Hormones: Biosynthesis, Signal Transduction, Action!, pp. 717–740. https://doi.org/10.1007/978-1-4020-2686-7_31.

López-Gresa, M.P., Maltese, F., Bellés, J.M., Conejero, V., Kim, H.K., Choi, Y.H., 2010. Metabolic response of tomato leaves upon different plant-pathogen interactions. Phytochem. Anal. 21, 89–94. https://doi.org/10.1002/pca.1179.

Lu, G.H., Chan, K., Liang, Y.Z., Leung, K., Chan, C.L., Jiang, Z.H., 2005. Development of high-performance liquid chromatographic fingerprints for distinguishing Chinese Angelica from related umbelliferae herbs. J. Chromatogr. A 1073, 383–392. Available at: http://www.ncbi.nlm.nih.gov/pubmed/15909545.

Lucio-Gutiérrez, J.R., Coello, J., Maspoch, S., 2011. Application of near infrared spectral fingerprinting and pattern recognition techniques for fast identification of *Eleutherococcus senticosus*. Food Res. Int. 44, 557–565. https://doi.org/10.1016/j.foodres.2010.11.037.

Ludwig, C., Viant, M.R., 2010. Two-dimensional J-resolved NMR spectroscopy: review of a key methodology in the metabolomics toolbox. Phytochem. Anal. 21, 22–32. https://doi.org/10.1002/pca.1186.

Luedemann, A., Strassburg, K., Erban, A., Kopka, J., 2008. TagFinder for the quantitative analysis of gas chromatography - mass spectrometry (GC-MS)-based metabolite profiling experiments. Bioinformatics 24, 732–737. https://doi.org/10.1093/bioinformatics/btn023.

Luengwilai, K., Saltveit, M., Beckles, D.M., 2012. Metabolite content of harvested Micro-Tom tomato (*Solanum lycopersicum* L.) fruit is altered by chilling and protective heat-shock treatments as shown by GC-MS metabolic profiling. Postharvest Biol. Technol. 63, 116–122. https://doi.org/10.1016/j.postharvbio.2011.05.014.

Lugan, R., Niogret, M.F., Leport, L., Guégan, J.P., Larher, F.R., Savouré, A., Kopka, J., Bouchereau, A., 2010. Metabolome and water homeostasis analysis of *Thellungiella salsuginea* suggests that dehydration tolerance is a key response to osmotic stress in this halophyte. Plant J. 64, 215–229. https://doi.org/10.1111/j.1365-313X.2010.04323.x.

Luo, Q., Wang, S., Sun, L., Wang, H., 2017. Metabolic profiling of root exudates from two ecotypes of *Sedum alfredii* treated with Pb based on GC-MS. Sci. Rep. 7, 39878. https://doi.org/10.1038/srep39878.

Luo, Q., Yue, G., Valaskovic, G.A., Gu, Y., Wu, S., Karger, B.L., 2007. On-line 1D and 2D porous layer open tubular/LC-ESI-MS using 10 μm poly (styrene-divinylbenzene) columns for ultrasensitive proteomic analysis. Anal. Chem. 79, 6174–6181.

Mandal, M.K., Chen, L.C., Hashimoto, Y., Yu, Z., Hiraoka, K., 2010. Detection of biomolecules from solutions with high concentration of salts using probe electrospray and nano-electrospray ionization mass spectrometry. Anal. Methods 2, 1905–1912. https://doi.org/10.1039/c0ay00530d.

Maher, A.D., Fonville, J.M., Coen, M., Lindon, J.C., Rae, C.D., Nicholson, J.K., 2012. Statistical total correlation spectroscopy scaling for enhancement of metabolic information recovery in biological NMR spectra. Anal. Chem. 84 (2), 1083–1091. https://doi.org/10.1021/ac202720f.

Mahrous, A., Farag, M.A., 2015. Two dimensional NMR spectroscopic approaches for exploring plant metabolome: a review. J. Adv. Res. 6, 3–15. https://doi.org/10.1016/j.jare.2014.10.003.

Matsuda, F., Hirai, M.Y., Sasaki, E., Akiyama, K., Yonekura-Sakakibara, K., Provart, N.J., Sakurai, T., Shimada, Y., Saito, K., 2010. AtMetExpress development: a phytochemical atlas of Arabidopsis development. Plant Physiol. 152, 566–578. https://doi.org/10.1104/pp.109.148031.

Melkus, G., Rolletschek, H., Fuchs, J., Radchuk, V., Grafahrend-Belau, E., Sreenivasulu, N., Rutten, T., Weier, D., Heinzel, N., Schreiber, F., Altmann, T., Jakob, P.M., Borisjuk, L., 2011. Dynamic 13C/1 H NMR imaging uncovers sugar allocation in the living seed. Plant Biotechnol. J. 9, 1022–1037. https://doi.org/10.1111/j.1467-7652.2011.00618.x.

Mirnezhad, M., Romero-Gonzalez, R.R., Leiss, K.A., Choi, Y.H., Verpoorte, R., Klinkhamera, P.G.L., 2010. Metabolomic analysis of host plant resistance to thrips in wild and cultivated tomatoes. Phytochem. Anal. 21, 110–117. https://doi.org/10.1002/pca.1182.

Mirsaleh-Kohan, N., Robertson, W.D., Compton, R.N., 2008. Electron ionization time-of-flight mass spectrometry: historical review and current applications. Mass Spectrom. Rev. 27, 237–285. https://doi.org/10.1002/mas.

Miura, D., Fujimura, Y., Wariishi, H., 2012. In situ metabolomic mass spectrometry imaging: recent advances and difficulties. J. Proteomics 75, 5052–5060. https://doi.org/10.1016/j.jprot.2012.02.011.

Miura, D., Tsuji, Y., Takahashi, K., Wariishi, H., Saito, K., 2010. A strategy for the determination of the elemental composition by Fourier transform ion cyclotron resonance mass spectrometry based on isotopic peak ratios. Anal. Chem. 82, 5887–5891. https://doi.org/10.1021/ac902931x.

Moseley, H.N., Lane, A.N., Belshoff, A.C., Higashi, R.M., Fan, T.W., 2011. A novel deconvolution method for modelling UDP-GlcNAc biosynthetic pathways based on ^{13}C mass isotopologue profiles under non steady-state conditions. BMC Biol. 9. https://doi.org/10.1186/1741-7007-9-37.

Motta, A., Paris, D., Melck, D., 2010. Monitoring real-time metabolism of living cells by fast two-dimensional NMR spectroscopy. Anal. Chem. 82, 2405–2411. https://doi.org/10.1021/ac9026934.

Munson, B., 2000. Chemical ionization mass spectrometry: theory and applications. Encycl. Anal. Chem. 1–22. http://dx.doi.org/10.1002/9780470027318.a6004.

Mungur, R., Glass, A.D., Goodenow, D.B., Lightfoot, D.A., 2005. Metabolite fingerprinting in transgenic *Nicotiana tabacum* altered by the *Escherichia coli* glutamate dehydrogenase gene. J. Biomed. Biotechnol. 198–214. https://doi.org/10.1155/JBB.2005.198.

Muscolo, A., Junker, A., Klukas, C., Weigelt-Fischer, K., Riewe, D., Altmann, T., 2015. Phenotypic and metabolic responses to drought and salinity of four contrasting lentil accessions. J. Exp. Bot. 66, 5467–5480. https://doi.org/10.1093/jxb/erv208.

Nagana Gowda, G.A., Raftery, D., 2014. Advances in NMR-based metabolomics. In: Comprehensive Analytical Chemistry. Elsevier B.V., pp. 187–211. https://doi.org/10.1016/B978-0-444-62651-6.00008-8.

Nakabayashi, R., Saito, K., 2013. Metabolomics for unknown plant metabolites. Anal. Bioanal. Chem. 405, 5005–5011. http://dx.doi.org/10.1007/s00216-013-6869-2.

Nakabayashi, R., Yamazaki, M., Saito, K., 2010. A polyhedral approach for understanding flavonoid biosynthesis in Arabidopsis. New Biotechnol. 27, 829–836. https://doi.org/10.1016/j.nbt.2010.03.004.

Nakanishi, Y., Fukuda, S., Chikayama, E., Kimura, Y., Ohno, H., Kikuchi, J., 2011. Dynamic omics approach identifies nutrition-mediated microbial interactions. J. Proteome Res. 10, 824–836. https://doi.org/10.1021/pr100989c.

Nguyen Hoai, N., Dejaegher, B., Tistaert, C., Nguyen Thi Hong, V., Rivière, C., Chataigné, G., Phan, V.K., Chau, V.M., Quetin-Leclercq, J., Vander, H.Y., 2009. Development of HPLC fingerprints for Mallotus species extracts and evaluation of the peaks responsible for their antioxidant activity. J. Pharm. Biomed. Anal. 50, 753–763. https://doi.org/10.1016/j.jpba.2009.06.016.

Ni, Y., Peng, Y., Kokot, S., 2008. Fingerprinting of complex mixtures with the use of high performance liquid chromatography, inductively coupled plasma atomic emission spectroscopy and chemometrics. Anal. Chim. Acta 616, 19–27. https://doi.org/10.1016/j.aca.2008.04.015.

Ni, Y., Mei, M., Kokot, S., 2012. One- and two-dimensional gas chromatography-mass spectrometry and high performance liquid chromatography-diode-array detector fingerprints of complex substances: a comparison of classification performance of similar, complex *Rhizoma Curcumae* samples with. Anal. Chim. Acta 712, 37–44. https://doi.org/10.1016/j.aca.2011.11.010.

Ni, J., Yang, X., Zhu, J., Liu, Z., Ni, Y., Wu, H., Zhang, H., Liu, T., 2015. Salinity-induced metabolic profile changes in *Nitraria tangutorum* Bobr. suspension cells. Plant Cell. Tissue Organ Cult. 122, 239–248. https://doi.org/10.1007/s11240-015-0744-0.

Nicholson, J.K., Lindon, J.C., 2008. Systems biology: metabonomics. Nature 455, 1054–1056. https://doi.org/10.1038/4551054a.

Novoa-Carballal, R., Fernandez-Megia, E., Jimenez, C., Riguera, R., 2011. NMR methods for unravelling the spectra of complex mixtures. Nat. Prod. Rep. 28, 78–98. https://doi.org/10.1039/c005320c.

Nyadong, L., Galhena, A.S., Fernández, F.M., 2009. Desorption electrospray/metastable-induced ionization: a flexible multimode ambient ion generation technique. Anal. Chem. 81, 7788–7794. https://doi.org/10.1021/ac9014098.

Obata, T., Fernie, A.R., 2012. The use of metabolomics to dissect plant responses to abiotic stresses. Cell. Mol. Life Sci. 69, 3225–3243. https://doi.org/10.1007/s00018-012-1091-5.

Oncescu, V., Mancuso, M., Erickson, D., 2013. Cholesterol testing on a smartphone. Lab. Chip 14, 759–763. https://doi.org/10.1039/C3LC51194D.

Oikawa, A., Nakamura, Y., Ogura, T., Kimura, A., Suzuki, H., Sakurai, N., Shinbo, Y., Shibata, D., Kanaya, S., Ohta, D., 2006. Clarification of pathway-specific inhibition by Fourier transform ion cyclotron resonance/mass spectrometry-based metabolic phenotyping studies. Plant Physiol. 142, 398–413. https://doi.org/10.1104/pp.106.080317.

Okazaki, Y., Kamide, Y., Hirai, M.Y., Saito, K., 2011. Plant lipidomics based on hydrophilic interaction chromatography coupled to ion trap time-of-flight mass spectrometry. Metabolomics 9, 121–131. https://doi.org/10.1007/s11306-011-0318-z.

Okazaki, Y., Saito, K., 2012. Recent advances of metabolomics in plant biotechnology. Plant Biotechnol. Rep. 6, 1–15. https://doi.org/10.1007/s11816-011-0191-2.

Pan, R., Guo, F., Lu, H., Feng, W., Liang, W., Zeng, Y., 2011. Development of the chromatographic fingerprint of *Scutellaria barbata* D. Don by GC-MS combined with Chemometrics methods. J. Pharm. Biomed. Anal. 55, 391–396. https://doi.org/10.1016/j.jpba.2011.01.016.

Pandey, S., Patel, M.K., Mishra, A., Jha, B., 2015. Physio-biochemical composition and untargeted metabolomics of cumin (*Cuminum cyminum* L.) make it promising functional food and help in mitigating salinity stress. PLoS One 10, 1–25. https://doi.org/10.1371/journal.pone.0144469.

Pang, Q., Zhang, A., Zang, W., Wei, L., Yan, X., 2016. Integrated proteomics and metabolomics for dissecting the mechanism of global responses to salt and alkali stress in *Suaeda corniculata*. Plant Soil 402, 379–394. https://doi.org/10.1007/s11104-015-2774-0.

Panagiotopoulos, N., Duschka, R., Ahlborg, M., Bringout, G., Debbeler, C., Graeser, M., Kaethner, C., Ludtke-Buzug, K., Medimagh, H., Stelzner, J., Buzug, T.M., Barkhausen, J., Vogt, F.M., Haegele, J., 2015. Magnetic particle imaging: current developments and future directions. Int. J. Nanomedicine 10, 3097–3114. https://doi.org/10.2147/IJN.S70488.

Passarelli, M.K., Winograd, N., 2011. Lipid imaging with time-of-flight secondary ion mass spectrometry (ToF-SIMS). Biochim. Biophys. Acta - Mol. Cell Biol. Lipids 1811, 976–990. https://doi.org/10.1016/j.bbalip.2011.05.007.

Patti, G.J., Woo, H.K., Yanes, O., Shriver, L., Thomas, D., Uritboonthai, W., Apon, J.V., Steenwyk, R., Manchester, M., Siuzdak, G., 2010. Detection of carbohydrates and steroids by cation-enhanced nanostructure-Initiator mass spectrometry (NIMS) for biofluid analysis and tissue imaging. Anal. Chem. 82, 121–128. https://doi.org/10.1021/ac9014353.

Patti, G.J., Yanes, O., Siuzdak, G., 2012. Innovation: metabolomics: the apogee of the omics trilogy. Nat. Publ. Gr 13, 263–269. https://doi.org/10.1038/nrm3314.

Pearson, H., 2007. Meet the human metabolome. J. Chromatogr. B Anal. Technol. Biomed. Life Sci. 446, 8. https://doi.org/10.1038/446008a.

Peng, J.B., Jia, H.M., Liu, Y.T., Zhang, H.W., Dong, S., Zou, Z.M., 2011. Qualitative and quantitative characterization of chemical constituents in Xin-Ke-Shu preparations by liquid chromatography coupled with a LTQ orbitrap mass spectrometer. J. Pharm. Biomed. Anal. 55, 984–995. https://doi.org/10.1016/j.jpba.2011.03.045.

Pereira, S.I., Figueiredo, P.I., Barros, A.S., Dias, M.C., Santos, C., Duarte, I.F., Gil, A.M., 2014. Changes in the metabolome of lettuce leaves due to exposure to mancozeb pesticide. Food Chem. 154, 291–298. https://doi.org/10.1016/j.foodchem.2014.01.019.

Porcari, A.M., Fernandes, G.D., Barrera-Arellano, D., Eberlin, M.N., Alberici, R.M., 2016. Food quality and authenticity screening via easy ambient sonic-spray ionization mass spectrometry. Analyst 141, 1172–1184. https://doi.org/10.1039/C5AN01415H.

Portolés, T., Sancho, J.V., Hernández, F., Newton, A., Hancock, P., 2010. Potential of atmospheric pressure chemical ionization source in GC-QTOF MS for pesticide residue analysis. J. Mass Spectrom. 45, 926–936. https://doi.org/10.1002/jms.1784.

Pourmortazavi, S.M., Hajimirsadeghi, S.S., 2007. Supercritical fluid extraction in plant essential and volatile oil analysis. J. Chromatogr. 1163, 2–24. http://dx.doi.org/10.1016/j.chroma.2007.06.021.

Putri, S.P., Yamamoto, S., Tsugawa, H., Fukusaki, E., 2013. Current metabolomics: technological advances. J. Biosci. Bioeng. 116, 9–16. https://doi.org/10.1016/j.jbiosc.2013.01.004.

Rabara, R.C., Tripathi, P., Reese, R.N., Rushton, D.L., Alexander, D., Timko, M.P., Shen, Q.J., Rushton, P.J., 2015. Tobacco drought stress responses reveal new targets for Solanaceae crop improvement. BMC Genomics 16, 484. http://dx.doi.org/10.1186/s12864-015-1575-4.

Raina, R., Hall, P., 2008. Comparison of gas chromatography-mass spectrometry and gas chromatography-tandem mass spectrometry with electron ionization and negative-ion chemical ionization for analyses of pesticides at trace levels in atmospheric samples. Anal. Chem. Insights 2008, 111–125.

Rai, R.K., Sinha, N., 2012. Fast and accurate quantitative metabolic profiling of body fluids by nonlinear sampling of ^{1}H–^{13}C two-dimensional nuclear magnetic resonance spectroscopy. Anal. Chem. 84, 10005–10011. https://doi.org/10.1021/ac302457s.

Ralston-Hooper, K., Hopf, A., Oh, C., Zhang, X., Adamec, J., Sepulveda, M.S., 2008. Development of GC9GC/TOF–MS metabolomics for use in eco-toxicological studies with invertebrates. Aqua. Toxicol. 88, 48–52. https://doi.org/10.1016/j.aquatox.2008.03.002.

Rodziewicz, P., Swarcewicz, B., Chmielewska, K., Wojakowska, A., Stobiecki, M., 2014. Influence of abiotic stresses on plant proteome and metabolome changes. Acta Physiol. Plant 36, 1–19.

Rosati, A., Cafiero, C., Paoletti, A., Alfei, B., Caporali, S., Casciani, L., Valentini, M., 2014. Effect of agronomical practices on carpology, fruit and oil composition, and oil sensory properties, in olive (*Olea europaea* L.). Food Chem. 159, 236–243. https://doi.org/10.1016/j.foodchem.2014.03.014.

Saito, K., Matsuda, F., 2010. Metabolomics for functional genomics, systems biology, and biotechnology. Annu. Rev. Plant Biol. 61, 463–489. http://dx.doi.org/10.1146/annurev.arplant.043008.092035.

Saritas, E.U., Goodwill, P.W., Croft, L.R., Konkle, J.J., Lu, K., Zheng, B., Conolly, S.M., 2013. Magnetic particle imaging (MPI) for NMR and MRI researchers. J. Magn. Reson 229, 116–126. https://doi.org/10.1016/j.jmr.2012.11.029.

Sana, T.R., Roark, J.C., Li, X., Waddell, K., Fischer, S.M., 2008. Molecular formula and METLIN personal metabolite database matching applied to the identification of compounds generated by LC/TOF-MS. J. Biomol. Tech. 19, 258–266.

Sanchez, D.H., Pieckenstain, F.L., Escaray, F., Erban, A., Kraemer, U., Udvardi, M.K., Kopka, J., 2011. Comparative ionomics and metabolomics in extremophile and glycophytic lotus species under salt stress challenge the metabolic pre-adaptation hypothesis. Plant, Cell Environ. 34, 605–617. https://doi.org/10.1111/j.1365-3040.2010.02266.x.

Sandasi, M., Kamatou, G.P.P., Viljoen, A.M., 2012. An untargeted metabolomic approach in the chemotaxonomic assessment of two Salvia species as a potential source of bisabolol. Phytochemistry 84, 94–101. https://doi.org/10.1016/j.phytochem.2012.08.009.

Sârbu, C., Nacu-Briciu, R.D., Kot-Wasik, A., Gorinstein, S., Wasik, A., Namieśnik, J., 2012. Classification and fingerprinting of kiwi and pomelo fruits by multivariate analysis of chromatographic and spectroscopic data. Food Chem. 130, 994–1002. https://doi.org/10.1016/j.foodchem.2011.07.120.

Sawada, Y., Hirai, M.Y., 2013. Integrated LC-MS/MS System for plant metabolomics. Comput. Struct. Biotechnol. J. 4, 1–6. https://doi.org/10.5936/csbj.201301011.

Schripsema, J., 2010. Application of NMR in plant metabolomics: techniques, problems and prospects. Phytochem. Anal. 21, 14–21. https://doi.org/10.1002/pca.1185.

Schanda, P., Ernst, M., 2016. Studying dynamics by magic-angle spinning solid-state NMR spectroscopy: principles and applications to biomolecules. Prog. Nucl. Magn. Reson. Spectrosc. 96, 1–122. https://doi.org/10.1016/j.pnmrs.2016.02.001.

Shi, H., Ye, T., Song, B., Qi, X., Chan, Z., 2015. Comparative physiological and metabolomic responses of four *Brachypodium distachyon* varieties contrasting in drought stress resistance. Acta Physiol. Plant 37, 1–12. https://doi.org/10.1007/s11738-015-1873-0.

Shulaev, V., 2006. Metabolomics technology and bioinformatics. Brief. Bioinform 7, 128–139. https://doi.org/10.1093/bib/bbl012.

Silvente, S., Sobolev, A.P., Lara, M., 2012. Metabolite adjustments in drought tolerant and sensitive soybean genotypes in response to water stress. PLoS One 7. https://doi.org/10.1371/journal.pone.0038554.

Simoh, S., Quintana, N., Kim, H.K., Choi, Y.H., Verpoorte, R., 2009. Metabolic changes in *Agrobacterium tumefaciens*-infected *Brassica rapa*. J. Plant Physiol. 166, 1005–1014. https://doi.org/10.1016/j.jplph.2008.11.015.

Sinija, V.R., Mishra, H.N., 2009. FT-NIR spectroscopy for caffeine estimation in instant green tea powder and granules. LWT - Food Sci. Technol. 42, 998–1002. https://doi.org/10.1016/j.lwt.2008.12.013.

Skirycz, A., De Bodt, S., Obata, T., De Clercq, I., Claeys, H., De Rycke, R., Andriankaja, M., Van Aken, O., Van Breusegem, F., Fernie, A.R., Inzé, D., 2010. Developmental stage specificity and the role of mitochondrial metabolism in the response of Arabidopsis leaves to prolonged mild osmotic stress. Plant Physiol. 152, 226–244. https://doi.org/10.1104/pp.109.148965.

Smith, C.A., Want, E.J., O'Maille, G., Abagyan, R., Siuzdak, G., Smith, C.A., Elizabeth, J., O'Maille, G., Abagyan, R., 2006. XCMS: processing mass spectrometry data for metabolite profiling using Nonlinear Peak Alignment, Matching,and Identification. ACS Publ. 78, 779–787. https://doi.org/10.1021/ac051437y.

Smith, L.M., Maher, A.D., Cloarec, O., Rantalainen, M., Tang, H., Elliott, P., Stamler, J., Lindon, J.C., Holmes, E., Nicholson, J.K., 2007. Statistical correlation and projection methods for improved information recovery from diffusion-edited NMR spectra of biological samples. Anal. Chem. 79, 5682–5689. https://doi.org/10.1021/ac0703754.

Sobhanian, H., Motamed, N., Jazii, F.R., Nakamura, T., Komatsu, S., 2010. Salt stress induced differential proteome and metabolome response in the shoots of *Aeluropus lagopoides* (Poaceae), a halophyte C4 plant. J. Proteome Res. 9, 2882–2897. https://doi.org/10.1021/pr900974k.

Soga, T., Baran, R., Suematsu, M., Ueno, Y., Ikeda, S., Sakurakawa, T., Kakazu, Y., Ishikawa, T., Robert, M., Nishioka, T., Tomita, M., 2006. Differential metabolomics reveals ophthalmic acid as an oxidative stress biomarker indicating hepatic glutathione consumption. J. Biol. Chem. 281, 16768–16776. https://doi.org/10.1074/jbc.M601876200.

Soga, T., Igarashi, K., Ito, C., Mizobuchi, K., Zimmermann, H.P., Tomita, M., 2009. Metabolomic profiling of anionic metabolites by capillary electrophoresis mass spectrometry. Anal. Chem. 81, 6165–6174. https://doi.org/10.1021/ac900675k.

Souto, U.T.C.P., Pontes, M.J.C., Silva, E.C., Galvão, R.K.H., Araújo, M.C.U., Sanches, F.A.C., Cunha, F., Oliveira, M., 2010. UV-Vis spectrometric classification of coffees by SPA-LDA. Food Chem. 119, 368–371. https://doi.org/10.1016/j.foodchem.2009.05.078.

Su, S., Hua, Y., Duan, J.A., Shang, E., Tang, Y., Bao, X., Lu, Y., Ding, A., 2008. Hypothesis of active components in volatile oil from a Chinese herb formulation, "Shao-Fu-Zhu-Yu decoction", using GC-MS and chemometrics. J. Sep. Sci. 31, 1085–1091. https://doi.org/10.1002/jssc.200700492.

Suhre, K., Shin, S., Petersen, A., Mohney, R.P., Wägele, B., Altmaier, E., Deloukas, P., Erdmann, J., Grundberg, E., Hammond, C.J., De, M.H., Meisinger, C., Meitinger, T., Mewes, H.W., Michael, V., 2013. Human metabolic individuality in biomedical and pharmaceutical research. Nature 477, 1–22. https://doi.org/10.1038/nature10354.

Sumner, L.W., Lei, Z., Nikolau, B.J., Saito, K., 2015. Modern plant metabolomics: advanced natural product gene discoveries, improved technologies, and future prospects. Nat. Prod. Rep. 32, 212–229. https://doi.org/10.1039/c4np00072b.

Sun, X., Zhang, J., Zhang, H., Ni, Y., Zhang, Q., Chen, J., Guan, Y., 2010. The responses of *Arabidopsis thaliana* to cadmium exposure explored via metabolite profiling. Chemosphere 78, 840–845. https://doi.org/10.1016/j.chemosphere.2009.11.045.

Sun, C., Gao, X., Chen, X., Fu, J., Zhang, Y., 2016. Metabolic and growth responses of maize to successive drought and re-watering cycles. Agric. Water Manag. 172, 62–73. https://doi.org/10.1016/j.agwat.2016.04.016.

Suzuki, H., Sasaki, R., Ogata, Y., Nakamura, Y., Sakurai, N., Kitajima, M., Takayama, H., Kanaya, S., Aoki, K., Shibata, D., Saito, K., 2008. Metabolic profiling of flavonoids in *Lotus japonicus* using liquid chromatography Fourier transform ion cyclotron resonance mass spectrometry. Phytochemistry 69, 99–111. https://doi.org/10.1016/j.phytochem.2007.06.017.

Tayyari, F., Gowda, G.A.N., Gu, H., Raftery, D., 2013. 15N-cholamine–a smart isotope tag for combining NMR- and MS-based metabolite profiling. Anal. Chem. 85, 8715–8721. https://doi.org/10.1021/ac401712a.

Tian, R.T., Xie, P.S., Liu, H.P., 2009. Evaluation of traditional Chinese herbal medicine: Chaihu (*Bupleuri Radix*) by both high-performance liquid chromatographic and high-performance thin-layer chromatographic fingerprint and chemometric analysis. J. Chromatogr. A 1216, 2150–2155. https://doi.org/10.1016/j.chroma.2008.10.127.

Tistaert, C., Dejaegher, B., Heyden, Y.V., 2011. Chromatographic separation techniques and data handling methods for herbal fingerprints: a review. Anal. Chim. Acta 690, 148–161. https://doi.org/10.1016/j.aca.2011.02.023.

Toh, D.F., New, L.S., Koh, H.L., Chan, E.C.Y., 2010. Ultra-high performance liquid chromatography/time-of-flight mass spectrometry (UHPLC/TOFMS) for time-dependent profiling of raw and steamed *Panax notoginseng*. J. Pharm. Biomed. Anal. 52, 43–50. https://doi.org/10.1016/j.jpba.2009.12.005.
Trygg, J., Holmes, E., Lundstedt, T., 2007. Chemometrics in metabonomics. J. Proteome Res. 6, 469–479. https://doi.org/10.1021/pr060594q.
Tschoep, H., Gibon, Y., Carillo, P., Armengaud, P., Szecowka, M., Nunes-Nesi, A., Fernie, A.R., Koehl, K., Stitt, M., 2009. Adjustment of growth and central metabolism to a mild but sustained nitrogen-limitation in Arabidopsis. Plant Cell Environ. 32, 300–318. https://doi.org/10.1111/j.1365-3040.2008.01921.x.
Tugizimana, F., Piater, L., Dubery, I., 2013. Plant metabolomics:A new frontier in phytochemical analysis. S. Afr. J. Sci. 109, 1–11. https://doi.org/10.1590/sajs.2013/20120005©.
Tulpan, D., Léger, S., Belliveau, L., Culf, A., Cuperlović-Culf, M., 2011. MetaboHunter: an automatic approach for identification of metabolites from ^{1}H-NMR spectra of complex mixtures. BMC Bioinforma 12, 400. https://doi.org/10.1186/1471-2105-12-400.
Urano, K., Maruyama, K., Ogata, Y., Morishita, Y., Takeda, M., Sakurai, N., Suzuki, H., Saito, K., Shibata, D., Kobayashi, M., Yamaguchi-Shinozaki, K., Shinozaki, K., 2009. Characterization of the ABA-regulated global responses to dehydration in Arabidopsis by metabolomics. Plant J. 57, 1065–1078. https://doi.org/10.1111/j.1365-313X.2008.03748.x.
Urbanczyk-Wochniak, E., Fernie, A.R., 2005. Metabolic profiling reveals altered nitrogen nutrient regimes have diverse effects on the metabolism of hydroponically-grown tomato (*Solanum lycopersicum*) plants. J. Exp. Bot. 56, 309–321. https://doi.org/10.1093/jxb/eri059.
Vagionas, K., Ngassapa, O., Runyoro, D., Graikou, K., Gortzi, O., Chinou, I., 2007. Chemical analysis of edible aromatic plants growing in Tanzania. Food Chem. 105, 1711–1717. https://doi.org/10.1016/j.foodchem.2007.05.029.
Van As, H., Van Duynhoven, J., 2013. MRI of plants and foods. J. Magn. Reson 229, 25–34. https://doi.org/10.1016/j.jmr.2012.12.019.
Van, Q.N., Issaq, H.J., Jiang, Q., Li, Q., Muschik, G.M., Waybright, T.J., Lou, H., Dean, M., Uitto, J., Veenstra, T.D., 2008. Comparison of ^{1}D and ^{2}D NMR spectroscopy for metabolic profiling. J. Proteome Res. 7, 630–639. https://doi.org/10.1021/pr700594s.
Verslues, P.E., Juenger, T.E., 2011. Drought, metabolites, and *Arabidopsis* natural variation: a promising combination for understanding adaptation to water-limited environments. Curr. Opin. Plant Biol. 14, 240–245. https://doi.org/10.1016/j.pbi.2011.04.006.
Vermaak, I., Hamman, J.H., Viljoen, A.M., 2011. Hoodia gordonii: an up-to-date review of a commercially important anti-obesity plant. Planta Med. 77, 1149–1160. https://doi.org/10.1055/s-0030-1250643.
Vincenti, M., 2001. The renaissance of desorption chemical ionization mass spectrometry: characterization of large involatile molecules and nonpolar polymers. Int. J. Mass Spectrom. 212, 505–518. https://doi.org/10.1016/S1387-3806(01)00492-4.
Vlachos, N., Skopelitis, Y., Psaroudaki, M., Konstantinidou, V., Chatzilazarou, A., Tegou, E., 2006. Applications of fourier transform-infrared spectroscopy to edible oils. Anal. Chim. Acta 573–574, 459–465. https://doi.org/10.1016/j.aca.2006.05.034.
Vogt, T., 2010. Phenylpropanoid biosynthesis. Mol. Plant 3, 2–20. https://doi.org/10.1093/mp/ssp106.
Wang, H., Sun, W., Zhang, J., Yang, X., Lin, T., Ding, L., 2010. Desorption corona beam ionization source for mass spectrometry. Analyst 135, 688–695. https://doi.org/10.1039/b922616h.
Wang, L., Wang, X., Kong, L., 2012. Automatic authentication and distinction of *Epimedium koreanum* and *Epimedium wushanense* with HPLC fingerprint analysis assisted by pattern recognition techniques. Biochem. Syst. Ecol. 40, 138–145. https://doi.org/10.1016/j.bse.2011.10.014.
Warren, C.R., Aranda, I., Cano, F.J., 2012. Metabolomics demonstrates divergent responses of two Eucalyptus species to water stress. Metabolomics 8, 186–200. https://doi.org/10.1007/s11306-011-0299-y.
Wen, H., Kang, S., Song, Y., Song, Y., Sung, S.H., 2010. Differentiation of cultivation sources of *Ganoderma lucidum* by NMR-based metabolomics approach. Phytochem. Anal. 21, 73–79. https://doi.org/10.1002/pca.1166.
Wen, H., Lee, M.Y., Song, Y., Moon, S., Park, S., 2011. Combined genomic à metabolomic approach for the differentiation of geographical origins of natural products : deer antlers as an Example. J. Agric. Food Chem. 59, 6339–6345. https://doi.org/10.1021/jf200048p.
Wenzel, A., Frank, T., Reichenberger, G., Herz, M., Engel, K.H., 2015. Impact of induced drought stress on the metabolite profiles of barley grain. Metabolomics 11, 454–467. https://doi.org/10.1007/s11306-014-0708-0.
Westerhuis, J.A., Hoefsloot, H.C.J., Smit, S., Vis, D.J., Smilde, A.K., Velzen, E.J.J., Duijnhoven, J.P.M., Dorsten, F.A., 2008. Assessment of PLSDA cross validation. Metabolomics 4, 81–89. https://doi.org/10.1007/s11306-007-0099-6.
Widodo, Patterson, J.H., Newbigin, E., Tester, M., Bacic, A., Roessner, U., 2009. Metabolic responses to salt stress of barley (*Hordeum vulgare* L.) cultivars, Sahara and Clipper, which differ in salinity tolerance. J. Exp. Bot. 60, 4089–4103. https://doi.org/10.1093/jxb/erp243.
Windt, C.W., Gerkema, E., Van As, H., 2010. Most water in the tomato truss is imported through the xylem, not the phloem: a nuclear magnetic resonance flow imaging study. Plant Physiol. 151, 830–842. https://doi.org/10.1104/pp.109.141044.
Wise, D.R., DeBerardinis, R.J., Mancuso, A., Sayed, N., Zhang, X.-Y., Pfeiffer, H.K., Nissim, I., Daikhin, E., Yudkoff, M., McMahon, S.B., Thompson, C.B., 2008. Myc regulates a transcriptional program that stimulates mitochondrial glutaminolysis and leads to glutamine addiction. Proc. Natl. Acad. Sci. U.S.A 105, 18782–18787. https://doi.org/10.1073/pnas.0810199105.
Wishart, D.S., 2007. Current progress in computational metabolomics. Brief. Bioinform 8, 279–293. https://doi.org/10.1093/bib/bbm030.
Wishart, D.S., 2011. Advances in metabolite identification. Bioanalysis 3, 1769–1782. https://doi.org/10.4155/bio.11.155.
Wishart, D.S., Knox, C., Guo, A.C., Eisner, R., Young, N., Gautam, B., Hau, D.D., Psychogios, N., Dong, E., Bouatra, S., Mandal, R., Sinelnikov, I., Xia, J., Jia, L., Cruz, J.A., Lim, E., Sobsey, C.A., Shrivastava, S., Huang, P., Liu, P., Fang, L., Peng, J., Fradette, R., Cheng, D., Tzur, D., Clements, M., Lewis, A., de souza, A., Zuniga, A., Dawe, M., Xiong, Y., Clive, D., Greiner, R., Nazyrova, A., Shaykhutdinov, R., Li, L., Vogel, H.J., Forsythei, I., 2009. HMDB: a knowledgebase for the human metabolome. Nucleic Acids Res. 37, 603–610. https://doi.org/10.1093/nar/gkn810.
Witt, S., Galicia, L., Lisec, J., Cairns, J., Tiessen, A., Araus, J.L., Palacios-Rojas, N., Fernie, A.R., 2012. Metabolic and phenotypic responses of greenhouse-grown maize hybrids to experimentally controlled drought stress. Mol. Plant 5, 401–417. https://doi.org/10.1093/mp/ssr102.
Worley, B., Powers, R., 2013. Multivariate analysis in metabolomics. Curr. Metabolomics 1, 92–107. https://doi.org/10.2174/2213235X11301010092.

Wu, D., Cai, S., Chen, M., Ye, L., Chen, Z., Zhang, H., Dai, F., Wu, F., Zhang, G., 2013. Tissue metabolic responses to salt stress in wild and cultivated barley. PLoS One 8. https://doi.org/10.1371/journal.pone.0055431.

Xiang, Z., Wang, X.Q., Caib, X.J., Zenga, S., 2011. Metabolomics study on quality control and discrimination of three *Curcuma* species based on gas chromatograph-mass spectrometry. Phytochem. Anal. 22, 411–418. https://doi.org/10.1002/pca.1296.

Xi, Y., de Ropp, J.S., Viant, M.R., Woodruff, D.L., Yu, P., 2008. Improved identification of metabolites in complex mixtures using HSQC NMR spectroscopy. Anal. Chim. Acta 614, 127–133. https://doi.org/10.1016/j.aca.2008.03.024.

Xu, Y., Goodacre, R., 2012. Multiblock principal component analysis: an efficient tool for analyzing metabolomics data which contain two influential factors. Metabolomics 8, 37–51. https://doi.org/10.1007/s11306-011-0361-9.

Yan, H., Han, B.X., Wu, Q.Y., Jiang, M.Z., Gui, Z.Z., 2011. Rapid detection of *Rosa laevigata* polysaccharide content by near-infrared spectroscopy. Spectrochim. Acta - Part a Mol. Biomol. Spectrosc. 79, 179–184. https://doi.org/10.1016/j.saa.2011.02.032.

Ye, T., Zhang, S., Nagana Gowda, G.A., Raftery, D., 2010. Nuclear magnetic resonance and statistical analysis. In: Meyers, R.A. (Ed.), Encyclopedia of Analytical Chemistry. John Wiley, Hoboken, NJ.

Yi, L.Z., Yuan, D.L., Liang, Y.Z., Xie, P.S., Zhao, Y., 2009. Fingerprinting alterations of secondary metabolites of tangerine peels during growth by HPLC-DAD and chemometric methods. Anal. Chim. Acta 649, 43–51. https://doi.org/10.1016/j.aca.2009.07.009.

Yi, L., Dong, N., Yun, Y., Deng, B., Liu, S., Zhang, Y., Liang, Y., 2014. Recent advances in chemometric methods for plant metabolomics: a review. Biotechnol. Adv. https://doi.org/10.1016/j.biotechadv.2014.11.008.

Zhang, J., Zhang, Y., Du, Y., Chen, S., Tang, H., 2011. Dynamic metabonomic responses of tobacco (*Nicotiana tabacum*) plants to salt stress. J. Proteome Res. 10, 1904–1914. https://doi.org/10.1021/pr101140n.

Zhang, L., Nie, L., 2010. Discrimination of geographical origin and adulteration of radix astragali using fourier transform infrared spectroscopy and chemometric methods. Phytochem. Anal. 21, 609–615. https://doi.org/10.1002/pca.1242.

Zhao, Y.Y., Zhang, Y., Lin, R.C., Sun, W.J., 2009. An expeditious HPLC method to distinguish *Aconitum kusnezoffii* from related species. Fitoterapia 80, 333–338. http://dx.doi.org/10.1016/j.fitote.2009.04.005.

Zhao, C., Chan, H.Y., Yuan, D., Liang, Y., Lau, T.Y., Chau, F.T., 2011. Rapid simultaneous determination of major isoflavones of *Pueraria lobata* and discriminative analysis of its geographical origins by principal component analysis. Phytochem. Anal. 22, 503–508. https://doi.org/10.1002/pca.1308.

Zhao, L., Huang, Y., Hu, J., Zhou, H., Adeleye, A.S., Keller, A.A., 2016. 1H NMR and GC-MS based metabolomics reveal defense and detoxification mechanism of cucumber plant under nano-Cu stress. Environ. Sci. Technol. 50, 2000–2010. https://doi.org/10.1021/acs.est.5b05011.

Zhu, H., Wang, Y., Liang, H., Chen, Q., Zhao, P., Tao, J., 2010. Identification of *Portulaca oleracea* L. from different sources using GC-MS and FT-IR spectroscopy. Talanta 81, 129–135. https://doi.org/10.1016/j.talanta.2009.11.047.

Zomer, S., Dixon, S.J., Xu, Y., Jensen, S.P., Wang, H., Lanyon, C.V., O'Donnell, A.G., Clare, A.S., Gosling, L.M., 2009. Consensus multivariate methods in gas chromatography mass spectrometry and denaturing gradient gel electrophoresis: MHC-congenic and other strains of mice can be classified according to the profiles of volatiles and microflora in their scent-marks. Analyst 134, 114–123. https://doi.org/10.1039/B807061J.

Zörb, C., Geilfus, C.M., Mühling, K.H., Ludwig-Müller, J., 2013. The influence of salt stress on ABA and auxin concentrations in two maize cultivars differing in salt resistance. J. Plant Physiol. 170, 220–224. https://doi.org/10.1016/j.jplph.2012.09.012.

Chapter 2

Metabolomics of Seaweeds: Tools and Techniques

Bhakti Tanna[1,2], Avinash Mishra[1,2]
[1]*CSIR-Central Salt and Marine Chemicals Research Institute, Bhavnagar, India;* [2]*Academy of Scientific and Innovative Research, Council of Scientific and Industrial Research, New Delhi, India*

1. INTRODUCTION

1.1 Seaweeds

It is well known that life originated in the sea. Owing to evolution, marine organisms have acquired very sophisticated biochemical and physiological systems. Marine algal polysaccharides are being used as food and pharmaceuticals from ancient times in Asian and European countries. The interest in food supplement research has grabbed attention because of consumers' awareness about health-promoting foods worldwide. Exotic foods with proved nutritional values are gaining obscurity even in developed countries as food supplements (Kumar et al., 2011). The marine environment is a very rich source of potent compounds that have activities against various conditions such as cancer, inflammation, pain, and allergy. The annual harvesting of wet seaweeds is about 8 million tons worldwide, which also includes stranded seaweeds on the beach. Harvested seaweeds are utilized for different purposes such as food, fertilizer, and in phytochemical production in industries as a raw material (Shiney et al., 2014).

Seaweeds include several species of multicellular, macroscopic, marine algae growing in the intertidal zone. Commonly, seaweed is divided in three groups: Phaeophyceae, Rhodophyceae, and Chlorophyceae as per their thallus pigmentation pattern, which is brown, red, and green, respectively. In addition to pigment coloration, there are also other factors that contribute to seaweed classification: photosynthetic pigments, storage compounds, cell wall composition, and presence or absence of flagella. In terms of botany, instead of stems, roots, and leaves, seaweeds have stiles, holdfast, and blades (of fronds). Thus they rather absorb nutrients directly from seawater than extract by roots as higher plants. They are a rich source of primary and secondary metabolites. Currently, seaweed consumption is higher in Asian countries, which is around 66.5% for brown algae followed by red (33%) and green algae (5%). Different species that are consumed now-a-days have great nutritional value as a source of various metabolites that were characterized by a wide range of biological activities (Jayalakshmi et al., 2014). Therefore seaweeds possess not only nutritional value but also nutraceutical potentials such as antioxidant, antimutagenic, anticoagulant, anticancerous, and antibacterial activities. Seaweeds have proven potential natural antioxidants because of the presence of a variety of metabolites, such as polyphenols, tocopherols, alkaloids, and carotenoids. Seaweeds have an annual market value of US $7 billion because of their enormous commercial importance for human food, animal feed, pharmaceuticals, nutraceuticals, biofuels, soil conditioner, and hydrocolloids production (Kumar et al., 2016).

1.2 Metabolomics: An Overview

Metabolites are biochemical intermediates or end products of metabolic reactions catalyzed by various enzymes that naturally occur within cells and their levels can be regarded as the ultimate response of biological systems to genetic or environmental changes. The complete set of metabolites is considered as metabolome and includes endogenous and exogenous metabolites. Endogenous metabolites are produced by the organism, whereas exogenous metabolites (such as chemical contaminants, food additives, and toxins) are not produced naturally by the organism. Endogenous metabolites are divided as primary metabolites that are responsible for growth, development, and reproduction, such as amino acids, lipids, and carbohydrates, and secondary metabolites, such as flavonoids, phenolics, terpenoids, and alkaloids, which are

Plant Metabolites and Regulation Under Environmental Stress. http://dx.doi.org/10.1016/B978-0-12-812689-9.00002-9

not directly linked to the growth. Metabolomics is a branch of "omics" that is defined as the comprehensive qualitative and quantitative analysis of all (or a subset of) metabolites in a biological system at a specific time point (Kumar et al., 2016; Young and Alfaro, 2016). Metabolomics can be used to study the geographic effect on bioactivity, effect of storage condition, and even response of an organism under different environmental stress conditions (Arbona et al., 2013; Arapitsas et al., 2016; Renault et al., 2016; Wei et al., 2016; Greff et al., 2017; Pilatti et al., 2017). Metabolomics is a fast evolving field and an integrated analysis of all the three omics studies are expected to provide association between metabolite and gene expression profiles. There is web-based system termed as "3Omics," which is used for analyzing, integrating, and visualizing human transcriptomics, proteomics, and metabolomics data (Kuo et al., 2013). The combination of metabolomics and genomics has already been used to discover secondary metabolites of good prospective and astonishing diversity in bacteria (Maansson et al., 2016; Viszwapriya and Pandian, 2017) Unlike transcriptomics and proteomics, which mainly rely on the genomic information, metabolomics is species independent. Metabolites are a true reflection of the phenotype of the organism, because their identities, concentrations, and fluxes are the results of a complex interplay of the gene expression, protein expression, and environment. Furthermore, metabolomics requires less time to reoptimize protocols for a new species; therefore it can be applied to widely diverse species (Schauer and Fernie, 2006). Thus metabolomics provides a functional snapshot of the cellular phenotype of a biological system at a given time. Metabolomics is generally categorized by three approaches. Nontargeted (untargeted) metabolic profiling is used for the identification and (relative) quantification of as many compounds as possible that change or are synthesized in response to a stress condition, environmental changes, or genetic perturbation with the ultimate goal of identifying discriminating metabolites (Mishra et al., 2015; Pandey et al. 2015; Patel et al., 2016a). Targeted metabolite analysis focuses on a limited number of metabolites chosen based on prior knowledge. Metabolite fingerprinting is normally performed to classify samples based on a comprehensive metabolite profiling without identification of individual peaks (Steinfath et al., 2008). A well-conceived method and analytical techniques are prerequisite for the sample preparation procedure and to avoid the exclusion of any metabolite. The chosen analytical method must contain high resolving power to maintain selectivity, universal applicability, sensitivity, and matrix independence. We require adequate tools to handle, store, normalize, and evaluate the acquired complex metabolic data to mark out the systemic response of the biological system. Furthermore, a true metabolic approach should have a systematic plan to identify unknown metabolites and should be able to correlate analytical results with theoretical biochemical pathway models. Different analytical methods are available to separate and quantitate the metabolite of interest (Fig. 2.1). Characterization of a large number of metabolites is done by nontargeted metabolic profiling using techniques such as nuclear magnetic resonance (NMR), chromatography and mass spectrometry (GC-MS and LC-MS), matrix-assisted laser desorption/ionization time-of-flight mass spectroscopy (MALDI-TOF-MS), and liquid chromatography (LC) with electrochemical detection. Combinations of these techniques can also be used to expand the analysis of information collected and/or to augment separations of the obtained peaks. The collected data are combined and analyzed by one or more software packages designed for use with large data sets (Kaddurah-Daouk et al., 2008).

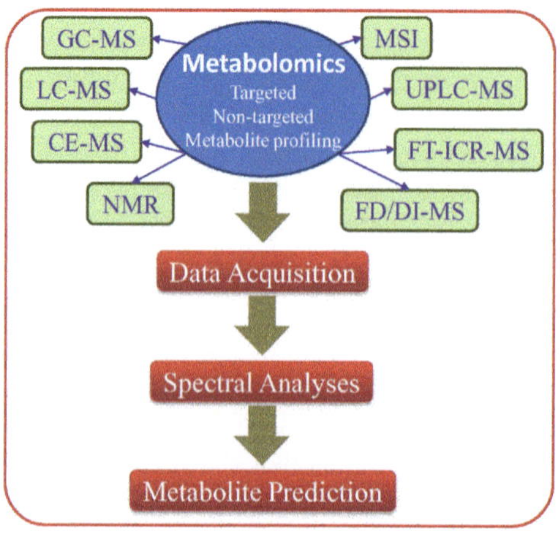

FIGURE 2.1 Metabolomics: analytical tools and techniques. *CE-MS*, capillary electrophoresis–mass spectrometry; *FD/DI-MS*, flow- or direct-injection/infusion MS; *FT-ICR-MS*, Fourier transform–ion cyclotron resonance–MS; *GC-MS*, gas chromatography–MS; *LC-MS*, liquid chromatography–MS; *MSI*, mass spectrometry imaging; *NMR*, nuclear magnetic resonance; *UPLC-MS*, ultra performance liquid chromatography–tandem MS. *(Adopted from Patel, M.K., Mishra A., Jha B., 2016b Untargeted metabolomics of halophytes. In: Kim S. (Ed.), Marine Omics: Principles and Applications, Taylor & Francis Group, CRC Press, Boca Raton, Florida, USA, pp. 309–325. https://doi.org/10.1201/9781315372303-18.)*

2. SAMPLING

2.1 Sampling and Sample Storage

When all metabolites from a biological system have to be studied, the data acquisition method is not the only technical question to be considered; one must also seriously consider adequate methods for sampling because the turnover of metabolome is very high in living organisms and it also changes very fast as per the surrounding environmental changes. Thus, considering both the factors, sampling should be as fast as possible to reduce any kind of biochemical activity occurring in cells. To increase the reproducibility of experiments, we need to rapidly inactivate all biochemical and enzymatic activities of the biological system; this process is termed as quenching. Different methods of quenching are available, such as the use of methanol (hot > 80°C or cold < 40°C), changes in pH using perchloric acid (PCA), or sudden temperature drops achieved by dipping the sample in liquid nitrogen (Mushtaq et al., 2014). Plants and algal samples are commonly stored in liquid nitrogen to stop any possibility of biochemical activities, which can alter the metabolite profile. Another method, drying, is also very crucial, as it inhibits enzymatic activity in the organism and facilitates longer shelf life of samples. One of the important factors is the water content, as presence of water may affect the solvation process and can also interfere with analytical instruments that are used for sample processing. For example, if NMR spectroscopy is being used, then traces of water present in the samples can distort the resolution of the spectra; it also creates loss of reproducibility in GC. There are different ways for drying but freeze-drying is the best, as it sublimates the water from the sample. Complete removal of water is not possible, as some water will be chemically bound; but it does not participate in chemical reactions or microbial growth, so it is not a threat for long-term storage. The completely dried samples should be stored at −80°C or in liquid nitrogen, but in any case, proper care should be taken so that reabsorption of water does not occur (Kim et al., 2010).

2.2 Sample Preparation

The rigid cell wall has to be broken to release metabolites and for that an effective extraction homogenization is very essential. There are different methods that disrupt the matrix or reduce the particle size. With the help of matrix disruption, the maximum amount of metabolites can be extracted, whereas solvent diffusion time can be decreased by smaller particle size (Bart and Pilz, 2011). There are many techniques for homogenizing and disrupting the matrix but mechanical shear seems to be the most efficient. It can be achieved by mortar and pestle but it is a laborious method. Many laboratories use automatic grinders, such as ball or cryomills, and acidic medium such as trichloroacetic acid or PCA for homogenization (Mushtaq et al., 2014). But care should be taken that whatever the method used the temperature should not exceed to activate enzymatic activities, as it can alter the metabolite profiling (Kim et al., 2010). Different extraction methods and solvent combinations are available and selection of an efficient method is key to obtain the maximum number of metabolites from the sample. The various extraction techniques include solvent extraction, supercritical fluid extraction, sonication, and solid phase extraction. Every extraction technique has its own advantages and disadvantages. Similarly, different solvents are available and certain properties such as toxicity, solubilization power, selectivity, dissolution rate, chemical reactivity, and pH should be considered while selecting a solvent (Schripsema, 2010). PCA is suitable for the extraction of polar, hydrophilic, or ionizable metabolites; in addition, it also denatures protein and inactivates all enzymatic reactions that can interfere in metabolomics (Kruger et al., 2008). Organic solvents are mainly used for moderately polar, nonpolar, or hydrophobic metabolites. The two-phase solvent systems, such as water–methanol–chloroform, are commonly used for not only polar, hydrophilic metabolites but also for nonpolar compounds, and have therefore been found suitable for plant metabolomics. Different solvents were reported for the plant metabolomics; Robbins et al. (2005) used 60% (v/v) aqueous methanol for the extraction of glucosinolates from *Brassica oleraceae* using high-performance (pressure) liquid chromatography (HPLC), whereas Moco et al. (2006) used 75% (v/v) aqueous methanol for the extraction of anthocyanins from *Solanum lycopersicum* using LC-MS. Similarly, hot 70% (v/v) methanol was used to extract phenolic acids from *B. oleraceae* using capillary electrophoresis (CE) LC-MS (Gratacós-Cubarsí et al., 2010). Till date, there is no comprehensive method that can give high recovery of all the metabolites with robustness and reproducibility. Yuliana et al. (2011) have developed a method with the ability to extract and preserve the integrity of a wide range of metabolites. They packed the fine plant tissue powder in kieselguhr, which is an anticlogging agent, and then a gradient of increasing polarity of solvents is then pumped through the column and all fractions were collected at different time points. It was claimed that the extract represents the total metabolome of the tissue.

3. ANALYTICAL TECHNIQUES

A number of different techniques have been reported to be used for metabolite profiling, such as CE, Fourier transform–ion cyclotron resonance–mass spectrometry (FT-ICR-MS or FTMS), HPLC, mass spectrometry (MS), liquid/gas chromatography (LC/GC), NMR, and MALDI-TOF-MS. Each technique is highly sophisticated but has an unavoidable intrinsic biased behavior to the specific metabolite profile (Hall et al., 2005). Selection of a technique should be based on its reproducibility for metabolite profiling and NMR is considered the most suitable in this context (Patel et al., 2016b). However, its sensitivity is quite low (1 μM–1 mM in an NMR tube) as compared with metabolic profiling by MS and chromatography-based platforms (Yuliana et al., 2011). Furthermore, NMR provides only a chemical fingerprint, whereas LC/ESI-MS resolves separate peaks of individual chemical components. Nonvolatile compounds that make a large portion of metabolites can be detected through NMR, whereas GC/MS can detect only volatile and thermostable compounds. The LC separation can reduce signal suppression and can be used to resolve isobaric compounds (Smith et al., 2006). For targeted metabolic profiling various methods are being used for different types of metabolites (Table 2.1). Each technique has its own advantages and disadvantages, but till date, no specific method has been employed for nontargeted metabolic profiling. Even the combination of analytical techniques can also be used to ensure the metabolite profiling reproducibility. Alsufyani et al., 2017 studied exometabolomic profiling in the green marine macroalga *Ulva* using multiple tools such as gas chromatography-mass spectrometry (GC-MS) and liquid chromatography-mass spectrometry (LC-MS) analysis as well as ultrahigh pressure liquid chromatography coupled to a time-of-flight mass spectrometer (TOF-MS).

3.1 Mass Spectrometry

The MS (or spectroscopy) is the primary detection technique widely used because of its sensitivity, speed, and broad applications. Selection of a separation method depends on the prepared extract and LC or GC is the most commonly used metabolite separation method that is coupled with the mass spectrometer (Sumner et al., 2003). CE separation is also gaining interest gradually (Sato et al., 2004).

TABLE 2.1 Analytical Techniques Used to Study Algal Metabolites

Algae	Targeted Metabolites	Techniques Employed	References
Grateloupia indica	Sulfated galactan	NMR spectroscopy	Chattopadhyay et al. (2007)
Iyengaria stellata	Sterols	Gas/liquid chromatography	Usmanghani et al. (1987)
Codium elongatum	Stigmasterol, hexadecanoic acid, and decortinone	NMR spectroscopy	Ali et al. (1999)
Enteromorpha intestinalis	β-Sitosterol	NMR spectroscopy	Ali et al. (1999)
Kappaphycus alvarezii	κ-Carrageenan	HPLC	Raman and Doble (2015)
Halimeda gracilis	β-Sitosterol and oleic acid	NMR	Hendri et al. (2016)
Acanthophora spicifera *Gracilaria foliifera* *Gracilaria edulis*	Fatty acids	GC and HPLC	Bhaskar et al. (2004)
Ulva fasciata *Dilophus fasciola* *Taonia atomaria* *Laurencia popillose* *Galaxoura cylindriea*	Phospholipids	LC-MS	El Baky et al. (2014)
Turbinaria conoides	Sulfated polysaccharides—fucoidan, alginic acid, laminaran (or laminarin)	GLC–MS	Chattopadhyay et al. (2010)
Porphyridium purpureum	Carotenoids—antheraxanthin, lutein, violaxanthin, xanthophylls, zeaxanthin	HPLC	Schubert et al. (2006)
Taonia atomaria	Phenolic compounds—stypodiol, isoepitaondiol, taondiol, stypoldione, sargaquinone, and sargaol	^{1}H and ^{13}C NMR GC-MS	Nahas et al. (2007)
Cystoseira sp.	Phenolic compounds—terpenoids	HRMS and ^{1}H NMR	Foti et al. (1994)
Gigartina skottsbergii, *Schizymenia binderi* *Lessonia vadosa*	Sulfated galactans—fucoidan	FT-IR	Barahona et al. (2011)

FTIR, Fourier-transform infrared spectroscopy; *GC*, gas chromatography; *GLC-MS*, gas liquid chromatography–mass spectrometry; *HPLC*, high-performance (pressure) liquid chromatography; *HRMS*, high-resolution mass spectrometry; *LC-MS*, liquid chromatography–mass spectrometry; *NMR*, nuclear magnetic resonance.

3.1.1 Gas Chromatography–Mass Spectrometry

This approach was used for metabolite profiling of Cabernet Sauvignon and Merlot cultivars during grapevine berry development, which showed a cultivar-dependent correlation between the primary metabolites and developmental stages (Cuadros-Inostroza et al., 2016). GC-MS is used for the separation and detection of naturally volatile or derivative volatile metabolites, such as alcohols, monoterpenes, and esters. The technique is commonly used for primary metabolites such as sugars and amino acids, which are nonvolatile and polar; by the use of chemical derivatization, these are converted into volatile and thermostable compounds. Single chromatogram can be developed for key primary metabolites by GC-MS. Because GC-MS is a robust separation and electron impact spectrometry technique it is the most popular global analysis method. GC-MS provides highly efficient metabolite separations, resolution, detection, and quantification of known metabolites even at low concentrations with remarkable reproducibility (Zhang et al., 2012). The identity of the detected metabolite is commonly confirmed by commercially available standards and databases (Desbrosses et al., 2005). The GC-MS is not suitable for thermolabile compounds such as di- and triphosphates, larger oligosaccharides, and high-molecular-weight (>500 Da) metabolites (Kopka, 2006).

One of the most important steps of this method is derivatization. Derivatization can be defined as a technical procedure in which a derivatizing reagent is used to produce a chemically related derivative of a chemical compound by modifying at one or more of its functional groups (Orata, 2012). A common derivatization procedure, used in routine GC-MS metabolomics occurs in two steps. The first step is methoxyamination, in which aldehyde or keto groups are converted to oxide with the use of methoxyamine; E- and Z-isomers of oximes get resolved with GC-MS and the second step is trimethylsilylation reaction to increase volatility of the derivative compound. Active hydrogens, such as $^{-}$OH, $^{-}$COOH, $^{-}$NH, or $^{-}$SH, react with derivatives and are converted into less polar, more thermostable and volatile trimethylsilyl (TMS) ether, TMS-ester, TMS-amine, or TMS-sulfide groups, respectively. *N*-methyl-trimethylsilyltrifluoroacetamide and *N,O*-bis-trimethylsilyltrifluoroacetamide are other silylation reagents, which are available and are the most common in metabolomics (Kopka et al., 2005).

The high reproducibility is partly achieved in GC-MS by the electron ionization (EI) method, in which sample molecules interact with kinetically activated electrons with an accepted average standard energy of 70 eV. Nowadays, quadrupole (Q) and orthogonal time of flight are the most commonly used mass analyzers but TOF-MS analyzers provide higher mass accuracy, higher duty cycles, and faster acquisition times than quadrupole mass analyzers. Thus this method is preferable for metabolic profiling, as it can accurately deconvolute overlapping GC peaks obtained from a complex extract; therefore time for comprehensive metabolome analysis is reduced (Villas-Bôas et al., 2005; Lei et al., 2011).

3.1.2 Liquid Chromatography–Mass Spectrometry

This approach has been used to study the effect of storage on red wine metabolite composition and also for studying bioprospection of marine macrophytes (Arapitsas et al., 2016; Maciel et al., 2016). LC-MS separates dissolved (liquid mobile phase) metabolites on the basis of their interaction with a solid stationary phase before their passage into a mass analyzer (Okazaki and Saito, 2012). LC-MS is suitable for heat-labile functional groups, chemically unstable structures, and high-vapor-point and high-molecular-weight secondary structures (>500 kDa), such as nonpolar metabolites, phospholipids, fatty acids, sterols, and steroids (Obata and Fernie, 2012). Nowadays, it has become possible to quickly perform a wide range of different analyses and obtain good efficiency in separations because of advances in particle sizes and column dimensions. Mostly reversed-phase (RP) stationary phases are used, which are composed of silica particles functionalized with hydrophobic alkyl chains; the most common phase is octade-cylsilane (i.e., phases with a C_{18} functional group bind covalently to the surface of the silica particle) (Chester, 2012). Mostly, organic solvent/aqueous mixed mobile phases, such as water:acetonitrile or water:methanol, are used in RP separations. Thus less polar compounds like primary metabolites such as carbohydrates and glycolytic intermediates will be retained in the nonpolar stationary phase (Dunn and Ellis, 2005).

Electron spray ionization (ESI) and atmospheric pressure ionization (API) are the most popular ionization methods that allow the analysis of biomolecules directly from the liquid phase and are thus suitable for LC coupling. ESI and API sources are soft ionization and they provide only limited structural information of molecules because they introduce little internal energy and generate a few fragments only (Holčapek et al., 2012). Detailed structural information is generally obtained through a number of fragments by collision of the analyte ions with inert gas molecules (e.g., He, Ar, or N_2), termed as the collision-induced dissociation (CID) process. CID is commonly used on a tandem MS instrument in which two or more sequential stages of mass spectrometric analysis can be allowed (tandem MS^n experiments). Tandem-in-time and tandem-in-space are two types of tandem MS^n instruments used in routine LC-MS metabolite analysis. Tandem-in-time includes ion-trapping instrument mass spectrometers, such as quadrupole ion traps (QIT-MS), the Orbitrap, and the Fourier-transform ion cyclotron resonance mass spectrometers (FT-ICR-MS). They can carry out fragmentation in different stages in which the first-generation product has significant power in structural elucidation, particularly of unknown metabolites, whereas other stages of fragmentation are needed to obtain structural information (MS^n analysis). The tandem-in-space instrument allows two sequential stages of mass spectrometric analysis (MS^2); it contains two mass analyzers separated by a collision cell (Schiffmann et al., 2014).

There are some limitations of LC-MS analysis, such as ion suppression in which compounds showing close resemblance get coeluted in a complex metabolite matrix (Trufelli et al., 2011). Multiple metabolites can compromise the accurate quantification of metabolites by ionization and the transfer of analytes from the liquid to gas phase (Wilson and Brinkman, 2003). Thus optimization is required in chromatographic separation to reduce the impact of matrix-induced effects in the mass spectrometer. Furthermore, LC-MS analysis requires authentic reference compounds for the identification and quantification of the measured metabolites (targeted metabolite), which limits the analysis of compounds that are not commercially available (Last et al., 2007).

3.1.3 Capillary Electrophoresis–Mass Spectrometry

CE has very high resolving power and efficient limits of detection but depends on the analyte charge to size ratio. However, the limit of detection depends on the class of metabolite to be analyzed, as not all classes of metabolites are equally amenable to CE-MS. CE is done in a fused silica capillary, the ends of which are dipped in buffer reservoirs across which high voltages (10–40 kV) are placed. Differences in the electrophoretic mobility of the different analytes are responsible for separation and occur over a bulk electro-osmotic flow toward the cathode (Nilsson et al., 2004).

CE-MS is faster and better than traditional LC-MS and GC-MS (Ramautar et al., 2009) and found suitable for the analysis of primary metabolites. A complicated interface was noticed between CE-MS and ESI ionization because of to the very low flow of the effluent from the capillary and applied voltage for electrophoretic separation. Electrical contact has to be maintained; in addition, in the running buffer, MS-compatible background electrolytes are needed. Owing to all these factors, reproducibility is poor for the CE-MS and a great deal of skill is required to use this technique effectively. Thus, owing to the lack of commercial solutions that are easy to set up and maintain, this technique is not widely accepted.

3.1.4 Fourier Transform–Ion Cyclotron Resonance–Mass Spectrometry

FT-ICR-MS is capable of very accurate mass measurements by ion trapping, which helps it to determine the empirical formula with the highest resolution and mass accuracy of any other mass spectrometers (Ohta et al., 2010). Orbitrap FTMS instruments have been recently introduced, which is not only a sensitive analytical technique (Hu et al., 2005) but also less expensive than the cyclotron FTMS machines (Sumner et al., 2003); thus it is predicted to have a bright future in the field of metabolomics (Dunn and Ellis, 2005). But the analytical method has a constraint, as it is unable to separate structural isomers with identical monoisomeric masses.

FT-ICR-MS with direct injection high-resolution mass spectrometry is a powerful ion-trapping system that has a very low detection limit. It also has good compatibility with MS^n analyzers. The high magnetic field associated with this instrument, the high cost, and complex ion–ion interactions are major obstacles for its widespread application and use in metabolomics researches (Jorge et al., 2016).

3.1.5 Ultra Performance Liquid Chromatography–Tandem Mass Spectrometer

Ultra performance liquid chromatography–tandem mass spectrometer (UPLC-MS) technology is a powerful technique, which consists of porous particles with internal diameters smaller than 2 mm and in conjunction with MS, it results in higher peak capacity, enhanced specificity, and high-throughput capabilities compared with conventional HPLC columns. Therefore it is suitable for metabolomics approaches. UPLC allows a more rapid analysis without loss of resolution because of the broader range of its optimum linear velocity. UPLC coupled with Q-TOF-MS can perform powerful and accurate mass measurement at less than 5 ppm level, with effective resolution, and thus can identify even trace components in complex mixtures. The UPLC-MS technique shows hyphenated microseparation platform in metabolomics, as a large number of primary metabolites are intrinsically polar (Zhang et al., 2012). This approach has already been used in the metabolomics study of marine compounds and to check their antitumor activity on cancer cell line (Gao et al., 2016).

3.1.6 Flow- or Direct-Injection/Infusion MS

This method is MS, in which a single mass spectrum is developed usually using the TOF (Verhoeven et al., 2006) or FTMS (Hirai et al., 2004) instrument. This method does not involve separation and so it is used not for quantification but as a rapid qualitative screening and also for comparative analysis screening for quality control, biodiversity analysis, mutant screening, etc. The run time can be kept as 1 min or even lesser (Goodacre et al., 2003).

3.1.7 Mass Spectrometry Imaging

Mass spectrometry imaging (MSI) helps to increase the understanding of the spatial distribution of various metabolites in different tissues and so provides more information about molecular processes occurring within these tissues (Lee et al., 2012; Matros and Mock, 2013). Sample extraction is not required but freeze-drying and cryosectioning are done by cooled (−20°C) cryomicrotome. Sample preparation is easy and analyses are normally performed in situ with ionization techniques that ionize and desorb the analytes directly from the tissue. Different ionization techniques are available, such as secondary ion mass spectrometry, in which primary ions (Ar^+, Ga^+, In^+) hit the sample surface inducing collision cascades with atoms and molecules to release secondary ions; MALDI, in which desorption of the analyte(s) from a matrix by a specific laser beam occurs; desorption electrospray ionization, in which impact interaction between electrospray formed charged droplets and the sample surface producing surface molecules dissolved in new charged droplets occurs; and laser ablation electrospray ionization, in which hybrid ionization method based on material ablation is done (van Hove et al., 2010; Wu et al., 2013). Among all these ionization techniques, MALDI is by far the most used method. Furthermore, MALDI-hybrid systems, such as MALDI-qTOF, MALDI-IT-Orbitrap, and MALDI-FT-ICR mass analyzers, are also considered as good options for metabolomics (Lee et al., 2012).

3.2 Nuclear Magnetic Resonance

NMR is being extensively used for metabolite fingerprinting, profiling, and metabolic flux analysis. The technique also has the ability to identify as well as quantify a number of organic compounds. NMR is a straightforward automated technique with a simple nondestructive sample preparation procedure and also does not require any chromatographic separation or derivatization. It was noticed that high-resolution NMR could provide an ideal mechanism for the profiling of metabolites but its sensitivity is less compared with MS methods; in addition, it is also pH sensitive (Wu et al., 2010). But because of the high quantitative accuracy, immense isotope editing potential, lack of standard requirement, structural identification, and direct isotopomer distribution analysis NMR has exceptional value for metabolic studies (Lane and Fan, 2017). ^{1}H NMR is fast and simple, so it has been the dominant profiling method but it suffers from signal overlapping in the profiling of complex metabolites. For such type of analysis, other advance NMR-based techniques, such as two-dimensional (2D) ^{1}H J-resolved NMR, heteronuclear single quantum coherence (HSQC) spectroscopy, total correlation spectroscopy, heteronuclear multiple quantum coherence and nuclear overhauser effect spectroscopy, and high-resolution magic angle spinning (HRMAS)-NMR spectroscopy is commonly used (Kim et al., 2011). The acquisition time can be longer for 2D NMR (20 min for 2D J-resolved spectroscopy) compared with one-dimensional (1D) NMR (only 1 min), but 2D NMR J-resolved spectroscopy spreads the resonances in a second dimension to overcome the spectral overlapping (as observed routinely in 1D NMR) for the identification and quantification of signals in the crowded spectral areas (Viant, 2003). In heteronuclear multiple quantum coherence (HMQC) and HSQC, any carbon atom directly attached to a proton is detected and the chemical shift of a ^{13}C atom connected to a specific proton can be obtained. HSQC is more sensitive than HMQC and thus the preferred method for the metabolic profiling, particularly for the determination of specific metabolites (Xia et al., 2008). In HRMAS-NMR, a solid lyophilized sample without extraction can directly be used. As in this method the sample is spun very fast (typically a few kilohertz) at a particular angle (magic angle) to the direction of the applied magnetic field, the line broadening is reduced and thus a high resolution is obtained. Intracellular low-molecular-weight metabolites and high-molecular-weight compounds, such as lipids and sugars, present in cell membranes and walls can be identified by this method (Beckonert et al., 2010). The 2D signal enhancement by spectral integration (SENSI) has been recently introduced to enhance the weak signals for a large metabolomics data set (Misawa et al., 2016) and by using advanced NMR tools and techniques, the total phytosterol profile of *Sargassum* seaweed has been obtained (Zhang et al., 2016).

4. SPECTRAL PROCESSING

Many software packages and databases are available for in silico data analysis of a vast amount of spectral data of metabolites generated by different analytical tools and techniques. These sophisticated bioinformatics tools are used for raw data processing, mining, integration, statistical analysis, management, as well as mathematical modeling for metabolomic networks. Generally, acquired data are processed for the adjustment of baseline shifts, reduction of background noise, peaks detection, alignment, and mass spectra deconvolution. After that, multivariate analyses are performed, such as principal components analysis, hierarchical clustering analysis, and partial least squares discriminant analysis (Alonso et al., 2015). The range of bioinformatic tools available for different analytical techniques is described in Table 2.2. Many of these data sets have been created by individual laboratories who are engaged in metabolomics studies; they share primary data sets from their own websites and also allow arbitrary laboratories to share their data sets publicly without even having to set up their own website, and thus serve as centralized metabolomics data repositories (Carroll, 2012).

After identification and quantification, the second crucial step is computational analysis of metabolites in a biological context and interpretation of metabolite data in terms of the underlying biochemical mechanisms and their phenotypic and physiological consequences (Wishart, 2010). There are also tools available for such kind of functional and biological interpretation of metabolome data set. These tools are mainly divided into two groups; the first creates graphical representation of the metabolic pathway of the metabolites by mapping and visualization, and the second comprises tools for the statistical analysis of metabolite annotations, commonly known as enrichment analysis. Automated and interactive tools are required because metabolism and other associated biological phenomena are very complex, and interactive tools make it easy to visualize the relation of metabolites to each other whether they are from the same metabolic pathway or close to each other in the metabolic network (Table 2.3). While in enrichment analysis (Table 2.4), targeted metabolites or a set of metabolites are annotated with the corresponding reference set and outline those annotations overrated in the interest set, as claimed with the accepted statistical test (Chagoyen and Pazos, 2013).

TABLE 2.2 Metabolome Databases

Database	Data Input	Source (URL)	References
ADAP-GC 2.0	GC/TOF-MS	http://www.du-lab.org	Ni et al. (2012)
AMDIS	GC-MS	http://www.amdis.net	Lu et al. (2008) Behrends et al. (2011)
Birmingham	NMR	http://www.bml-nmr.org	Ludwig et al. (2012)
BMRB	NMR	http://www.bmrb.wisc.edu	Markley et al. (2008)
COLMAR	NMR, TOCSY and HSQC	http://spin.ccic.ohio-state.edu/index.php/hsqc/	Bingol et al. (2014)
FiD	MS	https://www.cs.helsinki.fi/group/sysfys/software/fragid/	Kokkonen et al. (2008) Heinonen et al. (2008)
FiehnLib	GC-MS	http://fiehnlab.ucdavis.edu/projects/FiehnLib	Kind et al. (2009)
Golm	GC-MS	http://gmd.mpimp-golm.mpg.de	Hummel et al. (2007, 2010)
HMDB	NMR, LC-MS and GC-MS	http://www.hmdb.ca	Wishart et al. (2013)
KNApSAcK	MS	http://kanaya.naist.jp/KNApSAcK/	Afendi et al. (2012)
LMSD	MS	http://www.lipidmaps.org	Sud et al. (2007)
MMCD	NMR and MS	http://mmcd.nmrfam.wisc.edu	Cui et al. (2008)
MassBank	LC-, GC-, EI-, CE- and MALDI-MS	http://massbank.eu/MassBank/	Horai et al. (2010)
MassBase 1.0	MS	http://webs2.kazusa.or.jp/massbase	Carroll (2012)
MDL	1D and 2D NMR	http://www.liu.se/hu/mdl/main/	Lundberg et al. (2005)
MeltDB 2.0	GC-MS and LC-MS	https://meltdb.cebitec.uni-bielefeld.de/cgi-bin/login.cgi	Kessler et al. (2013)
MetaboAnalyst	NMR and LC/GC-MS	http://www.metaboanalyst.ca	Xia et al. (2009, 2015)
Metabolome Express	GC-MS	https://www.metabolome-express.org	Carroll et al. (2010)
MetaboMiner	2D NMR	http://wishart.biology.ualberta.ca/metabominer/	Xia et al. (2008)
MetaboSearch	m/z	http://omics.georgetown.edu/metabosearch.html	Zhou et al. (2012)
MET-IDEA	GC/LC-MS	http://bioinfo.noble.org/gateway/	Broeckling et al. (2006) Lei et al. (2012)
METLIN	MS	https://metlin.scripps.edu	Sana et al., 2008 Tautenhahn et al. (2012a)

TABLE 2.2 Metabolome Databases—cont'd

Database	Data Input	Source (URL)	References
MolFind	HPLC/MS	http://metabolomics.pharm.uconn.edu/Software.html	Menikarachchi et al. (2012)
Mzcloud	MS/MS	https://www.mzcloud.org	Mistrik et al. (2013)
MZmine	LC-MS	http://mzmine.github.io	Pluskal et al. (2010)
NAPROC-13	^{13}C NMR	http://c13.usal.es	López-Pérez et al. (2007)
NIST	GC/LC–MS	http://www.nist.gov/srd/nist1a.cfm	Naumkin et al. (2012)
NMR-ShiftDB	NMR	http://nmrshiftdb.nmr.uni-koeln.de	Steinbeck et al. (2003) Steinbeck and Kuhn (2004)
OpenMS	LC-MS	http://open-ms.sourceforge.net/about/	Sturm et al. (2008)
PMDB	Raw query sequence	https://bioinformatics.cineca.it/PMDB/	Castrignanò et al. (2006)
PRIMe	NMR, GC/LC-MS, and CE/MS	http://prime.psc.riken.jp	Akiyama et al. (2008) Matsuda et al. (2010) Matsuda et al. (2011) Sawada et al. (2012) Sakurai et al. (2013) Tsugawa et al. (2015)
SDBS	FTIR, NMR, and ESR	http://sdbs.db.aist.go.jp/sdbs/cgi-bin/cre_index.cgi	Yamaji et al. (2014)
XCMS	LC/GC-MS	https://xcmsonline.scripps.edu	Tautenhahn et al. (2012b)

CE, capillary electrophoresis; *EI*, electron ionization; *ESR*, electron spin resonance; *GC*, gas chromatography; *HPLC*, high-performance (pressure) liquid chromatography; *HSQC*, heteronuclear single quantum coherence; *LC*, liquid chromatography; *MALDI*, matrix-assisted laser desorption/ionization; *MS*, mass spectrometry; *m/z*, mass-to-charge ratio; *NMR*, nuclear magnetic resonance; *TOCSY*, total correlation spectroscopy; *TOF*, time of flight.

TABLE 2.3 Pathway Oriented Database

Database	URL	References
"Cyc" family MetaCyc BioCyc EcoCyc AraCyc 8.0 YeastCyc PlantCyc/PMN HumanCyc	http://metacyc.org http://biocyc.org http://ecocyc.org https://www.arabidopsis.org/biocyc/ http://pathway.yeastgenome.org http://www.plantcyc.org http://humancyc.org	Mueller et al. (2003) Carroll (2012) Caspi et al. (2014)
KappaView	http://kpv.kazusa.or.jp	Sakurai and Shibata (2006)
KEGG	http://www.genome.jp/kegg/	Kanehisa et al. (2014)
SMPDB	http://smpdb.ca	Jewison et al. (2013)

TABLE 2.4 Database for Metabolite Enrichment Analysis

Name	URL
IMPaLA	http://impala.molgen.mpg.de
MBRole	http://csbg.cnb.csic.es/mbrole/
MPEA	http://ekhidna.biocenter.helsinki.fi/poxo/mpea/
MSEA	http://www.metaboanalyst.ca

5. BIOLOGICAL ACTIVITY OF ALGAL METABOLITES

Seaweeds contain several biologically active compounds, which may be helpful in the treatment of various diseases. There are several reports indicating the beneficial effects of algal metabolites with respect to the treatment of many diseases. Furthermore, as they are extracted from natural sources, the harmful effect will be negligible and in the near future these will be a preferred source for treatments. The antibacterial activity of marine macro algae *Amphiroa anceps* using three solvent extracts, aqueous, methanol, and ethyl acetate, was studied and the highest zone of inhibition was detected with ethyl acetate extract against the bacteria *Enterobacter* sp. (Shiney et al., 2014). The phytochemical analysis of *A. anceps* showed the presence of flavonoids, amino acids, and terpenoids in the methanolic extract and protein, phenol, flavonoids, and terpenoids in the ethyl acetate extract, whereas only tannins were observed in the aqueous extract. Selected seaweeds, i.e., *Ulva fasciata*, *Chaetomorpha antennina*, *Spyridia hypnoides*, *Amphyroa anceps*, *Sargassum wightii*, *Chnoospora maxima* from Kanyakumari (India) coast, were extracted with methanol to study their biochemical composition and in vitro antioxidant properties (Jayalakshmi et al., 2014). Among these six seaweeds, the highest protein content was found in *S. wightii*, whereas the maximum carbohydrate and lipid content were recorded in *C. maxima*. The maximum phenolic content was also detected in *C. maxima*. Furthermore, high antioxidant activities were reported for brown algae *S. wightii* and *C. maxima*, which complements the fact that brown algae contains high phenolic contents. *Grateloupia filicina* polysaccharide was observed to improve immune-related gene expression in *Litopenaeus vannamei* (AiYing and LinYuan, 2014). Researchers fed different concentrations of *G. filicina* polysaccharide to shrimps and observed the differential expression of seven immune-related genes in hepatopancreas tissues using real-time quantitative polymerase chain reaction. The results showed that the expression level of hemocyanin, prophenoloxidase, and heat shock protein 70 significantly increased in the treatment groups and thus they concluded that *G. filicina* polysaccharide supplementation can be used as a substitute for chemical drugs for *L. vannamei* treatments. Sulfated polysaccharides were extracted from a different class of algae; red seaweeds (Rhodophyta) *Grateloupia indica*, *Scinaia hatei*, and *Gracilaria corticata*; brown seaweeds (Phaeophyta) *Stoechospermum marginatum* and *Cystoseira indica*; and green seaweed (Chlorophyta) *Caulerpa racemosa* and their antiviral activities were studied against dengue virus (Pujol et al., 2012). Antiviral activities were observed for all algal polysaccharides against DENV-2 with IC_{50} in the range 0.12–20 µg/mL. The highest activity was observed with sulfated galactan from *G. indica* with IC_{50} values around 0.1 µg/mL. Sulfated galactan from *G. indica* also showed antiviral activity for other viruses such as herpes simplex virus (Chattopadhyay et al., 2007). Different antiinflammatory activities, such as *MMP-9* expressions, mucin synthesis, and eosinophil infiltration, were found with *Eucheuma cottonii*, an edible red alga, on rat model (Anyanji et al., 2015). The reverse-phase HPLC and LC-MS repeatedly showed the absence of phlorotannins, phloroglucinol, rutin, or quercetin in aqueous methanol (50%v/v) extract, whereas Fourier-transform infrared spectroscopy results showed predominant bioactive compounds, simple proteins, or peptides. The study demonstrated that total white blood cells, eosinophil, band neutrophil, segmented neutrophil, and monocyte counts were significantly decreased. Furthermore, 300 mg extract/kg eased lung inflammation and airway smooth muscle constriction in rat compared with the 10 mg/kg loratadine treatment. Other activities of *E. cottonii* have also been studied, such as tumor suppressor activity by polyphenolic rich extract (Namvar et al., 2012). In vitro assays showed antiproliferative activity of the extract against estrogen-dependent MCF-7 and estrogen-independent MB-MDA-231 human breast cancer cells (IC_{50} values of 20 and 42 µg/mL, respectively) but was nontoxic to normal cell lines. Its extract also inhibited tumor development and erythrocyte lipid peroxidation in the cancer-induced rats in as dose-dependent manner. Sugiura et al. (2007) purified antiallergic phlorotannins (eckol, 0,0-bieckol, 0,2-bieckol, 2,2-bieckol, phlorofucofuroeckol-A, and phlorofucofuroeckol-B) from *Eisenia arborea*, an edible brown alga, using HPLC. All of the isolated phlorotannins showed inhibition of b-Hex release from RBL-H cells. *S. marginatum* extract was studied for biological activities and chemical composition by GC-MS (Esmaeili and Khakpoor, 2012). Polysaccharide κ-carrageenan has been isolated from marine red algae, *Kappaphycus alvarezii*, and checked against human colon cancer cells (Raman and Doble, 2015). Exposure of human colon cancer cells, HCT116, resulted in apoptotic cell death; apoptosome formation; nuclear fragmentation; upregulation of the mRNA expression levels of Bcl-xL, Bcl-2, and caspase3; and downregulation of PARP-1 and XIAP. *Solieria robusta* methanolic extract has shown activity against oral cancer Ca9-22 cells in terms of apoptosis and oxidative stress (Yen et al., 2014). The extract showed the presence of sugar moiety and fatty acid in NMR spectra. Annexin V/PI assay of apoptosis by using flow cytometry showed that the viability of oral cancer Ca9-22 cells decreased in a dose-responsive manner. They performed propidium iodide (PI)-based assay, which proved that cell cycle distribution was dramatically changed, mainly for subG1 accumulation. For oxidative stress evaluation, reactive oxygen species was found to be increased in a dose- and time-responsive manner, which leads to mitochondrial depolarization. The hypolipidemic potential of some 13 selected seaweeds has been studied by taking normal, triton-induced rats for examination (Ruqqia et al., 2015). Eight seaweed extracts, *Dictyota dichotoma* var. *intrica*, *Jolyna laminarioides*, *Melanothamnus afaqhusainii*, *Sargassum binderi*, *Sargassum swartzii*, *Halimeda tuna*, *Spatoglossum variabile*, and *S. robusta*, were studied for their efficacy and finally three potential seaweeds, *J. laminarioides*, *S. binderi*, and *M. afaqhusainii*, were further evaluated in rats with high-fat diet–induced hyperlipidemia (Ruqqia et al., 2015). Even some of the seaweed hypolipidemic activity was comparable with that of drugs such as bezafibrate (Lipocor) and fenofibrate. *Ulva intestinalis* collected from the coast of Zhejiang, China, has shown algicidal activity against toxic red tide microalgae *Heterosigma akashiwo* and *Prorocentrum micans* (Sun et al., 2015). EtOAc extracts gave three algicidal compounds, which were analyzed by NMR spectroscopy: (1) 15-ethoxy-(6z,9z,12z)-hexadecatrienoic acid, (2) (6E,9E,12E)-(2-acetoxy -β-D-glucose)-octadecatrienoic acid ester, and (3) hexadecanoic acid. From these, compound 2 is the most active against microalgae. Sulfated polysaccharides from *U. fasciata, Gloiopeltis furcata*, and *Sargassum henslouianum* have been studied for the relation between the sulfate content and in vitro antioxidant and antitumor activity (Shao et al., 2013). *U. fasciata* with a lower sulfate content showed excellent antioxidant activities in (2,2'-azino-bis(3-ethylbenzothiazoline-6-sulphonic acid)) assay, (2,2-diphenyl-1-picrylhydrazyl) assay, and superoxide radical assay and inhibitory activity on different cell lines such as MKN45 and DLD cancer cells. *S. henslouianum* was observed to have the lowest sulfate content and gave very less activity.

6. CONCLUSION AND FUTURE PROSPECTS

Marine algae are a rich source of distinct metabolites that have nutritional, nutraceutical, and pharmaceutical applications. Metabolomics, an emerging field of "omics," is rapidly evolving and becoming the major tool for fast and accurate analysis of high-throughput data to unravel new knowledge on diverse metabolites produced from different sources. METLIN and XCMS are considered the most useful databases for the identification of metabolites, whereas KEGG is one of the most widely used databases for elucidating and correlating metabolic pathways. Today, different types of analytical tools are available and day-by-day computational and bioinformatics technology is evolving and thus metabolomics studies will be boosted in the near future. However, for untargeted metabolomics major improvements are still needed. There are some limitations that should be taken care of while using this approach, such as external influence, which may cause sensitive effects while collecting sample or biological interpretation, which need complex bioinformatics procedures. Metabolite identification, annotation, and spectral deconvolution are the major bottleneck of untargeted metabolomics. The scientific community requires tools and methods that can be integrated efficiently in all different sources of "omics" fields, such as genomics, proteomics, and metabolomics, to develop a better understanding of an overall biological system. Some developments such as advances in metabolite biomarker libraries and robust analytical platforms are required. Furthermore, sensitivity improvement, metabolite coverage, as well as manageable bioinformatics packages are also needed to deduce a logical result. The database of metabolites from seaweeds is expected to integrate with transcriptomics and proteomics, which will be useful in future for the discovery of novel metabolic pathways. The advancement in "omics" will be able to allow new approaches to drug discovery. Advances in these studies will soon be able to produce high-quality data of secondary metabolites of seaweeds at a very high speed. Owing to the presence of various biologically active metabolites, seaweeds can be used in different ways compared with synthetic drugs, as naturally derived metabolites are safe to our body. Algal metabolites can be considered as the breakthrough in pharmaceutical research as well as nutraceutical applications as a functional food. Thus we can say metabolomics of seaweeds will be able to contribute a lot to scientific research fields such as medical science, food science, as well as agriculture.

ACKNOWLEDGMENTS

CSIR-CSMCRI Communication No.: PRIS-002/2017.

REFERENCES

Afendi, F.M., Okada, T., Yamazaki, M., Hirai-Morita, A., Nakamura, Y., Nakamura, K., Ikeda, S., Takahashi, H., Altaf-Ul-Amin, M., Darusman, L.K., Saito, K., 2012. KNApSAcK family databases: integrated metabolite–plant species databases for multifaceted plant research. Plant Cell Physiol. 53 (2), e1–e12.

AiYing, L., LinYuan, J., 2014. Effects of *Grateloupia filicina* polysaccharide on immune-related gene expression in *Litopenaeus vannamei*. J. South. Agric. 45 (4), 682–686.

Akiyama, K., Chikayama, E., Yuasa, H., Shimada, Y., Tohge, T., Shinozaki, K., Hirai, M.Y., Sakurai, T., Kikuchi, J., Saito, K., 2008. PRIMe: a Web site that assembles tools for metabolomics and transcriptomics. In Silico Biol. 8 (3, 4), 339–345.

Ali, M.S., Ahmad, V.U., Mahzar, F., Azhar, I., Usmanghani, K., 1999. Some chemical constituents from marine algae of Karachi coast (Arabian Sea). Turkish J. Chem. 23 (2), 181–184.

Alonso, A., Marsal, S., Julià, A., 2015. Analytical methods in untargeted metabolomics: state of the art in 2015. Front. Bioeng. Biotechnol. 3, 23.

Alsufyani, T., Weiss, A., Wichard, T., 2017. Time course exo-metabolomic profiling in the green marine macroalga *Ulva* (Chlorophyta) for identification of growth phase-dependent biomarkers. Mar. drugs 15 (1), 14.

Anyanji, V.U., Mustapha, N.M., Lim, S.L., Mohamed, S., 2015. Seaweed (*Eucheuma cottonii*) reduced inflammation, mucin synthesis, eosinophil infiltration and MMP-9 expressions in asthma-induced rats compared to Loratadine. J. Funct. Foods 19, 710–722.

Arapitsas, P., Della Corte, A., Gika, H., Narduzzi, L., Mattivi, F., Theodoridis, G., 2016. Studying the effect of storage conditions on the metabolite content of red wine using HILIC LC–MS based metabolomics. Food Chem. 197, 1331–1340.

Arbona, V., Manzi, M., Ollas, C.D., Gómez-Cadenas, A., 2013. Metabolomics as a tool to investigate abiotic stress tolerance in plants. Int. J. Mol. Sci. 14 (3), 4885–4911.

Barahona, T., Chandía, N.P., Encinas, M.V., Matsuhiro, B., Zúñiga, E.A., 2011. Antioxidant capacity of sulfated polysaccharides from seaweeds, A kinetic approach. Food Hydrocoll. 25 (3), 529–535.

Bart, H.J., Pilz, S., 2011. Industrial Scale Natural Products Extraction. John Wiley & Sons.

Beckonert, O., Coen, M., Keun, H.C., Wang, Y., Ebbels, T.M., Holmes, E., Lindon, J.C., Nicholson, J.K., 2010. High-resolution magic-angle-spinning NMR spectroscopy for metabolic profiling of intact tissues. Nat. Protoc. 5 (6), 1019–1032.

Behrends, V., Tredwell, G.D., Bundy, J.G., 2011. A software complement to AMDIS for processing GC-MS metabolomic data. Anal. Biochem. 415 (2), 206–208.

Bhaskar, N., Kinami, T., Miyashita, K., Park, S.B., Endo, Y., Fujimoto, K., 2004. Occurrence of conjugated polyenoic fatty acids in seaweeds from the Indian Ocean. Z. Naturforsch. C 59 (5–6), 310–314.

Bingol, K., Li, D.W., Bruschweiler-Li, L., Cabrera, O.A., Megraw, T., Zhang, F., Brüschweiler, R., 2014. Unified and isomer-specific NMR metabolomics database for the accurate analysis of 13C–1H HSQC spectra. ACS Chem. Biol. 10 (2), 452–459.

Broeckling, C.D., Reddy, I.R., Duran, A.L., Zhao, X., Sumner, L.W., 2006. MET-IDEA: data extraction tool for mass spectrometry-based metabolomics. Anal. Chem. 78 (13), 4334–4341.

Carroll, A.J., Badger, M.R., Millar, A.H., 2010. The Metabolome Express Project: enabling web-based processing, analysis and transparent dissemination of GC/MS metabolomics datasets. BMC Bioinforma. 11, 376.

Carroll, A.J., 2012. Online Metabolomics Databases and Pipelines. In Tech, Rijeka, Croatia, pp. 47–73.

Caspi, R., Altman, T., Billington, R., Dreher, K., Foerster, H., Fulcher, C.A., Holland, T.A., Keseler, I.M., Kothari, A., Kubo, A., Krummenacker, M., 2014. The MetaCyc database of metabolic pathways and enzymes and the BioCyc collection of Pathway/Genome Databases. Nucleic Acids Res. 42 (D1), D459–D471.

Castrignanò, T., De Meo, P.D., Cozzetto, D., Talamo, I.G., Tramontano, A., 2006. The PMDB protein model database. Nucleic Acids Res. 34 (Suppl. 1), D306–D309.

Chagoyen, M., Pazos, F., 2013. Tools for the functional interpretation of metabolomic experiments. Briefings Bioinforma. 14 (6), 737–744.

Chattopadhyay, K., Mateu, C.G., Mandal, P., Pujol, C.A., Damonte, E.B., Ray, B., 2007. Galactan sulfate of *Grateloupia indica*: isolation, structural features and antiviral activity. Phytochemistry 68 (10), 1428–1435.

Chattopadhyay, N., Ghosh, T., Sinha, S., Chattopadhyay, K., Karmakar, P., Ray, B., 2010. Polysaccharides from *Turbinaria conoides*: structural features and antioxidant capacity. Food Chem. 118 (3), 823–829.

Chester, T.L., 2012. Recent developments in high-performance liquid chromatography stationary phases. Anal. Chem. 85 (2), 579–589.

Cuadros-Inostroza, A., Ruíz-Lara, S., González, E., Eckardt, A., Willmitzer, L., Peña-Cortés, H., 2016. GC–MS metabolic profiling of *Cabernet Sauvignon* and *Merlot* cultivars during grapevine berry development and network analysis reveals a stage-and cultivar-dependent connectivity of primary metabolites. Metabolomics 12 (2), 39.

Cui, Q., Lewis, I.A., Hegeman, A.D., Anderson, M.E., Li, J., Schulte, C.F., Westler, W.M., Eghbalnia, H.R., Sussman, M.R., Markley, J.L., 2008. Metabolite identification via the madison metabolomics consortium database. Nat. Biotechnol. 26 (2), 162–164.

Desbrosses, G.G., Kopka, J., Udvardi, M.K., 2005. *Lotus japonicus* metabolic profiling. Development of gas chromatography-mass spectrometry resources for the study of plant-microbe interactions. Plant Physiol. 137 (4), 1302–1318.

Dunn, W.B., Ellis, D.I., 2005. Metabolomics: current analytical platforms and methodologies. TrAC Trends Anal. Chem. 24 (4), 285–294.

El Baky, H.H.A., El Baz, F.K., El Baroty, G.S., Asker, M.M.S., Ibrahim, E.A., 2014. Phospholipids of some marine macroalgae: identification, antivirus, anticancer and antimicrobial bioactivities. Pharm. Chem. 6, 370–382.

Esmaeili, A., Khakpoor, M., 2012. Biological activities and chemical composition of solvent extracts of *Stoechospermum marginatum* (C. Agardh). Acta Biochim. Pol. 59, 581–585.

Foti, M., Piattelli, M., Amico, V., Ruberto, G., 1994. Antioxidant activity of phenolic meroditerpenoids from marine algae. J. Photochem. Photobiol. B Biol. 26 (2), 159–164.

Gao, D., Wang, Y., Xie, W., Yang, T., Jiang, Y., Guo, Y., Guan, J., Liu, H., 2016. Metabolomics study on the antitumor effect of marine natural compound flexibilide in HCT-116 colon cancer cell line. J. Chromatogr. B 1014, 17–23.

Goodacre, R., York, E.V., Heald, J.K., Scott, I.M., 2003. Chemometric discrimination of unfractionated plant extracts analyzed by electrospray mass spectrometry. Phytochemistry 62 (6), 859–863.

Gratacós-Cubarsí, M., Ribas-Agustí, A., García-Regueiro, J.A., Castellari, M., 2010. Simultaneous evaluation of intact glucosinolates and phenolic compounds by UPLC-DAD-MS/MS in *Brassica oleracea* L. var. botrytis. Food Chem. 121 (1), 257–263.

Greff, S., Zubia, M., Payri, C., Thomas, O.P., Perez, T., 2017. Chemogeography of the red macroalgae *Asparagopsis*: metabolomics, bioactivity, and relation to invasiveness. Metabolomics 13 (4), 33.

Hall, R.D., de Vos, C.R., Verhoeven, H.A., Bino, R.J., 2005. Metabolomics for the assessment of functional diversity and quality traits in plants. In: Metabolome Analyses: Strategies for Systems Biology. Springer, Berlin, Heidelberg, pp. 31–44.

Heinonen, M., Rantanen, A., Mielikäinen, T., Kokkonen, J., Kiuru, J., Ketola, R.A., Rousu, J., 2008. FiD: a software for ab initio structural identification of product ions from tandem mass spectrometric data. Rapid Commun. Mass Spectrom. 22 (19), 3043–3052.

Hendri, J.S., Prayitno, B., Radjasa, O.K., Elvita, M., 2016. The isolation of metabolite compounds from seaweed (*Halimeda Gracillis*) in the waters of Teluk Lampung as a source of antibacterial compounds. Glob. J. Sci. Front. Res. 15 (2), 22–32.

Hirai, M.Y., Yano, M., Goodenowe, D.B., Kanaya, S., Kimura, T., Awazuhara, M., Arita, M., Fujiwara, T., Saito, K., 2004. Integration of transcriptomics and metabolomics for understanding of global responses to nutritional stresses in *Arabidopsis thaliana*. Proc. Natl. Academy Sci. U. S. A. 101 (27), 10205–10210.

Holčapek, M., Jirásko, R., Lísa, M., 2012. Recent developments in liquid chromatography–mass spectrometry and related techniques. J. Chromatogr. A 1259, 3–15.

Horai, H., Arita, M., Kanaya, S., Nihei, Y., Ikeda, T., Suwa, K., Ojima, Y., Tanaka, K., Tanaka, S., Aoshima, K., Oda, Y., 2010. MassBank: a public repository for sharing mass spectral data for life sciences. J. Mass Spectrom. 45 (7), 703–714.

Hu, Q., Noll, R.J., Li, H., Makarov, A., Hardman, M., Graham Cooks, R., 2005. The Orbitrap: a new mass spectrometer. J. Mass Spectrom. 40 (4), 430–443.

Hummel, J., Selbig, J., Walther, D., Kopka, J., 2007. The Golm Metabolome Database: a database for GC-MS based metabolite profiling. In: Metabolomics. Springer, Berlin, Heidelberg, pp. 75–95.

Hummel, J., Strehmel, N., Selbig, J., Walther, D., Kopka, J., 2010. Decision tree supported substructure prediction of metabolites from GC-MS profiles. Metabolomics 6 (2), 322–333.

Jayalakshmi, J., Subramanian, V., Anantharaman, P., 2014. Evaluation of biochemical composition and in vitro antioxidant properties of selected seaweeds from Kanyakumari coast, Tamil Nadu, India. Adv. Appl. Sci. Res. 5 (4), 74–81.

Jewison, T., Su, Y., Disfany, F.M., Liang, Y., Knox, C., Maciejewski, A., Poelzer, J., Huynh, J., Zhou, Y., Arndt, D., Djoumbou, Y., 2013. SMPDB 2.0: big improvements to the small molecule pathway database. Nucleic Acids Res. 6, gkt1067.

Jorge, T.F., Rodrigues, J.A., Caldana, C., Schmidt, R., van Dongen, J.T., Thomas-Oates, J., António, C., 2016. Mass spectrometry-based plant metabolomics: metabolite responses to abiotic stress. Mass Spectrom. Rev. 35 (5), 620–649.

Kaddurah-Daouk, R., Kristal, B.S., Weinshilboum, R.M., 2008. Metabolomics: a global biochemical approach to drug response and disease. Annu. Rev. Pharmacol. Toxicol. 48, 653–683.

Kanehisa, M., Goto, S., Sato, Y., Kawashima, M., Furumichi, M., Tanabe, M., 2014. Data, information, knowledge and principle: back to metabolism in KEGG. Nucleic Acids Res. 42 (D1), D199–D205.

Kessler, N., Neuweger, H., Bonte, A., Langenkämper, G., Niehaus, K., Nattkemper, T.W., Goesmann, A., 2013. MeltDB 2.0–advances of the metabolomics software system. Bioinformatics 29 (19), 2452–2459.

Kim, H.K., Choi, Y.H., Verpoorte, R., 2010. NMR-based metabolomic analysis of plants. Nat. Protoc. 5 (3), 536–549.

Kim, H.K., Choi, Y.H., Verpoorte, R., 2011. NMR-based plant metabolomics: where do we stand, where do we go? Trends Biotechnol. 29 (6), 267–275.

Kind, T., Wohlgemuth, G., Lee, D.Y., Lu, Y., Palazoglu, M., Shahbaz, S., Fiehn, O., 2009. FiehnLib: mass spectral and retention index libraries for metabolomics based on quadrupole and time-of-flight gas chromatography/mass spectrometry. Anal. Chem. 81 (24), 10038–10048.

Kokkonen, J., Kiuru, J., Ketola, R.A., Rousu, J., 2008. FiD: a software for ab initio structural identification of product ions from tandem mass spectrometric data. Mass Spectrom. 22, 3043.

Kopka, J., Schauer, N., Krueger, S., Birkemeyer, C., Usadel, B., Bergmüller, E., Dörmann, P., Weckwerth, W., Gibon, Y., Stitt, M., Willmitzer, L., 2005. GMD@ CSB. DB: the Golm metabolome database. Bioinformatics 21 (8), 1635–1638.

Kopka, J., 2006. Current challenges and developments in GC–MS based metabolite profiling technology. J. Biotechnol. 124 (1), 312–322.

Kruger, N.J., Troncoso-Ponce, M.A., Ratcliffe, R.G., 2008. 1H NMR metabolite fingerprinting and metabolomic analysis of perchloric acid extracts from plant tissues. Nat. Protoc. 3 (6), 1001–1012.

Kumar, M., Kumari, P., Trivedi, N., Shukla, M.K., Gupta, V., Reddy, C.R., Jha, B., 2011. Minerals, PUFAs and antioxidant properties of some tropical seaweeds from Saurashtra coast of India. J. Appl. Phycol. 23 (5), 797–810.

Kumar, M., Kuzhiumparambil, U., Pernice, M., Jiang, Z., Ralph, P.J., 2016. Metabolomics: an emerging frontier of systems biology in marine macrophytes. Algal Res. 16, 76–92.

Kuo, T.C., Tian, T.F., Tseng, Y.J., 2013. 3Omics: a web-based systems biology tool for analysis, integration and visualization of human transcriptomic, proteomic and metabolomic data. BMC Syst. Biol. 7 (1), 64.

Lane, A.N., Fan, T.W., 2017. NMR-based stable isotope resolved metabolomics in systems biochemistry. Arch. Biochem. Biophys. http://dx.doi.org/10.1016/j.abb.2017.02.009.

Last, R.L., Jones, A.D., Shachar-Hill, Y., 2007. Towards the plant metabolome and beyond. Nat. Rev. Mol. Cell Biol. 8 (2), 167–174.

Lee, Y.J., Perdian, D.C., Song, Z., Yeung, E.S., Nikolau, B.J., 2012. Use of mass spectrometry for imaging metabolites in plants. Plant J. 70 (1), 81–95.

Lei, Z., Huhman, D.V., Sumner, L.W., 2011. Mass spectrometry strategies in metabolomics. J. Biol. Chem. 286 (29), 25435–25442.

Lei, Z., Li, H., Chang, J., Zhao, P.X., Sumner, L.W., 2012. MET-IDEA version 2.06; improved efficiency and additional functions for mass spectrometry-based metabolomics data processing. Metabolomics 8 (1), 105–110.

López-Pérez, J.L., Therón, R., del Olmo, E., Díaz, D., 2007. NAPROC-13: a database for the dereplication of natural product mixtures in bioassay-guided protocols. Bioinformatics 23 (23), 3256–3257.

Lu, H., Liang, Y., Dunn, W.B., Shen, H., Kell, D.B., 2008. Comparative evaluation of software for deconvolution of metabolomics data based on GC-TOF-MS. TrAC Trends Anal. Chem. 27 (3), 215–227.

Ludwig, C., Easton, J.M., Lodi, A., Tiziani, S., Manzoor, S.E., Southam, A.D., Byrne, J.J., Bishop, L.M., He, S., Arvanitis, T.N., Günther, U.L., 2012. Birmingham Metabolite Library: a publicly accessible database of 1-D 1H and 2-D 1H J-resolved NMR spectra of authentic metabolite standards (BML-NMR). Metabolomics 8 (1), 8–18.

Lundberg, P., Vogel, T., Malusek, A., Lundquist, P.O., Cohen, L., Dahlqvist, O., 2005. MDL—the magnetic resonance metabolomics database (mdl. imv. liu. se). In: 22nd Annual Meeting of the European Society for Magnetic Resonance in Medicine and Biology, Magnetic Resonance Materials in Physics, Biology and Medicine.

Maansson, M., Vynne, N.G., Klitgaard, A., Nybo, J.L., Melchiorsen, J., Nguyen, D.D., Sanchez, L.M., Ziemert, N., Dorrestein, P.C., Andersen, M.R., Gram, L., 2016. An integrated metabolomic and genomic mining workflow to uncover the biosynthetic potential of bacteria. mSystems 1 (3) e00028–15.

Maciel, E., Leal, M.C., Lillebø, A.I., Domingues, P., Domingues, M.R., Calado, R., 2016. Bioprospecting of marine macrophytes using MS-based lipidomics as a new approach. Mar. Drugs 14 (3), 49.

Markley, J.L., Ulrich, E.L., Berman, H.M., Henrick, K., Nakamura, H., Akutsu, H., 2008. BioMagResBank (BMRB) as a partner in the Worldwide Protein Data Bank (wwPDB): new policies affecting biomolecular NMR depositions. J. Biomol. NMR 40 (3), 153–155.

Matros, A., Mock, H.P., 2013. Mass spectrometry based imaging techniques for spatially resolved analysis of molecules. Front. Plant Sci. 4, 89.

Matsuda, F., Hirai, M.Y., Sasaki, E., Akiyama, K., Yonekura-Sakakibara, K., Provart, N.J., Sakurai, T., Shimada, Y., Saito, K., 2010. AtMetExpress development: a phytochemical atlas of *Arabidopsis* development. Plant Physiol. 152 (2), 566–578.

Matsuda, F., Nakabayashi, R., Sawada, Y., Suzuki, M., Hirai, M.Y., Kanaya, S., Saito, K., 2011. Mass spectra-based framework for automated structural elucidation of metabolome data to explore phytochemical diversity. Front. Plant Sci. 2, 40.

Menikarachchi, L.C., Cawley, S., Hill, D.W., Hall, L.M., Hall, L., Lai, S., Wilder, J., Grant, D.F., 2012. MolFind: a software package enabling HPLC/MS-based identification of unknown chemical structures. Anal. Chem. 84 (21), 9388–9394.

Misawa, T., Wei, F., Kikuchi, J., 2016. Application of two-dimensional nuclear magnetic resonance for signal enhancement by spectral integration using a large data set of metabolic mixtures. Anal. Chem. 88 (12), 6130–6134.

Mishra, A., Patel, M.K., Jha, B., 2015. Non-targeted metabolomics and scavenging activity of reactive oxygen species reveal the potential of *Salicornia brachiata* as a functional food. J. Funct. Foods 13, 21–31.

Mistrik, R., Lutisan, J., Huang, Y., Suchy, M., Wang, J., Raab, M., 2013. mzCloud: a key conceptual shift to understand 'Who's Who'in untargeted metabolomics. In: Metabolomics Society Conference, Glasgow, July 2013, pp. 1–4.

Moco, S., Bino, R.J., Vorst, O., Verhoeven, H.A., de Groot, J., van Beek, T.A., Vervoort, J., De Vos, C.R., 2006. A liquid chromatography-mass spectrometry-based metabolome database for tomato. Plant Physiol. 141 (4), 1205–1218.

Mueller, L.A., Zhang, P., Rhee, S.Y., 2003. AraCyc: a biochemical pathway database for *Arabidopsis*. Plant Physiol. 132 (2), 453–460.

Mushtaq, M.Y., Choi, Y.H., Verpoorte, R., Wilson, E.G., 2014. Extraction for metabolomics: access to the metabolome. Phytochem. Anal. 25 (4), 291–306.

Nahas, R., Abatis, D., Anagnostopoulou, M.A., Kefalas, P., Vagias, C., Roussis, V., 2007. Radical-scavenging activity of aegean sea marine algae. Food Chem. 102 (3), 577–581.

Namvar, F., Mohamed, S., Fard, S.G., Behravan, J., Mustapha, N.M., Alitheen, N.B., Othman, F., 2012. Polyphenol-rich seaweed (*Eucheuma cottonii*) extract suppresses breast tumour via hormone modulation and apoptosis induction. Food Chem. 130 (2), 376–382.

Naumkin, A.V., Kraut-Vass, A., Gaarenstroom, S.W., Powell, C.J., 2012. NIST Standard Reference Database 20.

Ni, Y., Qiu, Y., Jiang, W., Suttlemyre, K., Su, M., Zhang, W., Jia, W., Du, X., 2012. ADAP-GC 2.0: deconvolution of coeluting metabolites from GC/TOF-MS data for metabolomics studies. Anal. Chem. 84 (15), 6619–6629.

Nilsson, S.L., Bylund, D., Jörntén-Karlsson, M., Petersson, P., Markides, K.E., 2004. A chemometric study of active parameters and their interaction effects in a nebulized sheath-liquid electrospray interface for capillary electrophoresis-mass spectrometry. Electrophoresis 25 (13), 2100–2107.

Obata, T., Fernie, A.R., 2012. The use of metabolomics to dissect plant responses to abiotic stresses. Cell. Mol. Life Sci. 69 (19), 3225–3243.

Ohta, D., Kanaya, S., Suzuki, H., 2010. Application of Fourier-transform ion cyclotron resonance mass spectrometry to metabolic profiling and metabolite identification. Curr. Opinion Biotechnol. 21 (1), 35–44.

Okazaki, Y., Saito, K., 2012. Recent advances of metabolomics in plant biotechnology. Plant Biotechnol. Rep. 6 (1), 1–5.

Orata, F., 2012. Derivatization Reactions and Reagents for Gas Chromatography Analysis. In Tech, Rijeka, Croatia, pp. 83–108.

Pandey, S., Patel, M.K., Mishra, A., Jha, B., 2015. Physio-biochemical composition and untargeted metabolomics of cumin (*Cuminum cyminum* L.) make it promising functional food and help in mitigating salinity stress. PLoS One 10 (12), e0144469.

Patel, M.K., Mishra, A., Jha, B., 2016a. Non-targeted metabolite profiling and scavenging activity unveil the nutraceutical potential of psyllium (*Plantago ovata* Forsk). Front. Plant Sci. 7, 431.

Patel, M.K., Mishra, A., Jha, B., 2016b. Untargeted metabolomics of halophytes. In: Kim, S. (Ed.), Marine Omics: Principles and Applications. Taylor & Francis Group, CRC Press, Boca Raton, Florida, USA, pp. 309–325. http://dx.doi.org/10.1201/9781315372303-18.

Pilatti, F.K., Ramlov, F., Schmidt, E.C., Costa, C., de Oliveira, E.R., Bauer, C.M., Rocha, M., Bouzon, Z.L., Maraschin, M., 2017. Metabolomics of *Ulva lactuca* Linnaeus (Chlorophyta) exposed to oil fuels: fourier transform infrared spectroscopy and multivariate analysis as tools for metabolic fingerprint. Mar. Pollut. Bull. 114 (2), 831–836.

Pluskal, T., Castillo, S., Villar-Briones, A., Orešič, M., 2010. MZmine 2: modular framework for processing, visualizing, and analyzing mass spectrometry-based molecular profile data. BMC Bioinforma. 11 (1), 395.

Pujol, C.A., Ray, S., Ray, B., Damonte, E.B., 2012. Antiviral activity against dengue virus of diverse classes of algal sulfated polysaccharides. Int. J. Biol. Macromol. 51 (4), 412–416.

Raman, M., Doble, M., 2015. κ-Carrageenan from marine red algae, *Kappaphycus alvarezii*–a functional food to prevent colon carcinogenesis. J. Funct. Foods 15, 354–364.

Ramautar, R., Somsen, G.W., de Jong, G.J., 2009. CE-MS in metabolomics. Electrophoresis 30 (1), 276–291.

Renault, D., Lombard, M., Vingère, J., Laparie, M., 2016. Comparative salinity tolerance in native flies from the subantarctic Kerguelen Islands: a metabolomic approach. Polar Biol. 39 (1), 47–56.

Robbins, R.J., Keck, A.S., Banuelos, G., Finley, J.W., 2005. Cultivation conditions and selenium fertilization alter the phenolic profile, glucosinolate, and sulforaphane content of broccoli. J. Med. Food 8 (2), 204–214.

Ruqqia, K., Sultana, V., Ara, J., Ehteshamul-Haque, S., Athar, M., 2015. Hypolipidaemic potential of seaweeds in normal, triton-induced and high-fat diet-induced hyperlipidaemic rats. J. Appl. Phycol. 27 (1), 571–579.

Sakurai, N., Shibata, D., 2006. KaPPA-View for integrating quantitative transcriptomic and metabolomic data on plant metabolic pathway maps. J. Pesticide Sci. 31 (3), 293–295.

Sakurai, T., Yamada, Y., Sawada, Y., Matsuda, F., Akiyama, K., Shinozaki, K., Hirai, M.Y., Saito, K., 2013. PRIMe update: innovative content for plant metabolomics and integration of gene expression and metabolite accumulation. Plant Cell Physiol. 54 (2), e5.

Sana, T.R., Roark, J.C., Li, X., Waddell, K., Fischer, S.M., 2008. Molecular formula and METLIN Personal Metabolite Database matching applied to the identification of compounds generated by LC/TOF-MS. J. Biomol. Tech. 19 (4), 258–266.

Sato, S., Soga, T., Nishioka, T., Tomita, M., 2004. Simultaneous determination of the main metabolites in rice leaves using capillary electrophoresis mass spectrometry and capillary electrophoresis diode array detection. Plant J. 40 (1), 151–163.

Sawada, Y., Nakabayashi, R., Yamada, Y., Suzuki, M., Sato, M., Sakata, A., Akiyama, K., Sakurai, T., Matsuda, F., Aoki, T., Hirai, M.Y., 2012. RIKEN tandem mass spectral database (ReSpect) for phytochemicals: a plant-specific MS/MS-based data resource and database. Phytochemistry 82, 38–45.

Schauer, N., Fernie, A.R., 2006. Plant metabolomics: towards biological function and mechanism. Trends Plant Sci. 11 (10), 508–516.

Schiffmann, C., Hansen, R., Baumann, S., Kublik, A., Nielsen, P.H., Adrian, L., Von Bergen, M., Jehmlich, N., Seifert, J., 2014. Comparison of targeted peptide quantification assays for reductive dehalogenases by selective reaction monitoring (SRM) and precursor reaction monitoring (PRM). Anal. Bioanal. Chem. 406 (1), 283–291.

Schripsema, J., 2010. Application of NMR in plant metabolomics: techniques, problems and prospects. Phytochem. Anal. 21 (1), 14–21.

Schubert, N., García-Mendoza, E., Pacheco-Ruiz, I., 2006. Carotenoid composition of marine red algae. J. Phycol. 42 (6), 1208–1216.

Shao, P., Chen, X., Sun, P., 2013. In vitro antioxidant and antitumor activities of different sulfated polysaccharides isolated from three algae. Int. J. Biol. Macromol. 62, 155–161.

Shiney, E., Reginald, M., Wilsy, J.I., 2014. Antibacterial activity and phytochemical screening of marine macro algae *Amphiroa anceps* using three solvent extracts. Int. J. Pharmacogn. 1 (9), 605–608.

Smith, C.A., Want, E.J., O'Maille, G., Abagyan, R., Siuzdak, G., 2006. XCMS: processing mass spectrometry data for metabolite profiling using nonlinear peak alignment, matching, and identification. Anal. Chem. 78 (3), 779–787.

Steinbeck, C., Krause, S., Kuhn, S., 2003. NMRShiftDB constructing a free chemical information system with open-source components. J. Chem. Inf. Comput. Sci. 43 (6), 1733–1739.

Steinbeck, C., Kuhn, S., 2004. NMRShiftDB–compound identification and structure elucidation support through a free community-built web database. Phytochemistry 65 (19), 2711–2717.

Steinfath, M., Groth, D., Lisec, J., Selbig, J., 2008. Metabolite profile analysis: from raw data to regression and classification. Physiol. Plant. 132 (2), 150–161.

Sturm, M., Bertsch, A., Gröpl, C., Hildebrandt, A., Hussong, R., Lange, E., Pfeifer, N., Schulz-Trieglaff, O., Zerck, A., Reinert, K., Kohlbacher, O., 2008. OpenMS–an open-source software framework for mass spectrometry. BMC Bioinforma. 9 (1), 163.

Sud, M., Fahy, E., Cotter, D., Brown, A., Dennis, E.A., Glass, C.K., Merrill, A.H., Murphy, R.C., Raetz, C.R., Russell, D.W., Subramaniam, S., 2007. Lmsd: lipid maps structure database. Nucleic Acids Res. 35 (Suppl. 1), D527–D532.

Sugiura, Y., Matsuda, K., Yamada, Y., Nishikawa, M., Shioya, K., Katsuzaki, H., Imai, K., Amano, H., 2007. Anti-allergic phlorotannins from the edible brown alga, *Eisenia arborea*. Food Sci. Technol. Res. 13 (1), 54–60.

Sumner, L.W., Mendes, P., Dixon, R.A., 2003. Plant metabolomics: large-scale phytochemistry in the functional genomics era. Phytochemistry 62 (6), 817–836.

Sun, X., Jin, H., Zhang, L., Hu, W., Li, Y., Xu, N., 2015. Screening and isolation of the algicidal compounds from marine green alga *Ulva intestinalis*. Chin. J. Oceanol. Limnol. 34 (4), 781–788.

Tautenhahn, R., Cho, K., Uritboonthai, W., Zhu, Z., Patti, G.J., Siuzdak, G., 2012a. An accelerated workflow for untargeted metabolomics using the METLIN database. Nat. Biotechnol. 30 (9), 826–828.

Tautenhahn, R., Patti, G.J., Rinehart, D., Siuzdak, G., 2012b. XCMS Online: a web-based platform to process untargeted metabolomic data. Anal. Chem. 84 (11), 5035–5039.

Trufelli, H., Palma, P., Famiglini, G., Cappiello, A., 2011. An overview of matrix effects in liquid chromatography–mass spectrometry. Mass Spectrometry Rev. 30 (3), 491–509.

Tsugawa, H., Cajka, T., Kind, T., Ma, Y., Higgins, B., Ikeda, K., Kanazawa, M., Vander-Gheynst, J., Fiehn, O., Arita, M., 2015. MS-DIAL: data-independent MS/MS deconvolution for comprehensive metabolome analysis. Nat. Methods 12 (6), 523–526.

Usmanghani, K., Shameel, M., Siddiqui, S., Alam, M.A., 1987. Studies on the sterols of a brown seaweed *Iyengaria stellata* from Pakistan. Pak. J. Bot. 19 (2), 249–252.

van Hove, E.R., Smith, D.F., Heeren, R.M., 2010. A concise review of mass spectrometry imaging. J. Chromatogr. A 1217 (25), 3946–3954.

Verhoeven, H.A., de Vos, C.R., Bino, R.J., Hall, R.D., 2006. Plant metabolomics strategies based upon quadrupole time of flight mass spectrometry (QTOF-MS). In: Plant Metabolomics. Springer, Berlin, Heidelberg, pp. 33–48.

Viant, M.R., 2003. Improved methods for the acquisition and interpretation of NMR metabolomic data. Biochem. Biophys. Res. Commun. 310 (3), 943–948.

Villas-Bôas, S.G., Mas, S., Åkesson, M., Smedsgaard, J., Nielsen, J., 2005. Mass spectrometry in metabolome analysis. Mass Spectrom. Rev. 24 (5), 613–646.

Viszwapriya, D., Pandian, S.K., 2017. Metagenomic approaches for novel active metabolites. In: Bioresources and Bioprocess in Biotechnology. Springer, Singapore, pp. 275–302.

Wei, D., Chen, D., Lou, Y., Ye, Y., Yang, R., 2016. Metabolomic profile characteristics of *Pyropia haitanensis* as affected by harvest time. Food Sci. Technol. Res. 22 (4), 529–536.

Wilson, I.D., Brinkman, U.T., 2003. Hyphenation and hypernation: the practice and prospects of multiple hyphenation. J. Chromatogr. A 1000 (1), 325–356.

Wishart, D.S., 2010. Computational approaches to metabolomics. Bioinforma. Methods Clin. Res. 593, 283–313.

Wishart, D.S., Jewison, T., Guo, A.C., Wilson, M., Knox, C., Liu, Y., Djoumbou, Y., Mandal, R., Aziat, F., Dong, E., Bouatra, S., 2013. HMDB 3.0—the human metabolome database in 2013. Nucleic Acids Res. 41, D801–D817.

Wu, C., Dill, A.L., Eberlin, L.S., Cooks, R.G., Ifa, D.R., 2013. Mass spectrometry imaging under ambient conditions. Mass Spectrom. Rev. 32 (3), 218–243.

Wu, J., An, Y., Yao, J., Wang, Y., Tang, H., 2010. An optimised sample preparation method for NMR-based faecal metabonomic analysis. Analyst 135 (5), 1023–1030.

Xia, J., Bjorndahl, T.C., Tang, P., Wishart, D.S., 2008. MetaboMiner–semi-automated identification of metabolites from 2D NMR spectra of complex biofluids. BMC Bioinforma. 9 (1), 507.

Xia, J., Psychogios, N., Young, N., Wishart, D.S., 2009. MetaboAnalyst: a web server for metabolomic data analysis and interpretation. Nucleic Acids Res. 37 (Suppl. 2), W652–W660.

Xia, J., Sinelnikov, I.V., Han, B., Wishart, D.S., 2015. MetaboAnalyst 3.0—making metabolomics more meaningful. Nucleic Acids Res. 43 (W1), W251–W257.

Yamaji, T., Saito, T., Hayamizu, K., Yanagisawa, M., Yamamoto, O., Wasada, N., Someno, K., Kinugasa, S., Tanabe, K., Tamura, T., Tanabe, K., 2014. Spectral Database for Organic Compounds, SDBS. Electronic resource http://sdbs.db.aist.go.jp/sdbs/cgi-bin/cre_index.cgi.

Yen, Y.H., Farooqi, A.A., Li, K.T., Butt, G., Tang, J.Y., Wu, C.Y., Cheng, Y.B., Hou, M.F., Chang, H.W., 2014. Methanolic extracts of *Solieria robusta* inhibits proliferation of oral cancer Ca9-22 cells via apoptosis and oxidative stress. Molecules 19 (11), 18721–18732.

Young, T., Alfaro, A.C., 2016. Metabolomic strategies for aquaculture research: a primer. Rev. Aquac. http://dx.doi.org/10.1111/raq.12146.

Yuliana, N.D., Khatib, A., Verpoorte, R., Choi, Y.H., 2011. Comprehensive extraction method integrated with NMR metabolomics: a new bioactivity screening method for plants, adenosine A1 receptor binding compounds in *Orthosiphon stamineus* Benth. Anal. Chem. 83 (17), 6902–6906.

Zhang, A., Sun, H., Wang, P., Han, Y., Wang, X., 2012. Modern analytical techniques in metabolomics analysis. Analyst 137 (2), 293–300.

Zhang, X.L., Wang, C., Chen, Z., Zhang, P.Y., Liu, H.B., 2016. Development and validation of quantitative 1H NMR spectroscopy for the determination of total phytosterols in the marine seaweed *Sargassum*. J. Agric. Food Chem. 64 (31), 6228–6232.

Zhou, B., Wang, J., Ressom, H.W., 2012. MetaboSearch: tool for mass-based metabolite identification using multiple databases. PLoS One 7 (6), e40096.

Chapter 3

Environmental Stresses and Metabolomics—Deciphering the Role of Stress Responsive Metabolites

Mohammad A. Ahanger[1], Fahima Gul[2], Parvaiz Ahmad[2,3], Nudrat A. Akram[4]

[1]*PG College Rajouri, Jammu, India;* [2]*SP College, Srinagar, India;* [3]*King Saud University, Riyadh, Saudi Arabia;* [4]*GC University, Faisalabad, Pakistan*

1. INTRODUCTION

Environmental stresses affect growth and development thereby restricting the yield potential of maximum food crops worldwide. It has been estimated that abiotic stresses including drought, salinity, and temperature (high and low) influence the survival rate, biomass, and yield production of staple food crops up to 70% (Thakur et al., 2010) therefore making the overall effect obvious on global food security. Particular plant species can either be exposed to one type of stress at a time or to a combination of more than one stresses, e.g., water stress is often accompanied by increase in temperature. Drought, extreme salinity, and temperatures lead to dehydration stress (Thakur et al., 2010). For counteracting the stress-induced alterations in the normal growth and metabolic patterns, several indigenously occurring tolerance mechanisms are initiated (Ahanger et al., 2014, 2015). As we know stress tolerance is multigenic in nature and the cumbersome challenge for plant scientists remains the understanding of the molecular, genetic, and biochemical interface of these mechanisms. No doubt considerable progress has been made in studying the tolerance mechanisms employing the molecular, genetic, and biochemical techniques (Ahmad et al., 2014; Ahanger et al., 2017a). Strides made to achieve stress tolerance employing conventional breeding methods have been marginally successful. Contrary to this, several efficient techniques including functional genomics backed by high-throughput sequencing have proved much affective in understanding the tolerance mechanism from perception to transduction to initiation with the involvement of key regulatory molecules like transcription factors, signaling molecules, etc. (Ray et al., 2010; Ahanger et al., 2017b). However, the activation of a stress response is considered as crucial as the perception and transmission, and these are of critical importance in introducing and improving the breeding and transgenic approach for attaining stress tolerance in plants. Whatever it shall be the ultimate identification of the molecules mediating tolerance at physiological, biochemical, or molecular level is essential, and in connection to this "omics" has contributed enormously (Liu et al., 2011).

One of the most accepted implications of environmental stresses involve the redox adjustments or alterations often linked to increased synthesis and accumulation of reactive oxygen species (ROS). Excess ROS accumulation induces oxidative damage to important sensitive organelles and structures. Progression of different developmental processes requires ROS-triggered signaling events, which are often evident when stress intensity is less or is at the initial (Mittler, 2017). However, with the increase in intensity and duration of stress the oxidative stress produces considerable alterations in the net cellular redox state having wide-ranging potential consequences on the metabolic processes of plants regulating normal growth (Ahmad et al., 2010; Ahanger et al., 2017c). Such conditions usually occur under conditions of extreme temperatures (cold or heat), water deficit, salinity, metal accumulation, pathogens or microbe infestation etc. However, to counteract these ill effects and to maintain the cellular functioning plants have developed the potential to synthesize and accumulate an array of different metabolic products that directly or indirectly contribute to growth maintenance and protection under extreme environmental conditions. For the investigation of such metabolic alterations and modifications for improved stress tolerance, metabolomics has emerged as a great tool. Metabolomics in the broader sense involves the ability to measure the existing metabolites at a time. Using techniques like mass spectrometry, metabolomics has allowed researchers to investigate the metabolites in a nontargeted fashion (Fernie et al., 2004). However, it should be noted here that there is no technique that could measure the all the existing metabolites in a single run, and this is attributed to the chemical diversity of metabolites based on the hydrophilic, hydrophobic or incomplete nature of any reference database.

Plant Metabolites and Regulation Under Environmental Stress. http://dx.doi.org/10.1016/B978-0-12-812689-9.00003-0

This limitation for measuring the majority of the metabolites in plant extract is in striking difference with that of other omics techniques like transcriptomics, ionomics, etc., which often provide quite comprehensive information. Instead of having such obvious drawbacks the metabolite alterations in the tissues, under normal or stressful conditions, are considered and counted among the powerful informative techniques due to the direct relation of the metabolome and phenotype. Therefore it could be concluded here that nontargeted metabolome profiling of plants while responding to environmental cues is somewhat interesting (Shulaev et al., 2008; Arbona et al., 2013).

The *omic* techniques including *proteomics, genomics, ionomics, transcriptionomics,* and *metabolomics,* studying proteins, genes, mineral ions, transcription factors, and metabolites, respectively, have revolutionized the progressing science. Nowadays the focus of metabolic studies has shifted from the traditional focus on certain known key metabolites serving particular functions under particular or different conditions to whole metabolome profiling. With the advent of efficient metabolome-quantifying instruments, quantifications of large numbers of metabolites has become possible in many instances. In addition, it has become possible to discern a combination of metabolite changes in response to particular types of environmental stresses. In the available literature, quiet often metabolomics has been discussed focusing on the physiological, biochemical, phytochemical, or mechanistic perspectives; however, metabolomics of stress-induced oxidative damage has been rarely visited. The present review addresses the role of key cellular metabolites in regulating the growth of plants under stressful conditions with focus on the recent advances and future requirements in improving our understanding.

2. STRESSES: REACTIVE OXYGEN SPECIES PRODUCTION AND SIGNIFICANCE

It has been observed quite often that plants exposed to stressful conditions exhibit alteration in normal metabolism, and one of the main causes for such hampered metabolism has been attributed to stress-induced increased generation of toxic ROS (Sirhindi et al., 2016; Ahmad et al., 2016, 2017; Ahanger and Agarwal, 2017a,b; Ahanger et al., 2017c). ROS are generated in several metabolically active cellular compartments like mitochondria (complexes I and III), chloroplast, and peroxisomes. ROS include singlet oxygen, superoxide, hydroxyl radical, hydrogen peroxide (H_2O_2), etc., which are basically derived through the modifications of molecular oxygen. These ROS can significantly modify the normal progression of most cellular pathways (Fischer et al., 2013). Photosystem (PS) II in chloroplasts serves as a prime source of singlet oxygen, and contrary to this the generation of superoxide or H_2O_2 involves the reduction of oxygen during the chloroplast or mitochondrial electron transport or by enzyme-mediated processes in peroxisomes by glycolate oxidase and at the cell surface/apoplast by NADPH oxidases and peroxidases (Foyer and Noctor, 2011; Ahmad et al., 2010; see Ahanger et al., 2017c). In addition, the more toxic hydroxyl radical is generated by reductive cleavage of H_2O_2 mediated by iron or related transition metals (Mittler, 2002). Nowadays it has been accepted that ROS-mediated signaling events can be exploited to develop mechanisms for enhancing the stress tolerance and normal growth and developmental events like cellular expansion (Schmidt et al., 2016; Mittler, 2017). For example, Verma et al. (2015) has demonstrated involvement of ROS, primarily H_2O_2, in seed germination after observing their involvement in mediating the signaling triggering the mobilization of stored reserves from the cotyledons in *Vigna radiata.* Quite before it has been suggested that the nature as well as the location of ROS plays key role in determining the response (Gadjev et al., 2006), therefore it can be inferred that the ROS molecules in mitochondria, chloroplast, or peroxisomes can initiate different signaling pathways. In this connection the variations displayed may be due to the potentiality of protective mechanisms including antioxidant and the redox system (Laloi et al., 2004; Ahmad et al., 2010; del Rio, 2015; Noctor and Foyer, 2016). Besides, the intra- and extracellular concentration of ROS is also considered as the key determinant of the metabolic outcome resulting from the intricate signaling cross-talk (Laloi et al., 2004; Richards et al., 2015; Noctor and Foyer, 2016).

Now moving toward the most consolidated role of ROS— growth restriction and hampered metabolism. ROS-induced lipid peroxidation results in leakage of key cellular constituents and hampers the structural integrity of nucleic acids, proteins, etc. However, for averting such heavy damages of ROS molecules cells have key indigenous protective enzymes, metabolites, and molecules that coordinately work for elimination of ROS (Ahmad et al., 2010). It shall be noted here that scavenging mechanisms of ROS show a clear contrasting trend in halophytes and glycophytes (Bose et al., 2014). The role and importance of the enzymatic defense systems can be seen elsewhere; however, here we shall focus on some of the key metabolites in mediating the scavenging of ROS and hence protecting against the oxidative damage.

3. METABOLITES AND PLANT GROWTH

Normal metabolism in plants is driven through the energy-transducing processes occurring in the cellular organelles including chloroplasts and mitochondria wherein adenosine triphosphate (ATP), NADPH, and associated metabolites are used as the chief energy sources for generating and exchanging of the energy. A plethora of metabolites including sugars, carbohydrates, amino acids, proteins, lipids, alkaloids, terpenoids, and other polyphenols needed for the maintenance of growth are synthesized by the plants. However, it shall be noted here that the requirement of metabolic products for developmental and growth progression via mediation of defense processes displays dynamic changes from the existing/changing environment and time, quite often triggering the short- and long-term regulation events (Griffiths et al., 2016). For example, the vibrant regulation of key metabolic processes like photosynthesis and respiration lies in the exchange of metabolites driven by the fine-tuned delivery of key raw skeletons. Contrarily, the understanding has increased about the coordinated regulation of primary carbon and nitrogen assimilatory pathways adding taste to decipher the unraveled mechanism; however, knowledge about the regulation of synthesis of metabolites, their transport (primary and secondary) within and between different cellular compartments and organs, and their involvement in protection against the environmental stresses needs much attention (Foyer et al., 2011). Sugars, amino acids, and other associated metabolites including the ones that control the redox homeostasis have key roles in bringing coordination among cellular pathways between different compartments (Lunn, 2007; Valluru and den Ende, 2008). Moreover, several metabolites initiate signaling events resulting in optimization of regulatory pathways via altered gene expression often determined by the prevailing environmental situation (Couee et al., 2006).

4. METABOLITES AND ANTIOXIDANT ACTIVITY—REVISITING THE REDOX BUFFERS

In cells, indigenously existing compounds can interact with the ROS molecules leading to their quick neutralization and hence can help to maintain their optimal required levels. These molecules leading to such ROS stability are referred to as antioxidants and more important in this context is that in a reaction mediated by an antioxidant molecule the oxidized product should be stable enough or even arbitrarily recyclable to the reduced form so that unsystematic initiation of potentially oxidizing reactions can be checked (Noctor and Foyer, 2016). Few antioxidants get modified by ROS into irreversible products. In addition to this, metabolically compatible organic solutes are also involved in the scavenging of toxic radicals (Bolouri-Moghaddam et al., 2010; Keunen et al., 2013). Understanding the possible involvement of important metabolites in improving the stress tolerance mechanisms and strengthening the stress-mitigating potential is important for drawing and designing future research toward sustainable crop production.

4.1 Carotenoids

Carotenoids a class of hydrocarbons, mostly C_{40} terpenoids, which actively participate in several biological processes including photosynthesis and photoprotection, photomorphogenesis, phytohormone biosynthesis, and development in plants, also act as colorants and critical components of human diet as antioxidants and provitamin A (Nisar et al., 2015). In plants, both carotenoids and their oxidative and enzymatic cleavage products, apocarotenoids, are very important for the progression and protection of numerous biological processes including assembly of photosystems and the light-harvesting antenna complexes for photosynthesis and photoprotection and regulation of growth and development (Havaux, 2014; Nisar et al., 2015). During the exposure of plants to stressful environments apocarotenoids have been proposed to mediate signaling and hence the plant–environment interactions (Cazzonelli, 2011). Like other metabolites environmental stress exposure alters the plant carotenoid profile. Therefore keeping in view the importance of the carotenoids for plant metabolic and developmental progression improvement of carotenoid synthesis via various techniques has been introduced in several crops.

Chen et al. (2015) isolated sweet potato *IbGGPS* gene from its storage roots, and its overexpression in *Arabidopsis* elevated the carotenoid content and tolerance to osmotic stress obvious in increased transcript levels of *AtPSY1*, *AtPDS*, *AtZDS*, and *AtBCH1*—the key genes for carotenoid biosynthesis. It is widely accepted that being synthesized in plastids carotenoids accumulate in the chloroplast and chromoplast, and within they chloroplast form a pigment–protein complex along with chlorophyll-binding protein contrary to the chromoplast where they combine with polar lipids and certain carotenoid-associated proteins giving carotenoid-lipoprotein-sequestering structure (Farre et al., 2010; Kang et al., 2017). From such evidences it can be inferred that the concentrations of carotenoids is maintained high in these organelles. It has been demonstrated that *Orange* (*Or*) is responsible for inducing the accumulation of carotenoids in *Brassica oleracea* var. botrytis and its possible involvement in the formation of carotenoid-sequestering structure has been inferred (Li et al., 2001). The *Or* gene encodes a DnaJ cysteine-rich-domain-containing protein resulting in improved accumulation of β-carotene and hence the regulation of plant growth (Lu et al., 2006). The involvement of *Or* protein in the regulation of cellular proliferation and growth has been demonstrated to result from the interaction with eukaryotic release factor 1-2 (Zhou et al., 2011).

Being hydrophobic in nature carotenoids form complexes with proteins thereby allowing their cytoplasmic localization, and in blue–green algae, these complexes known as orange carotenoid proteins protect the PSII and PSI by preventing their overexcitation through phycobilisomes (Zakar et al., 2016). In addition to this β-carotene functions as a redox component in alternate pathway of electron transfer within PSII, and such redox role of carotenoid in PSII stems from the highly oxidizing intermediates generated during the water oxidation process (Tracewell et al., 2001). More essentially, β-carotene regulates the assembly of the D1 protein into the functional PSII and high light–induced bleaching of β-carotene in PSII destabilizes and degrades D1 protein (Trebst and Depka, 1997). Under stressful conditions, particularly high irradiance, ROS-induced photoinhibition is often followed by destruction of carotenoids, bleaching of chlorophylls, and increased lipid peroxidation. Under such conditions carotenoids quench triplet chlorophyll and scavenge singlet oxygen and other toxic ROS. The interconversion of violaxanthin into zeaxanthin in light-harvesting complexes regulates photon harvesting and hence the dissipation of energy (Choudhury and Behera, 2001). The net effect of this interconversion is the direct quenching of chlorophyll fluorescence by singlet–singlet energy transfer from chlorophyll to zeaxanthin and transthylakoid membrane–mediated pH-dependent indirect quenching of chlorophyll fluorescence (Choudhury and Behera, 2001; Jahns and Holzwarth, 2012). Such photoprotective role of carotenoids has been demonstrated in *Arabidopsis thaliana* using szl1 (zeaxanthin-less) mutants displaying reduced lycopene-β-cyclase activity affecting the carotene content, and imparting reduced chlorophyll fluorescence quenching—the rapidly reversible component of nonphotochemical quenching (Cazzaniga et al., 2012). In addition to their obvious positive effects on photosynthesis by acting as antioxidants, accessory light-harvesting pigments, attractants for pollinators, and seed dispersers carotenoids can trigger stress signaling. Bioactive oxidative products of carotenoids during stresses like volatile β–cyclocitral, derived from the oxidation of β-carotene ($^1O_2$-responsive genes), can reprogram gene expression changes resulting in stress acclimation (Havaux, 2014). However, dissecting such mechanisms needs thorough research to reach a clear conclusion.

4.2 Tocochromanols—Vitamin E Family

Tocochromanols, a group constituted by four each of tocopherols and tocotrienols collectively constitute vitamin E, an essential vitamin. Only oxygenic photosynthetic organisms have the potential to synthesize tocochromanols (Sies and Stahl, 1995; DellaPenna, 2005). It has been observed that they display almost similar antioxidant activities under in vitro studies contrary to the in vivo conditions where the individual tocochromanols vary in vitamin E activities. From the past years, research studies using mutant and transgenic approaches in promising model organisms like *A. thaliana* and *Synechocystis* sp. have resulted in successful isolation of the whole complement of the core genes responsible for the biosynthesis of tocochromanols in addition to providing insight into molecular genetic and biochemical control. Such kind of knowledge made available is exploited to manipulate the levels of this essential vitamin in key staple food crops for world populace in both developed and developing countries (DellaPenna, 2005). α-Tocopherol supplementation at nanomolar levels protects the kinase and dehydrogenase activities of enzymes from H_2O_2-induced oxidative damage (Zakharova et al., 2012). Biosynthetically derived from the geranylgeranyl pyrophosphate in the terpenoid pathway the only difference between tocopherols and tocotrienols lies in the presence and absence of double bonds in side chains (Sies and Stahl, 1995). Both are recognized for their enormous contribution in ROS scavenging particularly in the protection of photosynthesis from the singlet oxygen. It shall be noted here that among these tocochromanols α-tocopherol has been immensely researched for its protective role against stress-induced oxidative damage. α-Tocopherol is a lipophilic antioxidant involved actively in a number of functions in plants and is synthesized from the homogentisic acid and isopentenyl diphosphate in the chloroplast envelope. It plays an essential role in the maintenance of the integrity of photosynthetic membranes by being involved in photo- and antioxidant protection (Munne-Bosch, 2007).

The concentration of tocopherol varies with the type of existing environmental condition. Within chloroplasts α-tocopherol is the major vitamin E found in the chloroplast envelope, thylakoid membranes, and plastoglobuli, leading to the deactivation of photosynthetically generated ROS like $^1O_2$ and OH, hence preventing the proliferation of lipid peroxidation and protection membranes by mediating scavenging of lipid peroxyl radicals in thylakoid (Munne-Bosch, 2005, 2007). α-Tocopherol physically quenches and chemically reacts with singlet oxygen. In addition to this α-tocopherol maintains membrane rigidity, and depending on its concentration and the concentration of other membrane components the production of α-tocopherol may be regulated to afford ample fluidity for optimal membrane function (Munne-Bosch, 2007). It has been suggested that α-tocopherol optimizes cellular signaling in plants, which is linked directly to redox homeostasis. α-Tocopherol regulates the cellular signaling events by controlling the progression of lipid peroxidation in chloroplasts, offering a way for the formation of oxylipins including jasmonic acid (Munne-Bosch, 2007; Miret and Munne-Bosch, 2015). Sattler et al. (2004) while analyzing the role of α-tocopherol in seed longevity mutated two vitamin E loci (VTE1 and VTE2) in *Arabidopsis* disrupting tocopherol cyclase activity and homogentisate phytyl transferase activity, respectively, exhibited reduced seed longevity despite the accumulation of VTE1, a redox-active biosynthetic intermediate.

It has been reported that transgenic *Brassica juncea* plants exhibited greater synthesis of α-tocopherol, reflecting an increased tolerance to a variety of abiotic stresses including salinity and metal and osmotic stress by protecting seed germination, shoot growth, and photosynthesis concomitant with the prevention of leaf senescence (Yusuf et al., 2010). The ill effects of silencing tocopherol synthesis have been reviewed in detail by Falk and Munne-Bosch (2010). α-Tocopherol contributes to the maintenance of redox homeostasis and expression of genes by bringing modulations in the peroxidation of lipids in leaves and hence to oxylipin levels; in addition, it neutralizes radicals like singlet oxygen in association with carotenoids leading to protection of photosynthesis. The interplay between α-tocopherol and carotenoids in ameliorating the singlet oxygen–induced deleterious effects on PSII functioning has been observed in *Chlamydomonas reinhardtii* (Trebst et al., 2002). Studies carried out using mutants, e.g., vte1 and vte2 in *A. thaliana* and sxd1 in maize, and sucrose export defective1 (SXD1)-RNAi-silenced potato plants have shown the possible involvement of tocopherols in the regulation of export of photoassimilates in leaves, reflecting on growth by affecting the carbohydrate metabolism and source–sink relation. In this context, Sattler et al. (2003) identified SXD1-encoded tocopherol cyclase in maize involved in photoassimilate export and growth regulation. Maize seedlings mutant for sxd1 exhibited occluded sucrose transport from leaves accompanied by callose plugging of plasmodesmata between bundle sheath and vascular parenchyma cells usually characterized by growth retardation and accumulation of anthocyanins and starch in leaves (Russin et al., 1996; Provencher et al., 2001). Existence of such events has been proved by silencing using RNA interference approach (Hofius et al., 2004). Considerable part of tocopherols is synthesized from the free phytol, therefore making it clear that the phytol released after stress- or senescence-induced chlorophyll breakdown may be accumulated in chloroplasts as tocopherols (Dormann, 2007). However, it has been confirmed using StSXD1:RNAi that stresses like salinity do not induce callose plugging and that the altered redox regulation of phloem loading by SUT1 contributes to the attenuation of sucrose export in salt-stressed SXD:RNAi plants (Asensi-Fabado et al., 2015).

4.3 Glutathione

It is an important feature of the antioxidant molecules to pace the interaction between ROS resulting in the formation of more stable species or to initiate the oxidation of reducing metabolites to relatively stable forms that do not promulgate stress. The differences in these mechanisms lie in their requirement of the accessory reductants, and the best studied pathway for the latter is the ascorbate–glutathione pathway (Ahmad et al., 2010; Hasanuzzaman et al., 2017a).

Glutathione is among the key metabolites involved in defense against the stresses in plants and animals. Glutathione, a tripeptide (γ-glutamyl-cysteinyl-glycine) constitutes most of the organic sulfur in plant cells. Existing in both reduced (GSH) and oxidized (GSSG) forms it has been adjudged as a promising molecule in the stress mitigation process. GSH is a low-molecular-weight thiol, and the enzyme glutathione reductase contributes to its regeneration or interconversion from oxidized to reduced forms (Mittler, 2002; Bose et al., 2014; Ahanger et al., 2017c). It has been observed that GSH accumulation under different stresses improves plant performance significantly. Both reduced and oxidized forms of glutathione play a key role in the maintenance and regulation of the cell cycle under stress (Bashandy et al., 2010; Diaz-Vivancos et al., 2015). In plant species the synthesis and accumulation of glutathione in different organelles like chloroplast and mitochondria prevents the ill effects of oxidative damage. The plant researchers are now working on genetic modifications of glutathione biosynthetic pathway for its exploitation in plant protection against stress (Jozefczak et al., 2012). Due to the high affinity of thiol group toward metal ions glutathione is considered as an efficient metal scavenger in addition to its role as a precursor for phytochelatins—the potential metal-scavenging molecules (Mittler et al., 2004; Smeets et al., 2009; Galano and Alvarez-Idaboy, 2011; Noctor et al., 2012; Das and Roychoudhury, 2014). Therefore it could be inferred that glutathione promotes growth under stressful conditions mainly by sequestration or chelation of toxic metal ions, protecting enzyme structures, etc. Nowadays research studies focus on the exogenous application of glutathione in strengthening plant tolerance mechanisms for enhancing yield productivity (Cao et al., 2015). In addition, the development and maturation of embryos is influenced by the endogenous concentration of glutathione (Cairns et al., 2006; Lim et al., 2011). Broadly speaking glutathione is involved in sulfur metabolism and defense in plants, regulating the sulfur metabolism at the root level. During detoxification of xenobiotics, glutathione is used by glutathione S-transferases (Anjum et al., 2012).

Exogenous application of GSH significantly increased the activity of Rubisco activase and Rubisco reflecting in improved photosynthesis under Cd stress (Son et al., 2014). Glutathione enhances plant stress tolerance by improving the expression of stress-responsive genes like phytohormone genes and proteins that code for signaling and ion transport. For example, in *Arabidopsis*, Cheng et al. (2015) has demonstrated increased polysome-bound messenger RNAs due to GSH treatment resulting in enhanced drought and salinity stress tolerance, and the identified genes include the ones coding for the regulation of cell cycle progression, phytohormone synthesis, root and embryonic development, sulfur metabolism, cell wall biogenesis, proline metabolism, and ion transport. GSH protects plants by acting as a cofactor to another key radical-detoxifying pathway, glyoxylase system, thereby mediating protection to cellular growth progression and the structural stability to DNA and thus preventing any chance of mutations (Hasanuzzaman et al., 2017b). Application of glutathione to

Oryza sativa L. improved the activity of GSH-dependent antioxidant enzymes leading to significant amelioration of oxidative damage induced by Cu treatment by maintaining the levels of other antioxidant molecules like ascorbate and concomitantly reducing the accumulation of free radicals (Mostafa et al., 2014). GSH acts as a key signaling molecule during stress mitigation process. Transgenic tobacco (NtGB lines) plants synthesizing more GSH have enhanced tolerance to biotic stress by activating the NPR1-dependent salicylic acid (SA)-mediated pathway (Ghanta et al., 2011). Transgenic plants overexpressing GSH biosynthetic enzymes exhibit greater stress tolerance; e.g., overexpression of shoot-specific bacterial γ-glutamylcysteine synthetase (ECS) gene, *S1ptTECS*, in ECS-deficient *A. thaliana* improved the cadmium, mercury, and arsenate tolerance by increasing the synthesis of ECS, GSH, and phytochelatins in both shoot and root tissues, indicating the involvement of long-distance transport mechanisms in inducing the selective absorption of ions at the root (Li et al., 2006). Choe et al. (2013) have demonstrated that rice plants overexpressing ECS protected the membrane and photosynthetic functioning under stresses induced by NaCl and methyl viologen by enhancing the GSH content, indicating its role in germination, biomass accumulation, and yield. Studies carried by Liedschulte et al. (2010) concluded that increased GSH accumulation after overexpressing GSH synthetase (GS) gene significantly improved stress tolerance without imparting any change in plant growth. However, Park et al. (2017) have shown that GS overexpression enhanced the redox homeostasis of rice seedlings thereby increasing photosynthesis, membrane functioning, and yield. During glutathione deficiency transcriptional regulatory networks are initiated so as to control the biosynthetic pathways that control the synthesis of radical-scavenging metabolites working independent of glutathione (Fukushima et al., 2017).

4.4 Ascorbate

Ascorbic acid (AsA) is synthesized in mitochondria and is transported to other cellular components either through proton electrochemical gradient or facilitated diffusion (Horemans et al., 2000). Four AsA biosynthetic pathways operate in plants, viz., (1) D-galacturonate pathway (2) the L-glucose pathway, (3) the myoinositol pathway, and (4) the L-galactose pathway (Huang et al., 2014; Wang et al., 2015). L-galactose pathway is also known as the Smirnoff–Wheeler pathway and is the dominant pathway contributing bulk of the AsA (Huang et al., 2014).

Metabolism of this key metabolite, AsA, contributing to the maintenance of redox homeostasis is correlated with oxidative stress defense. Accumulation of AsA in plant tissues has been reported to be altered by normal physiological events like senescence, cellular expansion, and development in addition exposure to environmental stresses (Davey et al., 2006; Ahanger and Agarwal, 2017a,b). The versatile role of this key antioxidant metabolite is ascribed to its potentiality to act as a reducing agent. It is widely accepted that AsA mediates the detoxification of H_2O_2 generated during photosynthesis (Smirnoff, 2011). AsA forms a ubiquitous part of the ROS scavenging pathway—the ascorbate–glutathione cycle (Foyer and Noctor, 2011; Ahanger et al., 2017c). For the progression of this ROS scavenging cycle AsA is required for the optimization of the activity of ascorbate peroxidase (APX), monodehydroascorbate reductase (MDHAR), dehydroascorbate reductase (DHAR), glutathione reductase (GR), leading to the elimination of H_2O_2 concomitant with the recycling and maintenance of the AsA pool (Noctor and Foyer, 1998; Ahmad et al., 2010). Synthesis of AsA in plants during germination depends on the ROS production rates, as has been demonstrated in rice treated with abscisic acid (ABA) (Ye et al., 2012), where ABA treatment brings down ROS levels accompanied with significant reduction of AsA concentration reflecting in the downregulation of the antioxidant system, amylase, and the seed germination rate. In addition to controlling the potentiality of antioxidant system, endogenous AsA influences the levels of other antioxidant molecules like GSH leading to the initiation of signaling events independent of SA (Pavet et al., 2005). For improving growth, plants growing under either normal or stressful conditions must maintain optimal concentrations of AsA to thrive under existing conditions; e.g., *Limonium stocksii*, a salt-tolerant halophyte when treated with AsA maintained increased expression of AsA-dependent antioxidant enzymes, leading to better salinity adaptation (Hameed et al., 2015). Water-stressed *Solanum lycopersicon* L. exhibited a significant increase in the AsA content affecting the transcript levels of different antioxidant enzymes like APX, GR, MDHAR, and DHAR (Murshed et al., 2013). Besides regulating the ROS metabolism under stress, concentrations of AsA bring modulations in root growth by enhancing the expression of expansion and xyloglucan endotransglycosylase genes in roots, reflecting on its involvement in cell wall loosening and cellular expansion (Xu et al., 2015). Till date numerous studies have reported the induction of stress tolerance after genetic manipulation of AsA biosynthetic pathways (reviewed in Venkatesh and Park, 2014; Akram et al., 2017).

AsA triggers signal transduction pathway for the integration of several developmental events like time of flowering, developmental senescence, programmed cell death, and responses to pathogen infestation, and such events have been found to be associated with the regulation of the expression of phytohormones and the interconversion of AsA (Barth et al., 2006). It has been speculated that low AsA affects the long-day pathway either altering the circadian rhythms or photoperiod or elevating the SA concentrations therefore inducing early flowering and senescence. Contrarily, AsA deficiency delays flowering and senescence under short days by inhibiting gibberellic acid pathway. In addition, high ABA level under short days delays flowering and senescence by inhibiting the autonomous pathways (see Barth et al., 2006).

4.5 Osmotic Constituents

Certain metabolites widely found in plants, regulate growth and helps them to withstand the stress and are commonly called compatible osmolytes (Ahanger et al., 2014). One of the best observed functions of these solutes when plants are exposed to episodes of water, salinity, metal, nutrient, or temperature stress is to maintain cell turgor so that turgor-dependent processes can keep going at optimal pace (Ahanger and Agarwal, 2017a,b; Ahmad et al., 2017). Widely identified compatible solutes have been reported to be organic as well as inorganic in nature. Among organic osmolytes are included sugars; amino acids such as free proline; quaternary amino acid derivatives including glycine betaine (GB) or proline betaine; sugar alcohols like mannitol, pinitol, and other low-molecular-weight metabolites; as well as tertiary sulfonium compounds (choline-o-sulfate, dimethyl sulfonium propironate) while ions like K^+ are inorganic osmolytes (Agarwal et al., 1999; Ahanger et al., 2014). Osmolyte-induced osmotic adjustment maintains cell enlargement and plant growth during severe stress conditions and allows stomata to remain partially open and CO_2 assimilation to continue at low water potentials, which can otherwise be inhibitory, and their relatively greater accumulation does not have any deleterious impact on the structural and functional stability of macromolecules and their dependent processes (Hayat et al., 2012; Ahanger et al., 2014). Other key functions regulated by osmolytes include the maintenance of cellular volume, processes like protein folding and protein–protein interactions, and more importantly, the increase of the unfolded state free energy thereby shifting the equilibrium toward the natively folded conformations. Balanced interactions of osmolytes with protein backbone assist in determining the fate of protein folding (Khan et al., 2010). At physiological pH osmolytes behave as neutral ions (zwitterion) and never interfere with the hydration shells of the important macromolecules like proteins. However, ions like Na^+ and Cl^- disrupt the protein hydration shell and interfere with the noncovalent interactions, an important aspect of the structural stability of proteins. More importantly, osmolytes reduce the efflux of ions including K^+ from cellular compartments by checking the ROS-induced activation of K efflux channels (Cuin and Shabala, 2007). Greater synthesis of osmotic constituents mediates proper root development so that access to water in the deep soils is increased (Gonzalez et al., 2008). The importance of osmolytes in the regulation of yield of major crop plants has been reported by several workers (reviewed in Agarwal et al., 2009; Ahanger et al., 2017d; Ahmad et al., 2017).

4.5.1 Amino Acids

Amino acids can be either proteinogenic or nonproteinogenic, where proteinogenic amino acids include natural amino acids, the syntheses of which are genetically controlled, and nonprotein amino acids can arise from the modification of existing proteins, amino acids and can also be the intermediates of the biosynthetic pathways, or post-translational incorporation into the protein. Of the 700 nonprotein amino acids, 300 exist in plants. Nonprotein amino acids have been reported in several legumes and grasses. Together with secondary metabolites amino acids are implicated in defense against herbivores and insects. Nonprotein amino acids like L-canavanine, γ-aminobutyric acid, DL-β-aminobutyric acid, L-3, 4-dihydroxyphenylalanine, 5-hydroxy-L-tryptophan, 5- hydroxytryptamine (serotonin), indole-3-acetyl-aspartic acid, *p*-aminophenylalanine, and L-azetidine-2-carboxylic acid have been identified in plants. Most of these nonprotein amino acids have structures similar to natural amino acids (Fellos and Bell, 1970; Huang et al., 2011). Misincorporation of non-protein amino acids into the proteins, obstruction of primary metabolism and mimicking, as well as interfering with insect neurological processes can have direct toxic effects on plants. In addition, under herbivore attack, nonprotein amino acids allow plants to store nitrogen in metabolically inaccessible form and may also act as signals for plant defense responses (Vranova et al., 2011; Huang et al., 2011).

Amino acids, in addition to their well-established role in protein synthesis, perform key functions in both primary and secondary metabolism in plants. Certain amino acids are involved in assimilating nitrogen and its subsequent transport from source to sink, whereas others serve as precursors for the synthesis of important secondary products such as hormones and other compounds that are involved in plant defense. Unraveling the regulatory mechanisms leading to the incorporation of inorganic nitrogen into amino acids and the identification of factors that regulate the biosynthesis of secondary metabolites from the amino acid precursors can be very handy in enhancing plant tolerance to adverse conditions. Amino acids act as osmolytes for maintaining the cellular water content and hence turgor-dependent processes like stomatal movements, ion transport, and tolerance to stresses (Rai, 2002). Moreover, the internal amino acid pool has been identified to serve as a signal for regulating the uptake of nitrate from the soil (Miller et al., 2007). Amino acids including proline, cysteine, methionine, glutamate, and arginine are actively involved in stress tolerance. Proline is a widely accepted osmolyte; however, amino acids serve as precursors for the synthesis of other important metabolites including hormones that have important role under consitions of stress. For example, ethylene is synthesized from methionine. whereas cysteine together with glutamate and glycine forms an important antioxidant glutathione required for optimal functioning of the ascorbate–glutathione cycle and for chelating heavy metals, in addition to its other important roles (Noctor et al., 2012). Enhanced synthesis of methionine in salt-stressed *V. radiata* seedlings resulted in enhanced synthesis of ethylene and hence imparted stress tolerance (Khan et al., 2014). Plants maintaining higher contents of amino acids have the potential to withstand stressed conditions (Tomar and Agarwal, 2013; Jatav et al., 2014; Ahanger and Agarwal, 2017a,b).

It has been observed that stress-induced accumulation of metabolites serving as osmolytes, e.g., proline, results from the up- and downregulation of their synthesis and catabolism, respectively (Iqbal et al., 2015). So it means that the delicate balance between synthesis and catabolism determines the resulting accumulation of a particular osmolyte. Upon exposure to environmental stresses, increased accumulation of proline has been reported in several plant species. In *Zea mays,* accumulated proline protects complex II of the mitochondrial electron transport chain from salt stress, thereby stabilizing mitochondrial respiration (Hamilton and Heckathorn, 2001). Proline helps in maintaining the cellular homeostasis, including redox as well as the energy status. In addition, it can also act as a signaling molecule to bring modulations in the mitochondrial functions, influence cell proliferation or cell death, and can also trigger the expression of specific genes (Banu et al., 2009). In salt-stressed tobacco, supplementation of growth media with proline reduced the production and accumulation of ROS and hence lipid peroxidation and suppressed nuclear deformation and chromatin condensation in addition to increasing the gene expression of different antioxidant enzymes (Banu et al., 2009). In addition, proline has been reported to regulate protein synthesis and the cell cycle progression. Plants showing defects in the accumulation of proline display altered transcriptional regulation of cyclin genes, resulting in cell cycle arrest and suppression of cell proliferation. Furthermore, it has also been reported that this negative affect can be reversed by the addition of proline (Wang et al., 2014). Plants accumulating higher contents of proline show better growth due to the substantial increment in the photosynthetic attributes and the activity of antioxidant enzymes affecting water use efficiency (WUE) and the carboxylation potential (Reddy et al., 2015). Optimizing the pathways of amino acid synthesis via molecular interventions can be helpful in improving crop quality. Simon-Sarkadi et al. (2006) has demonstrated significant increase in the total amino acid profile of soybean-overexpressing biosynthetic genes, and proline was reported to be the abundant amino acid under stress conditions. Engineering amino acids must focus on improving the levels of high-quality stress-responsive proteins. Moreover, stress-tolerant plants have the potential to maintain greater amounts of amino acids (Kovacs et al., 2012).

4.5.2 Glycine Betaine

GB levels are determined by the rate of biosynthesis and its transport to the growing tissues. GB protects PSII activity by stabilizing the extrinsic proteins with the photosynthetic oxygen-evolving complex proteins, maintaining membrane integrity, and helping to combat oxidative stress (Chen and Murata, 2011). Under normal conditions, plants capable of accumulating GB show improved morphological and yield characters like increased flower size and fruit and seed number. Introducing GB biosynthetic pathways in nonsynthesizing plant species can induce the expression of stress-responsive genes like ROS-scavenging enzymes (Chen and Murata, 2011). Park et al. (2007) while comparing transgenic tomato seedlings showing potential of accumulating GB in different cellular organelles observed that cultivars accumulating larger amounts of GB in the chloroplast are more hard than those accumulating it in the cytosol. Positive correlation was observed between the accumulation of GB and the amelioration of oxidative stress through its active involvement in the prevention of free radical production and protection of membranes and photosynthetic attributes. Exogenous application of GB significantly affected the growth by enhancing the antioxidant potential and osmolyte accumulation (Malekzadeh, 2015). Nuclear magnetic resonance analysis of *Arabidopsis*-overexpressing ALDH10A8 and ALDH10A9 genes showed increased GB and other important metabolites including sugars, amino acids, and organic acids, thereby regulating stress responses (Missihoun et al., 2015). The overexpression of *BADH* gene from *Atriplex hortensis* L. improved the photosynthetic efficiency of *Triticum aestivum* L. under salt stress, and more importantly, the levels of lipids and associated fatty acids of the thylakoid membranes were observed to increase contributing to greater chlorophyll and carotenoid synthesis, increased Hill reaction, and higher Ca^{2+}-ATPase activity (Tian et al., 2017).

4.5.3 Carbohydrates

Accumulation of carbohydrates like sugars (glucose, fructose, sucrose), starch, trehalose, and polyols plays an important role in bringing osmotic adjustment, carbon storage, and free radical scavenging. Exogenous application of glucose to wheat seedlings imparted salt tolerance by enhancing the antioxidant defense system and mediating efficient ion compartmentalization causing an increase in K/Na ratio and membrane stability (Hu et al., 2012). Several studies have reported that accumulation of higher amounts of sugars is an important protective response to abiotic stress (Pandey et al., 2004; Tomar and Agarwal, 2013; Masood et al., 2013; Jatav et al., 2014; Benzarti et al., 2014; Tomar et al., 2015; Hu et al., 2015;

Ahanger and Agarwal, 2017a,b). Sugars accumulated during stressed conditions function as osmolytes to maintain the cell turgor and protect membrane and protein structure (Kaplan and Guy, 2004).

Carbohydrates such as starch and fructans are accumulated as storage substances and are utilized during periods when their supply is perturbed and energy demands are enhanced. Most of the plants use starch as the stored carbohydrate, and during the freezing or dry periods fructans may also contribute a good amount (Hendry, 1993). High water solubility of fructans makes them important cryoprotectants under such conditions in addition to their role in stabilizing membranes and contributing to osmoregulation by releasing hexose sugars during dry periods (Spollen and Nelson, 1994; Valluru and den Ende, 2008). Fructans are branched polymers of fructose synthesized in vacuole by fructosyltransferases, which mediate transfer of fructose from sucrose to the growing fructan chain (Vijn and Smeekens, 1999; Livingston et al., 2009; Xiang et al., 2010). Introduction and overexpression of the fructan biosynthetic pathway in tobacco (Pilon-Smits et al., 1995) and rice (Li et al., 2011) plants enhanced the production of fructans leading to enhanced tolerance to stresses like drought and chilling. During the periods of drought or freezing fructans contribute to maintain membrane stability either by direct hydrogen bonding interaction with the phosphate or choline groups of membrane lipids (Valluru and den Ende, 2008) or by feeding part of polysaccharide into the lipid head group region of the membranes, thereby reducing the water efflux from the membranes (Livingston et al., 2009). Randomly coiled fructans (mostly inulin type) existing in different conformations insert themselves deep into the membranes and protect them for prolonged periods due to their low crystallizations rates (Valluru and den Ende, 2008).

Starch, an osmotically inert molecule, is composed of glucose polymers and is synthesized in plastids. It serves as the main carbohydrate store in plants and can be mobilized rapidly to produce soluble sugars. Plant mutants having less ability to synthesize or degrade starch show impaired growth. Starch metabolism is greatly influenced by the changes in the environment. Besides prominent diurnal fluctuations, stresses like high salinity, temperature, and water deficit induce enhanced cleavage of starch, resulting in the accumulation of sugars (Kaplan and Guy, 2004; Pandey et al., 2004). Starch may be degraded into simple soluble sugars either through phosphorylation by glucan-water dikinase and phosphoglucan water dikinase or by amylases (α and β) and α-glycosidase (Tetlow et al., 2004; Kotting et al., 2010; Stitt and Zeeman, 2012). In rice, osmotic and water stress conditions upregulated amylase activity resulting in increased accumulation of sugars (Pandey et al., 2004). Cold stress caused significant increase in the activity of starch-degrading enzymes and hence the accumulation of sugars in potato leaves resulting in enhanced stress tolerance (Sitnicka and Orzechowski, 2014).

Trehalose, a nonreducing disaccharide, is another important osmoprotectant that accumulates to high amounts during stressful environmental conditions. Optimal accumulation of trehalose can stabilize membranes and protein structures from stress-induced denaturation (Elbein et al., 2003; Paul et al., 2008). Three different pathways exist for the biosynthesis of trehalose, but the most studied and widely distributed is the one involving transfer of glucose from uridine 5-diphosphate glucose (UDP)-glucose to glucose-6-phosphate to form trehalose-6-phosphate (T6P) and UDP. Finally, T6P is converted to free trehalose (Elbein et al., 2003). Accumulation of trehalose in rice seedlings enhanced tolerance to drought and salt stress considerably. Transgenic plants showing higher accumulation of trehalose show better stress adaptation when compared with the wild counterparts (Redillas et al., 2012). Velazquez-Marquez et al. (2015) have observed enhanced accumulation of trehalose in *Z. mays* L. seedlings exposed to lower water potential. The beneficial role of trehalose in circumventing stress-induced changes can be exploited by maintaining the fine-tuning between its biosynthesis and hydrolysis (Delorge et al., 2014). T6P, an intermediate in trehalose biosynthesis, is now confirmed to act as a sensor for sucrose availability and helps in maintaining the availability of sucrose within optimal range therefore reflecting to its bidirectional role. Trehalose acts as a potential signal metabolite in plant–pathogen or plant–microbe interactions besides its involvement in the regulation of stomatal conductance and WUE under stressful conditions (Lunn et al., 2014). T6P inhibits sucrose nonfermenting 1(SNF1)-related protein kinase1 (SnRK1; AKIN10/AKIN11) leading to maintenance of the expression of genes involved in biosynthetic reactions like those of amino acid, protein, and nucleotide synthesis; the tricarboxylic acid cycle; and the mitochondrial electron transport, which are normally downregulated by SnRK1 (Zhang et al., 2009). It should be noted here that supra-accumulation of T6P leads to the development of certain undesirable phenotypes like poor seed set, reduced apical dominance, hastened flowering, and reduced rosette and lancet-shaped leaves (Delorge et al., 2014; Lunn et al., 2014). However, equilibrating the T6P levels by overexpressing the trehalase activity improves stress tolerance by increasing the water retaining capacity (Houtte et al., 2013).

4.6 Pyridine Nucleotides—Versatile Role of Small Molecules

Pyridine nucleotides are profusely occurring soluble coenzymes that undergo reversible oxidation and reduction during various biological electron transfer reactions. They are composed of two mononucleotides, i.e., adenosine monophosphate and nicotinamide mononucleotide, existing as oxidized or reduced nicotinamide adenine dinucleotides either in their phosphorylated ($NADP^+$ and NADPH) or unphosphorylated (NAD^+ and NADH) forms. In early times the primary function of pyridine nucleotides was considered as electron-shuttling agents supporting the activity of enzymes that catalyze redox reactions. Recent research has confirmed that pyridine nucleotides and the balance between their oxidized and reduced forms regulate several cellular functions including maintenance of redox status, regulation of ion channels, cell survival and death, and cell signaling under normal as well as extreme environments (Nakamura et al., 2012). Therefore targeting pyridine nucleotides can provide potential avenues for combating stresses. Pyridine nucleotides like NAD and NADP have vital roles in the metabolic conversions as signal transducers and more importantly in cellular defense systems. These coenzymes behave as electron carriers during energy transduction and several biosynthetic processes. Oxidized forms of pyridine nucleotides (NAD^+ and $NADP^+$) form the key elements in the regulatory pathways; e.g., NAD^+ serves as a substrate during ribosylation of adenosine diphosphate (ADP) and for the Sir2 family of NAD^+-dependent protein deacetylases, and as precursor of cyclic ADP ribose involved in calcium mobilizing. In addition, conversion of $NADP^+$ into the 2′-phosphorylated form of cyclic ADP ribose or to nicotinic acid derivative, NaADP, also results in the generation of powerful intracellular calcium signaling agents (Pollak et al., 2007). Additionally, during posttranslational protein modification transfer of the ADP ribose units from NAD to target proteins is very important (Ying et al., 2003).

During ROS-mediated signaling pyridine nucleotides play an important role as they form crucial components in the regulation of both ROS-generating and ROS-neutralizing pathways in plants (Apel and Hirt, 2004; Mittler et al., 2004). NADP/NADPH ratio in the chloroplast and NADP status in the mitochondria and NADPH oxidases at the plasma membranes influence ROS generation significantly. Similarly, the scavenging of ROS depends on ascorbate and glutathione concentrations with redox flux through these pools being mostly maintained by NAD(P)H (Ahanger et al., 2017c). More importantly, the pyridine nucleotides get involved in defensive and signaling reactions like generation of nitric oxide and metabolism of reactive lipid derivatives (Mano et al., 2005). Therefore it can be inferred that besides acting as cofactors in various energy-generating and biosynthetic reactions pyridine nucleotides are crucial players during the redox signaling linked to stress and development (Noctor et al., 2006). Several studies have witnessed significant reduction in the ratio of NADP/NADPH leading to alterations in the normal physiology and metabolism of plants (Valderrama et al., 2006; Fatma et al., 2016).

5. CONCLUSION AND FUTURE PROSPECTS

Stresses alter the growth by bringing modifications in the metabolome of plants. Minimizing the plant potentiality for synthesizing the metabolites hampers plant functioning through alteration in the redox homeostasis, antioxidant potential, and related tolerance mechanisms. Identifying the stress-responsive metabolite–coding genes and developing markers for improvement of the metabolomic profile of plant can be beneficial for yield improvement. Besides displaying antioxidant properties for ROS neutralization plant metabolites can assist in bringing osmotic adjustment therefore providing extra strength to the tolerance mechanisms. Genome-assisted optimization of metabolite synthesis can control growth by maintaining phytohormone concentration therefore integrating themselves in the signaling events during growth and developmental progression. Unraveling the involvement of the metabolic products of metabolites in stress signaling can help in enhancing the selective metabolites for their exploitation in stress tolerance.

REFERENCES

Agarwal, R.M., Pandey, R., Gupta, S., 1999. Certain aspects of water stress induced changes and tolerance mechanisms in plants. J. Indian Bot. Soc. 78, 255–269.

Agarwal, R.M., Tomar, N.S., Singh, K., Sharma, G.L., 2009. Potassium induced changes in flowering plants. In: Flower Retrospect and Prospect (Professsor Vishwambhar Puri Birth Centenary Volume). SR Scientific Publication, Delhi 110553, pp. 158–186.

Ahanger, M.A., Agarwal, R.M., Tomar, N.S., Shrivastava, M., 2015. Potassium induces positive changes in nitrogen metabolism and antioxidant system of oat (*Avena sativa* L. cultivar Kent). J. Plant Int. 10 (1), 211–223.

Ahanger, M.A., Agarwal, R.M., 2017a. Potassium improves antioxidant metabolism and alleviates growth inhibition under water and osmotic stress in wheat (*Triticum aestivum* L.). Protoplasma 254 (4), 1471–1486.

Ahanger, M.A., Agarwal, R.M., 2017b. Salinity stress induced alterations in antioxidant metabolism and nitrogen assimilation in wheat (*Triticum aestivum* L.) as influenced by potassium supplementation. Plant Physiol. Biochem. 115, 449–460.

Ahanger, M.A., Akram, N.A., Ashraf, M., Alyemeni, M.N., Wijaya, L., Ahmad, P., 2017b. Signal transduction and biotechnology in response to environmental stresses. Biol. Plant 61 (3), 401–416.

Ahanger, M.A., Akram, N.A., Ashraf, M., Alyemni, M.N., Wijaya, L., Ahmad, P., 2017a. Plant responses to environmental stresses – from gene to biotechnology. AoB Plants 9 (4), PLX025. https://doi.org/10.1093/aobpla/plx025.

Ahanger, M.A., Tittal, M., Mir, R.A., Agarwal, R.M., 2017d. Alleviation of water and osmotic stress-induced changes in nitrogen metabolizing enzymes in *Triticum aestivum* L. cultivars by potassium. Protoplasma 254 (5), 1953–1963.

Ahanger, M.A., Tomar, N.S., Tittal, M., Argal, S., Agarwal, R.M., 2017c. Plant growth under water/salt stress: ROS production; antioxidants and significance of added potassium under such conditions. Physiol. Mol. Biol. Plants. https://doi.org/10.1007/s12298-017-0462-7.

Ahanger, M.A., Tyagi, S.R., Wani, M.R., Ahmad, P., 2014. Drought tolerance: roles of organic osmolytes, growth regulators and mineral nutrients. In: Ahmad, P., Wani, M.R. (Eds.). Ahmad, P., Wani, M.R. (Eds.), Physiological Mechanisms and Adaptation Strategies in Plants under Changing Environment, vol. I. Springer Science+Business Media, Inc., pp. 25–56.

Ahmad, P., Abdel Latef, A.A., Abd Allah, E.F., Hashem, A., Sarwat, M., Anjum, N.A., Gucel, S., 2016. Calcium and potassium supplementation enhanced growth, osmolyte secondary metabolite production, and enzymatic antioxidant machinery in cadmium-exposed chickpea (*Cicer arietinum* L.). Front. Plant Sci. 7, 513. https://doi.org/10.3389/fpls.2016.00513.

Ahmad, P., Ahanger, M.A., Alam, P., Alyemeni, M.N., 2017. Modification of osmolytes and antioxidant enzymes by 24-epibrassinolide in chickpea seedlings under mercury (Hg) toxicity. J. Plant Growth Regul. https://doi.org/10.1007/s00344-017-9730-6.

Ahmad, P., Ashraf, M., Hakeem, K.R., Azooz, M.M., Rasool, S., Chandna, R., Akram, N.A., 2014. Potassium starvation-induced oxidative stress and antioxidant defense responses in *Brassica juncea*. J. Plant Int. 9 (1), 1–9.

Ahmad, P., Jaleel, C.A., Sharma, S., 2010. Antioxidant defense system, lipid peroxidation, proline metabolizing enzymes, and biochemical activities in two *Morus alba* genotypes subjected to NaCl stress. Russ. J. Plant Physiol. 57 (4), 509–517.

Akram, N.A., Shafiq, F., Ashraf, M., 2017. Ascorbic acid-a potential oxidant scavenger and its role in plant development and abiotic stress tolerance. Front. Plant Sci. 8, 613. https://doi.org/10.3389/fpls.2017.00613.

Anjum, N.A., Ahmad, I., Mohmood, I., Pacheco, M., Duarte, A.C., Pereira, E., Umar, S., Ahmad, A., Khan, N.A., Iqbal, M., Prasad, M.N.V., 2012. Modulation of glutathione and its related enzymes in plants' responses to toxic metals and metalloids–a review. Environ. Exp. Bot. 75, 307–324.

Apel, K., Hirt, H., 2004. Reactive oxygen species: metabolism, oxidative stress, and signal transduction. Annu. Rev. Plant Biol. 55, 373–399.

Arbona, V., Manzi, M., de Ollas, C., Gomez-Cadenas, A., 2013. Metabolomics as a tool to investigate abiotic stress tolerance in plants. Int. J. Mol. Sci. 14, 4885–4911.

Asensi-Fabado, M.A., Ammon, A., Sonnewald, U., Munné-Bosch, S., Voll, L.M., 2015. Tocopherol deficiency reduces sucrose export from salt-stressed potato leaves independently of oxidative stress and symplastic obstruction by callose. J. Exp. Bot. 66 (3), 957–971.

Banu, M.N.A., Hoque, M.A., Watanabe-Sugimoto, M., Matsuoka, K., Nakamura, Y., Shimoishi, Y., Murata, Y., 2009. Proline and glycine betaine induce antioxidant defense gene expression and suppress cell death in cultured tobacco cells under salt stress. J. Plant Physiol. 166, 146–156.

Barth, C., De Tullio, M., Conklin, P.L., 2006. The role of ascorbic acid in the control of flowering time and the onset of senescence. J. Exp. Bot. 57 (8), 1657–1665.

Bashandy, T., Guilleminot, J., Vernoux, T., Caparros-Ruiz, D., Ljung, K., Meyer, Y., Reichheld, J.P., 2010. Interplay between the NADP-linked thioredoxin and glutathione systems in Arabidopsis auxin signaling. Plant Cell 22, 376–391.

Benzarti, M., Rejeb, K.B., Messedi, D., Mna, A.B., Hessini, K., Ksontini, M., Abdelly, C., Debez, A., 2014. Effect of high salinity on *Atriplex portulacoides*: growth, leaf water relations and solute accumulation in relation with osmotic adjustment. S. Afr. J. Bot. 95, 70–77.

Bolouri-Moghaddam, M.R., Roy, K.L., Xiang, L., Rolland, F., den Ende, W.V., 2010. Sugar signalling and antioxidant network connections in plant cells. FEBS J. 277, 2022–2037.

Bose, J., Ana Rodrigo-Moreno, A., Shabala, S., 2014. ROS homeostasis in halophytes in the context of salinity stress tolerance. J. Exp. Bot. 65 (5), 1241–1257.

Cairns, N.G., Pasternak, M., Wachter, A., Cobbett, C.S., Meyer, A.J., 2006. Maturation of Arabidopsis seeds is dependent on glutathione biosynthesis within the embryo. Plant Physiol. 141, 446–455.

Cao, F., Cai, Y., Liu, L., Zhang, M., He, X., Zhang, G., Wu, F., 2015. Differences in photosynthesis, yield and grain cadmium accumulation as affected by exogenous cadmium and glutathione in the two rice genotypes. Plant Growth Regul. 75, 715–723.

Cazzaniga, S., Li, Z., Niyogi, K.K., Bassi, R., Dall'Osto, L., 2012. The Arabidopsis szl1 mutant reveals a critical role of β-carotene in photosystem I photoprotection. Plant Physiol. 159 (4), 1745–1758.

Cazzonelli, C.I., 2011. Carotenoids in nature: insights from plants and beyond. Funct. Plant Biol. 38, 833–847.

Chen, T.H., Murata, N., 2011. Glycine betaine protects plants against abiotic stress: mechanisms and biotechnological applications. Plant Cell Environ. 34, 1–20.

Chen, W., He, S., Liu, D., Patil, G.B., Zhai, H., Wang, F., Stephenson, T.J., Wang, Y., Wang, B., Valliyodan, B., Nguyen, H.T., Liu, Q., 2015. A sweet potato geranylgeranyl pyrophosphate synthase gene, *IbGGPS*, increases carotenoid content and enhances osmotic stress tolerance in *Arabidopsis thaliana*. PLoS One 10, e0137623.

Cheng, M.C., Ko, K., Chang, W.L., Kuo, W.C., Chen, G.H., Lin, T.P., 2015. Increased glutathione contributes to stress tolerance and global translational changes in *Arabidopsis*. Plant J. 83, 926–939.

Choe, Y.K., Kim, I., Bae, M.J., Lee, E.J., Kim, Y.H., Park, H.M., Yoon, H.S., 2013. Homologous expression of γ-glutamylcysteine synthetase increases grain yield and tolerance of transgenic rice plants to environmental stresses. J. Plant Physiol. 170 (6), 610–618.

Choudhury, N.K., Behera, R.K., 2001. Photoinhibition of photosynthesis: role of carotenoids in photoprotection of chloroplast constituents. Photosynthetica 39 (4), 481–488.

Couee, I., Sulmon, C., Gouesbet, G., El Amrani, A., 2006. Involvement of soluble sugars in reactive oxygen species balance and responses to oxidative stress in plants. J. Exp. Bot. 57 (3), 449–459.

Cuin, T.A., Shabala, S., 2007. Compatible solutes reduce ROS-induced potassium efflux in *Arabidopsis* roots. Plant Cell Environ. 30, 875–885.

Das, K., Roychoudhury, A., 2014. Reactive oxygen species (ROS) and response of antioxidants as ROS-scavengers during environmental stress in plants. Front. Environ. Sci. https://doi.org/10.3389/fenvs.2014.00053.

Davey, M.W., Kenis, K., Keulemans, J., 2006. Genetic control of fruit vitamin C contents. Plant Physiol. 142, 343–351.

del Rio, L.A., 2015. ROS and RNS in plant physiology: an overview. J. Exp. Bot. 66 (10), 2827–2837.

DellaPenna, D., 2005. Progress in the dissection and manipulation of vitamin E synthesis. Trends Plant Sci. 10 (12), 574–579.

Delorge, I., Janiak, M., Carpentier, S., Dijck, P.V., 2014. Fine tuning of trehalose biosynthesis and hydrolysis as novel tools for the generation of abiotic stress tolerant plants. Front. Plant Sci. https://doi.org/10.3389/fpls.2014.00147.

Diaz-Vivancos, P., Simone, A., Kiddle, G., Foyer, C.H., 2015. Glutathione – linking cell proliferation to oxidative stress. Free Radic. Biol. Med. 89, 1154–1164.

Dormann, P., 2007. Functional diversity of tocochromanols in plants. Planta 225 (2), 269–276.

Elbein, A.D., Pan, Y.T., Pastuszak, I., Carroll, D., 2003. New insights on trehalose: a multifunctional molecule. Glycobiology 13 (4), 17–27.

Falk, J., Munne-Bosch, S., 2010. Tocochromanol functions in plants: antioxidation and beyond. J. Exp. Bot. 61 (6), 1549–1566.

Farre, G., Sanahuja, G., Naqvi, S., Bai, C., Capell, T., Zhu, C., Christou, P., 2010. Travel advice on the road to carotenoids in plants. Plant Sci. 179, 28–48.

Fatma, M., Masood, A., Per, T.S., Khan, N.A., 2016. Nitric oxide alleviates salt stress inhibited photosynthetic performance by interacting with sulfur assimilation in mustard. Front. Plant Sci. 7, 521. https://doi.org/10.3389/fpls.2016.00521.

Fellos, L.E., Bell, E.A., 1970. 5-hydroxy-l-tryptophan, 5-hydroxytryptamine and l-tryptophan-5-hydroxylase in *Griffonia simplicifolia*. Phytochem 9 (11), 2389–2396.

Fernie, A.R., Trethewey, R.N., Krotzky, A.J., Willmitzer, L., 2004. Metabolite profiling: from diagnostics to systems biology. Nat. Rev. Mol. Cell Biol. 5, 763–769.

Fischer, B.B., Hideg, E., Krieger-Liszkay, A., 2013. Production, detection, and signaling of singlet oxygen in photosynthetic organisms. Antioxid. Redox Signal 18, 2145–2162.

Foyer, C.H., Noctor, G., Hodges, M., 2011. Respiration and nitrogen assimilation: targeting mitochondria-associated metabolism as a means to enhance nitrogen use efficiency. J. Exp. Bot. 62, 1467–1482.

Foyer, C.H., Noctor, G., 2011. Ascorbate and glutathione: the heart of the redox hub. Plant Physiol. 155, 2–18.

Fukushima, A., Iwasa, M., Nakabayashi, R., Kobayashi, M., Nishizawa, T., Okazaki, Y., Saito, K., Kusano, M., 2017. Effects of combined low glutathione with mild oxidative and low phosphorus stress on the metabolism of *Arabidopsis thaliana*. Front. Plant Sci. 8, 1464. https://doi.org/10.3389/fpls.2017.01464.

Gadjev, I., Vanderauwera, S., Gechev, T.S., Laloi, C., Minkov, I.N., Shulaev, V., Apel, K., Inze, D., Mittler, R., Van Breusegem, F., 2006. Transcriptomic footprints disclose specificity of reactive oxygen species signaling in Arabidopsis. Plant Physiol. 141, 436–445.

Galano, A., Alvarez-Idaboy, J.R., 2011. Glutathione: mechanism and kinetics of its non-enzymatic defense action against free radicals. RSC Adv. 1, 1763–1771.

Ghanta, S., Bhattacharyya, D., Chattopadhyay, S., 2011. Glutathione signaling acts through NPR1-dependent SA-mediated pathway to mitigate biotic stress. Plant Signal. Behav. 6 (4), 607–609.

Gonzalez, A., Martin, I., Ayerbe, L., 2008. Yield and osmotic adjustment capacity of barley under terminal water stress conditions. J. Agron. Crop Sci. 194, 81–91.

Griffiths, C.A., Paul, M.J., Foyer, C.H., 2016. Metabolite transport and associated sugar signalling systems underpinning source/sink interactions. Biochim. Biophys. Acta 1857, 1715–1725.

Hameed, A., Gulzar, S., Aziz, I., Hussain, T., Gul, B., Khan, M.A., 2015. Effects of salinity and ascorbic acid on growth, water status and antioxidant system in a perennial halophyte. AoB Plants 7, plv004. https://doi.org/10.1093/aobpla/plv004.

Hamilton 3rd, E.W., Heckathorn, S.A., 2001. Mitochondrial adaptations to NaCl complex I is protected by anti-oxidants and small heat shock proteins, whereas complex II is protected by proline and betaine. Plant Physiol. 126, 1266–1274.

Hasanuzzaman, M., Nahar, K., Anee, T.I., Fujita, M., 2017a. Glutathione in plants: biosynthesis and physiological role in environmental stress tolerance. Physiol. Mol. Biol. Plants 23 (2), 249–268.

Hasanuzzaman, M., Nahar, K., Hossain, M.S., Mahmud, J.A., Rahman, A., Inafuku, M., Oku, H., Fujita, M., 2017b. Coordinated actions of glyoxalase and antioxidant defense systems in conferring abiotic stress tolerance in plants. Int. J. Mol. Sci. 18 (1). https://doi.org/10.3390/ijms18010200.

Havaux, M., 2014. Carotenoid oxidation products as stress signals in plants. Plant J. 79, 597–606.

Hayat, S., Hayat, Q., Alyemeni, M.N., Wani, A.S., Pichtel, J., Ahmad, A., 2012. Role of proline under changing environments A review. Plant Signal. Behav. 7 (11), 1–11.

Hendry, G.A.F., 1993. Evolutionary origins and natural functions of fructans: a climatological, biogeography and mechanistic appraisal. New Phytol. 123, 3–14.

Hofius, D., Hajirezaei, M., Geiger, M., Tschiersch, H., Melzer, M., Sonnewald, U., 2004. RNAi-mediated tocopherol deficiency impairs photoassimilate export in transgenic potato plants. Plant Physiol. 135, 1256–1268.

Horemans, N., Foyer, C.H., Asard, H., 2000. Transport and action of ascorbate at the plant plasma membrane. Trends Plant Sci. 5, 263–267.

Houtte, H.V., Vandesteene, L., Lopez-Galvis, L., Lemmens, L., Kissel, E., Carpentier, S., Feil, R., Avonce, N., Beckman, T., Lunn, J.E., Dijck, P.V., 2013. Overexpression of the trehalase Gene *AtTRE1* leads to increased drought stress tolerance in *Arabidopsis* and is involved in abscisic acid-induced stomatal closure. Plant Physol. 161 (3), 1158–1171.

Hu, M., Shi, Z., Zhang, Z., Zhang, Y., Li, H., 2012. Effects of exogenous glucose on seed germination and antioxidant capacity in wheat seedlings under salt stress. Plant Growth Regul. 68, 177–188.

Hu, Y., Wang, B., Hu, T., Chen, H., Li, H., Zhang, W., Zhong, Y., Hu, Y., 2015. Combined action of an antioxidant defence system and osmolytes on drought tolerance and post-drought recovery of *Phoebe zhennan* S. Lee saplings. Acta Physiol. Plant 37, 84. https://doi.org/10.1007/s11738-015-1831-x.

Huang, M., Xu, Q., Deng, X.X., 2014. L-Ascorbic acid metabolism during fruit development in an ascorbate-rich fruit crop chestnut rose (*Rosa roxburghii* Tratt). J. Plant Physiol. 171, 1205–1216.

Huang, T., Jander, G., de Vos, M., 2011. Non-protein amino acids in plant defense against insect herbivores: representative cases and opportunities for further functional analysis. Phytochem 72, 1531–1537.

Iqbal, N., Umar, S., Khan, N.A., 2015. Nitrogen availability regulates proline and ethylene production and alleviates salinity stress in mustard (*Brassica juncea*). J. Plant Physiol. 178, 84–91.

Jahns, P., Holzwarth, A.R., 2012. The role of the xanthophyll cycle and of lutein in photoprotection of photosystem II. Biochim. Biophys. Acta 1817 (1), 182–193.

Jatav, K.S., Agarwal, R.M., Tomar, N.S., Tyagi, S.R., 2014. Nitrogen metabolism, growth and yield responses of wheat (*Triticum aestivum* L.) to restricted water supply and varying potassium treatments. J. Indian Bot. Soc. 93 (3&4), 177–189.

Jozefczak, M., Remans, T., Vangronsveld, J., Cuypers, A., 2012. Glutathione is a key player in metal-induced oxidative stress defenses. Int. J. Mol. Sci. 13, 3145–3175.

Kang, L., Park, S.C., Yoon, C., Kim, H.S., Lee, H.S., Kwak, S.S., 2017. Metabolic engineering of carotenoids in transgenic sweet potato. Breed. Sci. 67, 27–34.

Kaplan, F., Guy, C.L., 2004. β-Amylase induction and the protective role of maltose during temperature shock. Plant Physiol. 135, 1674–1684.

Keunen, E., Peshev, D., Vangronsveld, J., den Ende, W.V., Cuypers, A., 2013. Plant sugars are crucial players in the oxidative challenge during abiotic stress: extending the traditional concept. Plant Cell Environ. 36, 1242–1255.

Khan, M.I.R., Asgher, M., Khan, N.A., 2014. Alleviation of salt-induced photosynthesis and growth inhibition by salicylic acid involves glycine betaine and ethylene in mungbean (*Vigna radiata* L.). Plant Physiol. Biochem. 80, 67–74.

Khan, S.H., Ahmad, N., Ahmad, F., Kumar, R., 2010. Naturally occurring organic osmolytes: from cell physiology to disease prevention. Life 62 (12), 891–895.

Kotting, O., Kossmann, J., Zeeman, S.C., Lloyd, J.R., 2010. Regulation of starch metabolism: the age of enlightenment? Curr. Opin. Plant Biol. 13, 321–329.

Kovacs, Z., Simon-Sarkadi, L., Vashegyi, I., Kocsy, G., 2012. Different accumulation of free amino acids during short- and long-term osmotic stress in wheat. Sci. World J. 2012, 216–521.

Laloi, C., Apel, K., Danon, A., 2004. Reactive oxygen signalling: the latest news. Curr. Opin. Plant Biol. 7, 323–328.

Li, H.W., Zang, B.S., Deng, X.W., Wang, X.P., 2011. Overexpression of the trehalose-6-phosphate synthase gene OsTPS1 enhances abiotic stress tolerance in rice. Planta 234, 1007–1018.

Li, L., Paolillo, D.J., Parthasarathy, M.V., DiMuzio, E.M., Garvin, D.F., 2001. A novel gene mutation that confers abnormal patterns of β-carotene accumulation in cauliflower (*Brassica oleracea* var. *botrytis*). Plant J. 26, 59–67.

Li, Y., Dankher, O.P., Carreira, L., Smith, A.P., Meagher, R.B., 2006. The shoot-specific expression of γ-glutamylcysteine synthetase directs the long-distance transport of thiol-peptides to roots conferring tolerance to mercury and arsenic. Plant Physiol. 141, 288–298.

Liedschulte, V., Wachter, A., Zhigang, A., Rausch, T., 2010. Exploiting plants for glutathione (GSH) production: uncoupling GSH synthesis from cellular controls results in unprecedented GSH accumulation. Plant Biotechnol. J. 8, 807–820.

Lim, B., Meyer, A.J., Cobbett, C.S., 2011. Development of glutathione-deficient embryos in Arabidopsis is influenced by the maternal level of glutathione. Plant Biol. 13, 693–697.

Liu, J., Litt, L., Segal, M.R., Kelly, M.J.S., Pelton, J.S., Kim, M., 2011. Metabolomics of oxidative stress in recent studies of endogenous and exogenously administered intermediate metabolites. Int. J. Mol. Sci. 12, 6469–6501.

Livingston, D.P., Hincha, D.K., Heyer, A.G., 2009. Fructan and its relationship to abiotic stress tolerance in plants. Cell. Mol. Life Sci. 66, 2007–2023.

Lu, S., Van Eck, J., Zhou, X., Lopez, A.B., O'Halloan, D.M., Cosman, K.M., Conlin, B.J., Paolillo, D.J., Vrebalov, J., Kochian, L.V., Kupper, H., Earle, E.D., Cao, J., Li, L., 2006. The cauliflower or gene encodes a DnaJ cysteine-rich domain-containing protein that mediates high levels of beta-carotene accumulation. Plant Cell 18 (12), 3594–3605.

Lunn, J.E., 2007. Compartmentation in plant metabolism. J. Exp. Bot. 58, 35–47.

Lunn, J.E., Delorge, I., Figueroa, C.M., Dijck, P.V., Stitt, M., 2014. Trehalose metabolism in plants. Plant J. 79, 544–567.

Malekzadeh, P., 2015. Influence of exogenous application of glycine betaine on antioxidative system and growth of salt-stressed soybean seedlings (*Glycine max* L.). Physiol. Mol. Biol. Plants 21 (2), 225–232.

Mano, J., Belles-Boix, E., Babiychuk, E., Inze, D., Torii, Y., Hiraoka, E., Takimoto, K., Slooten, L., Asada, K., Kushnir, S., 2005. Protection against photooxidative injury of tobacco leaves by 2-alkenal reductase. Detoxification of lipid peroxide-derived reactive carbonyls. Plant Physiol. 139, 1773–1783.

Masood, A., Iqbal, N., Asgher, M., Khan, M.I.R., Fatma, M., Khan, N.A., 2013. Variation in carbohydrate accumulation in two cultivars of mustard and its association with salt tolerance. J. Func. Environ. Bot. 3 (2), 94–102.

Miller, A.J., Fan, X., Shen, Q., Smith, S.J., 2007. Amino acids and nitrate as signals for the regulation of nitrogen acquisition. J. Exp. Bot. 59 (1), 111–119.

Miret, J.A., Munne-Bosch, S., 2015. Redox signaling and stress tolerance in plants: a focus on vitamin E. Ann. N. Y. Acad. Sci. 1340, 29–38.

Missihoun, T.D., Willee, E., Guegan, J.P., Berardocco, S., Shafiq, M.R., Bouchereau, A., Bartels, D., 2015. Overexpression of ALDH10A8 and ALDH10A9 genes provides insight into their role in glycine betaine synthesis and affects primary metabolism in *Arabidopsis thaliana*. Plant Cell Physiol. 56 (9), 1798–1807.

Mittler, R., Vanderauwera, S., Gollery, M., Van Breusegem, F., 2004. Reactive oxygen gene network of plants. Trends Plant Sci. 9, 490–498.

Mittler, R., 2017. ROS are good. Trends Plant Sci. 22 (1), 11–19.

Mittler, R., 2002. Oxidative stress, antioxidants and stress tolerance. Trends Plant Sci. 7, 405–410.

Mostafa, M.G., Seraj, Z.I., Fujita, M., 2014. Exogenous sodium nitroprusside and glutathione alleviate copper toxicity by reducing copper uptake and oxidative damage in rice (*Oryza sativa* L.) seedlings. Protoplasma 251 (6), 1373–1386.

Munne-Bosch, S., 2005. The role of alpha-tocopherol in plant stress tolerance. J. Plant Physiol. 62 (7), 743–748.

Munne-Bosch, S., 2007. α-tocopherol: a multifaceted molecule in plants. Vitam. Horm. 76, 375–392.

Murshed, R., Lopez-Lauri, F., Sallanon, H., 2013. Effect of water stress on antioxidant systems and oxidative parameters in fruits of tomato (*Solanum lycopersicon* L., cv. Micro-tom). Physiol. Mol. Biol. Plants 19 (3), 363–378.

Nakamura, M., Bhatnagar, A., Sadoshima, A., 2012. Overview of pyridine nucleotides review series. Circ. Res. 604–610. https://doi.org/10.1161/circresaha.111.247924.

Nisar, N., Li, L., Lu, S., Khin, N.C., Pogson, B.J., 2015. Carotenoid metabolism in plants. Mol. Plant 8, 68–82.

Noctor, G., Foyer, C.H., 2016. Intracellular redox compartmentation and ROS-related communication in regulation and signaling. Plant Physiol. 171, 1581–1592.

Noctor, G., Foyer, C.H., 1998. Ascorbate and glutathione: keeping active oxygen under control. Annu. Rev. Plant Physiol. Plant Mol. Biol. 49, 249–279.

Noctor, G., Mhamdi, A., Chaouch, S., Han, Y., Neukermans, J., Marquez-Garcia, B., Queval, G., Foyer, C.H., 2012. Glutathione in plants: an integrated overview. Plant Cell Environ. 35, 454–484.

Noctor, G., Queval, G., Gakiere, B., 2006. NAD(P) synthesis and pyridine nucleotide cycling in plants and their potential importance in stress conditions. J. Exp. Bot. 57 (8), 1603–1620.

Pandey, R., Agarwal, R.M., Jeevaratnam, K., Sharma, G.L., 2004. Osmotic stress-induced alterations in rice (*Oryza sativa* L.) and recovery on stress release. Plant Growth Regul. 42, 79–87.

Park, E.J., Jeknic, Z., Pino, M.T., Murata, N., Chen, T.H.H., 2007. Glycine betaine accumulation is more effective in chloroplasts than in the cytosol for protecting transgenic tomato plants against abiotic stress. Plant Cell Environ. 30, 994–1005.

Park, I.M., Kim, Y.S., Kim, J.J., Mok, J.E., Kim, J.E., Kim, Y.H., Park, H.M., Kim, I.S., Yoon, H.S., 2017. Improved stress tolerance and productivity in transgenic rice plants constitutively expressing the *Oryza sativa* glutathione synthetase *OsGS* under paddy field conditions. J. Plant Physiol. 215, 39–47.

Paul, M.J., Primavesi, L.F., Jhurreea, D., Zhang, Y.H., 2008. Trehalose metabolism and signaling. Annu. Rev. Plant Biol. 59, 417–441.

Pavet, V., Olmos, E., Kiddle, G., Mowla, S., Kumar, S., Antoniv, J., Foyer, C.H., 2005. Ascorbic acid deficiency activates cell death and disease resistance responses in *Arabidopsis*. Plant Physiol. 139, 1291–1303.

Pilon-Smits, E.A.H., Ebskamp, M.J.M., Paul, M.J., Jeuken, M.J.W., Weisbeek, P.J., Smeekens, S.C.M., 1995. Improved performance of transgenic fructan-accumulating tobacco under drought stress. Plant Physiol. 107, 125–130.

Pollak, N., Dolle, C., Ziegler, M., 2007. The power to reduce: pyridine nucleotides – small molecules with a multitude of functions. Biochem. J. 402, 205–218.

Provencher, L.M., Miao, L., Sinha, N., Lucas, W.J., 2001. Sucrose export defective1 encodes a novel protein implicated in chloroplast-to-nucleus signaling. Plant Cell 13, 1127–1141.

Rai, V.K., 2002. Role of amino acids in plant responses to stresses. Biol. Plant 45 (4), 481–487.

Ray, S., Dansana, P.K., Bhaskar, A., Giri, J., Kapoor, S., Khurana, J.P., Tyagi, A.K., 2010. Emerging trends in functional genomics for stress tolerance in crop plants. In: Hirt, H. (Ed.), Plant Stress Biology: From Genomics to Systems Biology. Wiley-VCH Verlag GmbH and Co. KGaA, Weinheim, Germany.

Reddy, P.S., Jogeswar, G., Rasineni, G.K., Maheswari, M., Reddy, A.R., Varshney, R.K., Kishor, P.B.K., 2015. Proline over-accumulation alleviates salt stress and protects photosynthetic and antioxidant enzyme activities in transgenic sorghum [*Sorghum bicolor* (L.) Moench]. Plant Physiol. Biochem. 94, 104–113.

Redillas, M.C.F.R., Park, S.H., Lee, J., Kim, Y.C., Jeong, J.S., Jung, H., Bang, S.W., Hahn, T.R., Kim, J.K., 2012. Accumulation of trehalose increases soluble sugar contents in rice plants conferring tolerance to drought and salt stress. Plant Biotechnol. Rep. 6, 89–96.

Richards, S.L., Wilkins, K.A., Swarbreck, S.M., Anderson, A.A., Habib, N., Smith, A.G., McAinsh, M., Davies, J.M., 2015. The hydroxyl radical in plants: from seed to seed. J. Exp. Bot. 66 (1), 37–46.

Russin, W.A., Evert, R.E., Vanderveer, P.J., Sharkey, T.D., Briggs, S.P., 1996. Modification of a specific class of plasmodesmata and loss of sucrose export ability in a sucrose export defective1 maize mutant. Plant Cell 8, 645–658.

Sattler, S.E., Cahoon, E.B., Coughlan, S.J., DellaPenna, D., 2003. Characterization of tocopherol cyclases from higher plants and cyanobacteria. Evolutionary implications for tocopherol synthesis and function. Plant Physiol. 132, 2184–2195.

Sattler, S.E., Gilliland, L.U., Magallanes-Lundback, M., Pollard, M., DellaPennaa, D., 2004. Vitamin E is essential for seed longevity and for preventing lipid peroxidation during germination. Plant Cell 16, 1419–1432.

Schmidt, R., Kunkowska, A.B., Schippers, J.H.M., 2016. Role of reactive oxygen species during cell expansion in leaves. Plant Physiol. 172, 2098–2106.

Shulaev, V., Cortes, D., Miller, G., Mittler, R., 2008. Metabolomics for plant stress response. Physiol. Plant 132, 199–208.

Sies, H., Stahl, W., 1995. Vitamins E and C, a-carotene, and other carotenoids as antioxidants. Am. J. Clin. Nutr. 62, 1315S–1321S.

Simon-Sarkadi, L., Kocsy, G., Várhegyi, A., Galiba, G., De Ronde, J.A., 2006. Stress-induced changes in the free amino acid composition in transgenic soybean plants having increased proline content. Biol. Plant 50 (4), 793–796.

Sirhindi, G., Mir, M.A., Abd-Allah, E.F., Ahmad, P., Gucel, S., 2016. Jasmonic acid modulates the physio-biochemical attributes, antioxidant enzyme activity, and gene expression in glycine max under nickel toxicity. Front. Plant Sci. 7, 591. https://doi.org/10.3389/fpls.2016.00591.

Sitnicka, D., Orzechowski, S., 2014. Cold-induced starch degradation in potato leaves – intercultivar differences in the gene expression and activity of key enzymes. Biol. Plant 58 (4), 659–666.

Smeets, K., Opdenakker, K., Remans, T., van Sanden, S., van Belleghem, F., Semane, B., Horemans, N., Guisez, Y., Vangronsveld, J., Cuypers, A., 2009. Oxidative stress-related responses at transcriptional and enzymatic levels after exposure to Cd or Cu in a multipollution context. J. Plant Phys. 166, 1982–1992.

Smirnoff, N., 2011. Vitamin C: the metabolism and functions of ascorbic acid in plants. In: Fabrice, R.B., Roland, D. (Eds.), Advances in Botanical Research. Academic Press, pp. 107–177.

Son, J.A., Narayanankutty, D.P., Roh, K.S., 2014. Influence of exogenous application of glutathione on rubisco and rubisco activase in heavy metal-stressed tobacco plant grown *in vitro*. Saudi J. Biol. Sci. 21, 89–97.

Spollen, W.G., Nelson, C.J., 1994. Response of fructan to water-deficit in growing leaves of tall fescue. Plant Physiol. 106, 329–336.

Stitt, M., Zeeman, S.C., 2012. Starch turnover: pathways, regulation and role in growth. Curr. Opin. Plant Biol. 15, 282–292.

Tetlow, I.J., Morell, M.K., Emes, M.J., 2004. Recent developments in understanding the regulation of starch metabolism in higher plants. J. Exp. Bot. 55, 2131–2145.

Thakur, P., Kumar, S., Malik, J.A., Berger, J.D., Nayyar, H., 2010. Cold stress effects on reproductive development in grain crops: an overview. Environ. Exp. Bot. 67 (3), 429–443.

Tian, F., Wang, W., Liang, C., Wang, X., Wang, G., Wang, W., 2017. Over accumulation of glycine betaine makes the function of the thylakoid membrane better in wheat under salt stress. Crop J. 5 (1), 73–82.

Tomar, N.S., Agarwal, R.M., 2013. Influence of treatment of *Jatropha curcas* L. leachates and potassium on growth and phytochemical constituents of wheat (*Triticum aestivum* L.). Am. J. Plant Sci. 4 (5), 1134–1150.

Tomar, N.S., Sharma, M., Agarwal, R.M., 2015. Phytochemical analysis of *Jatropha curcas* L. during different seasons and developmental stages and seedling growth of wheat *Triticum aestivum* L.) as affected by extracts/leachates of *Jatropha curcas* L. Physiol. Mol. Biol. Plants 21 (1), 83–92.

Tracewell, C.A., Vrettos, J.S., Baustista, J.A., Frank, H.A., Brudvig, G.W., 2001. Carotenoid photooxidation in photosystem II. Arch. Biochem. Biophys. 385 (1), 61–69.

Trebst, A., Depka, B., Holländer-Czytko H, 2002. A specific role for tocopherol and Hof chemical singlet oxygen quenchers in the maintenance of photosystem II structure and function in *Chlamydomonas reinhardtii*. FEBS Lett. 43, 2157–2162.

Trebst, A., Depka, B., 1997. Role of carotene in the rapid turnover and assembly of photosystem II in *Chlamydomonas reinhardtii*. FEBS Lett. 400 (3), 359–362.

Valderrama, R., Corpas, F.J., Carreras, A., Gómez-Rodríguez, M.V., Chaki, M., Pedrajas, J.R., Fernández-Ocaña, A., Del Río, L.A., Barroso, J.B., 2006. The dehydrogenase-mediated recycling of NADPH is a key antioxidant system against salt-induced oxidative stress in olive plants. Plant Cell Environ. 29, 1449–1459.

Valluru, R., den Ende, W.V., 2008. Plant fructans in stress environments: emerging concepts and future prospects. J. Exp. Bot. 59 (11), 2905–2916.

Velazquez-Marquez, S., Conde-Martinez, V., Trejo, C., Delgado-Alvarado, A., Carballo, A., Suarez, R., Mascorro, J.O., Trujillo, A.R., 2015. Effects of water deficit on radicle apex elongation and solute accumulation in *Zea mays* L. Plant Physiol. Biochem. 96, 29–37.

Venkatesh, J., Park, S.W., 2014. Role of L-ascorbate in alleviating abiotic stresses in crop plants. Bot. Stud. 55, 38.

Verma, G., Mishra, S., Sangwan, N., Sharma, S., 2015. Reactive oxygen species mediate axis-cotyledon signaling to induce reserve mobilization during germination and seedling establishment in *Vigna radiata*. J. Plant Physiol. 184, 79–88.

Vijn, I., Smeekens, S., 1999. Fructan: more than a reserve carbohydrate? Plant Physiol. 120, 351–359.

Vranova, V., Rejsek, K., Skene, K.R., Formanek, P., 2011. Non-protein amino acids: plant, soil and ecosystem interactions. Plant Soil 342, 31–48.

Wang, G., Zhang, J., Wang, G., Fan, X., Sun, X., Qin, H., Xu, N., Zhong, M., Qiao, Z., Tang, Y., Songa, R., 2014. Proline responding-1 plays a critical role in regulating general protein synthesis and the cell cycle in maize. Plant Cell 26 (6), 2582–2600.

Wang, G.L., Xu, Z.S., Wang, F., Li, M.Y., Tan, G.F., Xiong, A.S., 2015. Regulation of ascorbic acid biosynthesis and recycling during root development in carrot (*Daucus carota* L.). Plant Physiol. Biochem. 94, 10–18.

Xiang, G., Mao-yun, S., Gui-xiang, Y., Yang, Y., Wei-hua, Q., Li-pu, D., Xing-guo, Y., 2010. Cloning and characterization of genes coding for fructan biosynthesis enzymes (FBEs) in *Triticeae* Plants. Agric. Sci. China 9 (3), 313–324.

Xu, Y., Xu, Q., Huang, B., 2015. Ascorbic acid mitigation of water stress-inhibition of root growth in association with oxidative defense in tall fescue (*Festuca arundinacea* Schreb.). Front. Plant Sci. 6, 807. https://doi.org/10.3389/fpls.2015.00807.

Ye, N., Zhu, G., Liu, Y., Zhang, A., Li, Y., Liu, R., Shi, L., Jia, L., Zhang, J., 2012. Ascorbic acid and reactive oxygen species are involved in the inhibition of seed germination by abscisic acid in rice seeds. J. Exp. Bot. 63 (5), 1809–1822.

Ying, W., Garnier, P., Swanson, R.A., 2003. NAD^+ repletion prevents PARP-1-induced glycolytic blockade and cell death in cultured mouse astrocytes. Biochem. Biophys. Res. Commun. 308, 809–813.

Yusuf, M.A., Kumar, D., Rajwanshi, R., Strasser, R.J., Tsimilli-Michael, M., Govindjee, Sarin, N.B., 2010. Overexpression of γ-tocopherol methyl transferase gene in transgenic Brassica juncea plants alleviates abiotic stress: physiological and chlorophyll a fluorescence measurements. Biochim. Biophys. Acta 1797, 1428–1438.

Zakar, T., Laczko-Dobos, H., Toth, T.N., Gombos, Z., 2016. Carotenoids assist in cyanobacterial photosystem II assembly and function. Front. Plant Sci. 7, 295. https://doi.org/10.3389/fpls.2016.00295.

Zakharova, I.O., Sokolova, T.V., Bayunova, L.V., Vlasova, Y.A., Rychkova, M.P., Avrova, N.F., 2012. α-Tocopherol at nanomolar concentration protects PC12 cells from hydrogen peroxide-induced death and modulates protein kinase activities. Int. J. Mol. Sci. 13, 11543–11568.

Zhang, Y., Primavesi, L.F., Jhurreea, D., Andralojc, P.J., Mitchell, R.A.C., Powers, S.J., Schluepmann, H., Delatte, T., Wingler, A., Paul, M.J., 2009. Inhibition of SNF1-related protein kinase1 activity and regulation of metabolic pathways by trehalose-6-phosphate. Plant Physiol. 149, 1860–1871.

Zhou, X., Sun, T.H., Wang, N., Ling, H.Q., Lu, S., Li, L., 2011. The cauliflower Orange gene enhances petiole elongation by suppressing expression of eukaryotic release factor 1. New Phytol. 190 (1), 89–100.

Chapter 4

Differential Responses of Plants to Biotic Stress and the Role of Metabolites

Maham Saddique[1], Muhammad Kamran[2], Muhammad Shahbaz[1]
[1]University of Agriculture Faisalabad, Faisalabad, Pakistan; [2]Government Postgraduate College of Science, Faisalabad, Pakistan

1. BIOTIC STRESS IN PLANTS

Biotic stresses are caused by various living organisms, such as bacteria, fungi, insects, and weeds (Lattanzio et al., 2006). They are very destructive for plant growth. Of the various biotic organisms, fungi create a more severe threat as compared with others (Waller et al., 2005). A tremendous number of fungi, i.e., about 8100 fungal species, have been reported to act as plant pathogens (Tarkowski and Vereecke, 2014). Similarly, viruses are also considered as dangerous pathogens destroying crops much more than fungi world over (Bai et al., 2002) These microorganisms may cause various types of symptoms in plants, such as wilting in plant, spots on leaves, rotting of roots, and damaging of seeds (Anderson et al., 2004). They can damage all plant parts, including stem, bark, flowers, and leaves (Walling, 2008) and also provide the source of transmission of bacteria and viruses from infected to uninfected plants (Mann et al., 2012). Weeds are wild, self-growing plant species that restrict or impede the growth of other plants either directly by damaging or indirectly by provoking competition for area and nutrient availability (Dorn et al., 2013; Dass et al., 2016). Weeds are rapidly growing plants and their yield is also high, so sometimes their faster growth helps them dominate the environment faster than the required crop plants (Dass et al., 2016).

1.1 Bacterial Infection

A majority of plant pathogenic bacteria are usually saprotrophic and do not damage the plant itself. On the other hand, about 100 known bacterial species have damaging ability for plants (Purcell and Hopkins, 1996). Bacterial diseases are more dominant in the subtropical and tropical areas of the world (Ashbolt, 2004). Rod-shaped bacteria mostly infect plants (Mansfield et al., 2012). Some specific pathogenicity factors enable these bacteria to spread disease in the plants. Five major types of factors are well known in this regard. These factors are effector proteins, cell wall–degrading enzymes, exopolysaccharides, phytohormones, and toxins (Gagne-Bourgue et al., 2013). *Agrobacterium* species cause the tumor formation through alteration in auxin levels (Lee et al., 2009). Similarly, bacteria generate exopolysaccharides, which may cause blockage in xylem vessels, and often even plant death (Thorne et al., 2006).

1.2 Viral Infection

Of the various biotic stresses, viruses are a big threat to yield in many crop plants. Two main groups of plant viruses are Geminiviridae and Reoviridae. These are small genome containing viruses (Reddy et al., 2009), which invade fields through infected seeds or virus carrier pests found in infected plants (Fereres and Moreno, 2009). Viral infection in plants usually spread through injured or damaged parts because they provide direct contact to the viral genome to enter in plant cells. Here duplication of genome and then further spread through the symplastic pathway and vascular tissues may cause more danger for plant growth (Taliansky et al., 2008). To reduce the viral infections, we need to control the viral vectors, using controlled plant material during crop trapping, using transgenic crops and increased agriculture practices (Prins et al., 2008).

1.3 Fungal Infection

Fungi are plant parasites and classified into two categories based on their lifestyle, biotrophs and necrotrophs (Kabbage et al., 2015). Biotrophs get their food directly from the living tissue of the host, whereas necrotrophs cause severe damage

Plant Metabolites and Regulation Under Environmental Stress. http://dx.doi.org/10.1016/B978-0-12-812689-9.00004-2

to the plant tissues and then dead tissues become a source of food for them (Campe et al., 2014). However, there are many fungi that are plant pathogens that use either mode depending on the environmental conditions and are known as hemibiotrophs (Mang et al., 2009). Initially, during the infection process, fungi act as biotrophs but as per previous studies they were considered as necrotrophs. They are now known as hemibiotrophs (Kabbage et al., 2015).

1.4 Herbivory: Insects and Nematodes

Many nematodes cause severe damage to many plants, including citrus. Of the various nematodes, the most famous is *Tylenchulus semipenetrans* (citrus nematode), which is found almost all over the world (Duncan, 2005). Slow decline disease of citrus is caused by *T. semipenetrans* (Safdar et al., 2013). *Radopholus similis* is also a citrus disease-causing nematode, which causes "vast dispersal decline disease" (Kaplan and Opperman, 2000). Some other diseases caused by nematodes are root-knot, stubby root, lesion, and sting, which are also involved in destruction of citrus (Etebu and Nwauzoma, 2014).

Coconut is also affected by many nematodes; one of the most devastating diseases that damage coconut is red ring disease (Mehlhorn et al., 2011). Red rings are formed on the root, stem, and leaf surfaces of coconut trees (Hunt et al., 2005). The palm weevil is one of the main carriers of nematodes from contaminated trees to normal trees (Magalhaes et al., 2008).

Lesion and corn cyst nematodes are considered as very harmful for corn (Saxena and Stotzky, 2001). The root-knot disease is caused by the root-knot nematode (Ramamoorthy et al., 2001). Other pathogenic nematodes of cotton are lance, reniform, dagger, lesion, and sting nematodes (Noe, 2004). Sting nematodes are found in sandy soils. Lance nematodes cause cell damage or necrosis in cotton (Goverse and Smant, 2014).

Several parasitic nematodes are harmful for legume crops (Engstrom et al., 2016). The pea cyst nematode damages the pea and broad bean plant tissues (Van-Dam and Bouwmeester, 2016). The stem nematode is another pathogen that is injurious to stem tissues of broad beans (Vovlas et al., 2011). Root-knot and reniform nematodes are injurious to chickpea and cowpea (Win et al., 2016).

Nematodes are also harmful for peanut plants growing all over the world (Johnson et al., 2004) and may cause about 12% loss annually (Koenning et al., 1999). Ring and stunt nematodes damage the peanut plants by causing root-knot and lesion diseases (Eschen et al., 2015). Potato cyst nematodes are the chief threat for destruction of potato plants (Rossi et al., 1998). The golden nematode disease is also known the world over, causing 80% of yield loss (Scurrah et al., 2005). Other nematodes involved in the destruction of potatoes are lesion, root-knot, and potato-rot nematodes (Balasubramanian et al., 2009). The quality of potato tubers is reduced by root-knot nematodes, which produce brown spots on the surface (Czajkowski et al., 2011). Root-lesion nematodes form lesions on tubers and the roots (George et al., 2016).

Many nematodes cause injuries in various plant parts of rice, including leaves, panicles, and stem (Bridge, 1996). One of the most devastating nematodes is the cyst nematode, which is the pest of soybean (Matthews et al., 2013). About 50 species of nematodes are known to act as pests of soybeans. The most famous are cyst and root-knot nematodes (Vuong et al., 2015). In addition, lance, reniform, lesion, and sting nematodes are also injurious for soybeans (Akhtar and Malik, 2000).

Lesion nematodes are the harmful pests of tobacco; others pests are tobacco reniform, spiral, and stunt nematodes (Reynolds et al., 2000). Root-knot nematodes are very dangerous for vegetables and they have markedly reduced the production of vegetables in the tropical and subtropical areas (Lamovsek et al., 2013). Other nematodes, such as reniform, sting, cyst, and stubby root, are well-known pests of vegetables (Grabau et al., 2017). Cereal cyst nematodes, the seed gall, root-knot, and lesion nematodes, harm wheat and barley (Nicol et al., 2011). In cereal nematodes, the seed gall nematode is familiar for "ear-cockle" disease (McDonald and Linde, 2002). Wheat gall nematode is an economically important parasite dispersed through contaminated seed to many wheat-producing regions of the world (Nicol and Rivoal, 2008). One of the most devastating impacts of nematodes is reduction in nutrient availability and metabolic processes, ultimately causing stressed growth and yield (Kerry, 2000).

1.5 Weeds

Weeds have many distinct growth habits and adaptations by which they can grow well in numerous ecological niches where other plants are unable to grow. Weeds compete not only with themselves but also with crop plants (Ramesh et al., 2017). Among the more important adaptations related to competitive benefit are properly coordinated germination, rapid establishment and growth of seedlings, tolerance to shading effects by the valuable plants or by other weeds at the time of rapid growth, rapid response to already present soil moisture and nutrients, adaptation to the most harsh climatic situations of the habitat, adaptations to the edaphic regimes, relative immunity to postseeding soil trouble, agriculture practices and resistance to herbicides (Sardana et al., 2017). During the initial stages of weeds invasion in ecological niches, a very tough competition was observed among weeds and crops for available resources. Even after completion of the establishment of the crop–weed connection, competition for the resources is more evident (Burger et al., 2015). Plant competition is a natural phenomenon whereby crop and weed plants tend to achieve a maximum correlated growth and yield (Benaragama et al., 2016).

1.5.1 Crop–Weed Competition

Weeds seem much more improved to agroecosystems than crop plants (Bidlack et al., 2006). Without intrusion by man, weeds would simply wipe out the crop plants (Jalloh et al., 2012). This is because of their struggle for nutrients, moisture, light, and area, which are the principle factors for crop production (Dunbabin, 2007). Commonly, a 1-kg increase of weed growth results in a 1-kg decrease of crop growth (Ashrafi et al., 2009).

Weeds usually absorb more mineral nutrients as compared with crop plants and accumulate a high amount of minerals in their tissues (Qasem, 1992). For example, the "N" uptake by *Echinochloa crusgalli* is higher than that of rice (Talbert and Burgos, 2007). Nutrient elimination by weeds may cause a massive deficiency of various nutrients for crop plants (Wortmann and Kaizzi, 1998).

Among the various plants, C_4 plants consume water more proficiently than C_3 plants and accumulate more biomass per unit of water. *Cynodon dactylon* had nearly twice as eminent transpiration rate as pearl millet (Polley et al., 1993). In weedy areas, soil moisture may be reduced by the weeds before maturity of crops, i.e., the stage where crops require more water for high yield, and deficiency of water caused substantial loss in crop production (Abouziena et al., 2015). Weed cover also reduced the availability of light for crop plants (Balandier et al., 2006). Weeds and crops also compete for the absorption of CO_2 (Dunbabin, 2007).

The leaf leachate extract of decomposed leaves of *Polygonum* comprises flavonoides, which are lethal to germination, hypocotyls and root growth of weeds as *Amaranthus spinosus*. Inhibitor produced by decaying rhizomes of *Sorghum halepense* reduces the growth of *Amaranthus* sp and *Digitaria sanguinalis* (Lajter et al., 2015).

2. RESPONSES OF PLANTS TO BIOTIC STRESSES

Plants are under the threat of harm and danger from many microbes, such as bacteria, fungi, and herbivore insects (Ramegowda and Senthil-Kumar, 2015). Plants have established a variety of resistance mechanisms for protection, several of which are generated as a result of pathogen attack (Durner et al., 1997). The microbial attack is sensed by the extracellular receptors of plants, which are found on the membrane, and they recognize pathogen-linked patterns (Chisholm et al., 2006) and causes of infection in the plant body before the harmful microbes spread on the whole body of the plant (Boller and He, 2009). However, disease-causing organisms have developed the mechanism to inhibit pathogen-associated molecular pattern (PAMP)-triggered immunity by producing specific proteins into the cytosol of the plant cell that change signaling for plant resistance, expression of defense responses (Chisholm et al., 2006). Once pathogens developed the ability to subdue the main resistances, plants established a more definite mechanism to sense microorganisms, named as effector triggered immunity (ETI) (Bilgin et al., 2010). In the ETI condition, resistance genes (R) expression, generally receptors identify the effector molecules expressed, brought by the pathogen that invades the host cell. The contact of the effector and receptors brought a change in plants, which may cause resistance (Miedes et al., 2015). Resistance resulting from R gene expression comes with an oxidative burst in the form of free oxygen radicals, which is essential for the next method of the protection, a kind of apoptosis supposed to stop the increase of pathogens from water and nutrients (Sato et al., 2010).

2.1 Plant Responses to Viral Infections

Viruses as pathogens cause damages to crop plants and yield loss (Anderson et al., 2004). The genome of viruses is short and because of its disease-causing ability ensure that the contact in plant and virus is inclusive, viral and plant proteins are responsible for the completion of the virus life cycle. In viral diseases, the molecular mechanisms and resistance remain mostly unfamiliar (Kang et al., 2005). *Nicotiana benthamiana* is suggested as a model plant to study many plant viruses as it is a host of many viruses (Goodin et al., 2008), the vehicle of *Arabidopsis* and Cherry leaf roll-virus to investigate the virus spreading by infected seed (Rumbou et al., 2009), and yeast as major host to study the plant viruses' relation with of eukaryotic cellular systems (Nagy, 2008). Proteomics uses subcellular fractionation to study plant and virus interaction (Haynes and Roberts, 2007). Pepper mottle virus affects *N. benthamiana* plants and proteomics studies described the changes in many proteins involved in the respiratory chain and the Calvin cycle (Pineda et al., 2010). Genetic resistance includes regulation of vectors, the use of plant materials that are always virus-free, and a decrease of the spreading methods (Wu et al., 2010). Most viral resistances are monogenic, whereas the rest show quantitative inheritance (Kang et al., 2005). Moreover, another family of host resistance has been recognized, and it has been found that many plant species with variety of pathogens suffer necrotic cell death (Van-Doorn et al., 2011), which results in decreased vulnerability to later pathogen attack (Durrant and Dong, 2004).

2.2 Plant Responses to Fungal Infection

Generally, plants show tolerance against pathogenic attack through systemic acquired resistance using salicylic acid (SA) signaling resistance against biotrophs. On the other hand, resistance against the necrotrophs is significantly by jasmonic acid (JA) and ethylene (ET) signaling (Caarls et al., 2015). In barley, gene-for-gene resistance reaction is evident against *Blumeria* fungus (Schulze-Lefert and Vogel, 2000). Therefore, SA-dependent protection methods are operative along with gene resistance throughout the biotrophic fungus attack (Thomma et al., 2001). JA-dependent gene responses are not started immediately but if these are initiated, they have been reported to be quite operative (Glazebrook, 2005). The involvement of nitrogen in the formation of photosynthetic proteins is reduced under biotic stress and most of the nitrogen is used to form the compounds that can be used in biotic defensive mechanism in a sensitive cultivar of pea against fungi (Amey et al., 2008). Proteomics study in the leaf of bean (*Phaseolus vulgaris*) revealed the constant change in protein concentration during the rust infection (Rodríguez-Moreno et al., 2008). It has been found that defense against the fungus can be achieved in plants by the expression of R gene, which is involved in the synthesis of plant defensive proteins (Collins et al., 2003). This gene expression is very important as a part of the plant natural defense system and actively participated in enhancing the defense system to establish resistance (Lee et al., 2009).

RNA synthesis study of different host–pest systems of fungi showed that fungal infection may cause regulation of many transcriptional factors (Pande et al., 2005). JA biosynthesis and other signaling are controlled by these transcriptional factors (Pieterse et al., 2009). Early response of chickpea was observed after infection with blight fungus and about 250 unique genes were identified (Pande et al., 2005). These genes were of different categories, i.e., genes involved in the regulation of stress, signaling, cell metabolic reactions, and several unknown processes (Jaiswal et al. 2014). Chitin in the cell wall of fungi acts as a membrane identifier; hence, it also plays a vital role in generating protection responses against pests of several plants (Petutschnig et al., 2010). Stress signaling of plant leaves showed that chitin has a stimulating function on several genetic factors because of responses shown during the defensive mechanism (Povero et al., 2011). It has also been found that some genes act as a vital moderator involved in the activation of the expression of genes, which are involved in JA and SA signaling (Birkenbihl and Somssich, 2011). Similarly, SA and JA signaling is induced by WRKY70 (zinc-finger-type), a transcription factor in plant (Li et al., 2004). Plants generate the mitogen-activated protein kinase (MAPK) signaling cascade after the identification of fungal PAMPs, which ultimately causes biosynthesis of tryptophan resulting in camalexin accumulation (Pitzschke et al., 2009). Proteins that are usually formed are some enzymes, which participate in the synthesis of oxygen free radicals, signaling controlled by their production, and signal transduction by various forms of auxin (Sharma et al., 2007). Therefore, studies about the role of proteins are helpful in understanding the reproduction and invading methods of the harmful fungi (Fernandes et al., 2014).

2.3 Plant Responses to Herbivory

In the parallel evolutionary history of insects and plants showed various approaches to develop tolerance mechanisms for each other (Verhage et al., 2010). This relationship led to the development of powerful protection mechanisms in plants, efficient enough to identify the stimuli from injured tissues by the pests' attack (Hare, 2011). To prevent infection, plants form specific body structures, metabolic products, and proteins that have piousness, harmful, and repulsive effects on the insects (War et al., 2011). Plants protect themselves from the pests directly by modification in reproductive behavior (direct defense) and indirectly by attracting other animals that are predators or enemies of the pests (indirect defense) (Dudareva et al., 2006). Direct defenses are mediated by plant characteristics that disturb the insects, such as plant surface safety by forming outgrowths (e.g., hairs, thorns, trichomes, and spines) or manufacture of poisonous substances such as anthocyanins, terpenoids, and phenols, which inhibit the growth of the pests (Sharma and Dietz, 2009). Indirect defenses involve the attraction of the pests' predators for food and accommodation (Arimura et al., 2009). However, in the race between plants and herbivores, plants have developed resistant genotypes (Sharma et al., 2009). Plants respond to wound produced by herbivores through a complex and active protection scheme that consists of physical fences and poisonous compounds (Karban, 2011). Plant response is a significant mechanism for the control of insects in crops, and infected for the control of pests (Sharma et al., 2009). These are of great importance if intended at mitigating the common stresses, these substances are formed to respond the pests' infection (Miranda et al., 2007). Variations in defensive ingredients of a plant against the attack of pests make the plants able to produce some toxins against insects; on the other hand, they also interrupt the health and performance of the insect pests (War et al., 2011). If induced response arises directly, it is of a big advantage to the crops, and declines the successive herbivore attack, also enhancing general suitability of valuable crops (Agrawal, 2011). Advancement in insect–plant communications has corrected our understanding about the development of self-protective methods, shown by plants against herbivory (Karban, 2010).

3. PLANT DIRECT DEFENSES

In tomato, enzyme inhibitors, alkaloids, and phenolic compounds when taken up distinctly caused reduction, but behave with each other's in an interactive way, involving the pests digestion (Carriere et al., 2015). Many changes occur in response to insects, which are as follows.

3.1 Morphological Structures

The initial method of defense against herbivory is the development of plant structures, for example, the formation of a waxy cuticle and the formation of trichomes, setae, and spines (Sharma et al., 2009). Physiological defenses consist of morphological and internal traits that give a physiological benefit to the crops by stopping the pests from infection directly (Agrawal and Konno, 2009), and microscopic variations in the composition of the cell wall, causing variations in the thickness because of lignification and suberization (He et al., 2011). Physical characters such as hardened leaves, thorns, and trichomes offer a significant role in crop defense against pest attack (Chamarthi et al., 2014). It has been found that shininess of leaf, plumule, and leaf pigmentation were a cause for shoot fly tolerance in *Sorghum bicolor* (Bloch and Richardson, 1991).

3.2 Trichomes

Trichomes play a crucial role in the protection of plants from many insects (Hanley et al., 2007). Furthermore, trichomes interfere with herbivory and affect the infection of other arthropods on the plant surface, thus inhibiting their contact with leaf epidermis (Agrawal and Konno, 2009). Secondary metabolites are released by glandular trichomes, which are in the form of terpenoids, flavonoids, and others that can be repulsive, harmful, and capture insects and other pests, thus creating a mixture of plant defense (Sharma et al., 2009). Pest infection may cause an increase in cell division of the outgrowth, whereas thickness remained uniform (Agrawal and Konno, 2009). It has been reported that the damages caused by leaf beetles resulted in a higher trichome thickness in the newly developed leaves (Dalin and Björkman, 2003). Trichome thickness distended in *Alnus incana* because the period of injury caused by the insects (Escobar-Bravo et al., 2017) varies from days to weeks (Dalin and Björkman, 2003). Trichome exudations also serve as extra floral nectar (EFN) for squash bugs (Swięczkowska and Kowalkowska, 2015).

4. PLANT METABOLITES

4.1 Primary Metabolites

Primary metabolites are essential for cell growth. They are produced during the various phases of cell growth and support basic metabolic processes, such as respiration and photosynthesis (Apel and Hirt, 2004). Well-known metabolites in most of the organisms are amino acids, tricarboxylic acids, sugars, building blocks, and energy sources (Zientz et al., 2004). In addition to the above-mentioned compounds, proteins, nucleic acids, and polysaccharides are also considered as primary metabolites (Tettelin et al., 2001).

4.2 Secondary Metabolites

Primary metabolites are involved in the synthesis of secondary metabolites (Hartmann, 1996). Secondary metabolites do not have uniform production (Bourgaud et al., 2001). They are the outputs of primary metabolites, such as phenolics, lignins, essential oils, resins, tannins, steroids, and alkaloids. They are usually not involved in the basic growth and developmental processes of an organism (Walker et al., 2003); their absence does not cause direct death, but threat to the organism's defense (Kumar et al., 2014). Secondary metabolites are a diverse group of compounds produced naturally by plants, algae, fungi, bacteria, and animals (Gribble, 2003). A majority of these metabolites, such as terpenes, alkaloids, and phenolics, are categorized based on their biosynthetic origin (Nicolaou et al., 2010). They have a short-term existence and plants can grow in their absence, showing that they are not essential for plant growth. Secondary metabolism furnishes a "genetic expression" that permits mutation to cause natural selection to fix new good traits through evolution (Firn and Jones, 2009). Secondary metabolism is also considered as a fundamental part of metabolism. It depends upon the basic metabolism to produce the cellular machinery, essential enzymes, and energy involved in the long-term survival of the plants (Roze et al., 2011).

4.2.1 Plant Defense and Secondary Metabolites

Secondary metabolites are not essentially involved in the normal growth and developmental processes in plants; their production decreases the palatability of the plant tissues (Boege and Marquis, 2005). They are also known as defensive metabolites, accumulated as inactive forms that become activated in response to the pest infection (Howe and Jander, 2008). Initially they were known as phytoanticipins, at present as phytoalexins. During herbivory, the β-glucosidase stimulates phytoanticipins, which control the production of different types of metabolites (Morant et al., 2008). Of the various metabolites, phytoalexins (alkaloids, terpenoids, and others) are important in the production of antiinsect compounds (Walling, 2008). They protect the plants from different stresses and increase the capability of the plants (Kessler et al., 2002). It has been studied that corn is defended from corn-worm by the production of the secondary metabolites phenylpropanoid product and chlorogenic acid (Nuessly et al., 2007). Lignin (phenolic in nature) provides support to plants against insect pests (War et al., 2012). Its synthesis is induced by infection by pests and its fast deposition affects pathogen fertility (Barakat et al., 2010). Quinones, derived from phenolic compounds, are very beneficial for the plant, as they produce some compounds by binding with proteins and stop the digestion of proteinic substances by the herbivores (Bhonwong et al., 2009). Accumulation of these compounds in amino acids decreases the digestion of proteins in the insects, which affects pest growth(Chen et al., 2005). Phenols also reduce the hydroxide radicals and singlet oxygen involved in a series of responses, which stimulate the initiation of defensive enzymes (Sharma et al., 2012). Salicylates are simple phenolics that serve as antifeedant in *Salix* leaves for insect herbivores; SA is very significant as preventive agent (Hardie et al., 2016).

4.2.2 Principal Groups

These are categorized into three different groups, viz: terpenes, nitrogen- and sulfur-containing compounds, and phenolic compounds.

4.2.2.1 Terpenes

Terpenes constitute one of the principal groups of compounds which are similar for their biosynthetic origin from glycolytic products. Most of the terpenes formed by plants are secondary metabolites that act as toxins against insects and other herbivores.

Terpenes are classified as follows.

4.2.2.1.1 Monoterpenes (C_{10}) Many chemical products secreted by plants are important mediators of insect toxicity. For example, leaves and flowers of *Chrysanthemum* species produce the pyrethroids (monoterpenes esters), which express strong pesticide responses to pests such as moths and are a widespread element in viable pesticides because of little stability in the surroundings and less poisonousness for mammals (Taft et al., 2015). In conifers, monoterpenes are released by resin-secreting cells present in the stem specifically as myrecene and limonene. All monoterpenes are poisonous to many pests, including beetles and other harmful pests of conifer species all the world over (Sarria et al., 2014).

4.2.2.1.2 Sesquiterpenes (C_{15}) Sesquiterpenes are well known for their role in plant defense as antiherbivore agents. They consist of a five-membered lactone ring and serve as a strong feeding repellent to several herbivorous insects. Abscisic acid is a sesquiterpene involved in controlling roles in the sustainment of seed and bud dormancy and plants show response to water stress by changing the properties of membrane; and it acts as an activator for transcription (Jiang et al., 2016). Furthermore, it increases the cytosolic calcium content (Cheval et al., 2013).

4.2.2.1.3 Diterpenes (C_{20}) Leguminous plants produce abietic acid, which is a diterpene. It is found in resins secreted from resin canals of the tree trunk; when these canals are penetrated by pests, the release of resin may significantly block feeding. One diterpene called phorbol, produced in plants of family Euphorbiaceae, acts as epithelium irritations and inner toxins to pest insects and mammals. Moreover, gibberellin is also a diterpene that plays numerous roles in plant developmental stages, such as germination, leaf expansion, and fruit setting (Ouzounidou and Ilias, 2005).

4.2.2.1.4 Triterpenes (C_{30}) Some of these triterpenes are a significant part of plant cell membranes, controlling protein channels and sustaining absorptivity to little particles by reducing the lipid globules. Phytoecdysones, a group of triterpenes, have some self-protective characters from insects by altering in developing procedures along with some side effects (Fatope et al., 1990). Limnoid, which is another triterpene in citrus, acts as a harmful compound against pests in plants of Rutaceae family (Mordue and Blackwell, 1993).

4.2.2.1.5 Polyterpenes (C_5) These are numerous high-molecular-weight terpenes synthesized in plants, including the tetra- and the polyterpenes (Silvestre and Gandini, 2008). Carotenoids are principal tetraterpenes, which is a family of pigments; another one is rubber, which is present in laticifers; it offers defense as a process for infection repair and as a resistance against pests (Eisner et al., 2000).

4.2.2.2 Phenolic Compounds

Plants synthesize some secondary products that consist of a phenol group, a $^-$OH functional group attached with benzene ring, which are the significant portion of defense scheme in plants from herbivore pests (Wuyts et al., 2006).

4.2.2.2.1 Coumarin Against insect herbivores and fungi, vascular plants produce simple phenolic compounds in various capacities in different plant defense processes, through the shikimic acid pathway (Ghasemzadeh et al., 2011). They are a very active group of compounds with antipathogenic action against bacteria and fungi (Brooker et al., 2008). It is found that natural pesticidal defense is produced and a range of antifungal activity was amended by these cyclic compounds in plants. Coumarin, derived from halogenated compounds, acts very efficiently in vitro to stop fungal growth. For example, parasitism of *Orobanche cernua* is prevented by the production of 7-hydroxylated simple coumarins for effective germination, interruption and connects to the transport system of host plants (Serghini et al., 2001). Against soil-borne plant pathogenic fungi, some coumarin by-products are considered as more stable compared to the usual coumarin (Brooker et al., 2008).

4.2.2.2.2 Furano-coumarins Furano-coumarins are a characteristic feature of the plants of the family Umbelliferae, with major interest because of phytotoxicity. They may be toxic when stimulated by ultraviolet light by stopping the process of transcription and repair of DNA and finally leading to cell death (Sajjadi et al., 2013). Psoralin, a very important furano-coumarin, is usually used by the plants for protection against fungal toxicity (Ali and Sharma, 2003).

4.2.2.2.3 Lignin Phenylpropanoid groups for a large branched polymer consist of coniferyl, sinapyl, and coumaryl. These are three different alcohols, which may be the source to produce lignin (Hatfield and Vermerris, 2001). The physical toughness and chemical strength of lignin prevents it from herbivorous feeding and penetration, respectively (Walters, 2011).

4.2.2.2.4 Flavonoids About 5000 types of flavonoids have been described in plants (Yao et al., 2004). Flavonoids protect plants against several stresses, including insect pests. They produce plant tolerance against the herbivory and regulate the reactive oxygen species (ROS), and decline in formation by metals chelating (Treutter, 2006). They are divided into various groups, anthocyanins, chalcones, aurones, flavonols, flavanones, dihydroflavonols, flavone, and proanthocyanidins. Several flavones have been studied as feeding deterrents for several insects (Harborne et al., 2000). Flavonoids such as flavones are present as feeding deterrents for *Spodoptera exempta* (Da-Silva et al., 2017). Flavonoid synthesis in *Arabidopsis* has been observed to produce resistance against pests (Johnson et al., 2004). Isoflavonoids obtained from the chickpea act as antifeedant against herbivore insect at a concentration of 100ppm (Simmonds and Stevenson, 2001).

4.2.2.2.5 Isoflavonoids Isoflavonoids are derivatives of flavonone, mostly found in plants, important in plant development, synthesized by the legumes, and play a vital role in encouraging the formation of root nodules by symbiotic bacteria (Sreevidya et al., 2006). Formation of these flavonoids by the plants is an active tactic against ROS (Posmyk et al., 2009).

4.2.2.2.6 Tannins Tannins are another type of phenolic polymers with defending properties found in plants. These are common toxins that prominently decrease the normal growth and sustainability of numerous pests and play a significant role as feeding repellents to pest animals (Barbehenn et al., 2008). Mammals typically avoid those plants that have high amount of tannin (Crozier et al., 2006). Chlorogenic and protocatechuic acids usually have a distinct role in disease resistance of various plants. It is considered that chlorogenic acid is produced and oxidized into operative quinones against the fungus by several disease-tolerant varieties of plants (Haldar et al., 2015). Tannins disrupt the growth of insect and because of the linking ability with proteins, decreases the efficiency of nutrient assimilation, and is also a source of lesions in the mid gut (Barbehenn and Constabel, 2011). Tannins are polyphenols that have an unpleasant smell and act as irritant during feeding by insects (Eneh et al., 2016). Furthermore, it also chelates the metal ions, decreasing their access to herbivores (Scalbert et al., 1999). When tannins were consumed by herbivores, the digestibility of proteins was decreased by tannins and so there was a decrease in the nutritive worth of plants to herbivores (Teferedegne, 2000). In Alaska paper birch, condensed tannins decreased the pupal volume and persistence of larvae of *Rheumaptera hastate* (L.) (Bryant et al., 1993). PtMYB134 genes are used for the formation of tannins, expressed when plants encounter wounding. Moreover, concentratio of tannin is also affected by stress in hybrid poplar (Mellway et al., 2009).

4.2.2.2.7 Plant Defensive Proteins It has been shown based on microarray and proteomic methods that an extensive range of PRPs is essential in plant protection against attack by insects (Chen et al., 2005). Because of different feeding behaviors of arthropods, numerous signaling pathways are switched on, causing the production of JA, SA, and ET, controlling arthropod-inducible proteins (Arimura et al., 2009).

4.2.2.3 Sulfur-Containing Secondary Metabolites

This group comprises glutathione (GSH), phytoalexins, glycosphingolipid (GSL), thionins, allinin, and defensins, which have been connected directly or indirectly to plants for the protection from microbial pathogens (Halkier and Gershenzon, 2006).

4.2.2.3.1 Glutathione GSH is an organic compound. S is present in different amino acids, cysteine and methionine, and in the soluble part and GSH acts as a transportable pond of S in development and growth settlement, and in stress it acts as an antioxidant in the cell by strengthening the defense system of the plants (Kang et al., 2007). It was described as an indication of plant S abundance that depressed the regulation of S absorption and S consumption by roots (Lappartient et al., 1999). Specific cells such as trichomes showed an increased production of enzymes for GSH synthesis and some other compounds required for heavy metal stress defense in plants (Choi et al., 2001). About 3–10 mM concentration of GSH in the major cellular sections in plants has been found (Leustek and Saito, 1999). GSH not only works to alleviate oxidative stress by removing ROS, but also acts as a reducing agent for some antioxidants, such as ascorbic acid (Nocito et al., 2002).

4.2.2.3.2 GSL GSLs are a class of S- and N-containing plant glucosides, have low molecular mass, and are formed by higher plants to develop tolerance against parasites and predators. Their by-products with poisonous or repellent features are emitted as volatile compounds; common examples are mustard plant oil glucosides and Allium allylcys sulfoxides (Van-de-Mortel et al., 2008). The volatiles with odor belong to the GSL group stimulated by myrosinase enzyme, split glucose from its S–C bond, results aglycon reorganizes with deficiency of the sulfate to produce pungent chemically reactive products, which includes nitriles and isothiocyanates, role in defense from poisons of herbivorous (Grubb and Abel, 2006). When plant tissue is wounded, GSL is produced because of myrosinase, an enzyme that eliminates the β-glucose, directing the synthesis of an intermediate that are nitriles and isothiocyanates, which have a role in defense by poisoning the herbivore insects during damage and as feeding repellents (Geu-Flores et al., 2009). Omirou et al. (2013) found that the amount of GSL produced inside may help to describe the distinct patchy spreading of many cruciferous plants, which indicates the opposite relationship, that GSLs are products of evolutionary reactions of patchy spreading important for enhancing plant yield.

4.2.2.3.3 Phytoalexins During biotic stress, plants produce phytoalexins that support the prevention of the spreading of the pathogens in the plant body by amassing around the infection site; this seems to be a usual method of defense against insect pests in numerous plants (Jeandet et al., 2013). Many of the variations are related to a fast, suitable response, which results in plant death, called as a hypersensitive response (Hammond-Kosack and Jones, 1996). Many plant families produce phytoalexins of organic nature, e.g., isoflavonoids are associated with Leguminosae and sesquiterpenoids with Solanaceae, whereas Crucifereae is the only plant family that produces these S metabolites, which are different from the other GSL compounds (Harborne, 1999). Cruciferous crops are very important; many phytoalexins are characteristic of this family called cruciferous phytoalexins (Monde et al., 2003). Moreover, they are involved in defense against pests' activity (Mehta et al., 1995); e.g., derivatives of furano-acetylenic produced in *Vicia fava* during pest attack (Soylu et al., 2002).

4.2.2.3.4 Defensins, Thionins, and Lectins S-rich proteins are produced under biotic stress, most of which provide resistance in plants against fungal infection (Thomma et al., 2002). Furthermore, defensins genes are partially induced by pathogens (Gu et al., 2002).

Lectins are proteins that bind with carbohydrate, which complete many protective functions against wound produced by several pests (Vandenborre et al., 2011). They are insecticides naturally produced by the plants (Saha et al., 2006). Their presence in the gut of pests gives them a powerful insecticidal potential; they line the digestive tract and result in a series of injurious systemic responses (Chakraborty et al., 2009). Lectins resist a high pH range and harm the epithelium, thus the problem with the digestion and absorption of nutrients (Ohno et al., 2006). Disturbance in the lipid, carbohydrate, and protein metabolism produces damage in specific tissues; on the other hand, it varies the hormonal and defensive position, inhibiting the growth of insect pests (Saha et al., 2006). When coleopteran, lepidopteran, and homopteran insects attack the plants, lectins are produced, initiating the defense mechanism (Macedo et al., 2007). Agglutinin was the first discovered plant lectin that acts against insects (Li et al., 2014). The effectiveness of agglutinin in wheat germ has been considered against numerous insect pests (Saha et al., 2006). When sucking insects start invading, they communicate with a specific carbohydrate and produce mannose-binding lectins to initiate defense mechanism (Stoger et al., 1999). In *Arum maculatum* L. lectins have been proved to be active against the aphids (Majumder et al., 2005). Plant lectins are stimulated by elicitors as a response brought against numerous stresses (Lannoo et al., 2007). In monocots, such as barley, maize, rye, rice, and wheat, many lectins are produced in leaf tissues (Jiang et al., 2006).

4.2.2.4 Nitrogen-Containing Secondary Metabolites

Secondary metabolites consist of alkaloids, cyanogenic glucosides, and nonprotein amino acids, which are of great importance for humans because of their antiherbivore defense (Wink and Mohamed, 2003).

4.2.2.4.1 Alkaloids Alkaloids are a group of secondary metabolites produced in almost 20% of the vascular plant species (Croteau et al., 2000). Pyrrolizidine alkaloids are the most abundant, poisonous to a little extent, and seem to help primarily in protection against microbial attack and herbivore competition. Livestock death is caused by the intake of alkaloids found in plants (Croteau et al., 2000). In animals, their mode of action is variable; when they enter pests and mammals, they disturb the components of the defense system, mainly the chemical transmitters, and affect protein synthesis, transport through membrane, and several enzyme activities (Creelman and Mullet, 1997).

4.2.2.4.2 Cyanogenic Glucosides These are also a group of N-containing compounds produced by the plants against pathogens for defensive purpose; these produce the toxin HCN and commonly occur in plants of families Rosaceae and Graminae. They are not typically so poisonous but when the plant is injured, slugs and snails are avoided because of these compounds in plants. Likewise, the existence of glucosides in cassava increases its storage life without the threat of insects (Adamolekun, 2010).

4.2.2.4.3 Nonprotein Amino Acids Many crop plants also contain amino acids that are uncommon, called nonprotein amino acids, which are present in the free form or in combination with proteins and work as protective compounds (Glover et al., 2014). After plants are damaged, they produce canavanine, which is recognized by the herbivore enzyme that frequently binds arginine to the arginine tRNA molecule, which binds it on specific position for the synthesis of proteins (Cornara et al., 2016). Arginine is used as an active source of nitrogen in *Arabidopsis* and proline as a component under water-deficit conditions. On identification of some pathogenic elements essential for the development o f defense responses inside plant species, individual plants often differ greatly in their resistance against pests. It has been shown that many genes found in plants function in defense against biological stress; these translate receptors that identify and attach to some molecules from pests and equip the plant for the pests' attack (Boller and He, 2009). Some of these are polysaccharides that are produced by the plants by the attack of bacteria and fungi on the cell wall of plant, whereas some others are few-carbon carbohydrates made by cell wall breakdown of fungus by plant enzymes, which are the result of the fungal attack (Cosgrove et al., 2003). Apparently, such extracellular substances are sensed by certain proteins in the membrane, which then stimulate the production of mRNA for enzymes that are the source of phytoalexins (Vidhyasekaran, 2007). Resistant gene products are mostly proteins that have a leucine-rich domain, which often has a uniform amino acid sequence; likewise, β-aminobutyric acid can produce tolerance in *Arabidopsis* against bacterial and fungal damage (Van-der-Ent et al., 2009).

4.2.2.5 Proteinase Inhibitors

Proteinase inhibitors (PIs) include one of the important groups of protective proteins; these are produced at a higher concentration in storage organs and stop different types of enzymatic reactions and play a major role in defense (Lawrence and Koundal, 2002). PIs attach with the enzymes in insect gut and decrease protein digestion, causing amino acid deficiency and disturbing the development of the insects (Azzouz et al., 2009). When lepidopteran and hemipteran insects harm the plant parts, PIs against insect pests are produced to create defensive function in many plants (Dunse et al., 2010). However, some insects respond by producing PI-insensitive proteases, thus stopping them from attachment to proteases; this function may result in even more damage to the infected plants (Zhu-Salzman et al., 2008).

4.2.2.6 Enzymes

These are the electrophiles and enzymes that include ascorbate peroxidases, polyphenol oxidases (PPOs), and peroxidases (PODs). Their accumulation inhibits the insect's digestion (Bhonwong et al., 2009). Some other significant antioxidative enzymes include phenylalanine ammonia lyase, superoxide dismutase, and lipoxygenases (LOXs). Production of antioxidative enzymes in plants following herbivory got great attention in recent years (Vandenborre et al., 2011).

4.2.2.6.1 Peroxidases The oxidative state of host plants has been associated with *N*-(4-Hydroxyphenyl) retinamide (HPR) to the insects, which affects the production ROS, which is consequently removed by antioxidative enzymes (Zhao et al., 2009). POD is one of the important enzymes that reduce the ROS production in maize, wheat, and corn under biotic stress (Fondevilla et al., 2011). PODs are a major part of the immediate reaction of plants against the damage by pests (Gulsen et al. 2010). PODs are monomeric hemoproteins propagated as a soluble form, attached to cell wall, membrane bound. Many processes are modulated by PODs for the defense against the pests' attack (War et al., 2011). The action of PODs on phenols forms phenoxy radicals, which inhibit the nourishing by insects and increase toxins that decrease the plant digestibility by the pests (Zhang et al., 2008). In addition to this, PODs also cause direct toxicity in guts of pests (Zhu-Salzman et al., 2008).

4.2.2.6.2 Polyphenol Oxidases These are important enzymes that affect the insects' growth and reproduction and play a significant role in plant defense against the biotic stress (Bhonwong et al., 2009). PPO-generated quinones may alkylate important amino acids and reduce the nutritional value of the plant; quinones may develop oxidative stress in the gut of pest by redox cycling and have lethal effects on herbivores (Zhang et al., 2008). Quinines are more harmful and toxic to pests than the common phenols (Bhonwong et al., 2009). In addition to their function, melanin production by PPOs improves the cell wall immunity to pests (Zhao et al., 2009). The PPO genes are activated by signaling molecules and cause harm because of wounding and pathogen (Zhao et al., 2009). A correlation between the start of PPO activity and infection has been observed in tomato and lettuce (Sethi et al., 2009). Because PPOs are stored in leaves, stems, and roots of the plants, newly formed tissues have more .vulnerability to pests' attack demonstrate greater induction (Tscharntke et al., 2001). Nevertheless, PPO levels had no or partial influence on *Blissus occiduus* (Heng-Moss et al., 2004).

4.2.2.6.3 Lipoxygenases LOXs are a group of antioxidative enzymes that are required in plant defense against many types of pests (Bruinsma et al., 2009). LOXs are the enzymes that cause the oxidation of unsaturated fatty acids by which hydroperoxides production is affected that are converted to JA and traumatin which make the plants protective to pests (Wang et al., 2008).

5. INDIRECT DEFENSES

Production of volatiles and the exudation of EFN enable interactions of plants with insect pests, which dynamically decrease the number of herbivore insects (Maffei, 2010).

5.1 Herbivore-Induced Plant Volatiles

Herbivore-induced plant volatiles (HIPVs) have an important role in plant defense. They protect the plants in two ways, by attracting the natural enemies of herbivores or by preventing feeding and/or oviposition (Arimura et al., 2009). Their production varies depending on the type of herbivore and plant species, growth stage, and plant condition (Maffei, 2010). Volatile compounds are generally expelled in large amounts by the plants into the environment, although a different mixture of these compounds is formed in response to insect attack (Arimura et al., 2009). These are specific for a specific insect–plant organization, comprising natural enemies and adjacent plants (Engelberth et al., 2004). To understand the function of C_6-aldehydes and their related by-products in plant defense, the green leaf volatiles (GLVs) intensity has been changed either by use of elicitors, or by controlling genetically. GLVs demonstrate an important function in plant defense mechanism by attracting natural enemies (Hare, 2011). MeSA is a frequent component of many leaf and floral mixtures and MeSA lures sticky cards attract numerous insect predators, including the big-eyed bug, ladybird beetle (James, 2003), and other enemies. Ulland et al. (2008) studied the oviposition inhibition of cabbage moths (*Mamestra brassicae* L.) by

MeSA expelled during attack, proposing that MeSA can also be identified by the harmful herbivores (Chen et al., 2003). Products of shikimic acid and LOX metabolic and terpenoid pathways play a major role in protection of plants under biotic stress, either directly or indirectly (Gill and Tuteja, 2010). After invasion of caterpillars in maize plants, (Z)-3-hexanol accelerate production is started, which appeals to the natural enemies (Ruther and Kleier, 2005). By production of volatile organic compounds, many signal transduction pathways begin to protect the plants under biotic stress (Kessler et al., 2002). *Diuraphis noxia* is a root nourishing insect, and controls the production of a monoterpene volatile (1, 8-cineole), which is toxic and repulsive to some insects (Ro et al., 2006). Sesquiterpene ((E)-β-caryophyllene) is synthesized by maize roots in reaction to nourishing by the larvae of *Diabrotica virgifera* attract the nematode (Rasmann et al., 2005).

5.2 Defense Elicitors (Insect Oral Secretion)

It is commonly found that plants infected by herbivores are controlled by pests' secretions (Pauwels et al., 2009). Fatty acid–amino acid conjugates (FACs) are the main constituents in the oral discharges of insects (Hettenhausen et al., 2014). In tobacco plants the FACs have also been described to increase the production of JA, which causes the expression of herbivore-responsive genes (Falk et al., 2014).

5.2.1 Jasmonic Acid

Many phytohormones contribute to plant protection from pests; JA is one of the significant phytohormones involved in the defense against herbivore insects and controls the initiation of direct and indirect protection responses (Shivaji et al., 2010). It is a derivative of linoleic acid, and damage of plant parts by insects causes defense signal transduction pathways (Itoh et al., 2002). Plant defense responses include the production of JA that comprise antioxidative enzymes and development of trichome (Pauwels et al., 2009). Various genes involved in plant defense against biotic stress are regulated by JAs (Shivaji et al., 2010). The production of jasmonates induces the defensive compound indole glucosinolate in plants (Ribot et al., 2008), which modifies the intracellular levels of calcium and redox status of cells (Walter et al., 2007). Calcium-dependent protein kinases (CDPKs) are activated by JA and are important in plants and protect the plants by initiating various signal transduction pathways under stress (Ludwig et al., 2004). However, JA also makes the plant resistant by producing the defensive compounds in them against the pests (Barbehenn et al., 2008).

5.2.2 Salicylic Acid

The precursor for the synthesis of SA is benzoic acid, one of the major phytohormones that contribute to maintaining the plant defense (War et al., 2011). It is a chief growth regulator that produces various defense reactions in addition to the effect on growth and development in plant. SA provides more defense against insects that penetrate and suck the plant phloem juice than against the chewing insects (Zhao et al., 2009; Maffei, 2010).

5.2.3 Ethylene

Ethylene is one of the important phytohormones that protect the plants from the attack of different types of insects (van Loon et al., 2006). It involves both direct and indirect defense against the pathogens and herbivore. It is reported that its indirect role is by HIPV production. Its signaling is related to the JA in expression of plant defense mechanisms against plant pathogen and herbivore insects. PI expression in response to the herbivore insects is observed in tomato by the ET- and JA-related signaling (Kunkel and Brooks, 2002). 1-Amino-cyclopropane-1-carboxylic acid, a precursor of ET, has been described to increase the production of unstable compounds from the leaves treated with JA (Horiuchi et al., 2003). Ethylene caused an increased production of volatiles in maize under biotic stress (Ruther and Kleier, 2005).

5.2.4 Calcium Ions (Ca^{2+})

When herbivores attack the plants, various signal transduction pathways are started in them to protect them under this biotic stress (Blechert et al., 1995). Production of CDPKs is seen in many plants under stress conditions, and they have a beneficial aspect to cope with biotic and abiotic stresses in plants (Kanchiswamy et al., 2010). It has been reported that in tobacco CDPKs activate MAPKs under biotic stress (Ludwig et al., 2004). Damage by pest larvae in *Phaseolus lunatus* L. resulted in the accumulation of Ca^{2+} not only in the cells near the wound site, but also in the entire leaf (Mithofer et al., 2005). Plant resistance signaling enhanced significantly in *Triticum aestivum* infected by pests by the expression of calmodulin-binding proteins (Reddy et al., 2013).

REFERENCES

Abouziena, H.F., Radwan, S.M., Eldabaa, M., 2015. Comparison of potato yield, quality and weed control obtained with different plastic mulch colours. Middle East J. Appl. Sci. 5 (2), 374–382.

Adamolekun, B., 2010. Etiology of Konzo, epidemic spastic paraparesis associated with cyanogenic glycosides in cassava: role of thiamine deficiency. J. Neurol. Sci. 296 (1), 30–33.

Agrawal, A.A., 2011. Current trends in the evolutionary ecology of plant defence. Funct. Ecol. 25 (2), 420–432.

Agrawal, A.A., Konno, K., 2009. Latex: a model for understanding mechanisms, ecology, and evolution of plant defense against herbivory. Annu. Rev. Ecol. Evol. Syst. 40, 311–331.

Akhtar, M., Malik, A., 2000. Roles of organic soil amendments and soil organisms in the biological control of plant-parasitic nematodes: a review. Bioresour. Technol. 74 (1), 35–47.

Ali, S.S., Sharma, S.B., 2003. Nematode survey of chickpea production areas in Rajasthan, India. Nematol. Med. 31 (2), 322–331.

Amey, R.C., Schleicher, T., Slinn, J., Lewis, M., Macdonald, H., Neill, S.J., Spencer-Phillips, P.T., 2008. Proteomic analysis of a compatible interaction between *Pisum sativum* (pea) and the downy mildew pathogen *Peronospora viciae*. Eur. J. Plant Pathol. 122 (1), 41–55.

Anderson, P.K., Cunningham, A.A., Patel, N.G., Morales, F.J., Epstein, P.R., Daszak, P., 2004. Emerging infectious diseases of plants: pathogen pollution, climate change and agrotechnology drivers. Trends Ecol. Evol. 19 (10), 535–544.

Apel, K., Hirt, H., 2004. Reactive oxygen species: metabolism, oxidative stress, and signal transduction. Annu. Rev. Plant Biol. 55, 373–399.

Arimura, G.I., Matsui, K., Takabayashi, J., 2009. Chemical and molecular ecology of herbivore-induced plant volatiles: proximate factors and their ultimate functions. Plant Cell Physiol. 50 (5), 911–923.

Ashbolt, N.J., 2004. Microbial contamination of drinking water and disease outcomes in developing regions. Toxicology 198 (1), 229–238.

Ashrafi, Z.Y., Mashhadi, H.R., Sadeghi, S., Blackshaw, R.E., 2009. Study Effects of planting methods and tank mixed herbiciedes on weeds controlling and wheat yield. J. Agric. Sci. 1 (1), 101.

Azzouz, N., Panasenko, O.O., Deluen, C., Hsieh, J., Theiler, G., Collart, M.A., 2009. Specific roles for the Ccr4-Not complex subunits in expression of the genome. RNA 15 (3), 377–383.

Bai, F.W., Yan, J., Qu, Z.C., Zhang, H.W., Xu, J., Ye, M.M., Shen, D.L., 2002. Phylogenetic analysis reveals that a dwarfing disease on different cereal crops in China is due to rice black streaked dwarf virus (RBSDV). Virus Gene. 25 (2), 201–206.

Balandier, P., Collet, C., Miller, J.H., Reynolds, P.E., Zedaker, S.M., 2006. Designing forest vegetation management strategies based on the mechanisms and dynamics of crop tree competition by neighbouring vegetation. Forestry 79 (1), 3–27.

Balasubramanian, N., Hao, Y.J., Toubarro, D., Nascimento, G., Simões, N., 2009. Purification, biochemical and molecular analysis of a chymotrypsin protease with prophenoloxidase suppression activity from the entomopathogenic nematode *Steinernema carpocapsae*. Int. J. Parasitol. 39 (9), 975–984.

Barakat, A., Bagniewska-Zadworna, A., Frost, C.J., Carlson, J.E., 2010. Phylogeny and expression profiling of CAD and CAD-like genes in hybrid *Populus* (*P. deltoids* × *P. nigra*): evidence from herbivore damage for sub-functionalization and functional divergence. BMC Plant Biol. 10 (1), 1–11.

Barbehenn, R.V., Constabel, C.P., 2011. Tannins in plant–herbivore interactions. Phytochemistry 72 (13), 1551–1565.

Barbehenn, R.V., Jaros, A., Yip, L., Tran, L., Kanellis, A.K., Constabel, C.P., 2008. Evaluating ascorbate oxidase as a plant defense against leaf-chewing insects using transgenic poplar. J. Chem. Ecol. 34 (10), 1331–1340.

Benaragama, D., Shirtliffe, S.J., Johnson, E.N., Duddu, H.S.N., Syrovy, L.D., 2016. Does yield loss due to weed competition differ between organic and conventional cropping systems? Weed Res. 56 (4), 274–283.

Bhonwong, A., Stout, M.J., Attajarusit, J., Tantasawat, P., 2009. Defensive role of tomato polyphenol oxidases against cotton bollworm (*Helicoverpa armigera*) and beet armyworm (*Spodoptera exigua*). J. Chem. Ecol. 35 (1), 28–38.

Bidlack, J.E., Middick, A., Shantz, D., MacKown, C.T., Williams, R.D., Rao, S.C., 2006. Weed control in a pigeon pea–wheat cropping system. Field Crops Res. 96 (1), 63–70.

Bilgin, D.D., Zavala, J.A., Zhu, J.I.N., Clough, S.J., Ort, D.R., Delucia, E., 2010. Biotic stress globally downregulates photosynthesis genes. Plant. Cell Environ. 33 (10), 1597–1613.

Birkenbihl, R.P., Somssich, I.E., 2011. Transcriptional plant responses critical for resistance towards necrotrophic pathogens. Front. Plant Sci. http://dx.doi.org/10.3389/fpls.2011.00076. Available from:.

Blechert, S., Brodschelm, W., Hölder, S., Kammerer, L., Kutchan, T.M., Mueller, M.J., Zenk, M.H., 1995. The octadecanoic pathway: signal molecules for the regulation of secondary pathways. Proc. Natl. Acad. Sci. 92 (10), 4099–4105.

Bloch, C., Richardson, M., 1991. A new family of small (5 kDa) protein inhibitors of insect α-amylases from seeds or sorghum (*Sorghum bicolor* (L) Moench) have sequence homologies with wheat γ-purothionins. FEBS Lett. 279 (1), 101–104.

Boege, K., Marquis, R.J., 2005. Facing herbivory as you grow up: the ontogeny of resistance in plants. Trends Ecol. Evol. 20 (8), 441–448.

Boller, T., He, S.Y., 2009. Innate immunity in plants: an arms race between pattern recognition receptors in plants and effectors in microbial pathogens. Science 324 (5928), 742–744.

Bourgaud, F., Gravot, A., Milesi, S., Gontier, E., 2001. Production of plant secondary metabolites: a historical perspective. Plant Sci. 161 (5), 839–851.

Bridge, J., 1996. Nematode management in sustainable and subsistence agriculture. Annu. Rev. Phytopathol. 34 (1), 201–225.

Brooker, R.W., Maestre, F.T., Callaway, R.M., Lortie, C.L., Cavieres, L.A., Kunstler, G., Armas, C., 2008. Facilitation in plant communities: the past, the present, and the future. J. Ecol. 96 (1), 18–34.

Bruinsma, M., Posthumus, M.A., Mumm, R., Mueller, M.J., van Loon, J.J., Dicke, M., 2009. Jasmonic acid-induced volatiles of *Brassica oleracea* attract parasitoids: effects of time and dose, and comparison with induction by herbivores. J. Exp. Bot. http://dx.doi.org/10.1093/jxb/erp101.

Bryant, P.J., Watson, K.L., Justice, R.W., Woods, D.F., 1993. Tumor suppressor genes encoding proteins required for cell interactions and signal transduction in *Drosophila*. Development 119, 239–249.

Burger, J., Darmency, H., Granger, S., Guyot, S.H., Messéan, A., Colbach, N., 2015. Simulation study of the impact of changed cropping practices in conventional and GM maize on weeds and associated biodiversity. Agric. Syst. 137, 51–63.

Caarls, L., Pieterse, C.M., Van Wees, S., 2015. How salicylic acid takes transcriptional control over jasmonic acid signaling. Front. Plant Sci. http://dx.doi.org/10.3389/fpls.2015.00170.

Campe, R., Loehrer, M., Conrath, U., Goellner, K., 2014. *Phakopsora pachyrhizi* induces defense marker genes to necrotrophs in *Arabidopsis thaliana*. Physiol. Mol. Plant Pathol. 87, 1–8.

Carriere, Y., Crickmore, N., Tabashnik, B.E., 2015. Optimizing pyramided transgenic Bt crops for sustainable pest management. Nat. Biotechnol 33 (2), 161–168.

Chakraborty, M., Karun, A., Mitra, A., 2009. Accumulation of phenylpropanoid derivatives in chitosan-induced cell suspension culture of *Cocos nucifera*. J. Plant Physiol. 166 (1), 63–71.

Chamarthi, S.K., Kumar, K., Gunnaiah, R., Kushalappa, A.C., Dion, Y., Choo, T.M., 2014. Identification of fusarium head blight resistance related metabolites specific to doubled-haploid lines in barley. Eur. J. Plant Pathol. 138 (1), 67–78.

Chen, F., D'auria, J.C., Tholl, D., Ross, J.R., Gershenzon, J., Noel, J.P., Pichersky, E., 2003. An *Arabidopsis thaliana* gene for methylsalicylate biosynthesis, identified by a biochemical genomics approach, has a role in defense. Plant J. 36 (5), 577–588.

Chen, H., Tang, W., Xu, C., Li, X., Lin, Y., Zhang, Q., 2005. Transgenic indica rice plants harboring a synthetic cry2A* gene of *Bacillus thuringiensis* exhibit enhanced resistance against lepidopteran rice pests. Theor. Appl. Genet. 111 (7), 1330–1337.

Cheval, C., Aldon, D., Galaud, J.P., Ranty, B., 2013. Calcium/calmodulin-mediated regulation of plant immunity. Acta Mol. Cell Res. 1833 (7), 1766–1771.

Chisholm, S.T., Coaker, G., Day, B., Staskawicz, B.J., 2006. Host-microbe interactions: shaping the evolution of the plant immune response. Cell 124 (4), 803–814.

Choi, Y.E., Harada, E., Wada, M., Tsuboi, H., Morita, Y., Kusano, T., Sano, H., 2001. Detoxification of cadmium in tobacco plants: formation and active excretion of crystals containing cadmium and calcium through trichomes. Planta 213 (1), 45–50.

Collins, N.C., Thordal-Christensen, H., Lipka, V., Bau, S., 2003. SNARE-protein-mediated disease resistance at the plant cell wall. Nature (London) 425 (6961), 973–977.

Cornara, L., Xiao, J., Burlando, B., 2016. Therapeutic potential of temperate forage legumes: a review. Crit. Rev. Food Sci Nutri. 56, 149–161.

Cosgrove, S.E., Sakoulas, G., Perencevich, E.N., Schwaber, M.J., Karchmer, A.W., Carmeli, Y., 2003. Comparison of mortality associated with methicillin-resistant and methicillin-susceptible *Staphylococcus aureus* bacteremia: a meta-analysis. Clin. Infect. Dis. 36 (1), 53–59.

Creelman, R.A., Mullet, J.E., 1997. Oligosaccharins, brassinolides, and jasmonates: nontraditional regulators of plant growth, development, and gene expression. Plant Cell 9 (7), 1211–1223.

Croteau, R., Kutchan, T.M., Lewis, N.G., 2000. Natural products (secondary metabolites). Biochem. Mol. Biol. Plants 24, 1250–1319.

Crozier, A., Yokota, T., Jaganath, I.B., Marks, S.C., Saltmarsh, M., Clifford, M.N., 2006. Secondary metabolites in fruits, vegetables, beverages and other plant based dietary components. In: Plant Secondary Metabolites: Occurrence, Structure and Role in the Human Diet. http://dx.doi.org/10.1002/9780470988558.ch7.

Czajkowski, R., Perombelon, M.C., van Veen, J.A., van der Wolf, J.M., 2011. Control of blackleg and tuber soft rot of potato caused by *Pectobacterium* and *Dickeya* species: a review. Plant Pathol. 60 (6), 999–1013.

Dalin, P., Björkman, C., 2003. Adult beetle grazing induces willow trichome defence against subsequent larval feeding. Oecologia 134 (1), 112–118.

Da-Silva, E.M., Roel, A.R., Porto, K.R., Falco, M.E., Matias, R., 2017. Insecticidal effect of the ethanol extract of *Baccharis dracunculifolia* (Asterales: Asteraceae). Rev. Biol. Trop. 65 (2), 517–523.

Dass, A., Shekhawat, K., Choudhary, A.K., Sepat, S., Rathore, S.S., Mahajan, G., Chauhan, B.S., 2016. Weed management in rice using crop competition-a review. Crop Prot. 95, 45–52.

Dorn, K.M., Fankhauser, J.D., Wyse, D.L., Marks, M.D., 2013. De novo assembly of the pennycress (*Thlaspi arvense*) transcriptome provides tools for the development of a winter cover crop and biodiesel feedstock. Plant J. 75 (6), 1028–1038.

Dudareva, N., Negre, F., Nagegowda, D.A., Orlova, I., 2006. Plant volatiles: recent advances and future perspectives. Crit. Rev. Plant Sci. 25 (5), 417–440.

Dunbabin, V., 2007. Simulating the role of rooting traits in crop-weed competition. Field Crops Res. 104 (1), 44–51.

Duncan, L.W., 2005. Nematode parasites of citrus. In: Luc, M., Sikora, R.A. (Eds.), Plant Parasitic Nematodes in Subtropical and Tropical Agriculture. CABI Bioscience, Egham, UK (Chapter 11).

Dunse, K.M., Kaas, Q., Guarino, R.F., Barton, P.A., Craik, D.J., Anderson, M.A., 2010. Molecular basis for the resistance of an insect chymotrypsin to a potato type II proteinase inhibitor. Proc. Natl. Acad. Sci. 107 (34), 15016–15021.

Durner, J., Shah, J., Klessig, D.F., 1997. Salicylic acid and disease resistance in plants. Trends Plant Sci. 2 (7), 266–274.

Durrant, W.E., Dong, X., 2004. Systemic acquired resistance. Annu. Rev. Phytopathol. 42, 185–209.

Eisner, T., Eisner, M., Aneshansley, D.J., Wu, C.L., Meinwald, J., 2000. Chemical defense of the mint plant, *Teucrium marum* (Labiatae). Chemoecology 10 (4), 211–216.

Eneh, F.U., Nwankwo, E.N., Okwuenu, P., 2016. Phytochemical analysis and toxicity of *Casuarina equisetifola* (whistling pine) to *Sitophilus zeamais* Motschulsky (Coleoptera: Curculionidae). Phytochem. Anal. 1 (3), 433–441.

Engelberth, J., Alborn, H.T., Schmelz, E.A., Tumlinson, J.H., 2004. Airborne signals prime plants against insect herbivore attack. Proc. Natl. Acad. Sci. U. S. A. 101 (6), 1781–1785.

Engstrom, M.T., Karonen, M., Ahern, J.R., Baert, N., Payré, B., Hoste, H., Salminen, J.P., 2016. Chemical structures of plant hydrolyzable tannins reveal their in vitro activity against egg hatching and motility of haemonchus contortus nematodes. J. Agric Food Chem 64 (4), 840–851.

Eschen, R., Roques, A., Santini, A., 2015. Taxonomic dissimilarity in patterns of interception and establishment of alien arthropods, nematodes and pathogens affecting woody plants in Europe. Divers Distrib. 21 (1), 36–45.

Escobar-Bravo, R., Klinkhamer, P.G., Leiss, K.A., 2017. Induction of jasmonic acid-associated defenses by thrips alters host suitability for conspecifics and correlates with increased trichome densities in tomato. Plant Cell Physiol. 58 (3), 622–634.

Etebu, E., Nwauzoma, A.B., 2014. A review on sweet orange (*Citrus sinensis* L Osbeck): health, diseases and management. Am. J. Res. Commun. 2 (2), 33–70.

Falk, K.L., Kästner, J., Bodenhausen, N., Schramm, K., Paetz, C., Vassão, D.G., Gershenzon, J., 2014. The role of glucosinolates and the jasmonic acid pathway in resistance of *Arabidopsis thaliana* against molluscan herbivores. Mol. Ecol. 23 (5), 1188–1203.

Fatope, M.O., Takeda, Y., Yamashita, H., Okabe, H., Yamauchi, T., 1990. New cucurbitane triterpenoids from *Momordica charantia*. J. Nat. Prod. 53 (6), 1491–1497.

Fereres, A., Moreno, A., 2009. Behavioural aspects influencing plant virus transmission by homopteran insects. Virus Res. 141 (2), 158–168.

Fernandes, I., Alves, A., Correia, A., Devreese, B., Esteves, A.C., 2014. Secretome analysis identifies potential virulence factors of *Diplodia corticola*, a fungal pathogen involved in cork oak (*Quercus suber*) decline. Fung Biol. 118 (5), 516–523.

Firn, R.D., Jones, C.G., 2009. A Darwinian view of metabolism: molecular properties determine fitness. J. Exp. Bot. 60 (3), 719–726.

Fondevilla, S., Küster, H., Krajinski, F., Cubero, J.I., Rubiales, D., 2011. Identification of genes differentially expressed in a resistant reaction to *Mycosphaerella pinodes* in pea using microarray technology. BMC Genom. 12 (1), 28.

Gagne-Bourgue, F., Aliferis, K.A., Seguin, P., Rani, M., Samson, R., Jabaji, S., 2013. Isolation and characterization of indigenous endophytic bacteria associated with leaves of switchgrass (*Panicum virgatum* L.) cultivars. J. App Microbiol 114 (3), 836–853.

George, C., Kohler, J., Rillig, M.C., 2016. Biochars reduce infection rates of the root-lesion nematode *Pratylenchus penetrans* and associated biomass loss in carrot. Soil Biol. Biochem. 95, 11–18.

Geu-Flores, F., Nielsen, M.T., Nafisi, M., Møldrup, M.E., Olsen, C.E., Motawia, M.S., Halkier, B.A., 2009. Glucosinolate engineering identifies a γ-glutamyl peptidase. Nat. Chem. Biol. 5 (8), 575–577.

Ghasemzadeh, A., Jaafar, H.Z., Rahmat, A., 2011. Effects of solvent type on phenolics and flavonoids content and antioxidant activities in two varieties of young ginger (*Zingiber officinale Roscoe*) extracts. J. Med. Plants Res. 5 (7), 1147–1154.

Gill, S.S., Tuteja, N., 2010. Reactive oxygen species and antioxidant machinery in abiotic stress tolerance in crop plants. Plant Physiol. Biochem. 48 (12), 909–930.

Glazebrook, J., 2005. Contrasting mechanisms of defense against biotrophic and necrotrophic pathogens. Annu. Rev. Phytopathol. 43, 205–227.

Glover, W.B., Mash, D.C., Murch, S.J., 2014. The natural non-protein amino acid N-[beta]-methylamino-l-alanine (BMAA) is incorporated into protein during synthesis. Amino Acids 46 (11), 2553.

Goodin, M.M., Zaitlin, D., Naidu, R.A., Lommel, S.A., 2008. *Nicotiana benthamiana*: its history and future as a model for plant–pathogen interactions. Mol. Plant Microbe Interact. 21 (8), 1015–1026.

Goverse, A., Smant, G., 2014. The activation and suppression of plant innate immunity by parasitic nematodes. Ann. Rev. Phytopathol. 52, 243–265.

Grabau, Z.J., Maung, Z.T.Z., Noyes, D.C., Baas, D.G., Werling, B.P., Brainard, D.C., Melakeberhan, H., 2017. Effects of cover crops on pratylenchus penetrans and the nematode community in carrot production. J. Nematol. 49 (1), 114.

Gribble, G.W., 2003. The diversity of naturally produced organohalogens. Chemosphere 52 (2), 289–297.

Grubb, C.D., Abel, S., 2006. Glucosinolate metabolism and its control. Trends Plant Sci. 11 (2), 89–100.

Gu, Y.Q., Wildermuth, M.C., Chakravarthy, S., Loh, Y.T., Yang, C., He, X., Martin, G.B., 2002. Tomato transcription factors Pti4, Pti5, and Pti6 activate defense responses when expressed in *Arabidopsis*. Plant Cell 14 (4), 817–831.

Gulsen, O., Eickhoff, T., Heng-Moss, T., Shearman, R., Baxendale, F., Sarath, G., Lee, D., 2010. Characterization of peroxidase changes in resistant and susceptible warm-season turfgrasses challenged by *Blissus occiduus*. Arthropod Plant Interact 4 (1), 45–55.

Haldar, A.K., Foltz, C., Finethy, R., Piro, A.S., Feeley, E.M., Pilla-Moffett, D.M., Coers, J., 2015. Ubiquitin systems mark pathogen-containing vacuoles as targets for host defense by guanylate binding proteins. Proc. Natl. Acad. Sci. 112 (41), E5628–E5637.

Halkier, B.A., Gershenzon, J., 2006. Biology and biochemistry of glucosinolates. Annu. Rev. Plant Biol. 57, 303–333.

Hammond-Kosack, K.E., Jones, J.D., 1996. Resistance gene-dependent plant defense responses. Plant Cell 8 (10), 1773.

Hanley, M.E., Lamont, B.B., Fairbanks, M.M., Rafferty, C.M., 2007. Plant structural traits and their role in anti-herbivore defense. Perspect. Plant Ecol. Evol. Syst. 8 (4), 157–178.

Harborne, J.B., 1999. The comparative biochemistry of phytoalexin induction in plants. Biochem. Syst. Ecol. 27 (4), 335–367.

Harborne, A., Fenner, D., Barnes, A., Beger, M., Harding, S., Roxburgh, T., 2000. Status report on the coral reefs of the east coast of Peninsula Malaysia. Report Prepared to Department of Fisheries Malaysia 361–369 Kuala Lumpur, Malaysia.

Hardie, D.G., Schaffer, B.E., Brunet, A., 2016. AMPK: an energy-sensing pathway with multiple inputs and outputs. Trends Cell Biol. 26 (3), 190–201.

Hare, P., 2011. Antimalarial fungal pesticide. Nat. Biotechnol. 29 (4), 330–331.

Hartmann, T., 1996. Diversity and variability of plant secondary metabolism: a mechanistic view. Entomol. Exp. Appl. 80 (1), 177–188.

Hatfield, R., Vermerris, W., 2001. Lignin formation in plants. The dilemma of linkage specificity. Plant Physiol. 126 (4), 1351–1357.

Haynes, P.A., Roberts, T.H., 2007. Subcellular shotgun proteomics in plants: looking beyond the usual suspects. Proteomics 7 (16), 2963–2975.

He, J., Chen, F., Chen, S., Lv, G., Deng, Y., Fang, W., He, C., 2011. Chrysanthemum leaf epidermal surface morphology and antioxidant and defense enzyme activity in response to aphid infestation. J. Plant Physiol. 168 (7), 687–693.

Heng-Moss, T., Sarath, G., Baxendale, F., Novak, D., Bose, S., Ni, X., Quisenberry, S., 2004. Characterization of oxidative enzyme changes in buffalograsses challenged by *Blissus occiduus*. J. Econ. Entomol. 97 (3), 1086–1095.

Hettenhausen, C., Heinrich, M., Baldwin, I.T., Wu, J., 2014. Fatty acid-amino acid conjugates are essential for systemic activation of salicylic acid-induced protein kinase and accumulation of jasmonic acid in *Nicotiana attenuata*. BMC Plant Biol. 14 (1), 326.

Horiuchi, K., Matsuzaki, H., Kobayashi, K., Goldberg, E.L., Shibata, Y., 2003. 10Be record and magnetostratigraphy of a Miocene section from Lake Baikal: Re-examination of the age model and its implication for climatic changes in continental Asia. Geophys Res. Lett. 30 (12), 1011–1022.

Howe, G.A., Jander, G., 2008. Plant immunity to insect herbivores. Annu. Rev. Plant Biol. 59, 41–66.

Hunt, D.J., Luc, M., Manzanilla-Lopez, R.H., 2005. Identification, morphology and biology of plant parasitic nematodes. In: Luc, M., Sikora, R.A. (Eds.), Plant Parasitic Nematodes in Subtropical and Tropical Agriculture. CABI Bioscience, Egham, UK.

Itoh, A., Schilmiller, A.L., McCaig, B.C., Howe, G.A., 2002. Identification of a jasmonate-regulated allene oxide synthase that metabolizes 9-hydroperoxides of linoleic and linolenic acids. J. Biol. Chem. 277 (48), 46051–46058.

Jaiswal, S., Fontanillas, P., Flannick, J., Manning, A., Grauman, P.V., Mar, B.G., Lindsley, R.C., Mermel, C.H., 2014. Age-related clonal hematopoiesis associated with adverse outcomes. N. Engl. J. Med 371, 2488–2498.

Jalloh, A., Roy-Macauley, H., Sereme, P., 2012. Major agro-ecosystems of West and Central Africa: brief description, species richness, management, environmental limitations and concerns. Agric. Ecosyst. Environ. 157, 5–16.

James, D.G., 2003. Field evaluation of herbivore-induced plant volatiles as attractants for beneficial insects: methyl salicylate and the green lacewing, *Chrysopa nigricornis*. J. Chem. Ecol. 29 (7), 1601–1609.

Jeandet, P., Clément, C., Courot, E., Cordelier, S., 2013. Modulation of phytoalexin biosynthesis in engineered plants for disease resistance. Int. J. Mol. Sci. 14 (7), 14136–14170.

Jiang, J.F., Han, Y., Xing, L.J., Xu, Y.Y., Xu, Z.H., Chong, K., 2006. Cloning and expression of a novel cDNA encoding a mannose-specific jacalin-related lectin from *Oryza sativa*. Toxicon 47 (1), 133–139.

Jiang, Y., Liu, Y., Guo, Q., Xu, C., Zhu, C., Shi, J., 2016. Sesquiterpene glycosides from the roots of *Codonopsis pilosula*. Acta Pharm. Sin. 6 (1), 46–54.

Johnson, J.M., Hough-Goldstein, J.A., Vangessel, M.J., 2004. Effects of straw mulch on pest insects, predators, and weeds in watermelons and potatoes. Environ. Entomol. 33 (6), 1632–1643.

Kabbage, M., Yarden, O., Dickman, M.B., 2015. Pathogenic attributes of *Sclerotinia sclerotiorum*: switching from a biotrophic to necrotrophic lifestyle. Plant Sci. 233, 53–60.

Kanchiswamy, C.N., Takahashi, H., Quadro, S., Maffei, M.E., Bossi, S., Bertea, C., Boland, W., 2010. Regulation of *Arabidopsis* defense responses against *Spodoptera littoralis* by CPK mediated calcium signaling. BMC Plant Biol. 10 (1), 97.

Kang, K.W., Lee, S.J., Kim, S.G., 2005. Molecular mechanism of nrf2 activation by oxidative stress. Antioxid. Redox Signal 7 (11–12), 1664–1673.

Kang, S.H., Cho, H., Cheong, H., Ryu, C.M., Kim, J.F., Park, S.H., 2007. Two bacterial entophytes eliciting both plant growth promotion and plant defense on pepper (*Capsicum annuum* L.). J. Microbiol. Biotechnol 17 (1), 96.

Kaplan, D.T., Opperman, C.H., 2000. Reproductive strategies and karyotype of the burrowing nematode, *Radopholus similis*. J. Nematol. 32 (2), 126.

Karban, R., 2010. Neighbors affect resistance to herbivory–a new mechanism. New Phytol. 186 (3), 564–566.

Karban, R., 2011. The ecology and evolution of induced resistance against herbivores. Funct. Ecol. 25 (2), 339–347.

Kerry, B.R., 2000. Rhizosphere interactions and the exploitation of microbial agents for the biological control of plant-parasitic nematodes. Annu. Rev. Phytopathol. 38 (1), 423–441.

Kessler, R.C., Andrews, G., Colpe, L.J., Hiripi, E., Mroczek, D.K., Normand, S.L., Zaslavsky, A.M., 2002. Short screening scales to monitor population prevalences and trends in non-specific psychological distress. Psychol. Med. 32 (06), 959–976.

Koenning, S.R., Overstreet, C., Noling, J.W., Donald, P.A., Becker, J.O., Fortnum, B.A., 1999. Survey of crop losses in response to phytoparasitic nematodes in the United States for 1994. J. Nematol. 31 (4S), 587.

Kunkel, B.N., Brooks, D.M., 2002. Cross talk between signaling pathways in pathogen defense. Curr. Opin. Plant Biol. 5 (4), 325–331.

Kumar, S., Bhanjana, G., Sharma, A., Sidhu, M.C., Dilbaghi, N., 2014. Synthesis, characterization and on field evaluation of pesticide loaded sodium alginate nanoparticles. Carbohydr. Polym 101, 1061–1067.

Lajter, I., Kusz, N., Hohmann, J., Vasas, A., 2015. Isolation, characterisation and chemotaxonomic significance of secondary metabolites from *Polygonum persicaria* L. Planta Medica 81 (16), 118.

Lamovsek, J., Urek, G., Trdan, S., 2013. Biological control of root-knot nematodes (*Meloidogyne* spp.): microbes against the pests/BIOTIcNO ZATIRANJE OGORcIC KORENINSKIH SISK (*Meloidogyne* spp.): MIKROORGANIZMI PROTI SKODLJIVCEM. Acta Agric Slov. 101 (2), 263.

Lannoo, N., Vandenborre, G., Miersch, O., Smagghe, G., Wasternack, C., Peumans, W.J., Van Damme, E.J., 2007. The jasmonate-induced expression of the *Nicotiana tabacum* leaf lectin. Plant Cell Physiol. 48 (8), 1207–1218.

Lappartient, A.G., Vidmar, J.J., Leustek, T., Glass, A.D., Touraine, B., 1999. Inter-organ signaling in plants: regulation of ATP sulfurylase and sulfate transporter genes expression in roots mediated by phloem-translocated compound. Plant J. 18 (1), 89–95.

Lattanzio, V., Lattanzio, V.M., Cardinali, A., 2006. Role of phenolics in the resistance mechanisms of plants against fungal pathogens and insects. Phytochem. Adv. Res. 661, 23–67.

Lawrence, P.K., Koundal, K.R., 2002. Plant protease inhibitors in control of phytophagous insects. Electron J. Biotechnol. 5 (1), 5–6.

Lee, C.W., Efetova, M., Engelmann, J.C., Kramell, R., Wasternack, C., Ludwig-Müller, J., Deeken, R., 2009. *Agrobacterium tumefaciens* promotes tumor induction by modulating pathogen defense in *Arabidopsis thaliana*. Plant Cell 21 (9), 2948–2962.

Leustek, T., Saito, K., 1999. Sulfate transport and assimilation in plants. Plant Physiol. 120 (3), 637–644.

Li, J., Brader, G., Palva, E.T., 2004. The WRKY70 transcription factor: a node of convergence for jasmonate-mediated and salicylate-mediated signals in plant defense. Plant Cell 16 (2), 319–331.

Li, X.T., Ju, R.J., Li, X.Y., Zeng, F., Shi, J.F., Liu, L., Lu, W.L., 2014. Multifunctional targeting daunorubicin plus quinacrine liposomes, modified by wheat germ agglutinin and tamoxifen, for treating brain glioma and glioma stem cells. Oncotarget 5 (15), 6497.

Ludwig, A.A., Romeis, T., Jones, J.D., 2004. CDPK-mediated signalling pathways: specificity and cross-talk. J. Exp. Bot. 55 (395), 181–188.

Macedo, M.L.R., Freire, M.D.G.M., da Silva, M.B.R., Coelho, L.C.B.B., 2007. Insecticidal action of *Bauhinia monandra* leaf lectin (BmoLL) against *Anagasta kuehniella, Zabrotes subfasciatus* and *Callosobruchus maculatus*. Comp. Biochem. Physiol. Part A Mol. Integr. Physiol. 146 (4), 486–498.

Maffei, M.E., 2010. Sites of synthesis, biochemistry and functional role of plant volatiles. South Afr. J. Bot. 76 (4), 612–631.

Magalhaes, J.A.S., De-Moraes-Neto, A.H.A., Miguens, F.C., 2008. Nematodes of *Rhynchophorus palmarum*, L, vector of the Red Ring disease in coconut plantations from the north of the Rio de Janeiro State. Parasitol. Res. 102 (6), 1281–1287.

Majumder, P., Mondal, H.A., Das, S., 2005. Insecticidal activity of *Arum maculatum* tuber lectin and its binding to the glycosylated insect gut receptors. J. Agric. Food Chem. 53 (17), 6725–6729.

Mang, H.G., Laluk, K.A., Parsons, E.P., Kosma, D.K., Cooper, B.R., Park, H.C., Chilosi, G., 2009. The Arabidopsis RESURRECTION1 gene regulates a novel antagonistic interaction in plant defense to biotrophs and necrotrophs. Plant Physiol. 151 (1), 290–305.

Mann, R.S., Ali, J.G., Hermann, S.L., Tiwari, S., Pelz-Stelinski, K.S., Alborn, H.T., Stelinski, L.L., 2012. Induced release of a plant-defense volatile 'deceptively' attracts insect vectors to plants infected with a bacterial pathogen. PLoS Pathog. 8 (3), 111–124.

Mansfield, J., Genin, S., Magori, S., Citovsky, V., Sriariyanum, M., Ronald, P., Toth, I.A.N., 2012. Top 10 plant pathogenic bacteria in molecular plant pathology. Mol. Plant Pathol. 13 (6), 614–629.

Matthews, B.F., Beard, H., MacDonald, M.H., Kabir, S., Youssef, R.M., Hosseini, P., Brewer, E., 2013. Engineered resistance and hypersusceptibility through functional metabolic studies of 100 genes in soybean to its major pathogen, the soybean cyst nematode. Planta 237 (5), 1337.

McDonald, B.A., Linde, C., 2002. The population genetics of plant pathogens and breeding strategies for durable resistance. Euphytica 124 (2), 163–180.

Mehlhorn, H., Al-Quraishy, S., Al-Rasheid, K.A., Jatzlau, A., Abdel-Ghaffar, F., 2011. Addition of a combination of onion (*Allium cepa*) and coconut (*Cocos nucifera*) to food of sheep stops gastrointestinal helminthic infections. Parasitol. Res. 108 (4), 1041–1046.

Mehta, R.G., Liu, J., Constantinou, A., Thomas, C.F., Hawthorne, M., You, M., Moriarty, R.M., 1995. Cancer chemopreventive activity of brassinin, a phytoalexin from cabbage. Carcinogenesis 16 (2), 399–404.

Mellway, R.D., Tran, L.T., Prouse, M.B., Campbell, M.M., Constabel, C.P., 2009. The wound-, pathogen and ultraviolet B-responsive MYB134 gene encodes an R2R3 MYB transcription factor that regulates proanthocyanidin synthesis in poplar. Plant Physiol. 150 (2), 924–941.

Miedes, E., Vanholme, R., Boerjan, W., Molina, A., 2015. The role of the secondary cell wall in plant resistance to pathogens. Plant Cell Wall Pathog. Parasit. Symb. 78, 213–221.

Miranda, M., Ralph, S.G., Mellway, R., White, R., Heath, M.C., Bohlmann, J., Constabel, C.P., 2007. The transcriptional response of hybrid poplar (*Populus trichocarpa* × *P. deltoids*) to infection by *Melampsora medusae* leaf rust involves induction of flavonoid pathway genes leading to the accumulation of proanthocyanidins. Mol. Plant Microbe Interact. 20 (7), 816–831.

Mithofer, A., Wanner, G., Boland, W., 2005. Effects of feeding *Spodoptera littoralis* on lima bean leaves. II. Continuous mechanical wounding resembling insect feeding is sufficient to elicit herbivory-related volatile emission. Plant Physiol. 137 (3), 1160–1168.

Monde, K., TaniguchiT, M.N., Nishimura, S.I., Harada, N., Dukor, R.K., Nafie, L.A., 2003. Preparation of cruciferous phytoalexin related metabolites, (−)-dioxibrassinin and (−)-3-cyanomethyl-3-hydroxyoxindole, and determination of their absolute configurations by vibrational circular dichroism (VCD). Tetrahedron Lett. 44 (32), 6017–6020.

Morant, A.V., Jørgensen, K., Jørgensen, C., Paquette, S.M., Sánchez-Pérez, R., Møller, B.L., Bak, S., 2008. β-Glucosidases as detonators of plant chemical defense. Phytochemistry 69 (9), 1795–1813.

Mordue, A.J., Blackwell, A., 1993. Azadirachtin: an update. J. Insect Physiol. 39 (11), 903–924.

Nagy, P.D., 2008. Recombination in plant RNA viruses. In: Plant Virus Evolution. vol. 26. Springer, Berlin Heidelberg, pp. 133–156.

Nicol, J.M., Rivoal, R., 2008. Global knowledge and its application for the integrated control and management of nematodes on wheat. Integrated Management and Biocontrol of Vegetable and Grain Crops Nematodes vol. 2, 251–294.

Nicol, J.M., Turner, S.J., Coyne, D.L., Den-Nijs, L., Hockland, S., Maafi, Z.T., 2011. Current nematode threats to world agriculture. In: Genomics and Molecular Genetics of Plant-nematode Interactions. vol. 45. Springer, Netherlands, pp. 21–43.

Nicolaou, C., Kang, Q., Ng, S.Y., Chen, D.Y.K., 2010. A comparative view of metabolite and substrate stress and tolerance in microbial bioprocessing: from biofuels and chemicals, to biocatalysis and bioremediation. Metabol. Eng. 5 (1), 6–15.

Nocito, F.F., Pirovano, L., Cocucci, M., Sacchi, G.A., 2002. Cadmium-induced sulfate uptake in maize roots. Plant Physiol. 129 (4), 1872–1879.

Noe, J.P., 2004. Plant-parasitic nematodes. In: Trigiano, R.N. (Ed.), Plant Pathology: Concepts and Laboratory Exercises. CRC Press, Taylor & Francis.

Nuessly, G.S., Scully, B.T., Hentz, M.G., Beiriger, R., Snook, M.E., Widstrom, N.W., 2007. Resistance to *Spodoptera frugiperda* and *Euxesta stigmatias* in sweet corn derived from exogenous and endogenous genetic systems. J. Econ. Entomol. 100 (6), 1887–1895.

Ohno, Y., Naganuma, T., Ogawa, T., Muramoto, K., 2006. Effect of lectins on the transport of food factors in Caco-2 cell monolayers. J. Agric. Food Chem. 54 (2), 548–553.

Omirou, M., Karpouzas, D.G., Papadopoulou, K.K., Ehaliotis, C., 2013. Dissipation of pure and broccoli-released glucosinolates in soil under high and low moisture content. Eur. J. Soil Biol. 56, 49–55.

Ouzounidou, G., Ilias, I., 2005. Hormone-induced protection of sunflower photosynthetic apparatus against copper toxicity. Biol. Plant 49 (2), 223.

Pande, S., Siddique, K.H.M., Kishore, G.K., Bayaa, B., Gaur, P.M., Gowda, C.L.L., Crouch, J.H., 2005. Ascochyta blight of chickpea (*Cicer arietinum* L.): a review of biology, pathogenicity, and disease management. Crop Past Sci 56 (4), 317–332.

Pauwels, L., Inzé, D., Goossens, A., 2009. Jasmonate-inducible gene: what does it mean? Trends Plant Sci. 14 (2), 87–91.

Petutschnig, E.K., Jones, A.M., Serazetdinova, L., Lipka, U., Lipka, V., 2010. The lysin motif receptor-like kinase (LysM-RLK) CERK1 is a major chitin-binding protein in *Arabidopsis thaliana* and subject to chitin-induced phosphorylation. J. Biol. Chem. 285 (37), 28902–28911.

Pieterse, C.M., Leon-Reyes, A., Van der Ent, S., Van-Wees, S.C., 2009. Networking by small-molecule hormones in plant immunity. Nat. Chem. Biol. 5 (5), 308–316.

Pineda, M., Sajnani, C., Barón, M., 2010. Changes induced by the Pepper mild mottle tobamovirus on the chloroplast proteome of *Nicotiana benthamiana*. Photosynth Res. 103 (1), 31–42.

Pitzschke, A., Schikora, A., Hirt, H., 2009. MAPK cascade signalling networks in plant defence. Curr. Opin. Plant Biol. 12 (4), 421–426.

Polley, H.W., Johnson, H.B., Marinoj, B.D., Mayeux, H.S., 1993. Increase in C3 plant water-use efficiency and biomass over Glacial to present C0. Nature 361, 7.

Posmyk, M.M., Kontek, R., Janas, K.M., 2009. Antioxidant enzymes activity and phenolic compounds content in red cabbage seedlings exposed to copper stress. Ecotoxicol. Environ. Saf. 72 (2), 596–602.

Povero, G., Loreti, E., Pucciariello, C., Santaniello, A., Di Tommaso, D., Di Tommaso, G., Perata, P., 2011. Transcript profiling of chitosan-treated *Arabidopsis* seedlings. J. Plant Res. 124 (5), 619–629.

Prins, M., Laimer, M., Noris, E., Schubert, J., Wassenegger, M., Tepfer, M., 2008. Strategies for antiviral resistance in transgenic plants. Mol. Plant Pathol. 9 (1), 73–83.

Purcell, A.H., Hopkins, D.L., 1996. Fastidious xylem-limited bacterial plant pathogens. Annu. Rev. Phytopathol. 34 (1), 131–151.

Qasem, J.R., 1992. Nutrient accumulation by weeds and their associated vegetable crops. J. Hortic. Sci. 67 (2), 189–195.

Ramamoorthy, V., Viswanathan, R., Raguchander, T., Prakasam, V., Samiyappan, R., 2001. Induction of systemic resistance by plant growth promoting rhizobacteria in crop plants against pests and diseases. Crop Prot. 20 (1), 1–11.

Ramegowda, V., Senthil-Kumar, M., 2015. The interactive effects of simultaneous biotic and abiotic stresses on plants: mechanistic understanding from drought and pathogen combination. J. Plant Physiol. 176, 47–54.

Ramesh, K., Rao, A.N., Chauhan, B.S., 2017. Role of crop competition in managing weeds in rice, wheat, and maize in India: a review. Crop Prot. 95, 14–21.

Rasmann, S., Köllner, T.G., Degenhardt, J., Hiltpold, I., Toepfer, S., Kuhlmann, U., Turlings, T.C., 2005. Recruitment of entomopathogenic nematodes by insect-damaged maize roots. Nature 434 (7034), 732–737.

Reddy, D.V.R., Sudarshana, M.R., Fuchs, M., Rao, N.C., Thottappilly, G., 2009. Genetically engineered virus-resistant plants in developing countries: status and future prospects. Adv. Virus Res. 75, 185–220.

Reddy, S.K., Weng, Y., Rudd, J.C., Akhunova, A., Liu, S., 2013. Transcriptomics of induced defense responses to greenbug aphid feeding in near isogenic wheat lines. Plant Sci 212, 26–36.

Reynolds, L.B., Potter, J.W., Ball-Coelho, B.R., 2000. Crop Rotation with sp. is an alternative to chemical fumigation for control of root-lesion nematodes. Agron. J. 92 (5), 957–966.

Ribot, C., Hirsch, J., Balzergue, S., Tharreau, D., Nottéghem, J.L., Lebrun, M.H., Morel, J.B., 2008. Susceptibility of rice to the blast fungus, Magnaporthe grisea. J. Plant Physiol. 165 (1), 114–124

Ro, D.K., Paradise, E.M., Ouellet, M., Fisher, K.J., Newman, K.L., Ndungu, J.M., Chang, M.C., 2006. Production of the antimalarial drug precursor artemisinic acid in engineered yeast. Nature 440 (7086), 940–943.

Rodríguez-Moreno, L., Pineda, M., Soukupová, J., Macho, A.P., Beuzón, C.R., Barón, M., Ramos, C., 2008. Early detection of bean infection by *Pseudomonas syringae* in asymptomatic leaf areas using chlorophyll fluorescence imaging. Photosynth Res. 96 (1), 27–35.

Rossi, M., Goggin, F.L., Milligan, S.B., Kaloshian, I., Ullman, D.E., Williamson, V.M., 1998. The nematode resistance gene Mi of tomato confers resistance against the potato aphid. Proc. Natl. Acad. Sci. 95 (17), 9750–9754.

Roze, L.V., Chanda, A., Linz, J.E., 2011. Compartmentalization and molecular traffic in secondary metabolism: a new understanding of established cellular processes. Fungal Genet. Biol. 48 (1), 35–48.

Rumbou, A., von Bargen, S., Büttner, C., 2009. A model system for plant-virus interaction—infectivity and seed transmission of Cherry leaf roll virus (CLRV) in *Arabidopsis thaliana*. Eur. J. Plant Pathol. 124 (3), 527–532.

Ruther, J., Kleier, S., 2005. Plant–plant signaling: ethylene synergizes volatile emission in *Zea mays* induced by exposure to (Z)-3-hexen-1-ol. J. Chem. Ecol. 31 (9), 2217–2222.

Safdar, A., Javed, N., Khan, S.A., Safdar, H., Haq, I.U., Abbas, H., Ullah, Z., 2013. Synergistic effect of a fungus, Fusarium semitectum, and a nematode, *Tylenchulus semipenetrans*, on citrus decline. Pak. J. Zool. 45 (3), 643–651.

Saha, S., Enugutti, B., Rajakumari, S., Rajasekharan, R., 2006. Cytosolic triacylglycerol biosynthetic pathway in oilseeds. Molecular cloning and expression of peanut cytosolic diacylglycerol acyltransferase. Plant Physiol. 141 (4), 1533–1543.

Sajjadi, S.E., Shokoohinia, Y., Mehramiri, P., 2013. Isolation and characterization of steroids, phthalide and essential oil of the fruits of *Kelussia odoratissima* Mozaff., an endemic mountain celery. Res. Pharma Sci 8 (1), 35.

Sardana, V., Mahajan, G., Jabran, K., Chauhan, B.S., 2017. Role of competition in managing weeds: an introduction to the special issue. Crop Prot. 95, 1–7.

Sarria, S., Wong, B., Martín, H.G., Keasling, J.D., Peralta-Yahya, P., 2014. Microbial synthesis of pinene. ACS Synth. Biol. 3 (7), 466–475.

Sato, K., Naito, M., Yukitake, H., Hirakawa, H., Shoji, M., McBride, M.J., Nakayama, K., 2010. A protein secretion system linked to bacteroidete gliding motility and pathogenesis. Proc. Natl. Acad. Sci. 107 (1), 276–281.

Saxena, D., Stotzky, G., 2001. *Bacillus thuringiensis* (Bt) toxin released from root exudates and biomass of Bt corn has no apparent effect on earthworms, nematodes, protozoa, bacteria, and fungi in soil. Soil Biol. Biochem. 33 (9), 1225–1230.

Scalbert, A., Mila, I., Expert, D., Marmolle, F., Albrecht, A.M., Hurrell, R., Tomé, D., 1999. Polyphenols, metal ion complexation and biological consequences. Plant Polyphenols 2, 545–554.

Schulze-Lefert, P., Vogel, J., 2000. Closing the ranks to attack by powdery mildew. Trends Plant Sci. 5 (8), 343–348.

Scurrah, M.I., Niere, B., Bridge, J., 2005. Nematode parasites of Solanum and sweet potatoes. Plant parasitic nematodes in subtropical and tropical agriculture. CAB Int. 97, 193–219 Wallingford.

Serghini, K., de Luque, A.P., Castejón-Muñoz, M., García-Torres, L., Jorrín, J.V., 2001. Sunflower (*Helianthus annuus* L.) response to broomrape (*Orobanche cernua* Loefl.) parasitism: induced synthesis and excretion of 7-hydroxylated simple coumarins. J. Exp. Bot. 52 (364), 2227–2234.

Sethi, A., McAuslane, H.J., Rathinasabapathi, B., Nuessly, G.S., Nagata, R.T., 2009. Enzyme induction as a possible mechanism for latex-mediated insect resistance in romaine lettuce. J. Chem. Ecol. 35 (2), 190–200.

Sharma, H.C., Sujana, G., Rao, D.M., 2009. Morphological and chemical components of resistance to pod borer, *Helicoverpa armigera* in wild relatives of pigeonpea. Arthropod Plant Interact. 3 (3), 151–161.

Sharma, P., Jha, A.B., Dubey, R.S., Pessarakli, M., 2012. Reactive oxygen species, oxidative damage, and antioxidative defense mechanism in plants under stressful conditions. J. Bot. 123, 211–222.

Sharma, S.K., Bryan, G.J., Winfield, M.O., Millam, S., 2007. Stability of potato (*Solanum tuberosum* L.) plants regenerated via somatic embryos, axillary bud proliferated shoots, microtubers and true potato seeds: a comparative phenotypic, cytogenetic and molecular assessment. Planta 226 (6), 1449–1458.

Sharma, S.S., Dietz, K.J., 2009. The relationship between metal toxicity and cellular redox imbalance. Trends Plant Sci. 14 (1), 43–50.

Shivaji, R., Camas, A., Ankala, A., Engelberth, J., Tumlinson, J.H., Williams, W.P., Luthe, D.S., 2010. Plants on constant alert: elevated levels of jasmonic acid and jasmonate-induced transcripts in caterpillar-resistant maize. J. Chem. Ecol. 36 (2), 179–191.

Silvestre, A.J., Gandini, A., 2008. Terpenes: major sources, properties and applications. In: Monomers. Polymers and Composites from Renewable Resources vol. 1, 17–38.

Simmonds, M.S., Stevenson, P.C., 2001. Effects of isoflavonoids from Cicer on larvae of *Heliocoverpa armigera*. J. Chem. Ecol. 27 (5), 965–977.

Soylu, S., Bennett, M.H., Mansfield, J.W., 2002. Induction of phytoalexin accumulation in broad bean (*Vicia faba* L.) cotyledons following treatments with biotic and abiotic elicitors. Turk. J. Agric. For 26 (6), 343–348.

Sreevidya, V.S., Rao, C.S., Sullia, S.B., Ladha, J.K., Reddy, P.M., 2006. Metabolic engineering of rice with soybean isoflavone synthase for promoting nodulation gene expression in rhizobia. J. Exp. Bot. 57 (9), 1957–1969.

Stoger, E., Williams, S., Christou, P., Down, R.E., Gatehouse, J.A., 1999. Expression of the insecticidal lectin from snowdrop (*Galanthus nivalis* agglutinin; GNA) in transgenic wheat plants: effects on predation by the grain aphid *Sitobion avenae*. Mol. Breed. 5 (1), 65–73.

Swięczkowska, E., Kowalkowska, A.K., 2015. Floral nectary anatomy and ultrastructure in mycoheterotrophic plant, *Epipogium aphyllum* Sw. (Orchidaceae). Sci. World J. 2015.

Taft, A.S., Ozawa, M., Fitch, A., Depasse, J.V., Halfmann, P.J., Hill-Batorski, L., Ghedin, E., et al., 2015. Identification of mammalian-adapting mutations in the polymerase complex of an avian H5N1 influenza virus. Nat. Commun 6.

Talbert, R.E., Burgos, N.R., 2007. History and management of herbicide-resistant barnyardgrass (*Echinochloa crusgalli*) in Arkansas rice. Weed Technol. 21 (2), 324–331.

Taliansky, M., Torrance, L., Kalinina, N.O., 2008. Role of plant virus movement proteins. Methods Mol. Biol. 451, 33–54.

Tarkowski, P., Vereecke, D., 2014. Threats and opportunities of plant pathogenic bacteria. Biotechnol. Adv. 32 (1), 215–229.

Teferedegne, B., 2000. New perspectives on the use of tropical plants to improve ruminant nutrition. Proc. Nutr. Soc. 59 (02), 209–214.

Tettelin, H., Nelson, K.E., Paulsen, I.T., Eisen, J.A., 2001. Complete genome sequence of a virulent isolate of *Streptococcus pneumoniae*. Science 293 (5529), 498.

Thomma, B.P., Cammue, B.P., Thevissen, K., 2002. Plant defensins. Planta 216 (2), 193–202.

Thomma, B.P., Tierens, K.F., Penninckx, I.A., Mauch-Mani, B., Broekaert, W.F., Cammue, B.P., 2001. Different micro-organisms differentially induce *Arabidopsis* disease response pathways. Plant Physiol. Biochem. 39 (7), 673–680.

Thorne, E.T., Young, B.M., Young, G.M., Stevenson, J.F., Labavitch, J.M., Matthews, M.A., Rost, T.L., 2006. The structure of xylem vessels in grapevine (Vitaceae) and a possible passive mechanism for the systemic spread of bacterial disease. Am. J. Bot. 93 (4), 497–504.

Treutter, D., 2006. Significance of flavonoids in plant resistance: a review. Environ. Chem. Lett. 4 (3), 147–157.

Tscharntke, T., Thiessen, S., Dolch, R., Boland, W., 2001. Herbivory, induced resistance, and interplant signal transfer in *Alnus glutinosa*. Biochem. Syst. Ecol. 29 (10), 1025–1047.

Ulland, S., Ian, E., Mozuraitis, R., Borg-Karlson, A.K., Meadow, R., Mustaparta, H., 2008. Methyl salicylate, identified as primary odorant of a specific receptor neuron type, inhibits oviposition by the moth *Mamestra brassicae* L. Chem. Senses 33 (1), 35–46.

van Loon, L.C., Rep, M., Pieterse, C.M., 2006. Significance of inducible defense-related proteins in infected plants. Annu. Rev. Phytopathol. 44, 135–162.

Van-Dam, N.M., Bouwmeester, H.J., 2016. Metabolomics in the rhizosphere: tapping into belowground chemical communication. Trends Plant Sci. 21 (3), 256–265.

Van-de-Mortel, J.E., Schat, H., Moerland, P.D., Van-Themaat, E.V.L., Van-Der-Ent, S.J., Blankestijn, H., Aarts, M.G., 2008. Expression differences for genes involved in lignin, glutathione and sulphate metabolism in response to cadmium in *Arabidopsis thaliana* and the related Zn/Cd-hyperaccumulator *Thlaspi caerulescens*. Plant Cell Environ. 31 (3), 301–324.

Vandenborre, G., Smagghe, G., Van Damme, E.J., 2011. Plant lectins as defense proteins against phytophagous insects. Phytochemistry 72 (13), 1538–1550.

Van-der-Ent, S., Van Wees, S.C., Pieterse, C.M., 2009. Jasmonate signaling in plant interactions with resistance-inducing beneficial microbes. Phytochemistry 70 (13), 1581–1588.

Van-Doorn, W.G., Beers, E.P., Dangl, J.L., Franklin-Tong, V.E., Gallois, P., Hara-Nishimura, I., Mur, L.A.J., 2011. Morphological classification of plant cell deaths. Cell Death Differ. 18 (8), 1241–1246.

Verhage, A., van Wees, S.C., Pieterse, C.M., 2010. Plant immunity: it's the hormones talking, but what do they say? Plant Physiol. 154 (2), 536–540.

Vidhyasekaran, P., 2007. Fungal Pathogenesis in Plants and Crops: Molecular Biology and Host Defense Mechanisms. CRC Press.
Vovlas, N., Troccoli, A., Palomares-Rius, J.E., De Luca, F., Liébanas, G., Landa, B.B., Castillo, P., 2011. *Ditylenchus gigas* sp. parasitizing broad bean: a new stem nematode singled out from the *Ditylenchus dipsaci* species complex using a polyphasic approach with molecular phylogeny. Plant Pathol. 60 (4), 762–775.
Vuong, T.D., Sonah, H., Meinhardt, C.G., Deshmukh, R., Kadam, S., Nelson, R.L., Nguyen, H.T., 2015. Genetic architecture of cyst nematode resistance revealed by genome-wide association study in soybean. BMC Genom. 16 (1), 593.
Walker, T.S., Bais, H.P., Grotewold, E., Vivanco, J.M., 2003. Root exudation and rhizosphere biology. Plant Physiol. 132 (1), 44–51.
Waller, F., Achatz, B., Baltruschat, H., Fodor, J., Becker, K., Fischer, M., Franken, P., 2005. The endophytic fungus *Piriformospora indica* reprograms barley to salt-stress tolerance, disease resistance, and higher yield. Proc. Natl. Acad. Sci. U. S. A. 102 (38), 13386–13391.
Walling, L.L., 2008. Avoiding effective defenses: strategies employed by phloem-feeding insects. Plant Physiol. 146 (3), 859–866.
Walters, D., 2011. Plant Defense: Warding Off Attack by Pathogens, Herbivores and Parasitic Plants. John Wiley & Sons.
Walter, M.G., Wamser, C.C., Ruwitch, J., Zhao, Y., Braden, D., Stevens, M., Pessiki, P.J., 2007. Syntheses and optoelectronic properties of amino/carboxy-phenylporphyrins for potential use in dye-sensitized TiO_2 solar cells. J. Porphyr. Phthalocyanines 11 (08), 601–612.
Wang, L., Allmann, S., Wu, J., Baldwin, I.T., 2008. Comparisons of lipoxygenase3-and jasmonate-resistant4/6-silenced plants reveal that jasmonic acid and jasmonic acid-amino acid conjugates play different roles in herbivore resistance of *Nicotiana attenuata*. Plant Physiol. 146 (3), 904–915.
War, A.R., Paulraj, M.G., Ahmad, T., Buhroo, A.A., Hussain, B., Ignacimuthu, S., Sharma, H.C., 2012. Mechanisms of plant defense against insect herbivores. Plant Signal Behav. 7 (10), 1306–1320.
War, A.R., Sharma, H.C., Paulraj, M.G., War, M.Y., Ignacimuthu, S., 2011. Herbivore induced plant volatiles: their role in plant defense for pest management. Plant Signal Behav. 6 (12), 1973–1978.
Win, P.P., Kyi, P.P., Maung, Z.T.Z., Myint, Y.Y., Cabasan, M.T.N., Waele, D., 2016. Host status of rotation crops in Asian rice-based cropping systems to the rice root-knot nematode *Meloidogyne graminicola*. Trop. Plant Pathol. 5 (41), 312–319.
Wink, M., Mohamed, G.I., 2003. Evolution of chemical defense traits in the Leguminosae: mapping of distribution patterns of secondary metabolites on a molecular phylogeny inferred from nucleotide sequences of the rbcL gene. Biochem. Syst. Ecol. 31 (8), 897–917.
Wortmann, C.S., Kaizzi, C.K., 1998. Nutrient balances and expected effects of alternative practices in farming systems of Uganda. Agric. Ecosyst. Environ. 71 (1), 115–129.
Wu, Q., Luo, Y., Lu, R., Lau, N., Lai, E.C., Li, W.X., Ding, S.W., 2010. Virus discovery by deep sequencing and assembly of virus-derived small silencing RNAs. Proc. Natl. Acad. Sci. 107 (4), 1606–1611.
Wuyts, N., Swennen, R., De Waele, D., 2006. Effects of plant phenylpropanoid pathway products and selected terpenoids and alkaloids on the behaviour of the plant-parasitic nematodes *Radopholus similis, Pratylenchus penetrans* and *Meloidogyne incognita*. Nematology 8 (1), 89–101.
Yao, L., Jiang, Y., Singanusong, R., D'Arcy, B., Datta, N., Caffin, N., Raymont, K., 2004. Flavonoids in Australian *melaleuca, guioa, lophostemon, banksia* and *helianthus* honeys and their potential for floral authentication. Food Res. Int. 37 (2), 166–174.
Zhang, H., Chen, L., Nieto, J.J., 2008. A delayed epidemic model with stage-structure and pulses for pest management strategy. Nonlinear Anal. Real World Appl. 9 (4), 1714–1726.
Zhao, Y., Wei, W., Lee, M., Shao, J., Suo, X., Davis, R.E., 2009. Construction of an interactive online phytoplasma classification tool, iPhyClassifier, and its application in analysis of the peach X-disease phytoplasma group (16SrIII). Int. J. Syst. Evol. Microbiol. 59 (10), 2582–2593.
Zhu-Salzman, K., Luthe, D.S., Felton, G.W., 2008. Arthropod-inducible proteins: broad spectrum defenses against multiple herbivores. Plant Physiol. 146 (3), 852–858.
Zientz, E., Dandekar, T., Gross, R., 2004. Metabolic interdependence of obligate intracellular bacteria and their insect hosts. Microbiol. Mol. Biol. Rev. 68 (4), 745–770.

Chapter 5

Metabolomics-Guided Elucidation of Abiotic Stress Tolerance Mechanisms in Plants

Asish K. Parida[1,2], Ashok Panda[1,2], Jaykumar Rangani[1,2]

[1]*Division of Plant Omics, CSIR-Central Salt and Marine Chemicals Research Institute (CSIR-CSMCRI), Bhavnagar, India;* [2]*Academy of Scientific and Innovative Research, Council of Scientific and Industrial Research (CSIR), CSIR-Central Salt and Marine Chemicals Research Institute (CSIR-CSMCRI), Bhavnagar, India*

1. INTRODUCTION

Metabolomics emerges as the most advanced "omics" study to characterize the metabolic profile of living organisms (Dubery et al., 2013). As a final downstream product of genome, the metabolome is composed of a large number of small-sized molecules (molecular mass less than 2000 Da) exhibiting vast diversity in chemical composition and structure. Metabolomics is the detection, identification, and quantification of all the metabolites (or subset) in biological samples (Kumar et al., 2016). Metabolomic studies are not only confined to the endogenous metabolites but also can ascertain the metabolites from exogenous sources such as drugs and diets (Kosmides et al., 2013). Hence metabolomics encompasses a broad array of small molecules or metabolic intermediates such as amino acids, peptides, lipids, nucleic acids, carbohydrates, organic acids, ketones, aldehydes, amines, steroids, and vitamins, hormones, signaling molecules, and secondary metabolites like polyphenols and flavonoids (Collino et al., 2013). In a wide-ranging term we can say that metabolomics is an all-inclusive qualitative and quantitative exploration of the metabolites present within the living organisms under certain environmental circumstances (Freund and Hegeman, 2017). Compared with transcriptomics and proteomics, metabolomics reproduces more thorough information about the biological regulatory mechanisms (Osorio et al., 2012; Dos Santos et al., 2017). Various metabolomic techniques such as nuclear magnetic resonance (NMR) spectroscopy and mass spectrometry (MS) and their advancements have immense contributions to the study of metabolite profiling of the plants for the elucidation of their stress tolerance mechanisms. During the last 15 years, different aspects of metabolomics have been implemented in studying the plants and their interaction with the environment. Metabolic responses of the plants to various stresses are currently gaining importance in mitigating the losses in agriculture due to imposed stresses. Several reports have been documented regarding targeted research studying the metabolites at the level of a single metabolic pathway. These studies set the foundation for the broader research that has been carried out during last 20 years.

Metabolomics is gaining importance in the field of plant science for quality improvement of crop plants, for discovery of vital natural products, and most importantly, to assess the impact of various abiotic stresses on plant's growth and physiology (Genga et al., 2011). Metabolomics is a fast-growing technology that helps in the phenotyping of plants, which can be helpful for increasing the nutritional value of the crop plants (Alla et al., 2012). Compared with animals, plants contains remarkably diverse metabolites that make plant metabolomics a potent tool, implemented in diverse areas such as fingerprinting of geno- and ecotypes, comparison between mutants with their wild types, activity determination of bioactive compounds of medicinal plants, and the environmental effects on the physiology of plants (Wolfender et al., 2013). Plants produce an array of metabolites that are either essential for regular metabolic processes such as respiration and photosynthesis (primary metabolites) or that are needed for survival in a certain environment (secondary metabolites). Secondary metabolites are not essential for growth, development, and reproduction of the plants but are indispensable for the survival and fecundity of plants especially for signaling and defense (Wolfender et al., 2013). The metabolite contents of plants are so numerous that the total number of metabolites in plants is found to be 2,00,000 among which 7000–15,000 are found in each individual species and around 3000–5000 are present in the leaves (Kim et al., 2010). This enormous discovery, identification, and documentation of metabolites aids in obtaining imperative and comprehensive knowledge about

Plant Metabolites and Regulation Under Environmental Stress. http://dx.doi.org/10.1016/B978-0-12-812689-9.00005-4

plant signaling and various regulatory pathways (Carreno-Quintero et al., 2013). The plant metabolomics has upgraded to a promising concept since last 20 years and intelligently implemented in the field of plant biotechnology and physiology (Cusido et al., 2014). Metabolomics approaches such as MS combined with gas chromatography (GC) or liquid chromatography(LC) (GC–MS and LC-MS), capillary electrophoresis-mass spectrometry (CE-MS), direct injection mass spectrometry (DIMS), high-performance liquid chromatography (HPLC) with photodiode array detection, thin layer chromatography with ultraviolet (UV) detection, and NMR are successfully employed in the field of plant science (Wolfender et al., 2013). Among these techniques, GC-MS, LC-MS, DIMS, and NMR are frequently used for the study of plant metabolomics. MS is more advantageous over NMR because of its high sensitivity and resolution. In MS the analytes are first ionized and then transported through the mass analyzers. There are a number of ionization techniques such as electron impact ionization (EI), fast atom bombardment, electrospray ionization (ESI), atmospheric pressure chemical ionization, and matrix-assisted laser desorption/ionization (MALDI). To study the diverse metabolites in plants, two types of metabolomic approaches can be taken, namely, targeted and nontargeted metabolomics. Targeted metabolomics is considered as the most promising tool that measures a subset of chemically characterized and interpreted metabolites (Li et al., 2014). However, the successful implementation of targeted metabolomics is hindered by unavailability and/or extremely high cost of isotopically labeled standards (Li et al., 2014). Identification of thousands of metabolites that are detected in untargeted metabolomics is difficult because of the presence of different isomeric forms of the same metabolite that cannot be differentiated solely on the basis of their *m/z* ratio. The number and concentration of metabolites that are usually synthesized in plants are significantly enhanced during stress encounter. The up- and down-regulations of metabolites play important roles in getting an insight into the stress responses of plants. Many reviews have been documented since the years 2005–16 explaining the alternations of different metabolites in plants under abiotic stress. However, the reviews comprehended the metabolite response of the plants in response to specific abiotic stress like salinity (Chandna et al., 2013), drought (Seki et al., 2007), heat (Bokszczanin and Fragkostefanakis, 2013), or heavy metal stress (Sytar et al., 2013). In this review, we have encompassed coordinated and integrated aspects of different metabolites and their useful roles in stress tolerance.

2. ADVANCES IN THE TECHNIQUES FOR METABOLOMICS STUDY

One of the primary purpose behind the emergence of studies like transcriptomics, proteomics, and metabolomics is to decipher how the genetic information is expressed so precisely and comprehensively. Unlike transcriptomics and proteomics, the precise interpretation of metabolites in metabolomics is quite difficult due to the lack of direct correlation between proteome and metabolome (Sumner et al., 2015). Noteworthy efforts have been made in the past 10 years for the successful integration of genomics and metabolomics for the enhancement of our knowledge of plant physiology and metabolism. Metabolomics not only represents the analytical investigation for the identification of myriad small molecules but also provides a relative quantification and emphasizes the correlation between their cellular concentration, physiological condition of plants, and the residing environment (Nicholson and Lindon, 2008). For the comprehensive identification of a vast range of metabolites present in biological samples, a combination of separation and identification techniques is required. NMR spectroscopy and MS represents the most powerful tools, taking the concept of metabolomics to a new level. In the process of NMR the metabolites are detected without any prior separation step by using specific atomic nuclei such as ^{1}H, ^{13}C, ^{31}P, or ^{15}N (Freund and Hegeman, 2017). NMR is able to detect a particular metabolite noninvasively with respect to other metabolites, and the detected signal intensity/area varies linearly with the concentration of the metabolites identified. The number of NMR-detectable metabolites in a typical biological sample ranges from approximately 20–60, although advancement in NMR technique have pushed this limit to over 100 (Nagana Gowda and Raftery, 2014). One and two-dimensional NMR, microcoil NMR, isotype-labeled NMR, and NMR imaging are the improvements of NMR that make it a more appreciated tool in plant metabolomics. New approaches like dynamic nuclear polarization made by combining NMR and electron paramagnetic resonance acquire hypersensitive NMR spectra for the identification and characterization of metabolites (Frydman and Blazina, 2007). The analysis of intact tissues has become possible by the implementation of high-resolution solid-state magic angle spinning NMR (Pereira et al., 2014). Advancement of metabolomics is possible due to the advancement of multidimensional NMR, which plays important roles in present-day spectroscopy. Accelerating multidimensional NMR spectroscopy includes non-Fourier-transform schemes through which acquisition of multiple NMR spectra in a single experiment and "single scan" is possible. Multidimensional NMR spectroscopy, also called ultrafast NMR spectroscopy (Herrera et al., 2009) has taken NMR spectroscopy to the next level. The fundamental principle behind ultrafast techniques is that they either decrease the t_{scan} repetition delay or reduce the overall number of N_{scan}, i.e., scans needed to retrieve the spectra. One of the most efficient ultrafast technique, selective optimized flip-angle short transient heteronuclear multiplequantum correlation spectroscopy (SOFAST-HMQC) reduces the t_{scan} time down to ≤100 while preserving high sensitivity (Gal et al., 2007). Better identification of potential biomarkers in metabolomics study can be achieved by implementation of the statistical total correlation spectroscopy (STOCSY) analysis method. STOCSY generates a pseudo-two-dimensional NMR spectrum that displays the correlation among the intensities of the various peaks across the whole sample by taking advantage of multicollinearity of the intensity variables in a set of spectra. The implication of STOCSY is further extended to enhance the recovery of latent biochemical information and improves the biological classification and interpretation.

In the process of MS, the metabolites are detected using the mass-to-charge ratio (m/z) of the molecular ions. Various separation techniques are coupled with MS to make it more useful for metabolite detection. Compared with NMR, MS can detect hundreds of metabolites routinely, offering high sensitivity. Technical upgradation in NMR and MS make these techniques more potent toward the advancement of metabolomics. Ionization is an essential step in MS wherein the analytes are ionized before injection into mass analyzers. Ionization techniques for MS include laser desorption ionization, MALDI, and desorption ESI. The improved MS techniques coupled with separation tools include GC-MS, LC-MS, Fourier-transform ion cyclotron resonance-MS (FT-ICR-MS), and CE-MS. LC-MS is suitable for a wide array of metabolites with heat-labile functional groups, chemically unstable substructures, and high-molecular-weight metabolites, as it does not require the compound to be volatile. Since 2000, considerable advancements have been made both in chromatography and MS techniques for efficient analysis of complex biological samples. Among the chromatographic methods, HPLC has undergone rapid development. The introduction of nano-ion pairing reverse-phase HPLC hyphenated with nano-ESI high-resolution MS significantly reduces the limit of detection thereby allowing to have a comprehensive knowledge about the central metabolism from a minimal amount of cell extract (Kiefer et al., 2011). Supercritical fluid chromatography (SFC) is one of the advancement of LC since 2005, that seems to be a promising alternative to traditional LC methods. SFC methods enhance the kinetic performance of the mobile phase by using low-viscosity supercritical fluid resulting in fast diffusion of the compounds in the mobile phase. Supercritical fluids possess features such as low viscosity and high diffusivity and their temperature and pressure are above the critical values.

Apart from the quantification of metabolites, technologies such as mass spectrometry imaging (MSI) and NMR imaging have been developed to provide spatial and temporal localization of metabolites in particular cell and subcellular compartments (Sumner et al., 2015). For MSI of the metabolites at cellular levels, ionization techniques using two-dimensional and three-dimensional coordinates of a biological tissue have been developed since last 10 years (Fletcher et al., 2011). NMR imaging is another powerful and chemically accurate tool that measures the resonance intensity of paramagnetic nuclei of the metabolites present in biological samples and generates images. Both MSI and NMR imaging enrich our knowledge about the differential and dynamic metabolism occurring in plants at different growth periods and in various environmental conditions (Sumner et al., 2015). Increased sensitivity and specificity of advanced analytical tools offer the opportunities of obtaining a large amount of data with a single approach. However, two different approaches can be used to analyze the data collected, namely, (1) the chemometric approach in which the chemical compounds are not identified, but their spectral patterns are statistically analyzed to identify relevant spectral features that could differentiate samples and (2) the targeted profiling approach in which most of the chemical compounds are first identified, and then various statistical methods are applied for the identification of valuable biomarkers (Genga et al., 2011). Statistical analysis can be divided into two classes: (1) monovariate or univariate analysis and (2) multivariate analysis. Because of the complexity of the plant metabolite profile, multivariate data analysis is more promising and productive to analyze the data obtained from analytical tools. As metabolomics is the study of the unknown, therefore distinctive precaution must be taken regarding confounding factors such as time of sampling, conditions of growth of the plants (e.g., sunlight or airing, nutrient source), age, and gender as they play important roles in defining the metabolite profile of the plants. Sample collection and preparation is a vital and imperative step in metabolomics. In contrast to most of the sample preparation approaches that aim to remove the metabolite interface, in metabolomics, sample preparation should treat all the low-molecular-mass metabolites that form the sample components equally and avoid alternation in sample composition.

3. METABOLOMICS WORKFLOW

Metabolomic analysis consists of four basic steps: (1) sample harvesting and extraction, (2) analysis, (3) data reduction, and (4) identification of metabolites (Choi and Verpoorte, 2014). Significant attention is given to the analytical and statistical analyses of the data neglecting sample preparation and extraction. A single organism contains enormous number of metabolites with different polarities, stabilities, and chemical properties. Therefore, preanalytical methods must be selected considering the type of samples and the targeted metabolites. Metabolomics basically deals with all the metabolites in an organism, therefore efficient extraction methods must be employed to extract all the metabolites. Otherwise the analysis will provide unreliable information resulting an alternative view of the metabolome. There are number of factors that must be taken into consideration while preparing samples for metabolomics study. To preserve the integrity of the metabolome at a given point of time factors like time of collection, contamination, light and temperature exposure, and quenching of chemicals and enzymatic reaction must be taken into consideration (Choi and Verpoorte, 2014). There are three essential aspects of any extraction methods: solute–solvent interaction, solute–solute interaction, and most importantly, dissociation of analytes from the biological matrix. Moreover, the dissociation of the analytes from the matrix is significantly affected by the chemical characteristics of the matrix, solvent, and metabolites (Choi and Verpoorte, 2014). Among the extraction methods quenching methods are the most focused extraction process in metabolomic analysis. Apart from quenching, mechanical and ultrasound extraction methods also seem to be promising in the field of metabolomics. Additionally, high extract yield and chemical diversity can be obtained by combining mechanical grinding and ultrasound extraction methods (Gong et al., 2017). Apart from the extraction methods, the wise selection of solvent is also important, as one solvent cannot feasible to extract diverse array of metabolites. Along with the solvent selection of other factors like solvent volume, sample–solvent ratio, and injection of sample into the analyzer can affect the outcomes of the metabolomics study. Therefore, a comprehensive extraction protocol using a gradient of extraction solvents was developed by Yuliana et al. (2011). The extraction procedure of metabolites is mainly centered around the solubility of the metabolites and is primarily categorized into three groups: (1) lipophilic, (2) medium polar, and (3) hydrophilic. Apart from the chemical aspects various physical aspects like temperature and pressure are known to have greater effects on the extraction efficiency. Many reports have been documented comparing the advantages and limitations of analytical tools that are used in metabolomics such as MS, NMR, infrared spectroscopy (IR), and UV spectroscopy with or without combination with chromatography (Kim et al., 2010; Wolfender et al., 2013). The large amount of data thus generated by the analytical instruments require statistical treatment to obtain maximum useful information out of them (Choi and Verpoorte, 2014). In the final step of metabolomics workflow, the selected signals must be interpreted for the identification of metabolites using computational spectral analysis along with various NMR and MS databases (Ludwig et al., 2012).

4. DATA ANALYSIS

Identification of the eluting compounds after separation with high-degree resolution is the next challenge. Natural product identification or LC peak annotation is generally based on database searches. Due to the higher sensitivity, MS detection is commonly used for the determination of the metabolites. Historically, FiehnLib was the first identification database that was developed for EI-MS due to higher rate of reproducible spectral pattern (Kind et al., 2009). Several other mass spectral databases such as METLIN (Smith et al., 2005), Lipid Search (Taguchi et al., 2007), and the Golm Metabolite Database (Yang et al., 2017) are used in the field of metabolomics. It has been reported that the analytical tools that are employed in the area of metabolomics are purely based on chemometric methods (Brown et al., 2005). It has also been argued that metabolomics study involved the combination of chemoinformatics and bioinformatics (Wishart, 2007). Before going for the analysis, preprocessing of the acquired data is essential. This preprocessing includes noise filtering, normalization, peak detection, deisotoping, alignment, and identification (Ernst et al., 2014). Similarly, for the processing of raw data obtained from LC-MS and GC-MS several open programs are available such as MZmine2 (Katajamaa et al., 2006), MetAlign (Vorst et al., 2005), MET-IDEA (Broeckling et al., 2006), and XCMS (Benton et al., 2008). After peak processing by these programs, mass spectral database is required for the interpretation of peaks. Apart from the aforementioned spectral databases, several compound databases such as Kyoto Encyclopedia of Genes and Genomes (Kanehisa et al., 2008), KNApSAcK (Shinbo et al., 2006), LipidBank, LIPIDMAPS (Fahy et al., 2007), and PubChem (Wheeler et al., 2008) are also engaged in metabolomics study. Sensational analytical strategies have been developed to increase the amount of information extracted from the data obtained from the instrumentation (Ernst et al., 2014). Metabolomics experimental design aims to compare the metabolome of two different groups of samples, i.e., control and treated. As the data of metabolomics are multivariate, predictive models based on latent variables, e.g., partial least squares (PLS) regression, are well suited to handle these data for discriminating analysis. Orthogonal partial least squares (OPLS) is an extension of PLS regression that integrates an orthogonal signal correction for the prediction of a quantitative response (Boccard and Rutledge, 2013). The main advantage of OPLS is that it focuses on discriminant information that helps in easier interpretation of the data. Among other multivariate analysis methods, principal component analysis (PCA) is commonly used for the analysis of metabolite data. The main advantage of PCA is that it compares all the data obtained from each sample emphasizing the holistic difference between the samples. Besides PCA, hierarchical cluster analysis, PLS discriminant analysis, and batch-learning self-organizing map analysis are the statistical methods that are often used for metabolomics data analysis (Ku et al., 2010). Classical analysis of variance (ANOVA) has wide applicability to unravel the effects of each factor on the data collected by decomposing the variance. Moreover, a multivariate extension of ANOVA is implemented to understand the multiple independent variables, and the covariance that describes different sources of variation (Boccard and Rudaz, 2016).

5. ROLE OF METABOLITES IN PLANT'S TOLERANCE TO VARIOUS ABIOTIC STRESSES

The abiotic stress can be defined as any variability in the plant's natural habitat that negatively alters plant's regular growth, physiology, and metabolism. Major dominating abiotic stress that affects crop productivity adversely are high salinity, drought, and waterlogging. Apart from these plants also encounter various abiotic stresses like high light intensity, heat, chilling, and heavy metal contamination and sometimes a combination of more than one stress. Salinity and drought adversely affect all fundamental processes of plants starting from germination to all vital physiological aspects (Zhang et al., 2011). Plants thriving in high saline and drought conditions encounter osmotic stress by disrupting ion homeostasis and balance (Guo et al., 2015). In high-salinity conditions, ion toxicity is mainly due to an excessive influx of Na^+ ion into the cells, which impairs the water and mineral uptake by the root (Wu et al., 2013). Plants adopt various strategies to acclimatize in high-salinity conditions. One strategy is to synthesize various compatible solutes to minimize osmotic stress, and the other is the efficient compartmentalization of Na^+ within vacuoles. Likewise, the acclimatory response of plants to drought stress is categorized as escape, avoidance, and tolerance (Yıldırım and Kaya, 2017). Escape is the strategy by which plants complete their reproductive phase (flowering to seed set) before the onset of summer drought. Avoidance is the strategy in which plants maximize their water uptake and minimize water loss by increasing the root–shoot ratio and by decreasing the transpiration rate. Tolerance is the mechanism by which plants maintain their regular homeostasis and cellular metabolism through osmotic adjustment. Apart from salinity and drought, soil waterlogging is another important abiotic stress that limits the crop yield by affecting plant's growth and physiology. Waterlogging for prolonged period creates a hypoxic condition (especially anoxic condition) that is directly felt by the roots and inhibits carbon assimilation and photosynthate utilization in the shoot (Wei et al., 2013). Acclimation to wet stress proceeds through three crucial stages: signal transduction, metabolic adaptation, and morphological modifications (Jackson and Colmer, 2005). To withstand the adverse effects of the aforementioned environmental alternation, plants adopt various regulatory strategies for their successful establishment in the altered environment. Plants' response to stress are dynamic and involve complex networking of various regulatory pathways including metabolite adjustment leading to physiological and morphological acclimation (Muscolo et al., 2015). Acclimation is a phenotypic response to changes in the environment, which include reprogramming of developmental and metabolic processes providing better fitness and survival (Sanchez et al., 2012). These alternations assist them to cope up with the changed environmental conditions that are usually referred to as acclimatization. Modifications like alteration in osmotic potential, changes in cell wall plasticity, reactive oxygen species (ROS) scavenging by the upregulation of enzymatic and/or nonenzymatic antioxidant systems, and controlled energy dissipation by regulated pigment biosyntheses are developed by plants in the course of adaptation.

A vast array of primary and secondary metabolites are employed by the plants to accomplish all the aforementioned alternations for successful stress adaptation. Primary metabolites are those that are directly involved in the growth, development, and reproduction of the plants. Primary metabolites are universal and present in many organism and cells, e.g., glutamic acid, aspartic acid, lactic acid, and isoascorbic acid. On the other hand, secondary metabolites are those that are not directly involved in the growth, development, and reproduction of plants. Secondary metabolites in plants are generally not essential for survival. However, they are required for protection, competition, and species interaction. Among the secondary compounds, three types of compounds are significant from the standpoint of metabolomics and stress tolerance: (1) compounds that are involved in acclimatization process such as antioxidants and osmoprotectants, (2) signal transduction molecules, and (3) by-products of disrupted normal homeostasis due to stress. The metabolites that confer osmoprotection include small-molecular-weight organic osmoprotectants, such as amino acids (proline, asparagine, serine, and glutamine), amines (polyamines and glycine betaine [GB]), soluble carbohydrates, and γ-amino-*N*-butyric acid (GABA); GB decreases the entropy levels and maintains the proteins in their native tertiary structures along with their primary role of osmoprotection (Fraire-Velazquez and Emmanuel, 2013). The antioxidative metabolites include nonenzymatic antioxidants (ascorbate and glutathione) and the molecules that help to reduce ROS production by efficient dissipation of excitation energy (intermediates of xanthophyll cycle). Alternation of these metabolites may be common to different stresses or specific to particular stress (Muscolo et al., 2015). Metabolomics has a profound role in discovering novel metabolites that enrich our understanding of the mechanisms underlying the abiotic stress management of plants, which can either be stress tolerance or stress avoidance. A schematic diagram showing the metabolomics workflow for the elucidation of abiotic stress tolerance in plants is given in Fig. 5.1.

The severity of stress depends on the plant's developmental phase at which the exposure to stress occurs along with its intensity and the period of exposure (Lafitte et al., 2007). Agriculture is the most severely affected by abiotic stresses and is worsening day by day mostly due to anthropogenic activities. To overcome the reduction of crop productivity abiotic stress-resistant crop varieties should be developed, and for this a better understanding of the mechanisms of plant stress tolerance is very important. Therefore the study of plant stress response has now shifted toward crop plants like wheat, rice, barley, maize, and other economically significant species. Knowledge about the important proteins and metabolites that are essential to the plant's growth and survival, their modulation upon the onset of various stresses, and the precise

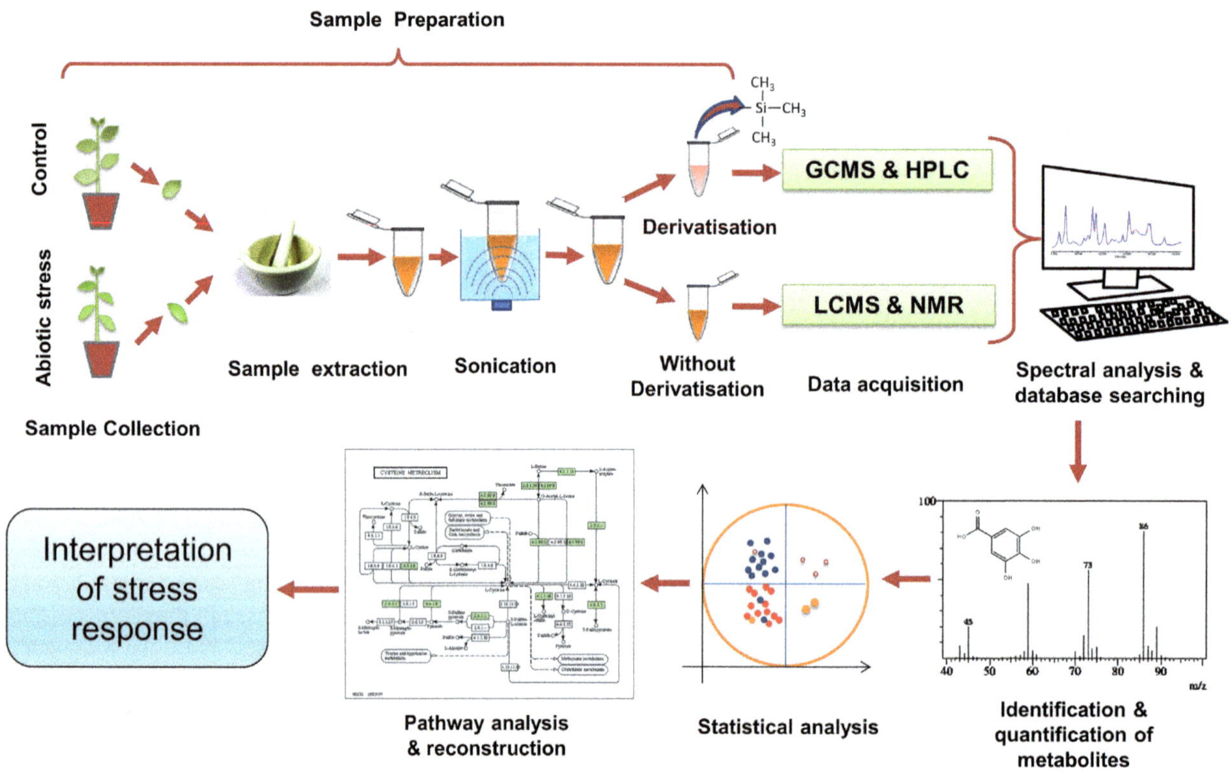

FIGURE 5.1 Schematic diagram showing the metabolomics workflow for elucidation of abiotic stress tolerance in plants. *HPLC*, high-performance liquid chromatography; *LC*, liquid chromatography; *MS*, mass spectrometry; *NMR*, nuclear magnetic resonance.

interpretation of the stress tolerance mechanisms offer scope for the identification of viable biomarkers that are important for abiotic stress tolerance of plants. The specific modulations of various metabolites in plants for the adaptation under different environmental stresses were given in Fig. 5.2 and are discussed in the following sections.

5.1 Salinity-Induced Alteration in Various Metabolites

Salinity is an abiotic stress that prevails worldwide and adversely affects the productivity of crop plants (Negrão et al., 2017). High levels of salinity cause ion toxicity and decrease in nutrient uptake, creating osmotic imbalance and metabolic disorders that lead to disturbances in various physiological activities and decrease in overall growth of the plants (Meng et al., 2016). To cope up with these adverse effects, plants respond by changing gene expression, hormonal balance, and metabolite pool at the cellular level. When a plant is exposed to a stress condition, it reduces energy utilization for the growth and development and shifts toward adjustment under harsh conditions. In consort with the changes in the levels of transcriptome and proteome, alternation of metabolites is also regarded as an adaptive mechanism adopted by the plants. The best described metabolic response is the increase in cellular accumulation of compatible solutes, which includes a range of soluble and neutral organic compounds. Accumulation of compatible solutes can contribute to reduction of the water potential of cytoplasm, balancing with the Na^+ accumulation in the vacuoles and extracellular spaces. To establish the tolerance, three interrelated aspects are important: (1) damage prohibition, (2) reestablishment of homeostatic condition under stress, and (3) continuation of growth maybe in reduced rate (Sobhanian et al., 2010). The metabolites whose levels are altered due to salinity may be osmoprotectants, ROS scavengers, and/or metabolites involved in energy metabolism. Storage, transport, and synthesis of various primary and secondary metabolites are affected upon exposure to salinity (Fraire-Velazquez and Emmanuel, 2013). The metabolic changes in response to salinity are well documented, whereas changes in the overall metabolome in response to abiotic stresses are yet to be elucidated (Widodo et al., 2009). Metabolites that act as a basis for developing stress tolerance in plants can be discussed as the metabolites of different metabolic and regulatory pathways like photosynthesis, amino acid biosynthesis, ROS scavengers, and tricarboxylic acid (TCA) cycle because these pathways are more adversely affected by salinity. Salinity-induced alterations of different classes of metabolites are given in Table 5.1 and discussed under the following headings.

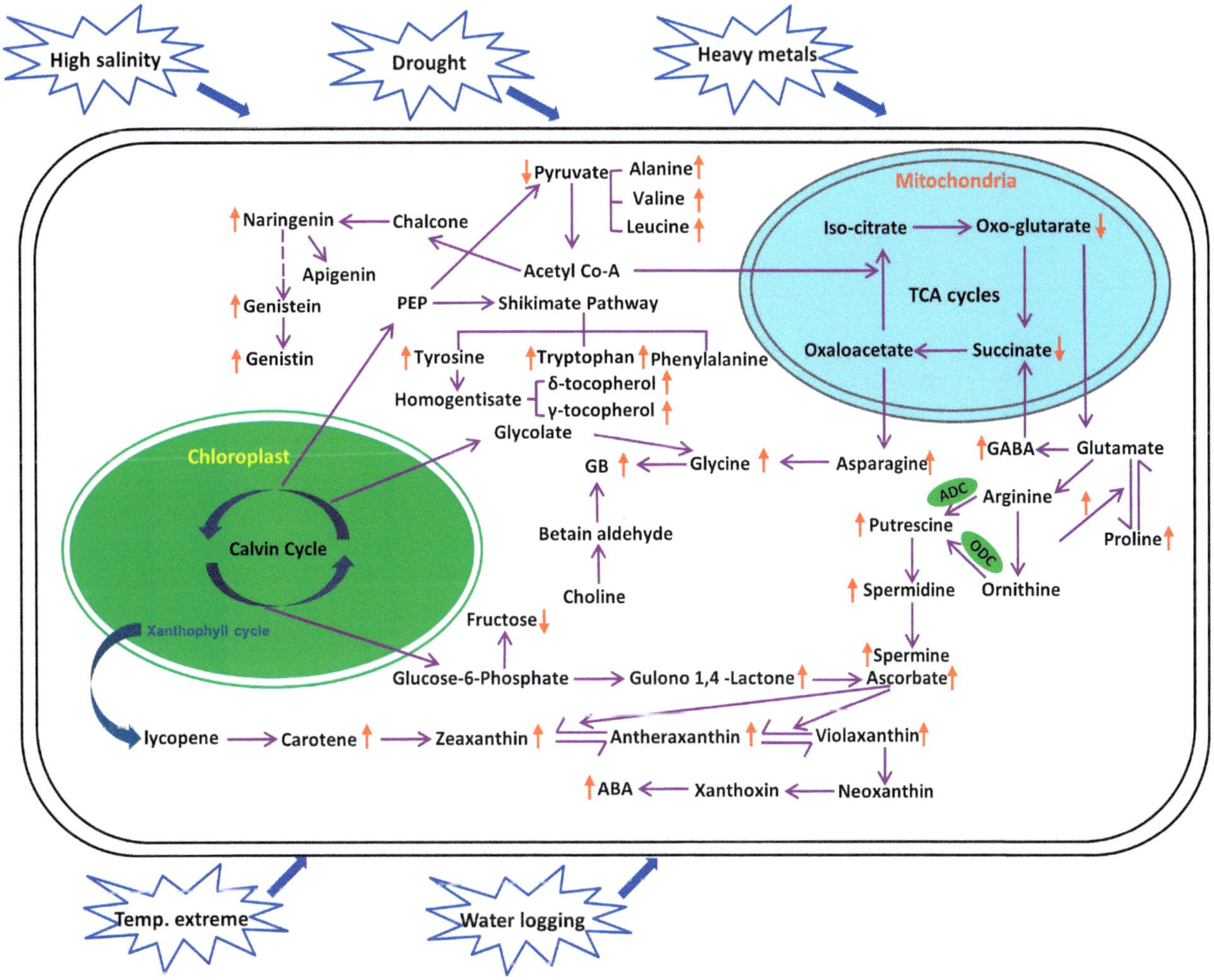

FIGURE 5.2 Schematic representation of alternations of different metabolites under various abiotic stresses and the biosynthetic pathways involved. ↑ And ↓ indicate up- and downregulated metabolites, respectively. *ABA*, abscisic acid; *ADC*, arginine decarboxylase; *GB*, glycine betaine; *GABA*, γ-amino-*N*-butyric acid; *ODC*, ornithine decarboxylase; *PEP*, phosphoenol pyruvate; *TCA*, tricarboxylic acid.

5.1.1 Metabolites Associated With Photosynthesis

Plants' responses to salinity stress involve various metabolic processes such as ion transport, photosynthesis, osmotic solute accumulation, and hormone synthesis. Upon exposure to stress conditions like salinity, photosynthesis is the most affected physiological process. Concurrently with the alternation of photosynthesis, the metabolites of the photosynthetic pathways are also altered under high salinity stress. Under high-salinity conditions, the photosynthetic ability of the plants are decreased due to unavailability of raw materials as a consequence of decline in the mineral and water uptake by the roots (Grewal, 2010). In addition, the reduction in photosynthesis may be attributed to the closure of stomata, which results in decrease in stomatal conductance and hence reduction of the CO_2 diffusion rate. Reduction in transpiration rate can also be considered as an adaptation of the plants to reduce the mobilization of salt to the leaf tissues. As reported by Wu et al. (2013) metabolites related to photosynthesis, i.e., glucose-6-phosphate, fructose-6-phosphate, and 3-phosphoglyceraldehyde (3-PGA), significantly accumulate in both wild and cultivated varieties of barley subjected to high-salinity treatment. Significant accumulation of 3-PGA signifies enhanced Calvin cycle under high salinity (Wu et al., 2013). A similar study conducted on *Thellungiella halophila* shows that the intermediates of Calvin cycle like ribulose-5-phosphate, ribose-5-phosphate, and glucose-6-phosphate decline (Wang et al., 2013; Kumari et al., 2015). These changes in metabolites can be attributed to the fact that salinity stress alters the activity of major Calvin cycle–related enzymes like aldolase, phosphoribulokinase, transketolase, ribose-5-phosphate isomerase, and fructose bisphosphatase (Chaves et al., 2009). Phosphoribulokinase and sedoheptulose-1, 7-bisphosphatase are the two most important enzymes of Calvin cycle whose regulation in dark/light transition is responsible for the regulation of the cycle. Upregulation of these enzymes under saline

TABLE 5.1 Differential Regulations of Important Metabolites in Various Plant Species in Response to Salt Stress

Abiotic Stress	Class of metabolites	Metabolite	Plant Species	Regulation	Function	References
Salinity	Amino acids	Alanine	*Nitraria tangutorum*	↑	Precursor for protein and nitrogen-containing compounds and acts as a signaling molecule	Ni et al. (2015)
			Arabidopsis thaliana	↑		Kim et al. (2007)
			Hordeum vulgare	↓		Widodo et al. (2009)
			Nicotiana tabacum	↑		Zhang et al. (2011)
		Glutamate	*Thellungiella halophila*	↑	Precursor for other amino acids	Gong et al. (2005)
			H. vulgare	↑		Widodo et al. (2009)
			Limonium latifolium	↑		Gagneul et al. (2007)
			H. vulgare	↑		Widodo et al. (2009)
			N. tangutorum	↑		Ni et al. (2015)
		GABA	*Zea mays*	↑	Osmoprotectant, growth regulators, nitrogen and carbohydrate metabolism	Gavaghan et al. (2011)
			N. tabacum	↓		Zhang et al. (2011)
			Nicotiana sylvestris	↑		Akçay et al. (2012)
		Leucine	*H. vulgare*	↑	Osmoprotectant	Wu et al. (2013)
		Aspartate	*T. halophila*	↑	Precursor for other amino acids	Gong et al. (2005)
			L. latifolium	↑		Gagneul et al. (2007)
	Glycolysis metabolites	Sucrose, fructose and glucose	*T. halophila*	↑	ROS scavenging and osmoprotectant	Gong et al. (2005)
		Pyruvate	*Aeluropus lagopoides*	↓	Energy metabolism	Sobhanian et al. (2010)
			H. vulgare (XZ16)	↑		Wu et al. (2013)
	Polyols	Cyclitols	*L. latifolium*	↑	Acts as an osmoticum	Gagneul et al. (2007)
		β-AB	*L. latifolium*	↑	Osmolyte	Gagneul et al. (2007)
	Organic acids		*L. latifolium*	↓	Precursor for many compounds	Gagneul et al. (2007)
			H. vulgare (Sahara genotype)	↑		Widodo et al. (2009)
		Allantoin	*N. tabacum*	↑	Nitrogen transport and assimilation	Zhang et al. (2011)
		Acetamide	*N. tangutorum*	↓	Intermediate of auxin biosynthesis	Ni et al. (2015)
	Polyphenols		*Z. mays*	↑	Antioxidant activity	Hichem et al. (2009)
			A. thaliana	↑		Kim et al. (2007)
	Sugar alcohol	Myoinositol	*Porteresia coarctata*	↓	Osmotic adjustment	Sengupta et al. (2008)
			H. vulgare	↑		Chen et al. (2007)
		D-Pinitol	*Porteresia coarctata*	↑	Osmoprotectant	Sengupta et al. (2008)

TABLE 5.1 Differential Regulations of Important Metabolites in Various Plant Species in Response to Salt Stress—cont'd

Abiotic Stress	Class of metabolites	Metabolite	Plant Species	Regulation	Function	References
	Cyclic acid	Phytate	*Oryza sativa* (during germination)	↓	Phosphate storage	Sung et al. (2015a)
			N. tabacum	↓		Zhang et al. (2011)
		Hypoxanthine	*N. tabacum*	↓	Purine metabolism	Zhang et al. (2011)
	TCA cycle metabolites	Succinate	*H. vulgare* (Sahara genotype)	↑	Growth and energy metabolism	Widodo et al. (2009)
		Citrate	*L. latifolium*	↓	Provide carbon skeletons for nitrogen assimilation and reducing equivalents for biosynthetic reactions	Gagneul et al. (2007)
			H. vulgare (Clipper genotype)	↓		Widodo et al. (2009)
			L. latifolium	↓		Gagneul et al. (2007)
		Malate	*N. tangutorum*	↓	Essential role in nitrogen fixation and phosphorus acquisition	Ni et al. (2015)
			Suaeda corniculata	↓		Pang et al. (2016)
			N. tangutorum	↑		Ni et al. (2015)

↑ and ↓ indicate up- and downregulation of different metabolites, respectively.
β-AB, β- alanine betaine; *GABA*, γ-amino-*N*-butyric acid; *TCA*, tricarboxylic acid.

conditions leads to the accumulation of the metabolites synthesized by the activity of these enzymes. Another study on *Arabidopsis* shows that the levels of sugar phosphates like glucose-6-phosphate and fructose-6-phosphate deplete when subjected to salt stress. This reduction in monophosphate levels under salinity may be due to the migration of the monosaccharides to other acid metabolism pathways (Sanchez et al., 2008A).

5.1.2 Metabolites Associated With Energy and Carbohydrate Metabolism

In addition to photosynthesis, pathways related to energy metabolism such as respiration and carbohydrate metabolism are also adversely affected by high salinity. It has been widely documented that plants accumulate different types of soluble carbohydrates like sucrose, hexoses, raffinose, trehalose, mannobiose, and sugar alcohols to counter osmotic stress generated due to salinity. Soluble carbohydrates have an indispensable role in plant metabolism and act as carbon and energy sources within the cells. The levels of soluble carbohydrates are altered by salt stress, as the carbohydrate content is related to various important physiological processes like photosynthesis and respiration. Soluble sugars act as osmoprotectants during osmotic stress and also help in turgor maintenance, cell membrane stabilization against ROS effects, and protection of proteins from degradation in stress condition. Furthermore, salinity-induced responses of metabolite level are different in tolerant and sensitive varieties of the plants. For example, the salt-tolerant variety (W05) and salt-sensitive variety (C08) of *Glycine max* show differential response to salinity with regard to sugar content. The major sugars found in the leaves of soybean seedlings under nonsaline condition are glucose, fructose, maltose, and sucrose. When the seedlings are subjected to salinity treatment, levels of sucrose decrease with no substantial alternation in the levels of fructose and glucose (Lu et al., 2013). Alteration in carbohydrate metabolism, i.e., glycolysis and sucrose metabolism pathway, is regarded as a long-term response to salinity (Kim et al., 2007). When a halophytic relative of *Arabidopsis thaliana*, i.e., *T. halophila*, is subjected to salt treatment, it shows diverse metabolic responses that are related to its glycophytic counterpart to some extent. Major differences between metabolic responses of these two species are the higher accumulation of hexoses, disaccharides, and additional sugars in *T. halophila*. The metabolomic study in *Suaeda salsa* under salinity shows that there is a decrease in glucose and fructose levels (Wu et al., 2012). Significant accumulation of sucrose was observed in the leaves of *Atriplex halimus* in high-salinity condition, which may be attributed to the impairment of sucrose transport outside the leaves (Bendaly et al., 2016). Being a precursor, elevated levels of sugars help to maintain the functionality of various nonenzymatic antioxidant compounds like ascorbate, carotenoids, and glutathione in *A. halimus* (Bendaly et al., 2016). A similar study conducted by Zhang et al. (2011) shows that short-term low-dose salt stress in *Nicotiana tabacum* induces an increase in the rate of gluconeogenesis thereby accumulating more glucose, sucrose, and fructose. Pathway analysis of salt-tolerant species *Populus euphratica* and salt-sensitive species *Populus canescens* shows the higher production of primary sugars in salt-tolerant *P. euphratica* species as a control mechanism of osmotic adjustment (Janz et al., 2010). Study on the effects of salinity on metabolite alternations in two cultivars of barley (Sahara and Clipper) shows some metabolic alternations 24h after treatment (Widodo et -al., 2009). The alternations include a decrease in fructose, galacturonate, glucose, maltose, sorbitol, and xylose in cultivar Sahara under salinity. However, in Clipper the metabolites that are increased in response to salinity include sugars, galactinol, gluconate, raffinose, and trehalose, and the metabolites that are decreased include fructose, galacturonate, glucose, and xylose (Widodo et al., 2009). However, the changes in metabolites are more immense in terms of number and magnitude in both the cultivars after 3h of treatment. This study suggests that the metabolic alternations under salt stress also depend on the period of stress exposure. It has been reported that the increase in sugar content in plants may be due to the decrease in the starch biosynthesis or increase in starch degradation due to Na^+ toxicity (Parida and Das, 2005). In plants in which salinity does not affect photosynthesis, the level of gluconeogenesis increases with increased levels of sugars to avoid osmotic stress (Guo et al., 2015).

Sugar alcohols are organic compounds that have polyhydroxy functional groups that participate in an array of metabolic reactions. Sugar alcohols are further categorized into two groups, namely, cyclic, which includes pinitol, and acyclic, which includes mannitol (Gupta and Huang, 2014). During the prevailing salinity conditions the cellular levels of sugar alcohols such as pinitol, myoinositol, mannitol, and sorbitol increase assisting the stress tolerance of plants. Sugar alcohols mostly function as low-molecular-weight chaperons and/or compatible solutes along with their ROS scavenging activities under abiotic stress. In addition to this, sugar alcohols also provide enzyme stabilization, protect cellular and subcellular structures, and mediate stress signaling. Inositols are used by many eukaryotes as protective compounds under stress conditions. Glucose-6-phosphate is the precursor molecule of inositol biosynthesis and is catalyzed by the enzymes 1-myoinositol-1-phosphate and inositol-O-methyltransferase. Inositol contents significantly increase in the halophyte *A. halimus* under salinity as reported by Daly et al. (2016). Among the halophytes, the majority of mangrove plants have pinitol as the source of compatible solute, and its concentration ranges from 40 to 230mM (Parida and Jha, 2010). Activation of sugar alcohol biosynthesis pathway is found to be more prominent in salt-tolerant species than the salt-sensitive ones (Janz et al., 2010). In salt-stressed chickpea, sugar alcohols like erythritol, xylitol, arabitol, mannitol, galactitol, and inositol are supposed to have important roles in salinity tolerance (Dias et al., 2015). Similarly, accumulation of sugar alcohols (lactitol or maltitol) is observed in *G. max* under salinity stress (Lu et al., 2013). Apart from pinitol and myoinositol, accumulation of galactinol due to hyperactivity of galactinol synthase has also been reported in *P. euphratica* in saline conditions. Wild and cultivated barely subjected to saline treatment show different stress responses in leaves and roots. Salt-responsive accumulation of mannitol, inositol, and xylitol are observed in the roots of barley imposed to salinity, whereas in leaves the concentrations are found to be decreased (Wu et al., 2013). The decrease in mannitol utilization in leaves may be attributed to the decreased activity of nicotinamide adenine dinucleotide (NAD)-dependent mannitol dehydrogenase in leaf tissues (Wu et al., 2013). From this, it can be conferred that mannitol and inositol can be regarded as root-specific metabolites. Similarly, the contents of polyols like sorbitol, mannitol, and pinitol are found to be higher in roots than in leaves. Upon exposure to salinity the contents of sorbitol, mannitol, and pinitol are decreased in the roots of chickpea (Dias et al., 2015). However, in response to salinity the concentrations of sorbitol and pinitol are increased in the leaves of chickpea, whereas the level of mannitol remains constant (Dias et al., 2015). The alternations in metabolites related to energy and carbohydrate metabolism under unfavorable environmental conditions like salinity suggest an important adaptive role of these pathway-related metabolites in conferring salinity tolerance.

5.1.3 Metabolites Associated With Amino Acid Metabolism

Amino acids are the most important metabolites required for the plants exposed to saline environment, for balancing their cellular osmotic concentration and ROS scavenging. It has been reported that the levels of various amino acids increase under high-salinity conditions (Sanchez et al., 2008a; Zhao et al., 2014). Amino acids such as threonine, leucine, and valine are taken as an index for salt tolerance in plants (Sanchez et al., 2008b). The decline in the content of arginine, methionine, and cysteine, which are the major constituents of total free amino acids, is observed in many plants exposed to high salinity. The accumulation of proline is an adaptive measure of plants after the onset of salinity stress (Kumari et al., 2015). Proline is a key amino acid that accumulates under salinity stress and serves as an osmolyte as well as a scavenger of ROS, thereby protecting the cells from salinity-induced damage. Proline also acts as a molecular chaperone to maintain enzyme integrity and stability (Hoque et al., 2007). Glutamate is the precursor molecule of proline biosynthesis and is catalyzed by the enzymes pyrroline-5-carboxylate synthetase and pyrroline-5-carboxylate reductase. In addition to proline, GB also participates in the maintenance of protein integrity by reducing protein carbonylation and enhancing the activity of glutathione peroxidase, glutathione-S-transferase, and glyoxalase thereby protecting the plants from salt-induced oxidative damage (Hoque et al., 2007). Widodo et al. (2009) have reported that the barley plants treated with high salinity accumulate amino acids like proline and GABA. The levels of various amino acids such as alanine, GABA, glutamate, and asparagine increase in maize plants exposed to salt stress, whereas the levels of cysteine, arginine, and methionine decrease (Gavaghan et al., 2011). Modulations in amino acid levels may be attributed to the activity of alanine/aspartate aminotransferase and glutamine synthetase, the up- and downregulations of which alter the ultimate cellular concentration of various amino acids (Wang et al., 2008). Long-term salinity with high dose of salt induced the accumulation of various osmolytes such as proline, changed the GABA shunt, and also induced the shikimate-mediated secondary metabolisms with increased levels of aromatic amino acids such as tyrosine, tryptophan, and phenylalanine (Zhang et al., 2011). Increased levels of the amino acid may be attributed to inhibited protein synthesis or enhanced protein degradation as the plant is growth inhibited during high-salinity condition. Metabolite profilings of four barley genotypes (CM72, Gairdner, XZ16, and XZ169) show a remarkable coordination between metabolite modulation and stress tolerance (Wu et al., 2013). The concentration of various amino acids like asparagine, glycine, isoleucine, leucine, proline, and serine are immensely increased, with the maximum increase observed in proline content in all the genotypes (Wu et al., 2013). However, the concentration of some amino acids like alanine, threonine, and valine are decreased only in CM72, whereas the concentration of aspartate and glutamate are decreased in both CM72 and XZ16 (Wu et al., 2013). In a study conducted on two genotypes of *G. max* (W05 and C08) a total of 98 metabolites were identified in the control plants and the alternation of these metabolites were studied in response to salt stress (Lu et al., 2013). The decrease in the amino acids such as alanine and glutathione are observed in both the genotypes under salinity stress, with an increase in the levels of glycine and serine in W05 genotype only (Lu et al., 2013). The metabolomic study in halophytic species, *Aeluropus lagopoides,* shows the upregulation of amino acids like alanine, asparagine, lysine, glutamine, arginine, glycine, proline, histidine, phenylalanine, serine, valine, and tyrosine (Sobhanian et al., 2010). A metabolic study of the xerohalophyte *A. halimus* shows an increase in the concentration of proline, α-alanine, tryptophan, threonine, leucine, phenylalanine, valine, glycine, and arginine with increased levels of salinity (Bendaly et al., 2016). However, decrease in the content of amino acids such as valine, leucine, isoleucine, glutamate, tyrosine, and phenylalanine and an increase in proline content are documented in *S. salsa* subjected to salt treatment (Wu et al., 2012). From this it can be assumed that the alternation in amino acid levels can be considered as an adaptive response of the plants to combat the adverse effects of salinity.

Apart from amino acids, quaternary ammonium compounds (QACs) also play an essential role in osmoprotection and ROS scavenging. Among the QACs, GB is the one that occurs most frequently in plants exposed to salinity (Lokhande and Suprasanna, 2012). Salinity-induced accumulation of GB has been documented in many halophytes as well as in glycophytes. Like proline, GB reduces protein carbonylation and enhances the activity of glutathione peroxidase, glutathione-S-transferase, and glyoxalase, thereby providing protection against salt-induced oxidative damage (Hoque et al., 2007). Most importantly, GB stabilizes the protein structure and minimizes stress-induced lipid peroxidation (Hoque et al., 2007). GB in chloroplast plays a significant role in maintaining the osmotic environment and stabilization of the thylakoid membrane and photosynthetic machinery (Ashraf and Foolad, 2007). GB accumulates in many halophytic species like *Suaeda maritima*, *Atriplex nummularia*, and *Salicornia europaea* and mangrove species like *Avicennia marina,* but is eventually absent in many crop species (e.g. rice) and model plant species like tobacco (Flowers and Colmer, 2008). The increased concentration of GB has also observed in salt-treated maize (Gavaghan et al., 2011). The accumulation of GB in *A. marina* may be due to the upregulation of glycine betaine aldehyde dehydrogenase (Chen and Murata, 2011). Salt-responsive accumulation of GB plays a central role in cytosolic osmotic adjustment in the halophyte *A. halimus* (Daly et al., 2016). Based on the correlative study on the pattern of metabolite concentration it is inferred that the methylation pathway and GB biosynthesis pathways are systematically coordinated as the early response to salt stress in barley genotypes (Chen et al., 2007). Salt-sensitive barley genotype shows higher accumulation of GB (over twofolds) when compared with the tolerant variety (Chen et al., 2007). In roots of the barley plants GB is undetected, which implies that the primary site of GB accumulation is the chloroplast (Chen et al., 2007). Chen et al. (2007) have reported that most of the cytoplasmic osmolarity is contributed by K^+ ion in salt-tolerant varieties. However, the levels of K^+ is lower in the sensitive variety of barley due to NaCl-induced K^+ efflux, leading to the higher synthesis of GB (Chen et al., 2007).

5.1.4 Metabolites Related to ROS Scavenging

ROS are often generated in plants by photosystems through photoexcited electrons. Under extreme environmental conditions such as salinity, drought, and high light intensity, the production of ROS in the chloroplast and mitochondria is essential for stress signaling and maintaining normal energy and metabolic fluxes. However, excess production of ROS can disrupt membrane integrity and cellular metabolism of plants under stress conditions. To restrict ROS generation within the tolerance limit of plants, a high degree of coordination is required between signaling and metabolic pathways in different cellular compartments. Plants synthesize an array of secondary metabolites to mitigate the effects of excess ROS generated due to the prevailing stress conditions. These metabolites include phenolic compounds, flavonoids, various nonenzymatic antioxidants (ascorbate, reduced glutathione [GSH], and reduced and oxidized glutathione [GSSG]), carotenoids, etc. Phenolic compounds are potent ROS scavengers that have H_2O_2 scavenging activity thereby protecting the cellular structure from oxidative damage. Significant accumulation of polyphenols in response to salinity has been reported in *A. halimus* along with other halophytes like *Thymelaea microphylla, Marrubium deserti, Helianthemum lippii,* and higher accumulation of polyphenols may be assigned to ROS scavenging activities (Belyagoubi-Benhammou et al., 2014; Daly et al., 2016). It has been reported that imposing salt treatment on the medicinal crop safflower (*Carthamus tinctorius* L.) induces the biosynthesis of flavonoids (Forieri et al., 2016). In addition to polyphenols and flavonoids, carotenoids also act as ROS scavenging molecules, and apart from their regular function as accessory light harvesting pigment, carotenoids stabilize the chloroplast membrane by scavenging or quenching triplet chlorophyll (Kang et al., 2017). The level of carotenoids is considered as a marker for oxidative stress tolerance in many plant species, e.g., in *A. halimus* (Daly et al., 2016). The antioxidative machinery of plants is developed to maintain optimum levels of ROS. The enzymatic antioxidants include superoxide dismutase, ascorbate peroxidase, catalase, and glutathione reductase (GR). Various metabolites like ascorbate, glutathione, proline, polyphenols, tocopherols, and carotenoids act as nonenzymatic antioxidants (Grace, 2005; Ozgur et al., 2013). Several studies have reported the increased levels of numerous antioxidants under salinity stress (Sudhakar et al., 2015). The ascorbate–glutathione cycle is an important biochemical pathway that detoxifies the toxic compounds that are generated under abiotic stress conditions. The ascorbate–glutathione cycle detoxifies methylglyoxal (MG), which is a highly reactive cytotoxic intracellular α-oxoaldehyde compound. In plants, MG can accumulate in normal growth condition, but higher accumulation is observed under abiotic stress conditions. Ascorbate acts as a potent ROS scavenger by donating electrons in various enzymatic and nonenzymatic reactions. Ascorbate is considered as the first cellular molecule that encounters stress-induced ROS. Antioxidant defense is provided by the changes in levels of GSH and GSSG, and the cellular redox potential is maintained by the appropriate ratio of GSH/GSSG, thereby playing a central role in the adaptation of plants to stressed environment (Basu et al., 2010). The levels of ascorbate decrease in rapeseed seedlings subjected to salt stress, but when selenium is applied along with the salt, a significant increase in ascorbate content is observed (Hasanuzzaman et al., 2011). GSH, a low-molecular-weight thiol present in most plant tissues, plays multiple roles in cellular metabolism and antioxidative defense. Elevated levels of GSH in rapeseed seedlings subjected to salt stress indicate its defensive role in salt tolerance (Hasanuzzaman et al., 2011).

5.1.5 Metabolites Associated With Mitochondrial Respiration

In plants exposed to salinity stress mitochondrial respiration pathways, like the TCA cycle, are severely affected leading to alternation of intermediate metabolites. The alternations include decrease in the levels of metabolites like shikimate, succinate, threonate, L-threonate-1,4-lactone, 1-monohexadecanoglycerol, and 1-monooctodecanoglycerol (Widodo et al., 2009). It has been observed that salinity stress results in decrease of TCA cycle intermediates like α-ketoglutaric acid, citric acid, malic acid, succinic acid, and fumaric acid in both wild and cultivated varieties of barley (Wu et al., 2013). The decrease in TCA cycle metabolites suggests increased requirement of sugars by plants for salt adaptation. In rice, it has also been documented that salt-sensitive varieties tend to have more metabolites related to TCA cycle than the tolerant ones (Zuther et al., 2007). However, the TCA cycle intermediates like isocitrate, *cis*-aconitate, citrate, and malate showed a salinity-induced increase in the seedlings of salt-sensitive chickpea (Rupali genotype), whereas there was no significant changes observed in tolerant genotypes (Genesis 836) except for turanose, which shows an increasing trend (Dias et al., 2015). Salinity-induced decline of citrate and oxoglutarate in the shoot of maize indicates the inhibition of TCA cycle in maize in response to salinity (Gavaghan et al., 2011). However, the levels of succinate and malate are increased in the root tissue of maize in response to salinity. The reason behind the increase in succinate and malate in root is still unclear (Gavaghan et al., 2011). Since malate acts to balance the charge and osmotic pressure, the accumulation of malate can be considered as an adaptive response to salinity (Gavaghan et al., 2011). Salinity-induced downregulations of TCA cycle–related metabolites in *A. lagopoides* have been observed, and this may be due to the unavailability of sugars for energy metabolism due to the utilization of sugars for other activities such as osmoprotection and ROS scavenging during stress adaptation (Sobhanian et al., 2010). Similarly, the difference in the metabolite levels of TCA cycle such as malate and citrate is observed in *A. thaliana* and its halophytic counterpart *T. halophila* (Gong et al., 2005). Likewise, the halophyte *S. salsa* experiencing salinity has increased levels of malate and fumarate (Wu et al., 2012). However, significant difference is observed in response to salinity between the shoot and root of *S. salsa*. In the root no significant change in malate, fumarate, and allantoin is observed during salinity treatment (Wu et al., 2012). In *Lotus japonicus*, salinity-induced depletion of TCA cycle intermediates like pyruvic acid, citric acid, aconitic acid, malic acid, 2-oxoglutaric acid, shikimic acid, and quinic acid is correlated with the increase in the amino acid content, which suggests that most of the TCA cycle intermediates act as precursors for amino acid biosynthesis (Sanchez et al., 2008b). Metabolites related to TCA cycle respond differentially to salinity in different plant species, but their modulations are aimed only to adapt in the stressed environment.

5.1.6 Phytohormones Involved in Salinity Tolerance

Phytohormones, also known as plant growth regulators are considered as the most potent signaling molecules, and they are critical in modulating the physiological processes for stress adaptation of the plants (Fahad et al., 2015). Abscisic acid (ABA) is considered as an important phytohormone in mediating stress signaling as ABA signaling regulates the expression of genes that are responsible for stress tolerance (Sharma et al., 2005). ABA acts as an indispensable messenger in maintaining the water status of the plants in stressed condition. Salinity stress induces the accumulation of ABA both in root and leaf tissues, and it is transported throughout the plants. However, the compartmentation of ABA is influenced by the pH of the xylem/apoplast and hence the amount of ABA received by the stomata is regulated. Therefore ABA influences the stomatal opening and closing in stressed conditions thereby controlling the transpirational water loss. ABA-induced stomatal closure under salinity is related to H_2O_2 production, which serves as an intermediate signal for stomatal closure and is also related to the concentration of Ca^{2+} in the cytoplasm of the guard cell (Kim and Wang, 2010). Salt-induced accumulation of ABA has been reported in many plants, and the increase of ABA is associated with leaf water potential (Shahzad et al., 2017). It has also been reported that ABA also helps in foliar accumulation of Na^+ and Cl^- and causes an increase in leaf succulence and leaf area expansion (Cabot et al., 2009). Reduction in Na^+ levels in root and its translocation to shoot is among the most beneficial role of ABA in mitigating the salt effect (Khadri et al., 2007). Apart from the important roles in stress signaling and regulation of stomatal opening/closing, ABA also plays an important role in the synthesis of osmoprotectants like proline and dehydrins to mitigate the ROS effects under high-salinity conditions (Iqbal et al., 2014). In addition to ABA, auxins (indole acetic acid [IAA]) also play a significant role in plant growth and survival under stress environments (Eyidogan et al., 2012). The impacts of salinity on seed germination are mediated by auxin signaling, which is facilitated by the membrane-bound transcription factor NTM2 (Park et al., 2011). It has been reported that wheat seeds treated with auxin prior to salt exposure are able to mitigate the inhibitory effects of salt on germination by inducing the synthesis of salicylic acid (Iqbal and Ashraf, 2007). Cytokinins are also known to facilitate plant growth and development under saline conditions by alleviating the salt effects (Iqbal et al., 2006; Fahad et al., 2014). As reported by Iqbal et al. (2006) wheat seeds treated with cytokinin prior to salt treatment tend to tolerate high salt concentration. Cytokinins are known for their important role in growth regulation and tend to govern the salinity adaptation of the plants (Hadiarto and Tran, 2011). In addition to ABA, auxin, and cytokinin, gibberellic acid (GA) also imparts salt tolerance to plants. Significant and rapid increase in the levels of GA is typical in plants exposed to various abiotic stress like salinity, and it regulates the signaling of sugars and antioxidants, which in turn is advantageous for the physiology and metabolism of the plants under stress (Iqbal et al., 2011). It has been reported that the reduction in the levels of GA in *Arabidopsis* is correlated with reduced growth under salinity stress. It has also been reported that GA reduces the salinity effects by improving water status and water use efficiency and by decreasing stomatal resistance (Maggio et al., 2010).

5.2 Drought-Induced Alteration in Different Metabolites

Irregular and inadequate rainfall and unsustainable groundwater use for irrigated crops are putting extreme pressure on the global agriculture and food crop production. Water deficit or drought is mainly of two types; one is physical drought in which water availability in the soil is low due to insufficient rainfall and/or low ground water level and another is physiological drought in which water is available in the soil but plants are unable to take it up due to other environmental conditions such as salinity. The condition of drought imposes various physiological and biochemical alternations in plants, including reduction in photosynthesis and respiration and hence growth and development. At the onset of water deficit stress, the levels of stress-responsive proteins and metabolites are modulated to counter the prevailing stress condition. Under such conditions the reconfiguration of metabolites is one of the major defense strategies employed by the plants. The first and foremost effect of water deficit is the accumulation of ABA in the leaf tissues and reduction in transpiration rate by stomatal closure. Apart from stomatal adjustment another important strategy of plants to cope up with water deficit is osmotic adjustment. Plants maintain positive turgor pressure by accumulating compatible solutes like amino acids and sugars. In addition to the primary metabolites, plants synthesize many secondary metabolites like flavonoids and polyamines that not only act as an osmoprotectants but also act as ROS scavengers. Elevation of the cellular concentrations of proline, valine, β-alanine, glucose, malate, sugars, and inositol plays important roles in maintaining osmotic concentration during water deficit condition (Shanker et al., 2014). Metabolomics is a powerful tool to obtain comprehensive knowledge about the mechanism of regulation of metabolic pathways and regulation of different stress-responsive metabolites. Drought-induced alternations of different class of metabolites in plants are given in Table 5.2 and discussed in the following sections.

TABLE 5.2 Differential Regulations of Important Metabolites in Various Plant Species in Response to Drought and Water Logging

Abiotic Stress	Class of Metabolites	Metabolite	Plant Species	Regulation	Function	References
Drought		Organic acid	*Phaselous vulgaris*	↓	Precursor for many compounds	Silvente et al. (2012)
			Hordeum vulgare	↑		Wenzel et al. (2015)
			Oryza sativa	↑		Farooq et al. (2010)
		Spermidine	*Thellungiella salsuginea*	↓	Membrane integrity, signal transduction	Lugan et al. (2010)
			O. sativa	↓		Do et al. (2013)
			Haberlea rhodopensis	↑		Gechev et al. (2013)
			Craterostigma plantagineum	↑		Alcázar et al. (2014)
			Arabidopsis thaliana	↑		Alet et al. (2011)
		Putrescine	*H. vulgare*	↑	Membrane integrity, signal transduction	Lanzinger et al. (2015)
			Solanum tuberosum	↑		Evers et al. (2010)
	Polyols		*T. salsuginea*	↓	Signal transduction	Lugan et al. (2010)
			H. vulgare	↓		Wenzel et al. (2015)
			Pinus pinaster	↑		de Miguel et al. (2016)
			Eucalyptus pauciflora	↑		Warren et al. (2012)
		Galactinol	*H. vulgare*	↑	Signal transduction	Wenzel et al. (2015)
			H. rhodopensis	↓		Gechev et al. (2013)
			Eucalyptus dumosa	↑		Warren et al. (2012)
		Vibo-quercitol	*E. dumosa*	↑	Osmoprotectant and antioxidant	Warren et al. (2012)
	Sugar alcohols	Myoinositol	*S. tuberosum* (tolerant variety)	↓	Osmotic adjustment	Silvente et al. (2012)
			Drought-sensitive *Glycine max*	↑		Sun et al. (2016)
			Zea mays	↑		De Miguel et al. (2016)
			P. pinaster	↑		Silvente et al. (2012)
			Drought-sensitive *G. max*	↑		Alla et al. (2012)
		Pinitol	*Atriplex halimus*	↑	Osmoprotectant	de Miguel et al. (2016)
			P. pinaster	↑		Warren et al. (2012)
			E. pauciflora	↑		Wenzel et al. (2015)
		Galactinol	*H. vulgare*	↑	Osmoprotectant	Gechev et al. (2013)
			H. rhodopensis	↓		Warren et al. (2012)
			E. pauciflora	↓		de Miguel et al. (2016)
	Sugars	Altrose	*Nicotiana tabacum*	↑	Osmoprotectant	Lanzinger et al. (2015)
		Trehalose	*H. vulgare*	↑	Osmoprotectant	Shi et al. (2015a)
		Raffinose	*Brachypodium distachyon*	↑	Osmoprotectant	Shi et al. (2015a)
		Mannobiose	*B. distachyon*	↑	Osmoprotectant	Evers et al. (2010)

TABLE 5.2 Differential Regulations of Important Metabolites in Various Plant Species in Response to Drought and Water Logging—cont'd

Abiotic Stress	Class of Metabolites	Metabolite	Plant Species	Regulation	Function	References
	Amino acids	Proline	*S. tuberosum* (tolerant variety)	↑	Protein structure stabilization, promotes activity of antioxidant enzymes	Krasensky and Jonak (2012)
			A. thaliana	↑		Alla et al. (2012)
			A. halimus	↑		Gechev et al. (2013)
			H. rhodopensis	↓		Parida and Jha (2013)
			Salicornia brachiata	↑		Shi et al. (2015a)
			B. distachyon	↑		Sun et al. (2016)
			Z. mays	↑		Lugan et al. (2010)
			T. salsuginea	↓		Alla et al. (2012)
		Serine	*A. halimus*	↑	Protein amino acid and plays a catalytic function in many enzymes	Shi et al. (2015a)
			B. distachyon	↑		Alla et al. (2012)
			A. halimus	↑		Warren et al. (2012)
		Methionine	*E. dumosa*	↑		Alla et al. (2012)
		Hydroxyproline	*A. halimus*	↑	Structural integrity of proteins	Sun et al. (2016)
		Tyrosine	*Z. mays*	↑	Osmoprotectant, cell signaling	Alla et al. (2012)
			A. halimus	↑		Sun et al. (2016)
		Valine	*Z. mays*	↑	Protein amino acid	Muscolo et al. (2015)
			Lens culinaris	↑		Muscolo et al. (2015)
		Asparagine	*L. culinaris*	↑	Amino group donor for biosynthesis of glycine and indicator of stress	de Miguel et al. (2016)
		Ornithine	*P. pinaster*	↑	Osmoprotectant	Gechev et al. (2013)
		GABA	*N. tabacum*	↑	Osmoprotectant, growth regulators, nitrogen and carbohydrate metabolism	Lanzinger et al. (2015)
			H. vulgare	↑		Witt et al. (2012)
			L. culinaris	↓		Lanzinger et al. (2015)
	TCA cycle metabolites	Succinate	*H. vulgare*	↑	Growth and energy metabolism	Griesser et al. (2015)
			Vitis vinifera	↑		Sun et al. (2016)
			Z. mays	↑		Silvente et al. (2012)
			G. max	↓		Alla et al. (2012)
			A. halimus	↑		Lanzinger et al. (2015)
		Malate	*H. vulgare*	↑	Essential role in nitrogen fixation and phosphorus acquisition	Griesser et al. (2015)
			V. vinifera	↑		de Miguel et al. (2016)
			P. pinaster	↑		Warren et al. (2012)
			E. pauciflora	↑		Warren et al. (2012)
		2-Oxoglutarate	*E. pauciflora*	↓	TCA cycle intermediate	Grimplet et al. (2009)
		Shikimic acid	*V. vinifera*	↑	Maintenance of osmotic pressure, precursor to aromatic amino acids and other secondary metabolites	Griesser et al. (2015)

Continued

TABLE 5.2 Differential Regulations of Important Metabolites in Various Plant Species in Response to Drought and Water Logging—cont'd

Abiotic Stress	Class of Metabolites	Metabolite	Plant Species	Regulation	Function	References
	Phenols	Catechin	*V. vinifera*	↑	Antioxidant	Griesser et al. (2015)
		Kaempferol-3-O-glucoside	*Vigna radiata*	↑	Antioxidant	Kreuzwieser et al. (2009)
	Hormones	Abscisic acid	*Gray poplar*	↑	Regulation of growth	Kumutha et al. (2008)
Waterlogging	Carbohydrate	Starch	*Cajanus cajan*	↓	Storage, osmotic adjustments, structural integrity	Pearson et al. (2013)
		Fructose	*Pinus sylvestris*	↑	Superoxide scavenging, osmoprotectant	Pearson et al. (2013)
	Polyamines	Spermidine	*Prunus persica*	↑	Membrane integrity, signal transduction	Tuo et al. (2015)

↑ and ↓, respectively, indicate up- and dowregulation of different metabolites.
GABA, γ-amino-*N*-butyric acid; *TCA*, tricarboxylic acid.

5.2.1 Metabolites Related to Photosynthesis

Being an important process of primary metabolism photosynthesis plays a vital role in the plant's performance under drought condition (Pinheiro and Chaves, 2011). Under water deficit condition, photosynthesis and turgor can be maintained by the plants for a short period. However, with an increase in the extent of stress the growth of the plants ceases with decrease in photosynthesis. Acclimation of plants under drought conditions is governed by the regulation of many small molecules or metabolites, which are mostly intermediates of photosynthesis. It has been reported that the enrichment of photosynthesis in *Brachypodium distachyon* under drought stress is mainly related to modulations of primary metabolites (Shi et al., 2015a). Higher accumulation of ribose reported in leaves of drought-stressed grapevine behaves as a stress indicator (Griesser et al., 2015). Similarly, increased levels of sucrose and ribose have been reported in drought-stressed *Pinus pinaster*, which play significant roles in the allocation of carbon resources and plant development (de Miguel et al., 2016). Glucose-6-phosphate dehydrogenase plays a protective role against ROS and provides an alternative way for the production of nicotinamide adenine dinucleotide phosphate (NADPH) (Liu et al., 2013). Although several reports of metabolomics studies have shown the modulations of different metabolites related to photosynthetic intermediates, their regulatory roles in drought tolerance are yet to be elucidated.

5.2.2 Metabolites Associated With Energy and Carbohydrate Metabolism

Metabolomics plays a major role in studying the effects of drought as an isolated stress factor in plant species including crops, vegetables, and the model plant *A. thaliana* (Obata and Fernie, 2012). Sensitive and tolerant varieties tend to respond differently to the prevailing stress conditions by differential regulations of their metabolite profile. It is evident from many reports that the metabolic profiles of drought-sensitive varieties are more affected than the drought-resistant varieties, as reported in eucalyptus, soybean, wheat, and lentils (Warren et al., 2012; Silvente et al., 2012; Muscolo et al., 2015). Several metabolites associated with energy and carbohydrate metabolism are altered in response to drought stress. It has been reported that the sucrose level in the leaves of drought-tolerant variety of soybean does not change significantly, whereas it decreases significantly in the sensitive variety (Evers et al., 2010). In contrast, the level of galactose is elevated considerably in the tolerant variety compared with the sensitive variety of soybean (Evers et al., 2010. From the study conducted in tomato fruit, it is evident that the metabolites of fruit surface (pericarp) can also contribute to the drought tolerance in this plant (Luengwilai et al., 2012). The study on maize suggests a significant accumulation of sucrose, glucose, and fructose in response to drought (Sun et al., 2016). Metabolic profiling of *P. pinaster* under drought stress suggests an increase in the levels of compatible solutes like malate, aspartate, α-tocopherol, luteolin-7-O-glucoside, and sugars (de Miguel et al., 2016). Plant responses to abiotic stresses like high salinity, cold, and drought depend on the stress-induced linear alteration of the threshold levels of various metabolites (Sanchez et al., 2008b, 2012). Similar to glycophytes, halophytes also synthesize various carbohydrates like sucrose, mannose, and trehalose for adjustment of the osmotic environment of the cells

(Wijayasinghe et al., 2017). The xero-halophyte *A. halimus* of coastal and inland areas responds differently to drought stress (Evers et al., 2010). For example, the enrichment of soluble sugars is higher in the inland arid variety than in the coastal one (Evers et al., 2010). The alteration of metabolites depends upon the intensity and duration of exposure to the stress condition, i.e., no significant accumulation of sugars is observed in *A. halimus* after 3 or 6 days of treatment, but the accumulation is immense after 10 days of water withholding (Alla et al., 2012). All the aforestated information indicates that the metabolites can be valuable and potential biomolecules to be considered for the production of drought-resistant crops. Cellular levels of polyols are also modulated in response to drought for conferring stress tolerance (Silvente et al., 2012). Drought-sensitive genotype of soybean shows a decrease in myoinositol level with no change in tolerant variety (Silvente et al., 2012). In contrast, pinitol content in sensitive variety increases under drought, but it is found to be decreased in the tolerant variety (Silvente et al., 2012). It has been reported that when both drought-tolerant and drought-sensitive genotypes of potato are subjected to drought, increased levels of galactinol and inositol are observed in the drought-tolerant genotype than the sensitive one (Evers et al., 2010). Likewise, drought-stressed *P. pinaster* shows an enhanced accumulation of inositol and pinitol (de miguel et al., 2016). Increased synthesis of green leaf volatile (GLV)-like (Z)-3-hexenal and its related aldehydes has been reported in *Punica granatum,* and it has been suggested that GLV has a signaling role during stress. Similarly increased endogenous levels of (E)-2-pentenal and (E)-2-hexenal have also been documented in *Arabidopsis* and tobacco plants (Yamauchi et al., 2014). By analyzing the data available in the literature, it can be confirmed that under stressful environment like drought judicious reconfigurations of the metabolic profile take place in the plants to downregulate the energy-consuming process.

5.2.3 Metabolites Associated With Amino Acid Metabolism

Plants synthesize an array of different amino acids in response to various abiotic stresses including drought. Up- and downregulation of the enzymes responsible for amino acid metabolism regulate the levels of the amino acids. For example, overexpression of D-1-pyrroline-5-carboxylate synthetase, which is an enzyme of the proline biosynthetic pathway, increases proline accumulation and enhances drought tolerance (Yamada et al., 2005). Apart from the upregulation of the synthesizing enzymes, the downregulation of the enzymes that are responsible for proline degradation also help in accumulation of proline under stress condition. Asparagine acts as an amino group donor in the biosynthesis of glycine, which is a photorespiration intermediate. It also serves as an indicator of drought stress in drought-tolerant and drought-sensitive plants (Shi et al., 2014). An increase in the levels of amino acids such as phenylalanine, tyrosine, ornithine, and asparagine has been detected under drought conditions in different plant species (Muscolo et al., 2015). Under stress condition, the plant's priority is to survive rather than to grow and develop. Therefore the changes in metabolites occur accordingly to support the survival of the plants. For example, the levels of amino acids increase in the drought-sensitive genotype of lentil, whereas the levels of organic acids decline (Muscolo et al., 2015). Similarly, the transient increase in L-3,4-dihydroxyphenylalanine (L-DOPA) has been reported in the sensitive genotypes of grass and lettuce (Soares et al., 2014). L-DOPA is a nonprotein amino acid having antioxidant properties that protect cellular structures from the damaging effects of ROS under stress condition. Drought-induced accumulations of proline and GB have also been reported in the coastal and inland arid varieties of *A. halimus* (Hassine et al., 2008). Accumulation of proline is found to be higher in the inland arid variety than in the coastal variety, whereas the accumulation of GB is more in the coastal variety. Higher accumulation of proline and proline analogs has also been reported in drought-tolerant lentil, conferring drought tolerance by maintaining cellular osmotic concentration (Evers et al., 2010). These reports suggest that the plants can simultaneously use different osmolytes for osmoregulation and protection of cellular structures. However, the kinetics of accumulation depends partly on the environmental factors and the habitat of the species (Hassine et al., 2008). In another report on *A. halimus* it has been inferred that polyethylene glycol (PEG)-induced drought could not enhance GB accumulation in *A. halimus* but that the addition of NaCl induces the accumulation of GB under water stress condition (Martinez et al., 2005). In contrasts, PEG-mediated drought had no impact on proline accumulation in *A. halimus* (Martinez et al., 2005). As discussed earlier the elevated cellular concentration of proline may be due to the increased activity of synthesizing enzymes and/or the decreased activity of degrading enzymes. Slama et al. (2007) have reported that the increased activity of ornithine aminotransferase and reduced activity of proline dehydrogenase may be the reason for the enhanced accumulation of proline in *Sesuvium portulacastrum.* In addition to osmoregulation, accumulation of proline may be attributed to other functions like enzyme protection and membrane stabilization (Parida and Jha, 2013). Drought-stressed maize exhibits higher accumulation of proline, isoleucine, leucine, and valine, whereas the levels of tryptophan and tyrosine decrease in maize under drought conditions (Sun et al., 2016). In the leaves of drought-stressed soybean, declines in the levels of alanine and glutamate have been reported in both drought-tolerant and drought-sensitive genotypes, whereas the GABA content declines in the tolerant variety only and aspartate content increases in the sensitive genotype (Silvente et al., 2012). The halophytic plants that grow in saline and arid environments frequently experience salt and drought stress. Like glycophytic species, the first and foremost action that halophyte take to counter the prevailing water stress condition is the adjustment of the osmotic concentration by synthesizing compatible solutes like proline and QACs such as GB, proline betaine, hydroxyproline betaine, and alanine betaine (Parida and Jha, 2013). Drought-induced accumulation of total free amino acid and proline has also been reported in the halophyte *Salicornia brachiata* (Parida and Jha, 2013), *S. portulacastrum* (Slama et al., 2007) and *Limonium linifolium* (Tabot and Adams, 2014).

5.2.4 Metabolites Related to Reactive Oxygen Species Scavenging

Plants synthesize an array of secondary metabolites to counter the effects of excess ROS that is generated due to the stressed conditions. Important ROS scavenging secondary metabolites include polyamines, flavonoids, ascorbate, and glutathione. Among the polyamines, spermidine, putrescine, and spermine are most common in higher plants and have an imperative role in plant stress management (Alcazar et al., 2011). The levels of polyamines are directly proportional to the stress tolerance of plants, conferring membrane integrity in adverse oxidative stress conditions (Hussain et al., 2011). It has been reported that higher levels of putrescine in response to stress-induced activation of arginine decarboxylase confer drought tolerance in *Arabidopsis*. Similarly, induction of the enzyme spermidine synthase under drought condition lead to the accumulation of higher amount of spermidine, which in turn confers drought tolerance in *Arabidopsis* (Krasensky and Jonak, 2012). It has been reported that being a precursor of spermidine and spermine, the level of putrescine decreases, whereas the level of spermine increases in drought-treated rice. Arginine, which is the main substrate for polyamine biosynthesis, is found to be accumulated increasingly during drought condition in rice (Do et al., 2013). However, the level of spermidine decreases in the cultivars of rice subjected to drought (Do et al., 2013). In contrast, the level of spermidine is found to be increased in oat and wheat in response to drought. Polyphenols are categorized as potent nonenzymatic ROS scavengers in plants. The increased accumulation of polyphenols has been reported in *S. brachiata*, conferring protection from ROS-induced damage to the plant (Parida and Jha, 2013). The antioxidant activity of polyphenols may be attributed to its utilization as a substrate for peroxidase-scavenging H_2O_2. The accumulation of polyphenols in plants is a common strategy for stress adaptation. However, the extent of accumulation depends on the plant species, tissue, age, severity, and duration of stress (Parida and Jha, 2013; Yildiz-Aktas et al., 2009). Apart from *S. brachiata*, drought-induced accumulations of polyphenols have also been reported in many other plant species (Warren et al., 2012; Griesser et al., 2015). Ascorbate is one of the important metabolites that is abundant in green leaves, and it plays an important role in ROS scavenging during stress condition. The accumulation of ascorbate under stress depends on the activity of the enzyme L-galactono-1,4-lactone dehydrogenase and varies from species to species and even among the cultivars of the same species (Bartoli et al., 2005). The ascorbate–glutathione cycle is one of the important nonenzymatic antioxidant system that mitigates the effect of ROS under stress condition. A study conducted on drought-resistant wheat shows a decrease in ascorbate content, whereas the application of salicylic acid elevates the ascorbate content up to 50% (Kang et al., 2013). On the other hand, the GSH level increases under drought condition and application of exogenous salicylic acid further accelerates the accumulation of GSH and hence increasing drought tolerance (Kang et al., 2013). The increased levels of GSH and ascorbate are the results of increased transcription of ascorbate–GSH cycle–related genes and thus provide drought tolerance (Kang et al., 2013).

5.2.5 Metabolites Associated With Mitochondrial Respiration

Mitochondrial respiratory pathway, like the TCA cycle, is very sensitive to drought stress. Drought stress affects the TCA cycle more adversely by decreasing the levels of TCA cycle intermediates like citrate, 2-oxoglutarate, and succinate in many plant species (Muscolo et al., 2015). In addition to the conventional osmolytes like proline, soluble sugars like sucrose, organic acids such as 2-oxoglutaric acid, and allantoin can also act as osmoprotectants (Silvente et al., 2012). The study conducted on maize seedlings suggests increased levels of TCA cycle metabolites like malate and succinate and decreased levels of shikimate and fumarate in response to drought (Sun et al., 2016). The metabolic profiling of *P. pinaster* suggests the upregulation of TCA cycle with an increased accumulation of malate and aspartate under drought stress (de Miguel et al., 2016). The xerohalophyte *A. halimus* shows upregulation of methyltetrahydropteroyl glutamate, fumarate, maleate, glutarate semialdehyde, ketoisovalerate, and aspartate in response to drought stress (Alla et al., 2012). In a study conducted on potato, an increase the levels of five TCA cycle intermediates, i.e. citrate, malate, fumarate, succinate, and isocitrate, has been reported, and these metabolites may have some important role in conferring drought tolerance in potato (Vasquez-Robinet et al., 2008). Among the aforementioned five metabolites, citrate is the most abundant TCA cycle intermediate in drought-stressed potato (Vasquez-Robinet et al., 2008). The higher TCA cycle rate suggests higher mitochondrial metabolic activity generating more reductant and adenosine triphosphate (ATP) during stressed condition (Vasquez-Robinet et al., 2008).

5.2.6 Phytohormones Involved in Drought Tolerance

Hormonal regulation is one of the most significant response to water deficit. Among the phytohormones, ABA is most important in triggering various endogenous signals for acclimation during drought (Li et al., 2017). The endogenous level of ABA is determined by the ratc of its biosynthesis and catabolism. ABA is synthesized from xanthoxin, which in turn is derived from the direct cleavage of C40 carotenoids (Seki et al., 2007). The cleavage is catalyzed by the cleavage enzyme 9-cis-epoxycarotenoid dioxygenase (NCED). The vital role of NCED in conferring drought tolerance by inducing drought-responsive biosynthesis of ABA is well established in *Arabidopsis* (Seki et al., 2007). ABA concentration remains same in control and drought-acclimated laurel seedlings, which suggests that low leaf water potential (−1.5 MPa) in control plants is enough to induce ABA synthesis and the same level of ABA is also synthesized in drought-treated seedlings at a lower leaf water potential of −2.5 MPa (Yildiz-Aktas et al. 2009). However, IAA concentration is found to be decreased in drought-acclimated seedlings of laurel, and this may be related to reducing growth and leaf area expansion in drought-treated seedlings. The decline in IAA content has also been reported by Wang et al. (2008). Yildiz-Aktas et al. (2009) have reported that increased level of GA_3 and zeatin in drought-acclimated seedlings may be attributed to the more negative water potential of leaf (−2.5 MPa). It has been suggested that ABA and GA have antagonistic role in many processes. Nemhauser et al. (2006) have reported that ABA opposed growth promotion by GA in *Arabidopsis*. It has also been reported that drought resistance is conferred by inhibiting GA production (Yuan et al., 2017). In the model plant *Arabidopsis*, it has been explained that biosynthesis of GA promotes degradation of DELLA proteins, which in turn improves stress resistance by coordinating the salicylic acid biosynthetic pathway (Alonso-Ramírez et al., 2009). The coordination of salicylic acid with other hormones in drought condition is still unknown. However, reduced drought tolerance has also been observed in maize seedlings to which salicylic acid has been externally applied (Nemeth et al., 2002). Therefore salicylic acid–induced drought tolerance relies on the relative levels of endogenous and exogenous salicylic acid (Horvath et al., 2007).

5.3 Waterlogging-Induced Alternation in Different Metabolites

Water stress is not only related to the deficit of water but also associated with water logging or excess irrigation that impairs water uptake by the plants. Soil waterlogging above the field capacity depletes O_2 level in the soil faster because of lower diffusion of this gas in water in addition to the consumption of O_2 by the plant root (Mittler et al., 2012). This depletion of O_2 level is directly proportional to the biomass ratio of root and soil microbiota (Mittler et al., 2012). Waterlogging induces hypoxic condition within a few hours, and prolonged waterlogging produces anoxic condition (Kumutha et al., 2008). Root hypoxia and anoxia are naturally caused by flooding and excess rainfall, causing lack of oxygen in the root (Barnawal et al., 2012). Waterlogging affects many vital physiological functions of the plants such as photosynthesis, stomatal closure, and growth inhibition of leaves, stem, and roots leading to reduction in crop growth and yield. Waterlogging induces alteration of various plant metabolites that are related to important cellular processes like energy and carbohydrate metabolism, amino acid metabolism, and phytohormone biosynthesis, which are mentioned in Table 5.2 and discussed under the following headings.

5.3.1 Metabolites Associated With Energy and Carbohydrate Metabolism

Waterlogging affects the energy metabolism and translocation of carbohydrates. The effects of waterlogging are also associated with temperature as high temperature during waterlogging is more detrimental than low temperature (Voesenek et al., 2014). However, low-temperature waterlogging is associated with carbohydrate crisis and may hinder energy dissipation and cellular homeostasis. Waterlogging tends to disturb vital physiological functions of the plants, which ultimately results in decreased crop growth and productivity. It tends to reduce gas exchange and induces acute energy crisis (Voesenek et al., 2014). Waterlogging stress induces the replacement of aerobic respiration with glycolysis and fermentation by breaking down sucrose using sucrose synthase to glucose-6-phosphate and finally to pyruvate, which can be fermented by alternative fermentation pathways (Brownstein et al., 2013). The study on rice coleoptiles has shown the enhanced activity of pyruvate phosphate dikinase and phosphoenol pyruvate carboxykinase and inhibition of pyruvate kinase (Lasanthi-Kudahettige et al., 2007). The principal end products of photosynthesis, i.e., sucrose and starch, are mostly altered by waterlogging. It is evident that the waterlogging-tolerant cultivars have more carbohydrates than the sensitive cultivars. The accumulation of starch increased in the leaf chloroplast during waterlogging in the sensitive variety because of the blocked phloem transport of photosynthates (Ferner et al., 2012). However, upregulated carbohydrate transporters, enhanced phloem loading, and better soluble carbohydrate partitioning are the key mechanisms behind waterlogging tolerance (Ferner et al., 2012). The enzymes related to sucrose metabolism are affected by waterlogging. The activities of sucrose synthase, sucrose phosphate synthase, and acid invertase are enhanced with the decrease in the rubisco activity in the waterlogged anaerobic condition

(Harada and Ishizawa, 2003). The enhanced activity of sucrose synthase suggests the important role of this enzyme in sucrose metabolism in anoxic condition (Harada and Ishizawa, 2003). Sucrose synthase reversibly catalyzes the conversion of sucrose and uridine diphosphate (UDP) to UDP-glucose and fructose, which provides adequate sugar supply during anoxic condition. Many workers have revealed that the catalysis of sucrose cleavage by invertase in roots is substituted by sucrose synthase under waterlogging condition (Kreuzwieser et al., 2009). It has been reported that as a result of higher fatty acid and starch catabolism, increased accumulation of soluble sugar is observed in waterlogged *Arabidopsis* (Hsu et al., 2011). Depletion in adenylate energy charge (AEC), which represents the depletion of usable energy (ATP) in the cells, is attributed to the hypoxia/anoxia-related injury. The maintenance of AEC in tolerant species of *Cajanus cajan* is ensured by the adequate supply of the substrates for glycolytic pathway and higher concentration of reducing and nonreducing sugars (Kumutha et al., 2008).

5.3.2 Metabolites Associated With Amino Acid Metabolism and Reactive Oxygen Species Scavenging

Waterlogging generally induces amino acid accumulation to enhance osmoprotection during hypoxic and anoxic conditions. However, inhibition of amino acid metabolism has been reported during prolonged waterlogging in *L. japonicus,* which suggests that inhibition of amino acid metabolism is an adaptive response of plants to the hypoxic condition (Rocha et al., 2010). The inhibition of amino acid metabolism can be attributed to the decreased rate of nitrogen fixation due to hypoxia-induced energy deprivation. It has been reported that the transgenic lotus overexpressing nitrogenase gene tends to accumulate higher amino acid in the shoot and that it might be due to the increased nitrogenase activity in oxygen-deficient waterlogged condition (Rocha et al., 2010). However, in roots of transgenic lotus, the concentration of amino acid that is derived from the intermediates of TCA cycle declines considerably with the increase in the concentration of three amino acids, i.e., alanine, glutamate, and GABA (Rocha et al., 2010). It is interesting that the accumulation of alanine occurs at the cost of other amino acids, which suggests the importance of alanine for survival of the plants under waterlogged condition. It has been reported that waterlogging stress induces a proline burst in the leaves of peach (Tuo et al., 2015). Similarly, waterlogging-induced accumulation of proline has also been reported in *Ocimum sanctum* (Barnawal et al., 2012). Higher concentration of proline helps to mitigate the cellular redox potential and scavenge the hydroxyl radicals generated due to water logging (Barnawal et al., 2012). In waterlogged *Arabidopsis* accumulation of soluble sugars has been observed, which may be attributed to the enhanced starch and fatty acid catabolism (Hsu et al., 2011). As discussed earlier, polyamines (spermidine, spermine, and their precursor putrescine) have important roles in conferring tolerance to drought, but their role in waterlogging is yet to be elucidated. Some workers have shown that the exogenous application of putrescine increases the plant's tolerance to waterlogging (Yiu et al., 2009; Gill and Tuteja, 2010b). It has also been reported that free spermidine and spermine increase in Scots pine seedlings under waterlogged condition (Pearson et al., 2013). Unlike polyamines the role proline in waterlogging stress tolerance is well established.

5.3.3 Metabolites Associated With Mitochondrial Respiration

Plants adopt different strategies to improve the consequences of oxygen deficiency due to flooding or waterlogging. Among the different types of adaptations are changes in the anatomical and morphological features like formation of aerenchyma, stem elongation, and gas film formation surrounding the flooded leaves (Yamauchi et al., 2014). Apart from the anatomical alternations, other adaptations includes downregulation of energy-consuming metabolism under hypoxic condition (Rocha et al., 2010). Such metabolic adaptation includes downregulation of storage metabolism and inhibition of mitochondrial respiration. Flooding or waterlogging affect the activity of many TCA cycle–related enzymes like pyruvate dehydrogenase, aconitase, NAD^+-dependent isocitrate dehydrogenase, 2-oxoglutarate dehydrogenase, NAD^+-dependent malate dehydrogenase, and NAD^+-dependent malic enzyme (Rocha et al., 2010). Metabolite profiling and enzyme assay of waterlogged *L. japonicus* root suggest a link between glycolysis and TCA cycle fluxes during hypoxia, and the link is established by alanine aminotransferase. The fluxes are particularly from oxaloacetate to malate and 2-oxoglutarate to succinate. Accumulation of alanine and succinate under waterlogged condition provides additional source for ATP by substrate-level phosphorylation through the action of succinate CoA ligase (Rocha et al., 2010). Accumulation of succinate during hypoxia and reduction of fumarate back to succinate by the activity of succinate dehydrogenase in the presence of reduced ubiquinone are also evident (Fan et al., 2003). Oxygen deficiency under waterlogging condition severely affects mitochondrial respiration leading to decrease in ATP synthesis for energy-demanding processes. Oxygen deficiency occurs due to multiple factors like replenishment of oxygen through photosynthesis and cellular consumption of oxygen through metabolic activity. However, plants encounter this energy crisis by increasing ATP synthesis by substrate-level phosphorylation. Alternative metabolism of pyruvate to alanine provides an nondetrimental product of anaerobic respiration such as 2-oxoglutarate, which can be further converted to succinate, providing additional ATP for energy-consuming metabolic processes (Bailey-Serres et al., 2012).

5.3.4 Phytohormones Involved in Waterlogging Tolerance

Plant hormones have significant roles in the morphological and anatomical changes that enable the plants to tolerate waterlogging. Ethylene plays an important role in aerenchyma and adventitious root formation along with the regulation of shoot elongation (Hu et al., 2016). The biosynthesis of ethylene is mediated by two enzymes, namely, 1-aminocyclopropane-1-carboxylic acid (ACC) synthase, which converts S-adenosylmethionine to ACC, and ACC oxidase, which converts ACC to ethylene. The activity of these enzymes is rapidly enhanced in tissues under waterlogging condition. Ethylene either when endogenously produced or when exogenously applied is involved in root aerenchyma formation in waterlogged plants belonging to the family Gramineae such as maize, rice, and wheat (Yamauchi et al., 2014). It has also been verified by the application of inhibitors of ethylene biosynthesis and action, e.g., aminoethoxyvinylglycine (Hu et al., 2016). However, the ethylene or ethylene inhibitors do not affect the aerenchyma formation if applied at higher concentrations, as reported in *Juncus effusus* (Visser and Bögemann, 2006). Auxin is a major phytohormone in initiating adventitious root formation because of its regulatory function in redifferentiation of shoot and root apical meristems (Verstraeten et al., 2013). Adventitious root formation is inhibited in waterlogged tomato plant treated with 1-naphthylphthalamic acid (NPA), an inhibitor of auxin transport (Vidoz et al., 2010). Along with auxin, the positive role of ethylene in adventitious root development and emergence under waterlogging condition has also been well established (Verstraeten et al., 2013). However, the coordination between auxin and ethylene signaling is important for adventitious root formation under flooding stress.

5.4 Alternation of Metabolites in Response to Temperature Extremes

Any alternation in the optimum temperature required for the growth and development of the plants is harmful to the plants, causing catastrophic effects on growth and development. The alternation may include high temperature leading to heat stress or low temperature leading to chilling and freezing stress. Plants belonging to the tropical zones, where high temperature prevails throughout the year, or to frigid zones, with prevailing low temperature, have to maintain the osmotic environment of their cells by altering their cellular metabolites. Temperatures above the normal or optimum for the growth and development of the plants are recognized as stress by the plants. The high-temperature condition disturbs cellular homeostasis and can severely affect the growth and development of the plants (Zandalinas et al., 2017). Under high temperature plants accumulate an array of secondary metabolites that include alanine, allantoin, arachidic acid, 2-ketoisocaproic acid, myoinositol, putrescine, and rhamnose (Luengwilai et al., 2012).

Plants in temperate zone frequently experience a low temperature that has an adverse impact on growth and development (Megha et al., 2017). The low-temperature stress includes (1) chilling (below 20°C) and (2) freezing stress (below 0°C). These low-temperature stresses cause disturbances in cellular homeostasis, membrane damage, inhibition of phloem transport, respiration, and plastid synthesis (Shi et al., 2012). Plants have developed numerous biochemical and physiological mechanisms to survive in this condition. During cold seasons, plants synthesize many cryoprotectant metabolites such as soluble sugars, various amino acids, polyamines, and sugar alcohol to increase their tolerance to low-temperature stress (Pathak et al., 2014). In *Arabidopsis* out of 434 metabolites identified by GC-MS, the levels of 325 metabolites increase in response to chilling condition (Cook et al., 2004). These metabolites are involved in urea cycle (ornithine, citrulline, and precursors of polyamine biosynthesis) and TCA cycle (malate, ketoglutarate, fumarate, and citrate), which suggest that upregulation of urea cycle and citric acid cycle occur in *Arabidopsis* plants exposed to chilling stress (Cook et al., 2004). The chilling adaptation mechanism includes the enhanced activity of various stress-related genes, which in turn lead to the accumulation of stress-responsive metabolites. The modulations of important metabolites in various temperature extreme conditions have been mentioned in Table 5.3 and are discussed in the following sections.

5.4.1 Alternation of Photosynthesis-Related Metabolites

Optimum temperature is essential for proper functioning of the photosynthetic machinery and the activity of photosynthesis-related enzymes. With any fluctuation in optimal temperature, the rate of photosynthesis is affected and so are the related metabolites. Alteration of the biochemical reaction of photosynthesis is considered as the primary effect of heat. Depending upon the duration and intensity of the exposure, heat can irreversibly damage ribulose 1,5-bisphosphate

TABLE 5.3 Differential Regulations of Important Metabolites in Various Plant Species in Response to Temperature Extreme and High Light Intensity

Abiotic Stress	Class of Metabolites	Metabolite	Plant Species	Regulation	Function	References
Heat stress	Amino acids	Alanine	*Solanum lycopersicum*	↑	Precursor for protein and nitrogen-containing compounds and acts as a signaling molecule	Luengwilai et al. (2012)
		Proline	*Nicotiana tabacum*	↑	Protein structure stabilization, promotes activity of antioxidant enzymes	Cvikrová et al. (2012)
			S. lycopersicon	↑		Rivero et al. (2014)
			Glycine max	↑		Chebrolu et al. (2016)
	Organic acids	Allantoin	*S. lycopersicum*	↑	Nitrogen transport and assimilation	Luengwilai et al. (2012)
	Fatty acid	Arachidic acid	*S. lycopersicum*	↑	Reconstruction of cell ultrastructure, isoprenoid biosynthesis	Luengwilai et al. (2012)
		2-Ketoisocaproic acid	*S. lycopersicum*	↑	Branched-chain amino acid catabolism	Luengwilai et al. (2012)
	Polyamines	Putrescine	*Lycopersicon esculentum*	↑	Acid-neutralizing and antioxidant properties	Cheng et al. (2009)
			S. lycopersicum	↑		Luengwilai et al. (2012)
			N. tabacum	↑		Cvikrová et al. (2012)
		Spermine	*L. esculentum*	↑	Membrane integrity, signal transduction	Cheng et al. (2009)
			N. tabacum	↑		Cvikrová et al. (2012)
	Sugars	Fructose	*S. lycopersicum*	↓	Superoxide scavenging, osmoprotectant	Luengwilai et al. (2012)
			S. lycopersicon	↓		Rivero et al. (2014)
			G. max	↑		Chebrolu et al. (2016)
	Flavonoids		*S. lycopersicum*	↑	Structural integrity, signaling, ROS scavenging	Gill and Tuteja (2010b)
			G. max	↑		Chebrolu et al. (2016)
		Tocopherols	*G. max*	↑	Membrane integrity, scavenge lipid peroxy radicals	Chebrolu et al. (2016)

TABLE 5.3 Differential Regulations of Important Metabolites in Various Plant Species in Response to Temperature Extreme and High Light Intensity—cont'd

Abiotic Stress	Class of Metabolites	Metabolite	Plant Species	Regulation	Function	References
Low temperature	Carbohydrate	Trehalose	*Arabidopsis rosettes*	↑	Membrane protection	Sowiński et al. (2005)
		Sucrose	*Oryza sativa*	↑	Cryoprotectant	Maruyama et al. (2014)
	Amino acids	Ornithine	*Arabidopsis thaliana*	↑	Metabolic adjustment	Cook et al. (2004)
	Galactolipid	MGDG	*A. thaliana*	↓	Membrane stabilization	Degenkolbe et al. (2012)
	Phospholipid	PE	*A. thaliana*	↑	Maintain membrane fluidity	Degenkolbe et al. (2012)
	Carotinoid	Zeaxanthin	*Spinacia oleracea*	↑	Energy dissipation	Latowski et al. (2011)
	Flavonoids	Quercetin	*Braseica oleracae*	↑	Antioxidant	Neugart et al. (2016)
		Luteolin-7-glucuronide	*Salix* sp.	↑	UV absorbent	Tegelberg and Julkunen-Tiitto (2001)
		Quercetin 3-(6″-malonyl-glucoside)	*Lactuca sativum*	↑	UV absorbent	DuPont et al. (2000)

↑ and ↓, respectively, indicate up- and downregulation of different metabolites.
MGDG, monogalactosyldiacylglycerol; *PE*, phosphotidylethanolamine; *ROS*, reactive oxygen species; *UV*, ultraviolet.

carboxylase/oxygenase, the thylakoid membranes, and the chloroplast ultrastructure (Chen et al., 2012). High temperature has greater adverse effects on the photosynthesis of C_3 plants than of C_4 plants (Hasanuzzaman et al., 2013). The decrease in photosynthesis under heat stress is mainly attributed to degradation of photosynthetic pigments and decrease in photosystem II activity (Marchand et al., 2005). In soybean, heat stress significantly decreases the chlorophyll content and photosynthetic rate, as a result of which a decline in sucrose level is observed (Tan et al., 2011).

5.4.2 Metabolites Associated With Energy and Carbohydrate Metabolism

High temperature affects many metabolites that are related to energy and carbohydrate metabolism. A study on soybean seed metabolome reports that higher accumulation of carbohydrate derivatives such as glycolate, ribose 1,3-dihydroxyacetone takes place in tolerant variety of soybean than in the sensitive variety. In contrast the levels of other carbohydrate derivatives such as arabitol, erythritol, pinitol, and chiroinositol are found to be lower in tolerant variety of soybean than sensitive variety (Chebrolu et al., 2016). The glucose and fructose levels are found to be consistently higher in both tolerant and sensitive varieties of soybean, whereas the levels of sugar alcohols such as fucitol and mannitol are found to be lower in both the genotypes (Chebrolu et al., 2016). It has been reported that the accumulation of sucrose, glucose, and fructose in the leaves in response to combined effects of salinity and heat stress can lead to the decrease in photosynthetic rate by feedback inhibition (Rivero et al., 2014). Adam and Murthy (2014) have reported that cryoprotectant carbohydrate molecules interact with cell membrane and protect the plants from damage caused by dehydration and ice-crystal formation under chilling stress. In rice plants exposed to low temperature stress, the pathways related to sucrose metabolism, starch degradation, and glyoxylate cycle are upregulated (Maruyama et al., 2014). The upregulation of these pathways can be correlated with the increased levels of sucrose, fructose, and glucose (Maruyama et al., 2014).

The comparative metabolic profiling of *Arabidopsis* grown in both heat and cold stress shows the overlapping trend of major metabolites (Maruyama et al., 2014). The combination of heat and other abiotic stress like salinity induce the accumulation of sucrose, fructose, glucose, starch, and trehalose as reported in tomato subjected to combination of salinity and high-temperature stress (Rivero et al., 2014).

5.4.3 Metabolites Associated With Amino Acid Metabolism

The accumulation of proline is widely reported in many plants in response to heat stress (Cvikrová et al., 2012; Rivero et al., 2014; Chebrolu et al., 2016). Proline accumulation is mainly due to the activation of synthesizing enzymes (D-1-pyrroline-5-carboxylate synthetase) or due to deactivation of degrading enzymes like proline dehydrogenase (Cvikrová et al., 2012). Plants grown under heat or in the combination of heat and salt accumulate GB more preferably than proline, whereas plants grown under saline conditions tend to accumulate more proline than GB (Rivero et al., 2014). The preferable accumulation of GB under high temperature suggests some significant role of this metabolite in protecting photosynthetic proteins under heat stress (Rivero et al., 2014). The high temperature–induced accumulation of various amino acids like asparagine, proline, methionine, and GABA has also been reported in perennial grass *Poa pratensis* (Du et al., 2011). Abiotic stresses like high and low temperature induce protein degradation leading to internal ammonium accumulation in the cells. To prevent cellular ammonium toxicity, the synthesis of amino acids, especially asparagine, is stimulated under temperature stress conditions (Du et al., 2011). The response of heat stress in tobacco and the correlation to proline accumulation in both wild and transformant has been studied by Cvikrová et al. (2012). A relatively higher stress tolerance of transformants has been reported with respect to the wild counterpart due to the enhanced levels of proline in nonstressed condition (Cvikrová et al., 2012). Heat stress induced the activity of arginine decarboxylase and ornithine decarboxylase (two main enzymes in the putrescine biosynthetic pathway) and resulted in the accumulation of putrescine in the early stage of stress in tobacco (Cvikrová et al., 2012). The metabolism of nitrogenous compound is also affected by low temperature or chilling stress. It has been reported that the transcription levels of the enzymes associated with amino acid biosynthesis are significantly altered in *A. thaliana* in response to temperature stress (Usadel et al., 2008). The upregulated enzymes are associated with proline, glutamine, and ornithine biosynthesis, and the downregulated enzymes are those that are associated with branched-chain amino acid degradation (Usadel et al., 2008).

5.4.4 Metabolites Associated With Reactive Oxygen Species Scavenging

The high-temperature stress leads to the generation of ROS, which functions as stress signaling molecules; however, overaccumulation of ROS can result in adverse impact on the cellular metabolic pathway (Szymańska et al., 2017). Unquestionable deleterious effects of global warming on crop productivity worldwide need to be addressed by making crop plants that are resistant to increasing atmospheric temperature and that will increase crop productivity. Plants respond to elevated ROS levels by elevated synthesis of enzymatic and nonenzymatic antioxidative machinery for the detoxification of ROS and maintaining steady cellular homeostasis. As the time course of stress increases, the levels of spermidine and spermine enhance with the decrease in putrescine content suggesting the conversion of putrescine to spermidine and spermine with the increase in heat intensity (Goyal and Asthir, 2009; Cvikrová et al., 2012). It has been reported that increasing accumulations of spermidine and spermine have pronounced effects on ROS detoxification by enhancing the antioxidative defense system (Cheng et al., 2009). Apart from common polyamines like spermidine and spermine, the levels of uncommon polyamines like norspermidine and norspermine are also enhanced in response to heat stress. Various polyamines like putrescine, spermidine, and spermine are also found in high levels under cold conditions (Pathak et al., 2014). Polyamines increase the tolerance of the plants to environmental stresses by regulating the level of stress-responsive signaling molecule ABA (Pathak et al., 2014). It has been reported that the polyamine metabolism is enhanced in transgenic tobacco with higher accumulation of spermidine and spermine when compared with the wild counterpart when subjected to high-temperature stress (Cheng et al., 2009). In wild plants the spermine and spermidine contents increase by 40% and 35% under heat stress when compared with controls, whereas in transgenic line it increases by 145% and 93%, respectively (Cheng et al., 2009). The upregulation of polyamine biosynthetic genes during the stress period may increase stress signaling thereby conferring stress tolerance in transgenic plants (Cheng et al., 2009). In addition, transgenic plants overexpressing polyamine biosynthesis genes show a decline in photosynthetic assimilation rate when compared with the wild type in response to high temperature (Cheng et al., 2009). Polyamines also play an antitranspirant role thereby helping in the conservation of water in heat-stressed plants (Cheng et al., 2009). Polyamines can increase the photosynthetic efficiency by maintaining the integrity of the thylakoid membrane under heat stress (Mitra and Bhatia, 2008). The elevated levels of polyamines help in the synthesis of heat shock proteins and ensure membrane stability and integrity upon the stress encounter (Hasanuzzaman et al., 2013). Spermidine induces the closure of stomata and thereby decreases water loss and increases the use of internal CO_2. Flavonoids also play a major role in mitigating the adverse effects of heat stress. The role of flavonoids in extenuating the negative effects of heat stress have been well documented (Gill and Tuteja, 2010a). It has been reported that high temperature tends to increase the concentration as well as the number of flavonoids in stressed soybean seeds (Chebrolu et al., 2016). In soybean seed, flavonoids are present mainly in the form of isoflavones. In control condition, a total of five flavonoids are detected, whereas in heat-treated samples a total of 10 flavonoids are identified (Chebrolu et al., 2016). In addition to the flavonoids, tocopherols also have an important role in heat stress tolerance of the plants. Tocopherols are potent lipophilic antioxidant molecules that have some significant contribution toward membrane stabilization by reducing ROS levels generated due to heat stress. The heat stress-responsive accumulation of gulono-1,4-lactone has been documented in many reports (Chebrolu et al., 2016). Being a precursor of L-ascorbic acid, an elevated level of gulono-1,4-lactone maintains a constant ascorbate level during stress condition (Lisko et al., 2013). L-Ascorbic acid has an important role in biomass accumulation and cell enlargement beside its well-known scavenging activity (Lisko et al., 2013). Threonic acid and galacturonic acid both are related to ascorbic acid biosynthesis. Since ascorbic acid is an important antioxidant molecule against oxidative stress, the accumulation of these two metabolites under heat stress can be considered as a defense strategy adapted by the plants to counter the oxidative stress (Du et al., 2011).

Lipid metabolism plays an important protective role during low-temperature stress (Badea and Basu, 2009). In many studies it has been reported that low temperature causes an increase in the levels of unsaturated fatty acids in cell membrane as it is an important factor that modulates the chilling and freezing tolerance in plants (Badea and Basu, 2009; Degenkolbe et al., 2012). Increased levels of unsaturation in fatty acids of phosphatidylglycerol in membrane of thylakoid help to stabilize the photosynthetic system in plants and also contributes to maintain fluidity of membrane at low temperature (Badea and Basu, 2009). Lipidomic analysis of *Arabidopsis* and rice plants shows that there is accumulation of massive amount of storage lipid triacylglycerides in cold condition; although the total amount of membrane lipid remains almost unchanged, the composition of membrane lipid greatly changes with increased proportions of phosphatidylcholine, phosphatidylethanolamine, and ceramide with decreased levels of glucosylceramide and monogalactosyldiacylglycerol (Degenkolbe et al., 2012). From the soybean seed metabolome data, it is conferred that the levels of glycerol-3-phosphate, glycerophosphorylcholine, azelate, and pantothenate are greater in the tolerant variety than in the sensitive one in response to high-temperature stress (Chebrolu et al., 2016).

5.4.5 Metabolites Associated With Mitochondrial Respiration

Mitochondria is a major cell organelle that is affected by abiotic stress not only because it is a major site for ROS production but also because heat stress disrupts the energy supply for the repair mechanism. Several enzymes involved in the TCA cycle are altered in plants grown in high-temperature conditions. The high-temperature-induced accumulation of fumarate, malate, and oxaloacetate and the low-temperature-stress-induced accumulation of α-ketoglutarate, fumarate, malate, and citrate have been reported in *Arabidopsis,* suggesting the upregulation of TCA cycle under temperature stress condition (Guy et al., 2008). Heat stress–induced higher accumulation of TCA cycle intermediates like malic acid, citric acid, isocitric acid, threonic acid, and galacturonic acid have also been reported in perennial grass *P. pratensis* (Du et al., 2011). Higher levels of TCA cycle intermediates induce mitochondrial activity, which is beneficial for generating more reducing agents and ATP or for amino acid biosynthesis (Vasquez-Robinet et al., 2008). However, knowledge about the metabolic responses of the plants to heat and chilling stress is limited and more research is needed to gain more knowledge and implement it for the production of heat- and chilling-resistant crop varieties.

5.4.6 Induction of Phytohormones in Response to Temperature Extremes

Temperature stress causes many physiological and biochemical alterations including changes in hormone homeostasis. One of the heat-induced process occurs in plants at the cellular and organelle level is hormone regulation. ABA is one of the most important phytohormone that mediates intracellular dehydration signaling pathway. In the field condition where drought and heat conditions occur simultaneously, induction of ABA biosynthesis is an important component of thermotolerance. However, the role of gibberellins in thermotolerance is antagonistic to that of ABA (Maestri et al., 2002). It has been proved by applying GA on barley, which changes heat-resistant barley to heat-sensitive barley; in contrast, the application of gibberellin-antagonist compound confers thermotolerance to barley (Maestri et al., 2002). The short-term heat stress significantly induces the accumulation of ABA in alga *Chlorella vulgaris* as ABA is an essential component of thermotolerance (Bajguz, 2009). It has also been observed that the application of brassinosteroid induces ABA accumulation, and it is pronounced in high-temperature condition (Bajguz, 2009). Apart from ABA, auxin is also an important phytohormone that coordinates many physiological mechanisms under heat stress. The endogenous concentration of auxin is reduced upon the exposure to high temperature. In *Arabidopsis* and barley, it has been reported that the endogenous levels of auxin are significantly declined upon exposure to high temperatures along with the expression of YUCCA auxin biosynthesis gene (Qu et al., 2013). In addition to ABA and auxin, salicylic acid acts as an important negotiator of heat stress response. Salicylic acid mitigates the effects caused by ROS by successful detoxification of superoxide radicals, enhancing the capacity of antioxidants, and provides protection from membrane damage (Sakata et al., 2010). As reported in rice and *Arabidopsis* application of salicylic acid increases the antioxidant capacity (Sakata et al., 2010).

Among the phytohormones, ABA is very important stress hormone that is involved in the regulation of expression of many stress-responsive genes under cold stress (Shi and Yang, 2014). ABA also plays a protective role in cold stress defense especially membrane stabilization and improvement of water status by elevation of root hydraulic conductivity and stomatal closure (Vishwakarma et al., 2017). It has been reported that in wheat seedlings treated with cold stress, ABA and related metabolites increase from day 1 of the treatment. The early function of ABA in cold acclimation is the stabilization of water potential (Shibasaki et al., 2009). However, there is also a decrease in the level of IAA observed in response to cold stress. This reduction in IAA level may be attributed to the inhibition of intracellular trafficking of auxin efflux carriers PIN2 and PIN3 (Shibasaki et al., 2009). Cytokinin and GA_1 levels also increase in wheat during acclimation period to cold stress. However, the level of ABA decreases in wheat during the acclimation period, which coincides with the elevation of other stress hormones during this period (Kosova et al., 2012).

5.5 Metabolic Alterations in Response to High Light Intensity

Light is a key environmental factor for maintaining plant growth and physiology. Plant growth, development, morphology, and productivity are determined by the intensity and duration of light (Ashraf and Harris, 2013). The photosynthetic reactions taking place in the thylakoid membrane include absorption of light, excitation of the chlorophyll molecule, release and transport of electrons, oxygen evolution, and energy transduction, all of which are sensitive to light (Ashraf and Harris, 2013). However, when the amount of light absorbed by chlorophyll is more than that required for photosynthesis it is called excess light. Excess light is harmful to the plants because it causes photooxidative damage to the plants by the generation of ROS and other harmful by-products. The light in excess of the photosynthetic requirement induces reorganization of the photosynthetic apparatus to facilitate light harvesting and to minimize the photooxidative damage. Simultaneously, synthesis of protective compounds like flavonoids, tocopherols, and carotenoids are enhanced, which helps in mitigating the stress effects (DellaPenna and Pogson, 2006). To acclimatize to various light conditions, plants have evolved various adaptive mechanisms including morphological and physiological changes at the leaf level. For example, in excess light plants tend to reduce the specific leaf area and increase the leaf thickness due to the increase in the layer of palisade tissue. These adaptations prevent the plants from damage caused by the excess light energy and ensure the proceedings of photosynthesis (Fan et al., 2013).

5.5.1 Metabolites Associated With Reactive Oxygen Species Scavenging Under Light Stress

Overexcitation of photosystem by high light intensity is responsible for the overproduction of superoxide radicals and also leads to increased lifetime of an excited chlorophyll molecule, resulting in enhanced chances of ROS production. One of the most efficient mechanisms adopted by the plants to mitigate the effects of photodamage is the xanthophyll cycle (Latowski et al., 2011). Some of the carotenoids of this cycle act as a quencher of the singlet chlorophyll reducing the formation of ROS (Gruszecki et al., 2006). Among the six xanthophyll cycles, Vx cycle is the most studied and is also called the xanthophyll cycle. The main product of this cycle is the light-stimulated production of zeaxanthin by de-epoxidation (Latowski et al., 2011). A better correlation between the increased level of zeaxanthin and nonphotochemical quenching (NPQ) has been established in isolated chloroplasts of *Spinacia oleracea* (Latowski et al., 2011). However, the xanthophyll cycle is working concomitantly with the lutein cycle, which facilitates the rapid engagement of NPQ with concurrent zeaxanthin accumulation inducing strong energy dissipation in plants (García-Plazaola et al., 2007). To control the ROS level, plants change their metabolism to synthesize various antioxidant molecules. The low-molecular-weight antioxidant molecules like ascorbate and glutathione are the primary antioxidant molecules whose levels increase during mild to moderate light stress but decrease in severe stress conditions (Fini et al., 2011). Therefore, in severe stress conditions other secondary antioxidant molecules come into play, i.e., synthesis of isoprenoid and flavonoids takes place in severe stress conditions to complement the primary antioxidant system (Brunetti et al., 2015). Neugart et al. (2016) have studied the response of *Brassica oleracea* to the high intensity of light, and it has been observed that plants grown at high light intensity have a higher level of total flavonoids with increased levels of quercetin glycosides, caffeic acid, monoacylated kaempferol triglycoside, and disinapoylgentiobiose. The accumulation of flavonoids has also been observed in the leaves of *Ligustrum vulgare* during excess sun light exposure (Tattini et al., 2004). These studies showing the higher accumulation of flavonoids during excess light suggests the role of flavonoids as an antioxidant molecule.

The ozone layer depletion due to increased production of chlorofluorocarbon has led to increased exposure of plants to UV radiation. UV light within the range of 280–320 (UV-B) is most damaging and causes damage to DNA, RNA, and proteins and also generates ROS in plants. The biochemical and physiological changes occur in plants, to minimize the negative effects of UV-radiation-induced oxidative damage in plants. The flavonoids and anthocyanins are the major UV-absorbing metabolites, they selectively absorb UV radiation and permit photosynthetically active radiation to pass through the mesophyll cell layer (Hidema and Kumagai, 2006). The major compounds that are upregulated during UV stress are the shikimate pathway–related metabolites. Luteolin-7-glucuronide is a flavonoid compound associated with UV tolerance in *Salix* sp. (Tegelberg and Julkunen-Tiitto, 2001). Similarly, quercetin 3-6′-malonyl-glucoside and quercetin-3-glucuronide are the flavonoid compounds whose concentrations increase during UV stress in *Lactuca sativa* and strawberry fruits, respectively (DuPont et al., 2000). In addition, ascorbate, tocopherol, vitamin B_6, and carotenoids also provide protection against UV stress (Harvaux and Kloppstech, 2001). Polyamines also play an important role in maintaining the structure and function of the photosynthetic apparatus. It has been reported that polyamines like putrescine, spermine, and spermidine are associated with the light harvesting complex and PSII complex, which are mostly affected by high light intensity and UV exposure (Lütz et al., 2005). The effects of positive charge of polyamines in protecting PSII against excessive illumination have been investigated in isolated thylakoids, and it is has been observed that the application of polyamines like spermine and spermidine improves the photosynthetic efficiency of the plant (Hamdani et al., 2011).

5.5.2 Metabolites Associated With Mitochondrial Respiration Under Light Stress

The alternations in light intensity, photoperiod, and exposure to UV radiations affect the mitochondrial respiration pathways like the TCA cycle. The metabolic study on coffee plant (*Coffea arabica* L.) shows the downregulation of intermediates of the TCA cycle accompanied by the increasing levels of sugars, various amino acids, sugar alcohols, polyamines, and flavonoids. The study on *Arabidopsis* shows the upregulation of metabolites involved in glycolysis, TCA, and pentose phosphate pathways (Wulff-Zottele et al., 2010). Plants adapt to UV radiation by producing UV-absorbing metabolites by structural changes and DNA damage repair (Hanns and Dorothee, 2003). The short- and long-term responses of *Arabidopsis* plants to UV- B exposure demonstrate that the short-term response largely includes changes in primary metabolism especially upregulation of the intermediates of the TCA cycle, so that it can divert energy and carbon flux toward the synthesis of UV-protective secondary metabolites as part of long-term protection from UV stress (Kusano et al., 2011). Cellular accumulation of pyruvate and malate upon exposure to high light intensity has been reported in pea, and it is thought to have a major role in activating alternative oxidative pathways for optimizing photosynthesis under high illumination (Dinakar and Bartels, 2013).

5.6 Alternations of Different Metabolites Associated With Nutrient Deficiency

Plants need various mineral nutrient elements for their growth and development, and the levels of these nutrients should be maintained at an appropriate range in plant cells (Chiou, 2007). Carbon (C), nitrogen (N), sulfur (S), and phosphorus (P) are the major structural elements of plants' metabolites, and carbon is the most abundant element in cells. Therefore the deficiency of these elements directly alters the normal growth and metabolism of the plants (Obata and Fernie, 2012). It has been reported that carbon-starved seedlings of *Arabidopsis* show downregulation of carbohydrates (raffinose, myoinositol), organic acids (pyruvate, succinate, fumarate, and malate), central amino acids (glutamate, asparagine, alanine, and other aromatic amino acids), and intermediates of shikimate pathway (Osuna et al., 2007). Increased levels of minor amino acids indicate increased proteolysis during carbon starvation (Osuna et al., 2007). The study on metabolic changes in tomato (*Solanum lycopersicum* L.) plants grown in nitrogen-, phosphorus-, and potassium-deficient medium has shown that N deficiency causes decreased levels of amino acids and other organic acids, whereas increase in levels of soluble sugars (Sung et al. (2015a)). In contrast, P deficiency has opposite effects than N deficiency, causing increased levels of amino acids and organic acids and downregulation of soluble sugars (Sung et al., 2015a). It is observed that N, P, and K limitations affect the xylem sap composition and cause an accumulation of glutamine, GABA, organic acids (i.e., malate and citrate), and soluble sugars (Sung et al., 2015b). Nitrogen is a primary element having structural role in a wide range of cellular metabolites like proteins, amino acids, and nucleic acids and in various secondary metabolites (Foito et al., 2012). The availability of N has a significant role in growth, metabolism, and development of plants. The deficiency of N causes increased levels of compounds derived from carbohydrates such as inositol, glycerate, and threonate and other soluble sugars (Comadira et al., 2015). In addition, the upregulation of intermediates of the TCA cycle is also a part of response to N deficiency (Comadira et al., 2015). In chamomile plants, higher accumulation of flavones and decreased levels of flavonols occur in response to nitrogen deficiency (Kováčik and Klejdus, 2014). Glucosinolates play important roles in sulfur (S) metabolism, and it has been reported that under sulfur deficiency condition breakdown of glucosinolates occurs in cells (Falk et al., 2007). In *Arabidopsis* due to the unavailability of sulfur a decrease in the levels of sulfur-containing metabolites like cysteine and glutathione occurs, whereas there is increased accumulation of putrescine (Nikiforova et al., 2005). Phosphorus is a major constituent of DNA, RNA, and ATP and many other metabolites that are related to energy metabolism. Phosphate starvation causes an induction of alternative pathways to glycolysis and electron transport chain allowing the cells to continue respiration even in the absence of ATP and inorganic phosphates (Pi) (Theodorou and Plaxton, 1993). A metabolic study on phosphorus-starved barley revealed that the phosphorus-deficient plants accumulate di- and trisaccharides in shoots, whereas the level of phosphorus- containing intermediates (i.e., glucose-6-phosphate, inositol phosphate, and glycerol-3-phosphate) and organic acid decrease (Huang et al., 2008). These results suggest that phosphorus-deficient plants modify their metabolism to reduce the utilization of phosphorus and that during extreme P starvation conditions they retrieve phosphorus from small phosphorus-containing metabolites (Huang et al., 2008). Although potassium (K) has no structural role in metabolites, it is required for ion homeostasis, enzyme activation, and maintenance of transmembrane potential of the plant cells (Marschner, 2011). The metabolic profile of shoot and root of *Arabidopsis* reveals that nitrate, organic acids, intermediates of glycolysis pathway, and negatively charged amino acids (Asp and Glu) are downregulated, whereas the levels of positively charged amino acids increase during potassium-starved conditions (Armengaud et al., 2009). Zinc deficiency in barley causes decreased levels of malate, fumarate, and succinate indicating downregulation of the TCA cycle (Rose et al., 2012). This information indicates that alteration of the nutrient pool available for plants induces metabolic adjustment, which helps the plants to cope up with the nutrient-deficit condition (Table 5.4).

TABLE 5.4 Differential Regulations of Important Metabolites in Various Plant Species in Response to Nutrient Deficiency and Heavy Metals Stress

Abiotic Stress	Class of metabolites	Metabolite	Plant Species	Regulation	Function	References
Nutrient deficiency	Peptides	Glutathione	*Arabidopsis thaliana*	↓	Metabolic adjustment toward S deficiency	Nikiforova et al. (2005)
	Polyamines	Putrescine	*A. thaliana*	↑	Membrane integrity, signal transduction	Nikiforova et al. (2005)
	Sugar alcohol	Myoinositol	*A. thaliana*	↑	Metabolic adjustment toward C starvation	Osuna et al. (2007)
		Inositol-phosphate	*Hordeum vulgare* L.	↓	Metabolic adjustment toward P deficiency	Huang et al. (2008)
	Organic acids		*A. thaliana*	↓	Precursor for many compounds	Osuna et al. (2007)
			Solanum lycopersicum	↓		Sung et al. (2015a)
	Sugar acid	Threonate	*H. vulgare*	↑	Metabolic adjustment toward N deficiency	Comadira et al. (2015)

TABLE 5.4 Differential Regulations of Important Metabolites in Various Plant Species in Response to Nutrient Deficiency and Heavy Metals Stress—cont'd

Abiotic Stress	Class of metabolites	Metabolite	Plant Species	Regulation	Function	References
Heavy metals	Alcohol	Antheraxanthin	*Secale cereale*	↑	Cellular photoprotection	Janik et al. (2010)
	Peptides	Glutathione	*Pisum sativum*	↑	Antioxidant	Metwally et al. (2005)
			Chlamydomonas reinhardtii	↑		Sharma and Dietz (2006)
			Bruguiera gymnorrhiza	↑		Huang and Wang (2010)
		Phtytochelatins	*C. reinhardtii*	↑	Metal chelators	Sharma and Dietz (2006)
			B. gymnorrhiza	↑		Huang and Wang (2010)
		Carbohydrate	*Triticum aestivum*	↓	Energy storage, structural component	Manivasagaperumal et al. (2011)
		Xanthophyll	*Pistia stratiotes*	↑	Protection from photodamage	Singh and Pandey (2011)
			Pinus sylvestris	↓		Sytar et al. (2013)
			Duckweed	↑		Sytar et al. (2013)
	Amino acids	Threonine	*P. stratiotes*	↓	Osmoprotectant and signaling	Vesely et al. (2012)
		Glutamine	*S. lycopersicum*	↑	Synthesis of phytochelatins	Hediji et al. (2010)
			P. stratiotes	↓		Vesely et al. (2012)
		Aspartate	*P. stratiotes*	↓	Synthesis of polyamines	Vesely et al. (2012)
		Proline	*Hydrilla*	↓	Osmoprotectant, integrity of protein structure	Dhir et al. (2004)
			B. gymnorrhiza	↑		Huang and Wang (2010)
			S. lycopersicum	↑		Hediji et al. (2010)
			Brassica juncea	↑		Ahmad and Gupta (2013)
			Drimia elata	↑		Okem et al. (2015)
		Glutamate	*S. lycopersicum*	↓	Osmoprotection and signaling	Hediji et al. (2010)
			T. aestivum	↓		Gajewska et al. (2013)
			Oryza sativa	↑		Begum et al. (2016)
		Lysine	*O. sativa*	↑	Protein amino acid and precursor of glutamate	Begum et al. (2016)
	Phenolics		*Raphanus sativus*	↑	Antioxidant. ROS scavenging	Sytar et al. (2013)
			Hypoxis hemerocallidea	↑		Okem et al. (2015)
			D. elata	↓		Okem et al. (2015)
	Flavonoids		*H. hemerocallidea*	↑	Structural integrity, signaling, ROS scavenging	Okem et al. (2015)
			D. elata	↓		Okem et al. (2015)
	Phytochelatins		*Rubia tinctorum*	↑	Metal chelation	Pal and Rai (2010)

↑ and ↓, respectively, indicate up and downregulation of different metabolites.
ROS, reactive oxygen species.

5.7 Heavy Metal–Induced Alternation in Different Metabolites

Metals having density higher than 5 gcm^{-3} are considered as heavy metals. Among the naturally occurring elements, 53 elements are recognized as heavy metals. Of the identified heavy metals 17 are available to living cells due to their solubility at physiological conditions and are essential for organisms and the environment. Some of the metals like Cu, Ni, Zn, Co, Cr, V, and W are regarded as trace elements because at higher concentrations these metals are lethal to living cells. Other metals like Hg, Ag, Sb, As, Cd, and Pb have no direct involvement in plant physiology and metabolism, and these metals appear to be lethal for the plants (Nazar et al., 2012). Metals like Fe, Mn, and Cu are essential for the plants in one or another way, because of their significant involvement in important biological functions. Fe, Mn, Cu, Mn, and Ni function as cofactors for the enzymes in plants and animals. Heavy metals are categorized depending upon their concentration in soil and are produced by decaying and pedogenic processes of the rocks. Apart from the natural processes, industrialization, rapid population growth, anthropogenic biosolid production, and agrochemical wastes increase the threat of soil pollution with heavy metals (Kashem et al., 2009). One of the significant factors that accounts for the toxicity of heavy metals is the form in which the metal is present in the soil. Metals found in soluble and bioexchangeable forms have greater bioavailability and have more negative effects on the growth and physiology of different plant species (Table 5.4).

5.7.1 Metabolites Associated With Energy and Carbohydrate Metabolism

Photosynthesis is the most adversely affected and most studied biological process under abiotic stress condition. The accumulation of heavy metals beyond the tolerance limit of the plants interacts with the photosystems leading to a variety of adverse consequences including photooxidative damage. In response to heavy metal contamination a reduction in photosynthetic pigments (chlorophyll and carotenoid) has been observed in many plants, which may be attributed to the adaptation mechanism of the plants to the prevailing stress environment (Vesely et al., 2012). The reduction in photosynthetic pigments helps in reducing the photoexcitation of photosystems thus minimizing the photooxidative damage to the plants (Ventrella et al., 2009). Heavy metals have deleterious effects on metabolism, growth, and reproduction of the plants by disturbing the absorption of essential mineral nutrients from the soil. Compared with other heavy metals, cadmium has a higher solubility and mobility in soil and therefore is most important in terms of its catastrophic effects. To cope up with the stress exerted by heavy metals, plants need multifaceted physiological and biochemical coordination, modifications of protein structure, and alternation of metabolite profile for appropriate signaling and tolerance to stress. The effects of heavy metals can be assessed by altered accumulation of carbohydrates in leaf tissues (Manivasagaperumal et al., 2011). It has been reported that the levels of carbohydrates are significantly reduced in leaf of wheat and cluster bean with increasing concentration of Cu and Zn (Manivasagaperumal et al., 2011). The decrease in total sugar content in cluster bean in response to Zn stress suggests the role of sugars in the enzymatic reaction of carbohydrate catabolic pathway (Manivasagaperumal et al., 2011). Cell wall is the crucial site for Cd accumulation in plants under Cd stress, and it has been reported that exogenous glucose application alleviates Cd toxicity in *Arabidopsis* by increasing fixation of cadmium in root cell wall and sequestration into vacuole (Shi et al., 2015b). In *S. salsa*, Cd exposure causes a decrease in sucrose and fumarate (Liu et al., 2011). Cu is also an essential micronutrient for many important physiological processes of the plants like photosynthesis, respiration, and carbohydrate distribution. However, excess amount of Cu can adversely affect the vital physiological processes, leading to undesirable physiological consequences (Guzel and Terzi, 2013).

5.7.2 Metabolites Associated With Amino Acid Metabolism and Reactive Oxygen Species Scavenging

Heavy metals at their toxic concentrations affect the plants by changing their biochemical processes, altering membrane integrity, and exerting oxidative stress by the generation of ROS through Haber–Weiss reaction. Superoxide radical ($O_2^{\cdot -}$), hydrogen peroxide (H_2O_2), and hydroxyl radical ($OH^{\cdot}$) are the ROS that have deleterious effects on cellular metabolism because of their negative effects on the integrity of lipids, proteins, and nucleic acids. Apart from carbohydrates, amino acids like proline act as indicators of heavy metal stress. The enrichment of proline level may be attributed to enhanced de novo synthesis and/or decreased catabolism of proline under heavy metal stress. The accumulation of proline in heavy metal stress is quite high and may have an active role in osmotic adjustment, stabilization of enzyme structure and function, and also stabilization of organelles and cellular biomolecules (Ahmad et al., 2011). Several reports have been documented suggesting elevated levels of proline upon exposure to Cd^{2+}stress in different plant species like *Brassica juncea*, *Triticum aestivum*, *Vigna radiata*, sunflower, and *B. oleracea* (Theriappan et al., 2011). However, in hydrophytes like *Ceratophyllum*, *Wolffia*, and *Hydrilla*, a decrease in proline content has been observed in response to Cd exposures (Dhir et al., 2004). Proline along with GB does not directly act as metal chelator, but these organic osmolytes stabilize the naive state of protein and hence help in preserving the conformation and integrity of proteins. It has also been reported that exogenous application of proline and betaine increase metal toxicity tolerance by inducing antioxidant defense system and glyoxalase pathway (Irfan et al., 2014). It has been reported that the elevated level of proline accumulation is accompanied by the increased antioxidant defense system with the increased activity of enzymes such as phytochelatin synthase, GR, and glutathione synthetase under As stress in *B. juncea* (Ahmad and Gupta, 2013). Apart from proline, the cellular concentrations of other amino acids like arginine, proline, leucine, valine, serine, and glycine are also increased in response to heavy metal stress. Glycine is involved in the synthesis of glutathione and phytochelatins, and arginine is involved in the synthesis of polyamines (Vesely et al., 2012). It has been reported that in *S. salsa*, Cd exposure causes an increase in branched-chain amino acids such as valine, leucine, and isoleucine and decrease in betaine and fumarate (Liu et al., 2011).

Plant's phenolic compounds such as flavonoids and lignin are important in combating metal stress tolerance because of their ability to neutralize the ROS by donating electrons to hydrogen atoms. The phytophenolics can act as antioxidants for detoxification of H_2O_2 generated in response to heavy metal stress. In addition, phytophenolics also acts as a pro-oxidants under certain stress conditions. It has been reported that the content of phenolic acid is increased in many plants treated with high levels of Cd (Sytar et al., 2013). Flavonoids are included in the large family of secondary metabolites, and their role as antioxidants is well established. Among the flavonoids anthocyanins, vacuole, and membrane-associated tocopherols are the major flavonoid antioxidant components present in plants for protection against ROS (Hernandez et al., 2009). The increase in accumulation of flavonoids and phenolics has been reported in *Hypoxis hemerocallidea* treated with a moderate concentration of Cd (Okem et al., 2015). The antioxidant role of tocopherols is established with many reports of its cellular accumulation in response to oxidative stress (Aslam et al., 2010). The tocopherol provides protection to plants from oxidative damage by removing ($O_2^{\cdot-}$) from chloroplasts by physical quenching. It is estimated that one tocopherol molecule can quench about 220 ($O_2^{\cdot-}$) molecules before its degradation (Sytar et al., 2013). Under toxic metal contamination, plants synthesize low-molecular-weight thiols that show high affinity for heavy metals. Among the thiols, glutathione, a sulfur-containing tripeptide thiol, and cysteine are the most important metabolites produced by the plants upon exposure to heavy metals.

Phytochelatins are heavy-metal-inducible and heavy-metal-binding cysteine-rich polypeptides that help in metal chelation. Several reports have been documented showing a positive correlation between heavy metal accumulation and phytochelatin synthesis. The phytochelatins are thiolic peptides synthesized by the activity of the phytochelatin synthase (activated by heavy metals) mainly from glutamate, cysteine, and glycine and form complexes with the toxic metal ions and are subsequently transported to the vacuole (Komal et al., 2014). Being the precursor of phytochelatins the decrease in GSH content is well documented in many plants. However, a simultaneous increase in both phytochelatins and GSH has also been reported in some plants (Huang and Wang, 2010). Metwally et al. (2005) reported a good correlation between the glutathione concentration and the tolerance index to Cd toxicity in genotypes of pea. The accumulation of phytochelatins and GSH has been reported in heavy metal hyperaccumulator plants like *Thlaspi goesingense* and *Aspergillus* species (Sharma and Dietz, 2006). Phytochelatins are instantaneously produced after exposure to heavy metals in many plants. Higher accumulations of phytochelatins have been documented in several plant species like *Rubia tinctorum*, *Brassica napus* and *Sedum alfredii* in response to heavy metal stress (Pal and Rai, 2010). Plants also adapt some photoprotective mechanisms under heavy metal stress to dissipate excess excitation energy, and this mechanism depends on the levels of zeaxanthin and antheraxanthin, which are components of the xanthophyll cycle (Latowski et al., 2011). The accumulation of antheraxanthin has been observed in the leaves of *Secale cereale* due to the reduction of violaxanthin de-epoxidation under the stimulation of heavy metals (Janik et al., 2010). Similarly, significant accumulation of xanthophyll cycle pigments has been observed in the Cd-treated duckweed plants, which suggests that xanthophylls play important roles in quenching the free radical–induced reaction, thus preventing generation of ROS under heavy metal stress (Sytar et al., 2013).

5.7.3 Metabolites Associated With Mitochondrial Respiration

The intermediates of TCA cycle are critical for mitigating the effects of heavy metals on the physiology of the plants. Metal exclusion is one of the best defense strategy adopted by plants to survive in heavy metal–contaminated soil. There is some specialized mechanism adopted by the plants by which the transport of heavy metals to the roots can be prevented or reduced. The roots secret vital chelating agents like citrate, malate, and oxalate, and these metabolites are involved in heavy metal tolerance of plants. The higher accumulation of TCA cycle intermediates has been reported in lettuce under heavy metal stress (Zorrig et al., 2010). Oxalate, malate, and citrate are low-molecular-weight organic acid intermediates of TCA cycle having at least one carboxylic group, and these metabolites act as oxygen-donating metal ligands (Viehweger, 2014). It has been reported that malate is an important metabolite having potential role in ion homeostasis, potassium, and iron uptake, along with Al and Zn tolerance in every cell type. Malate also functions as a cytoplasmic Zn chelator in Zn-tolerant plants (Viehweger, 2014).

In the cytoplasm, Zn is integrated with malate and transported to the vacuole where malate is dissociated and retranslocated to the cytoplasm (Viehweger, 2014). Compared with malate and oxalate, citrate has greater affinity for metal ions. The major role of citrate is the Fe^{2+} chelation, although it has affinity for other divalent cations like Zn^{2+}, Cd^{2+}, and Ni^{2+} (Schneider et al., 2013). In plants, oxalate is present in small amounts. However, significant accumulation of oxalate has been reported in several plant species belonging to the family Caryophyllaceae, Chenopodiaceae, and Polygonaceae (Norman et al., 2013). The role of oxalate in chelation of Al and conferring Al tolerance has been reported in many plants (Yang et al., 2013).

5.7.4 Phytohormones Involved in Heavy Metal Tolerance

Phytohormones like auxin, cytokinin, and gibberellin can alleviate stress symptoms by decreasing heavy metal absorption and restoring growth and primary metabolite levels (Piotrowska-Niczyporuk et al., 2012). Phytohormones stimulate the activities of antioxidative enzymes and the levels of nonenzymatic antioxidants like ascorbate and glutathione thereby increasing the antioxidant capacity of the plants growing under heavy metal stress. Phytohormones are important members of signal transduction cascade involved in initiating plant stress response. The inhibitory effects of phytohormones on heavy metal bioaccumulation have been documented in the green alga *C. vulgaris* (Piotrowska-Niczyporuk et al., 2012). It has been observed that cytokinin has the highest inhibitory effecton metal bioaccumulation followed by other hormones, with auxin treatment having the least effect (Piotrowska-Niczyporuk et al., 2012). Among the auxins, IAA has the highest inhibitory effect on metal bioaccumulation. By inhibiting the bioaccumulation of metals, these growth regulators minimize the stress effects generated by the toxic metals (Piotrowska-Niczyporuk et al., 2012). These phytohormones exhibit a protective function against metal toxicity leading to restoration of growth of the plants. Apart from the inhibitory effects on metal absorption cytokinin also inhibits carotenoid degradation induced by heavy metals stress and helps in increasing carotenoid level. Along with cytokinin, auxin (especially IAA) and gibberellin have stimulatory effects on the carotenoid content of plants. However, jasmonate has a stimulatory effects on heavy metal accumulation and inhibitory effects on carotenoid content. In contrast to the aforementioned reports, it has been reported that cytokinin induces the accumulation of heavy metals (Pb and Zn) in *Helianthus annuus* (Tassi et al., 2008). The changes in levels of various metabolites in different plant species in response to heavy metal stress are given in Table 5.4.

6. CONCLUSION AND FUTURE PROSPECTS

Metabolome analysis with the advanced metabolomics techniques has become a valuable strategy to study the metabolic alternations of the plants to adapt under prevailing stress conditions. The combination of different techniques has made the metabolomic approach more promising in deciphering the mechanisms that the plants adapt for their survival in stress environments. Metabolomics is a fundamental biotechnological approach that offers an array of opportunities in basic research for interpreting solitary or combined environmental effects on cellular processes. Plants in the natural habitat are experienced to the combination of various abiotic stresses rather than an individual stress. For example, plants experiencing salt stress simultaneously experience osmotic and ionic stresses, and flooding results in hypoxia and shading stresses. However, most of the studies have been conducted to analyze the effects of solitary abiotic stress on plant metabolism. There are only a few studies reported on the effects of combined stresses on plant metabolism. The metabolic profile of the *Arabidopsis* has been analyzed under combined drought and heat stress (Rizhsky et al., 2004). The response of the plants towards the combined stress of drought and heat is quite similar to that of the plants subjected to only drought treatment, and the plants treated with heat stress alone. Plants subjected to the combined stress of drought and heat accumulate higher amounts of sucrose and other sugars as compared to proline. However, proline accumulation is higher in plants subjected to drought only but not in those exposed to combination of stresses. The combined effects of high light irradiance and sulfur depletion simultaneously occur in natural habitat (Wulff-Zottele et al., 2010). The combination of high light intensity and sulfur depletion exerts similar modification in the plant's whole metabolome as high light intensity alone (Wulff-Zottele et al., 2010). Higher proline accumulation has been observed in many plants subjected to high light intensity, but under the combination of high light intensity stress with other stresses, proline is replaced with sugars, which is quite similar to that observed in the combined treatment of drought and heat. The data integration of metabolomics with other "omics" field such as genomics, transcriptomics, and proteomics is a prerequisite to understand the networking and coordinations of metabolic pathways. The integrated study of two omics, i.e, transcriptomics and metabolomics, in *Arabidopsis* in response to eight environmental factors including the combinations of changing light (darkness, high light intensity) and/or temperature (cold and heat) has been reported by Caldana et al. (2011). It has been demonstrated that darkness and high temperature have synergistic effects thus exerting more complex response. Similarly, integrated transcriptomics, ionomics, and metabolomics study has been conducted in barley inoculated with endophytic fungus *Piriformospora indica* under salt stress (Ghaffari et al., 2016). The response of *Astragalus membranaceus* to drought stress has been analyzed by multiple omics approaches such as transcriptomics and metabolomics (Jia et al., 2016). This review offers important information about the specific alternations of various metabolites in response to certain stress conditions. Taken together, all these information indicate that the study of metabolome with developed identification and interpretation techniques paves the way for better understanding of complex metabolic pathways and their regulation during abiotic stress conditions. The metabolomics study can pave the way for getting the information of metabolites as biomarkers of different environmental stresses. The identification of metabolite biomarkers can be helpful for plant biologists in two major ways: (1) to understand the regulatory factors perturbing plant metabolism under various abiotic stress conditions and (2) to develop strategies using knowledge/data of metabolomics for improvement of stress tolerance in crops by metabolic/genetic engineering via reverse genetics.

ACKNOWLEDGMENT

The grant of SERB (SB/SO/PS-14/2014), DST, Government of India, New Delhi, to AKP is duly acknowledged. JR is thankful to Council of Scientific and Industrial Research (CSIR), New Delhi, India, for financial support in the form of Junior Research Fellowship. The manuscript has been assigned CSIR-CSMCRI registration number 113/2016.

REFERENCES

Adam, S., Murthy, S.D.S., 2014. Effect of cold stress on photosynthesis of plants and possible protection mechanisms. In: Gaur, R.K., Sharma, P. (Eds.), Approaches to Plant Stress and Their Management, pp. 281–295. https://doi.org/10.1007/978-81-322-1620-9_16.

Ahmad, M.A., Gupta, M., 2013. Exposure of *Brassica juncea* (L.) to arsenic species in hydroponic medium: comparative analysis in accumulation and biochemical and transcriptional alterations. Environ. Sci. Pollut. Res. 20, 8141–8150. https://doi.org/10.1007/s11356-013-1632-y.

Ahmad, P., Nabi, G., Ashraf, M., 2011. Cadmium-induced oxidative damage in mustard [*Brassica juncea* (L.) Czern. & Coss.] plants can be alleviated by salicylic acid. S. Afr. J. Bot. 77, 36–44. https://doi.org/10.1016/j.sajb.2010.05.003.

Akçay, N., Bor, M., Karabudak, T., Özdemir, F., Türkan, I., 2012. Contribution of Gamma amino butyric acid (GABA) to salt stress responses of *Nicotiana sylvestris* CMSII mutant and wild type plants. J. Plant Physiol. 169, 452–458. https://doi.org/10.1016/j.jplph.2011.11.006.

Alcázar, R., Bitrián, M., Bartels, D., Koncz, C., Altabella, T., Tiburcio, A.F., 2014. Polyamine metabolic canalization in response to drought stress in Arabidopsis and the resurrection plant *Craterostigma plantagineum*. Plant Signal Behav. 6, 243–250. https://doi.org/10.4161/psb.6.2.14317.

Alcazar, R., Cuevas, J.a., Planas, J., Zarza, X., Bortolotti, C., Carrasco, P., Salinas, J., Tiburcio, A.F., Altabella, T., 2011. Integration of polyamines in the cold acclimation response. Plant Sci. 180, 31–38. https://doi.org/10.1016/j.plantsci.2010.07.022.

Alet, A.I., Sanchez, D.H., Cuevas, J.C., Del Valle, S., Altabella, T., Tiburcio, A.F., Marco, F., Ferrando, A., Espasandín, F.D., González, M.E., Ruiz, O.A., Carrasco, P., 2011. Putrescine accumulation in *Arabidopsis thaliana* transgenic lines enhances tolerance to dehydration and freezing stress. Plant Signal. Behav. 6, 278–286. https://doi.org/10.4161/psb.6.2.14702.

Alla, M.M.N., Khedr, A.H.A., Serag, M.M., Abu-Alnaga, A.Z., Nada, R.M., 2012. Regulation of metabolomics in *Atriplex halimus* growth under salt and drought stress. Plant Growth Regul. 67, 281–304. https://doi.org/10.1007/s10725-012-9687-1.

Alonso-Ramírez, A., Rodríguez, D., Reyes, D., Jiménez, J.A., Nicolás, G., López-Climent, M., 2009. Evidence for a role of gibberellins in salicylic acid-modulated early plant responses to abiotic stress in Arabidopsis seeds. Plant Physiol. 150, 1335–1344. https://doi.org/10.1104/pp.109.139352.

Armengaud, P., Sulpice, R., Miller, A.J., Stitt, M., Amtmann, A., Gibon, Y., 2009. Multilevel analysis of primary metabolism provides new insights into the role of potassium nutrition for glycolysis and nitrogen assimilation in Arabidopsis roots. Plant Physiol. 150, 772–785. https://doi.org/10.1104/pp.108.133629.

Ashraf, M., Foolad, M.R., 2007. Roles of glycine betaine and proline in improving plant abiotic stress resistance. Environ. Exp. Bot. 59, 206–216. https://doi.org/10.1016/j.envexpbot.2005.12.006.

Ashraf, M., Harris, P.J.C., 2013. Photosynthesis under stressful environments : an overview. Photosynthetica 51, 163–190. https://doi.org/10.1007/s11099-013-0021-6.

Aslam, M., Kumar, D., Rajwanshi, R., Jörg, R., Tsimilli-michael, M., Bhalla, N., 2010. Overexpression of γ -tocopherol methyl transferase gene in transgenic *Brassica juncea* plants alleviates abiotic stress : physiological and chlorophyll a fluorescence measurements. Biochim. Biophys. Acta 1797, 1428–1438. https://doi.org/10.1016/j.bbabio.2010.02.002.

Badea, C., Basu, S.K., 2009. The effect of low temperature on metabolism of membrane lipids in plants and associated gene expression. Plant Omi. J. 2, 78–84.

Bailey-Serres, J., Fukao, T., Gibbs, D.J., Holdsworth, M.J., Lee, S.C., Licausi, F., 2012. Making sense of low oxygen sensing. Trends Plant Sci. 17, 129–138. https://doi.org/10.1016/j.tplants.2011.12.004.

Bajguz, A., 2009. Brassinosteroid enhanced the level of abscisic acid in *Chlorella vulgaris* subjected to short-term heat stress. J. Plant Physiol. 166, 882–886. https://doi.org/10.1016/j.jplph.2008.10.004.

Barnawal, D., Bharti, N., Maji, D., Chanotiya, C.S., Kalra, A., 2012. 1-Aminocyclopropane-1-carboxylic acid (ACC) deaminase-containing rhizobacteria protect *Ocimum sanctum* plants during waterlogging stress via reduced ethylene generation. Plant Physiol. Biochem. 58, 227–235. https://doi.org/10.1016/j.plaphy.2012.07.008.

Bartoli, C.G., Guiamet, J.J., Kiddle, G., Pastori, G.M., Di Cagno, R., Theodoulou, F.L., Foyer, C.H., 2005. Ascorbate content of wheat leaves is not determined by maximal L-galactono-1, 4-lactone dehydrogenase (GalLDH) activity under drought stress. Plant Cell Environ. 28, 1073–1081. https://doi.org/10.1111/j.1365-3040.2005.01338.x.

Basu, S., Roychoudhury, A., Saha, P.P., Sengupta, D.N., 2010. Comparative analysis of some biochemical responses of three indica rice varieties during polyethylene glycol-mediated water stress exhibits distinct varietal differences. Acta Physiol. Plant 32 (3), 551–563. https://doi.org/10.1007/s11738-009-0432-y.

Begum, M.C., Islam, M.S., Monirul Islam, R.A., Parvez, A.H., Kabir, M.S., 2016. Biochemical and molecular responses underlying differential arsenic tolerance in rice (*Oryza sativa* L.). Plant Physiol. Biochem. https://doi.org/10.1016/j.toxicon.2008.05.012.

Belyagoubi-Benhammou, N., Belyagoubi, L., Bekkara, F.A., 2014. Phenolic contents and antioxidant activities in vitro of some selected Algerian plants. J. Med. Plants Res. 8, 1198–1207. https://doi.org/10.5897/JMPR2014.5554.

Bendaly, A., Messedi, D., Smaoui, A., Ksouri, R., Bouchereau, A., Abdelly, C., 2016. Physiological and leaf metabolome changes in the xerohalophyte species *Atriplex halimus* induced by salinity. Plant Physiol. Biochem. 103, 208–218. https://doi.org/10.1016/j.plaphy.2016.02.037.

Benton, H.P., Wong, D.M., Trauger, S.A., Siuzdak, G., 2008. XCMS2: processing tandem mass spectrometry data for metabolite identification and structural characterization. Anal. Chem. 80, 6382–6389. https://doi.org/10.1021/ac800795f.

Bokszczanin, K.L., Fragkostefanakis, S., 2013. Perspectives on deciphering mechanisms underlying plant heat stress response and thermotolerance. Front. Plant Sci. 4, 315. https://doi.org/10.3389/fpls.2013.00315.

Broeckling, C.D., Reddy, I.R., Duran, A.L., Zhao, X., Sumner, L.W., 2006. MET-IDEA: data extraction tool for mass spectrometry-based metabolomics. Anal. Chem. 78, 4334–4341. https://doi.org/10.1021/ac052 1596.

Brown, M., Dunn, W.B., Ellis, D.I., Goodacre, R., Handl, J., Knowles, J.D., 2005. A metabolome pipeline: from concept to data to knowledge. Metabolomics 1, 39–51. https://doi.org/10.1007/s11306-005-1106-4.

Brownstein, G., Wilson, J.B., Burritt, D.J., 2013. Waterlogging tolerance on a New Zealand saltmarsh. J. Exp. Mar. Bio. Ecol. 446, 202–208. https://doi.org/10.1016/j.jembe.2013.05.025.

Brunetti, C., Guidi, L., Sebastiani, F., Tattini, M., 2015. Isoprenoids and phenylpropanoids are key components of the antioxidant defence system of plants facing severe excess light stress. Environ. Exp. Bot. 119, 54–62. https://doi.org/10.1016/j.envexpbot.2015.04.007.

Boccard, J., Rudaz, S., 2016. Exploring omics data from designed experiments using analysis of variance multiblock orthogonal partial least squares. Anal. Chim. Acta 920, 18–28. https://doi.org/10.1016/j.aca.2016.03.042.

Boccard, J., Rutledge, D.N., 2013. A consensus orthogonal partial least squares discriminant analysis (OPLS-DA) strategy for multiblock Omics data fusion. Anal. Chim. Acta 769, 30–39.

Cabot, C., Sibole, J.V., Barceló, J., Poschenrieder, C., 2009. Abscisic acid decreases leaf Na+ exclusion in salt-treated *Phaseolus vulgaris* L. J. Plant Growth Regul. 28, 187–192. https://doi.org/10.1007/s00344-009-9088-5.

Caldana, C., Degenkolbe, T., Cuadros-Inostroza, A., Klie, S., Sulpice, R., Leisse, A., Steinhauser, D., Fernie, A.R., Willmitzer, L., Hannah, M.A., 2011. High-density kinetic analysis of the metabolomic and transcriptomic response of *Arabidopsis* to eight environmental conditions. Plant J. 67, 869–884. https://doi.org/10.1111/j.1365-313X.2011.04640.x.

Carreno-Quintero, N., Bouwmeester, H.J., Keurentjes, J.J.B., 2013. Genetic analysis of metabolome-phenotype interactions: from model to crop species. Trends Genet. 29, 41–50. https://doi.org/10.1016/j.tig.2012.09.006.

Chandna, R., Azooz, M.M., Ahmad, P., 2013. Recent advances of metabolomics to reveal plant response during salt stress. In: Ahmad, P., Azooz, M.M., Prasad, M.N.V. (Eds.), Salt Stress in Plants. Springer, New York, pp. 1–14.

Chaves, M.M., Flexas, J., Pinheiro, C., 2009. Photosynthesis under drought and salt stress: regulation mechanisms from whole plant to cell. Ann. Bot. 103, 551–560. https://doi.org/10.1093/aob/mcn125.

Chebrolu, K.K., Fritschi, F.B., Ye, S., Krishnan, H.B., Smith, J.R., Gillman, J.D., 2016. Impact of heat stress during seed development on soybean seed metabolome. Metabolomics 12, 1–14. https://doi.org/10.1007/s11306-015-0941-1.

Chen, L., Jia, H., Tian, Q., Du, L., Gao, Y., Miao, X., Liu, Y., 2012. Protecting effect of phosphorylation on oxidative damage of D1 protein by down-regulating the production of superoxide anion in photosystem II membranes under high light. Photosynth. Res. 112, 141–148. https://doi.org/10.1007/s11120-012-9750-9.

Chen, Z., Cuin, T.A., Zhou, M., Twomey, A., Naidu, B.P., Shabala, S., 2007. Compatible solute accumulation and stress-mitigating effects in barley genotypes contrasting in their salt tolerance. J. Exp. Bot. 58, 4245–4255. https://doi.org/10.1093/jxb/erm284.

Chen, T.H.H., Murata, N., 2011. Glycinebetaine protects plants against abiotic stress: mechanisms and biotechnological applications. Plant Cell Environ. 34 (1), 1–20. https://doi.org/10.1111/j.1365-3040.2010.02232.x.

Cheng, L., Zou, Y., Ding, S., Zhang, J., Yu, X., Cao, J., Lu, G., 2009. Polyamine accumulation in transgenic tomato enhances the tolerance to high temperature stress. J. Integr. Plant Biol. 51, 489–499. https://doi.org/10.1111/j.1744-7909.2009.00816.x.

Chiou, T.-J., 2007. The role of microRNAs in sensing nutrient stress. Plant Cell Environ. 30, 323–332. https://doi.org/10.1111/j.1365-3040.2007.01643.x.

Choi, Y.H., Verpoorte, R., 2014. Metabolomics: what you see is what you extract. Phytochem. Anal. 25 (4), 289–290. https://doi.org/10.1002/pca.2513.

Collino, S., Martin, F.P., Rezzi, S., 2013. Clinical metabolomics paves the way towards future healthcare strategies. Br. J. Clin. Pharmacol. 75, 619–629. https://doi.org/10.1111/j.1365-2125.2012.04216.x.

Comadira, G., Rasool, B., Karpinska, B., Morris, J., Verrall, S.R., Hedley, P.E., 2015. Nitrogen deficiency in barley (*Hordeum vulgare*) seedlings induces molecular and metabolic adjustments that trigger aphid resistance. J. Exp. Bot. 66, 3639–3655. https://doi.org/10.1093/jxb/erv276.

Cook, D., Fowler, S., Fiehn, O., Thomashow, M.F., 2004. A prominent role for the CBF cold response pathway in configuring the low-temperature metabolome of Arabidopsis. PNAS 101, 15243–15248. https://doi.org/10.1073/pnas.0406069101.

Cusido, R.M., Onrubia, M., Sabater-Jara, A.B., Moyano, E., Bonfill, M., Goossens, A., 2014. A rational approach to improving the biotechnological production of taxanes in plant cell cultures of *Taxus* spp. Biotechnol. Advt. 32, 1157–1167. https://doi.org/10.1016/j.biotechadv.2014.03.002.

Cvikrová, M., Gemperlová, L., Dobrá, J., Martincová, O., Prásil, I.T., Gubis, J., Vanková, R., 2012. Effect of heat stress on polyamine metabolism in proline-over-producing tobacco plants. Plant Sci. 182, 49–58. https://doi.org/10.1016/j.plantsci.2011.01.016.

Daly, B.A., Messedi, D., Smaoui, A., Ksouri, R., Bouchereau, A., Abdelly, C., 2016. Physiological and leaf metabolome changes in the xerohalophyte species *Atriplex halimus* induced by salinity. Plant Physiol. Biochem. 103, 208–218. https://doi.org/10.1016/j.plaphy.2016.02.037.

de Miguel, M., Guevara, M.Á., Sánchez-Gómez, D., de María, N., Díaz, L.M., Mancha, J.A., 2016. Organ-specific metabolic responses to drought in *Pinus pinaster* Ait. Plant Physiol. Biochem. 102, 17–26. https://doi.org/10.1016/j.plaphy.2016.02.013.

Degenkolbe, T., Giavalisco, P., Zuther, E., Seiwert, B., Hincha, D.K., Willmitzer, L., 2012. Differential remodelling of the lipidome during cold acclimation in natural accessions of *Arabidopsis thaliana*. Plant J. 72,972–982. https://doi.org/10.1111/tpj.12007.

DellaPenna, D., Pogson, B.J., 2006. Vitamin synthesis in plants: tocopherols and carotenoids. Annu. Rev. Plant Biol. 57, 711–738. https://doi.org/10.1146/annurev.arplant.56.032604.144301.

Dhir, B., Sharmila, P., Saradhi, P.P., 2004. Hydrophytes lack potential to exhibit cadmium stress induced enhancement in lipid peroxidation and accumulation of proline. Aquat. Toxicol. 66, 141–147. https://doi.org/10.1016/j.aquatox.2003.08.005.

Dias, D.A., Hill, C.B., Jayasinghe, N.S., Atieno, J., Sutton, T., Roessner, U., 2015. Quantitative profiling of polar primary metabolites of two chickpea cultivars with contrasting responses to salinity. J. Chromatogr. B1000, 1–13. https://doi.org/10.1016/j.jchromb.2015.07.002.

Dinakar, C., Bartels, D., 2013. Desiccation tolerance in resurrection plants: new insights from transcriptome, proteome and metabolome analysis. Front. Plant Sci. 4, 482. https://doi.org/10.3389/fpls.2013.00482.

Do, P.T., Degenkolbe, T., Erban, A., Heyer, A.G., Kopka, J., Köhl, K.I., 2013. Dissecting rice polyamine metabolism under controlled long-term drought stress. PLoS One 8. https://doi.org/10.1371/journal.pone.0060325.

Dos Santos, V.S., Macedo, F.A., do Vale, J.S., Silva, D.B., Carollo, C.A., 2017. Metabolomics as a tool for understanding the evolution of Tabebuia sensu lato. Metabolomics 13, 72.

Du, H., Wang, Z., Yu, W., Liu, Y., Huang, B., 2011. Differential metabolic responses of perennial grass *Cynodon transvaalensis*× *Cynodon dactylon* (C_4) and *Poa pratensis* (C_3) to heat stress. Physiol. Plant 141, 251–264. https://doi.org/10.1111/j.1399-3054.2010.01432.x.

Dubery, I., Tugizimana, F., Piater, L., Dubery, I., 2013. Plant metabolomics: a new frontier in phytochemical analysis. S. Afr. J. Sci. 109, 1–11. https://doi.org/10.1590/sajs.2013/20120005.

DuPont, M.S., Mondin, Z., Williamson, G., Price, K.R., 2000. Effect of variety, processing, and storage on the flavonoid glycoside content and composition of lettuce and endive. J. Agric. Food Chem. 48, 3957–3964. https://doi.org/10.1021/jf0002387.

Ernst, M., Silva, D.B., Silva, R.R., Vêncio, R.Z.N., Lopes, N.P., 2014. Mass spectrometry in plant metabolomics strategies: from analytical platforms to data acquisition and processing. Nat. Prod. Rep. 31, 784–806. https://doi.org/10.1039/c3np70086k.

Evers, D., Lefèvre, I., Legay, S., Lamoureux, D., Hausman, J.-F., Rosales, R.O.G., 2010. Identification of drought-responsive compounds in potato through a combined transcriptomic and targeted metabolite approach. J. Exp. Bot. 61, 2327–2343. https://doi.org/10.1093/jxb/erq060.

Eyidogan, F., Oz, M.T., Yucel, M., Oktem, H.A., 2012. Signal transduction of phytohormones under abiotic stresses. In: Khan, N.A. (Ed.), Phytohormones and Abiotic Stress Tolerance in Plants. Springer-Verlag, Berlin Heidelberg, pp. 1–49. https://doi.org/10.1007/978-3-642-25829-9.

Fahad, S., Hussain, S., Bano, A., Saud, S., Hassan, S., Shan, D., 2014. Potential role of phytohormones and plant growth-promoting rhizobacteria in abiotic stresses: consequences for changing environment. Environ. Sci. Pollut. Res. 22, 4907–4921. https://doi.org/10.1007/s11356-014-3754-2.

Fahad, S., Hussain, S., Matloob, A., Khan, F.A., Khaliq, A., Saud, S., 2015. Phytohormones and plant responses to salinity stress: a review. Plant Growth Regul. 75, 391–404. https://doi.org/10.1007/s10725-014-0013-y.

Fahy, E., Sud, M., Cotter, D., Subramaniam, S., 2007. LIPID MAPS online tools for lipid research. Nucleic Acids Res. 35, W606–W612. https://doi.org/10.1093/nar/gkm324.

Falk, K.L., Tokuhisa, J.G., Gershenzon, J., 2007. The effect of sulfur nutrition on plant glucosinolate content: physiology and molecular mechanisms. Plant Biol. 9, 573–581. https://doi.org/10.1055/s-2007-965431.

Fan, T.W.M., Lane, A.N., Higashi, R.M., 2003. In vivo and in vitro metabolomic analysis of anaerobic rice coleoptiles revealed unexpected pathways. Russ. J. Plant Physiol. 50, 787–793. https://doi.org/10.1023/B:RUPP.0000003276.14947.16.

Fan, X.X., Xu, Z.G., Liu, X.Y., Tang, C.M., Wang, L.W., Han, X., 2013. Effects of light intensity on the growth and leaf development of young tomato plants grown under a combination of red and blue light. Sci. Hortic. 153, 50–55. https://doi.org/10.1016/j.scienta.2013.01.017.

Farooq, M., Wahid, A., Lee, D.-J., Cheema, S.A., Aziz, T., 2010. Drought stress: comparative time course action of the foliar applied glycinebetaine, salicylic acid, nitrous oxide, brassinosteroids and spermine in improving drought resistance of rice. J. Agron. Crop. Sci. 196, 336–345. https://doi.org/10.1111/j.1439-037X.2010.00422.x.

Ferner, E., Rennenberg, H., Kreuzwieser, J., 2012. Effect of flooding on C metabolism of flood-tolerant (*Quercus robur*) and non-tolerant (*Fagus sylvatica*) tree species. Tree Physiol. 32, 135–145. https://doi.org/10.1093/treephys/tps009.

Fini, A., Brunetti, C., Di Ferdinando, M., Ferrini, F., Tattini, M., 2011. Stress-induced flavonoid biosynthesis and the antioxidant machinery of plants. Plant Signal. Behav. 6, 709–711. https://doi.org/10.4161/psb.6.5.15069.

Fletcher, J.S., Lockyer, N.P., Vickerman, J.C., 2011. Developments in molecular SIMS depth profiling and 3D imaging of biological systems using polyatomic primary ions. Mass Spectrom. Rev. 30, 142–174. https://doi.org/10.1002/mas.20275.

Flowers, T.J., Colmer, T.D., 2008. Salinity tolerance in halophytes. New Phytol. 179, 945–963. https://doi.org/10.1111/j.1469-8137.2008.02531.x.

Foito, A., Byrne, S.L., Hackett, C.A., Hancock, R.D., Stewart, D., Barth, S., 2012. Short-term response in leaf metabolism of perennial ryegrass (*Lolium perenne*) to alterations in nitrogen supply. Metabolomics 9, 145–156. https://doi.org/10.1007/s11306-012-0435-3.

Forieri, I., Hildebrandt, U., Rostas, M., 2016. Salinity stress effects on direct and indirect defence metabolites in maize. Environ. Exp. Bot. 122, 68–77. https://doi.org/10.1016/j.envexpbot.2015.09.007.

Fraire-Velazquez, S., Emmanuel, V., 2013. Abiotic stress in plants and metabolic responses. Abiotic Stress - Plant Responses Appl. Agric. https://doi.org/10.5772/54859.

Freund, D.M., Hegeman, A.D., 2017. Recent advances in stable isotope-enabled mass spectrometry-based plant metabolomics. Curr. Opin. Biotechnol. 43, 41–48. https://doi.org/10.1016/j.copbio.2016.08.002.

Frydman, L., Blazina, D., 2007. Ultrafast two-dimensional nuclear magnetic resonance spectroscopy of hyperpolarized solutions. Nat. Phys. 3, 415–419. https://doi.org/10.1038/nphys597.

Gagneul, D., Aïnouche, A., Duhazé, C., Lugan, R., Larher, F.R., Bouchereau, A., 2007. A reassessment of the function of the So-called compatible solutes in the halophytic *Plumbaginaceae Limonium latifolium*. Plant Physiol. 144, 1598–1611. https://doi.org/10.1104/pp.107.099820.

Gajewska, E., Niewiadomska, E., Tokarz, K., Słaba, M., Skłodowska, M., 2013. Nickel-induced changes in carbon metabolism in wheat shoots. J. Plant Physiol. 170, 369–377. https://doi.org/10.1016/j.jplph.2012.10.012.

Gal, M., Schanda, P., Brutscher, B., Frydman, L., 2007. UltraSOFAST HMQC NMR and the repetitive acquisition of 2D protein spectra at Hz rates. J. Am. Chem. Soc. 129 (5), 1372–1377. https://doi.org/10.1021/ja066915g.

García-Plazaola, J.I., Matsubara, S., Osmond, C.B., 2007. The lutein epoxide cycle in higher plants: its relationships to other xanthophyll cycles and possible functions. Funct. Plant Biol. 34, 759. https://doi.org/10.1071/FP07095.

Gavaghan, C.L., Li, J.V., Hadfield, S.T., Hole, S., Nicholson, J.K., Wilson, I.D., 2011. Application of NMR-based metabolomics to the investigation of salt stress in maize (*Zea mays*). Phytochem. Anal. 22, 214–224. https://doi.org/10.1002/pca.1268.

Gechev, T.S., Benina, M., Obata, T., Tohge, T., Sujeeth, N., Minkov, I., 2013. Molecular mechanisms of desiccation tolerance in the resurrection glacial relic *Haberlea rhodopensis*. Cell. Mol. Life Sci. 70, 689–709. https://doi.org/10.1007/s00018-012-1155-6.

Genga, A., Mattana, M., Coraggio, I., 2011. Plant Metabolomics: a characterization of plant responses to abiotic stresses. Abiotic Stress Plants - Mech. Adapt. 1–43. https://doi.org/10.5772/23844.

Ghaffari, M.R., Ghabooli, M., Khatabi, B., Hajirezaei, M.R., Schweizer, P., Salekdeh, G.H., 2016. Metabolic and transcriptional response of central metabolism affected by root endophytic fungus *Piriformospora indica* under salinity in barley. Plant Mol. Biol. 90, 699–717. https://doi.org/10.1007/s11103-016-0461-z.

Gill, S.S., Tuteja, N., 2010a. Reactive oxygen species and antioxidant machinery in abiotic stress tolerance in crop plants. Plant Physiol. Biochem. 48, 909–930. https://doi.org/10.1016/j.plaphy.2010.08.016.

Gill, S.S., Tuteja, N., 2010b. Polyamines and abiotic stress tolerance in plants. Plant Signal. Behav. 5, 26–33.

Gong, Q., Li, P., Ma, S., Indu Rupassara, S., Bohnert, H.J., 2005. Salinity stress adaptation competence in the extremophile *Thellungiella halophila* in comparison with its relative *Arabidopsis thaliana*. Plant J 44, 826–839. https://doi.org/10.1111/j.1365-313X.2005.02587.x.

Gong, Z.-G., Hu, J., Wu, X., Xu, Y.-J., 2017. The recent developments in sample preparation for mass spectrometry-based metabolomics. Crit. Rev. Anal. Chem. 8347, 1–7. https://doi.org/10.1080/10408347.2017.1289836.

Goyal, M., Asthir, B., 2009. Polyamine catabolism influences antioxidative defense mechanism in shoots and roots of five wheat genotypes under high temperature stress. Plant Growth Regul. 60, 13–25. https://doi.org/10.1007/s10725-009-9414-8.

Grace, S.C., 2005. Phenolics as Antioxidants. In: Antioxidants and Reactive Oxygen Species in Plants. Blackwell Publishing Ltd., pp. 141–168. https://doi.org/10.1002/9780470988565.ch6.

Grewal, H.S., 2010. Water uptake, water use efficiency, plant growth and ionic balance of wheat, barley, canola and chickpea plants on a sodic vertosol with variable subsoil NaCl salinity. Agric. Water Manag. 97 (1), 148–156. https://doi.org/10.1016/j.agwat.2009.09.002.

Griesser, M., Weingart, G., Schoedl-Hummel, K., Neumann, N., Becker, M., Varmuza, K., 2015. Severe drought stress is affecting selected primary metabolites, polyphenols, and volatile metabolites in grapevine leaves (*Vitis vinifera* cv. Pinot noir). Plant Physiol. Biochem. 88, 17–26. https://doi.org/10.1016/j.plaphy.2015.01.004.

Grimplet, J., Wheatley, M.D., Jouira, H.B., Deluc, L.G., Cramer, G.R., Cushman, J.C., 2009. Proteomic and selected metabolite analysis of grape berry tissues under well-watered and water-deficit stress conditions. Proteomics 9, 2503–2528. https://doi.org/10.1002/pmic.200800158.

Gruszecki, W.I., Grudzinski, W., Gospodarek, M., Patyra, M., Maksymiec, W., 2006. Xanthophyll-induced aggregation of LHCII as a switch between light-harvesting and energy dissipation systems. Biochim. Biophys. Acta 1757, 1504–1511. https://doi.org/10.1016/j.bbabio.2006.08.002.

Guo, R., Yang, Z., Li, F., Yan, C., Zhong, X., Liu, Q., 2015. Comparative metabolic responses and adaptive strategies of wheat (*Triticum aestivum*) to salt and alkali stress. BMC Plant Biol. 15, 1–13. https://doi.org/10.1186/s12870-015-0546-x.

Gupta, B., Huang, B., 2014. Mechanism of salinity tolerance in plants: Physiological, biochemical, and molecular characterization. Int. J. Genomics 1–19. https://doi.org/10.1155/2014/701596.

Guy, C., Kaplan, F., Kopka, J., Selbig, J., Hincha, D.K., 2008. Metabolomics of temperature stress. Physiol. Plant 132, 220–235. https://doi.org/10.1111/j.1399-3054.2007.00999.x.

Guzel, S., Terzi, R., 2013. Exogenous hydrogen peroxide increases dry matter production, mineral content and level of osmotic solutes in young maize leaves and alleviates deleterious effects of copper stress. Bot. Stud. 54, 26. https://doi.org/10.1186/1999-3110-54-26.

Hadiarto, T., Tran, L.-S.P., 2011. Progress studies of drought-responsive genes in rice. Plant Cell Rep. 30, 297–310. https://doi.org/10.1007/s00299-010-0956-z.

Hamdani, S., Gauthier, A., Msilini, N., Carpentier, R., 2011. Positive charges of polyamines protect PSII in isolated thylakoid membranes during photo inhibitory conditions. Plant Cell Physiol. 52, 866–873. https://doi.org/10.1093/pcp/pcr040.

Hanns, F., Dorothee, S., 2003. Ultraviolet-B radiation-mediated responses in plants. Balancing damage and protection. Plant Physiol. 133, 1420–1428. https://doi.org/10.1104/pp.103.030049.

Harada, T., Ishizawa, K., 2003. Starch degradation and sucrose metabolism during anaerobic growth of pondweed (*Potamogeton distinctus* A. Benn.) turions. Plant Soil 253,125–135. https://doi.org/10.1023/A:1024585015697.

Harvaux, M., Kloppstech, K., 2001. The protective functions of carotenoid and flavonoid pigments against excess visible radiation at chilling temperature investigated in Arabidopsis npq and tt mutants. Planta 213, 953–966.

Hasanuzzaman, M., Hossain, M.A., Fujita, M., 2011. Selenium-induced up-regulation of the antioxidant defense and methylglyoxal detoxification system reduces salinity-induced damage in rapeseed seedlings. Biol. Trace Elem. Res. 143, 1704–1721. https://doi.org/10.1007/s12011-011-8958-4.

Hasanuzzaman, M., Nahar, K., Alam, M.M., Roychowdhury, R., Fujita, M., 2013. Physiological, biochemical, and molecular mechanisms of heat stress tolerance in plants. Int. J. Mol. Sci. 14, 9643–9684. https://doi.org/10.3390/ijms14059643.

Hassine, A.B., Ghanem, M.E., Bouzid, S., Lutts, S., 2008. An inland and a coastal population of the Mediterranean xero-halophyte species *Atriplex halimus* L. differ in their ability to accumulate proline and glycinebetaine in response to salinity and water stress. J. Exp. Bot. 59, 1315–1326. https://doi.org/10.1093/jxb/ern040.

Hediji, H., Djebali, W., Cabasson, C., Maucourt, M., Baldet, P., Bertrand, A., 2010. Effects of long-term cadmium exposure on growth and metabolomic profile of tomato plants. Ecotoxicol. Environ. Saf. 73, 1965–1974. https://doi.org/10.1016/j.ecoenv.2010.08.014.

Hernandez, I., Alegre, L., Van Breusegem, F., Munne-Bosch, S., 2009. How relevant are flavonoids as antioxidants in plants? Trends Plant Sci 14, 125–132. https://doi.org/10.1016/j.tplants.2008.12.003.

Herrera, A., Fernández-Valle, E., Martínez-Álvarez, R., Molero, D., Pardo, Z.D., Sáez, E., Gal, M., 2009. Real-time monitoring of organic reactions with two-dimensional ultrafast TOCSY NMR spectroscopy. Angew. Chem. - Int. Ed. 48, 6274–6277.

Hichem, H., Mounir, D., Naceur, E.A., 2009. Differential responses of two maize (*Zea mays* L.) varieties to salt stress: changes on polyphenols composition of foliage and oxidative damages. Ind. Crops Prod. 30, 144–151. https://doi.org/10.1016/j.indcrop.2009.03.003.

Hidema, J., Kumagai, T., 2006. Sensitivity of rice to ultraviolet-B radiation. Ann. Bot. 97, 933–942. https://doi.org/10.1093/aob/mcl044.

Hoque, M.A., Banu, M.N.A., Nakamura, Y., Shimoishi, Y., Murata, Y., 2007. Proline and glycinebetaine enhance antioxidant defence and methylglyoxal detoxification systems and reduce NaCl-induced damage in cultured tobacco cells. J. Plant Physiol. 165, 813–824. https://doi.org/10.1016/j.jplph.2007.07.013.

Horvath, E., Szalai, G., Janda, T., 2007. Induction of abiotic stress tolerance by salicylic acid signaling. J. Plant Growth Regul. 26, 290–300. https://doi.org/10.1007/s00344-007-9017-4.

Hsu, F.C., Chou, M.Y., Peng, H.P., Chou, S.J., Shih, M.C., 2011. Insights into hypoxic systemic responses based on analyses of transcriptional regulation in Arabidopsis. PLoS One 6, e28888. https://doi.org/10.1371/journal.pone.0028888.

Hu, Z., Liu, A., Bi, A., Amombo, E., Gitau, M.M., Huang, X., Chen, L., Fu, J., 2016. Identification of differentially expressed proteins in bermudagrass response to cold stress in the presence of ethylene. Environ. Exp. Bot. 139, 67–78. https://doi.org/10.1016/j.envexpbot.2017.04.001.

Huang, C.Y., Roessner, U., Eickmeier, I., Genc, Y., Callahan, D.L., Shirley, N., 2008. Metabolite profiling reveals distinct changes in carbon and nitrogen metabolism in phosphate-deficient barley plants (*Hordeum vulgare* L.). Plant Cell Physiol. 49, 691–703. https://doi.org/10.1093/pcp/pcn044.

Huang, G.Y., Wang, Y.S., 2010. Physiological and biochemical responses in the leaves of two mangrove plant seedlings (*Kandelia candel* and *Bruguiera gymnorrhiza*) exposed to multiple heavy metals. J. Hazard. Mater 182, 848–854. https://doi.org/10.1016/j.jhazmat.2010.06.121.

Hussain, S.S., Ali, M., Ahmad, M., Siddique, K.H., 2011. Polyamines: natural and engineered abiotic and biotic stress tolerance in plants. Biotechnol. Adv. 29, 300–311. https://doi.org/10.1016/j.biotechadv.2011.01.003.

Iqbal, M., Ashraf, M., 2007. Seed treatment with auxins modulates growth and ion partitioning in salt-stressed wheat plants. J. Integr. Plant Biol. 49, 1003–1015. https://doi.org/10.1111/j.1672-9072.2007.00488.x.

Iqbal, M., Ashraf, M., Jamil, A., 2006. Seed enhancement with cytokinins: changes in growth and grain yield in salt stressed wheat plants. Plant Growth Regul. 50, 29–39. https://doi.org/10.1007/s10725-006-9123-5.

Iqbal, N., Nazar, R., Khan, M.I.R., Masood, A., Khan, N.A., 2011. Role of gibberellins in regulation of source- sink relations under optimal and limiting environmental conditions. Curr. Sci. 100, 998–1007.

Iqbal, N., Umar, S., Khan, N.A., Khan, M.I.R., 2014. A new perspective of phytohormones in salinity tolerance: regulation of proline metabolism. Environ. Exp. Bot. 100, 34–42. https://doi.org/10.1016/j.envexpbot.2013.12.006.

Irfan, M., Ahmad, A., Hayat, S., 2014. Effect of cadmium on the growth and antioxidant enzymes in two varieties of *Brassica juncea*. Saudi J. Biol. Sci. 21, 125–131. https://doi.org/10.1016/j.sjbs.2013.08.001.

Jackson, M.B., Colmer, T.D., 2005. Response and adaptation by plants to flooding stress. Ann. Bot. 96, 501–505. https://doi.org/10.1093/aob/mci205.

Janik, E., Maksymiec, W., Mazur, R., Garstka, M., Gruszecki, W.I., 2010. Structural and functional modifications of the major light-harvesting complex II in cadmium-or copper-treated *Secale cereale*. Plant Cell Physiol. 51, 1330–1340. https://doi.org/10.1093/pcp/pcq093.

Janz, D., Behnke, K., Schnitzler, J.-P., Kanawati, B., Schmitt-Kopplin, P., Polle, A., 2010. Pathway analysis of the transcriptome and metabolome of salt sensitive and tolerant poplar species reveals evolutionary adaption of stress tolerance mechanisms. BMC Plant Biol. 10, 1–17. https://doi.org/10.1186/1471-2229-10-150.

Jia, X., Sun, C., Zuo, Y., Li, G., Li, G., Ren, L., Chen, G., 2016. Integrating transcriptomics and metabolomics to characterise the response of *Astragalus membranaceus* Bge. var. mongolicus (Bge.) to progressive drought stress. BMC Genom. 188, 1–16. https://doi.org/10.1186/s12864-016-2554-0.

Kanehisa, M., Araki, M., Goto, S., Hattori, M., Hirakawa, M., Itoh, M.M., 2008. KEGG for linking genomes to life and the environment. Nucleic Acids Res. 36, D480–D484. https://doi.org/10.1093/nar/gkm882.

Kang, G.Z., Li, G.Z., Liu, G.Q., Xu, W., Peng, X.Q., Wang, C.Y., 2013. Exogenous salicylic acid enhances wheat drought tolerance by influence on the expression of genes related to ascorbate-glutathione cycle. Biol. Plant 57, 718–724. https://doi.org/10.1007/s10535-013-0335-z.

Kang, L., Ji, C.Y., Kim, S.H., Ke, Q., Park, S.-C., Kim, H.S., et al., 2017. Suppression of the β-carotene hydroxylase gene increases β-carotene content and tolerance to abiotic stress in transgenic sweet potato plants. Plant Physiol. Biochem. 117, 24–33.

Kashem, M.A., Kawai, S., Kikuchi, N., Takahashi, H., Sugawara, R., Singh, B.R., 2009. Effect of Lherzolite on chemical fractions of Cd and Zn and their uptake by plants in contaminated soil. Water. Air. Soil Pollut. 207, 241–251. https://doi.org/10.1007/s11270-009-0132-7.

Katajamaa, M., Miettinen, J., Oresic, M., 2006. MZmine: toolbox for processing and visualization of mass spectrometry based molecular profile data. Bioinformatics 22, 634–636. https://doi.org/10.1093/bioinformatics/btk039.

Khadri, M., Tejera, N.A., Lluch, C., 2007. Sodium chloride-ABA interaction in two common bean (*Phaseolus vulgaris*) cultivars differing in salinity tolerance. Environ. Exp. Bot. 60, 211–218. https://doi.org/10.1016/j.envexpbot.2006.10.008.

Kiefer, P., Delmotte, N., Vorholt, J.A., 2011. Nanoscale ion-pair reversed-phase HPLC-MS for sensitive metabolome analysis. Anal. Chem. 83 (3), 850–855. https://doi.org/10.1021/ac102445r.

Kim, H.K., Choi, Y.H., Verpoorte, R., 2010. NMR-based metabolomic analysis of plants. Nat. Protoc. 5, 536–549. https://doi.org/10.1038/nprot.2009.237.

Kim, J.K., Bamba, T., Harada, K., Fukusaki, E., Kobayashi, A., 2007. Time-course metabolic profiling in *Arabidopsis thaliana* cell cultures after salt stress treatment. J. Exp. Bot. 58, 415–424. https://doi.org/10.1093/jxb/erl216.

Kim, T.-W., Wang, Z.-Y., 2010. Brassinosteroid signal transduction from receptor kinases to transcription factors. Ann. Rev. Plant Biol. 61, 681–704. https://doi.org/10.1146/annurev.arplant.043008.092057.

Kind, T., Wohlgemuth, G., Lee do, Y., Lu, Y., Palazoglu, M., Shahbaz, S., Fiehn, O., 2009. FiehnLib: mass spectral and retention index libraries for metabolomics based on quadrupole and time-of flight gas chromatography/mass spectrometry. Anal. Chem. 81, 10038–10048. https://doi.org/10.1021/ac9019522.

Komal, T., Mustafa, M., Kazi, A.G., 2014. Heavy metal induced adaptation strategies and repair mechanisms in plants. J. Endocytobiosis Cell Res. 25, 33–41.

Kosmides, A.K., Kamisoglu, K., Calvano, S.E., Corbett, S.A., Androulakis, I.P., 2013. Metabolomic fingerprinting: challenges and opportunities. Crit. Rev. Biotechnol. 205–221.

Kosova, K., Prasilil, I.T., Vitamvas, P., Dobrev, P., Motyka, V., Flokova, K., 2012. Complex phytohormone responses during the cold acclimation of two wheat cultivars differing in cold tolerance, winter Samanta and spring Sandra. J. Plant Physiol. 169, 567–576. https://doi.org/10.1016/j.jplph.2011.12.013.

Kováčik, J., Klejdus, B., 2014. Induction of phenolic metabolites and physiological changes in chamomile plants in relation to nitrogen nutrition. Food Chem. 142, 334–341. https://doi.org/10.1016/j.foodchem.2013.07.074.

Krasensky, J., Jonak, C., 2012. Drought, salt, and temperature stress-induced metabolic rearrangements and regulatory networks. J. Exp. Bot. 63, 1593–1608. https://doi.org/10.1093/jxb/err460.

Kreuzwieser, J., Hauberg, J., Howell, K.A., Carroll, A., Rennenberg, H., Millar, A.H., Whelan, J., 2009. Differential response of *gray Poplar* leaves and roots underpins stress adaptation during hypoxia. Plant Physiol. 149, 461–473. https://doi.org/10.1104/pp.108.125989.

Ku, K.M., Choi, J.N., Kim, J., Kim, J.K., Yoo, L.G., Lee, S.J., 2010. Metabolomics analysis reveals the compositional differences of shade grown tea (*Camellia sinensis* L.). J. Agric. Food Chem. 58, 418–426. https://doi.org/10.1021/jf902929h.

Kumar, M., Kuzhiumparambil, U., Pernice, M., Jiang, Z., Ralph Peter, J., 2016. Metabolomics: an emerging frontier of systems biology in marine macrophytes. Algal Res. 16, 76–92. https://doi.org/10.1016/j.algal.2016.02.033.

Kumari, A., Das, P., Parida, A.K., Agarwal, P.K., 2015. Proteomics, metabolomics, and ionomics perspectives of salinity tolerance in halophytes. Front. Plant Sci. 6, 1–20. https://doi.org/10.3389/fpls.2015.00537.

Kumutha, D., Sairam, R.K., Ezhilmathi, K., Chinnusamy, V., Meena, R.C., 2008. Effect of waterlogging on carbohydrate metabolism in pigeon pea (*Cajanus cajan* L.): Upregulation of sucrose synthase and alcohol dehydrogenase. Plant Sci. 175, 706–716. https://doi.org/10.1016/j.plantsci.2008.07.013.

Kusano, M., Tohge, T., Fukushima, A., Kobayashi, M., Hayashi, N., Otsuki, H., 2011. Metabolomics reveals comprehensive reprogramming involving two independent metabolic responses of Arabidopsis to UV-B light. Plant J. 67, 354–369. https://doi.org/10.1111/j.1365-313X.2011.04599.x.

Lafitte, H.R., Yongsheng, G., Yan, S., Li, Z.K., 2007. Whole plant responses, key processes, and adaptation to drought stress: the case of rice. J. Exp. Bot. 58, 169–175. https://doi.org/10.1093/jxb/erl101.

Lanzinger, A., Frank, T., Reichenberger, G., Herz, M., Engel, K.H., 2015. Metabolite profiling of barley grain subjected to induced drought stress: Responses of free amino acids in differently adapted cultivars. J. Agric. Food Chem. 63, 4252–4261. https://doi.org/10.1021/acs.jafc.5b01114.

Lasanthi-Kudahettige, R., Magneschi, L., Loreti, E., Gonzali, S., Licausi, F., Novi, G., 2007. Transcript profiling of the anoxic rice coleoptile. Plant Physiol. 144, 218–231. https://doi.org/10.1104/pp.106.093997.

Latowski, D., Kuczyńska, P., Strzałka, K., 2011. Xanthophyll cycle- a mechanism protecting plants against oxidative stress. Red. Rep. 16, 78–90. https://doi.org/10.1179/174329211X13020951739938.

Li, Z., Yu, J., Peng, Y., Huang, B., 2017. Metabolic pathways regulated by abscisic acid, salicylic acid and γ-aminobutyric acid in association with improved drought tolerance in creeping bentgrass (*Agrostis stolonifera*). Physiol. Planta 159, 42–58.

Li, K., Wang, X., Pidatala, V.R., Chang, C.-P., Cao, X., 2014. Novel quantitative metabolomic approach for the study of stress responses of plant root metabolism. J. Proteome Res 13, 5879–5887. https://doi.org/10.1021/pr5007813.

Lisko, K.A., Torres, R., Harris, R.S., Belisle, M., Vaughan, M.M., Jullian, B., 2013. Elevating vitamin C content via overexpression of myo-inositol oxygenase and l-gulono-1, 4-lactone oxidase in Arabidopsis leads to enhanced biomass and tolerance to abiotic stresses. In vitro cellular & developmental biology. Plant J. Tissue Cult. Assoc. 49, 643–655. https://doi.org/10.1007/s11627-013-9568-y.

Liu, J., Wang, X., Hu, Y., Hu, W., Bi, Y., 2013. Glucose-6-phosphate dehydrogenase plays a pivotal role in tolerance to drought stress in soybean roots. Plant Cell Rep. 32, 415–429. https://doi.org/10.1007/s00299-012-1374-1.

Liu, X., Yang, C., Zhang, L., Li, L., Liu, S., Yu, J., 2011. Metabolic profiling of cadmium-induced effects in one pioneer intertidal halophyte *Suaeda salsa* by NMR-based metabolomics. Ecotoxicol 20, 1422–1431. https://doi.org/10.1007/s10646-011-0699-9.

Lokhande, H.V., Suprasanna, P., 2012. Prospects of halophytes in understanding and managing abiotic stress tolerance. In: Ahmad, P., Prasad, M.N.V. (Eds.), Environmental Adaptations and Stress Tolerance of Plants in the Era of Climate Change. Springer, pp. 1–515. https://doi.org/10.1007/978-1-4614-0815-4.

Lu, Y., Lam, H., Pi, E., Zhan, Q., Tsai, S., Wang, C., 2013. Comparative metabolomics in *Glycine max* and *Glycine soja* under salt stress to reveal the phenotypes of their offspring. J. Agric. Food Chem. 61, 8711–8721. https://doi.org/10.1021/jf402043m.

Ludwig, C., Easton, J.M., Lodi, A., Tiziani, S., Manzoor, S.E., Southam, A.D., 2012. Birmingham Metabolite Library: a publicly accessible database of 1-D 1H and 2-D 1H J-resolved NMR spectra of authentic metabolite standards (BML-NMR). Metabolomics 8 (1), 8–18. https://doi.org/10.1007/s11306-011-0347-7.

Luengwilai, K., Saltveit, M., Beckles, D.M., 2012. Metabolite content of harvested Micro-Tom tomato (*Solanum lycopersicum* L.) fruit is altered by chilling and protective heat-shock treatments as shown by GC–MS metabolic profiling. Postharvest Biol. Technol. 63, 116–122. https://doi.org/10.1016/j.postharvbio.2011.05.014.

Lugan, R., Niogret, M.F., Leport, L., Guégan, J.P., Larher, F.R., Savouré, A., 2010. Metabolome and water homeostasis analysis of *Thellungiella salsuginea* suggests that dehydration tolerance is a key response to osmotic stress in this halophyte. Plant J. 64, 215–229. https://doi.org/10.1111/j.1365-313X.2010.04323.x.

Lütz, C., Navakoudis, E., Seidlitz, H.K., Kotzabasis, K., 2005. Simulated solar irradiation with enhanced UV-B adjust plastid- and thylakoid-associated polyamine changes for UV-B protection. Biochim. Biophys. Acta - Bioenerg. 1710, 24–33. https://doi.org/10.1016/j.bbabio.2005.09.001.

Maestri, E., Klueva, N., Perrotta, C., Gulli, M., Nguyen, H.T., Marmiroli, N., 2002. Molecular genetics of heat tolerance and heat shock proteins in cereals. Plant Mol. Biol. 48, 667–681. https://doi.org/10.1023/A:1014826730024.

Maggio, A., Barbieri, G., Raimondi, G., de Pascale, S., 2010. Contrasting effects of GA3 treatments on tomato plants exposed to increasing salinity. J. Plant Growth Regul. 29, 63–72. https://doi.org/10.1007/s00344-009-9114-7.

Manivasagaperumal, R., Balamurugan, S., Thiyagarajan, G., Sekar, J., 2011. Effect of zinc on germination, seedling growth and biochemical content of cluster bean (*Cyamopsis tetragonoloba* (L.) Taub). Curr. Bot. 2, 11–15.

Marchand, F.L., Mertens, S., Kockelbergh, F., Beyens, L., Nijs, I., 2005. Performance of high Arctic tundra plants improved during but deteriorated after exposure to a simulated extreme temperature event. Glob. Change Biol. 11, 2078–2089. https://doi.org/10.1111/j.1365-2486.2005.01046.x.

Martinez, J.P., Kinet, J.M., Bajji, M., Lutts, S., 2005. NaCl alleviates polyethylene glycol-induced water stress in the halophyte species *Atriplex halimus* L. J. Exp. Bot. 56, 2421–2431. https://doi.org/10.1093/jxb/eri235.

Maruyama, K., Urano, K., Yoshiwara, K., Morishita, Y., Sakurai, N., Suzuki, H., 2014. Integrated analysis of the effects of cold and dehydration on rice metabolites, phytohormones, and gene transcripts. Plant Physiol. 164, 1759–1771. https://doi.org/10.1104/pp.113.231720.

Marschner, H., 2011. Marschner's Mineral Nutrition of Higher Plants. Academic Press, London, p. 672.

Megha, S., Basu, U., Kav, N.N.V., 2017. Regulation of low temperature stress in plants by microRNAs. Plant Cell Environ. 1–15. https://doi.org/10.1111/pce.12956.

Meng, F., Luo, Q., Wang, Q., Zhang, X., Qi, Z., Xu, F., 2016. Physiological and proteomic responses of diploid and tetraploid black locust (*Robinia pseudoacacia* L.) subjected to salt stress. Sci. Rep. 14, 20299–20325. https://doi.org/10.3390/ijms141020299.

Metwally, A., Safronova, V.I., Belimov, A.A., Dietz, K.-J., 2005. Genotypic variation of the response to cadmium toxicity in *Pisum sativum* L. J. Exp. Bot. 56, 167–178. https://doi.org/10.1093/jxb/eri017.

Mittler, R., Finka, A., Goloubinoff, P., 2012. How do plants feel the heat? Trends Biochem. Sci. 37, 118–125. https://doi.org/10.1016/j.tibs.2011.11.007.

Mitra, R., Bhatia, C.R., 2008. Bioenergetic cost of heat tolerance in wheat crop. Curr. Sci. 94, 1049–1053.

Muscolo, A., Junker, A., Klukas, C., Weigelt-Fischer, K., Riewe, D., Altmann, T., 2015. Phenotypic and metabolic responses to drought and salinity of four contrasting lentil accessions. J. Exp. Bot. https://doi.org/10.1093/jxb/erv208.

Nagana Gowda, G.A., Raftery, D., 2014. Advances in NMR-based metabolomics. Comp. anal. Chem. 63, 187–211. https://doi.org/10.1016/B978-0-444-62651-6.00008-8.

Nazar, R., Iqbal, N., Masood, A., Khan, M.I.R., Syeed, S., Khan, N.A., 2012. Cadmium toxicity in plants and role of mineral nutrients in its alleviation. Am.J. Plant Sci. 3, 1476–1489. https://doi.org/10.4236/ajps.2012.310178.

Negrão, S., Schmöckel, S.M., Tester, M., 2017. Evaluating physiological responses of plants to salinity stress. Ann. Bot. 119, 1–11. https://doi.org/10.1093/aob/mcw191.

Nemeth, M., Janda, T., Horvath, E., Paldi, E., Szalai, G., 2002. Exogenous salicylic acid increases polyamine content but may decrease drought tolerance in maize. Plant Sci. 162, 569–574. https://doi.org/10.1016/S0168-9452(01)00593-3.

Nemhauser, J.L., Hong, F., Chory, J., 2006. Different plant hormones regulate similar processes through largely non overlapping transcriptional responses. Cell 126, 467–475. https://doi.org/10.1016/j.cell.2006.05.050.

Neugart, S., Angelika, K., Zrenner, R., 2016. Influence of light and temperature on gene expression leading to accumulation of specific flavonol glycosides and hydroxycinnamic acid derivatives in Kale (*Brassica oleracea* var. sabellica). Front. Plant Sci. 7. https://doi.org/10.3389/fpls.2016.00326.

Ni, J., Yang, X., Zhu, J., Liu, Z., Ni, Y., Wu, H., 2015. Salinity-induced metabolic profile changes in *Nitraria tangutorum* Bobr. suspension cells. Plant Cell Tissue Organ Cult. 122, 239–248. https://doi.org/10.1007/s11240-015-0744-0.

Nicholson, J.K., Lindon, J.C., 2008. Systems biology: Metabonomics. Nature 455, 1054–1056. https://doi.org/10.1038/4551054a.

Nikiforova, V.J., Kopka, J., Tolstikov, V., Fiehn, O., Hopkins, L., Hawkesford, M.J., 2005. Systems rebalancing of metabolism in response to sulfur deprivation, as revealed by metabolome analysis of *Arabidopsis* plants. Plant Physiol. 138, 304–318. https://doi.org/10.1104/pp.104.053793.304.

Norman, H.C., Masters, D.G., Barrett-Lennard, E.G., 2013. Halophytes as forages in saline landscapes: Interactions between plant genotype and environment change their feeding value to ruminants. Environ. Exp. Bot. 92, 96–109. https://doi.org/10.1016/j.envexpbot.2012.07.003.

Obata, T., Fernie, A.R., 2012. The use of metabolomics to dissect plant responses to abiotic stresses. Cell. Mol. Life Sci. 69, 3225–3243. https://doi.org/10.1007/s00018-012-1091-5.

Okem, A., Stirk, W.A., Street, R.A., Southway, C., Finnie, J.F., Van Staden, J., 2015. Effects of Cd and Al stress on secondary metabolites, antioxidant and antibacterial activity of *Hypoxis hemerocallidea* Fisch. & C.A. Mey. Plant Physiol. Biochem. 97, 147–155. https://doi.org/10.1016/j.plaphy.2015.09.015.

Osorio, S., Do, P.T., Fernie, A.R., 2012. Profiling primary metabolites of tomato fruit with gas chromatography/mass spectrometry. Methods Mol. Biol. 860, 101–109. https://doi.org/10.1007/978-1-61779-594-7_7.

Osuna, D., Usadel, B., Morcuende, R., Gibon, Y., Bläsing, O.E., Höhne, M., 2007. Temporal responses of transcripts, enzyme activities and metabolites after adding sucrose to carbon-deprived *Arabidopsis* seedlings. Plant J 49, 463–491. https://doi.org/10.1111/j.1365-313X.2006.02979.x.

Ozgur, R., Uzilday, B., Sekmen, A.H., Turkan, I., 2013. Reactive oxygen species regulation and antioxidant defence in halophytes. Funct. Plant Biol. 40, 832–847. https://doi.org/10.1071/FP12389.

Pal, R., Rai, J.P.N., 2010. Phytochelatins: Peptides involved in heavy metal detoxification. Appl. Biochem. Biotechnol. 160, 945–963. https://doi.org/10.1007/s12010-009-8565-4.

Pang, Q., Zhang, A., Zang, W., Wei, L., Yan, X., 2016. Integrated proteomics and metabolomics for dissecting the mechanism of global responses to salt and alkali stress in *Suaeda corniculata*. Plant Soil 402, 379–394. https://doi.org/10.1007/s11104-015-2774-0.

Parida, A.K., Das, A.B., 2005. Salt tolerance and salinity effects on plants: A review. Ecotoxicol. Environ. Saf. 60, 324–349. https://doi.org/10.1016/j.ecoenv.2004.06.010.

Parida, A.K., Jha, B., 2010. Salt tolerance mechanisms in mangroves: A review. Trees - Struct. Funct. 24, 199–217. https://doi.org/10.1007/s00468-010-0417-x.

Parida, A.K., Jha, B., 2013. Physiological and biochemical responses reveal the drought tolerance efficacy of the halophyte *Salicornia brachiata*. J. Plant Growth Regul. 32, 342–352. https://doi.org/10.1007/s00344-012-9303-7.

Park, J., Kim, Y.-S., Kim, S.-G., Jung, J.-H., Woo, J.-C., Park, C.-M., 2011. Integration of auxin and salt signals by the NAC transcription factor NTM2 during seed germination in *Arabidopsis*. Plant Physiol. 156, 537–549. https://doi.org/10.1104/pp.111.177071.

Pathak, M.R., Teixeira da Silva, J.A., Wani, S.H., 2014. Polyamines in response to abiotic stress tolerance through transgenic approaches. GM Crops food 5, 87–96. https://doi.org/10.4161/gmcr.28774.

Pearson, M., Saarinen, M., Nummelin, L., Heiskanen, J., Roitto, M., Sarjala, T., Laine, J., 2013. Tolerance of peat-grown Scots pine seedlings to waterlogging and drought: Morphological, physiological, and metabolic responses to stress. For. Ecol. Manage 307, 43–53. https://doi.org/10.1016/j.foreco.2013.07.007.

Pereira, S.I., Figueiredo, P.I., Barros, A.S., Dias, M.C., Santos, C., Duarte, I.F., Gil, A.M., 2014. Changes in the metabolome of lettuce leaves due to exposure to mancozeb pesticide. Food Chem. 154, 291–298. https://doi.org/10.1016/j.foodchem.2014.01.019.

Pinheiro, C., Chaves, M.M., 2011. Photosynthesis and drought: can we make metabolic connections from available data? J. Exp. Bot. 62, 869–882. https://doi.org/10.1093/jxb/erq340.

Piotrowska-Niczyporuk, A., Bajguz, A., Zambrzycka, E., Godlewska-Zylkiewicz, B., 2012. Phytohormones as regulators of heavy metal biosorption and toxicity in green alga *Chlorella vulgaris* (Chlorophyceae). Plant Physiol. Biochem. 52, 52–65. https://doi.org/10.1016/j.plaphy.2011.11.009.

Qu, A.-L., Ding, Y.-F., Jiang, Q., Zhu, C., 2013. Molecular mechanisms of the plant heat stress response. Biochem. Biophys. Res. Commun. 432, 203–207. https://doi.org/10.1016/j.bbrc.2013.01.104.

Rivero, R.M., Mestre, T.C., Mittler, R., Rubio, F., Garcia-Sanchez, F., Martinez, V., 2014. The combined effect of salinity and heat reveals a specific physiological, biochemical and molecular response in tomato plants. Plant Cell Environ. 37, 1059–1073. https://doi.org/10.1111/pce.12199.

Rizhsky, L., Liang, H., Shuman, J., Shulaev, V., Davletova, S., Mittler, R., Rizhsky, L., Liang, H., Shuman, J., Shulaev, V., Davletova, S., Mittler, R., 2004. When defense pathways collide. The response of *Arabidopsis* to a combination of drought and heat stress. Plant Physiol. 134, 1683–1696. https://doi.org/10.1104/pp.103.033431.1.

Rocha, M., Licausi, F., Araújo, W.L., Nunes-Nesi, A., Sodek, L., Fernie, A.R., van Dongen, J.T., 2010. Glycolysis and the tricarboxylic acid cycle are linked by alanine aminotransferase during hypoxia induced by waterlogging of *Lotus japonicus*. Plant Physiol. 152, 1501–1513. https://doi.org/10.1104/pp.109.150045.

Rose, M.T., Rose, T.J., Pariasca-Tanaka, J., Yoshihashi, T., Neuweger, H., Goesmann, A., 2012. Root metabolic response of rice (*Oryza sativa* L.) genotypes with contrasting tolerance to zinc deficiency and bicarbonate excess. Planta 236, 959–973. https://doi.org/10.1007/s00425-012-1648-4.

Sakata, T., Oshino, T., Miura, S., Tomabechi, M., Tsunaga, Y., Higashitani, N., 2010. Auxins reverse plant male sterility caused by high temperatures. Proc. Natl. Acad. Sci. 107, 8569–8574. https://doi.org/10.1073/pnas.1000869107.

Sanchez, D.H., Schwabe, F., Erban, A., Udvardi, M.K., Kopka, J.O., 2012. Comparative metabolomics of drought acclimation in model and forage legume species. Plant Cell Environ. 35, 136–149. https://doi.org/10.1111/j.1365-3040.2011.02423.x.

Sanchez, D.H., Siahpoosh, M.R., Roessner, U., Udvardi, M., Kopka, J., 2008a. Plant metabolomics reveals conserved and divergent metabolic responses to salinity. Physiol. Plant 132, 209–219. https://doi.org/10.1111/j.1399-3054.2007.00993.x.

Sanchez, D.H., Lippold, F., Redestig, H., Hannah, M.A., Erban, A., Krämer, U., 2008b. Integrative functional genomics of salt acclimatization in the model legume *Lotus japonicus*. Plant J 53, 973–987.

Schneider, T., Persson, D.P., Husted, S., Schellenberg, M., Gehrig, P., Lee, Y., 2013. A proteomics approach to investigate the process of Zn hyperaccumulation in *Noccaea caerulescens* (J & C. Presl) F.K. Meyer. Plant J. Cell Mol. Biol. 73, 131–142. https://doi.org/10.1111/tpj.12022.

Seki, M., Umezawa, T., Urano, K., Shinozaki, K., 2007. Regulatory metabolic networks in drought stress responses. Curr. Opin. Plant Biol. 10, 296–302. https://doi.org/10.1016/j.pbi.2007.04.014.

Sengupta, S., Patra, B., Ray, S., Majumder, A.L., 2008. Inositol methyl transferase from a halophytic wild rice, *Porteresia coarctata* Roxb. (Tateoka): regulation of pinitol synthesis under abiotic stress. Plant Cell Environ. 31, 1442–1459. https://doi.org/10.1111/j.1365-3040.2008.01850.x.

Shahzad, R., Khan, A.L., Bilal, S., Waqas, M., Kang, S.M., Lee, I.J., 2017. Inoculation of abscisic acid-producing endophytic bacteria enhances salinity stress tolerance in *Oryza sativa*. Environ. Exp. Bot. 136, 68–77.

Shanker, A.K., Maheswari, M., Yadav, S.K., Desai, S., Bhanu, D., Attal, N.B., Venkateswarlu, B., 2014. Drought stress responses in crops. Funct. Integr. Genomics 14, 11–22. https://doi.org/10.1007/s10142-013-0356-x.

Sharma, N., Abrams, S.R., Waterer, D.R., 2005. Uptake, movement, activity, and persistence of an abscisic acid analogue (8' acetylene ABA methyl ester) in marigold and tomato. J. Plant Growth Regul. 24, 28–35. https://doi.org/10.1007/s00344-004-0438-z.

Sharma, S.S., Dietz, K.J., 2006. The significance of amino acids and amino acid-derived molecules in plant responses and adaptation to heavy metal stress. J. Exp. Bot. 57, 711–726. https://doi.org/10.1093/jxb/erj073.

Shi, H., Jiang, C., Ye, T., Tan, D., Reiter, R.J., Zhang, H., 2014. Comparative physiological, metabolomic, and transcriptomics analyses reveal mechanisms of improved abiotic stress resistance in bermudagrass [*Cynodon dactylon* (L). Pers.] by exogenous melatonin. J. Exp. Bot. 66, 681–694. https://doi.org/10.1093/jxb/eru373.

Shi, H., Ye, T., Song, B., Qi, X., Chan, Z., 2015a. Comparative physiological and metabolomic responses of four *Brachypodium distachyon* varieties contrasting in drought stress resistance. Acta Physiol. Plant 37, 1–12. https://doi.org/10.1007/s11738-015-1873-0.

Shi, Y.Z., Zhu, X.F., Wan, J.X., Li, G.X., Zheng, S.J., 2015b. Glucose alleviates cadmium toxicity by increasing cadmium fixation in root cell wall and sequestration into vacuole in Arabidopsis. J. Integr. Plant Biol. https://doi.org/10.1111/jipb.12312.

Shi, Y., Yang, S., 2014. Abscisic acid: Metabolism, transport and signaling. In: Zhang, D.-P. (Ed.), Abscisic Acid: Metabolism, Transport and Signaling. Springer, Dordrecht, Netherlands, pp. 337–363. https://doi.org/10.1007/978-94-017-9424-4.

Shi, Y., Tian, S., Hou, L., Huang, X., Zhang, X., Guo, H., Yang, S., 2012. Ethylene Signaling Negatively Regulates Freezing Tolerance by Repressing Expression of CBF and Type-A ARR Genes in Arabidopsis. Plant Cell 24, 2578–2595. https://doi.org/10.1105/tpc.112.098640.

Shinbo, Y., Nakamura, Y., Altaf-Ul-Amin, M., Asahi, H., Kurokawa, K., Arita, M., 2006. KNApSAcK: a comprehensive species-metabolite relationship database. In: Saito, K., Dixon, R., Willmitzer, L. (Eds.), Biotechnology in Agriculture and Forestry. Springer, Heidelberg, pp. 165–184.

Shibasaki, K., Uemura, M., Tsurumi, S., Rahman, A., 2009. Auxin response in *Arabidopsis* under cold stress: underlying molecular mechanisms. Plant Cell 21, 3823–3838. https://doi.org/10.1105/tpc.109.069906.

Silvente, S., Sobolev, A.P., Lara, M., 2012. Metabolite adjustments in drought tolerant and sensitive soybean genotypes in response to water stress. PLoS One 7. https://doi.org/10.1371/journal.pone.0038554.

Singh, K., Pandey, S.N., 2011. Effect of nickel-stresses on uptake, pigments and antioxidative responses of water lettuce, *Pistia stratiotes* L. J. Environ. Biol. 32, 391–394.

Slama, I., Ghnaya, T., Hessini, K., Messedi, D., Savouré, A., Abdelly, C., 2007. Comparative study of the effects of mannitol and PEG osmotic stress on growth and solute accumulation in *Sesuvium portulacastrum*. Environ. Exp. Bot. 61, 10–17. https://doi.org/10.1016/j.envexpbot.2007.02.004.

Smith, C.A., O'Maille, G., Want, E.J., Qin, C., Trauger, S.A., Brandon, T.R., 2005. METLIN: a metabolite mass spectral database. Ther. Drug Monit. 27, 747–751.

Soares, A.R., Marchiosi, R., de Cassia Siqueira-Soares, R., Barbosa de Lima, R., Dantas dos Santos, W., Ferrarese-Filho, O., 2014. The role of l-DOPA in plants. Plant Signal. Behav. 9 e28275-1–;e28275–e28279.

Sobhanian, H., Motamed, N., Jazii, F.R., Nakamura, T., Komatsu, S., 2010. Salt stress induced differential proteome and metabolome response in the shoots of *Aeluropus lagopoides* (Poaceae), a halophyte C4 plant. J. Proteome Res. 9, 2882–2897. https://doi.org/10.1021/pr900974k.

Sowiński, P., Rudzińska-Langwald, A., Adamczyk, J., Kubica, I., Fronk, J., 2005. Recovery of maize seedling growth, development and photosynthetic efficiency after initial growth at low temperature. J. Plant Physiol. 162, 67–80. https://doi.org/10.1016/j.jplph.2004.03.006.

Sudhakar, C., Veeranagamallaiah, G., Nareshkumar, A., Sudhakarbabu, O., Sivakumar, M., Pandurangaiah, M., 2015. Polyamine metabolism influences antioxidant defence mechanism in foxtail millet (*Setaria italica* L.) cultivars with different salinity tolerance. Plant Cell Rep. 34, 141 156. https://doi.org/10.1007/s00299-014-1695-3.

Sumner, L.W., Lei, Z., Nikolau, B.J., Saito, K., 2015. Modern plant metabolomics: advanced natural product gene discoveries, improved technologies, and future prospects. Nat. Prod. Rep. 32, 212–229. https://doi.org/10.1039/c4np00072b.

Sun, C., Gao, X., Chen, X., Fu, J., Zhang, Y., 2016. Metabolic and growth responses of maize to successive drought and re-watering cycles. Agric. Water Manag. 172, 62–73. https://doi.org/10.1016/j.agwat.2016.04.016.

Sung, J., Lee, S., Lee, Y., Ha, S., Song, B., Kim, T., 2015a. Metabolomic profiling from leaves and roots of tomato (*Solanum lycopersicum* L.) plants grown under nitrogen, phosphorus or potassium-deficient condition. Plant Sci. 241, 55–64. https://doi.org/10.1016/j.plantsci.2015.09.027.

Sung, J., Sonn, Y., Lee, Y., Kang, S., Ha, S., Krishnan, H.B., Oh, T.-K., 2015b. Compositional changes of selected amino acids, organic acids, and soluble sugars in the xylem sap of N, P, or K-deficient tomato plants. J. Plant Nut. Soil Sci. 178, 792–797. https://doi.org/10.1002/jpln.201500071.

Sytar, O., Kumar, A., Latowski, D., Kuczynska, P., Strzałka, K., Prasad, M.N.V., 2013. Heavy metal-induced oxidative damage, defense reactions, and detoxification mechanisms in plants. Acta Physiol. Plant 35, 985–999. https://doi.org/10.1007/s11738-012-1169-6.

Szymańska, R., Ślesak, I., Orzechowska, A., Kruk, J., 2017. Physiological and biochemical responses to high light and temperature stress in plants. Environ. Exp. Bot. https://doi.org/10.1016/j.envexpbot.2017.05.002.

Tabot, P.T., Adams, J.B., 2014. Salt secretion, proline accumulation and increased branching confer tolerance to drought and salinity in the endemic halophyte *Limonium linifolium*. S. Afr. J. Sci. 94, 64–73. https://doi.org/10.1016/j.sajb.2014.05.009.

Taguchi, R., Nishijima, M., Shimizu, T., 2007. Basic analytical systems for lipidomics by mass spectrometry in Japan. Methods Enzym. 432, 185–211. https://doi.org/10.1016/S0076-6879(07)320089.

Tan, W., Meng, Q.W., Brestic, M., Olsovska, K., Yang, X., 2011. Photosynthesis is improved by exogenous calcium in heat-stressed tobacco plants. J. Plant Physiol. 168, 2063–2071. https://doi.org/10.1016/j.jplph.2011.06.009.

Tassi, E., Pouget, J., Petruzzelli, G., Barbafieri, M., 2008. The effects of exogenous plant growth regulators in the phytoextraction of heavy metals. Chemosphere 71, 66–73. https://doi.org/10.1016/j.chemosphere.2007.10.027.

Tattini, M., Galardi, C., Pinelli, P., Massai, R., Remorini, D., Agati, G., 2004. Differential accumulation of flavonoids and hydroxycinnamates in leaves of *Ligustrum vulgare* under excess light and drought stress. New Phytol. 163, 547–561. https://doi.org/10.1111/j.1469-8137.2004.01126.x.

Tegelberg, R., Julkunen-Tiitto, R., 2001. Quantitative changes in secondary metabolites of dark-leaved willow (*Salix myrsinifolia*) exposed to enhanced ultraviolet-B radiation. Physiol. Plant 113, 541–547. https://doi.org/10.1034/j.1399-3054.2001.1130413.x.

Theodorou, M.E., Plaxton, W.C., 1993. Metabolic adaptations of plant respiration to nutritional phosphate deprivation. Plant Physiol. 101, 339–344. https://doi.org/10.1104/pp.101.2.339.

Theriappan, P., Gupta, A.K., Dhasarathan, P., 2011. Accumulation of proline under salinity and heavy metal stress in cauliflower seedlings. J. Appl. Sci. Environ. Manag. 15, 251–255. https://doi.org/10.4314/jasem.v15i2.68497.

Tuo, X.Q., Li, S., Wu, Q.S., Zou, Y.N., 2015. Alleviation of waterlogged stress in peach seedlings inoculated with Funneliformis mosseae: Changes in chlorophyll and proline metabolism. Sci. Hortic. 197, 130–134. https://doi.org/10.1016/j.scienta.2015.09.022.

Usadel, B., Bläsing, O.E., Gibon, Y., Poree, F., Höhne, M., Günter, M., et al., 2008. Multilevel genomic analysis of the response of transcripts, enzyme activities and metabolites in *Arabidopsis* rosettes to a progressive decrease of temperature in the non-freezing range. Plant Cell Environ. 31, 518–547. https://doi.org/10.1111/j.1365-3040.2007.01763.x.

Vasquez-Robinet, C., Mane, S.P., Ulanov, A.V., Watkinson, J.I., Stromberg, V.K., De Koeyer, D., 2008. Physiological and molecular adaptations to drought in Andean potato genotypes. J. Exp. Bot. 59, 2109–2123. https://doi.org/10.1093/jxb/ern073.

Ventrella, A., Catucci, L., Piletska, E., Piletsky, S., Agostiano, A., 2009. Interactions between heavy metals and photosynthetic materials studied by optical techniques. Bioelectrochemistry 77, 19–25. https://doi.org/10.1016/j.bioelechem.2009.05.002.

Verstraeten, I., Beeckman, T., Geelen, D., 2013. Adventitious root induction in *Arabidopsis thaliana* as a model for in vitro root organogenesis. Methods Mol. Biol. 959, 159–175. https://doi.org/10.1007/978-1-62703-221-6_10.

Vesely, T., Neuberg, M., Trakal, L., Szakova, J., Tlustoa, P., 2012. Water lettuce *Pistia stratiotes* L. response to lead toxicity. Water. Air. Soil Pollut. 223, 1847–1859. https://doi.org/10.1007/s11270-011-0989-0.

Vidoz, M.L., Loreti, E., Mensuali, A., Alpi, A. and Perata, P., 2010. Hormonal interplay during adventitious root formation in flooded tomato plants. Plant Journal, 63, 551-562. https://doi:10.1111/j.1365-313X.2010.04262.x.

Vishwakarma, K., Upadhyay, N., Kumar, N., Yadav, G., Singh, J., Mishra, R.K., et al., 2017. Abscisic acid signaling and abiotic stress tolerance in plants: a review on current knowledge and future prospects. Front. Plant Sci. 8, 1–12.

Viehweger, K., 2014. How plants cope with heavy metals. Bot. Stud. 55. https://doi.org/10.1186/1999-3110-55-35.

Visser, E.J.W., Bögemann, G.M., 2006. Aerenchyma formation in the wetland plant *Juncus effusus* is independent of ethylene. New Phytol 171, 305–314. https://doi.org/10.1111/j.1469-8137.2006.01764.x.

Voesenek, L.A.C.J., Van Veen, H., Sasidharan, R., 2014. Learning from nature: the use of non-model species to identify novel acclimations to flooding stress. AoB Plants 10, 1093. https://doi.org/10.1093/aobpla/plu016.

Vorst, O., de Vos, C.H., Lommen, A., Staps, R.V., Visser, R.G., Bino, R.J., 2005. A non-directed approach to the differential analysis of multiple LC–MS-derived metabolic profiles. Metabolomics 1, 169–180. https://doi.org/10.1007/s11306-005-4432-7.

Wang, X., Chang, L., Wang, B., Wang, D., Li, P., Wang, L., Yi, X., Huang, Q., Peng, M., Guo, A., 2013. Comparative proteomics of *Thellungiella halophila* leaves from plants subjected to salinity reveals the importance of chloroplastic starch and soluble sugars in halophyte salt tolerance. Mol. Cell. Proteom. 12, 2174–2195. https://doi.org/10.1074/mcp.M112.022475.

Wang, X., Yang, P., Gao, Q., Liu, X., Kuang, T., Shen, S., He, Y., 2008. Proteomic analysis of the response to high-salinity stress in *Physcomitrella patens*. Planta 228, 167–177. https://doi.org/10.1007/s00425-008-0727-z.

Warren, C., Aranda, I., Cano, F., 2012. Metabolomics demonstrates divergent responses of two Eucalyptus species to water stress. Metabolomics 8, 186–200.

Wei, W., Li, D., Wang, L., Ding, X., Zhang, Y., Gao, Y., Zhang, X., 2013. Morpho-anatomical and physiological responses to waterlogging of sesame (*Sesamum indicum* L.). Plant Sci. 208, 102–111. https://doi.org/10.1016/j.plantsci.2013.03.014.

Wenzel, A., Frank, T., Reichenberger, G., Herz, M., Engel, K.H., 2015. Impact of induced drought stress on the metabolite profiles of barley grain. Metabolomics 11, 454–467. https://doi.org/10.1007/s11306-014-0708-0.

Wheeler, D.L., Barrett, T., Benson, D.A., Bryant, S.H., Canese, K., Chetvernin, V., 2008. Database resources of the National Center for Biotechnology Information. Nucleic Acids Res. 36, D13–D21. https://doi.org/10.1093/nar/gkm1000.

Widodo, Patterson, J.H., Newbigin, E., Tester, M., Bacic, A., Roessner, U., 2009. Metabolic responses to salt stress of barley (*Hordeum vulgare* L.) cultivars, Sahara and Clipper, which differ in salinity tolerance. J. Exp. Bot. 60, 4089–4103. https://doi.org/10.1093/jxb/erp243.

Wijayasinghe, Y.S., Tyagi, A., Poddar, N.K., 2017. Regulation of Cell Volume by Osmolytes. In: Singh, L.R., Dar, T.A. (Eds.), Cellular Osmolytes, pp. 195–228. https://doi.org/10.1007/978-981-10-3707-8.

Wishart, D.S., 2007. Current progress in computational metabolomics. Brief. Bioinform 8, 279–293. https://doi.org/10.1093/bib/bbm030.

Witt, S., Galicia, L., Lisec, J., Cairns, J., Tiessen, A., Araus, J.L., 2012. Metabolic and phenotypic responses of greenhouse-grown maize hybrids to experimentally controlled drought stress. Mol. Plant 5, 401–417. https://doi.org/10.1093/mp/ssr102.

Wolfender, J.-L., Rudaz, S., Choi, Y.H., Kim, H.K., 2013. Plant metabolomics: from holistic data to relevant biomarkers. Curr. Med.Chem. 20, 1056–1090.

Wu, D., Cai, S., Chen, M., Ye, L., Chen, Z., Zhang, H., 2013. Tissue metabolic responses to salt stress in wild and cultivated barley. PLoS One 8. https://doi.org/10.1371/journal.pone.0055431.

Wu, H., Liu, X., You, L., Zhang, L., Zhou, D., Feng, J., 2012. Effects of salinity on metabolic profiles, gene expressions, and antioxidant enzymes in halophyte *Suaeda salsa*. J. Plant Growth Regul. 31, 332–341. https://doi.org/10.1007/s00344-011-9244-6.

Wulff-Zottele, C., Gatzke, N., Kopka, J., Orellana, A., Hoefgen, R., Fisahn, J., Hesse, H., 2010. Photosynthesis and metabolism interact during acclimation of *Arabidopsis thaliana* to high irradiance and sulphur depletion. Plant Cell. Environ. 33, 1974–1988. https://doi.org/10.1111/j.1365-3040.2010.02199.x.

Yamada, M., Morishita, H., Urano, K., Shiozaki, N., Yamaguchi- Shinozaki, K., Shinozaki, K., Yoshiba, Y., 2005. Effects of free proline accumulation in petunias under drought stress. J. Exp. Bot. 56, 1975–1981. https://doi.org/10.1093/jxb/eri195.

Yamauchi, T., Watanabe, K., Fukazawa, A., Mori, H., Abe, F., Kawaguchi, K., 2014. Ethylene and reactive oxygen species are involved in root aerenchyma formation and adaptation of wheat seedlings to oxygen-deficient conditions. J. Exp. Bot. 65, 261–273. https://doi.org/10.1093/jxb/ert371.

Yang, L.-T., Qi, Y.-P., Jiang, H.-X., Chen, L.-S., Yang, L.-T., Qi, Y.-P., 2013. Roles of organic acid anion secretion in aluminium tolerance of higher plants. Biomed. Res. Int. 2013, 173682. https://doi.org/10.1155/2013/173682.

Yang, X., Feng, L., Zhao, L., Liu, X., Hassani, D., Huang, D., 2017. Effect of glycine nitrogen on lettuce growth under soilless culture: a metabolomics approach to identify the main changes occurred in plant primary and secondary metabolism. J. Sci. Food Agri 0–2. https://doi.org/10.1002/jsfa.8482.

Yıldırım, K., Kaya, Z., 2017. Gene regulation network behind drought escape, avoidance and tolerance strategies in black poplar (*Populus nigra* L.). Plant Physiol. Biochem. 115, 183–199. https://doi.org/10.1016/j.plaphy.2017.03.020.

Yildiz-Aktas, L., Dagnon, S., Gurel, A., Gesheva, E., Edreva, A., 2009. Drought tolerance in cotton: Involvement of non-enzymatic ROS-scavenging compounds. J. Agr. Crop Sci. 195, 247–253. https://doi.org/10.1111/j.1439-037X.2009.00366.x.

Yiu, J.C., Juang, L.D., Fang, D.Y.T., Liu, C.W., Wu, S.J., 2009. Exogenous putrescine reduces flooding-induced oxidative damage by increasing the antioxidant properties of Welsh onion. Sci. Hortic. 120, 306–314.

Yuan, C., Ai, J., Chang, H., Xiao, W., Liu, L., Zhang, C., et al., 2017. CKB1 is involved in abscisic acid and gibberellic acid signaling to regulate stress responses in *Arabidopsis thaliana*. J. Plant Res. 130, 587–598.

Yuliana, N.D., Khatib, A., Verpoorte, R., Choi, Y.H., 2011. Comprehensive extraction method integrated with NMR metabolomics: A new bioactivity screening method for plants, adenosine A1 receptor binding compounds in *Orthosiphon stamineus* Benth. Anal. Chem. 83 (17), 6902–6906. https://doi.org/10.1021/ac201458n.

Zandalinas, S.I., Mittler, R., Balfagon, D., Arbona, V., Gomez-Cadenas, A., 2017. Plant adaptations to the combination of drought and high temperatures. Physiol. Planta. https://doi.org/10.1111/ppl.12540.

Zhang, J., Zhang, Y., Du, Y., Chen, S., Tang, H., 2011. Dynamic metabolomic responses of tobacco (*Nicotiana tabacum*) plants to salt stress. J. Proteome Res. 10, 1904–1914. https://doi.org/10.1021/pr101140n.

Zhao, X., Wang, W., Zhang, F., Deng, J., Li, Z., Fu, B., 2014. Comparative metabolite profiling of two rice genotypes with contrasting salt stress tolerance at the seedling stage. PLoS One 9, 1–7. https://doi.org/10.1371/journal.pone.0108020.

Zorrig, W., Rouached, A., Shahzad, Z., Abdelly, C., Davidian, J.-C., Berthomieu, P., 2010. Identification of three relationships linking cadmium accumulation to cadmium tolerance and zinc and citrate accumulation in lettuce. J. Plant Physiol. 167, 1239–1247. https://doi.org/10.1016/j.jplph.2010.04.012.

Zuther, E., Koehl, K., Kopka, J., 2007. Comparative metabolome analysis of the salt response in breeding cultivars of rice. In: Jenks, M.A., Hasegawa, P.M., Jain, S.M. (Eds.), Advances in Molecular Breeding toward Drought and Salt Tolerance Crops. Springer-Verlag, New York and Heidelberg, Berlin, pp. 285–300.

Chapter 6

Metabolomic Approach to Understand Plant Adaptations to Water and Salt Stress

Analía Llanes[a], Andrea Andrade[a], Sergio Alemano, Virginia Luna
Facultad de Ciencias Exactas, Físico-Químicas y Naturales, Universidad Nacional de Río Cuarto, Río Cuarto, Argentina

1. INTRODUCTION

Plants are unable to escape from unfavorable environmental conditions, e.g., biotic and abiotic stresses; thus their responses are manifested through physiological and metabolomic changes (Maritim et al., 2015). Salt and water stresses are the major abiotic environmental conditions that reduce plant growth and result in significant yield losses. Although plants have a wide spectrum of mechanisms to adapt to adverse environmental conditions, the current understanding of mechanisms associated with the ability of plants to maintain their growth under abiotic stresses are poorly understood. New advances in "omic" technologies are providing opportunities leading to the identification of pathways and mechanisms involved in plant stress responses (Ahuja et al., 2010). As most studies focused on model plants, here we consider plant species such as halophytes and other economically important glycophyte species such as sunflower. In response to salt and water stresses disruptions in plant metabolism occur, resulting in metabolic reconfiguration to maintain cell homeostasis and to assure plant survival (Hamanishi et al., 2015). There are numerous reports about the use of metabolomics to dissect plant responses to abiotic stresses, especially in the model plant *Arabidopsis* (Obata and Fernie, 2012). This technology is a powerful approach to achieve an integrated knowledge about the metabolic networks that regulate plant stress responses (Urano et al., 2010).

Metabolomics can be defined as the study of "metabolome," and the metabolome comprises all the chemical molecules with low molecular weight (<1800 Da) or metabolites present in a plant cell or tissue under a specified range of conditions (reviewed by Hall, 2011). Therefore the metabolites are the end products of cellular functions, and their levels can be considered as the plant responses to environmental or genetic manipulation. In plants, metabolomic studies aim to identify and quantify the set of primary and secondary metabolites involved in biological processes. Plant primary metabolites are implicated in the normal plant growth, development, and reproduction, whereas the secondary metabolites are crucial to plant survival under unfavorable conditions by maintaining a fine tuning with the environment. In addition, primary metabolites are highly conserved in their structures and abundances, whereas those of secondary metabolites vary widely across the plant kingdom (Scossa et al., 2016). The diversity of plant metabolites and their complicated regulatory mechanism highlight the necessity to investigate the biochemical nature of these compounds. Plant metabolome reported to date is composed of approximately 30,000 endogenous metabolites that mainly comprise carbohydrates, amino acids, organic acids, and lipids. Also, small molecules such as plant hormones and signaling molecules are vital for plant growth and development. Plant metabolomic research depends largely on its methodologies and instrumentation to comprehensively identify, quantify, and localize every metabolite. Thus several strategies for the analysis of metabolites are being developed rapidly (Hegeman, 2010): (1) *metabolite profiling*: identification and quantification of a number of predefined metabolites, which are related to a specific metabolic pathway(s); (2) *metabolic fingerprinting*: global screening of samples to discriminate among samples of different biological status or origin; (3) *metabolite target analysis*: qualitative and quantitative analysis of one or a few metabolites related to a specific metabolic reaction; (4) *metabonomics*: analysis of tissues and biological fluids for changes in endogenous metabolite contents resulting from disease or therapeutic treatments.

The metabolic networks in plants are highly complex and comprise numerous biochemical steps. Therefore the employment of appropriate analytical methods capable of identifying and quantifying changes in metabolite composition is crucial. At present, a set of integrated technologies and methodologies are used, such as nondestructive nuclear magnetic

a. These authors contributed equally to this chapter.

Plant Metabolites and Regulation Under Environmental Stress. https://doi.org/10.1016/B978-0-12-812689-9.00006-6

resonance spectroscopy (NMR) (Schripsema, 2010) and mass spectrometry (MS)-based methods, including GC-MS (gas chromatography-MS), LC-MS (liquid chromatography-MS) and CE-MS (capillary electrophoresis-MS) (Baran et al., 2009; Kaspar et al., 2011; Lee et al., 2012), and FI-ICR-MS (Fourier transform ion cyclotron resonance-MS) (Junot et al., 2010) (Fig. 6.1). Particularly, the liquid phase separations, high performance liquid chromatography (HPLC) and ultrahigh performance liquid chromatography (UPLC), provide a useful tool for the analysis of metabolites in several applications, and are a powerful analytical technique that allows the detection of a wide range of plant metabolites when they are coupled with MS (Theodoridis et al., 2012). GC-MS is excellent for the detection of biological samples with very complex matrices, offering highly efficient separation and resolution because this technology uses an electronic impact (EI) ionization source. On the other hand, LC-MS is useful in handling thermolabile, polar metabolites and high-molecular-weight compounds without derivatization, using soft ionization source, such as electrospray ionization (ESI) or atmospheric pressure chemical ionization. Indeed, MS with LC is the most frequently applied analytical technique in plant metabolomics studies because of its high sensitivity and extensive coverage of biological information relevant to the metabolism of organisms (Jorge et al., 2016). Metabolomic analysis of plant samples is now regarded as a viable counterpart for protein and transcript profiling technologies. The comprehensive identification and quantification of metabolites through the integration of methods based on MS allows metabolomics to be a robust technology comparable with or even better than conventional genomic or proteomic technologies. Indeed, metabolomic analyses have shown their potential in plant metabolomics studies in several food crops, among others. Nevertheless, assessing their robustness on plant metabolomic studies for more crops other than rice (Hu et al., 2014), maize (Rao et al., 2014), soybean (Kusano et al., 2015), and wheat (Francki et al., 2016) is needed.

In this chapter, we summarize the metabolomics technology to dissect the responses of plant species to water and salt stresses. To establish the similarity and specificity of the metabolic responses to both types of abiotic stresses, data available in the literature as well as our own data were used. The understanding of these stress responses through changes occurring at the metabolite level is a fundamental contribution to plant stress improvement programs.

2. METABOLIC RESPONSES OF PLANTS TO ABIOTIC STRESS

Plant metabolism is notably perturbed under abiotic stress conditions. In the last years, metabolomics has been employed for the identification of putative metabolites responsible for phenotypes tolerant/sensitive to several environmental stressors. In general, the metabolic changes that have been observed in plants subjected to stress may have different causes; thus they differ in their significance and are expected to differently correlate with tolerance/sensitivity phenotypes (Fig. 6.1). The main goal of studying metabolic changes during stress responses is to identify metabolites that allow the reestablishment

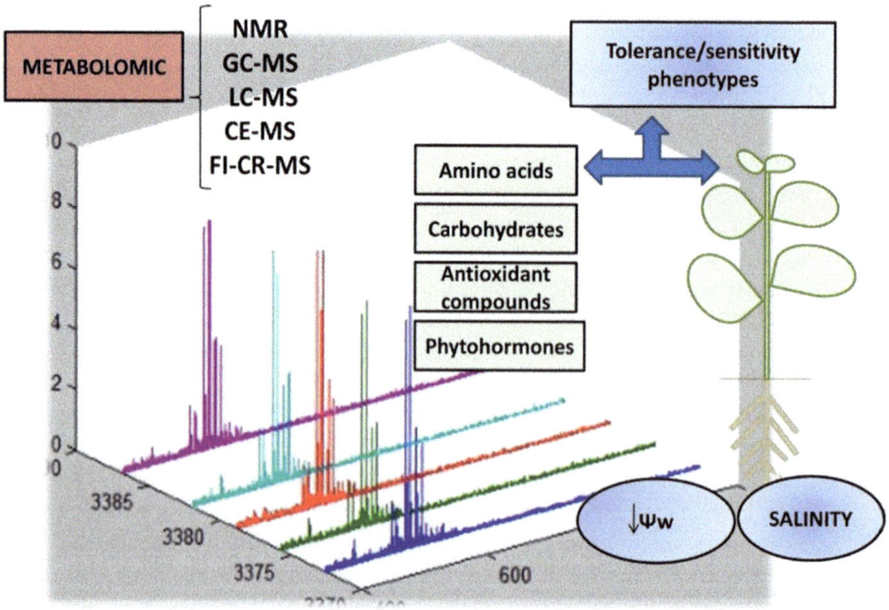

FIGURE 6.1 Integrated technologies for metabolomic analysis and identified metabolites in plants. *CE-MS*, capillary electrophoresis–mass spectrometry; *FI-ICR-MS*, Fourier transform ion cyclotron resonance-MS; *GC-MS*, gas chromatography-MS; *LC-MS*, liquid chromatography-MS; *NMR*, nuclear magnetic resonance spectroscopy.

of homeostasis and normal metabolic fluxes as well as to detect the accumulation of groups of compounds involved in mediating the stress tolerance.

A set of primary metabolites (osmolytes, osmoprotectants) and secondary metabolites (defense metabolites) accumulate to enhance plant stress tolerance. Among them, primary metabolites are the most important metabolites affected by stress, usually as a result of impairment in CO_2 assimilation. Although an increased accumulation of osmolytes by plants exposed to abiotic stresses has been reported, not all plant species synthesize all kinds of osmolytes; some species synthesize and accumulate very low quantities of some of these compounds, whereas some others do not do so at all (Slama et al., 2015). In some cases, the flux through a metabolic pathway rather than the accumulation of a specific metabolite per se might contribute to stress tolerance. Also, changes in secondary metabolite levels cannot be inferred only from variations in their primary metabolite precursors but are usually a result of a complex regulatory process.

Here we will attempt a summary of several metabolomic studies with the main associated metabolic changes in plant responses to two abiotic stresses: water stress and salinity. Furthermore, we discuss commonalities between both stress responses.

2.1 Water Stress

Water limitation is one of the major threats in crop production and it is projected to get considerably worse in the coming decades (Cominelli et al., 2009). Plants have developed numerous strategies in response to this environmental adverse factor (Ogbaga et al., 2014), most of them compatible with maintaining crop productivity. Morphophysiological responses can include reduction in leaf area, leaf abscission, stimulation of root growth by mobilization of nutrients to the underground parts of the plant, and biosynthesis of abscisic acid (ABA) to stomatal closure thereby reducing water loss through transpiration, among others. In addition, another strategy implies a precise metabolic regulation through the regulation of photosynthesis and accumulation of compounds such as osmolytes (Slama et al., 2015). As a consequence of an increase in osmolytes the osmotic potential of the cell is lowered, and water is drawn into the cell to maintain turgor pressure. Osmolytes include compounds such as soluble carbohydrates (e.g., glucose, sucrose, trehalose) and carbohydrate alcohols (e.g., mannitol, sorbitol); oligosaccharides (raffinose, stachyose, and verbascose); different amino acids, such as proline, betaine, valine, leucine, isoleucine, and agmatine (as precursor of polyamines); quaternary ammonium compounds (e.g., glycine betaine, b-alanine betaine, proline betaine, choline-O-sulfate); as well as polyamines (e.g., putrescine, spermidine, and spermine) (Ruan and Teixeira da Silva, 2011; Arbona et al., 2013). These compounds have an osmoprotective role during stress, repair damaged tissues, and support growth. In general, sugar alcohols are classified as acyclic (e.g., mannitol) and cyclic (e.g., pinitol) polyols. The accumulation of polyols may have dual functions: facilitating osmotic adjustment and supporting redox control. The up- and downregulation of specific primary and secondary metabolites constitute a defense mechanism to better tolerate the stress, and a number of them can be analyzed to assess the stress status of plants (Conde et al., 2014; Patel and Williamson, 2016).

A characterization of metabolic changes induced by water depletion in the roots of *Salvia miltiorrhiza* Bunge plants was performed by using two combined analytical techniques: NMR and LC-MS. The results showed that the NMR method was effective to detect both, the primary and secondary metabolites, whereas the LC-MS method was excellent for selectively detecting the secondary metabolites. Among the primary metabolites, five carbohydrates, galactose, glucose, sucrose, melibiose, and raffinose, were detected (Dai et al., 2010). In another study, nontargeted metabolomic approach by Gas Chromatography/Electron Impact-Time of Flight Mass Spectrometry (GC/EI-TOF-MS) was used to explore relative changes of the metabolite pool of gradually water-stressed *Lotus japonicus* ecotype Gifu plants. As a general metabolic trend in response to water stress, significant increases of fructose, glucose, galactose, and maltose were reported (Sánchez et al., 2012). GC-MS-based metabolite profiling was conducted to analyze the metabolites in leaf blades of 10 hybrids of maize exposed to drought (Obata et al., 2015). This study showed a large cluster that included several sugars and sugar alcohols, which decreased (e.g., maltose and erythritol) or did not change (e.g., raffinose and xylitol). In addition, a wide variety of amino acids, e.g., tryptophan, alanine, serine, glycine, leucine, and γ-aminobutyric acid (GABA) were accumulated in plants under drought stress (Obata et al., 2015). In other study, Sun et al. (2016) examined the metabolic responses of maize plants grown under drought through NMR metabolomic techniques. The authors found that stressed plants accumulated high levels of organic acids involved in the tricarboxylic acid (TCA) cycle (malate and citrate), amino acids such as proline and alanine, and sugars, particularly fructose, among other metabolites. In general, sucrose and particularly glucose act as osmolytes to maintain cell homeostasis. Nyarukowaa et al. (2016) identified and quantified metabolites in tolerant and drought-sensitive *Camellia sinensis* cultivars under water stress conditions using targeted analysis by GC–MS/MS and UPLC-MS. There was a significant variation in metabolite levels between drought tolerant and drought susceptible tea cultivars. Asparagine, glycine, valine, isoleucine, proline, and leucine were significantly higher in the DT (drought-tolerant) cultivars than in the DS (drought-sensitive) cultivars. The high amino acid levels observed in DT cultivars support their role as osmolytes (Shi and Chan, 2014). In addition, these elevated levels of amino acids were also attributed to a reduction in protein synthesis and an increase in the breakdown of proteins present. In regard to carbohydrates, an increase in fructose level was observed in the DS tea cultivars, whereas a decrease was detected in the DT cultivars. This could be because DT cultivars utilize fructose at a faster rate than the DS cultivars, ensuring their survival under drought stress. Under water deficit, glucose concentrations have also been documented to increase in DS plant varieties (Iordachescu and Imai, 2008). The glucose accumulation resulted in the increase of trehalose. This metabolite is a sugar reserve, supplying the plant with energy to cope with stress, as well as a stress protectant. This carbohydrate is also responsible for protein and membrane stabilization (Yoshida and Sakamoto, 2009). Furthermore, a significant increase in the levels of polyols, e.g., mannitol, was observed. Nyarukowaa et al. (2016) also observed that mannitol levels were lower in the DT than in the DS tea cultivars. This compound confers stress tolerance through actively scavenging hydroxyl radicals and is also found in lower concentrations in DT cultivars than in DS wheat crops. The accumulation of sugars in this study was accompanied by an increase in the concentrations of the organic acid citrate, although there was no statistically significant difference between the DT and DS tea cultivars. An increase in citrate leads to an increase in other Krebs cycle metabolites. Alternative column chemistries (e.g., hydrophilic interaction chromatography, porous graphitic carbon [PGC], and anion exchange chromatography) must be applied for the target analysis of the wide range of polar metabolites typically found in the plant metabolome. Among them, a less-laborious PGC-ESI-QIT-MS/MS (porous graphitic carbon liquid chromatography–electrospray ionization mass spectrometry) method has been developed, validated, and reported for the analysis of neutral sugars in a wide range of plant metabolomics studies related to plant environment and agriculture, including the quantification of glucose, sucrose, and raffinose from *Lupinus albus* cortex and stele stem in response to water stress and subsequent recovery treatments (António et al., 2008), and raffinose family oligosaccharides (RFOs) from *Haberlea rhodopensis* leaves in response to severe desiccation (Gechev et al., 2013). Overall, these examples show that as primary metabolites involved in plant response, amino acids and soluble sugars play important roles in cell responses to water deficit because they are directly linked to osmoprotection and photosynthetic performance.

The production of reactive oxygen species (ROS) is one of the primary responses to stress following the decline in photosynthesis. Therefore plants respond to this induced oxidative stress by overproducing antioxidant compounds, such as ascorbic acid, glutathione, and polyphenols (Das and Roychoudhury, 2014). Plants produce a huge variety of secondary metabolites that have diverse physical and chemical properties. This enables to them to play a role in coordinating ecology and defense strategies of plants (Wink, 2010). Among these metabolites, polyphenols are commonly involved in the defense mechanisms toward different abiotic stresses, including those related to drought (Di Ferdinando et al., 2014), because of their function as hydrogen donors and singlet oxygen scavengers (Labanowska et al., 2013). Ancillotti et al. (2015) analyzed metabolic changes by HPLC-MS/MS in *Nicotiana langsdorffii* plants under water stress. A decrease of antiradical activity and phenolics (represented by coumarins, such as esculetin and especially scopoletin) was observed. The complex information obtained from phenolic secondary metabolism data is in agreement with studies that reported different phenolic profiles in response to water deficit, depending on the species analyzed and even within the same species as a function of the more or less sensitive/tolerant genotype under investigation. Moreover, inconstant trends were found for different phenols in the same genotype. For example, a significant decrease of TP (polyphenols totals) was evidenced by Tejavathi et al. (2010) in in vitro plants of *Macrotyloma uniflorum* L. as a function of the increase of polyethylene glycol (PEG) percentage in the growing medium. Decreasing trends of TP and individual phenolic compounds were also observed in one tolerant and one sensitive cultivar of *Gossypium hirsutum* L. cultivated in vivo under water deficit conditions (Yildiz-Aktas et al., 2009). Conversely, TP were found to increase in other tolerant and sensitive genotypes of *G. hirsutum* cultivated in vivo with decreasing the watering level (Shah et al., 2011), which demonstrated the complex response of plant phenolic metabolism to water stress. Scalabrin et al. (2015) reported an untargeted metabolomic analysis using LC-MS in *N. langsdorffii* grown under water deficit, which showed an enhanced antioxidant level (e.g., polyamines) and damage of lipids, probably as a consequence of ROS production. These studies were highlighted by the results described by Sánchez-Rodríguez (2011), who reported either increasing or decreasing trends for hydroxycinnamic acids and flavonoids in different species as observed in leaves of different *Solanum lycopersicum* cultivars in response to water deficit.

Since the response to drought and salinity stresses is controlled by the hormonal network, the presence and relative concentration of phytohormones may be regarded as a good indicator of a plant's physiological condition. ABA is a phytohormone that plays an essential role in the dehydration-stress response. Studies of the molecular processes of the dehydration-stress response pathways have revealed both ABA-dependent and ABA-independent pathways (Yoshida et al., 2014). In Arabidopsis wild-type and knockout mutant of the NCED3 gene (nc3-2) plants subjected to dehydration stress, a metabolite profiling was performed using two types of MS systems, GC/TOF-MS and CE-MS. This study revealed that a total of 64 dehydration-increased metabolites were detected in the wild type and nc3-2 mutant. The accumulation of amino acids (valine, leucine, isoleucine, tyrosine, and tryptophan, among others) depended on ABA production, but the level of RFOs, citrate, malate, and succinate (TCA cycle) were regulated by ABA independently under dehydration stress (Urano et al., 2009). The analysis of wild and transgenic *N. langsdorffii* plants exposed to water deficit was performed by a sensitive HPLC-HRMS (high-resolution mass spectrometry) quantitative method to assure the simultaneous and sensitive determination of two relevant plant metabolites: salicylic acid (SA) and jasmonic acid (JA), which proved to be useful in the assessment of the plant metabolic and biochemical status. Particularly in this study the levels of these two phytohormones were not directly affected by water stress; however, SA and JA could take part in the activation of early plant defenses, to be progressively depleted during stress exposure. Some studies indicate that high concentrations of SA could determine oxidative stress, thus leading to reduced abiotic stress tolerance, whereas, at moderate levels, SA is able to alleviate drought damages (Miura and Tada, 2014); in *Arabidopsis thaliana* increased SA levels were related to an enhanced tolerance toward drought stress as a consequence of the induction of H_2O_2 as signaling molecule (He et al., 2014). JA seems to be involved in ascorbate and glutathione metabolism, which is a key element in the cellular redox balance, and therefore could protect plants from water deficit–induced oxidative stress, as highlighted by the malondialdehyde content and electrolyte leakage (Shan and Liang, 2010). Mahouachi et al. (2007) observed a transient accumulation of JA in papaya seedlings after 15 days of water deficit treatment, suggesting its role as a triggering signal for ABA accumulation.

In addition, other studies investigated the effect of moderate water stress on the hormonal and metabolic responses of two inbred lines of sunflower with different abilities to tolerate water stress: B59 (sensitive) and B71 (tolerant) during the early growth stage. Thus the hormonal (JA, ABA, SA, and indole-3-acetic acid [IAA]) and primary and nontargeted secondary metabolite analyses were carried out using GC and UPLC-qTOF-MS/MS. In both genotypes, the aerial organs showed a hormone accumulation higher than that in roots, irrespective of the relative tolerance. In stressed B59 seedlings, leaves showed SA levels higher than in controls, whereas in the B71 line both ABA and SA showed a high accumulation in stressed seedlings. In both sunflower lines, SA accumulation was a common response to water stress, whereas ABA significantly accumulated only in the tolerant line B71. In addition, JA and IAA levels were also positively altered in leaves of the tolerant line. Regarding metabolomic characterization, the shoots of the B59 line showed 1034 chromatographic signals, which corresponded to 50 potential metabolites. The B71 line presented a total of 1871 signals; they accounted for 160 metabolites. A total of 233 signals were common between both lines; among them, 41 were potential metabolites that had a higher accumulation under water stress (Fig. 6.2). In roots, for the B59 line a total of 779 chromatographic signals were obtained showing correspondence with 59 potential metabolites. The B71 line recorded 1088 signals, and they were equivalent to 40 potential metabolites. A total of 106 signals were common to both lines, of which 20 were for potential metabolites, 15 were induced and 5 repressed by water stress (Fig. 6.3) (Andrade et al., 2015).

So far, 25 primary metabolites were identified in the aerial part and root of sunflower seedlings, most being related to carbohydrate metabolism, and their profile was modified significantly in response to water stress in the tolerant genotype.

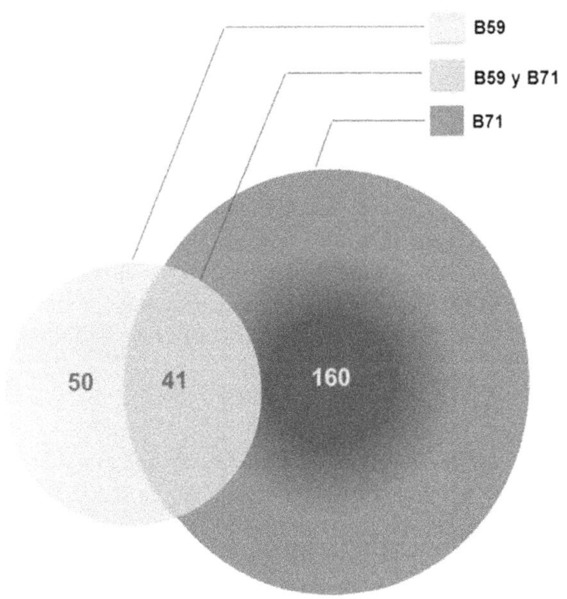

FIGURE 6.2 Number of potential common and differential metabolites in shoots of both B59 and B71 sunflower lines.

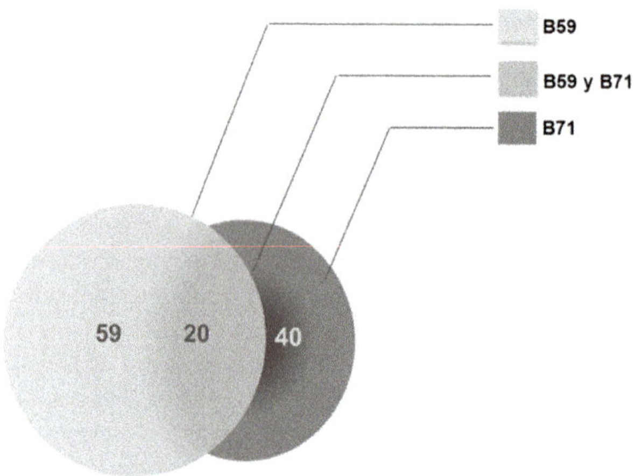

FIGURE 6.3 Number of potential common and differential metabolites in roots of both B59 and B71 sunflower lines.

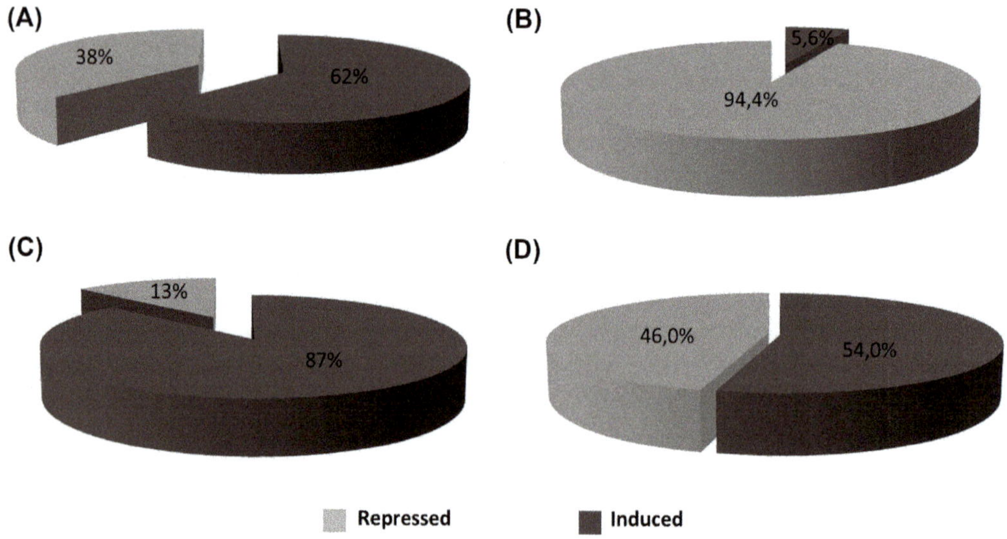

FIGURE 6.4 Percentage of potential metabolites induced and repressed by water stress. (A) B59 shoots; (B) B59 roots; (C) B71 shoots; (D) B71 roots.

Nontargeted metabolite profiling analysis showed that 62% of total altered metabolites accumulated under water stress conditions in leaves of the sensitive B59 line, whereas the rest reduced their concentration (Fig. 6.4A). In roots, the most altered metabolites (94%) were repressed by stress treatment (Fig. 6.4B). Interestingly, levels of most of the altered secondary metabolites (87%) increased under stress in shoots of the tolerant line (Fig. 6.4C). In roots, an opposite response was observed and approximately half of the altered metabolites were repressed (Fig. 6.4D). Some of these metabolites have been annotated. Preliminary results indicate that metabolic pathways in leaves were more altered by water stress conditions and this was more evident in B71 seedlings. This suggests that the tolerant line could have a differential hormonal and metabolic setup, responsible for its superior performance under water stress conditions (Andrade et al., 2015).

2.2 Salt Stress

When plants are subjected to salt stress, osmotic stress affects cell functions immediately leading to growth retardation and even tissue damage. Excessive sodium accumulation in plant cells causes a secondary stress that inhibits essential cellular processes, including photosynthesis, potassium ion absorption, and protein biosynthesis (Deinlein et al., 2014). The biosynthesis and accumulation of osmolytes is an important adaptive mechanism that enables the maintenance of cell turgor and restoration of cell water status by regulating the cellular water potential; these compounds can also be membrane stabilizing agents and/or ROS scavengers (Kao, 2015). Overproduction of osmolytes has been extensively used in an attempt to improve salt tolerance in several crops. Indeed, several studies using metabolomic techniques revealed a salt-induced accumulation of only few primary metabolites, such as proline or primary products in the N metabolism, depending on the species and organ (allantoin, urea, and glutamine, among others) (Slama et al., 2015).

The metabolite profile of perennial ryegrass (*Lolium perenne*) determined by GC-MS in response to salt stress (255 mM NaCl) showed that carbohydrates were sharply increased in stressed plants and accounted for more than 71.9% of the total metabolite content in leaves (Hu et al., 2016). In this study, sucrose represented approximately 42.2% of the total carbohydrates, which correlates with its crucial functions as a major signal molecule in source–sink regulation and in plant structure and metabolism. Similarly, previous studies reported that treatment with 200 mM NaCl increased sucrose and glucose levels in shoots of ryegrass plants (Hu et al., 2013). Higher levels of other carbohydrates such as tagatose, psicose, glucoheptose, idose, allose, talose, lactulose, and cellobiose and one sugar alcohol (inositol) were also observed in leaves of plants exposed to 255 mM NaCl for 12h or 4days. Therefore carbohydrate accumulation induced by salt stress might be critical in plant growth responses. Sugar alcohols could function as ROS scavengers to prevent cell peroxidation of lipids (Keunen et al., 2013).

Metabolic fingerprinting studies in two species with different sensitivity to salinity, the salt-tolerant plant *Thellungiella halophyte* and the salt-sensitive *A. thaliana*, were analyzed by Sanchez et al. (2008). These species exhibited similar NaCl dose-dependent patterns, including a gradual shifting of metabolome with increasing salt contents in the medium until a breakpoint at higher stress intensities, triggering a significant metabolic reconfiguration. Both species accumulated soluble sugars (fructose, glucose, sucrose, raffinose) and other unknown putative complex glycans under salt stress; however, levels of these metabolites were more pronounced in *Thellungiella* (Lugan et al., 2010). Likewise, *Thellungiella* plants had a higher proline content than *Arabidopsis* plants. In addition, *Thellungiella* had a higher detoxification efficiency and showed a lower mobilization of starch reserves during exposure to salt stress than *Arabidopsis*. This differential accumulation of stress-related compounds could be due to the photosynthetic rate, which was less affected in the salt-tolerant *Thellungiella* (Lugan et al., 2010). Similar results have been observed in the halophyte *Prosopis strombulifera* grown hydroponically in iso-osmotic solutions of NaCl and Na_2SO_4 at −1.0, −1.8, and −2.6MPa (250, 500, and 700mM NaCl and 190, 380, and 530 mM Na_2SO_4, respectively). This species is highly tolerant to NaCl and much less tolerant to Na_2SO_4. Metabolite profiles were analyzed by GC and UPLC-qTOF-MS/MS (Llanes et al., 2016). Several compounds, including amino acids, carbohydrates, and organic acids, among others, were salt induced. Amino acid accumulation in Na_2SO_4-treated plants was associated with cell damage, possibly due to an increase in the respiration caused by sulfate anion in the medium. On the contrary, specific amino acids, such as leucine, in roots or peptides, such as cysteinyl-glycine, in leaves were registered only in NaCl-treated plants, which could be a response induced to maintain cell homeostasis during salt stress. The high level of accumulation of carbohydrates observed only in salt-treated plants of *P. strombulifera* could reflect a high level of regulation of several metabolites from glycolysis/gluconeogenesis pathways. Some carbohydrates seem to be involved in the NaCl tolerance of *P. strombulifera*, such as maltose, maltotriitol, sucrose, and cellobiose (Llanes et al., 2016).

Phytohormones regulate every aspect of plant growth, development, and plant responses to environmental cues. One of the primary plant responses to abiotic stresses is ABA accumulation, which results in stomatal closure and reduced water loss via transpiration, eventually restricting cellular growth (Sah et al., 2016). Several studies have demonstrated that exogenous ABA application was able to offset the osmotic and ion stress effects from salt stress in plants by reducing the sodium concentration and enhancing the osmotic adjustment (Sripinyowanich et al., 2013; Gurmani et al., 2013).

Wang et al. (2015) analyzed the metabolomic profiles using GC-MS in the shoots and roots of two related rice genotypes with contrasting salt tolerance under salt and ABA treatments and their combinations. This study reported an organ-specific transcriptome reprogramming and primary metabolite changes in the two genotypes in their response to the treatments. Several primary metabolites in both genotypes were induced in response to salinity, which were partially ABA mediated and partially ABA independent. The analysis of metabolite profiles with public transcript data allowed demonstrating that ABA could trigger a readjustment of carbohydrate metabolism and carbon mobilization occurring during salinity. However, there is little information at the level of the metabolome profile regarding the relationship between ABA and salt stress responses in plants.

Alterations in the ABA pathway can affect the pathways of other hormones related to salt stress–induced responses. One of these hormonal pathways currently attracting much attention is jasmonate (JA) signaling, conveyed by JA and its derivatives (Riemann et al., 2015). JAs exogenous application diminished the damage by salinity in some species, such as rice seedlings (Kurotani et al., 2015). JA derives from the 13-LOX pathway and is one of the main oxylipins with a signaling role in the growth and development of plants growing under stress conditions (Wu and Baldwin, 2010). In the metabolomic analysis of *P. strombulifera* plants the intensity of the oxylipin 13 hydroxyoctadecadienoic acid detected by UPLC-qTOF-MS/MS decreased in the roots of salt-treated plants. Similarly, jasmonic acid methylester identified in this species by GC-MS varied with organ and salt treatment (Llanes et al., 2016). This low content of oxylipin acids in plants under salinity could be associated with an increase in the activity of the β-oxidation pathway to produce JAs with a possible role in stress tolerance responses.

It has been reported that plants accumulate excessive ROS inducing oxidative damage and cell death when exposed to salt stress (Dietz et al., 2016). The excessive ROS production during stress could result in serious damage to membranes and macromolecules, but this stress-induced damage could be attenuated by inhibiting ROS accumulation. However, it was proposed that an appropriate ROS level is required for optimal growth responses (Mittler, 2017). Enzymatic and nonenzymatic compounds are necessary for the scavenging or avoidance of ROS formation; compounds such as ascorbic acid, glutathione, α-tocopherols, some amino acids, carbohydrates, carotenoids, quinic acid derivatives, saponins, glucosinolates, phenolamides, phenylpropanoids, flavonoids are antioxidants in plants (reviewed by Demidchik, 2015; Noctor et al., 2015). Leaf metabolome changes in the xerohalophyte species *Atriplex halimus* induced by salinity (400 mM NaCl) were analyzed by Gas Chromatography - Flame Ionization Detector (GC-FID) (Bendaly et al., 2016). Results showed an increase in the total antioxidant activity, which was correlated with several compounds such as sucrose, threonine, phenylalanine, α-alanine, and proline. Salinity did not affect carotenoid levels in *A. halimus* leaves, which suggests their implication in oxidative stress resistance with a role as stabilizer of chloroplast membranes and reducing the formation of ROS directly by scavenging or indirectly by quenching the excited triplet state of chlorophyll (Li et al., 2010). Similarly, results about polyphenol levels in *A. halimus* leaves showed that these metabolites could be partly responsible for the antioxidant activity in leaves but evidently there were other factors involved (Bendaly et al., 2016). Metabolomic studies were also performed on the roots of *Salicornia herbacea* plants at different NaCl levels (0–400 mM NaCl) using UPLC–MS (Lee et al., 2016). Structural and functional analyses showed that 53 metabolites were accumulated when the NaCl concentration increased in the medium. These metabolites were linked to primary oxidative stress responses that increase under salt treatments, and were classified as polyols, alkaloids, steroids, shikimic acid derivates, and indole-3-carboxylic acids, which could protect against ROS generated by salt stress.

3. COMMON METABOLIC PLANT RESPONSES TO DROUGHT AND SALT STRESSES

The signaling components of water and salt stress display cross talk because both stress conditions result in cell dehydration, osmotic imbalance, and associated oxidative stress (Golldack et al., 2014). Comparison of the metabolic profiles of species with different tolerance to water and salt stresses (Muscolo et al., 2015) could show not only overlapping of crucial biochemical pathways, but also specificity in metabolic adjustments to these different environmental conditions. Several authors have evaluated the combined effects of water and salt stress, and the results suggest that both conditions likely elicit an additive effect on plant development (Suzuki et al., 2014). Some studies demonstrated that specific salinity levels might improve the ability of plants to cope with water stress. Sodium in plants might positively induce responses to water stress (Sucre and Suarez, 2011; Alla et al., 2012). In *P. strombulifera* plants, for example, Devinar et al. (2013) demonstrated that different salts at different concentrations and different relative humidity conditions caused particular responses of phytohormone production and growth parameters. Low humidity amplified the effects of salinity by modifying the biosynthesis of ABA and SA. Thus low-humidity conditions accentuated the stressing effect of the salt proposing that, when three stressing factors are applied simultaneously, they interact bringing about a response that is different from that triggered by any of them individually. However, further studies should be conducted to elucidate the exact mechanisms through which plants show improved characteristics when grown under simultaneous water and salt stresses compared with those subjected to individual stresses (Devinar et al., 2013).

The changes in plant metabolism summarized in this chapter show that accumulation of certain metabolites such as proline, soluble carbohydrates, glycine betaine, and γ-aminobutyric acid (GABA) is common to both water and salt stress. These metabolites are involved in maintaining the osmotic adjustment within the cell. Among the variety of quaternary ammonium compounds, glycine betaine is one of the most abundantly occurring in plants exposed to dehydration due to salinity and drought, among other abiotic stresses (Lokhande and Suprasanna, 2012). Carbohydrates such as the oligosaccharides raffinose and stachyose accumulate in different plant species in response to a broad range of abiotic stress conditions, such as drought or salinity (ElSayed et al., 2014). These compounds have been associated with a reduction in oxidative membrane damage and ROS scavenging. Also, the accumulation of other enzymatic and nonenzymatic compounds contributes to substantial reduction of the harmful effect of ROS generated by both abiotic stresses, whereas ROS themselves might act as an important messenger during stress responses (Suzuki et al., 2012). Phytohormones, mainly ABA, SA, and jasmonates (JAs), produced as a result of stress can also serve as signaling molecules activating systemic defense in plant responses under drought and salt stress. Taken together, the combined effects of water and salt stress could have a positive effect on plant metabolism by reducing the susceptibility to stress. The interaction between both types of stress points to a cross talk between their specific signaling pathways. This cross talk could be synergistic and/or antagonistic and include among others the involvement of amino acids, carbohydrates, ROS, and phytohormones.

4. CONCLUSION

Considering the studies described earlier, we may conclude that metabolomics is a powerful tool to advance the knowledge of plant physiology, because of its comprehensive information content concerning dynamic metabolic networks. The discrepancy and large number of compounds detected in each study may make difficult the understanding of the role of each one of them in the abiotic stress responses of plants. Notwithstanding, the diversity of compounds has to be interpreted as a picture of the metabolic pathways moving on in the tissue under a specific condition. Moreover, it can be seen that several metabolites are repeatedly found in plant stress responses to drought and salt stresses. Comparisons of metabolite profiling among species and/or cultivars differing in their stress tolerance, as well as the metabolite content of wild-type plants with that of mutants or transgenics, were reported to highlight qualitative and/or quantitative differences correlating with the phenotypes.

5. FUTURE PROSPECTS

The diversity of metabolites putatively providing stress protection that accumulate in different species upon salt and drought stress indicates that their production is not controlled by one specific pathway. Regulations of several pathways are involved in the stress responses in different species of plants. Nevertheless, the identification of specific metabolites can also represent good predictors for stress tolerance phenotypes in the screening of varieties as in plant breeding programs (Gauthier et al., 2015). Currently, more studies that include metabolomics as part of the integral approaches to study plant responses to a variety of stress conditions are needed. In this regard, the combination of metabolomics, proteomics, transcriptomics, and mathematical modeling will provide us a whole view of how plants respond to stresses and enable us to develop strategies to enhance drought and salt stress tolerance in different plants. In addition, the incorporation of these traits of tolerance to abiotic stresses in breeding programs of crops with agricultural interest is a challenge for the future.

REFERENCES

Ahuja, I., de Vos, R.C., Bones, A.M., Hall, R.D., 2010. Plant molecular stress responses face climate change. Trends Plant Sci. 15 (12), 664–674.

Alla, M.M.N., Khedr, A.H.A., Serag, M.M., Abu-Alnaga, A.Z., Nada, R.M., 2012. Regulation of metabolomics in *Atriplex halimus* growth under salt and drought stress. Plant Growth Reg. 67, 281–304.

Ancillotti, C., Bogani, P., Biricolti, S., Calistri, E., Checchini, L., Ciofi, L., Gonnelli, C., Del Bubba, M., 2015. Changes in polyphenol and sugar concentrations in wild type and genetically modified *Nicotiana langsdorffii Weinmann* in response to water and heat stress. Plant Physiol. Biochem. 97, 52–61.

Andrade, A., Arbona, V., Gómez-Cádenas, A., Alemano, S., 2015. Hormonal and metabolic profile of two sunflower inbred lines (*Helianthus annuus* L.) in response to water stress. In: Current Research in Plant Physiology, XXI Reunión de la Sociedad Española de Fisiología Vegetal. XIV Congreso-Hispano Luso de Fisiología Vegetal. Toledo, España, p. 935.

António, C., Pinheiro, C., Chaves, M.M., Ricardo, C.P., Ortuno, M.F., Thomas-Oates, J., 2008. Analysis of carbohydrates in *Lupinus albus* stems on imposition of water deficit, using porous graphitic carbon liquid chromatography-electrospray ionization mass spectrometry. J. Chromatogr. A 1187, 111–118.

Arbona, V., Manzi, M., Ollas, C., Gómez-Cadenas, A., 2013. Metabolomics as a tool to investigate abiotic stress tolerance in plants. Int. J. Mol. Sci. 14, 4885–4911.

Baran, R., Reindl, W., Northen, T.R., 2009. Mass spectrometry based metabolomics and enzymatic assays for functional genomics. Curr. Opin. Microbiol. 12 (5), 547–552.

Bendaly, A., Messedi, D., Smaoui, A., Ksouri, R., Bouchereau, A., Abdelly, C., 2016. Physiological and leaf metabolome changes in the xerohalophyte species *Atriplex halimus* induced by salinity. Plant Physiol. Biochem. 103, 208–218.

Cominelli, E., Galbiati, M., Tonelli, C., Bowler, C., 2009. Water: the invisible problem. EMBO Rep. 10, 671–676.

Conde, A., Regalado, A., Rodrigues, D., Costa, J.M., Blumwald, E., Chaves, M., Gerós, H., 2014. Polyols in grape berry: transport and metabolic adjustments as a physiological strategy for water-deficit stress tolerance in grapevine. J. Exp. Bot. 66, 889–906.

Dai, H., Xiao, C., Liu, H., Hao, F., Tang, H., 2010. Combined NMR and LC-DAD-MS analysis reveals comprehensive metabonomic variations for three phenotypic cultivars of *Salvia miltiorrhiza Bunge*. J. Proteome Res. 9, 1565–1578.

Das, K., Roychoudhury, A., 2014. Reactive oxygen species (ROS) and response of antioxidants as ROS-scavengers during environmental stress in plants. Front. Environ. Sci. 2, 53.

Deinlein, U., Stephan, A.B., Horie, T., Luo, W., Xu, G., Schroeder, J.I., 2014. Plant salt-tolerance mechanisms. Trends Plant Sci. 19, 371–379.

Demidchik, V., 2015. Mechanisms of oxidative stress in plants: from classical chemistry to cell biology. Environ. Exp. Bot. 109, 212–228.

Devinar, G., Llanes, A., Luna, V., 2013. Abscisic acid and salicylic acid levels induced by different relative humidity and salinity conditions in the halophyte *Prosopis strombulifera*. Plant Growth Regul. 70, 247–256.

Dietz, K.J., Mittler, R., Noctor, G., 2016. Recent progress in understanding the role of reactive oxygen species in plant cell signaling. Plant Physiol. 171, 1535–1539.

Di Ferdinando, M., Brunetti, C., Agati, G., Tattini, M., 2014. Multiple functions of polyphenols in plants inhabiting unfavorable Mediterranean areas. Environ. Exp. Bot. 103, 107–116.

ElSayed, A.I., Rafudeen, M.S., Golldack, D., 2014. Physiological aspects of raffinose family oligosaccharides in plants: protection against abiotic stress. Plant Biol. 16, 1–8.

Francki, M.G., Hayton, S., Gummer, J., Rawlinson, C., Trengove, R.D., 2016. Metabolomic profiling and genomic analysis of wheat aneuploid lines to identify genes controlling biochemical pathways in mature grain. Plant Biotechnol. J. 14 (2), 649–660.

Gauthier, L., Atanasova-Penichon, V., Chéreau, S., Richard-Forget, F., 2015. Metabolomics to decipher the chemical defense of cereals against *Fusarium graminearum* and deoxynivalenol accumulation. Int. J. Mol. Sci. 16 (10), 24839–24872.

Gechev, T., Benina, M., Obata, T., Tohge, T., Sujeeth, N., Minkov, I., Hille, J., Temanni, R., Marriott, A.S., Thomas-Oates, J., Antonio, C., Mueller-Roeber, B., Schippers, J.H.M., Fernie, A., Toneva, V., 2013. Molecular mechanisms of desiccation tolerance in the resurrection glacial relic *Haberlea rhodopensis*. Cell Mol. Life Sci. 70, 689–709.

Golldack, D., Li, C., Mohan, H., Probst, N., 2014. Tolerance to drought and salt stress in plants: unraveling the signaling networks. Front. Plant Sci. 5.

Gurmani, A.R., Bano, A., Najeeb, U., Zhang, J., Khan, S.U., Flowers, T.J., 2013. Exogenously applied silicate and abscisic acid ameliorates the growth of salinity stressed wheat (*Triticum aestivum* L) seedlings through Na+ exclusion. Aust. J. Crop Sci. 7 (8), 1123.

Hall, R.D., 2011. Plant metabolomics in a nutshell: potential and future challenges. Ann. Plant Rev. Biol. Plant Metab. 43, 1–24.

Hamanishi, E.T., Barchet, G.L., Dauwe, R., Mansfield, S.D., Campbell, M.M., 2015. Poplar trees reconfigure the transcriptome and metabolome in response to drought in a genotype-and time-of-day-dependent manner. BMC Genomics 16 (1), 329.

He, Q., Zhao, S., Ma, Q., Zhang, Y., Huang, L., Li, G., Hao, L., 2014. Endogenous salicylic acid levels and signaling positively regulate *Arabidopsis* response to polyethylene glycol-simulated drought stress. Plant Growth Regul. 33, 871–880.

Hegeman, A.D., 2010. Plant metabolomics—meeting the analytical challenges of comprehensive metabolite analysis. Brief. Funct. Genomics 9 (2), 139–148.

Hu, C., Shi, J., Quan, S., Cui, B., Kleessen, S., Nikoloski, Z., Wang, J., 2014. Metabolic variation between japonica and indica rice cultivars as revealed by non-targeted metabolomics. Sci. Rep. 4, 5067.

Hu, T., Chen, K., Hu, L., Amombo, E., Fu, J., 2016. H_2O_2 and Ca^{2+} signals integrated to enhance perennial ryegrass acclimation to salt stresses. Sci. Rep. 6. https://doi.org/10.1038/srep36396.

Hu, Y., Chen, L., Wang, H., Zhang, L., Wang, F., Yu, D., 2013. *Arabidopsis* transcription factor WRKY8 functions antagonistically with its interacting partner VQ9 to modulate salinity stress tolerance. Plant J. 74 (5), 730–745.

Iordachescu, M., Imai, R., 2008. Trehalose biosynthesis in response to abiotic stress. J. Integr. Plant Biol. 50 (10), 1223–1229.

Jorge, F., Rodrigues, J., Caldana, C., Schmidt, R., van Dongen, J., Thomas-Oates, J., António, C., 2016. Mass spectrometry-based plant metabolomics: metabolite responses to abiotic stress. Mass Spectrom. Rev. 35, 620–649.

Junot, C., Madalinski, G., Tabet, J.C., Ezan, E., 2010. Fourier transform mass spectrometry for metabolome analysis. Analyst 135 (9), 2203–2219.

Kao, C.H., 2015. Mechanisms of salt tolerance in rice plants: compatible solutes and aquaporins. Crop Environ. Bioinform. 12, 73–82.

Kaspar, S., Peukert, M., Svatos, A., Matros, A., Mock, H.P., 2011. MALDI-imaging mass spectrometry–An emerging technique in plant biology. Proteomics 11 (9), 1840–1850.

Keunen, E.L.S., Peshev, D., Vangronsveld, J., Van Den Ende, W.I.M., Cuypers, A.N.N., 2013. Plant sugars are crucial players in the oxidative challenge during abiotic stress: extending the traditional concept. Plant Cell Environ. 36 (7), 1242–1255.

Kurotani, K.I., Hayashi, K., Hatanaka, S., Toda, Y., Ogawa, D., Ichikawa, H., Hattori, T., 2015. Elevated levels of CYP94 family gene expression alleviate the jasmonate response and enhance salt tolerance in rice. Plant Cell Physiol. pcv006.

Kusano, M., Yang, Z., Okazaki, Y., Nakabayashi, R., Fukushima, A., Saito, K., 2015. Using metabolomic approaches to explore chemical diversity in rice. Mol. Plant 8 (1), 58–67.

Labanowska, M., Filek, M., Kurdziel, M., Bidzińska, E., Miszalski, Z., Hartikainen, H., 2013. EPR spectroscopy as a tool for investigation of differences in radical status in wheat plants of various tolerances to osmotic stress induced by NaCl and PEG-treatment. J. Plant Physiol. 170, 136–145.

Lee, S.J., JeongEM, Ki, A.Y., Oh, K.S., Kwon, J., Jeong, J.H., Chung, N.J., 2016. Oxidative defense metabolites induced by salinity stress in roots of *Salicornia herbacea*. J. Plant Physiol. 206, 133–142.

Lee, Y.J., Perdian, D.C., Song, Z., Yeung, E.S., Nikolau, B.J., 2012. Use of mass spectrometry for imaging metabolites in plants. Plant J. 70 (1), 81–95.

Li, J., Pandeya, D., Nath, K., Zulfugarov, I.S., Yoo, S.C., Zhang, H., Paek, N.C., 2010. ZEBRA-NECROSIS, a thylakoid-bound protein, is critical for the photoprotection of developing chloroplasts during early leaf development. Plant J. 62 (4), 713–725.

Llanes, A., Arbona, V., Gómez-Cadenas, A., Luna, V., 2016. Metabolomic profiling of the halophyte *Prosopis strombulifera* shows sodium salt-specific response. Plant Physiol. Bioch. 108, 145–157.

Lokhande, V.H., Suprasanna, P., 2012. Prospects of halophytes in understanding and managing abiotic stress tolerance. In: Ahmad, P., Prasad, M.N.V. (Eds.), Environmental Adaptations and Stress Tolerance of Plants in the Era of Climate Change. Springer, New York, pp. 29–56.

Lugan, R., Niogret, M.F., Leport, L., Guegan, J.P., Larher, F., Savoure, A., Kopka, J., Bouchereau, A., 2010. Metabolome and water homeostasis analysis of *Thellungiella salsuginea* suggests that dehydration tolerance is a key response to osmotic stress in this halophyte. Plant J. 64 (215), 229.

Mahouachi, J., Arbona, V., Gómez-Cadenas, A., 2007. Hormonal changes in papaya seedlings subjected to progressive water stress and re-watering. Plant Growth Regul. 53, 43–51.

Maritim, T.K., Kamunya, S.M., Mireji, P., Mwendia, V., Muoki, R.C., Cheruiyot, E.K., Wachira, F.N., 2015. Physiological and biochemical responses of tea (*Camellia sinensis* L. O Kuntze) to water-deficit stress. J. Hortic. Sci. Biotech. 90 (4), 395–400.

Mittler, R., 2017. ROS are good. Trends Plant Sci. 22, 11–19.

Miura, K., Tada, Y., 2014. Regulation of water, salinity, and cold stress responses by salicylic acid. Front. Plant Sci. 5, 1–12.

Muscolo, A., Junker, A., Klukas, C., Weigelt-Fischer, K., Riewe, D., Altmann, T., 2015. Phenotypic and metabolic responses to drought and salinity of four contrasting lentil accessions. J. Exp. Bot. 66 (18), 5467–5480.

Noctor, G., Lelarge-Trouverie, C., Mhamdi, A., 2015. The metabolomics of oxidative stress. Phytochem 112, 33–53.

Nyarukowaa, C., Koecha, R., Lootsc, T., Apostolidesa, Z., 2016. SWAPDT: a method for short-time withering assessment of probability for drought tolerance in *Camellia sinensis* validated by targeted metabolomics. J. Plant Physol. 198, 39–48.

Obata, T., Fernie, A.R., 2012. The use of metabolomics to dissect plant responses to abiotic stresses. Cell Mol. Life Sci. 69 (19), 3225–3243.

Obata, T., Witt, S., Lisec, J., Palacios-Rojas, N., Florez-Sarasa, I., Yousfi, S., Araus, J.L., Cairns, J.E., Fernie, A.R., 2015. Metabolite profiles of maize leaves in drought, heat, and combined stress field trials reveal the relationship between metabolism and grain yield. Plant Physiol. 169, 2665–2683.

Ogbaga, C.C., Stepien, P., Johnson, G.N., 2014. Sorghum (*Sorghum bicolor*) varieties adopt strongly contrasting strategies in response to drought. Physiol. Plant 152 (2), 389–401.

Patel, T.K., Williamson, J.D., 2016. Mannitol in plants, fungi, and plant–fungal interactions. Trends Plant Sci. 21, 486–497.

Rao, A., Ahmad, S.D., Sabir, S.M., Awan, S.I., Shah, A.H., Abbas, S.R., Shafique, S., Khan, F., Chaudhary, A., 2014. Potential antioxidant activities improve salt tolerance in ten varieties of wheat (*Triticum aestivum* L.). Am. J. Plant Sci. 4, 69–76.

Riemann, M., Dhakarey, R., Hazman, M., Miro, B., Kohli, A., 2015. Exploring jasmonates in the hormonal network of drought and salinity responses. Front. Plant Sci. 6, 1077.

Ruan, C.J., Teixeira da Silva, J.A., 2011. Metabolomics: creating new potentials for unraveling the mechanisms in response to salt and drought stress and for the biotechnological improvement of xero-halophytes. Crit. Rev. Biotechnol. 31 (2), 153–169.

Sah, S.K., Reddy, K.R., Li, J., 2016. Abscisic acid and abiotic stress tolerance in crop plants. Front. Plant Sci. 7, 571–578.

Sanchez, D.H., Siahpoosh, M.R., Roessner, U., Udvardi, M., Kopka, J., 2008. Plant metabolomics reveals conserved and divergent metabolic responses to salinity. Physiol. Plant 132 (2), 209–219.

Sánchez, F.S., Erban, A., Udvardi, M.K., Kopka, J., 2012. Comparative metabolomics of drought acclimation in model and forage legumes. Plant Cell Environ. 35, 136–149.

Sánchez-Rodríguez, E., Rubio-Wilhelmi, M., del, M., Ríos, J.J., Blasco, B., Rosales, M.Á., Melgarejo, R., Romero, L., Ruiz, J.M., 2011. Ammonia production and assimilation: its importance as a tolerance mechanism during moderate water deficit in tomato plants. J. Plant Physiol. 168, 816–823.

Scalabrin, E., Radaelli, M., Rizzato, G., Bogani, P., Buiatti, M., Gambaro, A., Capodaglio, G., 2015. Metabolomic analysis of wild and transgenic *Nicotiana langsdorffii* plants exposed to abiotic stresses: unraveling metabolic responses. Anal. Bioanal. Chem. 407 (21), 6357–6368.

Schripsema, J., 2010. Application of NMR in plant metabolomics: techniques, problems and prospects. Phytochem. Anal. 21 (1), 14–21.

Scossa, F., Brotmana, Y., de Abreu e Lima, F., Willmitzer, L., Nikoloski, Z., Tohge, T., Fernie, A.R., 2016. Genomics-based strategies for the use of natural variation in the improvement of crop metabolism. Plant Sci. 242, 47–64.

Shah, A.R., Khan, T.M., Sadaqat, H.A., Chatha, A.A., 2011. Alterations in leaf pigments in cotton (*Gossypium hirsutum*) genotypes subjected to drought stress conditions. Int. J. Agric. Biol. 13, 902–908.

Shan, C., Liang, Z., 2010. Jasmonic acid regulates ascorbate and glutathione metabolism in *Agropyron cristatum* leaves under water stress. Plant Sci. 178, 130–139.

Shi, H., Chan, Z., 2014. Improvement of plant abiotic stress tolerance through modulation of the polyamine pathway. J. Integr. Plant Biol. 56, 114–121.

Slama, I., Abdelly, C., Bouchereau, A., Flower, T., Savouré, A., 2015. Diversity, distribution and roles of osmoprotective compounds accumulated in halophytes under abiotic stress. Ann. Bot. 115, 433–447.

Sripinyowanich, S., Klomsakul, P., Boonburapong, B., Bangyeekhun, T., Asami, T., Gu, H., Chadchawan, S., 2013. Exogenous ABA induces salt tolerance in indica rice (*Oryza sativa* L.): the role of OsP5CS1 and OsP5CR gene expression during salt stress. Environ Exp. Bot. 86, 94–105.

Sucre, B., Suárez, N., 2011. Effect of salinity and PEG-induced water stress on water status, gas exchange, solute accumulation, and leaf growth in *Ipomoea pes-caprae*. Environ. Exp. Bot. 70, 192–203.

Sun, C.X., Li, M.Q., Gao, X.X., Liu, L.N., Wu, X.F., Zhou, J.H., 2016. Metabolic response of maize plants to multi-factorial abiotic stresses. Plant Biol. 18, 120–129.

Suzuki, N., Koussevitzky, S., Mittler, R., Miller, G., 2012. ROS and redox signaling in the response of plants to abiotic stress. Plant Cell Environ. 35, 259–270.

Suzuki, N., Rivero, R.M., Shulaev, V., Blumwald, E., Mittler, R., 2014. Abiotic and biotic stress combinations. New Phytol. 203, 32–43.

Tejavathi, D.H., Devaraj, V.R., Murthy, S.M., Anitha, P., Nijagunaiah, R., 2010. Regeneration of multiple shoots from the callus cultures of *Macrotyloma uniflorum* (Lam.) Verdc. Indian J. Biotech. 9, 101–105.

Theodoridis, G., Gika, H., Franceschi, P., Caputi, L., Arapitsas, P., Scholz, M., Mattivi, F., 2012. LC-MS based global metabolite profiling of grapes: solvent extraction protocol optimisation. Metabolomics 8 (2), 175–185.

Urano, K., Kurihara, Y., Seki, M., Shinozaki, K., 2010. 'Omics' analyses of regulatory networks in plant abiotic stress responses. Curr. Opin Plant Biol. 13 (2), 132–138.

Urano, K., Maruyama, K., Ogata, Y., Morishita, Y., Takeda, M., Sakurai, N., Kobayashi, M., 2009. Characterization of the ABA-regulated global responses to dehydration in *Arabidopsis* by metabolomics. Plant J. 57 (6), 1065–1078.

Wang, C., Liu, Y., Li, S.S., Han, G.Z., 2015. Insights into the origin and evolution of the plant hormone signaling machinery. Plant Physiol. 167 (3), 872–886.

Wink, M., 2010. Functions and Biotechnology of Plant Secondary Metabolites, second ed. Biochemistry of Plant Secondary Metabolism, vol. 40. Wiley-Blackwell, Oxford.

Wu, J., Baldwin, I.T., 2010. New insights into plant responses to the attack from insect herbivores. Annu. Rev. Genet. 44, 1–24.

Yildiz-Aktas, L., Dagnon, S., Gurel, A., Gesheva, E., Edreva, A., 2009. Drought tolerance in cotton: involvement of non-enzymatic ROS scavenging compounds. J. Agron. Crop Sci. 195, 247–253.

Yoshida, T., Sakamoto, T., 2009. Water stress induced trehalose accumulation and control of trehalase in the cyanobacterium *Nostoc punctiforme* IAM M-15. J. Gen. Appl. Microbiol. 55 (2), 135–145.

Yoshida, T., Mogami, J., Yamaguchi-Shinozaki, K., 2014. ABA-dependent and ABA-independent signaling in response to osmotic stress in plants. Curr. Op Plant Biol. 21, 133–139.

FURTHER READING

Fichman, Y., Gerdes, S.Y., Kovács, H., Szabados, L., Zilberstein, A., Csonka, L.N., 2015. Evolution of proline biosynthesis: enzymology, bioinformatics, genetics, and transcriptional regulation. Biol. Rev. 90, 1065–1099.

Renault, H., El Amrani, A., Berger, A., Mouille, G., Soubigou-Taconnat, L., Bouchereau, A., Deleu, C., 2013. γ-Aminobutyric acid transaminase deficiency impairs central carbon metabolism and leads to cell wall defects during salt stress in *Arabidopsis* roots. Plant Cell Environ. 36, 1009–1018.

Rontein, D., Basset, G., Hanson, A.D., 2002. Metabolic engineering of osmoprotectant accumulation in plants. Metab. Eng. 4, 49–56.

Witt, S., Galiciab, L., Liseca, J., Cairnsc, J., Tiessend, A., Arause, J.L., Palacios-Rojas, N., Ferniea, A.R., 2012. Metabolic and phenotypic responses of greenhouse-grown maize hybrids to experimentally controlled drought stress. Mol. Plant 5 (2), 401–417.

Chapter 7

Vitamins and Environmental Stresses in Plants

David Jiménez-Árias[1], Francisco J. García-Machado[1], Saray Morales-Sierra[1], Cristina Garrido-Orduña[1], Andrés A. Borges[1], Francisco V. González[2], Juan Cristo Luis Jorge[2]
[1]Instituto de Productos Naturales y Agrobiología-CSIC, La Laguna, Spain; [2]University of La Laguna, La Laguna, Spain

1. INTRODUCTION

In 1912, Casimir Funk first isolated the vitamin thiamin (vitamin B_1) and showed the important role of this new natural compound in beriberi disease (Smith et al., 2007). Over the following years, isolation and identification of all vitamins were established, thus acknowledging that vitamins, which play a central role in human health, can be obtained by a balanced diet rich in fruits and vegetables that are a rich source of such essential compounds (Alissa and Ferns, 2017). Although the significance of these molecules in plant metabolism has been ignored for a long time, nowadays vitamins are recognized as a diverse group of organic molecules traditionally classified as fat-soluble and water-soluble vitamins (Table 7.1). The fat-soluble vitamins are molecules with important antioxidant properties, whereas the water-soluble vitamins, or substances derived from them, commonly act as enzyme cofactors that carry out chemical reactions unmanageable for some cellular enzymes (Hanson et al., 2016). These essential compounds sometimes have the ability to play different roles in plant cells. For example, vitamin C (ascorbic acid) is essential in the ascorbate–glutathione pathway, acting as an antioxidant, and also in the protection of enzymes with a prosthetic transition metal ion (Hanson et al., 2016). In addition, it is a cofactor of many enzymes, including those in the synthesis of cell walls, especially in the hydroxylation of the proline residues (Macknight et al., 2017). In fact, the inability to make hydroxyl proline residues in collagen is manifested in the human disease scurvy, caused by a deficiency of vitamin C.

2. VITAMIN BIOSYNTHESIS, DEVELOPMENTAL AND ENVIRONMENTAL REGULATION

Vitamins, which are considered primary metabolites, are synthesized in small amounts as many secondary metabolites in plants, making the study of their metabolic pathways very difficult. Only in the past 20 years, the molecular analysis of the vitamin biosynthetic pathways, the genomic sequence information, and the interest in increasing the vitamin levels in plants, all improved our understanding on this matter (Giovanni, 2017). It has to be said that the elucidation of some vitamin biosynthetic pathways in plants was possible only because of our knowledge of bacterial pathways. Examples are biotin (vitamin B_7), pantothenic acid (vitamin B_5), and thiamin (vitamin B_1) (Colinas and Fitzpatrick, 2015). However, not all vitamins are synthesized by prokaryotes; in fact, vitamin C is exclusively from eukaryotes (Bulley and Laing, 2017).

Fortunately, today vitamin biosynthetic pathways have been completely elucidated, showing the complex allocation of metabolites due to the subcellular compartmentalization of eukaryotic cells, which sometimes involves one compartment (chloroplasts; examples are vitamins A, K_1, and riboflavin) or more compartments (cytosol, plastids, mitochondrion; examples are vitamins B_2, B_9, B_7, and C) in their biosynthesis (Miret and Munné-Bosch, 2014). In addition, many upstream precursors of vitamins come from the carbohydrate metabolism, which regulates the pool of pentoses, hexoses, and trioses in the plastids and the cytosol. Moreover, metabolites in the carbohydrate pool can be interconverted by a range of enzymatic activities, which keeps the cytosolic and plastid pools in equilibrium. Consequently, the carbohydrate pools and the metabolic pathways that feed and drain them could constitute the vital point for determining the fluxes of precursors toward the different branches that lead to downstream vitamin biosynthetic pathways (Asensi-Fabado and Munné-Bosch, 2010).

Plant Metabolites and Regulation Under Environmental Stress. http://dx.doi.org/10.1016/B978-0-12-812689-9.00007-8

TABLE 7.1 Functions of Some Vitamins Found in Plants and Algae

Vitamin	Biologically Active Form	Biological Function	Deficiency Diseases
Thiamin (B_1)	Thiamin pyrophosphate	Essential for several enzymes in central carbon metabolism	Beriberi
Riboflavin (B_2)	Flavin mononucleotide, flavin adenine dinucleotide	Used ubiquitously throughout the cell	Ariboflavinosis
Niacin (B_3)	Nicotinamide adenine dinucleotide	Used ubiquitously throughout the cell	Pellagra
Pantothenic acid (B_5)	Phosphopantetheine moiety in coenzyme A and acyl carrier protein	Cofactor for enzymes in lipid biosynthesis and catabolism and secondary metabolite biosynthesis	No known disease
Pyridoxine (B_6)	Pyridoxal phosphate	Used ubiquitously in amino acid transformations and in many secondary metabolite pathways	Risk factor in cardiovascular disease
Biotin (B_7)	Biotin	Cofactor for a small number of enzymes, including acetyl CoA carboxylase, propionyl CoA carboxylase, and pyruvate carboxylase	Basal ganglia disease
Folic acid (B_9)	Di- and tetra-hydrofolate polyglutamates	Required for several biosynthetic enzymes	Pernicious anemia
Cobalamin (B_{12})	Adenosylcobalamin, methylcobalamin	Cobalamin is not present in land plants, but in algae it is required, for example, for cobalamin-dependent methionine synthase	Pernicious anemia
Ascorbate (C)	Ascorbate	Essential antioxidant and substrate and/or cofactor in several reactions such as ascorbate oxidase and ascorbate peroxidases	Scurvy
Provitamin A (β-carotene)	Vitamin A	Gene expression by retinoic acids (all-trans and 9-cis) and by the ability of 11-cis retinal to absorb photons in rhodopsin	Xerophthalmia, corneal ulceration, blindness
Vitamin K	2-Methyl-3-phytyl-1, 4-naphthoquinone	Cofactor for γ-carboxylation of specific Glu residues, on certain proteins that must bind calcium to be biologically active	Bruising, hemorrhages, high risk of bone fracture

Following the carbohydrate metabolism, several branching points have been identified in plastids, which are involved in the biosynthesis of vitamins A, E, and K_1; however, other water-soluble vitamins (examples are B_5 and B_6) do not share precursors of the carbohydrate pools (Colinas et al., 2016). In addition, some vitamins are required as cofactors in the biosynthesis of other vitamins, for example, vitamin B_1 participates in the metabolic route of lipid-soluble vitamins and vitamin B_6, acting as a cofactor in the B_7 biosynthetic pathway (Asensi-Fabado and Munné-Bosch, 2010). Finally, the amino acid metabolism also plays a role in vitamin biosynthesis because many amino acids (glutamine, glutamic acid, valine, aspartic acid, alanine, glycine, cysteine) are precursors of several water-soluble and fat-soluble vitamins (Miret and Munné-Bosch, 2014).

Vitamin biosynthesis is also subjected to a developmental control. During the embryo development in maize, for example, there is an increase in gene transcripts leading to the biosynthesis of vitamin B_1 (Guan et al., 2014). Similarly, the enzymes involved in the folate biosynthesis showed a constant increase as the cell metabolism is engaged during the seed germination and seedling establishment (Gorelova et al., 2017). In addition, light is recognized as an important factor in vitamin biosynthesis, probably associated with the strong metabolic activity of the photosynthetic tissues. In fact, plants growing under high-light environments showed high levels of vitamin C in their tissues. Moreover, vitamin C is associated with increasing gene transcripts involved in the plant response to biotic and abiotic stress (Akram et al., 2017; Fujiwara et al., 2016). Similarly, the vitamin B_1 metabolic pathway is induced during plant development and when plants are exposed to light; however, these genes are also upregulated by stress conditions, such as hypoxia, high salinity, and sugar deprivation (Dong et al., 2015). These observations might reflect a fine-tuning of plant metabolism in response to environmental stress.

3. ANTIOXIDANT ACTION

Vitamin C is a well-known cellular antioxidant that plays a leading role in plants' defense antioxidant network owing to its excellent reactive oxygen species (ROS) scavenging properties (Asensi-Fabado and Munné-Bosch, 2010). It acts in chloroplasts, mitochondria, peroxisomes, and cytosol (ascorbate–glutathione cycle), adjusting the amount of hydrogen peroxide formed within the plant cells. Therefore it plays an important role in cellular signaling and in oxidative damage prevention (Akram et al., 2017). This vitamin is also found in the cell's apoplast where, with other antioxidants, it plays a role in controlling the cell's redox state and coordinates multiple stress responses (Foyer and Noctor, 2016). Provitamin A is involved in controlling ROS levels in chloroplasts, reducing singlet oxygen formation either by intercepting chlorophyll triplet states or by scavenging the existing singlet oxygen (Miret and Munné-Bosch, 2015). Vitamin E gives protection to membranes mainly by reacting with lipid peroxy radicals, reducing the extent of lipid peroxidation in seeds (Sattler et al., 2004; Rajjou et al., 2012).

Although all these vitamins showed an overlapping antioxidant role, their function in plants is fully complementary. For example, vitamin C is ubiquitously present in plant cells, being able to act as an excellent redox state sensor capable of ROS control and at the same time protecting the photoantioxidant machinery presumably by recycling the oxidized vitamin E (Akram et al., 2017). Provitamin A and other carotenoids are very effective in preventing the singlet oxygen accumulation, a role carried out also by vitamin E. They act at different locations in chloroplasts, at the site of production in photosystem II, and also inside the lipid matrix of thylakoids. Furthermore, vitamin E seems to play an essential role in preventing the propagation of lipid peroxidation. Vitamin B_6 also showed an efficient singlet oxygen scavenging role, but its presence outside the chloroplasts and its role as a cofactor clearly show that the function of this vitamin is more complex (Asensi-Fabado and Munné-Bosch, 2010).

Vitamin B_1 might also be a potent antioxidant in plants. Previous studies in *Zea mais* L. plants treated with vitamin B_1 showed a reduction of oxidative stress (Kaya et al., 2015). In fact, *Arabidopsis thaliana* L. seedlings exposed to high light, low temperatures, osmotic and salt stress, and paraquat exhibited high levels of vitamin B_1. These increases were correlated with the high gene transcript levels in the vitamin B_1 biosynthetic pathway. Therefore some authors suggested that vitamin B_1 could play an indirect role as an antioxidant in plant cells by providing NADH and NADPH to combat oxidative stress (Tunc-Ozdemir et al., 2009). Vitamin B_1 is a potent scavenger of superoxide anions and hydroxyl radicals with the ability to protect membranes from lipid peroxidation (Lukienko et al., 2000); however, it is unclear if these effects are the result of a direct antioxidant effect.

Vitamin B_9, especially its reduced forms, showed similar antioxidant properties in vitro when compared with vitamins C and E and also inhibited lipid peroxidation (Gliszczynska-Swiglo, 2007). However, vitamin B_9 and its reduced forms also act as cofactors in the transfer and use of one-carbon groups. They donate one-carbon groups in the biosynthesis of purine, pyrimidine, and DNA and play a key role in the regeneration of methionine and in the biosynthesis of vitamin B_5 (Gorelova et al., 2017). They are indirectly involved in the biosynthesis of *S*-adenosyl methionine and carotenoids and vitamin E. Consequently, it is difficult to discern between the direct and indirect effects of their antioxidant function in vivo. Finally, other vitamins are involved in the regulation of the cellular redox state and can therefore play a role in the plant antioxidant defense network. Examples are vitamin B_2 participating in multiple redox reactions (Sandoval et al., 2008), vitamin B_3 playing a key role in cellular redox chemistry, and vitamin K_1 acting in redox reactions at the plasma membrane of plants (Miret and Munné-Bosch, 2014). However, further research is needed to clarify the roles of vitamins B_3 and K in the antioxidant defense network of plants (Table 7.2).

4. VITAMINS PROTECTION UNDER BIOTIC STRESS

All organisms are constantly exposed to a wide range of pathogens and plants are not an exception. To cope with this threat, they have developed a wide range of strategies and complex molecular mechanisms. Therefore pathogens have to face physical, structural, and chemical defense barriers that can preexist or be developed after pathogen attack (Boubakri et al., 2016). Conserved microbial structures can be recognized by plants through transmembrane receptors resulting in intracellular defense mechanisms. This process described by some authors as "pathogen associated molecular patterns triggered immunity" is generally manifested by extracellular alkalinization, protein phosphorylation, defense genes upregulation, and ROS generation. The aim of these responses is limiting pathogen growth; however, pathogens can neutralize these responses through the secretion of specific proteins, resulting in a successful infection (Hann and Rathjen, 2010). Pathogen control in these cases can be achieved only by human intervention, which could be a serious risk to the environment and human health (Komárek et al., 2010).

Today's alternative in plant disease control arises from the use of elicitors of disease resistance through the manipulation of the natural host-defense strategies, also known as induced resistance (IR) (Walters et al., 2012). IR generally increases

TABLE 7.2 Antioxidant Capacity and Presence of Vitamin in Plants

Vitamin	Solubility	Antioxidant Capacity	Presence
Thiamin (B_1)	W.S.	O_2^-/OH^- scavenger	
Riboflavin (B_2)	W.S.	None described	Seeds: high levels; fruits, leaves, flowers, roots, tubers, and bulbs: small amounts
Niacin (B_3)	W.S.	None described	
Pyridoxine (B_6)	W.S.	$O_2^-/^1O_2$ scavenger	
Pantothenic acid (B_5)	W.S.	None described	Seeds and leaves: high levels; fruits, flowers, roots, tubers, and bulbs: small amounts
Folic acid (B_9)	W.S.	$O_2^-/^1O_2$ scavenger	
Biotin (B_7)	W.S.	None described	Seeds and fruits: high levels; leaves, flowers, roots, tubers, and bulbs: small amounts
Ascorbate (C)	W.S.	$H_2O_2/OH^-/^1O_2$ scavenger	Fruits and leaves: high levels; seeds, flowers, roots, tubers, and bulbs: small amounts
Provitamin A	L.S.	$^1O_2$ quencher	Seeds, fruits, leaves, flowers, roots, tubers, and bulbs: small amounts
Vitamin K	L.S.	Unknown	

L.P., lipid soluble; *W.S.*, water soluble.

plant innate defense responses against pathogens, which are triggered by the application of external factors. These defense responses could be established either before the pathogen attack (a process called elicitation) or after (also known as priming). The defense responses established during the IR led to plant resistance against a wide range of pathogens, such as fungi, bacteria, viruses, and also insect herbivores (Walters et al., 2013). IR is regulated by a complex signal transduction network. Two strategies have been described in plants; first, the systemic-acquired resistance (induced by chemical compounds and relies on salicylic acid defense signaling pathway), and second, the induced systemic resistance (an enhanced resistance developed after the colonization by beneficial soil-borne microorganisms) (Pieterse et al., 2013).

Several molecules have been studied as inducers of disease resistance, for example, aminobutyric acid, cyclodextrins, methyl jasmonate, and benzothiadiazole, with significant results against a wide range of pathogens; however, as a side effect they reduced plant growth (Boubakri et al., 2016). This unwanted consequence known as "allocation fitness cost" or "trade-off," is characterized by a growth reduction in treated plants because they use their energy in the development of IR (Cipollini and Lieurance, 2012). For some authors the molecular mechanisms behind this effect are the overexpression of phytotoxic chemicals (phenols and derivative molecules), the downregulation of genes involved in cell wall modifications and microtubule-driven movement, and the downregulation of translation-related transcripts, such as a set of ribosomal proteins, which led to the block of the cell cycle. It seems that such behavior in the elicited plants may contribute to energy saving (Walters et al., 2013; Reglinski et al., 2013; Almagro et al., 2014).

At present, work in under way to identify compounds capable of inducing resistance without reducing plant growth. Vitamins have received some attention; in fact, some were identified as inducers of disease resistance and were cost-effective (Song et al., 2013). Moreover, different vitamins were shown to induce resistance against a wide range of pathogens (fungus, bacteria, and viruses) in *A. thaliana* L. and different crop species when applied exogenously. Examples are vitamin B_1, vitamin B_2, vitamin K_3 synthetic derived molecule menadione sodium bisulfite (MSB), and vitamin B_9. Others such as vitamin C and vitamin E were linked to the molecular events associated to the IR process when elicited by the conventional inducers (Boubakri et al., 2016; Miret and Munné-Bosch, 2014).

4.1 Vitamin B_1

Vitamin B_1 showed high antioxidant activity in vivo. Previous studies described this antioxidant protection when *A. thaliana* L. plants were treated with paraquat and also exposed to abiotic stresses, such as low temperatures, high light, osmotic stress, and salinity (Tunc-Ozdemir et al., 2009). Against biotic stress, vitamin B_1 also induced resistance against a wide

TABLE 7.3 Vitamins and Vitamin-Derived Compounds That Induce Resistance in Plants

Vitamin	Plant/Crop	Pathogen	References
Thiamin (B_1)	Rice	*Xanthomonas oryzae*	Ahn et al. (2005)
	Cucumber	*Colletotrichum lagenarium*	Ahn et al. (2005)
	Arabidopsis	*Pseudomonas syringae*	Ahn et al. (2005)
	Tobacco	Pepper mild mottle virus	Ahn et al. (2005)
	Soybean	*Macrophomina phaseolina*	Abdel-Monaim (2011)
	Barley	Aphids	Hamada and Jonsson (2013)
	Pea	Aphids	Hamada and Jonsson (2013)
Riboflavin (B_2)	Arabidopsis	*P. syringae*	Zhang et al. (2009)
	Tobacco	*Alternaria alternate*	Zhang et al. (2009)
	Rice	*Rhizoctonia solani*	Taheri and Tarighi (2010)
	Chickpea	Fusarium wilt	Saikia et al. (2006)
	Bean	*Botrytis cinerea*	Azami-Sardooei et al. (2010)
	Soybean	*M. phaseolina*	Abdel-Monaim (2011)
	Grapevine	*Plasmopara viticola*	Boubakri et al. (2013)
Folic acid (B_9)	Arabidopsis	*P. syringae*	Wittek et al. (2015)
Vitamin-Derived Compounds			
Para-aminobenzoic acid	Cucumber	Cucumber mosaic virus	Song et al. (2013)
Menadione sodium bisulfite	Arabidopsis	*P. syringae*	Borges et al. (2009)
	Banana	*Fusarium oxysporum*	Borges et al. (2004)
	Oilseed rape	*Leptosphaeria maculans*	Borges et al. (2003),
	Citrus	*Candidatus liberibacter*	Liu et al. (2006)
	Pearl millet	Downy mildew	Borges-Rodríguez and Borges-Pérez (2010)
			Pushpalatha et al. (2007)

range of pathogens (see Table 7.3). In crop plants, such as rice, tobacco, and cucumber, vitamin B_1 treatments enhanced resistance against *Xanthomonas oryzae*, pepper mild mottle virus, *Colletotrichum lagenarium*, and powdery mildew (Ahn et al., 2005). Similarly, when applied under field conditions, vitamin B_1 significantly reduced the severity of the effects caused by *Macrophomina phaseolina* in soybeans (Abdel-Monaim, 2011) and enhanced the resistance against *Plasmopara viticola* in grapevine (Boubakri et al., 2012, 2013). Other studies pointed out a reduction of aphid infestations in barley and pea (Hamada and Jonsson, 2013).

4.2 Vitamin B_2

Vitamin B_2, when applied to *A. thaliana* L. plants induced protection against *Pseudomonas syringae* and *Peronospora parasitica*. Similarly, treated tobacco plants showed protection against tobacco mosaic virus and *Alternaria alternata* (Dong and Beer, 2000; Zhang et al., 2009). Moreover, vitamin B_2 was able to induce resistance against Fusarium wilt and charcoal rot diseases in chickpea and enhanced resistance to *Phytophthora parasitica* and *Ralstonia solanacearum* in treated tobacco seedlings (Saikia et al., 2006; Liu et al., 2010). In beans, vitamin B_2 treatments reduced *Botrytis cinerea* effects by approximately 25% compared with control plants (Azami-Sardooei et al., 2010). When applied to seeds under field conditions, vitamin B_2 reduced the damage caused by *M. phaseolina* in two consecutive seasons when compared with control seeds (Abdel-Monaim, 2011).

4.3 Menadione Sodium Bisulfite

MSB, which is a chemical modification of vitamin K_3, was initially studied for its effects on free indole-3-acetic acid in several crops, but in tomato plants it showed a correlation with an increased fruit yield (Borges et al., 2009). Later, MSB was studied as a plant IR activator against several pathogens, such as *Fusarium oxysporum f.* sp. *cubense* (Borges et al., 2003, 2004), downy mildew (Pushpalatha et al., 2007), and the virulent *P. syringae* pv. tomato DC3000 in *A. thaliana* L.

(Borges et al., 2009). In addition, insect growth reduction has been shown after MSB treatments, and more recently, this molecule was shown to protect citrus plants against both *Triozaerytreae* and *Diaphorinacitri* and also to repel mollusk pests in tomato plants (Borges et al., 2014; Carrillo-Perdomo et al., 2016).

4.4 Folic Acid

A study using *A. thaliana* L. described for the first time vitamin B_9 precursors 7,8-dihydropteroate and 4-amino-4-deoxychorismate (Wittek et al., 2015). In this report, vitamin B_9 and the DHP precursor 7,8-dihydroneopterin were able to induce resistance against *P. syringae* in infected *A. thaliana* L. plants. In addition, vitamin B_9 increased *A. thaliana* L. susceptibility to the necrotrophic fungus *Alternaria brassicicola* when it was locally applied.

5. VITAMIN BIOFORTIFICATION

Vitamin biofortification (breeding for enhanced vitamin content by conventional or transgenic selective approaches) could be a suitable strategy in terms of plants stress resistance and human nutrition (Díaz-Gómez et al., 2017). In fact, from the nutritional point of view, several attempts in enhancing the vitamin content have been published over the last 20 years in numerous crops with a large potential nutritional impact (Fudge et al., 2017; Guiliano, 2017; Strobbe and Van Der Straeten, 2017; Vanderschuren et al., 2013). Moreover, transgenic plants with high levels of vitamin E or provitamin A showed a decrease in their susceptibility to abiotic-induced oxidative stress (Guiliano, 2017; Jin and Daniell, 2014). Transgenic biofortification with vitamins B_1 and B_6 under different stress situations has been studied in *A. thaliana* L. (Raschke et al., 2011; Pourcel et al., 2013; Vanderschuren et al., 2013; Dong et al., 2015). *A. thaliana* L. leafs showed a higher increase in vitamin B_1 than control plants (Dong et al., 2015), and a similar result was obtained with vitamin B_6-engineered plants (Raschke et al., 2011). The vitamin B_6 biofortified plants exhibited a higher vegetative growth and were more resistant to oxidative stress when paraquat was applied (Raschke et al., 2011); however, vitamin B_1 biofortified plants showed no differences from control plants when exposed to salinity, cold, osmotic stress, drought, or oxidative stress (Dong et al., 2015).

6. CONCLUSION AND FUTURE PROSPECTS

Using natural compounds to control biotic or abiotic stress effects on plants could be a safer strategy than the use of pesticides or other hazardous molecules. Several compounds have been tested and promising results were shown when these were exogenously applied, despite some undesirable side effects on plant growth. Vitamins can be used as stress control agents, as several studies have demonstrated their beneficial effects on fruit yields and plant growth. Despite their multiple biological roles in plant metabolism, vitamins and also vitamin-derived compounds capable of inducing stress defense activities can be used as model systems that allow researchers to study the stress signal perception and transduction involved in stress tolerance or defense triggering. In this regard, biofortification has been demonstrated to increase vitamin levels in laboratory models and several crops with large potential nutritional impacts (for example, Golden Rice, canola, and tomatoes). Future findings of these studies may provide tools or practices that are environmentally safer than those used today in plant stress or disease treatments. In addition, the fact that some vitamins are essential components of plants' responses to different stresses could open new opportunities for future work on developing metabolically engineered crops overexpressing specific vitamin biosynthetic genes. In fact, several reports have demonstrated this possibility (overproducing vitamin B_6), whereas others did not show any beneficial effects. Finally, we believe that future strategies in crop plant stress responses must come by implementing integrated management programs from different disciplines (for example, agronomy, plant physiology, plant genetics) apart from other control strategies.

REFERENCES

Abdel-Monaim, M.F., 2011. Role of riboflavin and thiamine in induced resistance against charcoal rot disease of soybean. Afr. J. Biotechnol. 10, 10842–10855.

Ahn, I.P., Kim, S., Lee, Y.H., 2005. Vitamin B_1 functions as an activator of plant disease resistance. Plant Physiol. 138, 1505–1515.

Akram, N.A., Shafiq, F., Ashraf, M., 2017. Ascorbic acid-a potential oxidant scavenger and its role in plant development and abiotic stress tolerance. Front. Plant Sci. 8, 1–17.

Alissa, E.M., Ferns, G.A., 2017. Dietary fruits and vegetables and cardiovascular diseases risk. Crit. Rev. Food Sci. Nutr. 57, 1950–1962.

Almagro, L., Carbonell-Bejerano, P., Belchí-Navarro, S., Bru, R., Martínez-Zapater, J.M., Lijavetzky, D., Pedreño, M.A., 2014. Dissecting the transcriptional response to elicitors in *Vitis vinifera* cells. PLoS One 9, e109777.

Asensi-Fabado, M.A., Munné-Bosch, S., 2010. Vitamins in plants: occurrence, biosynthesis and antioxidant function. Trends Plant Sci. 15, 582–592.

Azami-Sardooei, Z., França, S.C., De Vleesschauwer, D., Höfte, M., 2010. Riboflavin induces resistance against *Botrytis cinerea* in bean, but not in tomato, by priming for a hydrogen peroxide-fueled resistance response. Physiol. Mol. Plant Pathol. 75, 23–29.

Borges, A.A., Borges-Pérez, A., Fernández-Falcón, M., 2004. Induced resistance to *Fusarial wilt* of banana by menadione sodium bisulphite treatments. Crop Prot. 23, 1245–1247.

Borges, A.A., Cools, H., Lucas, J.A., 2003. Menadione sodium bisulphite: a novel plant defence activator which enhances local and systemic resistance to infection by *Leptosphaeria maculans* in oilseed rape. Plant Pathol. 52, 429–436.

Borges, A.A., Dobon, A., Expósito-Rodríguez, M., Jiménez-Arias, D., Borges-Pérez, A., Casañas-Sánchez, V., Pérez, J.A., Luis, J.C., Tornero, P., 2009. Molecular analysis of menadione-induced resistance against biotic stress in Arabidopsis. Plant Biotechnol. J. 7, 744–762.

Borges, A.A., Jiménez-Arias, D., Expósito-Rodríguez, M., Sandalio, L.M., Pérez, J.A., 2014. Priming crops against biotic and abiotic stresses: MSB as a tool for studying mechanisms. Front. Plant Sci. 5, 642.

Borges-Rodríguez, A.A., Borges-Pérez, A., 2010. Compositions for Controlling the Psyllid *Trioza erytreae* and *Diaphorina citri*, Vectors of Bacteria of the Genus Candidatus Liberibacter Causal Agents of the Most Serious Citrus Disease Known as Huanglong-bing (HLB). Patent WO2012/045901.

Boubakri, H., Gargouri, M., Mliki, A., Brini, F., Chong, J., Jbara, M., 2016. Vitamins for enhancing plant resistance. Planta 244, 529–543.

Boubakri, H., Poutaraud, A., Wahab, M.A., Clayeux, C., Baltenweck-Guyot, R., Steyer, D., Marcic, C., Mliki, A., Soustre-Gacougnolle, I., 2013. Thiamine modulates metabolism of the phenylpropanoid pathway leading to enhanced resistance to *Plasmopara viticola* in grapevine. BMC Plant. Biol. 13, 31.

Boubakri, H., Wahab, M.A., Chong, J., Bertsch, C., Mliki, A., Soustre-Gacougnolle, I., 2012. Thiamine induced resistance to *Plasmopara viticola* in grapevine and elicited host-defense responses, including HR like-cell death. Plant Physiol. Biochem. 57, 120–133.

Bulley, S., Laing, W., 2017. The regulation of ascorbate biosynthesis. Curr. Opin. Plant Biol. 33, 15–22.

Carrillo-Perdomo, A., Jiménez-Arias, D., Aller, A., Borges, A.A., 2016. Menadione Sodium Bisulphite (MSB) enhances the resistance response of tomato leading to repel mollusc pests. Pest Manag. Sci. 72, 950–960.

Cipollini, D., Lieurance, D.M., 2012. Expression and costs of induced defense traits in *Alliaria petiolata*, a widespread invasive plant. Basic Appl. Ecol. 13, 432–440.

Colinas, M., Eisenhut, M., Tohge, T., Pesquera, M., Fernie, A.R., Weber, A.P.M., Fitzpatrick, T.B., 2016. Balancing of B6 vitamers is essential for plant development and metabolism in Arabidopsis. Plant Cell 28, 439–453.

Colinas, M., Fitzpatrick, T.B., 2015. Natures balancing act: examining biosynthesis de novo, recycling and processing damaged vitamin B metabolites. Curr. Opin. Plant Biol. 25, 98–106.

Díaz-Gómez, J., Twyman, R.M., Zhu, C., Farré, G., Serrano, J.C.E., Portero-Otin, M., Muñoz, P., Sandmann, G., Capell, T., Christou, P., 2017. Biofortification of crops with nutrients: factors affecting utilization and storage. Curr. Opin. Biotechnol. 44, 115–123.

Dong, H., Beer, S.V., 2000. Riboflavin induces disease resistance in plants by activating a novel signal transduction pathway. Phytopathology 90, 801–811.

Dong, W., Stockwell, V.O., Goyer, A., 2015. Enhancement of thiamin content in *Arabidopsis thaliana* by metabolic engineering. Plant Cell Physiol. 56, 2285–2296.

Foyer, C., Noctor, G., 2016. Stress-triggered redox signalling: what's in prospect? Plant Cell Environ. 39, 951–964.

Fudge, J., Mangel, N., Gruissem, W., Vanderschuren, H., Fitzpatrick, T.B., 2017. Rationalising vitamin B6 biofortification in crop plants. Curr. Opin. Biotechnol. 44, 130–137.

Fujiwara, A., Togawa, S., Hikawa, T., Matsuura, H., Masuta, C., Inukai, T., 2016. Ascorbic acid accumulates as a defense response to Turnip mosaic virus in resistant *Brassica rapa* cultivars. J. Exp. Bot. 67, 4391–4402.

Giovanni, G., 2017. Provitamin A biofortification of crop plants: a gold rush with many miners. Curr. Opin. Biotechnol. 44, 169–180.

Gliszczynska-Swiglo, A., 2007. Folates as antioxidants. Food Chem. 101, 1480–1483.

Gorelova, V., Ambach, L., Rébeillé, F., Stove, C., Van Der Straeten, D., 2017. Folates in plants: research advances and progress in crop biofortification. Front. Chem. 5, 1–21.

Guan, J.C., Hasnain, G., Garrett, T.J., Chase, C.D., Gregory, J., Hanson, A.D., McCarty, D.R., 2014. Divisions of labor in the thiamin biosynthetic pathway among organs of maize. Front. Plant Sci. 5, 1–11.

Guiliano, G., 2017. Provitamin A biofortification of crop plants: a gold rush with many miners. Curr. Opin. Biotechnol. 44, 169–180.

Hamada, A.M., Jonsson, L.M., 2013. Thiamine treatments alleviate aphid infestations in barley and pea. Phytochemistry 94, 135–141.

Hann, D.R., Rathjen, J.P., 2010. The long and winding road: virulence effector proteins of plant pathogenic bacteria. Cell Mol. Life Sci. 67, 3425–3434.

Hanson, A.D., Beaudoin, G.A., McCarty, D.R., Gregory III, J.F., 2016. Does abiotic stress cause functional B vitamin deficiency in plants? Plant Physiol. 172, 2082–2097.

Jin, S., Daniell, H., 2014. Expression of γ-tocopherol methyltransferase in chloroplasts results in massive proliferation of the inner envelope membrane and decreases susceptibility to salt and metal-induced oxidative stresses by reducing reactive oxygen species. Plant Biotechnol. J. 12, 1274–1285.

Kaya, C., Ashraf, M., Sonmez, O., Tuna, A.L., Polat, T., Aydemir, S., 2015. Exogenous application of thiamin promotes growth and antioxidative defense system at initial phases of development in salt-stressed plants of two maize cultivars differing in salinity tolerance. Acta Physiol. Plant. 37, 1–12.

Komárek, M., Cadková, E., Chrastny, V., Bordas, F., Bollinger, J.C., 2010. Contamination of vineyard soils with fungicides: a review of environmental and toxicological aspects. Environ. Int. 36, 138–151.

Liu, F., Wei, F., Wang, L., Liu, H., Zhu, X., Liang, Y., 2010. Riboflavin activates defense responses in tobacco and induces resistance against *Phytophthora parasitica* and *Ralstonia solanacearum*. Physiol. Mol. Plant Pathol. 74, 330–336.

Liu, S., Liu, Z., Fitt, B.D., Evans, N., Foster, S., Huang, Y., Latunde-Dada, A., Lucas, J., 2006. Resistance to *Leptosphaeria maculans* (phoma stem canker) in *Brassica napus* (oilseed rape) induced by *L. biglobosa* and chemical defence activators in field and controlled environments. Plant Pathol. 55, 401–412.

Lukienko, P.I., Melnichenko, N.G., Zverinskii, I.V., Zabrodskaya, S.V., 2000. Antioxidant properties of thiamine. Bull. Exp. Biol. Med. 130, 874–876.

Macknight, R.C., Laing, W.A., Bulley, S.M., Broad, R.C., Johnson, A.A.T., Hellens, R.P., 2017. Increasing ascorbate levels in crops to enhance human nutrition and plant abiotic stress tolerance. Curr. Opin. Biotechnol. 44, 153–160.

Miret, J.A., Munné-Bosch, S., 2014. Plant amino acid-derived vitamins: biosynthesis and function. Amino Acids 46, 809–824.

Miret, J.A., Munné-Bosch, S., 2015. Redox signaling and stress tolerance in plants: a focus on vitamin E. Ann. N.Y. Acad. Sci. 1340, 29–38.

Pieterse, C.M.J., Poelman, E.H., Van Wees, S.C.M., Dicke, M., 2013. Induced plant responses to microbes and insects. Front. Plant Sci. 4, 475.

Pourcel, L., Moulin, M., Fitzpatrick, T.B., 2013. Examining strategies to facilitate vitamin B_1 biofortification of plants by genetic engineering. Front. Plant Sci. 4, 160.

Pushpalatha, H., Mythrashree, S., Shetty, R., Geetha, N., Sharathchandra, R., Amruthesh, K., Shetty, H.S., 2007. Ability of vitamins to induce downy mildew disease resistance and growth promotion in pearl millet. Crop. Prot. 26, 1674–1681.

Rajjou, L., Duval, M., Gallardo, K., Catusse, J., Bally, J., Job, C., Claudette Job, C., Job, D., 2012. Seed germination and vigor. Annu. Rev. Plant Biol. 63, 507–533.

Raschke, M., Boycheva, S., Crèvecoeur, M., Nunes-Nesi, A., Witt, S., Fernie, A.R., Amrhein, N., Fitzpatrick, T.B., 2011. Enhanced levels of vitamin B_6 increase aerial organ size and positively affect stress tolerance in Arabidopsis. Plant J. 66, 414–432.

Reglinski, T., Vanneste, J.L., Wurms, K., Gould, E., Spinelli, F., Rikkerink, E., 2013. Using fundamental knowledge of induced resistance to develop control strategies for bacterial canker of kiwifruit caused by *Pseudomonas syringae* pv actinidiae. Front. Plant Sci. 4, 24.

Saikia, R., Yadav, M., Varghese, S., Singh, B.P., Gogoi, D.K., Kumar, R., Arora, D.K., 2006. Role of riboflavin in induced resistance against Fusarium wilt and charcoal rot diseases of Chickpea. Plant Pathol. J. 22, 339–347.

Sandoval, F.J., Zhang, Y., Roje, S., 2008. Flavin nucleotide metabolism in plants monofunctional enzymes synthesize FAD in plastids. J. Biol. Chem. 283, 30890–30900.

Sattler, S.E., Gilliland, L.U., Magallanes-Lundback, M., Pollard, M., DellaPenna, D., 2004. Vitamin E is essential for seed longevity, and for preventing lipid peroxidation during germination. Plant Cell 16, 1419–1432.

Smith, A.G., Croft, M.T., Moulin, M., Webb, M.E., 2007. Plants need their vitamins too. Curr. Opin. Plant Biol. 10, 266–275.

Song, G.C., Choi, H.K., Ryu, C.M., 2013. The folate precursor para-aminobenzoic acid elicits induced resistance against *Cucumber mosaic* virus and *Xanthomonas axonopodis*. Ann. Bot. 111, 925–934.

Strobbe, S., Van Der Straeten, D., 2017. Folate biofortification in food crops. Curr. Opin. Biotechnol. 44, 202–211.

Taheri, P., Tarighi, S., 2010. Riboflavin induces resistance in rice against *Rhizoctonia solani* via jasmonate-mediated priming of phenylpropanoid pathway. J. Plant Physiol. 167, 201–208.

Tunc-Ozdemir, M., Miller, G., Song, L., Kim, J., Sodek, A., Koussevitzky, S., Misra, A.N., Mittler, R., Shintani, D., 2009. Thiamin confers enhanced tolerance to oxidative stress in Arabidopsis. Plant Physiol. 151, 421–432.

Vanderschuren, H., Boycheva, S., Li, K.T., Szydlowski, N., Gruissem, W., Fitzpatrick, T.B., 2013. Strategies for vitamin B_6 biofortification of plants: a dual role as a micronutrient and a stress protectant. Front. Plant Sci. 4, 143.

Walters, D., Avrova, A., Bingham, I., Burnett, F., Fountaine, J., Havis, N., Hoad, S., Hughes, G., Looseley, M., Oxley, S.P., Renwick, A., Topp, C.E., Newton, A., 2012. Control of foliar diseases in barley: towards an integrated approach. Eur. J. Plant Pathol. 133, 33–73.

Walters, D.R., Ratsep, J., Havis, N.D., 2013. Controlling crop diseases using induced resistance: challenges for the future. J. Exp. Bot. 64, 1263–1280.

Wittek, F., Kanawati, B., Wenig, M., Hoffmann, T., Franz-Oberdorf, K., Schwab, W., Schmitt-Kopplin, P., Vlot, A.C., 2015. Folic acid induces salicylic acid-dependent immunity in Arabidopsis and enhances susceptibility to *Alternaria brassicicola*. Mole. Plant Pathol. 16, 616–622.

Zhang, S., Yang, X., Sun, M., Sun, F., Deng, S., Dong, H., 2009. Riboflavin-induced priming for pathogen defense in *Arabidopsis thaliana*. J. Integr. Plant Biol. 51, 167–174.

FURTHER READING

Fitzpatrick, T.B., Basset, G.J.C., Borel, P., Carrari, F., DellaPenna, D., Fraser, P.D., Hellmann, H., Osorio, S., Rothan, C.J., Valpuesta, V., Caris-Veyrat, C., Fernie, A.R., 2012. Vitamin deficiencies in humans: can plant science help? Plant Cell 24, 395–414.

Chapter 8

Environmental Stress and Secondary Metabolites in Plants: An Overview

Muhammad A. Ashraf[1], Muhammad Iqbal[1], Rizwan Rasheed[1], Iqbal Hussain[1,2], Muhammad Riaz[1], Muhammad Saleem Arif[1]

[1]*Government College University Faisalabad, Faisalabad, Pakistan;* [2]*Graduate School of Life Sciences, Tohoku University, Sendai, Japan*

1. INTRODUCTION

Plant secondary metabolites (SMs) are derivatives of primary metabolites produced by plants because of diverse physiological changes. The SMs significantly improve plant growth and survival under different environmental stresses and thereby act as important primary metabolites (Kliebenstein, 2013; Zandalinas et al., 2017). The SMs exert long-term effects on plant growth and survival under stressful environments (Agostini-Costa et al., 2012; Kurepin et al., 2017). About 100,000 SMs are present in the plant kingdom confined to specific taxonomic groups. There are three major groups of SMs in plants based on their biosynthetic pathway. These groups include nitrogen-containing compounds (cyanogenic glycosides, alkaloids, and glucosinolates), phenolic compounds (flavonoids and phenylpropanoids), and terpenes (isoprenoids) (Fang et al., 2011). The important secondary metabolites (SMs) and their variation in plants under environmental stress has been discussed in detail.

2. NITROGEN-CONTAINING COMPOUNDS

Cyanogenic glycosides, glucosinolates, and alkaloids comprise an important class of nitrogen-containing SMs. Alkaloids are low-molecular-weight compounds, and more than 12,000 nitrogen-containing compounds are included in this family (Facchini, 2001; Khan and ur Rahman, 2017). The biosynthesis of alkaloids is associated with the availability of a few amino acids such as tryptophan, tyrosine, and lysine (Taiz and Zeiger, 2006; Khalil, 2017). Alkaloids further include nicotine, tropane alkaloids, purine alkaloids, terpenoid indole alkaloids, and benzylisoquinoline alkaloids. Approximately 3000 compounds represent terpenoid indole alkaloids, including camptothecin and vinblastine (antineoplastic agents), strychnine (poison for rats), and quinine (antimalarial drug). Tryptophan is a primary amino acid involved in the biosynthesis of terpenoid indole alkaloids. A number of terpenoid indole alkaloids protect plants from biotic stresses (Facchini, 2001). Terpenoid indole alkaloids are also medicinally important alkaloids (Akhgari et al., 2017) as vinblastine and ajmaline have been used to cure cancer and cardiovascular diseases, respectively. Some plant species such as *Rauwolfia serpentina* and *Catharanthus roseus* are rich in terpenoid indole alkaloids of medicinal values. Tyrosine amino acid undergoes condensation to form benzylisoquinoline alkaloids. A number of compounds of pharmacological importance are included in this category such as antimicrobial agents (berberine and sanguinarine), muscle relaxants (papaverine), cough suppressants (codeine), and analgesics (morphine). These alkaloids are commonly found in plant families, namely, Papaveraceae, Ranunculaceae, Berberidaceae, and Menispermaceae. Scopolamine and hyoscyamine are tropane alkaloids obtained from *Datura*, *Atropa*, and *Hyosyamus* plant species. Convolvulaceae and Solanaceae plant families are a rich source of calystegines (widespread tropane alkaloids) that act as glucosidase inhibitor. Another amino acid ornithine is involved in the biosynthesis of nicotine, which causes addiction through is influence on the receptor for nicotinic acetylcholine. Purine nucleotides are the precursors for the biosynthesis of purine alkaloids (Ziegler and Facchini, 2008). Plants also contain different nitrogenous protecting compounds other than alkaloids such as cyanogenic glycosides and glucosinolates (Patra et al., 2013). Hydrogen cyanide is a poisonous gas (Ning et al., 2013) released by cyanogenic glycosides. Production of hydrogen cyanide occurs in plants that can produce cyanogenic glycosides and enzymes required for the hydrolysis of sugars. Hydrogen cyanide is a lethal gas that inhibits the activity of key metalloproteins such as cytochrome oxidase. The tubers of *Manihot esculenta* may cause limb paralysis because they contain significant amounts of cyanogenic glycosides.

Plant Metabolites and Regulation Under Environmental Stress. http://dx.doi.org/10.1016/B978-0-12-812689-9.00008-X

Glucosinolates also known as mustard oil glycosides are an important class of glycosides of plant origin and produce volatile substances upon hydrolysis. Particular taste and smell in vegetables, including radish, broccoli, and cabbage, is due to the presence of glucosinolates (Taiz and Zeiger, 2006).

2.1 Phenolics

Phenolics constitute an essential class of aromatic plant SMs (Malcovska et al., 2014) with the phenyl ring having one or more acidic hydroxyl groups attached (Achakzai et al., 2009; Ali et al., 2013). Approximately 10,000 compounds, both soluble in water and organic solvents, make phenolics a class of heterogeneous molecules (Taiz and Zeiger, 2006). The biosynthesis of phenolic compounds involves shikimic and malonic acid pathways (Verpoorte and Alfermann, 2013; Świeca et al., 2014). Biosynthesis of phenolics through malonic acid pathway is widespread in microorganisms (Cheynier et al., 2013) such as fungi and bacteria, and this pathway is not common in plants. In contrast, the biosynthesis of phenolics from shikimic acid pathway is more common in plants, fungi, and bacteria and is not present in animals (Ayoub et al., 2016; Puthur, 2016; Butkhup et al., 2016). Shikimic acid pathway uses precursors derived from pentose phosphate pathway (Bartwal et al., 2013) and glycolysis to produce diverse aromatic amino acids with phenylalanine as a frequent intermediate (Taiz and Zeiger, 2006). A wide range of compounds constitute phenolics, which includes arylpyrones and styrylpyrones, stibenes, tannins, coumarins, flavonoids, lignins, and lignans (Fang et al., 2011). The cinnamic acid, *p*-coumaric acid and their derivatives constitute simple phenolic compounds known as phenylpropanoids (Marienhagen and Bott, 2013). In contrast, a complex class of phenolics is formed by conspicuously branched polymers' phenylpropanoid groups. Lignin is an example of complex phenolic compound that is extensively found in plants after cellulose (Taiz and Zeiger, 2006). Phenolics are known to perform different important functions in plants (Chaparro et al., 2013) such as growth, reproduction, and tolerance of plants against various abiotic and biotic stresses (Achakzai et al., 2009; Giménez et al., 2014). Plants producing phenolics with allelopathic activity may suppress the growth of surrounding plants (Taiz and Zeiger, 2006). Phenolic compounds have redox properties (Vallverdú-Queralt et al., 2014) that enable them to act as antioxidants and thereby detoxify singlet oxygen (Huang and Bie, 2010). Phenolics are also reported to significantly reduce the threat of cancer in humans (Verma and Shukla, 2015). Phenolics also play other essential functions in plants such as nutrient uptake, enzyme activity, photosynthesis, and protein synthesis (Sharma et al., 2012; Fayez and Bazaid, 2014). Phenols also act as indicators of stress since their production is maximal in plants exposed to various stresses and lethal chemicals (Achakzai et al., 2009; Verma and Shukla, 2015). Effects of heavy metals and temperature on the production of SMs are summarized in Table 8.1. Plants also produce phenolics as antigrazing factors because phenolics impart a significant aroma and taste that prohibits insects and animals from feeding on them (Dey and Harborne, 1997). Isoflavones, flavonols, flavones, and anthocyanins are the main phenolic compounds that mediate plant growth and defense response against insects and microbes (Taiz and Zeiger, 2006). Petal colors are the result of certain phenolic compounds such as anthocyanins (flavonoids), which attract pollinators (Khatiwora et al., 2010).

2.2 Terpenes

The biosynthesis of terpenes in plants involves two distinct pathways occurring in plastids and cytoplasm. The methylerythritol 4-phosphate (MEP) and mevalonate (MVA) pathways mediate the biosynthesis of terpenes. The precursors are supplemented by the MEP pathway for the synthesis of carotenoids, isoprene, mono- and diterpenes, plant hormones [abscisic acid (ABA), gibberellins], phytol, the side chain of chlorophyll, tocopherols, phylloquinones, plastoquinones, etc. In contrast, the MVA pathway delivers isopentenyl diphosphate, which is involved in the biosynthesis of polyphenols, brassinosteroids, sterols, and sesquiterpenes. Furthermore, isopentenyl diphosphate is also consumed for prenylated proteins (Verma and Shukla, 2015). Isoprenoids or terpenoids constitute the major group of plant SMs and are insoluble in water. Acetyl coenzyme A also serves as the precursor for the biosynthesis of these compounds. Isopentenyl diphosphate forms as an intermediate compound in both the MEP and MVA pathways, leading to the formation of terpenes (Taiz and Zeiger, 2006; Thimmappa et al., 2014).

There exists different classes of terpenes based on the isoprene unit (five-carbon unit), including polyterpenoids ([C5]n), tetraterpenes (eight C5 units), triterpenes (six C5 units), diterpenes (four C5 units), sesquiterpenes (three C5 units), and monoterpenes (two C5 units) (Verma and Shukla, 2015). Sesquiterpenes (ABA), diterpenes (gibberellins), tetraterpenes (carotenoids), and triterpenes (sterole) constitute important terpenes that play a critical role in plant growth. Terpenes also mediate plant stress response because these are toxic to mammals and insects. Some plants possess saponins as terpene glycosides and essential oils as volatile terpenes. The biosynthesis of saponins involves triterpenes or steroids (lipids) and sugar (water-soluble elements). Saponins may be toxic (Tomar et al., 2014) because they damage cell membranes and interact with sterol uptake (Taiz and Zeiger, 2006).

TABLE 8.1 Changes in the Biosynthesis and Accumulation of SMs in Plants Under Temperature and Heavy Metal Stress

Plant Species	Stress Level	Effects on the Concentration of SMs	References
Dulce maravilla, Citrullus lanatus, Lycopersicum esculentum	15, 25, and 35°C temperature	High-temperature stress induced significant changes in plant biomass, activities of peroxidase, polyphenol oxidase and phenylalanine ammonia lyase, and total phenolics in *C. lanatus, D. maravilla,* and tomato plants. Temperature ranges include 15, 25, and 35°C. There existed a marked variation in the response of plant species to different temperature ranges, and this was more evident in watermelon and tomato. Onset of heat stress in tomato started at 35°C, whereas 15°C induced chilling stress in watermelon. Researchers observed decreased plant biomass, induced accumulation of phenolic contents, and enhanced activity of phenylalanine ammonia lyase. They further recorded decrease in the activities of polyphenol oxidase and peroxidase. It was suggested that thermal stress induced the production of phenolics and thereby induced thermal tolerance in plants	Rivero et al. (2001)
Vitis vinifera L.	10/7°C day/night	*V. vinifera* plants when subjected to constant or prolonged cold stress had significant changes in the accumulation of some SMs. It was proposed that tolerant varieties contained higher reducing power, antioxidant activity and phenolic contents. Cold stress resulted in a marked decrease in the concentration of phenolics and antioxidant activity in grapevine varieties. Ferulic acid, p-coumaric acid, caffeic acid, and caffeic acid derivatives were the major phenolic acids determined from leaf extracts of two grapevine varieties. Among various phenolic acids, caffeic acid was present in greater concentration. It was further recorded that plants facing prolonged cold stress had shown decrease in concentration of phenolic acids. It was found that SM concentration varied markedly in response to the intensity of the stress	Król et al. (2015)
Capsicum annuum L.	4°C for 3 days	Exposure of pepper seedlings to the cold stress resulted in a significant increase in the endogenous levels of phenolic compound, proline, and total soluble proteins, whereas chlorophyll contents decreased in plants in response to cold stress	Esra et al. (2010)
Sorghum bicolor L.	38/21 °C	Sorghum grains are a rich source of polyphenols. An experiment was conducted to appraise the concentration of polyphenols in different sorghum genotypes subjected to high-temperature stress. About 23 different phenolic compounds were detected with help of HPLC method. White sorghum contained simple phenolic profile because it contained fewer polyphenol contents. In contrast, brown sorghum possessed rich profile of phenolics in terms of ferulic and caffeic acid. However, luteolin and apigenin were not detected in this genotype. High-temperature stress increased the concentration of apigeninidin and luteolinidin in all sorghum genotypes. Furthermore, authors correlated the rich phenolic profile with greater temperature tolerance in sorghum	Wu et al. (2016)
Camellia sinensis L.	10, 20, 30 °C	Exposure of *C. sinensis* resulted in a marked increase in phenolic compounds	Upadhyaya (2012)
Matricaria chamomilla L.	60 μM and 120 μM of cadmium and copper	Influence of copper and cadmium stress on mineral nutrients, accumulation of phenolics, and phenolic metabolism was evaluated in *M. chamomilla* under 60 μM and 120 μM concentrations of both trace metals for 7 days. Plants facing copper stress had more prominent effects on ascorbate, polyphenol oxidase peroxidase, cinnamyl alcohol dehydrogenase, and shikimate dehydrogenase activities. Whereas, plants facing cadmium stress were rich in cinnamic acid derivatives	Kováčik et al. (2009)

Continued

TABLE 8.1 Changes in the Biosynthesis and Accumulation of SMs in Plants Under Temperature and Heavy Metal Stress—cont'd

Plant Species	Stress Level	Effects on the Concentration of SMs	References
Glycyrrhiza uralensis L.	0, 0.05, 0.1, 0.2, and 0.4 mmol/L of cadmium acetate	*G. uralensis* grown in varying concentrations of cadmium acetate had significant changes in growth and activities of polyphenol oxidase, phenylalanine ammonia lyase, catalase, peroxidase, and superoxide dismutase. Activities of different antioxidant enzymes exhibited dose-dependent increase in their activities as maximal activities were evident at higher dose of cadmium acetate. Elevated concentration of cadmium stress triggered increase in the activity of polyphenol oxidase, whereas the increase in phenylalanine ammonia lyase activity was concentration dependent. It was suggested that plants counteracted the inhibitory effects of cadmium stress through increased antioxidant and phenylalanine ammonia lyase activities	Zheng et al. (2010)
Vaccinium corymbosum L.	50 and 100 μM cadmium	An experiment was conducted to appraise the influence of cadmium on antioxidant potential and accumulation of phenolic compounds in *V. corymbosum* L. antioxidant activity was determined in the form of DPPH activity. Cadmium stress triggered marked changes in the production of phenolic compounds. It was recorded that blueberry plants were rich in chlorogenic acid, which had shown maximal accumulation at higher cadmium level. It was suggested that phenolic compound accumulation was positively correlated with the antioxidant activity of plants	Manquián-Cerda et al. (2016)

DPPH, 2-diphenyl-1-picrylhydrazyl; *HPLC*, high-performance liquid chromatography; *SM*, secondary metabolites.

2.3 Other Plant Secondary Metabolites

Plants are able to modulate their developmental programs through integration of external abiotic signals. Plants being sessile in nature cannot escape adverse environmental conditions and counteract the stress by provoking diverse biochemical, physiological, and molecular processes. These processes result in alterations in endogenous levels of SMs, ion homeostasis, gene expression, and regulation of protein activity. In biological systems, major control points are the regulation of genes at transcription levels, which is mediated by a number of regulators and transcription factors. Plant hormones are a group of small quantities of growth regulators, including salicylic acid (SA), ethylene, jasmonic acid (JA), ABA, gibberellic acid (GA), cytokinin, and auxins (Wani et al., 2016). Phytohormones are central for plastic growth and developmental response of plants. Phytohormones not only mediate developmental processes in plants but also significantly determine plant responses under environmental stresses (Wolters and Jurgens, 2009). A number of signaling pathways related to stress have been identified with the help of molecular genetics approach. Our current knowledge in plant hormone perception and signal transduction is based on the studies on defects in some metabolic events in *Arabidopsis*. Plant hormones such as SA, ethylene, JA, and ABA play substantial part in mediating plant responses to biotic and abiotic stresses. Besides the primary function of cytokinins, GAs, and auxins in plant developmental events, their involvement in regulating plant responses to stress has also been reported in the literature. There has been a rapid growth in the list of plant hormones, which has been extended to include strigolacones, nitric oxide, brassinosteroids, and polyamines (Gray, 2004). Plants treated with supplemented plant hormones exhibited strong changes in transcription levels (Chapman and Estelle, 2009). For instance, in a study on *Arabidopsis*, short exposure of plants to plant hormone induced significant alteration in gene expression, which included 10–300 genes that were overexpressed and approximately the same number of genes that were repressed (Paponov et al., 2008). Similarly, plants treated with plant hormones for a longer period may show wider changes in gene expression (Eyidogan et al., 2012). A variety of *cis*- or *trans*-acting factors are involved in phytohormone-induced regulation of abiotic stresses (Mahmoody and Noori, 2014).

Plant tissues treated with auxins show a number of alterations that entails cell differentiation, expansion, cell division, changes in transcriptional levels, and electrophysiology. Auxin induces the accumulation of a wide spectrum of transcripts, including LBD, small auxin-up RNAs, GH3, auxin/indole-3-acetic acid, early response genes, auxin response factors, and mediator of auxin response genes (Yang et al., 2004; Eyidogan et al., 2012). The function of such transcription factors is not fully known; however, it is speculated that these factors might also mediate plant defense responses to environmental stresses (Jain and Khurana, 2009). Examination of auxin-treated cells revealed that alteration in gene expression in part mediated the auxin response, which did not involve the biosynthesis of proteins. Within 5–60 min of auxin treatment, three major gene families related to early response genes (SAUR, GH3, and Aux/IAA) were expressed (Tromas and Perrot-Rechenmann, 2010). Plants are able to respond to environmental stresses and auxin signals through the active integration of these genes and thereby maintain plant growth and development (Eyidogan et al., 2012).

An enormous group of tetracyclic, diterpenoid plant hormones constitutes GAs that mediate plant growth and development. The literature indicates the active involvement of GAs in diverse plant developmental events such as pollen maturation, stem elongation, and seed germination (Olszewski et al., 2002). Environmental stresses cease plant growth and development. In this regard, the constituents of GA signaling pathways are important putative determinants of growth and defense response. Furthermore, cross talk of GA with signaling events of other plant hormones mediating plant defense response confers an important function to GAs in plants facing environmental constraints. Changes in the expression of components involved in GA signaling initiates its cross talk with the signaling pathways of other plant hormones (Achard et al., 2006; Eyidogan et al., 2012). Likewise, signaling pathways of cytokinins are similar to the two-component signaling transduction pathways of some fungi and bacteria. Researchers also counteract the inhibitory effects of environmental stresses through exogenous application of various plant hormones (Shakirova et al., 2012).

3. FUNCTIONS OF PLANT SECONDARY METABOLITES

Plant SMs are important compounds that add color, taste, and odor to plants and also mediate plant responses to adverse environmental conditions (Verma and Shukla, 2015). A number of factors cause significant perturbations in the production of SM in plants. The endogenous levels of different SMs vary among different plant species and also within the same plant species (Barton, 2007). A number of cellular and biochemical factors influence the storage and transportation of SM. Developmental factors influence the initiation and subsequent differentiation of particular cellular structures involved in the biosynthesis and storage of SM (Broun et al., 2006). Furthermore, the endogenous levels of SM are also influenced by a number of environmental stresses such as nutrient deficiencies, wounding, metal ions, ultraviolet (UV) radiation, light, circadian rhythm, seasonality, salinity, drought, and temperature (Gouvea et al., 2012; Verma and Shukla, 2015). Apart from this, the endogenous concentration of SM is associated with the metabolic pathway of the particular SM and growth conditions (Akula and Ravishankar, 2011). The concentration of SM is also affected by biotic factors (pathogen attack) and thereby mediates plant defense mechanism. For instance, there exists a significant variation in the levels of phenolics in plants in response to environmental stresses such as light intensity and nutrient availability (Verma and Shukla, 2015).

The endogenous levels and storage of SM is also influenced by genetic factors. A number of factors influence the expression of genes related to the biosynthesis of SMs. These factors play an essential role in mediating the biosynthesis, storage, and endogenous concentration of various SMs. In vitro tissue culture may also trigger an increase in the biosynthesis of different SMs via signaling molecules and plant growth regulators. This clearly indicated that the endogenous levels of SM can be changed. Overall, factors that trigger alterations in SMs fall into four categories, viz., environmental factors, morphogenetic factors, ontogenetic factors, and genetic factors. Among these factors, environmental factors are the key determinants for the fluctuations in plant SMs (Verma and Shukla, 2015).

4. ROLE OF SECONDARY METABOLITES UNDER ENVIRONMENTAL STRESSES

Plants show variation in the biosynthesis and accumulation of SMs in response to abiotic and biotic factors, which mediate the biosynthesis of SMs (Zhi-lin et al., 2007). The concentration of SMs may also vary in plants of the same species growing under contrasting environmental conditions (Radušienė et al., 2012). Plants growing under natural conditions are prone to the damages of biotic and abiotic stresses. These environmental stresses trigger the synthesis of certain SMs in plants, and thereby these SMs enable the plants to counteract the adverse effects of biotic and abiotic stresses. Therefore environmental stresses are the crucial determinants for the production of SMs in plants. Plants face abiotic stresses in terms of composition and type of soil, temperature stress, light intensity and drought, etc., which significantly influence plant growth and productivity. In contrast, reduction in plant growth and yield due to living organisms is recognized as a biotic factor (Radušienė et al., 2012). Abiotic stress is a broad-spectrum term that entails stress due to radiation, nutrient deficiency, pesticides, heavy metals, pollutants, toxic gases such as ozone, and salinity (Akula and Ravishankar, 2011).

4.1 Biotic Stresses

Biotic stresses in plants are caused by diverse living organisms that include nematodes, bacteria, virus, and fungi. Plants cannot change their position so as to avoid the stressful environment. However, plants exhibit a great degree of tolerance to pathogen attacks mainly through the production of SMs. Phytoalexins are such SMs that mediate plant defense response to pathogen attack due to their antimicrobial activities (Taiz and Zeiger, 2006). Plants under pathogen attack show enhanced biosynthesis of SMs. Plants undergoing fungal infections by strains of *Lupinus angustifolius* exhibit significant variation in the endogenous levels of phenolics (Verma and Shukla, 2015). An innate immune system is developed in plants due to pathogen attack. The innate immune system of plants is mediated by effector-triggered immunity and basal immunity. In basal immunity system, infected cells identify pathogens through microbe-associated molecular patterns that are perceived by pattern recognition receptors present on host cells. Besides, host cells also recognize pathogen attack via effector-triggered immunity in response to effectors or damages due to toxins of pathogens. Plants perceive signals from such effectors and stimulate a number of metabolic pathways to produce SMs. The concentration of SMs is significantly decreased during stress recovery (Wojakowska et al., 2013).

4.2 Abiotic Stresses

Plants are often exposed to a variety of abiotic stresses such as chemical fertilizers, soil type and composition, temperature stress, light intensity, availability of water, and salinity during different ontogenic phases. Plants need apposite quantities of abiotic components, and excess or deficiency of these components trigger variation in the biosynthesis of SMs, which determine plant growth and productivity (Verma and Shukla, 2015). The concentrations of phenylpropanoids (important SMs) is often increased in plants in response to environmental stresses such as temperature, nutrient deficiency, wounding, and UV radiation (Dixon and Paiva, 1995). Plants also show alterations in phenolic concentrations due to nutrient deficiency (Chalker-Scott and Fuchigami, 1989).

4.2.1 Chemical Stress

A variety of chemicals including plant SMs, growth regulators, elicitors, and nutrients are required in apposite amounts to maintain plant growth and development. However, the deficiency of any of these compounds results in chemical stress in plants. Chemical stress could be induced by various chemicals that entail elicitors, growth regulators, pesticides, ozone, pollutants, minerals, trace metals, etc. For example, supplementation of micronutrients in *Cassia angustifolia* increased the concentration of primary metabolites, which in turn resulted in greater production of SMs (Verma and Shukla, 2015). The endogenous levels of phenols, proteins, and chlorophyll also undergo variation in response to $CuSO_4$, $ZnSO_4$, and $FeSO_4$ (Shitole and Dhumal, 2012). Similarly, application of phosphorous and nitrogen fertilizers induced alterations in flavonoid concentration in St. John's wort. Phosphorous and nitrogen are important determinants of plant growth and development (Azizi and Dias, 2004). In addition, application of nitrogen as fertilizers may trigger the expression of certain genes in *Arabidopsis* and some other plant species (Vidal and Gutiérrez, 2008). Plants possess various mechanisms to respond to varying levels of nitrogen. A number of approaches including bioinformatics and genetics are being applied to unveil the metabolic pathways that mediate plant responses to different nitrogen levels in the environment. Plants undergo alterations in gene expression due to accessibility of nitrogen that results in developmental, physiological, and metabolic changes (Vidal and Gutiérrez, 2008). Phosphorous is also an important component that influences the biosynthesis of SMs and hence improves plant growth and development. Application of phosphorous did not influence the quantity and quality of essential oils in *Salvia officinalis* but caused a marked increase in the levels of rosmarinic acid and total phenolics and leaf biomass (Nell et al., 2009). Apart from the nutrient factors, plants also require certain gases such as oxygen and carbon dioxide to synthesize SMs and accomplish important biological mechanisms. For instance, variation in endogenous levels of SMs is associated with CO_2 in *Taxus baccata*, *Echinacea purpure*, and *Hypericum perforatum* (Save et al., 2007). This review of the literature clearly indicated that the chemicals needed by plants for their growth and development also determine the biosynthesis of SMs and thereby these chemical possess the potential to cause alterations in the concentrations of SMs (Verma and Shukla, 2015).

4.2.2 Radiation Stress

Plants require appropriate intensity of light for photosynthesis, which in turn influences the quality and accumulation of SMs. For instance, in *Mikania glomerata* the accumulation of coumarins is favored by sunlight. The endogenous level of coumarins is significantly decreased in different plant parts due to shorter light duration. In contrast, longer light period markedly improved the concentration of coumarins (de Castro et al., 2006). Therefore the literature indicates the conspicuous effect of photoperiod and light intensity on the biosynthesis and storage of SMs (Verma and Shukla, 2015).

Light is an essential abiotic component required by the plants for photosynthesis and growth, which influences the accumulation and quality of plant SMs. Sunlight favors the accumulation of coumarins in *M. glomerata*. A shorter light period decreases the level of coumarin in the leaves and stems of plants, whereas a longer light period significantly increases the level of coumarin. The coumarin content was significantly influenced by the photoperiod in leaves and stems (de Castro et al., 2006). So the light intensity and photoperiod have significant effects on the accumulation of plant SMs.

4.2.3 Temperature Stress

Plant growth and development is directly linked with apposite temperature ranges. Low and high temperature ranges may have negative impact on plant growth and productivity (Yadav, 2010). Plants growing under high temperatures have heat stress. Decrease in plant growth and productivity is associated with heat stress–induced decline in stomatal conductance and net CO_2 fixation. The biosynthesis of SMs is also correlated with heat stress in plants (Verma and Shukla, 2015). Plants growing under heat stress exhibit a decline in the photochemical efficiency of photosystem II. A literature survey indicated that high-temperature stress usually increased the production of SMs, whereas some reports are available in the literature that highlighted the decrease in SMs in plants under heat stress. For instance, *Panax quinquefolius* grown under high-temperature stress exhibited increase in ginsenosides (Jochum et al., 2007).

Plant growth and productivity is significantly hampered as a result of low-temperature stress. Cellular dehydration, water uptake, and metabolic reactions in plants are inhibited due to low-temperature stress (Chinnusamy et al., 2007). Exposure of *Capsicum annuum* plants to low temperature resulted in inhibited photosynthesis, which in turn suppressed plant growth and yield. Plants growing under low temperature exhibit significant alterations in various physiochemical and molecular processes that enable plants to survive low temperature stress, a phenomenon known as cold acclimation. Information is also available in the literature that indicated the decrease in chlorophyll and total soluble protein in plants under low-temperature stress (Esra et al., 2010). Low-temperature stress also caused a conspicuous decrease in the biosynthesis and storage of SMs (Verma and Shukla, 2015) (Table 8.1).

4.2.4 Salinity Stress

Greater levels of salts in the growth medium induce osmotic stress whereby plants face physiological drought, a condition in which plants are unable to take up water despite the availability of water. Soils with elevated levels of salts exhibit decline in photosynthesis, growth, and uptake of nutrients in plants (Ashraf et al., 2015). Plant SMs may undergo increase or decrease in their concentration in response to salinity-induced osmotic stress or specific ion toxicity (Akula and Ravishankar, 2011). For instance, *C. roseus* and *Rauvolfia tetraphylla* had shown substantial accumulation of vincristine alkaloids and reserpine when exposed to salinity stress in the growth medium (Said-Al Ahl and Omer, 2011). Salinity stress resulted in a significant rise in alkaloid concentration in *Achillea fragrantissima*, *C. roseus,* and *Solanum nigrum*. Exposure of *A. fragrantissima* to salinity stress increased the endogenous levels of phenolics (Verma and Shukla, 2015). When *Matricaria chamomilla* was subjected to salinity stress, there was a significant accumulation of different phenolics including caffeic, chlorogenic, and protocatechuic acids (Kováčik et al., 2009). In another study, salinity stress in *Mentha pulegium* and *Nigella sativa* resulted in a marked increase in phenols (Bourgou et al., 2009; Oueslati et al., 2010). There are contrasting reports of salinity stress on the production of essential oils. For instance, salinity stress caused decrease in essential oils in *Mentha suaveolens*, *M. pulegium*, *S. officinalis*, *M. chamomilla*, *Majorana hortensis*, *Origanum vulgare*, *Thymus maroccanus*, *Mentha piperita,* and *Trachyspermum ammi*. Contrarily, salinity stress triggered increase in the contents of essential oil in *S. officinalis*, *Satureja hortensis,* and *Matricaria recutita* (Said-Al Ahl and Omer, 2011). Salinity stress also cause increase in alkaloids. Likewise, salinity stress caused increase in an antihypertension alkaloid (ajmalicine) and antioxidant potential in roots of *C. roseus* (Jaleel, 2009). Similarly, salinity stress also increased the levels of saponins, flavonoids, and proline in *Plantago ovata* (Haghighi et al., 2012). The effects of salinity on plant SM production are summarized in Table 8.2. The levels of primary metabolites and SMs also vary in response to the availability of different nutrients (Verma and Shukla, 2015).

4.2.5 Drought Stress

Water serves as an essential factor for plants due to its direct involvement in the transportation of nutrients and metabolites in different plant parts. Unavailability of appropriate amounts of water or greater transpiration rates induces drought stress in plants and alters SM production (Table 8.3). Drought stress causes decline in turgor and water potentials in plants, which thereby negatively influence various physiological processes (Lisar et al., 2012). Drought-induced suppression in photosynthesis and growth induces significant alterations in important biochemical processes of plants (Aimar et al., 2011). Furthermore, decrease in photosynthesis due to drought stress is associated with disrupted activities of enzymes, loss of

TABLE 8.2 Influence of Salinity Stress on the Biosynthesis and Accumulation of Different SMs in Plants

Plant Species	Salinity Level	Effects on the Concentrations of SMs	References
Cotton	50, 100, 150, and 200 mM	There was a marked increase in the concentration of tannic acid (15.1%–24.3%), flavonoids (22.5%–37.6%), and gossypol (26.8%–51.4%) in cotton plants subjected to salinity stress. Researchers have further declared that salt-induced biosynthesis and accumulation of SMs significantly contributed to decrease in aphid population	Wang et al. (2015)
Safflower	0, 50, 100, and 150 mM	Safflower plants grown in hydroponics culture with varying levels of salinity (0, 50, 100, and 150 mM NaCl) exhibited a significant alteration in plant growth, nutrient accumulation, antioxidant enzyme activities, and flavonoid contents. Authors found that safflower tolerated <100 mM salinity because there was nonsignificant effect of salinity on plant biomass. However, the levels of Na^+ increased due to salinity along with concomitant higher levels of Ca^{2+} and K^+ in different plant parts. Interestingly, plants growing under <100 mM of NaCl under shown a conspicuous rise in the endogenous levels of flavonoids. Salinity stress also triggered increase in the levels of proteins and soluble sugars. Authors concluded that safflower plants facing moderate salinity levels showed significant accumulation of SMs, which in turn caused alterations in important physiological processes in plants	Gengmao et al. (2015)
Sugarcane	12.5 and 6.8 dS/m	Two sugarcane clones, namely, CP-4333 (salinity tolerant) and HSF-240 (salinity sensitive), exhibited a significant alteration in the number of SMs, namely, flavones, anthocyanins, and soluble phenolics, under salinity stress. Exposure of two clones to salinity stress markedly decreased plant biomass, photosynthesis, chlorophyll content, and SMs. However, this decrease in SMs and photosynthesis was more in the sensitive clone HSF-240 than in CP-4333. Salinity tolerance in sugarcane was positively associated with these SMs, which appeared to prevent oxidative damage due to ionic toxicity. Carotenoid is an important component of light harvesting complex and thereby induces tolerance to chloroplast under salinity stress	Wahid and Ghazanfar (2006)
Buckwheat	0, 50, 100, and 200 mM	When buckwheat plants were subjected to different salinity levels, there was a marked effect on nutritional quality in terms of carotenoids, phenolic compounds, and antioxidant activity. Plants growing under different salinity doses exhibit significant increase in the concentration of phenolic compounds when compared with plants growing under nonsaline conditions. Furthermore, four major compounds, namely, vitexin, rutin, orientin, and isoorientin, were the major contributors to phenolic compounds under salinity stress. Similarly, salt-treated plants exhibited significantly higher carotenoid contents when compared with control plants. Authors concluded that salinity treatment improved the nutritional quality in buckwheat in terms of enhanced carotenoids, antioxidant activity, and phenolic compounds	Lim et al. (2012)
Artichoke	1.5–20.6 dS/m in pots and 13–29 dS/m in field	Artichoke plants occupy important pharmaceutical values because they contain substantial amounts of phenolic compounds. Effect of salinity on the phenolic concentration was observed in artichoke in field and pot experiments. Moderate levels of salinity induced increase in the flavonoid and phenolic concentration in artichoke grown in pot or field experiments. Antioxidant activity of plants was also improved by higher salinity levels. There was also an increase in the contents of chlorogenic and caffeic acids in response to moderate salinity, whereas higher concentration of salinity limited the contents of caffeic and chlorogenic acids. Authors concluded that salinity stress markedly enhanced polyphenol contents in artichoke plants	Rezazadeh et al. (2012)
Lepidium sativum	30, 60, 90 and 120 mM	An experiment was conducted to evaluate the effect salinity stress on the accumulation of flavonols, flavonoids, and phenolic compounds in *L. sativum*. It was observed that plants grown under moderate salinity levels possessed greater concentrations of flavonols, flavonoids, and phenolic compounds	Ahmed et al. (2012)
Rosmarinus officinalis	0, 50, 100 and 150 mM	A significant increase in the phenolic concentration was recorded in *R. officinalis* in response to salinity stress. Increase in the phenolic contents was greater in plants exposed to 100 mM	Kiarostami et al. (2010)

TABLE 8.2 Influence of Salinity Stress on the Biosynthesis and Accumulation of Different SMs in Plants—cont'd

Plant Species	Salinity Level	Effects on the Concentrations of SMs	References
Salvia mirzayanii	4.5, 6.8 and 9.1 dS/m	Antioxidant activity, phenolic concentration, and volatile components in *S. mirzayanii* were examined under salinity stress. Antioxidant activity in terms of DPPH scavenging activity increased with increase in salinity stress, whereas the phenolic contents and volatile components increased in response to moderate salinity levels. Volatile compounds mainly included bicyclogermacrene, 1,8-cineole, and α-terpinyl acetate	Valifard et al. (2014)
Cherry tomato	0, 150, 300, and 450 mM	Exposure of tomato plants to varying levels of salinity stress resulted in differential effect on carotenoids, phenolics, and flavonoids. Elevated levels of salinity triggered increase in the endogenous levels of the aforementioned mentioned SMs	Al Hassan et al. (2015)
Maize	0 and 100 mM	Total phenolic and flavonoid contents increased in maize plants subjected to salinity stress	Salama et al. (2015)
Lentil	0 and 100 mM	Salinity stress in lentil plants resulted in marked alterations in total flavonoids, total phenolics, and antioxidant activity. It was observed that phenolics such as resveratrol, gallic acid, kaempferol, quercetin, *p*-coumaric acid, rutin, epicatechin, and catechin were the major contributor in improving antioxidant activity in lentils. Salinity stress decreased the endogenous levels of phenolics in lentil plants	Giannakoula et al. (2012)
Phaseolus vulgaris	0, 10, 30 and 50 mM	Salinity stress (moderate levels) in *P. vulgaris* decreased phenols, whereas increase in flavonoid contents was erratic	Telesiñski et al. (2008)
Catharanthus roseus	0 and 100 mM	Salinity stress caused increase in vincristine in *C. roseus* when subjected to salinity stress	Misra and Gupta (2006)
Rauvolfia tetraphylla	0, 25, 50, 75 and 100 mM	Salinity stress increased the endogenous levels of reserpine in *R. tetraphylla*	Anitha and Kumari (2006)
Rice	0 and 25 mM	Salinity stress induced a significant increase the nutritional quality in grains in terms of antioxidant activities, anthocyanins, proanthocyanins, and total phenolics	Chunthaburee et al. (2015)
Calendula	3.5–7 dS/m	Plants growing in saline alkaline and saline soils , calendula, undergo changes in riboflavin, phenolics, catalase activity, and total proteins. Salinity stress in calendula did not significantly influence phenolics and total proteins. All parts of plants contained substantial amount of riboflavin contents. However, plants grown under saline–alkaline soils had greater riboflavin concentration in flowers	Iqbal et al. (2015)
Solanum nigrum	0, 50, 100 and 150 mM	An experiment was conducted to appraise the effect of salinity on flavonoid and carotenoid content in *S. nigrum*. Salinity stress enhanced the expression of flavonoid genes encoding flavonol synthase, chalcone synthase, and phenylalanine ammonia lyase, which in turn increased the synthesis of quercetin 3-β-d-glucoside, lutein, and lutein. Likewise, some genes related to carotenoids entailing β-lycopene cyclase and phytoene synthase 2 were also overexpressed in response to salinity stress. Authors suggested that antioxidant potential of flavonoids and carotenoids enabled the plants to overcome salt-induced oxidative damage	Ben Abdallah et al. (2016)
Wheat	0 and 150 mM	Hydroponics experiment was conducted to appraise growth stage–based variation in phenolics in wheat under varying salinity levels. It was observed that salinity stress induced increase in phenolics, which were significantly higher at the boot stage than in the vegetative or reproductive stage. Authors suggested that phenolics contributed to salinity tolerance in wheat in terms of improved plant biomass	Ashraf et al. (2010)
Pea	75, 120, 150, and 200 mM	There was significant increase in total phenolics in three different pea species (*Pisum fulvum, Pisum sativum,* and *Pisum arvense*) plants under moderate salinity levels, whereas decrease in total phenolics was evident due to elevated levels of salinity	Miljus-Djukic et al. (2013)

DPPH, 2-diphenyl-1-picrylhydrazyl; *SM*, secondary metabolites.

TABLE 8.3 Drought-Induced Alterations in the Biosynthesis and Storage of Plant SMs

Plant Species	Drought Level	Effects of the Concentrations of SMs	References
Grapevine	70% and 30% soil moisture	Grapevine plants had significant alterations in the biosynthesis and storage of phenolics that mainly include ferulic acid, *p*-coumaric acid, and caffeic acid. Amount of phenolics was estimated from different plant parts (leaves and roots). It was observed that all acids existed in ester-bound form in both leaves and roots except for leaf caffeic acid, which was present either as ester bound or free form. Among different phenolic compounds, the concentration of the caffeic acid was maximal, whereas ferulic acid was detected in lesser quantity in both plant parts. The concentrations of mentioned phenolic compounds decreased markedly in plants under drought stress. Antioxidant property was exhibited by both plant parts, but root extracts were inferior with respect to antioxidant activity when compared with control plants. Authors concluded that plants facing drought for longer period undergo significant decline in certain compounds of secondary metabolism in contrasting plant parts	Król et al. (2014)
Achillea species	100, 75, 50% and 25% FC (field capacity)	An experiment was undertaken to appraise the influence of drought on oxidant production, antioxidant activity, proline, and SM accumulation (total flavonoids and phenolic contents) in different *Achillea* species. Total phenolic and flavonoid contents were greater in *Achillea filipendulina* under both moderate and severe drought stress. Antioxidant activity determined in terms of DPPH was associated with drought-induced changes in SMs	Gharibi et al. (2016)
Triticale	70% and 25% FC	Drought-induced changes in ferulic acid and phenolic compounds were carried out in spring triticale during flowering stage in drought-sensitive (CHD 247, CHD 220) and drought-resistant (CHD 147, CHD 12) genotypes. Interestingly, drought conditions caused sharp increase in the levels of ferulic acid and phenolic compounds in drought-resistant genotypes. Authors declared that accumulation of phenolics and ferulic acid could be used as a selection criterion to screen plants for drought resistance. In contrast, in drought-sensitive genotypes, decrease in ferulic acid was linked with a concomitant increase in phenolic compounds	Hura et al. (2009)
Rice	85, 65, 45%, and 25% soil moisture	Influence of drought stress on SM accumulation and antioxidant activity was determined in 20 rice genotypes. It was found that greater antioxidant activity in terms of DPPH and accumulation of SMs is correlated with drought tolerance in rice plants. Authors declared Q2 as the sensitive and Q8 as the tolerant rice genotype. Drought-induced increase in antioxidant activity and SMs (flavonoids and phenolics) was greater in tolerant Q8genotype, whereas rice genotype Q2 was inferior in this context. Presence of *p*-hydroxybenzoic acid was only detected in Q8 genotype, whereas vanillic acid accumulation was evident in both genotypes under drought stress. It was suggested that vanillic acid and *p*-hydroxybenzoic acid (phenolic compounds) were the major determinants of drought tolerance or sensitivity in rice genotypes	Quan et al. (2016)
Maize	PEG-induced stress (−0.6 MPa)	PEG-induced drought conditions caused decrease in phenolic compounds that was accompanied by decrease in plant biomass in maize plants	Asgari and Shiri (2015)
Safflower	100, 50%, and 25% water deficit	Drought stress in *Carthamus tinctorius* reduced carotenoids, phenolics, and hence growth. It was observed that moderate drought stress increased phenolic compound, whereas a significant decrease in phenolics was evident in plants facing severe drought conditions. Likewise, significantly higher carotenoids were recorded in plants under moderate drought conditions	Salem et al. (2014)
Maize	30%–35% FC	Imposition of drought stress in maize ceased plant growth, reduced relative water contents, and increased phenolics and soluble proteins.	Latif et al. (2016)
Maize	70 and 30%–35% FC.	An experiment was conducted to appraise whether or not drought tolerance in maize is determined by the accumulation of phenolics. There existed a significant alteration in phenylalanine ammonia lyase, ferulic acid, and total phenolics. In drought-tolerant genotypes, low water potential was accompanied by the accumulation of ferulic acid and total phenolics. Total phenolics absorb light and transform into blue fluorescence and thereby protect mesophyll tissues. It was suggested that phenolics acted as photoprotectors because these compounds limit chlorophyll excitation under drought conditions	Hura et al. (2008)
Soybean	−15 to −20 (control) and drought (−90 and −100 kPa)	Onset of drought conditions in soybean plants resulted in a significant increase in total phenolics and lignin	Nacer (2012)

DPPH, 2-diphenyl-1-picrylhydrazyl; *FC*, field capacity; *PEG*, polyethylene glycol; *SM*, secondary metabolite.

membrane integrity, and stomatal closure. Productivity of important cereals is also negatively influenced by drought-induced osmotic stress (Valentovic et al., 2006). A number of SMs produced in plants are also helpful in the induction of drought tolerance (Verma and Shukla, 2015). The rise in endogenous levels of plant SMs in response to drought stress was recorded in various medicinal plants such as *C. roseus*, *H. perforatum*, and *Artemisia annua*. For instance, drought stress caused increase in phenolics and photosynthetic pigments and decrease in plant fresh and dry biomass in *T. ammi* (Azhar et al., 2011). In another study, imposition of drought stress resulted in improved quality of key SMs such as rutin, quercetin, and betulinic acid in *Hypericum brasiliense*, and that of artemisinin in *Artemisia* (Verma and Shukla, 2015). Likewise, exposure of St. John's wort plants to water-limited conditions resulted in a significant decrease in photosynthesis along with concomitant increase in the concentrations of SMs including pseudohypericin, hypericin, and hyperforin (Zobayed et al., 2007). Likewise, increase in total flavonoids was observed in *Glechoma longituba* grown under drought conditions (Zhang et al., 2012). Water-limited conditions in *Ocimum americanum* and *Ocimum basilicum* significantly affected concentrations of different macronutrients, proline, carbohydrates, and essential oils (Khalid, 2006).

5. CONCLUSION AND FUTURE PROSPECTS

It is clear from the literature cited in this chapter that plant SMs play a significant role in plant adaptation to various environmental constraints that not only influence plant growth but also alter the biosynthesis of SMs. Plant growth and development is precisely regulated by secondary metabolism that acts as a reservoir of key phytochemicals protecting plants against multiple environmental constraints. It is also documented in the literature that these plant SMs are important candidates for human nutrition. A survey of the literature cited earlier in the chapter also indicates that biosynthesis of plant SMs is not fully known. More research is needed to completely understand the regulatory proteins and genes involved in the biosynthesis of plant SMs so that these may be manipulated for improving plant tolerance to environmental stresses. A number of plant SMs possess antioxidant properties that act as first line of defense against oxidative damage induced by different environmental constraints (salinity, drought, temperature, heavy metal, and hypoxia stress). Oxidative damage is usually the result of excessive generation of reactive oxygen species (ROS) due to environmental stress. Membrane integrity is hampered by elevated cellular levels of ROS. Plant SMs have not been fully studied with respect to their relevant functions in plants under environmental stress. How plant SMs enable plants to avoid oxidative damage is an area of research that must be studied in detail. There is need to explore the functions of various plant SMs in plants grown under stressful conditions. Biotic and abiotic factors that influence production of plant SMs should be studied. Strategies must be developed to improve the potential of plants for SM production.

REFERENCES

Achakzai, A.K.K., Achakzai, P., Masood, A., Kayani, S.A., Tareen, R.B., 2009. Response of plant parts and age on the distribution of secondary metabolites on plants found in Quetta. Pak J. Bot. 41 (5), 2129–2135.

Achard, P., Cheng, H., De Grauwe, L., Decat, J., Schoutteten, H., Moritz, T., Van Der Straeten, D., Peng, J., Harberd, N.P., 2006. Integration of plant responses to environmentally activated phytohormonal signals. Science 311 (5757), 91–94.

Agostini-Costa, TdS., Vieira, R.F., Bizzo, H.R., Silveira, D., Gimenes, M.A., 2012. Secondary metabolites. In: Dhanarasu Dr., S. (Ed.), Chromatography and Its Applications Brazil, pp. 131–164.

Ahmed, A., Gabr, A., AL–Sayed, H., Smetanska, I., 2012. Effect of drought and salinity stress on total phenolic, flavonoids and flavonols contents and antioxidant activity in in vitro sprout cultures of garden cress (*Lepidium sativum*). J. Appl. Sci. Res. 8 (8), 3934–3942.

Aimar, D., Calafat, M., Andrade, A., Carassay, L., Abdala, G., Molas, M., 2011. Drought tolerance and stress hormones: from model organisms to forage crops. Plants Environ. 137–164.

Akhgari, A., Oksman-Caldentey, K.-M., Rischer, H., 2017. Biotechnology of the medicinal plant *Rhazya stricta*: a little investigated member of the Apocynaceae family. Biotechnol. Lett. 6 (39), 829–840.

Akula, R., Ravishankar, G.A., 2011. Influence of abiotic stress signals on secondary metabolites in plants. Plant Signal. Behav. 6 (11), 1720–1731.

Al Hassan, M., Fuertes, M.M., Sánchez, F.J.R., Vicente, O., Boscaiu, M., 2015. Effects of salt and water stress on plant growth and on accumulation of osmolytes and antioxidant compounds in cherry tomato. Not. Bot. Horti Agrobot. Cluj-Napoca 43 (1), 1.

Ali, M., Abbasi, B.H., Ihsan-ul-haq, 2013. Production of commercially important secondary metabolites and antioxidant activity in cell suspension cultures of *Artemisia absinthium* L. Ind. Crops Prod. 49, 400–406. http://dx.doi.org/10.1016/j.indcrop.2013.05.033.

Anitha, S., Kumari, B.R., 2006. Reserpine accumulation in NaCl treated calli of *Rauvolfia tetraphylla* L. Sci. Asia 32, 417–419.

Asgari, A., Shiri, M., 2015. Proline, glycine betaine, total phenolics and pigment contents in response to osmotic stress in maize seedlings.

Ashraf, M.A., Iqbal, M., Hussain, I., Rasheed, R., 2015. Physiological and Biochemical Approaches for Salinity Tolerance. Managing Salt Tolerance in Plants: Molecular and Genomic Perspectives, p. 79.

Ashraf, M.A., Ashraf, M., Ali, Q., 2010. Response of two genetically diverse wheat cultivars to salt stress at different growth stages: leaf lipid peroxidation and phenolic contents. Pak J. Bot. 42 (1), 559–565.

Ayoub, M., de Camargo, A.C., Shahidi, F., 2016. Antioxidants and bioactivities of free, esterified and insoluble-bound phenolics from berry seed meals. Food Chem. 197, 221–232. http://dx.doi.org/10.1016/j.foodchem.2015.10.107.

Azhar, N., Hussain, B., Ashraf, M.Y., Abbasi, K.Y., 2011. Water stress mediated changes in growth, physiology and secondary metabolites of desi ajwain (*Trachyspermum ammi* L.). Pak. J. Bot. 43 (1), 15–19.

Azizi, M., Dias, A., 2004. Nitrogen and phosphorus fertilizers affect flavonoids contents of St. John's wort (*Hypericum perforatum* L.). In: Proc. 4th Int. Iran & Russia Conf. Shahrkurd, Iran, pp. 8–10.

Barton, K.E., 2007. Early ontogenetic patterns in chemical defense in Plantago (Plantaginaceae): genetic variation and trade-offs. Am. J. Bot. 94 (1), 56–66.

Bartwal, A., Mall, R., Lohani, P., Guru, S.K., Arora, S., 2013. Role of secondary metabolites and brassinosteroids in plant defense against environmental stresses. J. Plant Growth Regul. 32 (1), 216–232. http://dx.doi.org/10.1007/s00344-012-9272-x.

Ben Abdallah, S., Aung, B., Amyot, L., Lalin, I., Lachâal, M., Karray-Bouraoui, N., Hannoufa, A., 2016. Salt stress (NaCl) affects plant growth and branch pathways of carotenoid and flavonoid biosyntheses in *Solanum nigrum*. Acta Physiol. Plant. 38 (3), 72. http://dx.doi.org/10.1007/s11738-016-2096-8.

Bourgou, S., Kchouk, M., Bellila, A., Marzouk, B., 2009. Effect of salinity on phenolic composition and biological activity of *Nigella sativa*. In: International Symposium on Medicinal and Aromatic Plants-SIPAM2009 853, pp. 57–60.

Broun, P., Liu, Y., Queen, E., Schwarz, Y., Abenes, M.L., Leibman, M., 2006. Importance of transcription factors in the regulation of plant secondary metabolism and their relevance to the control of terpenoid accumulation. Phytochem. Rev. 5 (1), 27–38. http://dx.doi.org/10.1007/s11101-006-9000-x.

Butkhup, L., Chowtivannakul, S., Gaensakoo, R., Prathepha, P., Samappito, S., 2016. Study of the phenolic composition of Shiraz red grape cultivar (*Vitis vinifera* L.) cultivated in north-eastern Thailand and its antioxidant and antimicrobial activity. South Afr. J. Enol. Vitic. 31 (2), 89–98.

Chalker-Scott, L., Fuchigami, L., 1989. The Role of Phenolic Compounds in Plant Stress Responses.

Chaparro, J.M., Badri, D.V., Bakker, M.G., Sugiyama, A., Manter, D.K., Vivanco, J.M., 2013. Root exudation of phytochemicals in Arabidopsis follows specific patterns that are developmentally programmed and correlate with soil microbial functions. PLoS One 8 (2), e55731.

Chapman, E.J., Estelle, M., 2009. Mechanism of auxin-regulated gene expression in plants. Annu. Rev. Genet. 43, 265–285.

Cheynier, V., Comte, G., Davies, K.M., Lattanzio, V., Martens, S., 2013. Plant phenolics: recent advances on their biosynthesis, genetics, and ecophysiology. Plant Physiol. Biochem. 72, 1–20. http://dx.doi.org/10.1016/j.plaphy.2013.05.009.

Chinnusamy, V., Zhu, J., Zhu, J.-K., 2007. Cold stress regulation of gene expression in plants. Trends Plant Sci. 12 (10), 444–451. http://dx.doi.org/10.1016/j.tplants.2007.07.002.

Chunthaburee, S., Sanitchon, J., Pattanagul, W., Theerakulpisut, P., 2015. Effects of salt stress after late booting stage on yield and antioxidant capacity in pigmented rice grains and alleviation of the salt-induced yield reduction by exogenous spermidine. Plant Prod. Sci. 18 (1), 32–42. http://dx.doi.org/10.1626/pps.18.32.

de Castro, E.M., Pinto, J.E.B.P., Bertolucci, S.K.V., Malta, M.R., Cardoso, M.G., Silva, F.A.M., 2006. Coumarin contents in young *Mikania glomerata* plants (guaco) under different radiation levels and photoperiod. Acta Farm. Bonaer. 25 (3), 387–392.

Dey, P.M., Harborne, J.B., 1997. Plant Biochemistry. Academic press.

Dixon, R.A., Paiva, N.L., 1995. Stress-induced phenylpropanoid metabolism. Plant Cell 7 (7), 1085.

Esra, K., Işlek, C., Üstün, A.S., 2010. Effect of cold on protein, proline, phenolic compounds and chlorophyll content of two pepper (*Capsicum annuum* L.) varieties. Gazi Univ. J. Sci. 23 (1), 1–6.

Eyidogan, F., Oz, M.T., Yucel, M., Oktem, H.A., 2012. Signal transduction of phytohormones under abiotic stresses. In: Khan, N.A., Nazar, R., Iqbal, N., Anjum, N.A. (Eds.), Phytohormones and Abiotic Stress Tolerance in Plants. Springer Berlin Heidelberg, Berlin, Heidelberg, pp. 1–48. http://dx.doi.org/10.1007/978-3-642-25829-9_1.

Facchini, P.J., 2001. Alkaloid biosynthesis in plants: biochemistry, cell biology, molecular regulation, and metabolic engineering applications. Annu. Rev. Plant Biol. 52 (1), 29–66.

Fang, X., Yang, C.-Q., Wei, Y.-K., Ma, Q.-X., Yang, L., Chen, X.-Y., 2011. Genomics grand for diversified plant secondary metabolites. Plant Div. Res. 33 (1), 53–64.

Fayez, K.A., Bazaid, S.A., 2014. Improving drought and salinity tolerance in barley by application of salicylic acid and potassium nitrate. J. Saudi Soc. Agric. Sci. 13 (1), 45–55. http://dx.doi.org/10.1016/j.jssas.2013.01.001.

Gengmao, Z., Yu, H., Xing, S., Shihui, L., Quanmei, S., Changhai, W., 2015. Salinity stress increases secondary metabolites and enzyme activity in safflower. Ind. Crops Prod. 64, 175–181. http://dx.doi.org/10.1016/j.indcrop.2014.10.058.

Gharibi, S., Tabatabaei, B.E.S., Saeidi, G., Goli, S.A.H., 2016. Effect of drought stress on total phenolic, lipid peroxidation, and antioxidant activity of *Achillea* species. Appl. Biochem. Biotechnol. 178 (4), 796–809. http://dx.doi.org/10.1007/s12010-015-1909-3.

Giannakoula, A., Ilias, I., Dragišić-Maksimović, J.J., Maksimović, V., Živanović, B.D., 2012. Does overhead irrigation with salt affect growth, yield, and phenolic content of lentil plants? Arch. Biol. Sci. 64 (2), 539–547.

Giménez, M.J., Valverde, J.M., Valero, D., Guillén, F., Martínez-Romero, D., Serrano, M., Castillo, S., 2014. Quality and antioxidant properties on sweet cherries as affected by preharvest salicylic and acetylsalicylic acids treatments. Food Chem. 160, 226–232. http://dx.doi.org/10.1016/j.foodchem.2014.03.107.

Gouvea, D.R., Gobbo-Neto, L., Sakamoto, H.T., Lopes, N.P., Lopes, J.L.C., Meloni, F., Amaral, J.G., 2012. Seasonal variation of the major secondary metabolites present in the extract of *Eremanthus mattogrossensis* Less (Asteraceae: Vernonieae) leaves. Quím. Nova 35 (11), 2139–2145.

Gray, W.M., 2004. Hormonal regulation of plant growth and development. PLoS Biol. 2 (9), e311.

Haghighi, Z., Modarresi, M., Mollayi, S., 2012. Enhancement of compatible solute and secondary metabolites production in *Plantago ovata* Forsk. by salinity stress. J. Med. Plants Res. 6 (18), 3495–3500.

Huang, X., Bie, Z., 2010. Cinnamic acid-inhibited ribulose-1,5-bisphosphate carboxylase activity is mediated through decreased spermine and changes in the ratio of polyamines in cowpea. J. Plant Physiol. 167 (1), 47–53. http://dx.doi.org/10.1016/j.jplph.2009.07.002.

Hura, T., Hura, K., Grzesiak, S., 2008. Contents of total phenolics and ferulic acid, and PAL activity during water potential changes in leaves of maize single-cross hybrids of different drought tolerance. J. Agron. Crop Sci. 194 (2), 104–112. http://dx.doi.org/10.1111/j.1439-037X.2008.00297.x.

Hura, T., Hura, K., Grzesiak, S., 2009. Leaf dehydration induces different content of phenolics and ferulic acid in drought-resistant and -sensitive genotypes of spring triticale. Z. für Naturforsch. C 64. http://dx.doi.org/10.1515/znc-2009-1-215.

Iqbal, M., Hussain, I., Habib, A., Ashraf, M.A., Rasheed, R., 2015. Effect of semiarid environment on some nutritional and antinutritional attributes of calendula (*calendula officinalis*). J. Chem. 2015.

Jain, M., Khurana, J.P., 2009. Transcript profiling reveals diverse roles of auxin-responsive genes during reproductive development and abiotic stress in rice. FEBS J. 276 (11), 3148–3162. http://dx.doi.org/10.1111/j.1742-4658.2009.07033.x.

Jaleel, C.A., 2009. Changes in non enzymatic antioxidants and ajmalicine production in *Catharanthus roseus* with different soil salinity regimes. Molecules 2, 2.

Jochum, G.M., Mudge, K.W., Thomas, R.B., 2007. Elevated temperatures increase leaf senescence and root secondary metabolite concentrations in the understory herb *Panax quinquefolius* (Araliaceae). Am. J. Bot. 94 (5), 819–826. http://dx.doi.org/10.3732/ajb.94.5.819.

Khalid, K.A., 2006. Influence of water stress on growth, essential oil, and chemical composition of herbs (*Ocimum* sp.). Int. Agrophys 20 (4), 289–296.

Khalil, A., 2017. Role of biotechnology in alkaloids production. In: Naeem, M., Aftab, T., Khan, M.M.A. (Eds.), *Catharanthus roseus*: Current Research and Future Prospects. Springer International Publishing, Cham, pp. 59–70. http://dx.doi.org/10.1007/978-3-319-51620-2_4.

Khan, S., ur Rahman, L., 2017. Pathway modulation of medicinal and aromatic plants through metabolic engineering using *Agrobacterium tumefaciens*. In: Jha, S. (Ed.), Transgenesis and Secondary Metabolism. Springer International Publishing, Cham, pp. 431–462. http://dx.doi.org/10.1007/978-3-319-28669-3_15.

Khatiwora, E., Adsul, V.B., Kulkarni, M.M., Deshpande, N., Kashalkar, R., 2010. Spectroscopic determination of total phenol and flavonoid contents of Ipomoea carnea. Int. J. Chem. Tech. Res. 2 (3), 1698–1701.

Kiarostami, K., Mohseni, R., Saboora, A., 2010. Biochemical changes of *Rosmarinus officinalis* under salt stress. J. Stress Physiol. Biochem. 6 (3).

Kliebenstein, D.J., 2013. Making new molecules — evolution of structures for novel metabolites in plants. Curr. Opin. Plant Biol. 16 (1), 112–117. http://dx.doi.org/10.1016/j.pbi.2012.12.004.

Kováčik, J., Klejdus, B., Hedbavny, J., Bačkor, M., 2009. Salicylic acid alleviates NaCl-induced changes in the metabolism of *Matricaria chamomilla* plants. Ecotoxicology 18 (5), 544–554. http://dx.doi.org/10.1007/s10646-009-0312-7.

Król, A., Amarowicz, R., Weidner, S., 2014. Changes in the composition of phenolic compounds and antioxidant properties of grapevine roots and leaves (*Vitis vinifera* L.) under continuous of long-term drought stress. Acta Physiol. Plant. 36 (6), 1491–1499. http://dx.doi.org/10.1007/s11738-014-1526-8.

Król, A., Amarowicz, R., Weidner, S., 2015. The effects of cold stress on the phenolic compounds and antioxidant capacity of grapevine (*Vitis vinifera* L.) leaves. J. Plant Physiol. 189, 97–104. http://dx.doi.org/10.1016/j.jplph.2015.10.002.

Kurepin, L.V., Ivanov, A.G., Zaman, M., Pharis, R.P., Hurry, V., Hüner, N.P.A., 2017. Interaction of Glycine betaine and plant hormones: protection of the photosynthetic apparatus during abiotic stress. In: Hou, H.J.M., Najafpour, M.M., Moore, G.F., Allakhverdiev, S.I. (Eds.), Photosynthesis: Structures, Mechanisms, and Applications. Springer International Publishing, Cham, pp. 185–202. http://dx.doi.org/10.1007/978-3-319-48873-8_9.

Latif, F., Ullah, F., Mehmood, S., Khattak, A., Khan, A.U., Khan, S., Husain, I., 2016. Effects of salicylic acid on growth and accumulation of phenolics in *Zea mays* L. under drought stress. Acta Agric. Scand. Sect. B — Soil & Plant Sci. 66 (4), 325–332. http://dx.doi.org/10.1080/09064710.2015.1117133.

Lim, J.-H., Park, K.-J., Kim, B.-K., Jeong, J.-W., Kim, H.-J., 2012. Effect of salinity stress on phenolic compounds and carotenoids in buckwheat (*Fagopyrum esculentum* M.) sprout. Food Chem. 135 (3), 1065–1070. http://dx.doi.org/10.1016/j.foodchem.2012.05.068.

Lisar, S.Y., Rahman, I.M., Hossain, M.M., Motafakkerazad, R., 2012. Water Stress in Plants: Causes, Effects and Responses. INTECH Open Access Publisher.

Mahmoody, M., Noori, M., 2014. Effect of gibberellic acid on growth and development plants and its relationship with abiotic stress. IJFAS 3, 717–721.

Malcovska, S.M., Ducaiova, Z., Maslaáková, I., Backor, M., 2014. Effect of silicon on growth, photosynthesis, oxidative status and phenolic compounds of maize (*Zea mays* L.) grown in cadmium excess. Water, Air Soil Pollut. 225 (8), 1.

Manquián-Cerda, K., Escudey, M., Zúñiga, G., Arancibia-Miranda, N., Molina, M., Cruces, E., 2016. Effect of cadmium on phenolic compounds, antioxidant enzyme activity and oxidative stress in blueberry (*Vaccinium corymbosum* L.) plantlets grown in vitro. Ecotoxicol. Environ. Saf. 133, 316–326. https://doi.org/10.1016/j.ecoenv.2016.07.029.

Marienhagen, J., Bott, M., 2013. Metabolic engineering of microorganisms for the synthesis of plant natural products. J. Biotechnol. 163 (2), 166–178. http://dx.doi.org/10.1016/j.jbiotec.2012.06.001.

Miljus-Djukic, J., Stanisavljevic, N., Radovic, S., Jovanovic, Z., Mikic, A., Maksimovic, V., 2013. Differential response of three contrasting pea (‘*Pisum arvense*, and *P. fulvum*) species to salt stress: assessment of variation in antioxidative defence and miRNA expression. Aust. J. Crop Sci. 7 (13), 2145.

Misra, N., Gupta, A.K., 2006. Effect of salinity and different nitrogen sources on the activity of antioxidant enzymes and indole alkaloid content in *Catharanthus roseus* seedlings. J. Plant Physiol. 163 (1), 11–18. https://doi.org/10.1016/j.jplph.2005.02.011.

Nacer, B., 2012. Soybean seed phenol, lignin, and isoflavones and sugars composition altered by foliar boron application in soybean under water stress. Food Nutr. Sci. 2012.

Nell, M., Vötsch, M., Vierheilig, H., Steinkellner, S., Zitterl-Eglseer, K., Franz, C., Novak, J., 2009. Effect of phosphorus uptake on growth and secondary metabolites of garden sage (*Salvia officinalis* L.). J. Sci. Food Agric. 89 (6), 1090–1096. http://dx.doi.org/10.1002/jsfa.3561.

Ning, P., Qiu, J., Wang, X., Liu, W., Chen, W., 2013. Metal loaded zeolite adsorbents for hydrogen cyanide removal. J. Environ. Sci. 25 (4), 808–814. https://doi.org/10.1016/S1001-0742(12)60138-7.

Olszewski, N., Sun, T.-P., Gubler, F., 2002. Gibberellin signaling biosynthesis, catabolism, and response pathways. Plant Cell 14 (Suppl. 1), S61–S80.

Oueslati, S., Karray-Bouraoui, N., Attia, H., Rabhi, M., Ksouri, R., Lachaal, M., 2010. Physiological and antioxidant responses of *Mentha pulegium* (Pennyroyal) to salt stress. Acta Physiol. Plant. 32 (2), 289–296. http://dx.doi.org/10.1007/s11738-009-0406-0.

Paponov, I.A., Paponov, M., Teale, W., Menges, M., Chakrabortee, S., Murray, J.A.H., Palme, K., 2008. Comprehensive transcriptome analysis of auxin responses in Arabidopsis. Mol. Plant. 1 (2), 321–337. https://doi.org/10.1093/mp/ssm021.

Patra, B., Schluttenhofer, C., Wu, Y., Pattanaik, S., Yuan, L., 2013. Transcriptional regulation of secondary metabolite biosynthesis in plants. Biochim. Biophys. Acta 1829 (11), 1236–1247. http://dx.doi.org/10.1016/j.bbagrm.2013.09.006.

Puthur, J.T., 2016. Antioxidants and cellular antioxidation mechanism in plants. South Indian J. Biol. Sci. 2 (1), 14–17.

Quan, N.T., Anh, L.H., Khang, D.T., Tuyen, P.T., Toan, N.P., Minh, T.N., Bach, D.T., Ha, P.T.T., Elzaawely, A.A., Khanh, T.D., 2016. Involvement of secondary metabolites in response to drought stress of rice (*Oryza sativa* L.). Agriculture 6 (2), 23.

Radušienė, J., Karpavičienė, B., Stanius, Ž., 2012. Effect of external and Internal factors on secondary metabolites accumulation in St. John's worth. Bot. Lith. 18. http://dx.doi.org/10.2478/v10279-012-0012-8.

Rezazadeh, A., Ghasemnezhad, A., Barani, M., Telmadarrehei, T., 2012. Effect of salinity on phenolic composition and antioxidant activity of artichoke (*Cynara scolymus* L.) leaves. Res. J. Med. Plant 6, 245–252.

Rivero, R.M., Ruiz, J.M., García, P.C., López-Lefebre, L.R., Sánchez, E., Romero, L., 2001. Resistance to cold and heat stress: accumulation of phenolic compounds in tomato and watermelon plants. Plant Sci. 160 (2), 315–321. http://dx.doi.org/10.1016/S0168-9452(00)00395-2.

Said-Al Ahl, H., Omer, E., 2011. Medicinal and aromatic plants production under salt stress. A review. Herba Polonica 57 (2), 72–87.

Salama, Z.A., Gaafar, A.A., El Fouly, M.M., 2015. Genotypic variations in phenolic, flavonoids and their antioxidant activities in maize plants treated with Zn (II) HEDTA grown in salinized media. Agric. Sci. 6 (3), 397.

Salem, N., Msaada, K., Dhifi, W., Sriti, J., Mejri, H., Limam, F., Marzouk, B., 2014. Effect of drought on safflower natural dyes and their biological activities. EXCLI J. 13, 1–18.

Save, R., de Herralde, F., Codina, C., Sanchez, X., Biel, C., 2007. Effects of atmospheric carbon dioxide fertilization on biomass and secondary metabolites of some plant species with pharmacological interest under greenhouse conditions. Afinidad 64 (528), 237–241.

Shakirova, F.M., Avalbaev, A.M., Bezrukova, M.V., Fatkhutdinova, R.A., Maslennikova, D.R., Yuldashev, R.A., Allagulova, C.R., Lastochkina, O.V., 2012. Hormonal intermediates in the protective action of exogenous phytohormones in wheat plants under salinity. In: Phytohormones and Abiotic Stress Tolerance in Plants. Springer, pp. 185–228.

Sharma, A., Gangwar, M., Chtaurvedi, A.P., Sinha, A.S.K., Tripathi, Y.B., 2012. Comparative analysis of phenolic and flavonoid content of *Jatropha curcas* L. Plant Arch. 12, 823–826.

Shitole, S., Dhumal, K., 2012. Influence of foliar application of micronutrients on photosynthetic pigments and organic constituents of medicinal plant *Cassia agustifolia* Vahl. Ann. Biol. Res. 3 (1), 520–526.

Świeca, M., Sęczyk, Ł., Gawlik-Dziki, U., 2014. Elicitation and precursor feeding as tools for the improvement of the phenolic content and antioxidant activity of lentil sprouts. Food Chem. 161, 288–295. http://dx.doi.org/10.1016/j.foodchem.2014.04.012.

Taiz, L., Zeiger, E., 2006. Secondary metabolites and plant defense. Plant Physiol. Fourth Ed. Sinauer Assoc. Inc. Capítulo 13, 125.

Telesiñski, A., Nowak, J., Smolik, B., Dubowska, A., Skrzypiec, N., 2008. Effect of soil salinity on activity of antioxidant enzymes and content of ascorbic acid and phenols in bean [*Phaseolus vulgaris* L.] plants. J. Elem. 13 (3).

Thimmappa, R., Geisler, K., Louveau, T., O'Maille, P., Osbourn, A., 2014. Triterpene biosynthesis in plants. Annu. Rev. Plant Biol. 65, 225–257.

Tomar, N.S., Ahanger, M.A., Agarwal, R.M., 2014. Jatropha curcas: an overview. In: Ahmad, P., Wani, M.R. (Eds.). Ahmad, P., Wani, M.R. (Eds.), Physiological Mechanisms and Adaptation Strategies in Plants under Changing Environment, vol. 2. Springer New York, New York, NY, pp. 361–383. http://dx.doi.org/10.1007/978-1-4614-8600-8_13.

Tromas, A., Perrot-Rechenmann, C., 2010. Recent progress in auxin biology. C R. Biol. 333 (4), 297–306. http://dx.doi.org/10.1016/j.crvi.2010.01.005.

Upadhyaya, H., 2012. Changes in antioxidative responses to low temperature in tea [*Camellia sinensis* (L) O. Kuntze] cultivars. Int. J. Mod. Bot. 2 (4), 83–87.

Valentovic, P., Luxova, M., Kolarovic, L., Gasparikova, O., 2006. Effect of osmotic stress on compatible solutes content, membrane stability and water relations in two maize cultivars. Plant Soil Environ. 52 (4), 184.

Valifard, M., Mohsenzadeh, S., Kholdebarin, B., Rowshan, V., 2014. Effects of salt stress on volatile compounds, total phenolic content and antioxidant activities of *Salvia mirzayanii*. S Afr. J. Bot. 93, 92–97. http://dx.doi.org/10.1016/j.sajb.2014.04.002.

Vallverdú-Queralt, A., Regueiro, J., Martínez-Huélamo, M., Rinaldi Alvarenga, J.F., Leal, L.N., Lamuela-Raventos, R.M., 2014. A comprehensive study on the phenolic profile of widely used culinary herbs and spices: rosemary, thyme, oregano, cinnamon, cumin and bay. Food Chem. 154, 299–307. http://dx.doi.org/10.1016/j.foodchem.2013.12.106.

Verma, N., Shukla, S., 2015. Impact of various factors responsible for fluctuation in plant secondary metabolites. J. Appl. Res. Med. Aromatic Plants 2 (4), 105–113. http://dx.doi.org/10.1016/j.jarmap.2015.09.002.

Verpoorte, R., Alfermann, A.W., 2013. Metabolic Engineering of Plant Secondary Metabolism. Springer Science & Business Media.

Vidal, E.A., Gutiérrez, R.A., 2008. A systems view of nitrogen nutrient and metabolite responses in Arabidopsis. Curr. Opin. Plant Biol. 11 (5), 521–529. http://dx.doi.org/10.1016/j.pbi.2008.07.003.

Wahid, A., Ghazanfar, A., 2006. Possible involvement of some secondary metabolites in salt tolerance of sugarcane. J. Plant Physiol. 163 (7), 723–730. http://dx.doi.org/10.1016/j.jplph.2005.07.007.

Wang, Q., Eneji, A.E., Kong, X., Wang, K., Dong, H., 2015. Salt stress effects on secondary metabolites of cotton in relation to gene expression responsible for aphid development. PLoS One 10 (6), e0129541.

Wani, S.H., Kumar, V., Shriram, V., Sah, S.K., 2016. Phytohormones and their metabolic engineering for abiotic stress tolerance in crop plants. Crop J. 4 (3), 162–176. http://dx.doi.org/10.1016/j.cj.2016.01.010.

Wojakowska, A., Muth, D., Narożna, D., Mądrzak, C., Stobiecki, M., Kachlicki, P., 2013. Changes of phenolic secondary metabolite profiles in the reaction of narrow leaf lupin (*Lupinus angustifolius*) plants to infections with *Colletotrichum lupini* fungus or treatment with its toxin. Metabolomics 9 (3), 575–589. http://dx.doi.org/10.1007/s11306-012-0475-8.

Wolters, H., Jurgens, G., 2009. Survival of the flexible: hormonal growth control and adaptation in plant development. Nat. Rev. Genet. 10 (5), 305–317. http://www.nature.com/nrg/journal/v10/n5/suppinfo/nrg2558_S1.html.

Wu, G., Johnson, S.K., Bornman, J.F., Bennett, S.J., Clarke, M.W., Singh, V., Fang, Z., 2016. Growth temperature and genotype both play important roles in sorghum grain phenolic composition. Sci. Rep. 6, 21835. http://dx.doi.org/10.1038/srep21835.

Yadav, S.K., 2010. Cold stress tolerance mechanisms in plants. A review. Agron. Sustainable Dev. 30 (3), 515–527.

Yang, G.-X., Jan, A., Shen, S.-H., Yazaki, J., Ishikawa, M., Shimatani, Z., Kishimoto, N., Kikuchi, S., Matsumoto, H., Komatsu, S., 2004. Microarray analysis of brassinosteroids- and gibberellin-regulated gene expression in rice seedlings. Mol. Genet. Genomics 271 (4), 468–478. http://dx.doi.org/10.1007/s00438-004-0998-4.

Zandalinas, S.I., Mittler, R., Balfagón, D., Arbona, V., Gómez-Cadenas, A., 2017. Plant adaptations to the combination of drought and high temperatures. Physiol. Plant. http://dx.doi.org/10.1111/ppl.12540.

Zhang, L., Wang, Q., Guo, Q., Chang, Q., Zhu, Z., Liu, L., Xu, H., 2012. Growth, physiological characteristics and total flavonoid content of *Glechoma longituba* in response to water stress. J. Med. Plants Res. 6 (6), 1015–1024.

Zheng, G., Lv, H., Gao, S., Wang, S., 2010. Effects of cadmium on growth and antioxidant responses in *Glycyrrhiza uralensis* seedlings. Plant Soil Environ. 56 (11), 508–515.

Zhi-lin, Y., Chuan-chao, D., Lian-qing, C., 2007. Regulation and accumulation of secondary metabolites in plant-fungus symbiotic system. Afr. J. Biotechnol. 6 (11).

Ziegler, J., Facchini, P.J., 2008. Alkaloid biosynthesis: metabolism and trafficking. Annu. Rev. Plant Biol. 59, 735–769.

Zobayed, S.M.A., Afreen, F., Kozai, T., 2007. Phytochemical and physiological changes in the leaves of St. John's wort plants under a water stress condition. Environ. Exp. Bot. 59 (2), 109–116. http://dx.doi.org/10.1016/j.envexpbot.2005.10.002.

Chapter 9

Engineering of Biomass Accumulation and Secondary Metabolite Production in Plant Cell and Tissue Cultures

Pinar Nartop
Namık Kemal University, Tekirdağ, Turkey

1. INTRODUCTION

Biotechnology is the use of living cells, tissues, organs, and their derivatives (enzymes, genes, etc.) to enhance the quality of their products. These biosystems and derivatives can be used in many areas such as pharmaceutical, food, agricultural, cosmetic, and textile industries. To obtain better quality from bioproducts, human, animal, and plant cells, tissues, and organs, and even bacteria and viruses and their derivatives are exposed to a wide range of effective biotechnological procedures. These procedures comprise multistep applications and the whole course is called "bioprocess."

Bioprocess needs highly educated and experienced researchers for keeping up with multistep applications and interpretation of the results because of the complexity of the study with living systems. Thus bioengineers, biomedical engineers, pharmacists, chemists, biologists, biochemists, and genetic, molecular, and agricultural engineers are contemporarily educated through multidisciplined and applied education programs in both undergraduate and graduate degrees.

Plant biotechnology is one of the most utilized branches of biotechnological bioprocesses. Plant organs, tissues, and cells are used as biofactories to obtain high-tech bioproducts in plant biotechnology. Moreover, plant cell organelles and enzymes are also used in bioprocess to harvest products with high added value. Bacteria, viruses, and human and animal cells, tissues, organs and their derivatives are also frequently used in these bioprocesses.

Plants along with their parts (leaves, stems, roots, fruits, etc.) and metabolites have been used since ancient times in many different areas for various purposes. Plant biotechnology focuses on obstacles and bottlenecks of plant-derived bioproducts and can help to find solutions for impossibilities faced during bioprocesses.

Secondary metabolite production via plant cell, tissue, and organ cultures is one of the most studied areas in plant biotechnology processes. Plant secondary metabolites are very valuable and unique organic compounds produced only by plant cells. Most of these metabolites have pharmacological effects. Thus many of them are being used for their anticarcinogenic, antioxidant, antibacterial, antiviral, antiseptic, and tonic bioactivities.

The aim of the plant biotechnology process is to enhance both the biomass of in vitro cultures and rates of production of secondary metabolite biomass to gain bioproducts at higher qualities. A number of implications are being applied to in vitro cultures of plants. In some cases, biomass and/or metabolite enhancing techniques cannot be the only solution to solve the problems and they should be combined. Therefore techniques, processes, and results must be known in detail. In this chapter, these implications and their results will be discussed.

2. CLASSIFICATION OF SECONDARY METABOLITES

Higher plants (vascular plants, tracheophytes) are unique sources of some biochemicals called plant secondary metabolites. Each of these organic compounds has very special characteristics owing to their chemical properties. Secondary metabolites have enormous chemical diversity (Table 9.1) and every plant family has its own type; however, relatively few building blocks are routinely employed in their biosynthesis.

Secondary metabolites are generally classified according to their biosynthetic pathways. There are four main classes of secondary metabolites (Dewick, 2002):

- *The acetate pathway*—fatty acids and polyketides (combinations of acetate unit's biosynthesis of saturated and unsaturated fatty acids together with prostaglandins, thromboxanes, and leukotrienes).

Plant Metabolites and Regulation Under Environmental Stress. http://dx.doi.org/10.1016/B978-0-12-812689-9.00009-1

TABLE 9.1 Numbers of Secondary Metabolites in Higher Plants

Components	Number of Chemical Structures	References
Terpenes	30,000	Degenhardt et al. (2009)
Flavonoids	6,000	Stobiecki and Kachlicki (2006)
Alkaloids	5,000	Pelletier (2001)
Glucosinolates	120	Fahey et al. (2001)

TABLE 9.2 Secondary Metabolite Classification in Plants

Phenylpropanoids	• Polyphenols
	  • Phenolic acids
	  • Flavonoids
	  • Stilbenes
	  • Lignans
	  • Phenolenones
	• Anthocyanins
	• Coumarins
	• Hydroxycinnamoylderivatives
	• Isoflavonoids
	• Proanthocyanidines
	• Tannins

Alkaloids	• Acridines • Betalains • Quinolizidines • Furonoquinones • Harringtonines • Isoquinolines indols • Purines • Pyridines • Tropane alkaloids
Terpenoids	• Carotenes • Monoterpenes • Sesquiterpenes • Diterpenes • Triterpenes
Quinones	• Anthraquinones • Benzoquinones • Naphthoquinones
Steroids	• Cardiac glycosides • Pregnenolone derivatives

- *The shikimate pathway*—aromatic amino acids and phenylpropanoids (formation of aromatic amino acids, benzoic acids, and cinnamic acids, lignans, lignin, phenylpropenes, and coumarins; combinations of the shikimate pathway and the acetate pathway are responsible for the biosynthesis of styrylpyrones, flavonoids and stilbenes, flavonolignans, and isoflavonoids; terpenoid quinones are formed by a combination of the shikimate pathway with the terpenoid pathway).
- *The mevalonate and deoxyxylulose phosphate pathways*—terpenoids and steroids (terpenoids, hemiterpenes, monoterpenes, iridoids, sesquiterpenes, diterpenes, sesterterpenes, triterpenes, tetraterpenes, higher terpenoids, steroids, cardioactive glycosides, phytosterols, corticosteroids, progestogens, estrogens, and androgens).
- *Alkaloid pathway* (indolizidine, pyridine, tetrahydroisoquinoline, isoquinoline, hyoscyamine, hyoscine, atropine, catecholamines, galanthamine, serotonin, ellipticine, camptothecin, Solanum alkaloids, caffeine, theobromine, and theophylline).

According to Rao and Ravishankar (2002) and Mulabagal and Tsay (2004), another classification (similar to Dewick's classification) is also given in Table 9.2.

3. STEPS OF PLANT SECONDARY METABOLITE PRODUCTION IN IN VITRO CONDITIONS

Plant cells are biofactories used in the production of plant secondary metabolites. High biomass accumulation is needed for obtaining successful results. To enhance biomass accumulation, media compositions and culture conditions should be optimized first (Fig. 9.1). Concentrations of macro- and microelements, growth regulators, and vitamins in the medium play a crucial role in cell division and culture growth. Moreover, temperature, light intensity, and pH of the medium are also very effective in both biomass accumulation and secondary metabolite production. Without optimization, biomass accumulation cannot be achieved and without biofactories, secondary metabolites cannot be produced in sufficient amounts. Thus, subsequent to optimization of media composition and culture conditions, high-yield cell lines or tissues in terms of biomass accumulation and secondary metabolite production should be selected.

There are some advantages of using plant cells as biofactories (Ishihara et al., 2003);

- They can be grown in laboratories under controlled conditions. They are homogenous and can independently be cultured from external conditions.
- Cultured tissues, cells, and plantlets can produce secondary metabolites at high amounts. Their growth cycle is approximately 1 2 weeks.
- Infinite volume of cultured biomass can be obtained in in vitro conditions.

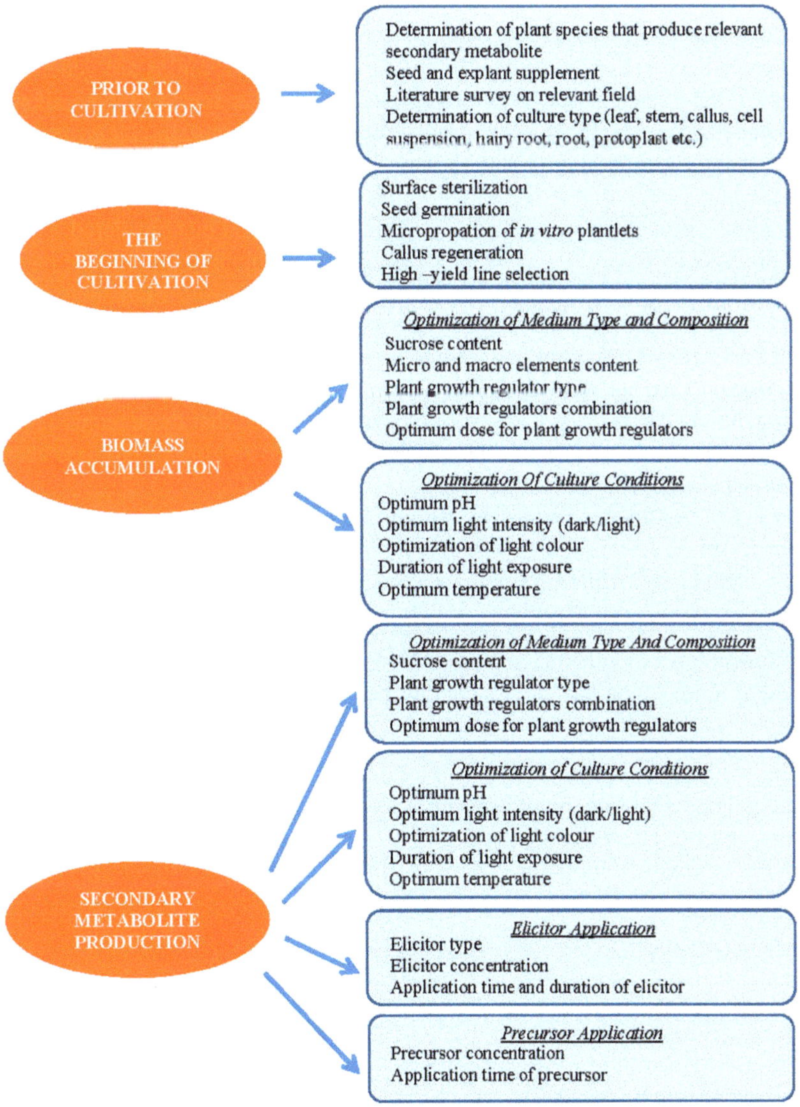

FIGURE 9.1 Steps of biomass accumulation and secondary metabolite production.

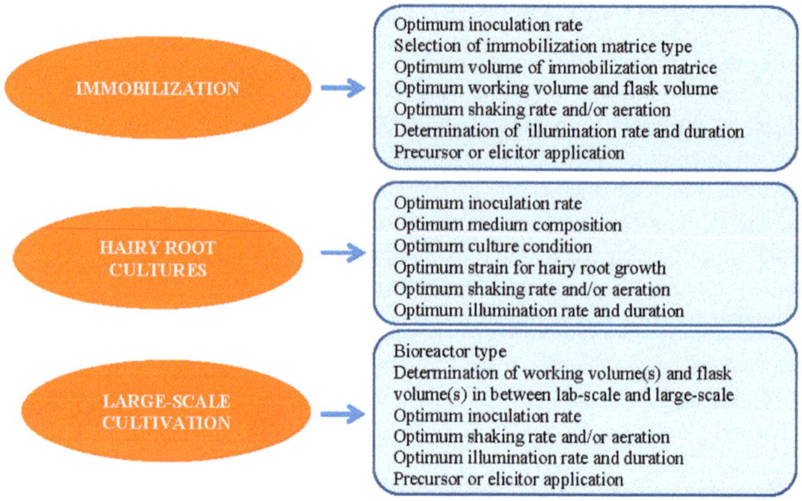

FIGURE 9.1 Cont'd.

Enzymes, intermediates, and end products of metabolic pathways are situated in membranous organelles (vacuoles, mitochondrion, etc.) in specialized cells. For example, flavonoids and anthocyanins are generally formed in the colorful parts (flowers and fruits) of the plants. Aerial parts of *Lupinus polyphyllus* formed lupine alkaloids in the chloroplasts. Moreover, secondary metabolites can be transported to the other parts of plants. In *Catharanthus roseus*, monoterpenoid indole alkaloids' enzymes were found in roots, young leaves, stems, and flower buds, whereas they were accumulated in idioblasts and laticifers (Seigler, 2002). Before starting bioproduction in in vitro conditions, the basics about identifications of biochemical and physiologic properties of the secondary metabolite should be considered.

Researchers should first specify the culture type to obtain high-yield bioproduction. The part of the studied plant that produces and accumulates the secondary metabolite should be cultivated. Tissues (leaf, stem, node, root, and hairy root), cells (callus and suspended cells), and the whole plant can be used in in vitro culture procedures.

Cells, tissues, and plantlets should be cultured in vitro to optimize medium composition and culture conditions after the determination of culture type. When sufficient biomass is obtained, secondary metabolite enhancing techniques, such as elicitation and precursor addition, can be implemented to biomass. Provided that these techniques are implemented before obtaining a high biomass accumulation, they may cause a decrease in the amount of biomass accumulation and results of the bioprocess will not be satisfactory.

Elicitors and precursors can be added to the growth medium to induce secondary metabolism. However, in some cases, adding an elicitor or a precursor directly to the growth medium may not affect secondary metabolite production positively. At this stage, a two-step method (two different medium compositions) can also be implemented to enhance secondary metabolite production:

1. Optimization of medium composition and culture conditions for biomass accumulation (biofactory building step),
2. Optimization of medium composition and culture conditions with elicitor or precursor treatment (high-yield production step).

As mentioned before, medium composition and culture conditions are the most effective factors in cell and tissue behaviors, whereas settling the second step might be as arduous as the first step. Therefore, another optimization for a new condition might be a solution for high productivity.

To summarize, secondary metabolite production steps should be:

- Determining the culture type
- Optimization of medium composition and culture conditions
- Biomass accumulation
- Implementation of secondary metabolite enhancing techniques

There is another way, which combines the third and fourth steps in cell suspension cultures. The immobilization technique causes both higher biomass accumulation and secondary metabolite production in suspended cells (Nartop et al., 2013).

4. OPTIMIZATION OF MEDIUM COMPOSITION AND CULTURE CONDITIONS

4.1 Medium Type

Medium composition is the most effective factor along with culture condition for biomass accumulation and secondary metabolite production. There are many medium compositions used for different purposes. Mostly used media are Murashige–Skoog (MS), Gamborg (B5), White (WH), Nitsch and Nitsch (NN), Linsmaier and Skoog (LS), Schenk and Hildebrandt (SH), McCown and Lloyd (woody plant medium, WPM). Their effects are distinctive because of their different major and minor element compositions (Table 9.3) (Sathyanarayana and Varghese, 2007).

TABLE 9.3 Composition of Some Plant Cell and Tissue Culture Media

Compounds	Murashige and Skoog (MS)	Gamborg (B5)	White (WH)	Chu (N6)
Macro Elements (mg/L)				
NH_4NO_3	1.650	–	–	
KNO_3	1.900	2.500	80	2.830
$CaCl_2 \cdot 2H_2O$	440	150	–	125.33
$MgSO_4 \cdot 7H_2O$	370	250	720	90.27
KH_2PO_4	170	–	–	400
$(NH_4)_2SO_4$	–	134	–	463
$NaH_2PO_4 \cdot H_2O$	–	150	16.5	–
Na_2SO_4	–	–	200	–
KCl	–	–	65	–
$Ca(NO_3)_2 \cdot 4H_2O$	–	–	300	–
Micro Elements (mg/L)				
KI	0.83	0.75	0.75	0.8
H_3BO_3	6.2	3	1.5	1.6
$MnSO_4 \cdot 4H_2O$	22.3	–	7	3.33
$MnSO_4 \cdot H_2O$	–	10	–	–
$ZnSO_4 \cdot 7H_2O$	8.6	2	3	1.5
$Na_2MoO_4 \cdot 2H_2O$	0.25	0.25	–	–
$CuSO_4 \cdot 5H_2O$	0.025	0.025	–	–
$CoCl_2 \cdot 6H_2O$	0.025	0.025	–	–
$Na_2 \cdot EDTA$	37.3	37.3	–	–
$FeSO_4 \cdot 7H_2O$	27.8	27.8	–	–
$Fe_2(SO_4)_3$	–	–	2.5	–
Fe_2Na EDTA	–	–	–	36.7
Organic Compounds (mg/L)				
Myoinositol	100	100	–	–
Nicotinic acid	0.5	1	0.5	0.5
Pyridoxin-HCl	0.5	1	0.1	0.5
Thiamine-HCl	0.1	10	0.1	1
Glisin	2	–	3	2
Sucrose	30.000	20.000	20.000	–

Continued

TABLE 9.3 Composition of Some Plant Cell and Tissue Culture Media—cont'd

Compounds	Nitsch and Nitsch (NN)	Linsmaier and Skoog (LS)	Schenk and Hilderbrandt (SH)	McCown and Lloyd (Woody Plant Medium - WPM)
Macro Elements (mg/L)				
NH_4NO_3	720	1.650	–	400
KNO_3	950	1.900	2.500	–
K_2SO_4	–	–	–	990
$CaCl_2{\cdot}2H_2O$	220	440	200	–
$CaCl_2$	–	–	–	72.5
$MgSO_4{\cdot}7H_2O$	185	370	400	368.5
KH_2PO_4	68	170	–	170
$NH_4H_2PO_4$	–	–	300	–
$Ca(NO_3)_2$	–	–	–	386.8
Micro Elements (mg/L)				
KI	–	0.83	1	–
H_3BO_3	10	6.2	5	6.2
$MnSO_4{\cdot}4H_2O$	25	–	–	–
$MnSO_4{\cdot}H_2O$	–	16.89	10	22.3
$ZnSO_4{\cdot}7H_2O$	10	10.58	1	8.6
$Na_2MoO_4{\cdot}2H_2O$	0.25	0.25	0.1	0.25
$CuSO_4{\cdot}5H_2O$	0.025	0.025	0.2	0.25
$CoCl_2{\cdot}6H_2O$	–	0.025	0.1	–
$Na_2{\cdot}EDTA$	37.3	–	20	37.3
$FeSO_4{\cdot}7H_2O$	27.8	27.85	15	27.8
$Fe_2(SO_4)_3$	–	–	–	–
Organic Compounds (mg/L)				
Myoinositol	100	–	1.000	100
L-Inositol	–	100	–	–
Nicotinic acid	5	–	5	0.5
Pyridoxine-HCl	0.5	–	0.5	0.5
Thiamin-HCl	0.5	0.4	5	1
Glycine	2	–	–	2
Folic acid	0.5	–	–	–
Biotin	0.05	–	–	–
Sucrose	20,000	–	30,000	20,000

Rosmarinus officinalis leaves and stems were cultured in both WPM and MS media with different naphthalene acetic acid (NAA) concentrations (1, 3, and 5 mg/L). In leaf cultures, the highest callus regeneration rate (8%) was obtained in WPM medium supplemented with 5 mg/L NAA, whereas MS medium supplemented with 1 mg/L NAA had the highest callus regeneration rate 65%) in stem cultures (Nartop, 2005). In *Lithospermum erythrorhizon* cell cultures, LS, B5, WH,

NN, and Blades media were used for the production of shikonin and its derivatives. The LS medium was represented as the optimum one for cell growth; however, shikonin and its derivatives cannot be produced. White medium did not trigger cell growth, but secondary metabolite production was found higher in this medium. In this case, the above-mentioned two-step process was used: (1) The cell growth was induced in LS medium; (2) shikonin and its derivatives production was triggered in White medium (Yamada and Fujita, 1983).

4.2 Sucrose Content

Owing to their heterotrophic features, most in vitro cultures consume sucrose as a carbon source. The sucrose content affects the culture's productivity. In *Colues blumei* cell cultures, 2.5% and 7.5% sucrose were consumed and rosmarinic acid was determined at 0.8 and 3.3 g/L, respectively. To optimize the concentration, sucrose was used at 4%–12% concentration in *C. roseus* cell cultures and the optimum ratio was found to be 8% for alkaloid accumulation. Benzophenanthridine alkaloids yield was enhanced 10 times and reached 150 mg/L in *Eschscholtzia californica* suspension cultures, whereas the sucrose concentration was 8%. Sucrose can also be used up as an osmotic agent to stress out the cells, tissues, and plantlets. Osmotic stress induced by sucrose in *Vitis vinifera* cell cultures demonstrated a regulatory effect on anthocyanin production. In *Aralia cordata* cell cultures, anthocyanin accumulation was enhanced with 5% sucrose, but decreased with 3% concentration in dark conditions (Sakamoto et al., 1993; Yağcı et al., 2008; Rao and Ravishankar, 2002). In *Astragalus trojanus* shoot cultures, when the sucrose concentration was doubled from 15 to 30 g/L in WPM medium, the rooting percentage increased fourfold, and the average root length increased up to 2.26 times, whereas the root regeneration ratio increased 9.3 times (Nartop et al., 2015a).

4.3 Nitrate and Phosphate Content

The nitrogen concentration affects protein and amino acid production rates in suspension cultures. MS, LS, and B5 media are rich in both nitrate and ammonium. However, the ammonium/nitrate ratio and total nitrogen are both important in secondary metabolite production. Low concentrations of NH_4^+ and NO_3^- enhanced shikonin and betacyanin production; higher concentrations induced berberine and ubiquinone productions. Low total nitrogen amount triggered productions of capsaicin in *Capsicum frutescens*, anthraquinones in *Morinda citrifolia*, and anthocyanins in *V. vinifera* (Rao and Ravishankar, 2002).

Phosphate concentration is another factor that influences secondary metabolite production. High amounts of phosphate generally affect cell growth and secondary metabolite production in positive and negative ways, respectively. Low phosphate concentrations enhanced ajmalicin and phenolics in *C. roseus*, cafeoil putrecins in *Nicotiana tabacum*, and harman alkaloids in *Perganum harmala*. However, high amounts triggered digitoxin in *Digitalis purpurea* and betacyanin in *Chenopodium rubrum* and *Phytolacca americana* cell cultures (Rao and Ravishankar, 2002; Yağcı et al., 2008).

4.4 Plant Growth Regulators

Both the type of plant growth regulators and their concentrations have a very complicated role in secondary metabolite accumulation. Not only auxin and cytokinin types and their concentrations but also auxin/cytokinin ratio causes remarkable alterations in cell growth and product accumulation. 2,4-Dichlorophenoxyacetic acid (2,4-D) generally inhibits secondary metabolite production. When 2,4-D was eliminated or replaced with NAA and indole acetic acid (IAA), anthocyanin production was enhanced in *Populus* cell cultures. Nevertheless, 2,4-D stimulated the biosynthesis of carotenoids in *Daucus carota* suspension cultures and anthocyanins in *Oxalis linearis* callus cultures. The effects of cytokinins depend on the metabolite and plant types. Kinetin triggered anthocyanin production in *Haplopappus gracilus* cell cultures; apart from this, it inhibits anthocyanins in *Populus* cell cultures. Gibberellic acid and abscisic acid mostly inhibited anthocyanin production in many cell cultures (Yağcı et al., 2008). Stem and leaf explants of *A. trojanus* were cultivated for callus regeneration in both MS and WPM media supplemented with different concentrations of kinetin, NAA, 2,4-D, thidiazurone, and IAA. The highest astragaloside IV accumulations were detected in MS medium supplemented with 0.1 mg/L 2,4-D at 35.03 μg/mg DW in stem-derived callus and 14.48 μg/mg DW in MS medium supplemented with 1 mg/L 2,4-D in leaf-derived callus (Nartop et al., 2015b).

The concentration of plant growth regulators and balance between them affect secondary metabolism as well. Implementations that cause structural differentiation such as shoot and root regeneration from callus tissue may change the biochemical profile of cultures (Yeoman and Yeoman, 1996).

5. ELICITATION AND PRECURSOR FEEDING

5.1 Elicitation

Secondary metabolites are mostly defensive products that are generally induced after an attack to the plant. In 1960, Condon and Kuc first isolated a secondary metabolite, 6-methoxymellein (antifungal isocoumarin), which accumulates in carrot roots and causes pathogen resistance to microbial infections. By the accumulation of 6-methoxymellein, the growth of fungi, yeasts, and bacteria was inhibited; on the other hand, this metabolite caused an important defense for carrot plant. In 1970s, the first biotic elicitors were described (Baenas et al., 2014); after that, Keen described a biotic elicitor, hydroxyphaseollin, in soybean (Keen, 1975).

Elicitors consisting of microbial molecules that are derived from bacterial cell walls are introduced as signaling compounds of insect oral secretions and molecules of the damaged plant cell walls. They can be classified into two groups depending on their origins, exogenous elicitors (substances of pathogenic origin) and endogenous elicitors (chemicals released from plant by the effect of pathogen). Elicitors can also be classified according to their nature: (1) abiotic and (2) biotic (Table 9.4). Abiotic elicitors are nonbiological factors and can be divided into three groups, physical, chemical, and hormonal elicitors. Physical elicitors mostly comprise environmental conditions. Chemical and hormonal elicitors are generally the result of media compositions. Biotic elicitors have a biological origin, as they are generally originated from bacteria, virus, fungi, and herbivore infections. Complex preparations can also be used as biotic elicitors (Naik and Al-Khayri, 2016) (Table 9.5).

Elicitors have various effects on plant primary metabolism; their effect over cell growth and plant productivity has been represented adversely. They cause changes in photosynthesis; growth cycle; respiration; nutrient uptake; homeostasis; tissue differentiation; biosynthesis of carbohydrates, lipids, and protein; and gene expressions (Baenas et al., 2014). These changes cause low biomass yield in in vitro cultures of plant cells and tissues.

All culture types (cell, callus, root, hairy root, whole plant, shoots, and seedling) can be exposed to elicitor treatments. Methyl jasmonate, jasmonic acid, and salicylic acid are the most commonly used elicitors. Moreover, polysaccharides, such as pectin, dextran, chitin, chitosan, and alginate, are frequently utilized as elicitors for triggering secondary metabolism in plant cell and tissue cultures.

In *A. trojanus* cell cultures, methyl jasmonate, jasmonic acid, salicylic acid, and pectin are used as elicitors to trigger astragaloside IV and cycloastragenol production. The highest astragaloside IV concentration was detected in methyl jasmonate–elicitated culture, whereas the highest cycloastrogenol concentration was detected in jasmonic acid–elicitated culture (Nartop, 2013).

Generally, the first response to elicitors starts at the cellular level, since cell membranes of plants have specific receptors that bind to elicitors. This signal is carried to the nucleus via cytosol receptors and causes phytoalexin formation. The response of the plant varies according to plant species, characteristics, physiological state, elicitor concentration, duration of elicitor application, age of culture, and nutrient composition of the medium. It has been reported that elicitors were more effective during the exponential phase of the growth cycle where the primary metabolism is at a maximum. Elicitation can be blocked in the absence of growth regulators. Therefore, adding growth regulators to the medium together with elicitors may be helpful to enhance secondary metabolite production (Naik and Al-Khayri, 2016).

In recent years, nanoparticles explanting surface sterility and elicitors have been used in plant cell and tissue cultures. Silver nanoparticles synthesized via green synthesis have been used for successful surface sterilization of *R. officinalis* and *Pyracantha coccinea* stem explants (Nartop, 2016a,b). Gold, zinc, and copper nanoparticles have been used as elicitors; however, the most frequently used ones are silver nanoparticles (Table 9.6). These studies about nanoparticles associate plant biotechnology and nanotechnology for better, more practical, and beneficial applications.

Callus proliferation, total protein content, dry weight, root length, leaf area, and shoot production rates have been induced in elicitation with silver nanoparticles (Fazal et al., 2016; Homaee and Ehsanpour, 2015; Gorczyca et al., 2015; Aghdaei et al., 2012). Moreover, superoxide dismutase and peroxide enzyme activities were found to be higher in *Prunella vulgaris* L. callus cultures (Fazal et al., 2016). Higher catalase activity was observed in hairy root cultures of *Artemisia annua* L. (Zhang et al., 2013). The production of secondary metabolites, such as phenolics, flavonoids, glycosides, and essential oils, was also triggered by the use of nanoparticles as elicitors (Table 9.6).

Specific compounds can also be induced by silver nanoparticles. Suspension cultures of *Capsicum* species and hairy root cultures of *A. annua* L. were elicitated with silver nanoparticles and capsaicin and artemisinin accumulations were effectively increased, respectively (Bhat and Bhat, 2016; Zhang et al., 2013). Aloin production was also increased in *Aloe vera* L. suspension cultures (Raei et al., 2014).

In brief, silver nanoparticles may be more beneficial than classical elicitors. They did not suppress primary metabolism while inducing secondary metabolism. Therefore, they can be the new generation of elicitors for plant biotechnology in the near future.

TABLE 9.4 Classification of Elicitors Based on Their Nature (Naik and Al-Khayri, 2016; Baenas et al., 2014)

Abiotic Elicitors	Physical elicitors	Injuries	Wind, insects, etc.
		Light (UV radiation)	UV-A, UV-B, UV-C
		Osmotic stress	Glucose, sucrose, fructose
			Mannitol
			Proline
			Polyethylene glycol
		Salinity	NaCl, KCl
		Drought	–
		Thermal stress	–
	Chemical elicitors	Metal ions and heavy metals	Ni, Ag, Fe, Cu, Co, Zn, Mn
		Mineral salts	Calcium chloride
			Silver nitrate
			Vanadium sulfate
		Gases	CO_2
	Hormonal elicitors	Jasmonates	Methyl jasmonate
			Jasmonic acid
		Salicylates	Salicylic acid
			Trifluoroethyl salicylate
		Ethylene	–
		Cytokinins	Adenine-type (kinetin, zeatin, and 6-benzylaminopurine)
			Phenylurea-type (thidiazuron)
		Gibberellic acid (GA)	GA_1, GA_2, GA_3, GA_4
Biotic elicitors	Defined compositions	Carbohydrates (polysaccharides, oligosaccharides, peptides)	Oligogalacturonic acid
			Agropectin
			Chitosan
			Chitin
			Mannuronate
			Guluronate
			Mannan
			Galacturonides
			Glutathione
			Hepta-β-glucoside
			β-Glucan
		Proteins	Cellulase
			Oligandrin
		Lipids	Lipopolysaccharides
	Complex compositions	Yeast extract	–
		Fungal elicitors	Necrotrophic pathogens (*Botrytis* sp.)
			Biotrophic pathogens (*Fusarium* sp., *Phoma* sp.)
			Agrobacterium rhizogenes *Agrobacterium tumefaciens*
			Bacillus cereus, *Bacillus aminovorans*
			Rhizobacterium
			Pseudomonas syringae

TABLE 9.5 Elicitor Applications and Secondary Metabolite Production in Plant Cell and Tissue Cultures

Elicitor	Plant	Culture Type	Secondary Metabolite	References
Coronatine Methyl jasmonate	*Taxus media*	Cell	Paclitaxel and related taxanes	Onrubia et al. (2013)
Chitin Pectin Dextran	*Hypericum perforatum*	Shoot	Phenylpropanoids (phenolics and flavonoids) Naphtodianthrones (hypericin and pseudohypericin)	Simic et al. (2014)
Ag^{+} Yeast extract Methyl jasmonate	*Salvia miltiorrhiza* Bunge	Hairy root	Tanshinones (cryptotanshinone, dihydrotanshinone I)	Cheng et al. (2013)
Extract of mycelium and the polysaccharide fraction of *Trichoderma atroviride* D16	*S. miltiorrhiza*	Hairy root	Tanshinone I Tanshinone IIA	Ming et al. (2013)
Dimethyl-β-cyclodextrin	*Vitis vinifera* *Vitis amurensis* *Vitis riparia* *Vitis berlandieri*	Cell	Stilbene	Zamboni et al. (2006)
Methyl jasmonate Coronatine Cyclodextrins	*Taxus* spp.	Cell	Taxanes	Cusido et al. (2014)
Strain N5.18 (Rhizobacteria)	*Hypericum perforatum* L.	Seedlings shoots	Hypericin Pseudohypericin	Manero et al. (2012)
Salicylic acid Coronalon	*Erythroxylum coca*	Cell	Cocaine Cinnamoylcocaine	Docimo et al. (2015)
Arginine Glutamate Proline Phenylalanine	*E. coca*	Cell	Tropane alkaloids	Docimo et al. (2015)
Folic acid	*S. miltiorrhiza*	Hairy root	Rosmarinic acid Salvianolic acid Caffeic acid Phenolics	Yang et al. (2016)
Yeast extract Methyl jasmonate $AgNO_3$ Sorbitol	*Perovskia abrotanoides* Karel	Adventitious root	Cryptotanshinone Tanshinone IIA	Zaker et al. (2015)
Ozone	*Melissa officinalis*	Shoot	Phenolics Rosmarinic acid	Tonelli et al. (2015)
Pseudomonas brassicacearum *Rhizobium radiobacter* *Pseudomonas thivervalensis* *P. frederiksbergensis*	*S. miltiorrhiza*	Hairy root	Tanshinone Rosmarinic acid Salvionic acid B	Yan et al. (2014)
Yeast extract Silver nitrate	*Morus alba* L.	Hairy root	Betulin Betulinic acid	Park et al. (2017)

5.2 Precursor Feeding

Plant primary and secondary metabolisms consume some organic compounds as building blocks to provide cell growth, biomass establishment, and homeostasis. These organic compounds called precursors can be self-biosynthetic or external. If the metabolism is inadequate for producing self-biosynthetic compounds, subsequent steps of the related pathway

TABLE 9.6 Nanoparticle Use in Plant Cell and Tissue Cultures in Terms of Biomass Accumulation and Secondary Metabolite Production

Plant Species	Nanoparticle Type	Results		References
Prunella vulgaris L.	AgNP (30 μg/L) AgNP–AuNP (1:2) AgNP–AuNP (2:1)	Callus proliferation	100%	Fazal et al. (2016)
		Total protein content	855 μg/mg (fw)	
		Superoxide dismutase and peroxidase	0.54 nM/min/mg (fw)	
			0.39 nM/min/mg (fw)	
		Phenolics	9.57 mg/g (dw)	
		Flavonoids	6.71 mg/g (dw)	
Capsicum sp.	AgNP	Capsaicin accumulation in suspension culture		Bhat and Bhat (2016)
Solanum tuberosum	AgNP, $AgNO_3$ (2 mg/L)	Dry weight, root length, and leaf area		Homaee and Ehsanpour (2015)
		Flavonoids		
		Total phenolics		
Calendula officinalis L.	$AgNO_3$ (0.3, 0.6, 0.9, and 1.2 mg/L)	Essential oil accumulation in in vitro plants		Mohammed Al-oubaidi and Mohammed-Ameen (2014)
Olea europaea L.	AgNP (0.5, 1.0, 1.5, 2.0 mg/L)	Glycosides compounds accumulation in in vitro shoot tips		Al-Sowaidi and Al-oubaidi (2015)
Artemisia annua L.	AgNP (900 mg/L)	Artemisinin accumulation in hairy root cultures	2.86 mg/g dw	Zhang et al. (2013)
		Catalase activity		
A. annua L.	AgNP (0.4, 0.8, and 1.2 mM)	Camphor	6.61%	Ghanati and Bakhtiarian (2016)
		[beta]-selinene	23.02%	
		Artemisia ketone	9.68%	
Wheat seedlings	AgNP	Higher total dry weight		Gorczyca et al. (2015)
Aloe vera L.	AgNP	Aloin content in cell suspension	127% higher than control	Raei et al. (2014)
Brassica nigra	ZnNP (1–20 mg/L)	Phenolics	Up to 0.15 μg/mg FW	Zafar et al. (2016)
		Flavonoids	Up to 0.22 μg/mg FW	
Eruca sativa	AgNP, CuNP AuNP (30 μg/mL)	Total protein, total flavonoids, and total phenolic content in in vitro plants		Zaka and Abbasi (2016)
Tecomella undulata (Roxb)	AgNP	Fresh shoots/explants		Aghdaei et al. (2012)
		Shoot production/explant		

dw, dry weight; *fw*, fresh weight.

cannot work efficiently; therefore, bioprocess cannot lead to high-yield production; in this case, the biosystem should be fed externally.

Absence of specific precursors may sometimes lead to lower yield of secondary metabolites. Biosynthetic pathways can be fed by providing precursor molecules both in vivo and in vitro. Amino acids (phenylalanine, proline, glutamine, tyrosine, tryptophan, cinnamic acid, naringenin, arginine) are the most used precursors in plant cell and tissue cultures. They can be incorporated into the medium at different concentrations in different culture types (Table 9.7). Similar to elicitors, the response differs according to the plant species, characteristics, physiological state, precursor concentration, duration of precursor application, age of culture, and nutrient composition of the medium.

TABLE 9.7 Secondary Metabolite-Enhancing Precursor Applications in Plant Cell and Tissue Cultures

Precursor	Plant	Culture Type	Secondary Metabolite	References
Phenylalanine (20 mg/L)	*Sauropus androgynus*	Callus	Phenolics, Flavonoids, Naringenin, Quercetin, Kaempferol	Wee (2015)
Proline (50–150 mg/L)	*Verbascum thapsus* L.	Callus	Coumarin, Eugenol	Al-Jibouri et al. (2016)
Tryptophan (150 mg/L)	*V. thapsus* L.	Callus	Thymol	Al-Jibouri et al. (2016)
Cinnamic acid Catalpol (2.5 mg/100 mL)	*Picrorhiza kurroa*	Shoot	Picroside-I	Kumar et al. (2016)
Progesterone (200–300 mg/L)	*Digitalis purpurea* L.	Shoot	Digitoxin, Digoxin	Patil et al. (2013)
Arginine	*Erythroxylum coca*	Cell	Tropane alkaloids	Docimo et al. (2015)
Glutamate				
Proline				
Phenylalanine				
Amyrin-enriched resins	*Centella asiatica*	Cell	Centelloside	Hidalgo et al. (2017)
Ferulic acid	*Capsicum frutescens*	Plant	Vanillin	Chee et al. (2017)
Phenylalanine (10 mM)	*Juniperus virginiana* L.	Callus-Cell	Podophyllotoxin	Kasparova et al. (2017)

Phenylalanine, introduced as one of the most used precursors among amino acids. This compound is the final product of the shikimate pathway and plays an effective role in the phenylpropanoid pathway. Phenylalanine produces phenolics and flavonoids by using enzymes such as phenylalanine ammonia-lyase, cinnamic acid 4-hydroxylase, and 4-coumarate-CoA ligase (Wee, 2015).

In secondary metabolite production, the precursor application approach can be very beneficial in the case the cheaper precursor is available (Mulabagal and Tsay, 2004). However, utilization of a common precursor induces a group of biochemicals; biosynthesis may not lead to the accumulation of a specific compound. In this case, enzyme inhibitors can be used to block the unwanted biosynthetic pathways. Intermediate product of a certain pathway can also be consumed as precursor to produce the final product of the relevant pathway.

6. BIOTRANSFORMATION AND METABOLIC ENGINEERING

All organic compounds are generated from precursor molecules and can be modified by live cells and their enzymes. This process is known as biotransformation. Metabolic engineering, a relatively new and multidisciplinary branch of plant biotechnology, comprises biochemical and genetic engineering principles. It is in charge of endogenous metabolic pathway modulations to induce metabolic flux to desirable molecules. Shortly, metabolic engineering investigates biotransformation during bioprocess. Modulation of metabolic pathways consists of enzyme expressions, enzyme inhibitions, enhancing the activity of enzymes, modulating pathway branch points to prevent the loss of flux, and promoting the development of sink compartments to store desired compounds (Zorrilla-López et al., 2013).

By means of genetic and molecular biotechnology, the genomes of some plant species, such as *Arabidopsis thaliana*, *Oryza sativa*, *A. annua*, *Panax ginseng*, and *Saussurea involucrate*, have been sequenced; moreover, some information about manipulation in biosynthesis of the target compound has been provided. A target gene can be introduced into another plant; furthermore, higher levels of relevant metabolite accumulation can be obtained (Gaosheng and Jingming, 2012).

Precursor application in plant cell and tissue cultures has a potential to produce new compounds that do not exist in nature. This opportunity is priceless for researchers who are in pursuit of new natural products. Instead of the original substrate, an analogue substrate can be added as a precursor into the culture medium. This approach may lead to the biosynthesis of a new molecule in vitro.

7. IMMOBILIZATION

Immobilization of plant cells has been used since 1979 to alter plant cell metabolism. Brodelius et al. (1979) reported their study on the immobilization of plant cells, belonging to *M. citrifolia*, *C. roseus*, and *Digitalis lanata*, in calcium alginate beads. This was the first report about plant cell immobilization and they concluded that the immobilized plant cells did not behave differently from immobilized microorganisms. After this publication, the cells of different plant species were immobilized with different matrices and it became obvious that immobilization of the plant cells has many advantages over freely suspended plant cells. Various matrices, such as calcium alginate, reticulate polyurethane foam, polyester and glass fibers, agar and agarose, hybrid sol–gel, and lignocellulosic (loofa, jute, and sisal) materials are commonly used in plant cell immobilization. These systems serve solutions to some important problems (low yield of biomass and metabolite production, shear stress, aeration) that are faced in cell suspension cultures and provide higher amounts of secondary metabolites under controlled conditions.

Most reports indicated that immobilized cultures may induce the yields of secondary metabolite production, even up to 16 times, compared with cell suspension cultures. Owing to the secretion of metabolites into the culture medium, the cells are not disturbed and may continue secondary metabolite production (Table 9.8).

Agitation is an obligation in plant cell cultures, as cells are often in the form of aggregates and cling to each other and surfaces. Agitation is detrimental to the plant cells because of their low tolerance to shear stress. Cell-to-cell contact is another important point in cell metabolism to transfer the metabolites from one cell to another. Secondary metabolite production is mostly higher in slow-growing or nongrowing cultures.

The immobilization of plant cells combines all of these factors and may be the best approach for increasing secondary metabolite production. This technique offers the potential for dealing with the limitations in secondary metabolite production in liquid plant cell cultures. When the cell is immobilized, cell replication is sometimes repressed. As a consequence of this situation, the genetic instability of the suspended cells may be reduced. Aggregation is a main problem and is not controllable in suspension cultures; however, aggregate sizes can be controlled by the help of immobilization techniques (Mavituna and Park, 1985).

The steps of the immobilization of plant cells are (1) obtaining a cell culture with defined and desirable properties (e.g., a cell line with high product yield and low growth rate), (2) determining a suitable immobilization method that maintains cell viability and product biosynthesis, (3) obtaining product release from the cells into the medium while maintaining the survival of cultures, and (4) designing a suitable vessel for the culture of the immobilized cells and operating conditions that maintain cellular characteristics (Chawla, 2002).

There are mainly three types of immobilization systems explained for plant cell cultures.

- *Physical entrapment system* uses a boundary, such as hollow fiber tube, paraffin, and silicon oil, to separate the cells from the media.
- *Entrapment in a gel matrix system* is also known as gel encapsulation technique and it is the most widely studied method for the immobilization of plant cells. Agar, alginate, agarose, carrageenan, collagen, pectate, gelatine, polyacrylamide, and a combination of these substances can also be utilized for entrapment of these substances.
- *Surface immobilization system* is used to attach the cells to a solid support, such as glass or polyurethane foam, loofa, sisal, and jute (Fig. 9.2).

In plant cell cultures, calcium alginate and reticulate polyurethane foam are utilized frequently. Agar, agarose, polyester and glass fibers, hybrid sol–gel matrices, loofa sponge, and jute and sisal fibers were also used as matrices for immobilization (Table 9.8). The matrices used for immobilization may act as a barrier for cells. Although they all have benefits, there are some disadvantages caused by the matrix itself and technique. Researchers must choose the immobilization matrix and technique that is most suitable to the purpose of their study to make immobilization of plant cells feasible for industrial production.

For plant cell cultures, there are no ideal immobilization matrices; however, cell carriers should have some features to ease the production of biomass and secondary metabolites. They should be (1) biocompatible, (2) nontoxic, (3) autoclavable, (4) reusable, (5) suitable for a large number of cells, and (6) available for large-scale bioreactor systems. They should also have (1) long-term mechanical and chemical stability, (2) high surface/volume ratio, (3) efficient diffusion of the nutrient medium into the center of the carriers, (4) macroporous structure for high cell density and shear stress protection, (5) uniform size distribution to obtain uniform cultures, (6) high batch-to-batch consistency, and (7) functional groups to bind the cells. Moreover, these matrices should induce cell activity and facilitate immobilization and harvesting of cells (Mavituna, 2004).

TABLE 9.8 Immobilization Matrices Utilized for Secondary Metabolite Production in Some Plant Species

Plant Species	Secondary Metabolites	Immobilization Matrices	References
Morinda citrifolia	Anthraquinones	Calcium alginate	Brodelius et al. (1979) and Brodelius and Mosbach (1982)
Catharanthus roseus	Indole alkaloids		
Digitalis lanata	Digoxin		
C. roseus	Ajmalicine	Calcium alginate	Brodelius et al. (1979) and Brodelius and Mosbach (1982)
C. roseus	Ajmalicine, Secologanin	Agarose entrapment	Brodelius and Nilsson (1983)
Daucus carota			
Datura innoxia			
Capsicum frutescens Mill cv. annuum	Capsaicin	Reticulated polyurethane	Lindsey and Yeoman (1984a,b) and Lindsey (1985)
C. frutescens	Capsaicin	Reticulated polyurethane	Mavituna and Park (1985)
Papaver somniferum	Codeine	Reticulated polyurethane	Corchete and Yeoman (1989)
Ginkgo biloba	Ginkgolide A and Ginkgolide B	Surface immobilization	Carrier et al. (1990)
Papaver somniferum	Sanguinarine, Dihydrosanguinarine	Polyester fibers	Kurz et al. (1990)
Mucuna pruriens L.	L-DOPA	Calcium alginate	Pras (1990)
Podophyllum hexandrum	Podophyllotoxin		
Linum flavum L.	5-Metoxypodophyllotoxin		
Thalictrum rugosum	Protoberberine	Surface immobilization	Archambault et al. (1990)
C. roseus	Catharantine, Ajmalicine	Glass fiber	Facchini and DiCosmo (1991a)
Thalictrum rugosum	Protoberberine		Facchini and DiCosmo (1991b)
C. frutescens	Capsaicin, Vanillin	Calcium alginate	Sudhakar Johnson et al. (1990, 1991, 1996) and Sudhakar Johnson (1993)
Coleus blumei	Rosmarinic acid	Loofa sponge	Park and Martinez (1994)
Taxus cuspidata	Taxol	Calcium alginate	Park et al. (1995)
Coronilla vaginalis Lam.	Umbelliferone, Marmesin	Hybrid sol–gel matrices	Campostrini et al. (1996)
Haplophyllum patavinum (L.) *G. Don Fil.*	Coumarin compounds		
C. roseus (L.) *G. Don.*	Vinblastine and Vincristine		
C. frutescens	Capsaicin, Dihydrocapsaicin	Calcium alginate	Sudhakar Johnson and Ravishankar (1996)
Nicotiana tabacum	Scopolin	Calcium alginate	Gillet et al. (2000)
Solanum chrysotrichum	Spirostanol saponin	Calcium alginate	Charlet et al. (2000)
Juniperus chinensis	Podophyllotoxin	Calcium alginate	Premjet and Tachibana (2004)
Taxus baccata	Paclitaxel, Baccatin III	Calcium alginate	Bentebibel et al. (2005)
Linum usitatissimum	Dehydrodiconiferyl alcohol-4-β-D-glucoside	Calcium alginate	Attoumbre et al. (2006)

TABLE 9.8 Immobilization Matrices Utilized for Secondary Metabolite Production in Some Plant Species—cont'd

Plant Species	Secondary Metabolites	Immobilization Matrices	References
Galphimia glauca	Galphimine-B	Calcium alginate	Osuna et al. (2008)
Dioscorea bulbifera L.	Diosgenin	Calcium alginate	Jyothishwaran and Seetharam (2008)
Scoparia dulcis Linn.	Scopadulcic acid B	Loofa sponge	Mathew and Jayachandran (2009)
Morus bombycis K.	Rutin, GABA	Calcium alginate	Han et al. (2012)
Rubia tinctorum L.	Alizarin, Purpurin	Loofa, Jute, and Sisal	Nartop et al. (2013)
G. biloba	Bilobalide, Ginkgolide	Jute	Sukito and Tachibana (2016)

DOPA, dihydroxyphenylalanine; *GABA*, gamma-aminobutyric acid.

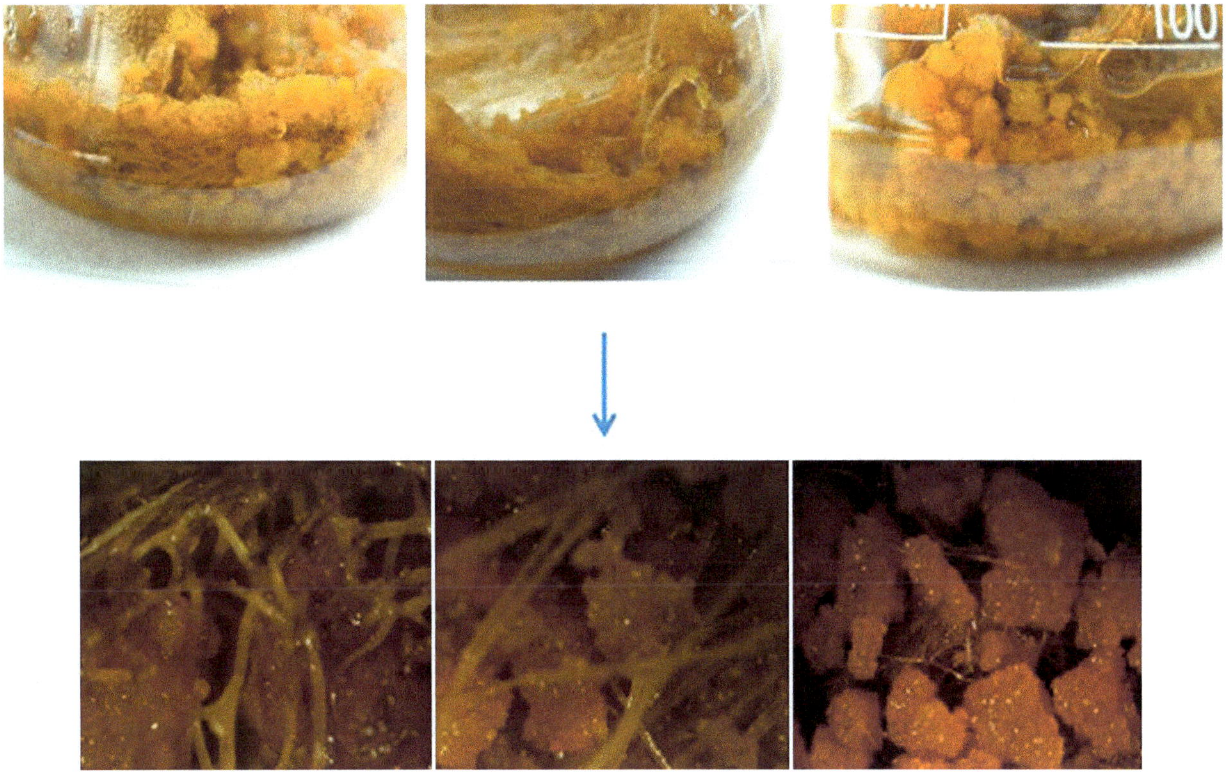

FIGURE 9.2 Immobilization of *Rubia tinctorum* cell cultures on loofa, sisal, and jute, respectively (Nartop et al., 2013).

Immobilization techniques can be combined with elicitor (methyl jasmonate, jasmonic acid, salicylic acid, and pectin), precursor (amino acids, tropane, and indole alkaloids), and feeding applications to gain more valuable biomass with higher metabolite composition. Sukito and Tachibana (2016) reported that bilobalide and ginkgolide A, B, and C accumulations of *Gingko biloba* cell suspension cultures were enhanced 1.78, 1.95, 2.05, and 2.95 times, respectively, when immobilization with jute and elicitor treatment (0.1 mM methyl jasmonate + 0.1 mM salicylic acid) were combined.

8. HAIRY ROOT CULTURES

Agrobacterium rhizogenes is a soil-borne gram-negative bacterium that belongs to the *Rhizobiacea* family. It causes hairy root disease in infected dicotyledonous plants. *Agrobacterium*, also known as natural genetic engineers, have begun to attract interest in gene transfer to plants in the mid-1980s. The integration of the T-DNA region of the Ri (root-inducing) plasmid of *A. rhizogenes* into the plant DNA results in infection in plants and reveals the hairy root phenotype. Hairy roots have some characteristic features, such as frequent branching, high density of capillary roots, high biomass production potential, and rapid development. *Agrobacterium* is routinely used in gene transfer studies because of its ease of application, economic viability, high rate of transformation, and rapid root growth. In hairy root cultures, secondary metabolite production rates can be as high as that of the parent plant. Moreover, it allows genetic manipulations to be performed in the metabolic pathway of the synthesis. The reason for choosing hairy roots is to provide rapid development without the need for any auxin source from outside. Usually, no light is needed for incubation. In addition to their genetic stability, metabolite production is also stable. Because of all these advantages, many root-derived plant products that are thought not to be suitable for cell culture have been produced via hairy root culture technology (Giri and Narasu, 2000). Reports on hairy root cultures established for secondary metabolite production in 2016 are listed in Table 9.9.

TABLE 9.9 Secondary Metabolite Production in Hairy Root Cultures

Plant Species	Secondary Metabolite	References
Tylophora indica	Tylophorine	Da Silva and Jha (2016)
Salvia miltiorrhiza	Rosmarinic acid, Salvianolic acid B, Caffeic acid, Phenolics	Yang et al. (2016)
Linum album	Lignan, Phenolic acids	Tashackori et al. (2016)
S. miltiorrhiza	Tanshinone, Salvianolic acids	Wang et al. (2016a,b)
Plumbago rosea L.	Plumbagin-An	Jose et al. (2016)
Bacopa monnieri (L.)	Bacoside A	Largia et al. (2016)
Prunella vulgaris L.	Rosmarinic acid	Ru et al. (2016)
Ocimum basilicum	Rosmarinic acid	Srivastava et al. (2016a,b)
Panax vietnamensis	Ocotillol-type ginsenosides	Ha et al. (2016)
Catharanthus roseus L.	Indole alkaloids	Hanafy et al. (2016)
S. miltiorrhiza	Rosmarinic acid, Salvianolic acid B	Wu et al. (2016)
Artemisia carvifolia Buch.	Caffeic acid, Quercetin, Isoquercetin, Rutin	Dilshad et al. (2016)
S. miltiorrhiza	Tanshinones	Wang et al. (2016a,b)
Lithospermum erythrorhizon	Shikonin	Fang et al. (2016)
Rauwolfia serpentina	Ajmaline, Ajmalicine	Srivastava et al. (2016a,b)
Solanum khasianum	Solasodine, Alpha-solanine	
Valeriana officinalis	Sesquiterpenoid	Ricigliano et al. (2016)
Artemisia dubia	Artemisinin and derivatives	Amanullah et al. (2016)
Brassica oleracea	Glucosinolates	Lee et al. (2016)
Fagopyrum tataricum Gaertn	Anthocyanins, Rutin	Thwe et al. (2016)
Arnebia hispidissima	Alkannin, Shikonin	Singh and Sharma (2016)
Fagopyrum tataricum Gaertn.	Rutin, Quercetin	Huang et al. (2016)
Datura sp.	Hyoscyamine	Harfi et al. (2016)
C. roseus	Catharanthine, Ajmalicine	Benyammi et al. (2016)
Datura stramonium	Hyoscyamine	Belabbassi et al. (2016)
Ajuga bracteosa	Phytoecdysteroid	Kayani et al. (2016)
Salvia castanea	Tanshinones	Li et al. (2016)

Hairy root cultures are immobilized cell culture systems. To trigger the production of secondary metabolites of the entrapped cells, elicitor is also applied to the hairy root cultures. Although there is little information about biosynthetic pathways of secondary metabolism in hairy roots, metabolite production can be increased by culturing the most commonly used elicitors, such as jasmonic acid and methyl jasmonate. NaCl, $CaCl_2$, and KCl can also be used as elicitors. They are cheaper than jasmonic acid derivatives and can be easily supplied. In *Rauwolfia serpentina* hairy root cultures, ajmalicine production was stimulated up to 14.8-fold at 100 mM concentration of NaCl after 1 week of treatment (Srivastava et al., 2016a,b). Potassium chloride (KCl) and calcium chloride ($CaCl_2$) were also used as elicitors in some *Datura* species. The highest hyoscyamine levels obtained for *Datura stramonium*, *Datura tatula*, and *Datura innoxia* hairy root cultures were 2.32-, 1.99-, and 1.85-fold the control levels with KCl elicitation and 2.08-, 2.07-, and 1.85-fold the control levels with $CaCl_2$ elicitation, respectively (Harfi et al., 2016).

Elicitors can be used alone or in combination with heavy metal ions (cobalt, silver, cadmium), polysaccharides (yeast extracts, chitosan), and signal compounds (salicylic acid). Coadministration of biotic and abiotic elicitors provides a synergistic effect that can lead to better results than single use of an elicitor (Sudha, 2014). Wang et al. (2016a,b) reported that 18-day-old *Salvia miltiorrhiza* hairy root cultures were exposed to ultraviolet B irradiation and methyl jasmonate application and the maximum production of tanshinone reached to 28.21 mg/L, a 4.9-fold increase over the control.

The effects of gene expression on secondary metabolism have been studied on hairy root cultures. Dilshad et al. (2016) reported that flavonoid accumulation was affected by rol genes in *Artemisia carvifolia* Buch hairy root cultures. Among the transgenic plants, those harboring rol B were slightly more active than the rol C-transformants in flavonoid biosynthesis. In *Artemisia dubia*, rolA transgenic hairy root cultures were reported to be cultivable and viable and an efficient source for artemisinin and its derivatives (Amanullah et al., 2016).

9. BIOREACTORS

Bioreactors (or fermenters) are tanks of large volumes designed for large-scale cultivation and they are used for high-yield products of plants, mammals, and bacteria. In plant biotechnology, bioreactors are advanced large-scale liquid culture systems that are used for mass propagation and/or secondary metabolite production. They provide improved developments for the plant cell and tissue culture industry.

Most plant species grow better in liquid than on solid media. Culture conditions in bioreactors can precisely be controlled (gaseous exchange, light intensity, agitation rate, temperature, pH changes, etc.) and can generate high multiplication and growth rates and biomass accumulation with high secondary metabolite content. Moreover, scaling up of a bioreactor system is easier. Once the system is set, because of reducing space and energy and labor requirements, it is more productive and profitable for commercial production than semisolid micropropagation techniques. However, they have some disadvantages: (1) contamination problem, (2) lack of protocols and production procedures, (3) increased hyperhydricity, (4) foaming and shear stress, and (5) accumulation of inhibiting metabolites (Akin-Idowu et al., 2009; Levin and Tanny, 2004). Except the fourth one, these factors are also disadvantages of semisolid systems. Thus bioreactors are still popular in plant biotechnology studies (Table 9.10).

For high-yield production at the commercial scale, before working with bioreactors, all parameters affecting the culture growth and productivity should be determined. Cultures should be started at small scale in 100–250 mL Erlenmayer flasks with 20–50 working volumes. If the growth is successful, cultures can be transferred to 500–1000–2000 mL Erlenmayer flasks (or bottles) with 100–200–500 mL working volumes (Figs. 9.3 and 9.4). In this step, after determination of all requirements of the culture, scaling up can be taken into consideration. According to laboratory-scale data, a bioreactor can be designed or an available bioreactor can be used for large-scale cultivation. If the progress goes well in 1- to 10-L bioreactors, larger ones (50–100 to 1000 L) can be taken into consideration.

Some bioreactor types, which are frequently used in microbiological studies, are favorable in plant biotechnology. In terms of secondary metabolite production, cell suspension cultures come to the mind first, because of some similarities with bacterial cultures. Bubble column, airlift, and stirred tank reactors are the most commonly used ones for plant cell suspension cultures. For adventitious and hairy root cultures, bubble column and airlift bioreactors are more suitable. However, temporary immersion systems and mist bioreactors come into prominence for mass propagation. In vitro mass propagation is commonly used in agricultural studies to get higher proliferation (multiple bud and shoot formations, somatic embryogenesis, plantlet regeneration, shoot induction and elongation) rates than ex vitro conditions. Higher biomass accumulation cannot be obtained without higher proliferation rates. As mentioned before, biomass, which consists of plant cells, is the biofactory of secondary metabolites. Thus, because of the importance of biomass accumulation, this gives an option to researchers; if sufficient biomass accumulation and secondary metabolite production cannot be obtained from

TABLE 9.10 Bioreactor Types Used in Plant Biotechnology and Their Productivity

Bioreactor Type	Plant	Culture Type	Production		References
Balloon-type airlift bioreactor	*Oplopanax elatus*	Adventitious roots	Polysaccharides	1913.9 g/L	Jiang et al. (2015)
			Phenolics	459.7 g/L	
			Flavonoids	713.7 g/L	
5-L stirred hybrid airlift bioreactor	*Catharanthus roseus*	Cambial meristematic cells	Vindoline	7.45 mg/L	Zhou et al. (2015)
			Catharanthine	1.76 mg/L	
			Ajmalicine	58.98 mg/L	
3 L airlift balloon-type bioreactor (continuous immersion culture)	Onion	Mature zygotic embryos	Plantlet regeneration		Wu et al. (2015)
Aeroponics and temporary immersion bioreactor	*Dioscorea rotundata* Poir.	Tuber pieces	Production of seed tubers		Aighewi et al. (2015)
Balloon-type bubble bioreactor	*Eurycoma longifolia*	Adventitious roots	Phenolics and flavonoids		Lulu et al. (2015)
60 L stirred-tank bioreactor	Tobacco	Transgenic cells	Intimin (adhesin protein, an oral cattle vaccine)		O'Neill et al. (2015)
Mist bioreactor	*Artemisia annua*	Single nodal explants	Rooted plantlets		Fei and Weathers (2016)
Airlift bioreactors	*Eleutherococcus koreanum* Nakai	Adventitious root	Bioactive compounds (total phenolics and flavonoids)	303.93 mg/L	Lee et al. (2015)
Stirred-tank bioreactor	*Boerhaavia diffusa* L.	Hairy root	Boeravinone B		Gupta et al. (2016)
			Eupalitin		
Temporary immersion bioreactor	*Artocarpus altilis*	Plantlets	Seedlings		Shandil and Tuia (2015)
Airlift bioreactor	*Coleus blumei*	Calli, normal roots, hairy roots	Rosmarinic acid		Bauer et al. (2015)
5-L tank bioreactor	*Vinca minor*	Transgenic cell suspension	Total alkaloid content	2.7 ± 0.3% dw	Verma et al. (2015)
			Vincamine	0.005 ± 0.001% dw	
Airlift bioreactor	*Castanea dentata* x *C. mollissima*	Small cell clumps	Transgenic embryogenic tissue proliferation		Kong et al. (2014)
Mist bioreactor	Carrot	Embryogenic cells	Rooted plantlets		Fei and Weathers (2014)
Temporary immersion system	*Leucojum aestivum*	Somatic embryos	Galanthamine	0.05% dw	Ptak et al. (2013)
2-L bioreactor	Rice	Transgenic suspension cell system	Mouse granulocyte-macrophage colony-stimulating factor	24.6 mg/L	Liu et al. (2012)
Bubble column, airlift, and stirred tank reactor (10–200 L)	*Bupleurum falcatum* L.	Root	Sailcosaponin-a Saikosaponin-d	500–600 mg/L/56 days (draft tube, 200 L scale)	Kusakari et al. (2012)
5-L balloon-type bubble bioreactor	*Astragalus membranaceus*	Adventitious roots	Biomass	30 g/inoculum	Wu et al. (2011)
			Total polysaccharide	30.0 mg/g dw	
			Total saponin	3.4 mg/g dw	
			Total flavonoid	6.4 mg/g dw	

TABLE 9.10 Bioreactor Types Used in Plant Biotechnology and Their Productivity—cont'd

Bioreactor Type	Plant	Culture Type	Production		References
Roller bottle system	*Stevia rebaudiana*	Root tip explants	Adventitious roots		Reis et al. (2011)
3-L stirred tank reactor and a 1-L glass-column bioreactor	*Harpagophytum procumbens*	Cell	Verbascoside	165.42 mg/L/day	Georgiev et al. (2011)
Temporary immersion system	*Curcuma zedoaria* *Zingiber zerumbet*	In vitro shoots	Multiple shoot induction		Stanly et al. (2010)
Temporary immersion system	*Capsicum chinense* Jacq.	Nodal segments	Shoot induction and elongation		Bello-Bello et al. (2010)
Temporary immersion system	*Cedrela odorata* L.	Juvenile and mature tissues	Multiple adventitious shoot formation		Pena-Ramirez et al. (2010)

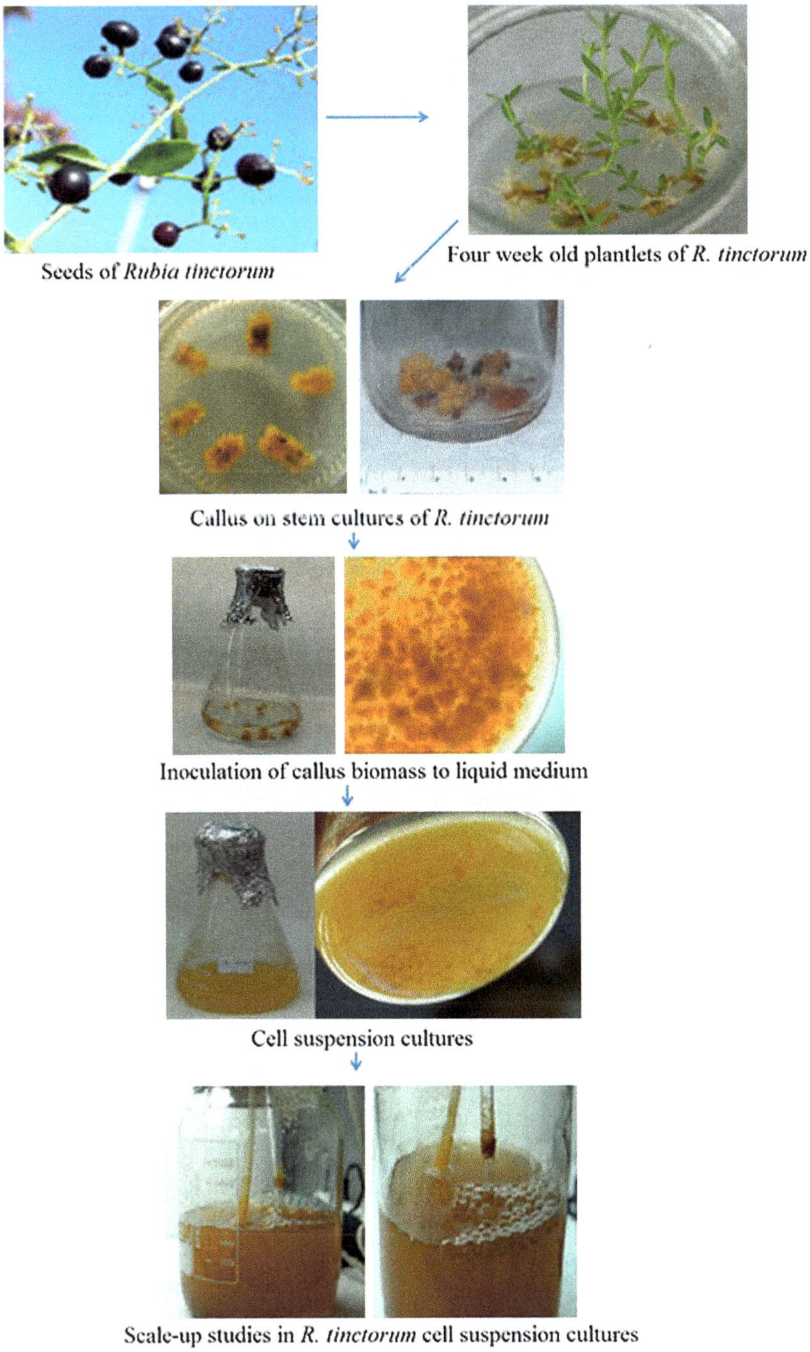

FIGURE 9.3 Establishment of *Rubia tinctorum* cell cultures and scale-up studies (Naziri, 2014).

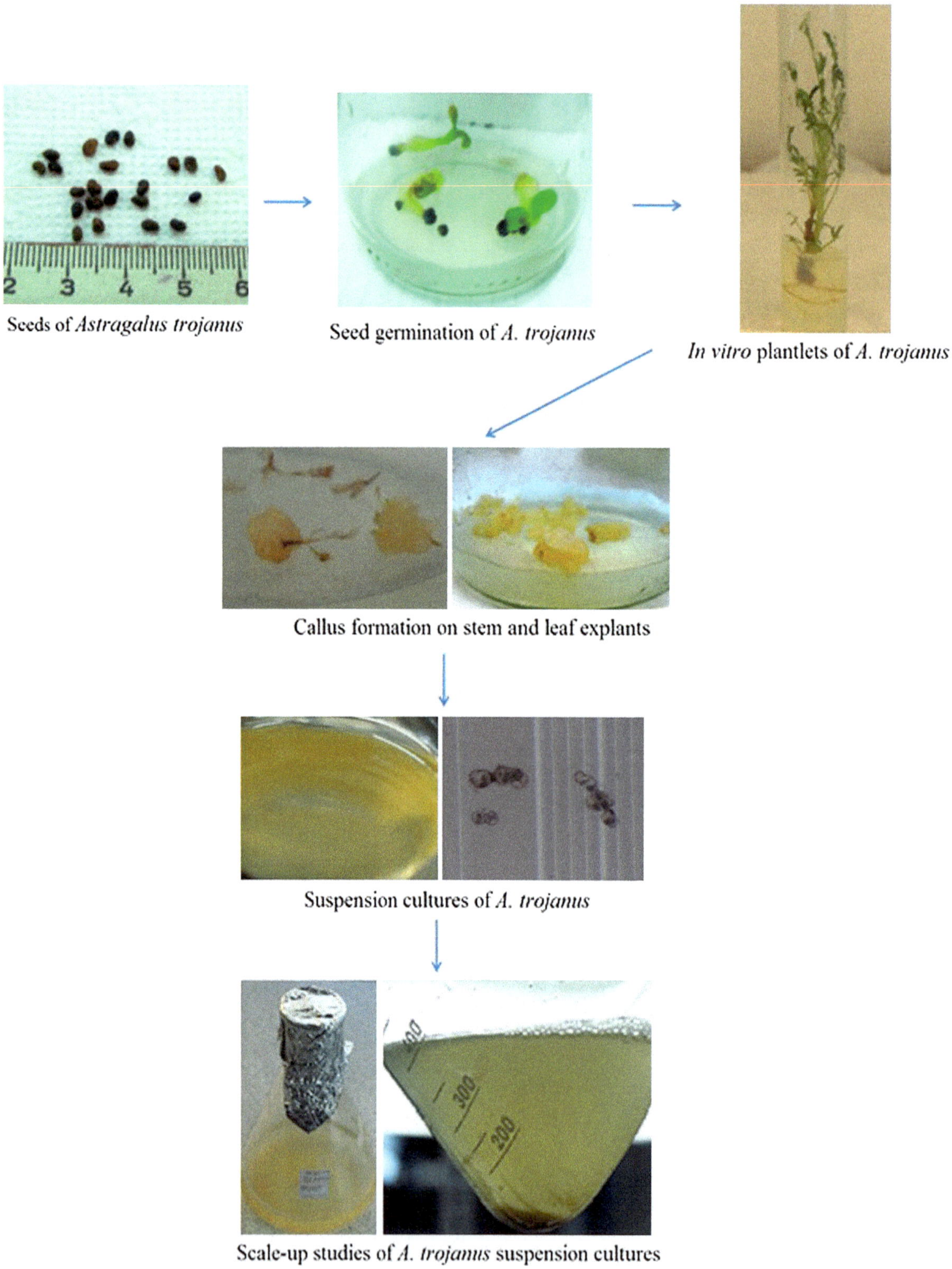

FIGURE 9.4 Establishment of *Astragalus trojanus* cell cultures and scale-up studies (Nartop, 2013).

cell suspension cultures, biomass at higher volumes for extraction of relevant metabolite can be obtained via temporary immersion or mist bioreactor systems of tissue and organ cultures.

Production at the industrial scale is possible in bioreactors. Ginseng (*P. ginseng* C. A. Meyer) adventitious root cultures are a good example for biomass accumulation and metabolite production for bioreactor scale-up. About 500 g fresh weight was obtained with 60 g of inoculum in a 5-L bioreactor, whereas 2.2 kg fresh weight was obtained in a 20-L bioreactor in 6 weeks of culture. At the eighth week of culture, fresh weight reached 2.8 kg in a 20-L airlift bioreactor. Maximum ginsenoside accumulations were detected 43, 34, 45, and 40 mg/g dry weight in airlift bioreactors with 4, 15, 500, and 1000 L working volumes, respectively (Paek et al., 2008).

10. COMBINATORIAL BIOSYNTHESIS

Combinatorial biosynthesis is a new method to produce valuable plant secondary metabolites in different species. Bacteria such as *Escherichia coli* and *Saccharomyces cerevisiae* have been used for the production of plant-specific metabolites. These bacteria have been preferred for their mutation, culture, transformation, and expression skills.

Related genes from *Malus domestica*, *Anthurium andraeanum*, and *Petunia hybrida* were cloned and transferred to *E. coli*. By the time the culture was fed with precursor of anthocyanins, naringenin and eriodictyol productions were detected in the culture media (Yan, 2005). For artemisinin production, amorphadiene synthase and cytochrome P450 monooxygenase were isolated and transferred to yeast chromosome. The yield of artemisinic acid was detected at 115 mg/L (Ro et al., 2006; Gaosheng and Jingming, 2012).

11. CONCLUSION

Plant biotechnology serves many ways for secondary metabolite production. Plant species as a starting material, explant type, culture type, and culture conditions are all changeable and replaceable. Moreover, advanced techniques such as immobilization and elicitor and precursor treatments render the bioprocess multiple choices. In some cases, this may cause confusion for researchers. However, by examining previous studies, confusion can be turned into clarity. New materials can be used in classic applications and new techniques can even be discovered.

REFERENCES

Aghdaei, M., Salehi, H., Sarmast, M.K., 2012. Effects of silver nanoparticles on *Tecomella undulata* (Roxb) Seem, micropropagation. Adv. Hort. Sci. 26 (1), 21–24.

Aighewi, B.A., Asiedu, R., Maroya, N., Balogun, M., 2015. Improved propagation methods to raise the productivity of yam (*Dioscorea rotundata* Poir). Food Secur. 7 (4), 823–834.

Akin-Idowu, P.E., Ibitote, D.O., Ademoyegun, O.T., 2009. Tissue culture as a plant production technique for horticultural crops. Afr. J. Biotech. 8 (16), 3782–3788.

Al-Jibouri, A.M.J., Abed, A.S., Ali, A.-J.A., Majeed, D.M., 2016. Improvement of phenols production by amino acids in callus cultures of *Verbascum thapsus* L. Am. J. Plant. Sci. 7, 84–91.

Al-Sowaidi, W.M.M., Al-oubaidi, H.K.M., 2015. Increasing (glycosides compounds) of *Olea europaea* L. from shoot tips using $AgNO_3$ nanoparticle *in vitro*. Int. J. Phytopharm. 6 (1), 31–35.

Amanullah, M.B., Rizvi, Z.F., Zia, M., 2016. Production of artemisinin and its derivatives in hairy roots of *Artemisia dubia* induced by rolA gene transformation. Pak. J. Bot. 48 (2), 699–706.

Archambault, J., Volesky, B., Kurz, W.G.W., 1990. Production of indole alkaloids by surface immobilized *C. roseus* cell. Biotechnol. Bioeng. 35 (7), 660–667.

Attoumbre, J., Charlet, S., Baltora-Rosset, S., Hano, C., Grandic, S.R.-L., Gillet, F., Bensaddek, L., Mesnard, F., Fliniaux, M.-A., 2006. High accumulation of dehydrodiconiferyl alcohol-4-β-D-glucoside in free and immobilized *Linum usitatissimum* cell cultures. Plant Cell Rep. 25, 859–864.

Baenas, N., García-Viguera, C., Moreno, D.A., 2014. Elicitation: a tool for enriching the bioactive composition of foods. Molecules 19, 13541–13563.

Bauer, N., Vukovic, R., Likic, S., Jelaska, S., 2015. Potential of different *Coleus blumei* tissues for rosmarinic acid production. Food Tech. Biotech. 53 (1), 3–10.

Belabbassi, O., Khelifi-Slaoui, M., Zaoui, D., Benyammi, R., Khalfallah, N., Malik, S., Makhzoum, A., Khelifi, L., 2016. Synergistic effects of polyploidization and elicitation on biomass and hyoscyamine content in hairy roots of *Datura stramonium*. Biotechnol. Agron. Soc. Environ. 20 (3), 408–416.

Bello-Bello, J.J., Canto-Flick, A., Balam-Uc, E., Gomez-Uc, E., Robert, M.L., Iglesias-Andreu, L.G., Santana-Buzzy, N., 2010. Improvement of *in vitro* proliferation and elongation of habanero pepper shoots (*Capsicum chinense* Jacq.) by temporary immersion. Hortscience 45 (7), 1093–1098.

Bentebibel, S., Moyano, E., Palazon, J., Cusido, R.M., Bonfill, M., Eibl, R., Pinol, M.T., 2005. Effects of immobilization by entrapment in alginate and scale-up on paclitaxel and baccatin III production in cell suspension cultures of *Taxus baccata*. Biotechnol. Bioeng. 89 (6), 647–655.

Benyammi, R., Paris, C., Khelifi-Slaoui, M., Zaoui, D., Belabbassi, O., Bakiri, N., Aci, M.M., Harfi, B., Malik, S., Makhzoum, A., 2016. Screening and kinetic studies of catharanthine and ajmalicine accumulation and their correlation with growth biomass in *Catharanthus roseus* hairy roots. Pharm. Biol. 54 (10), 2033–2043.

Bhat, P., Bhat, A., 2016. Silver nanoparticles for the enhancement of accumulation of capsaicin in suspension culture of *Capsicum* sp. J. Exp. Sci. 7 (2), 1–6.

Brodelius, P., Deus, B., Mosbach, K., Zenk, M.H., 1979. Immobilized plant cells for the production and transformation of natural products. FEBS Lett. 103 (1), 93–97.

Brodelius, P., Mosbach, K., 1982. Immobilized plant cells. Appl. Biochem. Biotechnol. 7, 31–34.

Brodelius, P., Nilsson, K., 1983. Permeabilization of immobilized plant cells, resulting in release of intracellularly stored products with preserved cell viability. Eur. J. Appl. Microbiol. Biotechnol. 17, 275–280.

Campostrini, R., Carturan, G., Caniato, R., Piovan, A., Filippini, R., Innocenti, G., Cappelletti, E.M., 1996. Immobilization of plant cells in hybrid sol-gel materials. J. Sol-Gel Sci. Technol. 7, 87–97.

Carrier, D.J., Coulombe, P., Mancini, M., Neufeld, R., Weber, M., Archambault, J., 1990. Immobilized *Ginkgo biloba* cell cultures. In: Nijkamp, H.J.J., Van Der Plas, L.H.W., Van Aartrijk, J. (Eds.), Progress in Plant Cellular and Molecular Biology. Proceedings of the VIIth International Congress on Plant Tissue and Cell Culture, Amsterdam, The Netherlands, 24–29 June, Kluwer Academic Publishers, Dordrecht/Boston/London, pp. 614–618.

Charlet, S., Gillet, F., Villarreal, M.-L., Barbotin, J.-N., Filiniaux, M.-A., Nava-Saucedo, J.-E., 2000. Immobilisation of *Solanum chrysotrichum* plant cells within Ca-alginate gel beads to produce an antimycotic spirostanol saponin. Plant Physiol. 38, 875–880.

Chawla, H.S., 2002. Introduction to Plant Biotechnology, second ed. Science Publishers, USA.

Chee, M.J.Y., Lycett, G.W., Khoo, T.J., Chin, C.F., 2017. Bioengineering of the plant culture of *Capsicum frutescens* with vanillin synthase gene for the production of vanillin. Mol. Biotechnol. 59 (1), 1–8.

Cheng, Q., He, Y., Li, G., Liu, Y., Gao, W., Huang, L., 2013. Effects of combined elicitors on tanshinone metabolic profiling and *SmCPS* expression in *Salvia miltiorrhiza* hairy root cultures. Molecules 18 (7), 7473–7485.

Corchete, P., Yeoman, M.M., 1989. Biotransformation of (–)-codeinone to (–)-codeine by *Papaver somniferum* cells immobilized in reticulate polyurethane foam. Plant Cell Rep. 8 (2), 128–131.

Cusido, R.M., Onrubia, M., Sabater-Jara, A.B., Moyano, E., Bonfill, M., Goossens, A., Pedreno, M.A., Palazon, J., 2014. A rational approach to improving the biotechnological production of taxanes in plant cell cultures of *Taxus* spp. Biotechnol. Adv. 32 (6), 1157–1167.

Da Silva, J.A.T., Jha, S., 2016. Micropropagation and genetic transformation of *Tylophora indica* (Burm. f.) Merr. a review. Plant Cell Rep. 35 (11), 2207–2225.

Degenhardt, J., Köllner, T.G., Gershenzon, J., 2009. Monoterpene and sesquiterpene synthases and the origin of terpene skeletal diversity in plants. Phytochemistry 70, 1621–1637.

Dewick, P.M., 2002. Medicinal Natural Products, a Biosynthetic Approach, second ed. John Wiley & Sons, Ltd., Baffins Lane, Chichester, West Sussex, England.

Dilshad, E., Ismail, H., Haq I.U., Cusido R.M., Palazon, J., Ramirez-Estrada, K., Mirza, B., 2016. Rol genes enhance the biosynthesis of antioxidants in *Artemisia carvifolia* Buch. BMC Plant Biol. 16 (1), 125.

Docimo, T., Davis, A.J., Luck, K., Fellenberg, C., Reichelt, M., Gershenzon, J., D'Auria, J.C., 2015. Influence of medium and elicitors on the production of cocaine, amino acids and phytohormones by *Erythroxylum coca* calli. Plant Cell Tiss. Org. Cult. 120 (3), 1061–1075.

Facchini, P.J., DiCosmo, F., 1991a. Secondary metabolite biosynthesis in cultured cells of *Catharanthus roseus* (L.) G. Don immobilized by adhesion to glass fibres. Appl. Microb. Biotechnol. 35, 382–392.

Facchini, P.J., DiCosmo, F., 1991b. Plant cell bioreactor for the production of protoberberine alkaloids from immobilized *Thalictrum rugosum* cultures. Biotech. Bioeng. 37 (5), 397–403.

Fahey, J.W., Zalcmann, A.T., Talalay, P., 2001. The chemical diversity and distribution of glucosinolates and isothiocyanates among plants. Phytochemistry 56 (1), 5–51.

Fang, R.J., Zou, A.L., Zhao, H., Wu, F.Y., Zhu, Y., Zhao, H., Liao, Y.H., Tang, R.J., Pang, Y.J., Yang, R.W., 2016. Transgenic studies reveal the positive role of LeEİL–1 in regulating shikonin biosynthesis in *Lithospermum erythrorhizon* hairy roots. BMC Plant Biol. 16, 121.

Fazal, H., Abbasi, B.H., Ahmad, N., Ali, M., 2016. Elicitation of medicinally important antioxidant secondary metabolites with silver and gold nanoparticles in callus cultures of *Prunella vulgaris* L. Appl. Biochem. Biotechnol. 180 (6), 1076–1092.

Fei, L.W., Weathers, P., 2016. From leaf explants to hanging rooted plantlets in a mist reactor. Plant Cell Tiss. Org. Cult. 124 (2), 265–274.

Fei, L.W., Weathers, P.J., 2014. From cells to embryos to rooted plantlets in a mist bioreactor. Plant Cell Tiss. Org. Cult. 116 (1), 37–46.

Gaosheng, H., Jingming, J., 2012. Production of useful secondary metabolites through regulation of biosynthetic pathway in cell and tissue suspension culture of medicinal plants. In: Leva, A., Rinaldi, L.M.R. (Eds.), Recent Advances in Plant In Vitro Culture. InTech. ISBN: 978-953-51-0787-3, pp. 197–210 (Chapter 11).

Georgiev, M., Ludwig-Muller, J., Weber, J., Stancheva, N., Bley, T., 2011. Bioactive metabolite production and stress-related hormones in Devil's claw cell suspension cultures grown in bioreactors. Appl. Microbiol. Biotechnol. 89 (6), 1683–1691.

Ghanati, F., Bakhtiarian, S., 2016. Changes of natural compounds of *Artemisia annua* L. by methyl jasmonate and silver nanoparticles. Adv. Environ. Biol. 2251.

Gillet, F., Roisin, C., Fliniaux, M.A., Jacquin-Dubreuil, A., Barbotin, J.N., Nava-Saucedo, J.E., 2000. Immobilization of *Nicotiana tabacum* plant cell suspensions within calcium aljinate gel beads for the production of enhanced amounts of scopolin. Enzyme Microb. Technol. 26, 229–234.

Giri, A., Narasu, L.M., 2000. Transgenic hairy roots: recent trends and applications. Biotechnol. Adv. 18, 1–22.

Gorczyca, A., Pociecha, E., Kasprowicz, M., Niemiec, M., 2015. Effect of nanosilver in wheat seedlings and *Fusarium culmorum* culture systems. Eur. J. Plant Pathol. 142 (2), 251–261.

Gupta, R., Pandey, P., Singh, S., Singh, D.K., Saxena, A., Luqman, S., Bawankule, D.U., Banerjee, S., 2016. Advances in *Boerhaavia diffusa* hairy root technology: a valuable pursuit for identifying strain sensitivity and up-scaling factors to refine metabolite yield and bioactivity potentials. Protoplasma 253 (4), 1145–1150.

Ha, L.T., Pawlicki-Jullian, N., Pillon-Lequart, M., Boitel-Conti, M., Duong, H.X., Gontier, E., 2016. Hairy root cultures of *Panax vietnamensis*, a promising approach production ocotillol-type ginsenosides. Plant Cell Tiss. Org. Cult. 126 (1), 93–103.

Han, K.-L., Lee, Y., Song, J.-H., Hwang, Y.-S., Lee, W.S., Kim, S.-H., 2012. Enhanced production and secretion of rutin and GABA in immobilized cells of mulberry tree (*Morus bombycis* K.). Plant Cell Tiss. Org. Cult. 108, 513–520.

Hanafy, M.S., Matter, M.A., Asker, M.S., Rady, M.R., 2016. Production of indole alkaloids in hairy root cultures of *Catharanthus roseus* L. and their antimicrobial activity. South Afr. J. Bot. 105 (9), 9–18.

Harfi, B., Khelifi-Slaoui, M., Bekhouche, M., Benyammi, R., Hefferon, K., Makhzoum, A., Khelifi, L., 2016. Hyoscyamine production in hairy roots of three *Datura* species exposed to high-salt medium. In Vitro Cell. Dev. Biol. Plant 52 (1), 92–98.

Hidalgo, D., Steinmetz, V., Brossat, M., Tournier-Couturier, L., Cusido, R.M., Corchete, P., Palazon, J., 2017. An optimized biotechnological system for the production of centellosides based on elicitation and bioconversion of *Centella asiatica* cell cultures. Eng. Life Sci. 17 (4), 413–419.

Homaee, M.B., Ehsanpour, A.A., 2015. Physiological and biochemical responses of potato (*Solanum tuberosum*) to silver nanoparticles and silver nitrate treatments under *in vitro* conditions. Indian J. Plant Physiol. 20 (4), 353–359.

Huang, X., Yao, J.W., Zhao, Y.Y., Xie, D.F., Jiang, X., Xu, Z.Q., 2016. Efficient rutin and quercetin biosynthesis through flavonoids-related gene expression in *Fagopyrum tataricum* Gaertn. hairy root cultures with UV-B irradiation. Front. Plant Sci. 7, 63.

Ishihara, K., Hamada, H., Hirata, T., Nakajima, N., 2003. Biotransformation using plant cultured cells. J. Mol. Catal. B Enzym. 23, 145–170.

Jiang, Y.J., Piao, X.C., Liu, J.S., Jiang, J., Lian, Z.X., Kim, M.J., Lian, M.L., 2015. Bioactive compound production by adventitious root culture of *Oplopanax elatus* in balloon-type airlift bioreactor systems and bioactivity property. Plant Cell Tiss. Org. Cult. 123 (2), 413–425.

Jose, B., Silja, P.K., Pillai, D.B., Satheeshkumar, K., 2016. *In vitro* cultivation of hairy roots of *Plumbago rosea* L. in a customized reaction kettle for the production of plumbagin - an anticancer compound. Ind. Crops Prod. 87, 89–95.

Jyothishwaran, G., Seetharam, Y.N., 2008. Biotransformation of cholesterol to diosgenin by freely suspended and immobilised cells of *Dioscorea bulbifera* L. J. Asian Nat. Prod. Res. 10 (2), 139–145.

Kasparova, M., Martin, J., Tumova, L., Spilkova, J., 2017. Production of podophyllotoxin by plant tissue cultures of *Juniperus virginiana*. Nat. Prod. Comm. 12 (1), 101–103.

Kayani, W.K., Palazon, J., Cusido, R.M., Mirza, B.H., 2016. The effect of rol genes on phytoecdysteroid biosynthesis in *Ajuga bracteosa* differs between transgenic plants and hairy roots. RSC Adv. 6 (27), 22700–22708.

Keen, N.T., 1975. Specific elicitors of plant phytoalexin production: determinants of race specificity in pathogens? Science 187, 74–75.

Kong, L.S., Holtz, C., Nairn, C., Houke, H., Powell, W., Baier, K., Merkle, S., 2014. Application of airlift bioreactors to accelerate genetic transformation in American chestnut. Plant Cell Tiss. Org. Cult. 117 (1), 39–50.

Kumar, V., Sharma, N., Sood, H., Chauhan, R.S., 2016. Exogenous feeding of immediate precursors reveals synergistic effect on picroside-I biosynthesis in shoot cultures of *Picrorhiza kurroa* Royle ex Benth. Sci. Rep. 6, 29750.

Kurz, W.G.W., Paiva, N.L., Tyler, R.T., 1990. Biosynthesis of sanguinarine by elisitation of surface immobilized cells of *Papaver somniferum*. In: Nijkamp, H.J.J., Van Der Plas, L.H.W., Van Aartrijk, J. (Eds.), Progress in Plant Cellular and Molecular Biology. Proceedings of the VIIth International Congress on Plant Tissue and Cell Culture, Amsterdam, The Netherlands, 24–29 June, Kluwer Academic Publishers, Dordrecht/Boston/London, pp. 682–688.

Kusakari, K., Yokoyama, M., Inomata, S., Gozu, Y., Katagiri, C., Sugimoto, Y., 2012. Large-scale production of saikosaponins through root culturing of *Bupleurum falcatum* L. using modified airlift reactors. J. Biosci. Bioeng. 113 (1), 99–105.

Largia, M.J.V., Satish, L., Johnsi, R., Shilpha, J., Ramesh, M., 2016. Analysis of propagation of *Bacopa monnieri* (L.) from hairy roots, elicitation and Bacoside A contents of Ri transformed plants. World J. Microbiol. Biotechnol. 32 (8), 131.

Lee, E.J., Park, S.Y., Paek, K.Y., 2015. Enhancement strategies of bioactive compound production in adventitious root cultures of *Eleutherococcus koreanum* Nakai subjected to methyl jasmonate and salicylic acid elicitation through airlift bioreactors. Plant Cell Tiss. Org. Cult. 120 (1), 1–10.

Lee, S.Y., Bong, S.J., Kim, J.K., Park, S.U., 2016. Glucosinolate biosynthesis as influenced by growth media and auxin in hairy root cultures of kale (*Brassica oleracea* var. acephala). Emir. J. Food Agric. 28 (4), 277–282.

Levin, R., Tanny, G., 2004. Bioreactors as a low cost option for tissue culture, Low cost options for tissue culture technology in developing countries. In: Proceedings of a Technical Meeting Organized by the Joint FAO/IAEA Division of Nuclear Techniques in Food and Agriculture, pp. 47–54, Vienna, Austria.

Li, B., Wang, B.Q., Li, H.Y., Peng, L., Ru, M., Liang, Z.S., Yan, X.J., Zhu, Y.H., 2016. Establishment of *Salvia castanea* Diels f. tomentosa Stib. hairy root cultures and the promotion of tanshinone accumulation and gene expression with Ag+, methyl jasmonate, and yeast extract elicitation. Protoplasma 253 (1), 87–100.

Lindsey, K., 1985. Manipulation, by nutrient limitation, of the biosynthetic activity of immobilized cells of *Capsicum frutescens* Mill. cv. annuum. Planta 165, 126–133.

Lindsey, K., Yeoman, M.M., 1984a. The synthetic potential of immobilised cells of *Capsicum frutescens* Mill cv. annuum. Planta 162, 495–501.

Lindsey, K., Yeoman, M.M., 1984b. The viability and biosynthetic activity of cells of *Capsicum frutescens* Mill. cv. annuum immobilized in reticulate polyurethane. J. Exp. Bot. 35 (11), 1684–1696.

Liu, Y.K., Huang, L.F., Ho, S.L., Liao, C.Y., Liu, H.Y., Lai, Y.H., Yu, S.M., Lu, C.A., 2012. Production of mouse granulocyte-macrophage colony-stimulating factor by gateway technology and transgenic rice cell culture. Biotech. Bioeng. 109 (5), 1239–1247.

Lulu, T., Park, S.Y., Ibrahim, R., Paek, K.Y., 2015. Production of biomass and bioactive compounds from adventitious roots by optimization of culturing conditions of *Eurycoma longifolia* in balloon-type bubble bioreactor system. J. Biosci. Bioeng. 119 (6), 712–717.

Manero, F.J.G., Algar, E., Gomez, M.S.M., Sierra, M.D.S., Solano, B.R., 2012. Elicitation of secondary metabolism in *Hypericum perforatum* by rhizosphere bacteria and derived elicitors in seedlings and shoot cultures. Pharm. Biol. 50 (10), 1201–1209.

Mathew, A.J., Jayachandran, K., 2009. Production of scopadulcic acid B from *Scoparia dulcis* Linn. using a luffa sponge bioreactor. Plant Cell Tiss. Org. Cult. 98, 197–203.

Mavituna, F., Park, J.M., 1985. Growth of immobilised plant cells in reticulate polyurethane foam matrices. Biotechnol. Lett. 7 (9), 637–640.

Mavituna, F., 2004. Pre-formed carriers for cell immobilization. In: Fundamentals of Cell Immobilisation Biotechnology Focus on Biotechnology, vol. 8A, pp. 121–139.

Ming, Q., Su, C., Zheng, C., Jia, M., Zhang, Q., Zhang, H., Rahman, K., Han, T., Qin, L., 2013. Elicitors from the endophytic fungus *Trichoderma atroviride* promote *Salvia miltiorrhiza* hairy root growth and tanshinone biosynthesis. J. Exp. Bot. 64 (18), 5687–5694.

Mohammed Al-Oubaidi, H.K., Mohammed-Ameen, A.S., 2014. The effect of ($AgNO_3$) NPs on increasing of secondary metabolites of *Calendula officinalis* L. *in vitro*. Int. J. Pharm. Ther. 5 (4), 267–272.

Mulabagal, V., Tsay, H.-S., 2004. Plant cell cultures - An alternative and efficient source for the production of biological important secondary metabolites. Int. J. Appl. Sci. Eng. 2 (1), 29–48.

Naik, P.M., Al-Khayri, J.M., 2016. Impact of abiotic elicitors on *in vitro* production of plant secondary metabolites: a review. J. Adv. Res. Biotechnol. 1 (1), 7.

Nartop, P., 2005. Identifying the Effective Factors in Callus and Cell Cultures in *Rosmarinus Officinalis* L. and Secondary Metabolite Production (M.S. thesis). Ege University, Graduate School of Applied Science, Izmir, Turkey.

Nartop, P., 2013. Enhancing Secondary Metabolite Production in *Astragalus trojanus* Stev. Cell Suspension Cultures (Ph.D. thesis). Ege University, Graduate School of Applied Science, İzmir, Turkey.

Nartop, P., 2016a. Green sterilization of *Rosmarinus officinalis* L. stem surfaces with silver nanoparticles synthesized using *Rubia tinctorum* L. cell culture extracts. Iran. J. Sci. Technol. Trans. A: Sci. 1–4 (in press).

Nartop, P., 2016b. Use of biosynthetic silver nanoparticles in surface sterilization of *Pyracantha coccinea* stem explants. Pamukkale Univ. Muh Bilim Derg. PAJES–04809 (in press).

Nartop, P., Akay, Ş., Gürel, A., 2013. Immobilization of *Rubia tinctorum* L. suspension cultures and its effects on alizarin and purpurin accumulation and biomass production. Plant Cell Tiss. Org. Cult. 112 (1), 123–128.

Nartop, P., Gurel, A., Akgun, I.H., Bedir, E., 2015b. Astragaloside IV and cycloastragenol production capacity of *Astragalus trojanus* calli. Rec. Nat. Prod. 9 (1), 49–61.

Nartop, P., Gürel, A., Akgün, I.H., Bedir, E., 2015a. Astragaloside IV and cycloastragenol production of *Astragalus trojanus* Stev. *in vitro* conditions. Ind. J. Biotech. 14, 540–546.

Naziri, E., 2014. Effects of Different Stress Factors on Biomass and Secondary Metabolite Content in Madder (*Rubia Tinctorum* L.) Cell Suspension Cultures (Ph.D. thesis). Ege University, Graduate School of Applied Science, İzmir, Turkey.

O'Neill, K.M., Schilthuis, A.M., Leiter, C.A., Neihaus, K.M., Judge, N.A., Twiddy, E., O'Brien, A.D., Curtis, W.R., 2015. Scale-up of transgenic tobacco cells that express intimin of enterohemorrhagic *Escherichia coli* O157: H7 for use as a transitional platform for an oral cattle vaccine. In Vitro Cell. Dev. Biol. Plant 51 (3), 315–323.

Onrubia, M., Moyano, E., Bonfill, M., Cusidó, R.M., Goossens, A., Palazón, J., 2013. J. Plant Physiol. 170 (2), 211–219.

Osuna, L., Moyano, E., Mangas, S., Bonfill, M., Cusidó, R.M., Piñol, M.T., Zamilpa, A., Tortoriello, J., Palazón, J., 2008. Immobilization of *Galphimia glauca* plant cell suspensions for the production of enhanced amounts of Galphimine-B. Planta Med. 74 (1), 94–99.

Paek, K.Y., Murthy, H.N., Hahn, E.J., Zhong, J.J., 2008. Large scale culture of ginseng adventitious roots for production of ginsenosides. Adv. Biochem. Eng. Biotechnol. 113, 151–176.

Park, C.H., Martinez, B., 1994. Growth and production characteristics of permeabilized *Coleus blumei* cells in immobilized fed-batch culture. Plant Cell Rep. 13 (8), 459–463.

Patil, J.G., Ahire, M.L., Nitnaware, K.M., Panda, S., Bhatt, V.P., Kishor, P.B.K., Nikam, T.D., 2013. *In vitro* propagation and production of cardiotonic glycosides in shoot cultures of *Digitalis purpurea* L. by elicitation and precursor feeding. Appl. Microbiol. Biotechnol. 97 (6), 2379–2393.

Park, C.H., Zhao, S.C., Yee, H.J., Park, Y.E., Baska, T.B., Arasu, M.V., Al-Dhabi, N.A., Park, S.U., 2017. Comparison of different strains of *Agrobacterium rhizogenes* for hairy root induction and betulin and betulinic acid production in *Morus alba*. Nat. Prod. Comm. 12 (4), 479–482.

Park, J.H., Kim, C.H., Chung, I.S., Lee, S.W., 1995. Production of taxol from *in vitro* cell culture of *Taxus cuspidata*: international symposium on medicinal and aromatic plants. ACTA Acta Hortic. 390, 127–132.

Pelletier, S.W., 2001. The Nature and Definition of an Alkaloid, Alkaloids: Chemical and Biological Perspectives. Wiley-Interscience, New York, pp. 2–31 (Chapter 1).

Pena-Ramirez, Y.J., Juarez-Gomez, J., Gomez-Lopez, L., Jeronimo-Perez, J., Garcia-Shesena, I., Gonzalez-Rodriguez, J., Robert, M.L., 2010. Multiple adventitious shoot formation in Spanish Red Cedar (*Cedrela odorata* L.) cultured *in vitro* using juvenile and mature tissues: an improved micropropagation protocol for a highly valuable tropical tree species. In Vitro Cell. Dev. Biol. Plant 46 (2), 149–160.

Pras, N., 1990. Bioconversion of precursors occurring in plants and of related synthetic compounds. In: Nijkamp, H.J.J., Van Der Plas, L.H.W., Van Aartrijk, J. (Eds.), Progress in Plant Cellular and Molecular Biology. Proceedings of the VIIth International Congress on Plant Tissue and Cell Culture, Amsterdam, The Netherlands, 24–29 June, Kluwer Academic Publishers, Dordrecht/Boston/London, pp. 640–649.

Premjet, D., Tachibana, S., 2004. Production of podophyllotoxin by immobilized cell cultures of *Juniperus chinensis*. Pak. J. Biol. Sci. 7 (7), 1130–1134.

Ptak, A., Simlat, M., Kwiecien, M., Laurain-Mattar, D., 2013. *Leucojum aestivum* plants propagated in *in vitro* bioreactor culture and on solid media containing cytokinins. Eng. Life Sci. 13 (3), 261–270.

Raei, M., Angaji, S.A., Omidi, M., Khodayari, M., 2014. Effect of abiotic elicitors on tissue culture of *Aloe vera*. Int. J. Biosci. 5 (1), 74–81.

Rao, R.S., Ravishankar, G.A., 2002. Plant cell cultures: chemical factories of secondary metabolites. Biotech. Adv. 20, 101–153.

Reis, R.V., Borges, A.P.P., Chierrito, T.P.C., de Souto, E.R., de Souza, L.M., Iacomini, M., de Oliveira, A.J.B., Goncalves, R.A.C., 2011. Establishment of adventitious root culture of *Stevia rebaudiana* Bertoni in a roller bottle system. Plant Cell Tiss. Org. Cult. 106 (2), 329–335.

Ricigliano, V., Kumar, S., Kinison, S., Brooks, C., Nybo, S.E., Chappell, J., Howarth, D.G., 2016. Regulation of sesquiterpenoid metabolism in recombinant and elicited *Valeriana officinalis* hairy roots. Phytochemistry 125, 43–53.

Ro, D.K., Paradise, E.M., Ouellet, M., 2006. Production of the antimalarial drug precursor artemisinic acid in engineered yeast. Nature 440 (13), 940–943.

Ru, M., An, Y.Y., Wang, K.R., Peng, L., Li, B., Bai, Z.Q., Wang, B.Q., Liang, Z.S., 2016. *Prunella vulgaris* L. hairy roots: culture, growth, and elicitation by ethephon and salicylic acid. Eng. Life Sci. 16 (5), 494–502.

Sathyanarayana, B.N., Varghese, D.B., 2007. Instrumentation, Plant Tissue Culture, Practices and New Experimental Protocols. International Publishing, House Pvt. Ltd, New Delhi, India.

Seigler, D.S., 2002. Plant Secondary Metabolism. Second Printing. Kluwer Academic Publishers, pp. 8–9.

Sakamoto, K., Lida, K., Sawamura, K., Hijaro, K., Asada, Y., Yoshikawa, T., Furuya, T., 1993. Effects of nutrients on anthocyanin production in cultured cells of *Aralia cordata*. Phytochemistry 33 (2), 357–360.

Shandil, A.S., Tuia, V.S., 2015. Micropropagation of breadfruit (*A. altilis*) enhanced using a bioreactor system. In: Jaenicke, H., Ashmore, S.E., Dulloo, M.E., Guarino, L., Taylor, M. (Eds.), XXIX International Horticultural Congress on Horticulture: Sustaining Lives, Livelihoods and Landscapes (IHC2014)IV International Symposium on Plant Genetic Resources Book Series: Acta Horticulturae, vol. 1101, pp. 159–163.

Simic, S.G., Tusevski, O., Maury, S., Delaunay, A., Joseph, C., Hagege, 2014. Effects of polysaccharide elicitors on secondary metabolite production and antioxidant response in *Hypericum perforatum* L. shoot cultures. Sci. World J.:609649 10 pp.

Singh, B., Sharma, R.A., 2016. Yield enhancement of phytochemicals by *Azotobacter chroococcum* biotization in hairy roots of *Arnebia hispidissima*. Ind. Crops Prod. 81, 169–175.

Srivastava, M., Sharma, S., Misra, P., 2016a. Elicitation based enhancement of secondary metabolites in *Rauwolfia serpentina* and *Solanum khasianum* hairy root cultures. Pharmacogn. Mag. 12 (46), 315–320.

Srivastava, S., Conlan, X.A., Adholeya, A., Cahill, D.M., 2016b. Elite hairy roots of *Ocimum basilicum* as a new source of rosmarinic acid and antioxidants. Plant Cell Tiss. Org. Cult. 126 (1), 19–32.

Stanly, C., Bhatt, A., Keng, C.L., 2010. A comparative study of *Curcuma zedoaria* and *Zingiber zerumbet* plantlet production using different micropropagation systems. Afr. J. Biotech. 9 (28), 4326–4333.

Stobiecki, M., Kachlicki, P., 2006. Isolation and identification of flavonoids. In: Grotewold, E. (Ed.), The Science of Flavonoids. Springer, United States of America, pp. 47–69.

Sudha, C.G., 2014. Progress and prospects of *Agrobacterium rhizogenes* mediated transformed root cultures for the production of therapeutically active secondary metabolites. In: Govil, J.N., Kumar, P.A. (Eds.), Recent Progress in Medicinal Plants. Biotechnology and Genetic Engineering II, vol. 39. Studium Press LLC, USA (Chapter 5).

Sudhakar Johnson, T., 1993. Studies in Production of Capsaicin in Immobilized Cells and Placental Tissues of *Capsicum Annuum* and *C. Frutescens* (Ph.D. thesis). University of Mysore, Mysore.

Sudhakar Johnson, T., Ravishankar, G.A., 1996. Precursor biotransformation in immobilized placental tissues of *Capsicum frutescens* Mill: I. Influence of feeding intermediate metabolites of the capsaicinoid pathway on capsaicin and dihydrocapsaicin accumulation. J. Plant Physiol. 147 (5), 481–485.

Sudhakar Johnson, T., Ravishankar, G.A., Venkataraman, L.V., 1996. Biotransformation of ferulic acid and vanillylamine to capsaicin and vanillin in immobilized cell cultures of *Capsicum frutescens*. Plant Cell Tiss. Org. Cult. 44, 117–121.

Sudhakar Johnson, T., Ravishankar, G.A., Venkataraman, L.V., 1990. *In vitro* capsaicin production by immobilized cells and placental tissues of *Capsicum annuum* L. grown in liquid medium. Plant Sci. 70, 223–229.

Sudhakar Johnson, T., Ravishankar, G.A., Venkataraman, L.V., 1991. Elicitation of capsaicin production in freely suspended cells and immobilized cell cultures of *Capsicum frutescens* Mill. Food Biotechnol. 5, 197–205.

Sukito, A., Tachibana, S., 2016. Effect of methyl jasmonate and salicylic acid synergism on enhancement of bilobalide and ginkgolide production by immobilized cell cultures of *Ginkgo biloba*. Bioresour. Bioprocess. 3 (24), 1–11.

Tashackori, H., Sharifi, M., Chashmi, N.A., Safaie, N., Behmanesh, M., 2016. İnduced-differential changes on lignan and phenolic acid compounds in *Linum album* hairy roots by fungal extract of *Piriformospora indica*. Plant Cell Tiss. Org. Cult. 127 (1), 187–194.

Thwe, A., Arasu, M.V., Li, X.H., Park, C.H., Kim, S.J., Al-Dhabi, N.A., Park, S.U., 2016. Effect of different *Agrobacterium rhizogenes* strains on hairy root induction and phenylpropanoid biosynthesis in Tartary Buckwheat (*Fagopyrum tataricum* Gaertn). Front. Microbiol. 7, 318.

Tonelli, M., Pellegrini, E., D'Angiolillo, F., Petersen, M., Nali, C., Pistelli, L., Lorenzini, G., 2015. Ozone-elicited secondary metabolites in shoot cultures of *Melissa officinalis* L. Plant Cell Tiss. Org. Cult. 120 (2), 617–629.

Verma, P., Sharma, A., Khan, S.A., Shanker, K., Mathur, A.K., 2015. Over-expression of *Catharanthus roseus* tryptophan decarboxylase and strictosidine synthase in rol gene integrated transgenic cell suspensions of *Vinca minor*. Protoplasma 252 (1), 373–381.

Wang, C.H., Zheng, L.P., Tian, H., Wang, J.W., 2016a. Synergistic effects of ultraviolet-B and methyl jasmonate on tanshinone biosynthesis in *Salvia miltiorrhiza* hairy roots. J. Photochem Photobiol. B Biol. 159, 93–100.

Wang, Y.J., Shen, Y., Shen, Z., Zhao, L., Ning, D.L., Jiang, C., Zhao, R., Huang, L.Q., 2016b. Comparative proteomic analysis of the response to silver ions and yeast extract in *Salvia miltiorrhiza* hairy root cultures. Plant Physiol. Biochem. 107, 364–373.

Wee, S.L., 2015. The Effects of Elicitors and Precursor on in Vitro Cultures of *Sauropus androgynus* for Sustainable Metabolite Production and Antioxıdant Capacity Improvement (Ph.D. thesis). University of Nottingham.

Wu, C.F., Karioti, A., Rohr, D., Bilia, A.R., Efferth, T., 2016. Production of rosmarinic acid and salvianolic acid B from callus culture of *Salvia miltiorrhiza* with cytotoxicity towards acute lymphoblastic leukemia cells. Food Chem. 201, 292–297.

Wu, S.Q., Lian, M.L., Gao, R., Park, S.Y., Piao, X.C., 2011. Bioreactor application on adventitious root culture of *Astragalus membranaceus*. In Vitro Cell. Dev. Biol. Plant 47 (6), 719–724.

Wu, X., Yang, F., Piao, X.C., Li, K.H., Lian, M.L., Dai, Y., 2015. High-frequency plantlet regeneration by somatic embryogenesis from mature zygotic embryos of onion. N.Z. J. Crop. Hortic. Sci. 43 (4), 249–260.

Yağcı, C., Toker, M.C., Toker, G., 2008. Bitki doku kültürü yoluyla üretilen flavonoidler. Türk Bilimsel Derlemeler Derg. 1 (1), 47–58.

Yamada, Y., Fujita, Y., 1983. Production of Useful Compounds in Culture. In: Evans, D.A., Sharp, W.R., Ammirato, P.V., Yamada, Y. (Eds.), Handbook of Plant Cell Culture. Techniques for Propagation and Breeding, vol. 1. Macmillan Publishing Co, New York (Chapter 23).

Yan, Y., Zhang, S., Zhang, J., Ma, P., Duan, J., Liang, Z., 2014. Effect and mechanism of endophytic bacteria on growth and secondary metabolite synthesis in *Salvia miltiorrhiza* hairy roots. Acta Physiol. Plant 36 (5), 1095–1105.

Yan, Y.J., 2005. Metabolic engineering of anthocyanin biosynthesis in *Escherichia coli*. Appl. Environ. Microb. 71 (7), 3617–3623.

Yang, D., Huang, Z., Xing, B., Jin, W., Yan, X., Guo, Z., Liang, Z., 2016. Regulation of folic acid on phenolic acids production in *Salvia miltiorrhiza* hairy roots. Plant Cell Tiss. Org. Cult. 127 (1), 175–185.

Yeoman, M.M., Yeoman, C.L., 1996. Manipulating secondary metabolism in cultured plant cells. New Phytol. 134, 553–559.

Zafar, H., Ali, A., Ali, J.S., Ihsan, U., Haq, I.U., Zia, M., 2016. Effect of ZnO nanoparticles on *Brassica nigra* seedlings and stem explants: growth dynamics and antioxidative response. Front. Plant Sci. 7, 535.

Zaka, M., Abbasi, B.H., 2016. Effects of bimetallic nanoparticles on seed germination frequency and biochemical characterisation of *Eruca sativa*. IET Nanobiotechnol. http://dx.doi.org/10.1049/iet-nbt.2016.0004.

Zaker, A., Sykora, C., Gössnitzer, F., Abrishamchi, P., Asili, J., Mousavi, S.H., Wawrosch, C., 2015. Effects of some elicitors on tanshinone production in adventitious root cultures of *Perovskia abrotanoides* Karel. Indus. Crops Prod. 67, 97–102.

Zamboni, A., Vrhovsek, U., Kassemeyer, H.-H., Mattivi, F., Velasco, R., 2006. Elicitor-induced resveratrol production in cell cultures of different grape genotypes (*Vitis* spp.). J. Grapevine Res. 45 (2), 63–68.

Zhang, B., Zheng, L.P., Ping, Y.L., Wen Wang, J., 2013. Stimulation of artemisinin production in *Artemisia annua* hairy roots by Ag-SiO_2 core-shell nanoparticles. Curr. Nanosci. 9 (3), 363–370 (8).

Zhou, P.F., Yang, J.Z., Zhu, J.H., He, S.J., Zhang, W.J., Yu, R.M., Zi, J.C., Song, L.Y., Huang, X.S., 2015. Effects of beta-cyclodextrin and methyl jasmonate on the production of vindoline, catharanthin, and ajmalicine in *Catharanthus roseus* cambial meristematic cell cultures. Appl. Microb. Biotechnol. 99 (17), 7035–7045.

Zorrilla-López, U., Masip, G., Arjó, G., Bai, C., Banakar, R., Bassie, L., Berman, J., Farré, G., Miralpeix, B., Pérez-Massot, E., Sabalza, M., Sanahuja, G., Vamvaka, E., Twyman, R.M., Christou, P., Zhu, C., Capell, T., 2013. Engineering metabolic pathways in plants by multigene transformation. Int. J. Dev. Biol. 57, 565–576.

FURTHER READING

Masoumian, M., Arbakariya, A., Syahida, A., Maziah, M., 2011. Effect of precursors on flavonoid production by *Hydrocotyle bonariensis* callus tissues. Afr. J. Biotech. 10 (32), 6021–6029.

Mathur, S., Goswami, A., 2012. Effect of precursor β-phenylalanine on production of flavonoids of *Maytenus emarginata* in vitro. Int. J. Sci. Res. 3 (7), 333–335.

Nartop, P., Gurel, A., 2016. Immobilization Systems in Plant Cell Cultures: An Approach to Enhance Biomass Accumulation and Secondary Metabolite Production. Recent Progress in Medicinal Plants, Biotechnology and Genetic Engineering II, vol. 39. Studium Press, LLC, USA. 1-933699-99-X, Series ISBN: 0-9656038-5-7 (Chapter 7).

Saw, N.M.M.T., Riedel, H., Kütük, O., Ravichandran, K., Smetanska, I., 2010. Effect of elicitors and precursors on the synthesis of anthocyanin in grape *Vitis vinifera* cell cultures. Energy Res. J. 1 (2), 189–192.

Chapter 10

Metabolic Responses of Plants Upon Different Plant–Pathogen Interactions

Miguel A. Matilla
Estación Experimental del Zaidín, Consejo Superior de Investigaciones Científicas, Granada, Spain

1. INTRODUCTION

The great majority of plant species are regularly attacked by pathogens including viruses, fungi, and bacteria. Some of these invading organisms colonize the rhizosphere, the area of soil surrounding the root system that supports high levels of microbial activity (Matilla and Krell, 2017, and references therein). The number and diversity of microorganisms existing in a particular rhizosphere is a reflection of both the complex interactions between rhizosphere-colonizing microorganisms and selective pressures derived from colonized plants (Raaijmakers et al., 2009; Laren et al., 2016; Bandyopadhyay et al., 2017; Lugtenberg et al., 2017). Hence the influence of plant species and their developmental stage on rhizosphere-associated microbial communities has been documented in several studies (Mougel et al., 2006; Micallef et al., 2009; Inceoglu et al., 2013).

To protect themselves from microbial pathogens, plants have developed different mechanisms. Among them is the secretion by plant roots to the rhizosphere in the form of root exudates different biologically active secondary metabolites, some of which may act against pathogens by different mechanisms (Badri et al., 2009; Doornbos et al., 2012; van Dam and Bouwmeester, 2016; Massalha et al., 2017). Thus, plant phytoalexins (PHALs) are low-molecular-mass antimicrobial secondary metabolites synthesized de novo after a pathogen attack (Lanoue et al., 2010; Ahuja et al., 2012; Baetz and Martinoia, 2014). Contrary to PHALs, phytoanticipins are described as low-molecular-mass antimicrobial compounds that are either present in plants before being challenged by microorganisms or produced after infection from preexisting precursors (van Etten et al., 1994; Pedras and Yaya, 2015). Alternatively to the synthesis and secretion of bioactive secondary metabolites, plants have evolved additional sophisticated defense strategies to perceive attack and to translate this perception into an appropriate and effective defense response. Thus, plants have evolved two branches in their immune system to defend themselves against pathogen infections. The primary innate immune response is triggered by the detection of evolutionarily conserved microbial- or pathogen-associated molecular patterns (MAMPs or PAMPs, respectively) and is known as PAMP-triggered immunity (PTI). The second plant innate immune response is triggered by the recognition of specific pathogen effector proteins and is referred to as effector-triggered immunity (ETI) (Jones and Dangl, 2006; Gao et al., 2014; Cui et al., 2015). The perception of PAMPs or MAMPs such as flagellin (Flg), lipopolysaccharides (LPS), peptidoglycans (PGNs), or elongation factor Tu (Macho and Zipfel, 2014) is carried out by pattern recognition receptors (PRRs) present at the plant cell surface (Boller and Félix, 2009; Dalio et al., 2017; Ranf, 2017). These PAMPs/MAMPs-induced defense responses are regulated by a network of interconnecting signal transduction pathways in which the hormones salicylic acid (SA), jasmonic acid (JA), and ethylene (ET) play key roles (Robert-Seilaniantz et al., 2011; Pieterse et al., 2012, 2014; Yang et al., 2015). Thus, in response to pathogen infection, plants either accumulate or increase their sensitivity to these plant hormones, resulting in the activation of distinct sets of defense-related genes, as confirmed by the phenotypic characterization of mutant and transgenic plants (revised in Matilla-Vazquez and Matilla, 2014; Pieterse et al., 2014; Bigeard et al., 2015). Depending on the signaling pathway activated, the induction of resistance can be divided into (1) induced systemic resistance (ISR), generally activated by nonpathogenic microorganisms and mediated by JA and ET, and (2) systemic acquired resistance (SAR), an SA-dependent signaling cascade mainly triggered by pathogenic microorganisms (Pieterse et al., 2001; Derksen et al., 2013; Zhang et al., 2015). Additionally, studies in the last decade have identified diverse mechanisms whereby plant pathogens can manipulate auxin biosynthesis, signaling cascades, and transport pathways to promote host susceptibility (Mutka et al., 2013; Naseem et al., 2015 and references cited therein). For example, the phytopathogenic bacterium *Pseudomonas syringae* alters auxin physiology of *Arabidopsis* plants through the synthesis of indole acetic acid and the production of virulence factors that alter auxin responses in host cells (Cui et al., 2013).

Plant Metabolites and Regulation Under Environmental Stress. http://dx.doi.org/10.1016/B978-0-12-812689-9.00010-8

In this chapter, recent progress focused on unraveling plant–pathogen interactions is reviewed, with special attention on the role of 14-3-3 proteins, oxidative-burst, Ca^{2+} signaling, and root exudates in pathogenicity processes. This field is of great relevance for plant health and has notable socioeconomic and environmental implications.

2. PLANT–PATHOGEN INTERACTIONS

2.1 Recognition Process

Recognition of PAMPs/MAMPs by PRRs represents a crucial layer of innate immunity in plants (Choi and Klessig, 2016; Boutrot and Zipfel, 2017; Dalio et al., 2017). Plants are constantly exposed to a wide range of potential pathogenic microorganisms, but, in response to these pathogens, plants can activate a set of complicated signaling pathways that can finally result in: (1) the accumulation of antimicrobial compounds (e.g., PHAL), (2) induction of plant resistance genes, (3) synthesis of pathogenesis-related proteins (PRP), and induction of systemic resistance (SR) (Moghaddam et al., 2016). To activate these responses, individual plant cells are capable of recognizing pathogens and turning on an effective defense response consisting of a combination of preformed defenses and an innate immune system (Fig. 10.1). These immune responses are mainly achieved through the specific interaction between plant-derived protein receptors (PPR or PRR), mostly localized in the plasma membrane (Nürnberger et al., 2004), and the corresponding pathogen elicitor or a virulence protein (i.e., MAMPs/PAMPs; Jones and Dangl, 2006; Trdá et al., 2015; Boutrot and Zipfel, 2017). In addition, certain proteins interact with PRRs forming immune receptor complexes necessary for a normal MAMP/PAMP perception and transduction of the signal (Macho and Zipfel, 2014; Macho et al., 2014). Thus, upon MAMP/PAMP perception by its cognate PPR, plants activate their innate defense responses (Gómez-Gómez and Boller, 2000; Kim et al., 2013; Trdá et al., 2015). The first identified MAMP/receptor pair was that of Flg and its cognate receptor Flg sensing 2 (FLS2). The characterization of the FLS2 receptor revealed that (1) it has an N-terminal signal peptide, 28 leucine-rich repeats, a transmembrane domain, and a cytoplasmic kinase domain (Gómez-Gómez and Boller, 2000); (2) it interacts with a 22-amino acid motif of the bacterial Flg (flg22; Chinchilla et al., 2007); and (3) mutants deficient in *FLS2* show enhanced susceptibility of bacterial pathogens (Zipfel et al., 2004). Another well-characterized PRR is EFR (EF-Tu receptor), a receptor that recognizes the minimal epitope of the elongation factor Tu (Zipzel et al., 2006).

In addition to ET, JA, and SA, some studies indicate that brassinosteroids (BRs) are also implicated in plant responses to pathogen attack (i.e., resistance against cucumber mosaic virus in *Arabidopsis*; Nakashita et al., 2003; Zhang et al., 2015). However, although a rapid progress was achieved in elucidating the signaling pathway of BRs in *Arabidopsis* (Guo et al., 2013),

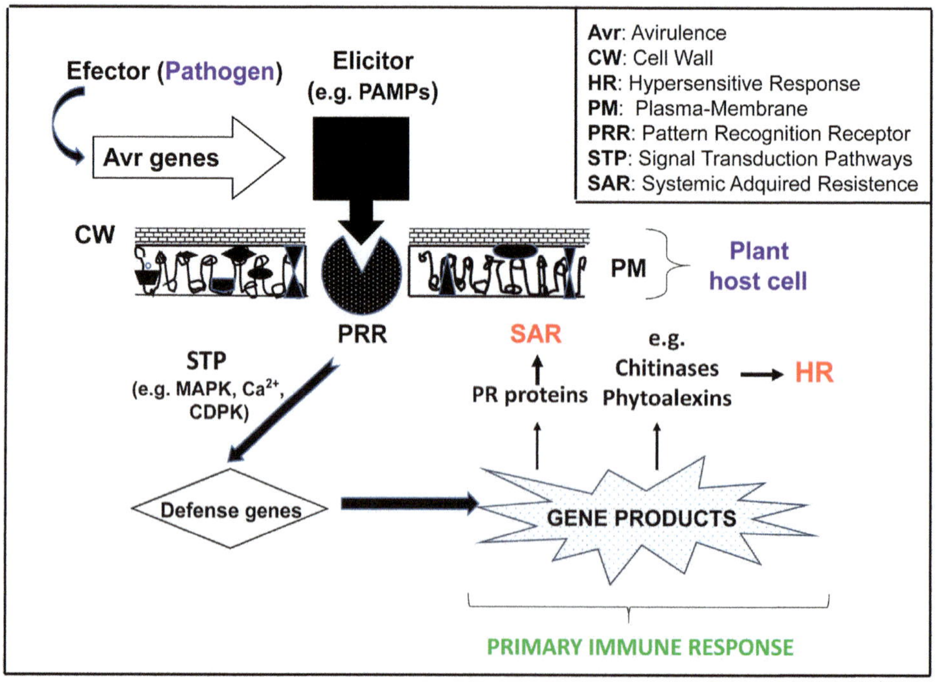

FIGURE 10.1 Overview of defense signaling pathways activated by the recognition of pathogen elicitors and effectors.

their involvement in plant defense is currently not fully understood. However, the receptor-like kinase (RLK) BAK1 (revised in Kim et al., 2013) has been found to be involved in the regulation of BR-dependent growth, development, innate immunity, and cell death control by directly interacting with the BR receptor BRI1 and immune receptors such as FLS2 (He et al., 2013 and references therein). Importantly, mutants deficient in *BAK1* are impaired in responses mediated by BRs and PAMPs/MAMPs such as Flg (revised in Postel and Kemmerling, 2009). BAK1 is negatively regulated by a specific protein phosphatase 2A, an isoform whose activity is inhibited upon PAMP perception (Segonzac et al., 2014).

Finally, damage-associated molecular patterns (DAMPs) are also perceived by PRRs in plants. In contrast to MAMPs/PAMPs, DAMPs are plant endogenous elicitors generated as a result of cell damage and that are able to initiate and perpetuate innate immune responses (revised by Choi and Klessig, 2016; Boutrot and Zipfel, 2017). Thus, for example, to infringe the structural barriers of plant tissues, the lytic activity of many plant pathogens can release DAMPs such as cell wall (CW) fragments, cutin monomers, and peptides (Boller, 2005; Huffaker et al., 2006). These molecules generally emerge in the apoplast and serve as danger signals to induce innate immunity (revised in Muthamilarasan and Prasad, 2013). As described earlier for MAMPs/PAMPs, DAMPs appear to activate innate immune signaling via BAK1 and BKK1 (closest paralog of BAK1) (Choi and Klessig, 2016).

2.2 Elicitors and Plant Innate Immunity

Although plants lack the circulating immune cells found in vertebrates, they possess an innate immune system that detects and limits pathogen colonization (Asai and Shirasu, 2015). Thus, plants have evolved an inducible immune system to face pathogen threats (Jones and Dangl, 2006). Elicitors (i.e., MAMPs/PAMPs, DAMPs) are defined as extra- or intracellular molecules from several sources, including bacteria, fungi, viruses, plant CW components, as well as chemicals released by plants upon pathogen attack (Postel and Kemmerling, 2009; Newman et al., 2013; Boutrot and Zipfel, 2017). Originally, the term elicitor was used for molecules capable of inducing the production of PHAL but it is currently used for those compounds that stimulate any type of plant defense response (Boller, 1995; Nürnberger, 1999; Zhao et al., 2005). MAMP/PAMP-triggered defense plays a key role in the basal resistance of plants and prevents the attack of phytopathogens. Molecules shown to possess elicitor activities include Flg, polysaccharides, glycoproteins, lipids, LPS, oligosaccharides, and enzymes (Boller and He, 2009; Zipfel, 2008). Within them, poly- and oligosaccharides are the best studied signal molecules in elicitor signal transduction pathways, but the molecular mechanisms that are used by plants to recognize and transduce oligosaccharide-mediated signals remain unclear (Trouvelot et al., 2014). However, five steps are currently known to intervene in this process: (1) recognition of oligosaccharide signals by the PRR on the plant cell membrane, (2) signal transfer and magnification, (3) activation of target genes and accumulation of responding proteins, (4) induction of defense-related secondary metabolites, and (5) defense reaction (Yin et al., 2016, and references therein).

Elicitors act as signal compounds at low concentrations, providing information for the plant to trigger a defense response. Unlike elicitors, toxins generally act at high concentrations and affect the plant detrimentally without activating plant metabolism (Boller, 1995). As indicated in Section 2.1, the elicitor is perceived by its cognate PRR situated in the plasma membrane, initiating thus a signal transduction cascade that finally results in the triggering of the defense response in the host plant (Fig. 10.1). In others words, MAMP/PAMP elicitors can be considered as evolutionarily conserved and essential building blocks of microbial surfaces that directly bind to PRRs and induce defense responses (Thomma et al., 2011). All currently known plant PRRs are plasma membrane–localized RLKs or receptor-like proteins (RLPs), and their inactivation leads to loss of immunity. RLKs possess a cytosolic kinase domain and a ligand-binding domain. After signal perceptions, these PRRs normally associate with additional proteins in the plasma membrane to form large protein complexes (Newman et al., 2013; Bohm et al., 2014; Zipfel, 2014). Importantly, the combination of genetic approaches, functional genomics, and biochemical analyses allowed the identification of regulatory mechanisms involved in the transcription of *PRR* genes and PRR biogenesis, providing therefore insights into the composition of PRR complexes at the plasma membrane and highlighting the roles of long-known components involved in immune signaling (Segonzac and Zipfel, 2011).

Plant perception of MAMPs/PAMPs triggers a phosphorylation relay leading to PTI (Monaghan et al., 2014). Data describing molecular interactions and protein modifications that occur between PRRs and their regulatory proteins have provided important mechanistic insight into how plants avoid infection and achieve immunity (Monaghan and Zipfel, 2012). The *RLK* and *RLP* gene families have greatly expanded in plants compared with mammals, indicating a greater MAMP/PAMP sensing capacity, possibly to compensate for the lack of adaptive immunity (Spoel and Dong, 2012). As described earlier, pathogens can be perceived by two different recognition systems. The first phase of defense induction is the so-called PAMP/MAMP-triggered immunity (PTI/MTI), or basal resistance (Zipfel, 2014; Bigeard et al., 2015), whereas the second phase is known as effector-triggered immunity (ETI; Jones and Dangl, 2006). Both systems are accompanied by a set of induced defenses that usually repel pathogen attacks. Data describing some characterized elicitor receptors can also be found in Table 10.1 from

TABLE 10.1 Known Phytoalexins Produced by Different Plant Families

Plant Families	Types of Phytoalexins
Amaryllidaceae	Flavans
Brassicaceae (Cruciferae)	Indole phytoalexins, sulfur-containing phytoalexins
Compositae	Polyacetylenes
Convolvulaceae	Furanosesquiterpenes
Chenopodiaceae	Flavanones, isoflavones
Euphorbiaceae	Diterpenes
Leguminosae	Isoflavones, pterocarpans, phaseolin, glyceolin, furanoacetylenes, stilbenes, pterocarpens
Linaceae	Phenylpropanoids
Malvaceae	Terpenoids
Moraceae	Furanopterocarpans
Orchidaceae	Dihydrophenanthrenes, iterpenoids, momilactones, oryzalexins, zealexins, phytocassanes, kauralexins
Poaceae	Diterpenoids, oryzalexins, zealexines, phytocassanes, kauralexins, deoxyanthocyanidins, flavanones, phenylamides
Rosaceae	Biphenyls, dibenzofurans
Solanaceae	Phenylpropanoid-related compounds, steroid glycoalkaloids, sesquiterpenoids, coumarins, polyacetylenic derivatives
Rutaceae	Methylated phenolic compounds
Umbelliferae	Polyacetylenes, phenolics, 6-methoxymellein
Vitaceae	Stilbenes

Adapted from Jeandet, P., Hébrard, C., Marie-Alice Deville, M.-A., Cordelier, S., Dorey, S., Aziz, A., Crouze, J., 2014. Deciphering the role of phytoalexins in plant-microorganism interactions and human health. Molecules 19, 18033–18056 and Jeandet, P., 2015. Phytoalexins: current progress and future prospects. Molecules 20, 2770–2774 with permission.

Zhao et al. (2005). In many cases, PTI prevents further pathogen colonization. However, some pathogens can overcome PTI by producing effectors that interfere with defense signaling. These effectors can trigger ETI, which is highly pathogen specific and involves the activation of further resistance (*R*) genes often accompanied by SAR (Dangl and Jones, 2001; Chisholm et al., 2006; Jones and Dangl, 2006). Thus R proteins activate ETI, a rapid response that amplifies PTI signaling events and limits pathogen spread (Gururani et al., 2012; Dangl et al., 2013). At this point, it is important to highlight that the first *R* gene cloned from a plant was *Pto*, which encodes an intracellular Ser/Thr protein kinase (Martin et al., 1993).

Both PTI and ETI have been shown to share many signaling elements (Dangl et al., 2013). However, ETI induces these defenses more rapidly and intensely than PTI and is considered an accelerated and amplified PTI/MTI response that, in most cases, leads to localized cell death known as the hypersensitive response (HR). Current hypothesis indicates that ETI has evolved from PTI to specifically recognize, mediated by R proteins, pathogen type III secreted effector (T3SEs). Thus although T3SEs block PTI, they show avirulence functions by triggering ETI (Hou et al., 2011). HR is programmed cell death (PCD) of plant cells surrounding the site of infection, which prevents further spread of the pathogen in plant tissue (Coll et al., 2011). HR is frequently accompanied by the cellular production of reactive oxygen species (ROS). Elicitors that trigger PTI and ETI can be classified as biotic and abiotic (Fig. 10.2). Biotic elicitors of pathogen origin are considered as exogenous elicitors, whereas those produced by the plant as a consequence of the interaction with the pathogenic organisms are considered as endogenous (Yamaguchi and Huffaker, 2011). General elicitors are able to trigger defense responses in both host and nonhost plants, whereas race-specific elicitors induce defense responses leading to disease resistance only in specific hosts. Elicitors of biological origin, especially those that induce SR in plants, are being intensively investigated due to their great potential for crop protection and environmental compatibility (Wiesel et al., 2014 and references therein). However, new strategies to combat plant diseases include the design of synthetic elicitors, small drug-like molecules that are able to induce defense responses in plants (Bektas and Eulgem, 2015, and references therein). These synthetic elicitors

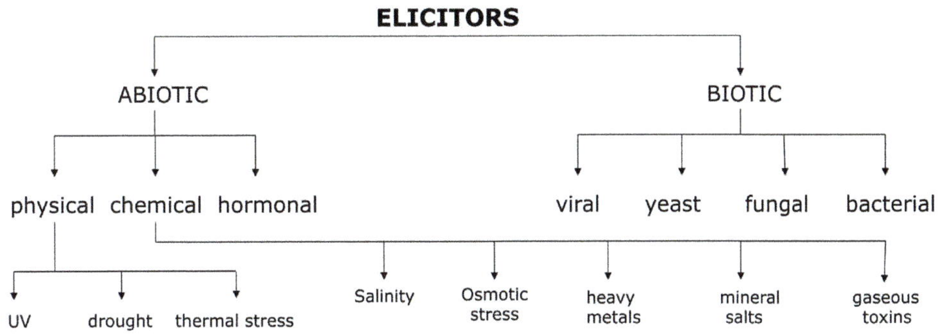

FIGURE 10.2 Classification of elicitors responsible for triggering plant immunity.

can be environment friendly alternatives to conventional chemical pesticides since they can protect crops from diseases without being harmful for the environment, farmers, and consumers.

An overlapping set of downstream immune responses results from the PTI/ETI continuum (Thomma et al., 2011). These responses include the activation of multiple MAMP-induced signaling pathways involving mitogen-activated protein kinases (MAPKs), production of ROS [also called the oxidative burst (OB)] and reactive nitrogen species such as nitric oxide (NO), alterations in the plant CW, synthesis of antimicrobial compounds (e.g., PHAL), alterations in the synthesis and sensitivity to defense hormones (e.g., SA, JA, and ET), and the synthesis of PRP (e.g., defensins and hydrolytic enzymes such as β-1,3-glucanases and chitinases) (Jones and Dangl, 2006; Zipfel, 2009; Stael et al., 2015; Dempsey and Klessig, 2017, and references therein). Some of these responses will be considered here.

2.2.1 Elicitors in Bacteria

Bacterial MAMPs (bMAMPs) are generally essential molecules for bacteria that can be recognized by plants (Ausubel, 2005; Trdá et al., 2015). Plant perception of bMAMPs activates diverse signaling events including the generation of Ca^{2+} signals and OB, synthesis of hormones, and the activation of phosphorylation cascades (Zipfel, 2014). Well-studied bMAMPs include PGN, LPS, or flg22 and elf18 peptides derived from the bacterial Flg and elongation factor Tu (EF-Tu), respectively (Albert, 2013; Newman et al., 2013; Gao et al., 2014 and references therein). Importantly, plants can detect bMAMPs at subnanomolar levels.

The bacterial flagellum is a complex nanomachine that provides motility and often increases adhesion of bacteria to their hosts. Flg is the main structural component of bacterial flagella, and the Flg peptide flg22 is the best known bacterial MAMP. This bacterial elicitor is a highly conserved domain in the N-terminal end of Flg that is constituted by 22 amino acids. The cognate receptor of flg22 is FLS2, a PRR belonging to the RLK family (revised in Newman et al., 2013). The specific interaction between flg22 with FLS2 in *Arabidopsis* was first demonstrated by Chinchilla et al. (2006). Thus, in this model plant, flg22 induces the activity of multiple Ca^{2+}-dependent protein kinases (CDPKs). Recently, it was found that glycosylation of the most externally located sites on the surface of the Flg is required for virulence in *P. syringae* pv. *tabaci* 6605 (Taguchi et al., 2010).

Bacterial translation EF-Tu is one of the most abundant proteins in the bacterial cells (Furano, 1975). Although it is not a surface exposed protein, plants recognize the N-terminal end of EF-Tu. Thus, an *N*-acetylated peptide having the first 18 amino acids (elf18) of this elongation factor is recognized by plant hosts triggering ion fluxes, Ca^{2+} influx, and MAPK activation (Zipfel et al., 2006). The PRR EFR, a close homolog of FLS2, has been identified as the receptor of EF-Tu, and it has only been found to elicit innate immunity in members of the family Brassicaceae (Zipfel et al., 2006). Thus mutants of *Arabidopsis* defective in *efr* are more susceptible to pathogen attack (i.e., *Agrobacterium tumefaciens*).

PGN, a molecule never found in eukaryotes, is an essential and unique membrane envelope component of all bacteria, making it an excellent target for the eukaryotic innate immune system. PGNs from both Gram-positive and Gram-negative bacteria were found to act as elicitors of plant innate immunity in *Arabidopsis*, tobacco, and rice (Gust et al., 2007; revised in Gust, 2015; Newman et al., 2013). Although the exact fragment size required for the immunogenic activity still remains to be determined (Gust, 2015), it has been shown that muropeptides from Gram-positive bacteria are more effective in inducing defense responses than the intact PGN molecule (Erbs et al., 2008). All plant receptors described for PGNs belong to the family of lysin motif proteins (Buist et al., 2008). Willmann et al. (2011) isolated and characterized the first three PGN receptors in *Arabidopsis,* and Ao et al. (2014) have done the same in rice. Experimental data demonstrated that exogenous PGNs provoke an increase in cytoplasmic Ca^{2+} concentration, ROS accumulation, MAPK activation, and transcriptional reprogramming (Gust, 2015; Bigeard et al., 2015).

Finally, LPS and its derivatives were found to act as bMAMPs, therefore inducing innate immune responses in plants. LPS are abundantly present in the outer membrane of Gram-negative bacteria and consist of three functionally distinct species domains: a lipid, a core oligosaccharide, and an O-polysaccharide part (Ranf, 2016 and references therein). Bacteria lacking O-polysaccharide are frequently nonvirulent and have difficulties to survive within plant hosts (Slipo et al., 2010). LPS-sensing systems in plants remained genetically unidentified, and no PRR for LPS and its derivatives have been characterized. Interestingly, the O-antigen part of the LPS molecule is believed to be responsible for triggering ISR in *Arabidopsis* (Newman et al., 2013). Recently, it has been found that a surface-localized lectin S-domain receptor kinase named RLK LORE (*l*ipo*o*ligosaccharide-specific *r*educed *e*licitation) mediates sensitive perception of bMAMPs in *Arabidopsis thaliana* and other crucifers, finally resulting in characteristic PTI responses (Newman et al., 2013; Lore, 2015; Ranf et al., 2015). The activity of LPS in plants has mostly been described in dicots; but studies in rice cells have revealed that LPS from various pathogenic and nonpathogenic bacteria induce the generation of ROS and defense-related gene expression, therefore indicating that the machinery recognizing LPS is evolutionary conserved in monocots and dicots (Desaki et al., 2006, 2012). Future research focused to understanding the LPS immune sensing and the virulence functions of LPS from economically relevant phytopathogens will help to develop antibacterial tools and to improve disease resistance in crop plants.

2.2.2 Elicitors in Fungal Pathogens

Fungal CW β-glucans are the most abundant fungal CW polysaccharides, and they have been proved to act as MAMPs, although plant perception mechanisms of β-glucans remains largely unknown (Fesel and Zuccaro, 2015). After cellulose, chitin is the second most abundant natural biopolymer and can also be found in the exoskeletons of arthropods. In fungi, β-glucans are cross-linked to chitin, and chitin and its fragments [chitin oligosaccharides (CTOS)] have been shown to trigger defense responses in both monocots and dicots (Newman et al., 2013). Thus plants can detect infecting fungi through the perception of chitin and CTOS, and CEBiP (*c*hitin-*e*licitor *bi*nding *p*rotein) and CERK1 (*c*hitin-*e*licitor *r*eceptor *k*inase-1) were identified as the cell surface lysin motif receptors for chitin (Kaku et al., 2006; Wan et al., 2008). In the case of rice, it has also been shown that CEBiP and CERK1 form a hetero-oligomeric receptor complex ligand (Shimizu et al., 2010). Alternatively, in *Arabidopsis*, the receptor kinase CERK1 is required for both chitin and PGN signaling (Willmann et al., 2011). Interestingly, pathogenic fungi are able to secrete effectors and change their CW composition to escape from the chitin-mediated immune system (Shinya et al., 2015 and references therein). Recently, to overcome the low solubility of chitin, soluble CTOS have been developed from the original polysaccharide and proved to act as potent MAMPs (Ying et al., 2016, and references therein).

3. METABOLIC ASPECTS INVOLVED IN PATHOGENESIS

3.1 Participation of 14-3-3 Proteins

14-3-3 Proteins function by binding to phosphorylated proteins to modulate their functions (Denison et al., 2011). These proteins are eukaryotic-specific evolutionary conserved proteins and are encoded by multigene families. Increasing evidence supports a prominent role of 14-3-3 proteins in regulating plant defense and immunity against pathogens at various levels (Bigeard et al., 2015, and references therein). Within the wide range of proteins modulated by 14-3-3 are PRP (Robert et al., 2001; Lozano-Durán and Robatzek, 2015). Thus, several 14-3-3 proteins are involved in *R*-gene-mediated plant disease resistance (e.g., tobacco N protein; Konagaya et al., 2004). Upon pathogen infection or perception, 14-3-3 proteins have been shown to display changes at transcript levels and, in some instances, in protein abundance or properties (revised in Lozano-Durán and Robatzek, 2015). This is exemplified in 14-3-3 proteins including TFT1 from tomato (Robb et al., 2007), GRF2 from soybean (Klink et al., 2009), Hv 14-3-3b from wheat (Manickavelu et al., 2010), GF14b from rice (Manosalva et al., 2011), ZmGF14-6 from maize (Campo et al., 2012), and GRF6 from *Arabidopsis* (Carrasco et al., 2014). Additionally, it has been also described that: (1) several 14-3-3 proteins have been found to interact with components of the plant immune system (Elmayan et al., 2007; Chang et al., 2009; Yang et al., 2009; Oh et al., 2010; Oh and Martin, 2011); (2) 14-3-3 proteins can be targets of bacterial effectors (Nomura et al., 2006; Kim et al., 2009; Taylor et al., 2012; Teper et al., 2014); (3) the 14-3-3 protein TFT7 can interact with the tomato MAPKKKα protein kinase and its downstream kinase, MAPKK (Oh et al., 2010; Oh and Martin, 2011); and (4) 14-3-3 proteins play important roles in modulating how plants respond to a variety of biotic stresses (Li et al., 2014; Teper et al., 2014) by binding to phosphorylated target proteins (Obsil and Obsilova, 2011).

The MAPK pathway is involved in plant stress and defense responses, and it has been shown to positively regulate PCD in response to the pathogen *P. syringae* (Zhang and Klessig, 2001; del Pozo et al., 2004). It has been suggested that the 14-3-3 protein TFT7 may act by stabilizing the MAPKKKα protein so that it can activate downstream MAPK cascades leading to PCD (Oh et al., 2010). Thus, a mutant version of TFT7 that does not interact with the MAPKKKα protein abolished its ability to efficiently induce PCD (Oh et al., 2010). Recent data indicate that 14-3-3 proteins may also act by bringing the MAPKKKα and MAPKK proteins together in the cell (Denison et al., 2011).

3.2 Oxidative-Burst

Perception of MAMPs by plants triggers a rapid transient accumulation of ROS (Matilla-Vázquez and Matilla, 2012; Ranf et al., 2011; O'Brien et al., 2013). This OB is thus defined as a ubiquitous rapid production of high levels of ROS in response to external stimuli, and it is currently considered one of the earliest events following elicitation. Two major mechanisms have been proposed for generating ROS: plasma membrane NAD(P)H-dependent oxidases (also known as *r*espiratory *b*urst *o*xidase *h*omologs; RBOHs) and class III CW peroxidases (Almagro et al., 2009; Chaouch et al., 2012; Choi et al., 2007; Daudi et al., 2012; Hemetsberger et al., 2012; revised in Görlach et al., 2015; Künstler et al., 2015; references cited therein; Marino et al., 2012; Suzuki et al., 2012). In accordance with this, the absence of expression of NADPH oxidase (*RBOH*)–encoding genes confers enhanced susceptibility to biotrophic pathogens (Berrocal-Lobo et al., 2010; Proels et al., 2010). However, NADPH oxidases are not the only source for ROS in plant cells since it is becoming obvious that a complex spatial and temporal coordination occurs in plants between ROS sources and between ROS and other signals (Baxter et al., 2014). Thus, many features confirm the involvement of GTP-binding proteins, ion channels (especially Ca^{2+}), protein kinases, phosphatases, and phospholipases in the signaling transduction cascade leading to ROS production (Wojtaszek, 1997).

The rapid apoplastic production of ROS upon pathogen attack is generally considered a defense mechanism for microbial killing and an initiation of host defense responses in plants (Torres, 2010; Ding et al., 2011; Rojas et al., 2014). However, this defense response also causes: (1) cytoskeletal reorganization (Higaki et al., 2011); (2) the reinforcement of the plant CW through the papilla (i.e., CW appositions by deposition of callose and lignification) (Collinge, 2009; Hückelhoven, 2007; Hardham et al., 2007) and the cross-linking of proline-rich glycoproteins present in the CW (Bradley et al., 1992); (3) the increase in synthesis of PHALs (Ahuja et al., 2012); and (4) the upregulation of the primary metabolism (Moghaddam and Van den Ende, 2012, 2013; Rojas et al., 2014). Several lines of research have shown the existence of a correlation between H_2O_2, compounds of papilla, Golgi vesicles containing CW components in transit to papilla, and functionality of plasmodesmata (Collins et al., 2003; An et al., 2006; Shetty et al., 2008). It is generally accepted that upon recognition of a pathogen, a cascade of signaling events is initiated, leading to an increase in extracellular pH, release of a reductant, activation of class III apoplastic CW peroxidases, and mitochondrial ROS production. As a consequence, the activation of these mechanisms ultimately results in an H_2O_2-mediated signal that activates defense response pathways (O'Brien et al., 2013). Interestingly, the PHAL resveratrol has been also shown to stimulate OB (Chang et al., 2011).

The magnitude and amplitude of ROS production vary depending on the plant–pathogen interaction and is positively correlated with plant immunity. Likewise, accumulation of sphingolipids, phospholipids, or linolenic acid triggers ROS production during defense responses (Kachroo and Kachroo, 2009; Berkey et al., 2012; Rojas et al., 2014). ROS accumulation can also be triggered in plants by antimicrobial peptides (AMPs) such as defensins, thionins, or lipid transfer proteins (Hwang et al., 2011; Rahnamaeian, 2011; Moghaddam et al., 2016). This AMP-triggered OB was supported by the action of defensins HsAFP1, RsAFP2, NaD1, and PvD1, which induce the accumulation of ROS in response to the pathogenic fungi *Candida albicans* and *Fusarium oxysporum,* therefore modulating intracellular signaling cascades and resulting in the permeabilization of the fungal membrane and growth arrest (Aerts et al., 2009, 2011; Mello et al., 2011; van der Weerden et al., 2008).

It is well established that ROS control NO production (and vice versa) and that ROS and NO play a critical role in plant immune responses (Perchepied et al., 2010; Scheler et al., 2013; Wang et al., 2013). NO cooperates with ROS and Ca^{2+} to mediate innate immune responses against plant pathogens (Bellin et al., 2013; Trapet et al., 2015). Therefore, the induction of NO accumulation by AMPs provides a stronger signal for the innate immune system, causing dangerous levels of oxidative stress in the pathogen and cooperating with scavenging agents to modulate the cellular redox status and other defense signaling pathways. Taken together, it is now well documented that NO produced in plant cells challenged by PAMS/MAMPs and DAMPs is integrated in signaling cascades leading to the expression of defense-related genes, production of secondary metabolites, and, ultimately, HR (Bellin et al., 2013; Trapet et al., 2015 and references cited therein).

3.2.1 Programmed Cell Death Modulated by Reactive Oxygen Species Is Included in Plant Defense

PCD is an integral part of plant development, but it occurs in response to biotic and abiotic stresses (Dorn et al., 2011). It can be defined as an active and tightly genetically modulated cell dismantling process that finally leads to the selective removal of unwanted or damaged cells (Lam, 2004). That is, only the programmed cells are eliminated, without affecting the neighboring cells. In contrast to PCD, necrosis refers to accidental cell death in response to a harmful environmental factor and is typically found under abiotic stresses (Doorn et al., 2011). During PCD, the cell content is removed by a combination of an autophagy-like process and the release of hydrolases from collapsed lytic vacuoles. In this process, specific proteases and nucleases are also involved.

PCD is an essential and beneficial process for plants, as well as for a proper response to environmental conditions (Gadjev et al., 2008). PCD occurs at all stages of plants' life cycle. Among other metabolic aspects, PCD is important for plant immunity against biotrophic pathogens. Interestingly, H_2O_2 and other ROS have recently become recognized to be key modulators of PCD (Gadjev et al., 2008; Kumar et al., 2016). Thus, the mechanism of PCD induction usually includes an increase in the ROS levels, and such increase is used as mediators of the stress signal (de Pinto et al., 2012). In general, low doses of ROS are employed as plant signals to mediate in stress responses, whereas high ROS concentrations pose a significant threat that may eventually lead to PCD (Bhattacharjee, 2012).

There is a strict association between H_2O_2 and NO regarding PCD activation (de Pinto et al., 2002), but the mechanism of this interplay is still largely unknown (Yoshioka et al., 2011). As described earlier, plants generally react to the attack of both adapted and nonadapted microorganisms by inducing pathogenesis-related (PR) genes and localized cell death at the site of infection, a process known as hypersensitivity response. The involvement of NO in this HR has been shown to require H_2O_2 since NO itself is not effective in triggering cell death (Zago et al., 2006; Kulik et al., 2014). However, how H_2O_2 channels NO through the cell death pathway is currently unknown; but it is certainly clear that NO plays a central role in the immune signaling of plants by controlling, transducing, and probably coordinating Ca^{2+} and ROS signaling.

3.3 Ca^{2+} Burst

Transient and drastic changes in intracellular (mostly cytosolic) Ca^{2+} levels upon pathogen infection are an essential early signaling event for plant defense responses (Lecourieux et al., 2006; Ranf et al., 2011). Thus, Ca^{2+} signaling is essential in both plant PTI and ETI responses (Grant et al., 2000; Spalding and Harper, 2011). This Ca^{2+} burst is positively regulated by BIK1, a plasma membrane–localized Ser/Thr protein kinase that is a crucial component in host signaling events required to activate the resistance responses to pathogens such as *Botrytis cinerea* and *Alternaria brassicicola* (Veronese et al., 2006). The increased levels of Ca^{2+} after pathogen infection are subsequently sensed and translated into intracellular responses, mainly via different Ca^{2+} sensor proteins (De Falco et al., 2010). In accordance with this, changes in the cytosolic concentration of Ca^{2+} has been verified in *Nicotiana plumbaginifolia* cells following treatment with cryptogein, an elicitor secreted by the oomycete *Phytophthora cryptogea* (Lecourieux et al., 2006). Other elicitors such as the peptide Pep-13 from *Phytophthora megasperma* or an oligopeptide elicitor derived from a CW protein of *Phytophthora sojae* were also shown to induce Ca^{2+} transients (Blume et al., 2000).

CDPKs have emerged as important Ca^{2+} sensor proteins in transducing differential Ca^{2+} signatures triggered by PAMPs (or other effectors) and activating complex downstream responses (Gao et al., 2014, and references therein). CDPKs contain a Ca^{2+}-binding domain and represent a unique subclass of kinases found in plants, green algae, and some protists. Thus, CDPKs can directly transmit Ca^{2+} signals by Ca^{2+} binding to the elongation factor (EF)-hand domain at the C-terminus and through the phosphorylation of substrates by the catalytic kinase domain at the N-terminus (Boudsocq and Sheen, 2013; Liese and Romeis, 2013). The genome of *Arabidopsis* encodes 34 CDPKs (Hrabak et al., 2003), and this superfamily of protein kinases has been found to participate in hormone signaling, flg22-mediated ROS burst, flg22-mediated transcriptional reprogramming, flg22-induced resistance to the bacterial pathogen *P. syringae* and in modulating the complex gene expression network that regulates plant defense responses (Schultz et al., 2013; Romeis and Herde, 2014; Bigeard et al., 2015). Therefore, CDPKs participate in plant defense mainly by interacting and phosphorylating diverse substrates to finally regulate the subsequent immune responses. Thus, several CDPKs (i.e., CPK4 and CPK6) were shown to participate in host immunity by phosphorylating the NADPH oxidase RBOHD (Kadota et al., 2014). However, how CDPKs determine specific substrates remains enigmatic. The *Nicotiana tabaccum* NtCDPK2 was the first CDPK known to be involved in ETI triggered by the fungal elicitor Avr9 (Romeis et al., 2001). Since then, emerging evidences suggest that specific and overlapping CDPKs phosphorylate distinct substrates in PTI and ETI to regulate diverse plant immune responses, including ROS production, transcriptional reprogramming of immune genes, and the HR (reviewed in Gao et al., 2014). Importantly, ROS and Ca^{2+} are mutually interconnected. Thus, calcium can increase ROS production and ROS can significantly affect Ca^{2+} influx into the cell and intracellular Ca^{2+} stores (revised by Görlach et al., 2015).

It is well known that H_2O_2 can trigger increases in cytosolic Ca^{2+} by activating the Ca^{2+}-permeable channels (Matilla-Vázquez and Matilla, 2012, and references therein). However, H_2O_2 production during OB also depends on continuous Ca^{2+} influx, subsequently activating not only the NADPH oxidase but also the calmodulin (CaM)-binding NAD kinase, which supplies $NADP^+$ for ROS production through NADPH oxidase (Karita et al., 2004; Dubiella et al., 2013). On the other hand, the involvement of the activation of Ca^{2+} ion channels in the induction of OB and other defense responses has also been indicated (Hahlbrock et al., 1995; Mori and Schroeder, 2004). As a consequence of the advances in the knowledge of the role of Ca^{2+} in the defense processes in plants (Lecourieux et al., 2006), it is currently suggested that NO may be an important component in the mechanism coordinating and regulating Ca^{2+} and ROS signaling in plant immunity (Trapet et al., 2015).

Ca^{2+} influx is one of the earliest responses upon recognition of pathogens in plant innate immunity, and distinct Ca^{2+} signatures have been observed in plant PTI and ETI signaling. Thus, different PAMPs or elicitors such as flg22 and elf18 triggered the rapid and transient increase of cytosolic Ca^{2+} in the cytosol, nucleus, mitochondria, or chloroplasts (Ranf et al., 2011). These data suggest that Ca^{2+} signaling plays essential roles in modulating diverse aspects of innate immunity through differential Ca^{2+} sensory machineries (Grant et al., 2000; Ali et al., 2007). Altogether, the action of channels and pumps governing Ca^{2+} fluxes across the plasma membrane and tonoplast contributes to generating cytosolic Ca^{2+} signals elicited by pathogens and PAMPs, and the disruption of Ca^{2+} homeostasis affects the ability of plants to mount appropriate defense responses against pathogens (Ma et al., 2012). Recent findings also indicate that chloroplasts have an unanticipated function as a hub for ROS and Ca^{2+} signaling that affects immune responses at an early stage after pathogen attack (Stael et al., 2015, and references therein). Cyclic nucleotide–gated ion channels and glutamate receptors facilitate cytosolic Ca^{2+} elevation and subsequent events, including NO synthesis in response to pathogenic signals. However, the exact mechanisms of Ca^{2+}-induced NO production are still an open question, but, collectively, NO synthesis requires Ca^{2+}, and, in turn, NO acts as an agonist that can further mobilize Ca^{2+} (for review, see Trapet et al., 2015). Thus NO appears as a step in the signaling cascade initiated by Ca^{2+} and as an amplifier of Ca^{2+} signaling.

4. CALMODULIN IN PLANT DEFENSE

Research on plant defense strategies against pathogens contributed to identifying mechanisms involved in the generation and processing of Ca^{2+} signals, besides demonstrating the roles of these signals in plant immunity. Thus, plants possess certain proteins that are very sensitive to Ca^{2+} and its cytosolic alterations. Among them, calmodulins (CaMs) belong to a class of Ca^{2+} sensors in all eukaryotic cells that are able to perceive changes in cellular Ca^{2+} and transduce these alterations into a cellular response (Du et al., 2011; Perochon et al., 2011; Zeng et al., 2015). Thus, for example, CaMs can bind proteins named CaMBPs (CaM-binding proteins) causing specific physiological responses (Poovaiah et al., 2013; Zhang et al., 2014a,b). Importantly, accumulating evidence indicates that either downregulation of the expression of CaM/CaM-like proteins (CML) or their loss of function in mutated plants strongly affects immune responses (see Table 10.1 in Cheval et al., 2013, and references therein). In accordance with this, Ca^{2+} sensing by CML has been shown to modulate synthesis of NO after pathogen attack, therefore including CaM/CML proteins as key players in plant immune responses against pathogens (Ma and Berkowitz, 2011). Thus, it has been shown that the nitric oxide synthase AtNOS1 from *Arabidopsis* contains CaM-binding motifs and that Ca^{2+}/CaM is necessary for the complete activation of the enzyme (Guo et al., 2003; Hong et al., 2008). Interestingly, CaMs can be *S*-nitrosylated at their Cys-27, a residue belonging to the EF-1 domain located in the N-terminal part of protein. This posttranslational modification has been suggested to alter the interaction of CaMs with protein partners, consequently modulating the downstream signaling cascade (Jeandroz et al., 2013).

As mentioned earlier, pathogen-derived molecules provoke elevations of free Ca^{2+} levels in different compartments of plant cells (Dodd et al., 2010). It has been recently demonstrated that AtSR1, one of the best-studied CaM-binding transcription factors (TFs) involved in plant defense (Du et al., 2009, 2011), has positive effects on plant resistance since a mutant in *AtSR1* is more susceptible to pathogen infection (Nie et al., 2012). Interestingly, another CaM-binding TF, named CBP60, has also been shown to be involved in plant defense by regulating SA biosynthesis and playing a positive role in SA-mediated defense responses (Zhang et al., 2010). CBP60 is believed to activate the expression of the SA biosynthesis gene *ICS1*, therefore providing a direct channel for Ca^{2+} signal to activate defense responses. Collectively, CaM-binding TFs are key modulators of pathogen-induced SA production, but they are also responsible for preventing the overproduction of the hormone since these enhanced levels may result in unfavorable effects for the plant (Du et al., 2009; Zhang et al., 2010).

5. ROOT EXUDATES AND PATHOGENICITY IN PLANTS

Plants live in complex environments in which they intimately interact with a broad range of microbial pathogens with different lifestyles and infection strategies. It has become increasingly evident that root exudation is a response to the release of multiple biotic and abiotic stressors and is an important strategy for the protection of plants against pathogens

(Badri et al., 2009; Chaparro et al., 2013, 2014; Weston et al., 2012, 2015; Zhu et al., 2016), and the quantity and quality of root exudates has been determined in the presence of multiple microorganisms. Thus composition of root exudates generates a selective pressure that favors different microbial communities (i.e., plant specific microbiome; Huang et al., 2014). Although the in vivo functions of most components of exudates have not been determined, multiple research data provided clear evidences that root exudates play an important role in plant defense and environmental adaptation (Bais et al., 2005, 2006; Bednarek and Osbourne, 2009; Baetz and Martinoia, 2014). As low-M_r (relative molecular mass) components of root exudates are commonly assimilated by microorganisms, it was proposed that exudates have a primary role in regulating rhizosphere microbial dynamics, abundance, and activity (Walker et al., 2003; Kapoor and Mukerji, 2006; Bais et al., 2006; Weisskopf et al., 2008; Goyal and Mattoo, 2016). In consequence, the use of artificial root exudates has demonstrated their importance in the establishment of microbial communities and activities in the rhizosphere (Beaudoin et al., 2003; Landi et al., 2006; Shi et al., 2011). Importantly, bacteria have developed chemosensory systems (i.e., chemotaxis signaling cascades) that sense components of root exudates thus guiding them to root-secreted components to provide them enough nutrient sources for survival (Corral-Lugo et al., 2016; Scharf et al., 2016).

5.1 Root Exudates Constitute a Convoluted Cocktail

Living roots accumulate and release multiple compounds through root exudates, including bioactive plant secondary metabolites responsible for creating barriers for plant pathogen penetration. Root exudation represents a high cost for the plant since it is currently estimated that around 5%–21% of the carbon fixed in photosynthesis is released to the rhizosphere as root exudates (Walker et al., 2004; Newmann and Römheld, 2007; Uren, 2007; Badri and Vivanco, 2009; Hamel and Beaudoin, 2010). Root exudates are considerably influenced by biotic and abiotic factors (Matilla et al., 2010) and have usually been clustered into compounds with low and high M_r (e.g., amino acids, sugars, organic acids, phenolic compounds, proteins, and polysaccharides; Bertin et al., 2003; Moe, 2013). The composition of root exudates was shown to vary depending on the plant species and the stage of plant development (Bais et al., 2006; Neumann and Römheld, 2007; Uren, 2007; Badri and Vivanco, 2009). Importantly, the defense signaling molecules SA, NO, and methyl jasmonate were demonstrated to independently elicit a differential genomewide transcription profile in roots that resulted in enhanced rhizosecretion (Badri et al., 2008; Badri and Vivanco, 2009; Ruiz-May et al., 2009).

Plants produce thousands of different natural products with low M_r. Some of these secondary metabolites act as antimicrobials, working in defense against plant pathogens. These substances can be divided into two major groups: phytoanticipins, which are constitutive in plants (González-Lamothe et al., 2009), and PHALs, which increase considerably in response to pathogen attack (Ahuja et al., 2012). Consequently, the plants that produce these must also have mechanisms to defend itself against autotoxicity (Weston and Duke, 2003). However, although these autotolerance mechanisms have been studied in isolated cells (Yazaki et al., 2008), the mechanisms of resistance of the radical system are far to be understood. Interestingly, several root exudates contain proteins involved in root resistance or likely related with root–microbe or plant–plant communication (Wen et al., 2007; De-la-Peña et al., 2008; Badri et al., 2012). These proteins are differentially secreted depending on the presence of compatible or incompatible bacteria. Similarly, depending on the colonized plant, proteins secreted by specific bacteria can also vary significantly (De-la-Peña et al., 2008). Taking together the aforementioned facts, root exudates are responsible for generating in the rhizosphere a wide and flexible barrier of chemical defense that may act as messengers that allow the initiation of diverse physical and molecular interactions between plant roots and soil organisms (Walker et al., 2003; Wen et al., 2007; De-la-Peña et al., 2008; Hawes et al., 2011, 2012; Baetz and Martinoia, 2014).

5.2 Exudation Mechanisms

Secretion of root exudates to the rhizosphere environment depends on root structure, M_r, polarity, and molecular modifications (i.e., glycosylation, acylation or hydroxylation) of the exuded compounds (Walker et al., 2003; Sirikantaramas et al., 2008; Weston et al., 2012). Plant root cells involved in exudation possess specific transport mechanisms to secrete these compounds outside the cell. Initially, it was suggested that exudates were passively released by diffusion, using ion channels and vesicle transport for high-M_r compounds (Bertin et al., 2003; Badri and Vivanco, 2009). However, research mainly in the last decade has revealed that primary and secondary active transport systems are becoming key players in these secretion processes. Thus, four families of transport proteins have been mainly shown to release organic substrates from cells (Weston et al., 2012, and references cited therein), including the ATP-binding cassette family (ABC; primary active transporters) (Rea, 2007; Kang et al., 2011; Song et al., 2014), the major facilitator superfamily (Reddy et al., 2012), the multidrug and toxic compounds extrusion family (MATE)

(Yazaki et al., 2008; Wu et al., 2014), and the aluminum-activated malate transporter family (Weston et al., 2012). Within them, ABC and MATE transporters are the best characterized.

ABC proteins are responsible for transporting from the cytosol to the apoplast or to internal organelles a wide range of compounds (i.e., peptides, steroids, lipids, phytohormones [i.e., abscisic acid, ABA; auxins and SA], carbohydrates, inorganic acids, and xenobiotics) using ATP hydrolysis as the energy source (Goossens et al., 2003; Kang et al., 2010, 2011; Weston et al., 2012; revised in Jeandet et al., 2014). However, there are consistent evidences involving ABC proteins in the secretion of compounds to the rhizosphere (Sugiyama et al., 2007; Badri et al., 2008, 2009). For example, the ABC protein *Np*PDR1 from *N. plumbaginifolia* was firstly described as a plant ABC transporter directly implicated in pathogen defense and active transport of terpenoids (Jasinski et al., 2001). These results were later confirmed by silencing of *NpPDR1,* which caused an increase of susceptibility to oomycete pathogen invasion (Bultreys et al., 2009). Additionally, roots of *Medicago truncatula* were quickly infected by *F. oxysporum* when *MtABCG10* (homolog of *NtPDR1*) was silenced (Banasiak et al., 2013). Recently, research on *Ph*PDR1 of *Petunia* concluded that there is a direct correlation between this strigolactone ABC transporter present in root cells and soil microbiota (Kretschmar et al., 2012). Interestingly, the ABC transporter PEN3 (*p*enetration *r*esistance 3; *AtABCG36*) from *Arabidopsis* was the only ABC protein reported to be involved in both auxin transport and plant resistance to bacterial, fungal, or oomycete pathogens, both effects being linked (Stein et al., 2006; Strader and Bartel, 2009, 2011; Xin et al., 2013). Importantly, PEN3 recruitment to sites of pathogen infection was shown to be triggered by PAMPs such as chitin and Flg (Stein et al., 2006). Additional data on the implication of ABC transporters on root exudate secretion resulted from the mutation of the ABC transporter–encoding gene *abcg30* (*Atpdr2*) from *Arabidopsis*, since its mutation caused changes in the composition of root exudates and markedly influenced the soil microbial community compared with the wild type (Badri et al., 2009).

The involvement of ABC proteins in the transport and tolerance to ABA has been largely described (Borghi et al., 2015, and references cited therein). However, the relationship between these two roles (i.e., ABA transport and tolerance) and resistance to bacterial infection has been recently demonstrated by Ji et al. (2014). In this study, the authors showed that the ABC transporter AtABCG16 from *Arabidopsis* is involved in plant resistance against *P. syringae* pv. *tomato* DC3000, besides playing a role in the tolerance to the deleterious effects of ABA after pathogen infection. Future work will help to decipher the complex regulatory network that modulates the expression of transporter proteins and their effect on the plant defense system.

5.3 Phytoalexins

In addition to constitutive root exudation (Matilla and Krell, 2017), induced defense-related compounds named phytoalexins can be generated after pathogen infection (Thakur and Sohal, 2013; Zhang et al., 2013). As indicated previously, PHALs are antimicrobial secondary metabolites of low M_r produced temporarily by plants as a response to biotic or abiotic stresses (Flores et al., 1999; Hasegawa et al., 2010; Jeandet et al., 2014; Jeandet, 2015 and references included). Thus, PHALs take part in an intricate defense system which enables plants to control invading microorganisms, and, as such, production of PHALs constitutes one of the several mechanisms that plants possess to combat pathogen infections. However, some PHALs such as momilactone A can be constitutively produced and exuded from root (Toyomasu et al., 2014).

PHALs show great chemical diversity, and more than 350 types have been chemically characterized, including coumarins, terpenoids, indoles, flavonoids, alkaloids, sulfur-containing compounds, or phenolic compounds (Jeandet, 2015). PHALs have been identified in families such as Brassicaceae, Fabaceae, Solanaceae, Vitaceae, and Poaceae in response to infection by pathogens or treatment with isolated elicitors (Table 10.1; Jeandet et al., 2014). PHALs are mostly lipophilic and therefore have the ability to cross the plasma membrane and act inside the invasive cell. The synthesis of PHALs, mainly those not detected in healthy plants, is subjected to a tight regulation and constitutes a temporally dynamic process regulated by several endogenous and exogenous stimuli (De-la-Peña et al., 2008; Baetz and Martinoia, 2014). Pisatin, originally isolated from pea plants, was the first chemically characterized and crystallized PHAL (Cruickshank and Perrin, 1960). However, nowadays, some of the most studied PHALs include phaseolin in bean; pisatin in pea; glyceolin in soybean, alfalfa, and clover; rishitin in potato; gossypol in cotton; and capsidiol in pepper. Importantly, most of the isolated PHALs derive from the phenylpropanoid and isoprenoid biosynthetic pathways (Iriti and Faoro, 2009). PHAL toxicity in plants occurs as a function of their acidic character, the high number of hydroxyl groups, and molecular substituents.

Interestingly, the regulatory networks involved in biosynthesis and production of camalexin, the main PHAL from Brassicaceae, has been investigated in *A. thaliana* (Ren et al., 2008). The indolic ring of camalexin derives from tryptophan (Trp), which in turn arises from chorismate. Indole-3-acetaldoxime, indole-3-acetonitrile, and dihydrocamalexic acid, direct precursors of camalexin, are three notable intermediates in this well-known indole PHAL pathway (Ahuja et al., 2012). Another well-characterized PHAL is resveratrol (*trans*-3,4′,5-trihydroxystilbene). This secondary metabolite shows antifungal activity against grapevine pathogens. It is well known that the transcription of stilbene synthases–encoding genes (*StSy*), key enzymes for resveratrol synthesis, is induced by pathogens and by the activation of the JA pathway (Tassoni et al., 2005). Engineering the *StSy* gene into plants of agronomic interest results in resveratrol accumulation and elevates pathogen resistance (Chang et al., 2011, and references therein). On the other hand, resveratrol stimulates reorganization of the actin cytoskeleton and induction of defense genes such as *PR*5 (Chang et al., 2011). Interestingly, analysis of the genomes of cereals such as wheat and barley allowed the identification of candidate PHAL biosynthesis genes, but no PHALs have been isolated from these plants (Ejike et al., 2013). This may be explained by the absence of a specific elicitor for the activation of the biosynthetic pathway(s).

6. CONCLUDING REMARKS

Plant defense responses require a notable supply of energy preferably derived from primary metabolic pathways (revised in Bolton, 2009; and Rojas et al., 2014). However, several studies have also shown an upregulation of primary metabolism after plant infection by phytopathogens, as an indication that primary metabolism may regulate signaling pathways leading to plant resistance against pathogens (Rojas et al., 2014). However, additional research is needed to understand aspects such as the implications of certain carbohydrates and amino acids in defense responses, or the routes by which some lipids trigger ROS accumulation following pathogen-induced HR.

Most of the current knowledge on both plant–pathogen and plant–microbiome derives from approaches under laboratory conditions. Nevertheless, under field conditions, pathogenicity can be affected by multitude of environmental factors. Thus, for example, plant diseases caused by phyllosphere bacteria often occur after rainfall or during periods of high humidity, and recent research showed that water-soaked areas favor bacteria multiplication, besides demonstrating the importance of high-humidity conditions for the virulence capabilities of bacterial pathogens inside plants (Xin et al., 2016). Therefore, this recent study illustrates the current need for future consideration of natural climate conditions to fully understand host–microbe interactions.

The world total population is constantly increasing, and the demands for food supplies are growing exponentially. Unfortunately, as described in this chapter, plant infections caused by pathogenic bacteria, fungi, oomycetes, viruses, or nematodes are causing major crop losses and challenging our capacity to maintain and enhance crop production. In consequence, modern agriculture should be oriented to the identification or development of resistant cultivars. In accordance with this, current research on plant disease management is allowing to decipher the molecular mechanisms involved in plant resistance against pathogens, with the final objective of generating plants with increased resistance to both biotic and abiotic stresses.

ACKNOWLEDGMENTS

Miguel A. Matilla was supported by the Spanish Ministry of Economy and Competitiveness Postdoctoral Research Program, Juan de la Cierva (JCI-2012-11815). We thank Prof. Pilar Carbonero (Centro de Biotecnología y Genómica de Plantas (UPM-INIA), Escuela Técnica Superior de Ingenieros Agrónomos, Universidad Politécnica de Madrid, Campus de Montegancedo, 28223-Pozuelo de Alarcón, Madrid, Spain) for the critical review, advice, and assistance in editing the manuscript. No conflict of interest declared. The author thanks Dr. Philippe Jeandet (Laboratory of Stress, Defenses & Plant Reproduction; University of Reims, France) for providing Table 10.1 to be published in this chapter.

REFERENCES

Aerts, A.M., Carmona-Gutiérrez, D., Lefevre, S., Govaert, G., François, I.E., Madeo, F., Santos, R., Cammue, B.P., Thevissen, K., 2009. The antifungal plant defensin RsAFP2 from radish induces apoptosis in a metacaspase independent way in *Candida albicans*. FEBS Lett. 583, 2513–2516.

Aerts, A.M., Bammens, L., Govaert, G., Carmona-Gutierrez, D., Madeo, F., Cammue, B.P.A., Thevissen, K., 2011. The antifungal plant defensin HsAFP1 from *Heuchera sanguínea* induces apoptosis in *Candida albicans*. Front. Microbiol. 2, 47.

Ahuja, I., Kissen, R., Bones, A.M., 2012. Phytoalexins in defense against pathogens. Trends Plant Sci. 17, 73–90.

Albert, M., 2013. Peptides as triggers of plant defence. J. Exp. Bot. 64, 5269–5279.

Ali, R., Ma, W., Lemtiri-Chlieh, F., Tsaltas, D., Leng, Q., Bodman, S., Berkowitz, G.A., 2007. Death don't have no mercy and neither does calcium: *Arabidopsis* cyclic nucleotide gated channel2 and innate immunity. Plant Cell 19, 1081–1095.

Almagro, L., Gómez Ros, L.V., Belchi-Navarro, S., Bru, R., Ros Barceló, A., Pedreño, M.A., 2009. Class III peroxidases in plant defence reactions. J. Exp. Bot. 60, 377–390.

An, Q., Ehlers, K., Kogel, K.H., van Bel, A.J.E., Hükelhoven, R., 2006. Multivesicular compartments proliferate in susceptible and resistant MLA12-barley leaves in response to infection by the biotrophic powdery mildew fungus. New Phytol. 172, 563–576.

Ao, Y., Li, Z., Feng, D., Xiong, F., Liu, J., Li, F.J., Wang, M., Wang, J., Liu, B., Wang, H.B., 2014. OsCERK1 and OsRLCK176 play important roles in peptidoglycan and chitin signaling in rice innate immunity. Plant J. 80, 1072–1084.

Asai, S., Shirasu, K., 2015. Plant cells under siege: plant immune system versus pathogen effectors. Curr. Opin. Plant Biol. 28, 1–8.
Ausubel, F.M., 2005. Are innate immune signaling pathways in plants and animals conserved? Nat. Immunol. 6, 973–979.
Badri, D.V., Loyola-Vargas, V.M., Boechling, C.D., De-la-Pena, C., Jasinsky, M., Santelia, D., Martinoia, E., Sumner, L.W., Banta, L.M., Stermitz, F., Vivanco, J.M., 2008. Transcriptome analysis of *Arabidopsis* roots treated with signaling compounds: a focus of signaling transduction, metabolic regulation and secretion. New Phytol. 179, 209–223.
Badri, D.V., Vivanco, J.M., 2009. Regulation and function of root exudates. Plant Cell Environ. 32, 666–681.
Badri, D.V., Weir, T.L., Lelie, D., Vivanco, J.M., 2009. Rhizosphere chemical dialogues: plant–microbe interactions. Curr. Opin. Biotechnol. 20, 642–650.
Badri, D.V., De-la-Peña, C., Lei, Z., Manter, D.K., Chaparro, J.M., Guimarães, R.L., Sumner, L.W., Vivanco, J.M., 2012. Root secreted metabolites and proteins are involved in the early events of plant-plant recognition prior to competition. PLoS One 7, e46640.
Baetz, U., Martinoia, E., 2014. Root exudates: the hidden part of plant defense. Trends Plant Sci. 19, 90–98.
Bais, H.P., Prithiviraj, B., Jha, A.K., Ausubel, F.M., Vivanco, J.M., 2005. Mediation of pathogen resistance by exudation of antimicrobials from roots. Nature 434, 217–221.
Bais, H.P., Weir, T.L., Perry, L.G., Gilroy, S., Vivanco, J.M., 2006. The role of root exudates in rhizosphere interactions with plants and other organisms. Ann Rev. Plant Biol. 57, 233–266.
Banasiak, J., Biala, W., Staszkow, A., Swarcewicz, B., Kepczynska, E., Figlerowicz, M., Jasinski, M., 2013. A *Medicago truncatula* ABC transporter belonging to subfamily G modulates the level of isoflavonoids. J. Exp. Bot. 64, 1005–1015.
Bandyopadhyay, P., Bhuyan, S.K., Yadava, P.K., Varma, A., Tuteja, N., 2017. Emergence of plant and rhizospheric microbiota as stable interactomes. Protoplasma 254, 617–626.
Baxter, A., Mittler, R., Suzuki, N., 2014. ROS as key players in plant stress signaling. J. Exp. Bot. 65, 1229–1240.
Beaudoin, E., Benizri, E., Guckert, A., 2003. Impact of artificial root exudates on the bacterial community structure in bulk soil and maize rhizosphere. Soil Biol. Biochem. 35, 1183–1192.
Bednareck, P., Osbourne, A., 2009. Plant-microbe interactions: chemical diversity in plant defense. Science 324, 746–748.
Bektas, Y., Eulgem, T., 2015. Synthetic plant defense elicitors. Front. Plant Sci. 5, 804.
Bellin, D., Asai, S., Delledonne, M., Yoshioka, H., 2013. Nitric oxide as a mediator for defense responses. Mol. Plant–Microbe Interact. 26, 271–277.
Berkey, R., Bendigeri, D., Xiao, S., 2012. Sphingolipids and plant defense/disease: the "death" connection and beyond. Front. Plant Sci. 3, 68.
Berrocal-Lobo, M., Stone, S., Yang, X., Antico, J., Callis, J., Ramonell, K.M., Somerville, S., 2010. ATL9, a RING zinc finger protein with E3 ubiquitin ligase activity implicated in chitin- and NADPH oxidase-mediated defense responses. PLoS One 5, e14426.
Bertin, C., Yang, X., Weston, L.A., 2003. The role of root exudates and allelochemicals in the rhizosphere. Plant Soil 256, 67–83.
Bhattacharjee, S., 2012. The language of reactive oxygen species signaling in plants. J. Bot.:3985298.
Bigeard, J., Colcombet, J., Hir, H., 2015. Signaling mechanisms in pattern triggered immunity (PTI). Mol. Plant 8, 521–539.
Blume, B., Nurnberger, T., Nass, N., Scheel, D., 2000. Receptor-mediated increase in cytoplasmic free calcium required for activation of pathogen defense in parsley. Plant Cell 12, 1425–1440.
Bohm, H., Albert, I., Fan, L., Reinhard, A., Nurnberger, T., 2014. Immune receptor complexes at the plant cell surface. Curr. Opin. Plant Biol. 20C, 47–54.
Bolton, M.D., 2009. Primary metabolism and plant defense–fuel for the fire. Mol. Plant Microbe Interact. 22, 487–497.
Boller, T., 1995. Chemoperception of microbial signals in plant cells. Ann. Rev. Plant Physiol. Plant Mol. Biol. 46, 189–214.
Boller, T., 2005. Peptide signalling in plant development and self/non self-perception. Curr. Opin. Cell Biol. 17, 116–122.
Boller, T., Felix, G., 2009. A renaissance of elicitors: perception of microbe-associated molecular patterns and danger signals by pattern-recognition receptors. Ann. Rev. Plant Biol. 60, 379–406.
Boller, T., He, S.Y., 2009. Innate immunity in plants: an arms race between pattern recognition receptors in plants and effectors in microbial pathogens. Science 324, 742–744.
Borghi, L., Kang, J., Ko, D., Lee, Y., Martinoia, E., 2015. The role of ABCG-type ABC transporters in phytohormone transport. Biochem. Soc. Trans. 43, 924–930.
Boudsocq, M., Sheen, J., 2013. CDPKs in immune and stress signaling. Trends Plant Sci. 18, 30–40.
Boutrotm, F., Zipfel, C., 2017. Function, discovery, and exploitation of plant pattern recognition receptors for broad-spectrum disease resistance. Annu. Rev. Phytopathol. http://dx.doi.org/10.1146/annurev-phyto-080614-120106. (Epub ahead of print).
Bradley, D.J., Kjellbom, P., Lamb, C.J., 1992. Elicitor- and wound-induced oxidative cross-linking of a proline-rich plant cell wall protein – a novel, rapid defense response. Cell 70, 21–30.
Buist, G., Steen, A., Kok, J., Kuipers, O.R., 2008. LysM, a widely distributed protein motif for binding to (peptido)glycans. Mol. Microbiol. 68, 838–847.
Bultreys, A., Trobik, T., Drozak, A., Boutry, M., 2009. *Nicotiana plumbaginifolia* plants silenced for the ATP-binding cassette transporter gene *NpPDR1* show increased susceptibility to a group of fungal and oomycete pathogens. Mol. Plant Pathol. 10, 651–663.
Campo, S., Peris-Peris, C., Montesinos, L., Penas, G., Messeguer, J., San Segundo, B., 2012. Expression of the maize ZmGF14-6 gene in rice confers tolerance to drought stress while enhancing susceptibility to pathogen infection. J. Exp. Bot. 63, 983–999.
Carrasco, J.L., Castello, M.J., Naumann, K., Lassowskat, I., Navarrete-Gómez, M., Scheel, D., Vera, P., 2014. Arabidopsis protein phosphatase DBP1 nucleates a protein network with a role in regulating plant defense. PLoS One 9, e90734.
Coll, N.S., Epple, P., Dangl, J.L., 2011. Programmed cell death in the plant immune system. Cell Death Differ. 18, 1247–1256.
Collinge, D.B., 2009. Cell wall appositions: the first line defence. J. Exp. Bot. 60, 351–352.
Collins, N.C., Thordal-Christensen, H., Lipka, V., Bau, S., Kombrink, E., Qiu, J.L., Hückelhoven, R., Stein, M., Freialdenhoven, A., Somerville, S.C., Schulze-Lefert, P., 2003. SNARE-protein-mediated disease resistance at the plant cell wall. Nature 425, 973–977.

Corral-Lugo, A., De la Torre, J., Matilla, M.A., Fernández, M., Morel, B., Espinosa-Urgel, M., Krell, T., 2016. Assessment of the contribution of chemoreceptor-based signalling to biofilm formation. Environ. Microbiol. 18, 3355–3372.

Cruickshank, I.A., Perrin, D.R., 1960. Isolation of a phytoalexin from *Pisum sativum* L. Nature 187, 799–800.

Cui, H., Tsuda, K., Parker, J.E., 2015. Effector-triggered immunity: from pathogen perception to robust defense. Annu. Rev. Plant Biol. 66, 487–511.

Cui, F., Wu, S., Sun, W., Coaker, G., Kunkel, B., He, P., Shan, L., 2013. The *Pseudomonas syringae* type III effector AvrRpt2 promotes pathogen virulence via stimulating Arabidopsis auxin/indole acetic acid protein turnover. Plant Physiol. 162, 1018–1029.

Chang, I.F., Curran, A., Woolsey, R., Quilici, D., Cushman, J.C., Mittler, R., Harmon, A., Harper, J.F., 2009. Proteomic profiling of tandem affinity purified 14-3-3 protein complexes in *Arabidopsis thaliana*. Proteomics 9, 2967–2985.

Chang, X., Heene, E., Qiao, F., Peter Nick, P., 2011. The phytoalexin resveratrol regulates the initiation of hypersensitive cell death in *Vitis* cell. PLoS One 6, e26405.

Chaouch, S., Queval, G., Noctor, G., 2012. AtRbohF is a crucial modulator of defence-associated metabolism and a key factor in the interplay between intracellular oxidative stress and pathogenesis responses in *Arabidopsis*. Plant J. 69, 613–627.

Chaparro, J.M., Badri, D.V., Bakker, M.G., Sugiyama, A., Manter, D.K., Vivanco, J.M., 2013. Root exudation of phytochemicals in Arabidopsis follows specific patterns that are developmentally programmed and correlate with soil microbial functions. PLoS One 8, e55731.

Chaparro, J.M., Badri, D.V., Vivanco, J.M., 2014. Rhizosphere microbiome assemblage is affected by plant development. ISME J. 8, 790–803.

Cheval, C., Aldon, D., Galaud, J.-P., Ranty, B., 2013. Calcium/calmodulin-mediated regulation of plant immunity. BBA Mol. Cell Res. 1833, 1766–1771.

Chinchilla, D., Bauer, Z., Regenass, M., Boller, T., Felix, G., 2006. The *Arabidopsis* receptor kinase FLS2 binds flg22 and determines the specificity of flagellin perception. Plant Cell 18, 465–476.

Chinchilla, D., Zipfel, C., Robatzek, S., Kemmerling, B., Nürnberger, T., Jones, J.D., Felix, G., Boller, T., 2007. A flagellin-induced complex of the receptor FLS2 and BAK1 initiates plant defence. Nature 448, 497–500.

Chisholm, S.T., Coaker, G., Day, B., Staskawicz, B.J., 2006. Host-microbe interactions: shaping the evolution of the plant immune response. Cell 124, 803–814.

Choi, H.W., Kim, Y.J., Lee, S.C., Hong, J.K., Hwang, B.K., 2007. Hydrogen peroxide generation by the pepper extracellular peroxidase CaPO2 activates local and systemic cell death and defense response to bacterial pathogens. Plant Physiol. 145, 890–904.

Choi, H.W., Klessig, D.F., 2016. DAMPs, MAMPs, and NAMPs in plant innate immunity. BMC Plant Biol. 16, 232.

Dalio, R.J.D., Magalhaes, D.M., Rodrigues, C.M., Arena, G.D., Oliveira, T.S., Souza-Neto, R.R., Picchi, S.C., Martins, P.M.M., Santos, P.J.C., Maximo, H.J., Pacheco, I.S., De Souza, A.A., Machado, M.A., 2017. PAMPs, PRRs, effectors and R-genes associated with citrus-pathogen interactions. Ann. Bot. 119, 749–774.

Dangl, J.L., Jones, J.D.G., 2001. Plant pathogens and integrated defence responses to infection. Nature 411, 826–833.

Dangl, J.L., Horvath, D.M., Staskawicz, B.J., 2013. Pivoting the plant immune system from dissection to deployment. Science 341, 746–751.

Daudi, A., Cheng, Z., O'Brien, J.A., Mammarella, N., Khan, S., Ausubel, F.M., Bolwell, G.P., 2012. The apoplastic oxidative burst peroxidase in Arabidopsis is a major component of pattern triggered immunity. Plant Cell 24, 275–287.

De Falco, T.A., Bender, K.W., Snedden, W.A., 2010. Breaking the code: Ca^{2+} sensors in plant signaling. Biochem. J. 425, 27–40.

De la Peña, C., Lei, Z., Watson, B.S., Sumner, L.W., Vivanco, J.M., 2008. Root-microbe communication through protein secretion. J. Biol. Chem. 283, 25247–25255.

de Pinto, M.C., Tommasi, F., De Gara, L., 2002. Changes in the antioxidant systems as part of the signaling pathway responsible for the programmed cell death activated by nitric oxide and reactive oxygen species in tobacco Bright-Yellow 2 cells. Plant Physiol. 130, 698–708.

de Pinto, M.C., Locato, V., De Gara, L., 2012. Redox regulation in plant programmed cell death. Plant Cell Environ. 35, 234–244.

Del Pozo, O., Pedley, K., Martin, G., 2004. MAPKKKα is a positive regulator of cell death associated with both plant immunity and disease. EMBO J. 23, 3072–3082.

Dempsey, M.A., Klessig, D.F., 2017. How does the multifaceted plant hormone salicylic acid combat disease in plants and are similar mechanisms utilized in humans? BMC Biol. 15, 23.

Denison, F.C., Paul, A.-L., Zupanska, A.K., Ferl, R.J., 2011. 14-3-3 proteins in plant physiology. Sem. Cell Dev. Biol. 22, 720–727.

Derksen, H., Rampitsch, C., Daayf, F., 2013. Signaling cross-talk in plant disease resistance. Plant Sci. 207, 79–87.

Desaki, Y., Miya, A., Venkatesh, B., Tsuyumu, S., Yamane, H., Kaku, H., Minami, E., Shibuya, N., 2006. Bacterial lipopolysaccharides induce defense responses associated with programmed cell death in rice cells. Plant Cell Physiol. 47, 1530–1540.

Desaki, Y., Otomo, I., Kobayashi, D., Jikumaru, Y., Kamiya, Y., Venkatesh, B., Tsuyumu, S., Kaku, H., Shibuya, N., 2012. Positive crosstalk of MAMP signaling pathways in rice cells. PLoS One 7, e51953.

Ding, L., Xu, H., Yi, H., Yang, L., Kong, Z., Zhang, L., Xue, S., Jia, H., Ma, Z., 2011. Resistance to hemi-biotrophic *F. graminearum* infection is associated with coordinated and ordered expression of diverse defense signaling pathways. PLoS One 6, e19008.

Dodd, A.N., Kudla, J., Sanders, D., 2010. The language of calcium signaling. Annu. Rev. Plant Biol. 61, 593–620.

Doorn, W.G., Beers, E.P., Dangl, J.L., Franklin-Tong, V.E., Gallois, P., Hara-Nishimura, I., Jones, A.M., Kawai-Yamada, M., Lam, E., Mundy, J., Mur, L.A.J., Petersen, M., Smertenko, A., Taliansky, M., Breusegem, F., Wolpert, T., Woltering, E., Zhivotovsky, B., Bozhkov, P.V., 2011. Morphological classification of plant cell deaths. Cell Death Differ. 18, 1241–1246.

Doornbos, L., Van Loon, L., Bakker, P.A.H.M., 2012. Impact of root exudates and plant defense signaling on bacterial communities in the rhizosphere. A review. Agron. Sustain Dev. 32, 227–243.

Du, L., Ali, G.S., Simons, K.A., Hou, J., Yang, T., Reddy, A.S., Poovaiah, B.W., 2009. Ca^{2+}/calmodulin regulates salicylic-acid-mediated plant immunity. Nature 457, 1154–1158.

Du, L., Yang, T., Puthanveettil, S.V., Poovaiah, B.W., Luan, S., 2011. Decoding of calcium signal through calmodulin: calmodulin-binding proteins in plants coding and decoding of calcium signals in plants. In: Luan, S. (Ed.), Coding and Decoding of Calcium Signals in Plants. Springer, Berlin, pp. 177–233.

Dubiella, U., Seybold, H., Durian, G., Komander, E., Lassig, R., Witte, C.P., Schulze, W.X., Romeis, T., 2013. Calcium-dependent protein kinase/NADPH oxidase activation circuit is required for rapid defense signal propagation. Proc. Natl. Acad. Sci. USA 110, 8744–8749.

Ejike, E.C.C., Gong, M., Udenigwe, Ch.C., 2013. Phytoalexins from the Poaceae: biosynthesis, function and prospects in food preservation. Food Rev. Int. 52, 167–177.

Elmayan, T., Fromentin, J., Riondet, C., Alcaraz, G., Blein, J.P., Simon-Plas, F., 2007. Regulation of reactive oxygen species production by a 14-3-3 protein in elicited tobacco cells. Plant Cell Environ. 30, 722–732.

Erbs, G., Silipo, A., Aslam, S., De Castro, C., Liparoti, V., Flagiello, A., Pucci, P., Lanzetta, R., Parrilli, M., Molinaro, A., Newman, M.-A., 2008. Peptidoglycan and muropeptides from pathogens *Agrobacterium* and *Xanthomanas* elicit innate immunity: structure and activity. Chem. Biol. 15, 438–448.

Fesel, P.H., Zuccaro, A., 2015. β-glucan: crucial component of the fungal cell wall and elusive MAMP in plants. Fungal Genet. Biol. 90, 53–60.

Flores, H.E., Vivanco, J.M., Loyola-Vargas, V.M., 1999. "Radicle" biochemistry: the biology of root-specific metabolism. Trends Plant Sci. 4, 220–222.

Furano, A.V., 1975. Content of elongation factor Tu in *Escherichia coli*. Proc. Natl. Acad. Sci. USA 72, 4780–4784.

Gadjev, I., Stone, J.M., Gechev, T.S., 2008. Programmed cell death in plants: new insights into redox regulation and the role of hydrogen peroxide. Int. Rev. Cell Mol. Biol. 270, 87–144.

Gao, X., Cox Jr., K.L., He, P., 2014. Functions of calcium-dependent protein kinases in plant innate immunity. Plants 3, 160–176.

Gómez-Gómez, L., Boller, T., 2000. FLS2: and LRR receptor-like kinase involved in the perception of the bacterial elicitor flagellin in Arabidopsis. Mol. Cell 5, 1003–1011.

González-Lamothe, R., Mitchell, G., Gattuso, M., Diarra, M.S., Malouin, F., Bouarab, K., 2009. Plant antimicrobial agents and their effects on plant and human pathogens. Int. J. Mol. Sci. 10, 3400–3419 Jones, J.D.G., Dangl, J.L., 2006. The plant immune system. Nature 444, 323–329.

Goossens, A., Häkkinen, S.T., Laakso, I., Oksman-Caldentey, K.M., Inzé, D., 2003. Secretion of secondary metabolites by ATP-binding cassette transporters in plant cell suspension cultures. Plant Physiol. 131, 1161–1164.

Görlach, A., Bertram, K., Hudecova, S., Krizanova, O., 2015. Calcium and ROS: a mutual interplay. Redox Biol. 6, 260–271.

Goyal, R., Mattoo, A., 2016. Plant Antimicrobial Peptides. . http://dx.doi.org/10.1007/978-3-319-32949-9_5.

Grant, M., Brown, I., Adams, S., Knight, M., Ainslie, A., Mansfield, J., 2000. The RPM1 plant disease resistance gene facilitates a rapid and sustained increase in cytosolic calcium that is necessary for the oxidative burst and hypersensitive cell death. Plant J. 23, 441–450.

Guo, F.Q., Okamoto, M., Crawford, N.M., 2003. Identification of a plant nitric oxide synthase gene involved in hormonal signaling. Science 302, 100–103.

Guo, H., Li, L., Aluru, M., Aluru, S., Yin, Y., 2013. Mechanisms and networks for brassinosteroid regulated gene expression. Curr. Opin. Plant Biol. 16, 545–553.

Gururani, M.A., Venkatesh, J., Upadhyaya, C.P., Nookaraju, A., Pandey, S.K., Park, S.W., 2012. Plant disease resistance genes: current status and future directions. Physiol. Mol. Plant Pathol. 78, 51–65.

Gust, A.A., Biswas, R., Lenz, H.D., Rauhut, T., Ranf, S., Kemmerling, B., Gotz, F., Glawischnig, E., Lee, J., Felix, G., Nürnberger, T., 2007. Bacteria-derived peptidoglycans constitute pathogen-associated molecular patterns triggering innate immunity in. Arab. J. Biol. Chem. 282, 32338–32348.

Gust, A.A., 2015. Peptidoglycan perception in plants. PLoS Pathog. 11, e1005275.

Hahlbrock, K., Scheel, D., Logemann, E., Nurberger, T., Parniske, M., Reinold, S., Sacks, W.R., Schmelzer, E., 1995. Oligopeptide elicitor-mediated defense gene activation in cultured parsley cells. Proc. Natl. Acad. Sci. USA 92, 4150–4157.

Hamel, L.P., Beaudoin, N., 2010. Chitooligosaccharide sensing and downstream signaling: contrasted outcomes in pathogenic and beneficial plant-microbe interactions. Planta 232, 787–806.

Hardham, A.R., Jones, D.A., Takemoto, D., 2007. Cytoskeleton and cell wall function in penetration resistance. Curr. Opin. Plant Biol. 10, 342–348.

Hasegawa, M., Mitsuhara, I., Seo, S., Imai, T., Koga, J., Okada, K., Yamane, H., Ohashi, Y., 2010. Phytoalexin accumulation in the interaction between in the rice and the blast fungus. Mol. Plant Microbe Interact. 23, 1000–1011.

Hawes, M.C., Curlango-Rivera, G., Wen, F., White, G.J., Vanetten, H.D., Xiong, Z., 2011. Extracellular DNA: the tip of root defenses? Plant Sci. 180, 741–745.

Hawes, M.C., Curlango-Rivera, G., Xiong, Z., Kessler, J.O., 2012. Roles of root border cells in plant defense and regulation of rhizosphere microbial population by extracellular DNA "trapping". Plant Soil 355, 1–16.

He, K., Xu, S., Li, J., 2013. BAK1 directly regulates brassinosteroid perception and BRI1 activation. J. Integr. Plant Biol. 55, 1264–1270.

Hemetsberger, C., Herrberger, C., Zechmann, B., Hillmer, M., Doehlemann, G., 2012. The *Ustilago maydis* effector Pep1 suppresses plant immunity by inhibition of host peroxidase activity. PLoS Pathog. 8, e1002684.

Higaki, T., Kurusu, T., Hasezawa, S., Kuchitsu, K., 2011. Dynamic intracellular reorganization of cytoskeletons and the vacuole in defense responses and hypersensitive cell death in plants. J. Plant Res. 124, 315–324.

Hong, J.K., Yun, B.W., Kang, J.G., Raja, M.U., Kwon, E., Sorhagen, K., Chu, C., Wang, Y., Loake, G.J., 2008. Nitric oxide function and signalling in plant disease resistance. J. Exp. Bot. 59, 147–154.

Hou, S., Yang, Y., Wu, D., Zhang, C., 2011. Plant immunity: evolutionary insights from PBS1, Pto, and RIN4. Plant Signal Behav. 6, 794–799.

Hraback, E.M., Chan, C.W., Gribskov, M., Harper, J.F., Choi, J.H., Halford, N., Kudla, J., Luan, S., Nimmo, H.G., Sussman, M.R., Thomas, M., Walker-Simmons, K., Zhu, J.K., Harmon, A.C., 2003. The *Arabidopsis* CDPK-SnRK superfamily of protein kinases. Plant Physiol. 132, 666–680.

Huang, X.F., Chaparro, J.M., Reardon, K.F., Zhang, R.F., Shen, Q.R., Vivanco, J.M., 2014. Rhizosphere interactions: root exudates, microbes, and microbial communities. Botany 92, 267–275.

Hückelhoven, R., 2007. Cell wall–associated mechanisms of disease resistance and susceptibility. Annu. Rev. Phytopathol. 45, 101–112.

Huffaker, A., Pearce, G., Ryan, C.A., 2006. An endogenous peptide signal in *Arabidopsis* activates components of the innate immune response. Proc. Natl. Acad. Sci. USA 103, 10098–10103.

Hwang, B., Hwang, J.-S., Lee, J., Lee, D.G., 2011. The antimicrobial peptide, psacotheasin induces reactive oxygen species and triggers apoptosis in *Candida albicans*. Biochem. Biophys. Res. Commun. 405, 267–271.

Inceoglu, Ö., Overbeek, L.S., Salles, J.F., Elsas, J.D., 2013. Normal operating range of bacterial communities in soil used for potato cropping. Appl. Environ. Microbiol. 79, 1160–1170.

Iriti, M., Faoro, F., 2009. Chemical diversity and defence metabolism: how plants cope with pathogens and ozone pollution. Int. J. Mol. Sci. 10, 3371–3399.

Jasiński, M., Stukkens, Y., Degand, H., Purnelle, B., Marchand-Brynaert, J., Boutry, M., 2001. A plant plasma membrane ATP binding cassette-type transporter is involved in antifungal terpenoid secretion. Plant Cell 13, 1095–1107.

Jeandet, P., Hébrard, C., Marie-Alice Deville, M.-A., Cordelier, S., Dorey, S., Aziz, A., Crouze, J., 2014. Deciphering the role of phytoalexins in plant-microorganism interactions and human health. Molecules 19, 18033–18056.

Jeandet, P., 2015. Phytoalexins: current progress and future prospects. Molecules 20, 2770–2774.

Jeandroz, S., Lamotte, O., Astier, J., Rasul, S., Trapet, P., Besson-Bard, A., Bourque, S., Nicolas-Francès, V., Ma, W., Berkowitz, G.A., Wendehenne, D., 2013. There's more to the picture than meets the eye: nitric oxide cross talk with Ca2+ signaling. Plant Physiol. 163, 459–470.

Ji, H., Peng, Y., Meckes, N., Allen, S., Stewart Jr., C.N., Traw, M.B., 2014. ATP-dependent binding cassette transporter G family member 16 increases plant tolerance to abscisic acid and assists in basal resistance against *Pseudomonas syringae* DC3000. Plant Physiol. 166, 879–888.

Jones, J.D.G., Dangl, J.L., 2006. The plant immune system. Nature 444, 323–329.

Kachroo, A., Kachroo, P., 2009. Fatty acid-derived signals in plant defense. Annu. Rev. Phytopathol. 47, 153–176.

Kadota, Y., Sklenar, J., Derbyshire, P., Stransfeld, L., Asai, S., Ntoukakis, V., Jones, J.D., Shirasu, K., Menke, F., Jones, A., Zipfel, C., 2014. Direct regulation of the NADPH oxidase RBOHD by the PRR-associated kinase BIK1 during plant immunity. Mol. Cell 54, 43–55.

Kaku, H., Nishizawa, Y., Ishii-Minami, N., Akimoto-Tomiyama, Ch., Dohmae, N., Takio, K., Minami, E., Shibuya, N., 2006. Plant cells recognize chitin fragments for defense signaling through a plasma membrane receptor. Proc. Nat. Acad. Sci. USA 103, 11086–11091.

Kang, J., Hwang, J.U., Lee, M., Kim, Y.Y., Assmann, S.M., Martinoia, E., Lee, Y., 2010. PDR-type ABC transporter mediates cellular uptake of the phytohormone abscisic acid. Proc. Natl. Acad. Sci. USA 107, 2355–2360.

Kang, J., Park, J., Choi, H., Burla, B., Kretzschmar, T., Lee, Y., Martinoia, E., 2011. Plant ABC transporters. Arab. Book 9, e0153.

Kapoor, R., Mukerji, K.G., 2006. In: Mukerji, K.G., Manoharachary, C., Singh, J. (Eds.), Rhizosphere Microbial Community Dynamics. Microbial Activity in the Rhizosphere. Springer, Berlin, pp. 55–66.

Karita, E., Yamakawa, H., Mitsuhara, I., Kuchitsu, K., Ohashi, Y., 2004. Three types of tobacco calmodulins characteristically activate plant NAD kinase at different Ca^{2+} concentrations and pHs. Plant Cell Physiol. 45, 1371–1379.

Kim, J.G., Li, X., Roden, J.A., Taylor, K.W., Aakre, C.D., Su, B., Lalonde, S., Kirik, A., Chen, Y., Baranage, G., 2009. *Xanthomonas* T3S effector XopN suppresses PAMP-triggered immunity and interacts with a tomato atypical receptor-like kinase and TFT1. Plant Cell 21, 1305–1323.

Kim, B.H., Kim, S.Y., Nam, K.H., 2013. Assessing the diverse functions of BAK1 and its homologs in *Arabidopsis*, beyond BR signaling and PTI responses. Mol. Cells 35, 7–16.

Klink, V.P., Hosseini, P., Matsye, P., Alkharouf, N.W., Matthews, B.F., 2009. A gene expression analysis of syncytia laser microdissected from the roots of the *Glycine max* (soybean) genotype PI 548402 (Peking) undergoing a resistant reaction after infection by *Heterodera glycines* (soybean cyst nematode). Plant Mol. Biol. 71, 525–567.

Konagaya, K., Matsushita, Y., Kasahara, M., Nyunoya, H., 2004. Members of 14-3-3 protein isoforms interacting with the resistance gene product N and the elicitor of Tobacco mosaic virus. J. Gen. Plant Pathol. 70, 221–231.

Kretzschmar, T., Kohlen, W., Sasse, J., Borghi, L., Schlegel, M., Bachelier, J.B., Reinhardt, D., Bours, R., Bouwmeester, H.J., Martinoia, E., 2012. A petunia ABC protein controls strigolactone-dependent symbiotic signalling and branching. Nature 483, 341–344.

Kulik, A., Noirot, E., Grandperret, V., Bourque, S., Fromentin, J., Salloignon, P., Truntzer, P., Dobrowolska, G., Simón-Plas, F., Wendehenne, D., 2014. Interplays between nitric oxide and reactive oxygen species in cryptogein signaling. Plant Cell Environ. 38, 331–348.

Kumar, S.R., Mohanapriya, G., Sathishkumar, R., 2016. Abiotic stress-induced redox changes and programmed in cell death in plants-a path to survival or death? In: Gupta, D.K., et al. (Ed.), Redox State as a Central Regulator of Plant-Cell Stress Responses. Springer International Publishing, Switzerland. http://dx.doi.org/10.1007/978-3-319-44081-1-11.

Künstler, A., Bacso, R., Hafez, Y.M., Kiraly, L., 2015. Reactive oxygen species and plant disease resistance. In: Gupta, D.K., et al. (Ed.), Reactive Oxygen Species and Oxidative Damage in Plants Under Stress. Springer International Publishing, Switzerland, pp. 269–303. http://dx.doi.org/10.1007/978-3-319-20421-5_11.

Lam, E., 2004. Controlled cell death, plant survival and development. Nat. Rev. Mol. Cell Biol. 5, 3015–3315.

Landi, L., Valori, F., Ascher, J., Renella, G., Falchini, L., Nannipieri, P., 2006. Root exudate effects on the bacterial communities, CO_2 evolution, nitrogen transformations and ATP content of rhizosphere and bulk soils. Soil Biol. Biochem. 38, 509–516.

Lanoue, A., Burlat, V., Henkes, G.J., Koch, I., Schurr, U., Röse, U.S.R., 2010. De novo biosynthesis of defense root exudates in response to Fusarium attack in barley. New Phytol. 185, 577–588.

Lareen, A., Burton, F., Schäfer, P., 2016. Plant root-microbe communication in shaping root microbiomes. Plant Mol. Biol. 90, 575–587.

Lecourieux, D., Ranjeva, R., Pujin, A., 2006. Calcium in plant defence-signaling pathway. New Phytol. 171, 249–269.

Li, H., Liu, D., He, H., Zhang, N., Ge, F., Chen, C., 2014. Molecular cloning of a 14-3-3 protein gene from *Lilium regale* Wilson and overexpression of this gene in tobacco increased resistance to pathogenic fungi. Sci. Hortic. 168, 9–16.

Liese, A., Romeis, T., 2013. Biochemical regulation of in vivo function of plant calcium-dependent protein kinases (CDPK). Biochim. Biophys. Acta 1833, 1582–1589.

Lore, C., 2015. A new receptor for LPS. Nat. Immunol. 16, 340–341.

Lozano-Durán, R., Robatzek, S., 2015. 14-3-3 proteins in plant-pathogen interactions. Mol. Plant Microbe Interact. 5, 511–518.

Lugtenberg, B., Rozen, D.E., Kamilova, F., 2017. Wars between microbes on roots and fruits. F1000Res. 6, 343.

Ma, W., Berkowitz, G.A., 2011. Ca^{++} conduction by plant cyclic nucleotide gated channels and associated signaling components in pathogen defense signal transduction cascades. New Phytol. 190, 566–572.

Ma, Y., Walker, R.K., Zhao, Y., Berkowitz, G.A., 2012. Linking ligand perception by PEPR pattern recognition receptor to cytosolic Ca^{++} elevation and downstream immune signaling in plants. Proc. Nat. Acad. Sci. USA 109, 19852–19857.

Macho, A.P., Zipfel, C., 2014. Plant PRRs and the activation of innate immune signaling. Mol. Cell 54, 263–272.

Macho, A.P., Schwessinger, B., Ntoukakis, V., Brutus, A., Segonzac, C., Roy, S., Kadota, Y., Oh, M.H., Sklenar, J., Derbyshire, P., et al., 2014. A bacterial tyrosine phosphatase inhibits plant pattern recognition receptor activation. Science 343, 1509–1512.

Manickavelu, A., Kawaura, K., Oishi, K., Shin, I.T., Kohara, Y., Yahiaoui, N., Keller, B., Suzuki, A., Yano, K., Ogihara, Y., 2010. Comparative gene expression analysis of susceptible and resistant near-isogenic lines in common wheat infected by *Puccinia triticina*. DNA Res. 17, 211–222.

Manosalva, P.M., Bruce, M., Leach, J.E., 2011. Rice 14-3-3 protein (GF14e) negatively affects cell death and disease resistance. Plant J. 68, 777–787.

Marino, D., Dunand, C., Puppo, A., Pauly, N., 2012. A burst of plant NADPH oxidases. Trends Plant Sci. 17, 9–15.

Martin, G.B., et al., 1993. Map-based cloning of a protein kinase gene conferring disease resistance in tomato. Science 262, 1432–1436.

Massalha, H., Korenblum, E., Tholl, D., Aharoni, A., 2017. Small molecules below-ground: the role of specialized metabolites in the rhizosphere. Plant J. 90, 788–807.

Matilla, M.A., Ramos, J.L., Barker, P., Doornbos, R.D., Badri, D.V., Vivanco, J.M., Ramos-González, M.I.R., 2010. *Pseudomonas putida* KT2440 causes induced systemic resistance and changes in *Arabidopsis* root exudation. Environ. Microb. Rep. 2, 381–388.

Matilla, M.A., Krell, T., 2017. Plant growth promotion and biocontrol mediated by plant-associated bacteria. In: Egamberdieva, D., Ahmad, P. (Eds.). Plant microbiome: Stress response. Springer, New York.

Matilla-Vázquez, M.A., Matilla, A.J., 2012. Role of H_2O_2 as signaling molecule in plants. In: Ahmad, P., Prasad, M.N.V. (Eds.), Environmental Adaptations and Stress Tolerance of Plants in the Era of Climate Change, Springer Science+Business Media, LLC.

Matilla-Vázquez, M.A., Matilla, A.J., 2014. Ethylene: role in plants under environmental stress. In: Ahmad, P., Wani, M.R. (Eds.). Physiological Mechanisms and Adaptation Strategies in Plants under Changing Environment, vol. 2. Springer Science+Business Media, New York. http://dx.doi.org/10.1007/978-1-4614-8600-8_7.

Mello, E.O., Ribeiro, S.F.F., Carvalho, A.O., Santos, I.S., Da Cunha, M., Santa-Catarina, C., Gomes, V.M., 2011. Antifungal activity of PvD1 defensin involves plasma membrane permeabilization, inhibition of medium acidification, and induction of ROS in fungi cells. Curr. Microbiol. 62, 1209–1217.

Micallef, S.A., Shiaris, M.P., Colon-Carmona, A., 2009. Influence of *Arabidopsis thaliana* accessions on rhizobacterial communities and natural variation in root exudates. J. Exp. Bot. 60, 1729–1742.

Moe, L.A., 2013. Amino acids in the rhizosphere: from plants to microbes. Am. J. Bot. 100, 1692–1705.

Moghaddam, M.R.B., Van den Ende, W., 2012. Sugars and plant innate immunity. J. Exp. Bot. 63, 3989–3998.

Moghaddam, M.R.B., Van den Ende, W., 2013. Sweet immunity in the plant circadian regulatory network. J. Exp. Bot. 64, 1439–1449.

Moghaddam, M.R.B., Vilcinskas, A., Rahnamaeian, M., 2016. Cooperative interaction of antimicrobial peptides with the interrelated immune pathways in plants. Mol. Plant Pathol. 17, 464–471.

Monaghan, J., Zipfel, C., 2012. Plant pattern recognition receptor complexes at the plasma membrane. Curr. Opin. Plant Biol. 15, 349–357.

Monaghan, J., Matschi, S., Shorinola, O., Rovenich, H., Matei, A., Segonzac, C., Malinovsky, F.G., Rathjen, J.P., MacLean, D., Romeis, T., Zipfel, C., 2014. The calcium-dependent protein kinase CPK28 buffers plant immunity and regulates BIK1 turnover. Cell Host Microbe 16, 605–615.

Mori, I.C., Schroeder, J.I., 2004. Reactive oxygen species activation of plant Ca^{2+} channels. A signaling mechanism in polar growth, hormone transduction, stress signaling, and hypothetically mechanotransduction. Plant Physiol. 135, 702–708.

Mougel, C., Offre, P., Ranjard, L., Corberand, T., Gamalero, E., Robin, C., Lemanceau, P., 2006. Dynamic of the genetic structure of bacterial and fungal communities at different developmental stages of *Medicago truncatula* Gaertn. cv. Jemalong line J5. New Phytol. 170, 165–175.

Muthamilarasan, M., Prasad, M., 2013. Plant innate immunity: an updated insight into defense mechanism. J. Biosci. 38, 433–449.

Mutka, A.M., Fawley, S., Tsao, T., Kunkel, B.N., 2013. Auxin promotes susceptibility to *Pseudomonas syringae* via a mechanism independent of suppression of salicylic acid-mediated defenses. Plant J. 74, 746–754.

Nakashita, H., Yasuda, M., Nitta, T., Asami, T., Fujioka, S., Arai, Y., Sekimata, K., Takatsuto, S., Yamaguchi, I., Yoshida, S., 2003. Brassinosteroid functions in a broad range of disease resistance in tobacco and rice. Plant J. 33, 887–898.

Naseem, M., Kaltdorf, M., Dandekar, T., 2015. The nexus between growth and defence signalling: auxin and cytokinin modulate plant immune response pathways. J. Exp. Bot. 66, 4885–4896.

Newmann, G., Römheld, V., 2007. The release of root exudates as affected by the plant physiological status. In: Pinton, R., Varanini, Z., Nannipieri, P. (Eds.), The Rhizosphere: Biochemistry and Organic Substances at the Soil-Plant Interface. CRC Press, New York, pp. 23–72.

Newman, M.-A., Sundelin, T., Nielsen, J.T., Erbs, G., 2013. MAMP (microbe-associated molecular pattern) triggered immunity in plants. Front. Plant Sci. 4, 139.

Nie, H., Zhao, C., Wu, G., Wu, Y., Chen, Y., Tang, D., 2012. SR1, a calmodulin-binding transcription factor, modulates plant defense and ethylene-induced senescence by directly regulating NDR1 and EIN3. Plant Physiol. 158, 1847–1859.

Nomura, K., Debroy, S., Lee, Y.H., Pumplin, N., Jones, J., He, S.Y., 2006. A bacterial virulence protein suppresses host innate immunity to cause plant disease. Science 313, 220–223.

Nürnberger, T., 1999. Signal perception in plant pathogen defense. Cell Mol. Life Sci. 55, 167–182.

Nürnberger, T., Brunner, F., Kemmerling, B., Piater, L., 2004. Innate immunity in plants and animals: striking similarities and obvious differences. Immunol. Rev. 198, 249–266.

O'Brien, J.A., Arsalan Daudi, A., Finch, P., But, V.S., Whitelegge, J.P., Souda, P., Ausubel, F.M., Bolwell, G.P., 2013. A peroxidase-dependent apoplastic oxidative burst in cultured Arabidopsis cells functions in MAMP-elicited defense. Plant Physiol. 158, 2013–2027.

Obsil, T., Obsilova, V., 2011. Structural basis of 14-3-3 protein functions. Semin. Cell Dev. Biol. 22, 663–672.

Oh, C.S., Pedley, K.F., Martin, G.B., 2010. Tomato 14-3-3 protein 7 positively regulates immunity-associated programmed cell death by enhancing protein abundance and signaling ability of MAPKKKα. Plant Cell 22, 260–272.

Oh, C.S., Martin, G.B., 2011. Tomato 14-3-3 protein TFT7 interacts with a MAP kinase kinase to regulate immunity-associated programmed cell death mediated by diverse disease resistance proteins. J. Biol. Chem. 286, 14129–14136.

Pedras, M.S., Yaya, E.E., 2015. Plant chemical defenses: are all constitutive antimicrobial metabolites phytoanticipins? Nat. Prod. Commun. 10, 209–218.

Perchepied, L., Balague, C., Riou, C., Claudel-Renard, C., Riviere, N., Grezes-Besset, B., Roby, D., 2010. Nitric oxide participates in the complex interplay of defense-related signaling pathways controlling disease resistance to *Sclerotinia sclerotiorum* in *Arabidopsis thaliana*. Mol. Plant Microbe Interact. 23, 846–860.

Perochon, A., Aldon, D., Galaud, J.P., Ranty, B., 2011. Calmodulin and calmodulin-like proteins in plant calcium signaling. Biochim 93, 2048–2053.

Pieterse, C.M.J., et al., 2001. Rhizobacteria-mediated induce systemic resistance: triggering, signaling, and expression. Eur. J. Plant Pathol. 107, 51–61.

Pieterse, C.M., Van der Does, D., Zamioudis, C., León-Reyes, A., Van Wees, S.C., 2012. Hormonal modulation of plant immunity. Annu. Rev. Cell Dev. Biol. 28, 489–521.

Pieterse, C.M., Zamioudis, C., Berendsen, R.L., Weller, D.M., Van Wees, S.C., Bakker, P.A., 2014. Induced systemic resistance by beneficial microbes. Annu. Rev. Phytopathol. 52, 347–375.

Poovaiah, B.W., Du, L., Wang, H., Yang, T., 2013. Recent advances in calcium/calmodulin-mediated signaling with an emphasis on plant-microbe interactions. Plant Physiol. 163, 531–542.

Postel, S., Kemmerling, B., 2009. Plant systems for recognition of pathogen-associated molecular patterns. Semin. Cell Dev. Biol. 20, 1025–1031.

Proels, R.K., Oberhollenzer, K., Pathuri, I.P., Hensel, G., Kumlehn, J., Hückelhoven, R., 2010. RBOHF2 of barley is required for normal development of penetration resistance to the parasitic fungus *Blumeria graminis* f. sp. *hordei*. Mol. Plant Microbe Interact. 23, 1143–1150.

Raaijmakers, J.M., Paulitz, T.C., Steinberg, Ch., Alabouvette, C., Moënne-Loccoz, Y., 2009. The rhizosphere: a playground and battlefield for soilborne pathogens and beneficial microorganisms. Plant Soil 321, 341–361.

Rahnamaeian, M., 2011. Antimicrobial peptides: modes of mechanism, modulation of defense responses. Plant Signal Behav. 6, 1325–1332.

Ranf, S., Eschen-Lippold, L., Pecher, P., Lee, J., Scheel, D., 2011. Interplay between calcium signalling and early signalling elements during defence responses to microbe- or damage-associated molecular patterns. Plant J. 68, 100–113.

Ranf, S., Gisch, N., Schäffer, M., Illig, T., Westphal, L., Knirel, Y.A., et al., 2015. A lectin S-domain receptor kinase mediates lipopolysaccharide sensing in *Arabidopsis thaliana*. Nat. Immunol. 16, 426–433.

Ranf, S., 2016. Immune sensing of lipopolysaccharide in plants and animals: same but different. PLoS Pathog. 12, e1005596.

Ranf, S., 2017. Sensing of molecular patterns through cell surface immune receptors. Curr. Opin. Plant Biol. 38, 68–77.

Rea, P.A., 2007. Plant ATP-binding cassette transporters. Annu. Rev. Plant Biol. 58, 347–375.

Reddy, V.S., Shlykov, M.A., Castillo, R., Sun, E.I., Saier, M.H., 2012. The major facilitator superfamily (MFS) revisited. FEBS J. 279, 2022–2035.

Ren, D., Liu, Y., Yang, K.Y., Han, L., Mao, G., Glazebrook, J., Zhang, S., 2008. A fungal-responsive MAPK cascade regulates phytoalexin biosynthesis in *Arabidopsis*. Proc. Natl. Acad. Sci. USA 105, 5638–5643.

Robb, J., Lee, B., Nazar, R.N., 2007. Gene suppression in a tolerant tomato-vascular pathogen interaction. Planta 226, 299–309.

Robert, N., Ferran, J., Breda, C., Coutos-Thévenot, P., Boulay, M., Buffard, D., Esnault, R., 2001. Molecular characterization of the incompatible interactions of *Vitis vinifera* leaves with *Pseudomonas syringae* pv. pisi: expression of genes coding for stilbene synthase and class 10 PR protein. Eur. J. Plant Pathol. 107, 249–261.

Robert-Seilaniantz, A., Grant, M., Jones, J.D., 2011. Hormone crosstalk in plant disease and defense: more than just jasmonate-salicylate antagonism. Annu. Rev. Phytopathol. 49, 317–343.

Rojas, C.M., Senthil-Kumar, M., Tzin, V., Mysore, K.S., 2014. Regulation of primary plant metabolism during plant-pathogen interactions and its contribution to plant defense. Front. Plant Sci. 5, 17.

Romeis, T., Ludwig, A.A., Martin, R., Jones, J.D., 2001. Calcium-dependent protein kinases play an essential role in a plant defence response. EMBO J. 20, 5556–5567.

Romeis, T., Herde, M., 2014. From local to global: CDPKs in systemic defense signaling upon microbial and herbivore attack. Curr. Opin. Plant Biol. 20, 1–10.

Ruiz-May, E., Galaz-Ávalos, R.M., Loyola-Vargas, V.M., 2009. Differential secretion and accumulation of terpene indole alkaloids in hairy roots of *Catharanthus roseus* treated with methyl jasmonate. Mol. Biotechnol. 41, 278–285.

Scharf, B.E., Hynes, M.F., Alexandre, G.M., 2016. Chemotaxis signaling systems in model beneficial plant-bacteria associations. Plant Mol. Biol. 90, 549–559.

Scheler, C., Durner, J., Astier, J., 2013. Nitric oxide and reactive oxygen species in plant biotic interactions. Curr. Opin. Plant Biol. 16, 534–539.

Schulz, P., Herde, M., Romeis, T., 2013. Calcium-dependent protein kinases: hubs in plant stress signaling and development. Plant Physiol. 163, 523–530.

Segonzac, C., Zipfel, C., 2011. Activation of plant pattern-recognition receptors by bacteria. Curr. Opin. Microbiol. 14, 54–61.

Segonzac, C., Macho, A.P., Sanmartín, M., Ntoukakis, V., Sánchez-Serrano, J.J., Zipfel, C., 2014. Negative control of BAK1 by protein phosphatase 2A during plant innate immunity. EMBO J. 33, 2069–2079.

Shetty, N.P., Jørgensen, H.J.L., Jensen, J.D., Collinge, D.B., Shetty, H.S., 2008. Roles of reactive oxygen species in interactions between plants and pathogens. Eur. J. Plant Pathol. 121, 267–280.

Shi, S., Richardson, A.E., O'Callaghan, M., De Angelis, K.M., Jones, E.E., Stewart, A., Firestone, M.K., Condron, L.M., 2011. Effects of selected root exudate components on soil bacterial communities. FEMS Microbiol. Ecol. 77, 600–661.

Shimizu, T., Nakano, T., Takamizawa, D., Desaki, Y., Ishii-Minami, N., Nishizawa, Y., Minami, E., Okada, K., Yamane, H., Kaku, H., Shibuya, N., 2010. Two lysM receptor molecules, CEBiP and OsCERK1, cooperatively regulate chitin elicitor signaling in rice. Plant J. 64, 204–214.

Shinya, T., Nakagawa, T., Kaku, H., Shibuya, N., 2015. Chitin-mediated plant–fungal interactions: catching, hiding and handshaking. Curr. Opin. Plant Physiol. 26, 64–71.

Silipo, A., Erbs, G., Shinya, T., Dow, J.M., Parrilli, M., Lanzetta, R., Shibuya, N., Newman, M.-A., Molinaro, A., 2010. Glyco-conjugates as elicitors or suppressors of plant innate immunity. Glycobiol 20, 406–419.

Sirikantaramas, S.M., Yazaki, K.A., Saito, K., 2008. Mechanisms of resistance to self-produced toxic secondary metabolites in plants. Phytochem. Rev. 7, 467–477.

Stael, S., Kmiecik, P., Willems, P., Van Der Kelen, K., Coll, N.S., Teige, M., Breusegem, F., 2015. Plant innate immunity – sunny side up? Trends Plant Sci. 20, 3–11.

Stein, M., Dittgen, J., Sánchez-Rodríguez, C., Hou, B.H., Molina, A., Schulze-Lefert, P., Lipka, V., Somerville, S., 2006. Arabidopsis PEN3/PDR8, an ATP binding cassette transporter, contributes to nonhost resistance to inappropriate pathogens that enter by direct penetration. Plant Cell 18, 731–746.

Song, W.-Y., Yamaki, T., Yamaji, N., Ko, D., Jung, K.-H., Fujii-Kashino, M., An, G., Martinoia, E., Lee, Y., Ma, J.F., 2014. A rice ABC transporter, OsABCC1, reduces arsenic accumulation in the grain. Proc. Nat. Acad. Sci. USA 111, 15699–15704.

Spalding, E.P., Harper, J.F., 2011. The ins and outs of cellular Ca^{2+} transport. Curr. Opin. Plant Biol. 14, 715–720.

Spoel, S.H., Dong, X., 2012. How do plants achieve immunity? Defence without specialized immune cells. Nat. Rev. Immunol. 12, 89–100.

Strader, L.C., Bartel, B., 2009. The *Arabidopsis* pleiotropic drug resistance8/ABCG36 ATP binding cassette transporter modulates sensitivity to the auxin precursor indole-3-butyric acid. Plant Cell 21, 1992–2007.

Strader, L.C., Bartel, B., 2011. Transport and metabolism of the endogenous auxin precursor indole-3-butyric acid. Mol. Plant 4, 477–486.

Sugiyama, A., Shitan, N., Yazaki, K., 2007. Involvement of a soybean ATP-binding cassette-type transporter in the secretion of genistein, a signal flavonoid in legume–Rhizobium symbiosis. Plant Physiol. 144, 2000–2008.

Suzuki, N., Miller, G., Morales, J., Shulaev, V., Torres, M.A., Mittler, R., 2012. Respiratory burst oxidases: the engines of ROS signaling. Curr. Opin. Plant Biol. 14, 691–699.

Taguchi, F., Yamamoto, M., Ohnishi-Kameyama, M., Iwaki, M., Yoshida, M., Ishii, T., Konishi, T., Ichinose, Y., 2010. Defects in flagellin glycosylation affect the virulence of *Pseudomonas syringae* pv. *tabaci* 6605. Microbiol 156, 72–80.

Tassoni, A., Fornalè, S., Franceschetti, M., Musiani, F., Micheal, A.J., Perri, B., Bagni, N., 2005. Jasmonates and Na-orthovanadate promote resveratrol production in *Vitis vinifera* L. cv. Barbera cell cultures. New Phytol. 166, 895–905.

Taylor, K.W., Kim, J.G., Su, X.B., Aakre, C.D., Roden, J.A., Adams, C.M., Mudgett, M.B., 2012. Tomato TFT1 is required for PAMP-triggered immunity and mutations that prevent T3S effector XopN from binding to TFT1 attenuate Xanthomonas virulence. PLoS Pathog. 8, e1002768.

Teper, D., Salomon, D., Sunitha, S., Kim, J.G., Mudgett, M.B., Sessa, G., 2014. *Xanthomonas euvesicatoria* type III effector XopQ interacts with tomato and pepper 14-3-3 isoforms to suppress effector-triggered immunity. Plant J. 77, 297–309.

Thakur, M., Sohal, B.S., 2013. Role of elicitors in inducing resistance in plants against pathogen infection: a Review. ISRN Biochem.:762412. http://dx.doi.org/10.1155/2013/762412.

Thomma, B.P.H.J., Nürnberger, T., Joosten, M.H.A.J., 2011. Of PAMPs and effectors: the blurred PTI-ETI dichotomy. Plant Cell 23, 4–15.

Torres, M.A., 2010. ROS in biotic interactions. Physiol. Plant 138, 414–429.

Toyomasu, T., Kagahara, T.K., Okada, K., Koga, J., Hasegawa, M., Mitsuhashi, W., Sassa, T., Yamane, H., 2014. Diterpene phytoalexins are biosynthesized in and exuded from the roots of rice seedlings. Biosci. Biotech. Biochem. 72, 562–567.

Trdá, L., Boutrot, F., Claverie, J., Brulé, D., Dorey, S., Poinssot, B., 2015. Perception of pathogenic or beneficial bacteria and their evasion of host immunity: pattern recognition receptors in the frontline. Front. Plant Sci. 6, 219.

Trapet, P., Kulik, A., Lamotte, O., Jeandroz, S., Bourque, S., Nicolas-Francès, V., Rosnoblet, C., Besson-Bard, A., Wendehenne, D., 2015. NO signaling in plant immunity: a tale of messengers. Phytochem 112, 72–79.

Trouvelot, S., Héloir, M.C., Poinssot, B., Gauthier, A., Paris, F., Guillier, C., Combier, M., Trdá, L., Daire, X., Adrian, M., 2014. Carbohydrates in plant immunity and plant protection: roles and potential application as foliar sprays. Front. Plant Sci. 5, 592.

Uren, N.C., 2007. Types, amounts and possible functions of compounds released into rhizosphere by soil-grown plants. In: Pinton, R., Varanini, Z., Nannipieri, P. (Eds.), The Rhizosphere: Biochemistry and Organic Substances at the Soil-Plant Interface. CRC Press, New York, pp. 1–21.

van Dam, N.M., Bouwmeester, H.J., 2016. Metabolomics in the rhizosphere: tapping into belowground chemical communication. Trends Plant Sci. 21, 256–265.

van der Weerden, N.L., Lay, F.T., Anderson, M.A., 2008. The plant defensin, NaD1, enters the cytoplasm of *Fusarium oxysporum* hyphae. J. Biol. Chem. 283, 14445–14452.

Van Etten, H.D., Mansfield, J.W., Bailey, J.A., Farmer, E.E., 1994. Two classes of plant antibiotics: phytoalexins versus "phytoanticipins". Plant Cell 6, 1191–1192.

Veronese, P., Nakagami, H., Bluhm, B., Abuqamar, S., Chen, X., Salmeron, J., Dietrich, R.A., Hirt, H., Mengiste, T., 2006. The membrane-anchored Botrytis-induced kinase1 plays distinct roles in *Arabidopsis* resistance to necrotrophic and biotrophic pathogens. Plant Cell 18, 257–273.

Walker, T.S., Bais, H.P., Grotewold, E., Vivanco, J.M., 2003. Root exudation and rhizosphere biology. Plant Physiol. 132, 44–51.

Walker, T.S., Bais, H.P., Deziel, E., Schweizer, H.P., Rahme, L.G., Fall, R., Vivanco, J.M., 2004. *Pseudomonas aeruginosa*-plant root interactions. Pathogenicity, biofilm formation, and root exudation. Plant Physiol. 134, 320–331.

Wan, J., Zhang, X-Ch., Neece, D., Ramonell, K.M., Clough, S., Kim, S.-Y., Stacey, M.G., Stacey, G., 2008. A lysM receptor-like kinase plays a critical role in chitin signaling and fungal resistance in *Arabidopsis*. Plant Cell 20, 471–481.

Wang, Y., Loake, G.L., Chu, C., 2013. Cross-talk of nitric oxide and reactive oxygen species in plant programed cell death. Front. Plant Sci. 4, 314.

Weisskopf, L., Le Bayon, R.C., Kohler, F., Page, V., Jossi, M., Gobat, J.M., Martinoia, E., Aragno, M., 2008. Spatio-temporal dynamics of bacterial communities associated with two plant species differing in organic acid secretion: a one-year microcosm study on lupin and wheat. Soil Biol. Biochem. 40, 1772–1780.

Wen, F., VanEtten, H.D., Tsaprailis, G., Hawes, M.C., 2007. Extracellular proteins in pea root tip and border cell exudates. Plant Physiol. 143, 773–783.

Weston, L.A., Duke, S.O., 2003. Weed and crop allelopathy. Crit. Rev. Plant Sci. 22, 367–389. http://dx.doi.org/10.1080/713610861.

Weston, L.A., Ryan, P.R., Watt, M., 2012. Mechanisms for cellular transport and release of allelochemicals from plant roots into the rhizosphere. J. Exp. Bot. 63, 3445–3454.

Weston, L.A., Skoneczny, D., Weston, P.A., Weidenhamer, J.D., 2015. Metabolic profiling: an overview – new approaches for the detection and functional analysis of biologically active secondary plant products. J. Allelochem. Interact. 2, 15–27.

Wiesel, L., Newton, A.C., Elliott, I., Booty, D., Gilroy, E.M., Birch, P.R.J., Hein, I., 2014. Molecular effects of resistance elicitors from biological origin and their potential for crop protection. Front. Plant Sci. 5, 655.

Willmann, R., Lajunen, H.M., Erbs, G., Newman, M.A., Kolb, D., Tsuda, K., Katagiri, F., Fliegmann, J., Bono, J.-J., Cullimore, J.V., Jehle, A.K., Götz, F., Kulik, A., Molinaro, A., Lipka, V., Gust, A.A., Nürnberger, T., 2011. Arabidopsis lysin-motif proteins LYM1 LYM3 CERK1 mediate bacterial peptidoglycan sensing and immunity to bacterial infection. Proc. Natl. Acad. Sci. USA 108, 19824–19829.

Wojtaszek, P., 1997. Oxidative burst: an early plant response to pathogen infection. Biochem. J. 322, 681–692.

Wu, X., Li, R., Shi, J., Wang, J., Sun, Q., Zhang, H., Guo, Y.-D., 2014. *Brassica oleracea* MATE encodes a citrate transporter, and enhances aluminum tolerance in *Arabidopsis thaliana*. Plant Cell Physiol. 55, 1426–1436.

Xin, X.F., Nomura, K., Underwood, W., He, S.Y., 2013. Induction and suppression of PEN3 focal accumulation during *Pseudomonas syringae* pv. tomato DC3000 infection of *Arabidopsis*. Mol. Plant Microbe Interact. 26, 861–867.

Xin, X.-F., Nomura, K., Aung, K., Velásquez, A., Yao, J., Boutrot, F., Chang, J.H., Zipfel, C., He, S.Y., 2016. Bacteria establish an aqueous living space in plants crucial for virulence. Nature 539, 524–530.

Yamaguchi, Y., Huffaker, A., 2011. Endogenous peptide elicitors in higher plants. Curr. Opin. Plant Biol. 14, 351–357.

Yang, X., Wang, W., Coleman, M., Orgil, U., Feng, J., Ma, X., Ferl, R., Turner, J.G., Xiao, S., 2009. Arabidopsis 14-3-3 lambda is a positive regulator of RPW8-mediated disease resistance. Plant J. 60, 539–550.

Yang, Y.X., Ahammed, G.J., Wu, C., Fan, S.Y., Zhou, Y.H., 2015. Crosstalk among jasmonate, salicylate and ethylene signaling pathways in plant disease and immune responses. Curr. Protein Pept. Sci. 16, 450–461.

Yazaki, K., Sugiyama, A., Morita, M., Shintan, N., 2008. Secondary transport as an efficient membrane transport mechanism for plant secondary metabolites. Phytochem. Rev. 7, 513–524.

Yin, H., Du, Y., Dong, Z., 2016. Chitin oligosaccharide and chitosan oligosaccharide: two similar but different plant elicitors. Front. Plant Sci. 7, 522.

Yoshioka, H., Mase, K., Yoshioka, M., Kobayashi, M., Asai, S., 2011. Regulatory mechanisms of nitric oxide and reactive oxygen species generation and their role in plant immunity. Nitric Oxide 25, 216–221.

Zago, E., Morsa, S., Datt, J.F., Alard, P., Ferrarini, A., Inzè, D., Delledonne, M., Van Breusegem, F., 2006. Nitric oxide- and hydrogen peroxide-responsive gene regulation during cell death induction in tobacco. Plant Physiol. 141, 404–411.

Zeng, H., Xu, L., Singh, A., Huizhong Wang, H., Du, L., Poovaiah, B.W., 2015. Involvement of calmodulin and calmodulin-like proteins in plant responses to abiotic stresses. Front. Plant Sci. 6, 600.

Zhang, S., Klessig, D.F., 2001. MAPK cascades in plant defense signaling. Trends Plant Sci. 6, 520–527.

Zhang, Y., Xu, S., Ding, P., Wang, D., Cheng, Y.T., He, J., Gao, M., Xu, F., Li, Y., Zhu, Z., Li, X., Zhang, Y., 2010. Control of salicylic acid synthesis and systemic acquired resistance by two members of a plant-specific family of transcription factors. Proc. Natl. Acad. Sci. USA 107, 18220–18225.

Zhang, L., Du, L., Poovaiah, B.W., 2014a. Calcium signaling and biotic defense responses in plants. Plant Signal Behav. 9 (11), e973818.

Zhang, J., Li, W., Xiang, T., Liu, Z., Laluk, K., Ding, X., Zou, Y., Gao, M., Zhang, X., Chen, S., Mengiste, T., Zhang, Y., Zhou, J.M., 2010. Receptor-like cytoplasmic kinases integrate signaling from multiple plant immune receptors and are targeted by a *Pseudomonas syringae* effector. Cell Host Microbe 7, 290–301.

Zhang, Y., Lubberstedt, T., Xu, M., 2013. The genetic and molecular basis of plant resistance to pathogens. J. Genet. Genomics 40, 23–35.

Zhang, L., Du, L., Shen, C., Yang, Y., Poovaiah, B.W., 2014b. Regulation of plant immunity through ubiquitin-mediated modulation of Ca^{2+}-calmodulin-AtSR1/CAMTA3 signaling. Plant J. 78, 269–281.

Zhang, D.W., Deng, X.G., Fu, F.Q., Lin, H.H., 2015. Induction of plant virus defense response by brassinosteroids and brassinosteroid signaling in *Arabidopsis thaliana*. Planta 241, 875–885.

Zhao, B., Lin, X., Poland, J., Trick, H., Leach, J., Hulbert, S., 2005. A maize resistance gene functions against bacterial streak disease in rice. Proc. Nat. Acad. Sci. USA 102, 15383–15388.

Zhu, X., Skoneczny, D., Weidenhamer, J.D., Mwendwa, J.M., Weston, P.A., Gurr, G.M., Callaway, R.M., Weston, L.A., 2016. Identification and localization of bioactive naphthoquinones in the roots and rhizosphere of Paterson's curse (*Echium plantagineum*), a noxious invader. J. Exp. Bot. 67, 3777–3788.

Zipfel, C., Robatzek, S., Navarro, L., Oakeley, E.J., Jones, J.D., Félix, G., Boller, T., 2004. Bacterial disease resistance in *Arabidopsis* through flagellin perception. Nature 428, 764–767.

Zipfel, C., Kunze, G., Chinchilla, D., Caniard, A., Jones, J.D.G., Boller, T., Felix, G., 2006. Perception of the bacterial PAMP EF-Tu by the receptor EFR restricts *Agrobacterium*-mediated transformation. Cell 125, 749–760.

Zipfel, C., 2008. Pattern-recognition receptors in plant innate immunity. Curr. Opin. Immunol. 20, 10–16.

Zipfel, C., 2009. Early molecular events in PAMP-triggered immunity. Curr. Opin. Plant Biol. 12, 414–420.

Zipfel, C., 2014. Plant pattern-recognition receptors. Trends Immunol. 35, 345–351.

Chapter 11

Primary Plant Metabolism During Plant–Pathogen Interactions and Its Role in Defense

Belur S. Kumudini, Nellickal S. Jayamohan, Savita V. Patil, Mallu Govardhana
Jain University, Bangalore, India

1. INTRODUCTION

The present-day food production requires ecologically sound and compatible agricultural strategies for sustainable crop production. This is one of the major challenges for a very rapidly growing population to meet the demand for adequate food production. Biotic factors or phytopathogens affect plants right from the stage of seed germination till storage of the produce. These phytopathogens are mainly grouped into three, based on their modes of plant infection: necrotrophs, biotrophs, and hemibiotrophs (Pandey et al., 2016). Microbes that attack and kill their host and feed on the host's nutritive materials are described as necrotrophs (e.g., the gray mold fungus *Botrytis cinerea* and the bacterial soft rot pathogen *Erwinia carotovora*), whereas pathogens that live on their host to continue their life cycle as they continue to feed on its nutritive material are referred to as biotrophs (e.g., powdery mildew fungus *Blumeria graminis* and *Cladosporium fulvum* and the bacterial rice pathogen *Xanthomonas oryzae*). Pathogens that require a living host initially, but kill it at later stage of infection, are described as hemibiotrophs (e.g., *Magnaporthe grisea*, the causative agent of rice blast disease). In general, bacteria and fungi adopt either a biotrophic or necrotrophic mode of infection, whereas viruses infect plants as "perfect biotrophs," although viral infection can consequently result in host cell death (Thomma et al., 2001).

However emerging and reemerging plant diseases continuously challenge this ability, and managing plant diseases seems to be an uphill task. For plants to grow and survive, the harmful effects of those antagonistic interactions must be overcome. Consequently, plants have evolved several different defense mechanisms by which they defend themselves against diseases or pathogenic infections (Jones and Dangl, 2006; Freeman and Beattie, 2008; Jan et al., 2011). These defense mechanisms of plants are called basal resistance, which refers to two distinct aspects of plant–pathogen interaction: constitutive (wax layers, rigid cell walls, cuticular lipids, cutin, bark, etc., are physical barriers to pathogens) and induced (lignification, deposition of gum and callose, suberization, formation of tyloses and abscission layer, etc., are nonspecific pathogen-elicited) defense mechanisms (van Baarlen et al., 2007; Anderson et al., 2010). The innate immunity of plants protects them from pathogen threats. Such innate immunity is based on preformed and induced defense responses. Preformed defense responses are nonspecific and include compounds with antimicrobial properties or structural barriers, such as the cell wall and the cytoskeleton, that deter pathogens and pests (Mysore and Ryu, 2004; Uma and Podile, 2015). Induced defenses are activated by the recognition of pathogen-associated molecular patterns (PAMPs) present on the pathogen surface or by recognition of proteins (effectors) translocated by the pathogen to the host cell (Bonardi and Dangl, 2012). Early induced defense responses include cytoskeletal reorganization, cell wall fortification, generation of reactive oxygen species (ROS), and synthesis of phytoalexins (Torres, 2010). Later events during defense responses include transcription of pathogenesis-related (PR) proteins and the development of a type of programmed cell death known as the hypersensitive response (HR) that limits pathogen spread (Ahuja et al., 2012). Although considerable progress has been made to understand plant defense responses, very little is known about the role of primary metabolic pathways required for the growth and development in regulating plant defense responses.

The induced defense response in the host plant depends upon the mode of nutrition and lifestyle of plant pathogen (Pandey et al., 2016). Plants respond to pathogens with the help of natural structural barriers as well as physiological changes such as respiration, photosynthesis, nutrient translocation, transpiration, growth, and development using primary

Plant Metabolites and Regulation Under Environmental Stress. http://dx.doi.org/10.1016/B978-0-12-812689-9.00011-X

carbon metabolism (Berger et al., 2007; Zeilinger et al., 2016). The first step in the infection process is the cell-to-cell communication between the host plant and pathogens. Plants of species or varieties may not be infected by pathogen if their epidermis lack specific recognition factors (Doughari, 2015). If the pathogen is unable to recognize the plant as one of its host, then it may not adhere to the host surface or produce elicitors such as enzymes (chitinase, lipase, cellulase) or structures such as appressoria and haustoria. These recognition molecules include various types of oligosaccharides, polysaccharides, and glycoproteins (van Baarlen et al., 2007).

For a pathogen to successfully infect a plant, it must

1. find the host and the appropriate site of entry
2. develop specialized infection structures
3. overcome preformed and induced host defenses
4. take up nutrients, grow, and colonize the host tissue

In addition, morphological changes are essential for pathogen penetration (van Baarlen et al., 2007). Access into the plant tissues can be either by direct penetration of plant surfaces or through physical injuries or natural openings such as the stomata (Zeilinger et al., 2016). Interestingly, the lignin barrier defense system in plant roots is much stronger than in the leaves (Underwood, 2012).

During plant–pathogen interactions, the first line of plant defense consists primarily of the PAMP/MAMP (membrane associated molecular pattern) recognition system that has been shown to confer the so-called basal resistance to a wide range of pathogens (Jones and Dangl, 2006). MAMPs include a broad category of compounds- typically essential for microbial life including the conserved bacterial cell surface structures like lipopolysaccharides, elongation factor Tu (EF-Tu), flagellin, cold-shock proteins, peptidoglucans, or fungal cell wall polysaccharide structures like glucan or chitin (Pieterse et al., 2009; Staal and Dixelius, 2009). Based on its localization, there are two types of receptors recognizing PAMPs: surface and intracellular receptors (Staal and Dixelius, 2009). These PAMPs are recognized by the pattern recognition receptor system of plants resulting in the activation of the host's basal immunity. This is known as PAMP-triggered immunity (PTI), which is the first line of defense (Jones and Dangl, 2006; Bigeard et al., 2015). A well-studied example is of *Botrytis*-induced kinase 1 (BIK1), which has been identified as a critical component in the PTI against the fungal necrotrophic pathogens *B. cinerea* and *Alternaria brassicicola*. BIK1 functions independently of the currently known defense-related mitogen-activated protein kinase (MAPK) cascades in inducing defense responses in the early stages of pathogen recognition (Veronese et al., 2006).

However, many of the pathogens are able to suppress this PTI by inducing effector molecules. Cui et al. (2015) showed that plants have a different way of tackling this invasion using R-proteins, which can recognize the effector molecules, called effector-triggered immunity (ETI), otherwise called gene-for-gene immunity as the role of R-proteins is to specifically detect and recognize particular pathogenic effectors (Chisholm et al., 2006). This *R*-gene-mediated resistance results in host defense activation, including HR or apoptosis at the site of infection, in an effort to limit the pathogen's spread throughout the plant and hence the development of the disease (Hwang et al., 2015).

Usually in plant–fungal interaction, PTI plays a major role before the invasion of appressorium, whereas ETI acts postinvasion (Rojas et al., 2014). The important feature of PTI is its ability to sense infectious self and nonself, whereas for ETI, it is the ability to sense microbe-mediated modifications inferred on points of vulnerability in the host (Malinovsky et al., 2014).

In addition to the two level-specific receptors, there are a few nonspecific elicitors that may lead to immunity. These include mostly abiotic elicitors such as damage-associated molecular pattern, elicitins, and *R*-gene-specific, insect-specific, and volatile organic compounds (Doughari, 2015).

The plant responses on pathogen infection include (Pieterse et al., 2009; Rojas et al., 2014):

- Photosynthesis-associated changes
- Synthesis of PR proteins
- Cytoskeleton reorganization
- Production of ROS
- Cell apoptosis

2. VARIATIONS IN PRIMARY METABOLITES

2.1 Photosynthesis Reduction During Plant Defense

During photosynthesis, plants harvest light energy to generate ATP and reducing power in the form of NADPH, which can be utilized in the production of assimilates for various biological processes (Tanaka and Makino, 2009). As plant defense responses may alter the pool size of a range of metabolic intermediates, the photosynthetic metabolism is likely to be influenced as it adjusts to meet the demands of the cell. Two hypotheses could be proposed as to the relationship between photosynthesis and energy for the defense response.

First, from a supply-and-demand standpoint, the rates of photosynthesis could increase to supply the carbon skeletons, energy, and reducing equivalents required to support the plant defense.

Second, if disposable cellular activities are reduced during the resistance response, the production of defense-related compounds becomes "top priority," with photosynthetic rates reduced until pathogenic growth has been terminated (Somssich and Hahlbrock, 1998).

Indeed, most studies on photosynthesis and plant defense have shown that the rates of photosynthesis are reduced locally when challenged by virulent or avirulent pathogens, herbivores, or wounding, or with hormone treatment (Horst et al., 2008; Schwachtje and Baldwin, 2008). Although such decreases are typically localized near the infection area, studies have shown that the rate of photosynthesis increases in the uninfected leaves of plants being attacked by virulent pathogens. Studies using chlorophyll *a* fluorescence imaging techniques showed that the rate of photosynthesis was nearly zero at the periphery of initial infection sites before the final steps of defense and cell death were initiated during the incompatible *Phytophthora parasitica* var. *nicotianae*–tobacco interaction (Swarbrick et al., 2006). Photosynthesis reduced both at the center of attempted pathogen penetration site and in the surrounding cells during the resistance response mediated by Mla12 (race specific) and mlo (broad spectrum) in barley against powdery mildew. This decrease is not only due to the elimination of green (photosynthetic) leaf area as a consequence of the HR but also due to an alteration in host metabolism.

Localized ribulose bisphosphate carboxylase (*RbcS*) and chlorophyll *a*/*b*-binding (*Cab*) gene repression may not be detectable in whole leaves due to incompatible interaction between *Arabidopsis* and *Pseudomonas syringae*; therefore plants avoid systemic downregulation to assimilate high production, especially in an incompatible interaction, as most cells do not encounter the pathogen (Bonfig et al., 2006). In another study of the same pathosystem, *RbcS* and *Cab* transcripts were both reduced more than 20-fold in the incompatible interaction. For example, during a compatible interaction between tomato and its pathogens *P. syringae* and *B. cinerea*, the expected zone of inhibited photosynthesis is in the direct vicinity of the infection sites, but there is also a distinct stimulation of photosynthesis surrounding the infection sites (Berger et al., 2007). It is not clear if the enhanced photosynthesis is due to the plant's attempt to fuel defense responses to limit pathogen growth or fungal manipulation of the plant to co-opt resources for fungal nutrition. Some studies have been carried out to show the mechanistic alteration of photosynthesis for suppression of plant defense mechanism and reprogramming of plant's metabolism to the pathogen's own benefit (Garavaglia et al., 2010). Indeed most of the studies reveal the considerable reduction in photosynthetic activity locally at the early stages during pathogen infestation, after herbivore attack or wounding. The downregulation of photosynthesis gene expression, most probably through hexose signaling, is a central feature following pathogen attack (Chou et al., 2000; Bonfig et al., 2006; Swarbrick et al., 2006; Bischof et al., 2011).

Studies by Rodríguez-Herva et al. (2012) on bacterial pathogen *P. syringae* indicate the effective role of HopN1 (cysteine protease), which targets the PsbQ protein involved in the oxygen evolving complex, mediating its effects by deteriorating electron transport in photosystem (PS) II. Studies by Jones et al. (2006) on *Arabidopsis* show increasing levels of PsbO1 and Psbp in thylakoid lumen when infected by an avirulent AvrRpm1-expressing strain of *P. syringae*. Photosynthetic regulation of defense mechanism during plant–pathogen interactions have been substantiated by many studies on plant–virus interactions (Lehto et al., 2003; Jang et al., 2013). Chloroplastic isochorismate synthase 1 (ICS1), contributes to the biosynthesis of PSI electron transfer component phylloquinone and salicylic acid (SA). This was evinced by *Arabidopsis* ICS1 mutants, increased plastoquinone reduction state, and altered thylakoid structure, linked with plant immune signaling (Gawroński et al., 2013; Kangasjärvi et al., 2013). Disparity in photosynthetic electron transfer induces various redox reactions, which induce an array of defense-related genes through chloroplast retrograde signaling pathways that share an intimate cross-talk with hormonal regulation. These complex interactions allow modulation of both local and systemic responses that promote resistance against light stress and plant pathogens (Karpinski et al., 1999, 2013; Miller et al., 2009).

Sugars regulate cellular activity at multiple levels, from transcription and translation to protein stability and activity. However, the classic pathogen signaling classes, i.e., MAMP/PAMP and ETI, completely exclude the role of sugar moieties in plant immunity. Recent evidences clearly show the significant role of sugar in plant defense (Morkunas and Ratajczak, 2014) owing to a new concept called "sweet priming"—it predicts specific key roles of saccharides in perceiving, mediating, and counteracting both biotic and abiotic stresses (Bolouri Moghaddam and Van den Eden, 2012; Morkunas and Ratajczak, 2014). Also, sugars are involved in multiple steps of the host immunity regulation—activation of PR proteins (Herbers et al., 1996; Salzman et al., 1998; Thibaud et al., 2004; Essmann et al., 2008; Kocal et al., 2008), swing in transcription of enzymes during biosynthesis of secondary metabolites (Herbers et al., 1996; Solfanelli et al., 2006; Ferri et al., 2011; Morkunas et al., 2011), production of reaction species, and maintainenance of the cell redox state (Kim et al., 2006; Essmann et al., 2008).

The lessened photosynthetic activity in infected tissue augments metabolic source–sink transitions, thereby inducing majority of the defense-related genes (Herbers et al., 1996; Gomez et al., 2007). Studies on induction of secondary metabolite synthesis have evidence overexpression of an yeast invertase in the vacuole or apoplast of tobacco leaves resulting in constitutive induction of defense-related genes, elevated callose content, and peroxidase activities (Herbers et al., 1996). Resistance to potato virus Y (PVY) in tobacco plants is due to inhibition of sugar export (Herbers et al., 2000), which leads to a model of sugar-enhanced plant defense.

Differential expression studies have been done on photosynthetic genes during the *Fusarium oxysporum* f.sp. *ciceri* race 1 (Foc1) infection in chickpea (*Cicer arietinum* L.). The results demonstrate that the infection of root tissues of chickpea by Foc1 dramatically increases the ROS levels in shoot tissues of susceptible plants, which hampers the photosynthetic stability by downregulating the key photosynthetic genes. On the contrary, resistant chickpea lines are grossly devoid of such instances with few behavioral irregularities at later time points (Bhar et al., 2016).

2.2 Changes Associated with Carbohydrate Molecules

Glucose and sucrose function as signaling molecules that enhance the expression of various defense-specific genes (Morkunas and Ratajczak, 2014). Trehalose is another disaccharide signal for growth and development processes of plants, as well as plant defense responses, whereas trehalose-6-phosphate is a powerful signaling molecule in plant cells (Delatte et al., 2011). *In vitro* studies using leaf disks of tobacco showed that sugar solutions induced PR transcripts PR-Q and PAR1 photoassimilate responding gene 1 (Rojas et al., 2014). Recent studies on sucrose-induced receptor kinase in *Arabidopsis* by Wu et al. (2013) substantiates sucrose acting as a signaling molecule. Glucose and sucrose increase the level of cell wall invertase (Cw-Inv) in *Pisum sativum* (Zhang et al., 1997), *Chenopodium rubrum* (Roitsch et al., 1995), and *Solanum lycopersicum* (Proels and Roitsch, 2009).

Cw-Inv plays a pivotal role in sugar sensing and metabolic reprogramming during plant–microbe interactions (Proels and Huckelhoven, 2014). Pathogen challenge induces expression or derepression of Cw-Inv, mediated by the downregulation of proteinaceous Cw-Inv inhibitors, resulting in enhanced sucrose breakdown into glucose and fructose. Cw-Inv localizes at the integration point of hormonal, sugar, defense, and diurnal/circadian regulation. Hexose breakdown further supports respiration and synthesis of secondary metabolites, namely, callose and phenolic compounds, which act as signaling molecules.

Hexokinase (HXK1) is the best investigated enzyme, catalyzing the first step of glycolysis (Smeekens et al., 2010), whose isoforms have been found in the cytosol, chloroplast, mitochondria, and nucleus (Hanson and Smeekens, 2009), which reflects its roles in a variety of cellular processes. Mitochondria-associated HXKs play a role in the control of programmed cell death (Morkunas and Ratajczak, 2014). In mutant HXK, gene expression studies in *Arabidopsis* plants by Bolouri-Moghaddam et al. (2010) reveal that nuclear HXK signaling integrates nutrient and hormone signals to regulate gene expression during plant growth and development. One of the earlier reports by Sheen (1990) on regulation of gene expression by sugars was indicated as HXK, which involves the repression of photosynthesis proteins (chlorophyll *a*/*b*-binding protein).

Although, extensive studies have been carried out on the involvement of sugars in plant defense mechanism, clear evidence of specific sugars involved in antagonism or in coordination with pathogenesis remains to be elucidated (Lecompte et al., 2017).

2.3 Amino Acid Changes Accompanying a Pathogen Encounter

The evidence accumulated so far suggests that the accumulation of some amino acids or their metabolic by-products triggers resistance responses against pathogens that can be dependent or independent of SA- and ROS-mediated defense pathways.

Mutation in the homoserine kinase encoded by downy mildew resistant 1 caused accumulation of the amino acid homoserine and increased resistance against the oomycete *Hyaloperonospora arabidopsidis* by an unknown mechanism that was independent of the major defense signaling pathways such as jasmonic acid (JA), SA, and Ethylene (ET) (van Damme et al., 2009).

The literature shows the role of 14-3-3 proteins in plant–pathogen interaction in different plant systems. Various isoforms of 14-3-3 proteins in *Arabidopsis* and tobacco reportedly interact with defense-related proteins and lead to improved plant defense activity, whereas isoforms of 14-3-3 proteins in tomato lead to apoptosis during pathogen encounter (as reviewed by Lozano-Duran and Robatzek, 2015). Mass spectrometry studies identified more than 300 potential 14-3-3 target proteins, which makes them one of the most connected nodes in plants (de Boer et al., 2013).

A noted study by Dulermo et al. (2009) observed depletion of glutamate in healthy leaf region in sunflower during *B. cinerea* infection. A 50%–80% amino acid decrease and a 50% ammonium decrease were detected in leaves at 48 hpi.

Glutamate was transferred to the infected region from the healthy region, which was correlated by the enhanced sunflower glutamate dehydrogenase transcription level in the area of pathogen invasion.

Gene expression studies revealed the upregulation of several genes involved in amino acid biosynthesis in *Arabidopsis*, as a response against avirulent *P. syringae* pv. tomato (AvrRpt2; Scheideler et al., 2002), *P. syringae* pv. tomato (AvrRpm1), and *P. syringae* pv. tomato (Hrc), or the pathogen-derived elicitors Flg22 and HrpZ (Less et al., 2011).

Metabolomic studies indicate the accumulation of valine, leucine, and tyrosine in *Arabidopsis* upon treatment with avirulent and virulent pathogens (Rojas et al., 2014). Disparity in amino acid accumulation by flow injection electrospray ionization-mass spectrometry data in host plants (*Brachypodium distachyon, Hordeum vulgare,* and *Oryza sativa*) has been observed when treated with *M. grisea* during the initial 72h after challenge inoculation (Parker et al., 2009). This change in metabolic profile also involves various other biosynthetic mechanisms, viz., metabolisms of carbohydrate, aromatic carbons, and urea/proline and fatty acids. *B. distachyon* infestation varied with the phospholipid concentrations after interactions with rice blast fungus (Allwood et al., 2006), and in a biotrophic interaction between *Ustilago maydis* and maize variety, the targeted metabolomics study found significant increase in aromatic amino acids, shikimate, and hydroxycinnamic acid derivatives and hexoses 4days after infection (Doehlemann et al., 2008). There was a significant increase in metabolites' pathways connected to sugar and amino acid metabolism, the tricarboxylic acid cycle, the γ-aminobutyric acid shunt, ROS scavengers, and phenylpropanoids, relative to the control plants at the start of the strong viral infection against PVY infection in potato plants (Kogovsek et al., 2016).

Mutation studies on glutamine indicate the increase in callose accumulation and H_2O_2 production (Pilot et al., 2004), by increase in genes involved in glutamine, proline, and aspartate metabolism have been shown to be induced after exposure to avirulent pathogens and pathogen-derived elicitors (Rojas et al., 2014). These changes in various amino acid accumulations indicate the direct or indirect regulation of defense response signaling and secondary metabolite biosynthesis in the host plants. These signal mechanisms induce metabolic by-products and trigger the resistance response against the pathogen, which can be dependent or independent of SA- and ROS-mediated mechanisms.

3. IMPACT OF PRIMARY METABOLIC VARIATIONS IN PLANT PHYSIOLOGY

Earlier it was generally accepted that the primary metabolism of plants, including photosynthesis, was suppressed during a pathogen encounter and the saved energy was diverted toward defense mechanism. However, the latest theories suggest that the primary metabolism of plants also function to confront a pathogen (Rojas et al., 2014).

3.1 Generation of Reactive Oxygen Species

One of the most rapid responses in this context is the oxidative burst, leading to the transient production of ROS following successful pathogen recognition (Staal and Dixelius, 2009). There is a difference in the ROS production by plants in response to virulent and avirulent pathogen encounter (Torres et al., 2006). A drastic increase in ROS production in the host can induce cell death and restrict further penetration of pathogen through activation of various signaling pathways such as PTI and ETI. Generally, avirulent pathogens are successfully recognized through the action of disease-resistant *R*-gene products in plant immune system. This elicits a biphasic accumulation of ROS with a low-amplitude, transient first phase, followed by a sustained second phase of much higher degree, which correlates with disease resistance (Kunstler et al., 2015). However, virulent pathogens that avoid host recognition induce only the transient, low-amplitude first phase (Torres et al., 2006). Although ROS usually correlates with successful disease resistance responses in plants, some pathogens induce production of ROS in the host to their own advantage (Kuzniak and Sklodowska, 2005). Therefore, the ability to maintain ROS balance can be correlated with the disease tolerance of a plant (Kartashov et al., 2008).

Major ROS are singlet oxygen ($^1O_2$), hydrogen peroxide (H_2O_2), superoxide radical ($O_2^{\cdot -}$), and hydroxyl radical (OH). The ROS production leads to the oxidation of proteins, precisely a site-specific modification of amino acids (especially those that contain thiol groups), leading to susceptibility in plants (Das and Roychoudhary, 2014). The cellular protein undergoes various modifications, which may either be direct or indirect. During direct modifications, the activity of the protein becomes varied as a result of different chemical modifications such as nitrosylation, carboxylation, disulfide bond formation, and glutathionylation (Moller et al., 2007).

The ROS burst occurs due to pathogen entry and may lead to direct pathogen killing. In addition, ROS also acts as a diffusible messenger resulting in HR, structural cell wall glycoprotein oxidative cross-linking, and induction of systemic acquired resistance (SAR) (Wojtaszek, 1997).

H_2O_2 is predominantly generated in chloroplasts through energy transfer from photo-excited chlorophyll to the molecular oxygen electron. H_2O_2 reacts easily with double bonds and has affinity toward amino acids such as histidine, methionine, tryptophan, and cysteine. H_2O_2 accumulation can depend on two consecutive steps: reduction followed by spontaneous or catalyzed dismutation of $O_2 \rightarrow O_2^{\cdot-}$ followed by spontaneous or catalyzed dismutation of $O_2^{\cdot-}$ to H_2O_2 (Thakur and Sohal, 2013). H_2O_2 also leads to lipid peroxidation in the plasma membrane, which acts as a signal for JA production. Within the cytoplasm, H_2O_2 also helps in stabilizing messenger RNA, and within the nucleus they promote defense-specific gene activation/repression (Oliveira et al., 2016).

3.2 Cytoskeleton Manipulations

Infection by fungal necrotrophs is a complex process that includes conidial attachment, germination, host penetration, lesion formation, expansion, and tissue maceration followed by sporulation (Laluk and Mengiste, 2010). The penetration through cuticular layer is achieved by cutinases and lipases (Serrano et al., 2014). Once penetrated they start producing a set of cell wall degrading enzymes and polygalacturonases. As the primary strategy of inhibition of cell wall polysaccharide degradation by microbial enzymes, plants produce proteinaceous inhibitors, viz., cell wall–associated polygalacturonase-inhibiting proteins (Bellincampi et al., 2014).

Biotrophs and hemibiotrophs attack the plant surface and penetrate the external barriers by developing appressoria and exploiting the mechanical pressure. To weaken the first structural barrier, pathogens secrete oxidases, cutinases, esterases, and lipases (Feng et al., 2011; Zhao et al., 2013). Generally, plants respond to them by the formation of "papillae," or cell wall oppositions, or cell wall, early at the site of pathogen entry. In addition to callose, as the major constituent, papillae encompass cellulose, hemicelluloses, pectins, lignin, and structural proteins such as arabinogalactan proteins and hydroxyproline-rich glycoproteins (Bellincampi et al., 2014). Papillae are also the sites where antimicrobial peptides, toxic secondary metabolites, and ROS accumulate and contribute to plant resistance (Daudi et al., 2012) at the site of pathogen penetration too (Fig. 11.1).

Recent research has pointed out that plants employ an additional layer of defense that functions outside the cell wall, perhaps before the onset of PTI. Studies by Toyoda et al. (2016) in pea plants showed that the ecto-ATPase would recognize

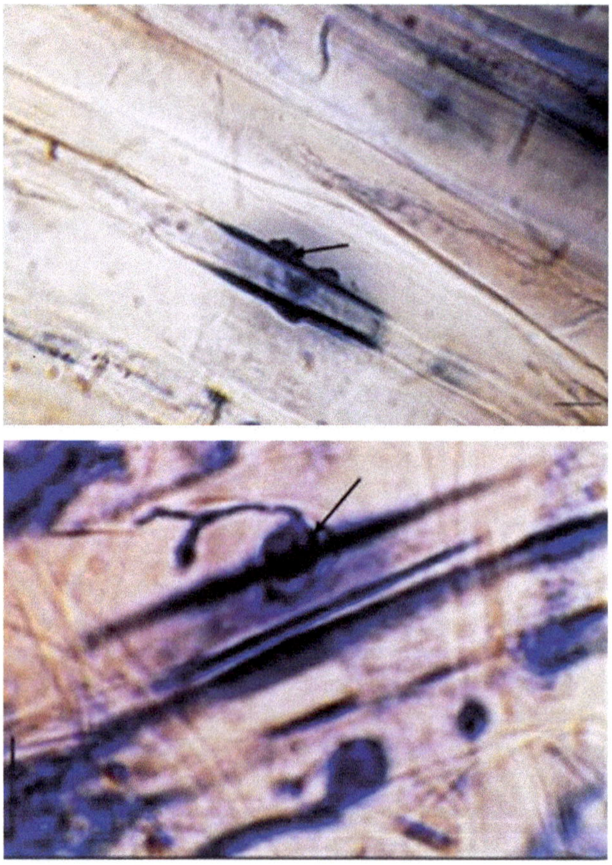

FIGURE 11.1 Formation of cell wall appositions at the site of pathogen entry (*arrow*) in an incompatible interaction between pearl millet and *Sclerospora graminicola* (Kumudini, 2004; unpublished data).

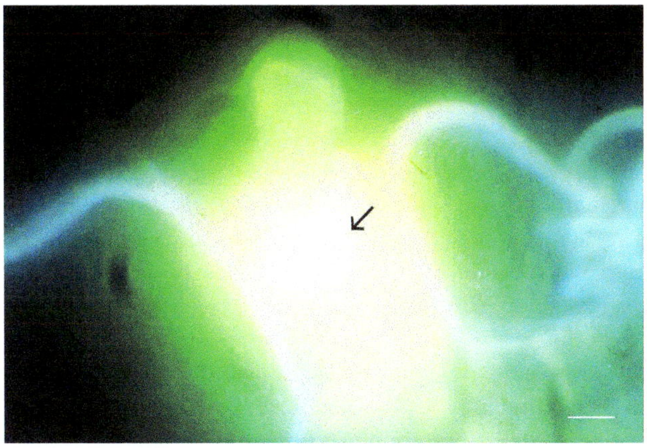

FIGURE 11.2 Deposition of callose to prevent pathogen entry (*arrow*) at the stomatal region in an incompatible interaction between pearl millet and *Sclerospora graminicola* (Kumudini, 2004; unpublished data).

the incoming pathogen-derived elicitors via direct binding and quickly initiate the oxidative burst, which eventually lead to production of $O_2^{\cdot-}$ and H_2O_2. Evidences show that the expression of *PR10-1* may cause temporary activation of JA-regulated cellular processes in the host plant.

Callose is an essential cell wall homopolymer, surrounding pollen mother cells, and encloses the four microspores after meiosis (Malinovsky et al., 2014). In response to pathogen attack, callose is deposited between the plasma membrane and the preexisting cell wall at sites of pathogen attack (Ellinger and Voigt, 2014). The cell wall thickenings due to callose formation at the sites of pathogen attack are a major physical barrier to slow pathogen invasion and are observed in all the plant species (Voigt, 2014). A time course histochemical study revealed that the accumulation of the polymers in cell walls of resistant (cv. IP 18293) seedlings was rapid and localized around the pathogen, as a defense strategy to restrict the downy mildew pathogen against *Sclerospora graminicola* by deposition of lignin and callose (Kumudini and Shetty, 2002) as shown in Fig. 11.2. Haustoria formation was prevented in resistant lines by JA-dependant defense mechanism induced by callose deposition during compatible and incompatible interactions in *Arabidopsis*-powdery mildew disease (Ellinger et al., 2013).

Plant defense is also established by lignification, which is another mechanical barrier to pathogen invasion, diminishing the diffusion of toxins into the host cells (Raj et al., 2012); lignin can also pose as a toxin to the pathogens (Nicholson and Hammerschmidt, 1992).

Various studies have implicated the role of cell wall/cytoskeleton rearrangements during plant–pathogen interactions, which have been directed by primary metabolite signaling during stress.

3.3 Signaling Associated with Cell Apoptosis

In a susceptible disease response, plant cells are killed directly by toxins produced by the pathogens, indirectly by a metabolic catastrophe due to secretion of virulence factors, or by an endogenous cell death program triggered the recognition of the specific signal molecules. Ceramide signaling pathway as a critical second messenger system has been studied in detail to understand apoptosis during both degenerative and proliferative disease expressions in animal systems (Khurana et al., 2005). The balance between the bioactive sphingolipid ceramide and its phosphorylated derivatives has been proposed to modulate the apoptosis not only in animals but also in plants (Liang et al., 2003). In compatible interactions, the toxin secreted by the fungus activates apoptosis through competitive inhibition of ceramide synthase, a key enzyme in sphingolipid biosynthesis, and this effect is modulated by codominant *Asc* alleles. When there is no inhibition of ceramide synthase by the dominant *Asc/Asc* alleles, it leads to a resistance response. The intermediate inhibition of ceramide synthase leads to partial apoptosis (intermediate response), whereas competitive inhibition of ceramide synthase leads to complete apoptosis, which is a susceptible response (Khurana et al., 2005).

ROS accumulate preceding cell death during HR. Neither SA nor ET can trigger HR; however, both the hormones are regarded as positive regulators of HR. JA is regarded as a negative regulator of plant cell apoptosis with the help of nitric oxide (NO) (Staal and Dixelius, 2009). In addition, brassinosteroids also function as hormones that promote senescence (Rao et al., 2002).

3.4 PR Proteins Dependent Transformations

PR proteins in plants were first discovered in tobacco infected with tobacco mosaic virus (Van Loon and Van Kammen, 1970). During a pathogen encounter, PR proteins are produced not only by the adjacent healthy cells but also in the infected plant cells (Oliveira et al., 2016).

Depending on their isoelectric points, PR proteins may be acidic or basic, although with similar functions. Most acidic PR proteins are located in the intercellular spaces, whereas basic PR proteins are predominantly located in the vacuole (Oliveira et al., 2016). PR proteins are categorized into 17 families according to their properties and functions, including β-1,3-glucanases, chitinases, thaumatin-like proteins, peroxidases, ribosome-inactivating protein defensins, thionins, nonspecific lipid transfer proteins, oxalate oxidase, and oxalate-oxidase-like proteins (Oliveira et al., 2016; Mahesh et al., 2017).

Among these PR proteins, chitinases and β-1,3-glucanases are two important hydrolytic enzymes that are abundant in many plant species after infection with different types of pathogens. They play a major role in the defense reactions against fungal pathogens by degrading their cell walls, containing chitin and β-1,3-glucan as major structural components. β-1,3-glucanases are expressed in coordination with chitinases after fungal infection (Ebrahim et al., 2011; Oliveira et al., 2016; Patil et al., 2016).

Lipoxygenase (LOX) is a major enzyme involved in plant growth and signals the developmental processes. It additionally takes part in the synthesis of regulatory metabolic signals, which help in their defense responses (Alvarez et al., 2016). LOX incorporates molecular oxygen to fatty acids to generate fatty acid hydroperoxides, which act as substrates for allene oxide synthase leading to the production of jasmonates, which are well-known signals in plant defense mechanisms (Lavanya et al., 2012).

4. VARIATIONS IN SECONDARY METABOLITE

4.1 SA- and JA-Centric Regulation of Plant Immunity

Hormones regulate plant immunity through various antagonistic or synergistic interactions under specific circumstances. Once the plant receptor genes identify pathogen-specific elicitors, a set of phytohormones are also activated, which may bind to the specific domains of nuclear proteins to activate downstream regulatory mechanism and R-genes. These chemical signal elicitor molecules that regulate plant defense include SA and JA (Pieterse et al., 2009; Kazan and Lyons, 2014; Siddaiah et al., 2017). SA is a phenolic compound hormone released in response to a wide number of pathogens. This is a major hormone for the establishment of local and systemic immunity in plants (Alazem and Lin, 2015). SA is mainly involved in resistance to biotrophs, whereas methyl jasmonate and ET play a significant role in resistance against necrotrophic pathogens (Grobkinsky et al., 2014; Kumudini et al., 2017).

Gene expression level studies in model plant *Arabidopsis* showed invasion of pathogens and insects resulting in mechanical wounding, generating signals/elicitors that activate a phosphorylation cascade and actually regulate JA signaling and synthesis (Kazan and Lyons, 2014). In the host plant JA is synthesized from the precursor α-linolenic acid, which is converted to 13-hydroperoxy linolenic acid, which is again converted to 12-oxo-phytodionic acid, in the chloroplast. This is transported to peroxisomes by ATP-binding cassette and subsequently reduced to 3-oxo-2 (2pentenyl) cyclopentane-1-octanoic acid, which is converted into JA through a series of β-oxidations (Lyons et al., 2013). The JA is converted to bioactive JA-Ile by jasmonate resistant 1, which encodes a jasmonate-amido synthetase. JAs activate a signaling cascade that leads to the activation of genes required for direct defense against phloem-feeding herbivores as well as fungal pathogens (Lyons et al., 2013).

In a noted study by Groszmann et al. (2015), it was demonstrated that *P. syringae* infection on *Arabidopsis* F1 hybrids led to reduced SA and auxin-regulated defense. A study on apple trees demonstrated that ET- and JA-mediated defense response are functional even in the root system of perennial tree species defending soil-borne pathogens. Exogenous application of ET and JA indicated that JA always acts as a positive regulator, whereas ET acts as either a positive or a negative feedback signal against *Pythium ultimum* in apple trees (Shin et al., 2014).

4.2 Role of Abscisic Acid (ABA) in Plant Defense

The plant hormone abscisic acid (ABA) plays a pivotal role in plant defense (Flors et al., 2005). ABA is another major plant hormone resulting from the cleavage of γ-carotene. This regulates a number of physiological processes including development and adaptive stress responses in plants (Alazem and Lin, 2015). The role of ABA reportedly varies in different plant systems. In some systems ABA positively regulates the disease incidence by priming callose deposition and stomatal closure (Ton and Mauch-Mani, 2004; He et al., 2006). However, a recent study shows that ABA acts a negative regulator in many plant systems. Studies on rice blast disease caused by *Magnaporthe oryzae* showed that increased ABA level leads to higher disease incidence. Similar results were observed in *X. oryzae* pv. *oryzae* in rice through suppression of SA (Xu et al., 2013).

Studies showed that ABA has a multifaceted role in plant immunity depending on the stage of infection. ABA can positively regulate plant defense at the early stages of infection through mediation of stomatal closure, or induction of callose deposition if the pathogen evades the first line of defense. However, if activated at later stages, ABA can suppress ROS production and SA or JA signaling, hence nullifying the defense response controlled by these two pathways (Ton et al., 2009).

A study also demonstrated cytokinin-mediated resistance through reduced ABA levels followed by enhanced ABA catabolism in tobacco–*P. syringae* interaction. ABA-impaired mutant tomato and *Arabidopsis* plants were shown to overexpress defensive signaling pathways, leading to enhanced resistance against bacterial and fungal pathogens (as reviewed by Denance et al., 2013). Studies on nematode pathogenicity (*Hirschmanniella oryzae*) showed that ABA plays a negative role in defense response in rice plants; however, the role of ABA is strongly correlated with the ET, JA, and SA biosynthesis pathways (Nahar et al., 2012). Although a negative regulator of plant immunity, ABA can also positively regulate the resistance to some pathogens such as *A. brassicicola*, *Rhizoctonia solanacearum,* and *Pythium irregulare*, in mutants (as reviewed by Denance et al., 2013). Earlier reports on the role of ABA in plant–virus interaction were made in various plant systems. *Phaseolus vulgaris* infected with tobacco necrosis virus showed that ABA restricts viral movement through callose deposition. However, ABA was not induced in response to white clover mosaic virus infection in the same host (Clarke et al., 1999).

JA and SA cross talk is initiated by *NPR1*, *JIN1/MYC2*, *MPK4*, *WRKY70*, and *SCF* genes simultaneously. JA–auxin cross talk is governed by *CSN*, *SGT1b*, *AFR*, and *AXR1* genes (Lyons et al., 2013; Kazan and Lyons, 2014). The *PDF1.2* gene, which encodes JA-inducible plant defensin in *Arabidopsis*, is a well-known marker for the induction of induced systemic resistance (Lyons et al., 2013).

4.3 Role of Ethylene

ET is one of the major volatile plant hormones that affect a myriad of developmental processes, such as seedling emergence, leaf and flower senescence, formation of the vascular system, ripening, organ abscission, and also responses to biotic and abiotic stresses (Bouchez et al., 2007). Its role in the interaction with necrotrophic and hemibiotrophic pathogens is characterized by two opposite effects: ET primarily promotes senescence and is involved in programmed cell death during HR (Van Loon et al., 2006). In vitro application of ET showed that a large burst of ET is produced after the early steps of HR initiation and can induce defense-related genes (Bouchez et al., 2007).

4.4 Role of Nitric Oxide in Plant Defense

Traditionally, plant defense has been described as SA- and SA/ET-centric pathway in response to pathogen attack in plants. However, the final response is mediated by a complex web of signal network composed of different phytohormones and other regulators.

NO is a volatile gas and also a free radical; it has recently been characterized as a coregulator of many plant processes including defense response. However, its regulatory role is often exerted in connection with other classical regulatory hormones such as ABA in the plant (Leon et al., 2013). NO and ABA together contribute to defense against phytopathogens both as alternative pathways to the major SA-related or JA/ET-related signaling events (Robert-Seilaniantz et al., 2011).

Evidently, both NO and ABA were demonstrated to play convincing roles in activating pattern-triggered immunity in plant systems (Leon et al., 2013; Nie et al., 2017). They have also been reported to modulate cell death typical of the HR in incompatible plant–pathogen interactions. Invariably, the role of NO in plant immunity has been associated with intracellular Ca^{2+} and cyclic nucleotide level changes (Ma, 2011). ABA negatively regulates resistance, whereas NO positively contributes to plant defense during ETI and necrotroph attack.

Many research groups with interest in NO mechanism in plant defense are concentrating on thiol oxidation by NO, generally referred as *S*-nitrosylation. *S*-nitrosylation comes about by the reaction of the oxidized form of NO, the nitrosonium ion (NO^+), which can electrophilically attack thiolate to produce *S*-nitrosylated thiols. This reaction can generate large pools of *S*-nitrosoglutathione (GSH + NO → GSNO + H+). GSNO itself can act as a nitrosylating agent and thus as a biochemical "memory," so that the effects of NO could persist after its generation has ceased or act as a mobile signal. Hence NO effects can be propagated throughout a plant as a component of SAR (Espunya et al., 2012). If the thiol group belongs to cysteine residues of proteins, the result is an *S*-nitrosoprotein that can affect protein function. A fascinating example of this is the control of ROS generation and cell death through protein *S*-nitrosylation of cysteine 890 residue flanking the flavin-binding domain of the NADPH oxidase, which is a major source of ROS generation during HR. When the concentrations of *S*-nitrosothiols were high, NO function also governed a negative feedback loop limiting the HR, mediated by *S*-nitrosylation of the NADPH oxidase, AtRBOHD, at Cys 890, abolishing its ability to synthesize ROS intermediates (Yun et al., 2011).

NO has emerged as a major signal influencing host resistance mediated by SA and ET signaling pathways. It acts as an inducer or suppressor of signaling in both pathways. NO initiates SA biosynthesis and nitrosylates key cysteines on TGACGTCA *cis*-element-binding protein (TGA)-class transcription factors to help in SA-dependent gene expression. Similarly, NO initiates the expression of JA biosynthetic enzymes and ET biosynthetic genes (Mur et al., 2013).

5. TRANSCRIPTIONAL REGULATION OF PATHOGENESIS

Over the last few years, it has become increasingly clear that histone modification, histone replacement, and chromatin remodeling can provide a memory that influences the later transcription of defense-related genes (Berr et al., 2011; Reimer-Michalski and Conrath, 2016).

Plant immune responses are also associated with the upregulation of a large number of WRKY transcripts and proteins. Expression studies showed that ~49 At WRKY genes exhibited differential regulation upon triggering of SA-dependent defenses (Dong et al., 2003). However, many WRKY factors also act as negative regulators of plant defense (Wang et al., 2006). A transcriptome profiling study in soybean during *F. oxysporum* encounter showed marked difference in gene expression against pathogenic and nonpathogenic strains, however, at a later stage of fungal challenge (Lanubile et al., 2015).

TGA-class transcription factors also appear to play a role in SA–JA interactions during pathogen encounter (Ren et al., 2008). Although many TGA factors regulate SA-responsive gene expression, such as TGA2, TGA5, and TGA6, they also induce JA and ET defense genes and crucially, also regulate SA-mediated antagonism (Zander et al., 2010). This suppressive mechanism includes a redox-regulated step catalyzed by glutaredoxins (GRX). GRX catalyze reactions such as oxidation of glutathione and reduction of cysteine residues, which influence the stability and/or activity of proteins. GRX interacts with TGA2, and overexpression of GRX480 countered the induction of the ORA59 promoter by EIN3 (Zander et al., 2012). In a parallel study, GCC box cis-elements such as those found in the promoter of the JA marker gene PDF1.2 were revealed as key sites through which SA–JA antagonism is affected. Focusing on two GCC-binding transcription factors it was found that ORA59 but not ERF1 was the key transcriptional target for the SA antagonistic mechanism (Van der Does et al., 2013).

Another detailed study on *F. oxysporum* f. sp. *conglutinans* in cabbage roots showed that the early defense systems, including the MAPK signaling pathway, calcium signaling, and SA-mediated-HR, were activated after pathogen infection. Lignin biosynthesis pathways, SA-dependent SAR, and ET- and JA-mediated pathways showed significant roles in plant resistance. Among the defense-related genes, such as genes encoding PR proteins, UDP glycosyltransferase, pleiotropic drug resistance, ATP-binding cassette transporters, myrosinase, transcription factors, and kinases showed differential expression (Xing et al., 2016).

6. CONCLUSION AND FUTURE PROSPECTS

Plants face consistent challenge from the surrounding environment, especially from pathogens, which is of great concern as human existence relies on agricultural productivity. Over the last few decades, there is a large amount of information pertaining to plant–pathogen interaction through the rationalization of high-precision techniques in cell biology and genetics. Convergent evolution of resistant pathogen strains is always a big concern for the plant biologists. A detailed understanding of the defense-specific signal cross talk has created the immense possibility to fine-tune the plant defense system to combat the new and evolving pathogen strains. However, focused upgradation of farming practices based on this vast data on plants, relevant to defense priming, is still trailing. However, it is not easy to alter a specific pathway/gene and expect a positive result, as these signal cross talk mechanisms are highly complex and interconnected. There is a need for integrated approach, considering various environmental interactions, which indeed holds great promise in the sustainability of crop productivity. Hence to exploit this rich knowledge so far achieved, researchers should concentrate more on altering the combination of mechanisms/genes. In addition, evolutionary studies are also necessary to check the ability of plants to evolve under specific pathogen elicitors. Indeed, this must contribute more toward identifying the critical mechanisms that can foster a defense against specific pathogen elicitors. The changes in the primary metabolic patterns are the valuable tools in understanding the initial changes in host defense.

ACKNOWLEDGMENTS

The authors wish to apologize to all those whose relevant work cannot be quoted in this book chapter due to space constraints. We acknowledge Jain University and DST-SERB, GoI, for the research funding (YSS/2015/001905).

REFERENCES

Ahuja, I., Kissen, R., Bones, A.M., 2012. Phytoalexins in defense against pathogens. Trends Plant Sci. 17, 1360–1385.

Alazem, M., Lin, N.-S., 2015. Roles of plant hormones in the regulation of host–virus interactions. Mol. Plant Pathol. 16, 529–540.

Allwood, J.W., Ellis, D.I., Heald, J.K., Goodacre, R., Mur, L.A., 2006. Metabolomic approaches reveal that phosphatidic and phosphatidyl glycerol phospholipids are major discriminatory non-polar metabolites in responses by *Brachypodium distachyon* to challenge by *Magnaporthe grisea*. Plant J. 46, 351–368.

Alvarez, A., Montesano, M., Schmelz, E., IPd, L., 2016. Activation of shikimate, phenylpropanoid and auxin pathways in *Pectobacterium carotovorum* elicitors treated moss. Front. Plant Sci. http://dx.doi.org/10.3389/fpls.2016.00328.

Anderson, J.P., Gleason, C.A., Foley, R.C., Thrall, P.H., Burdon, J.B., Singh, K.B., 2010. Plants versus pathogens: an evolutionary arms race. Funct. Plant Biol. 37, 499–512.

Bellincampi, D., Cervone, F., Lionetti, V., 2014. Plant cell wall dynamics and wall-related susceptibility in plant–pathogen interactions. Front. Plant Sci. http://dx.doi.org/10.3389/fpls.2014.00228.

Berger, S., Sinha, A.K., Roitsch, T., 2007. Plant physiology meets phytopathology: plant primary metabolism and plant–pathogen interactions. J. Exp. Bot. 58, 4019–4026.

Berr, A., Shafiq, A., Shen, W., 2011. Histone modifications in transcriptional activation during plant development. Biochim. Biophys. Acta 1809, 567–576.

Bhar, A., Gupta, S., Chatterjee, M., Sen, S., Das, S., 2016. Differential expressions of photosynthetic genes provide clues to the resistance mechanism during *Fusarium oxysporum* f.sp. *ciceri* race 1 (Foc1) infection in chickpea (*Cicer arietinum* L.). Eur. J. Plant Pathol. http://dx.doi.org/10.1007/s10658-016-1109-1.

Bigeard, J., Colcombet, J., Hirt, H., 2015. Signalling mechanisms in pattern triggered immunity (PTI). Mol. Plant 8, 521–539.

Bischof, M., Eichmann, R., Hückelhoven, R., 2011. Pathogenesis-associated transcriptional patterns in Triticeae. J. Plant Physiol. 168, 9–19.

Bolouri Moghaddam, M.R., Van den Eden, W., 2012. Sugars and plant innate immunity. J. Exp. Bot. 63, 3989–3998.

Bolouri-Moghaddam, M.R., Le Roy, K., Xiang, L., Rolland, F., Van den Ende, W., 2010. Sugar signalling and antioxidant network connections in plant cells. FEBS J. 277, 2022–2037.

Bonardi, V., Dangl, J.L., 2012. How complex are intracellular immune receptor signaling complexes? Front. Plant Sci. http://dx.doi.org/10.3389/fpls.2012.00237.

Bonfig, K.B., Schreiber, U., Gabler, A., Roitsch, T., Berger, S., 2006. Infection with virulent and avirulent *P. syringae* strains differentially affects photosynthesis and sink metabolism in Arabidopsis leaves. Planta 225, 1–12.

Bouchez, O., Huard, C., Lorrain, S., Roby, D., Balague, C., 2007. Ethylene is one of the key elements for cell death and defense response control in the Arabidopsis lesion mimic mutant vad1 [w]. Plant Physiol. 145, 465–477.

Chisholm, S.T., Coaker, G., Day, B., Staskawicz, B.J., 2006. Host-microbe interactions: shaping the evolution of the host immune response. Cell 124, 803–814.

Chou, H.M., Bundock, N., Rolfe, S.A., Scholes, J.D., 2000. Infection of *Arabidopsis thaliana* leaves with *Albugo candida* (white blister rust) causes a reprogramming of host metabolism. Mol. Plant Pathol. 1, 99–113.

Clarke, S.F., McKenzie, M.J., Burritt, D.J., Guy, P.L., Jameson, P.E., 1999. Influence of white clover mosaic potexvirus infection on the endogenous cytokinin content of bean. Plant Physiol. 120, 547–552.

Cui, H., Tsuda, K., Parker, J.E., 2015. Effector-triggered immunity: from pathogen perception to robust defence. Ann. Rev. Plant Pathol. 66, 487–511.

Das, K., Roychoudhary, A., 2014. Reactive oxygen species (ROS) and response of antioxidants as ROS-scavengers during environmental stress in plants. Front. Environ. Sci. http://dx.doi.org/10.3389/fenvs.2014.00053.

Daudi, A., Cheng, Y.Z., O'Brien, J.A., Mammarella, N., Khan, S., Ausubel, F.M., Bolwell, G.P., 2012. The apoplastic oxidative burst peroxidase in Arabidopsis is a major component of pattern-triggered immunity. Plant Cell 24, 275–287.

de Boer, A.H., van Kleeff, P.J., Gao, J., 2013. Plant 14-3-3 proteins as spiders in a web of phosphorylation. Protoplasm 250, 425–440.

Delatte, T.L., Sedijani, P., Kondou, Y., Matsui, M., de Jong, G.J., Somsen, G.W., Wiese-Klingenberg, A., Primavesi, L.F., Paul, M.J., Schluepmann, H., 2011. Growth arrest by trehalose-6-phosphate: an astonishing case of primary metabolite control over growth by way of the SnRK1 signaling pathway. Plant Physiol. 157, 160–174.

Denance, N., Sánchez-Vallet, A., Goffner, D., Molina, A., 2013. Disease resistance or growth: the role of plant hormones in balancing immune responses and fitness costs. Front. Plant Sci. http://dx.doi.org/10.3389/fpls.2013.00155.

Doehlemann, G., Wahl, R., Horst, R.J., Voll, L.M., Usadel, B., Poree, F., Sitt, M., Pons-Kuhnemann, J., Sonnewald, U., Kahmann, R., Kamper, J., 2008. Reprogramming a maize plant: transcriptional and metabolic changes induced by the fungal biotroph *Ustilago maydis*. Plant J. 56, 181–195.

Dong, J., Chen, C., Chen, Z., 2003. Expression profiles of the Arabidopsis *WRKY* gene superfamily during plant defense response. Plant Mol. Biol. 51, 21–37.

Doughari, J.H., 2015. An overview of plant immunity. J. Plant Pathol. Microbiol. http://dx.doi.org/10.4172/2157-7471.1000322.

Dulermo, T., Bligny, R., Gout, E., Cotton, P., 2009. Amino acid changes during sunflower infection by the necrotrophic fungus *B. cinerea*. Plant Signal Behav. 4, 859–861.

Ebrahim, S., Uhsh, K., Singh, B., 2011. Pathogenesis related (PR) proteins in plant defence mechanism. In: Méndez-Vilas, A. (Ed.), Science Against Microbial Pathogens: Communicating Current Research and Technological Advances. Formatec, pp. 1043–1054.

Ellinger, D., Voigt, A.C., 2014. Callose biosynthesis in arabidopsis with a focus on pathogen response: what we have learned within the last decade. Ann. Bot. http://dx.doi.org/10.1093/aob/mcu120.

Ellinger, D., Naumann, M., Falter, C., Zwikowics, C., Jamrow, T., Manisseri, C., et al., 2013. Elevated early callose deposition results in complete penetration resistance to powdery mildew in *Arabidopsis*. Plant Physiol. 161, 1433–1444.

Espunya, M.C., De Michele, R., Gomez-Cadenas, A., Martinez, M.C., 2012. S-nitrosoglutathione is a component of wound- and salicylic acid-induced systemic responses in *Arabidopsis thaliana*. J. Exp. Bot. 63, 3219–3227.

Essmann, J., Schmitz-Thom, I., Schoen, H., Sonnewald, S., Weis, E., Scharte, J., 2008. RNA interference-mediated repression of cell wall invertase impairs defense in source leaves of tobacco. Plant Physiol. 147, 1288–1299.

Feng, J., Wang, F., Hughes, G.R., Kaminskyj, S., Wei, Y.D., 2011. An important role for secreted esterase in disease establishment of the wheat powdery mildew fungus *Blumeria graminis* f. sp *tritici*. Can. J. Microbiol. 57, 211–216.

Ferri, M., Righetti, L., Tassoni, A., 2011. Increasing sucrose concentrations pro- mote phenylpropanoid biosynthesis in grapevine cell cultures. J. Plant Physiol. 168, 189–195.

Flors, V., Ton, J., Jakab, G., Mauch-Mani, B., 2005. Abscisic acid and callose: team players in defence against pathogens? J. Phytopathol. 153, 377–383.

Freeman, B.C., Beattie, G.A., 2008. An overview of plant defenses against pathogens and herbivores. Plant Health Instr. http://dx.doi.org/10.1094/PHI-I-2008-00226-01.

Garavaglia, B.S., Thomas, L., Gottig, N., Zimaro, T., Garofalo, C.G., Gehring, C., Ottado, J., 2010. Shedding light on the role of photosynthesis in pathogen colonization and host defense. Comm. Integr. Biol. 3, 382–384.

Gawroński, P., Górecka, M., Bederska, M., Rusaczonek, A., Ślesak, I., Kruk, J., Karpiński, S., 2013. Isochorismate synthase 1 is required for thylakoid organization, optimal plastoquinone redox status, and state transitions in *Arabidopsis thaliana*. J. Exp. Bot. 64, 3669–3679.

Gomez, L., Bancel, D., Rubio, E., Vercambre, G., 2007. The microplate reader: an efficient tool for the separate enzymatic analysis of sugars in plant tissues – validation of a micro-method. J. Sci. Food Agric. 87, 1893–1905.

Grobkinsky, D.K., van der Graaff, E., Roitsch, E., 2014. Abscisic acid-cytokinin antagonism modulates resistance against *Pseudomonas syringae* in tobacco. Phytopathology 104, 1283–1288.

Groszmann, M., Gonzalez-Bayon, R., Lyons, R.L., Greaves, I.K., Kazan, K., Peacock, W.J., Dennis, E.S., 2015. Hormone-regulated defense and stress response networks contribute to heterosis in Arabidopsis F1 hybrids. PNAS. http://dx.doi.org/10.1073/pnas.1519926112.

Hanson, J., Smeekens, S., 2009. Sugar perception and signaling—an update. Curr. Opin. Plant Biol. 12, 562–567.

He, S.Y., Melotto, M., Underwood, W., Koczan, J., Nomura, K., 2006. Plant stomata function in innate immunity against bacterial invasion. Cell 126, 969–980.

Herbers, K., Meuwly, P., Frommer, W.B., Métraux, J.P., Sonnewald, U., 1996. Systemic acquired resistance mediated by the ectopic expression of invertase: possible hexose sensing in the secretory pathway. Plant Cell 8, 793–803.

Herbers, K., Takahata, Y., Melzer, M., Mock, H.P., Hajirezaei, M., Sonnewald, U., 2000. Regulation of carbohydrate partitioning during the interaction of potato virus Y with tobacco. Mol. Plant Pathol. 1, 51–59.

Horst, R.J., Engelsdorf, T., Sonnewald, U., Voll, L.M., 2008. Infection of maize leaves with *Ustilago maydis* prevents establishment of C4 photosynthesis. J. Plant Physiol. 165, 19–28.

Hwang, E.E., Wang, M.B., Bravo, J.E., Banta, L.M., 2015. Unmasking host and microbial strategies in the *Agrobacterium*-plant defense tango. Front. Plant Sci. http://dx.doi.org/10.3389/fpls.2015.00200.

Jan, A.T., Azam, M., Ali, A., Haq, Q., 2011. Novel approaches of beneficial *Pseudomonas* in mitigation of plant diseases – an appraisal. J. Plant Interact. 6, 195–205.

Jang, C., Seo, E.Y., Nam, J., Bae, H., Gim, Y.G., Kim, H.G., et al., 2013. Insights into alternanthera mosaic virus TGB3 functions: interactions with *Nicotiana benthamiana* PsbO correlate with chloroplast vesiculation and veinal necrosis caused by TGB3 over-expression. Front. Plant Sci. http://dx.doi.org/10.3389/fpls.2013.00005.

Jones, A.M., Dangl, J.L., 2006. The plant immune system. Nature 444, 323–329.

Jones, A.M., Thomas, V., Bennett, M.H., Mansfield, J., Grant, M., 2006. Modifications to the *Arabidopsis* defense proteome occur prior to significant transcriptional change in response to inoculation with *Pseudomonas syringae*. Plant Physiol. 142, 1603–1620.

Kangasjärvi, S., Neukermans, J., Li, S., Aro, E.M., Noctor, G., 2012. Photosynthesis, photorespiration, and light signalling in defence responses. J. Exp. Bot. 63, 1619–1636.

Karpinski, S., Reynolds, H., Karpinska, B., Wingsle, G., Creissen, G., Mullineaux, P., 1999. Systemic signalling and acclimation in response to excess excitation energy in Arabidopsis. Science 284, 654–657.

Karpinski, S., Szechynska-Hebda, M., Wituszynska, W., Burdiak, P., 2013. Light acclimation, retrograde signalling, cell death and immune defences in plants. Plant Cell Environ. 36, 736–744.

Kartashov, A.V., Radyukina, N.L., Ivanov, Y.V., Pashkoviskii, P.P., Shevyakova, N.I., Kuznetsov, V.V., 2008. Role of antioxidant systems in wild plant adaptation to salt stress. Russ. J. Plant Physiol. 55, 463–468.

Kazan, K., Lyons, R., 2014. Intervention of phytohormone pathways by pathogen effectors. Plant Cell 26, 2285–2309.

Khurana, S.M.P., Pandey, S.K., Sarkar, D., Chanemougasoundharam, A., 2005. Apoptosis in plant disease response: a close encounter of the pathogen kind. Curr. Sci. 88, 740–752.

Kim, M., Lim, J.-H., Ahn, C.S., Park, K., Kim, G.T., Kim, W.T., Pai, H.S., 2006. Mitochondria-associated hexokinases play a role in the control of programmed cell death in *Nicotiana benthamiana*. Plant Cell 18, 2341–2355.

Kocal, N., Sonnewald, U., Sonnewald, S., 2008. Cell wall-bound invertase limits sucrose export and is involved in symptom development and inhibition of photosynthesis during compatible interaction between tomato and *Xanthomonas campestris* pv *vesicatoria*. Plant Physiol. 148, 1523–1536.

Kogovsek, P., Pompe-Novak, M., Petek, M., Fragner, L., Weckwerth, W., Gruden, K., 2016. Primary metabolism, phenylpropanoids and antioxidant pathways are regulated in potato as a response to potato virus Y infection. PLoS One. http://dx.doi.org/10.1371/journal.pone.0146135.

Kumudini, B.S., 2004. Induced Systemic Resistance in Pearl Millet against Downy Mildew Disease – a Histological Study on the Pattern of Pathogen Development and Host Responses. Submitted to University of Mysore, Mysuru.

Kumudini, B.S., Shetty, H.S., 2002. Association of lignification and callose deposition with host cultivar resistance and induced systemic resistance in pearl millet to *Sclerospora graminicola*. Australas. Plant Pathol. 31, 157–164.

Kumudini, B.S., Jayamohan, N.S., Patil, V.P., 2017. Integrated mechanisms of plant disease containment by rhizospheric bacteria: Unraveling the signal cross talk between plant and fluorescent *Pseudomonas*. In: Meena, V.S., et al. (Ed.), Agriculturally Important Microbes for Sustainable Agriculture. http://dx.doi.org/10.1007/978-981-10-5343-6_9.

Kunstler, A., Bacso, R., Hafez, Y.M., Kiraly, L., 2015. Reactive oxygen species and plant disease resistance. In: Gupta, D.K., et al. (Ed.), Reactive Oxygen Species and Oxidative Damage in Plants under Stress. Springer International Publishing, Switzerland.

Kuzniak, E., Sklodowska, M., 2005. Fungal pathogen-induced changes in the antioxidant system of leaf peroxisomes from infected tomato plants. Planta 222, 192–200.

Laluk, K., Mengiste, T., 2010. Necrotroph attacks on plants: Wanton destruction or covert extortion? Arab. Book. http://dx.doi.org/10.1199/tab.0136.

Lanubile, A., Muppirala, U.K., Severin, A.J., Marocco, A., Munkvold, G.P., 2015. Transcriptome profiling of soybean (Glycine max) roots challenged with pathogenic and non-pathogenic isolates of *Fusarium oxysporum*. BMC Genomics. http://dx.doi.org/10.1186/s12864-015-2318-2.

Lavanya, S.N., Raj, S.N., Udayashankar, A.C., Kini, K.R., Amruthesh, K.N., Niranjana, S.R., Shetty, H.S., 2012. Comparative analysis of activities of vital defence enzymes during induction of resistance in pearl millet against downy mildew. Arch. Phytopathol. Plant Protect. 45, 1252–1272.

Lecompte, F., Nicot, P.C., Ripoll, J., Abro, M.A., Raimbault, A.K., Lopez-Lauri, F., Bertin, N., 2017. Reduced susceptibility of tomato stem to the necrotrophic fungus *Botrytis cinerea* is associated with a specific adjustment of fructose content in the host sugar pool. Ann. Bot. http://dx.doi.org/10.1093/aob/mcw240.

Lehto, K., Tikkanen, M., Hiriart, J.B., Paakkarinen, V., Aro, E.M., 2003. Depletion of the photosystem II core complex in mature tobacco leaves infected by the flavum strain of tobacco mosaic virus. Mol. Plant-Microbe Interact. 16, 1135–1144.

Leon, J., Castillo, M.C., Coego, A., Lozano-Juste, J., Mir, R., 2013. Diverse functional interactions between nitric oxide and abscisic acid in plant development and responses to stress. J. Exp. Bot. http://dx.doi.org/10.1093/jxb/ert451.

Less, H., Angelovici, R., Tzin, V., Galili, G., 2011. Coordinated gene networks regulating Arabidopsis plant metabolism in response to various stresses and nutritional cues. Plant Cell 23, 1264–1271.

Liang, H., Yao, N., Song, J.T., Luo, S., Lu, H., Greenberg, J.T., 2003. Ceramides modulate programmed cell death in plants. Curr. Sci. 17, 2636–2641.

Lozano-Duran, R., Robatzek, S., 2015. 14-3-3 proteins in plant-pathogen interactions. Mol. Plant Microbe Interact. 28, 511–518.

Lyons, R., Manners, J.M., Kazan, K., 2013. Jasmonate biosynthesis and signaling in monocots: a comparative overview. Plant Cell Rep. http://dx.doi.org/10.1007/s00 299-013-1400-y.

Ma, W., 2011. Roles of Ca^{2+} and cyclic nucleotide gated channel in plant innate immunity. Plant Sci. 181, 342–346.

Mahesh, H.M., Murali, M., Chandra Pal, M.A., Melvin, P., Sharada, M.S., 2017. Salicylic acid seed priming instigates defense mechanism by inducing PR-proteins in *Solanum melongena* L. upon infection with *Verticillium dahliae* Kleb. Plant Physiol. Biochem. http://dx.doi.org/10.1016/j.plaphy.2017.05.012.

Malinovsky, F.G., Fangel, J.U., Willats, W.G.T., 2014. The role of the cell wall in plant immunity. Front. Plant Sci. http://dx.doi.org/10.3389/fpls.2014.00178.

Miller, G., Schlauch, K., Tam, R., Cortes, D., Torres, M.A., Shulaev, V., Dangl, J.L., Mittler, R., 2009. The plant NADPH oxidase RBOHD mediates rapid systemic signaling in response to diverse stimuli. Sci. Signal. http://dx.doi.org/10.1126/scisignal.2000448.

Moller, I.M., Jensen, P.E., Hansson, A., 2007. Oxidative modifications to cellular components in plants. Annu. Rev. Plant Biol. 58, 459–481.

Morkunas, I., Ratajczak, L., 2014. The role of sugar signaling in plant defense responses against fungal pathogens. Acta Physiol. Plant 36, 1607–1619.

Morkunas, I., Narozna, D., Nowak, W., Samardakiewicz, S., Remlein-Starosta, D., 2011. Cross-talk interactions of sucrose and *Fusarium oxysporum* in the phenylpropanoid pathway and the accumulation and localization of flavonoids in embryo axes of yellow lupine. J. Plant Physiol. 168, 424–433.

Mur, L.A.J., Prats, E., Pierre, S., Hall, M.A., Hebelstrup, K.H., 2013. Integrating nitric oxide into salicylic acid and jasmonic acid/ethylene plant defense pathways. Front. Plant Sci. http://dx.doi.org/10.3389/fpls.2013.00215.

Mysore, K.S., Ryu, C.M., 2004. Non-host resistance: how much do we know? Trends Plant Sci. 9, 97–104.

Nahar, K., Kyndt, T., Nzogela, Y.B., Gheysen, G., 2012. Abscisic acid interacts antagonistically with classical defense pathways in rice–migratory nematode interaction. New Phytol. 196, 901–913.

Nicholson, R.L., Hammerschmidt, R., 1992. Phenolic compounds and their role in disease resistance. Annu. Rev. Phytopathol. 30, 369–389.

Nie, P., Li, X., Wang, S., Guo, J., Zhao, H., Niu, D., 2017. Induced systemic resistance against *Botrytis cinerea* by *Bacillus cereus* AR156 through a JA/ET- and NPR1-dependent signaling pathway and activates PAMP-triggered immunity in *Arabidopsis*. Front. Plant Sci. http://dx.doi.org/10.3389/fpls.2017.00238.

Oliveira, M.D.M., Varanda, C.M.R., Félix, M.R.F., 2016. Induced resistance during the interaction pathogen x plant and the use of resistance inducers. Phytochem. Lett. 15, 152–158.

Pandey, D., Rajendran, S.R.C.K., Gaur, M., Sajeesh, P.K., Kumar, A., 2016. Plant defense signaling and responses against necrotrophic fungal pathogens. J. Plant Growth Regul. http://dx.doi.org/10.1007/s00344-016-9600-7.

Parker, D., Beckmann, M., Zubair, H., Enot, D.P., Caracuel-Rios, Z., Overy, D.P., Snowdon, S., Talbot, N.J., Draper, J., 2009. Metabolomic analysis reveals a common pattern of metabolic re-programming during invasion of three host plant species by *Magnaporthe grisea*. Plant J. 59, 723–737.

Patil, S.V., Jayamohan, N.S., Kumudini, B.S., 2016. Strategic assessment of multiple plant growth promotion traits for shortlisting of fluorescent *Pseudomonas* spp. and seed priming against ragi blast disease. Plant Growth Regul. 80, 47–58.

Pieterse, C.M.J., Leon-Reyes, A., SVd, E., Wees, S.C.M.V., 2009. Networking by small-molecule hormones in plant immunity. Nat. Chem. Biol. 5, 308–316.

Pilot, G., Stransky, H., Bushey, D.F., Pratelli, R., Ludewig, U., Wingate, V.P.M., Frommer, W.B., 2004. Overexpression of GLUTAMINE DUMPER1 leads to hypersecretion of glutamine from hydathodes of Arabidopsis leaves. Plant Cell 16, 1827–1840.

Proels, R.K., Huckelhoven, R., 2014. Cell-wall invertases, key enzymes in the modulation of plant metabolism during defence responses. Mol. Plant Pathol. 15, 858–864.

Proels, R.K., Roitsch, T., 2009. Extracellular invertase LIN6 of tomato: a pivotal enzyme for integration of metabolic, hormonal, and stress signals is regulated by a diurnal rhythm. J. Exp. Bot. 60, 1555–1567.

Raj, S.N., Lavanya, S.N., Amruthesha, K.N., Niranjana, S.R., Reddy, M.S., Shetty, H.S., 2012. Histo-chemical changes induced by PGPR during induction of resistance in pearl millet against downy mildew disease. Biol. Control 60, 90–102.

Rao, S.S.R., Vardhini, B.V., Sujatha, E., Anuradha, S., 2002. Brassinosteroids-A new class of phytohormones. Curr. Sci. 82, 1239–1245.

Reimer-Michalski, E.-M., Conrath, U., 2016. Innate immune memory in plants. Semin. Immunol. http://dx.doi.org/10.1016/j.smim.2016.05.006.

Ren, C.M., Zhu, Q., Gao, B.D., Ke, S.Y., Yu, W.C., et al., 2008. Transcription factor WRKY70 displays important but no indispensable roles in jasmonate and salicylic acid signaling. J. Integr. Plant Biol. 50, 630–637.

Robert-Seilaniantz, A., Grant, M., Jones, J.D., 2011. Hormone crosstalk in plant disease and defense: more than just jasmonate-salicylate antagonism. Ann. Rev. Phytopathol. 49, 317–343.

Rodríguez-Herva, J.J., Gonzalez-Melendi, P., Cuartas-Lanza, R., Antúnez- Lamas, M., Río-Alvarez, I., Li, Z., Lopez-Torrejon, G., Díaz, I., del Pozo, J.C., Chakravarthy, S., Collmer, A., Rodríguez-Palenzuela, P., Lopez-Solanilla, E., 2012. A bacterial cysteine protease effector protein interferes with photosynthesis to suppress plant innate immune responses. Cell Microbiol. 14, 669–681.

Roitsch, T., Bittner, M., Godt, D.E., 1995. Induction of apoplastic invertase of *Chenopodium rubrum* by D-glucose and a glucose analogue and tissue-specific expression suggest a role in sink–source regulation. Plant Physiol. 108, 285–294.

Rojas, C.M., Sentil-Kumar, M., Tzin, V., Mysore, K.S., 2014. Regulation of primary plant metabolism during plant-pathogen interactions and its contribution to plant defense. Front. Plant Sci. http://dx.doi.org/10.3389/fpls.2014.00017.

Salzman, R.A., Tikhonova, I., Bordelon, B.P., Hasegawa, P.M., Bressan, R.A., 1998. Coordinate accumulation of antifungal proteins and hexoses constitutes a developmentally controlled defense response during fruit ripening in grape. Plant Physiol. 117, 465–472.

Scheideler, M., Schlaich, N.L., Fellenberg, K., Beissbarth, T., Hauser, N.C., Vingron, M., Slusarenko, A.J., Hoheisel, J.D., 2002. Monitoring the switch from housekeeping to pathogen defense metabolism in *Arabidopsis thaliana* using cDNA arrays. J. Biol. Chem. 277, 10555–10561.

Schwachtje, J., Baldwin, I.T., 2008. Why does herbivore attack reconfigure primary metabolism? Plant Physiol. 146, 845–851.

Serrano, M., Coluccia, F., Torres, M., L'Harison, F., Metraux, J.-P., 2014. The cuticle and plant defence to pathogens. Front. Plant Sci. http://dx.doi.org/10.3389/fpls.2014.00274.

Sheen, J., 1990. Metabolic repression of transcription in higher plants. Plant Cell 2, 1027–1038.

Shin, S., Lv, J., Fazio, G., Mazzola, M., Zhu, Y., 2014. Transcriptional regulation of ethylene and jasmonate mediated defense response in apple (*Malus domestica*) root during *Pythium ultimum* infection. Horti Res. http://dx.doi.org/10.1038/hortres.2014.53.

Siddaiah, C.N., Satyanarayana, N.R., Mudili, V., Gupta, V.K., Gurunathan, S., Rangappa, S., Shekar, H.S., Srivastava, R.K., 2017. Elicitation of resistance and associated defense responses in *Trichoderma hamatum* induced protection against pearl millet downy mildew pathogen. Nat. Sci. Rep. http://dx.doi.org/10.1038/srep43991.

Smeekens, S., Ma, J., Hanson, J., Rolland, F., 2010. Sugar signals and molecular networks controlling plant growth. Curr. Opin. Plant Biol. 13, 274–279.

Solfanelli, C., Poggi, A., Loreti, E., Alpi, A., Perata, P., 2006. Sucrose-specific induction of the anthocyanin biosynthetic pathway in arabidopsis. Plant Physiol. 140, 637–646.

Somssich, I.E., Hahlbrock, K., 1998. Pathogen defence in plants–a paradigm of biological complexity. Trends Plant Sci. 3, 86–90.

Staal, J., Dixelius, C., 2009. Plant innate immunity. In: Encyclopedia of Life Sciences. John Wiley & Sons LTD, Chichester. http://dx.doi.org/10.1002/9780470015902.a0020114.

Swarbrick, P.J., Schulze-Leftert, P., Scholes, J.D., 2006. Metabolic consequences of susceptability and resistance (race-specific and broad-spectrum) in barley leaves challenged with powdery mildew. Plant Cell Environ. 29, 1061–1076.

Tanaka, A., Makino, A., 2009. Photosynthetic research in plant science. Plant Cell Physiol. 50, 681–683.

Thakur, M., Sohal, B.S., 2013. Role of elicitors in inducing resistance in plants against plant infection. ISRN Biochem. 3, 1–11.

Thibaud, M.C., Gineste, S., Nussaume, L., Robaglia, C., 2004. Sucrose increases pathogenesis-related *PR-2* gene expression in *Arabidopsis thaliana* through an SA-dependent but *NPR1*-independent signaling pathway. Plant Physiol. Biochem. 42, 81–88.

Thomma, B.P.H.J., Penninckx, I.A., Broekaert, W.T., Cammue, B.P., 2001. The complexity of disease signaling in Arabidopsis. Curr. Opin. Immunol. 13, 63–68.

Ton, J., Mauch-Mani, B., 2004. β-amino-butyric acid-induced resistance against necrotrophic pathogens is based on ABA-dependent priming for callose. Plant J. 38, 119–130.

Ton, J., Flors, V., Mauch-Mani, B., 2009. The multifaceted role of ABA in disease resistance. Trends Plant Sci. 14, 310–317.

Torres, M.A., 2010. ROS in biotic interactions. Physiol. Plant 138, 414–429.

Torres, M.A., Jones, J.D.G., Dangl, J.L., 2006. Reactive oxygen species signaling in response to pathogens. Plant Physiol. 141, 373–378.

Toyoda, K., Yao, S., Tkagi, M., Uchioki, M., Miki, M., Tanaka, K., Suzuki, T., Amano, M., Kiba, A., Kato, T., Takahashi, H., Ishiga, Y., Matsui, H., Noutoshi, Y., Yamamoto, M., Ichinose, Y., Shiraishi, T., 2016. The plant cell wall as a site for molecular contacts in fungal pathogenesis. Physiol. Mol. Plant Pathol. http://dx.doi.org/10.1016/j.pmpp.2016.02.006.

Uma, B., Podile, A.R., 2015. Apoplastic oxidative defenses during non-host interactions of tomato (*Lycopersicon esculentum* L.) with *Magnaporthe grisea*. Acta Physiol. Plant 2015, https://doi.org/10.1007/s11738-015-1779-xs.

Underwood, W., 2012. The plant cell wall: a dynamic barrier against pathogen invasion. Front. Plant Sci. http://dx.doi.org/10.3389/fpls.2012.00085.

van Baarlen, P., van Belkum, A., Summerbell, R.C., Crous, P.W., Thomma, B.P., 2007. Molecular mechanisms of pathogenicity: how do pathogenic microorganisms develop cross-kingdom host jumps? FEMS Microbiol. Rev. 31, 239–277.

van Damme, M., Zeilmakerm, T., Elberse, J., Andel, A., de Sain-van der Velden, M., van den Ackerveken, G., 2009. Downy mildew resistance in Arabidopsis by mutation of HOMOSERINE KINASE. Plant Cell 21, 2179–2189.

Van der Does, D., Leon-Reyes, A., Koornneef, A., Van Verk, M.C., Rodenburg, N., Pauwels, L., et al., 2013. Salicylic acid suppresses jasmonic acid signaling downstream of SCF^{COI1}-JAZ by targeting GCC promoter motifs via transcription factor ORA59. Plant Cell 25, 744–761.

Van Loon, L.C., Van Kammen, A., 1970. Polyacrylamide disc electrophoresis of the soluble leaf proteins from *Nicotiana tabacum* var *'Samsun'* and Samsun NN II. Changes in protein constitution after infection with tobacco mosaic virus. Virology 40, 199–211.

Van Loon, L.C., Geraats, B.P., Linthorst, H.J., 2006. Ethylene as a modulator of disease resistance in plants. Trends Plant Sci. 11, 184–191.

Veronese, P., Nakagami, H., Bluhm, B., AbuQamar, S., Chen, X., Salmeron, J., Dietrich, R.A., Hirt, H., Mengiste, T., 2006. The membrane anchored *Botrytis*-induced kinase1 plays distinct roles in Arabidopsis resistance to necrotrophic and biotrophic pathogens. Plant Cell 18, 257–273.

Voigt, C.A., 2014. Callose-mediated resistance to pathogenic intruders in plant defense-related papillae. Front. Plant Sci. http://dx.doi.org/10.3389/fpls.2014.00168.

Wang, D., Amornsiripanitch, N., Dong, X., 2006. A genomic approach to identify regulatory nodes in the transcriptional network of systemic acquired resistance in plants. PLoS Pathog. 2, e123.

Wojtaszek, P., 1997. Oxidative burst: an early plant response to pathogen infection. Biochem. J. 322, 681–692.

Wu, X.N., Rodriguez, C.S., Pertl-Obermeyer, H., Obermeyer, G., Schulze, W.X., 2013. Sucrose-induced receptor kinase SIRK1 regulates a plasma membrane aquaporin in Arabidopsis. Mol. Cell Proteomics 12, 2856–2873.

Xing, M., Lv, H., Ma, J., Xu, D., Li, H., Yang, L., Kang, J., Wang, X., Fang, Z., 2016. Transcriptome profiling of resistance to *Fusarium oxysporum* (*Brassica oleracea*) r. PLoS One. http://dx.doi.org/10.1371/journal.pone.0148048.

Xu, J., Kris, A., Monica, H., Vleesschauwer, D.D., 2013. Abscisic acid promotes susceptibility to the rice leaf blight pathogen *Xanthomonas oryzae* pv *oryzae* by suppressing salicylic acid-mediated defenses. PLoS One. http://dx.doi.org/10.1371/journal.pone.0067413.

Yun, B.-W., Feechan, A., Yin, M., Saidi, N.B.B., Bihan, T.L., Yu, M., Moore, J.W., Kang, J.-G., Kwon, E., Spoel, S.H., Pallas, J.A., Loake, G.J., 2011. S-nitrosylation of NADPH oxidase regulates cell death in plant immunity. Nature. http://dx.doi.org/10.1038/nature10427.

Zander, M., Chen, S., Imkampe, J., Thurow, C., Gatz, C., 2012. Repression of the *Arabidopsis thaliana* jasmonic acid/ethylene-induced defense pathway by TGA-interacting glutaredoxins depends on their C-terminal ALWL motif. Mol. Plant. http://dx.doi.org/10.1093/mp/ssr113.

Zander, M., La Camera, S., Lamotte, O., Métraux, J.P., Gatz, C., 2010. *Arabidopsis thaliana* class-II TGA transcription factors are essential activators of jasmonic acid/ethylene-induced defense responses. Plant J. 61, 200–210.

Zeilinger, S., Gupta, V.K., Dahms, T.E.S., Silva, R.N., Singh, H.B., Upadhyay, R.S., Gomes, E.V., Tsui, C.K.-M., Nayak, C.S., 2016. Friends or foes? Emerging insights from fungal interactions with plants. FEMS Microbiol. Rev. 40, 182–207.

Zhang, L., Cohn, N.S., Mitchell, J.P., 1997. A pea cell-wall invertase gene (*PsInv-1*) with tissue-specific expression. Plant Physiol. Biochem. 35, 751–760.

Zhao, Z.T., Liu, H.Q., Wang, C.F., Xu, J.R., 2013. Comparative analysis of fungal genomes reveals different plant cell wall degrading capacity in fungi. BMC Genomics. http://dx.doi.org/10.1186/1471-2164-14-274.

Chapter 12

Effect of Root-Associated Bacteria on Soluble Sugar Metabolism in Plant Under Environmental Stress

Yachana Jha, R.B. Subramanian
Sardar Patel University, Anand, India

1. INTRODUCTION

Plants growing on land experience a multitude of environmental stresses (Slama et al., 2015). These stresses include drought, salinity, waterlogging, extremes of temperature, radiation, pathogen, and mineral deficiency or excess. Plant responses to environmental stress can be analyzed by systematically identifying traits related to stress tolerance followed by analysis of the physiological, cellular, biochemical, and molecular basis of the trait. Environmental stress adversely affects many aspects of the plant's physiology, especially photosynthetic capacity. Plants, because of their immobility, are extremely sensitive and responsive to their surroundings and need to adjust the altered carbohydrate availability for their survival. Plants are autotrophic and photosynthetic organisms that both produce and consume sugars. In plants, sugars are not merely a carbohydrate metabolite or a photoassimilate of photosynthesis. Sugars are not only important energy sources and structural components but also central regulatory molecules controlling physiology, metabolism, cell cycle, development, and gene expression in plants as well as in root-associated bacteria. In plants, sugars affect growth and development throughout the life, from germination to flowering to senescence. The rhizosphere is a sugar/nutrient-rich habitat and harbors a huge variety of root-associated bacteria and fungi that having neutral, beneficial, or deleterious effects on the plant (Mendas et al., 2013). These root-associated bacteria also get associated with the plant to avail free available sugar/nutrient from them and these organisms can improve plant growth by different mechanisms (Braun et al., 2014). The stress affects the plant growth, and productivity is severely diminished. For this reason, plants have evolved complex physiological and biochemical adaptations to adjust and adapt to a variety of environmental stresses and the root-associated bacteria help in enhancing plant growth and yield under different types of stress. Plant growth-promoting rhizobacteria (root-associated bacteria) is one among the most effective and best studied soil microorganisms that can promote plant performance. As the plant and root-associated bacteria, both are related to each other through soluble sugar (Upadhyay et al., 2012). As the plant and root associated bacteria, both are related to each other through soluble sugar, so effect of soluble sugar is an important parameter for the study. The root-associated bacteria have been known to increase the growth and productivity of many commercial crops by modulating the soluble sugar concentration in the plant (Paul and Lade, 2014).

2. ISOLATION AND IDENTIFICATION OF ROOT-ASSOCIATED BACTERIAL ISOLATES

The rhizosphere is the most active ecological niche where inter- and intraspecies communication of microorganisms, such as fungi, bacteria, and protozoa, persists because of the occurrence of a rich and diverse microbial food source (Bais et al., 2006). Root-associated bacteria influence the inductions of genes in plants under various stresses and can also modulate the population of phytopathogen around the root of the plant. The rhizosphere bacterial inhabitants play a significant role in maintaining a healthy root and developing tolerance for environmental stress. Use of root-associated bacteria as biofertilizer can increase the plant nutrient status by associative nitrogen fixation, phosphorus and potassium solubilization (Sindhu et al., 2016), and siderophores production. It can enhance the bioavailability of such nutrients by changing the absorptivity and converting nutrients in the rhizosphere (Jha and Subramanian, 2014a). Two bacterial strains are isolated from the rhizosphere soil and root of *Suaeda nudiflora* wild mosque plant from khambhat near the seashore of Gujarat having high levels of salt concentration as per our previously published method (Jha and Subramanian, 2013a). The soil sample was tested in SICART (Sophisticated Instrumentation Centre for Applied Research and Testing) laboratory by the extracted water sample method. The soil possessed the following physiochemical properties: pH 6.58, electrical conductivity 1480 μS/cm, salinity

Plant Metabolites and Regulation Under Environmental Stress. https://doi.org/10.1016/B978-0-12-812689-9.00012-1

8.6%, nitrate 112.5 mg/kg, chloride 128 mg/kg, sulfate 155 mg/kg, ammonia nitrogen 23.3 mg/kg, CEC 3 cmol, and organic carbon 5500 mg/kg. Initially, the bacteria were isolated in the semisolid NFb medium and the white veil-like pellicle that formed below the surface of the semisolid NFb medium was transferred on NFb agar plates. That the pure culture has been maintained on the NFb agar plate indicates that the isolates have the ability for nitrogen fixation. NFb agar plates containing bromothymol blue, a pH indicator dye, is used for the isolation of bacteria. The change in plate color from green to blue indicates that the medium pH has shifted toward alkalinity because of the growth of bacteria, as such bacteria have the ability for nitrogen fixation.

Molecular identification of bacterial isolates has been done by isolation of the total genomic DNA from the isolates and amplification with 16S rDNA specific primers16S F: 5′AGAGTTTGATCCTGGCTCAG3′ and 16S R: 5′AGGTTACCTTGTTACGACTT3′ for polymerase chain reaction (PCR) amplification of the DNA followed by sequencing according to our published method (Jha and Subramanian, 2013b). PCR amplicons of 16S rDNA of about 1500 bp were obtained for both the isolates as discrete bands in agarose gel. The phylogenetic trees have been constructed using *Basic Local Alignment Search Tool (BLAST)* software by comparing the 16S rDNA sequence of isolates and related genera from a database using the neighbor-joining algorithm and maximum likelihood method. Nucleotide homology and phylogenetic analysis have identified the root isolate as *Pseudomonas aeruginosa* (GenBank Accession Number: JQ790515) and soil isolates have been identified as *Bacillus megaterium* (GeneBank Accession Number: JQ790514) (Fig. 12.1). *B. megaterium* and *P. aeruginosa* are nonsymbiotic bacteria and of two isolates, one is gram-positive endospore-forming bacteria; it is large, approximately 2.6 × 1.2 μm, noncapsulated, and motile by perichate flagella. The other is gram-negative short rods, motile, having polar flagella, and nonsporulating as shown in Table 12.1.

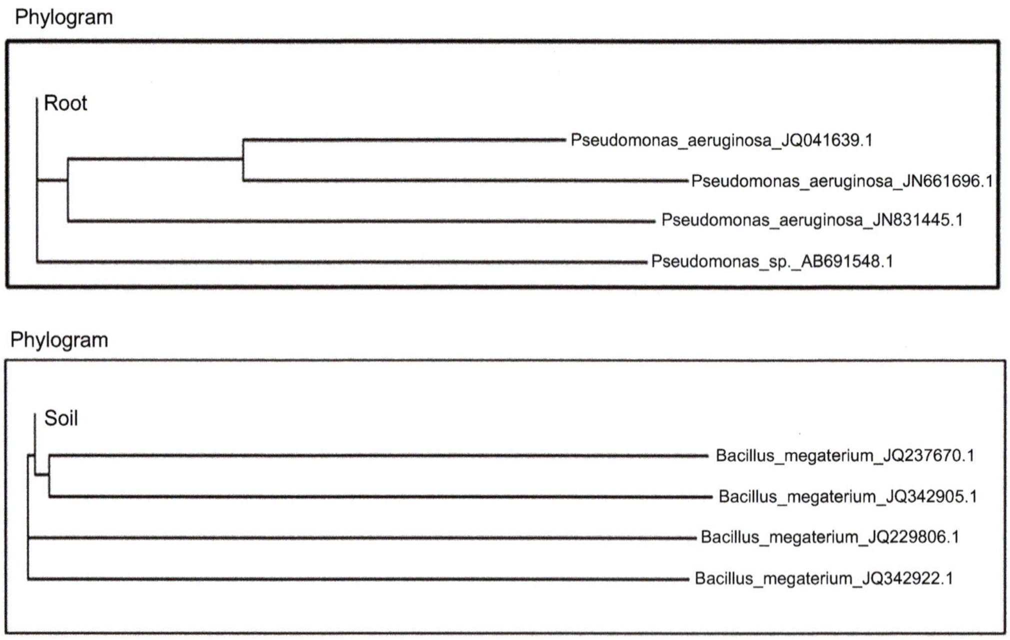

FIGURE 12.1 Phylogenetic tree for *Pseudomonas aeruginosa* and *Bacillus megaterium* made in MEGA 3.1 software using Neighbor Joining method.

TABLE 12.1 The Morphological Characteristics of the Final Short-Listed Isolates

Colony Characteristics	Size	Shape	Margin	Elevation	Consistency	Texture	Opacity	Pigmentation
Bacillus megaterium	Large	Round	Entire	Raised	Moist	Smooth	Semitransparent	Light yellow
Pseudomonas aeruginosa	Small	Irregular	Entire	Convex	Dry	Smooth	Opaque	Blue-green

3. EFFECT OF ROOT-ASSOCIATED BACTERIA ON SOLUBLE SUGAR IN PLANT GROWTH RESPONSE

Sugar plays a role as a signaling molecule that regulates a variety of genes. It probably affects various aspects of development in plants (Granot et al., 2013). High levels of sugar in the medium inhibit both hypocotyl elongation in the dark and light-induced cotyledon opening. In addition to these developmental aspects, flowering also seems to be influenced by the sugar (Lastdrager et al., 2014). Sugar also plays an important role in the intricate machinery of signal transduction. Sugar signaling is part of an ancient system for cellular adjustment to shifting environments and has been found to be crucial in responses to various stimuli, most importantly to the carbohydrate status of the plant (OHara et al., 2013). Gene responses to a changing carbohydrate status can vary greatly between plants and plant tissues. In general, carbohydrate depletion upregulates genes for photosynthesis, reserve mobilization, and export processes, whereas high carbohydrate levels induce genes involved in storage or growth (Ruan, 2014). In the life cycle of plants, seed germination and early seedling growth depend upon stored substances, sumainly carbohydrates, which are mobilized in the form of soluble sugars (sucrose, glucose and fructose) to provide energy for the differentiation of seed tissues in various organs like stem and radicle, and it also maintain the osmotic homeostasis of the differentiating cells (Santos and Buckeridge, 2004). Soluble sugars, like hormones, can act as primary messengers and regulate signals that control the expression of different genes involved in plant growth and metabolism (Chen, 2007). They regulate the growth and metabolism by modulation of gene expression and enzyme activities of both sugar exporting (source) and importing (sink) tissues. This ensures optimal synthesis and use of carbon and energy resources (Coruzzi and Bush, 2001). Accumulation of soluble sugars in source tissues downregulates photosynthesis thus maintaining homeostasis. Thus germinated seeds and growing seedlings seem to be the most vulnerable stages to soluble sugar fluctuations, and constitute an excellent material to study the effect of different environmental stresses. At the same time the optimum sugar concentration is required for bacterial growth, as a high concentration of sugar results in osmotic stress in the bacterial cell and reduces its growth, whereas a low concentration also reduced its growth (Lastdrager et al., 2014). The low concentration of sugar increases competition between the plant and root-associated bacteria, resulting in reduced plant growth (Li et al., 2016). However, a high concentration of sugar causes osmotic stress in both, i.e., plant and root-associated bacteria. Therefore the association of root-associated bacteria with the plant counterbalances the sugar concentration for the proper growth and development of both (Ma et al., 2017). Seeds of maize cultivar Pioneer 30 V92 have been inoculated with isolates as per our published methods with some modification (Jha and Subramanian, 2013c). The seeds are washed thoroughly with sterile distilled water and kept in a shaker for 5–6 h in autoclaved distilled water on a rotary shaker. Later the seeds are transferred to Petri dishes containing tryptone glucose yeast extract agar medium and incubated in the dark at 30°C to test for possible contamination. The germinated seedlings devoid of any contamination have been used for inoculation experiments. To study the effect of the isolated bacteria on the biochemical parameters selected, maize seedlings are transferred to culture tubes containing 400 mL Hoagland's nutrient medium, 400 mL micronutrients, and 1% agar in 40 mL distilled water. Before the transfer, bacterial inoculum of the isolated bacteria is added to the medium at a concentration of 6×10^8 cfu/mL. To obtain a mixture of both bacterial cultures, an equal volume of both the cultures is mixed into the medium to give a concentration of 6×10^8 cfu/mL. The tubes are incubated at 27°C in a 12-h light–dark cycle in a growth chamber. The pot study showed that both individual bacterial isolates and their mixture enhance plant height, root length, and dry weight. Among the two isolates, *P. aeruginosa* is more effective than *B. megaterium* and the mixture of both showed the best growth response of maize plant as shown in Table 12.2. Soluble sugars in the plant downregulate photosynthesis, resulting in lower plant growth, but inoculation with root-associated bacteria enhances the plant growth by enhancing photosynthesis. Thus root-associated bacteria are coordinated between production and accumulation of sugar in the plant and help the plant in better survival.

TABLE 12.2 Effect of Root-Associated Bacteria on Plant Growth Promoting Activity in Maize Plant

Treatment	Germination (%)	Shoot Length (cm)	Root Length (cm)	Total Plant Height (cm)	Total Biomass (g)
Control	66.66	14.50	15.00	29.50	0.295
Control + *Bacillus megaterium*	83.33	22.66	22.00	44.66	0.480
Control + *Pseudomonas aeruginosa*	83.33	24.00	22.50	49.50	0.540
Control + *B. megaterium* + *P. aeruginosa*	86.66	28.16	24.00	52.16	0.670

4. EFFECT OF ROOT-ASSOCIATED BACTERIA ON PLANT VASCULAR TISSUE AND SUGAR METABOLISM

Plant growth is impaired by severe environmental stress because of a decrease in stomatal opening, which limits CO_2 uptake and hence reduces photosynthetic activity. Reduced photosynthetic ability results in reduced production of sugar, required for the growth of root-associated bacteria and plants. In plants, altered photosynthesis also affects the vascular system at different levels:

- it results in less sugar available for export through the vascular system,
- the increased demand for sugar for the growth of plant, root-associated bacteria, and pathogens required a balance between plant and pathogens for the survival of plant and yield,
- the distribution of sugar between plant, root associated bacteria and plant pathogens results in impaired sugar delivery, and ultimately decides the winner, plant or pathogen. As the plant is host for the root associated bacteria, so the root associated bacteria help plants to become winner.

So the complexity of sugar exchanges among the host/pathogen interface depend on concentration of sugar in the cytoplasm of mesophyll cells of plant leaves and the export of sugar from source (leaves) to entire plant, which indicates source-limitation for both pathogen and plants (Chen, 2014).

Vascular plants transport energy in the form of sugars from the leaves where they are produced to sites of active growth. The mass flow of sugars through the phloem vascular system is determined by the sap flow rate and the sugar concentration (Zanon et al., 2015). The sugars produced by photosynthesis are transported in the phloem vascular system of plants. Transport is initiated in the leaves where sugars are either passively or actively loaded into the phloem (Lemoine et al., 2013). In active loading species, the process is driven by membrane transporters and sugar polymerization and occurs against a sugar concentration gradient. However, in passive loading species, sugars move into the phloem without the use of metabolic energy by travelling down a concentration gradient from the mesophyll to phloem (Rennie and Turgeon, 2009). In the phloem, an aqueous solution of sugars, amino acids, proteins, ions, and signaling molecules flows through a series of narrow, elongated cylindrical cells, known as sieve tube elements, lying end to end, forming a microfluidic network spanning the entire length of the plant (Péron et al., 2017). Sugar concentrations vary over a wide range in plant tissues, which provides plants with a broad range of signals (Peuke et al., 2015).

4.1 Effect on Lignin

Lignins are complex three-dimensional polymers that are embedded in specialized plant cell walls, such as the secondary wall of xylem cells. Lignins reinforce and rigidify plant cell walls and render them impermeable to water and resistant to microbial degradation. Lignin biosynthesis is a part of phenylpropanoid metabolism and generates the C_6–C_3 skeletons that are used to build a diverse array of compounds, including lignins, and lignification represents a significant, nonrecoverable investment of carbon, i.e., sugar and energy. Consequently, it is likely that plants monitor resource availability and regulate lignification so that the requirements for carbon skeletons and sugar to build lignins do not outpace availability (Amthor, 2003). The sugar-dependent accumulation of lignin could simply be a consequence of increased availability of sugar flowing through a pathway. Alternatively, sugar may function as a signal to induce lignin biosynthesis (Le Gall et al., 2015). The root-associated bacteria also affect the lignification of plant root cell. So in our study, the plants from each treatment after 35 days of sowing the seeds are carefully collected with the root and subjected to phloroglucinol staining. The cross-sections of dark-grown root samples are placed in a 1% phloroglucinol-HCl solution for 10 min, and mounted in 50% glycerol, 6N HCl and observed immediately. Bright-field illuminated sections of roots are examined under an image analyzer microscope (Carl Zeiss). In noninoculated control as well as in inoculated plants, no anatomical change has been observed in the xylem tissues, but inoculated plants show an increase in the root length, root hair development, as well as enhanced lignification (Fig. 12.2). The reports showed that, pretreatment with biological elicitors such as root associated bacteria like *Pseudomonas* species induce systemically acquired resistance in crops, result in enhanced lignin content in plants (Vidhyasekaran et al., 2001). Lignin solidifies the cell wall, providing a nondegradable barrier for pathogens, and is therefore thought to enhance protection against biotic stresses (Magalhaes Silva Moura et al., 2010). The carbohydrates have a direct relation to the extent of lignification, and sugar-mediated control of lignin deposition occurs through a hexokinase-independent sugar-signaling pathway as reported by Rogers et al. (2005). Hilal et al. (2004) also demonstrated that sugar is related to the synthesis of other carbohydrates such as C4 sugar (erythrose-4-phosphate), which acts as a substrate for lignin and phenolic compounds synthesis. Hence, the metabolism of soluble sugars is a dynamic process simultaneously involving degrading and synthetic reactions.

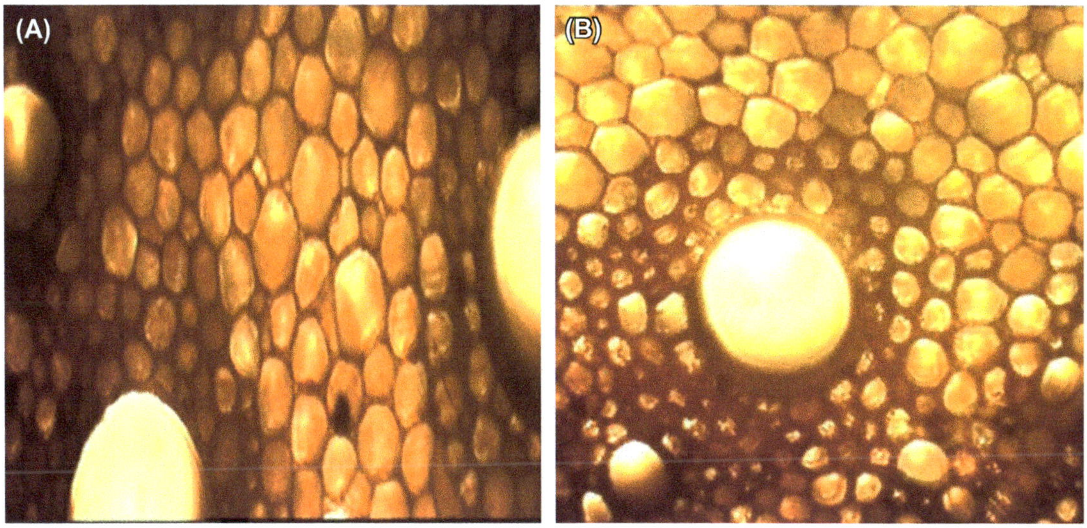

FIGURE 12.2 Effect of root-associated bacteria on lignin in maize cell under stress, where A= Control, B= Inoculated with root-associated bacteria.

5. EFFECT OF ROOT-ASSOCIATED BACTERIA ON SOLUBLE SUGAR IN PLANT UNDER BIOTIC STRESS

Plants possess a group of chemical and physical barriers to avoid nearly all unfavorable plant biotic stresses or interactions. The chemical constituent of the constitutive defense is to inhibit enzymatic functions of pathogens or rapid accumulation of secondary metabolites or protein, which inhibits the growth of the pathogen, or toxic metabolite that kills the pathogen. Plants also construct physical obstacles, such as cuticle and cell wall, to avoid harmful biotic interactions and depending on the species and the environment in which plants grow, there are different levels and forms of barriers. So for efficient response to biotic stress plants have developed a system that permits rapid and targeted responses. Root-associated bacteria are native to the soil and plant rhizosphere and play an important role in the plant as biocontrol agent for plant pathogens (Morkunas and Ratajczak, 2014). The sugar produced by the plant is one of the important parameters for the growth and development of root-associated bacteria as well as plant pathogen. Root associated bacteria reduce the frequency or severity of plant diseases, as free sugar liberated by the plant root is efficiently utilized by them than the plant pathogen. That is why root-associated bacteria are regularly mentioned as biocontrol agents. Sugars, owing to their regulatory function, affect all phases of the life cycle of both the plants and bacteria. In the plant it interacts with phytohormones, and thus controls the processes of growth and development of plants (Stokes et al., 2013). In root-associated bacteria also an optimum concentration is required for proper growth and development, as it is the sole source of energy. There are many reports on the importance of sugar levels in plant disease resistance caused by pathogens (Bolouri Moghaddam and Van den Eden, 2012), as high and low concentrations of sugar opposes the growth and development of the pathogen. Many plant responses to the attack of a pathogen are closely connected with the pathways regulating the level of sugar in the plant cell and ensuring energy homeostasis (Hey et al., 2010). The innate resistance of plants to pathogens and systemic resistance induced by signals originating from the infection site are well known. A significant role in such responses is played by sugars themselves, acting as signaling molecules and such defense response results in a substantial reprogramming of plant cells (Bolton, 2009). The types of carbohydrates (sugar) and their levels are studied in plant inoculated with root-associated bacteria under normal and stress conditions. The results show accumulation of different types of sugar, in inoculated plant under both conditions (Tables 12.3A and B). Sugars can be considered as signaling molecules rather than just providers of carbon and energy in plants (BolouriMoghaddam et al., 2010). Similar to plant defense hormones, they participate in various mechanisms, including immune and defense responses. Plants undergo large changes in their environment throughout their life and have developed many strategies to respond to these changes. Recent data on biotic stress have unraveled the pathways of sugar exchange, and the corresponding molecular players (https://www.ncbi.nlm.nih.gov/pmc/articles/PMC3721551/; Doidy et al., 2012). Unexpectedly, sucrose can be absorbed by both host and pathogen, whereas it is commonly admitted that glucose resulting from the invertase activity is the main host sugar taken up by pathogens (Tauzin et al., 2014). These new findings highlight the complexity of sugar exchanges at the host/pathogen interface.

Sugars regulate cellular activity at multiple levels, from transcription and translation to protein stability and activity (Rolland et al., 2006). This diversity of subcellular localizations of sugar plays an important role in the control of programmed cell death

TABLE 12.3A Different Soluble Carbohydrates in Noninoculated Plant Leaf Extracts Identified in GC-MS Analysis Under Salinity

Hit	Rev	For	Compound Name	M.W.	Formula	CAS
1.	823	682	D-Galactose 6-deoxy	184	$C_8H_{12}O_5$	3815-37-0
2.	788	588	Lactose	342	$C_{12}H_{22}O_{11}$	63-42-3
3.	774	625	D-Glucohexodialdose	178	$C_8H_{10}O_8$	900149-19-1
4.	772	608	D- Mannose	180	$C_6H_{12}O_6$	3458-28-4
5.	771	668	L-Mannose 6-deoxy	184	$C_6H_{12}O_6$	3815-41-8
6.	769	570	D-Glucose 4-D-α-D glucopyranosyl	342	$C_{12}H_{22}O_{11}$	69-79-4
7.	766	563	β-D-Glucopyranose 4-D-β-galactopyranosyl	342	$C_{12}H_{22}O_{11}$	5965-66-2
8.	765	558	D-Lyxose	150	$C_5H_{10}O_5$	1114-34-7

Where hit means attempt Number, Rev Reverse match of peak, For Forward match of Peak, *M.W.*, molecular weight of compound, and *CAS*, chemical abstract service.

TABLE 12.3B Different Soluble Carbohydrate in KSB Inoculated Plant Leaves Extract Identified in GC–MS Analysis Under Salinity

Hit	Rev	For	Compound Name	M.W.	Formula	CAS
1.	863	667	Ethyl α-D-Glucopyranoside	208	$C_8H_{16}O_6$	900127-29-4
2.	849	670	α-D-Galactopyranoside Methyl 6-deoxy	178	$C_7H_{14}O_5$	14887-15-1
3.	834	508	Ethyl β D-Riboside	178	$C_7H_{14}O_5$	900128-95-4
4.	798	691	L-Galactose 6-deoxy	184	$C_8H_{12}O_5$	2438-80-4
5.	796	637	2,3,4,5-Tetrahydroxypbtanal	150	$C_8H_{10}O_5$	53108-52-8
6.	796	585	1,2,3,4-Cyclopentanetetrol 1-α, 2-β, 3-β, 4-α	134	$C_8H_{10}O_4$	14003-71-5
7.	787	868	B-D Ribopyranoside Methyl	184	$C_8H_{12}O_5$	17289-61-1
8.	788	644	L-Lyxose	150	$C_8H_{10}O_5$	1994-78-6
9.	781	580	α-Rhamnopyranose	164	$C_8H_{12}O_5$	35810-58-1
10.	778	585	α-D-Glucopyranose 4-o-α-D-galactopyranosl	342	$C_{12}H_{22}O_{11}$	14641-83-1
11.	775	802	α-D-Glucopyranoside o-α-D-glucopyranosyl	504	$C_{18}H_{32}O_{16}$	597-12-8
12.	775	862	2,3,4,5-TetrahydroxyPentanal	150	$C_5H_{10}O_5$	5310-52-8

Where hit means attempt Number *Rev*, Reverse match of peak *For*, Forward match of Peak, *M.W.*, molecular weight of compound, and *CAS*, chemical abstract service.

(PCD) under biotic stress (Jha and Subramanian, 2014b). Kim et al. (2006) showed that PCD promotes the expression of many of the pathogenesis-related genes induced during the hypersensitive response cell death. In our study, caspase-like activity was assayed on the upcoming third youngest leaves from each treatment at the end of 35 days of stress. Plants under stress suffered a loss in cell viability, as indicated by an increase in the uptake of Evans blue (which is indicative of cell death). The loss of cell viability was higher in plant under stress compared with the corresponding controls. Inoculation with root-associated bacteria reduced the loss of cell viability by two times in normal condition and three times under stress condition compared with noninoculated plants. Inoculation with both the root-associated bacteria showed the highest cell viability compared with individual ones. The sugar signals are also believed to boost plant defense responses (Kocal et al., 2008), presumably through mitogen-activated protein kinase (MAPK) signaling cascades. The sugar-specific MAPK activation can function as an early signal to regulate various sugar-specific responses, including defense reactions in plant cells (Li et al., 2014).

6. EFFECT OF ROOT-ASSOCIATED BACTERIA ON ACCUMULATION OF SOLUBLE SUGAR IN PLANT UNDER ABIOTIC STRESS

Many environmental stresses, such as drought, cold, and salinity, lead to major alterations in carbohydrate metabolism and the sugar signaling pathways interact with stress pathways to modulate metabolism (Gong et al., 2015). Indirectly, the sugars play an important role during plant growth and development under abiotic stresses in regulating carbohydrate metabolism. A large number of stress-responsive genes have been reported to be induced by glucose, indicating the role of sugars in environmental responses (Yue et al., 2015). The sugars produced during photosynthesis are the substrates of carbon and energy metabolism and are used in the biosynthesis of polysaccharides, such as starch and cellulose, in plants (Ruan, 2014). Recent reports showed that sugars are physiological signals repressing reactivating plant genes involved in many essential processes, including photosynthesis, glyoxylate metabolism, respiration, starch and sucrose synthesis and degradation, nitrogen metabolism, pathogen defense, wounding response, cell cycle regulation, pigmentation, and even senescence (Das et al., 2017). Drought, salinity, low temperature, and flooding, in general, increased soluble sugar concentrations. Sugar acts either as a substrate for cellular respiration or as an osmolyte to maintain cell homeostasis (Gupta and Kaur, 2005). In our study, a significant difference was observed in the concentration of sugar accumulated in plants under nonsaline and saline conditions in the plants inoculated with root-associated bacteria. With the increase in the salinity level, the concentration of sugar increased gradually in all plants inoculated with root-associated bacteria and the highest accumulation of sugar was recorded in the case of the control plant leaves at the highest stress (Jha et al., 2011). Recent studies report increasing tolerance to environmental stresses, through metabolic engineering of compatible solutes, and show that increases in soluble sugars and/or other osmolytes provide optimism to increase plant tolerance to abiotic stresses (Harsh et al., 2016). Sassi-Aydi et al. (2014) also reported that mild and moderate osmotic stress decreased sugar in plant. A decreased sugar content could be the consequence of osmotic stress–induced reduction in photosynthesis and the subsequent shortage of photoassimilates in the aerial parts (Hussin and Koyro, 2014). Accumulation of sugars in source tissues downregulates photosynthesis thus maintaining homeostasis. The differential source–sink effects allow the adaptation of carbon metabolism to changing environmental conditions and to the availability of other nutrients (Rolland et al., 2002). The effect of abiotic stresses on metabolic events in source and sink tissues affect the accumulation of soluble sugars in source tissues, which down regulates photosynthesis for maintaining homeostasis. Differential source-sink effects on metabolism induced by unfavorable environmental factors lead to a differential expression of several proteins related to carbohydrate metabolism.

7. EFFECT OF ROOT-ASSOCIATED BACTERIA ON SUGAR SIGNALING FOR DIFFERENTIAL GENE EXPRESSION IN PLANT UNDER ABIOTIC STRESS

The survival of all living organisms on this planet directly depends on photosynthesis, i.e., fixation of carbon and light energy from the sun. As the sugar molecule is the prime carbon and energy source, sugars have acquired important regulatory functions in different organisms early during evolution. The daily life of photosynthetic plants revolves around sugar production, transport, storage, and utilization, and the complex sugar metabolic and signaling networks integrate internal regulators and environmental cues to govern and sustain plant growth and survival. Although diverse sugar signals have emerged as pivotal regulators from embryogenesis to senescence, glucose is the most ancient and conserved regulatory signal that controls gene and protein expression, cell-cycle progression, central and secondary metabolism, as well as growth and development programs. Glucose signals are perceived and transduced by two principal mechanisms: direct sensing through glucose sensors and indirect sensing via a variety of energy and metabolite sensors. Hence, sugar signaling plays an important role in controlling metabolism, stress resistance, growth, and development (Rolland et al., 2006). Sugar signaling can be divided into three steps, sugar sensing, signal transduction, and target gene expression. Further complicating matters are the dual function of sugars as nutrients and signaling molecules and the interaction between sugar signaling and hormonal networks. Hexoses, sucrose and trehalose, might serve as elicitors of plant sugar signaling (Jansson, 2005). Sugar sensing is the interaction between a sugar molecule and a sensor protein in such a way that a signal is generated. The signal then initiates signal transduction cascades that result in cellular responses such as altered gene expression and enzymatic activities. Sugars as signaling molecules affect the plants at all stages of growth starting from seed germination to seed development. Sugars, like hormones, can act as primary messengers and regulate signals that control the expression of various genes involved in sugar metabolism. Sucrose and hexoses both play dual functions in gene regulation as exemplified by the upregulation of growth-related genes and downregulation of stress-related genes. Although coordinately regulated by sugars, these growth- and stress-related genes are upregulated or downregulated through HXK-dependent and/or HXK-independent pathways. Very little is known about the effect that sugars have on the expression of genes involved in sugar signaling cascade. The sugar sensors feed into signal transduction cascades that result in various types of plant responses. The signal transduction cascades involve MAPKs, protein phosphatases, Ca^{2+}, and calmodulin (Barker et al., 2000). It has been reported that a putative sugar signaling component that belongs to the protein family is induced by sucrose, glucose, and fructose.

Previous studies indicate that most of the stress-inducible genes are also induced by RAB 18 (Shinozaki and Yamaguchi-Shinozaki, 2000), and speculatively, this gene family appears to be widely distributed throughout the plant kingdom and is expressed differently upon physiological stresses such as salinity, drought, and cold in various tissues. Inoculation of root-associated bacteria as osmoprotectant has been demonstrated by inoculating *Bacillus amyloliquefaciens* NBRISN13, which confers osmoprotectant ability by modulating differential transcription in a set of at least 14 genes, which underlines the role of the root-associated bacteria as a next-generation osmoprotectant (Nautiyal et al., 2013). To augment the previous conformation and to understand the relative differences in induced gene expression as a result of inoculation of root-associated bacteria in plants exposed to the stress, total mRNA has been isolated from rice inoculated with root-associated bacteria and exposed to different levels of salinity for seven consecutive days. For RT-PCR, the cDNA obtained from the control as well as treated samples for stress are PCR amplified using RAB18 specific primers and the sequence is submitted to the National Center for Biotechnology Information under the accession number JF495696. The induction of RAB18 is observed in inoculated plants as intense bands (data already communicated).

Earlier, it has been reported that the inoculation of other root-associated bacteria, such as *Poenibacillus polymyxa*, in other plants resulted in the induction of the RAB18 fourfold to sixfold in the RT-PCR experiment (Barriuso et al., 2008), which is in agreement with our observed results. Using the *Oryza sativa* as a model, it has been demonstrated that coinoculation of the root-associated bacteria can effectively change the gene expression profiles of the responsive genes under salt stress (Nautiyal et al., 2013). Soluble sugars are highly sensitive to environmental stresses, which act on the supply of carbohydrates from source organs to sink ones. The sugar status of plant cells is sensed by sensor proteins. The regulation is controlled at numerous levels in plants, for example, by allosteric regulation of metabolic enzymes, and tissue-specific or temporal-specific expression of genes (Rook and Bevan, 2003). Sugars, owing to their regulatory function, affect all phases of the life cycle of plants and, interacting within phytohormones, control the processes of growth and development of plants (Stokes et al., 2013). Research on molecular plant responses to abiotic stresses also provides information that is useful in the interpretation of reactions occurring in plants during stress (Hey et al., 2010). Most environmental changes are stressful, although some may be beneficial. To counteract stressful changes and grow successfully, a majority of plants launch resistance mechanisms to stressful environments by reprogramming sugar metabolism and gene expression, and acquiring a new equilibrium between development and stress (Yu et al., 2010). At the same time, attempts have been made to discuss the role of sugar level in resistance to abiotic stresses. Based on the results of ecological and agronomic studies, there is a strong correlation between the soluble sugar concentration and stress tolerance (Acosta-Motos et al., 2017). Root-associated bacteria coordinated between sugar production and utilization and induced different small protein molecules in plant under stress as well as control conditions to establish themselves in the host plant and to protect the plant under stress (Bharti et al., 2016). Mechanisms of root-associated bacteria–mediated phytostimulation would help us to find more capable rhizobacterial strains having the ability to function efficiently under different agroecological conditions for sustainable agriculture (Jha and Subramanian, 2016).

8. CONCLUSION AND FUTURE PROSPECTS

This study contributed to understand the interactions of root-associated bacteria and plants, which resulted in the induction of a series of morphological, anatomical, and enzymatic changes, including changes in sugar metabolism in plants. Such inductions enhance the capability of plant to overcome infection of phytopathogenic microorganism. It also affects the vascular tissues, lignin, and programmed cell death on the interaction of the root-associated bacteria and the plant under biotic stresses. The study showed that expression of the RAB18 gene in abiotic stress is influenced by sugar signals on inoculation with root-associated bacteria in the plant. Our results concur with the suggestion that inoculation with root-associated bacteria can effectively manage sugar concentration to act as an energy source, signaling molecule, or sensor molecule and protect plants against stresses by various mechanisms.

The need of today's world is high-output yield and enhanced production of the crop as well as fertility of soil in an ecofriendly manner. Plant growth promoting rhizobacteria (PGPR)-mediated agriculture is now gaining worldwide importance and acceptance as the safe method for an increasing the number of crops and managed ecosystems. PGPR-mediated crop improvement has untapped potential and is underused, underexploited, underestimated, often untried, and therefore unproven. The new tools of recombinant DNA technology, mathematical modeling, and computer technology combined with a continuation of the more classical approaches and field testing should quickly move PGPR research and technology into a new era. The major challenges in developing PGPR for commercial application include screening and selection of potential PGPR strain, microbial ecology and interaction, fermentation technology and shelf life of formulations, patent protection and prohibitive registration cost, awareness, training and education shortfalls, and lack of multidisciplinary approach.

Hence, the research has to be focused on the new concept of rhizoengineering based on favorable partitioning of the exotic biomolecules, which create a unique setting for the interaction between plant and PGPR. Future research in rhizosphere biology will rely on the development of molecular and biotechnological approaches to increase our knowledge of rhizosphere biology and to achieve an integrated management of soil microbial populations.

REFERENCES

Acosta-Motos, J.R., Ortuño, M.F., Bernal-Vicente, A., Diaz-Vivancos, P., Sanchez-Blanco, M.J., Hernandez, J.A., 2017. Plant responses to salt stress: adaptive mechanisms. Agronomy 7, 18.

Amthor, J.S., 2003. Efficiency and lignin biosynthesis: a quantitative analysis. Ann. Bot. 91 (6), 673–695.

Bais, H.P., Weir, T.L., Perry, L.G., Gilroy, S., Vivanco, J.M., 2006. The role of root exudates in rhizosphere interactions with plants and other organisms. Annu. Rev. Plant Biol. 57, 233–266.

Barker, L., Kuhn, C., Weise, A., Schulz, A., Gebhardt, C., Hirner, B., Hellman, H., Schulze, W., Ward, J.M., Frommer, W.B., 2000. SUT2, a putative sucrose sensor in sieve elements. Plant Cell 12, 1153–1164.

Barriuso, J., Solano, B.R., Gutiérrez Mañero, F.J., 2008. Protection against pathogen and salt stress by four plant growth-promoting rhizobacteria isolated from Pinus sp. on *Arabidopsis thaliana*. Phytopathology 98, 666–672.

Bharti, N., Pandey, S.S., Barnawal, D., Patel, V.K., Kalra, A., 2016. Plant growth promoting rhizobacteria *Dietzia natronolimnaea* modulates the expression of stress responsive genes providing protection of wheat from salinity stress. Sci. Rep. 6, 34768.

Bolouri-Moghaddam, M.R., Le Roy, K., Xiang, L., Rolland, F., Van den Ende, W., 2010. Sugar signalling and antioxidant network connections in plant cells. FEBS J. 277, 2022–2037.

Bolouri-Moghaddam, M.R., Van den Eden, W., 2012. Sugars and plant innate immunity. J. Exp. Bot. 63 (11), 3989–3998.

Bolton, M.D., 2009. Primary metabolism and plant defense-fuel for the fire. Mol. Plant Microbe Interact. 22, 487–497.

Braun, D.M., Wang, L., Ruan, Y.L., 2014. Understanding and manipulating sucrose phloem loading, unloading, metabolism, and signalling to enhance crop yield and food security. J. Exp. Bot. 65 (7), 1713–1735.

Chen, J.G., 2007. Sweet sensor, surprising partners. Sci. STKE 373, 7.

Chen, L.Q., 2014. SWEET sugar transporters for phloem transport and pathogen nutrition. New Phytol. 201, 1150–1155.

Coruzzi, G.M., Bush, D.R., 2001. Nitrogen and carbon nutrient and metabolite signaling in plants. Plant Physiol. 125, 61–64.

Das, A., Paul, J., Rushton, P.J., Rohila, J.S., 2017. Metabolomic profiling of soybeans (*Glycine max* L.) reveals the importance of sugar and nitrogen metabolism under drought and heat stress. Plants 6, 3–21.

Doidy, J., Grace, E., Kuhn, C., Simon-Plas, F., Casieri, L., Wipf, D., 2012. Sugar transporters in plants and in their interactions with fungi. Trends Plant Sci. 17, 413–422.

Gong, X., Liu, M., Zhang, L., Ruan, Y., Ding, R., Ji, Y., Zhang, N., Zhang, S., Farmer, J., Wang, C., 2015. *Arabidopsis AtSUC2* and *AtSUC4*, encoding sucrose transporters, are required for abiotic stress tolerance in an ABA dependent pathway. Physiol. Plant 153, 119 136.

Granot, D., David-Schwartz, R., Kelly, G., 2013. Hexose kinases and their role in sugar-sensing and plant development. Front. Plant Sci. 4, 44.

Gupta, A.K., Kaur, N., 2005. Sugar signalling and gene expression in relation to carbohydrate metabolism under abiotic stresses in plants. J. Biosci. 30, 761–776.

Harsh, A., Sharma, Y.K., Joshi, U., Rampuria, S., Singh, G., Kumar, S., Sharma, R., 2016. Effect of short-term heat stress on total sugars, proline and some antioxidant enzymes in moth bean (*Vigna aconitifolia*). Ann. Agric. Sci. 61, 57–64.

Hey, S.J., Byrne, E., Halford, N.G., 2010. The interface between metabolic and stress signaling. Ann. Bot. 105, 197–203.

Hilal, M., Parrado, M.F., Rosa, M., Gallardo, M., Orce, L., Massa, E.D., 2004. Epidermal lignin deposition in quinoa cotyledons in response to UV-B radiation. Photochem. Photobiol. 79, 205–210.

Hussin Geissler, N., Koyro, H.W., 2014. Effect of NaCl salinity on *Atriplex nummularia* with special emphasis on carbon and nitrogen metabolism. Acta Physiol. Plant 35, 1025–1038.

Jansson, C., 2005. Sugar signaling mutants in *Arabidopsis*. In: Esser, K., Lüttge, U., Beychlag, W., Murata, J. (Eds.), Progress in Botany. Springer-Verlag, Berlin Heidelberg.

Jha, Y., Subramanian, R.B., 2013a. Paddy physiology and enzymes level is regulated by rhizobacteria under saline stress. J. Appl. Bot. Food Qual. 85, 168–173.

Jha, Y., Subramanian, R.B., 2013b. Root associated bacteria from the rice antagonizes the growth of *Magnaporthe grisea*. J. Plant Pathol. Microbiol. 4, 164.

Jha, Y., Subramanian, R.B., 2013c. Paddy inoculated with PGPR show better growth physiology and nutrient content under salinity. Chil. J. Agric. Res. 73 (1), 213–219.

Jha, Y., Subramanian, R.B., 2014a. Characterization of root associated bacteria from paddy and its growth promotion efficacy. Biotech 4 (3), 325–330.

Jha, Y., Subramanian, R.B., 2014b. PGPR regulate caspaselike activity, programmed cell death, and antioxidant enzyme activity in paddy under salinity. Physiol. Mol. Biol. Plants 20 (2), 201–207.

Jha, Y., Subramanian, R.B., 2016. Rhizobacteria enhance oil content and physiological status of *Hyptis suaveolens* under salinity stress. Rhizosphere 1, 33–35.

Jha, Y., Subramanian, R.B., Patel, S., 2011. Combination of endophytic and rhizospheric plant growth promoting rhizobacteria in *Oryza sativa* shows higher accumulation of osmoprotectant against saline stress. Acta Physiol. Plant 33, 797–802.

Kim, M., Lim, J.H., Ahn, J.H., Park, K., Kim, G.T., Kim, W.T., Pai, H.S., 2006. Mitochondria-associated hexokinases play a role in the control of programmed cell death in *Nicotiana benthamiana*. Plant Cell 18, 2341–2355.

Kocal, N., Sonnewald, U., Sonnewald, S., 2008. Cell wall-bound invertase limits sucrose export and is involved in symptom development and inhibition of photosynthesis during compatible interaction between tomato and *Xanthomonas campestris* pv *vesicatoria*. Plant Physiol. 148, 1523–1536.

Lastdrager, J., Johannes Hanson, J., Smeekens, S., 2014. Sugar signals and the control of plant growth and development. J. Exp. Bot. 65, 799–807.

Le Gall, H., Philippe, F., Domon, J.M., Gillet, F., Pelloux, J., Rayon, C., 2015. Cell wall metabolism in response to abiotic stress. Plants 4 (1), 112–166.

Lemoine, R., La Camera, S., Atanassova, R., Dédaldéchamp, F., Allario, T., Pourtau, N., 2013. Source-to-sink transport of sugar and regulation by environmental factors. Front. Plant Sci. 4, 272.

Li, G., Ma, J., Tan, M., Mao, J., An, N., Sha, G., Zhang, D., Zhao, C., Han, H., 2016. Transcriptome analysis reveals the effects of sugar metabolism and auxin and cytokinin signaling pathways on root growth and development of grafted apple. BMC Genomics 17, 150.

Li, Y., Van Den Ende, W., Rolland, F., 2014. Sucrose induction of anthocyanin biosynthesis is mediated by della. Mol. Plant 7, 570–572.

Ma, Q., Cao, X., Xie, Y., Xiao, H., Tan, X., Wu, L., 2017. Effects of glucose on the uptake and metabolism of glycine in pakchoi (*Brassica chinensis* L.) exposed to various nitrogen sources. BMC Plant 17, 58.

Magalhaes Silva Moura, J.C., ValenciseBonine, C.A., FernandesViana, J.D.O., Dornelas, M.C., Mazzafera, P., 2010. Abiotic and biotic stresses and changes in the lignin content and composition in plants. J. Integr. Plant Biol. 52, 360–376.

Mendes, R., Garbeva, P., Raaijmakers, J.M., 2013. The rhizosphere microbiome: significance of plant beneficial, plant pathogenic, and human pathogenic microorganisms. FEMS 37 (5), 634–663.

Morkunas, I., Ratajczak, L., 2014. The role of sugar signaling in plant defense responses against fungal pathogens. Acta Physiol. Plant 4 (36), 1607–1619.

Nautiyal, C.S., Srivastava, S., Chauhan, P.S., Seem, K., Mishra, A., Sopory, S.K., 2013. Plant growth-promoting bacteria *Bacillus amyloliquefaciens* NBRISN13 modulates gene expression profile of leaf and rhizosphere community in rice during salt stress. Plant Physiol. Biochem. 66, 1–9.

OHara, L.E., Paul, M.J., Wingler, A., 2013. How do sugars regulate plant growth and Development? New insight into the role of trehalose-6-phosphate. Mol. Plant 6, 261–274.

Paul, D., Lade, H., 2014. Plant-growth-promoting rhizobacteria to improve crop growth in saline soils: a review. agronomy for sustainable development. Springer Verlag/EDP Sciences/INRA 34 (4), 737–752.

Péron, T., Candat, A., Montiel, G., Veronesi, C., Macherel, D., Delavault, P., Simier, P., 2017. New insights into phloem unloading and expression of sucrose transporters in vegetative sinks of the parasitic plant *Phelipanche ramosa* L. (Pomel). Front. Plant Sci. 7, 2048.

Peuke, A.D., Gessler, A., Trumbore, S., Windt, C.W., Homan, N., Gerkema, E., VAN As, H., 2015. Phloem flow and sugar transport in *Ricinus communis* L. is inhibited under anoxic conditions of shoot or roots. Plant Cell Environ. 38 (3), 433–447.

Rennie, E.A., Turgeon, R., 2009. A comprehensive picture of phloem loading strategies. Proc. Natl. Acad. Sci. USA 106, 14162–14167.

Rogers, L.A., Dubos, C., Surman, C., Willment, J., Cullis, I.F., Mansfield, S.D., 2005. Comparison of lignin deposition in three ectopic lignification mutants. New Phytol. 168, 123–140.

Rolland, F., Baena-Gonzalez, E., Sheen, J., 2006. Sugar sensing and signaling in plants: conserved and novel mechanisms. Ann. Rev. Plant Biol. 57, 675–709.

Rolland, F., Moore, B., Sheen, J., 2002. Plant sugar sensing and signaling. Plant Cell 14, S185–S205.

Rook, F., Bevan, M.W., 2003. Genetic approaches to understanding sugar-response pathways. J. Exp. Bot. 546 (1), 495–50.

Ruan, Y.L., 2014. Sucrose metabolism: gateway to diverse carbon use and sugar signaling. Annu. Rev. Plant Biol. 65, 33–67.

Santos, H.P., Buckeridge, M.S., 2004. The role of the storage carbon of cotyledons in the establishment of seedlings of *Hymenaea courbaril* under different light conditions. Ann. Bot. 94, 819–830.

Sassi-Aydi, S., Aydi, S., Abdelly, C., 2014. Inorganic nitrogen nutrition enhances osmotic stress tolerance in *Phaseolus vulgaris*: lessons from a drought-sensitive cultivar. Hort. Sci. 49 (5), 550–555.

Shinozaki, K., Yamaguchi-Shinozaki, K., 2000. Molecular responses to dehydration and low temperature: differences and cross-talk between two stress signaling pathways. Curr. Opin. Plant Biol. 3, 217–223.

Sindhu, S.S., Parmar, P., Phour, M., Sehrawat, A., 2016. Potassium-solubilizing microorganisms (KSMs) and its effect on plant growth improvement. In: Meena, V., Maurya, B., Verma, J., Meena, R. (Eds.), Potassium Solubilizing Microorganisms for Sustainable Agriculture. Springer, New Delhi.

Slama, I., Abdelly, C., Bouchereau, A., TFlowers, T., Savouré, A., 2015. Diversity, distribution and roles of osmoprotective compounds accumulated in halophytes under abiotic stress. Ann. Bot. 115 (3), 433–447.

Stokes, M.E., Chattopadhyay, A., Wilkins, O., Nambara, E., Campbell, M.M., 2013. Interplay between sucrose and folate modulates auxin signalling in *Arabidopsis*. Plant Physiol. 162, 1552–1565.

Tauzin, A.S., Giardina, T., 2014. Sucrose and invertases, a part of the plant defense response to the biotic stresses. Front. Plant Sci. 5, 293.

Upadhyay, S.K., Singh, J.S., Saxena, A.K., Singh, D.P., 2012. Impact of PGPR inoculation on growth and antioxidant status of wheat under saline conditions. Plant Biol. 14, 605–611.

Vidhyasekaran, P., Kamala, N., Ramanathan, A., Rajappan, K., Paranidharan, V., Velazhahan, R., 2001. Induction of systemic resistance by *Pseudomonas fluorescens* Pf1 against *Xanthomonas oryzae* pv. *oryzae* in rice leaves. Phytoparasitica 29, 155–166.

Yu, L., Nie, J., Cao, C., Jin, Y., Yan, M., Wang, F., 2010. Phosphatidic acid mediates salt stress response by regulation of MPK6 in *Arabidopsis thaliana*. New Phytol. 188, 762–773.

Yue, C., Cao, H.L., Wang, L., Zhou, Y.H., Huang, Y.T., Hao, X.Y., Wang, Y.C., Wang, B., Yang, Y.J., Wang, X.C., August 2015. Effects of cold acclimation on sugar metabolism and sugar-related gene expression in tea plant during the winter season. Plant Mol. Biol. 88 (6), 591–608.

Zanon, L., Falchi, R., Hackel, A., Kühn, C., Vizzotto, G., 2015. Expression of peach sucrose transporters in heterologous systems points out their different physiological role. Plant Sci. 238, 262–272.

Chapter 13

Metabolic Responses of Sugarcane Plants Upon Different Plant–Pathogen Interactions

María E. Legaz[1], Elena Sánchez-Elordi[1], Rocío Santiago[2], Roberto de Armas[3], Blanca Fontaniella[1], Ana M. Millanes[4], María Blanch[5], Carlos Vicente[1]

[1]*Complutense University of Madrid, Madrid, Spain;* [2]*Federal University of Pernambuco, Recife, Brazil;* [3]*Havana University, Havana, Cuba;* [4]*Universidad Rey Juan Carlos, Móstoles, Spain;* [5]*Institute of Food Science Technology and Nutrition (ICTAN-CSIC), Madrid, Spain*

1. INTRODUCTION

Within the large family of Poaceae, the genus *Saccharum* includes six species characterized by different degrees of polyploidy. One of them, *Saccharum officinarum,* is used in extensive cultures mainly to produce sucrose. Its ancestors are *Saccharum spontaneum*, *Miscanthus sinensis,* and *Erianthus arundinaceus*. Commercial varieties currently cultivated are complex hybrids of two or more species. A large number of commercial cultivars arise from interspecific hybridizations through genetic improvement processes.

The leaf of sugarcane plants presents a classical anatomy of C4 plants called Kranz anatomy. The vascular bundles of sugarcane leaf include xylem, phloem, and fibers, and they vary in size, from small to large vascular packages (Fig. 13.1A). The fibers consist of external sclereids surrounding the vascular bundle and an inner ring of cells that surround the phloem. Some sclereids serve as a coupling between the inner epidermal layer and the vascular bundle (docking cells), whereas a compact set of sclereids occupies the total internal volume of the leaf edge (Legaz et al., 2006). The vascular bundle is surrounded by a ring of large green cells known as bundle sheath cells. Near the small vascular bundles, bulliform cells often extend to the upper epidermis (Fig. 13.1B).

The entry of a pathogen into the leaf tissues causes serious anatomical and functional alterations. *Xanthomonas albilineans*, for example, occludes the xylem vessels when it secretes a xanthanlike gum and causes water stress, which is visually observed by the appearance of yellow files alternating with green lines along the leaf blade, parallel to the main vascular bundles (Lopes et al., 1998). An immediate mechanism of resistance to desiccation consists in the rolling up of the leaf, after the increase in volume of the bulliform cells (Fig. 13.1C). The leaf blade forms a kind of cylinder whose inner surface contains the majority of stomata. In this way, the atmosphere inside the rolled sheet is rapidly saturated with water vapor and transpiration decreases.

The case of smut, one of the most important diseases of sugarcane, is fundamentally different. The spores of *Sporisorium scitamineum* are deposited on the surface of leaves or on the stem internodes, where they germinate forming appressoria, mainly in the innermost scales of the buds or at the base of the younger leaves (Waller, 1970). However, some buds, highly protected by strong waxy scales laden with toxic flavonoids, can stop the infection and prevent the formation of a penetrating mycelium. After germination, the mycelium penetrates the meristematic tissues, preferably using intercellular spaces to advance, between 6 and 36h after infection. The growing mycelium invades not only the meristematic tissues of the buds but also the parenchymatous cells below the internode (Fig. 13.1D). Whips (sori with teliospores produced only at the base of this structure) usually are formed above the internodes (Marques et al., 2017).

Once the pathogen, whether fungal or bacterial, has settled in the host plant, the morphological changes described earlier are often accompanied by profound metabolic changes whose differences may define the susceptibility or resistance of sugarcane plants to the invasive agent (Ramesh Sundar et al., 2012; Karen et al., 2013). After whip formation, drastic changes in carbon partitioning, normally directed to sucrose storage (Wang et al., 2013) occur related to overexpression of invertases and trehalose-6-phosphate synthase (Schaker et al., 2016). These changes can be related to drastic modifications in the plant pathways for pathogen cell recognition and tolerance depending on sugar signaling (Ruan, 2014).

Plant Metabolites and Regulation Under Environmental Stress. http://dx.doi.org/10.1016/B978-0-12-812689-9.00013-3

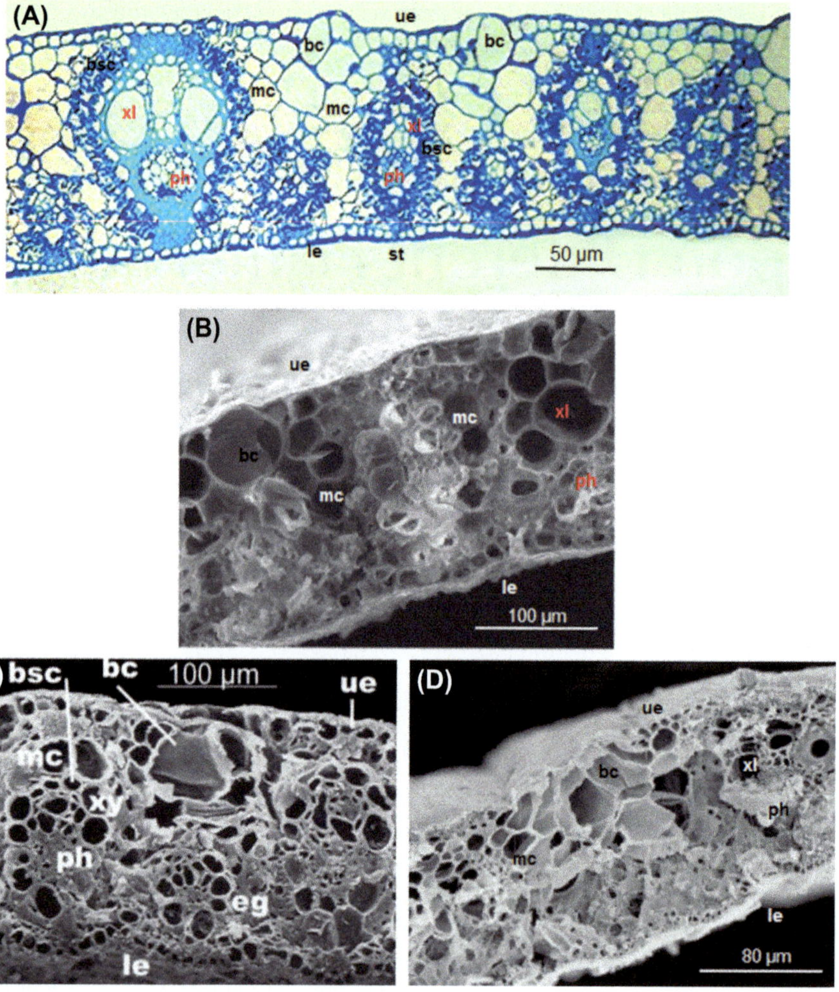

FIGURE 13.1 (A) Cross-section of a sugarcane leaf (cultivar My 55-14) showing the characteristic Kranz anatomy under light microscope. (B) Structure of a healthy sugarcane leaf, cultivar My 55-14, showing complete and unmodified bundle vessels under scanning electron microscopy. (C) Scanning electron micrographs of a cross-section of leaves from B42 231 sugarcane cultivar. (D) Scanning electron micrographs of a cross-section of the yellow zone of a scalded leaf of L55-5 sugarcane cultivar showing unmodified bundle sheath cells, but bulliform cells are enlarged and xylem vessels are often occluded. *bc*, Bulliform cell; *bsc*, bundle sheath cells; *le*, lower epidermis; *mc*, mesophyll cells; *ox*, occluded xylem; *ph*, phloem; *ue*, upper epidermis; *xl*, xylem.

1.1 Production of Soluble Polysaccharides by Sugarcane Plants as a Response to Stress, Mechanical Injuries, or Impairment

Sugar metabolism in *S. officinarum* is mainly directed to the synthesis and accumulation of sucrose in the parenchyma of the stems. Secondarily, some soluble polysaccharides may transiently accumulate, along with sucrose such as small amounts of arabinogalactans, starchlike glucopolysaccharides containing some of them branches consisting of 1,6-glucans as well as 1,4-glucans (Roberts et al., 1985).

However, under special stress conditions, sugarcane plants can synthesize several compounds of low molecular mass (Pacheco et al., 2013) as well as soluble heteropolysaccharides, consisting of fructose and galactitol, which have been tentatively classified into two groups according to their molecular mass: high-molecular-weight carbohydrates (HMMC), composed of more than 100 monosaccharide units, and midmolecular mass carbohydrates (MMMC), composed of 10 to 50 monosaccharide units (Legaz et al., 2005). The stress circumstances that provoke their accumulation in detriment of that of sucrose are usually conditions of deterioration of the cut stems: collection at the postripening stage (aged stems), number of cuts applied on stems, or low oxygen tension during transport and storage. Varietal differences in soluble polysaccharides have also been reported (Rodriguez et al., 1985). Other soluble polysaccharides, such as glucans or sarkaran, have been found after stalk impairment.

Fructose and galactitol were the only products that appeared after total acid hydrolysis of both HMMC and MMMC (Vicente et al., 1991). The quantitative composition of these polysaccharides after their complete acid hydrolysis proved to be [$fructose_2$:$galactitol_3$]$_n$ for MMMC and [$fructose_4$:$galactitol_5$]$_n$ for HMMC. The occurrence of galactitol in the hydrolysates from both polysaccharides was an unexpected result since, in the literature, only the occurrence of mannitol was reported for sugarcane juices. Since the detection of mannitol was always simultaneous with those of ethanol and lactic acid in highly deteriorated cane juice, their occurrence indicated that these three compounds were products of a fermentation process caused by *Leuconostoc mesenteroides* (Eggleston et al., 2004). To verify whether galactitol had the same origin, even as included in a polymeric heterofructan, *L. mesenteroides* and *Erwinia amylovora*, isolated from impaired juices, were grown on inorganic media containing HMCM or MMMC as sole sources of organic carbon. The absence of growth of both microorganisms indicated their inability to produce fructose and galactitol from these substrates (Martínez et al., 1990a).

The biosynthetic pathway of the sugarcane heterofructans is so far unknown. However, Legaz et al. (1990) were able to purify from sugarcane juices an enzymatic complex with glycosidase activity that hydrolyzed both HMMC and MMMC, although with different efficiency, by releasing three times more reducing sugars (fructose) than galactitol. From these results it can be deduced that the bonds through which polyalcohol residues were included in the sequence of the complete polysaccharide were stronger and, therefore, more resistant to enzymatic hydrolysis than the glycosidic bond between two consecutive fructose units. The glycosidase complex also released small polymers, of about 35–50 units, containing variable amounts of fructose and galactitol. The enzyme used Mn^{2+} as a cofactor, which could be partially replaced by Mg^{2+}. The transient binding of Mn^{2+} to the enzyme to form the enzyme–cofactor–substrate complex followed a hysteresis cycle as a function of cation concentration (Legaz et al., 1991). The purified enzyme showed an optimum pH of 6.0 and a molecular mass of 13.2 kDa. Logically, differences in affinity of the enzyme for HMMC or MMMC have been observed, probably derived from the different degrees of substrate polymerization.

The enzyme was also able to release discrete amounts of both HMMC and MMMC from a suspension of cell walls in aqueous buffer, isolated from parenchymatous cells of sugarcane stems (Martínez et al., 1990b). This ability implied that, under deterioration conditions of sugarcane stalks, the enzyme could dissolve several heterofructan fractions of different molecular sizes into the cytoplasm of the parenchymatous cells, or released them to the intercellular spaces, from polysaccharides located in the wall, which would imply their recovery in industrial juices destined to obtain sucrose. The overall reaction catalyzed by the glycosidase complex could be written as in Fig. 13.2.

1.2 Chemical Structure of Soluble Polysaccharides of Sugarcane Plants

The polysaccharides HMMC and MMMC were fully hydrolyzed by acid and partially hydrolyzed by the enzyme invertase. Prolonged treatment of both polymers with 6N HCl released only two classes of monomers, identified by high-performance liquid chromatography (HPLC) as fructose and galactitol. For this reason, both polysaccharides were defined as heterofructans, as opposed to those that, like inulin, were exclusively composed of fructose units, except for the initial glucose monomer (Van den Ende, 2013), a relict of sucrose used by sucrose:sucrose fructosyltransferase. However, hydrolysis with invertase only released fructose and oligosaccharides of variable molecular mass. The spectrophotometric quantitation of fructose obtained in both cases revealed that partial hydrolysis with invertase released only 62% of the amount of that fructose released by total acid hydrolysis (Legaz et al., 2005). From the results obtained by both procedures, it could be deduced that the fructose units linked to each other by a glycosidic β-(1→2)bond, hydrolyzed by invertase, whereas the galactitol units might bind to each other and the fructose molecules by stronger bonds, sensitive to the acid but insensitive to invertase. Hexitol–hexose bond was seen to be sensitive to acidic hydrolysis but unaccessible to invertase action (Fig. 13.3).

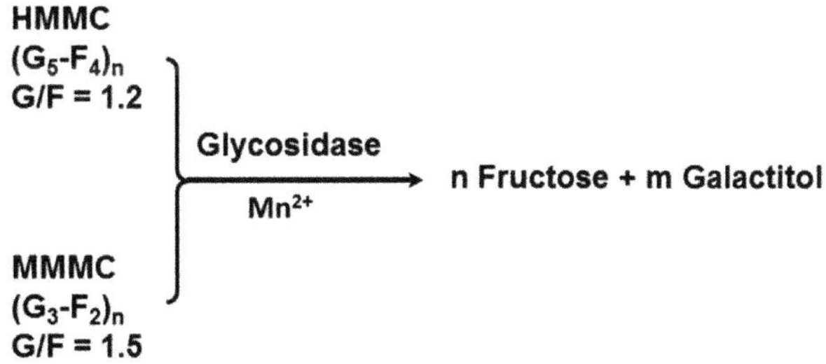

FIGURE 13.2 Schematic representation of the sugarcane glycosidase action on high-molecular-weight carbohydrates (HMMC) and midmolecular mass carbohydrates (MMMC) combining β-1,2-glycosidase and ether hydrolase enzymatic activities.

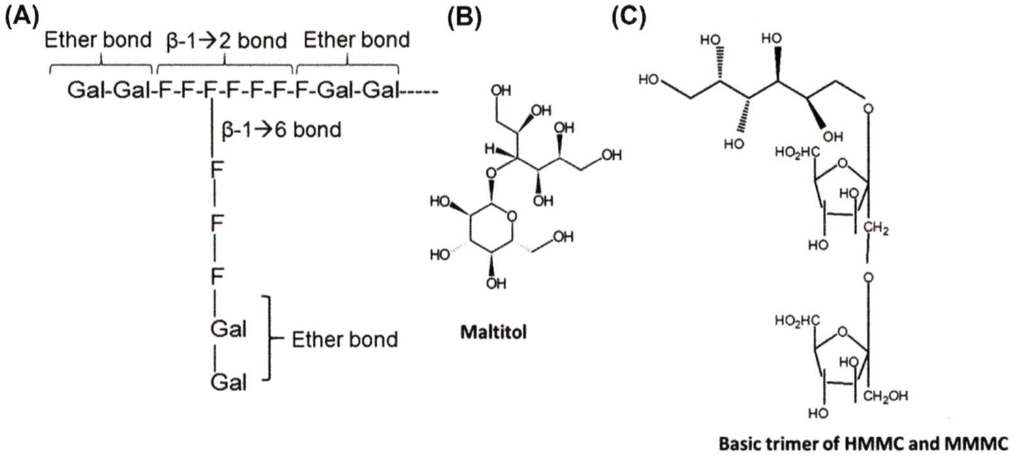

FIGURE 13.3 (A) Proposed structure of both HMMC and MMMC, where F is fructose and Gal, galactitol. (B) Chemical structure of matiltol. (C) Chemical structure of the basic trimer composing both HMMC and MMMC. *HMMC*, high-molecular-weight carbohydrates; *MMMC*, midmolecular mass carbohydrates.

Hexytol–hexose dimers are not very abundant in nature, but some of them, such as maltitol (4-O-α-D-glucopyranosyl-D-glucitol), can be obtained by chemical treatment of some vegetable juices and syrups. In addition, the glycoside β-(1→4) bond between galactose and glucitol produces lactitol (Fig. 13.3). Other dimers or trimers can be naturally produced, as occurs when an α-galactosidase from *Bifidobacterium breve* 203, which normally synthesizes a trisaccharide derived from melibiose, uses inositol as an acceptor of the second galactose unit (Schröder et al., 2004). However, fructose–galactitol dimer must involve a bond stronger than a glycosidic one. In fact, infrared spectra of both HMMC and MMMC showed a strong peak at 1100 cm^{-1} (de Armas et al., 1992) that did not appear for maltitol or lactitol molecules. This peak could reveal the occurrence of an ether bond between fructose and galactitol, which could not be hydrolyzed by invertase but broken by acidic hydrolysis. In addition, a peak for MMMC near 890 cm^{-1} clearly corroborates the fact that fructose units are linked in β-conformation.

1.3 Soluble Polysaccharides of Sugarcane Plants Are Usually Associated with Proteins

The absorption spectra of both HMMC and MMMC showed a maximum at 274 nm, with a shoulder at 250 nm exclusively for MMMC (Fig. 13.4A). On exciting both samples with light of 275 nm wavelength, HMMC emitted fluorescence at 334–336 nm, whereas MMMC did so with a maximum at 320 nm and a second maximum at 346 nm. Tryptophan emits fluorescence with a maximum at 346–347 nm, which would correspond to the secondary maximum detected in MMMC (hereinafter MMMG, mid molecular mass glycoproteins) fluorescence spectrum after excitation with a light of 275 nm (Fig. 13.4B). However, this maximum did not appear in the fluorescence emission spectrum of HMMC (hereinafter HMMG, high molecular mass glycoproteins), but was shown shifted to the 334–336 nm area. This difference could be explained if some of the tryptophan residues in MMMG glycoprotein were found in a domain of the protein highly exposed to the aqueous environment, whereas, in the HMMG, those same residues would be located in strongly hydrophobic domains, completely isolated from the hydrophilic zone in direct contact with the solvent. When solutions of both glycoproteins were excited with far-ultraviolet light (210 nm), the fluorescence emission corresponding to tryptophan residues did not change, but a new fluorescence peak appeared at 296–302 nm (Fig. 13.4B), corresponding to the occurrence of tyrosine. Quantitative analysis of the polysaccharide and protein domains revealed that HMMG contains identical proportion of both moieties, whereas MMMG contains 20% protein and 80% polysaccharide (Legaz et al., 1995).

The glycoprotein nature of MMMG and HMMG has also been evidenced by anion exchange chromatography and capillary electrophoresis. Anion exchange chromatography on diethyl aminoethyl-Sephadex A-60 can separate electrically neutral, cationic, or anionic peptides as different components of the total pool of sugarcane glycoproteins. The occurrence of both cationic and anionic glycoproteins was corroborated after separation by capillary electrophoresis. This separation procedure revealed similarities and differences between the different components of MMMG and HMMG. Identical glycoproteins in both pools corresponded to peaks whose migration time in the capillary electrophoresis presented values of 14.07, 22.26, 22.73, and 23.74 min. However, MMMG differed from HMMG in the existence of two glycoproteins with migration times of 16.57 and 21.28 min, which did not appear in HMMG samples (Legaz et al., 1998a,b). On the other

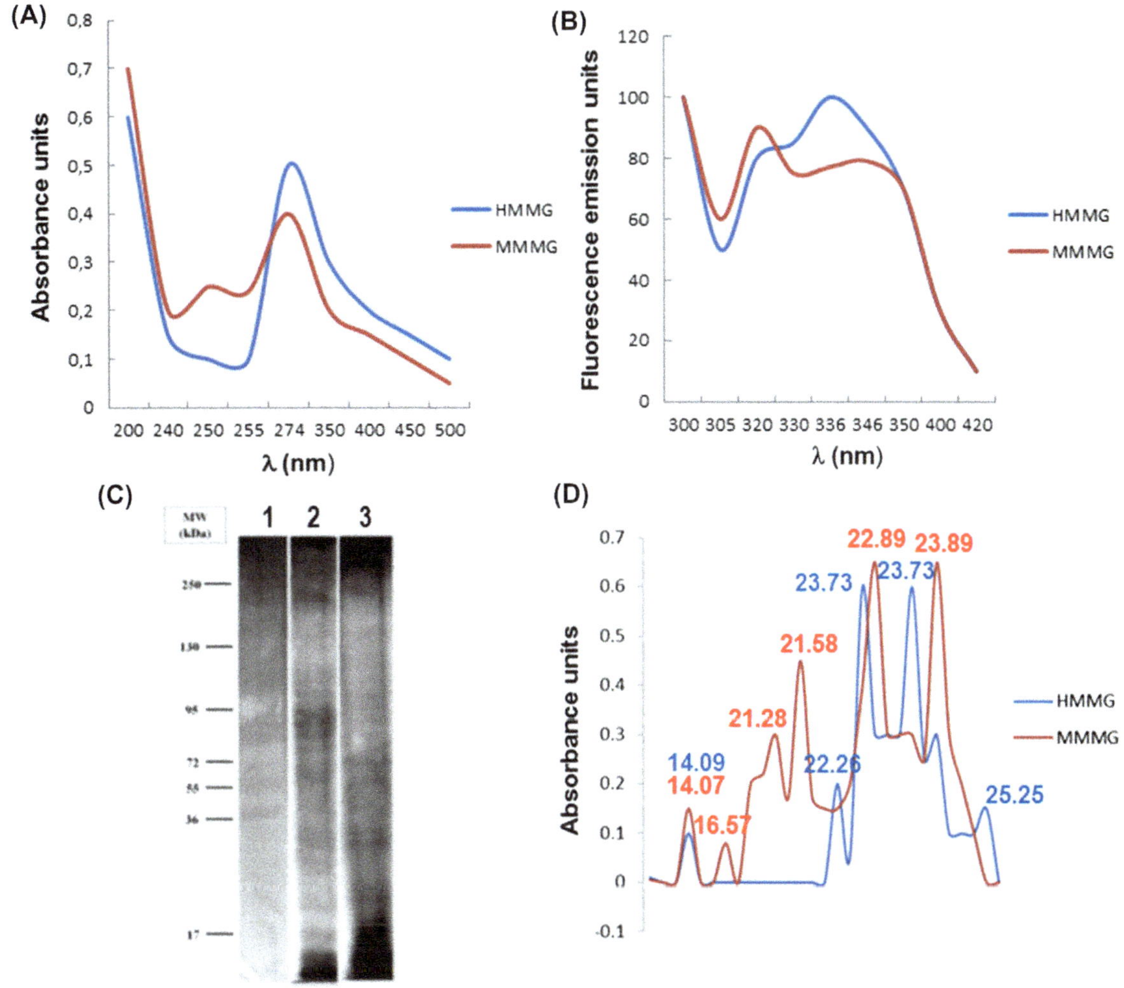

FIGURE 13.4 (A) Absorption spectra of high molecular mass glycoproteins (HMMG) and mid molecular mass glycoproteins (MMMG). (B) Spectra of fluorescence emission of HMMG and MMMG. (C) Polyacrylamide gel electrophoresis of HMMG and MMMG. Lane 1, molecular markers; lane 2, HMMG; lane 3, MMMG. (D) Electropherogram of the separation of HMMG and MMMG components by capillary electrophoresis. Numbers near the peaks indicate mobility time in minutes.

hand, HMMG contained a glycoprotein with migration time of 25.25 min, without any equivalent in MMMG (Fig. 13.4C and D). With the migration time of neutral marker estimated at 17.37 min, it can be concluded that the MMMG pool contained two cationic glycoproteins (only one from HMMC), whereas both MMMG and HMMG were composed of four anionic glycoproteins, albeit with different net charges.

The mobility of glycoproteins as a function of the net charge/mass ratio could be modified through two different processes: (1) the occurrence of a polysaccharide coverage of the protein, structured by ether and glycosidic bonds and (2) the conjugation of these glycoproteins with molecules possessing their own net charge. The shortening of the polysaccharide moieties by their enzymatic hydrolysis, catalyzed by a β-1,2-fructofuranosidase, causes the cationic glycopeptides of MMMG to move toward the neutral or anionic zone of the electropherogram, whereas the only HMMG cationic glycoprotein does not modify its electrophoretic mobility after hydrolysis of its polysaccharide moiety. These results indicate that these polysaccharide moieties, at least in MMMG, are able to modify the net charge of the protein, or the net charge/mass ratio, whereas this does not occur for HMMG. In addition, the analysis by HPLC of the hydrolysates of both HMMG and MMMG revealed the occurrence of conjugated polyamines (PAs) to the different glycoproteins, mainly putrescine (PUT), although cadaverine (CAV), spermidine (SPD), and spermine (SPM) also appeared in less amounts. Polycationic PAs bound to MMMG could provide the net positive charge to their cationic components. The displacement of these MMMG cationic components toward the neutral and anionic zones of the electropherogram after the hydrolysis catalyzed by a fructofuranosidase would demonstrate that these PAs preferably would conjugate to the heterofructan components of the glycoproteins.

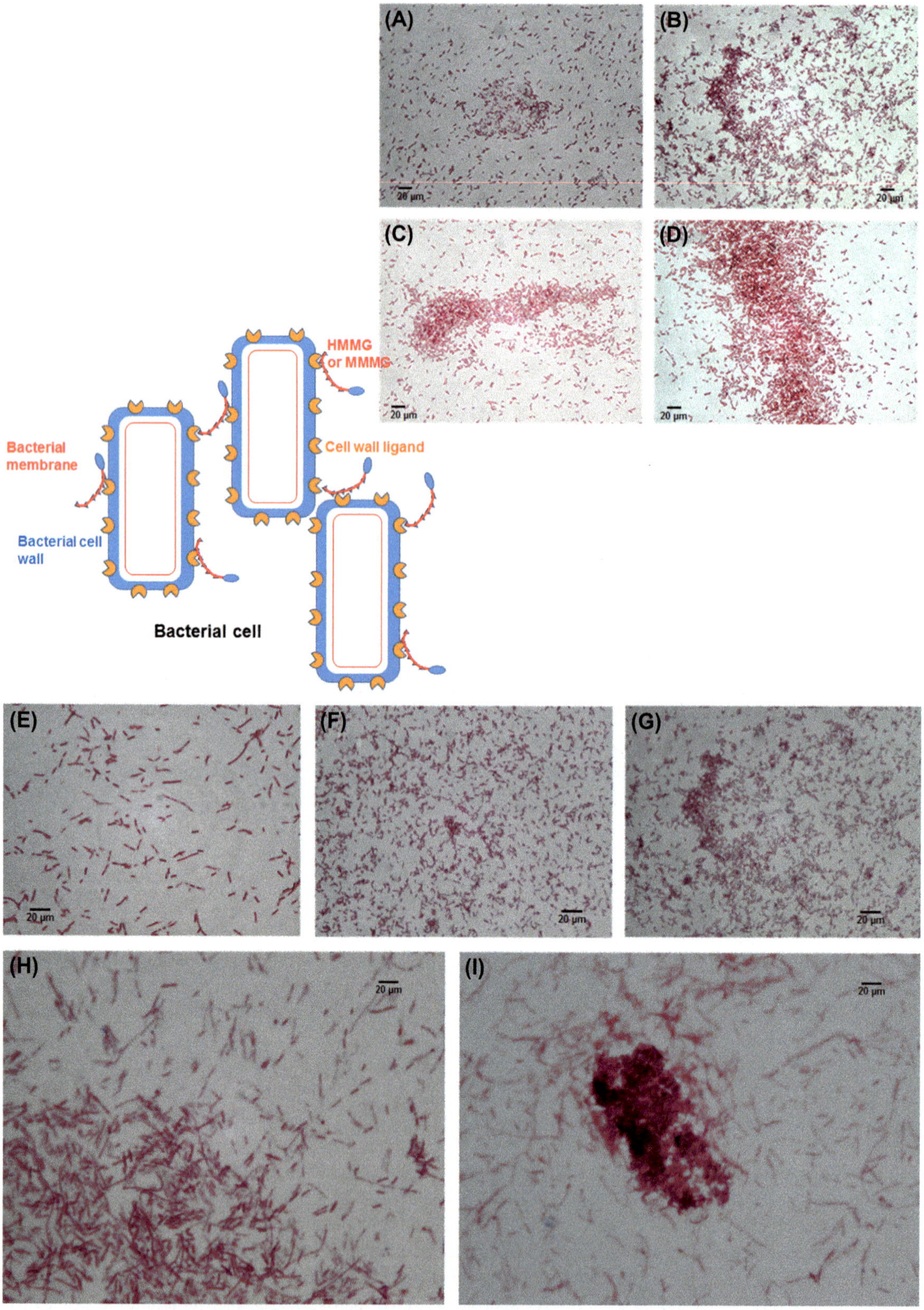

FIGURE 13.5 Glycoproteins from sugarcane stalks induce bacterial cytoagglutination. (A–D) Cell aggregation of *Gluconacetobater diazotrophicus* induced by high molecular mass glycoproteins (HMMG). (A) Untreated control, (B) after 30 min, (C) after 1 h, and (D) after 2 h incubation of a cell suspension with HMMG. (E–I) Cell aggregation of *Xanthomonas albilineans* induced by mid molecular mass glycoproteins (MMMG) and HMMG. (E) Untretated control, (F) after 30 min of incubation with MMMG, (G) after 30 min of incubation with HMMG, (H) after 2 h of incubation with MMMG, (I) after 2 h of incubation with HMMG. The inset drawing shows the use of sugarcane glycoproteins, shared by ligands of several bacterial cells, to promote cytogglutination.

2. SUGARCANE PLANTS PRODUCE GLYCOPROTEINS AS A RESPONSE TO INVASION BY SEVERAL ENDOSYMBIONTS: *GLUCONACETOBACTER, XANTHOMONAS, SPORISORIUM*

As mentioned previously, not only pathogenic microorganisms but also some beneficial ones, can penetrate into the tissues of sugarcane plants and settle into them. This is the case of *Gluconacetobacter diazotrophicus*, for example, a nitrogen-fixing endosymbiont (Cavalcante and Döbereiner, 1988). It could be thought then that any endophyte penetrating these tissues could cause, as a biological response, an increase in the production of HMMG and MMMG. However, experimental infection of sugarcane stalks with *G. diazotrophicus* did not increase the production of HMMG and MMMG, whereas *X. albilineans*, a pathogenic bacterium, increased by 2.5 times the production of these glycoproteins (Blanco et al., 2005a,b). On the other hand, the amount of glycoproteins did not show a significant increase when both microorganisms were simultaneously introduced in growing plants. This difference of action could be translated into a competition between both microorganisms inside the same host, in such a way that the presence of *G. diazotrophicus* impedes the action of the pathogen. Blanco et al. (2005a,b) demonstrated that parenchymal stem cells, containing the nitrogen fixer *G. diazotrophicus*, and experimentally infected with *X. albilineans* prevented the growth of the pathogenic bacterium. The endophyte synthesized a lysozyme-like bacteriocin that diffused into the tissue and lysed the pathogen cells (Piñón et al., 2002). This impeded the elicitation of HMMG and MMMG produced by the pathogen in the absence of its antagonist.

Infection of sugarcane plants with *S. scitamineum* teliospores also increased the production of some of these glycoproteins (Fontaniella et al., 2002) but in a different way. Experimental standardized (Singh et al., 2014) inoculation of My 55-14 cultivar, resistant to smut, or cultivar Jaronu 60-5, sensitive to smut, with teliospores of *S. scitamineum* preferentially increased the amount of HMMG but decreased the amount of MMMG, even when inoculated plants did not show external symptoms of smut disease (Millanes et al., 2005). The difference in the pattern of glycoprotein production has been explained on the basis of the simultaneous production of a glycosidase, the affinity of which for the polysaccharide moiety of MMMG is higher than that observed for HMMG. This gives greater role to HMMG than that suspected for MMMG in the biochemical defense of the sugarcane plants against the pathogen. The different degrees of resistance of the different cultivars to smut would be then a function of the amount of HMMG that the plant could accumulate after the entry of the pathogen (Legaz et al., 2005; Fontaniella et al., 2002). The use of sugars or glycoproteins as signaling processes against fungal pathogens is a common strategy in the plant kingdom (Morkunas and Ratajczak, 2014).

The glycosidase system also released small heterofructans from the cell wall of the parenchymatous cells of sugarcane stalks (Martínez et al., 1990b; Legaz et al., 2005). The occurrence of soluble polysaccharides, associated or not with peptides, in sugarcane juices causes a reduction in the amount of sucrose recovered during the industrial process of crystallization from Ja 60-5 cultivar, sensitive to smut, stalks (Fontaniella et al., 2003a,b) by forming molasses that affect the purity of the syrup, changing its physicochemical properties.

Accumulation of sugarcane glycoproteins as a response to a smut infection also seemed to be a function of the age of the plant. Thus the amounts of both HMMG and MMMG decrease when 22-month-old plants of cultivars were experimentally infected, as opposed to when 12- or 14-month-old plants were inoculated (Fontaniella et al., 2002). This will be further discussed in the following sections.

3. SUGARCANE PLANT GLYCOPROTEINS PRODUCE ENDOSYMBIONT CYTOAGGLUTINATION: IDENTIFICATION OF THEIR LIGANDS

3.1 Cytoagglutination of *Gluconacetobacter diazotrophicus*

To demonstrate whether sugarcane glycoproteins actually bind to the cells of their potential endophytes, such as *G. diazotrophicus*, both HMMG and MMMG can be labeled with a fluorophore, fluorescein isothiocyanate, for example, and incubated with these cells over a suitable period to be observed later by fluorescence microscopy. In this way, such binding can be easily visualized, since the fluorescence is peripherally and externally located on the bacterial cell walls. Even its desorption can be studied after binding, using sucrose as the releasing agent. The reason for this remotion is logical: the β-(1→2) glycosidic bond between glucose and fructose of sucrose molecule, added to fluorescent bacterial suspension, can displace both HMMG and MMMG by competition with the β-(1→2)-fructofuranosyl fructose link of the polysaccharide moiety of these glycoproteins. This implies that sugarcane glycoproteins use their β-(1→2)-fructofuranosyl fructose from the glycosidic moieties of HMMG and MMMG to bind to their bacterial ligands in the bacterial cell walls.

This chemical property provides a good tool for the isolation of such ligands from bacterial cell walls. HMMG and MMMG may be covalently bound to active agarose to retain bacterial ligands by affinity chromatography, ligands which could subsequently be desorbed with sucrose. The galactitol domains of sugarcane glycoproteins do not appear to be reactive, since any detectable desorption of bacterial glycoproteins is not achieved using this polyalcohol as eluent. In this way, at least six classes of *G. diazotrophicus* receptors for MMMG and another six for HMMG have been characterized by capillary electrophoresis. All of them behave as anionic proteins, although their charge-to-mass ratio is different for the ligands of the two classes of macromolecules. This difference can be explained on the basis of the different lengths of the polysaccharide chains of both classes of sugarcane glycoproteins (Blanco et al., 2005a,b). Since *G. diazotrophicus* behaves as an endosymbiont of sugarcane plants and moves from the infection points to the intercellular apoplastic spaces, their cell wall glycoproteins, identified as ligands, would interact in situ with the saccharide domain of both HMMG and MMMG, located in the cell walls of the parenchymatous cells of sugarcane stalks or even secreted from these cells to their intercellular spaces (Martínez et al., 1990b).

From the set of ligands of *X. albilineans* eluted from agarose with sucrose, nine different anionic glycoproteins are recognized by MMMG, although only six were recognized by HMMG. In this case, four of these anionic glycoproteins are identical for both classes of sugarcane glycoproteins, whose migration times in capillary electrophoresis were 1.6, 1.9, 2.6, and 2.8 min (Blanco et al., 2006). Three of these ligands were also desorbed with galactitol (migration time values of 1.7, 1.9, and 2.6 min), although differing in the different amounts of sugar required to be eluted. This effect could probably be related to a different number of binding points in the peptide for the saccharide domain of sugarcane glycoproteins. Two other anionic ligands, also eluted with the polyalcohol, were glycoproteins different from those previously identified.

The binding of sugarcane glycoproteins to their cell wall ligands in *X. albilineans* results in cell recruitment, which probably increases the infectivity of the microorganism. In addition, production of bacterial polysaccharide is also affected by the synthesis of bacterial quorum-sensing molecules (QSMs) (Mensi et al., 2016) that included a diffusion signaling factor synthesized by a diffusible signal factor synthase (Rott et al., 2013). Relationships between HMMG and MMMG and their bacterial receptors seem to be related to a bacterial recruitment, which probably improves plant infection rather than the defense mechanism. By promoting such binding, MMMG and HMMMG cannot be considered as resistance (R) proteins or avirulence factors (Avr). MMMG and HMMG favor the penetration and establishment of the bacterium in the tissues of the host, whereas R and Avr stop the infection. In addition, the location of both kinds of proteins is distinct. HMMG and MMMG are cell wall glycoproteins (Martínez et al., 1990b), whereas R proteins are transmembrane or cytoplasmic factors.

Since both bacteria, *G. diazotrophicus* and *X. albilineans*, have sucrose-sensitive ligands for sugarcane glycoproteins, it could be thought that some competition or antagonism could be established between them that would have repercussions on bacterial development. In fact, this antagonism exists, but not on the basis of an impediment of cell recognition by cross-receptor occupation. The physiological basis of this antagonism has been studied using tissues of sugarcane stalks previously inoculated with the endosymbiont, then immobilized in calcium alginate, and then maintained after this in a culture medium for *Gluconacetobacter*. Under these conditions, bacterial endosymbionts occurring in immobilized tissues are able to secrete to the inert matrix a lysozyme-like bacteriocin that inhibits the growth of *X. albilineans* (Blanco et al., 2005a,b).

Similar results on cytoagglutination were obtained using *Herbaspirillum rubrisubalbicans* treated with MMMG and HMMG (Fig. 13.5). *H. rubrisubalbicans* is a nitrogen-fixing bacterium of the genus *Herbaspirillum* found in the roots and stems of sugarcane plants. *H. rubrisubalbicans* can cause symptoms of the mottled stripe disease in sugarcane (Kirchhof et al., 2001). MMMGs were preferentially desorbed with sucrose and galactitol, whereas HMMGs were mainly desorbed with glucose and mannose. This would indicate that, against this bacterium, MMMG behaves as signal molecules that bind to their receptor, or receptors, using their polysaccharide moiety, whereas, on the contrary, HMMG would use their peptide moiety for binding to different receptors, similar to the action mode of the lectin concanavalin A (Fontaniella et al., 2004), from *Canavalia ensiformis.*

Surprisingly, receptors for both HMMG and MMMG do not behave as the typical adhesion receptors containing polymeric monosaccharides that bind by affinity to a specific peptide domain in the signaling molecule, the recognition of which implies the binding of this to selected carbohydrate moieties in their ligands (Zihni et al., 2016). In this case, the carbohydrate moiety of the signal molecule seems to be of use to recognize a particular amino acid domain on the ligand (receptor) in an inverse way to that described for plant lectins and animal selectins. This fact suggests that HMMG and MMMG, with independence of their possible enzymatic activities (Su et al., 2016a), behave as true protein of resistance (PR), according to Su et al. (2016b), which would require ligands similar to Toll-like receptors, studied in animals (Kumar et al., 2009).

3.2 Cytoagglutination of Smut Teliospores

Cytoagglutination induced by sugarcane glycoproteins is not restricted to prokaryotic endosymbionts. Teliospores of the basidiomycete *S. scitamineum* also aggregate in the presence of both HMMG and MMMG fractions (Millanes et al., 2005; Sánchez-Elordi et al., 2016b).

To investigate this fact, samples of smut teliospores were floated on glycoprotein fractions obtained from both healthy and inoculated plants of three different sugarcane cultivars. In all cases, the adhesion of sugarcane glycoproteins to fungal teliospores produced aggregation, as shown in Fig. 13.6. Interestingly, in these preliminary experiments, the aggregation effect seemed to be unrelated to the degree of resistance of the cultivar from which the glycoproteins had been isolated. The level of cytoagglutination produced was therefore initially attributed to the concentration of glycoproteins in the fraction employed in the treatments rather than to the origin of the glycoprotein fraction. However, very low concentrations of glycoproteins isolated from a cultivar resistant to smut were enough to induce aggregation of smut teliospores. In contrast, higher concentrations were needed to induce cytoagglutination in suspensions of teliospores when glycoproteins from non-resistant cultivars were used (Millanes et al., 2005, 2008).

The cytoagglutinating effect was clearly reduced using invertase-digested HMMG or MMMG, which hydrolyzed the glycidic moiety of sugarcane glycoprotein. This suggested that this glycidic moiety, which contains fructose residues, polymerized by β-D-fructofuranosyl-1,2-β-D-fructose, could be involved in the process of cell aggregation since the extensive hydrolysis of β-(1→2) bond impeded cell adhesion. Thus the integrity of such β-(1→2) bond would also be essential to bind both HMMG and MMMG to a hypothetical cell-wall receptor of the teliospores.

To obtain experimental evidence of the presence of such cell wall receptor or receptors, we isolated glycoproteins, in this case from the cell wall of the fungal pathogen. These proteins were separated by affinity chromatography through activated agarose columns, to which sugarcane glycoproteins from different cultivars had been previously bound. Fungal cell wall receptors retained by sugarcane glycoproteins were recovered next, desorbed by certain monosaccharides introduced in the columns as eluents. In particular, we observed that sugarcane HMMG and MMMG fractions exhibited a high affinity for the monosaccharide N-acetyl-D-glucosamine component of the cell wall of filamentous fungi. Thus, cell wall receptors of *S. scitamienum* retained by sugarcane glycoproteins bound to the columns were competitively displaced by high concentrations of that monosaccharide (Fig. 13.6). Several plant lectins bind oligomers of N-acetyl-D-glucosamine. Sampietro et al. (2001) purified a new lectin from *Cyphomandra betacea* whose agglutinating ability was only inhibited by N-acetyl-D-glucosamine oligomers. Although we did not test the possible reversion of the cytoagglutinating effect by N-acetyl-D-glucosamine, we demonstrated that the sugarcane glycoprotein bound this monosaccharide with certain degree

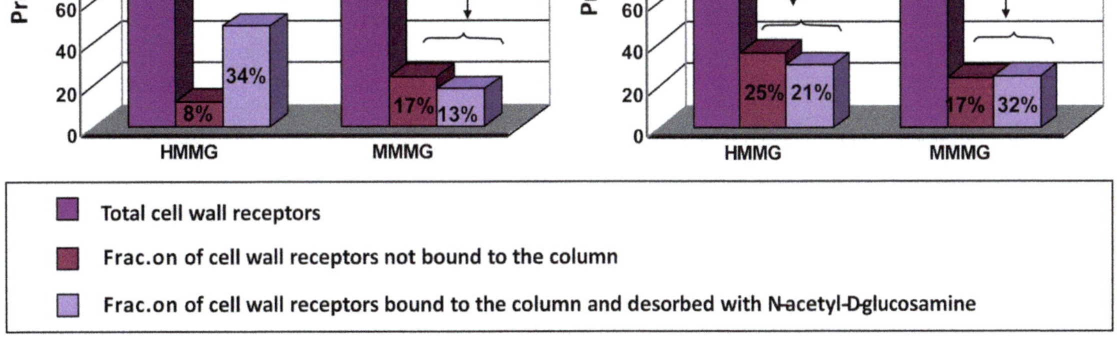

FIGURE 13.6 Proteins isolated from a sample of cell walls of *Sporisorium scitamineum* by filtration through an affinity chromatography column. Agarose-activated columns bound high molecular mass glycoproteins or mid molecular mass glycoproteins, from healthy (A) or inoculated (B) sugarcane plants isolated from a resistant cultivar, My 55-14. In each case the percentage of fungal protein irreversibly retained by the column is indicated. Results are the mean of three replicates. Standard error was never higher than 9%.

of affinity. Therefore at least part of the glycoproteins contained in HMMG and MMMG fractions behaved as lectins capable of binding fungal N-acetyl-D-glucosamine. Owing to the high heterogeneity of HMMG and MMMG fractions it is not strange that in different cultivars, or even within the same cultivar, there are several possibilities regarding specificity of the binding oligosaccharide. Moreover, the same lectin can have different binding sites for different sugars. Sugarcane glycoproteins could therefore contain a lectin with different affinity for different carbohydrates, or include different lectins each of them with affinity for a certain saccharide.

Interestingly, this binding mechanism differs, for example, from that described by Blanco et al. (2005a,b) for the cell wall receptors of *G. diazotrophicus*. In that case, glycoproteins bound through the β-(1→2)-fructofuranosyl fructose domain from its glycidic moiety to the bacterial cell wall receptors, which exhibited a binding site for this saccharide residue. Therefore in the cases that HMMG or MMMG bind to their ligands using their polysaccharide moiety, either to bacterial cells or to fungal teliospores, they did not behave as lectins but as recognition factors using monosaccharide units or glycosidic bonds to bind to a particular domain of their ligands (Aguilar-Cuenca et al., 2017). In addition, and as previously explained, Blanco et al. (2006) showed later that HMMG and MMMG fractions behaved differently in their binding mechanisms to cell walls of *H. rubrisubalbicans*. These differences in the recognition mechanism could be interpreted as discrimination factors between pathogens and endosymbionts.

4. CYTOAGGLUTINATION IS FOLLOWED BY MODIFICATIONS OF INFECTIVITY OF *XANTHOMONAS ALBILINEANS*

4.1 Sugarcane Can Regulate Xanthan Production by *Xanthomonas albilineans*

Leaf scald is a bacterial disease that mainly affects the vascular organs of sugarcane, and it is produced by *X. albilineans* (Ashby) Dowson. The initial visual symptom consists of white stripes on the leaf ("pencil line"), although it may also appear in foliar pods (Lopes et al., 1998), 1–2 mm wide, parallel to the main vascular bundles. These pencil lines may widen over time, and the leaf will gradually dry until the first symptoms of necrosis are observed. The pencil line may also be visible in the leaf pods. These symptoms can already be observed after the first buds sprout or in the young shoots of infected cane plants. From this moment, the described symptoms may disappear, although the plants remain diseased or the plants are actually infected. Alternatively, mature stems may wither and die suddenly, without previous onset of other symptoms.

The bacterium is transmitted from one plant to another by infected, although without appreciable symptoms, agamic seeds, by the culture instruments or instruments used to cut the stems, through the soil or the water of irrigation. The infected plants are isolated from healthy plants, and the rest of the population is kept in quarantine to prevent the spread of the disease. In those countries or areas where the disease is endemic, resistant cultivars are used to control the disease. In addition, the cane should be subjected to a cold bath/hot water treatment before planting.

The genus *Xanthomonas* comprises at least by 27 species and over 120 pathovars (variants of pathogeny) of Gram-negative bacteria, they are associated with both monocotyledonous (e.g., rice, sugarcane, or banana) and dicotyledonous (e.g., citrus, cauliflower, bean, pepper, cabbage, and tomato) plants. Many species from *Xanthomonas* genus produce fastidious diseases, but *X. albilineans* is the only xanthomonad that produces the phytotoxin albicidin causing the white foliar stripe symptoms characteristic of leaf scald disease of sugarcane.

Previous studies found that *X. albilineans* shows a reduced genome compared with other species of the same genus. Mainly, two loci, occurring in the genome of other pathogenic species of the same genus and required for pathogenicity, are lacking in *X. albilineans*: that for the biosynthesis of the pentameric xanthan unit and that for the secretion system of pathogenicity type 3 (Hrp-T3SS) (Pieretti et al., 2015a,b). Xanthans are extracellular polysaccharides (EPS) involved in the formation of biofilm and the filling of host vascular bundles. The main EPS produced by *Xanthomonas* is called xanthan. The best studied xanthan, synthesized by *X. campestris*, is the product of the action of 12 consecutively acting genes that result in a polymeric structure of 1.0 MDa molecular mass, which is formed by repeats of a pentamer consisting of two glucose units in β-(1→4)-D-glucopyranose bonds, dimers that are joined to form a β-(1→4)-glucan backbone similar to cellulose, and a (3→1)-α-linked D-mannopyranose-(2→1)-β-D-glucuronic acid-(4→1)-β side chain-D-mannopyranose appearing in the polymer on alternating residues of glucose. *X. albilineans*, however, contains genes involved in the biosynthesis of uridine diphosphate (UDP) glucose (Pieretti et al., 2012), a precursor of glucuronic acid, a constitutive of xanthan branches. This allows the pathogenic cane bacteria to produce a xanthanlike gum composed only of tetrameric units (Fig. 13.7A) including mannose and glucuronic acid as branching components (Blanch et al., 2008).

Besides, the genome of *X. albilineans* encodes specific genomic T3SS of the *Salmonella* pathogenicity island-1 family and six nonribosomal peptide synthase (NRPS) loci directed to albicidin biosynthesis. *X. albilineans* exhibits considerable intraspecies genetic variability, and numerous haplotypes have been described by using pulsed field gel electrophoresis and multilocus sequence analysis (Pieretti et al., 2015a,b).

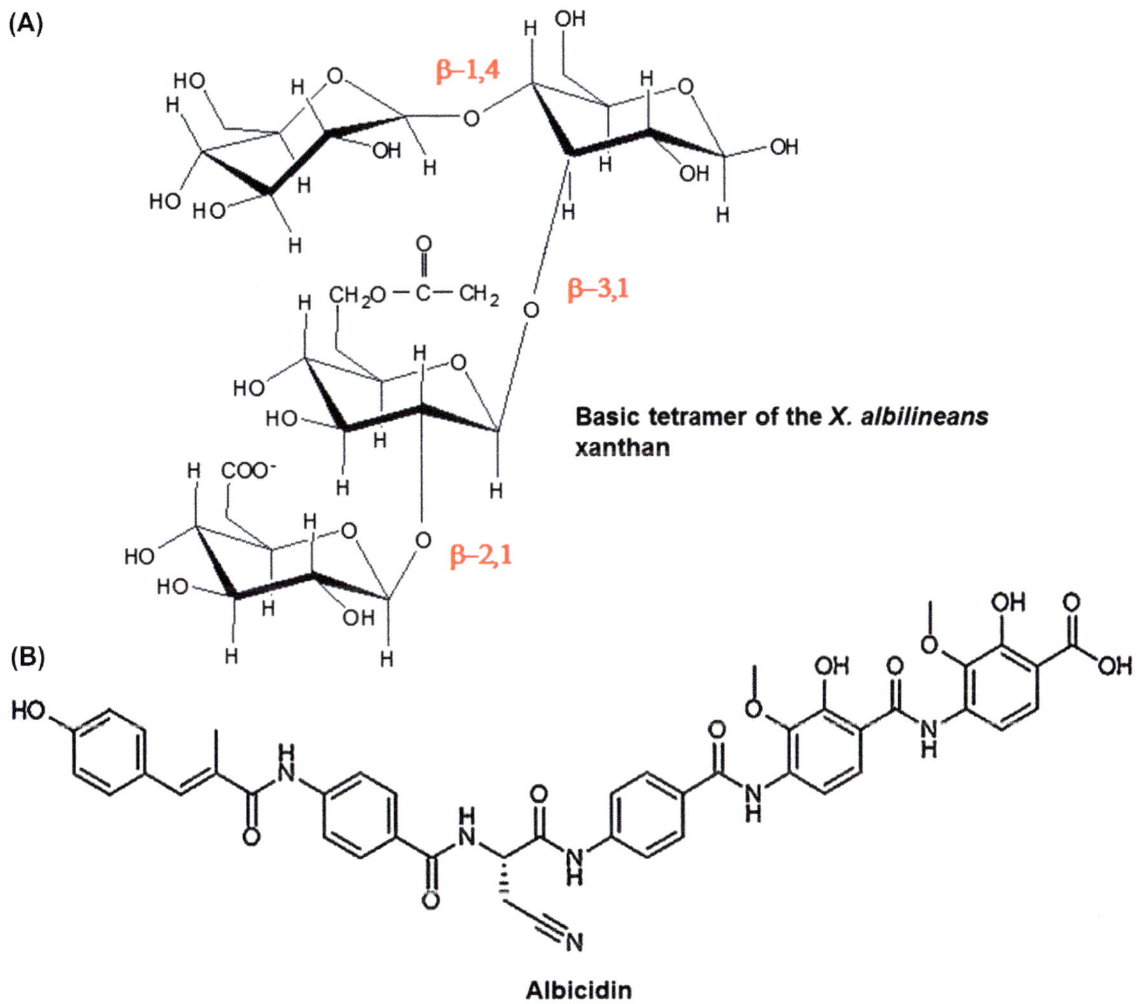

FIGURE 13.7 Chemical structures of the basic tetramer of xanthanlike gum from *Xanthomonas albilineans* (A) and of albicidin (B).

4.2 Albicidin: a Unique Feature of *Xanthomonas albilineans*

Albicidin is a phytotoxin related to the appearance of the characteristic pencil line in leaf scald disease. It behaves as an inhibitor of DNA gyrase, impeding then the differentiation of chloroplasts. It also acts on bacterial gyrase by a mechanism different from that described for other inhibitors' of DNA gyrase, such as coumarins and quinolones (Hashimi et al., 2007). Albicidin is the final product of action of a hybrid polyketide synthase (PKS)/NRPS enzyme complex. PKS and NRPS genes are often clustered together with a large set of regulatory, transport, or tailoring genes, as well as with genes involved in the biosynthesis of nonproteinogenic amino acids.

Albicidin (Fig. 13.7B) proved to be a linear pentapeptide composed of cyanoalanine and *p*-aminobenzoic acid, a p-coumaric acid derivative (Cociancich et al., 2015). Three genes involved in the synthesis of albicidin can be distinguished: xabA, xabB, and xabC. The xabB gene encodes a long protein with a modular architecture indicative of a multifunctional PKS bound to a NRPS (Huang et al., 2001). The xabC gene, located immediately after the xabB gene, encodes an O-methyltransferase-S-adenosyl-L-methionine-dependent. Finally, the xabA gene, located in another region of the genome, encodes a phosphopantetheinyl transferase required for the posttranslational enzymatic activation of PKS and NRPS. The first module of the PKS system and the NRPS module cooperate in the formation of the thiazole heterocycle from acetate and cysteine, by successive reactions of condensation, cyclodehydration, and dehydrogenation. Possibly, this latter dehydrogenation is a function of an enoyl reductase domain, located in the first PKS module, which until now had been considered as cryptic. The successive PKS modules contain the corresponding dehydrase except for module 4, which lacks it, and must have been responsible for the dehydration that builds a double bond between carbons 12 and 13. Therefore this dehydration could be catalyzed by the dehydratase domain of the next module, which would be atypical, or by enzymatic post-PKS-modifying structure.

Albicidin also behaves as a potent antibacterial agent, which inhibits the growth of competitive Gram-positive and Gram-negative pathogenic bacteria at nanomolar concentrations (Magnani et al., 2013) except two sugarcane-living bacteria containing an albicidin resistance gene: *Leifsonia xyli* (Monteiro-Vitorello et al., 2004) and *Pantoea dispersa* (Zhang and Birch, 1997).

Some albicidin-deficient mutants are unable to produce disease symptoms. This indicates that this toxin potentially plays an important role in leaf scald disease expression. However, other albicidin-deficient mutants are still able to produce severe leaf symptoms. This indicates that albicidin alone is not the sole molecule responsible of the bleaching mechanism. Albicidin production varies according to the strain of the pathogen, but there is no clear relationship between the amount of albicidin and the pathotypes used in the production of the phytotoxin (Renier et al., 2007). The biosynthesis of albicidin has been a metabolic process difficult to study, but this study was greatly facilitated after transferring the block of genes involved in the production of phytotoxin to *Xantjomonas axonopodis* pv. *vesicatoria*, which resulted in a high yield of albicidin (Vivien et al., 2007).

4.3 Purification and Properties of an Unusual UDP-Glucose Dehydrogenase, NADPH-Dependent, From *Xanthomonas albilineans*

This microorganism produces a xanthanlike polysaccharide that occludes both xylem and phloem producing desiccation of leaves. This xanthanlike polysaccharide consisted of a basal tetramer composed of two molecules of fructose, one mannose residue, and a final glucuronic acid (Fig. 13.7A) that is repeated many times to form the macromolecule (Fontaniella et al., 2002).

X. albilineans strain NCPPB887 in culture on Wilbrink media produces a dehydrogenase that catalyzes the oxidation of UDP-glucose, using reduced nicotinamide adenine dinucleotide phosphate (NADPH) as a cofactor (Blanch et al., 2006), according to the equation:

$$\text{UDP-glucose} + NADPH_2 + O_2 \Leftrightarrow \text{UDP-glucuronic acid} + NADP^+ + H_2O$$

$NADP^+$-dependent UDP-glucose dehydrogenase belongs to a small group of dehydrogenases that are able to carry out the twofold oxidation of an alcohol to an acid without the release of an aldehyde as intermediate. The mode of action of this dehydrogenase is through a tri–uni–uni–tri ping-pong mechanism, as proposed by Blanch et al. (2012), according to the scheme given below

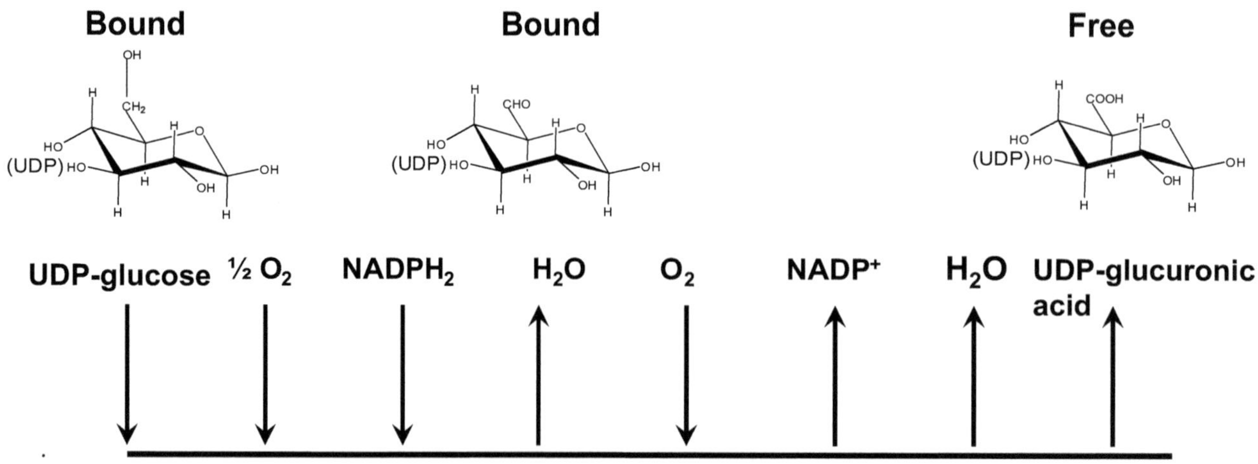

This enzyme has been purified from sugarcane plants infected by *X. albilineans*. Its N-terminal sequence was found to be IQPYNH (Ile-Gln-Pro-Tyr-Asn-His). The enzyme seemed to be extremely sensitive to bacterial proteases, synthesized in parallel to UDP-glucose dehydrogenase. The proteolytic hydrolysis of this dehydrogenase prevented the bacterial synthesis of glucuronic acid, and consequently, the production of the xanthanlike polysaccharide was impeded (Blanch et al., 2008). Other protease inhibitors have been characterized from sugarcane plants, such as canecystatin, a competitive and reversible inhibitor of cysteine proteases (Valadares et al., 2013), and some endoproteinases from the PR-7 family (Souza et al., 2017).

4.4 Sugarcane Glycoproteins Are Required for the Production of an Active UDP-Glucose Dehydrogenase by *Xanthomonas albilineans*

X. albilineans growing on Wilbrink media supplemented with HMMG or MMMG maintained the synthesis of active UDP-glucose dehydrogenase, whereas a net, although variable, loss of this protein was measured for control bacteria growing on the Wilbrink medium without any addition.

UDP-glucose dehydrogenase activity does not disappear by adding 0.1 mM 8-azaguanine to the culture media, even when HMMG or MMMG were simultaneously added, with the quantity of MMMG added being slightly higher. This could be interpreted as the absence of competition between sugarcane glycoproteins and 8-azaguanine to enhance transcription of the corresponding UDP-glucose dehydrogenase gene. A similar effect was observed by adding chloramphenicol to the Wilbrink media. Both inhibitors 8-azaguanine and chloramphenicol nullified the reported loss of UDP-glucose dehydrogenase activity in the absence of sugarcane glycoproteins (Blanch et al., 2007). *Xylella fastidiosa*, another pathogenic bacterium of plants producing xanthanlike gums, also synthesized extracellular proteases, belonging to the serine- and metalloprotease group, secreted to the culture medium (Fedatto et al., 2006).

The hydrolysis by proteases of bacterial UDP-glucose dehydrogenase explains why the xanthanlike polysaccharide is only produced by *X. albilineans in plants* but never in culture. Sugarcane glycoproteins are produced as a response to bacterial invasion, being then able to inhibit bacterial proteases and preventing degradation of the UDP-glucose dehydrogenase. As a consequence, this polysaccharide can be produced only in sugarcane plants infected by *X. albilineans* (Fig. 13.8).

These data support the hypothesis that the production of fibrous material, the xanthanlike polysaccharide, that invades leaf tissues of sugarcane plants that exhibit visible symptoms of leaf scald, could be due to the maintenance of active production of UDP glucuronic acid and subsequent incorporation of the glucuronate to the xanthan structure (Legaz et al., 2011).

4.5 Modifications of Sucrose Crystallization by Xanthans Produced by *Xanthomonas albilineans*

Leaf scald is a problem in over 60 tropical countries producing sugarcane. It causes yield losses of sucrose production in both field and factory (reduction of the quantity and quality of the juice extracted from the sugarcane stalks). Xanthanlike gum, derived from HMMG and MMMG action, is able to modify the crystallization sequence of sucrose, a product of primary commercial interest of this culture. Sucrose crystallization from the juices of the sugarcane Barbados 42231 cultivar stalks (sensible to the disease), inoculated with *X. albilineans,* showed notable differences when compared with that observed for juices obtained from stalks of healthy plants. The number of sucrose crystals obtained drastically decreased when inoculated juices were used as a source of sucrose. In addition, some of these crystals showed the typical agglomeration of prisms in complexes with a starlike form, although nuclei were amorphous, or, on the contrary, individual prisms showed a small degree of aggregation (four to five prisms for each complex). This would indicate that xanthans, secreted by the bacteria and recovered in juices obtained from infected stalks blocked the center of adhesion of individual crystals (Blanch et al., 2006).

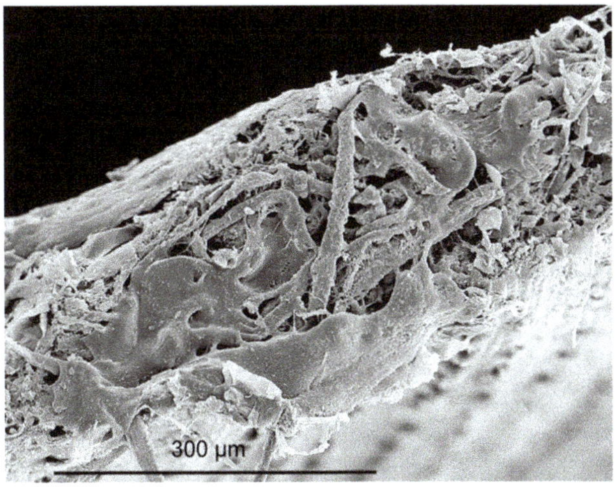

FIGURE 13.8 Scanning electron micrograph of a cross-section of a leaf-scald-diseased sugarcane blade completely occupied by mucilaginous material (xanthanlike gum) that masks the entire internal leaf structure.

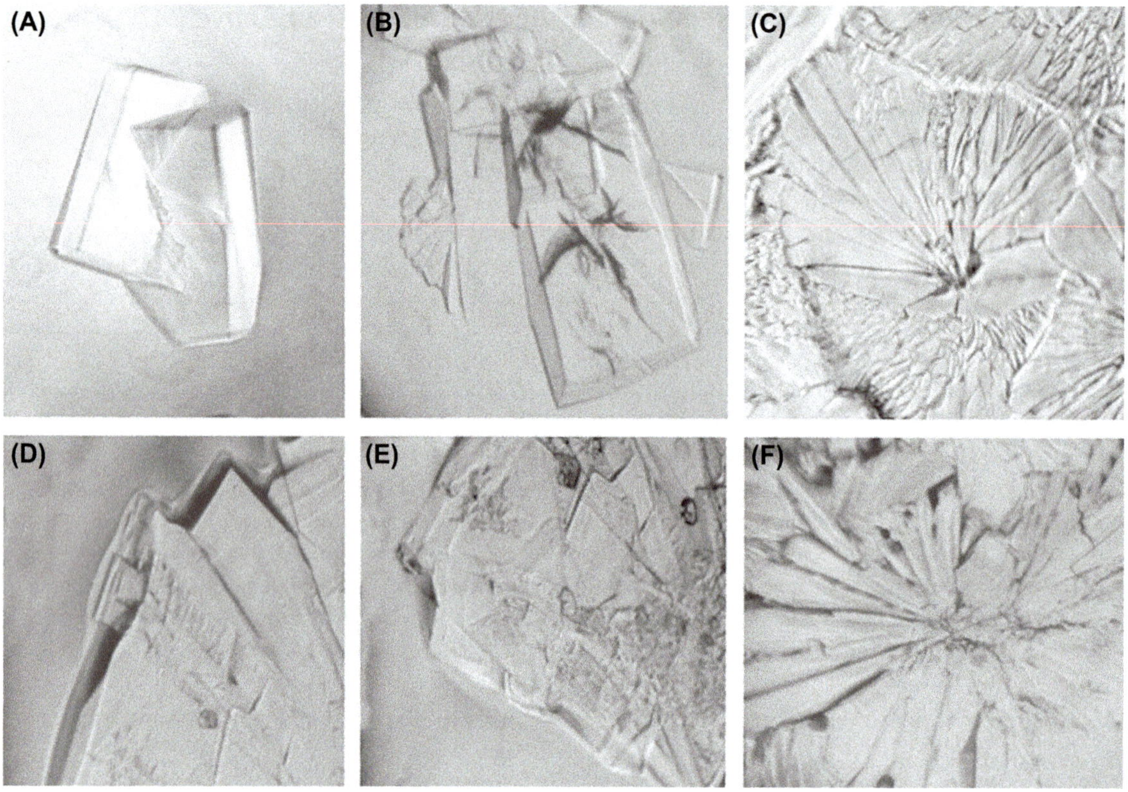

FIGURE 13.9 (A, B, C) Different sucrose crystals obtained at 24h from a saturated sucrose solution without added xanthan from *Xanthomonas albilineans*. Partial crystallization of sucrose at 4h (D) or 6h (E) from a saturated sucrose solution with added xanthan. (F) Complete crystallization of sucrose at 24h from a saturated sucrose solution with added xanthan.

By adding bacterial xanthans to saturated sucrose solutions, it is possible to simulate the in vitro effect of bacterial polysaccharides on sucrose crystallization. Adhesion of individual crystals in starlike aggregates failed to some extent, and several crystals were broken during the crystallization process (Fig. 13.9). However, in the absence of added xanthans, sucrose recrystallization is started by the formation of irregular prismatic crystals growing on their lateral surfaces. These crystals aggregate by contact between their small, lower surfaces to produce starlike crystals.

Fontaniella et al. (2003a,b) demonstrate that the colligative properties of the peptide moiety of both HMMG and MMMG enable the binding of these glycoproteins to the crystalline surface of sucrose, using the lower face of the prismatic crystals to adhere to it. This impedes the formation of starlike complexes. When polysaccharide and protein moieties separate, carbohydrate is unable to impede the formation of these complex crystals and heterofructans obtained from impaired sugar beet, that do not contain the peptide moiety.

4.6 Inhibition of the Germination of *Sporisorium* Teliospores

The actin cytoskeleton of fungal cells has an essential role in spore germination and the subsequent growth of mycelium. In *S. scitamineum* we have shown how, before the emergence of the germinative tube, actin concentrates close to the germinative pore to facilitate the transport of cell material toward that point (Fig. 13.10).

The mechanisms of cytoskeleton organization were well described in *Ustilago maydis*, a close species to *Sporisorium*. Millanes et al. (2005) described for the first time the reorganization of the actin microfilaments before the germination of *S. scitamineum* teliospores. Sugarcane glycoproteins were seen to be involved in the regulation of the actin cytoskeleton, at least during germination. In this case, the reaction of smut teliospores in response to different treatments with HMMG and MMMG varied enormously depending on the origin of these glycoproteins. Glycoprotein fractions isolated from resistant cultivars inhibited cytoskeleton polarization to a much higher degree, compared with the effect of glycoproteins from nonresistant cultivars (Fig. 13.11) (Millanes et al., 2005, 2008). However, our results pointed to a scenario in which different glycoproteins could act in different moments of the fungal growth. Thus HMMG fractions from the resistant cultivar inhibited polarization to an high extent, but some spores were still able to germinate and form mycelium.

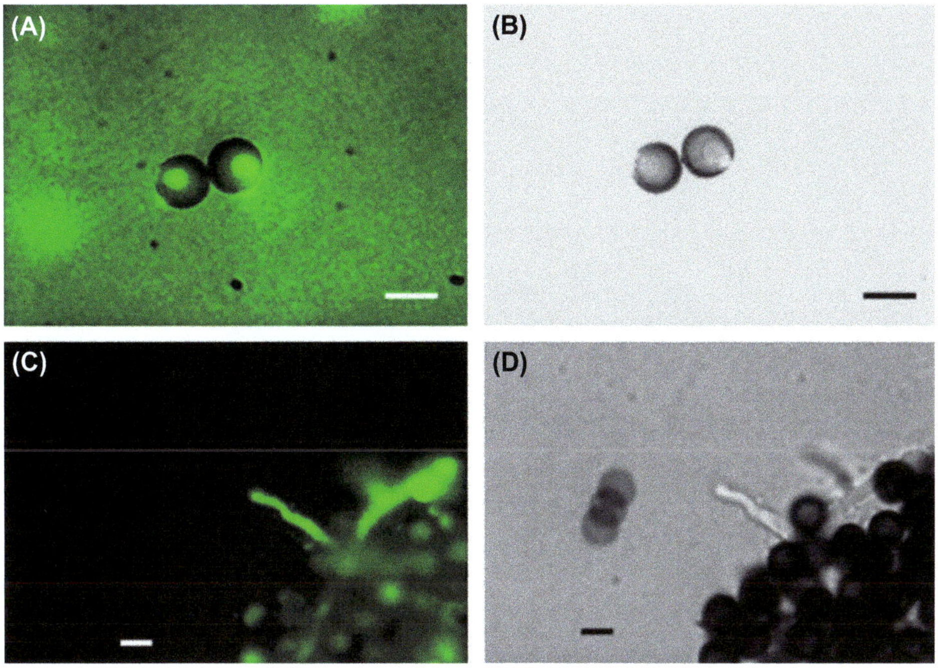

FIGURE 13.10 (A–B) Polarization of actin cytoskeleton before germination of smut teliospores. The highest concentration of actin, stained with fluorescein isothiocyanate, is situated next to the germination pore visualized by fluorescence microscopy (A), or (B) observed with light microscopy. Afterward, actin concentrates in the germinative tube, observed with fluorescent microscopy in (C) and light microscopy in (D). Scale bar = 5.0 μm.

Conversely, the MMMG fraction isolated from the same cultivar inhibited actin polarization and, in addition, completely prevented the formation of mycelium (Millanes et al., 2005). Considering the high heterogeneity of the proteins included in HMMG and MMMG, it was not strange to hypothesize that different proteins could alter different mechanisms related with pathogenicity.

A more detailed investigation of the components of HMMG and MMMG sheds some light on the mechanisms of action of both glycoprotein fractions. At the first stage, it felt logical to consider that chitinase activity of some glycoproteins from HMMG or MMMG fractions could hinder or impede teliospore polarization. This enzyme may produce structural damage in the teliospore cell wall preventing polarization. However, several glycoprotein fractions with no chitinase activity impeded polarization at a high percentage. Therefore we studied the possible effect of arginase on cell polarization. Sugarcane juice contains several enzymes related to PA biosynthesis, mainly ornithine decarboxylase (ODC) (Legaz et al., 2001), which uses ornithine produced by enzymatic hydrolysis of arginine as a substrate. Arginase is therefore the first enzyme in the synthesis of PUT and acts as a precursor. It has been shown that glycosylated arginases produced by lichenized and pathogenic ascomycetes bind to the algal partner or change the level of PUT produced by the invaded cell. We therefore compared polarization of smut teliospores treated with arginase or PUT with the controls maintained in distilled water. These controls showed teliospore polarization, whereas treatment with exogenous arginase or PUT completely inhibited cytoskeleton reorganization (Millanes et al., 2005). When using HMMG and MMMG from healthy plants, sucrose reversed the binding of arginase to teliospores, which is easily explained since the polysaccharide moiety of both HMMG and MMMG contain several domains composed of β-D-fructofuranosyl-1,2-β-D-fructose. However, sucrose failed to produce arginase desorption when HMMG and MMMG from inoculated plants were used. This could be explained by the internalization of arginase within the teliospores. Moreover, particular glycoproteins isolated from HMMG or MMMG fractions for different cultivars also showed the ability to promote actin polymerization (Millanes et al., 2005, 2008).

In the light of these results, we considered two possible mechanisms through which cytoskeleton reorganization could be affected by sugarcane glycoproteins. First, glycoproteins with arginase activity not retained by the cell wall receptors penetrated teliospores, increasing the levels of PUT, which would produce cell disorganization (Millanes et al., 2005). Alternatively, glycoproteins other than arginase bound different cell wall ligands, activating a signal that did not involve PUT, and that could promote or inhibit actin polymerization (Millanes et al., 2005, 2008). The ability to inhibit or promote actin polymerization was not related to the amount of glycoprotein adsorbed to the spores. This pointed to signal

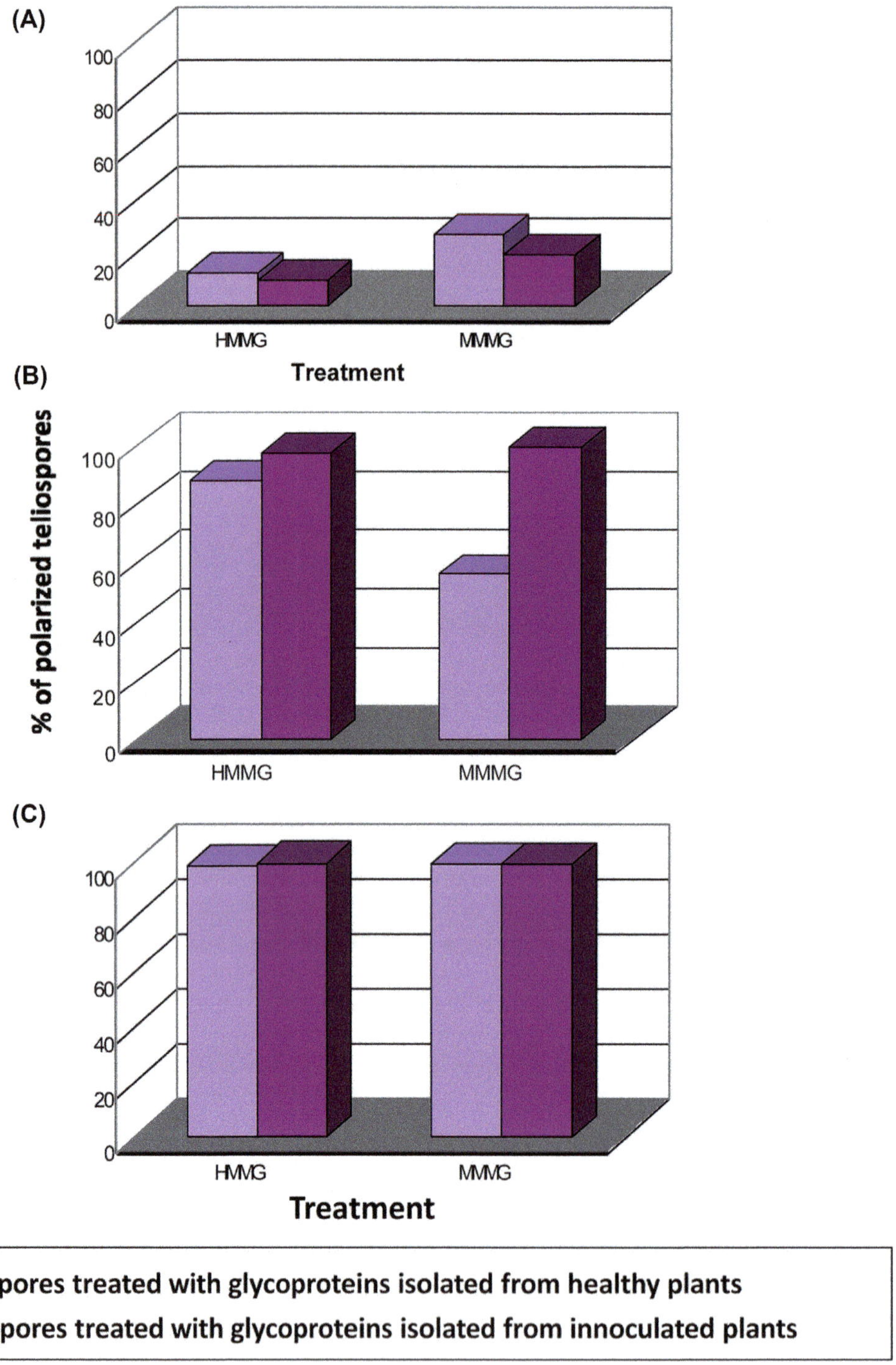

FIGURE 13.11 Percentage of polarized spores after 2 h incubation in 2.0 mL of phosphate buffer, containing 300 μg of high molecular mass glycoproteins or mid molecular mass glycoproteins isolated from one resistant cultivar (A), and two nonresistant cultivars (B–C). Values are the mean of three replicates. Standard error was never higher than 7%.

transduction mechanisms, triggered by the binding of glycoproteins to the fungal cell wall. There are several evidences showing that very low concentrations of signal molecules can launch highly intense responses in plants (Memon and Durakovik, 2014). The specific path of signal transduction modulating actin cytoskeleton reorganization has been largely found for *U. maydis*, a species closely phylogenetically related to *S. scitamineum* (Basse and Steinberg, 2004; Barnabas et al., 2017), but the particular biochemical mechanism acting in *Sporisorium* needs further investigations. The most common mechanism in other fungi involves the action of several Ras and Rho guanosine triphosphatases (GTPases), mediated by cyclic adenosine monophosphate, and connected to mitogen-activated protein kinases (MAPKs) that regulate the assembly of actin monomers.

The role of different enzymatic activities of sugarcane glycoproteins, and the involvement of fungal cytoskeleton in the complex fungal–plant interactions observed in the smut–sugarcane system will be discussed further in Sections 6 and 7.

5. ENZYMATIC NATURE OF THE SUGARCANE GLYCOPROTEINS

As mentioned in Section 1.2, HMMG and MMMG fractions contain between four and six different glycoproteins, which can be identified according to their different biological activities The concentration of both HMMG and MMMG glycoproteins, obtained from the resistance to smut cultivar Mayari 5514, clearly increases after inoculation of sugarcane plants with smut teliospores, although major symptoms of disease are not observed.

5.1 Arginase

Sugarcane glycoproteins, defined by HMMG and MMMG, not only induce homotypic adhesion of smut teliospores but also inhibit their germination and exhibit arginase activity, which is enhanced after plant inoculation (Santiago et al., 2008b). The desorption of purified arginase from healthy plants, adhered to the cell wall of the teliospores, is completely carried out by sucrose, but this desorbed arginase only accounts for 50% of the total extracted arginase and also loses its activity after binding (Millanes et al., 2005). In other words, soluble arginase from healthy sugarcane, when immobilized on its ligands in the cell wall of teliospores, is completely inactivated. However, the desorption pattern changes when the binding is performed using arginase obtained from inoculated plants. In this case, sucrose only desorbs 60% of the arginase bound to the ligands, whereas the remaining 40% is retained by the teliospore. These data are consistent with a model of the entry of an excess of arginase into teliospores. Arginase synthesized by sugarcane plants as a response to the experimental infection would increase the synthesis of PUT inside the teliospore, and this impedes polarization at concentrations higher than 0.05 mM. However, teliospores smut seems to be able to change the pattern of glycoprotein production by sugarcane plants, thereby promoting the synthesis of different glycoproteins that activate polarization after binding to their cell wall ligand (Millanes et al., 2005).

5.2 Fungal Cell Wall Hydrolytic Enzymes

Since some other enzymes, such as chinitases (Su et al., 2015), glucanases and peroxidases (POXs), and sometimes catalases (Su et al., 2014), have been traditionally related to defense mechanisms, these activities were assayed in different glycoprotein fractions. As reported by Millanes et al. (2005), chitinase activity was practically undetected from glycoprotein obtained from My 5514 cultivar, whereas POX was detected in juices, mainly in those obtained from healthy specimens (Santiago et al., 2009a). Sánchez-Elordi et al. (2016b) reported that many teliospores cytoaggregated by the action of HMMG were visualized, after 6 h, as "ghosts," or cell walls without protoplasts, probably by the action of degradative enzymes. Changes in the number and complexity of aggregates after being in contact for 6 h with HMMG also supported this hypothesis. Thus cell wall breakdown of smut teliospores and protoplast leakage by the action of HMMG suggested that this glycoprotein fraction may contain enzymatic activities that would promote the hydrolysis of the cell walls. In agreement with this rationale, it has been found that the fraction of HMMG contained chitinase and β-1,3- and β-1,4-glucanase activities (Santiago et al., 2012a). The first two enzymes increased their activity after 3 h of contact of sugarcane glycoproteins with teliospores, whereas the activity of β-1,4-glucanase decreased. Earlier we knew that the inhibition of the growth of several pathogenic fungi could be related to the production of chitinase by plants (Guan et al., 2008).

Other sugarcane pathogens produce similar increases in the hydrolytic activities in host plant tissues. For example, leaves of sugarcane plants, resistant or sensitive to leaf scald, 7 days after inoculation with a suspension of *X. albilineans*, were assayed for these hydrolytic activities. An increase of β-1,3-glucanase activity in leaves of the different cultivars was found, except for Mayarí cultivar (resistant to leaf scald), in which it significantly decreased. Nevertheless, the degree of increase is according to the cultivar. The major increases of activity were measured for sensitive cultivars, whereas lower increases were observed for moderately sensitive cultivars. (Santiago et al., 2012a).

Often, plant pathogens induce changes in certain enzymatic activities produced by the host cells by the action of elicitors (Su et al., 2013), of which salicylate and jasmonic acid are the most studied. The possible action of these inducers on β-1,3-glucanase activity, using healthy leaf disks from cultivar Jaronú 60-5, was tested by Santiago et al. (2012a). Salicylic acid at concentrations below 0.5 mM increased this glucanase activity, whereas higher concentrations of the elicitor decreased the activity. In contrast, jasmonic acid had no effect on glucanase activity. The use of laminarin (a β-1,3-glucan) as a possible enzyme inducer resulted in a significant loss of β-1,3-glucanase activity after 24 h of treatment.

Resistance to pathogens supported by pathogenesis-related proteins (PRP) can be induced by salicylic acid even in the absence of the pathogenic organism. Salicylic acid produced by pathogens at the primary site of infection may subsequently be exported to uninfected tissues. Lawton et al. (1995) characterized transgenic *Arabidopsis* plants able to express constitutively a bacterial gene, nahG, encoding salicylate hydroxylase, catabolic enzyme of salicylic acid, followed by activation of the acquired systemic resistance of *A. thaliana* plants against the pathogenic bacteria and the addition of exogenous salicylic acid. However, salicylic and jasmonic acids do not always act as inducers of plant defense responses. Hause et al. (1997) were unable to find any accumulation of jasmonate-inducible protein JIP-23 in a barley cultivar susceptible to *Erysiphe graminis* f. sp. hordea.

5.3 Dirigent Proteins

Lignin is a phenolic polymer that provides to the plant cell wall an additional structural resistance, hydrophobicity, and resistance to the entry of the pathogen. The polymer contains *p*-hydroxyphenyl, guaiacyl, and syringyl units derived from the monolignol precursors *p*-coumaryl, coniferyl, and sinapyl alcohol, respectively. The production of dimers and trimers from monolignols requires dirigent proteins (DIR proteins), a family of proteins, exclusive to plants, involved in the production of lignins and in the responses to the invasion by pathogenic organisms and abiotic stresses. These proteins are responsible for the stereoselective coupling of monolignol radicals to produce lignan or lignin (Fig. 13.12). The first DIR protein was described as a (+)-pinoresinol-forming DIR protein in *Forsythia intermedia* (Gang et al., 1999).

There are only few studies about sugarcane DIR, all of them in stems. Damaj et al. (2010) identify genes codifying for DIR protein and O-methyltransferase, highly expressed in the stem of sugarcane plants and potentially involved in lignification and defense responses. Jin-Long et al. (2012) reports a novel DIR protein gene in sugarcane stems, expressed as a response to drought, salt, and oxidative stress. Sánchez-Elordi et al. (submitted) were able to find that transcripts from *SofDIR16* gene in both sugarcane meristem and leaves of My 55-14 cultivar, resistant to smut, were higher than in B 42231 cultivar, susceptible to smut. *SofDIR16* expression was almost nullified in the resistant cultivar after inoculation with smut sporidia, whereas this expression was increased in the susceptible cultivar. It is proposed that lignans are involved in the defense response of the resistant cultivar, My 55-14 (see Section 8.3). It has been proposed that lignans are implicated in the defense response of the resistant cultivar My 55-14 since their levels would be increased when DIR activity decreases.

6. CHEMOTACTIC ROLE OF THE SUGARCANE GLYCOPROTEINS: FALSE QUORUM EFFECT ON *SPORISORIUM* TELIOSPORES

6.1 Quorum Sensing: Cell-to-Cell Communication

During plant–pathogen interactions, small diffusible signaling molecules are individually produced by several microorganisms to inform to the others about the cell density of the population. This communal behavior, which is detected in many phytopathogenic bacteria (Galloway et al., 2012) and fungi (Hogan, 2006), is necessary to promote the attraction of the members in such a way that they group a large number of potentially infective cells. If the host plant protects itself against potential invaders by production of defense molecules, pathogen cells could more likely survive if they conform to a cluster big enough to prevent, at least in part, an excessive percentage of cell death caused by the defensive response of the plant (Sánchez-Elordi et al., 2016a).

Intraspecific communication is possible via production and subsequent response to QSMs that pathogens accumulate in their extracellular environment (Hogan, 2006). These signal molecules cause variations in the regulation of gene expression as a response to cell density (Yajima, 2014) to promote invasion under optimal conditions. Usually, these molecules are lactones, cyclic dipeptides, or some methyl esters of fatty acids. However, QSMs can be of very different nature so that Sánchez-Elordi et al. (2016a) propose an enzyme as the quorum signal molecule responsible for cytoagglutination of pathogenic cells during sugarcane–*S. scitamineum* interaction.

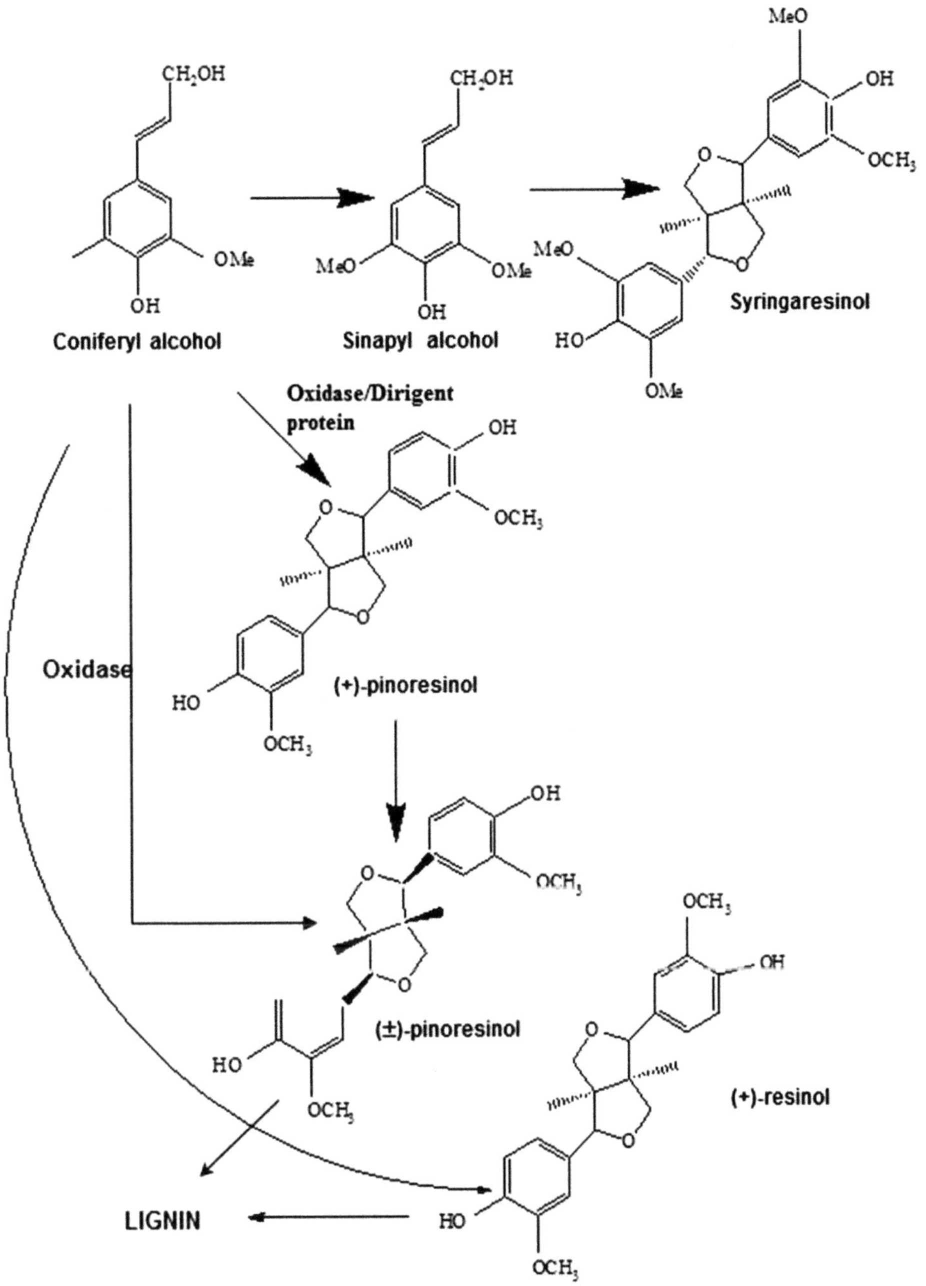

FIGURE 13.12 Role of dirigent proteins in the biosynthetic pathways of lignin from cinnamoyl alcohols.

6.2 "Unity Makes Strength": Cytoagglutination of Smut Teliospores by the Action of a Fungal Arginase

According to this, teliospores of *S. scitamineum* secrete an active arginase that causes agglutination of fungal cells when they bind to the teliospore cell wall. Fungal arginase causes aggregation of smut teliospores with the purpose that part of the group avoids sugarcane defense (Fig. 13.13). Interestingly, arginase produced by cells at first stages of fungal growth activates a signal transduction cascade that accelerates the germination of the rest of the members (Sánchez-Elordi et al., 2016b). Arginase activity is necessary to stimulate germination, probably by means of synthesis of catalytic amounts of urea, ornithine, or PUT.

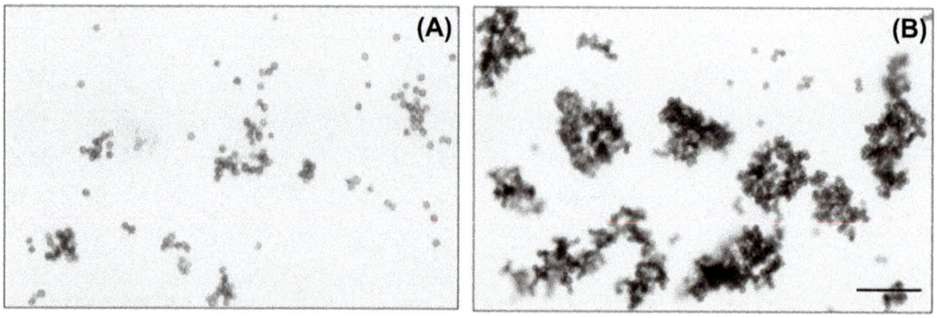

FIGURE 13.13 Cytoagglutinating effect of sugarcane arginase on smut teliospores. Images show aggregation after incubation of teliospores in water (A) or plant arginase purified from stalks of a resistant cultivar of sugarcane (B).

Many researchers have confirmed the correlation between fungal pathogenicity and PA production (Walters, 2003a). In fungi, variations in the concentration of PUT and SPD are closely associated with the onset of important physiological processes for the development of the organism. Thus it has been found that the addition of low concentrations (0.1–1 mM) of PUT and SPD to the culture media leads to increased growth rate of some phytopathogenic fungi, such as *Botrytis* sp, *Botrytis cinerea*, *Monilinia fructicola,* and *Rhizoctonia solani* (Rajam and Galston, 1985), suggesting that PAs play an important role in the regulation of fungal progress.

Ornithine is synthesized in the initial stages of PA production from arginine by hydrolysis catalyzed by arginase, and PUT appears immediately after by the action of an ODC. SPD and SPM are produced by the addition of aminopropyl groups, $[NH_2(CH_2)_3—]$, to PUT units by the action of SPM/SPM synthases (see Section 9.1).

SPD, specifically, seems to be the most important PA to stimulate fungal growth. PUT has only been related to SPD biosynthesis, acting as a simple intermediate in *U. maydis*. However, although PUT does not seem to be an essential factor for cell growth, it is involved in stress responses and in virulence of this last fungal species. Moreover, during sugarcane–smut interaction, the addition of PUT to germinating smut teliospores does not significantly alter the rate of germination for concentrations from 0 to 1.0 mM, although it is slightly inhibited by 5.0 mM diamine. However, the frequency of sporidia formation was considerably enhanced by 0.1 mM PUT, indicating that PUT, the product of combined arginase/ODC activities, increases the potential infectivity of the pathogen (Sánchez-Elordi et al., 2015).

In plants, arginase activity, coupled to an ODC, is also directed to the synthesis of PUT. However, PA biosynthesis in plants and bacteria differs from that in the rest of eukaryotic organisms by the existence of an alternative route of PA production by means of successive arginine decarboxylase (ADC), agmatine iminohydrolase, and N-carbamoyl PUT amidohydrolase activities (see Section 9.1). These enzymes are all derived from a common cyanobacterial ancestor (Fuell et al., 2010).

PAs are responsible for important physiological processes in plants, and they are involved in tolerance against abiotic stress (Gupta et al., 2013). Moreover, many secondary metabolites, which are so important in plant development, are PA derivatives (Fuell et al., 2010). PAs have a role in plant defense against pathogens (Legaz et al., 1998a,b; Piñón et al., 1999; Wojtasik et al., 2015), but little is known about this fact.

6.3 Arginase Role in Sugarcane Plant Defense

As previously stated, sugarcane glycoproteins mediate their effects through binding to specific receptors in the wall of the teliospores (Legaz et al., 1995, 1998a,b; Millanes et al., 2005). To understand the putative defensive mechanism of resistant plants of sugarcane against smut teliospores we must integrate the signals of (1) chemotaxis, (2) cytoagglutination, and (3) cell wall degradation as a defensive response triggered by PR HMMG produced by resistant sugarcane plants.

First, a chemoattractive protein recently described in sugarcane stimulates the displacement of teliospores toward the chemoattractive agent by successive episodes of cytoskeleton contraction and relaxation without depending on external motile structures. It means that glycoproteins activate a signal transduction cascade in teliospores to organize the actin cytoskeleton and to produce the movement of fungal cells (Sánchez-Elordi et al., 2016c). After chemotaxis, HMMG glycoproteins cause agglutination of cells that prevents their germination by disruption of microtubular polymerization. Finally, degradative enzymes, also contained in the pool of HMMG, induce the dispersal of teliospore aggregates and the loss of chemotaxis by breakage of the teliospore cell wall after prolonged contact with the chemoattractant (Sánchez-Elordi et al., 2016b).

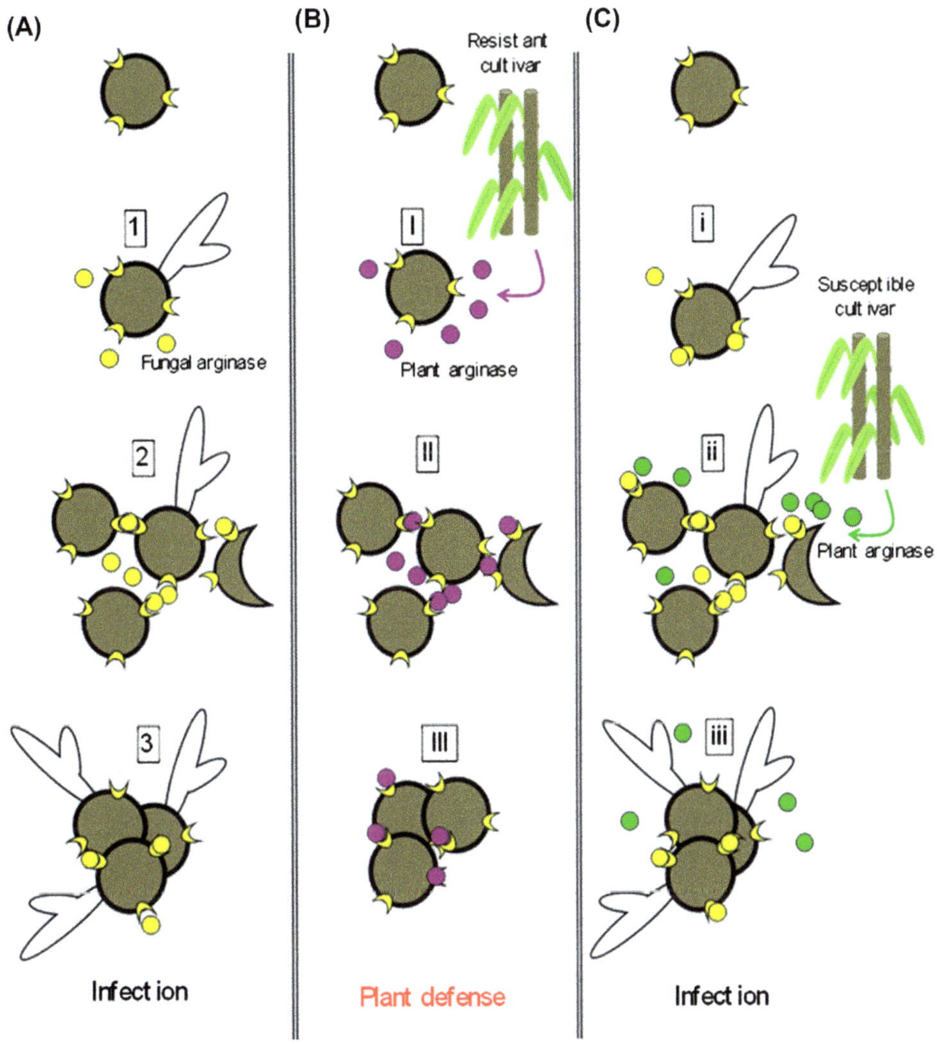

FIGURE 13.14 (A) Fungal arginase effect on quorum-sensing establishment. Teliospores produce arginase activity at first stages of germination (1). It causes spore agglutination (2) and germination (3), which leads to sugarcane infection. (B) Arginase from resistant varieties of sugarcane: effect on false-quorum establishment. Constitutive plant arginase production avoids fungal arginase release (I) by means of the cytoagglutination (II) without germination (III). (C) Arginase from nonresistant plants is produced after fungal arginase release (i). It presents more affinity for its ligand on the teliospore wall, and it cannot be displaced by plant arginase (ii). Teliospore germination takes place, and infection is observed in susceptible cultivars (iii).

Arginase activity contained in the HMMG is the protein responsible for agglutination of cells (Fig. 13.14), in the same way that aggregation occurred in the presence of fungal arginase. However, arignase produced by the pathogen stimulates spore germination, whereas sugarcane arginase (SCA) inhibits their germination when it binds to their cell wall ligand (Sánchez-Elordi et al., 2015) due to the inhibition of hyphae emergence by cytoskeleton disorganization (Fontaniella et al., 2002; Millanes et al., 2005) and nuclear decondensation (Sánchez-Elordi et al., 2016a). This suggests that small differences between arginases types seem to have important consequences in the defensive capability of the plant against infection (Sánchez-Elordi et al., 2016b). Moreover, arginase effect on germination seems to be related to the excessive increase of the amount of PAs in fungal cells, PAs that interact with the cytoskeleton, and other negative charged molecules (Savarin et al., 2010). It is due to the polycationic nature of PAs at physiological pH values (Gupta et al., 2013). The defensive role of plant arginase has been also found in *Arabidopsis* plants. Those containing overexpressed arginase genes are less susceptible to *B. cinerea*, whereas silencing lines remained as susceptible in a similar way to the wild type (Brauc et al., 2012).

Surprisingly, HMMG from nonresistant cultivars does not promote teliospore aggregation, the first stage of defensive response, which results in a lower inhibitory effect on germination (Sánchez-Elordi et al., 2016b). This could be probably related to the absence of arginase activity in the pool of glycoproteins produced by nonresistant cultivars (Sánchez-Elordi et al., 2016a), whereas it is produced constitutively in healthy plants of resistant varieties.

Then, the defense capacity of sugarcane against infection by *S. scitamineum* seems to be related to the establishment of a false quorum signal by arginase production that concludes with the destruction of the pathogenic cells and prevents their germination. Aggregation seems to be necessary to tramp the teliospores in a small region of the space in contact with the HMMG, which would increase their accessibility to the cell walls and facilitate the enzymatic activities of the HMMG to degrade the trapped teliospores (Sánchez-Elordi et al., 2016b). Therefore previous cytoagglutination (that is only produced by resistant varieties) appears to be essential for the plant to attack simultaneously the maximum number of invading cells (Fig. 13.14).

6.4 Both Plant and Fungal Arginases Compete for Binding to Teliospore Cell Wall Ligands

Chemoattraction toward agglutination points is also stimulated by teliospores at the first stages of germination. Binding of fungal arginase to smut teliospores activates a signal transduction cascade that enhances germination after cytoagglutination (Fig. 13.14a). This last effect can be related to quorum-sensing mechanisms described for yeasts and fungi (Wang et al., 2013). However, teliospore agglutination without germination, stimulated by SCA, must then be the result of a false quorum signal that prevents plant infection by smut teliospores (Fig. 13.14b).

Competition assays between sugarcane and fungal arginases are reported by Sánchez-Elordi et al. (2015). The affinity of active fungal arginase (AFA) for its ligand is greater than that found for SCA. In fact, AFA, which significantly enhances teliospore germination, removes previously bound SCA, whereas the inverse process is practically negligible.

Enzymatic activity of arginase is required for the binding to teliospore cell wall and to activate germination. Competition experiments were repeated by using fungal arginase previously inactivated with Hg^{2+}, which causes irreversible inhibition of the thiol-dependent functions of the enzyme. Binding of inactivated fungal arginase (IFA) to teliospore wall is very ineffective and is easily reversed by AFA. Moreover, arginase activity is necessary to stimulate teliospore germination since IFA is able to bind to the cell wall but, nevertheless, does not activate germination.

In fungi, cell polarity establishment and cytoskeleton polarization are essential processes for teliospore germination. PAs are proposed as a link between the defensive role of plant arginase and the cytoskeleton disorganization that impedes germination (Millanes et al., 2005). Although it is known that low levels of PAs stimulate fungal development (Sánchez-Elordi et al., 2015), an excessive and damaging accumulation of these compounds has been recently observed in teliospores after contact with HMMG isolated from resistant cultivars. This means that a transient and controlled depolarization of cytoskeleton, induced by low concentrations of PUT, could stimulate sporidia liberation but an increased PUT production, as a consequence of early SCA synthesis, could not be regulated by teliospores and the effect would become negative. Thus, if SCA binds to the teliospore wall before AFA production, the germination started by the fungal enzyme cannot take place. The binding of early produced SCA will avoid germination, whereas fungal arginase still has not been secreted to activate the germination signaling cascade. Agglutination not linked to germination is then the result of a false quorum signal triggered by SCA (Sánchez-Elordi et al., 2015).

This model is also consistent with the observations made on nonresistant cultivars, in which the lack of chemotactic, agglutinating, and hydrolytic effects of the sugarcane glycoproteins enables single teliospores to reach the surface of the plant, promoting its successful infection. Interestingly, nonresistant sugarcane varieties try to increase arginase activity in juices after detecting the presence of the pathogen (Sánchez-Elordi et al., 2016a). However, SCA will not be able to reach the cell wall when fungal arginase is already present in the system (Fig. 13.14c) because this last SCA exhibits the highest affinity for its cell wall ligand. This explains why the production of active arginase by susceptible plants "comes too late" for an effective defense response (Sánchez-Elordi et al., 2015) based on the inhibition of teliospore germination. If plant arginase appears after fungal enzyme, it will not able to stimulate defensive cytoagglutination.

Therefore, it has been concluded that resistant sugarcane plants produce a pool of glycoproteins that induce chemotaxis, cytoagglutination, and death of smut teliospores as the result of a perfectly orchestrated defense response consequence of a false quorum signal establishment (Sánchez-Elordi et al., 2016b).

7. ROLE OF THE CYTOSKELETON IN THE CHEMOTAXIS OF *SPORISORIUM SCITAMINEUM*

7.1 Chemotactic Strategy

The term chemotaxis defines a directional change of cell movement either toward or away from the source of a chemoattractant or repellent agent, respectively. When chemotaxis is positive, the movement of an organism toward high concentrations of a chemical stimulus is induced. Cells recognize external chemical gradients of the chemoattractant and translate a primary response to the perception of this signal inside them, which results in migration along the gradient (Wang et al., 2011).

Chemotaxis in mammals has many physiological and pathophysiological implications. It has a central role in reproduction since it has been related to sperm–egg interaction in a lot of species. Chemotaxis is required for development, organ patterning, mobilization, and homing of stem and progenitor cells. For example, during the development of central nervous system, neuroblasts proliferate within germinal zones of the neuroepithelium and thereafter they migrate to their final positions. Chemotaxis is also involved in inflammation episodes and tissue regeneration, pathogenesis, and dissemination of cancer cells (Vorotnikov, 2011).

In other cellular types, chemoattraction represents a social strategy for bacteria (Sourjik and Wingreen, 2012) or amoebae (Franca-Koh et al., 2006), allowing cells to follow gradients of nutrients. Chemotaxis is an effective mechanism to aggregate a larger number of pathogenic cells in the surface of the host during quorum--sensing signaling (Sánchez-Elordi et al., 2015, 2016b).

7.2 Role of the Cytoskeleton in Chemotaxis

Polarization is a key event to initiate cell migration, which results in morphological changes in the cell shape and ensures motile behavior (Vorotnikov, 2011). Chemoattractive migration has been widely related to reorganization of actin filaments in cells. The stimulus, by means of its binding to a ligand on cell surface, activates a guided organization of the cytoskeleton that causes the cell to move to the chemoattractant. Chemotaxis is also involved in the migration of cancer cells to other regions during metastasis by means of actin polymerization and pseudopodia formation. The same process is described in *Dictyostelium discoideum*, a soil-living amoeba, as a modular process composed of directional sensing, pseudopod extension, and polarization in which actin organization plays an essential role (Franca-Koh et al., 2006).

Thus, migration is accepted as a consequence of the extension of a leading-edge protrusion or lamellipodium, followed by the establishment of new adhesion sites at the front, cell body contraction, and the detachment of adhesions at the cell rear pole. A constant reorganization of the actin cytoskeleton occurs in each step of the process (Raftopoulou and Hall, 2004). Supramolecular complexes containing adhesion and adaptor proteins that are anchored to actin filaments (Hayakawa et al., 2012) act like traction points in several eukaryotic systems.

Myosin superfamily is also involved in motility. Myosins develop adenosine triphosphatase (ATPase) activity, being able to transform the chemical energy of ATP in motion through conformational changes of its structure (Aguilar-Cuenca et al., 2014). Phosphorylation of myosin II heavy chain plays a critical role in *Dictyostelium* cells for the maintenance of cell shape and for the persistent translocation in a spatial gradient of chemoattractant. Ca^{2+}-dependent myosin II activation is required for establishing a polarized pseudopod and for uropod retraction during neutrophil migration.

Polarized microtubular distribution in fibroblasts is also observed before cellular asymmetric establishment, and Keays et al. (2007) demonstrated that mutations of the GTP-binding sites in tubulin monomers avoided microtubular polymerization during neuronal migration in mammals. Smut teliospore displacement is also affected by nocodazole, an inhibitor of microtubule polymerization. These results indicate that microtubule polarization is involved in cell motility in the same way as actin organization. However, the effect of microtubular polarization on cell displacement could be mediated by interaction with the actin filaments (Rodriguez et al., 2003). Actin and microtubules collaborate in different and important processes such as motility, cortical flow, and cell division.

On the other hand, many intracellular signaling molecules are involved in cell migration, including MAPK cascades, lipid kinases, phospholipases, Ser/Thr and Tyr kinases, and scaffold proteins. GTP molecules play an essential role in both signal transduction and actin organization through Rho GTPases, a large family of Ras-related monomeric small G proteins that seem to be the most important members of signaling cascade involved in cell motility. They are best known as regulators of the organization of actin filaments, but also control cell polarity, gene expression, microtubule dynamics, and vesicular traffic (Raftopoulou and Hall, 2004). In the same way, Rho-GTPases coordinate transduction pathways that link cell surface receptors to a lot of intracellular responses in filamentous fungi. This group of proteins plays the main role in the establishment and maintenance of cell polarity and polarized hyphal growth (Wendland and Philippsen, 2000).

Specific models of chemotaxis have been described for highly invasive pathogens in which motility is likely to play a major role in the disease process. For spirochetes, translational motility requires asymmetrical rotation of the two internally located flagellar bundles. Coordinated flagellar rotation involves an efficient mechanism. Signal is transmitted along the length of the cell, which in some cases is over 100-fold greater than the cell diameter.

Chemotaxis is also described in plant–pathogen interactions where motility is an important trait for virulence and pathogenic fitness (Yao and Allen, 2006). The fact that microorganisms are able to penetrate in their plant hosts through opened stomata or wounds suggests that they must have developed motility mechanisms toward these entrance ways (Santiago et al., 2012b). The two most suggested strategies for the displacement are the presence of submicroscopical contractions of helically arranged fibrils within the cell walls and the occurrence of motile appendages in zoospores. Other species of pathogenic fungi produce spores that are able to glide in the presence of a substrate without any obvious external organ or change in cell shape (Brand and Gow, 2012).

Teliospores of pathogenic fungus *S. scitamineum* display a polarized distribution of actin after their contact with chemoattractants, as it has been revealed by using antiactin antibodies. This suggests that cytoskeleton also plays the main role in displacement of fungal cells during quorum-sensing establishment. However, smut teliospores cannot develop lamellipodia in the direction of movement because of the rigidity of the cell wall and because fungal cells do not "crawl" on a substrate, but "swim" in solution (Sánchez-Elordi et al., 2016c).

7.3 Motility of Smut Teliospores in Chemotaxis

A mechanism for teliospore motility consistent with a jellyfish-like "swimming" mechanism driven by polarized actin polymerization has been recently reported. Light and electron microscopies demonstrate the absence of motile external structures in cells able to produce movement. The only morphological feature consistent with chemotaxis of smut teliospores was the presence of membrane/wall invaginations, detected by transmission and scanning electron microscopies. Appendages were often observed, but their features would not likely support migration (Fig. 13.15).

A similar distribution of polymerized, myosin II-decorated actin at the rear pole of migrating cells has been observed in different eukaryotic models, including *Dictyostelium* amoebae, leukocytes, and fibroblasts (Vicente-Manzanares et al., 2008; Juanes et al., 2015). In these cellular lines, lamellipodia and filopodia formation entails the net movement of microfilaments toward the back of the protrusion to extend the cytoskeletal filaments (Choi et al., 2008).

In the same way, displacement of smut teliospores toward a chemoattractant source was necessarily related to some morphological changes consistent with cyclic episodes of polymerization–depolymerization of actin filaments, in collaboration with a myosin II-like protein and some anchorage proteins. In fact, chemotaxis assays were performed in the presence of several cytoskeleton inhibitors that clearly avoided cell migration by means of cytoskeleton disarrangement. Moreover, actin and myosin have been localized in fungal cells after immunofluorescence assays. Again, actin–myosin polarization was related to the presence of invaginations in teliospores (Fig. 13.16).

Events that cause smut teliospore displacement have been described. First, a translation of actin filaments toward its minus end, in the equatorial plane, seems to occur inward the smut teliospores after binding the chemoattractant to its receptors. Translocation of actin filaments has been described as a consequence of the interaction of an ATPase sensitive to blebbistatin, with contractile ability, with F-actin cytoskeleton. It leads to an increase in the cell volume at the front of advance and the retraction of the opposite pole, in which filaments must necessarily be anchored to cell membrane, since

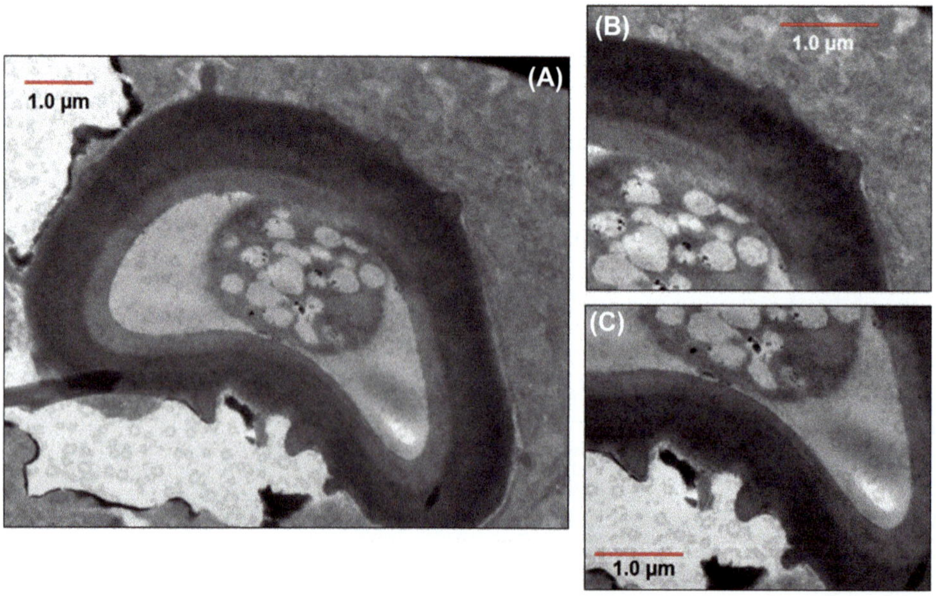

FIGURE 13.15 (A) Transmission electron micrograph of a smut teliospore after chemotaxis assay. (B) Zooming in the image to highlight the absence of external motile structures and (C) the presence of an invagination. Scale bar indicates 1 μm.

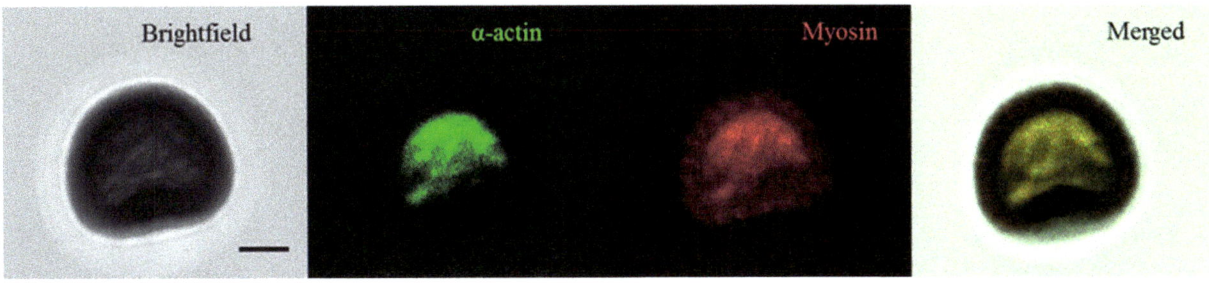

FIGURE 13.16 Confocal series (z-projection) of teliospores after chemotactic stimulation. Brightfield, fluorescence, and brightfield/fluorescence images are shown. Scale bar indicates 2 μm.

invaginations are observed. Finally, depolymerization of F-actin is achieved at the opposite pole, the repolymerization of which leads to cell advancement. Teliospore displacement arises as a consequence of repeated blows of force produced by myosin II, which "pushes" actin filaments toward the minus end in a similar way to that described for Chinese hamster ovary CHO.K1 cells (Juanes et al., 2015).

7.4 The Role of Cytoskeleton in *Sporisorium scitamineum* Chemoattraction During Fungus–Sugarcane Interaction

Chemotaxis of teliospores occurs during quorum establishment before sugarcane infection. However, chemotaxis has been also observed in the presence of defensive HMMG and MMMG produced by sugarcane. The protein moieties of both HMMG and MMMG induce chemoattraction and teliospore agglutination when they bind to their cell wall ligand, hindering the invasive capability of the pathogen. Chemotaxis induced by sugarcane glycoproteins has been explained as a false quorum signal since it simulates natural chemotaxis induced by the fungus (Sánchez-Elordi et al., 2015).

Interestingly, sugarcane glycoproteins increase the percentage of teliospores showing an asymmetric distribution of actin filaments, although it results in a minor number of polymerized filaments per cell (Sánchez-Elordi et al., 2016c). It suggests that sugarcane glycoproteins reorganize actin filaments involved in chemotaxis but disrupt the same involved in germination, probably by means of its interaction with microtubules (Sánchez-Elordi et al., 2016a).

Agglutination of smut cells occurs immediately after chemotaxis. Although chemotaxis and agglutination are independent events, the first one is previously required to stimulate the rest of the process. Aggregation depends on chemotaxis in *Dictyostelium* cells also. It has been demonstrated that mutants defective in chemotaxis are not able to form aggregates (Wang et al., 2011).

Aggregation induced by sugarcane glycoproteins has a defensive role since it finally results in germination blockage. That is so because HMMG fraction of smut-resistant cultivars contains degradative enzymatic activities (β-1,3-glucanase, β-1,4-glucanase, and chitinase) directed to the hydrolysis of the cell walls of teliospores (see Section 5) when they reach the agglutination points. It leads to cell wall rupture, to the disintegration of teliospore aggregates, and to the loss of chemotaxis (Sánchez-Elordi et al., 2016b).

8. SUGARCANE PLANTS DEVELOP ALTERNATIVE DEFENSE PATHWAYS: SYNTHESIS OF PHENOLS, LIGNANS, AND LIGNINS

Sugarcane plants need to defend themselves against attack from different biotrophic pathogens such as *S. scitamineum* and *X. albilineans*, which invade living cells. Plant–pathogen recognition causes the rapid activation of local and systemic appropriate defenses, and this involves cellular reprogramming. Because plants lack a circulatory system, each plant cell must possess a preformed and/or inducible defense capability against a particular pathogen (Sinha ey al. 2013). Signaling pathways activate a broad series of defense responses that curb or eliminate the pathogen.

Knowledge of disease resistance is a prerequisite to evolve breeding strategies to develop disease-resistant varieties of sugarcane. A comprehensive knowledge of factors governing sugarcane resistance is needed. Sugarcane cultivars with ability to resist infection by a particular pathogen depend on very different strategies that can be simultaneously activated and thus potentially enhance plant protection against diverse pathogen types. Therefore resistance of sugarcane plants to diseases seems to be a multifactorial process. The choice of the variety is one of the most important factors in sugarcane breeding and production.

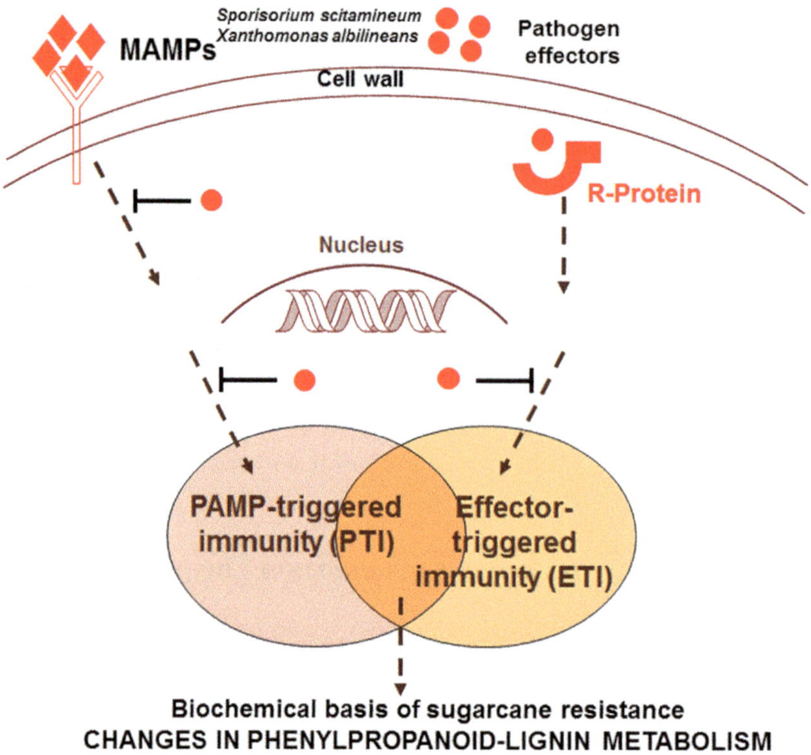

FIGURE 13.17 Partial diagram of pathogen recognition and signal transduction in sugarcane. Changes in phenylpropanoid lignin metabolism are activated as a part of the sugarcane immunity system.

Some of the sugarcane defense mechanisms (Fig. 13.17) include the accumulation of different compounds such as systemic enzymes that generate antimicrobial compounds (phenolic acids) and protective biopolymers (lignans and lignins) that restrict the spread of pathogens (Que at al. 2014a,b). In this way, sugarcane resistance triggered by early recognition of the invader leads to an efficient and nearly immune response that includes upregulation of phenylalanine ammonium lyase (PAL), a key enzyme in the phenylpropanoid pathway and sugarcane defense. Esh et al. (2014) observed significant variations in the levels of polyphenol oxides, PAL, POX, esterase, chitinase, and β-1.3-glucanase in sugarcane clones that are resistant or susceptible to smut.

The sensitivity or resistance against *S. scitamineum* and *X. albilineans* has been related with changes in sugarcane leaf PAL activity and levels of free phenolic compounds in response to microbe-associated molecular patterns (MAMPs), classically termed "elicitors" (de Armas et al., 2007; Santiago et al., 2008a, 2009a; Coutinho et al., 2016). This interaction was studied by using sugarcane leaf disks obtained from cultivars B 42-231 (susceptible to smut), L 55-5 (susceptible to scald), and My 55-14 (resistant to smut and scald). Water-soluble MAMPs isolated from *S. scitamineum* mycelium enhanced the accumulation of free phenolics, mainly hydroxycinnamic acids such as *p*-coumaric acid, by activation of PAL in the resistant cultivar, and mainly hydroxybenzoic acids as syringic acid, in the susceptible cultivar (de Armas et al., 2007; Santiago et al., 2008a). B 42-231 showed low PAL activity related to the accumulation of caffeic acid that produced feedback inhibition of the enzyme (Santiago et al., 2008b). Smut MAMPs were separated by capillary electrophoresis and resolved as negatively charged proteins, peptides, or glycopeptides of mid molecular mass. To facilitate the infection, plant pathogens secrete numerous effector proteins into the plant apoplast or cytosol (Koeck et al., 2011).

Similarly, water-soluble scald MAMPs were resolved as proteins of high molecular mass and glycoproteins. Scald protein MAMPs develop a resistance response by increasing hydroxycinnamic acids synthesis, whereas scald carbohydrates MAMPs increase the phenolic content by accumulation of different hydroxybenzoic acids in susceptible cultivar, but antifungal resistance does not ensue (Santiago et al., 2009a). The pathway of hydroxybenzoic acids is activated, once the level of *p*-coumaric acid justifies the accumulation of hydroxycinnamic acids.

However, a decrease in the amount of free hydroxycinnamic acids was found in both pathosystems when the highest value of PAL activity was reached. Resistant cultivar needs to maintain a high level of PAL activity without accumulation of free hydroxycinnamic acids over time. This could be due to a very fast metabolic conversion of newly produced hydroxycinnamic acids to more structurally complicated compounds, lignans, and lignins, as a possible mechanical defense response against the potential entry of pathogens. This hypothesis was corroborated by using the same model of interaction in vitro (leaf disks-MAMPs). The level and organization of lignin in response to smut MAMPs was evaluated in cross-sections of sugarcane leaves stained with phloroglucinol/HCl. *S. officinarum* leaf shows the classical arrangement of cells that defines a C4 species (Santiago et al., 2009b). Fibers of the vascular bundle and the most external layer of edge of sclereids were strongly lignified (Legaz et al., 2006). Changes in the pattern of lignification were revealed. Smut MAMPs produce an increase of lignification of sclerenchyma as well as of the medium and small xylem vessels of the leaves from resistant cultivar, whereas delignification of sclereids seems to occur in the susceptible cultivar (Santiago et al., 2010a, 2012b; Vicente et al., 2017). Thus active penetration and progression of fungal hyphae through the cell wall of plant cells is also possible in susceptible cultivar, as described by Santiago et al. (2008b, 2012a). The cells with lower levels of lignin are the first to be hydrolyzed by the pathogen in several plant species, including sugarcane.

Induction (or activation) of POX by smut MAMPs was related to de novo biosynthesis of lignin in resistant sugarcane plants (de Armas et al., 2007; Santiago et al., 2008a), most probably using phenolic acids as substrate, with the formation of ester–ether bridges. It could be considered as hypotheses that laccases are able to carry out the oxidative breakdown of lignin in susceptible cultivar. This would not only arrest lignification but also induce a quantitative loss, more or less large, of previously deposited lignin. According to Eggert et al. (1996), the level of phenols (or one of them in particular) would displace the equilibrium toward depolymerizing oxidative reactions, which would be interpreted as a response of susceptibility, or of resistance when the equilibrium displaces to oxidative polymerization. The ability of laccases to degrade lignin increases when some phenols act as mediators, controlling in this way the system (Fig. 13.18).

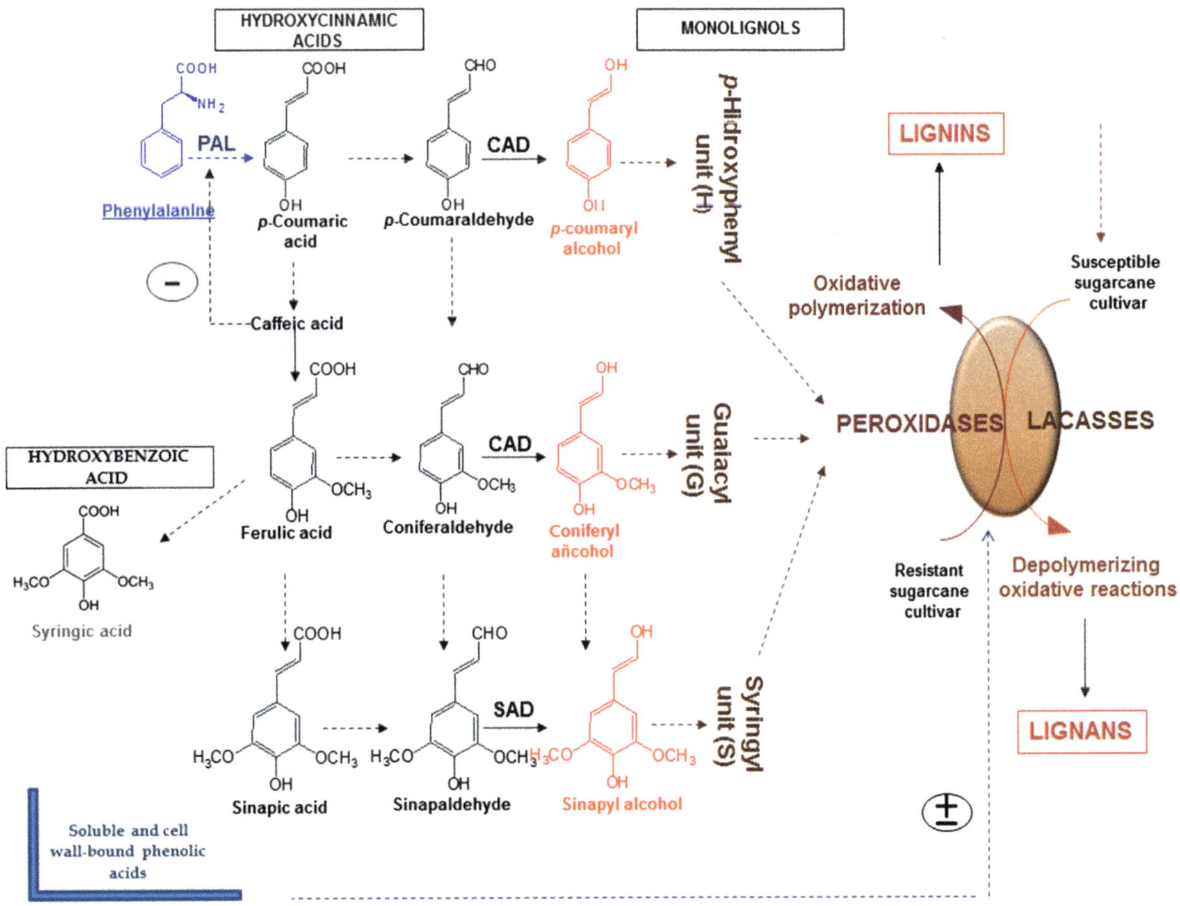

FIGURE 13.18 Schematic pathway of phenylpropanoid and lignin biosynthesis with a hypothetical equilibrium system of metabolites depending on sugarcane cultivar.

Therefore the results suggested that the leaf of sugarcane plants recognize and respond to smut and/or scald MAMPs to trigger pathogen-associated molecular pattern (PAMP)-triggered immunity (PTI), based on phenylpropanoid and lignin metabolism. However, according to the zigzag model of the plant immune system, the ultimate amplitude of disease resistance or susceptibility is proportional to:

MAMP-triggered immunity (PTI) – effector-triggered susceptibility (ETS) + effector-triggered immunity (ETI).

S. scitamineum seems to possess an arsenal of effectors that can potentially manipulate host metabolism (Soyer et al., 2014; Que et al., 2014a,b) as the most effective strategy used to defend itself from plant immune system. During the coevolution of fungal plant pathogens and their hosts, a seesawing interplay between pathogen virulence and host resistance has been developed.

For this reason, in vivo assays must be carried out to understand the amplitude of the response. For the first time, phenolic fractionation from stems of sugarcane plants with changes in the level of soluble and wall-bound phenolic acids, produced after inoculating young plants with smut sporidia, has been achieved. The ratio inoculated/noninoculated plants of some cell wall–bound phenolics, such as ferulic, caffeic, and syringic acids, increased in the resistant cultivar, whereas it was maintained more or less unchanged in the susceptible one (Santiago et al., 2009c). The increased flux through the phenylpropanoid pathway resulted in the synthesis of phenolic monomers that were eventually esterified and incorporated into the cell wall. Phenolic acid constituents of sugarcane cell wall include sinapic acid as well as *p*-coumaric and ferulic acids. Ferulic acid forms deposits at the very early stage of lignification, whereas *p*-coumaric and sinapic acids increase their deposit as lignification progresses. These new depositions result in a better capacity to cultivar My 55-14 for an increase in the frequency of bridges between lignin fragments through ester–ether linkages for reinforcing the cell wall and major resistance to the disease.

An increase of 29% of the lignin content in the resistant cultivar after inoculation but only 13% in the susceptible cultivar compared with uninoculated plants has been described (Santiago et al., 2012b). The final concentration of lignin after inoculation treatment is very similar in both cultivars. So the possible defense mechanism of the resistant cultivar would be based on the neosynthesis of the polyphenolic polymer. Sugarcane is known to respond to the presence of smut by modulating the expression of genes related to lignin pathways (You-Xiong et al., 2011). This can be justified on the basis of the recently complete genome sequencing of *S. scitamineum* (Barnabas et al., 2016). A comparative transcriptome profiling shows that approximately 13.5% of smut genes are related to several metabolic processes important for the fungal survival and protection inside the host tissues. Increased smut extracellular and nonsecreted laccase activities were detected, that must be regarded as potentially involved in lignin breakdown (Taniguti et al., 2015).

Structurally, lignin is a heteropolymer that results from the oxidative coupling of three cinnamyl alcohols (monolignols), *p*-coumaryl, coniferyl, and sinapyl alcohols. Cinnamyl alcohol dehydrogenase is the generic name of enzymes that catalyze the reduction of cinnamaldehydes into cinnamyl alcohols. This is the last step of monolignol biosynthesis before oxidative polymerization in the cell wall. Within the lignin, these alcohols give rise to the H (hydroxyphenyl), G (guaiacyl), and S (syringyl) units of the polymer, respectively.

Inoculation of sugarcane stems with smut sporidia produces significant increases of coniferyl alcohol dehydrogenase (CAD) and sinapyl alcohol dehydrogenase (SAD), enzymes able to catalyze the reduction of coniferyl and sinapyl aldehydes, respectively, to the corresponding alcohols (Santiago et al., 2012b). The initial concentration of lignin in both cultivars could be described as G type since only CAD activity was detected. Thereafter, synthesis of G and S lignin types can occur. Activities were chromatographically assayed by the oxidation of the corresponding alcohols (coniferyl alcohol and sinapyl alcohol) into the aldehydes (coniferaldehyde and sinapaldehyde). A metabolic compartmentalization would facilitate an independent regulation of two independent pathways, guaiacyl and syringyl lignin. The highest SAD activity is consistent with high levels of cell-wall-bound syringic acid as well as a time-dependent increase in the resistant cultivar than in the susceptible cultivar (Santiago et al., 2009c).

Biochemical resistance of sugarcane against *S. scitamineum* and *X. albilineans* involves the rapid (<24 h) production of cell wall appositions, physical barriers, and antimicrobial metabolites at the site of entry of the pathogen. This postinvasion mechanism probably requires elements to regulate it. Therefore metabolic regulation of PAL, CAD, SAD, and POX will be studied, since they are the key enzymes involved in phenylpropanoid–lignin metabolism. Biochemicals, genomics, and proteomics tools should be used (Ashwin et al., 2017). It is know, for example, that target genes of several microRNAs play essential roles in the MAPK signaling pathway, plant hormone signal transduction, and plant–pathogen interaction (Su et al., 2017). Continuous research efforts are required to unravel the enigmatic relationship between sugarcane and its important pathogens, so as to sustain the productivity of the released varieties for commercial cultivation (Gao et al., 2016).

8.1 Factors Affecting the Development of *Sporisorium* Mycelium: Growth of *Sporisorium scitamineum*

The sequence of nuclear events that characterize the life cycle of *S. scitamineum* is governed by homologous genes that regulate their sexuality and pathogenicity. There are three distinct phases during the fungal life cycle: diploid teliospores, haploid yeastlike sporidia, and dikaryotic mycelium (Fig. 13.19). At the teliospore germination time, spores extend in size, split its exosporidium (external wall of teliospores), and produce a probasidium from which, following meiosis, four sporidia emerge. Haploid sporidia represent the secondcell type in this life cycle. Germination of sporidia leads to the production of a germ tube or same sporidial cells. Heterothallism in *S. scitamineum* is bipolar, and sporidia belongs to two mating types: plus (+) and minus (−). Single haploid (n) sporidia continue to multiply by budding in a yeastlike way. Dikaryotic mycelium (n+n) phase, the third cell type, appears following three consecutive ways: (1) fusion of compatible sporidia, (2) promycelial anastomosis (the growth of two promycelium together, the lysis of their cell wall in crossed site, and finally the migration of the nucleus to the fused cell), and (3) hyphal anastomosis of two mycelia from different mating types arising from sporidial germination. Mating genes are crucial factors for smut establishment since a successful mating reaction is needed to form the infective dikaryotic mycelium. Results of cutting inoculations and detection of the fungus in plantlets using polymerase chain reaction indicated that, regardless of the existence of the smut fungus in plants, infection was only produced by dikaryotic forms (Moosawi-Jorf and Izadi, 2007). Mating and pathogenicity processes are regulated by two loci, *a* and *b*. The *a* locus is composed of genes encoding a pheromone receptor system necessary for cell–cell recognition and hyphal fusion, whereas the *b* locus is composed of two genes encoding subunits of a heterodimeric homodomain transcription factor regulating the maintenance of the dikaryon.

During *S. scitamineum* infection, the fungus grows within the meristematic tissue and induces the formation of reproductive structures to produce teliopores. The reproductive structures, usually typical grass panicles, are transformed into a *whip*like sorus that grows rapidly and protrudes out between the leaf sheaths and is responsible for disseminating smut propagules. Within these structures, millions of dark teliospores develop, which have an aerial dispersal mode. The *whip* is composed partly by host plant tissue and partly by fungus tissue, and its morphology differs from short to long, twisted, or multiple *whips* (Ramesh Sundar et al., 2012). Under appropriate conditions, germination of smut spores occurs on the internodal surface, which is followed by the formation of appressorium structures to penetrate plant tissues. Hyphae are found throughout the plant mostly in the parenchymatous cells toward the lower internodes. Infective mycelia penetrate through the buds at each node and systemically colonize the apical meristem.

8.2 Caffeic Acid as a Possible Mycotoxin

Infected sugarcane plants show a strong metabolic modification resulting in the synthesis of phenols, lignans, and lignins, changes that occur during the period between the appearance of stress on the plant from the invasion of smut and the subsequent

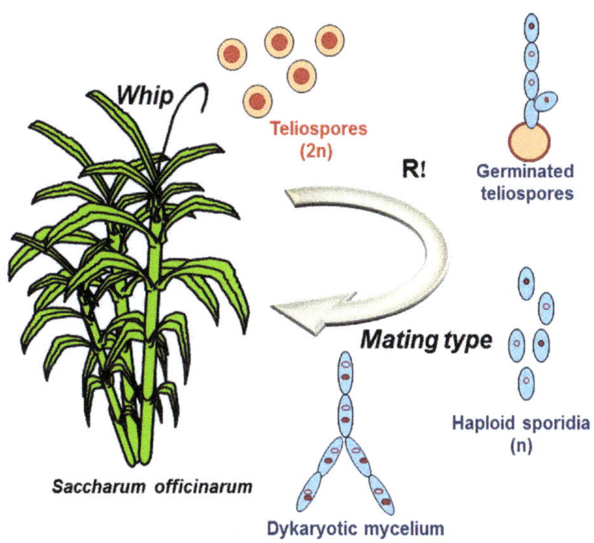

FIGURE 13.19 Developmental stages in the *S. scitamineum* life cycle: diploid teliospores (2n); haploid yeastlike sporidia (n) after meiosis (R!); dikaryotic mycelium after *mating type*.

plant–pathogen interaction. It is known that early sugarcane infection with *S. scitamineum* triggers changes of phenylpropanoid and lignin metabolism as a means of protection against smut infection. In addition to the physical barrier that strongly lignified tissues represent against teliospore penetration, some of the monolignols synthesized as lignin precursors show biological activity. Many intermediates and end products of the phenylpropanoid pathway play important roles in plants as phytoalexins, antiherbivory compounds, antioxidants, ultraviolet protectants, pigments, and aroma compounds. It is well known that phenolic acids are precursors of a variety of antimicrobial compounds that play an important role in plant defense responses. The synthesis of these compounds has been associated with the resistance of plants to fungi and viruses (Ralph et al., 2006).

Caffeic acid sometimes behaves as a potent phytotoxin affecting plant and fungi growth and physiology. Santiago et al. (2010b) studied the in vitro effects of commercial caffeic acid upon different phases of *S. scitamineum* growth cycle reproducing the amount of this phenolic detected on treated sugarcane. Inoculation of plants with smut MAMPs produced a significant increase of caffeic acid concentration to reach values about 5 mg/mL in the resistant cultivar leaves, whereas no effect was produced in the susceptible one. The same concentration range of soluble caffeic acid was detected in sugarcane stems inoculated with smut sporidia, but a significant decrease with time after treatment was achieved in a resistant cultivar (Santiago et al., 2009c). At 5.0 mg/mL, caffeic acid produced inhibition of teliospore germination, haploid sporidia production, and dikaryotic mycelium appearance. Inhibition was more evident 24 or 28 h after incubation of teliospores in caffeic acid solution than after 48 h. It seems that these cells were able to use caffeic acid as a carbon source for higher times of incubation (Santiago et al., 2010b). Higher caffeic acid concentrations, detected in susceptible cultivar, had no significant effect on dikaryotic mycelium appearance. Pathogenicity of *S. scitamineum* depends on the hyphal fusion and sexual cycle for disease establishment and dissemination.

With the extension of hyphal growth around the vascular tissues, there is the possibility that *S. scitamineum* spread quickly and systemically into the plant through the vascular system. However, if resistant cultivar is able to inhibit dikaryotic mycelium formation, infectivity would be reduced. Thus caffeic acid in sugarcane gives a better capacity for decreasing smut pathogenicity, contributing to modulate in part the defense response of resistant cultivars. However, during the coevolution of smut and sugarcane, *S. scitamineum* could use another protective strategy to detoxify the environment. So sugarcane could secrete various antimicrobial compounds into the apoplast to restrict pathogen growth.

8.3 Pinoresinol as a Possible Mycotoxin

The biotrophic interaction of *S. scitamineum* and sugarcane leads to modified lignin metabolism (Santiago et al., 2010a, 2012b; Vicente et al., 2017); sugarcane lignin precursors and/or intermediates may also exhibit antimicrobial properties. In this way, lignans constitute a group of important plant phenols that are structurally characterized by the coupling of two phenylpropanoid units. The biological role of lignans in plants is still under discussion, but it is generally assumed that they play a role in plant defense. Lignans have been found to possess a variety of biological properties (cytotoxic, antioxidant, antiviral, antimicrobial). França et al. (2001) built sugarcane complementary DNA sequences with similarity to a *F. intermedia* gene encoding (+)-pinoresinol/(+)-lariciresinol reductase, which regulates a pivotal branch point in lignan biosynthesis (Fig. 13.20A).

(+)-Pinoresinol behaves as a phytotoxin affecting smut growth. The effect of commercial (+)-pinoresinol upon *S. scitamineum* growth cycle depends on both lignan concentration and time incubation. At 20–40 µg/mL range concentration, (+)-pinoresinol inhibited teliospore germination after 3–24 h incubation but it had no significant effect for incubation times longer than a day. Additionally, it is important to notice that 40 µg/mL (+)-pinoresinol produces abnormal and irregular growth of smut sporidia, with migration of basic cellular content (Fig. 13.20B). Effect of appearance of yeastlike colonies was also corroborated in the presence of commercial (+)-pinoresinol and a lignan extract obtained from infected sugarcane cultivars.

It could be hypothesized that expression of the lignan pathway in sugarcane seems to be regulated as a response to smut attack. In fact, the induction of phenylpropanoid–lignan–lignin synthesis under conditions of stress must be the result of increased transcription of genes encoding the corresponding biosynthetic enzymes, as it has been confirmed in other plants. Further evidence as to which specific class of secondary compounds sugarcane defense products belong would contribute to the understanding of the evolution of plant defense mechanisms.

9. SUGARCANES DEVELOP ALTERNATIVE DEFENSE PATHWAYS: POLYAMINES

9.1 Polyamines: Nature, Metabolism, and Biological Functions

PAs constitute a group of ubiquitous and essential aliphatic polycations found in both eukaryotic and prokaryotic organisms with the exception of some *Archaea* (Kusano et al., 2007). In eukaryotic organisms, the most common PAs are the diamine PUT and the triamines SPD and SPM. Nevertheless, a large number of fungal species do not contain SPM

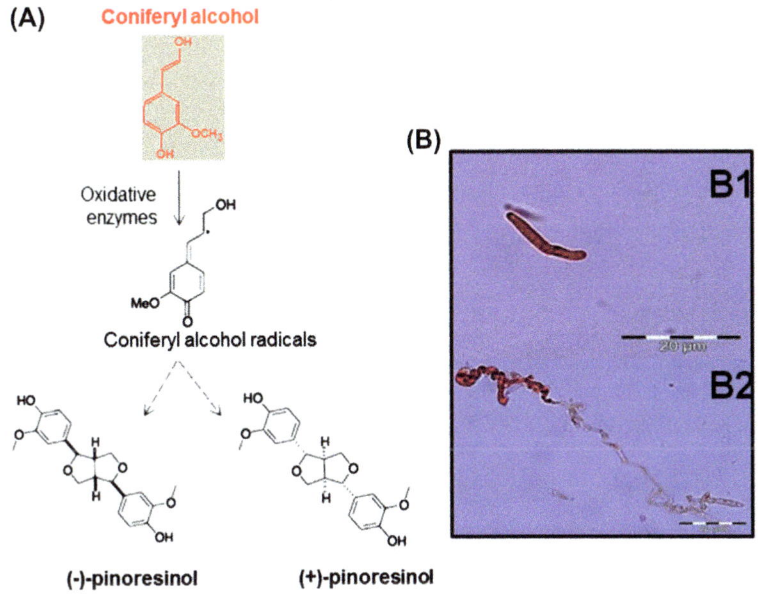

FIGURE 13.20 (A) Schematic biosynthetic pathway of (+)-pinoresinol and (–)-pinoresinol from coniferyl alcohol monomers. Lignan biosynthesis involves strict stereochemical controls. (B) Effect of (+)-pinoresinol upon growth of smut sporidia stained with safranin.

(Valdés-Santiago et al., 2012) whereas other PAs, such as the diamine CAV, are widely distributed. PAs are required for normal development of prokaryotic and eukaryotic organisms. In higher plants, PAs are involved in a wide range of developmental processes, e.g., root growth, somatic embryogenesis, floral initiation, and development of flowers and fruits. Research conducted on fungi also support their essential role in the regulation of developmental processes, including spore germination, dimorphism, appressorium formation, and conidiation. In addition, PAs are involved not only in fundamental cell processes but also in adaptive responses to abiotic and biotic stress of both plants and fungi (Walters, 2003a).

Plants have a natural array of defense mechanisms to protect them from the pathogenic organisms. Among chemical substances related to plant–pathogen interactions, PAs and phenolic acids have been widely recognized. It has been shown that PA levels can vary greatly in plants infected by pathogens of different nature (Walters, 2003b). These changes can be accounted for de novo biosynthesis, degradation, and conjugation or binding to molecules of diverse nature.

The pathways of PA biosynthesis have been established for many organisms. The first step is the formation of PUT. In mammals and fungi, PUT is produced by a single pathway catalyzed by ODC. In contrast, PUT in plants is usually produced by two alternative pathways. One is the ODC-catalyzed reaction, as in fungi; the other is from arginine, as a result of the action of ADC to produce agmatine (Fig. 13.21). Conversion of agmatine into PUT requires two enzymes, agmatine iminohydrolase and N-carbamoylputrescine amidohydrolase. In a subsequent step, PUT is converted into SPD and then into SPM by symmetrical addition of an aminopropyl residue from decarboxylated S-adenosylmethionine (AdoMet), which results from decarboxylation of AdoMet by AdoMet decarboxylase. These reactions are sequentially catalyzed by two closely related but distinct enzymes SPD synthase and SPM synthase. The diamine CAD can be formed from lysine either via the action of ODC or as a result of lysine decarboxylase (LDC) activity.

In addition to the biosynthetic pathway, in plants and fungi there is a retroconversion pathway for the oxidative transformation of SPM to SPD and SPD to PUT, via acetylated PAs, with the production of H_2O_2. Regarding catabolism of PAs, copper-containing diamine oxidases (CuAOs) and flavin-containing polyamine oxidases (PAOs) catalyze the oxidative deamination of PAs. CuAO oxidizes PUT and CAD, whereas PAO oxidizes SDP and SPM, producing 4-aminobutanal and N-(3-aminopropyl)-4-aminobutanal, respectively, in addition to 1,3-diaminopropane and H_2O_2 (Fig. 13.22).

In plant cells, PAs can occur as free molecules (S-PAs) or conjugated to small molecules, such as sugars or phenolic acids (SH-PAs), or bounded to larger molecules (PH-PAs), such as proteins, DNA, polysaccharides, or polymeric phenolic compounds (D'Agostino et al., 2005). PAs conjugated to phenolic acids are synthesized as a response to pathogen attack. These conjugated PAs have been correlated with reduced fungal or viral dispersion, reduced germination of fungal spores, or reinforcement of plant cell walls making them more resistant to pathogenic hydrolytic enzymes.

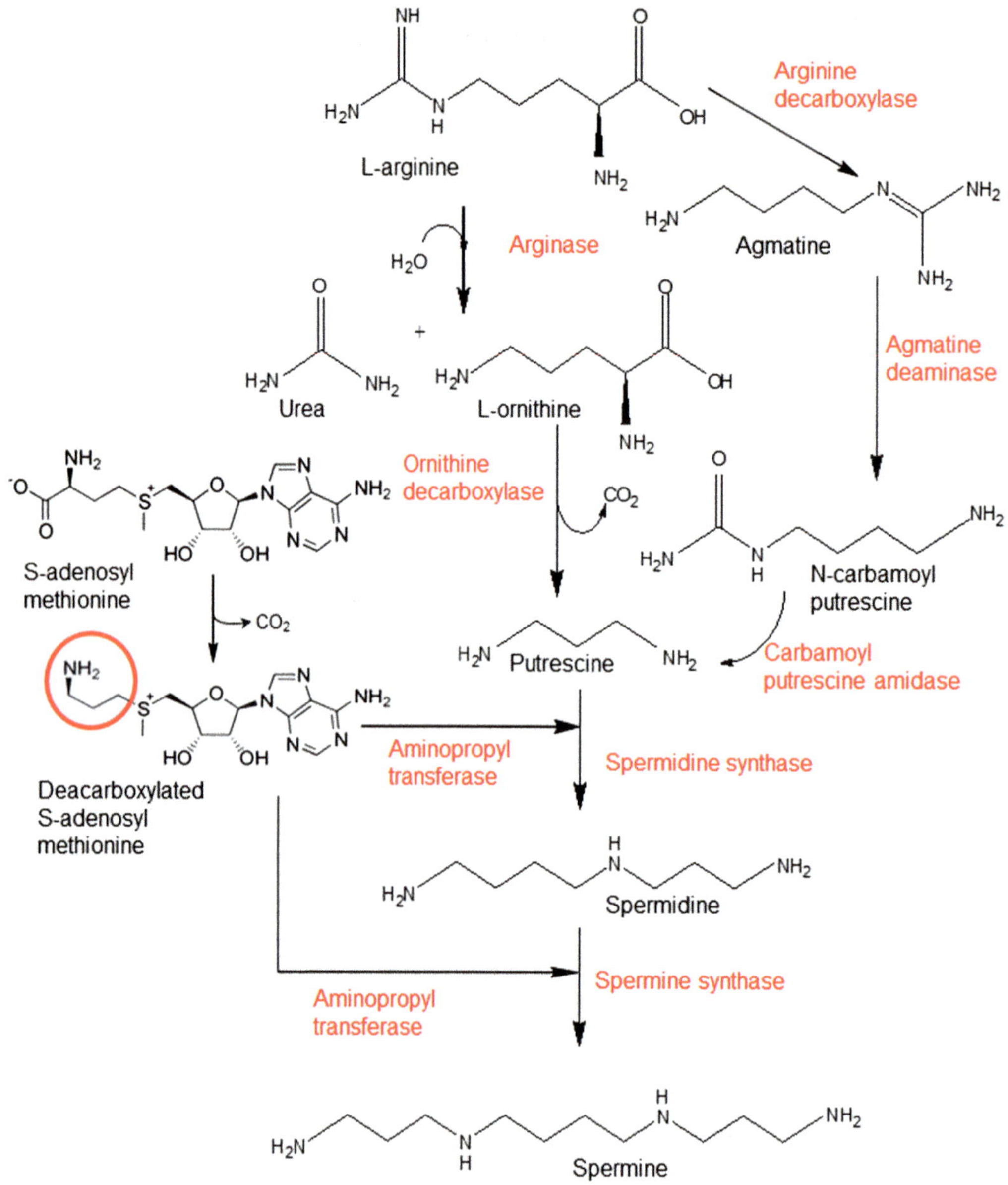

FIGURE 13.21 Biosynthetic pathways of polyamines in fungi and higher plants.

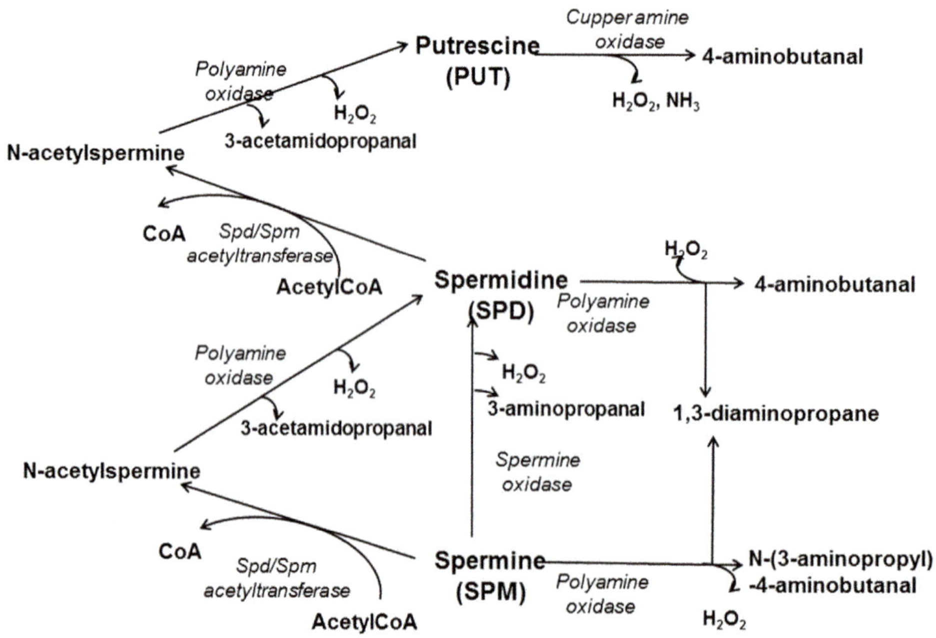

FIGURE 13.22 Retroconversion pathway for the oxidative transformation of spermine in other polyamines.

9.2 Polyamines in Sugarcane as Defense Against Pathogens

PAs and phenolic acids normally occur in sugarcane, and their titers and interconversion between free and conjugated forms are usually altered by pathogenic infections. The first research that studied PA levels in sugarcane plants as well as possible changes in response to pathogenic organisms was carried out by Legaz et al. (1998a,b). The authors infected buds of a cultivar (Ba 42231) susceptible to smut with *S. scitamineum* teliospores. After 2 months, the general symptoms of infected plantlets included the appearance of a sole, central whip in the meristematic region of the stalks and narrowed leaves giving the plants a herbaceous aspect. The fact that pathogen was able to complete its biological cycle (penetration, invasion, and spore formation) in that short period of time was related to the high susceptibility of the cultivar. In resistant cultivars, there is no whip appearance, which means that the fungal sporogenic phase remains unrealized.

Two-month-old healthy plants of sugarcane contained PUT, CAD, SPM, and SPD as free and conjugated (-SH and -PH) forms. PA levels and conjugation patterns differed according the tissue. Leaves contained free and conjugated PUT, SPD, and CAD but SPM was only present in the free form. Stalks showed high levels of free PUT, SPD, SPM, and PH-SPD, but the absence of any PA in the -SH fraction was noticed. Roots did not contain SPM but high levels of free and conjugated SPD.

Analysis of PAs in organs of infected plantlets revealed that the most dramatic changes after smut inoculation occurred in the stalks, where the total PA content and conjugation pattern were strongly affected. Free forms of PUT, SPD, SPM, and CAD, as well as PH conjugated forms of PUT, SPD, and CAD, completely disappeared, whereas -SH conjugate forms of PUT, SPD, SPM, and CAD appeared. Since a linear correlation exists between the concentration of phenolic compounds produced in sugarcane tissue and smut infection, the authors suggested that susceptibility of sugarcane to smut was the result of the conjugation of phenolics to PAs, thus reducing the fungus' toxic properties demonstrated for free phenols.

A subsequent research was carried out by the same authors to determine the possible existence of a correlation between pathogenicity and PA metabolism during the initial infection events (Piñón et al., 1999). To demonstrate this hypothesis, the authors provoked the experimental infection of buds of two sugarcane cultivars with teliospores of *S. scitamineum*, the sensitive cultivar Ba 42231 and the resistant cultivar Mayari 5514. Changes in PA levels along 20 days after infection were analyzed, and it was revealed that, in healthy buds from both cultivars, PA levels increased during growth, reaching a maximum value at 13–17 days after germination, and decreased later. Inoculation produced, in both cultivars, a remarkable increase in the levels of free and conjugated PAs and enhanced the activity of the biosynthetic enzymes ODC, arginase, and LDC. Although the level of PAs increased in both cultivars upon infection, changes varied according the cultivar, so that, infected buds from the resistant My 5514 buds mainly produced free PUT, whereas infected buds form the sensitive Ba 42231 produced acid-soluble conjugated (-SH) as well as acid-insoluble conjugated (-PH) PUT, SPD, and SPM. Since resistance has been previously explained as depending on phenolic accumulation in sugarcane tissues, differences in the titer of -SH and -PH fractions for both SPD and SPM between resistant and sensitive cultivars must be biologically significant. Thus cultivar Ba 42231 becomes sensitive to smut as a consequence of the high value of PA conjugates. Moreover, conjugation of SPD and SPM could prevent their degradation via PAO, thereby reducing activation of defense pathways against pathogens (Fujimura et al., 2009). SPM has been shown to be a salicylate-independent endogenous inducer of acidic PRP and conferred resistance to tobacco mosaic virus (TMV) infection (Yamakawa et al., 1998).

Mature sugarcane stalks are ripened and used for industrial production of sucrose, the main compound found in the juice of mature plants. For industry, juice quality is defined by high sucrose content, but low levels of other soluble compounds such as glycoproteins and phenolic acids, the presence of which can interfere with the industrial process of clarification and sucrose crystallization. Diverse abiotic and biotic stresses can provoke alterations in plant metabolites, and such alterations can be reflected in the juice quality. Changes in some saccharimetric parameters as well as in the levels of other compounds, such as phenolic acids and PAs, were studied in raw juices from sugarcane plants affected by diverse pathogenic organisms (Fontaniella et al., 2003a,b, 2007).

Sugarcane widely suffers in many countries a yellowing and stunting disease called sugarcane yellow leaf syndrome (YLS). The syndrome produced important economic losses, either by a significant yield decrease or by a decline in the productivity of the affected cultivar. Evidences about the association of two possible causal agents, a phytoplasma and a luteovirus, have been extensively described. Juices from YLS plants showed higher levels of reducing sugars, glucose index, and glycoproteins (HMMG and MMMG). High hexose levels in plant tissues infected by pathogens can be associated with the action of plant hydrolases that digest cell wall polysaccharides into monosaccharides to provide the energy required for plant resistance against pathogens. Analysis of PAs in raw sugarcane juices from different sugarcane cultivars revealed the existence of PUT, SPD, SPM, and CAD. They were found mainly either free (-S) or bound to macromolecules (-PH), although PA conjugated to small molecules (-SH) was rarely found in sugarcane juices.

Juices from healthy sugarcanes contained arginase, ODC, and LDC, whereas YLS juices did not contain LDC. Both arginase and ODC activities were almost three times higher in YLS juices than in healthy ones. Analysis of PAs revealed that juices from YLS plants contained an amount of PAs slightly higher than that obtained from healthy ones, a fact that could indicate the existence of an active catabolic process among biosynthesis. In YLS plants PUT, synthesized from arginine via arginase and ODC activities, was mainly converted to SPD, whereas a small percentage of it seemed to be conjugated to produce PH-PUT. The importance of this form of conjugated PUT seems to be derived from its binding to HMMG glycoproteins as well as PH-CAD, which increased in YLS juices. Another change observed in YLS juices was the disappearance of SH-SPM. As it was not recovered as free SPM, it could be possible that once deconjugated, SPM was converted into SPD by the activity of a retroconversion pathway for the oxidative transformation via acetylated PAs with the production of H_2O_2. Cell death caused by TMV infection or cryptogein, an elicitor secreted by the oomycete *Phytophthora cryptogea* that remodels phenylpropanoid metabolism (Amelot et al., 2011), was partially mediated by H_2O_2 generated through PA catabolism. These authors suggested that the PA substrate for H_2O_2 production is SPD since, during elicitation of hypersensitive reaction, SPD but not SPM accumulated in the apoplast. That plants employ PA catabolism-derived H_2O_2 as a defensive tool upon exposure to biotic stresses has been reviewed by other groups.

On the other hand, PAs and phenolic acid levels were also examined in two sugarcane cultivars differing in susceptibility to leaf scald, a disease caused by the bacterium *X. albilineans*. Raw juices obtained from infected plants of both cultivars showed a significant increase in the levels of free PUT and ODC activities. However, the pathogen induced different changes in subsequent metabolic steps. Acid-insoluble bound SPD (PH-SPD) completely disappeared from the juices of highly susceptible cultivar C 439-52, whereas an increase in the PAs bound to macromolecules (PH-PUT, PH-CAD, and PH-SPD) occurred in the moderately susceptible cultivar L55-5 (Fontaniella et al., 2007).

In planta, bacterial growth was strongly decreased in tobacco Tet/ADC transgenic plants or in wild-type plants infiltrated with SPM, an effect that was blocked by the PAO inhibitor 1,19-bis(ethylamino)5,10,15-triazanonadecane (Marina et al., 2008). The authors concluded that the accumulation and further oxidation of PAs in the leaf apoplast of tobacco plants is a common response against infection by microorganisms with contrasting pathogenesis strategies, ultimately leading to plant tissue cell death. They suggested that this response plays a beneficial role in defense against biotrophic pathogens that depend on living tissue for successful host colonization while being detrimental for plant defense against necrotrophic pathogens that feed on necrotic tissue.

10. CONCLUSIONS

Sugarcane plants have developed a complex battery of defenses against invasion by occasional pathogens. First, the host plant is able to discriminate beneficial or harmful endobionts by a simple cell signaling system based on the production of glycoproteins capable of binding to receptors on the invader cell wall. Some of these glycoproteins also behave as chemotactic factors, capable of guiding the movement of the pathogen in a chemical gradient, and to produce cytoagglutination, acting as a false quorum signal that prevents the development of the pathogen. Other glycoproteins show hydrolytic activities on pathogen wall polymers, inducing cell death. Finally, the production of DIR proteins allows the plant to reinforce its cell walls by means of new deposits of lignin, making it difficult for the invader to enter, while increases the levels of lignans, cytotoxic agents against eventual pathogens.

11. FUTURE PROSPECTS

Research in phytopathology, particularly in relation to diseases of sugarcane plants, has advanced in a prodigious way during the last 40 years. The diagnosis was increasingly accurate describing the morphological and functional alterations that resulted from the invasion of the plant by the pathogen: foliar chlorosis, desiccation, low root development, stunting and thinning of stems, losses in the sucrose accumulation capacity, floral malformations, appearance of abnormal or atypical organs, etc.

A next step forward was the study of the physiological dysfunctions underlying this symptomatology: chlorophyll degradation, xylem vessel occlusion, hormonal imbalance, drift of part of sucrose-P toward the synthesis of polysaccharides, and advances that manifest the influence of metabolites produced by the pathogen on the gene expression of the plant. To all this, the knowledge about defense responses that the plant developed against the invading agent was added, a circumstance that was favored by the previous existence of an enormous amount of cultivars, of greater or lesser resistance to a specific pathogen, all of them obtained through techniques of classic plant genetic improvement. Today, we are able to differentiate, thanks to the advanced techniques of proteomics, the sequence of an enzyme produced by a resistant cultivar, capable of inhibiting the germination of spores of pathogen, from that of the same enzyme produced by a sensitive cultivar, which does not alter this process, and even from that of the same enzyme produced by the pathogen, which activates the germination of its spores. The study of these genes, orthologs in the first case and paralogs in the second, and their cloning will allow, through genetic transformation techniques, to obtain transformed plants resistant to a specific disease with much greater efficiency, speed, and security.

REFERENCES

Aguilar-Cuenca, R., Juanes-García, A., Vicente-Manzanares, M., 2014. Myosin II in mechanotransduction: master and commander of cell migration, morphogenesis, and cancer. Cell Mol. Life Sci. 71, 479–492.

Aguilar-Cuenca, R., Llorente-Gonzalez, C., Vicente, C., Vicente-Manzanares, M., 2017. Microfilament-coordinated adhesion dynamics drives single cell migration and shapes whole tissues. F1000Research 6 (F1000 Faculty Rev), 160. http://dx.doi.org/10.12688/f1000research.10356.1.

Amelot, N., Carrouche, A., Danoun, S., Bourque, S., Haiech, J., Pugin, A., Ranjeva, R., Grima-Pettenati, J., Mazars, C., Briere, C., 2011. Cryptogein, a fungal elicitor, remodels the phenylpropanoid metabolism of tobacco cell suspension cultures in a calcium-dependent manner. Plant Cell Environ. 34, 149–156.

Ashwin, N.M.R., Barnabas, L., Amalraj, R., Malathi, P., Viswanathan, R., Masi, A., Agrawal, G.K., Rakwal, R., 2017. Advances in proteomic technologies and their scope of application in understanding plant–pathogen interactions. J. Plant Biochem. Biotechnol. http://dx.doi.org/10.1007/s13562-017-0402-1.

Barnabas, L., Ashwin, N.M.R., Kaverinathan, K., Trentin, A.R., Pivato, M., Sundar, A.R., Malathi, P., Viswanathan, R., Rosana, O.B., Neethukrishna, K., Carletti, P., Giorgio Arrigoni, G., Masi, A., Ganesh Agrawal, K., Rakwal, R., 2016. Proteomic analysis of a compatible interaction between sugarcane and *Sporisorium scitamineum*. Proteomics 16, 1111–1122.

Barnabas, L., Ashwin, N.M.R., Amalraj, R., Malathi, P., Viswanathan, R., 2017. Putative orthologs of *Ustilago maydis* effectors screened from the genome of sugarcane smut fungus - *Sporisorium scitamineum*. Austr. Plant Pathol. 46 (2). http://dx.doi.org/10.1007/s13313-017-0471-6.

Basse, C.W., Steinberg, G., 2004. *Ustilago maydis*, model system for analysis of the molecular basis of fungal pathogenicity. Mol. Plant Pathol. 5, 83–92.

Blanch, M., Legaz, M.E., Vicente, C., 2008. Purification and properties of an unusual UDP-glucose dehydrogenase, NADPH dependent, from *Xanthomonas albilineans*. Microbiol. Res. 163, 362–371.

Blanch, M., Legaz, M.E., Vicente, C., 2012. Structure and biosynthesis of a xanthan-like polysaccharide produced by *Xanthomonas albilineans*. Funct. Plant Sci. Biotechnol. 6, 85–90.

Blanch, M., Piñon, D., Vicente, C., Legaz, M.E., 2007. Sugar cane glycoproteins are required to the production of an active UDP-glucose dehydrogenase by *Xanthomonas albilineans*. Ann. Microbiol. 57, 217–221.

Blanch, M., Rodriguez, C.W., Legaz, M.E., Vicente, C., 2006. Modifications of sucrose crystallization by xanthans produced by *Xanthomonas albilineans*, a sugar cane pathogen. Sugar Tech. 8, 255–259.

Blanco, Y., Arroyo, M., Legaz, M.E., Vicente, C., 2005a. Isolation from *Gluconacetobacter diazotrophicus* cell walls of specific receptors for sugar cane glycoproteins, which act as recognition factors. J. Chromatogr. A 1093, 204–211.

Blanco, Y., Blanch, M., Piñón, D., Legaz, M.E., Vicente, C., 2005b. Antagonism of *Gluconacetobacter diazotrophicus* (a sugar cane endosymbiont) against *Xanthomonas albilineans* (pathogen) studied in alginate-immobilized sugar cane stalk tissues. J. Biosci. Bioeng. 99, 366–371.

Blanco, Y., Sacristán, M., Legaz, M.E., Vicente, C., 2006. Isolation of specific receptors from *Xanthomonas albilineans* cell walls for lectins-like glycoproteins of sugar cane. J. Plant Interact. 1, 107–114.

Brand, A., Gow, N.A.R., 2012. Tropic orientation responses of pathogenic fungi. In: Pérez-Martin, J., Di Pietro, A. (Eds.), Morphogenesis and Pathogenicity in Fungi. Springer Verlag, Berlin, pp. 21–41.

Brauc, S., De Vooght, E., Claeys, M., Geuns, J.M.C., Höfte, M., Angenon, G., 2012. Overexpression of arginase in *Arabidopsis thaliana* influences defence responses against *Botrytis cinerea*. Plant Biol. 14, 39–45.

Cavalcante, V.A., Döbereiner, J., 1988. A new acid-tolerant nitrogen fixing bacterium associated with sugar cane. Plant Soil 108, 23–31.

Choi, C.K., Vicente-Manzanares, M., Zareno, J., Whitmore, L.A., Mogilner, A., Horwitz, A.R., 2008. Actin and α-actinin orchestrate the assembly and maturation of nascent adhesions in a myosin II motor-independent manner. Nat. Cell Biol. 10, 1039–1050.

Cociancich, S., Pesic, A., Petras, D., Uhlmann, S., Kretz, J., Schubert, V., Vieweg, L., Duplan, S., Marguerettaz, M., Noëll, J., Pieretti, I., Hügelland, M., Kemper, S., Mainz, A., Rott, P., Royer, M., Süssmuth, R.D., 2015. The gyrase inhibitor albicidin consists of p-aminobenzoic acids and cyanoalanine. Nat. Chem. Biol. 11, 195–197.

Coutinho, I.D., Baker, J.M., Ward, J.L., Beale, M.H., Creste, S., Cavalheiro, A.J., 2016. Metabolite profiling of sugarcane genotypes and identification of flavonoid glycosides and phenolic acids. J. Agric. Food Chem. 64, 4198–4206.

D'Agostino, L., Di Pietro, M., Di Luccia, A., 2005. Nuclear aggregates of polyamines are supramolecular structures that play a crucial role in genomic DNA protection and conformation. FEBS J. 272, 3777–3787.

Damaj, M.B., Kumpatla, S.P., Emani, C., Beremand, P.D., Reddy, A.S., Rathore, K.S., Buenrostro-Nava, M.T., Curtis, I.S., Thomas, T.L., Mirkov, T.E., 2010. Sugar cane dirigent and o-methyltransferase promoters confer stem-regulated gene expression in diverse monocots. Planta 231, 1439–1458.

de Armas, R., Mateos, J.L., Caffaro, S.V., Legaz, M.E., Vicente, C., 1992. Changes in sucrose crystal shape induced by cane juice fructans. Intern Sugar J. 94, 141–143.

de Armas, R., Santiago, R., Legaz, M.E., Vicente, C., 2007. Levels of phenolic compounds and enzyme activity can be used to screen for resistance of sugar cane to smut (*Ustilago scitaminea*). Aust. Plant Pathol. 36, 32–38.

Eggleston, G., Legendre, B., Tew, T., 2004. Indicators of freeze-damaged sugar cane varieties which can predict processing problems. Food Chem. 87, 119–133.

Eggert, C., Temp, U., Dean, J.F.D., Eriksson, K.E.L., 1996. A fungal metabolite mediates degradation of nonlignin structures and synthetic lignin. FEBS Lett. 391, 144–148.

Esh, A.M.H., Guirgis, A., El-kholi, M.M.A., El-Absawy, E.A., Nasr, M.I., Hassanien, E.H., 2014. The activity of pathogenesis related proteins in smut resistant and susceptible sugar cane (GT54-9) mutants induced by gamma radiation. Adv. Plants Agric. Res. 1, 1–12.

Fedatto, L.M., Silva-Stenico, M.E., Etchegaray, A., Pacheco, F.T.H., Rodrigues, J.L.M., Tsai, S.M., 2006. Detection and characterization of protease secreted by the plant pathogen *Xylella fastidiosa*. Microbiol. Res. 161, 263–272.

Fontaniella, B., Rodriguez, C.W., Piñon, D., Vicente, C., Legaz, M.E., 2002. Identification of xanthans isolated from sugar cane juices obtained from scalded plants infected by *Xanthomonas albilineans*. J. Chromatogr. B 770, 275–281.

Fontaniella, B., Millanes, A.M., Piñón, D., Rodríguez, C.W., Vicente, C., Legaz, M.E., 2003a. Effect of leaf scald on the content of sucrose and polysaccharides of two sugar cane cultivars. Food Sci. Biotechnol. 12, 346–350.

Fontaniella, B., Millanes, A.M., Vicente, C., Legaz, M.E., 2004. Concanavalin A binds to a mannose-containing ligand in the cell wall of some lichen phycobionts. Plant Physiol. Biochem. 42, 773–779.

Fontaniella, B., Vicente, C., Legaz, M.E., de Armas, R., Rodríguez, C.W., Martínez, M., Piñón, D., Acevedo, R., Solas, M.T., 2003b. Yellow leaf síndrome modifies the composition of sugar cane juices in polysaccharides, phenols and polyamines. Plant Physiol. Biochem. 41, 1027–1036.

Fontaniella, B., Vicente, C., de Armas, R., Legaz, M.E., 2007. Effect of leaf scald (*Xanthomonas albilineans*) on polyamine and phenolic acid metabolism of two sugar cane cultivars. Eur. J. Plant Pathol. 119, 401–409.

França, S.C., Roberto, P.G., Marins, M.A., Puga, R.D., Rodrigues, A., Pereira, J.O., 2001. Biosynthesis of secondary metabolites in sugar cane. Gen. Mol. Biol. 24, 243–250.

Franca-Koh, J., Kamimura, Y., Devreotes, P., 2006. Navigating signaling networks: chemotaxis in *Dictyostelium discoideum*. Curr. Opin. Genet. Dev. 16, 333–338.

Fuell, C., Elliott, K.A., Hanfrey, C.C., Franceschetti, M., Michael, A.J., 2010. Polyamine biosynthetic diversity in plants and algae. Plant Physiol. Biochem. 48, 513–520.

Fujimura, K., Takahashi, H., Munemura, I., Uchimiya, H., Sano, H., 2009. Polyamines as a common source of hydrogen peroxide in host- and nonhost hypersensitive response during pathogen infection. Plant Mol. Biol. 70, 103–112.

Galloway, W.R.J.D., James, T., Hodgkinson, J.T., Bowden, S., Welch, M., Spring, D.R., 2012. Applications of small molecule activators and inhibitors of quorum sensing in Gram-negative bacteria. Trends Microbiol. 20, 449–458.

Gang, D.R., Costa, M.A., Fujita, M., Dinkova-Kostova, A.T., Wang, H.B., Burlat, V., Martin, W., Sarkanen, S., Davin, L.B., Lewis, N.G., 1999. Regiochemical control of monolignol radical coupling: a new paradigm for lignin and lignan biosynthesis. Chem. Biol. 6, 143–151.

Gao, S., Yang, Y., Wang, C., Guo, J., Zhou, D., Wu, Q., Su, Y., Xu, L., Que, Y., 2016. Transgenic sugarcane with a cry1Ac gene exhibited better phenotypic traits and enhanced resistance against sugarcane borer. PLoS One 11 (4), e0153929.

Guan, Y., Ramalingam, S., Nagegowda, D., Taylor, P.W., Chye, M.L., 2008. *Brassica juncea* chitinase BjCHI1 inhibits growth of fungal phytopathogens and agglutinates Gram-negative bacteria. J. Exp. Bot. 59, 3475–3484.

Gupta, K., Dey, A., Gupta, B., 2013. Plant polyamines in abiotic stress responses. Acta Physiol. Plant 35, 2015–2036.

Hashimi, S.M., Wall, M.K., Smith, A.B., Maxwell, A., Birch, R.G., 2007. The phytotoxin albicidin is a novel inhibitor of DNA gyrase. Antimicrob. Agents Chemother. 51, 181–187.

Hause, B., Kogel, K.H., Parthier, B.C., Wasternack, C., 1997. In barley leaf cells, jasmonates do not act as a signal during compatible or incompatible interactions with the powdery mildew fungus, *Erysiphe graminis* f. sp. hordei. J. Plant Physiol. 150, 127–132.

Hayakawa, K., Tatsumi, H., Sokabe, M., 2012. Mechano-sensing by actin filaments and focal adhesion proteins. Commun. Integr. Biol. 5, 572–577.

Hogan, D.A., 2006. Talking to themselves: autoregulation and quorum sensing in fungi. Eukaryot. Cell 5, 613–619.

Huang, G., Zhang, L., Birch, R.G., 2001. A multifunctional polyketide peptide synthetase essential for albicidin biosynthesis in *Xanthomonas albilineans*. Microbiology 147, 631–642.

Jin-Long, G., Li-Ping, X., Jing-Ping, F., Ya-Chun, S., Hua-Ying, F., You-Xiong, Q., Jing-Sheng, X., 2012. A novel dirigent protein gene with highly stem-specific expression from sugar cane, response to drought, salt and oxidative stresses. Plant Cell Rep. 31, 1801–1812.

Juanes, A., Chapman, J.R., Aguilar-Cuenca, R., Delgado-Arevalo, C., Hodges, J., Whitmore, L.A., Jeffrey Shabanowitz, J., Hunt, D.F., Horwitz, A.R., Vicente-Manzanares, M., 2015. A regulatory motif in nonmuscle myosin II-B regulates its role in migratory front–back polarity. J. Cell Biol. 209, 23–32.

Karen, S.A., Shamsul, B., Paul, J.B., Barry, C., Meredith, M., 2013. Investigation of the genetic mechanisms of resistance to smut in sugarcane. Plant Anim. Genome 1, 12–16.

Keays, D.A., Tian, G., Poirier, K., Huang, G.J., Siebold, C., Cleak, J., Oliver, P.L., Fray, M., Harvey, R.J., Molnár, Z., Piñon, M.C., Dear, N., Valdar, W., Brown, S.D., Davies, K.E., Rawlins, J.N., Cowan, N.J., Nolan, P., Chelly, J., Flint, J., 2007. Mutations in α-tubulin cause abnormal neuronal migration in mice and lissencephaly in humans. Cell 128, 45–57.

Kirchhof, G., Eckert, B., Stoffels, M.J., Ivo Baldani, J.I., Reis, V.M., Anton Hartmann, A., 2001. *Herbaspirillum frisingense* sp. nov., a new nitrogen-fixing bacterial species that occurs in C4-fibre plants. Intern J. Syst. Evol. Microbiol. 51, 157–168.

Koeck, M., Hardham, A.R., Dodds, P.N., 2011. The role of effectors of biotrophic and hemibiotrophic fungi in infection. Cell Microbiol. 13, 1849–1857.

Kumar, H., Kawai, T., Akira, S., 2009. Toll-like receptors and innate immunity. Biochem. Biophys. Res. Commun. 388, 621–625.

Kusano, T., Yamaguchi, T.B., Takahashi, Y., 2007. Advances in polyamine research in 2007. J. Plant Res. 120, 345–350.

Lawton, K., Weymann, K., Friedrich, L., Vernooij, B., Uknes, S., Ryals, J., 1995. Systemic acquired resistance in *Arabidopsis* requires salicylic acid but not ethylene. Mol. Plant Microbe Interact. 8, 863–870.

Legaz, M.E., de Armas, R., Martínez, M., Medina, I., Vicente, C., 1991. Binding studies of Mn^{2+} to a glycosidase system from sugar cane juice. Plant Physiol. Biochem. 29, 601–605.
Legaz, M.E., Pedrosa, M.M., Martínez, M., Vicente, C., 1995. Soluble glycoproteins from sugar cane juice analyzed by SEHPLC and fluorescence emission. J. Chromatogr. 697, 329–335.
Legaz, M.E., de Armas, R., Piñón, D., Vicente, C., 1998a. Relationships between phenolics-conjugated polyamines and sensitivity of sugar cane to smut (*Ustilago scitaminea*). J. Exp. Bot. 49, 1723–1728.
Legaz, M.E., Pedrosa, M.M., de Armas, R., Rodringuez, C.W., de los Rios, V., Vicente, C., 1998b. Separation of soluble glycoproteins from sugar cane juice by capillary electrophoresis. Anal. Chim. Acta 372, 201–208.
Legaz, M.E., Fontaniella, B., de Armas, R., Vicente, C., 2001. Determination by high performance liquid chromatography of ornithine and lysine decarboxylases in sugar cane juices. Chromatographia 53, S260–S265.
Legaz, M.E., de Armas, R., Millanes, A.M., Rodriguez, C.W., Vicente, C., 2005. Heterofructans and heterofructan-containing glycoproteins from sugar cane: structure and function. Recent Res. Devel Biochem. 6, 31–51.
Legaz, M.E., Millanes, A.M., Fontaniella, B., Piñón, D., de Armas, R., Rodríguez, C.W., Solas, M.T., Vicente, C., 2006. Ultrastructural alterations of sugar cane leaves caused by common sugar cane pathogens. Belg. J. Bot. 139, 14–26.
Legaz, M.E., Blanch, M., Piñón, D., Santiago, R., Fontaniella, B., Blanco, Y., Solas, M.T., Vicente, C., 2011. Sugar cane glycoproteins may act as signals for the production of xanthan in the plant-associated bacterium *Xanthomonas albilineans*. Plant Signal Behav. 6, 1132–1139.
Lopes, S.A., Damann, K.E., Grelen, L.B., 1998. Comparison of methods for identification of sugar cane pathogen *Xanthomonas albilineans*. Sum. Phytopathol. 24, 114–119.
Magnani, G.S., Cruz, L.M., Weber, H., Bespalhok, J.C., Daros, E., Baura, V., Yates, M.G., Monteiro, R.A., Faoro, H., Pedrosa, F.O., Souza, E.M., 2013. Culture-independent analysis of endophytic bacterial communities associated with Brazilian sugar cane. Genet. Mol. Res. 12, 4549–4558.
Marina, M., Maiale, S.J., Rossi, F.R., Romero, M.F., Rivas, E.I., Gárriz, A., Ruiz, O.A., Pieckenstain, F.L., 2008. Apoplastic polyamine oxidation plays different roles in local responses of tobacco to infection by the necrotrophic fungus *Sclerotinia sclerotiorum* and the biotrophic Bacterium *Pseudomonas viridiflava*. Plant Physiol. 147, 2164–2178.
Marques, J.P.R.M., Appezzato-da-Glória, B., Piepenbring, M., Massola, N.S., Monteiro-Vitorello, C.B., Carneiro Vieira, M.L., 2017. Sugarcane smut: shedding light on the development of the whip-shaped sorus. Ann. Bot. 119, 815–827.
Martínez, M., Legaz, M.E., Paneque, M., Domech, R., de Armas, R., Medina, I., Rodríguez, C.W., Vicente, C., 1990a. Glycosidase activities and polysaccharide accumulation in sugar cane stalks during post-collection impairment. Plant Sci. 72, 193–198.
Martínez, M., Legaz, M.E., Paneque, M., de Armas, R., Pedrosa, M.M., Medina, I., Rodríguez, C.W., Vicente, C., 1990b. The origin of soluble fructans in sugar cane juice. Intern Sugar J. 92, 155–159.
Memon, A.R., Durakovik, C., 2014. Signal perception and transduction in plants. PEN 2, 15–29.
Mensi, I., Daugrois, J.H., Pieretti, I., Gargani, D., Fleites, L.A., Noell, J., Bonnot, F., Gabriel, D.W., Rott, P., 2016. Surface polysaccharides and quorum sensing are involved in the attachment and survival of *Xanthomonas albilineans* on sugarcane leaves. Mol. Plant Pathol. 17, 236–246.
Millanes, A.M., Fontaniella, B., Legaz, M.E., Vicente, C., 2005. Glycoproteins from sugar cane plants regulate cell polarity of *Ustilago scitaminea* teliospores. J. Plant Physiol. 162, 253–265.
Millanes, A.M., Legaz, M.E., Vicente, C., 2008. Sugar cane bind to surface, specific ligands and modify cytoskeleton arrangement of *Ustilago scitaminea* teliospores. J. Plant Interact. 3, 95–110.
Monteiro-Vitorello, C.B., Camargo, L.A., Van Sluys, M.A., Kitajima, J.P., Truffi, D., do Amaral, A.M., Harakava, R., de Oliveira, J.C.F., Wood, D., de Oliveira, M.C., Miyaki, C., Takita, M.A., da Silva, A.C.R., Furlan, L.R., Carraro, D.M., Camarotte, G., Almeida, N.F., Carrer, H., Coutinho, L.L., El-Dorry, H.A., Ferro, M.I.T., Gagliardi, P.R., Giglioti, E., Goldman, M.H.S., Goldman, G.H., Kimura, E.T., Ferro, E.S., Kuramae, E.E., Lemos, E.G.M., Lemos, M.V.F., Mauro, S.M.Z., Machado, M.A., Marino, C.L., Menck, C.F., Nunes, L.R., Oliveira, R.C., Pereira, G.G., Siqueira, W., de Souza, A.A., Tsai, S.M., Zanca, A.S., Simpson, A.J.G., Brumbley, S.M., Setúbal, J.C., 2004. The genome sequence of the gram-positive sugar cane pathogen *Leifsonia xyli* subsp. *xyli*. Mol. Plant Microbe Interact. 17, 827–836.
Moosawi-Jorf, S.A., Izadi, M.B., 2007. In vitro detection of yeast-like and mycelial colonies of *Ustilago scitaminea* in tissue-cultured plantlets of sugar cane using polymerase chain reaction. J. Appl. Sci. 7, 3768–3773.
Morkunas, I., Ratajczak, L., 2014. The role of sugar signaling in plant defence responses against fungal pathogens. Acta Physiol. Plant 36, 1607–1619.
Pacheco, C.M., Pestana-Calsa, M.C., Gozzo, F.C., Mansur-Custodio-Nogueira, R.J., Menossi, M., Calsa, T., 2013. Differentially delayed root proteome responses to salt stress in sugar cane varieties. J. Proteome Res. 12, 5681–5695.
Pieretti, I., Cociancich, S., Bolot, S., Carrère, S., Morisset, A., Rott, P., Royer, M., 2015a. Full genome sequence analysis of two isolates reveals a novel *Xanthomonas* species close to the sugar cane pathogen *Xanthomonas albilineans*. Genes 6, 714–733.
Pieretti, I., Pesic, A., Petras, D., Royer, M., Süssmuth, R.D., Cociancich, S., 2015b. What makes *Xanthomonas albilineans* unique amongst xanthomonads? Front. Plant Sci. 6, 289.
Pieretti, I., Royer, M., Barbe, V., Carrere, S., Koebnik, R., Couloux, A., Darrasse, A., Gouzy, J., Jacques, M.A., Lauber, E., Manceau, C., Mangenot, S., Poussier, S., Segurens, B., Szurek, B., Verdier, V., Arlat, M., Gabriel, D.W., Rott, P., Cociancich, S., 2012. Genomic insights into strategies used by *Xanthomonas albilineans* with its reduced artillery to spread within sugar cane xylem vessels. BMC Genom. 13, 658–680.
Piñón, D., de Armas, R., Vicente, C., Legaz, M.E., 1999. Role of polyamines in the infection of sugar cane buds by *Ustilago scitaminea* spores. Plant Physiol. Biochem. 37, 57–64.
Piñón, D., Casas, M., Blanch, M., Fontaniella, B., Blanco, Y., Vicente, C., Solas, M.T., Legaz, M.E., 2002. *Gluconacetobacter diazotrophicus*, a sugar cane endosymbiont, produces a bacteriocin against *Xanthomonas albilineans*, a sugar cane pathogen. Res. Microbiol. 153, 345–351.

Que, Y., Su, Y., Guo, J., Wu, Q., Xu, L., 2014a. A Global view of transcriptome dynamics during *Sporisorium scitamineum* challenge in sugar cane by RNA-seq. PLoS One 9 (8), e106476.

Que, Y., Xu, L., Wu, Q., Liu, Y., Ling, H., Liu, Y., Zhang, Y., Guo, J., Su, Y., Chen, J., Wang, S., Zhang, C., 2014b. Genome sequencing of *Sporisorium scitamineum* provides insights into the pathogenic mechanisms of sugarcane smut. BMC Genom. 15, 996–1015.

Raftopoulou, M., Hall, A., 2004. Cell migration: Rho GTPases lead the way. Dev. Biol. 265, 23–32.

Rajam, M.V., Galston, A.W., 1985. The effects of some polyamine biosynthetic inhibitors on growth and morphology of phytopathogenic fungi. Plant Cell Physiol. 26, 683–692.

Ralph, S., Park, J.Y., Bohlman, N.J., Mansfield, S.D., 2006. Dirigent proteins in conifer defence: gene discovery, phylogeny, and differential wound-and insect-induced expression of a family of DIR and DIR-like genes in spruce (*Picea* spp.). Plant Mol. Biol. 60, 21–40.

Ramesh Sundar, A., Leonard Barnabas, E., Malathi, P., Viswanathan, R., 2012. A mini-review on smut disease of sugar cane caused by *Sporisorium scitamineum.* In: Mworia, J.K. (Ed.), Botany. InTech Publishers, Croatia, pp. 107–128.

Renier, A., Vivien, E., Cociancich, S., Letourmy, P., Perrier, X., Rott, P.C., Royer, M., 2007. Substrate specificity-conferring regions of the nonribosomal peptide synthetase adenylation domains involved in albicidin pathotoxin biosynthesis are highly conserved within the species *Xanthomonas albilineans*. Appl. Environ. Microbiol. 73, 5523–5530.

Roberts, E.J., Clarke, M.A., Godshall, M.A., Parris, F.W., 1985. A glucan from sugar cane. Intern Sugar J. 87, 227–231.

Rodriguez, O.C., Schaefer, A.W., Mandato, C.A., Forscher, P., Bement, W.M., Waterman-Storer, C.M., 2003. Conserved microtubule–actin interactions in cell movement and morphogenesis. Nat. Cell Biol. 5, 599–609.

Rodriguez, C.W., Valdes, P., Martinez, M., 1985. Formación de polisacáridos solubles de alto peso molecular en distintas variedades de caña de azúcar. Cienc. Agric. 22, 63–68.

Rott, P., Fleites, L.A., Mensi I,Sheppard, L., Daugrois, J.H., Dow, J.M., Gabriel, D.W., 2013. The RpfCG two-component system negatively regulates the colonization of sugar cane stalks by *Xanthomonas albilineans*. Microbiol 159, 1149–1159.

Ruan, Y.L., 2014. Sucrose metabolism: gateway to diverse carbon use and sugar signaling. Annu. Rev. Plant Biol. 65, 33–67.

Sampietro, A.R., Isla, M.I., Quiroga, E.N., Vattuone, M.A., 2001. An N-acetylglucosamine oligomer binding agglutinin (lectin) from ripe *Cyphomendra betacea* Sendt.fruits. Plant Sci. 160, 659–667.

Sánchez-Elordi, E., Baluška, F., Echevarría, C., Vicente, C., Legaz, M.E., 2016a. Defence sugar cane glycoproteins disorganize microtubules and prevent nuclear polarization and germination of *Sporisorium scitamineum* teliospores. J. Plant Physiol. 200, 111–123.

Sánchez-Elordi, E., de los Ríos, L.M., Díaz, E.M., Ávila, A., Legaz, M.E., Vicente, C., 2016b. Defensive glycoproteins from sugar cane plants induced chemotaxis, cytoagglutination and death of smut teliospores. J. Plant Pathol. 98, 493–501.

Sánchez-Elordi, E., de los Ríos, L.M., Vicente, C., Legaz, M.E., 2015. Sugar cane arginase competes with the same fungal enzyme as a false quorum signal against smut teliospores. Phytochem. Lett. 14, 115–122.

Sánchez-Elordi, E., Vicente-Manzanares, M., Díaz, E., Legaz, M.E., Vicente, C., 2016c. Plant–pathogen interactions: sugar cane glycoproteins induce chemotaxis of smut teliospores by cyclic contraction and relaxation of the cytoskeleton. S Afr. J. Bot. 105, 66–78.

Santiago, R., de Armas, R., Legaz, M.E., Vicente, C., 2008a. Separation from smut (*Ustilago scitaminea*) of different elicitors which modify the pattern of phenolic accumulation in sugar cane leaves. J. Plant Pathol. 90, 87–96.

Santiago, R., Fontaniella, B., Millanes, A.M., Vicente, C., Legaz, M.E., 2008b. Biochemical responses of sugar cane plants to smut (*Ustilago scitaminea*). In: Legaz, M.E., Vicente, C. (Eds.), Biochemical Responses of Plants to Invaders. Research SingPost, Trivandrum, pp. 125–138.

Santiago, R., de Armas, R., Legaz, M.A., Vicente, C., 2009a. Changes in phenolic acids content, phenylalanine ammonia-lyase and peroxidase activities in sugar cane leaves induced by elicitors isolated from *Xanthomonas albilineans*. Austr Plant Pathol. 38, 357–365.

Santiago, R., Díaz, E.M., Quintana, J., Rodríguez, S., Fontaniella, B., Vicente, C., Legaz, M.E., 2009b. On the nature and distribution of sclereids in sugar cane leaves. Res. Rev. Biosci. 3, 251–258.

Santiago, R., de Armas, R., Fontaniella, B., Vicente, C., Legaz, M.E., 2009c. Changes in soluble and cell-bound hydroxycinnamic and hydroxybenzoic acids in sugar cane cultivars inoculated with *Sporisorium scitamineum sporidia*. Eur. J. Plant Pathol. 124, 439–450.

Santiago, R., Quintana, J., Rodriguez, S., Díaz, E.M., Legaz, M.E., Vicente, C., 2010a. An elicitor isolated from smut teliospores (*Sporisorium scitamineum*) enhances lignin deposition on the cell-wall of both sclerenchyma and xylem in sugar cane leaves. Pak J. Bot. 42, 2867–2881.

Santiago, R., de Armas, R., Blanch, M., Vicente, C., Legaz, M.E., 2010b. In vitro effects of caffeic acid upon growth of the fungi *Sporisorium scitamineum*. J. Plant Interact. 5, 233–240.

Santiago, R., Millanes, A.M., Legaz, M.E., Vicente, C., 2012a. Measurement of β-1,3 glucanase activity in permeabilized discs of leaves of healthy and scald-diseased plants. J. Life Sci. 6, 175–181.

Santiago, R., Alarcón, B., de Armas, R., Vicente, C., Legaz, M.E., 2012b. Changes in cinnamyl alcohol dehydrogenases from sugar cane cultivars inoculated with *Sporisorium scitamineum sporidia*. Physiol. Plant 145, 245–259.

Savarin, P., Barbet, A., Delga, S., Joshi, V., Hamon, L., Lefevre, J., Nakib, S., de Bandt, J.P., Moinard, C., Curmi, P.A., Pastré, D., 2010. A central role for polyamines in microtubule assembly in cells. Biochem. J. 430, 151–159.

Schröder, S., Schmidt, U., Thiem, J., Kowalczyk, J., Kunz, M., Vogel, M., 2004. Synthesis of oligosaccharides as potential novel food components and upscaled enzymatic reaction employing the β-galactosidase from bovine testes. Tetrahedron 60, 2601–2608.

Schaker, P.D.C., Palhares, A.C., Taniguti, L.M., Peters, L.P., Creste, S., Aitken, K.S., Van Sluys, M.A., Kitajima, J.P., Vieira, M.L.C., Monteiro-Vitorello, C.B., 2016. RNAseq transcriptional profiling following whip development in sugarcane smut disease. PLoS One E 11 (9), e0162237.

Sinha, D., Gupta, M.K., Patel, H.K., Ranjan, A., Sonti, R.V., 2013. Cell wall degrading enzyme induced rice innate immune responses are suppressed by the type 3 secretion system effectors XopN, XopQ, XopX and XopZ of *Xanthomonas oryzae* pv. *oryzae*. PLoS One 8, e75867.

Singh, P., Kumar, B., Rani, R., Jindal, M.M., 2014. Standardization of inoculation technique of sugarcane smut (*Ustilago scitaminea*) for evaluation of resistance. Afr. J. Microbiol. Res. 8, 3108–3111.

Sourjik, V., Wingreen, N.S., 2012. Responding to chemical gradients: bacterial chemotaxis. Curr. Opin. Cell Biol. 24, 262–268.

Souza, T.P., Dias, R.O., Silva-Filho, M.C., 2017. Defense-related proteins involved in sugarcane responses to biotic stress. Gen. Mol. Biol. 40, 360–372.

Soyer, J.L., El Ghalid, M., Glaser, N., Ollivier, B., Linglin, J., Grandaubert, J., Balesdent, M.H., Connolly, L.R., Freitag, M., Rouxel, T., Fuda, I., 2014. Epigenetic control of effector gene expression in the plant pathogenic fungus *Leptosphaeria maculans*. PLoS Gen. 10, e1004227.

Su, Y.C., Xu, L.P., Xue, B.T., Wu, Q.B., Guo, J.L., Wu, L.G., Que, Y.X., 2013. Molecular cloning and characterization of two pathogenesis-related β-1,3-glucanase genes ScGluA1 and ScGluD1 from sugarcane infected by *Sporisorium scitamineum*. Plant Cell Rep. 32, 1503–1519.

Su, Y., Guo, J., Ling, H., Chen, S., Wang, S., Xu, L., Allan, A.C., Que, Y., 2014. Isolation of a novel peroxisomal catalase gene from Sugarcane, which is responsive to biotic and abiotic stresses. PLoS One 9 (1), e84426.

Su, Y., Xu, L., Wang, S., Wang, Z., Yang, Y., Chen, Y., Que, Y., 2015. Identification, phylogeny, and transcript of chitinase family genes in sugarcane. Nat. Sci. Rep. 5, 10708,. http://dx.doi.org/10.1038/srep10708.

Su, Y., Wang, Z., Xu, L., Peng, Q., Liu, F., Zhu Li, Z., Que, Y., 2016a. Early selection for smut resistance in sugarcane using pathogen proliferation and changes in physiological and biochemical indices. Front. Plant Sci 7. http://dx.doi.org/10.3389/fpls.2016.01133. art 1133.

Su, Y., Xu, L., Wang, Z., Peng, Q., Yang, Y., Chen, Y., Que, Y., 2016b. Comparative proteomics reveals that central metabolism changes are associated with resistance against *Sporisorium scitamineum* in sugarcane. BMC Genom. 17, 800–820.

Su, Y., Zhang, Y., Huang, N., Liu, F., Su, W., Xu, L., Ahmad, W., Wu, Q., Guo, J., Que, Y., 2017. Small RNA sequencing reveals a role for sugarcane miRNAs and their targets in response to *Sporisorium scitamineum* infection. BMC Genom. 18, 325–343.

Taniguti, L.M., Schaker, P.D.C., Benevenuto, J., Peters, L.P., Carvalho, G., Palhares, A., Quecine, M.C., Nunes, F.R.S., Kmit, M.C.P., Wai, A., Hausner, G., Karen, S., Aitken, K.S., Berkman, P.J., Fraser, J.A., Moolhuijzen, P.M., Coutinho, L.L., Creste, S., Maria, L.C., Vieira, M.L.C., Kitajima, J.P., Monteiro-Vitorello, C.B., 2015. Complete genome sequence of *Sporisorium scitamineum* and biotrophic interaction transcriptome with sugar cane. PLoS One 10, e0129318.

Valadares, N.F., de Oliveira-Silva, R., Cavini, I.A., Marques Ide, A., Pereira, H.D., Soares-Costa, A., Henrique-Silva, F., Kalbitzer, H.R., Munte, C.E., Garratt, R.C., 2013. X-ray crystallography and NMR studies of domain-swapped canecystatin-1. FEBS J. 280, 1028–1038.

Valdés-Santiago, L., Cervantes-Chávez, J.A., Geraldine León-Ramírez, C., Ruiz-Herrera, J., 2012. Polyamine metabolism in fungi with emphasis on phytopathogenic species. J. Amino Acids. http://dx.doi.org/10.1155/2012/837932.

Van den Ende, W., 2013. Multifunctional fructans and raffinose family oligosaccharides. Front. Plant Sci. 4, 247. http://dx.doi.org/10.3389/fpls.2013.00247.

Vicente, C., Mateos, J.L., Pedrosa, M.M., Legaz, M.E., 1991. High-performance liquid chromatography determination of sugars and polyols in extracts of lichens and sugar cane juice. J. Chromatogr. 553, 271–283.

Vicente, C., Alarcón, B., Santiago, R., Armas, R., Sánchez-Elordi, E., Legaz, M.E., 2017. An elicitor isolated from *Sporisorium scitamineum* multiplies xylem bridges between vascular neighboring bundles in sugarcane leaves. Res. Rev. Biosci. 12, 114–124.

Vicente-Manzanares, M., Koach MA,Whitmore, L., Lamers, M.L., Horwitz, A.F., 2008. Segregation and activation of myosin IIB creates a rear in migrating cells. J. Cell Biol. 183, 543–554.

Vivien, E., Pitorre, D., Cociancich, S., Pieretti, I., Gabriel, D.W., Rott, P.C., Royer, M., 2007. Heterologous production of albicidin: a promising approach to over producing and characterizing this potent inhibitor of DNA gyrase. Antimicrob. Agents Chemother. 51, 1549–1552.

Vorotnikov, A.V., 2011. Chemotaxis: movement, direction, control. Biochemistry 76, 1528–1555.

Waller, J.M., 1970. Sugar cane smut (*Ustilago scitaminea*) in Kenya. II. Infection and resistance. Trans Br. Mycol. Soc. 54, 405–414.

Walters, D.R., 2003a. Polyamines and plant disease. Phytochemistry 64, 97–107.

Walters, D., 2003b. Resistance to plant pathogens: possible roles for free polyamines and polyamine catabolism. New Phytol. 159, 109–115.

Wang, Y., Chen, C.L., Iijima, M., 2011. Signaling mechanisms for chemotaxis. Dev. Growth Differ. 53, 495–502.

Wang, J., Nayak, S., Karen Koch, K., Ray Ming, R., 2013. Carbon partitioning in sugarcane (*Saccharum* species). Front. Plant Sci. 4. http://dx.doi.org/10.3389/fpls.2013.00201. Art 201.

Wendland, J., Philippsen, P., 2000. Determination of cell polarity in germinated spores and hyphal tips of the filamentous ascomycete *Ashbya gossypii* requires a rhoGAP homolog. J. Cell Sci. 113, 1611–1621.

Wojtasik, W., Kulma, A., Namysł, K., Preisner, M., Szopa, J., 2015. Polyamine metabolism in flax in response to treatment with pathogenic and non-pathogenic *Fusarium* strains. Front. Plant Sci. 6, 291.

Yamakawa, H., Kamada, H., Satho, M., Ohashi, Y., 1998. Spermine is a salicylate independent endogenous inducer for both tobacco acidic pathogenesis-related proteins and resistance against tobacco mosaic virus infection. Plant Physiol. 118, 1213–1222.

Yajima, A., 2014. Recent progress in the chemistry and chemical biology of microbial signaling molecules: quorum-sensing pheromones and microbial hormones. Tetrahedron Lett. 55, 2773–2780.

Yao, J., Allen, C., 2006. Chemotaxis is required for virulence and competitive fitness of the bacterial wilt pathogen *Ralstonia solanacearum*. J. Bacteriol. 188, 3697–3708.

You-Xiong, Q., Jian-Wei, L., Xian-Xian, S., Li-Ping, X., Ru-Kai, C., 2011. Differential gene expression in sugar cane in response to challenge by fungal pathogen *Ustilago scitaminea* revealed by cDNA-AFLP. J. Biomed. Biotechnol. 2011, 1–10.

Zhang, L., Birch, R., 1997. The gene for albicidin detoxification from *Pantoea dispersa* encodes an esterase and attenuates pathogenicity of *Xanthomonas albilineans*. Proc. Natl. Acad. Sci. U.S.A. 94, 9984–9989.

Zihni, C., Mills, C., Matter, K., Balda, M.S., 2016. Tight junctions: from simple barriers to multifunctional molecular gates. Nat. Rev. Mol. Cell Biol. 17, 564–580.

FURTHER READING

Kretz, J., Kerwat, D., Schubert, V., Grätz, S., Pesic, A., Semsary, S., Cociancich, S., Royer, M., Süssmuth, R.D., 2015. Total synthesis of albicidin: a lead structure from *Xanthomonas albilineans* for potent antibacterial gyrase inhibitors. Angew. Chem. Int. Ed. Engl. 54, 1969–1973.

Legaz, M.E., de Armas, R., Barriguete, E., Vicente, C., 2000. Binding of soluble glycoproteins from sugar cane juice to cells of *Acetobacter diazotrophicus*. Int. Microbiol. 3, 177–182.

Vicente, C., Fontaniella, B., Legaz, M.E., 2000. Fructan-like polysaccharides produced by sugar beet during deterioration. Int. Sugar J. 102, 250–256.

Voznesenskaya, E.V., Franceschi, V.R., Kiirats, O., Freitag, H., Edwards, G.E., 2001. Kranz anatomy is not essential for terrestrial C4 plant photosynthesis. Nature 414, 543–546.

Walters, D., Cowley, T., Mitchell, A., 2002. Methyl jasmonate alters polyamine metabolism and induces systemic protection against powdery mildew infection in barley seedlings. J. Exp. Bot. 53, 747–756.

Chapter 14

Coordinated Regulation of Photosynthesis in Plants Increases Yield and Resistance to Different Types of Environmental Stress

Hans-Werner Koyro[1], Bernhard Huchzermeyer[2,3]

[1]*Justus-Liebig-Uniniversität, Giessen, Germany;* [2]*Northeastern University, Boston, MA, United States;* [3]*Leibniz Universität Hannover, Hannover, Germany*

1. INTRODUCTION

Feeding the rapidly growing human population is one of the biggest challenges in our planet today and in the future. Especially in arid and semiarid regions, the increasing population density is collaterally catalyzed by the consequences of global climate change, which is caused by anthropogenic emissions of trace gases such as CO_2 (IPCC, 2014). There is a general understanding that feeding an ever-growing population requires tapping alternative resources in terms of both crop species and agriculturally used area. In this context, using areas not productive any more due to desertification and salinization is under discussion (Rozema and Flowers, 2008). Global scarcity of water resources and salinization of water and soil are worsening the situation (www.fao.org/agl/agl1/terrastat). Thus salinity—already affecting 7% of the global total land area (Millennium Ecosystem Assessment, 2005) and 20%–50% of the global irrigated farmland (Koyro et al., 2008, 2014b)—will pose an increasingly serious threat to agriculture and human nutrition in the future. Abiotic stress already is a major factor limiting crop yield, and effected losses already are exceeding by far those caused by lack of fertilizers and herbivore and pathogen activities (www.fao.org/nr/aboutnr/nrl/en/). The awkward situation is aggravated by a competition for resources of food and feed production and the so-called sustainable use of biomass to produce bioenergy and fine chemicals.

Metabolite patterns of individual plant species vary when grown in different environments. A complex regulatory network is adjusting metabolic activities, thus allowing an individual plant to adapt to the environment and adopt a new homeostasis. The range of this flexibility is determined in the genome of a plant species. Resource use efficiency, especially the use of energy and water, is decreased if plants have to grow under suboptimal conditions (Munns, 2005). This adverse effect of the environment is expressed in terms of a reduced growth rate, a reduced fertility, and at harsh conditions, an interruption of their life cycle and death. Plants differ a lot in stress resistance depending on their anatomy and metabolic capacity (Flowers et al., 1977). Production of metabolites that protect from osmotic and ionic stresses or eventually from toxic compounds is an important aspect to develop stress resistance (Koyro and Huchzermeyer, 2008). Some of the compounds overproduced under mild stress are of pharmaceutical interest or are high-value fine chemicals. Therefore results from plant stress research are directly feeding in biotechnology research as well. Relevant information for identification of traits for respective projects can be found in the database eHaloph (Santos et al., 2016). Areas rendered not productive for regular cropping, due to water scarcity or salinization, may be used to grow such high-value plant species.

In this review, we will summarize information on stress effects on photosynthesis. Moreover, we will try to explain how these effects are linked to subsequent metabolic activity and what information is available concerning regulation. In parallel, we will comment on experimental methods and traits that may be used to monitor plant performance. With respect to clarity we have to focus on a limited number of examples.

The viability of plants in saline habitats depends on their ability to cope with (1) water deficit due to low water potential of the soil, (2) restriction of CO_2 uptake, (3) ion toxicity, and (4) nutrient imbalance. The first two constraints are directly related to photosynthesis and are prerequisites for biomass and crop production, respectively. In this chapter we will focus on information related to these constraints, thereby discussing the different levels of photosynthetic

Plant Metabolites and Regulation Under Environmental Stress. http://dx.doi.org/10.1016/B978-0-12-812689-9.00014-5

reactions. First, there are primary reactions of photosynthesis, namely, absorption of light energy and its conversion (1) to redox energy, conserved in the coenzyme NADPH, and (2) energy of chemical bounds, conserved in the coenzyme adenosine triphosphate (ATP). On the second level, we find reactions of the Calvin cycle, nitrate and sulfate reduction, as well as sugar, lipid, and amino acid metabolism. Typical reactions on the third level are transmembrane and intertissue transport of metabolites. The fourth level of photosynthesis relates to physiological aspects of gas exchange and water relations.

2. INFLUENCE OF STRESS ON WATER RELATIONS

In the face of global climate change that is increasing desertification and salinization, one has also to consider the influence of low water potential on photosynthesis, which we will do in this chapter. In C_3 plants, water deficiency leads to a reduction of stomatal conductance, resulting in an increase of photorespiration and oxidative stress (as these effects are much less pronounced in C_4 plants, it will be differentiate between these two plant types).

According to Munns, plants show a two-phase growth response to salinity (Munns, 2002). During the first phase of growth, reduction water or osmotic stress prevails and plant responses are presumably regulated by hormonal signals coming from the roots.

Aside from arid regions, terrestrial plants growing in saline habitats often are surrounded by low water potentials in the soil solution and atmosphere. For water to flow through the soil–plant–atmosphere continuum, a gradient of decreasing water potentials (Ψ) must be established. The Ψ of pure water is defined as 0 MPa; increasing the salinity or concentrations of other solutes will decrease Ψ. Thus any sharp rise in salinity could effectively hold water osmotically away from plants that lack any physiological or morphological modifications. To limit restrictions on water uptake, plants must generate increasingly lower Ψ to allow continued water flux into belowground structures (Koyro, 2003).

It is also important to prevent water loss by transpiration from being higher than the influx rate. This is only possible if the water potential in the plant is lower than that in the soil. However, experiments demonstrated that the leaf water potential of halophytes does not exclusively correlate as a single factor with salinity resistance. Plant species although having different levels of salt resistance such as *Aster tripolium*, *Beta vulgaris* ssp. maritima, *Chenopodium quinoa*, *Atriplex nummularia*, *Spartina townsendii*, and *Sesuvium portulacastrum*, exhibit a sufficient adjustment mechanism even at high salinities. Indeed, resistance in the form of osmotic adjustment plays an important role in halophytes residing in saline environments (Flowers and Colmer, 2008). However, osmotic potentials of these species (and many others) constitutively are sufficiently low to explain full turgescence of leaves up to sea water salinity level (results not shown). Except for differences in concentration and type of osmotica used in plant tissues (ions are more prevalent in halophytes), physiological plant responses to either hyperosmotic salinity or drought were remarkably similar. For plants exposed to periods of limited water availability, physiological adjustments often involve avoidance and tolerance, with most plants using some combination of the two (Romanello et al., 2008).

Assuming there is no interruption of the water supply, water can flow passively from the root to the shoot, and there seems to be no reason for growth reduction by water deficit for any of the studied species. However, by regulating the extent of apoplastic barriers and their chemical composition (long-distance response coordination), plants can effectively regulate the uptake or loss of water and solutes (by structures such as barriers in the hypo- or exodermis). This appears to be an additional or compensatory strategy of plants to acquire water and solutes. This way at the extremes of growth under saline or dry conditions, the exodermis becomes an absolute barrier for water and ions in the strict sense (Robbins et al., 2014).

Thus the rate of water supply to the shoot can be restricted due to the coupling between the flows of water and solutes (Na and Cl) even if the leaf water potential is low. Therefore, the balance between water flow (sum of water accumulation and transpiration) and (1) the decrease in the amounts of nutrients taken up or (2) unfavorable nutrient ratios (e.g., Na^+/K^+), is a basic cause for impaired leaf elongation and plant growth (Cramer, 2003).

3. IMPACT ON GAS EXCHANGE

In any case, plant water loss has to be minimized at low soil water potential. Biomass production depends mainly on the ability to keep high net photosynthesis and low transpiration simultaneously. In this field of tension, biomass production has always to be seen in connection with CO_2/H_2O gas exchange, which can be estimated by the water use efficiency (WUE) of photosynthesis. A critical point for plant physiology is reached when CO_2 fixation (apparent photosynthesis) falls below CO_2 production (compensation point).

Several halophytic plants such as *A. tripolium*, *B. vulgaris* ssp. maritima, *C. quinoa*, or *S. townsendii* reveal a combination of low (but positive) net photosynthesis, minimum transpiration, high stomatal resistance, and minimum internal CO_2

concentration at their threshold salinity resistance (Koyro and Huchzermeyer, 2004). However, there is a big bandwidth among halophytes, especially for succulent species such as *S. portulacastrum* or *Avicennia marina*, which have alternatives if the water balance (water uptake minus water loss) is still positive and not the limiting factor for photosynthesis. In case of *S. portulacastrum*, net photosynthesis and WUE increase but stomatal resistance decreases. These results show that it is quite important to describe the regulation of gas exchange at high salinity in correlation with other parameters (such as water relations). Water deficit is one major constraint at high salinity and can lead to restriction of CO_2 uptake and, as a final consequence, to the development of reactive oxygen species (ROS) (see later). Monitoring the balance between water loss and CO_2 uptake helps to find the weakest link in the mechanism of adjustment (of photosynthesis) to high salinity (Badawi et al., 2004).

4. IONIC AND OSMOTIC LOAD

In principle, resistance to ionic stress can be achieved by exclusion or inclusion of the harmful substance. Several physiological ion injury–avoiding (and the symplast-protecting) mechanisms described in the literature are known as the so-called major plant responses (Munns, 2002; Koyro, 2003).

It seems to be essential that resistance includes, at different parts of the root system and in younger and older leaf tissues, changes in morphology, growth, water relations, gas exchange, and composition of minerals (and compatible solutes). The measurement of such data at plant, organ, or tissue level will reveal general trends. However, as such data represent a mean response of several cell types, much information on adjustment at single cell level is lost. There will be no sufficient information about the compartmentation inside a cell, along a diffusion zone in a root apoplast, or about ultrastructural changes such as apoplastic barriers. To uncover such individual resistance mechanisms against ion toxicity, special physiological research at the single-cell level is needed, such as analysis of gene expression and its genetic basis (Ahn et al., 2017; Jung et al., 2017).

The general scientific data give an impression of various adaptation mechanisms to high NaCl salinity. Besides water stress and ion-specific toxic physiological disorders on tissue level, intracellular ionic imbalances (K^+, Ca^{2+}, and Mg^{2+}) can be caused by ion toxicity. The capacity of plants to maintain homeostasis and low concentrations of harmful substances in the cytoplasm appears to be an important determinant of resistance (Rahman et al., 2017). A possibility to find such limiting factors is the study of the relations inside single cells, such as (1) the compartmentation between cytoplasm and vacuole and (2) the distribution of elements in different cell types or along a diffusion zone in a root apoplast, and the related ultrastructural changes.

In case of hyperosmotic salinity, *B. vulgaris* ssp. maritima and *S. townsendii* keep Na and Cl concentrations low in their young growing tissues (such as juvenile leaves) and in their storage organs (such as tap root or rhizome). However, *B. vulgaris* ssp. maritima is a typical Cl includer and *S. townsendii* is a typical Cl excluder (although it is an excreter as well) with high Na accumulation in mature leaves. Both species seem to react similar to salinity with respect to changes of leaf water potential, gas exchange, and nutrients (Koyro and Huchzermeyer, 2004). However, this number of partially congruent or complementary results does not allow concluding analogical intracellular relations. The comparison of their intracellular ionic balance will be used to demonstrate the necessity of special physiological investigations (see later).

In contrast to water stress effects, occurring in the meristematic region of young leaves, salt toxicity predominantly occurs in adventitious roots and mature leaves. Furthermore, most of the Na and Cl are stored preferentially in the shoot of halophytes such as *B. vulgaris* ssp maritima and *S. townsendii*. This results in a significantly more pronounced growth reduction of the aboveground parts than the inhibition found in the roots (Koyro and Huchzermeyer, 2004). These changes can be interpreted as signs of a critical load. Therefore to distinguish between the individual mechanisms of salinity resistance, further investigations of the intracellular ionic balance were performed, initially at the epidermal leaf cells (the end of the transpiration stream) of both these species. The single-cell data of the vacuolar and cytoplastic composition of the cells of the upper leaf epidermis were measured for controls. High-salinity treatments, up to seawater salinity, have been performed first by Koyro and Stelzer (1988). The intracellular composition of leaf epidermal cytoplasm and vacuoles of control *B. vulgaris* ssp. maritima and *S. townsendii* show some more congruities of both species. The epidermal vacuoles of controls of both species contain most of the elements (with the exception of P) at higher concentrations as the cytoplasm, indicating the overall picture of a vacuolar buffer. The leaf vacuoles in their entirety can be described as voluminous potassium pools with high storage capacity for sodium and chloride. This pool is needed in case of high NaCl salinity for maintenance of K homeostasis in the cytoplasm. The dominant elements in the cytoplasm were P and K. In the epidermal cytoplasm of control plants K concentrations were at an ideal range for enzymatic reactions (Koyro and Stelzer, 1988).

It is obvious that seawater salinity leads to a decrease of P, S, Mg, and K in the epidermal vacuoles of both species. In *Spartina* only the remaining K, S, and Mg millimolar concentrations were two-digit values, and especially the K value was much higher as in *Beta*. The vacuolar buffer of K seemed to be exhausted.

NaCl salinity led to a significant decrease of the K and P concentrations especially in the cytoplasm of *Beta* and resulted in breakdown of homeostasis. This result indicates a deficiency for both elements in the cytoplasm. Additionally, at high NaCl salinity the concentrations of sodium and chloride were below 5 mol $\times$ m^{-3} in the cytoplasm of the epidermal cells, and the gradients of both ions between cytoplasm and vacuole were higher in experiments with *Beta* when compared with the results achieved with *Spartina*. In summary, these results support the hypothesis that the sea beet does not sustain ion toxicity but ion deficiency! It is hypothesized that such low K^+ levels as found in the cytoplasm of *Beta* will lead to a reduction of protein synthesis, which is of utmost importance in the process of leaf expansion. Both ion toxicity and ion deficiency can have a direct effect on photosynthesis.

The salt-induced reductions of the cytoplasmic K and P concentrations were much less pronounced in *Spartina* as in *Beta*. The results of *Spartina* point at a working system to keep ionic homeostasis. However, there was one important exception: the sodium concentration increased significantly in the epidermal cytoplasm. Sodium might substitute for potassium in its cytoplasmic functions, or it could be the first sign of intoxication.

These results and interpretations are in agreement with the hypothesis that plant growth is affected by ion imbalance and toxicity. This probably will lead to the growth differences between the salt-resistant and salt-sensitive species observed in long-term studies. However, *Beta* and *Spartina* are two excellent examples to demonstrate how important it can be to validate intracellular ionic imbalances (K^+, Ca^{2+}, and Mg^{2+}) at high salt concentrations, to uncover individual salt resistance mechanisms, and to understand the threshold levels of individual species.

5. STRESS EFFECTS ON STRUCTURE AND FUNCTION OF THYLAKOID MEMBRANES

Photosynthesis is a key phenomenon controlling plant growth and development and thus crop yield. Coordinated regulation of photosynthesis in plants can increase yield and resistance to environmental stress. Primary reactions of photosynthesis involve a sequence of reactions converting light energy into a usable chemical form of energy. Subsequently this energy is used for assimilation of carbon, nitrogen, and minerals. This is a key process occurring in all green plants. However, stressful environments, such as drought and soil salinity, considerably hamper this process. We had to learn that photosynthesis has to be understood as a strictly coupled sequence of reactions. Stress-induced inhibition of one of these reaction steps will result in the inhibition of the whole reaction chain. The analysis of stress effects is hampered by the fact that inhibition of some of the individual reaction steps may result in stimulation of side reactions leading to the production of toxic compounds. Moreover, it has been demonstrated that intermediates and products of photosynthesis are acting as messengers controlling gene activity (Rolland et al., 2006). Thus stress-induced changes in cellular metabolite patterns may result in a feedback reaction controlling changes of metabolic activity. Such effects may explain why we can observe sudden bursts of a signal, when stress is exceeding a certain (plant-specific) threshold level (Koyro and Huchzermeyer, 2008).

From this synopsis, it becomes obvious why stress responses have been characterized to be multifactorial traits (Munns, 2002). In the following discussion we will try to analyze these observations while discussing step by step the reaction sequence of photosynthesis.

5.1 Harvesting Light Energy

The process of photosynthesis starts with absorption of the energy of light quanta and transferring chlorophyll in an activated state. Chlorophylls are found in protein complexes of chloroplast thylakoid membranes. We can discriminate light-harvesting complexes from reaction centers of photosystem (PS) I and PSII (Anderson, 2012). Activation energy is transferred from light-harvesting complexes containing most of the chlorophyll *b* to the core of the reaction centers of the two photosystems that are formed by a special pair of chlorophyll *a*. Optimal yield of energy transfer is achieved when the pigments are placed in optimal order allowing energy transfer within overlapping orbitals of their pi electron systems. Energy use efficiency will be reduced if distances among pigments are enhanced or if activation energy is absorbed and subsequently released as heat by other pigments such as xanthophylls (Bennett et al., 2013) (see later).

The structure of thylakoid membranes is quite flexible, and the interaction of intrinsic protein complexes is controlled by modifications such as protein phosphorylation and redox state of cystein residues. [Moreover, interaction of protein moieties located at the surface of thylakoid vesicles controls the structures that have been described as grana stacks (Farooq et al., 2016)]. Light-harvesting complexes and PSII are preferentially involved in stack formation, whereas PSI and the ATP-synthesizing complex are mostly excluded from stacked regions (Huchzermeyer et al., 1986; Farooq et al., 2016).

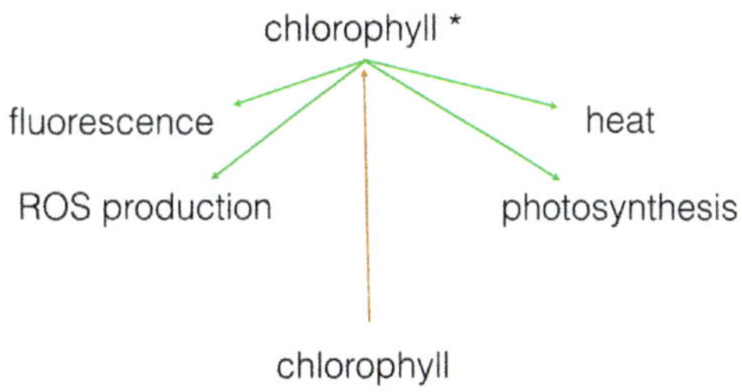

FIGURE 14.1 **Competition for absorbed light energy.** Several pathways are competing for energy from light-activated chlorophyll. Inhibition of one of the pathways will favor energy flow through the remaining ones. Under physiological conditions, probability of heat release is constant. Therefore monitoring of chlorophyll fluorescence can provide indirect information on energy flow through the photosynthetic pathway.

This way membrane stacking contributes to a heterogeneous pattern of protein complexes within thylakoid membranes. As pigments as well as redox systems of the photosynthetic electron transport chain are bound to these complexes, membrane stacking affects not only the efficiency of light absorption but also the subsequent electron transport (Anderson, 2012). At this stage salt stress comes into play, because readiness of protein complexes to interact as well as stability of membrane stacks is controlled by the availability of cations.

In its light activated state chlorophyll is highly reactive. While an overall potential difference between the donor and acceptor side of ~1V was found, a half-wave reduction potential of −1.1V in DMSO (dimethyl sulfoxide) versus the saturated calomel electrode can be measured (Caffarri et al., 2014; Shanmugam et al., 2015). Thus, upon illumination, it becomes an electron donor capable of reducing O_2 and form ROS under unfavorable conditions. Inside both photosystems activation of the special pair of chlorophyll *a* molecules initiates the release of an electron and its cross-membrane transfer within the respective protein complexes. Reactions occur in the time frame of picoseconds to milliseconds, and the acceptor is a quinone (plastoquinone and phylloquinone, respectively) that is bound to the core of the respective photosystem (PSII and PSI, respectively) (Anderson, 2012). By this reaction sequence absorbed light energy is converted to form energy of an electric membrane potential (positive on the inner surface of the thylakoid vesicles, because the special pair of chlorophyll *a* is located there, whereas quinones are bound near the outer surface).

The efficiency of this initial energy conversion is reduced by the fact that chlorophyll is unstable in its activated state. A high proportion of activation energy is lost by heat release when activated chlorophyll returns to its ground state. Due to its negative redox potential activated chlorophyll can transfer an electron to an oxygen molecule to form an oxygen radical that may be converted to other ROS (Miyake, 2010). As oxygen is produced by the water-splitting system located adjacent to PSII, ROS formation is a realistic risk and chloroplasts are equipped with ROS detoxification systems as described later. Another option is release of fluorescent light quanta. The probability of fluorescence is quite low (up to 5%), whereas heat release is the most probable reaction. As shown in Fig. 14.1, all mentioned reactions are competing for energy from light-activated chlorophyll. Heat release at physiological temperatures occurs at an almost constant rate, whereas the probability of ROS formation varies with the partial pressure of oxygen and probability of photosynthetic electron transport significantly depends on the availability of electron acceptors (Koyro et al., 2011; Murchie and Lawson, 2013). We will discuss this aspect later. This scheme of competing energy flows explains why chlorophyll fluorescence signals provide information on energy use efficiency of the photosynthetic apparatus. As discussed later, analysis of chlorophyll fluorescence is a reliable noninvasive method widely used to monitor plant performance (Murchie and Lawson, 2013).

5.2 Conversion of Harvested Energy to Drive Metabolism

Subsequent to the transfer of an electron to the photosynthetic redox chain, the special pair of PSII will receive an electron from the water-splitting system. After four such events an oxygen molecule is released, whereas the four electron transfer reactions are paralleled by the release of four protons into the aqueous inner phase of the thylakoid vesicles. The spatial arrangement of the photosynthetic electron transport system is depicted in Fig. 14.2- A pool of plastoquinone molecules functions as acceptor of electrons released from PSII (Allen, 2003). Plastoquinone will bind to PSII near the outer surface of the thylakoid vesicles (Trebst, 1974). Acceptance of two electrons paralleled by binding of two protons from the chloroplast matrix space turns plastoquinone to plastohydroquinone (plastoquinol). This molecule is released from the PSII-binding pocket. It will cross the thylakoid membrane and find a binding site at the cytochrome b_6/f complex near the inner surface of the thylakoid vesicles. The complex will accept electrons, but protons will be released into the inner aqueous phase of the thylakoid vesicles. In open chain photosynthetic electron transport electrons will be transferred further via

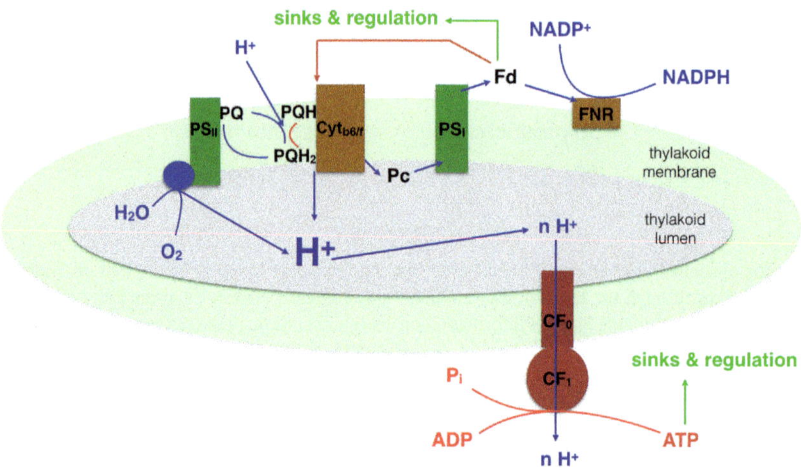

FIGURE 14.2 Energy conversion. *CF_1*, catalytic moiety of the chloroplast coupling factor; *CFo*, trans membrane proton channel of the chloroplast coupling factor; *Fd*, ferredoxin; *FNR*, *ferredoxin*-NADP+ reductase (EC 1.18.1.2); *Pc*, plastocyanin; *PQ*, plastoquinone; *PQH*, plastosemiquinone; *PQH_2*, plastoquinol; *PS_I*, photosystem I; *PS_{II}*, photosystem II. Absorbed light energy is converted to the energy of a chemiosmotic proton gradient and redox energy represented by NADPH. Adenosine triphosphate (ATP) is formed at the expense of protons from the pH gradient. In this figure alternative oxidases and reductases are not shown. With respect to various ferredoxin activities, only cyclic, photosystem (PS) I-driven, electron transport is indicated. There is no strict stoichiometry between formed NADPH and ATP synthesis.

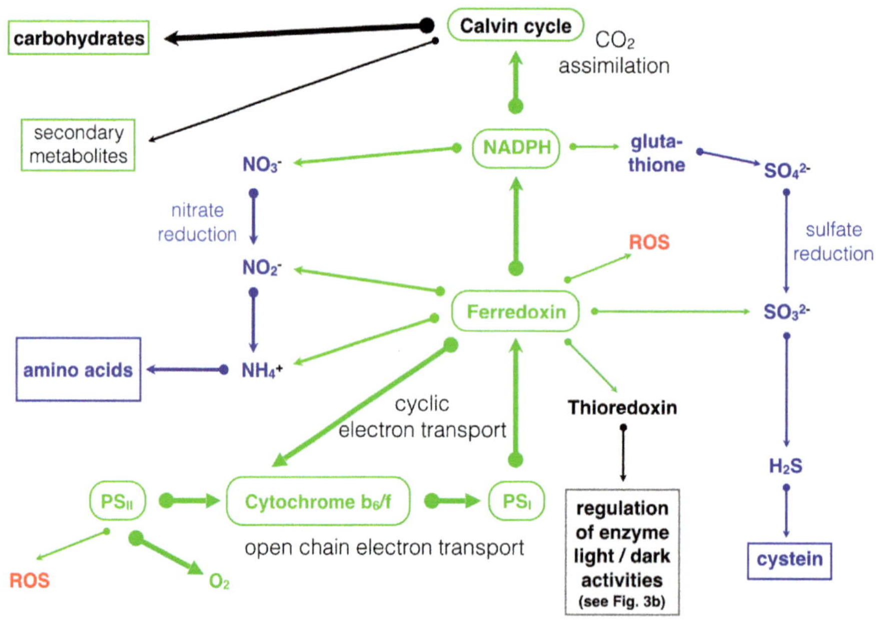

FIGURE 14.3A **Reactions depending on NADPH or ferredoxin$_{red}$.** Redox energy is not only used for CO_2 fixation. Assimilation of nitrate and sulfate are important sinks as well. Moreover, the chloroplast matrix contains important pathways leading to secondary metabolites. Although only a minor fraction of electrons are consumed by thioredoxins, this system regulating day/night activities of important enzymes, such as RuBisCO and the chloroplast adenosine triphosphatase, have to be mentioned as well. (see Fig. 14.3B for more details.). *ROS*, reactive oxygen species.

plastocyanin to PSI. Three electrons are needed to allow the special pair of chlorophyll *a* molecules to return to the ground state. Electrons released from PSI will be accepted by ferredoxin. This is a small soluble iron–sulfur protein that can serve several electron-consuming pathways (see Fig. 14.3A for an overview), such as nitrite reduction, cyclic photosynthetic electron transport, and open chain electron transport (Hanke and Mulo, 2013). In open chain electron transport, ferredoxin will transfer electrons to NADP+ reductase to form NADPH (Hill and Bendall, 1960).

Water-splitting and plastoquinone-mediated steps of electron transport result in the accumulation of protons inside the aqueous space of the thylakoid vesicles. Thus a pH gradient of 3–3.5 units is formed; pH values below 5 can be measured inside the vesicles, whereas in light the chloroplast matrix typically has a pH of 8.3 (Buchanan et al., 2015). As explained in great details by Mitchell and Jagendorf the absorbed light energy is used to form this gradient

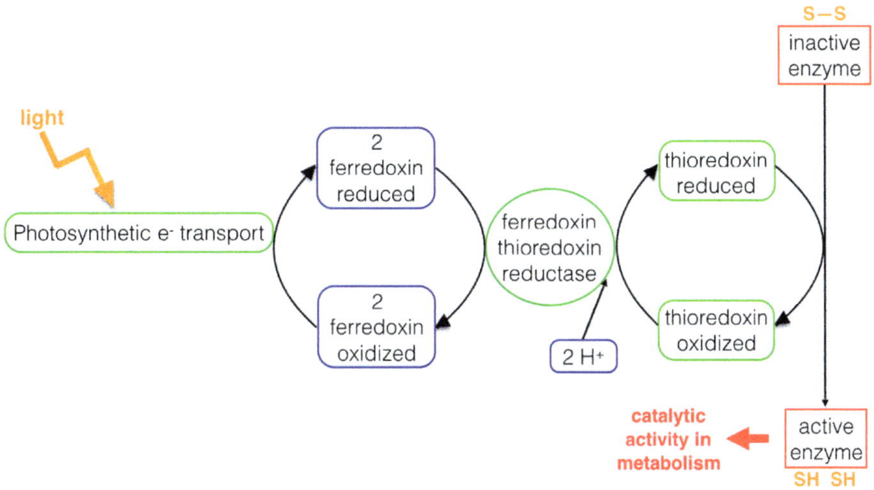

FIGURE 14.3B **Redox regulation of catalytic activity of enzymes.** Thioredoxin is reduced by ferredoxin. In subsequent reactions, reduced thioredoxin will reduce S–S-bounds of "light-regulated" target enzymes.

(Jagendorf, 2002; Mitchell, 2011). This implies that the gradient will reach its maximal value, when the energy necessary to release a proton into the inner phase of the thylakoid vesicles will equal the energy of a red light quantum, absorbed by chlorophyll to initiate this proton release.

Meanwhile we know that the proton gradient is used by the chloroplast ATPase complex to form ATP from adenosine diphosphate (ADP) plus phosphate (Mitchell, 2011). Protons will be allowed to pass the ATPase complex and transfer the energy of the pH gradient to drive catalysis. [The ATPase complex consists of a proton channel (CF_0) located inside the thylakoid membrane and an attached catalytic unit (CF_1) located on the outer surface of the thylakoids.] The amount of this ATPase complexes corresponds to a major share of the protein mass of the thylakoid membrane, as can be calculated from protein/chlorophyll ratios (Strotmann et al., 1973). Such ratios have been analyzed by several researchers, as the protein content of this membrane was observed to vary under stress and with the developmental stage of the plastids. It was found that interactions of membrane proteins including the ATPase complex are affecting the membrane structure, spatial organization of proteins, and energy use efficiency of chloroplasts (Anderson, 2012). Catalytic activity of the ATPase is strictly regulated to prevent the back reaction (ATP hydrolysis) during the night, when energy for ATP synthesis (the pH gradient) is missing. There are several levels of regulation allowing tuning of apparent activity of the enzyme population of a thylakoid vesicle. Among the parameters regulating catalytic activity are: (1) a threshold size of the pH gradient. The enzyme becomes activated when the energy quanta of passing protons are high enough to drive ATP synthesis. (2) The redox state of the complex (Huchzermeyer, 1988; Richter et al., 2000; Hisabori et al., 2013). Cysteine disulfide bridges have to be reduced at the expense of reduced ferredoxin to allow maximal turnover rate of the enzymes. Reduced ferredoxin is not available at night, and oxidation will transfer the enzyme to a resting state (Hisabori et al., 2013; Buchanan et al., 2015). (3) Threshold concentration of magnesium cations. Cations are needed as free nucleotides are not a substrate for the enzyme. The Mg–ADP complex was found to be the ideal substrate (He et al., 2000). Tuning the catalytic activity of enzymes via the ferredoxin/thioredoxin system is a common feature in chloroplasts (see Fig. 14.3B). It allows adapting the metabolic activity to the energy supply and the redox status to the availability of light (day/night situation). We will discuss the activation of the catalytic activity of RuBisCO as a second example.

As shown by Boyer and Walker (Walker, 1994; Boyer, 2000), the ATPase is special as it contains only a single proton channel that is rotating and serving three catalytic sites. During ATP synthesis the three sites are sequentially changing their conformational state and thus are stabilizing the intermediate states of ATP synthesis. It was found that the release of spontaneously formed tightly bound ATP is the energy-consuming step, while tight binding of ADP to the catalytic site will render the whole enzyme complex catalytically inactive. Based on this finding it could be shown that a fraction of the ATPase population is inactive in catalysis and the percentage of active enzymes depends on (light) energy supply (Huchzermeyer, 1988). Moreover, control of catalytic activity of the ATPase population is paralleled by rearrangements within the protein patterns of the thylakoid membrane. Clusters formed by photosystems and ATPase proteins vary in abundance as indicated by protonation–deprotonation reactions of amino acid residues exposed to the water phase (Theg et al., 1988).

6. STRESS EFFECTS ON CO_2 ASSIMILATION AND EXPORT OF CARBOHYDRATES

Within the chloroplast matrix the concentration of ADP and $NADP^+$ is about 1 mM. As there is no efficient transmembrane transport, especially of $NADP^+$, conversion of these coenzymes to ATP and NADPH requires recycling (consumption of fixed energy) at a rate resembling light-driven reactions described earlier (Koyro et al., 2011). It is generally understood that assimilation of CO_2 is the main energy sink in the chloroplasts, although other reactions consuming significant portions of redox energy, such as nitrate reduction, have to be mentioned as well (see Fig. 14.7 and the corresponding discussion). This becomes obvious when the photosynthetic activity of leaves (1) differing in developmental stage or (2) harvested from different plant species, respectively, is analyzed. Significant differences in photosynthetic activity per unit of leaf area can be observed. This may be explained by the differences in leaf anatomy when leaves of different plant species are compared. However, differences in photosynthetic activity have to be attributed to variations in catalytic activity of enzymes, when using leaves from identical plant species. Accordingly, it could be demonstrated that such differences can be found subsequent to shading periods, for instance (Sassenrathcole and Pearcy, 1994). Moreover, these authors have nicely demonstrated that the observed differences will vary with both the intensity and duration of shading stress, respectively. Induction of alternative sinks, such as nitrate assimilation, has a positive effect on the photosynthetic activity. This latter observation indicates that photosynthetic electron transport rate is limited by the turnover rate of sinks. Plants have the ability to modify their growth habit to a given light environment, and this is demonstrated by the changes in the anatomy and morphology observed in leaves grown in full sunlight and shade, respectively. This developmental acclimation occurs in a timescale of days and weeks (Sassenrathcole and Pearcy, 1994).

In all plant species the focal reaction of CO_2 assimilation, i.e., synthesis of carbohydrates, is RuBisCO-catalyzed binding of CO_2 to the acceptor molecule ribulose-1,5-bisphosphate (Ru-1,5-P). RuBisCO is the most abundant protein in leaves. The catalytic activity of this enzyme is strictly controlled; RuBisCO is activated by a sequence of reactions as depicted in Fig. 14.4. As the name indicates, the enzyme is capable of catalyzing both a carboxylation and an oxygenation reaction: in place of CO_2 it can catalyze the binding of O_2 to Ru-1,5-P as well (Buchanan et al., 2015). Instead of two molecules of 3-phosphoglycerate (3-PGA) formed by CO_2 binding one molecule of 3-PGA and one molecule of 2-phosphoglycolate (2-PG) are formed when using O_2 as a substrate. Thus CO_2 and O_2 compete at the catalytic site of RuBisCO. Based on substrate specificity of the enzyme CO_2 is the preferred substrate under physiological conditions. However, when gas exchange is limited under stress, CO_2/O_2 partial pressure ratio may increase, resulting in an increased proportion of 2-PG synthesis. Under such conditions there is a risk that the chloroplast 2-PG content may reach toxic concentrations characterized by the formation of 2-PG manganese complexes, resulting in inhibition of the water-splitting system. As other divalent cations competing with manganese for complex formation, the toxicity level of 2-PG varies with cation nutrition of plants under investigation.

Inside the chloroplast stroma no 2-PG turnover worth mentioning has been found, but detoxification occurs in a reaction sequence named photorespiration (Fig. 14.5) involving reactions taking place inside peroxisomes and mitochondria.

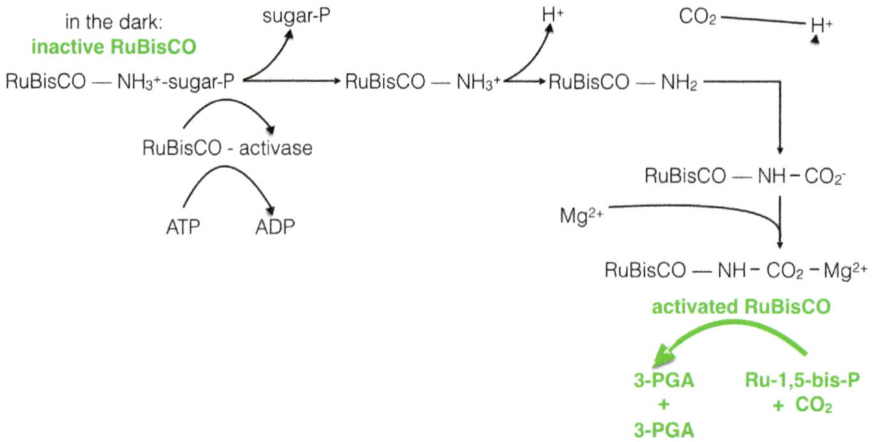

FIGURE 14.4 **Regulating catalytic activity of the enzyme RuBisCO.** Catalytic activity of RuBisCO is strictly regulated, and only a fraction of the enzymes is active at a time. Activation of the enzyme has to occur in an ordered sequence of reactions. Only if ATP, CO_2, and Mg^{2+} are available, activation to full activity is possible. In the dark the catalytic center of the enzyme is shielded by a bound phosphorylated sugar. The type of sugar varies among plant species. *ADP*, adenosine diphosphate; *ATP*, adenosine triphosphate.

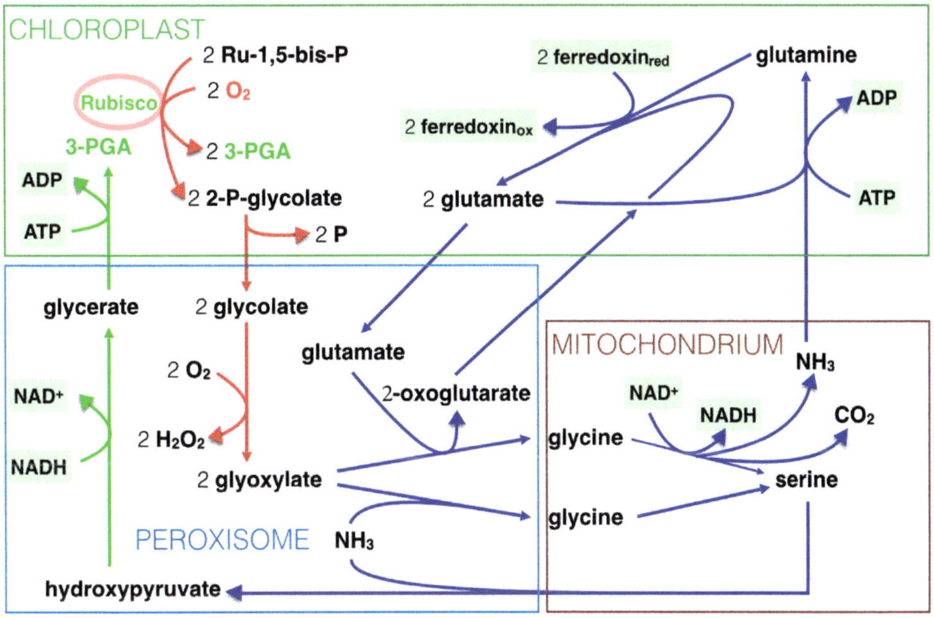

FIGURE 14.5 Photorespiration. Photorespiration is initiated when the enzyme RuBisCO is accepting O_2 instead of CO_2 as a substrate. Instead of two molecules of 3-PGA only one molecule of 3-PGA and one molecule of 2-PG will be formed. 2-PG has the potential to form a Mn^{2+} complex, thus inhibiting the water-splitting system. 2-PG is not a substrate of any plastidic enzyme, but it can be efficiently exported to peroxisomes to be turned over there. In a subsequent series of reactions involving the peroxisome and the mitochondria, two molecules of 2-PG will be turned over to form CO_2 and one molecule of 3-PGA. Photorespiration depends on the catalytic activity of RuBisCO. Therefore the reaction exclusively takes place in light. It competes with the Calvin cycle for the acceptor molecule, ribulose-1,5 bis-phosphate (Ru-1,5-bis-P). *2-PG*, 2-phosphoglycolate; *3-PGA*, 3-phosphoglycerate; *ADP*, adenosine diphosphate; *ATP*, adenosine triphosphate.

In total, two molecules of 2-PG are turned over to form one molecule of 3-PGA that is returned to the chloroplast and one molecule of CO_2 that is released during the reaction sequence inside the mitochondria. The name photorespiration has been chosen because catalytic activity of the enzyme RuBisCO is strictly light dependent and the reaction sequence is characterized by the consumption of oxygen and release of CO_2.

Recycling of the acceptor molecule Ru-1,5-P occurs in a reaction sequence named reductive pentose-P cycle or Calvin cycle. All substrates of this cycle are phosphorylated sugars, and ATP-dependent phosphorylation of ribulose-5-phosphate to form Ru-1,5-P is the last step of this reaction sequence (Buchanan et al., 2015). Several intermediates of the Calvin cycle are substrates of metabolic pathways leading to the formation of secondary plant compounds. The shikimate pathway leading to the synthesis of aromatic compounds including aromatic amino acids has to be mentioned in this context. However, export into the cytosol of 3-PGA (glycerate 3-phosphate) and DHAP (dihydroxyacetone phosphate) is the most significant trade-off from the Calvin cycle. This export is mediated in exchange for inorganic phosphate by the phosphate translocator of the inner chloroplast envelope membrane. This countertransport of phosphate and phosphorylated sugar is essential to keep a constant phosphate pool inside the chloroplast matrix (Buchanan et al., 2015).

To understand the observations made under stressful conditions it is important to know that (1) Cl^- anions are competitive inhibitors of phosphate binding to the phosphate translocator and (2) the maximal turnover rate of the translocator is significantly lower than the maximal turnover rate of the Calvin cycle. As indicated in Fig. 14.6, export into the cytosol of sugar phosphates is the rate-limiting step in assimilate supply (Stitt, 1986). This restriction of the rate of the photosynthetic chain reaction is compensated to some extent by starch production inside the chloroplasts. Starch production allows recycling of phosphate, and degradation of starch and subsequent sugar export allows a continuous assimilate supply to the cytosol during the night as well. As shown in Fig. 14.7A, amino acid synthesis, and other plastid-localized metabolic pathways, can partly compensate for the limitation in assimilate export.

Nevertheless, stress-mediated inhibition of CO_2 supply as well as inhibited consumption of assimilates (sugar export to sink organs) will result in the inhibition of photosynthetic activity (Stitt et al., 1990). This will include inhibition of photosynthetic electron transport. As a result, overreduction of components of the electron transport chain will be found. In laboratory experiments this tight coupling of reactions allows noninvasive and remote monitoring of photosynthetic activity of leaves (see discussion on chlorophyll fluorescence measurement later in the chapter for details).

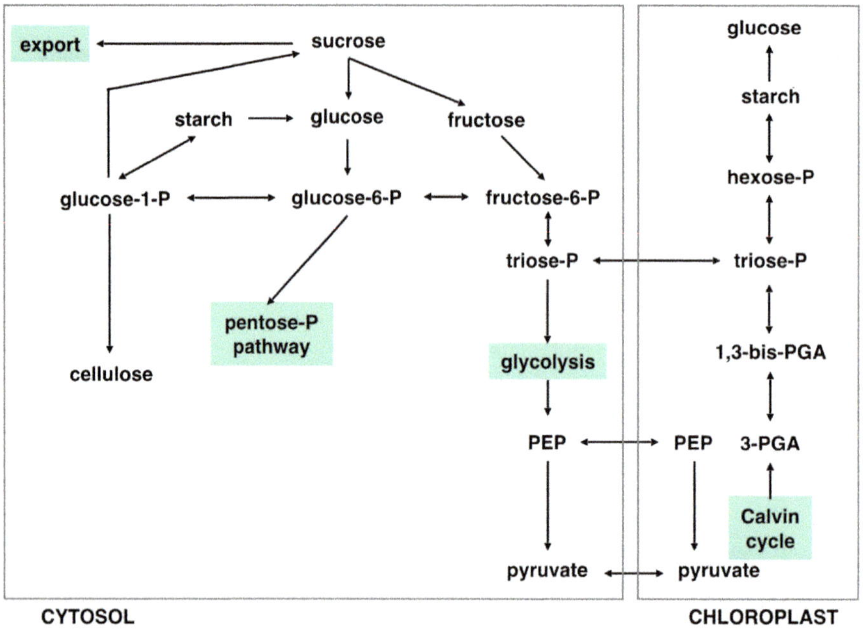

FIGURE 14.6 **Interactions of C_3 phosphate and hexose phosphate pools.** In light, triose phosphates are exported out of the chloroplast in exchange for inorganic phosphate. As a rule, this export is rate limiting for photosynthesis. Inside the chloroplast matrix, inorganic phosphate can be recycled from phosphorylated sugars by starch production. In the cytosol, exported triose phosphates are substrates for metabolism including hexose synthesis. Sucrose formed from glucose and fructose is used for carbohydrate export to sink organs.

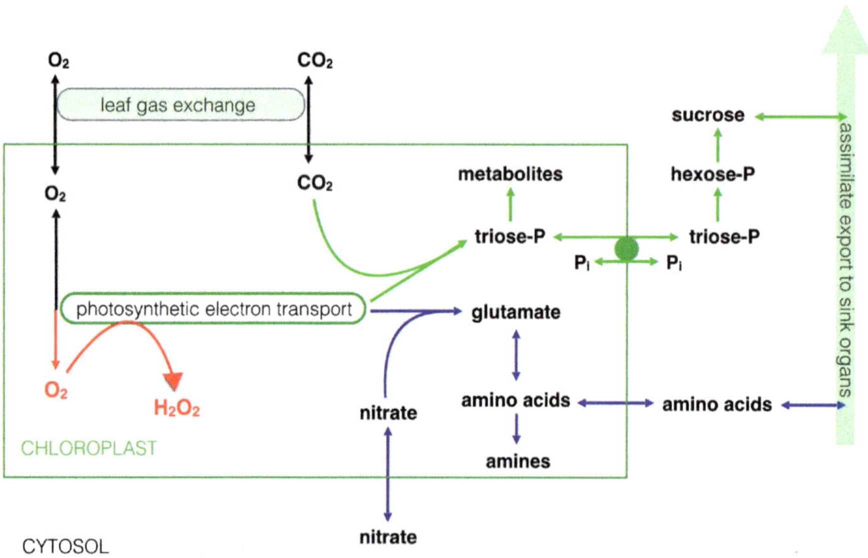

FIGURE 14.7A **Inhibition of assimilate use is stimulating ROS production.** Inside the chloroplast matrix the concentrations of the coenzymes ADP/ATP and of NAD^+/NADH are about 1 mM. This requires permanent recycling by assimilatory and metabolic activity. CO_2 assimilation and carbohydrate use in the cytosol are strictly coupled via the counterexchange mechanism at the inner chloroplast membrane. Therefore any inhibition of metabolic activity will provide feedback to the inhibition of primary events of photosynthesis. On the other hand, absorption of light energy is a continuous process. Assimilation and ROS production are competing for absorbed light energy. If the rate of carbohydrate use does not come up with the rate of energy capture, an overreduction of the electron transport chain will result in increased ROS production.

6.1 Adaptation to Light Supply

Open chain electron transport involves two photosystems differing in spectral properties. A red shift of actinic light will result in preferential activation of PSI (P_{700}), whereas PSII (P_{680}) will be preferentially activated by shorter wavelengths. This effect can be compensated by the allocation of light-harvesting complexes to the "less activated" photosystem (Buchanan et al., 2015). The effect is controlled by specific protein kinase activity that is regulated in turn by the redox state of the

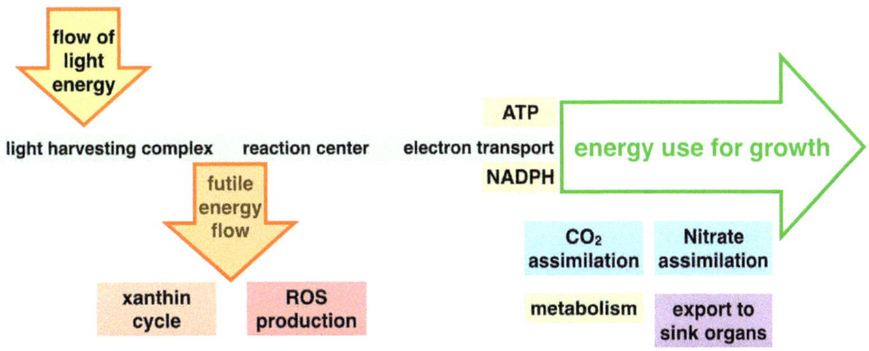

FIGURE 14.7B **Flow of redox-energy.** Energy-dissipating reactions are competing with assimilatory activity and ROS production for absorbed light energy. Activation energy transfer to xanthins and subsequent heat release is a very efficient safety valve for surplus energy. *ADP*, adenosine diphosphate; *ATP*, adenosine triphosphate; *ROS*, reactive oxygen species.

plastoquinone pool. Typically, spectral light quality does not significantly vary within short periods of time. Therefore this kind of adaptation is important for adaptation to local light conditions, such as open range or forest sites. The adaptation window of a plant species is influenced by factors such as thylakoid membrane fluidity (activity of membrane stacking and unstacking) and maximal kinase activity controlled by divalent cations.

As a rule, high light intensity results in bleaching of pigments and overreduction of components of the electron transport chain (To prevent damage, algae as well as higher plants have evolved two different mechanisms of light energy sensing and response. Excess light can be sensed by photoreceptors, such as phototropin, neochrome, and cryptochrome. Respective signals control slow responses such as chloroplast movement and gene activities (Li et al., 2009). These regulatory mechanisms are understood to match the photosynthetic activity to the general needs of the environment. Faster adaptation mechanisms are needed for fine-tuning of the photosynthetic machinery of individual leaves, which have to respond to varying light conditions during a day. Degradation especially of the PSII reaction center can be observed. This is paralleled by reduced abundance of light-harvesting complexes. Reactions related to this phenomenon are described in summary as photoinhibition (Murata et al., 2007). They result in reduced absorbance of light energy and, thus, reduced photosynthetic capacity. Avoidance of overreduction of the electron transport chain along with avoidance of adverse side reactions is an advantage of reduced energy use.

Photoinhibition occurs within minutes, whereas high-energy-induced quenching of activation energy transfer is an even faster reaction. This "safety valve" is activated when the transmembrane pH gradient exceeds a threshold value. Conformational changes in protein complexes of the thylakoid membrane modify distances among pigments. Activation energy will be directed to xanthophylls rather than the special pairs of chlorophyll *a* of the photosynthetic reaction centers. Energy from xanthophylls can be efficiently released as heat, whereas chlorophylls (due to the negative redox potential of the activated state) can transfer an electron to molecular oxygen (Zaks et al., 2012).

In summary, as can be seen from Fig. 14.7B, adaptation mechanisms are necessary (1) to improve absorption and use of light energy and (2) to reduce the risk of side reactions leading to the production of ROS and hence destruction. The probability of occurrence of such side reactions will increase if the rate of energy use (electron transport rate and subsequent assimilation) is significantly lower when compared with the absorption rate of light quanta.

6.2 Stress Effects Specific to C_4 Plants

As discussed earlier, using O_2 instead of CO_2 as a substrate hampers photosynthetic efficiency. Some plant species, such as maize and millet (C_4 plants), have developed CO_2-concentrating cycles to reduce this problem. Distinct sequences of these cycles are located in leaf mesophyll cells and bundle sheath cells, respectively. Two types of chloroplasts differing in their tendency to form grana stacks can be discriminated. Typically bundle sheath chloroplasts lack stacked thylakoid membranes (Buchanan et al., 2015).

C_4-type photosynthesis includes a CO_2-concentrating mechanism. In mesophyll cells the enzyme carbonic anhydrase uses CO_2 as a substrate to form HCO_3^-. Subsequently the formation of oxaloacetic acid (OAA) from HCO_3^- and phosphoenolpyruvate (PEP) is catalyzed in the cytosol of mesophyll cells by the enzyme PEP carboxylase (PEPCase). In all C_4 plants a C_4 compound is transported into the bundle sheath cells to be decarboxylated. The CO_2 released there will be used as a substrate by the enzyme RuBisCO for final CO_2 assimilation (Hatch, 1987). C_4 plants are characterized by a low photorespiratory activity, as competition of O_2 and CO_2 for binding to RuBisCO is low. This can be explained by the fact that (1) O_2 is not a substrate for the enzyme carbonic anhydrase and PEPCase, (2) gas diffusion to the bundle sheath is limited, and (3) the water-splitting system of bundle sheath chloroplasts is inactive or significantly reduced in activity. Due to the high substrate affinity of carbonic anhydrase, C_4 photosynthesis can even take place at high rates when CO_2 diffusion is limited by semiclosed stomata. Thus it may be concluded that C_4 plants have evolved to display higher photosynthetic efficiency than C_3 plants under elevated temperatures, high light intensities, and drought.

C_4 photosynthesis requires a complex interaction between metabolic activities in two different compartments, the mesophyll and the bundle sheath. It has to be mentioned that CAM plants (plants expressing the crassulacean acid metabolism) are special, because the two phases, CO_2 concentration and final fixation, are separated in time. Moreover, some bacteria, algae, and special higher plants such as *Bienertia* sp. perform the two phases of the C_4 cycle within specialized cell compartments. C_4 plants have to overcome two problems, (1) tuning substrate transport rates as well as metabolic activities in both compartments and (2) transport of reducing equivalents together with the "CO_2 carrier molecule," to compensate for inhibited NADPH supply. NADPH production is coupled to open chain electron transport. There will be no NADPH supply when the water-splitting system is inhibited inside bundle sheath chloroplasts. Reduction of the coenzyme has to be coupled to substrate consumption as there is no NADPH transporter inside the plasma membrane. C_4 plants are solving the latter problem by transferring to the bundle sheath reduced C_4 compounds rather than OAA. However, due to different abundance of transporters and different patterns of enzyme activities, different types of C_4 plants can be discriminated. In the literature, C_4 plants are classified as $NADP^+$ malic enzyme type ($NADP^+$-ME), NAD^+ malic enzyme type (NAD^+-ME), and PEP carboxykinase type (Buchanan et al., 2015; see Table 14.1 for description of the respective biochemistry).

TABLE 14.1 Distinguishing Marks of C_4 Photosynthesis Types

NADP-ME Type	
Typical representative	Maize (*Zea mays*)
Decarboxylation reaction Malate + $NADP^+$ — [NADP-ME] → pyruvate + CO_2 + NADPH + H^+	
Location of decarboxylation reaction	Chloroplast
C_4 acid preferentially transported to the bundle sheath	Malate
C_3 acid preferentially transported to the mesophyll	Pyruvate
Regional relationship with rainfall	Abundance of species is increasing with rainfall
NAD-ME Type	
Typical representative	Millet (*Panicum milaceum*)
Decarboxylation reaction Malate + NAD^+ — [NAD-ME] —> pyruvate + CO_2 + NADH + H^+	
Location of decarboxylation reaction	Mitochondria
C_4 acid preferentially transported to the bundle sheath	Aspartate
C_3 acid preferentially transported to the mesophyll	Alanine
Regional relationship with rainfall:	Abundance of species decreases with rainfall
PCK Type	
Typical representative	Guinea grass (*Panicum maximum*)
decarboxylation reaction OAA + ATP — [PCK] —> PEP + ADP + CO_2	
Location of decarboxylation reaction	Mitochondria
C_4 acid preferentially transported to the bundle sheath	Aspartate (+some malate)
C_3 acid preferentially transported to the mesophyll	Alanine, PEP, or pyruvate
Regional relationship with rainfall	Abundance of species does not correlate with rainfall

ME, malic enzyme; *NAD; NADP; NADH; OAA*, oxaloacetic acid; *PCK*, PEP carboxykinase; *PEP*, phosphoenolpyruvate.

C_4 plants differ in stress sensitivity, especially with respect to hyperosmotic salinity. An enhanced sensitivity of C_4 plants when compared with C_3 plants was found in shading experiments. It can be explained by cation requirements of enzyme activities involved in the respective C_4 cycle. Typically, K^+ requirement of C_4 plants correlates with K^+ stimulation of pyruvate kinase activity, and Mg^{++} requirement mirrors dependency on this cation of ATPases and other enzymes, such as protein kinases. Protein kinases are essential for coordination of enzyme activities, such as PEPCase and pyruvate phosphate dikinase, in mesophyll and bundle sheath cells (Vidal and Chollet, 1997). In general, hyperosmotic salinity sensitivity can be explained by adverse salt effects on membrane potentials. There are two strategies to keep cytosolic salt concentration at a tolerable level: (1) salt export into the apoplast and (2) sequestration of salt to the vacuole. In both cases, transport occurs at the expense of the respective membrane potential. Thus response to salt stress and transport activity of the C_4 cycle compete for the same energy source. Salt stress directly affects the CO_2-concentrating mechanism and thus inhibits photosynthesis.

Although C_4-type photosynthesis is apparently characterized by an improved WUE when compared with C_3-type photosynthesis, in the natural environment some populations of C_4 species are more vulnerable to water shortage than C_3 populations. Ibrahim et al. (2008) have tested C_3 and C_4 subspecies of *Alloteropsis semialata* accordingly. When compared with C_3 plants C_4 subspecies were observed to lose their photosynthetic advantage during natural drought events, despite greater WUE under irrigated conditions. Presented data confirm earlier observations indicating that drought stress might affect C_4 plants in a way different from the one well characterized in C_3 species. When comparing drought effects on C_3 and C_4 subspecies of the grass *Alloteropsis*, more severe stomatal and metabolic limitations have been found in the C_4 subspecies, indeed. More detailed analysis showed that reduced gas exchange rates of the C_4 plants were based on a severe inhibition of the C_4 metabolism rather than on stomatal effects. Authors therefore discussed whether drought sensitivity of C_4 metabolism can explain the observed decline in C_4 species relative to C_3 species in drought-affected areas.

Currently, this problem has not been sufficiently solved. CO_2 assimilation in NAD-ME-type bundle sheath chloroplasts depends on the supply of redox equivalents (malate–oxaloacetate shuttle) from mitochondria, whereas malate is supplied by mesophyll cells in NADP-ME-type C_4 plants. Moreover, NADP-ME- and NAD-ME-type bundle sheath chloroplasts differ in the abundance of ribulose-1,5-bis-phosphate (Rao and Dixon, 2016). In NADP-ME chloroplasts the Calvin cycle can be significantly stimulated by external supply, whereas no such stimulation was found in NAD-ME-type chloroplasts (Hatch and Kagawa, 1976). Bundle sheath mitochondria are not directly involved in NADP-ME-type C_4 metabolism, whereas NAD-ME-type metabolism depends on a well-tuned interaction of chloroplasts and mitochondria. Therefore it may be assumed that observed differences in stress resistance are based on specific reactions involved in the interplay of both cellular compartments of bundle sheath cells.

7. REACTIVE OXYGEN SPECIES PRODUCTION AND DETOXIFICATION

ROS are involved in pathogen defense not only in animals but also in plants (Floryszak-Wieczorek and Arasimowicz-Jelonek, 2016; Yadav and Sharma, 2016). The importance of ROS in control of development may explain why it was not possible yet to create mutants of C_4 plants completely lacking photorespiratory activity (Baxter et al., 2014). We will focus in this chapter on the adverse effects created by stress-induced ROS production.

As shown in Fig. 14.8 ROS are produced in several metabolic reactions, such as acetyl-CoA dehydrogenase reaction initiating fatty acid degradation, glycolate peroxidation during photorespiration, or reactions catalyzed by enzymes

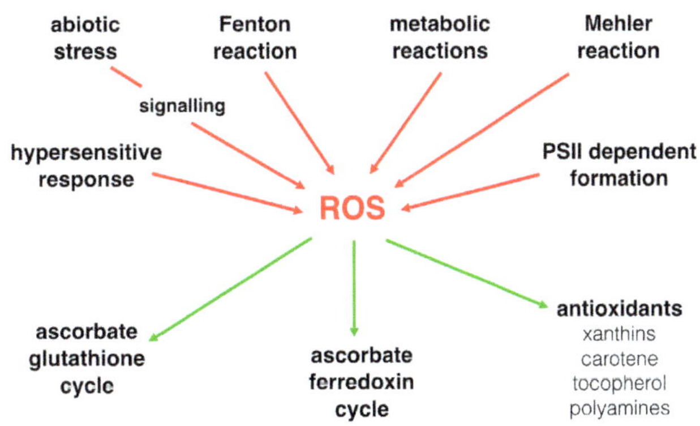

FIGURE 14.8 **Production, detoxication, and scavenging of reactive oxygen species (ROS).** Although ROS are needed for signaling and some polymerization reactions, plants have to prevent ROS from reaching toxic concentrations. Antioxidant molecules and enzymes catalyzing detoxication reactions control cellular ROS concentration.

of the cytochrome P450 family (Bak et al., 2011). Here we will focus on stress-induced ROS production inside chloroplasts.

As mentioned earlier (see Figs. 14.1 and 14.7), ROS production is at several levels competing for electrons with photosynthetic electron transport. (1) In its light-activated state, chlorophyll is an electron donor. It is hosted within the protein complex of a photosystem. Orientation within this protein complex preferentially allows photosynthetic energy transfer. However, due to its redox potential, activated chlorophyll may transfer an electron to molecular oxygen to create an oxygen radical (Pospísil, 2009). (2) Quinones involved in electron transfer reactions will be converted to unstable semiquinones by acceptance of an electron. Semiquinones are reactive radicals. They adopt the stable state of a quinol (hydroquinone) only by subsequent acceptance of a second electron. Similar to ROS semiquinones are reactive and can cause damage if no "physiological" reaction partner is available within a reasonable period of time (Song and Buettner, 2010). (3) Similar to activated chlorophyll reduced ferredoxin is sufficiently electronegative to form ROS as well. (4) Artificial redox partners, such as viologen-type herbicides, can be another source for ROS formation (Sharma et al., 2012).

The extent of ROS production can be estimated by measuring the peroxidation rate of membrane lipids, applying the malondialdehyde method, for instance (Sharma et al., 2012; Koyro et al., 2013). In addition to lipids, chloroplast pigments, proteins, and nucleic acids are targets of ROS damage. Such damage occurs in all plants during the day. Damage will be tolerated by plants as long as rates of repair mechanisms are exceeding the ones of damage (Sofo et al., 2015). In parallel to repair, scavenging and detoxification of ROS play an important role to build stress resistance. Scavengers are antioxidant molecules that typically are more reactive/more unstable when compared with molecules under risk of ROS damage that are essential for plant physiology. Tocopherols (vitamin E) and carotenoids are among the well-analyzed scavengers (Sharma et al., 2012) (see Fig. 14.8).

In addition to ROS scavengers, in all plants enzyme-catalyzed reaction sequences resulting in fast detoxification of ROS are found (Yadav and Sharma, 2016). In many cases the primary electron acceptor molecule will be recycled at the expense of redox equivalents. This way an ROS molecule will be removed, and the cycle will also function as an electron sink. Thus recycling of the acceptor prevents overreduction of the electron transport chain, and it competes with ROS production for electrons.

Anyways, we have to keep in mind that both ROS production and ROS scavenging cycles are competing for electrons with assimilation and growth rate. ROS scavenging prevents from ROS damage, but it will reduce the maximal growth capacity. Plants can thrive in the presence of surplus light energy offer, but breeding for stimulated use of assimilates or enhanced gas exchange rate would solve the problem as well and would lead to better yields (von Caemmerer et al., 2012).

Although activities of ROS scavenging mechanisms are regulated, there always is a minimal energy leakage through such pathways. Therefore the abundance of respective enzymes is low under optimal growth conditions. Genes are activated in response to stress perception, and it takes hours to days until a plant species will reach its specific maximal detoxification rate (Koyro et al., 2014a).

From such observation the advantage of a permanent antioxidant pool becomes obvious. Especially at locations characterized by a very variable stress level, antioxidants will function as the first level saving from ROS damage (Das and Roychoudhury, 2014; Koyro et al., 2014a). Cells will need some time to synthesize ROS scavenging enzymes and further develop the aforementioned detoxification sequences. In experiments it has been observed that respective gene expression was not detectable in plant species containing high concentrations of antioxidants until ROS stress exceeded a threshold level (Koyro et al., 2014a; Papenbrock et al., 2016).

Some of the antioxidants found in plants have been identified to be useful for medical treatments (Shebis et al., 2013). Therefore there is some interest in breeding for high antioxidant content or to stimulate the content by plant cultivation praxis. Due to the aforementioned complex regulation of such contents, respective approaches require monitoring of growth conditions as well as definition of an ideal time point for harvesting the plant material (Boestfleisch et al., 2014).

8. STRESS-INDUCED CHANGES IN METABOLITE PATTERNS

As mentioned earlier, synthesis of secondary metabolites is another important sink for energy, redox equivalents, and assimilated carbon. Two aspects have to be mentioned in context of this review: (1) synthesis of secondary metabolites can lessen constraints on the photosynthetic apparatus when energy use and efflux of assimilates are inhibited under stress and (2) an existing pool of secondary metabolites (antioxidants and compatible solutes) can absorb eventually toxic byproducts (such as ROS) or side reactions (osmotic and ionic effects) emerging under stress (Selmar and Kleinwächter, 2013; Boestfleisch et al., 2014; Gil et al., 2014). As indicated in Fig. 14.9, a pool of ROS scavengers may react as a first line

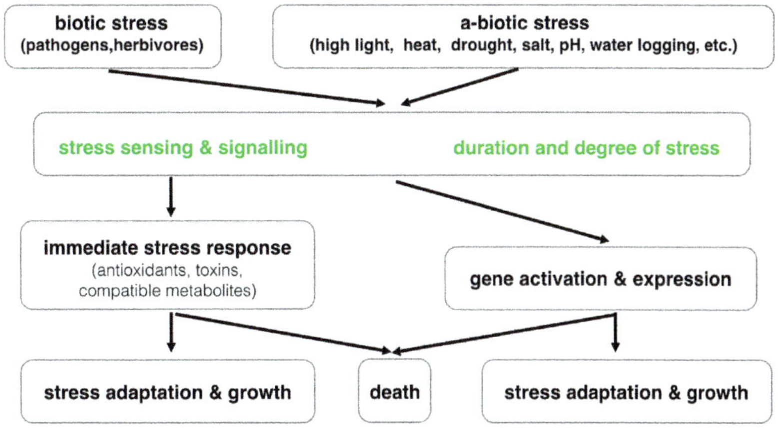

FIGURE 14.9 **Stress perception and response.** In the field, plants sense different types of stress at a time. Signaling at cellular level results in cross talk of messengers and an overlapping regulatory response that is only poorly understood to date.

of defense, but will be consumed during long-lasting periods of stress. In such a case, activity of stress-responsive genes has to initiate synthesis of antioxidant enzymes (second defense line). This may explain two lines of observations made in experiments. (1) Plants significantly differing in their secondary metabolite contents apparently differ as well in the degree of stress resistance. For instance, salt resistance of accessions of beans rich in antioxidants was found to be significantly higher when compared with accessions of beans with low antioxidant content (Ventura et al., 2014). (2) When plants were subjected to light, salt, or drought stress, significantly higher stress levels (in terms of light intensity, salt concentrations, and withdrawal of water) could be tolerated if stress was applied gradually (increasing salt concentration stepwise over a certain period of time, for instance). Putting plants from the shade into full sunlight or adding salt to a hydroponics culture to immediately reach a final salt concentration would kill plants at a much lower stress level (Koyro, 2003). These observations could be explained by the finding that plants developed a higher internal concentration of secondary metabolites during the respective adaptation period. This was achieved not only by stimulation of enzyme activity but also by overexpression of respective gene activities, thus by adapting a new metabolic homeostasis (Koyro et al., 2014a; citations therein; Bhaskara et al., 2015). Many of the enzymes involved in metabolic pathways leading to secondary metabolites are located inside the chloroplast stroma. Compounds are produced inside the compartment subjected to most of the early stress events. Nevertheless, not only de novo synthesis but also turnover rates and import and export events are issues to be analyzed (Bhaskara et al., 2015).

Plants differ a lot in their contents of secondary metabolites and, thus, in the extent of substrate fluxes leading to secondary metabolites. This results in differences in growth rate and light energy requirements (Agati et al., 2013; Cheynier et al., 2013; Caretto et al., 2015). Depending on control of enzymatic activities, this latter aspect can serve as a feedback to the respective light compensation point, i.e., the light intensity necessary to compensate for respiratory activity (metabolic energy demand) of a plant (Zandalinas et al., 2017). From this line of argument it may be concluded that under ideal growth conditions fast growing plants will be dominant in an ecosystem, whereas an area that is characterized by regularly occurring stress events will preferentially be inhabited by stress-tolerant plants containing high levels of secondary metabolites that are synthesized at the expense of a reduced growth rate (Caretto et al., 2015). Accordingly, it is well documented in the literature that high-yielding crops, such as cereals, are more stress sensitive when compared with accessions producing only moderate yields.

Many of the secondary metabolites have been characterized as compatible solutes, antioxidants, and compounds stabilizing structure and functions of enzymes and membranes. Thus actual concentrations and metabolite patterns correlate with an observed level of stress resistance, when using respective parameters as references (Agati et al., 2013; Cheynier et al., 2013). As enhanced ROS concentrations are harmful, most compounds of "beneficial" function are either ROS scavengers or help avoiding ROS production (as symbolized in Fig. 14.10A). The list in Fig. 14.10B shows that synthesis of these compounds starts with early products of photosynthesis. This line of argument implies that researchers may not find a direct correlation between the level of applied stress and the concentration of an individual metabolite or some other parameter. It is important to identify the bottleneck in stress resistance of the respective experimental plant. This applies not only to secondary metabolites but also to any other physiological parameter (Papenbrock et al., 2016 and citations therein).

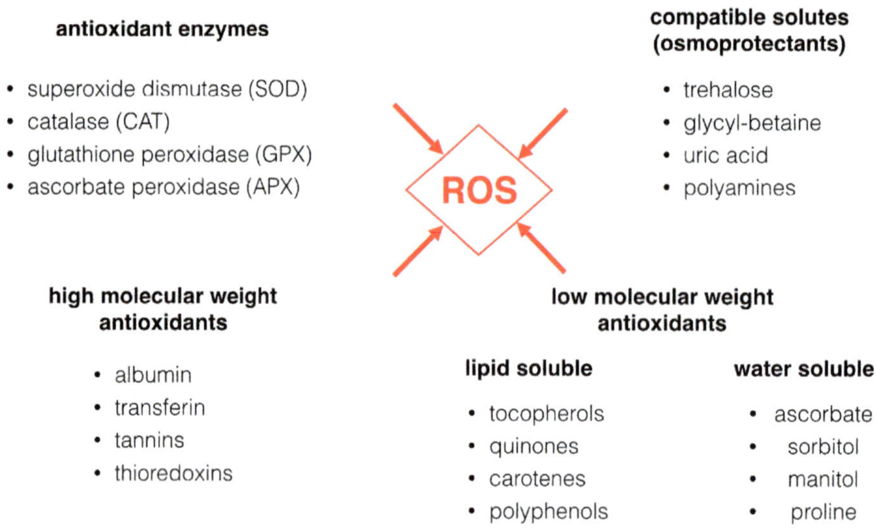

FIGURE 14.10A **Antioxidants and compatible solutes.** Most types of stress result in reactive oxygen species (ROS) production. Without knowing signaling and response reactions in detail, it can be difficult to discriminate in field trials reactions resulting in ROS detoxication from those preventing ROS production. This especially applies for salt and drought stress. Therefore many compatible solutes are termed antioxidants in the literature as well (some of them may have both functions, indeed).

compound	substrates
sorbitol, manitol, trehalose	glucose
proline	glutamate
glycyl-betaine	methionine —> choline
polyamines	arginine or ornithine
ascorbate	GDP-mannose
quinones, carotenes & tocopherol	pyruvate + glyceraldehyde 3-P
polyphenols & tannins	Acetyl-CoA / Malonyl-CoA

FIGURE 14.10B **Synthesis of antioxidants and compatible metabolites.** Substrates for the synthesis of common antioxidants and compatible solutes are listed.

8.1 Production of Compatible Solutes

Drought and hyperosmotic salinity are the most adverse conditions limiting crop production (Munns, 2011). Drought preferentially affects osmotic homeostasis, whereas hyperosmotic salinity is characterized by exerting ionic stress as well (Koyro et al., 2008). Both stresses are sensed by roots, and abscisic acid (ABA) signaling, for instance, will result in stress responses of the shoot as well. [In both cases, adaptation of the osmotic tissue potential to the new situation is essential to allow water and nutrient uptake (Koyro et al., 2008)]. In plants water potential has the highest preference. Therefore gas and water exchange at stomata level will be adapted to the water uptake capacity of the root system (Koyro et al., 2008). This will affect nutrient supply, as nutrient intake and transport is linked to water flow.

To adapt cellular osmotic potential there are two options: (1) import of inorganic ions (salt) from the soil water phase and (2) synthesis and storage of organic compounds, i.e., use of assimilates to compensate for the external osmotic potential and, thus, allow water and nutrient uptake in an adverse environment (Koyro et al., 2008). Uptake of ions may be risky, as inorganic ions will compete with proteins for water shells (Koyro et al., 2008; Munns, 2011). As we know from enzyme purification experiments, big protein complexes carrying a low charge density will be denatured and precipitated first. Respective catalytic activity of high-molecular-weight enzymes will be lost at quite low ionic strength. Increasing cytosolic ion concentration will reach its limits soon and is not suitable as the only strategy to compensate for ionic stress. In most cases ions were found to be sequestered to the vacuole compartment. However, with respect to complex analytical methods (ion-selective transmission electron microscopy), this aspect is only poorly documented in the current literature (Koyro and Stelzer, 1988).

When the vacuole volume is increasing under succulence-inducing stress the share of cytosolic volume will decrease. Together with an increasing Donan potential increasing concentrations of cytosolic solutes will contribute to balance the increased osmotic potential of the soil water phase (Höhner et al., 2016). Within the cytosol the osmotic potential can be further increased by production and storage of compatible solutes. Sugars, polyalcohols, amino acids, and polyamines are among the most important compatible solutes (Wani et al., 2013). In response to salinity and drought stress, and also under heat and light stress, significant increase in cytosolic proline concentration was observed in several plant species (Bhaskara et al., 2015). Proline concentration inside the leaf mesophyll can increase at a rate that requires a concerted action of proline synthesis and proline import from other plant tissues. It has been reported that proline concentrations may exceed 100 mM, but as proline apparently is excluded from the vacuole, cytosolic protein concentrations are not easy to measure (Bhaskara et al., 2015). Accessions characterized by a high protein level under stress at the same time show only little symptoms of ROS toxicity. Therefore, in addition to osmotic effects, the antioxidant capacity of proline is discussed in the literature (Hayat et al., 2012). Moreover, it has been suggested to breed for proline content to improve stress resistance of crops (Bhaskara et al., 2015). Polyamines resemble proline with respect to beneficial effects under stress, although the polyamine content of leaf cell is by far lower as compared to proline. Breeding for polyamine content may be an option for plant species that do not accumulate proline under stress. As shown in Fig. 14.10B, these compounds derive from early assimilates and the respective synthesis takes only very few catalytic steps. Thus the number of genes that need to be activated to overproduce compatible solutes is very low. It has been shown that plant species differ in the spectrum of respective stress-responsive genes (Koyro and Huchzermeyer, 2008). However, more detailed analysis has shown that there is a complex regulation system controlling stress response. Regulation of gene activity not only by sugars but also by polyamines has been documented (Kumar et al., 2013).

8.2 Production of Antioxidants

ROS formation is a by-product of metabolism in living cells (Yadav and Sharma, 2016). ROS formation has to be tightly regulated to prevent levels exceeding the antioxidative capacity of cells. Otherwise oxidative stress that may result in malfunctioning and ultimately death of the cell can be expected (Yadav and Sharma, 2016). Enzymes such as superoxide dismutase (SOD), catalase, ascorbate peroxidase (APX), glutathione peroxidase, glutathione-*S*-transferase, monodehydroascorbate reductase, and dehydroascorbate reductase (DHAR) catalyze ROS degradation reactions, whereas antioxidants such as ascorbate, glutathione (GSH), phenolic compounds, carotenoids, and tocopherols either directly interact with ROS or function as substrates in enzyme-catalyzed ROS-degrading reactions (Yadav and Sharma, 2016).

Some of the most common antioxidant compounds found in higher plants are listed in Fig. 14.10B along with information on the substrate of the respective metabolic pathway. It is obvious that antioxidant synthesis starts with early metabolites of CO_2 assimilation. Any limitation of assimilate export from leaves to sink organs therefore will result in increased concentrations of these substrates and thus will stimulate antioxidant synthesis. Moreover, additional stimulation of antioxidant synthesis is based on the activity of stress-responsive genes: Enzymes involved in pathways leading to antioxidant synthesis will be overexpressed under stress (see information provided in the following chapter). Therefore stress-resistant plants, such as spices and medicinal plants, are characterized by constitutive high internal antioxidant concentrations allowing to tolerate moderate stress–induced ROS production. Moreover, antioxidant content can be significantly enhanced to tolerate long-lasting periods of stress (Selmar and Kleinwächter, 2013). Stress signaling and regulation of gene and enzyme activities is not easy to analyze, due to a complex cross talk among signals released by different types of stresses.

Plant species differ in the concentrations as well as the patterns of antioxidants found in leaf tissues. Moreover, the antioxidant concentration found under optimal growth conditions can be significantly stimulated by drought, high light intensity, and other stresses. This observation especially applies to plant species that can be found in ecosystems under risk of regularly occurring stress events (Selmar and Kleinwächter, 2013). It has been nicely demonstrated that antioxidant production will be stimulated when leaf gas exchange is limited and consumption of NADPH by CO_2 assimilation is decreased under stress (Selmar and Kleinwächter, 2013).

Among the tocopherols, α-tocopherol is most important with respect to its pharmaceutical and nutritive use. Tocopherols are highly lipophilic, thus preferentially protecting the structure and function of membranes and membrane proteins, such as PSII (Das and Roychoudhury, 2014). α-Tocopherol can scavenge lipid peroxy radicals while yielding a tocopheroxyl radical that can be recycled back to α-tocopherol by reacting with ascorbate. Tocopherols as well as carotenes that are compounds typically found in plastids. Hydrophilic antioxidants, on the other hand, are not strictly located in only one cell compartment; this applies for ascorbate and GSH, for instance. In some experiments, antioxidant concentration has been used as a parameter to evaluate the stress perception of experimental plants. It therefore has to be pointed out that antioxidant concentrations can be tissue specific within individual plants. Moreover, antioxidant concentration may undergo a day-night fluctuation (García-Plazaola et al., 1999).

Ascorbate can be detected in most plant cell types and cell compartments. Therefore, and because of its pharmaceutical relevance, it is one of the most extensively studied antioxidants. Ascorbate is synthesized in the mitochondria and is transported to the other cell compartments. Similar to GSH, ascorbate not only directly interacts with ROS but also is a substrate of enzymes of antioxidant pathways. In the major pathway active in plant tissues GSH (reduced glutathione) is oxidized to GSSG (oxidized glutathione) and ascorbate is oxidized to monodehydroascorbate and dehydroascorbate. Subsequently these three oxidized compounds are reduced to form the initial substrates. Ascorbate is a substrate in a reaction cycle efficiently scavenging H_2O_2 at the expense of NADPH. As mentioned earlier, the photosynthetic apparatus is at high risk of ROS-mediated damage. In addition to the ascorbate–GSH cycle chloroplast membranes are receiving additional protection from oxidative damage by the exothermic production of zeaxanthin (Buchanan et al., 2015). In *Arabidopsis* zeaxanthin is not detectable in leaves of plants grown under optimal conditions, but it accumulates rapidly under excess light energy. Both the cycles, the xanthin cycle and the GSH–ascorbate cycle, were found to consume significant shares of absorbed light energy, thus preventing chloroplasts from ROS damage (Das and Roychoudhury, 2014). Plant species significantly differ in the expression of respective genes, resulting in different relevance of these protective mechanisms.

8.3 Reactive Oxygen Species and Metabolites Controlling Metabolic or Physiological Functions

In early experiments, sucrose export from leaf mesophyll cells was inhibited in transgenic tobacco plants. During the phase of maturation leaves of such plants developed bleached spots. Green areas of the leaves were characterized by a high rate of photosynthesis, a low rate of respiration, and low sugar content, whereas the opposite situation (low photosynthetic activity, high respiratory activity, and high soluble sugar concentrations) was found in bleached areas (Stitt et al., 1990). These experiments showed for the first time that metabolite concentrations not only affect catalytic enzyme activities but also can affect the abundance of respective enzyme proteins (Stitt et al., 1990). Moreover, in experiments of Stitt et al. (1990) it was shown that inhibition of sucrose export affects cytosolic enzymes as well as those of the plastids and the mitochondria. These experiments were extended by feeding sugars and sugar analogs to detached leaves as well as suspended cells of cell cultures, and subsequently monitoring the gene expression patterns. Two results were found: (1) photosynthetic genes (*rbcS*, *cab*, and *atp8*) were inhibited by addition of sugars and (2) inhibition occurred exclusively when added sugars could be metabolized. Stable sugar analogs did not result in inhibition of gene expression (Krapp et al., 1993). Inhibition of transcriptional activity by sugars was also observed in experiments with *Chenopodium* and with freshly prepared maize protoplasts (Krapp et al., 1993). In experiments further extending the spectrum of tested metabolites, metabolic inhibition of photosynthetic genes by acetate was also demonstrated, and the inhibitory effect was allocated to distinct upstream regulatory elements. Further experiments using other plant species led to the conclusion that the metabolic repression of photosynthetic genes is likely universal (Rolland et at., 2006). It overrides other regulation by light, tissue type, and developmental stage and serves potentially as the molecular basis of interactions between sink and source tissues (Krapp et al., 1993). Involvement of ROS in this regulatory network is a special aspect in tissues active in photosynthesis: Observed bleaching and photoinhibitory effects have to be attributed to direct ROS action.

ROS are important messengers involved in modifying important processes such as mitosis, cell elongation, and senescence. They also provide to the cell information on cellular redox state and regulate gene expression associated with biotic and abiotic stress responses to optimize defense and survival. We have to mention, though, that cellular redox state is determined by an interplay of several redox-active compounds, such as ascorbate, GSH, NAD(P)H, and proteins from the thioredoxin superfamily. Due to the instability of the compounds, measuring ROS concentrations is complicated. Therefore measuring the pool of ascorbic acid and its redox status can indicate its involvement in the cellular mechanisms by which plants respond to drought-induced oxidative stress. Anyway, analysis of ROS effects is complicated by the conflicting nature of these compounds. They are produced as a normal product of plant cellular metabolism. Whether ROS would serve as signaling molecules or could cause oxidative damage depends on a delicate equilibrium between their production and scavenging (You and Chan, 2015). Various environmental stresses lead to excessive ROS production that has to be balanced by antioxidant activity of respective tissues to prevent progressive oxidative damage and ultimately cell death (Sharma et al., 2012). Stress-induced production of ROS mostly results from pathways such as photorespiration, overreduction of the photosynthetic electron transport chain, and mitochondrial respiration (but as a stress response, additional ROS production by NADPH oxidases has been observed). In general, it can be observed that stress resistance of plant species correlates with antioxidative capacity, especially of leaves. Manipulating ROS levels provides an opportunity to enhance stress resistance of crop plants under a variety of unfavorable environmental conditions (You and Chan, 2015). Respective research is based on the finding that genes coding for antioxidant enzymes are among ROS-responsive genes of plants native to environments at risk of sudden stress events (drought, flooding, high light intensity) (Sharma et al., 2012). In the model plant *Arabidopsis*, a network of at least 152 genes involved in managing the level of ROS has been identified. This network is highly dynamic and redundant, and it encodes ROS-scavenging and ROS-producing proteins. When evaluating oxidative stress management in the leaves of two *Cerasus humilis* genotypes differing in drought resistance a direct correlation between the degree of ROS damage (measured in terms of malondialdehyde accumulation) and inhibition of growth was found. Stress resistance correlated with enhanced SOD activity, turnover rate of the ascorbate–GSH cycle, and maintenance of ascorbate levels. In agreement with this observation the expressions of cytosolic APX and DHAR increased. This indicates that drought resistance of *C. humilis* depends on the upregulation of these two enzymes at transcriptional level (Jing et al., 2016). In agreement with these data, experiments with the salt-tolerant plant species *A. marina* demonstrated that the genes *Sod1*, *Cat1*, and *Fer1* are upregulated upon oxidative stress, thus reducing accumulation of ROS (Jithesh et al., 2006).

Avoidance of ROS stress is a strategy commonly found in all plant species. This can be achieved, for instance, by the production of compatible solute, such as proline. Surprisingly it has been found that stress resistance correlates with proline synthesis and degradation activities rather than mere proline storage (Bhaskara et al., 2015). Proline is preferentially found in the cytosol, and no proline transporter has been found in the tonoplast membrane. Cytosolic proline concentration is regulated by control of both proline synthesis and proline degradation. Both pathways are under the control of ABA-responsive genes. Among the gene products a protein phosphatase has been identified. Among the catalytic activities of this enzyme are dephosphorylation of pyrroline-5-carboxylate synthetase (an enzyme involved in proline synthesis) and proline dehydrogenase 1, a key enzyme of proline degradation. Opposite effects of dephosphorylation on catalytic activities of these enzymes results in proline accumulation (Bhaskara et al., 2015).

It is a special challenge to keep cellular ion content balanced under drought and hyper-osmotic salinity. Losses in crop yield worldwide can be attributed to respective problems. Therefore elucidation of the salt overly sensitive (SOS) network gained high attention and has stimulated crop breeding activity. The system is composed of SOS3, SOS2, and SOS1. Initially it was proposed to mediate cellular signaling under salt stress, to maintain ion homeostasis. More recently, several reports have stated that in roots the SOS proteins may have novel roles in addition to their known functions. SOS3, for instance, plays a critical role in the development of lateral roots through modulation of auxin gradients and maxima in roots under mild salt conditions. The SOS proteins also play a role in the dynamics of cytoskeleton under stress (Ji et al., 2013, and citations therein).

Investigations on plant adaptation to a certain environment have started with analysis of hormone synthesis and hormone effects. As expected, stress-responsive hormones did play an important role in coordinately regulating growth responses under stress (Kohli et al., 2013). However, interpretation of experimental results is complicated by the finding that interactions among various phytohormones are the rule rather than exception in integrating the diverse input signals and readjusting growth as well as acquiring stress resistance (Kohli et al., 2013). It is generally agreed, meanwhile, that observed stress responses are plausibly being revealed to involve cross talk with energy signaling pathways as any growth-limiting factor alters the plant's energy status. Stress sensing is primarily reflected in cellular energy imbalance, which may be discussed in terms of imbalance in redox homeostasis and changes in cellular sugar level (Biswal et al., 2011). Signals generated by these changes bring about photochemical, metabolic, and molecular reprogramming for stress adaptation through different signal transduction pathways (Biswal et al., 2011). These statements are linking hormone research to findings on stress responses of metabolism as discussed earlier. This line of discussion leads to a demand for analysis of gene-to-metabolite networks in primary and secondary metabolism. In turn, this can lead to identification of gene function and subsequent improvement of production of useful compounds in plants.

Using transgenic experimental plants allowed elucidation of signaling pathways that are linking stress perception to control of gene expression. As expected, it was found that mitogen-activated protein kinases (MAPKs) are involved similar to signaling in other eukaryotes. This view is supported by the finding that all model plant genomes contain high numbers of MAPK genes (Nakagami et al., 2005). The view of similarity among eukaryotes is supported by the identification of conserved protein kinases KIN10 and KIN11 in *Arabidopsis*. Together with specific bZIP transcription factors, these two kinases are involved in reprogramming of transcription in response to several stresses, such as low light conditions and the aforementioned sugar stress.

9. DEVELOPING CONCEPTS TO IMPROVE STRESS RESISTANCE AND INCREASE CROP YIELD

Chloroplasts are known to accommodate catalytic reaction sequences summarized under the term photosynthesis. However, metabolic activities of chloroplasts are not only composed of the sequence of processes by which light energy is used to form glucose from CO_2 and water. Assimilation of nitrogen (i.e., nitrate reduction) and synthesis of many secondary plant metabolites also takes place in chloroplasts (Buchanan et al., 2015). In fact, most of the compounds (in animal nutrition) termed essential nutrients are produced by metabolic pathways located inside the chloroplasts.

There is a strictly regulated exchange of metabolites among metabolic pathways. Elucidation of reaction sequences leading to the production of individual metabolites was complicated by the finding that (1) catalytic activities of enzymes can be regulated by several factors, such as light (energy and redox status) (Buchanan et al., 2015) and feedback inhibition brought about by end products, and (2) different substrates may be used to build the carbon skeleton of the respective compounds (Buchanan et al., 2015). Branching and merging of metabolic pathways significantly aggravated the biochemical dissection of metabolic pathways and is still leading to misinterpretation of experimental results.

The need to improve productivity of agricultural land use requires the development of concepts to increase biomass production. Improving stress resistance of crops is a promising approach (Munns, 2005). This requires to identify breeding goals and traits that are easily and reliably to be measured. With this respect, three problems have to be solved: (1) transfer of results achieved in laboratory experiments to applications in the field (Poorter et al., 2016), (2) dissecting the tightly knit network of metabolic pathways, and (3) elucidating the interplay of plant architecture, biomass formation, and photosynthetic capacity (Tschiersch et al., 2017). Each of these problems requires to establish reliable protocols to monitor plant performance during a growth period. As outlined earlier, plant metabolism is a tightly coupled sequence of enzyme-controlled reactions fanning out from primary light reactions to sequences leading to carbohydrates and secondary metabolites (Buchanan et al., 2015). However, most of the individual reactions are well understood, enzymes involved have been isolated, and substrate affinities and catalytic activities have been measured. The situation may be compared to chess: all pieces and their individual moves are known. However, apparently there are several piled layers of chess sets, and some of the chess pieces have the option to move in a third dimension. This third dimension may symbolize regulatory activity exerted by intermediates (Rolland et al., 2006) as well as by stress perception and subsequent signaling (Ye et al., 2017).

Each plant species is characterized by a typical metabolite pattern. This has been the basis for selection of crop plants and breeding activities aimed at improving this pattern to overproduce metabolites of interest, such as starch or oil. Moderate stress may (1) generally inhibit all catalytic activities or (2) preferentially inhibit production of only a part of the metabolic spectrum. These two different types of stress response may be explained as follows. (1) A general inhibition of metabolic activity will be observed, if enzymes catalyzing early events of photosynthesis will be most stress sensitive. In this case all reactions downstream of the metabolic pathway will be inhibited due to lack of substrates. (2) Stress will result in a modified metabolite pattern, if enzymes located downstream of a branching point of metabolic pathways will be more stress sensitive than those catalyzing production of other metabolites. Stress-inhibited enzymes will not compete for substrates with enzymes of the other pathways, and respective end products will be generated at an enhanced rate. In some cases, initial stress-induced stimulation of individual metabolite production has been observed followed by a decrease under prolonged or high stress (Boestfleisch et al., 2014). Such effects already are used to "improve quality" of plants or plant organs with respect to sensory aspects or for subsequent pharmaceutical applications, for instance (de Matos Nunes et al., 2014).

These observations have led to the development of different strategies to identify breeding targets for improvement of stress resistance. The first ones are based on the fact that all plant activities depend on photosynthetic activity. This allowed to introduce methods for noninvasive monitoring of plant performance (Papenbrock et al., 2016 and citations therein). The inhibition of early events of photosynthesis will result in an overreduction of the photosynthetic electron transport chain and a reduced efficiency in light energy use. This may be observed by monitoring chlorophyll fluorescence parameters. Analyzing the kinetics of respective measurements will allow to estimate further data, such as redox state of quinones and ATP/ADP and $NADP^+$/NADPH ratios (Flexas et al., 2002).

Several researchers have reported that stress resistance of plants may vary during development (physiological parameters of seedling stage and mature plants may differ) (de Souza et al., 2016). This observation may be explained in terms of the aforementioned branching of metabolic pathways as follows. (1) Enhanced demand for energy (ATP and NADPH) will prevent overreduction of the electron transport chain and will reduce the risk of ROS production and subsequent oxidative damage. (2) If seeds preferentially depend on carbohydrates of storage organs, stress-induced inhibition of photosynthesis will escape from detection.

The aforementioned problems may be avoided by analyzing the activities of regulatory systems and subsequent monitoring of gene expression at the level of messenger RNA abundance or proteome analysis

(Chen and Hoehenwarter, 2015; Liang et al., 2016). In the following discussion we will comment on problems related to these approaches. All of the methods require a precise description of growth conditions and resulting physiology of experimental plants. Without such data, comparison of individual results among research teams is not possible (Pérez-Harguideguy et al., 2013).

9.1 Measuring Chlorophyll Fluorescence

Any stress reaction that inhibits the export or consumption of assimilates will affect primary reactions of photosynthesis. If the pool of ATP and NADPH is not recycled, overreduction of the electron transport chain will be observed. As depicted in Fig. 14.1 already, ROS production and emission from activated chlorophyll of fluorescent light quanta will be favored, if the rate of photosynthetic energy use does equal the absorption rate of light energy quanta. Although only a minor portion (1%–4%) of absorbed light energy will be released in the form of fluorescent light quanta, changes in photosynthetic activity will result in changes of chlorophyll fluorescence as well. Chlorophyll fluorescence measurement is widely used to analyze stress effects on photosynthesis (Rungrat et al., 2016). A characteristic pattern of chlorophyll fluorescence signals is observed when dark-adapted leaves are exposed to continuous light (see Fig. 14.11A for two sketches).

Optimized light absorption will allow plants to grow at low light intensity. However, in the presence of high light intensities, use of absorbed light energy is limited by the maximal turnover rate of CO_2 fixation or export and metabolic use of products. Excess absorbed light energy will result in ROS production, which then might damage cell components. Plants have two options to react: (1) scavenging of ROS at the expense of redox energy (water-to-water cycle, for instance) or (2) preventing excess light absorption [degradation of chlorophyll, photoinhibition (qI) mechanisms (Gururani et al., 2015; Kromdijk et al., 2016)]. Either of these options will lead to reduced growth rate, reproduction, and even death. To balance between optimal light absorption and maximal risk avoidance, plants have evolved short- and long-term acclimation mechanisms. Activation of any protective pathway will result in a typical signature of chlorophyll fluorescence (Lawson et al., 2012). Data analysis is based on the assumption that under physiological conditions the portion of heat released from light-activated chlorophyll is constant. Therefore the observed degree of quenching of chlorophyll fluorescence depends on the efficiency of competing pathways, i.e., it depends on the current rates of photosynthetic electron transport and energy dissipation by nonphotosynthetic reactions. This interpretation implies that photosynthetic efficiency can be described as the fraction of absorbed light energy converted to energy of chemical bonds during photosynthesis. Thus the ratio of Fv/Fm, the efficiency of excitation capture by "open" photosynthetic reaction centers (regular values: 0.74–0.85), is used to characterize plant performance. A decline in Fv/Fm indicates a decline in the quantum yield of photochemistry and a disturbance in or damage to the photosynthetic apparatus (Papenbrock et al., 2016). However, this is only a very simplified approach not taking into account the fact that modification of the fine structure of thylakoid membrane will affect the fluorescence signals as well. However, such changes can be analyzed by analysis of fluorescence signals at high time resolution as shown in Fig. 14.11B.

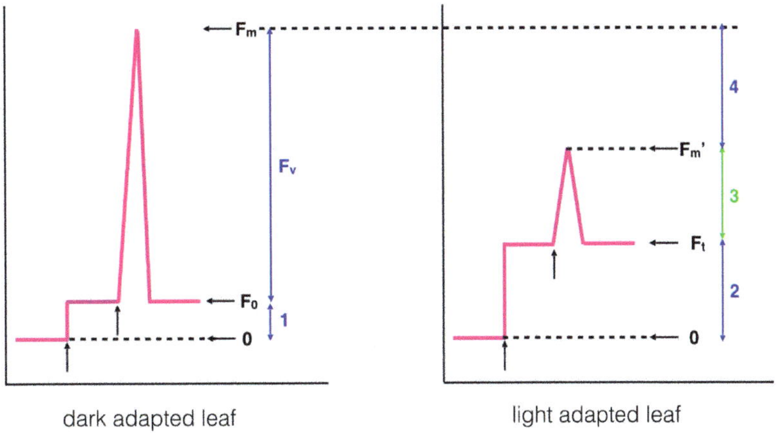

FIGURE 14.11A Sketches of Kautsky curves. Chlorophyll fluorescence of dark- and light-adapted leaves. (Left) Dark-adapted leaves: During dark adaptation the light-harvesting system and components of the photosynthetic electron transport chain adopt their respective steady state. Thus the whole system is not ready for immediate electron transport. Therefore a high-intensity light flash will induce maximal chlorophyll fluorescence. The respective fluorescence intensity is termed F_m. (Right) Light-adapted leaves: Such a high intensity of fluorescence will not be reached when light at the same intensity is flashed on illuminated (photosynthetically active) leaves. In such case only a reduced level of chlorophyll fluorescence, F_m', is achieved. Moreover, in the presence of actinic light, photosynthetically active leaves are continuously emitting fluorescent light at a level termed F_t. For data analysis the extent of flash-induced chlorophyll fluorescence (F_v in dark-adapted leaves and F_v' in light- adapted leaves, respectively) is calculated from measurements symbolized in this figure.

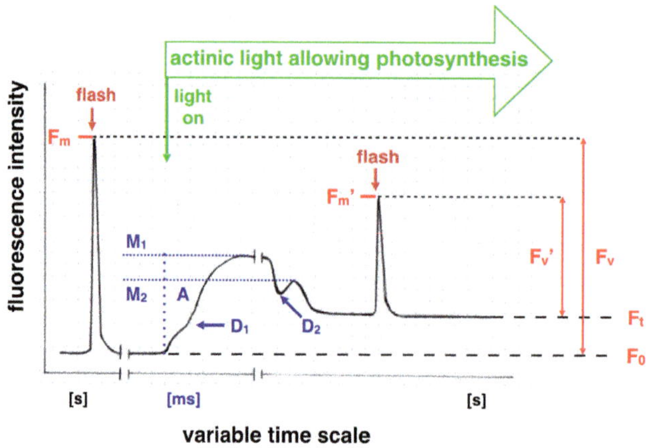

FIGURE 14.11B **Sketch of a Kautsky curve at high time resolution.** Take into account that there are two interruptions in the time scale (second–milliseconds–seconds). At low time resolution some noise in the fluorescence signal can be detected when actinic light is switched on and dark-adapted leaves will start photosynthesis. However, if the initial phase of photosynthetic activity is analyzed at millisecond time resolution, a characteristic fluorescence pattern becomes detectable as depicted in this figure. In the course of time the curve shows some dips (D_1, D_2, and may be even more) and two overshoot phases (M_1 and M_2) separated by a minimum (D_2) before reaching the steady-state level of continuous illumination. These observations are discussed based on the assumption that energy and electron transfer reactions involve pools of reaction partners. These pools have to be filled and buffered to allow optimal energy transfer rate. As individual pools differ in reaction kinetics, the observed dips and overshoots can be found. It is assumed, for instance, that dip D_1 can be attributed to the PSII plastoquinol pool Q_A. Overreduction of individual pools will be indicated by intermediate maxima (M_1 and M_2). Ratios of the pool sizes may be calculated from areas (A) marked by the curve of fluorescence intensity and lines, such as the dotted lines indicated in this figure. In the presence of continuous actinic light, chlorophyll fluorescence will reach a steady-state level F_t as indicated in the figure. This fluorescence intensity is indicating an equilibrium, characteristic for the plant species and the respective experimental conditions (including the developmental stage of the leaf).

As nonphotochemical pathways significantly differ in turnover rates, different types of nonphotochemical quenching (NPQ) can be discriminated by their kinetics (García-Plazaola et al., 2012).

qI has been identified as the fraction of NPQ occurring at the slowest rate. It has been attributed to damage, especially degradation of the D1 subunit of PSII, caused by abiotic stress, such as excess of light. qI can be calculated from a decrease in chlorophyll fluorescence at maximal quantum efficiency (Baker, 2008). Energy-dependent quenching (qE) is the most rapid fraction of NPQ. It is induced within seconds when the transthylakoid membrane pH gradient exceeds a threshold value. qE fluorescence quenching is caused by energy transfer from light-activated chlorophyll to zeaxanthin and subsequent heat release. This fraction of NPQ is controlled by the activity of the enzyme violaxanthin de-epoxidase (activity of this enzyme is under the control of the pH of the thylakoid lumen), and the protonation state of the PsbS subunit, which controls the three-dimensional orientation of the light-harvesting antenna of PSII and distances among bound pigments (Ware et al., 2015; Ikeuchi et al., 2016). At high light intensity qE is the most important fraction of NPQ. A third fraction (qZ) of NPQ occurring at an intermediate rate (in the range of minutes) is associated with enzyme activities, preferentially the synthesis of zeaxanthin (Nilkens et al., 2010).

At a first glance, chlorophyll fluorescence measurement appears to be a reliable tool to analyze and evaluate plant stress resistance and performance of crops in laboratories and in the field. Indeed, the technique has been routinely used for many years to monitor the photosynthetic performance of plants noninvasively (Murchie and Lawson, 2013). However, this success story does not always apply. The fact that *Fv*/*Fm* is a relatively inert ratio where stress-induced changes are detected rather late is among the reasons for failure to see early stress responses. Therefore improved methods of data analysis have been suggested early. However, another aspect is even more important: the biochemical reason for changes in the *Fv*/*Fm* ratio cannot be judged for sure. This reservation especially applies (1) when plants are grown under severe stress, (2) if physiology of plants allows alternating between different types (C_3, C_4, CAM) of photosynthesis, and (3) if there is an option to use alternative sinks for energy or assimilates (Papenbrock et al., 2016).

This problem becomes obvious when remembering basics of photorespiration (see earlier discussion). Photorespiration is initiated by binding of molecular oxygen to the acceptor Ru-1,5-P. This reaction is catalyzed by the light-activated enzyme RuBisCO, and subsequent reactions are driven at the expense of energy supplied by photosynthetic electron transport. Thus photorespiration is a major energy sink under conditions when CO_2 assimilation rate match the rate of absorption of light quanta. In contrast to an "energy wasting" reaction resulting in NPG, photorespiration uses energy provided by "photosynthetic electron transport", i.e., photorespiration activity is contributing to photosynthetic quenching. The same holds true for the water-to-water cycle that involves the Mehler reaction (Mehler, 1951). In facultative C_4 plants, on the other hand, the contribution of PSI-driven cyclic photophosphorylation may significantly vary with the degree of stress and subsequent stress adaptation (Wilhelm and Selmar 2011). As monitoring of chlorophyll fluorescence parameters has failed to evaluate the degree of stress exerted on some experimental plant species (Papenbrock et al., 2016) and interpretation may be uncertain due to the aforementioned arguments, it is advisable to apply additional methods. Chlorophyll fluorescence measurement allows noninvasive continuous monitoring of cultures, whereas most alternative methods require harvesting of plant samples.

9.2 Parameters to Monitor Plant Performance

Obviously monitoring chlorophyll fluorescence is ideal for evaluating the degree of stress a plant is exposed to (Kalaji et al., 2016; Tschiersch et al., 2017). As outlined earlier, any inhibition of growth and use of assimilates will serve as a feedback to primary reactions of photosynthesis and thus will affect a parameter detectable by chlorophyll fluorescence. Therefore chlorophyll fluorescence monitoring of plants became an established method over the years. However, there is an awkward aspect: chlorophyll fluorescence parameters per se are not reliable traits to plan a breeding project. In most cases respective measurements do not allow to precisely localize the site of stress attack. For breeding purpose it would be essential to identify the most vulnerable point in plant biochemistry/physiology. Therefore most current projects are based on "-omics" approaches; i.e., researchers are analyzing stress effects on gene expression, patterns of enzyme activities, and interaction of metabolite pools (Shulaev et al., 2008). Such approaches became feasible as respective databases of model plants and most crop plants now are available. Nevertheless, data analysis still is extremely laborious as stresses induce a range of physiological and biochemical responses in plants. Accordingly, gene networks rather than individual genes are involved in building stress resistance. Nevertheless, analysis of the complete metabolome could significantly contribute to the study of stress biology in plants by identifying products of stress metabolism or molecules that are part of the acclimation response of plants. Results could be further tested by correlation with changes in transcriptome, and subsequently could be confirmed by mutant analysis (Shulaev et al., 2008).

There are an ever-growing number of stress-responsive genes that are identified. However, plant species differ in the patterns of respective gene activities, and similar differences are found with respect to patterns of protein abundance, enzyme activities, and metabolite concentrations and its turnover rates. Some of the stress-inducible genes have been introduced via gene transfer into sensitive model plants, such as *Arabidopsis*, or crops, such as rice, and the degree of improved stress resistance has been evaluated (Ahn et al., 2017; Garg et al., 2017; Jung et al., 2017; Zhao et al., 2016). Aligning gene activity to respective biochemical and physiological functions will be the next step in such projects. By comparing salt-sensitive and salt-resistant inbred lines of sweet sorghum, Sui et al. (2015) found out that salt resistance of this plant species is linked to increased synthesis of sucrose synthetase. Resistant plants are able to keep a high level sucrose export to the phloem and thus prevent leaves from bleaching. This result is in accordance with the view discussed earlier that inhibition of assimilate use will enhance the risk of ROS production and peroxidative damage.

It is well documented in the literature, for instance, that drought stress triggers ABA signals, and physiological responses, such as stomata closure, are mentioned in every botany textbook. Accordingly, it has been shown that several drought-responsive genes are responding to exogenous ABA treatment (Zhao et al., 2016). With this information in mind the finding that some of the drought-responsive genes are not activated by ABA treatment was a surprise (Yamaguchi-Shinozaki and Shinozaki, 2005). This finding indicates that there may be more than one signaling pathway linking sensors of an individual stress type to genes responding to that respective stress. Indeed, meanwhile, several researchers have observed that cross talk among signaling pathways apparently is a common rule rather than an exception (Chinnusamy et al., 2004).

In many publications authors have classified stress-responsive genes to find clues to understand the respective plant species–specific mechanism leading to stress resistance. Shinozaki et al. (2003), for instance, have defined two groups of drought-inducible genes in rice. They have discriminated resistance-mediating proteins (chaperons, aquaporin, enzymes for osmolyte biosynthesis and transport) from regulatory proteins (transcription factors, protein kinases and phosphatases, and other signaling proteins). A special aspect was the finding that several of the drought-responsive transcription factors are responding to other stresses as well. This result applied not only for the model plant *Arabidopsis* but also was found to be true for other plants, such as rice. Moreover, the pattern of stress-inducible genes was found to be similar in plants as different as rice and *Arabidopsis*. Among the 73 genes identified as stress inducible in rice by Shinozaki and Yamaguchi-Shinozaki (2007), 51 had been identified in *Arabidopsis* as well.

It may be speculated that some of the responsive genes are under the control of stress-mediated changes in metabolite concentrations (see earlier chapter: genes may be under the control of metabolites) rather than regulated by stress-mediated signaling pathways. This question can be solved by analyzing gene-to-metabolite networks. Such networks may be constructed based on data mining of gene profiling and metabolite profiling data of respective stress experiments (Awlia et al., 2016; Zhang et al., 2016).

The observation that a multitude of genes is responding to an individual stress can be explained as well by the observation that several genes can be under the control of the same transcription factor. Meanwhile, several families of transcription factors are known to translate stress signals into changes in gene expression. Some of these families, such as heat shock transcription factors, are found not only in plants but also in other kingdoms. Among plant-specific families, the AP2/ERF family has been identified to be involved in cold and drought signaling (Mizoi et al., 2012; Ahn et al., 2017; Jung et al., 2017). It is composed of four subfamilies, AP2, RAV, ERF, and DREB, which share a well-conserved DNA-binding domain. DREB transcription factors (dehydration-responsive element binding) are regulating the expression of many stress-inducible genes mostly in an ABA-independent manner. They control gene expression by interacting with a DRE/CRT cis-element present in the promoter region of target genes (Lata and Prasad, 2011). NAC proteins are a second family of plant-specific transcription factors. Typically, they share a conserved N-terminal domain and a diversified C-terminal transcription regulatory region (see Puranik et al., 2012; for a review). Meanwhile, more than 100 NAC genes have been identified in *Arabidopsis* and rice. Some of them are stress responsive but are not necessarily ABA inducible (Nakashima et al., 2012). It has been shown in several experiments that transgenic *Arabidopsis* and rice plants overexpressing stress-responsive NAC genes have acquired an improved drought resistance. Based on these studies it may be expected that improved stress resistance of crop plants can be achieved via biotechnological approaches (Nakashima et al., 2012). This expectation is supported by successful improvement of wheat by overexpression of an NAC gene (Xue et al., 2011).

9.3 Improvement of the Habitat

Beside optimization of plant response further adjustments can be done in the complex interacting system of soil, plant, and atmosphere. Soil properties (such as conductivity, moisture, nutrients), terrestrial vegetation (including vegetation–albedo feedbacks), and atmospheric flows are parts of a complex interacting system, characterized by the presence of many feedback mechanisms between the various components (Baudena et al., 2008). The vegetation together with soil properties determines the flux of water from the soil to the atmosphere through evapotranspiration, i.e., the surface moisture flux. Feedback between plants and the soil frequently is invoked on the basis of evidence of mutual effects. Feedback can operate through pathways involving soil physical properties, chemical and biogeochemical properties and processes, and biological properties, including the community composition of the microbiota and soil fauna (Ehrenfeld et al., 2005). The evidence of feedback is strongest for plants growing in extreme environments and for plant–mutualist or plant–enemy interactions.

Sands hold little water in the range of water potentials available to plants because of their particle size distribution, whereas clays hold a large amount of water (Fig. 14.12). However, in the latter one much of the water is unavailable to plants because of the strong surface forces (fine colloidal nature). As an intermediate substrate loam holds more plant-available water than clays but dispose also a much lower saturated hydraulic conductivity than sand. However, even drainage and evaporation are comparably low. The portion of evaporation can further be reduced by amendments such as biochar

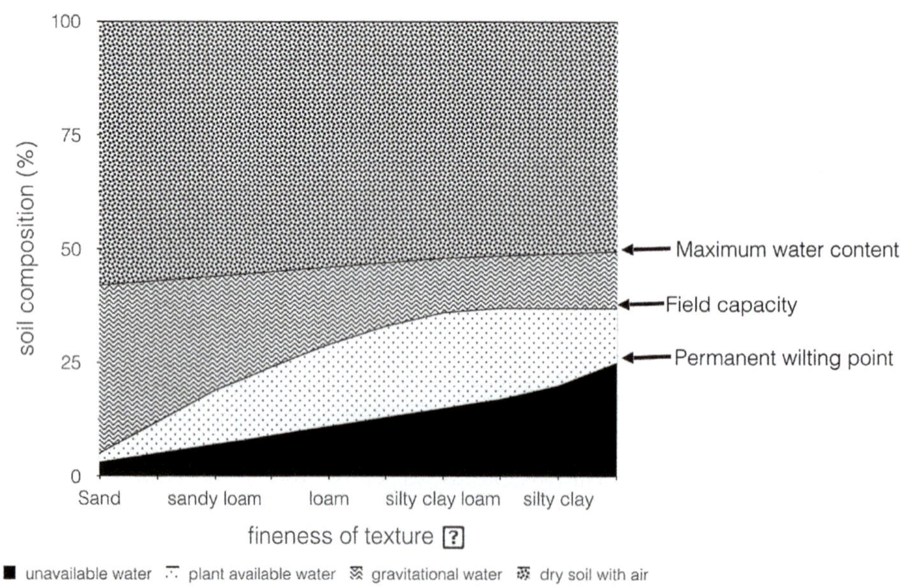

FIGURE 14.12 **Schematic illustration of the water-holding capacity of different soil materials.**

and/or compost. One important property is the surface area in the pores of the biochar (specific surface area up to 25.38 m^2/g), which helps to increase the water-holding capacity of the soil, the plant-available water, and organic matter content without much effect on evaporation. It also buffers heavy metal toxicity. Compost can additionally enhance nutrient supply and ion exchange capacity. If both are supplied together, they improve soil properties such as mineralization and microbial activity, and can reduce environmental stress by the reduction of soil evaporation and optimization of nutrient supply.

In the complex interacting system of soil, plant, and atmosphere further adjustments can be done in the last two spheres. There is abundant evidence for the mutual effects of plants on soil, and vice versa, acting through mechanisms involving all aspects of plant growth, morphology, and physiology, as well as all the physical, chemical, and biological components of the soil (Ehrenfeld et al., 2005). With respect to plants, an effective uptake of the plant-available water from the soil and high resistance to water loss to the atmosphere, including a high WUE of productivity and photosynthesis, are essential parts for a sustainable and productive survival in extreme habitats. These properties can reduce the risk of ion stress but may enhance the risk of exposure to oxidative stress.

The latter can be reduced by optimization of the atmospheric conditions to reduce the water or light stress of plants. In the last century several projects have started with the implementation of—or close adjustment of—local environmental conditions. The Sahara forest project (Jordan and Qatar), the Watergy greenhouse (Almería, Spain), or the Sundrop Farm in Adelaide are just some good examples with the aim of increasing atmospheric water and CO_2 content near the plant. Ehrenfeld et al. (2005) conclude with recommendations for a more critical appraisal of feedback and for new directions of research.

10. CONCLUSIONS

Plant stress research originates from experiments elucidating environmental effects on whole plant and cellular phenotype. High light, drought, and soil salinity stresses are among the conditions analyzed in much detail. Researchers were puzzled by the observation that drought and salinity, although targeting the roots, triggered inhibition of photosynthesis at an early phase of stress response. Vice versa, responses to high light stress were not limited to leaves. Meanwhile, it is generally accepted that plant stress response represents an active process aimed at establishing a homeostasis under an array of suboptimal environmental conditions. Based on this concept researchers have been and still are trying to generate better yielding crops. Recent developments of technologies in molecular breeding, biological control, and biotechnological applications in agriculture have been considered most powerful approaches in maintaining or even increasing crop yield.

Nevertheless, development of cultivars with higher yields, better product quality, and improved stress tolerance requires detailed information on the biological processes that enable plants to survive in stressful environments.

It is generally observed that most abiotic stresses cause ROS production in parallel to inhibition of photosynthesis. To prevent oxidative damage, plants are equipped with an antioxidative system composed of low-molecular-weight antioxidants and protective enzymes involved in futile, ROS-consuming reaction sequences. As a rule, there is a sequence of events leading from (1) stress perception; via (2) signaling at cellular (Ca^{++}; MAP kinases; etc.), tissue, or whole plant level (ABA; jasmonate, etc.); to (3) activation of transcription factors and corresponding stress-controlled genes; and (4) finally to metabolic and physiological responses. Monitoring of essential reactions, such as photosynthesis, is a good approach to gain general information on plant performance in growth and reproduction. In this context measuring chlorophyll fluorescence signals is a well-accepted method, because it allows nondestructive distant measurement without sampling of plant material. However, information gained by this method is not sufficient to exactly identify the bottlenecks in plant performance under stress and to develop cultivation strategies or breeding concepts.

REFERENCES

Agati, G., Brunetti, C., Di Ferdinando, M., Ferrini, F., Pollastri, S., Tattini, M., 2013. Functional roles of flavonoids in photoprotection: new evidence, lessons from the past. Plant Physiol. Biochem. 72, 35–45.

Ahn, H., Jung, I., Shin, S.-J., Park, J., Rhee, S., Kim, J.-K., Jung, W., Kwon, H.–B., Kim, S., 2017. Transcriptional network analysis reveals drought resistance mechanisms of AP2/ERF transgenic rice. Front. Plant Sci. http://dx.doi.org/10.3389/fpls.2017.01044.

Allen, J.F., 2003. Cyclic, pseudocyclic and noncyclic photophosphorylation: new links in the chain. Trends Plant Sci. 8, 15–19.

Anderson, J.M., 2012. Lateral heterogeneity of plant thylakoid protein complexes: early reminiscences. Phil. Trans. R. Soc. B 367, 3384–3388.

Awlia, M., Nigro, A., Fajkus, J., Schmoeckel, S.M., Negrao, S., Santelia, D., Trtilek, M., Tester, M., Julkowska, M.M., Panzarova, K., 2016. High-throughput non-destructive phenotyping of traits that contribute to salinity tolerance in *Arabidopsis thaliana*. Front. Plant Sci. http://dx.doi.org/10.3389/fpls.2016.01414.

Badawi, G.H., Yamauchi, Y., Shimada, E., Sasaki, R., Kawano, N., Tanaka, K., Tanaka, K., 2004. Enhanced tolerance to salt stress and water deficit by overexpressing superoxide dismutase in tobacco (*Nicotiana tabacum*) chloroplasts. Plant Sci. 166, 919–928.

Bak, S., Beisson, F., Bishop, G., Hamberger, B., Höfer, R., Paquette, S., Werck-Reichhart, D., 2011. Cytochromes P450. In: The *Arabidopsis* Book. American Society of Plant Biologists. http://dx.doi.org/10.1199/tab.0144.

Baker, N.R., 2008. Chlorophyll fluorescence: a probe of photosynthesis in vivo. Annu. Rev. Plant Biol. 59, 89–113.

Baudena, M., Andrea, F.D., Provenzale, A., 2008. A model for soil-vegetation-atmosphere interactions in water-limited ecosystems. Water Resour. Res. 44, W12429. http://dx.doi.org/10.1029/2008WR007172.

Baxter, A., Mittler, R., Suzuki, N., 2014. ROS as key players in plant stress signalling. J. Exp. Bot. 65, 1229–1240.

Bennett, D.I.G., Amarnath, K., Fleming, G.R., 2013. A structure-based model of energy transfer reveals the principles of light harvesting in photosystem II supercomplexes. J. Am. Chem. Soc. 135, 9164–9173.

Bhaskara, G.B., Yang, T.-H., Verslues, P.E., 2015. Dynamic proline metabolism: importance and regulation in water limited environments. Front. Plant Sci. 6, 484. http://dx.doi.org/10.3389/fpls.2015.00484.

Biswal, B., Joshi, P.N., Raval, M.K., Biswal, U.C., 2011. Photosynthesis, a global sensor of environmental stress in green plants: stress signalling and adaptation. Curr. Sci. 101, 47–56.

Boestfleisch, C., Wagenseil, N.B., Buhmann, A.K., Seal, C.E., Wade, E.M., Muscolo, A., Papenbrock, J., 2014. Manipulating the antioxidant capacity of halophytes to increase their cultural and economic value through saline cultivation. AoB Plants 6. http://dx.doi.org/10.1093/aobpla/plu046.

Boyer, P.D., 2000. Catalytic site forms and controls in ATP synthase catalysis. Biochim. Biophys. Acta 1458, 252–262.

Buchanan, B.B., Gruissem, W., Jones, R.L., 2015. Biochemistry & Molecular Biology of Plants, second ed. John Wiley & Sons, Ltd., Chichester, UK.

Caffarri, S., Tibiletti, T., Jennings, R.C., Santabarbara, S., 2014. A comparison between plant photosystem I and photosystem ii architecture and functioning. Curr. Protein Peptide Sci. 15, 296–331.

Caretto, S., Linsalata, V., Colella, G., Giovanni Mita, G., Vincenzo Lattanzio, V., 2015. Carbon fluxes between primary metabolism and phenolic pathway in plant tissues under stress. Int. J. Mol. Sci. 16, 26378–26394.

Chen, Y., Hoehenwarter, W., 2015. Changes in the phosphoproteome and metabolome link early signaling events to rearrangement of photosynthesis and central metabolism in salinity and oxidative stress response in *Arabidopsis*. Plant Physiol. 169, 3021–3033.

Cheynier, V., Comte, G., Davies, K.M., Lattanzio, V., Martens, S., 2013. Plant phenolics: recent advances on their biosynthesis, genetics, and ecophysiology. Plant Physiol. Biochem. 72, 1–20.

Chinnusamy, V., Schumaker, K., Zhu, J.-K., 2004. Molecular genetic perspectives on cross-talk and specificity in abiotic stress signalling in plants. Crosstalk in plant signal transduction special issue. J. Exp. Bot. 55, 225–236.

Cramer, G.R., 2003. Differential effects of salinity on leaf elongation kinetics of three grass species. Plant Soil 253, 233–244.

Das, K., Roychoudhury, A., 2014. Reactive oxygen species (ROS) and response of antioxidants as ROS-scavengers during environmental stress in plants. Front. Environ. Sci. http://dx.doi.org/10.3389/fenvs.2014.00053.

de Matos Nunes, J., Bertodo, L.O.O., daRosa, L.M.G., Von Poser, G.L., Rech, S.B., 2014. Stress induction of valuable secondary metabolites in *Hypericum polyanthemum* acclimatized plants. South Afr. J. Bot. 94, 182–189.

de Souza, T.C., Magalhães, P.C., de Castro, E.M., Duarte, V.P., Lavinsky, A.O., 2016. Corn root morpho-anatomy at different development stages and yield under water stress. Pesq. Agropec. Bras 51 (4) Brasília.

Ehrenfeld, J.G., Ravit, B., Elgersma, K., 2005. Feedback in the plan-soil system. Annu. Rev. Environ. Resour. 30, 75–115.

Farooq, S., Chmeliov, J., Trinkunas, G., Valkunas, L., van Amerongen, H., 2016. Is there excitation energy transfer between different layers of stacked photosystem-II-containing thylakoid membranes? J. Phys. Chem. Lett. 7, 1406–1410.

Flexas, J., Escalona, J.M., Evain, S., Gulías, J., Moya, I., Osmond, C.B., Medrano, H., 2002. Steady-state chlorophyll fluorescence (Fs) measurements as a tool to follow variations of net CO_2 assimilation and stomatal conductance during water-stress in C_3 plants. Physiol. Plant. 114 (2), 231–240.

Floryszak-Wieczorek, J., Arasimowicz-Jelonek, M., 2016. Contrasting regulation of no and ROS in potato defense-associated metabolism in response to pathogens of different lifestyles. PLoS One. http://dx.doi.org/10.1371/journal.pone.0163546.

Flowers, T.J., Colmer, T.D., 2008. Salinity tolerance in halophytes. New Phytol. 179, 945–963.

Flowers, T.J., Troke, P.F., Yeo, A.R., 1977. The mechanism of salt tolerance in halophytes. Annu. Rev. Plant Physiol. 28, 89–121.

García-Plazaola, J.I., Artetxe, U., Becerril, J.M., 1999. Diurnal changes in antioxidant and carotenoid composition in the Mediterranean schlerophyll tree *Quercus ilex* (L) during winter. Plant Sci. 143, 125–133.

García-Plazaola, J.I., Esteban, R., Fernández-Marín, B., Kranner, I., Porcar-Castell, A., 2012. Thermal energy dissipation and xanthophyll cycles beyond the *Arabidopsis* model. Photosynth Res. 113, 89–103.

Garg, B., Gill, S.S., Biswas, D.K., Sahoo, R.K., Kunchge, N.S., Tuteja, R., Tuteja, N., 2017. Simultaneous expression of PDH45 with EPSPS gene improves salinity and herbicide tolerance in transgenic tobacco plants. Front. Plant Sci. http://dx.doi.org/10.3389/fpls.2017.00364.

Gil, R., Bautista, I., Boscaiu, M., Lidón, A., Wankhade, S., Sánchez, H., Llinares, J., Vicente, O., 2014. Responses of five Mediterranean halophytes to seasonal changes in environmental conditions. AoB Plants 6. http://dx.doi.org/10.1093/aobpla/plu049.

Gururani, M.A., Venkatesh, J., Tran, L.-S.P., 2015. Regulation of photosynthesis during abiotic stress-induced photoinhibition. Mol. Plant 8, 1304–1320.

Hanke, G., Mulo, P., 2013. Plant type ferredoxins and ferredoxin-dependent metabolism. Plant Cell Environ. 36, 1071–1084.

Hatch, M.D., 1987. C_4 photosynthesis: a unique blend of modified anatomy and ultrastructure. Biochim. Biophys. Acta 895, 81–106.

Hatch, M.D., Kagawa, T., 1976. Photosynthetic activities of isolated bundle sheath cells in relation to differing mechanisms of C_4 pathway photosynthesis. Arch. Biochem. Biophys. 175, 39–53.

Hayat, S., Hayat, Q., Alyemeni, M.N., Wani, A.S., Pichtel, J., Ahmad, A., 2012. Role of proline under changing environments. Plant Signal. Behav. 7, 1456–1466.

He, X., Miginiac-Maslow, M., Sigalat, C., Keryer, E., Haraux, F., 2000. Mechanism of activation of the chloroplast ATP synthase. J. Biol. Chem. 275 (18), 13250–13258 Issue of May 5.

Hill, R., Bendall, F., 1960. Function of the two cytochrome components in chloroplasts – a working hypothesis. Nature 186, 136–137.

Hisabori, T., Sunamura, E.-I., Kim, Y., Konno, H., 2013. The chloroplast ATP synthase features the characteristic redox regulation machinery. Antioxidants Redox Signal. 19, 1846–1854.

Höhner, R., Aboukila, A., Kunz, H.–H., Venema, K., 2016. Proton gradients and proton-dependent transport processes in the chloroplast. Front. Plant Sci. 7, 218. http://dx.doi.org/10.3389/fpls. 016.00218.

Huchzermeyer, B., 1988. Nucleotide binding and ATPase activity of membrane bound chloroplast coupling factor (CF1). Z. Naturforsch. 43c, 133–139.

Huchzermeyer, B., Löhr, A., Willms, I., 1986. A direct interaction between photosystem I and the chloroplast coupling factor. Biochem. J. 234, 217–220.

Ibrahim, D.G., Gilbert, M.E., Ripley, B.S., Osborne, C.P., 2008. Seasonal differences in photosynthesis between the C_3 and C_4 subspecies of *Alloteropsis semialata* are offset by frost and drought. Plant Cell Environ. 31, 1038–1050.

Ikeuchi, M., Uebayashi, N., Sato, F., Endo, T., 2016. Physiological functions of PsbS-dependent and PsbS-independent NPQ under naturally fluctuating light conditions. Plant Cell Physiol. 55, 1286–1295.

IPCC, 2014. Climate Change 2014: Synthesis Report.

Jagendorf, A.T., 2002. Photophosphorylation and the chemiosmotic perspective. Photosynth. Res. 73, 233–241.

Ji, H., Pardo, J.M., Batelli, G., van Oosten, M.J., Bressan, R.A., Li, X., 2013. The salt overly sensitive (SOS) pathway: established and emerging roles. Mol. Plant 6, 275–286.

Jing, R., Sun, L.N., Zhang, Q.Y., Song, X.S., 2016. Drought tolerance is correlated with the activity of antioxidant enzymes in *Cerasus humilis* seedlings. BioMed Res. Int.:9851095. http://dx.doi.org/10.1155/2016/9851095.

Jithesh, M.N., Prashanth, S.R., Sivaprakash, K.R., Parida, A., 2006. Monitoring expression profiles of antioxidant genes to salinity, iron, oxidative, light and hyperosmotic stresses in the highly salt tolerant grey mangrove, *Avicennia marina* (Forsk.) Vierh. by mRNA analysis. Plant Cell Rep. 25, 865–876.

Jung, H., Chung, P.J., Park, S.–H., Reveche, M.C.F.R., Kim, Y.S., Suh, J.-W., Kim, J.-K., 2017. Overexpression of OsERF48 causes regulation of OsCML16, a calmodulin-like protein gene that enhances root growth and drought tolerance. Plant Biotechnol. J. http://dx.doi.org/10.1111/pbi.12716.

Kalaji, H.M., Jajoo, A., Oukarroum, A., Brestic, M., Zivcak, M., Samborska, I.A., Cetner, M.D., Lukasik, I., Goltsev, V., Ladle, R.J., 2016. Chlorophyll a fluorescence as a tool to monitor physiological status of plants under abiotic stress conditions. Acta Physiol. Plant 38, 102. http://dx.doi.org/10.10077/s11738-016-2113-y.

Kohli, A., Sreenivasulu, N., Lakshmanan, P., Kumar, P.P., 2013. The phytohormone crosstalk paradigm takes center stage in understanding how plants respond to abiotic stresses. Plant Cell Rep. 32, 945–957.

Koyro, H.–W., Huchzermeyer, B., 2008. Designing new plant genotypes with enhanced resistance to salinity. In: Neumann, K.-H., Kumar, A., Sopory, S.K. (Eds.), Recent Advances in Plant Biotechnology and its Applications: Prof. Dr. Karl-Hermann Neumann Commemorative Volume. I.K. International Pub. House, New Delhi, pp. 428–451.

Koyro, H.–W., Geißler, N., Hussin, S., Debez, A., Huchzermeyer, B., 2008. Strategies of halophytes to survive in a salty environment. In: Khan, N.A., Singh, S. (Eds.), Abiotic Stress and Plant Responses. I. K. International Publishing House, New Delhi, pp. 83–104.

Koyro, H.–W., Geissler, N., Seenivasan, R., Huchzermeyer, B., 2011. Plant stress physiology: physiological and biochemical strategies allowing plants/crops to thrive under ionic stress. In: Pessarakli, M. (Ed.), Handbook of Plant and Crop Stress, 3d edition. CRC Press, Boca Raton, FL, USA, pp. 1051–1093.

Koyro, H.–W., Huchzermeyer, B., 2004. Ecophysiological needs of the potential biomass crop *Spartina townsendii* GROV. Trop. Ecol. 45, 123–139.

Koyro, H.–W., Huchzermeyer, B., Zörb, C., 2014a. Effects of hyperosmotic salinity on protein patterns and enzyme activities. In: Pessarakli, M. (Ed.), Handbook of Plant and Crop Physiology, third ed. Taylor & Francis, NY, USA, pp. 487–507.

Koyro, H.–W., Hussain, T., Huchzermeyer, B., Khan, M.A., 2013. Photosynthetic and growth responses of a perennial halophytic grass *Panicum turgidum* to increasing NaCl concentrations. Environ. Exp. Bot. 91, 22–29.

Koyro, H.–W., Stelzer, R., 1988. Ion concentrations in the cytoplasm and vacuoles of rhizodermal cells from NaCl treated Sorghum, *S. partina* and *P. uccinellia* plants. J. Plant Physiol. 133, 441–446.

Koyro, H.–W., 2003. Study of potential cash crop halophytes by a quick check system: determination of the threshold of salinity tolerance and the ecophysiological demands. In: Lieth, H., Mochtchenko, M. (Eds.), Cash Crop Halophytes: Recent Studies. Tasks for Vegetation Science, vol. 38. Springer, Dordrecht. http://dx.doi.org/10.1007/978-94-017-0211-9_1.

Koyro, H.–W., Lieth, H., Gul, B., Ansari, R., Huchzermeyer, B., Abideen, Z., Hussain, T., Khan, M.A., 2014b. Importance of the diversity within the halophytes to agriculture and land management in arid and semiarid countries. In: Khan, M.A., Böer, B., Öztürk, M., Zahran, T., Abdessalaam, A., Clüsener-Godt, M., Gul, B. (Eds.), Sabkha Ecosystems Volume IV: Cash Crop Halophyte and Biodiversity Conservation. Tasks for Vegetation Science, vol. 47. Springer Dordrecht Heidelberg, New York London, pp. 175–198. http://dx.doi.org/10.1007/978-94-007-7411-7. ISBN 978-94-007-7410-0 ISBN 978-94-007-7411-7 (eBook).

Krapp, A., Hofmann, B., Schafer, C., Stitt, M., 1993. Regulation of the expression of rbcS and other photosynthetic genes by carbohydrates: a mechanism for the 'sink regulation' of photosynthesis? Plant J. 3, 817–828.

Kromdijk, J., Glowacka, K., Leonelli, L., Gabilly, S.T., Iwai, M., Niyogi, K.K., Long, S.P., 2016. Improving photosynthesis and crop productivity by accelerating recovery from photoprotection. Science 354, 857–861.

Kumar, M.N., Jane, W.-N., Verslues, P.E., 2013. Role of the putative osmosensor *Arabidopsis Histidine Kinase1* in dehydration avoidance and low-water-potential response. Plant Physiol. 161, 942–953.

Lata, C., Prasad, M., 2011. Role of DREBs in regulation of abiotic stress responses in plants. J. Exp. Bot. 62, 4731–4748.

Lawson, T., Kramer, D.M., Raines, C.A., 2012. Improving yield by exploiting mechanisms underlying natural variation of photosynthesis. Curr. Opin. Biotechnol. 23, 215–220.

Li, Z., Wakao, S., Fischer, B.B., Niyogi, K.K., 2009. Sensing and responding to excess light. Annu. Rev. Plant Biol. 60, 239–260.

Liang, C., Cheng, S., Zhang, Y., Sun, Y., Fernie, A.R., Kang, K., Panagiotou, G., Lo, C., Lim, B.L., 2016. Transcriptomic, proteomic and metabolic changes in *Arabidopsis thaliana* leaves after the onset of illumination. BMC Plant Biol. 16, 43. http://dx.doi.org/10.1186/s12870-016-0726-3.

Mehler, A.H., 1951. Studies on reactions of illuminated chloroplasts. I. Mechanism of the reduction oxygen and other Hill reagents. Arch. Biochem. Biophys. 34, 65–77.

Millennium Ecosystem Assessment, 2005. Ecosystems and Human Well-being: Desertification Synthesis. World Resources Institute, Washington, DC. http://www.maweb.org/documents/document.355.aspx.pdf.

Mitchell, P., 2011. Chemiosmotic coupling in oxidative and photosynthetic phosphorylation. Biochim. Biophys. Acta 1807, 1507–1538.

Miyake, C., 2010. Alternative electron flows (water–water cycle and cyclic electron flow around PSI) in photosynthesis: molecular mechanisms and physiological functions. Plant Cell Physiol. 51, 1951–1963.

Mizoi, J., Shinozaki, K., Yamaguchi-Shinozaki, K., 2012. AP2/ERF family transcription factors in plant abiotic stress responses. Biochim. Biophys. Acta 1819, 86–96.

Munns, R., 2002. Comparative physiology of salt and water stress. Plant Cell Environ. 25, 239–250.

Munns, R., 2005. Genes and salt tolerance: bring them together. New Phytol. 167, 645–663.

Munns, R., 2011. Plant adaptations to salt and water stress: differences and commonalities. Adv. Bot. Res. 57, 1–32.

Murata, N., Takahashi, S., Nishiyama, Y., Allakhverdiev, S.I., 2007. Photoinhibition of photosystem II under environmental stress. Biochim. Biophys. Acta 1767, 414–421.

Murchie, E.H., Lawson, T., 2013. Chlorophyll fluorescence analysis: a guide to good practice and understanding some new applications. J. Exp. Bot. 64, 3983–3998.

Nakagami, H., Pitzschke, A., Hirt, H., 2005. Emerging MAP kinase pathways in plant stress signalling. Trends Plant Sci. 10, 339–346.

Nakashima, K., Takasaki, H., Mizoi, J., Shinozaki, K., Yamaguchi-Shinozaki, K., 2012. NAC transcription factors in plant abiotic stress responses. Biochim. Biophys. Acta 1819, 97–103.

Nilkens, M., Kress, E., Lambrev, P., Miloslavina, Y., Müller, M., Holzwarth, A.R., Jahns, P., 2010. Identification of a slowly inducible zeaxanthin-dependent component of non-photochemical quenching of chlorophyll fluorescence generated under steady-state conditions in *Arabidopsis*. Biochim. Biophys. Acta 1797, 466–475.

Papenbrock, J., Churakova, Y., Huchzermeyer, B., Koyro, H.-W., 2016. Current methods in photosynthesis research. In: Pessarakli, M. (Ed.), Handbook of Plant and Crop Physiology, third ed. CRC Press, Boca Raton, Florida, pp. 447–464.

Pérez-Harguindeguy, N., Díaz, S., Garnier, E., Lavorel, S., Poorter, H., Jaureguiberry, P., Bret-Harte, M.S., Cornwell, W.K., Craine, J.M., Gurvich, D.E., Urcelay, C., Veneklaas, E.J., Reich, P.B., Poorter, L., Wright, I.J., Ray, P., Enrico, L., Pausas, J.G., de Vos, A.C., Buchmann, N., Funes, G., Quétier, F., Hodgson, J.G., Thompson, K., Morgan, H.D., ter Steege, H., van der Heijden, M.G.A., Sack, L., Blonder, B., Poschlod, P., Vaieretti, M.V., Conti, G., Staver, A.C., Aquino, S., Cornelissen, J.H.C., 2013. New handbook for standardised measurement of plant functional traits worldwide. Aust. J. Bot. 61, 167–234.

Poorter, H., Fiorani, F., Pieruschka, R., Wojciechowski, T., van der Putten, W.H., Kleyer, M., Schurr, U., Postma, J., 2016. Pampered inside, pestered outside? Differences and similarities between plants growing in controlled conditions and in the field. New Phytol. 212, 838–855.

Pospíšil, P., 2009. Production of reactive oxygen species by photosystem II. Biochim. Biophys. Acta 1787, 1151–1160.

Puranik, S., Sahu, P.P., Srivastava, P.S., Prasad, M., 2012. NAC proteins: regulation and role in stress tolerance. Trends Plant Sci. 17, 369–381.

Rahman, M.M., Rahman, M.A., Miah, M.G., Saha, S.R., Karim, M.A., Mostofa, M.G., 2017. Mechanistic insight into salt tolerance of *Acacia auriculiformis*: the importance of ion selectivity, osmoprotection, tissue tolerance, and Na+ exclusion. Front. Plant Sci. http://dx.doi.org/10.3389/fpls.2017.00155.

Rao, X., Dixon, R.A., 2016. The differences between NAD-ME and NADP-ME Subtypes of C_4 photosynthesis: more than Decarboxylating enzymes. Front. Plant Sci. 7, 1525. http://dx.doi.org/10.3389/fpls.2016.01525.

Richter, M.L., Hein, R., Huchzermeyer, B., 2000. Important subunit interactions in the chloroplast atp synthase. Biochim. Biophys. Acta 1458, 326–342.

Robbins, N.E., Trontin, C., Duan, L., Dinneny, J.R., 2014. Beyond the barrier: communication in the root through the Endodermis1. Plant Physiol. 166, 551–559.

Rolland, F., Baena-Gonzalez, E., Sheen, J., 2006. Sugar sensing and signaling in plants: conserved and novel mechanisms. Annu. Rev. Plant Biol. 57, 675–709.

Romanello, G.A., Chuchra-Zbytniuk, K.L., Vandermer, J.L., Touchette, B.W., 2008. Morphological adjustments promote drought avoidance in the wetland plant *Acorus americanus*. Aquat. Bot. 89, 390–396.

Rozema, J., Flowers, T.J., 2008. Crops for a salinized world. Science 322, 1478–1480.

Rungrat, T., Awlia, M., Brown, T., Cheng, R., Sirault, X., Fajkus, J., Trtilek, M., Furbank, B., Badger, M., Tester, M., Pogson, B.J., Borevitz, O.J., Wilson, P., 2016. Using phenomic analysis of photosynthetic function for abiotic stress response gene discovery. In: The *Arabidopsis* Book. The American Society of Plant Biologists. http://dx.doi.org/10.1199/tab.0185.

Santos, J., Al-Azzawi, M., Aronson, J., Flowers, T.J., 2016. eHALOPH a database of salt-tolerant plants: helping put halophytes to work. Plant Cell Physiol. 57 (1), e101–e110. http://dx.doi.org/10.1093/pcp/pcv155.

Sassenrathcole, G.F., Pearcy, R.W., 1994. Regulation of photosynthetic induction state by the magnitude and duration of low-light exposure. Plant Physiol. 105, 1115–1123.

Selmar, D., Kleinwächter, M., 2013. Stress enhances the synthesis of secondary plant products: the impact of stress-related over-reduction on the accumulation of natural products. Plant Cell Physiol. 54, 817–826.

Shanmugam, S., Xu, J., Boyer, C., 2015. Utilizing the electron transfer mechanism of chlorophyll a under light for controlled radical polymerization. Chem. Sci. 6, 1341–1349.

Sharma, P., Jha, A.B., Dubey, R.S., Pessarakli, M., 2012. Reactive oxygen species, oxidative damage, and antioxidative defense mechanism in plants under stressful conditions. Hindawi J. Bot 2012:217037. http://dx.doi.org/10.1155/2012/217037.

Shebis, Y., Iluz, D., Kinel-Tahan, Y., Dubinsky, Z., Yehoshua, Y., 2013. Natural antioxidants: function and sources. Food Nutr. Sci. 4, 643–649.

Shinozaki, K., Yamaguchi-Shinozaki, K., 2007. Gene networks involved in drought stress response and tolerance. J. Exp. Bot. 58, 221–227.

Shinozaki, K., Yamaguchi-Shinozaki, K., Seki, M., 2003. Regulatory network of gene expression in the drought and cold stress responses. Curr. Opin. Plant Biol. 6, 410–417.

Shulaev, V., Cortes, D., Miller, G., Mittler, R., 2008. Metabolomics for plant stress response. Physiol. Plant. 132, 199–208.

Sofo, A., Scopa, A., Nuzzaci, M., Vitti, A., 2015. Ascorbate peroxidase and catalase activities and their genetic regulation in plants subjected to drought and salinity stresses. Int. J. Mol. Sci. 16, 13561–13578.

Song, Y., Buettner, G.R., 2010. Thermodynamic and kinetic considerations for the reaction of semiquinone radicals to form superoxide and hydrogen peroxide. Free Radic. Biol. Med. 15, 919–962. http://dx.doi.org/10.1016/j.freeradbiomed.2010.05.009.

Stitt, M., 1986. Limitation of photosynthesis by carbon metabolism. I. Evidence for excess electron transport capacity in leaves carrying out photosynthesis in saturating light and CO_2. Plant Physiol. 81, 1115–1122.

Stitt, M., von Schaewen, A., Willmitzer, L., 1990. "Sink" regulation of photosynthetic metabolism in transgenic tobacco plants expressing yeast invertase in their cell wall involves a decrease of the Calvin-cycle enzymes and an increase of glycolytic enzymes. Planta 183, 40–50.

Strotmann, H., Hesse, H., Edelmann, K., 1973. Quantitative determination of coupling factor CF1 of chloroplasts. Biochim. Biophys. Acta 314, 202–210.

Sui, N., Yang, Z., Liu, M., Wang, B., 2015. Identification and transcriptomic profiling of genes involved in increasing sugar content during salt stress in sweet sorghum leaves. BMC Genomics 16, 534. http://dx.doi.org/10.1186/s12864-015-1760-5.

Theg, S.M., Chian, G., Dilley, R.A., 1988. Protons in the thylakoid membrane-sequestered domains can directly pass through the coupling factor during ATP synthesis in flashing light. J. Biol. Chem. 263, 673–681.

Trebst, A., 1974. Energy conservation in photosynthetic electron transport of chloroplasts. Annu. Rev. Plant Physiol. 25, 423–458.

Tschiersch, H., Junker, A., Meyer, R.C., Altmann, T., 2017. Establishment of integrated protocols for automated high throughput kinetic chlorophyll fluorescence analyses. Plant Methods 13, 54. http://dx.doi.org/10.1186/s13007-017-0204-4.

Ventura, Y., Myrzabayeva, M., Alikulov, Z., Omarov, R., Khozin-Goldberg, I., Sagi, M., 2014. Effects of salinity on flowering, morphology, biomass accumulation and leaf metabolites in an edible halophyte. AoB Plants 6. http://dx.doi.org/10.1093/aobpla/plu053.

Vidal, J., Chollet, R., 1997. Regulatory phosphorylation of C_4 PEP carboxylase. Trends Plant Sci. 2, 230–237.

von Caemmerer, S., Quick, W.P., Furbank, R.T., 2012. The Development of C Rice: current progress and future challenges. Science 336, 1671–1672.

Walker, J.E., 1994. The regulation of catalysis in ATP synthase. Curr. Opin. Struct. Biol. 4, 912–918.

Wani, S.H., Singh, N.B., Haribhushan, A., Mir, J.I., 2013. Compatible solute engineering in plants for abiotic stress tolerance - role of Glycine Betaine. Curr. Genomics 14, 157–165.

Ware, M.A., Belgio, E., Ruban, A.V., 2015. Comparison of the protective effectiveness of NPQ in *Arabidopsis* plants deficient in PsbS protein and zeaxanthin. J. Exp. Bot. 66, 1259–1270.

Wilhelm, C., Selmar, D., 2011. Energy dissipation is an essential mechanism to sustain the viability of plants: the physiological limits of improved photosynthesis. J. Plant Physiol. 168, 79–87.

Xue, G.P., Way, H.M., Richardson, T., Drenth, J., Joyce, P.A., McIntyre, C.L., 2011. Overexpression of TaNAC69 leads to enhanced transcript levels of stress up-regulated genes and dehydration tolerance in bread wheat. Mol. Plant 4, 697–712.

Yadav, N., Sharma, S., 2016. Reactive Oxygen Species, Oxidative stress and ROS scavenging system in plants. J. Chem. Pharm. Res. 8, 595–604.

Yamaguchi-Shinozaki, K., Shinozaki, K., 2005. Organization of cis- acting regulatory elements in osmotic- and cold-stress-responsive promoters. Trends Plant Sci. 10, 88–94.

Ye, Y., Ding, Y., Jiang, Q., Wang, F., Sun, J., Zhu, C., 2017. The role of receptor-like protein kinases (RLKs) in abiotic stress response in plants. Plant Cell Rep. 36, 235–242.

You, J., Chan, Z., 2015. ROS regulation during abiotic stress responses in crop plants. Front. Plant Sci. 6, 1092. http://dx.doi.org/10.3389/fpls.2015.01092.

Zaks, J., Amarnath, K., Kramer, D.M., Niyogi, K.K., Fleming, G.R., 2012. A kinetic model of rapidly reversible nonphotochemical quenching. Proc. Nat. Acad. Sci. USA. 109, 15757–15762.

Zandalinas, S.I., Sales, C., Beltrán, J., Gómez-Cadenas, A., Arbona, V., 2017. Activation of secondary metabolism in citrus plants is associated to sensitivity to combined drought and high temperatures. Front. Plant Sci. 7, 1954. http://dx.doi.org/10.3389/fpls.2016.01954.

Zhang, J., Yang, D., Li, M., Shi, L., 2016. Metabolic profiles reveal changes in wild and cultivated soybean seedling leaves under salt stress. PLoS One 11 (7), e0159622. http://dx.doi.org/10.1371/journal.pone.0159622.

Zhao, P., Cui, R., Xu, P., Wu, J., Mao, J.-L., Chen, Y., Zhou, C.-Z., Yu, L.–H., Xiang, C.B., 2016. ATHB17 enhances stress tolerance by coordinating photosynthesis associated nuclear gene and ATSIG5 expression in response to abiotic stress. Sci. Rep. 7, 45492. http://dx.doi.org/10.1038/srep45492.

Chapter 15

Physiological and Biochemical Mechanisms of Drought Stress Tolerance in the Argan Tree

Abdelghani Chakhchar[1], Mouna Lamaoui[1], Salama Aissam[1], Abderrahim Ferradous[2], Said Wahbi[1], Abdelhamid El Mousadik[3], Saad Ibnsouda-Koraichi[4], Abdelkarim Filali-Maltouf[5], Cherkaoui El Modafar[1]

[1]*Université Cadi Ayyad, Marrakech, Morocco;* [2]*Centre Régional de la Recherche Forestière Marrakech, Marrakech, Morocco;* [3]*Université Ibn Zohr, Agadir, Morocco;* [4]*Université Sidi Mohamed Ben Abdellah, Fèz, Morocco;* [5]*Université Mohammed V Agdal, Rabat, Morocco*

1. INTRODUCTION

Drought is the principal limiting factor for agricultural production. It has become an increasingly serious problem in many parts of the world and urgent issue for the international community (Passioura, 2007; He et al., 2014; Feller, 2016). In semiarid and arid areas, water loss through transpiration in plants often exceeds root absorption capacity, exposing the plants to water stress, which is accentuated by environmental factors. This constraint is responsible for a wide range of physiological and biochemical tolerance reactions in higher plants. Despite extensive literature on this topic, there is controversy regarding plants' tolerance mechanisms to drought. Responses to drought are multiple and interconnected (Osakabe et al., 2014; Liu et al., 2017). It has been established that drought stress induces a certain number of physiological, biochemical, and molecular reactions that regulate plant growth and productivity (Yoon et al., 2014; Wu et al., 2016; Fahad et al., 2017). However, this abiotic stress impairs numerous metabolic and physiological processes in plants (Khoyerdi et al., 2016; Pompeiano et al., 2016; Fahad et al., 2017). In the Mediterranean ecosystem, plants are subjected to continuous and severe drought stress (Nogués and Baker, 2000; Toscano et al., 2016). Some endemic species, such as the argan tree [*Argania spinosa* (L.) skeels], are well adapted to these conditions in Morocco (Msanda et al., 2005; Diaz-Barradas et al., 2010). The argan tree is ranked among the first forest massifs in Morocco, a category that includes the holm oak and Atlas cedar forests. The local forest area, which forms vast natural zones also called *arganeraie*, is currently estimated at 800,000 ha (Msanda et al., 2005). In southwestern Morocco, the main *A. spinosa* area is bounded to the north by Oued Tensift, to the south by Oued Noun and Oued Seyad, to the west by the Atlantic Ocean, and to the east by the surroundings of Aoulouz to the west of Jbel Siroua. It is the northernmost species in the Mediterranean region of the tropical family Sapotaceae and grows naturally and abundantly in arid and semiarid areas in southwestern Morocco. The climate of this region is arid Mediterranean, where rainfall is unequally distributed over the year. This plant formation belongs to the Mediterranean–Saharan transition zone (McGregor et al., 2009). It supports temperatures ranging from 3 to 50°C, and it receives very low rainfall. Indeed, mature argan trees can survive in very dry conditions for long periods (annual rainfall <100 mm) (Msanda et al., 2005). The tree plays important socioeconomic and ecological roles in this area and is even more significantly linked to the biodiversity of the forest's ecosystem (Lybbert et al., 2002; Msanda et al., 2005; Genin et al., 2017). About 1.3 million people are living in rural areas where traditional sylvo-pastoral systems are based on the argan tree (Chaussod et al., 2005). Potentially a very important tree species for the vegetable oil (Lybbert et al., 2010; AitAabd et al., 2014) extracted from its seed kernels, the argan tree is of considerable interest to the local horticultural industry.

Drought adaptation poses enormous challenges to plants. Drought is a complex physicochemical processes involving many biological macromolecules and small molecules, such as nucleic acids (DNA, RNA, micro-RNA, etc.), proteins, carbohydrates, lipids, plant hormones, ions, free radicals, and mineral elements (Shao et al., 2008; Ahmad et al., 2014; Katam et al., 2016; Kolenc et al., 2016; Landi et al., 2016). The study of drought has been a key disciplinary focus in global plant biology (Somerville and Dangl, 2000). However, although almost all plants are drought tolerant, the extent varies from one species to another and even within the same species. For this purpose to overcome the adverse effects of drought, plants are equipped with the ability to manage their water status through a number of physiological mechanisms and an arsenal

Plant Metabolites and Regulation Under Environmental Stress. https://doi.org/10.1016/B978-0-12-812689-9.00015-7

of enzymes, metabolites, antioxidants, and osmoprotectants, which are sufficient enough to face the reactive oxygen species (ROS) and other substances generated by different metabolisms (Król et al., 2014; Sewelam et al., 2016; Landi et al., 2017). Typically, the mechanisms of drought tolerance include ionic homeostasis, biosynthesis of osmolytes, scavenging of free radicals, and circulation of water and synergy of a long-distance signaling system, the "root-leaf" (Reddy et al., 2004; Golldack et al., 2014).

It is in this sense that this pot study is undertaken on the argan tree to study and characterize the physiological and biochemical mechanisms of tolerance involved in drought stress and recovery responses and to assess the differences in physiological and biochemical traits analyzed in four contrasting *A. spinosa* ecotypes to contribute to an assisted selection of the most tolerant ecotypes.

2. MATERIALS AND METHODS

2.1 Plant Materials and Experimental Design

A. spinosa seed sampling was conducted in four regions of the argan tree forest in southwestern Morocco (Fig. 15.1). We chose two contrasting coastal ecotypes (site: Rabia [Rab] and Admine [Adm]) and two contrasting inland ecotypes (site: Aoulouz [Alz] and Lakhssas [Lks]) for better interpretation of the mechanisms regulating biochemical and physiological processes. Climatic, geographical, and hydrological conditions of these four regions are markedly distinct (Table 15.1). The protocol of cultivation and experimental layout was the same as that used previously (Chakhchar et al., 2015a,b,c). Uniform young *A. spinosa* plants of similar height were selected for the experiment for each ecotype. The environmental conditions in the chamber during the experiment were maintained at 28 ± 1°C during day and 25 ± 1°C during night in a 16:8 photoperiod and the rate of relative humidity varied between 65% and 70%. The average maximum photosynthetically active radiation was approximately 400 μmol/m^{2}s, provided by a combination of incandescent and fluorescent lamps.

The experimental layout was completely randomized with regard to three factors (ecotype, watering regime, and time). Two approaches were adopted to characterize drought tolerance in the argan tree. In the first, plants were randomly exposed to two stress treatments: 50% and 25% field water capacity (FWC), which corresponds to medium and severe drought stress, respectively, for 2 months. Our second approach consisted of simulating the effect of prolonged drought stress by cessation of irrigation for 15 and 30 days followed by rehydration for 4 days. For the control and rehydrated plants, soil humidity was maintained at 75% FWC.

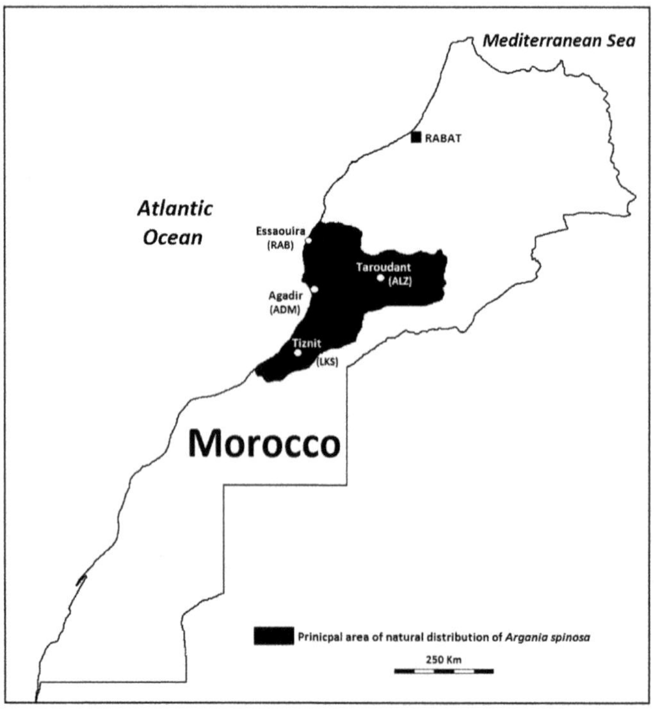

FIGURE 15.1 Biogeographical distribution of the four studied *Argania spinosa* ecotypes (Chakhchar et al., 2016).

TABLE 15.1 Geographical and Climatic Characteristics of the Sites of the Four *Argania spinosa* Ecotypes (Chakhchar et al., 2015a,b,c)

Origin Site	Province	Territory	Temperature (°C)	Rainfall (mm)	Altitude (m)	Humidity (%)	*m* (°C)	*M* (°C)
Rabia	Essaouira	Coastal	17–18	295	181–226	80–90	9.6	22.2
Admine	Agadir	Coastal	18–19	225	275–430	75–85	7.2	27.1
Aoulouz	Taroudant	Inland	19–20	232	700–850	60–70	5.6	35.7
Lakhssas	Tiznit	Inland	21–22	189	916–988	50–60	7.3	31.2

m, average minimum temperature of the coldest month of the year; *M*, average maximum temperature of the coldest month of the year.

TABLE 15.2 Physiological and Biochemical Traits Studied in the Four Contrasting Ecotypes of *Argania spinosa*

Physiological traits	- Stomatal conductance (g_s) - Predawn leaf water potential (Ψ_{pb}) - Leaf relative water content (RWC) - Leaf dry matter content (LDMC) - Leaf moisture (LM) - Residual transpiration rate - Electrolyte leakage (EL)	- Leaf concentrations of K^+, Ca^{2+} and Mg^{2+} - Maximum efficiency of PSII (F_v/F_m) - Intrinsic efficiency of open PSII (F_v'/F_m') - Coefficient of photochemical quenching (q_p) - Quantum yield of PSII electron transport (Φ_{PSII}) - Coefficient of non-photochemical quenching (NPQ)
Biochemical traits	- Total superoxide dismutase (SOD) and their isoenzymes Cu/Zn-SOD, Cu/Zn-SOD and Fe-SOD - Catalase (CAT) - Peroxidase (POD) - Polyphenoloxidase (PPO) - Lipoxygenase (LOX) - Ascorbate peroxidase (APX) - Monodehydro-ascorbate reductase (MDHAR) - Dehydroascorbate reductase (DHAR) - Glutathione reductase (GR) - Glutathione-*S*-transferase (GST) - Glutathione peroxidase (GP) - Ribulose-1,5-biphosphate carboxylase-oxygenase (Rubisco) - Glucose-6-phsphate dehydrogenase (G6PDH) - Sorbitol dehydrogenase (SDH) - Aldose-6-phosphate reductase (A6PR) - Acid invertase (AI) - Phenylalanine ammonium lyase (PAL)	- Shikimate dehydrogenase (SKDH) - Cinnamate 4-hydroxylase (C4H) - Ferredoxin-NADP$^+$-oxidoreductase (FNR) - Chlorophyllase - Malonyldialdehyde (MDA - Ascorbic acid (AA) - α-tocopherol (α-toc) - Reduced gluthatione (GSH) - Total polyphenols content - Leaf epicuticular wax loads (EWL) - Free proline content - Soluble sugars content - Thiol compounds content - Chlorophyll content - Anthocyanin content - Pheophytin contents - Hydrogenperoxide (H_2O_2) - Superoxide radical content (O_2^-) - Carbonyl groups

2.2 Physiological and Biochemical Traits

The study of the physiological and biochemical traits can positively contribute to understanding the behavior of *A. spinosa* species while also characterizing the mechanisms of tolerance to drought stress. At the physiological level, the incidence of drought stress was assessed by measuring some physiological traits related to the water status of the plant, ionic state, and photosynthesis (chlorophyll fluorescence) (Table 15.2) (Chakhchar, 2015; Chakhchar et al., 2015a,b, 2016, 2017a,b). From the biochemical perspective, the focus of the present study is a wide range of biochemical traits related to oxidative damage, enzymatic defense systems, osmoregulation, metabolism of carbohydrates, secondary metabolism, and certain pigments and enzymes of photosynthesis (Table 15.2) (Chakhchar, 2015; Chakhchar et al., 2015a,b,c, 2016).

2.3 Statistical Analysis

All data were subjected to analysis of variance to test the effect of ecotype, watering regime, time, and their interactions for each trait measured. The means were compared using Tukey's posthoc test after checking the normality and homoscedasticity of data. Pearson correlation analysis was conducted for determined variables for each ecotype after checking the assumptions of parametric tests. A canonical discriminant analysis (CDA), involving the simultaneous analysis of all independent variables into the equation, was performed on the four contrasting species of *A. spinosa* to determine which variables differed between them. Statistical analyses were conducted using SPSS 17.0.

3. RESULTS AND DISCUSSION

Intraspecific differences were observed for all physiological and biochemical traits studied. Variations in these traits depended on the intensity and duration of the drought stress treatments. The *Argania* ecotype also had significant effects on the set of traits. The results showed that, under drought stress conditions, the water status indices of *A. spinosa* leaves Ψ_{pb}, g_s, and relative water content (RWC) significantly decreased in stressed plants compared with controls. In addition, significant correlations between these three physiological traits were recorded in the four contrasting ecotypes. In fact, g_s was usually correlated with the leaf water potential for numerous species, suggesting the existence of chemical signals between the shoot and the root (Tardieu and Simonneau, 1998; Daszkowska-Golec and Szarejko, 2013). The changes in water potential and/or soil water content, due to drought stress treatments, were expected to cause significant stomatal closure, which is very effective in preventing water loss in *A. spinosa* plants (Diaz-Barradas et al., 2013; Chakhchar et al., 2015a,b, 2016, 2017c). Thus leaf damage at the cell membrane level was evaluated by monitoring the electrolytes. Indeed, a significant increase of electrolyte leakage (EL) was noted in stressed plants compared with the controls, indicating an injury and loss of membrane integrity of *A. spinosa* leaves related to chain reactions induced by ROS. Moreover, plants subjected to severe drought stress treatment showed a significant decrease in F_v/F_m, F_v'/F_m', q_p, and Φ_{PSII} compared with control ($P<.001$); however, the nonphotochemical quenching (NPQ) increased significantly. This finding implies that the photoprotective efficiency of the quenching processes in the photosystem (PS) II antenna seems to be affected. No significant differences were recorded for these five traits of chlorophyll fluorescence under moderate drought stress conditions. These changes in chlorophyll fluorescence traits indicate modifications in photosynthesis at photosystem II level in *A. spinosa* plants (Chakhchar et al., 2017b). Since photoinhibition is one of the major consequences of drought stress in plants, this phenomenon can be regarded as an adaptation to drought conditions or to injury to the PSII (Baker and Rosenqvist, 2004; Ping et al., 2015; Gururani et al., 2015). Thus significant correlations were recorded between $\Phi_{PSII} \times q_p$ on the one hand and between $\Phi_{PSII} \times F_v'/F_m'$ on the other hand. The decrease in F_v'/F_m' and q_p caused a decrease in the quantum yield of PSII electron transport. Indeed, the significant decrease of Φ_{PSII} may be due to a decreased capacity to use the excitation energy in leaf photosynthesis under drought conditions. In addition, the stomatal limitations noted can affect photosynthesis and be accompanied by a decrease in the rate of consumption of adenosine triphosphate and reduced nicotinamide adenine dinucleotide for C assimilation, which could result in decreases in the rate of linear electron transport and consequently in Φ_{PSII} (Baker and Rosenqvist, 2004; Ping et al., 2015).

According to Pearson correlation analysis, significant negative correlations between Residual transpiration rate (RTR) and leaf epicuticular wax loads (EWL) were obtained in all four studied ecotypes. With the increase in drought stress intensity, the RTR decreased significantly, whereas the EWL showed a significant increase. This relationship between both traits suggests that drought stress enhanced the biosynthesis and transport of cuticular wax at the leaf level, which can improve the ability of leaves to retain water by minimizing cuticular transpiration (Seo and Park, 2011; González and Ayerbe, 2010; Chakhchar et al., 2015a).

Biochemically, drought stress causes an increase of ROS at the cellular level, reflecting an oxidative stress. This kind of stress manifests as a serious oxidative damage and metabolic imbalance of the cell (Munemasa et al., 2013; Noctor et al., 2015; Landi et al., 2017). Drought stress induced a significant increase in the endogenous hydrogen peroxide concentration and the lipid peroxidation rate, assessed by the malonyldialdehyde (MDA) content. Under moderate and severe drought stress conditions, positive and significant correlations were found between H_2O_2 and MDA in *A. spinosa* leaves. These correlations confirmed the hypothesis that H_2O_2 brings about lipid peroxidation, leading to membrane damage. It would therefore appear that the plants in question have developed several protection mechanisms to prevent the damaging effect of ROS and maintain homeostasis through the stimulation of an effective antioxidant defense system. In *A. spinosa*, significant and high induction of both enzymatic and nonenzymatic defense systems that act as ROS scavengers, either in conjunction or independently, suggests a better ability of drought tolerance (Chakhchar et al., 2015c). Both drought stress treatments preferentially stimulated, in a time-dependent manner, the enzymatic activity of superoxide dismutase (SOD), catalase, peroxidase (POD), polyphenoloxidase (PPO), and lipoxygenase (LOX) in *A. spinosa* leaves. Thus positive and significant correlations were registered between LOX activity and MDA content. Given that polyunsaturated fatty acids, the major components of cell membranes phospholipids, are considered as the substrate of LOX, the enhancement of LOX activity has been observed to damage cell membranes causing lipid peroxidation. With regard to the comparison of ecotypes, the constitutive activities of SOD, POD, PPO, and LOX were higher in inland ecotypes than in coastal ecotypes. In addition, the effect of drought stress treatments on *A. spinosa* plants was marked by a significant increase in the activity of the ascorbate–glutathione cycle enzymes including ascorbate peroxidase, monodehydroascorbate reductase (MDHAR), dehydroascorbate reductase (DHAR), and glutathione reductase (GR), as well as the content of nonenzymatic antioxidants, namely Acid ascorbic (ASA), α-tocopherol, and reduced glutathione (GSH) in the four ecotypes and especially in the Lks ecotype (Chakhchar et al., 2015c). These nonenzymatic antioxidants serve as potent antioxidants for plant cell defense systems, while the ascorbate–glutathione cycle is an important pathway of indirect and direct ROS scavenging in cells (Foyer and Noctor, 2005a,b; Munemasa et al., 2013; Landi et al., 2017). Furthermore, our results showed that drought stress can induce the shikimate and phenylpropanoid pathways in *A. spinosa* plants. A significant induction of the activity of shikimate dehydrogenase (SKDH), phenylalanine ammonium lyase (PAL), and cinnamate-4-hydroxylase (C4H) and a significant increase in the total polyphenol content was recorded in all ecotypes. Under drought stress conditions, the Lks ecotype showed the strongest induction and accumulation compared with other ecotypes. The total polyphenol accumulation was positively correlated with the activities of SKDH, PAL, and C4H. Generally, biotic and abiotic stresses promote the synthesis of polyphenols, involved in plant protection mechanism (Ma et al., 2014; Di Ferdinando et al., 2014; Rodziewicz et al., 2014). The enzyme stimulation implied in flavonoid and phenylpropanoid synthesis would be key in ensuring protection and prevention against damage caused by drought stress, suggesting de novo synthesis of polyphenols under drought stress. The enhanced level of polyphenols may be instrumental in enhancing the degree of drought tolerance in *A. spinosa* (Chakhchar et al., 2015a). As for the enzymes involved in carbohydrate metabolism, drought stress treatments have significantly affected the activity of glucose-6-phsophate dehydrogenase, aldose-6-phosphate reductase, sorbitol dehydrogenase (SDH), and acid invertase. Compared with the control, drought stress induced a significant increase in the activity of these four enzymes. In parallel, a significant increase of both hexoses (glucose and fructose) was observed in the four contrasting ecotypes, especially under severe drought stress treatment. In fact, these enzymes had highly significant and positive direct effects on foliar glucose concentration during drought stress. Also, significant positive correlations between both enzymes SDH and acid invertase and foliar fructose concentration were recorded. Indeed, *A. spinosa* plants showed significant changes in carbohydrate concentrations, which are related to increased activities of enzymes associated with carbohydrate metabolism. In addition to the accumulation of soluble sugars in terms of osmoregulation, the proline content in the leaves was increased significantly in parallel with the severity of drought stress. Moreover, both levels of drought stress, namely, moderate and severe drought stress, significantly affected the concentrations of chlorophylls and carotenoids in all four ecotypes. Thus *A. spinosa* plants showed a significant decrease of Chl*a*/*b* ratio in response to drought stress. In fact, Chl*a*/*b* ratio can be considered as a key trait to judge environmental stress tolerance in higher plants (Parida et al., 2007). Positive and significant correlations were recorded between Chl*a*/*b* ratio and activity of chlorophyllase in leaves of *A. spinosa* plants, under both drought stress treatments, suggesting that the degradation of chlorophyll in leaves may be partly due to the chlorophyllase enzyme. This enzyme is considered preeminent in the pathway of chlorophyll degradation, i.e., the hydrolysis of chlorophyll to chlorophyllide and phytol (Arkus et al., 2005). Our results also showed a great increase in the percentage of pheophytinized chlorophyll suggesting that drought stress destabilizes the chloroplastic pigment–protein complexes, rendering the chlorophyll susceptible to pheophytinization in *A. spinosa* plants. However, the significant increase of anthocyanin content in stressed *A. spinosa* plants may help reduce oxidative damage caused by free radicals. In addition, the anthocyanins are thought to act as osmoregulators under drought stress, taking into account their biochemical nature. The increase recorded in the present study suggests that anthocyanin biosynthesis is actively induced in *A. spinosa* under drought conditions. Recent studies of the last decade have shown that anthocyanins in vegetative tissues are mainly involved in photoprotection of chlorophyll under direct drought stress–induced photoinhibition (Taulavuori et al., 2010; Landi et al., 2015). We have also studied the extent of limitation imposed by ribulose-1,5-bisphosphate carboxylase-oxygenase (Rubisco) on photosynthesis of *A. spinosa*. Under drought stress conditions, the highest decrease in Rubisco activity was recorded in the both coastal ecotypes compared with both inland ecotypes. Rubisco plays an important part in photosynthetic carbon assimilation and photorespiration (Kroth, 2015). Thus positive and significant correlations were recorded between Rubisco activity and stomatal conductance in the four ecotypes, suggesting that the decrease of Rubisco activity during drought stress may be related to the decrease in leaf g_s. In response to drought stress, the decrease in g_s restricts CO_2 entry into leaves (stomatal limitation) thereby decreasing CO_2 assimilation (Flexas et al., 2002; Ping et al., 2015; Feller, 2016; Kolenc et al., 2016).

However, neither drought stress treatment substantially affected the activity of ferredoxin-NADP$^+$-oxidoreductase enzyme (FNR) in the 4 *A. spinosa* ecotypes. Nonetheless, we found that both inland ecotypes have higher constitutive activity of FNR compared with the coastal ecotypes.

In terms of statistical analysis, ecotype × watering regime, ecotype × time, watering regime × time, and ecotype × watering regime × time were examined for all physiological and biochemical traits measured. According to the CDA for all traits, three discriminant functions (DF) have been established corresponding to 60.2%, 24.2%, and 15.6% of the total variance for physiological traits and 60.3%, 28.8%, and 10.9% for biochemical traits. The two-dimensional scatterplots of discriminant space obtained (Figs. 15.2 and 15.3) showed a clear separation of the four contrasting ecotypes taking into account

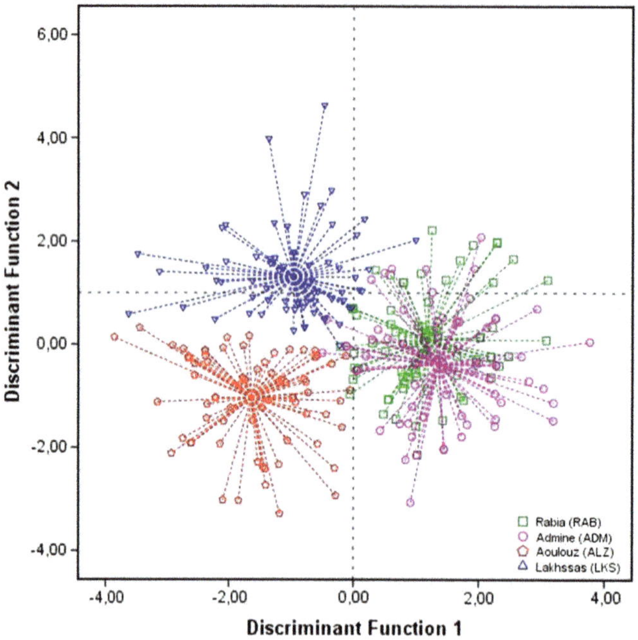

FIGURE 15.2 Two-dimensional scatterplot showing the separation of the four contrasting ecotypes of *Argania spinosa* according to the two discriminant function gradients obtained by canonical discriminant analysis for physiological traits under drought stress.

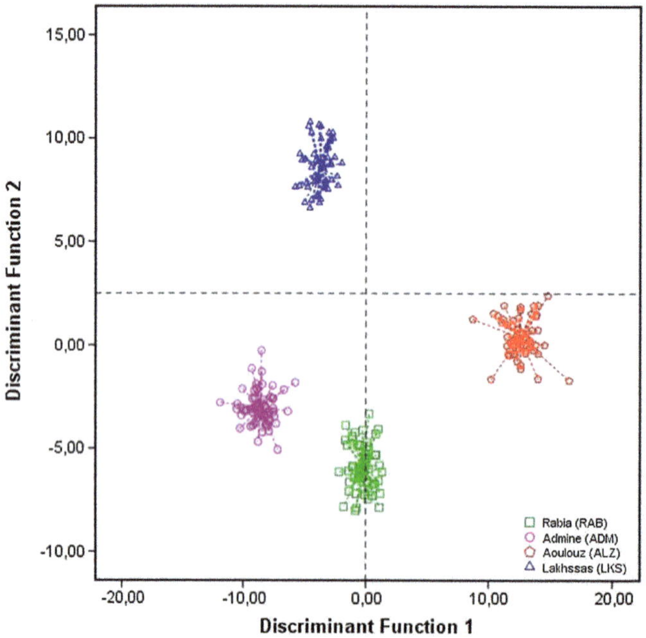

FIGURE 15.3 Two-dimensional scatterplot showing the separation of the four contrasting ecotypes of *Argania spinosa* according to the two discriminant function gradients obtained by canonical discriminant analysis for biochemical traits under drought stress.

only the first two functions for each category of studied traits. Based on the standardized coefficients of the canonical DF, interpretation of this separation according to the distribution of studied traits was possible. Regarding the physiological traits (Fig. 15.2), both coastal contrasting ecotypes (Adm, Rab) were particularly separated from the inland ecotypes (Lks and Alz), with regard to the first DF, with high decrease of the traits of fluorescence of chlorophyll such as Φ_{PSII}, q_p, and F_v'/F_m', and with a significant physiological difference in water status assessed by Ψ_{pb}, g_s, and RWC. With regard to the second DF, the Lks ecotype was slightly distinguished from other ecotypes by a high decrease of g_s and Ψ_{pb} and low decrease of q_p and F_v'/F_m'. Concerning the biochemical traits (Fig. 15.3), the differences were also evident between ecotypes under drought stress conditions. The first DF revealed that the Alz ecotype is clearly separated from other ecotypes, whereas the second DF separated the Lks ecotype from other ecotypes. The separation of each of these inland ecotypes from both coastal ecotypes was primarily due to the high stimulation of certain antioxidant enzyme activity, especially the ascorbate–glutathione cycle enzymes, high accumulation of certain nonenzymatic antioxidants including GSH and ASA, and low accumulation of ROS, suggesting that both these inland ecotypes have a very effective antioxidant defense system as an internal mechanism for drought tolerance.

Under edaphic drought simulated by cessation of irrigation for 15 and 30 days, we recorded a significant decrease in g_s and Ψ_{pb} in all ecotypes. These results have been illustrated by the significant correlations between g_s, Ψ_{pb}, and the decrease of soil moisture content (Chakhchar et al., 2016). Thus the induced drought stress yielded significant accumulation in leaf concentrations of inorganic ions Mg^{2+}, K^+, and Ca^{2+} ($P<.001$). This significant accumulation recorded in stressed argan plants may be interpreted as a physiological mechanism of tolerance to mitigate the negative effects of drought stress (Chakhchar et al., 2017a). Both inland ecotypes appear more tolerant than both coastal ecotypes, considering their constitutive high concentrations of these inorganic ions. Biochemically, we noted oxidative damage in the leaves of *A. spinosa* under edaphic drought. This oxidative damage has been assessed by estimating the content of superoxide radical (O_2^-), hydrogen peroxide (H_2O_2), and carbonyl groups (CG). Indeed, significant accumulation ($P<.001$) of these three oxidative products was recorded in the leaves of stressed plants compared with the control. Under drought stress conditions, we recorded significant and positive correlations between the endogenous concentration of H_2O_2 and the content of CG. When proteins are oxidized, the CG occur endogenously in cells. The oxidation of proteins, linked to the production of ROS, was induced in *A. spinosa* leaves subjected to drought stress highlighting the accumulation of H_2O_2. In parallel, a very significant increase was recorded in total thiol content and activity of isoenzymes of superoxide dismutase (Fe-SOD, Cu/Zn-SOD, Mn-SOD), glutathione peroxidase (GP), GR, and glutathione-S-transferase (GST) in stressed *A. spinosa* plants. Meanwhile, a substantial increase in the activity of the isoenzymes Cu/Zn-SOD, Mn-SOD, and Fe-SOD has been reported in drought stress conditions (Uzilday et al., 2012; Szőllősi, 2014). The upregulation of SOD and its isoenzymes are involved in the defense systems against oxidative stress due to biotic and abiotic stress, playing a crucial role in the survival of plants under environmental stresses (Badawi et al., 2004; Szőllősi, 2014). Significant and positive correlations were found between the concentration of both ROS (O_2^- and H_2O_2) and the activity of total SOD and its isoenzymes. This indicates that the antioxidant ability to eliminate ROS in the argan tree has increased considerably in response to drought stress. Moreover, the glutathione-related enzymes GP, GR, and GST showed a significant increase in activity, suggesting the stimulation of enzymatic antioxidant system in *A. spinosa*, under drought stress, to maintain homeostasis and prevent oxidation (Chakhchar et al., 2017a). Besides these enzymatic antioxidants, other nonenzymatic compounds are also considered to be relevant markers of oxidative stress, such as the thiol compounds (Deneke, 2000; Rouhier et al., 2015). Indeed, high levels of glutathione and other thiols have been associated with an increased tolerance to oxidative stress under abiotic stress conditions (Szalai et al., 2009; Nazar et al., 2011). According to our results, an increase in the total thiol content (protein thiols + nonprotein thiols) could have signified the occurrence of oxidative stress and reflected the antioxidant capacity of thiols in the argan tree (Chakhchar et al., 2017a). In terms of comparison, it was revealed that both inland ecotypes (Alz and Lks) have a very effective enzymatic and nonenzymatic antioxidant defense system compared with the other ecotypes.

However, after the rehydration phase, the set of studied physiological and biochemical traits significantly reached a level close to the values recorded in control plants. The kinetics of recovery of the argan tree as a result of rehydration was rapid, only after 4 days. The *A. spinosa* plants showed effective recovery capacity taking into account the interecotype differences.

According to the CDA for physiological and biochemical traits under drought stress (30 days) and rehydration conditions, we considered the first two DF: first under drought stress conditions, 84.3% and 10.1% of the total variance for physiological traits (Fig. 15.4A) and 87.0% and 10.9% for biochemical traits (Fig. 15.4B), then under rehydration conditions, 71.6% and 23.9% of the total variance for physiological traits (Fig. 15.5A) and 99.5% and 0.4% for biochemical traits (Fig. 15.5B). The canonical correlations for the first two functions in each model were highly significant. The canonical plots made using the first two functions showed a clear separation of the four contrasting ecotypes. Based on the standardized coefficients, the DF allowed us to make a connection between the studied traits and the registered differences among ecotypes in their tolerance to drought stress. Regarding the physiological traits, the first DF contributed mostly to

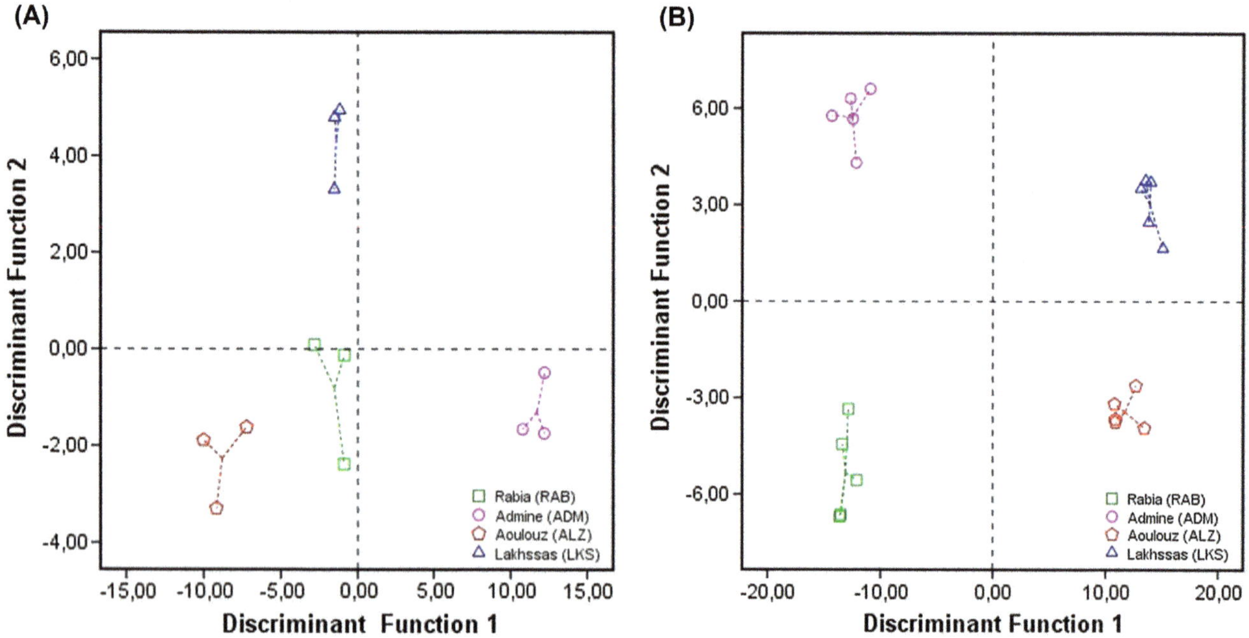

FIGURE 15.4 Two-dimensional scatterplot showing the separation of the four contrasting ecotypes of *Argania spinosa* according to the two discriminant function gradients obtained by canonical discriminant analysis under edaphic stress conditions. (A) Physiological traits and (B) biochemical traits.

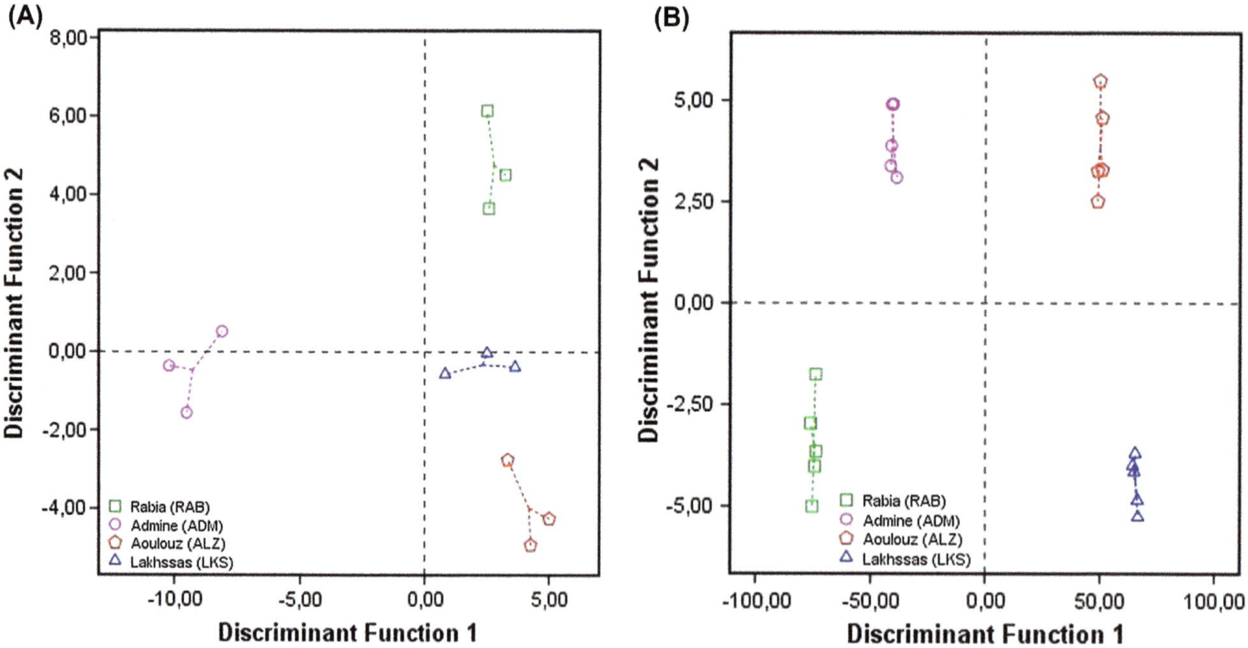

FIGURE 15.5 Two-dimensional scatterplot showing the separation of the four contrasting ecotypes of *Argania spinosa* according to the two discriminant function gradients obtained by canonical discriminant analysis under recovery conditions. (A) physiological traits and (B) biochemical traits.

distinguishing between the Adm ecotype and the other ecotypes under drought stress and rehydration conditions by the both physiological traits Ψ_{pb} and K^+. Considering the second DF, the Lks ecotype was clearly and mainly separated from the other ecotypes, under drought conditions, by high accumulation of Mg^{2+}, whereas the Rab ecotype was distinguished from the other ecotypes, under rehydration conditions, by stomatal conductance trait. Meanwhile, concerning biochemical traits, the first DF clearly separated both inland ecotypes (Lks and Alz) from the both coastal ecotypes (Rab and Adm) by GST and Mn-SOD enzymes under drought stress conditions and mainly by Mn-SOD and GR enzymes under rehydration conditions. The second DF contributed to distinguish among the inland ecotypes on the one hand, and, on the other

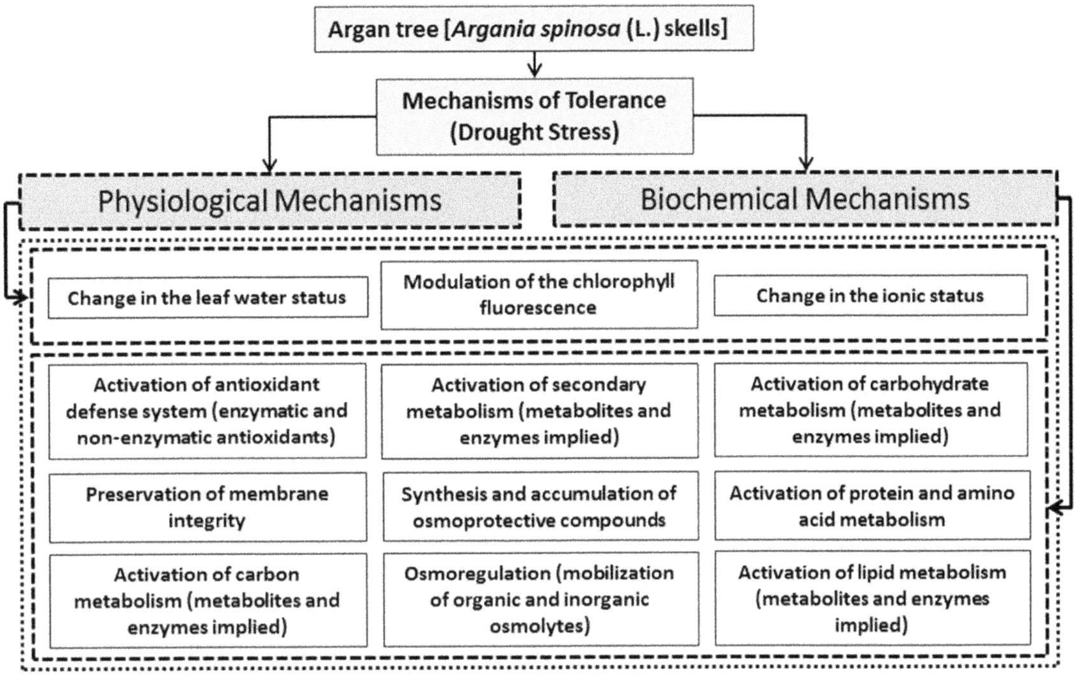

FIGURE 15.6 Different physiological and biochemical mechanisms involved in drought stress tolerance in *Argania spinosa.*

hand, among the coastal ecotypes under drought stress and rehydration conditions. Total SOD, Mn-SOD, GP, GR, and the nonprotein thiols were the biochemical traits having most power to discriminate between the 4 *A. spinosa* ecotypes under drought stress and recovery conditions.

Based on our multifactorial approach that involved the characterization of the physiological and biochemical tolerance mechanisms active in the argan tree under drought stress and recovery conditions, as well as the integration into a discriminating statistical model of all studied traits, we highlighted interecotypic differences in tolerance to drought stress and a variety of *A. spinosa's* effective adaptation mechanisms (Fig. 15.6).

4. CONCLUSION

Plant exposure to environmental stresses such as drought leads to changes in their physiological and biochemical status. To avoid such disturbances and to instead promote a state of equilibrium that ensures survival plants have developed physiological and biochemical mechanisms of tolerance, including various molecules and organs fulfilling their functions to counter the negative impact of environmental constraints. Our integrated multifactorial approach has combined physiological and biochemical analyses to study the reactions and characterize the behavior of *A. spinosa* under drought stress. It appears from our findings that there are significant differences between the four contrasting ecotypes of argan tree in terms of their physiological and biochemical responses to drought stress at different levels on the one hand and, to edaphic drought+recovery on the other hand. Multivariate analysis of all studied traits revealed that both the inland ecotypes (Lks and Alz) showed good ability to tolerate drought stress compared with both the coastal ecotypes (Adm and Rab). Thus the CDA has allowed to distinguish between these four contrasting ecotypes taking into account their physiological and biochemical traits of tolerance having the most discriminating power. At the physiological level, we noted that the traits Ψ_{pb}, g_s, q_p, and F_v'/F_m' are more discriminant among all the studied physiological traits. At the biochemical level, the ascorbate–glutathione cycle enzymes (GR, GP, MDHAR and DHAR), GSH, ascorbic acid, and C4H showed higher discriminating power among all the analyzed biochemical traits. These traits seem to be considered as heritable adaptive traits constituting the internal mechanism of tolerance to drought stress in *A. spinosa*. In the light of results from the statistical analysis, we suggest the following order of the four contrasting ecotypes in terms of tolerance to drought stress: Lks>Alz>Rab>Adm. Nonetheless, these results require confirmation and validation by field study to determine the effects of natural environmental conditions. Future research on the drought tolerance/adaptation degree of *A. spinosa* will be necessary to fully appreciate the agronomic performance of inland and coastal ecotypes and, consequently, the development of the argan sector within Morocco and the region.

ACKNOWLEDGMENTS

I thank all my coauthors who have contributed effectively toward the realization of this work. This research was financially supported by the Académie Hassan II des Sciences et Techniques and the Ministère Marocain de l'Enseignement Supérieur, de la Recherche Scientifique et de la Formation des Cadres, Morocco.

REFERENCES

Ahmad, P., 2014. Oxidative damage to plants: antioxidant networks and signaling. Academic Press, San Diego, pp. 672.

AitAabd, N., Msanda, F., El Mousadik, A., 2014. Evaluation of variability in argan oil content through different environments and preselection of elite genotypes. Euphytica 195, 157–167.

Arkus, K.A.J., Cahoon, E.B., Jez, J.M., 2005. Mechanistic analysis of wheat chlorophyllase. Arch. Biochem. 438, 146–155.

Badawi, G.H., Yamauchi, Y., Shimada, E., Sasaki, R., Kawano, N., Tanaka, K., Tanaka, K., 2004. Enhanced tolerance to salt stress and water deficit by overexpressing superoxide dismutase in tobacco (*Nicotiana tabacum*) chloroplasts. Plant Sci. 166, 919–928.

Baker, N.R., Rosenqvist, E., 2004. Applications of chlorophyll fluorescence can improve crop production strategies: an examination of future possibilities. J. Exp. Bot. 55, 1607–1621.

Chakhchar, A., 2015. Mécanismes physiologiques et biochimiques de la tolérance de l'arganier au stress hydrique. Université Cadi Ayyad.Thèse, p. 210.

Chakhchar, A., Lamaoui, M., Wahbi, S., Ferradous, A., El Mousadik, A., Ibnsouda-Koraichi, S., Filali-Maltouf, A., El Modafar, C., 2015a. Leaf water status, osmoregulation and secondary metabolism as a model for depicting drought tolerance in *Argania spinosa*. Acta Physiol. Plant 37, 1–16.

Chakhchar, A., Wahbi, S., Lamaoui, M., Ferradous, A., El Mousadik, A., Ibnsouda-Koraichi, S., Filali-Maltouf, A., El Modafar, C., 2015b. Physiological and biochemical traits of drought tolerance in *Argania spinosa*. J. Plant Interact. 10, 252–261.

Chakhchar, A., Lamaoui, M., Ferradous, A., Wahbi, S., El Mousadik, A., Ibnsouda-Koraichi, S., Filali-Maltouf, A., El Modafar, C., 2015c. Differential drought tolerance of four contrasting *Argania spinosa* ecotypes assessed by enzymatic and non-enzymatic antioxidant. Int. J. Recent Sci. Res. 6, 3002–3009.

Chakhchar, A., Lamaoui, M., Aissam, S., Ferradous, A., Wahbi, S., El Mousadik, A., Ibnsouda-Koraichi, S., Filali-Maltouf, A., El Modafar, C., 2016. Differential physiological and antioxidative responses to drought stress and recovery among four contrasting *Argania spinosa* ecotypes. J. Plant Interact. 11, 30–40.

Chakhchar, A., Lamaoui, M., Aissam, S., Ferradous, A., Wahbi, S., El Mousadik, A., Ibnsouda-Koraichi, S., Filali-Maltouf, A., El Modafar, C., 2017a. Electrolyte ions and glutathione enzymes as stressmarkers in *Argania spinosa* subjected to drought stress and recovery. Afr. J. Biotechnol. 16, 10–21.

Chakhchar, A., Lamaoui, M., Aissam, S., Ferradous, A., Wahbi, S., El Mousadik, A., Ibnsouda-Koraichi, S., Filali-Maltouf, A., El Modafar, C., 2017b. Using chlorophyll fluorescence, photosynthetic enzymes and pigment composition to discriminate drought tolerant ecotypes of *Argania spinosa*. Plant Biosyst. http://dx.doi.org/10.1080/11263504.2017.1297334.

Chakhchar, A., Haworth, M., El Modafar, C., Lauteri, M., Mattioni, C., Wahbi, S., Centritto, M., 2017c. An assessment of genetic diversity and drought tolerance in argan tree(*Argania spinosa*) populations:potential for the development of improved drought tolerance. Front. Plant Sci. 8, 276.

Chaussod, R., Adlouni, A., Christon, R., 2005. The argan tree and argan oil in Morocco: towards a deep change in a traditional agroforestry system. Economic and Scientific challenges. Cah. Agric. 14, 351–356.

Daszkowska-Golec, A., Szarejko, I., 2013. Open or close the gate – stomata action under the control of phytohormones in drought stress conditions. Front. Plant Sci. 4, 138.

Deneke, S.M., 2000. Thiol-based antioxidants. Curr. Top. Cell Regul. 36, 151–180.

Diaz-Barradas, M.C., Zunzunegui, M., Ain-Lhout, F., Jáuregui, J., Boutaleb, S., Álvarez-Cansino, L., Esquivias, M.P., 2010. Seasonal physiological responses of *Argania spinosa* tree from Mediterranean to semi-arid climate. Plant Soil 337, 217–231.

Diaz Barradas, M.C., Zunzunegui, M., Paz Esquivias, M., Boutaleb, S., Valera-Burgos, J., Tagma, T., Ain-Lhout, F., 2013. Some secrets of *Argania spinosa* water economy in a semiarid climate. Nat. Prod. Commun. 8, 11–14.

Di Ferdinando, M., Brunetti, C., Agati, G., Tattini, M., 2014. Multiple functions of polyphenols in plants inhabiting unfavorable Mediterranean areas. Environ. Exp. Bot. 103, 107–116.

Fahad, S., Bajwa, A.A., Nazir, U., Anjum, S.A., Farooq, A., Zohaib, A., Sadia, S., Nasim, W., Adkins, S., Saud, S., Ihsan, M.Z., Alharby, H., Wu, C., Wang, D., Huang, J., 2017. Crop production under drought and heat stress: plant responses and management options. Front. Plant Sci. 8, 1147.

Feller, U., 2016. Drought stress and carbon assimilation in a warming climate: reversible and irreversible impacts. J. Plant Physiol. 203, 84–94.

Flexas, J., Bota, J., Escalona, J.M., Sampol, B., Medrano, H., 2002. Effects of drought on photosynthesis in grapevines under field conditions: an evaluation of stomatal and mesophyll limitations. Funct. Plant Biol. 29, 461–471.

Foyer, C.H., Noctor, G., 2005a. Redox homeostasis and antioxidant signaling: a metabolic interface between stress perception and physiological responses. Plant Cell 17, 1866–1875.

Foyer, C.H., Noctor, G., 2005b. Oxidant and antioxidant signaling in plants: a re-evaluation of the concept of oxidative stress in a physiological context. Plant Cell Environ. 28, 1056–1071.

Genin, M., Alifriqui, M., Fakhech, A., Hafidi, M., Ouahmane, L., Genin, D., 2017. Back to forests in pre-Saharan Morocco? When prickly pear cultivation and traditional agropastoralism reduction promote argan tree regeneration. Silva Fenn. 51, 1–22.

Golldack, D., Li, C., Mohan, H., Probst, N., 2014. Tolerance to drought and salt stress in plants: unraveling the signaling networks. Front. Plant Sci. 5, 151.

González, A., Ayerbe, L., 2010. Effect of terminal water stress on leaf epicuticular wax load, residual transpiration and grain yield in barley. Euphytica 172, 341–349.

Gururani, M.A., Venkatesh, J., Tran, L.S., 2015. Regulation of photosynthesis during abiotic stress-induced photoinhibition. Mol. Plant 8, 1304–1320.

He, B., Cui, X., Wang, H., Chen, A., 2014. Drought: the most important physical stress of terrestrial ecosystems. Acta Ecol. Sin. 34, 179–183.

Katam, R., Sakata, K., Suravajhala, P., Pechan, T., Kambiranda, D.M., Naik, K.S., Guo, B., Basha, S.M., 2016. Comparative leaf proteomics of drought-tolerant and susceptible peanut in response to water stress. J. Proteomics 143, 209–226.

Khoyerdi, F.F., Shamshiri, M.H., Estaji, A., 2016. Changes in some physiological and osmotic parameters of several pistachio genotypes under drought stress. Sci. Hortic. 26, 44–51.

Kolenc, Z., Vodnik, D., Mandelc, S., Javornik, B., Kastelec, D., Čerenak, A., 2016. Hop (*Humulus lupulus* L.) response mechanisms in drought stress: proteomic analysis with physiology. Plant Physiol. Biochem. 105, 67–78.

Król, A., Amarowicz, R., Weidner, S., 2014. Changes in the composition of phenolic compounds and antioxidant properties of grapevine roots and leaves (*Vitis vinifera* L.) under continuous of long-term drought stress. Acta Physiol. Plant 36, 1491–1499.

Kroth, P.G., 2015. The biodiversity of carbon assimilation. J. Plant Physiol. 172, 76–81.

Landi, M., Tattini, M., Gould, K.S., 2015. Multiple functional roles of anthocyanins in plant-environment interactions. Environ. Exp. Bot. 119, 4–17.

Landi, S., De Lillo, A., Nurcato, R., Grillo, S., Esposito, S., 2017. In-field study on traditional Italian tomato landraces: the constitutive activation of the ROS scavenging machinery reduces effects of drought stress. Plant Physiol. Biochem. 118, 150–160.

Landi, S., Nurcato, R., De Lillo, A., Lentini, M., Grillo, S., Esposito, S., 2016. Glucose-6-phosphate dehydrogenase plays a central role in the response of tomato (*Solanum lycopersicum*) plants to short and long-term drought. Plant Physiol. Biochem. 105, 79–89.

Liu, Y., Li, P., Xu, G.C., Xiao, L., Ren, Z.P., Li, Z.B., 2017. Growth, morphological, and physiological responses to drought stress in *Bothriochloa ischaemum*. Front. Plant Sci. 8, 230.

Lybbert, T.J., Barrett, C.B., Narjisse, H., 2002. Market-based conservation and local benefits: the case of argan oil in Morocco. Ecol. Econ. 41, 125–144.

Lybbert, T.J., Magnan, N., Aboudrare, A., 2010. Household and local forest impacts of Morocco's argan oil bonanza. Environ. Dev. Econ. 15, 439–464.

Ma, D., Sun, D., Wang, C., Li, Y., Guo, T., 2014. Expression of flavonoid biosynthesis genes and accumulation of flavonoid in wheat leaves in response to drought stress. Plant Physiol. Biochem. 80, 60–66.

McGregor, H.V., Dupont, L., Stuut, J.B.W., Kuhlmann, H., 2009. Vegetation change, goats, and religion: a 2000-year history of land use in southern Morocco. Quat. Sci. Rev. 28, 1434–1448.

Msanda, F., El Aboudi, A., Peltier, J.P., 2005. Biodiversity and biogeography of Moroccan argan tree communities. Cah. Agric. 4, 357–364.

Munemasa, S., Muroyama, D., Nagahashi, H., Nakamura, Y., Mori, I.C., Murata, Y., 2013. Regulation of reactive oxygen species-mediated abscisic acid signaling in guard cells and drought tolerance by glutathione. Front. Plant Sci. 4, 472.

Nazar, R., Iqbal, N., Masood, A., Syeed, S., Khan, S.A., 2011. Understanding the significance of sulfur in improving salinity tolerance in plants. Environ. Exp. Bot. 70, 380–387.

Noctor, G., Lelarge-Trouverie, C., Mhamdi, A., 2015. The metabolomics of oxidative stress. Phytochemistry 112, 33–53.

Nogués, S., Baker, N.R., 2000. Effects of drought on photosynthesis in Mediterranean plants grown under enhanced UV-B radiation. J. Exp. Bot. 51, 1309–1317.

Osakabe, Y., Osakabe, K., Shinozaki, K., Tran, L.S.P., 2014. Response of plants to water stress. Front. Plant Sci. 5, 86.

Parida, A.K., Dagaonkar, V.S., Phalak, M.S., Umalkar, G.V., Aurangabadkar, L.P., 2007. Alterations in photosynthetic pigments, protein and osmotic components in cotton genotypes subjected to short-term drought stress followed by recovery. Plant Biotechnol. Rep. 1, 37–48.

Passioura, J.B., 2007. The drought environment: physical, biological and agricultural perspectives. J. Exp. Bot. 58, 113–117.

Ping, M.A., Tuan-hui, B.A.I., Feng-wang, M.A., 2015. Effects of progressive drought on photosynthesis and partitioning of absorbed light in apple trees. J. Integr. Agric. 14, 681–690.

Pompeiano, A., Remorini, D., Vita, F., Guglielminetti, L., Miele, S., Morini, S., 2016. Growth and physiological response of *Arundo donax* L. to controlled drought stress and recovery. Plant Biosyst. http://dx.doi.org/10.1080/11263504.2016.1249427.

Reddy, A.R., Chiatanya, K.V., Vivekanandan, M., 2004. Drought induced responses of photosynthesis and antioxidant metabolism in higher plants. J. Plant Physiol. 161, 1189–1202.

Rodziewicz, P., Swarcewicz, B., Chmielewska, K., Wojakowska, A., Stobiecki, M., 2014. Influence of abiotic stresses on plant proteome and metabolome changes. Acta Physiol. Plant 36, 1–19.

Rouhier, N., Cerveau, D., Couturier, J., Reichheld, J.P., Rey, P., 2015. Involvement of thiol-based mechanisms in plant development. Biochim. Biophys. Acta 1850, 1479–1496.

Seo, P.J., Park, C.M., 2011. New cuticular wax biosynthesis as a way of inducing drought resistance. Plant Signal Behav. 6, 1043–1045.

Sewelam, N., Kazan, K., Schenk, P.M., 2016. Global plant stress signaling: reactive oxygen species at the cross-road. Front. Plant Sci. 7, 187.

Shao, H.B., Chu, L.Y., Lu, Z.H., Kang, C.M., 2008. Primary antioxidant free radical scavenging and redox signaling pathways in higher plant cells. Int. J. Biol. Sci. 4, 8–14.

Somerville, C., Dangl, J., 2000. Plant biology in 2010. Science 290, 2077–2078.

Szalai, G., Kellos, T., Galiba, G., Kocsy, G., 2009. Glutathione as an antioxidant and regulatory molecule in plants under abiotic stress conditions. J. Plant Growth Regul. 28, 66–80.

Szőllősi, R., 2014. Superoxide dismutase (SOD) and abiotic stress tolerance in plants: an overview. In: Ahmad, P. (Ed.), Oxidative Damage to Plants. Academic Press, San Diego, pp. 89–129 (Chapter 3).

Tardieu, F., Simonneau, T., 1998. Variability among species of stomatal control under fluctuating soil water status and evaporative demand: modelling isohydric and anisohydric behaviours. J. Exp. Bot. 49, 419–432.

Taulavuori, E., Tahkokorpi, M., Laine, K., Taulavuori, K., 2010. Drought tolerance of juvenile and mature leaves of a deciduous dwarf shrub *Vaccinium myrtillus* L. in a boreal environment. Protoplasma 241, 19–27.

Toscano, S., Farieri, E., Ferrante, A., Romano, D., 2016. Physiological and biochemical responses in two ornamental shrubs to drought stress. Front. Plant Sci. 7, 645.

Uzilday, B., Turkan, I., Sekmen, A.H., Ozgur, R., Karakaya, H.C., 2012. Comparison of ROS formation and antioxidant enzymes in *Cleome gynandra* (C4) and *Cleome spinosa* (C3) under drought stress. Plant Sci. 182, 59–70.

Wu, X., Yuan, J., Luo, A., Chen, Y., Fan, Y., 2016. Drought stress and re-watering increase secondary metabolites and enzyme activity in dendrobium moniliforme. Ind. Crops Prod. 94, 385–393.

Yoon, S.K., Park, E.J., Choi, Y.I., Bae, E.K., Kim, J.H., Park, S.Y., Kang, K.S., Lee, H., 2014. Response to drought and salt stress in leaves of poplar (*Populus alba* × *Populus glandulosa*): expression profiling by oligonucleotide microarray analysis. Plant Physiol. Biochem. 26, 158–168.

Chapter 16

Dynamic Proline Metabolism: Importance and Regulation in Water-Limited Environments

Muhammad A. Ashraf[1], Muhammad Iqbal[1], Rizwan Rasheed[1], Iqbal Hussain[1,2], Shagufta Perveen[1], Saqib Mahmood[1]

[1]*Government College University Faisalabad, Faisalabad, Pakistan;* [2]*Graduate School of Life Sciences, Tohoku University, Sendai, Japan*

1. INTRODUCTION

Plants under drought conditions accumulate substantial amounts of proline. In this context, roots of maize seedlings accumulated more than 100 mM proline when subjected to limited water availability (Rejeb et al., 2014; Bhaskara et al., 2015; Hong et al., 2016; Zandalinas et al., 2017). Proline is one of the widely studied compatible solutes having zwitterionic structure and greater solubility (Yancey et al., 1982; Fariduddin et al., 2014; Filippou et al., 2014). Endogenous levels of proline are higher in organelles and the cytoplasm due to exclusion of this imino acid from vacuoles (Büssis and Heineke, 1998; Abbas et al., 2010; Li et al., 2013; Per et al., 2017). Why plants accumulate significantly greater levels of proline over other metabolites and how proline is involved in the induction of drought tolerance is a question of widespread interest in plant science discipline (Kavi Kishor and Sreenivasulu, 2014; Bhaskara et al., 2015; Dinesh et al., 2016). A review of the literature is presented in Table 16.1 where proline involvement in drought tolerance is enlisted. Enzymes such as P5CR (Δ1-pyrroline-5-carboxylate reductase) and P5CS (Δ1-pyrroline-5-carboxylate synthetase) synthesize proline from glutamate. Catabolism of proline into glutamate is carried out by proline dehydrogenase (ProDH) and P5CDH (Δ1-pyrroline-5-carboxylate dehydrogenase) (Szabados and Savouré, 2010; Verslues and Sharma, 2010). Gene expression studies for proline dehydrogenase 1 (ProDH1) and Δ1-pyrroline-5-carboxylate synthetase 1 (P5CS1) indicated that photosynthetic tissues had greater biosynthesis and suppressed catabolism of proline. However, the breakdown of proline is significantly greater in regions of roots and shoot with meristematic activity (Sharma et al., 2011). *Arabidopsis* plants had shown enhanced expression of P5CS1 in response to various environmental constraints including drought stress (Peng et al., 1996). The P5CS1 is a rate-limiting enzyme in the biosynthesis of proline (Székely et al., 2008; Surekha et al., 2014; Zarattini and Forlani, 2017). Limited expression of ProDH1 also determined proline accumulation in *Arabidopsis* plants under water-limited conditions (Sharma et al., 2011). Gene expression studies on ProDH1 and P5CS1 indicated that proline accumulation is not merely a symptom of stress injury, rather the metabolism of proline is precisely regulated (Ober and Sharp, 1994; Giberti et al., 2014; Kaur et al., 2017). Plants under stressful conditions show stimulated generation of reactive oxygen species (ROS) including H_2O_2, singlet oxygen, as well as superoxide and hydroxyl radicals. These species induce lipid peroxidation and react with important macromolecules thereby resulting in substantial damage. Plants possess an array of essential systems that keep ROS levels under control. In this context, detoxification of ROS is carried out by low-molecular-weight metabolites functioning as antioxidants, e.g., α-tocopherol, glutathione, ascorbic acid, flavonoids, phenolics, and proline (Rasheed et al., 2014). In addition, enzymatic antioxidants, namely, superoxide dismutase (SOD), catalase (CAT), and ascorbate peroxidase (APX) can efficiently scavenge ROS.

Metabolism of proline is different from that of other amino acids in a variety of ways. Among proteogenic amino acids, only proline contains α-amino group as secondary amine. Such distinctive features of proline must be known so as to completely understand the functions of proline in plants. An important characteristic of proline is its accumulation in greater levels in plants as well as microorganisms under stressful conditions. Accumulation of proline is more conspicuous in plants facing dehydration stress in the form of salinity, freezing, or drought (Verslues and Bray, 2006; Verslues and Sharma, 2010; Suzuki et al., 2014; Balestro et al., 2017). However, plants also accumulate proline, though in minimal concentration,

Plant Metabolites and Regulation Under Environmental Stress. http://dx.doi.org/10.1016/B978-0-12-812689-9.00016-9

TABLE 16.1 Some Examples From the Literature Representing Association of Proline Metabolism With Drought Tolerance in Plants

Plant Species	Drought Intensity	Proline Alters Key Physiochemical Attributes Linked With Drought Tolerance	References
Medicago saliva L.	Irrigation was restricted by withholding water supply for 8 days	Proline and total soluble protein accumulation in leaves and nodules was significant, which in turn enhanced water status of plants in these organs. Authors suggested that increase in the accumulation of total soluble proteins and proline stabilized protein, buffered cytosolic pH, and mediated osmotic adjustment	Irigoyen et al. (1992)
Petunia hybrida L.	Plants were grown in growth chamber without water in culture with the following conditions: 16/8 light/darkness at 25°C and 60% humidity	Proline functioned as an important osmoprotectant in *P. hybrid* L. plants under drought conditions. Plants exhibited conspicuous increase in the cellular levels of proline upon exposure to drought conditions. In the present experiment, *P. hybrid* L. plants were transformed with AtP5CS from *Arabidopsis thaliana* or with OsP5CS from rice. These transgenic plants accumulated more proline and correlation of proline was studied with drought tolerance. Proline accumulation accounted for 0.57%–1.01% of the total free amino acids in transformed plants. This proline accumulation was 1.5–2.6 times higher than in wild-type plants under nonstress conditions. Plants transformed with P5CS had shown tolerance to 14-day period of drought conditions, which showed the involvement of P5CS in enabling the plants to tolerate drought stress. Likewise, plants treated with supplemented proline have shown significant proline accumulation. Plants supplied with 5 mM exogenous proline showed greater proline accumulation (18 times) when compared with control plants. Supplemented proline suppressed the growth of *Petunia* plants (wild type) over *Arabidopsis* plants. This result clearly suggested that proline accumulation is related to the plant species. The alterations in growth of *Petunia* plants were related to proline accumulation as well as relative proline contents over total free amino acids	Yamada et al. (2005)
Cicer arietinum L.	Drought stress was imposed by withholding irrigations at the vegetative or anthesis stages	A field experiment on chickpea was conducted in complete block randomized design with three replicates per treatment to determine the influence of drought on transpiration, photosynthesis, stomatal conductance, yield, chlorophyll content, and proline accumulation. Drought tolerance cultivars, namely, ILC482 and Bivaniej, as well as drought-sensitive cultivar, Pirouz, were used in the experiment. Treatment breakup was control (normal irrigation), drought at the vegetative stage, at anthesis stage, and at both vegetative and anthesis stages. Drought stress had significant negative effects on key physiological attributes. Drought stress at the vegetative or anthesis stage caused considerable decline in chlorophyll contents. Accumulation of proline was less in Pirouz than in ILC482 under normal or drought stress conditions. Authors further concluded that drought stress at anthesis stage is more critical to plants than at the vegetative stage. It was recorded in this experiment that plants subjected to drought conditions at the anthesis stage had shown greater endogenous levels of proline over plants exposed to drought conditions at the vegetative stage. Greater yield losses were evident in plants facing drought at the anthesis stage. Therefore it could be concluded that proline accumulation at the anthesis stage is an important factor that might contribute to drought tolerance of plants	Mafakheri et al. (2010)
Olea europaea L.	Irrigation was restricted for 20 days	When 2-year-old olive plants were subjected to drought conditions for 20 days, a significant increase in proline contents corresponding to the intensity of drought was recorded. A marked rise in the activity of lipoxygenase and MDA contents were recorded in roots and leaves of plants. Authors suggested that proline accumulation might have determined drought tolerance in olive plants. Increase in MDA contents and LOX activity is a reflection of severity of drought stress	Sofo et al. (2004)

TABLE 16.1 Some Examples From the Literature Representing Association of Proline Metabolism With Drought Tolerance in Plants—cont'd

Plant Species	Drought Intensity	Proline Alters Key Physiochemical Attributes Linked With Drought Tolerance	References
Nicotiana tabacum L.	–	Plants under stress conditions could not have feedback inhibition for P5CS. In the present experiment, authors compared proline levels in transgenic tobacco plants (transformed with P5CS from *Vigna aconitifolia* and a mutated enzyme, namely, P5CSF129A, with lost ability to feedback inhibition by proline by site-directed mutagenesis). Plants transformed with P5CSF129A had shown twofold greater proline accumulation when compared with plants transformed with P5CS from *V. aconitifolia*. This variation in proline accumulation was further amplified in plants treated with 200 mM NaCl salinity. This study indicated that P5CS played an essential part in controlling the endogenous levels of proline under stress or normal conditions through feedback inhibition. Transgenic plants had shown greater tolerance to salinity stress due to the ROS scavenging potential of proline. Authors further concluded that proline played a critical role in protecting plants from oxidative damage under osmotic stress	Hong et al. (2000)
Helianthus annuus L.	Drought stress was imposed by maintaining 60% field capacity	Authors studied various variations in biomass allocation, growth, chlorophyll contents, important biochemical constituents, and proline metabolism in five sunflower cultivars. Authors have found a significant variation in proline metabolism along with other attributes in sunflower plants under drought conditions. Furthermore, drought stress induced increase in glycine-betaine and amino acids and proline accumulation in sunflower plants. Moreover, plants exposed to drought conditions had shown greater γ-glutamyl kinase, whereas the activity of proline oxidase enzyme was substantially decrease, which indicated the inherent potential of sunflower plants to accumulate proline in substantial amounts	Manivannan et al. (2007)
Zea mays L.	PEG 10% (−0.15–1.76 MPa), PEG 20% (−0.49–1.76 MPa), PEG 30% (−1.03–1.76 MPa) and PEG 40% (−1.76 MPa).	Drought stress caused a significant decrease in growth and production in maize plants. Drought stress was imposed using different PEG-6000 concentrations i.e., PEG 10% (−0.15–1.76 MPa), PEG 20% (−0.49–1.76 MPa), PEG 30% (−1.03–1.76 MPa) and PEG 40% (−1.76 MPa), and control (without PEG). Authors have found increase in the levels of proline and total soluble sugars in maize plants under PEG-induced water deficit conditions. Authors suggested that proline might have reduced the inhibitory effects of dehydration stress in plants	Mohammadkhani and Heidari (2008)
Populus przewalskii Maximowicz	100%, 50%, and 25% field capacity	Explants of *P. przewalskii* Maximowicz were exposed to various water regimes, i.e., 25, 50%, and 100% field capacity. The effect of onset of drought stress on important morphophysiological and biochemical attributes was determined. In this experiment, two different species of *P. przewalskii* (from dry and wet climates) were used. Drought had significant effect on plant biomass partitioning. Drought stress triggered decrease in chlorophyll contents and induced increase in the accumulation of total free amino acids and proline. In addition, drought stress resulted in a marked increase in the endogenous levels of ABA, superoxide radicals, and hydrogen peroxide, which functioned as secondary messenger molecules thereby activating the antioxidant defense system in plants. This defense system was composed of glutathione reductase, ascorbate peroxidase, guaiacol peroxidase, and superoxide dismutase. The population of *P. przewalskii* from dry land accumulated more proline and total free amino acids and exhibited more biomass partitioning to root system. Authors suggested that accumulation of more proline was positively associated with osmotic adjustment. The antioxidant system of the population from dry climate had shown better antioxidant system in terms of ascorbic acid contents, glutathione reductase, and ascorbate peroxidase than the population from wet climate	Lei et al. (2006)

Continued

TABLE 16.1 Some Examples From the Literature Representing Association of Proline Metabolism With Drought Tolerance in Plants—cont'd

Plant Species	Drought Intensity	Proline Alters Key Physiochemical Attributes Linked With Drought Tolerance	References
Gossypium hirsutum L.	Drought stress was induced by withholding irrigation for 14 days	Influence of drought on activities of antioxidant enzymes, RWC, and accumulation of free proline was studied in sensitive (Ca/H 148) and tolerant (Ca/H 680) cotton genotypes. These determinations on the aforementioned attributes were made during and after stress recovery. Induction of water stress resulted in increased activity of P5CS and accumulation of proline in both sensitive and tolerant genotypes. The activity of OAT did not change, whereas there was minimal decline in the activity of P5CR. Drought conditions induced decrease in ProDH activity in both cotton genotypes. Plants of tolerant genotype had shown greater RWC, photosynthesis rate, and proline contents, as well as greater P5CR and P5CS activities under drought conditions. Data after recovery were similar to those in control plants. The results of the experiment suggested that increased proline accumulation due to downregulation of ProDH and upregulation of P5CS might be critical for the development of drought resistance in plants	Parida et al. (2008)
A. thaliana L.	Drought stress was induced by withholding water for 10 days	The association among soluble sugars, proline, anthocyanin, and photosynthesis was apprised in *Arabidopsis* during drought onset, mild drought, and moderate drought. Authors suggested that proline had a played significant role in the induction of drought tolerance through the reduction of MDA contents in plants	Sperdouli and Moustakas (2012)
Triticum aestivum L.	Drought stress induced by restricting irrigation for 15 days	Plants accumulate proline as an adaptive strategy to tolerate drought conditions. Plants expressing genes for constitutive expression of proline had shown decrease in growth. Exposure of wheat plants to 15 days of drought conditions resulted in a marked accumulation in proline in transgenic plants. Authors suggested that drought tolerance in transgenic plants was correlated with the potential of proline to protect plants from oxidative damage rather than its involvement in osmotic adjustment	Vendruscolo et al. (2007)

ABA, abscisic acid; *AtP5CS*, Δ1-pyrroline-5-carboxylate synthetase genes; *LOX*, lipoxygenase; *MDA*, malondialdehyde; *OAT*, ornithine aminotransferase; *P5CR*, Δ1-pyrroline-5-carboxylate reductase; *P5CS*, Δ1-pyrroline-5-carboxylate synthetase; *ProDH*, proline dehydrogenase; *PEG*, polyethylene glycol; *ROS*, reactive oxygen species; *RWC*, relative water content.

under trace metal stress and other biotic and abiotic elicitors (Fabro et al., 2004; Sharma and Dietz, 2009; Verslues and Sharma, 2010; Slama et al., 2015). Proline performs diverse functions in plants, which explains its accumulation particularly under abiotic stress conditions (Nadeem et al., 2014). For instance, proline exists most of the time as zwitterion with negative and positive charges at nitrogen and carboxylic groups, respectively. Furthermore, proline is highly water soluble when compared with other amino acids. The feature of proline to exist in zwitterionic state is shared by other compatible solutes that are accumulated in significant amounts by organisms to perform osmotic adjustment (Farahani et al., 2013). Unlike other compatible solutes proline performs several other crucial roles in plants including the potential to act as cellular redox buffer (de Carvalho et al., 2013; Sekmen et al., 2014). Significance of proline in plant stress biology could be explained in part by the inherent characters of this biomolecule and in part through its metabolism (Liang et al., 2013; Li et al., 2014). Mechanisms mediating the regulation of proline metabolism are greatly variable when compared with those of other amino acids (Verslues and Sharma, 2010). A very interesting step forward in this context came from the microarray analysis of available data that revealed significant alterations in transcription levels of genes involved in catabolism, whereas genes encoding enzymes involved in biosynthesis pathway were only slightly affected due to abiotic stress (Less and Galili, 2008). An important exception to this research includes the substantial influence of abiotic stress on the regulation of biosynthetic enzymes in interlinked pathways of arginine and proline metabolism. Apart from the regulation of proline metabolism, compartmentalization of proline between cytoplasm, chloroplast, and mitochondria is critical and still not fully understood. Proline transport (inter- and intracellular), association of metabolic pathways of proline with other important cellular metabolism, and differential accumulation of proline in tissues are critical to completely understand the basics of proline metabolism, which may improve our understanding of proline functions under stressful conditions (Verslues and Sharma, 2010; Ali et al., 2013; Huang et al., 2013).

2. ENZYMES INVOLVED IN THE REGULATION OF PROLINE METABOLISM

Proline metabolism involves various enzymes. Biosynthesis of proline occurs in the chloroplast and/or cytoplasm where two enzymes carry out the biosynthesis of proline from glutamate. P5CS catalyzes the formation of 1-pyrroline-5-carboxylic acid (P5C) from glutamate where P5CR catalyzes the conversion of P5C to proline. P5C may also be yielded by ornithine, and this reaction is catalyzed by the ornithine aminotransferase (OAT) enzyme. The catabolism of proline occurs in mitochondria. Proline is first converted to P5C via enzyme ProDH. P5C is then converted to glutamate and ornithine in the presence of OAT and P5CDH enzymes, respectively. There is speculation that mitochondria may also provide P5C to mediate proline biosynthesis in chloroplast and/or cytoplasm. Likewise, glutamate formed in mitochondria is transported to cytoplasm and/or chloroplast. This speculated proline metabolism is known as proline cycle.

Stress-induced enhanced synthesis of proline from glutamate and suppressed catabolism of proline are the major control levels for proline. This process mostly holds good only with a few exceptions (Stines et al., 1999). This model is in line with the available literature that indicates the regulation of proline metabolism via gene transcriptomes of key enzymes. However, this is one part of the picture; the other part that involves posttranscriptional control over enzymes in proline metabolism and function of ornithine as precursor of proline biosynthesis remains unclear. Likewise, integrated regulation and metabolic changes induced by proline cycle need to be more clearly studied so as to completely understand the function and metabolism of proline (Verslues and Sharma, 2010).

Microorganisms under osmotic stress accumulate proline in substantial amounts (Kavi Kishor and Sreenivasulu, 2014). Under drought and saline conditions, proline protects plants from damage because it can scavenge ROS (Szabados and Savouré, 2010; Natarajan et al., 2012). Proline is considered as an ideal solute due to its greater solubility in water (Kavi Kishor and Sreenivasulu, 2014). Strong evidence is available in the literature that highlights a positive association between abiotic stress tolerance and accumulation of proline (Saradhi et al., 1995; Gururani et al., 2013; Kavi Kishor and Sreenivasulu, 2014; Vurukonda et al., 2016). Proline plays critical role in mediating plant growth under abiotic stress as evident from transgenic studies (Kavi Kishor and Sreenivasulu, 2014). Studies on enhanced or suppressed expression of different genes involved in proline metabolism revealed significant physiological functions of proline (Strizhov et al., 1997; Beltrano et al., 2013). Constitutive expression of genes regulating proline metabolism induced various pleiotropic effects (Kishor et al., 1995). It has been found that plants engineered to overexpress genes of proline biosynthesis had higher yield, better inflorescence architecture, and greater plant biomass (Hare et al., 1999; Kavi Kishor and Sreenivasulu, 2014). Likewise, studies on transgenic plants also indicated involvement of proline during development (Funck et al., 2012). There exists a consensus among most plant physiologists that proline functions as a key osmoticum and mediates a variety of developmental processes (Kavi Kishor and Sreenivasulu, 2014; Yu et al., 2016). There has been extensive research focused on determining the mechanisms by which proline regulates plant growth in response to stress. However, despite ample literature on proline involvement in stress tolerance, the relative contribution of spatial and temporal accumulation of proline toward stress tolerance is not fully known (Kavi Kishor and Sreenivasulu, 2014).

To completely understand proline homeostasis, a number of important questions should be given considerations. The first is how proline is synthesized in different plant parts and is then transported across the plant body at various ontogenic phases. How the accumulation of proline in response to stress influences plant growth, in particular during the transitions from vegetative to reproductive phases. The most important question is whether the accumulation of proline is per se enough to induce stress tolerance in plants or is proline homeostasis involved in the induction of stress tolerance (Kavi Kishor and Sreenivasulu, 2014; Khoyerdi et al., 2016). Kavi Kishor and Sreenivasulu (2014) suggested that expansion in proline sink when plants shift from vegetative to reproductive phase significantly determines the potential of plants to set seeds and thereby markedly affects yield stability (Kavi Kishor and Sreenivasulu, 2014).

3. PROLINE ACCUMULATION ALTERS HOMEOSTASIS IN PLANT TISSUES UNDER ENVIRONMENTAL CONSTRAINTS

Plants usually accumulate proline in substantial amount under abiotic stress (Dar et al., 2016; Martínez-Fernández et al., 2016; Zanella et al., 2016). However, proline accumulation is particularly significant during pollen and seed formation. The accumulation of proline in plants seems to be positively correlated with desiccation that occurs during natural developmental phases or that is induced (Sinay et al., 2016; Singh, 2016). In all cases, P5CS acts as a rate-limiting enzyme mediating proline accumulation (Székely et al., 2008). Two genes are involved in the formation of P5CS enzyme. One gene is P5CS1, which is more expressed in shoot and root, whereas the second gene is P5CS2, expressed significantly in dividing cells and poorly regulated by external elicitors (Székely et al., 2008). P5CS2 gene is also relatively more expressed in embryo suspensor, endosperm, and inflorescence (Kavi Kishor and Sreenivasulu, 2014). In addition, the expression of P5CS1 is also significantly induced in vegetative tissues in response to the supplemented abscisic acid (ABA) or abiotic stresses including drought. The significant alterations in the expression pattern of both enzymes (P5CS1 and P5CS2) suggest that proline plays a prominent role both in plant development and in mediating plant responses to abiotic stresses (Kavi Kishor and Sreenivasulu, 2014). The alternative splicing of P5CS1 in *Arabidopsis thaliana* suggests the inherent potential for changes in proline accumulation (Kesari et al., 2012). Moreover, plants also show epigenetic control over proline accumulation because DNA methylation improves proline accumulation through induced expression of OAT and P5CS (Zhang et al., 2013). Therefore proline is a key stress-related amino acid that is accumulated by plants under various environmental conditions. In addition, the accumulation of proline is also carried to the next generation as a signal of stress memory (Kavi Kishor and Sreenivasulu, 2014).

Plants subjected to abiotic constraints show induced expression of P5CS1, whereas the expression of ProDH is suppressed especially in plants under salinity, dehydration, or drought stress (Verslues and Sharma, 2010). The schematic proline metabolism is shown in Fig. 16.1. In addition, the expression of ProDH is enhanced upon stress recovery (Satoh et al., 2002). It could be concluded from the information in the literature that increased proline accumulation in different tissues is associated with the induced expression of P5CS1 and suppressed expression of ProDH (Kavi Kishor and Sreenivasulu, 2014). Under stressful conditions, proline homeostasis is disturbed due to increased expression of proline biosynthesis genes and downregulation of genes involved in proline catabolism and thereby results in enhanced accumulation of proline (Kavi Kishor and Sreenivasulu, 2014).

How proline biosynthesis and degradation are regulated via the interaction of hormones with ProDH and P5CS is an important factor in proline homeostasis. In this context, ABA is known to increase biosynthesis of proline (Ábrahám et al., 2003). However, ABA does not have any impact on the expression of PrDH (Verslues and Sharma, 2010). The proline biosynthesis pathway is significantly associated with the endogenous levels of ABA (Ábrahám et al., 2003). Transcription factors (group-S basic leucine zipper) linked with ABA-signaling pathways exerted a very poor influence on the regulation of ProDH (Satoh et al., 2004). Likewise, auxin also mediates the regulation of OAT and P5CS (You et al., 2012). Similarly, the expression of P5CS1 is downregulated by cytokinin and that of P5CS2 is upregulated by cytokinins (Yoshiba et al., 1995). The activity of P5CR is stimulated by salicylic acid that induces proline accumulation (Misra and Saxena, 2009). Likewise, the accumulation of proline is positively correlated with the endogenous levels of soluble sugars (Balibrea et al., 1997). Proline accumulation also increases when carbohydrates are supplemented externally (Larher et al., 1993). It could be inferred from the aforementioned literature that proline homeostasis may also be regulated via different sugars and hormones (Kavi Kishor and Sreenivasulu, 2014).

The energy requirements of dividing cells and conditions after stress recovery are significantly associated with proline homeostasis (Kavi Kishor and Sreenivasulu, 2014). For instance, genes involved in proline metabolism, viz., ProDH1, P5CS, and P5CR are activated in meristematic tissues (Sharma et al., 2011). The cycling of the substrate for proline biosynthesis determines metabolic homeostasis and maintains $NADP^+/NADPH$ through oxidative pentose phosphate pathway

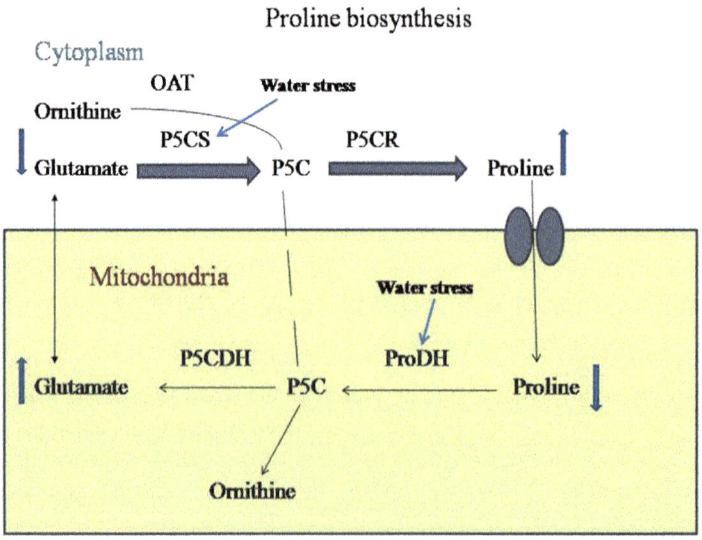

FIGURE 16.1 Schematic diagram of proline metabolism cycle under limited water regimes. *Upward arrow* shows higher concentration, whereas *downward arrows* show low concentration. *OAT*, ornithine aminotransferase; *P5C*, 1-pyrroline-5-carboxylic acid; *P5CDH*, Δ1-pyrroline-5-carboxylate dehydrogenase; *P5CR*, Δ1-pyrroline-5-carboxylate reductase; *P5CS*, Δ1-pyrroline-5-carboxylate synthetase; *ProDH*, proline dehydrogenase.

(OPPP) especially in root tip (Hare et al., 2003). For instance, plants mutant for AtP5CS1 had shown decrease in proline accumulation in germinating seedlings that exhibited delayed appearance of radicle, whereas wild-type plants did now show any impediment in proline accumulation and had successfully replenished $NADP^+$ via the OPPP pathway that is active in plastids (Hare et al., 2003). Thus there is a direct association of increased activity of OPPP with increased proline accumulation, and this association plays significant part in improving germination. Cell wall proteins of germinating seedlings, proteolytic activity of proline-rich reserves, and significant accumulation of proline serve as sufficient sources of proline. Interestingly, when endosperm was removed from the seed, there was a significant decline in proline accumulation. It was also observed that the scutellum acts as an important source of proline in young dividing roots (Raymond and Smirnoff, 2002). Moisture stress induces the accumulation of proline in root and shoot. However, proline porters (ProT2) may also transport proline through phloem to roots. Likewise, phloem becomes rich in proline particularly in plants subjected to moisture stress (Lee et al., 2009). Similarly, proline oxidation generates NADPH, which in turn improves root growth. Apart from the distant transport of proline, its compartmentalization also contributes significantly to proline homeostasis. Under stressful environment, plants synthesize proline in cytoplasm and chloroplast, whereas catabolism of proline is carried out in root mitochondria (Lehmann et al., 2010). Studies on transgenic lines indicated that enhanced accumulation of proline resulted in greater root growth in plants subjected to salinity and drought (Kishor et al., 1995). In a study on maize, plants accumulated up to 120 mM proline in roots and decline in cellular levels of ABA triggered caused decrease in endogenous levels of proline (Ober and Sharp, 1994). The increased cellular levels of ABA determine the increase in proline accumulation in maize roots, and proline also played an essential part in osmotic adjustment. The literature also indicates that plants subjected to low or high water potentials had more consumption of proline over its biosynthesis (Raymond and Smirnoff, 2002). Plants under abiotic stresses show increased shoot to root proline transport where proline degradation supplies energy and thereby increases root biomass. Such alterations in roots enable plants to take up more water from soil with limited moisture level (Kavi Kishor and Sreenivasulu, 2014).

3.1 Apoptosis Regulation Is Associated With the Pyrroline-5-carboxylic Acid Cycle

In plants, higher cellular levels of ROS are lethal, whereas optimal levels of ROS are needed in the modulation and signaling events in a number of developmental processes. Excessive levels of ROS induce significant damage to chlorophyll and nucleic acids and oxidize lipids and proteins (Ashraf et al., 2015). Plants depend on redox signals produced in mitochondria and chloroplast to interact with the changing environment and to maintain metabolic flux (Suzuki et al., 2012). Organisms facing osmotic stress usually downregulate protein catabolism so as to accumulate proline in substantial amount. Moreover, transport of P5C across the mitochondria determines proline metabolism (Miller et al., 2009). The existence of transporters for P5C is not known. However, if this P5C cycle is present, it might control the addition of P5C within mitochondria and cytoplasm. Furthermore, it is observed that ROS levels increase in mitochondria supplemented with exogenous proline (Miller et al., 2009). The generation of glutamate as a result of oxidation of P5C is carried out by P5CDH. The degradation pathways of proline are compromised due to downregulation of P5CDH activity, which in turn increases ROS generation due to excess proline levels (Miller et al., 2009). An intricate balance between mitochondrial and chloroplast ROS generation is required when oxidation of proline increases so as to escape proline-associated noxiousness. Proline metabolism must be precisely regulated so as to maintain redox balance and hence plant growth (Sharma et al., 2011). Proline metabolism generates NADPH and $NADP^+$, whose ratio is critical to redox balance and maintenance of plant growth under stressful conditions (Kavi Kishor and Sreenivasulu, 2014). Actively dividing cells are intricately associated with proline homeostasis, which is stimulated by catabolism of proline. Under dehydration stress, proline efficiently scavenges ROS and functions as an osmoticum. During stress recovery, proline is consumed via ProDH to resume growth by consuming the reservoirs of proline (Nanjo et al., 2003; Kavi Kishor and Sreenivasulu, 2014; Faës et al., 2015). Collapse of proline homeostasis in dividing cells encourages the cells to grow, whereas fractional proline catabolysis produces P5C and thereby initiates apoptosis (Deuschle et al., 2004). The dilemma whether proline accumulation damages or protects plants could be explained by studying the apoptosis initiated by proline-induced P5C (Maxwell and Davis, 2000). P5CDH is critical to the metabolism of P5C to glutamate so as to fulfill the energy requirements for germination, pollen tube development, for vegetative tissues, and for recommencement of growth upon stress recovery (Kavi Kishor and Sreenivasulu, 2014).

3.2 Proline Acts as an Antioxidant

Plants keep the cellular levels in check by employing enzymatic and nonenzymatic antioxidant systems. In this context, proline protected thylakoid membranes from photoinhibition and hence improved photochemical efficiency of thylakoid membrane (Alia and Saradhi, 1991). Likewise, proline significantly reduced lipid peroxidation in terms of malondialdehyde and conjugated dienes in response to high light intensity. Therefore proline is known to stabilize membranes, DNA, and proteins in plants exposed to stressful conditions (Rasheed et al., 2014). Likewise, proline increases the enzymatic activities of POD and APX and decreases the activities of CAT and SOD in grapevine (Ozden et al., 2009). Similarly, wheat plants with enhanced expression of P5CS showed negative association of proline with lipid peroxidation, which indirectly indicates the antioxidant potential of this biomolecule (Vendruscolo et al., 2007). Citrumelo plants having heterologous expression for VvP5CSF129A had greater endogenous contents of proline, which in turn improved the activities of APX and SOD. Proline did not affect the activity of CAT in plants under drought stress (de Campos et al., 2011). Probably proline protects plants from damages of osmotic stress through enhancing antioxidant potential (Kasote et al., 2015). Proline is also reported to scavenge singlet oxygen (Kavi Kishor and Sreenivasulu, 2014). However, this statement has been denied by Signorelli et al. (2013) who advocated that proline did not alter the cell's potential to quench singlet oxygen. Furthermore, it is observed that proline's function as antioxidants largely relies on its ability to quench OH radicals (Kavi Kishor and Sreenivasulu, 2014; Nakabayashi and Saito, 2015).

3.3 Transporters of Proline

Cellular homeostasis is significantly determined by the inter- and intracellular proline transport (Kavi Kishor and Sreenivasulu, 2014). Identification of intracellular transporters of proline needs further research. However, evidences could be traced from transcription levels of intracellular transporters from a variety of species (Lehmann et al., 2011). Studies on intracellular affinity, substrate selection, and localization of three *A. thaliana* proline transporters AtProT showed a marked resemblance. Despite striking resemblance, their expression pattern is variable among tissues, which suggests their divergent function. All these proline transporter genes (*AtProT1*, *AtProT2,* and *AtProT3*) are localized at the cell membrane, but these proline transporters had different expression pattern. For example, *AtProT1* is transcribed more in phloem, cortex and root epidermis showed greater expression for *AtProT2*, whereas *AtProT3* was greatly transcribed in leaf epidermis (Grallath et al., 2005). Plants engineered for overexpression of AtProT showed poor growth, implying the presence of an intricate equilibrium between proline transport, degradation, and biosynthesis (Lehmann et al., 2011). Tomato *LeProT1* was exclusively expressed in pollen grain, and barley *HvProT* was specifically expressed in root tip and thereby resulted in root elongation due to proline accumulation (Schwacke et al., 1999; Ueda et al., 2001). Similarly, *Arabidopsis* mutant lines *Atprot1*, *Atprot2,* and *Atprot3* were morphologically similar but had distinct proline accumulation when plant faces stressful conditions (Lehmann et al., 2011). Apart from the fact that increased proline content relied greatly on improved biosynthesis and suppressed catabolism of proline, proline transporters mediating long-distance transport of proline significantly contributed in the expansion of proline sink (Grallath et al., 2005; Kavi Kishor and Sreenivasulu, 2014).

3.4 Proline Mediates Flower Transition

A large amount of proline is transported to reproductive parts of plants under normal conditions (Mattioli et al., 2009; Biancucci et al., 2015). In this context, proline accounts for 26% of total amino acid reservoir in the reproductive parts of *A. thaliana* and it makes only 1%–3% of the total amino acids in the vegetative parts (Schwacke et al., 1999). Likewise, the amount of proline is 60-fold greater in tomato flowers than in vegetative tissue. Furthermore, proline accounts for more than 70% total free amino acids in pollen grains (Fujita et al., 1998). The significant accumulation of proline in reproductive parts implies that proline acts as an important signaling molecule (Kavi Kishor and Sreenivasulu, 2014; Zhang and Becker, 2015). For example, P5CS overexpression in *A. thaliana* and tobacco increases proline accumulation, alters floral architecture, accelerates the onset of flowering, and increases the number of flowers per plant (Kishor et al., 1995; Mattioli et al., 2009). Molecular studies are required to further explore the role of proline in flowering.

4. WHETHER OR NOT ORNITHINE AMINOTRANSFERASE MEDIATES NITROGEN RECOVERY

Arginine acts as a precursor in ornithine biosynthesis. The enzyme arginase catalyzes the formation of arginine to ornithine that plays a critical role in grain formation and panicle development in rice (Ma et al., 2013). The formation of P5C from ornithine is catalyzed by OAT (Funck et al., 2008). Transgenic plants with enhanced expression of OAT showed greater proline accumulation that effectively scavenged ROS and improved germination under osmotic stress (You et al., 2012). The increase in proline accumulation by these transgenic lines is associated with the increased cytoplasmic expression of OAT. In wild-type plants, OAT exists in mitochondrion that lacks proline biosynthesis mechanism. It is hypothesized that

OAT generates P5C that is used by P5CDH for energy release and synthesis of glutamate in the mitochondrion. *Arabidopsis* plant mutants with OAT signified the role of OAT in the synthesis of glutamate rather that its correlation with proline biosynthesis, and this depicts the essential role of OAT in nitrogen recycling (Funck et al., 2008; Winter et al., 2015). The glutamate generated due to proline oxidation is transported to chloroplast. However, the transporters mediating the transport of glutamate to chloroplast are still a matter of research (Kavi Kishor and Sreenivasulu, 2014). Mining of data from the available literature indicates the association of OAT with metabolism of glutamate and catabolism of glutamine. Furthermore, OAT also intermingles with floral regulators. Thus the degradation of ornithine is not concerned with proline synthesis but it influenced floral development through the production of glutamate/γ-aminobutyric acid (GABA) (Kavi Kishor and Sreenivasulu, 2014).

5. PROLINE PLAYS ESSENTIAL ROLE IN POLLEN DEVELOPMENT

Significant levels of proline have been found in ovules of faba beans and pollen grains of tomato and petunias (Hong-qi et al., 1982; Fujita et al., 1998; Venekamp and Koot, 1984). Significant levels of proline in reproductive parts are attributed to enhanced expression of ProT and P5CS in flower (Xue et al., 2009; Kishor et al., 2015). Higher concentration of proline in reproductive parts is an important ready source of nitrogen for pollen grain development and fertilization. Thus increased expression of P5Cs is needed in pollen development (Funck et al., 2012). Plants with heterozygous mutations for P5CS1 and P5CS2 produced smaller pollen grains than normal. Likewise, plants that were double mutants for p5cs1/p5cs2 produced sterile pollen grains (Funck et al., 2012). Furthermore, greater transcript levels of P5CR have been found in developing seeds, ovules, pollen grains, guard cells, and actively dividing cells (Kavi Kishor and Sreenivasulu, 2014). Pollen fertility and development in P5CR mutants are variable from wild type. Complete absence of P5CR did not show any influence on female or male fertility (Funck et al., 2012). When LeProT was transcribed in germinating and mature pollen, it facilitated prompt proline uptake (Schwacke et al., 1999). Increased expression of ProTs in pollen grains significantly altered proline levels under drought conditions (Lehmann et al., 2010). Male sterility due to high temperature is ascribed to altered expression of ProT1 in microspore that disturbs sugar metabolism and proline translocation during pollen development (Sato et al., 2006). Ample literature is available that highlights the positive impact of greater proline levels in pollen grains that in turn determine economic yield (Claussen, 2005; Špoljarević et al., 2011). Pollen grains with higher proline levels predict improved seed yield and seed set (Claussen, 2005). In another study, there was a positive correlation of proline with the number of seeds per panicle and seed set in sorghum hybrid under normal or low-temperature stress (Lansac et al., 1996). Likewise, at anthesis stage significantly higher proline levels in maize silk were positively linked with the number of kernels per ear (Špoljarević et al., 2011). Higher proline levels may determine better reproduction in certain plant species including maize wherein temperature or radiation stress markedly decreases pollen viability (Luna et al., 2001; Špoljarević et al., 2011). Since proline prevents water loss from stigma and pollen, its involvement in reproductive success is obvious (Kavi Kishor and Sreenivasulu, 2014). Apart from the fact that comprehensive mechanism demonstrating proline involvement in economic yield is obscure, the relation of proline with yield stability could not be denied. Proline (up to 2.02 mM) in floral nectars and extrafloral parts when compared with the other amino acids (present in the range of 114–547 mM) is suggested as a substantial factor that attracts pollinators (Carter et al., 2006). For instance, proline is preferred over serine and alanine by bees, which consume proline as energy source during flight (Bertazzini et al., 2010). Likewise, proline also has significant impact on nectar taste (Nepi et al., 2009). Therefore reproductive tissues rich in proline exert multiple benefits (Kavi Kishor and Sreenivasulu, 2014).

6. PROLINE MEDIATES SEED DEVELOPMENT

Supplemented proline improved somatic embryo formation in in vitro established cells in a number of plant species (Trigiano and Conger, 1987). Medium rich in proline not only increased the size of the embryo but also improved the generation of mature plants from the somatic embryo (Jariteh et al., 2015). In line with these suggestions, P5CS1 and P5CS2 showed greater transcription during embryogenesis in *A. thaliana* (Székely et al., 2008). Likewise, interruption in P5CR expression produced defects in embryos with ceased development after the embryo proper has undergone a second round of cell division (Meinke et al., 2008). It has been reported in the literature that early embryogenesis is highly associated with cellular proline levels, which are maintained via P5CS and P5CR activities, and thereby proline served as an essential factor for plant growth and development (Kavi Kishor and Sreenivasulu, 2014).

Proline accumulated in seeds during germination also contributes to tolerance of water scarcity environments. In this context, the underlying cause for the sensitivity of *Machilus thunbergii* seeds is the unavailability of oligosaccharides and proline (Lin and Chen, 1995). It has been shown in the literature that ABA induced the accumulation of proline (Yaish, 2015).

Furthermore, proline and ABA are also significant mediators of embryogenesis (Hare et al., 1999). Exposure of plants to prematuration desiccation may break down important biomolecules as well as trigger enhanced cellular production of ROS. Proline is an efficient scavenger of ROS, as we have mentioned earlier in pollen development. Proline contents are significantly higher in mature grape berries. However, the accumulation pattern of proline is variable over various developmental phases. For instance, accumulation of proline begins during the period of 4–6 weeks of fruit ripening when accumulation of protein and development are suppressed (Kavi Kishor and Sreenivasulu, 2014). Proline has been reported to interfere with maturation and ripening in grapes. Apart from the essential role of proline in embryogenesis, it also showed significant involvement in endosperm formation. The distinctive fragrance in basmati rice is due to the presence of the compound 2-acetyl-1-pyrroline. Glutamate and ornithine serve as precursors for the biosynthesis of 2-acetyl-1-pyrroline (Yoshihashi et al., 2002). Proteins present in small grain cereals are predominantly rich in proline. Therefore proline acts as an essential determinant for aroma and storage proteins in rice grain (Kavi Kishor and Sreenivasulu, 2014).

7. CONCLUSIONS AND FUTURE PERSPECTIVES

Plants tend to accumulate substantial levels of proline in response to various environmental stresses, in particular dehydration stress. We have provided a literature survey of the recent decade that indicates a positive association of proline and drought tolerance in crop plants. There is a school of thought that proline is merely a symptom of stress. However, we have documented compelling evidence from the literature that indicates significant involvement of this biomolecule in conferring drought tolerance in plants. In addition, researchers have also reported better tolerance against abiotic stress in plants treated with foliar application of proline. Therefore, there is need to completely understand the metabolism of proline. It is also important to know how abiotic stresses stimulate proline metabolic enzymes. A survey of the literature indicates that a particular set of proline metabolic enzymes is upregulated and/or downregulated in response to particular abiotic stress. It is known that proline enables plants to maintain plant growth under water-limited conditions. Plants mutant for proline biosynthetic enzymes P5CS and ProDH had conspicuous decrease in plant growth under drought stress. Proline not only protects subcellular compartments of cell but also buffers cytosolic pH. Proline also functions as an antioxidant, which in turn protects plants from ROS-induced oxidative damage under water-limited conditions.

REFERENCES

Abbas, W., Ashraf, M., Akram, N.A., 2010. Alleviation of salt-induced adverse effects in eggplant (*Solanum melongena* L.) by glycinebetaine and sugar-beet extracts. Sci. Hortic. 125 (3), 188–195. http://dx.doi.org/10.1016/j.scienta.2010.04.008.

Ábrahám, E., Rigó, G., Székely, G., Nagy, R., Koncz, C., Szabados, L., 2003. Light-dependent induction of proline biosynthesis by abscisic acid and salt stress is inhibited by brassinosteroid in *Arabidopsis*. Plant Mol. Biol. 51 (3), 363–372.

Ali, Q., Anwar, F., Ashraf, M., Saari, N., Perveen, R., 2013. Ameliorating effects of exogenously applied proline on seed composition, seed oil quality and oil antioxidant activity of maize (*Zea mays* L.) under drought stress. Int. J. Mol. Sci. 14 (1), 818–835.

Alia, Saradhi, P.P., 1991. Proline accumulation under heavy metal stress. J. Plant Physiol. 138 (5), 554–558. http://dx.doi.org/10.1016/S0176-1617(11)80240-3.

Ashraf, M.A., Iqbal, M., Hussain, I., Rasheed, R., 2015. Physiological and Biochemical Approaches for Salinity Tolerance. Managing Salt Tolerance in Plants: Molecular and Genomic Perspectives, vol. 79.

Balestro, G.C., Higashi, B., Lopes, S.M.S., Gonçalves, J.E., Vieira, L.G.E., de Oliveira, A.J.B., Gonçalves, R.A.C., 2017. Biochemical composition of symplastic sap from sugarcane genetically modified to overproduce proline. Plant Physiol. Biochem. 113, 133–140.

Balibrea, M.E., Rus-Alvarez, A.M., Bolarín, M.C., Pérez-Alfocea, F., 1997. Fast changes in soluble carbohydrates and proline contents in tomato seedlings in response to ionic and non-ionic iso-osmotic stresses. J. Plant Physiol. 151 (2), 221–226. http://dx.doi.org/10.1016/S0176-1617(97)80156-3.

Beltrano, J., Ruscitti, M., Arango, M., Ronco, M., 2013. Effects of arbuscular mycorrhiza inoculation on plant growth, biological and physiological parameters and mineral nutrition in pepper grown under different salinity and p levels. J. Soil Sci. Plant Nutrition 13 (1), 123–141.

Bertazzini, M., Medrzycki, P., Bortolotti, L., Maistrello, L., Forlani, G., 2010. Amino acid content and nectar choice by forager honeybees (*Apis mellifera* L.). Amino Acids 39 (1), 315–318.

Bhaskara, G.B., Yang, T.-H., Verslues, P.E., 2015. Dynamic proline metabolism: importance and regulation in water limited environments. Front. Plant Sci. 6 (484). http://dx.doi.org/10.3389/fpls.2015.00484.

Biancucci, M., Mattioli, R., Forlani, G., Funck, D., Costantino, P., Trovato, M., 2015. Role of proline and GABA in sexual reproduction of angiosperms. Front. Plant Sci. 6.

Büssis, D., Heineke, D., 1998. Acclimation of potato plants to polyethylene glycol-induced water deficit II. Contents and subcellular distribution of organic solutes. J. Exp. Bot. 49 (325), 1361–1370. http://dx.doi.org/10.1093/jxb/49.325.1361.

Carter, C., Shafir, S., Yehonatan, L., Palmer, R.G., Thornburg, R., 2006. A novel role for proline in plant floral nectars. Naturwissenschaften 93 (2), 72–79. http://dx.doi.org/10.1007/s00114-005-0062-1.

Claussen, W., 2005. Proline as a measure of stress in tomato plants. Plant Sci. 168 (1), 241–248. http://dx.doi.org/10.1016/j.plantsci.2004.07.039.

Dar, M.I., Naikoo, M.I., Rehman, F., Naushin, F., Khan, F.A., 2016. Proline accumulation in plants: roles in stress tolerance and plant development. In: Osmolytes and Plants Acclimation to Changing Environment: Emerging Omics Technologies. Springer, pp. 155–166.

de Campos, M.K.F., de Carvalho, K., de Souza, F.S., Marur, C.J., Pereira, L.F.P., Filho, J.C.B., Vieira, L.G.E., 2011. Drought tolerance and antioxidant enzymatic activity in transgenic 'Swingle' citrumelo plants over-accumulating proline. Environ. Exp. Bot. 72 (2), 242–250. http://dx.doi.org/10.1016/j.envexpbot.2011.03.009.

de Carvalho, K., Kaphan, M., de Campos, F., Domingues, D.S., Filipe, L., Pereira, P., Gonzaga, L., Vieira, E., 2013. The accumulation of endogenous proline induces changes in gene expression of several antioxidant enzymes in leaves of transgenic *Swingle citrumelo*. Mol. Biol. Rep. 40 (4), 3269.

Deuschle, K., Funck, D., Forlani, G., Stransky, H., Biehl, A., Leister, D., van der Graaff, E., Kunze, R., Frommer, W.B., 2004. The role of Δ1-pyrroline-5-carboxylate dehydrogenase in proline degradation. Plant Cell 16 (12), 3413–3425.

Dinesh, A., Muralidhara, B., Gangurde, S., Yogeshwar, M., 2016. Molecular Response of Plants to Drought, Cold and Heat Stress – A Review.

Fabro, G., Kovács, I., Pavet, V., Szabados, L., Alvarez, M.E., 2004. Proline accumulation and AtP5CS2 gene activation are induced by plant-pathogen incompatible interactions in *Arabidopsis*. Mol. Plant-Microbe Interact. 17 (4), 343–350.

Faës, P., Deleu, C., Aïnouche, A., Le Cahérec, F., Montes, E., Clouet, V., Gouraud, A.-M., Albert, B., Orsel, M., Lassalle, G., 2015. Molecular evolution and transcriptional regulation of the oilseed rape proline dehydrogenase genes suggest distinct roles of proline catabolism during development. Planta 241 (2), 403.

Farahani, A., Lebaschi, H., Hussein, M., Hussein, S.A., Reza, V.A., Jahanfar, D., 2013. Effects of arbuscular mycorrhizal fungi, different levels of phosphorus and drought stress on water use efficiency, relative water content and proline accumulation rate of Coriander (*Coriandrum sativum* L.). J. Med. Plants Res. 2 (6), 125–131.

Fariduddin, Q., Yusuf, M., Ahmad, I., Ahmad, A., 2014. Brassinosteroids and their role in response of plants to abiotic stresses. Biol. Plant 58 (1), 9–17.

Filippou, P., Bouchagier, P., Skotti, E., Fotopoulos, V., 2014. Proline and reactive oxygen/nitrogen species metabolism is involved in the tolerant response of the invasive plant species *Ailanthus altissima* to drought and salinity. Environ. Exp. Bot. 97, 1–10.

Fujita, T., Maggio, A., Garcia-Rios, M., Bressan, R.A., Csonka, L.N., 1998. Comparative analysis of the regulation of expression and structures of two evolutionarily divergent genes for Δ1-pyrroline-5-carboxylate synthetase from tomato. Plant Physiol. 118 (2), 661–674.

Funck, D., Stadelhofer, B., Koch, W., 2008. Ornithine-δ-aminotransferase is essential for arginine catabolism but not for proline biosynthesis. BMC Plant Biol. 8 (1), 40.

Funck, D., Winter, G., Baumgarten, L., Forlani, G., 2012. Requirement of proline synthesis during *Arabidopsis* reproductive development. BMC Plant Biol. 12 (1), 191.

Giberti, S., Funck, D., Forlani, G., 2014. Δ1-pyrroline-5-carboxylate reductase from *Arabidopsis thaliana*: stimulation or inhibition by chloride ions and feedback regulation by proline depend on whether NADPH or NADH acts as co-substrate. New Phytol. 202 (3), 911–919.

Grallath, S., Weimar, T., Meyer, A., Gumy, C., Suter-Grotemeyer, M., Neuhaus, J.-M., Rentsch, D., 2005. The AtProT family. Compatible solute transporters with similar substrate specificity but differential expression patterns. Plant Physiol. 137 (1), 117–126.

Gururani, M.A., Upadhyaya, C.P., Baskar, V., Venkatesh, J., Nookaraju, A., Park, S.W., 2013. Plant growth-promoting rhizobacteria enhance abiotic stress tolerance in *Solanum tuberosum* through inducing changes in the expression of ROS-scavenging enzymes and improved photosynthetic performance. J. Plant Growth Regul. 1–14.

Hare, P., Cress, W., Van Staden, J., 1999. Proline synthesis and degradation: a model system for elucidating stress-related signal transduction. J. Exp. Bot. 50 (333), 413–434.

Hare, P.D., Cress, W.A., van Staden, J., 2003. A regulatory role for proline metabolism in stimulating *Arabidopsis thaliana* seed germination. Plant Growth Regul. 39 (1), 41–50. http://dx.doi.org/10.1023/a:1021835902351.

Hong-qi, Z., Croes, A.F., Linskens, H.F., 1982. Protein synthesis in germinating pollen of *Petunia*: role of proline. Planta 154 (3), 199–203. http://dx.doi.org/10.1007/bf00387864.

Hong, Y., Zhang, H., Huang, L., Li, D., Song, F., 2016. Overexpression of a stress-responsive NAC transcription factor gene ONAC022 improves drought and salt tolerance in rice. Front. Plant Sci. 7.

Hong, Z., Lakkineni, K., Zhang, Z., Verma, D.P.S., 2000. Removal of feedback inhibition of Δ1-pyrroline-5-carboxylate synthetase results in increased proline accumulation and protection of plants from osmotic stress. Plant Physiol. 122 (4), 1129–1136.

Huang, C., Zhao, S., Wang, L., Anjum, S.A., Chen, M., Zhou, H., Zou, C., 2013. Alteration in chlorophyll fluorescence, lipid peroxidation and antioxidant enzymes activities in hybrid ramie ('*Boehmeria nivea*' L.) under drought stress. Aust. J. Crop Sci. 7 (5), 594.

Irigoyen, J.J., Einerich, D.W., Sánchez-Díaz, M., 1992. Water stress induced changes in concentrations of proline and total soluble sugars in nodulated alfalfa (*Medicago sativa*) plants. Physiol. Plant 84 (1), 55–60. http://dx.doi.org/10.1111/j.1399-3054.1992.tb08764.x.

Jariteh, M., Ebrahimzadeh, H., Niknam, V., Mirmasoumi, M., Vahdati, K., 2015. Developmental changes of protein, proline and some antioxidant enzymes activities in somatic and zygotic embryos of Persian walnut (*Juglans regia* L.). Plant Cell Tissue Organ Cult. (PCTOC) 122 (1), 101–115.

Kasote, D.M., Katyare, S.S., Hegde, M.V., Bae, H., 2015. Significance of antioxidant potential of plants and its relevance to therapeutic applications. Int. J. Biol. Sci. 11 (8), 982.

Kaur, D., Grewal, S., Kaur, J., Singh, S., 2017. Differential proline metabolism in vegetative and reproductive tissues determine drought tolerance in chickpea. Biol. Plant 61 (2), 359–366.

Kavi Kishor, P.B., Sreenivasulu, N., 2014. Is proline accumulation per se correlated with stress tolerance or is proline homeostasis a more critical issue? Plant Cell Environ. 37 (2), 300–311. http://dx.doi.org/10.1111/pce.12157.

Kesari, R., Lasky, J.R., Villamor, J.G., Des Marais, D.L., Chen, Y.-J.C., Liu, T.-W., Lin, W., Juenger, T.E., Verslues, P.E., 2012. Intron-mediated alternative splicing of *Arabidopsis* P5CS1 and its association with natural variation in proline and climate adaptation. Proc. Natl. Acad. Sci. 109 (23), 9197–9202.

Khoyerdi, F.F., Shamshiri, M.H., Estaji, A., 2016. Changes in some physiological and osmotic parameters of several pistachio genotypes under drought stress. Sci. Hortic. 198, 44–51. http://dx.doi.org/10.1016/j.scienta.2015.11.028.

Kishor, P.B.K., Kumari, P.H., Sunita, M., Sreenivasulu, N., 2015. Role of proline in cell wall synthesis and plant development and its implications in plant ontogeny. Front. Plant Sci. 6.

Kishor, P.K., Hong, Z., Miao, G.-H., Hu, C.-A.A., Verma, D.P.S., 1995. Overexpression of [delta]-pyrroline-5-carboxylate synthetase increases proline production and confers osmotolerance in transgenic plants. Plant Physiol. 108 (4), 1387–1394.

Lansac, A., Sullivan, C., Johnson, B., 1996. Accumulation of free proline in sorghum (Sorghum bicolor) pollen. Can. J. Botany 74 (1), 40–45.

Larher, F., Leport, L., Petrivalsky, M., Chappart, M., 1993. Effectors for the osmoinduced proline response in higher plants. Plant Physiol. Biochem. 31 (6), 911–922.

Lee, B.-R., Jin, Y.L., Avice, J.-C., Cliquet, J.-B., Ourry, A., Kim, T.-H., 2009. Increased proline loading to phloem and its effects on nitrogen uptake and assimilation in water-stressed white clover (*Trifolium repens*). New Phytol. 182 (3), 654–663. http://dx.doi.org/10.1111/j.1469-8137.2009.02795.x.

Lehmann, S., Funck, D., Szabados, L., Rentsch, D., 2010. Proline metabolism and transport in plant development. Amino Acids 39 (4), 949–962. http://dx.doi.org/10.1007/s00726-010-0525-3.

Lehmann, S., Gumy, C., Blatter, E., Boeffel, S., Fricke, W., Rentsch, D., 2011. In planta function of compatible solute transporters of the AtProT family. J. Exp. Bot. 62 (2), 787–796.

Lei, Y., Yin, C., Li, C., 2006. Differences in some morphological, physiological, and biochemical responses to drought stress in two contrasting populations of *Populus przewalskii*. Physiol. Plant 127 (2), 182–191. http://dx.doi.org/10.1111/j.1399-3054.2006.00638.x.

Less, H., Galili, G., 2008. Principal transcriptional programs regulating plant amino acid metabolism in response to abiotic stresses. Plant Physiol. 147 (1), 316–330.

Li, P., Zheng, X., Liu, Y., Zhu, Y., 2014. Pre-storage application of oxalic acid alleviates chilling injury in mango fruit by modulating proline metabolism and energy status under chilling stress. Food Chem. 142, 72–78.

Li, Z.-G., Ding, X.-J., Du, P.-F., 2013. Hydrogen sulfide donor sodium hydrosulfide-improved heat tolerance in maize and involvement of proline. J. Plant Physiol. 170 (8), 741–747.

Liang, X., Zhang, L., Natarajan, S.K., Becker, D.F., 2013. Proline mechanisms of stress survival. Antioxid. Redox Signaling 19 (9), 998–1011.

Lin, T.-P., Chen, M.-H., 1995. Biochemical characteristics associated with the development of the desiccation-sensitive seeds of *Machilus thunbergii* Sieb. & Zucc. Ann. Bot. 76 (4), 381–387.

Luna, V., Figueroa, M., Baltazar, M., Gomez, L., Townsend, R., Schoper, J., 2001. Maize pollen longevity and distance isolation requirements for effective pollen control. Crop Sci. 41 (5), 1551–1557.

Ma, X., Cheng, Z., Qin, R., Qiu, Y., Heng, Y., Yang, H., Ren, Y., Wang, X., Bi, J., Ma, X., Zhang, X., Wang, J., Lei, C., Guo, X., Wang, J., Wu, F., Jiang, L., Wang, H., Wan, J., 2013. OsARG encodes an arginase that plays critical roles in panicle development and grain production in rice. Plant J. 73 (2), 190–200. http://dx.doi.org/10.1111/j.1365-313x.2012.05122.x.

Mafakheri, A., Siosemardeh, A., Bahramnejad, B., Struik, P., Sohrabi, Y., 2010. Effect of drought stress on yield, proline and chlorophyll contents in three chickpea cultivars. Aust. J. Crop Sci. 4 (8), 580.

Manivannan, P., Jaleel, C.A., Sankar, B., Kishorekumar, A., Somasundaram, R., Lakshmanan, G.M.A., Panneerselvam, R., 2007. Growth, biochemical modifications and proline metabolism in *Helianthus annuus* L. as induced by drought stress. Colloids Surf. B Biointerfaces 59 (2), 141–149. http://dx.doi.org/10.1016/j.colsurfb.2007.05.002.

Martínez-Fernández, D., Barroso, D., Komárek, M., 2016. Root water transport of *Helianthus annuus* L. under iron oxide nanoparticle exposure. Environ. Sci. Poll. Res. Int. 23 (2), 1732.

Mattioli, R., Costantino, P., Trovato, M., 2009. Proline accumulation in plants. Plant Signal. Behav. 4 (11), 1016–1018. http://dx.doi.org/10.4161/psb.4.11.9797.

Maxwell, S.A., Davis, G.E., 2000. Differential gene expression in p53-mediated apoptosis-resistant vs. apoptosis-sensitive tumor cell lines. Proc. Natl. Acad. Sci. 97 (24), 13009–13014.

Meinke, D., Muralla, R., Sweeney, C., Dickerman, A., 2008. Identifying essential genes in *Arabidopsis thaliana*. Trends Plant Sci. 13 (9), 483–491. http://dx.doi.org/10.1016/j.tplants.2008.06.003.

Miller, G., Honig, A., Stein, H., Suzuki, N., Mittler, R., Zilberstein, A., 2009. Unraveling Δ1-pyrroline-5-carboxylate-proline cycle in plants by uncoupled expression of proline oxidation enzymes. J. Biol. Chem. 284 (39), 26482–26492.

Misra, N., Saxena, P., 2009. Effect of salicylic acid on proline metabolism in lentil grown under salinity stress. Plant Sci. 177 (3), 181–189. http://dx.doi.org/10.1016/j.plantsci.2009.05.007.

Mohammadkhani, N., Heidari, R., 2008. Drought-induced accumulation of soluble sugars and proline in two maize varieties. World Appl. Sci. J. 3 (3), 448–453.

Nadeem, S.M., Ahmad, M., Zahir, Z.A., Javaid, A., Ashraf, M., 2014. The role of mycorrhizae and plant growth promoting rhizobacteria (PGPR) in improving crop productivity under stressful environments. Biotechnol. Adv. 32 (2), 429–448.

Nakabayashi, R., Saito, K., 2015. Integrated metabolomics for abiotic stress responses in plants. Curr. Opin. Plant Biol. 24, 10–16.

Nanjo, T., Fujita, M., Seki, M., Kato, T., Tabata, S., Shinozaki, K., 2003. Toxicity of free proline revealed in an *Arabidopsis* T-DNA-tagged mutant deficient in proline dehydrogenase. Plant Cell Physiol. 44 (5), 541–548.

Natarajan, S.K., Zhu, W., Liang, X., Zhang, L., Demers, A.J., Zimmerman, M.C., Simpson, M.A., Becker, D.F., 2012. Proline dehydrogenase is essential for proline protection against hydrogen peroxide-induced cell death. Free Radic. Biol. Med. 53 (5), 1181–1191. http://dx.doi.org/10.1016/j.freeradbiomed.2012.07.002.

Nepi, M., von Aderkas, P., Wagner, R., Mugnaini, S., Coulter, A., Pacini, E., 2009. Nectar and pollination drops: how different are they? Ann. Bot. http://dx.doi.org/10.1093/aob/mcp124.

Ober, E.S., Sharp, R.E., 1994. Proline accumulation in maize (*Zea mays* L.) primary roots at low water potentials (I. Requirement for increased levels of abscisic acid). Plant Physiol. 105 (3), 981–987.

Ozden, M., Demirel, U., Kahraman, A., 2009. Effects of proline on antioxidant system in leaves of grapevine (*Vitis vinifera* L.) exposed to oxidative stress by H_2O_2. Sci. Hortic. 119 (2), 163–168. http://dx.doi.org/10.1016/j.scienta.2008.07.031.

Parida, A.K., Dagaonkar, V.S., Phalak, M.S., Aurangabadkar, L.P., 2008. Differential responses of the enzymes involved in proline biosynthesis and degradation in drought tolerant and sensitive cotton genotypes during drought stress and recovery. Acta Physiol. Plant. 30 (5), 619–627. http://dx.doi.org/10.1007/s11738-008-0157-3.

Peng, Z., Lu, Q., Verma, D.P.S., 1996. Reciprocal regulation of Δ1-pyrroline-5-carboxylate synthetase and proline dehydrogenase genes controls proline levels during and after osmotic stress in plants. Mol. General Genet. MGG 253 (3), 334–341. http://dx.doi.org/10.1007/pl00008600.

Per, T.S., Khan, N.A., Reddy, P.S., Masood, A., Hasanuzzaman, M., Khan, M.I.R., Anjum, N.A., 2017. Approaches in modulating proline metabolism in plants for salt and drought stress tolerance: phytohormones, mineral nutrients and transgenics. Plant Physiol. Biochem. 115, 126–140. http://dx.doi.org/10.1016/j.plaphy.2017.03.018.

Rasheed, R., Ashraf, M.A., Hussain, I., Haider, M.Z., Kanwal, U., Iqbal, M., 2014. Exogenous proline and glycinebetaine mitigate cadmium stress in two genetically different spring wheat (*Triticum aestivum* L.) cultivars. Braz. J. Bot. 37 (4), 399–406. http://dx.doi.org/10.1007/s40415-014-0089-7.

Raymond, M.J., Smirnoff, N., 2002. Proline metabolism and transport in maize seedlings at low water potential. Ann. Bot. 89 (7), 813–823.

Rejeb, K.B., Abdelly, C., Savouré, A., 2014. How reactive oxygen species and proline face stress together. Plant Physiol. Biochem. 80, 278–284.

Saradhi, P.P., AliaArora, S., Prasad, K.V.S.K., 1995. Proline accumulates in plants exposed to uv radiation and protects them against uv-induced peroxidation. Biochem. Biophys. Res. Commun. 209 (1), 1–5. http://dx.doi.org/10.1006/bbrc.1995.1461.

Sato, S., Kamiyama, M., Iwata, T., Makita, N., Furukawa, H., Ikeda, H., 2006. Moderate increase of mean daily temperature adversely affects fruit set of *Lycopersicon esculentum* by disrupting specific physiological processes in male reproductive development. Ann. Bot. 97 (5), 731–738.

Satoh, R., Fujita, Y., Nakashima, K., Shinozaki, K., Yamaguchi-Shinozaki, K., 2004. A novel subgroup of bZIP proteins functions as transcriptional activators in hypoosmolarity-responsive expression of the ProDH gene in *Arabidopsis*. Plant Cell Physiol. 45 (3), 309–317.

Satoh, R., Nakashima, K., Seki, M., Shinozaki, K., Yamaguchi-Shinozaki, K., 2002. ACTCAT, a novel cis-acting element for proline-and hypoosmolarity-responsive expression of the ProDH gene encoding proline dehydrogenase in *Arabidopsis*. Plant Physiol. 130 (2), 709–719.

Schwacke, R., Grallath, S., Breitkreuz, K.E., Stransky, E., Stransky, H., Frommer, W.B., Rentsch, D., 1999. LeProT1, a transporter for proline, glycine betaine, and γ-amino butyric acid in tomato pollen. Plant Cell 11 (3), 377–391.

Sekmen, A.H., Ozgur, R., Uzilday, B., Turkan, I., 2014. Reactive oxygen species scavenging capacities of cotton (*Gossypium hirsutum*) cultivars under combined drought and heat induced oxidative stress. Environ. Exp. Bot. 99, 141–149. http://dx.doi.org/10.1016/j.envexpbot.2013.11.010.

Sharma, S., Villamor, J.G., Verslues, P.E., 2011. Essential role of tissue-specific proline synthesis and catabolism in growth and redox balance at low water potential. Plant Physiol. 157 (1), 292–304.

Sharma, S.S., Dietz, K.-J., 2009. The relationship between metal toxicity and cellular redox imbalance. Trends Plant Sci. 14 (1), 43–50. http://dx.doi.org/10.1016/j.tplants.2008.10.007.

Signorelli, S., Arellano, J.B., Melø, T.B., Borsani, O., Monza, J., 2013. Proline does not quench singlet oxygen: evidence to reconsider its protective role in plants. Plant Physiol. Biochem. 64, 80–83. http://dx.doi.org/10.1016/j.plaphy.2012.12.017.

Sinay, H., Arumingtyas, E.L., Harijati, N., Indriyani, S., 2016. Proline content and yield components of local corn cultivars from Kisar Island, Maluku, Indonesia. Int. J. Plant Biol. 6 (1).

Singh, B., 2016. Response of Wheat (*Triticum aestivum* L.) Varieties to Different Sowing Dates and Growth Regulator. Punjab Agricultural University, Ludhiana.

Slama, I., Abdelly, C., Bouchereau, A., Flowers, T., Savouré, A., 2015. Diversity, distribution and roles of osmoprotective compounds accumulated in halophytes under abiotic stress. Ann. Bot. 115 (3), 433–447.

Sofo, A., Dichio, B., Xiloyannis, C., Masia, A., 2004. Lipoxygenase activity and proline accumulation in leaves and roots of olive trees in response to drought stress. Physiol. Plant 121 (1), 58–65. http://dx.doi.org/10.1111/j.0031-9317.2004.00294.x.

Sperdouli, I., Moustakas, M., 2012. Interaction of proline, sugars, and anthocyanins during photosynthetic acclimation of *Arabidopsis thaliana* to drought stress. J. Plant Physiol. 169 (6), 577–585. http://dx.doi.org/10.1016/j.jplph.2011.12.015.

Špoljarević, M., Agić, D., Lisjak, M., Gumze, A., Wilson, I.D., Hancock, J.T., Teklić, T., 2011. The relationship of proline content and metabolism on the productivity of maize plants. Plant Signal. Behav. 6 (2), 251–257. http://dx.doi.org/10.4161/psb.6.2.14336.

Stines, A.P., Naylor, D.J., Høj, P.B., Van Heeswijck, R., 1999. Proline accumulation in developing grapevine fruit occurs independently of changes in the levels of Δ1-pyrroline-5-carboxylate synthetase mRNA or protein. Plant Physiol. 120 (3), 923.

Strizhov, N., Ábrahám, E., Ökrész, L., Blickling, S., Zilberstein, A., Schell, J., Koncz, C., Szabados, L., 1997. Differential expression of two P5CS genes controlling proline accumulation during salt-stress requires ABA and is regulated by ABA1, ABI1 and AXR2 in *Arabidopsis*. Plant J. 12 (3), 557–569. http://dx.doi.org/10.1111/j.0960-7412.1997.00557.x.

Surekha, C., Kumari, K.N., Aruna, L., Suneetha, G., Arundhati, A., Kishor, P.K., 2014. Expression of the *Vigna aconitifolia* P5CSF129A gene in transgenic pigeon pea enhances proline accumulation and salt tolerance. Plant Cell Tissue Organ Cult. (PCTOC) 116 (1), 27–36.

Suzuki, N., Koussevitzky, S., Mittler, R.O.N., Miller, G.A.D., 2012. ROS and redox signalling in the response of plants to abiotic stress. Plant Cell Environ. 35 (2), 259–270. http://dx.doi.org/10.1111/j.1365-3040.2011.02336.x.

Suzuki, N., Rivero, R.M., Shulaev, V., Blumwald, E., Mittler, R., 2014. Abiotic and biotic stress combinations. New Phytol. 203 (1), 32–43.

Szabados, L., Savouré, A., 2010. Proline: a multifunctional amino acid. Trends Plant Sci. 15 (2), 89–97. http://dx.doi.org/10.1016/j.tplants.2009.11.009.

Székely, G., Ábrahám, E., Cséplő, Á., Rigó, G., Zsigmond, L., Csiszár, J., Ayaydin, F., Strizhov, N., Jásik, J., Schmelzer, E., Koncz, C., Szabados, L., 2008. Duplicated P5CS genes of *Arabidopsis* play distinct roles in stress regulation and developmental control of proline biosynthesis. Plant J. 53 (1), 11–28. http://dx.doi.org/10.1111/j.1365-313X.2007.03318.x.

Trigiano, R.N., Conger, B.V., 1987. Regulation of growth and somatic embryogenesis by proline and serine in suspension cultures of *Dactylis glomerata*. J. Plant Physiol. 130 (1), 49–55. http://dx.doi.org/10.1016/S0176-1617(87)80300-0.

Ueda, A., Shi, W., Sanmiya, K., Shono, M., Takabe, T., 2001. Functional analysis of salt-inducible proline transporter of barley roots. Plant Cell Physiol. 42 (11), 1282–1289.

Vendruscolo, E.C.G., Schuster, I., Pileggi, M., Scapim, C.A., Molinari, H.B.C., Marur, C.J., Vieira, L.G.E., 2007. Stress-induced synthesis of proline confers tolerance to water deficit in transgenic wheat. J. Plant Physiol. 164 (10), 1367–1376. http://dx.doi.org/10.1016/j.jplph.2007.05.001.

Venekamp, J.H., Koot, J.T.M., 1984. The distribution of free amino acids, especially of proline, in the organs of field bean plants, *Vicia faba* L., during development in the field. J. Plant Physiol. 116 (4), 343–349. http://dx.doi.org/10.1016/S0176-1617(84)80113-3.

Verslues, P.E., Bray, E.A., 2006. Role of abscisic acid (ABA) and *Arabidopsis thaliana* ABA-insensitive loci in low water potential-induced ABA and proline accumulation. J. Exp. Bot. 57 (1), 201–212.

Verslues, P.E., Sharma, S., 2010. Proline metabolism and its implications for plant-environment interaction. Arabidopsis Book e0140. http://dx.doi.org/10.1199/tab.0140.

Vurukonda, S.S.K.P., Vardharajula, S., Shrivastava, M., SkZ, A., 2016. Enhancement of drought stress tolerance in crops by plant growth promoting rhizobacteria. Microbiol. Res. 184, 13–24. http://dx.doi.org/10.1016/j.micres.2015.12.003.

Winter, G., Todd, C.D., Trovato, M., Forlani, G., Funck, D., 2015. Physiological implications of arginine metabolism in plants. Front. Plant Sci. 6.

Xue, X., Liu, A., Hua, X., 2009. Proline accumulation and transcriptional regulation of proline biosynthesis and degradation in *Brassica napus*. BMB Reports 42 (1), 28–34.

Yaish, M., 2015. Proline accumulation is a general response to abiotic stress in the date palm tree (*Phoenix dactylifera* L.). Genet. Mol. Res. 14 (3), 9943–9950.

Yamada, M., Morishita, H., Urano, K., Shiozaki, N., Yamaguchi-Shinozaki, K., Shinozaki, K., Yoshiba, Y., 2005. Effects of free proline accumulation in petunias under drought stress. J. Exp. Bot. 56 (417), 1975–1981. http://dx.doi.org/10.1093/jxb/eri195.

Yancey, P., Clark, M., Hand, S., Bowlus, R., Somero, G., 1982. Living with water stress: evolution of osmolyte systems. Science 217 (4566), 1214–1222. http://dx.doi.org/10.1126/science.7112124.

Yoshiba, Y., Kiyosue, T., Katagiri, T., Ueda, H., Mizoguchi, T., Yamaguchi-Shinozaki, K., Wada, K., Harada, Y., Shinozaki, K., 1995. Correlation between the induction of a gene for Δ1-pyrroline-5-carboxylate synthetase and the accumulation of proline in *Arabidopsis thaliana* under osmotic stress. Plant J. 7 (5), 751–760. http://dx.doi.org/10.1046/j.1365-313X.1995.07050751.x.

Yoshihashi, T., Huong, N.T.T., Inatomi, H., 2002. Precursors of 2-acetyl-1-pyrroline, a potent flavor compound of an aromatic rice variety. J. Agric. Food Chem. 50 (7), 2001–2004.

You, J., Hu, H., Xiong, L., 2012. An ornithine δ-aminotransferase gene OsOAT confers drought and oxidative stress tolerance in rice. Plant Sci. 197, 59–69. http://dx.doi.org/10.1016/j.plantsci.2012.09.002.

Yu, X., Liu, Y., Wang, S., Tao, Y., Wang, Z., Shu, Y., Peng, H., Mijiti, A., Wang, Z., Zhang, H., 2016. CarNAC4, a NAC-type chickpea transcription factor conferring enhanced drought and salt stress tolerances in *Arabidopsis*. Plant Cell Rep. 35 (3), 613.

Zandalinas, S.I., Mittler, R., Balfagón, D., Arbona, V., Gómez-Cadenas, A., 2017. Plant adaptations to the combination of drought and high temperatures. Physiol. Plant. http://dx.doi.org/10.1111/ppl.12540. n/a–n/a.

Zanella, M., Borghi, G.L., Pirone, C., Thalmann, M., Pazmino, D., Costa, A., Santelia, D., Trost, P., Sparla, F., 2016. β-amylase 1 (BAM1) degrades transitory starch to sustain proline biosynthesis during drought stress. J. Exp. Bot. 67 (6), 1819–1826.

Zarattini, M., Forlani, G., 2017. Toward unveiling the mechanisms for transcriptional regulation of proline biosynthesis in the plant cell response to biotic and abiotic stress conditions. Front. Plant Sci. 8.

Zhang, C., Wang, N., Zhang, Y., Feng, Q., Yang, C., Liu, B., 2013. DNA methylation involved in proline accumulation in response to osmotic stress in rice (Oryza sativa). Genet. Mol. Res. 12 (2), 1269–1277.

Zhang, L., Becker, D.F., 2015. Connecting proline metabolism and signaling pathways in plant senescence. Front. Plant Sci. 6.

Chapter 17

Mechanism of Proline Biosynthesis and Role of Proline Metabolism Enzymes Under Environmental Stress in Plants

Ágnes Szepesi, Réka Szőllősi
University of Szeged, Szeged, Hungary

1. INTRODUCTION

Proline (Pro), a multifunctional molecule in plant development and involved in the response to a wide range of plant abiotic and biotic stresses, can act as both a compatible osmolyte and radical scavenger or by its degradation as a supply of energy for regrowth after stress situations. In this chapter, we discuss the complexity of proline biosynthesis and catabolism, deciphering components involved in the regulation of proper plant growth and development as well as in the context of abiotic and biotic stress-induced processes. Although numerous excellent reviews have been written, we provide a broad view of the most recent approaches of how proline can take part in other signaling pathways and interconnect with some phytohormones. Investigating the posttranscriptional modification of Pro-mediated enzymes provides a new uncovered area for better understanding the efficiency of proline-regulated stress responses and plant developmental processes.

2. PROLINE BIOSYNTHESIS AND DEGRADATION IN PLANTS

Proline biosynthesis in plants takes place in the chloroplast and cytoplasm, whereas degradation occurs in the mitochondria. The pathways involved in proline synthesis and degradation are regulated by multiple enzymes. Proline dehydrogenase (ProDH) catalyzes the first and rate-limiting step in the transformation of proline (Pro) into glutamic acid (Glu) that takes place in the mitochondria. Some studies indicated that, as ProDH is bound to the inner mitochondrial membrane, it may have a role in transferring electrons to ubiquinone in the mitochondrial electron chain in higher plants (Elthon and Stewart, 1981; Kiyosue et al., 1996; Liang et al., 2013). This enzyme oxidizes Pro into delta-1-pyrroline 5-carboxylate (P5C) using FAD as cofactor, which can transform nonenzymatically into glutamate semialdehyde (GSA). P5C dehydrogenase (P5CDH) can generate Glu and nicotinamide adenine dinucleotide (NADH) from GSA.

Knowledge about the transporters of proline inside the plant cell is limited; there may be glutamate/proline antiporters. More investigations are needed to decipher the export of GSA/P5C into the cytoplasm or into the mitochondria from the cytoplasm.

The Pro-Glu pathway was confirmed to have an important role in supporting root growth, as the proline oxidation can generate reducing equivalents for mitochondrial oxidative phosphorylation (Verslues and Sharma, 2010; Liang et al., 2013). Signorelli (2016) has proposed a fermentation analogy for proline accumulation in stressed plants.

The rate-limiting enzyme of the biosynthesis of proline in higher plants is D1-pyrroline-5-carboxylate synthetase (P5CS), which is responsible for converting glutamate into pyrroline-5-carboxylate (P5C). Next, the second enzyme D1-pyrroline-5-carboxylate reductase (P5CR) can reduce P5C to proline, and it is encoded by only one gene in most plant species that have been investigated (Armengaud et al., 2004). P5CS is regulated by several factors such as abscisic acid (ABA) natural antisense small interfering RNAs and epigenetically or by alternative splicing (Abraham et al., 2003; Borsani et al., 2005; Kesari et al., 2012; Zhang et al., 2013).

The level of proline varies depending on the species involved or severity and duration of the stress situation (Delauney and Verma, 1993); however, the relation between proline accumulation and stress adaptation is contradictory (Hare and Cress, 1997; Chen et al., 2007). Kavi Kishor and Sreenivasulu (2014) suggest the importance of proline homeostasis over proline accumulation with stress tolerance. Some studies provide evidence that stress-induced proline accumulation could

Plant Metabolites and Regulation Under Environmental Stress. https://doi.org/10.1016/B978-0-12-812689-9.00017-0

coexist with elevation of catabolic activities (Fabro et al., 2004; Kaplan et al., 2007). The overexpression of key proline synthesis pathway genes influences the modulation of the shoot/root biomass ratio, the inflorescence architecture, and economic yield under stressful conditions (Kavi Kishor et al., 1995; Hare and Cress, 1997) and highlighting the importance of these genes by knockout transgenic *Arabidopsis* lines (Szekely et al., 2008; Funck et al., 2012).

P5CS1 and ProDH show different regulation profiles under stress; the former is upregulated under stress, whereas the transcription of the latter enzyme in several plant species is downregulated by moisture stress, dehydration, and salinity (Verslues et al., 2007; Sharma and Verslues, 2010) and upregulated when stress is relieved (Yoshiba et al., 1997; Satoh et al., 2002).

The Pro synthesized in shoots can be transported to roots for subsequent degradation, sustaining root growth under unfavorable conditions (Sharma et al., 2011).

Proline responding1, gene encoding a P5CS catalyzing the biosynthesis of proline from glutamic acid, plays a critical role in regulating general protein synthesis and the cell cycle in maize (Wang et al., 2014). Loss of function of Pro1 showed inhibited proline biosynthesis and decreased its accumulation in the pro1 mutant. Proline deficiency triggered the increase of the level of uncharged tRNApro AGG accumulation and the phosphorylation of eukaryotic initiation factor 2α (eIF2α) in the pro1 mutant, which could lead to a general reduction in protein synthesis in this mutant. Downregulation of major cyclin genes at the transcriptional level, which caused cell cycle arrest and the suppression of cell proliferation due to proline deficiency, was reversible when external proline is supplied to the mutant, suggesting a regulatory role of proline in the cell cycle transition. Based on these evidences, it can be demonstrated that proline plays an important role in the regulation of general protein synthesis and the cell cycle transition in plants.

Reproductive organs can show parallel induction of P5CS (Savouré et al., 1995), ProDH (Nakashima et al., 1998), and P5CDH (Deuschle et al., 2004), whereas the expression of P5CS2, P5CR, and ProDH1 increase in meristematic tissues, such as the root tip, shoot apex, and inflorescences (Kavi Kishor and Sreenivasulu, 2014). In *Brassica napus* genome, six BnaProDH1 and two BnaProDH2 genes were described; the BnaProDH1 genes are mainly expressed in pollen and roots, whereas BnaProDH2 gene expression is associated with leaf vascular tissues at senescence (Faes et al., 2015).

The inner mitochondrial membrane is the place of the two-step oxidation of proline in all eukaryotes. The first step is the consecutive action of proline dehydrogenase (ProDH) that produces delta(1)-pyrroline-5-carboxylate (P5C) and the second is the action of P5C dehydrogenase (P5CDH) that oxidizes P5C to glutamate. Once osmotic stress occurs, this catabolic route is downregulated, allowing free Pro accumulation in plants. The cellular Pro to P5C ratio under ambient and osmotic stress conditions did not change in case of overexpression of MsProDH in tobacco and *Arabidopsis* or impairment of P5C oxidation in the *Arabidopsis* p5cdh mutant, indicating that P5C excess was reduced to Pro in a mitochondrial-cytosolic cycle. This cycle, with ProDH and P5C reductase, exists in animal cells, and Miller et al. (2009) have described it in plants also. In this cycle, Pro oxidation by the ProDH-FAD complex delivers electrons to the mitochondrial electron transport chain and, in the case of hyperactivity of the cycle, e.g., when an excess of exogenous Pro is provided, can generate reactive oxygen species (ROS) by delivering electrons to O_2. After exogenous Pro treatment, the lack of P5CDH activity induced higher ROS production under dark and light conditions, and rendered plants hypersensitive to heat stress. To avoid Pro-mediated toxic effects, it is important to balance mitochondrial ROS production during increased Pro oxidation by P5CDH, which is the key enzyme to prevent P5C-Pro intensive cycling, and avoid ROS production from electron run-off.

Fichman et al. (2015) reviewed proline biosynthesis pathways in different domains of life. The most prevalent pathway of proline synthesis is assisted by bifunctional Δ1-pyrroline-5-carboxylate synthase, in higher plants and animals, whereas there are alternative pathways of proline formation using the initial steps of the arginine biosynthetic pathway to ornithine, which can be converted to Δ1-pyrroline-5-carboxylate by ornithine aminotransferase and then reduced to proline or converted directly to proline by ornithine cyclodeaminase. In some organisms, the latter pathways contribute to or could be fully responsible for the synthesis of proline. Fichman et al. (2015) provided novel data about proline biosynthesis in organisms contributing to elucidate the conservation of proline biosynthetic enzymes, significance of specific residues for catalytic activity and allosteric regulation are analyzed on the basis of protein structural data, multiple sequence alignments, and mutant studies.

3. PROLINE AND ABIOTIC STRESS FACTORS

It is well known that Pro has a diversified role in plant metabolism, especially in plant responses to several types of biotic and abiotic stresses, e.g., it not only acts as a nonenzymatic antioxidant but also may form complexes with heavy metals (HMs) or it can be a signaling molecule (reviewed by Hare and Cress, 1997; Hossain et al., 2014; Emamverdian et al., 2015). Numerous results have exhibited that Pro accumulation can be regarded as a general consequence of the stresses and its level depends on the degree and duration of the stress, as well as the ontogenetic stage of the plants (Kaur et al., 2011; Rejeb et al., 2014). Besides, other investigations executed on transgenic plants also confirmed that changes of Pro metabolism, mainly up- or downregulation of enzymes, improve the stress tolerance (review from Szabados and Savouré, 2010; Ku et al., 2011; Lv et al., 2011; Rejeb et al., 2014). Moreover, exogenous application of Pro may also alleviate the negative effects of various abiotic stresses (Xu et al., 2009; Kaur et al., 2011; Zouari et al., 2016a,b).

3.1 Salinity

Owing to climate change, salinity has become one of the most serious abiotic stress factors, often in combination with drought stress, which may result in the loss of the yield and the quality of agricultural products (Mittler, 2006; Shabala, 2013). Salt-tolerant plants, namely, halophytes (such as quinoa from Chenopodiaceae), have acclimated to the high salt concentration of the environment, whereas most plants (including the main crops) are nontolerant and usually show, e.g., dysfunctions in photosynthetic activity, oxidative stress, or other morphophysiological changes of the organs, as it has been presented by several investigations (Meloni et al., 2003; Szepesi et al., 2008; Shabala, 2013; Deinlein et al., 2014; Singh et al., 2016). All the results pronounce that supraoptimal salt concentration not only provokes oxidative stress and usually the upregulated activities of the enzymatic antioxidants, but also triggers Pro accumulation accompanied by the changes of the enzymes of Pro metabolism (Table 17.1). It was found that in the leaves of salt-stressed Indian mustard (*Brassica juncea* L. cv. Pusa Jai Kisan) or eggplant (*Solanum melongena* L. var. Neelam) the elevated level of Pro content was probably due to the enhanced activity of Δ1-pyrroline-5-carboxylate synthetase (P5CS) and dysfunction of proline oxidase or proline dehydrogenase (PDH) (Iqbal et al., 2015; Singh et al., 2016; Table 17.1). Moreover, transgenic potato plants with accrued level of P5CS and Δ1-pyrroline-5-carboxylate reductase (P5CR) activities expressed a much higher Pro content compared with nontransgenic (control) lines (Hmida-Sayari et al., 2005; Table 17.1).

3.2 Drought

Lack of water, beside salinization, may affect the fecundity of crops, especially in the arid fields. Therefore, the acclimation and the adaptation to water deprivation is a great challenge for the intolerant plants. Several researchers displayed the detrimental impact of water deficit resulting in numerous morphophysiological variations in plants, e.g., a decline in photosynthetic efficiency, ROS overproduction, or reduced development of organs (Alexieva et al., 2001; Arora et al., 2002; Karataş et al., 2014). Both wild-type and transformed tobacco (*Nicotiana tabacum* L. cv. M51) plants exposed to water retention showed an increment of Pro level in leaves and roots (Cvikrová et al., 2013). Dobrá et al. (2011) showed that Pro accumulation was in correlation with the higher activity of P5CS and the lower level of PDH activity (Table 17.1). Similarly, significant inhibition of Pro accumulation was observed in P5CS-silenced *Nicotiana benthamiana* plants (Ku et al., 2011).

3.3 Chilling and Heat Stress

Extremely high and low temperature values also might affect negatively the productivity of crops, although some plants have acclimated having specialized epidermal surface, e.g., waxy cuticle on the needles of pines or species covered with hairs can bear heat or cold stress. Most of the commercial species (vegetables, cereals, etc.) can hardly bear weather extremities, and several morphological alterations and/or physiological symptoms appear. Low temperature can cause dysfunction of the chloroplasts or mitochondria in sensitive plants, inducing oxidative stress. At the same time, chilling may also change the Pro content (mainly accumulation), providing stabilization of the cellular structure, protection of enzymes, and stress tolerance (Hare and Cress, 1997; Prasad et al., 1994; Rivero et al., 2001; Szabados and Savouré, 2010).

In chickpea (*Cicer arietinum* L. cv. GPF2), which is an important food legume in Asia, after a 1 week-long chilling stress, endogenous Pro level increased then decreased in the leaves at the reproductive phase, whereas after foliar application of Pro it was significantly higher compared with nontreated chilling-stressed plants (Kaur et al., 2011; Table 17.1). Kumar and Yadav (2009) also confirmed the protective effects of exogenous usage of Pro. In the topmost leaves of cold-stressed tea plant (*Camellia sinensis* (L.) O. Kuntze), lipid peroxidation diminished on Pro exposure. In the case of bamboo (*Phyllostachys praecox* f. prevernalis), induced activities of P5CS, PDH, and ornithine-Δ-aminotransferase (OAT) were found, which later attenuated (Liu et al., 2016; Table 17.1).

Several studies have shown that heat stress can dramatically destroy the plants at physiological and/or morphological level, especially combined with water deficit or salt stress (Mittler, 2006). Elevated temperature usually triggers the decline of the photosynthetic rate, oxidative stress, and Pro accumulation as well, as it was found in the leaves if wheat, eggplant, or chickpea (see Table 17.1; Hare and Cress, 1997; Kotak et al., 2007; Rejeb et al., 2014; Wu et al., 2014).

TABLE 17.1 Summary of Abiotic Stress Factors That Affect Proline Metabolism in Plants

Type of Abiotic Stress Factor	Plant Name	Duration of Precultivation	Concentration or Degree of the Stress	Time of Exposure	Plant Organ Investigated	Change in Pro content[a]	Activity of Enzymes of Pro metabolism[a]	References
Salinity (NaCl)	*Brassica juncea* L. cv. Pusa Jai Kisan	No data	100 mM	30 days	Leaves	↑	P5CS ↑, PO ↓	Iqbal et al. (2015)
	Cucumis sativus L. var. Jumbo	30 days	150 mM	3 days	Leaves	↑	No data	Fariduddin et al. (2013)
	C. sativus L. var. Rocket	30 days	150 mM	3 days	Leaves	↑	No data	
	Mesembryanthemum crystallinum L.	4 weeks	100 mM	24 h	Leaves	↑	No data	Shevyakova et al. (2009)
		9 weeks	100 mM	24 h	Leaves	↑	No data	
	Pisum sativum L. "Sprinter"	20 days	100 mM	No data	Leaves	↑	No data	Balal et al. 2016
	Solanum melongena L. var. Neelam	15 days	0.3 g/kg (sand)	12 days	Leaves	↑	P5CS ↑, PDH ↓	Singh et al. (2016)
			0.5 g/kg (sand)			↑	P5CS ↑, PDH ↓	
	Solanum tuberosum L. cv. Nicola nontransgenic line	4–6 weeks	100 mM	10 days	Leaves	↑ ns	Nondetectable	Hmida-Sayari et al. (2005)
			180 mM			↑	Nondetectable	
	S. tuberosum L. cv. Nicola transgenic line 15B	4–6 weeks	100 mM	10 days	Leaves	↑↑	P5CS and P5CR ↑	
			180 mM			↑↑	P5CS and P5CR ↑	
	S. tuberosum L. cv. Nicola transgenic line 17	4–6 weeks	100 mM	10 days	Leaves	↑↑	P5CS and P5CR ↑	
			180 mM			↑↑	P5CS and P5CR ↑	
	Spirodela polyrhiza L., clone DR	24 ± 2 days	100 mM	4 days	Frond (i.e., whole plant)	No change	P5CS ↑ ns, P5CR ↑ ns, GS ↓ ns	Cheng et al. (2013)
			200 mM	4 days	Frond (i.e., whole plant)	↓	P5CS ↑ ns, P5CR ↑, GS ↓	

Drought (water deficit)	*Nicotiana tabacum* L. cv. Wisconsin 38	6 weeks	No irrigation	1 day	Leaves	↑ in upper and lower leaves but ↓ ns in middle leaves	PDH ↓, P5CS ↑	Dobrá et al. (2011)
				6 days		↑	PDH ↓, P5CS ↑	
				1 day	Roots	↑	PDH ↓, P5CS ↑	
				6 days		No change	PDH ↓, P5CS ↑	
	N. tabacum L. cv. M51 (wild type)	6 weeks	No irrigation	5 and 10 days	Leaves	↑ and ↑↑	No data	Cvikrová et al. (2013)
					Roots	↑ ns	No data	
	N. tabacum L. M51-1 transformant			5 and 10 days	Leaves	↑ and ↑↑	No data	
					Roots	↑	No data	
Low temperature (chilling)	*Cicer arietinum* L. cv. GPF2	ca. 2 months	8.3–9.6/2.8–5.3°C (max./min.)	7 days	Leaves	↑, later ↓	No data	Kaur et al. (2011)
	Phyllostachys praecox f. prevernalis	14 days	1°C	42 days	Shoots	↑	OAT ↑, later ↓; P5CS ↑; PDH ↑, later ↓	Liu et al. (2016)
High temperature (heat)	*C. arietinum* L. cv. GPF2	No precultivation	35–45°C	10 days	Leaves	↑ and ↑↑	No data	Kaushal et al. (2011)
	Triticum aestivum L. cultivars (Unnat Halna, Halna, Raj3765, C306, NIAW34, WR544, WH730, HD2877, and PBW343)	4 weeks	38°C	4 h	Leaves	↑	No data	Mishra et al. (2017)
	Vigna aconitifolia L. varieties (e.g., Jwala, RMO 40, RMO 225, and RMO 423)	7 days	42°C	1 h	Leaves	↑ or ↓ ns	No data	Harsh et al. (2016)

PO, proline oxidase.

[a]*↑ indicates significant and ↑↑ significantly high increase, ↑ ns indicates nonsignificant increase, whereas ↓ refers to significant decrease and ↓ ns indicates nonsignificant reduction in Pro content and enzyme activities.*

3.4 Heavy Metals

The emission of HMs has increased owing to anthropogenic activities such as mining, industrial and agricultural works, and the output of municipal wastes, which may have disadvantageous consequences in plants. Although aluminum (Al) is not an HM because of its density (2.70 g/cm^3), owing to its toxic effects it is often mentioned among the classical HMs, such zinc (Zn), iron (Fe), copper (Cu), cadmium (Cd), or lead (Pb) (Matsumoto and Motoda, 2012). The main visual symptoms of HM stress and/or toxicity in plants are chlorotic leaves, decreased or inhibited germination of seeds, and disturbed development of vegetative parts (root and shoot) (Arora et al., 2002; Maksymiec, 2007; Farooq et al., 2009; Emamverdian et al., 2015). Beside toxic metals such as Cd, Pb, and Hg, which can cause oxidative stress in the plant cells, the essential HMs (e.g., Fe, Zn, Ni, or Cu) at supraoptimal concentration also might provoke the increment of ROS. Here we discuss the influence of excess HM on Pro metabolism in the case of the main essential and nonessential HMs.

Copper (Cu) is an essential micronutrient for plants, and its mean concentration in the soil is 2–40 mg/kg dry matter (DM), but it can reach 10- to 15-fold higher concentrations in agricultural soils because of the usage of sewage sludge, fertilizers, and pesticides, and may induce dramatic changes in photosynthetic efficiency accompanied by reduced shoot growth and leaf chlorosis (Hirt and Shinozaki, 2003; Janas et al., 2010; Yadav, 2010; Emamverdain). Cu being a redox-active HM can cause lipid peroxidation, catalyze •OH radical formation through Haber-Weiss and Fenton reaction and change the activity of several enzymes (Arora et al., 2002; reviewed by Schützendübel and Polle, 2002; Feigl et al., 2013; Sytar et al., 2013). In plants, after Cu exposure there was an increment of Pro content, which was in correlation with the elevated activity of P5CS, OAT, and GDH, but diminution of PDH activity (Table 17.2).

Zinc (Zn) is also an essential trace element for plants; its mean concentration is 10–80 mg/kg DM (in soil), but in polluted areas the range can be 150–300 mg/kg DM and it becomes phytotoxic (Hirt and Shinozaki, 2003; Emamverdian et al., 2015). Zn plays an important role in N-metabolism, chlorophyll biosynthesis, cell growth regulation, and root development and acts as a constituent of several enzymes such as oxidoreductases, transferases, hydrolases, ligases (Rout and Das, 2003; Broadley et al., 2007; Halušková et al., 2010; Emamverdian et al., 2015). At supraoptimal concentrations, it may cause growth inhibition and chlorosis owing to oxidative stress, lipid peroxidation, or disorders in various metabolic processes (Halušková et al., 2010; Janas et al., 2010; Stoláriková-Vaculíková et al., 2015). Zinc is a non-redox active HM; thus it evokes oxidative stress indirectly (Dat et al., 2000; Cuypers et al., 2002). Pro accumulation was reported in tomato shoots and the leaves and roots of wheat after Zn excess (Li et al., 2013b; Al Khateeb and Al-Qwasemeh, 2014; Table 17.2).

Nickel (Ni) is a micronutrient required by plants at a very low concentration (Gajewska and Skłodowska, 2005; Emamverdian et al., 2015). In natural soils its level ranges 10–1000 mg/kg DM, but at 20- to 30-fold levels it becomes toxic (Yadav, 2010; Emamverdian et al., 2015; Pavlovkin et al., 2016). Several studies have exhibited that excess Ni may cause growth inhibition, chlorosis, tissue damage, and functional disorders of plasma membrane or enhances ROS generation in the plant cells (Gajewska and Skłodowska, 2005; Kumar et al., 2007; Fiala et al., 2013; Pavlovkin et al., 2016). It was also proved that in crop plants, such as *Eleusine coracana* (mainly in Africa and Asia) or pea, high levels of Ni may provoke Pro accumulation in all vegetative plant organs (Gajewska and Skłodowska, 2005; Viswanath et al., 2016; Table 17.2).

Cobalt (Co), similarly to Ni, is a transitional element and the cofactor of several enzymes. Excess Co may affect plant development, the integrity of plant cell wall and membrane, drought tolerance, photosynthetic activity, or the uptake and translocation of other elements (Tewari et al., 2002; Jayakumar et al., 2007; Li et al., 2007; Arora et al., 2012). Although there are not much data about the relationship between Co toxicity and Pro metabolism, it seems that this HM also increases Pro content in plants (Table 17.2).

Cadmium (Cd), known as a nonessential toxic HM, can induce numerous morphophysiological disorders in plants. Typical signs of Cd stress are reduced plant development and photosynthesis, chlorotic leaves, or declined seed germination rate (Arora et al., 2002; Benavides et al., 2005; Halušková et al., 2010; Fiala et al., 2013). These symptoms are the probable consequences of oxidative stress and altered activity of several enzymes (Gallego et al., 1999; Szőllősi et al., 2009; Dietz et al., 2016). In natural soils the Cd content is estimated to be about 0.06–0.50 mg/kg DM, but owing to anthropogenic activities the Cd output to agricultural environment has increased. Since Cd is not redox active, it induces indirectly ROS overproduction and has a great affinity to bind to —SH groups of proteins, which generally results in the dysfunction of the enzymes (Arora et al., 2002; Emamverdian et al., 2015). Several data support that plants at various ages exposed to this toxic HM try to protect themselves by accumulating Pro (see Table 17.2).

The level of toxic lead (Pb) has increased in soils as a result of several anthropogenic activities. Some plants are tolerant to Pb and can accumulate it, but most of the species exhibit toxicity symptoms, such as damage of root cells, declined root length, or chlorotic leaves (Islam et al., 2007; Çavuşoğlu et al., 2009; reviewed by Nagajyoti et al., 2010). These visual

TABLE 17.2 Changes of Proline Metabolism During Heavy Metal Stress in Plants

Type of Abiotic Stress Factor	Plant Name	Duration of Precultivation	Concentration or Degree of the Stress	Time of Exposure	Plant Organ Investigated	Change in Pro content[a]	Activity of Enzymes of Pro metabolism[a]	References
Heavy Metals								
Cu	*Cucumis sativus* L. var. Jumbo	30 days	100 mg/kg (sand)	3 days	Leaves	↑	No data	Fariduddin et al. (2013)
	C. sativus L. var. Rocket	30 days	100 mg/kg (sand)	3 days	Leaves	↑	No data	
	Solanum nigrum L.	No data	50, 100, 150, and 200 μM	5 weeks	Shoot	↑	No data	Al Khateeb and Al-Qwasemeh (2014)
	Solanum lycopersicum L.	No data	50, 100, 150, and 200 μM	5 weeks	Shoot	↑	No data	Al Khateeb and Al-Qwasemeh (2014)
	Zea mays var. rugosa Bona	54 h	50 μM	4–24 h	Shoot	↑ns and ↑ at 12–24 h	P5CS ↑ at 12–24 h, GDH ↑ at 12–24 h, OAT ↑, PDH ↓ at 8–24 h	Wen et al. (2013)
Zn	*S. nigrum* L.	No data	0.25, 0.5, 0.75, and 1.0 mM	5 weeks	Shoot	↑	No data	Al Khateeb and Al-Qwasemeh (2014)
	S. lycopersicum L.	No data	0.25, 0.5, 0.75, and 1.0 mM	5 weeks	Shoot	↑ns and ↑ at 1.0 mM	No data	Al Khateeb and Al-Qwasemeh (2014)
	Triticum aestivum L. cv. Xihan 3	No data	0.5, 1, and 3 mM	6 days	Leaves	↑	OAT↑, PDH ↑	Li et al. (2013a,b)
					Roots	↑	OAT ↑ns and ↑ at 1 and 3 mM; PDH no change and ↓ at 3 mM	
Ni	*Eleusine coracana* L. (Sri Chaitanya VR-847)	7 days	0.5 mM	7 days	Shoot	↑	No data	Viswanath et al. (2016)
					Root	↑	No data	
	Pisum sativum L. cv. "Cud Kelvedonu"	14 days	10, 100, and 200 μM	1–9 days	Leaves	↑	No data	Gajewska and Skłodowska (2005)
					Roots	↑ and ↑ ns	No data	

Continued

TABLE 17.2 Changes of Proline Metabolism During Heavy Metal Stress in Plants—cont'd

Type of Abiotic Stress Factor	Plant Name	Duration of Precultivation	Concentration or Degree of the Stress	Time of Exposure	Plant Organ Investigated	Change in Pro content[a]	Activity of Enzymes of Pro metabolism[a]	References
	P. sativum L. "Sprinter"	20 days	100 μM	No data	Leaves	↑	No data	Balal et al. 2016
Co	*Glycine max* L. varieties (SL-688, PS-1347, and DS-9712)	No data	50, 100, and 150 μM	No data	Leaves	↑↑	No data	Imtiyaz et al. (2014)
Cd	*Avicennia marina* (Forsk.)	2 weeks	0.5 mg/kg	30 days	Leaves	↑	No data	Dai et al. (2017)
			5 mg/kg			↑	No data	
	Kandelia obovata (S., L.)	2 weeks	0.5 mg/kg	30 days	Leaves	↑ ns	No data	Dai et al. (2017)
			5 mg/kg			↑	No data	
	Olea europaea L. cv Chemlali	2 years	10 and 30 mg/kg soil	5 months	Leaves	↑	No data	Zouari et al. (2016a,b)
					Roots	↑	No data	
	S. nigrum L.	No data	100, 200, and 400 μM	5 weeks	Shoot	↑	No data	Al Khateeb and Al-Qwasemeh (2014)
	S. lycopersicum L.	No data	100, 200, and 400 μM	5 weeks	Shoot	↓ns and ↑ at 400 mM	No data	Al Khateeb and Al-Qwasemeh (2014)
	T. aestivum L. cv. Gerek-79	7 days	50, 100, 250, and 500 mg/L	6 days	Leaves	↑	No data	Öncel et al. (2000)
	T. aestivum L. cv. Bolal-2973	7 days	50, 100, 250, and 500 mg/L	6 days	Leaves	↑ ns	No data	
Pb	*Glycine max* L. varieties (SL-688, PS-1347, and DS-9712)	No data	50, 100, and 150 μM	No data	Leaves	↑	No data	Imtiyaz et al. (2014)
	Sesamum indicum	30 days	2 mM	48 h	Root	↑	No data	Amooaghaie and Enteshari (2017)

	T. aestivum L. cv. Gerek-79	7 days	50, 100, 250, and 500 mg/L	6 days	Leaves	No change	No data	Öncel et al. (2000)
	T. aestivum L. cv. Bolal-2973					No change	No data	
Al	*Lycopersicum esculentum* Mill. cv. Gowri (sensitive)	No data	500 μM	2 weeks	Leaves	↓	No data	Surapu et al. (2014)
	L. esculentum Mill. cv. Siri (resistant)					↓ ns	No data	
	Zea mays L. cv. Zea 769, (tolerant)	3 days	0.22 mM	7 days	Shoot	↑	No data	Khan et al. (2000)
					Roots	↑↑	No data	
	Zea mays L. cv. Bozm 1335 (tolerant)				Shoot	↑	No data	
					Roots	↑↑	No data	
	Zea mays L. cv. Reward (sensitive)				Shoot	↑ ns	No data	
					Roots	↑ ns	No data	
	Zea mays L. cv. Golden (sensitive)				Shoot	↑ ns	No data	
					Roots	↑ ns	No data	

[a]↑ indicates significant and ↑↑ significantly high increase, ↑ ns indicates nonsignificant increase, whereas ↓ refers to significant decrease and ↓ ns indicates nonsignificant reduction in Pro content and enzyme activities.

symptoms are caused by water imbalance, disturbed mineral nutrition, or malfunction of enzymes (Seregin and Ivanov 2001; Yadav, 2010). Crop plants, such as soya or wheat, showed elevated levels of Pro in the leaves or expressed no change after Pb stress (Öncel et al., 2000; Imtiyaz et al., 2014; Table 17.2).

Although aluminum (Al) is not an HM because of its density, it is regarded to be toxic. At acidic pH (5–5.5) it produces a trivalent cation (Al^{3+}) and becomes toxic for plants (Emamverdian et al., 2015). Several studies have assessed that excess Al results in root growth inhibition and severe damage to root epidermal cells as a consequence of lipid peroxidation and ROS overproduction (Dat et al., 2000; Matsumoto and Motoda, 2012). Experiments executed on maize exhibited that the accrual of Pro in both shoots and roots may be protective against Al stress, whereas higher Al concentrations can cause Pro degradation in both sensitive and resistant tomato plants (Khan et al., 2000; Surapu et al., 2014; Table 17.2).

Oh et al. (2017) investigated Pro accumulation and related gene expression during spring regrowth in three Rosaceae species, suggesting that Pro accumulation in response to higher temperatures during the spring regrowth is a more general phenomenon across Rosaceae fruit trees, and although peach shoots may accumulate Pro via the ornithine pathway during spring regrowth, there was no evidence for this in plum and apple.

Lipid signaling can be involved in the regulation of proline biosynthesis as a negative regulator (Thiery et al., 2004); however, during high salt stress (200 mM NaCl), calcium signaling and phospholipase C could trigger P5CS1 transcription and proline accumulation (Parre et al., 2007). Halophyte plants, such as *Thellungiella salsuginea*, function in an opposite manner in the case of lipid signaling (Ghars et al., 2012).

PLD (Phospholipase D) signaling for proline biosynthesis shows a process that is similar to RD29A gene expression and different from the abscisic acid-dependent RAB18 gene expression. Some studies revealed that PLDs play positive and negative roles in hyperosmotic stress signal transduction in plants, contributing to a precise regulation of ion homeostasis and plant salt tolerance. In cashew plants treated with NaCl, parallel increases of Orn and Pro have also been observed, which activated OAT and curiously repressed P5CDH (Da Rocha et al., 2012).

In addition, NADPH oxidases trigger proline accumulation in response to salt stress independently of the SOS pathway (Rejeb et al., 2015).

P5CS1 expression has also been confirmed to be stimulated by light (Hayashi et al., 2000) and nitric oxide (Zhao et al., 2009) in *Arabidopsis*. Under osmotic and salt stresses, ABA can induce AtP5CS1 expression through the cis-acting ABA responsive element (Strizhov et al., 1997). However, Savouré et al. (1997) showed that stress induction of P5CS1 occurs independently of ABA upon cold and osmotic stresses. Sharma and Verslues (2010) provided evidences supporting this by indicating significant ABA-independent expression of P5CS1 at low water potential demonstrated by comparing ABA-deficient (aba2-1), proline accumulation-deficient (p5cs1-4), and double (p5cs1-4/aba2-1) mutants.

4. PROLINE METABOLISM DURING PLANT BIOTIC STRESS

Pro metabolism seems to be tightly regulated by environmental and endogenous signals. During plant–pathogen interactions, there are increases in glutamate, proline, and γ-aminobutyric acid that can, directly or indirectly, exhibit ROS scavenging activity (Takahashi et al., 2008; Verslues and Sharma, 2010). ProDH has been implicated in defenses against pathogens. In *Arabidopsis* plants, Cecchini et al. (2011a) reported enhanced gene expression and enzyme activity on elicitation of the hypersensitive response in the case of treatment with *Pseudomonas syringae* pv. tomato (Pst) AvrRpm1 (Pst-AvrRpm1). Senthil-Kumar and Mysore (2012) stated that ProDH1/2 genes are also induced in *Nicotiana benthamiana* plants establishing a nonhost interaction with Pst T1. In both cases, silencing of ProDH genes compromises the accumulation of ROS (Cecchini et al., 2011a), the generation of cell death, and disease resistance (Senthil-Kumar and Mysore, 2012). Moreover, ProDH activation is believed to sustain the oxidative burst and reduce cell viability in other kingdoms (Cecchini et al., 2011b). ProDH1 and ProDH2 genes could be differentially regulated in infected tissues (Cecchini et al., 2011a). Rizzi et al. (2016) found that ProDH1 and ProDH2 are both necessary for achieving maximum resistance against Pst-AvrRpm1 and *Botrytis cinerea*, but ProDH2 has a more significant effect on early restriction of *B. cinerea* growth. It is important to note that, although ProDH1 is upregulated by salicylic acid (SA) and jasmonic acid (JA), ProDH2 is activated only by JA, and both genes show transcriptional interregulation at basal and infection conditions. Rizzi et al. (2016) provided the first evidence of the contribution of ProDH2 to disease resistance, showing the differential regulation and complementary function of both enzyme isoforms in infected tissues, deciphering the fundamental role of ProDH in the control of biotrophic and necrotrophic pathogens. Transient exposure to exogenous Pro or infection with *P. syringae* pv. tomato could trigger the increase of ProDH activity in *Arabidopsis* tissues (Rizzi et al., 2015). Wild-type (Col-0) and p5cdh *Arabidopsis* plants showed consecutively activated ProDH and Pro biosynthetic genes under both conditions by different coordination mechanisms between the biosynthetic pathways; in Col-0 plants, both Pro biosynthetic routes were induced, whereas in p5cdh mutant plants, the Orn route may preferentially be activated.

5. PROLINE AND REDOX SIGNALING IN PLANTS

Szabados and Savouré summarized in an excellent review (2010) that proline has multifunctional roles. There are some data about how proline can have an impact on ROS generation processes affecting the relationship between ROS and proline metabolism (Rejeb et al., 2014, 2015). Some studies, such as Zhang and Becker (2015), showed evidence that the proline metabolic ROS production seems to be a general phenomenon in diverse organisms potentially affecting cellular processes such as aging and plant senescence. Recent studies have highlighted the roles of proline as an important player in the redox signaling in plants. Giberti et al. (2014) studied the different regulation of P5CR activity by proline and chloride ions depending on the cofactor used, NADH or NADPH. Sharma et al. (2011) also confirmed that proline metabolism has a special effect on the ratio of NADP to NADPH (Sharma et al., 2011). Some studies also indicated the influence of redox-sensitive enzymes on proline accumulation, e.g., thioredoxins (Verslues et al., 2014) and mitochondrial NAD dehydrogenases (Lovell et al., 2015).

There is some evidence to prove that Pro has a role in controlling the redox status in plants. Shinde et al. (2016) observed that lipid metabolism mutants have higher proline accumulation under stress, and the greater proline accumulation is caused mainly by the altered redox status of these mutants, as ProDH can load electrons into the mitochondrial electron transport chain regulating the cellular redox state. Based on studies using ProDH inhibitors and exogenous Pro treatment as a ProDH inducer process, Fabro et al. (2016) demonstrated that *Arabidopsis* ProDH could modulate the ROS production by affecting RBOHD generating ROS and contributing to flagellin-mediated PAMP-triggered immunity. Ben Rejeb et al. (2015) reported that hydrogen peroxide produced by NADPH oxidases increases proline accumulation during salt or mannitol stress in *Arabidopsis thaliana*. Some potential mechanisms by which proline metabolism influences plant senescence were studied in the case of petal and leaf, suggesting that the proline content is influenced by the energy demands of senescing cells. Proline metabolism may influence ROS signal pathways in the flower and leaf to delay senescence progression. Mechanisms by which proline could regulate senescence should be elucidated, offering new methods to improve and preserve postharvest agricultural products (for review see Zhang and Becker, 2015). Filippou et al. (2016) reported enhanced proline accumulation after priming of *Medicago truncatula* plants by kresoxim-methyl, a type of strobilurins widely used as an important agricultural fungicide. Data highlighted the necessity of monitoring the proline level during priming processes and if controlling the effects of different biotic compounds are needed. The above-mentioned data can help decipher the complex regulation of proline-mediated redox status, and further investigations are needed to uncover novel steps in proline metabolism.

6. POSTTRANSCRIPTIONAL MODIFICATION PROCESSES OF PROLINE METABOLISM ENZYMES

Although there are a few interesting reports, our knowledge about the posttranslational modification of Pro metabolism enzymes is limited. Redox-sensitive modification is one possible type of modification, such as in the case of ProDH1 involved in electron transport (Servet et al., 2012; Schertl et al., 2014). Studies have revealed that enzymes involved in Pro metabolism can be affected at different levels (Fichman et al., 2014; Kavi Kishor and Sreenivasulu, 2014), e.g., OAT, P5CS, and ProDH can be affected at the transcriptional and posttranscriptional levels, whereas P5CS can be inhibited allosterically by its end product. P5CDH has posttranscriptional regulation by cis antisense transcripts, whereas P5CR shows sensitivity to redox regulation (Giberti et al., 2014). Although Bashkara et al. (2015) have shown immunoblot data for posttranslational modifications of both P5CS1 and ProDH1, our knowledge about these modifications is limited. Further studies to decipher these posttranslational modifications of Pro metabolism enzymes will advance our understanding of the Pro-mediated biotic and abiotic stress tolerance of plants, which is important for agriculture and food safety.

7. PROLINE AND OTHER SIGNAL PATHWAYS IN PLANTS

An interaction between proline metabolism and the phytohormone ABA under stress was reported by some studies (Savoure et al., 1997; Strizhov et al., 1997; Abraham et al., 2003), although ABA alone cannot duplicate drought-induced proline accumulation (Sharma and Verslues, 2010). Our current knowledge about how proline can coordinate and affect the cell redox status by other metabolic pathways is very limited. Shinde et al. (2016) proved that both proline and very-long-chain fatty acid synthesis can help buffer the cellular redox state and also produce some products, e.g., compatible osmolytes and cuticle lipids, for improving stress resistance.

8. FUTURE PERSPECTIVES

A review by Mansour and Ali (2017) provided an excellent evaluation of proline functions in saline conditions and concluded that comprehensive future research is needed to establish the exact mechanism by which proline enhances plant salt tolerance. Further investigations are needed to determine the effective proline concentration range for external proline application or the best mode of application, e.g., seed soaking, and to focus on the fruit quality of these plants, controlling food safety. Recent examples of exogenous proline application in the case of olive fruits and date palm (Zouari et al., 2016a,b) also provided evidence that exogenous treatment with proline improved mineral nutrition and enzymatic and nonenzymatic antioxidant systems, counteracting cadmium inhibitory effects in young date palm plants. Aksakal et al. (2017) described that proline applied exogenously could enhance the levels of gibberellic acid (GA) and indole acetic acid (IAA), the concentrations of soluble sugars and organic acids, and the expression of phenylalanine ammonia lyase (PAL), γ-tocopherol methyltransferase (γ-TMT), and proline dehydrogenase (ProDH) genes as compared with the control. The results obtained in this study suggest that pretreatment with exogenous Pro provides important contributions to the increase in the ultraviolet (UV)-B tolerance of lettuce by regulating the biochemical mechanisms of UV-B response. There are some contradictory data about the effectiveness of exogenous Pro treatment, e.g., Wani et al. (2016) reported that proline was not highly effective in alleviating the undesirable effects of higher degrees of salt stress in either *B. juncea* cultivar investigated. As in the study of Aliferis et al. (2014) that provides metabolic data of plant–pathogen pathosystems, metabolomics of proline homeostasis in plants may uncover novel possible strategies for improving crop management and food safety.

LIST OF ABBREVIATIONS

ABA Abscisic acid
bZIP Basic leucine zipper transcription factor
Ca^{2+} Calcium
GABA Gamma aminobutyric acid
GSA Glutamic γ-semialdehyde
H_2O_2 Hydrogen peroxide
OAT Ornithine D-aminotransferase
P5C Pyrroline-5-carboxylate
P5CDH P5C dehydrogenase
P5CR Pyrroline-5-carboxylate reductase
P5CS Pyrroline-5-carboxylate synthetase
PLC Phospholipase C
PLD Phospholipase D
ProDH Proline dehydrogenase
SiRNA Small interfering RNA
SNAC Stress responsive NAC transcription factor
SRO5 Similar to radical-induced cell death one, which shares a 3′ overlapping region with P5CDH

REFERENCES

Abraham, E., Rigo, G., Szekely, G., Nagy, R., Koncz, C., Szabados, L., 2003. Light-dependent induction of proline biosynthesis by abscisic acid and salt stress is inhibited by brassinosteroid in *Arabidopsis*. Plant Mol. Biol. 51, 363–372. https://doi.org/10.1023/A:1022043000516.

Aksakal, O., Tabay, D., Esringu, A., Aksakal, F.I., Esim, N., 2017. Effect of proline on biochemical and molecular mechanisms in lettuce (*Lactuca sativa* L.) exposed to UV-B radiation. Photochem. Photobiol. Sci. 16, 246–254. https://doi.org/10.1039/C6PP00412A.

Al Khateeb, W., Al-Qwasemeh, H., 2014. Cadmium, copper and zinc toxicity effects on growth, proline content and genetic stability of *Solanum nigrum* L., a crop wild relative for tomato; comparative study. Physiol. Mol. Biol. Plants 20 (1), 31–39.

Alexieva, V., Sergiev, I., Mapelli, S., Karanov, E., 2001. The effect of drought and ultraviolet radiation on growth and stress markers in pea and wheat. Plant Cell Environ. 24, 1337–1344.

Aliferis, K.A., Faubert, D., Jabaji, S., 2014. A metabolic profiling strategy for the dissection of plant defense against fungal pathogens. PLoS One. https://doi.org/10.1371/journal.pone.0111930.

Amooaghaie, R., Enteshari, S., 2017. Role of two-sided crosstalk between NO and H_2S on improvement of mineral homeostasis and antioxidative defense in *Sesamum indicum* under lead stress. Ecotoxicol. Environ. Saf. 139, 210–218.

Armengaud, P., Thiery, L., Buhot, N., Grenier-de March, G., Savoure, A., 2004. Transcriptional regulation of proline biosynthesis in *Medicago truncatula* reveals developmental and environmental specific features. Physiol. Plant. 120, 442–450.

Arora, A., Sairam, R.K., Srivastava, G.C., 2002. Oxidative stress and antioxidative system in plants. Curr. Sci. 82, 1227–1238.

Arora, P., Bhardwaj, R., Kanwar, M.K., 2012. Effect of 24-epibrassinolide on growth, protein content and antioxidative defense system of *Brassica juncea* L. subjected to cobalt ion toxicity. Acta Physiol. Plant. 34, 2007–2017.

Balal, R.M., Shahid, M.A., Javaid, M.M., Anjum, M.A., Ali, H.H., Mattson, N.S., Garcia-Sanchez, F., 2016. Foliar treatment with Lolium perenne (Poaceae) leaf extract alleviates salinity and nickel-induced growth inhibition in pea. Braz. J. Bot. 39 (2), 453–463.

Bashkara, G.B., Yang, T.H., Verslues, P.E., 2015. Dynamic proline metabolism: importance and regulation in water limited environments. Front. Plant Sci. 6, 1–7.

Benavides, M.P., Gallego, S.M., Tomaro, M.L., 2005. Cadmium toxicity in plants. Braz. J. Plant Physiol. 17, 21–34.

Broadley, M.R., White, P.J., Hammond, J.P., Zelko, I., Lux, A., 2007. Zinc in plants. New Phytol. 173, 677–702.

Borsani, O., Zhu, J., Verslues, P.E., Sunkar, R., Zhu, J.-K., 2005. Endogenous siRNAs derived from a pair of natural cis-antisense transcripts regulate salt tolerance in Arabidopsis. Cell 123, 1279–1291.

Çavuşoğlu, K., Ergene, A., Yalçin, E., Tan, S., Yapar, K., 2009. Cytotoxic effects of lead (Pb) and mercury (Hg) ions on root tip cells of *Cicer arietinum* L. Fresenius Environ. Bull. 18 (9), 1654–1661.

Cecchini, N.M., Monteoliva, M.I., Alvarez, M.E., 2011a. Proline dehydrogenase contributes to pathogen defense in *Arabidopsis*. Plant Physiol. 155, 1947–1959.

Cecchini, N.M., Monteoliva, M.I., Alvarez, M.E., 2011b. Proline dehydrogenase is a positive regulator of cell death in different kingdoms. Plant Signal. Behav. 6, 1195–1197.

Chen, Z., Cuin, T.A., Zhou, M., Twomey, A., Naidu, B.P., Shabala, S., 2007. Compatible solute accumulation and stress-mitigating effects in barley genotypes contrasting in their salt tolerance. J. Exp. Bot. 58, 4245–4255.

Cheng, T.S., Hung, M.J., Cheng, Y.I., Cheng, L.J., 2013. Calcium-induced proline accumulation contributes to amelioration of NaCl injury and expression of glutamine synthetase in greater duckweed (*Spirodela polyrrhiza* L.). Aquat. Toxicol. 144, 265–274.

Cuypers, A., Vangronsveld, J., Clijsters, H., 2002. Peroxidases on roots and primary leaves of *Phaseolus vulgaris* copper and zinc phytotoxicity: a comparison. J. Plant Physiol. 159, 869–876.

Cvikrová, M., Gemperlová, L., Martincová, O., Vanková, R., 2013. Effect of drought and combined drought and heat stress on polyamine metabolism in proline-over-producing tobacco plants. Plant Physiol. Biochem. 73, 7–15.

Da Rocha, I.M., Vitorello, V.A., Silva, J.S., Ferreira Silva, S.L., Viegas, R.A., Silva, E.N., et al., 2012. Exogenous ornithine is an effective precursor and the D-ornithine amino transferase pathway contributes to proline accumulation under high N recycling in salt-stressed cashew leaves. J. Plant Physiol. 169, 41–49.

Dai, M., Lu, H., Liu, W., Jia, H., Hong, H., Liu, J., Yan, C., 2017. Phosphorus mediation of cadmium stress in two mangrove seedlings *Avicennia marina* and *Kandelia obovata* differing in cadmium accumulation. Ecotoxicol. Environ. Saf. 139, 272–279.

Dat, J., Vandenabeele, S., Vranova, E., Van Montagu, M., Inzé, D., Van Breusegem, F., 2000. Dual action of the active oxygen species during plant stress responses. Cell. Mol. Life Sci. 57, 779–795.

Deinlein, U., Stephan, A.B., Horic, T., Luo, W., Xu, G., Schroeder, J.I., 2014. Plant salt tolerance mechanisms. Trends Plant Sci. 19 (6), 371–379.

Delauney, A.J., Verma, D.P.S., 1993. Proline biosynthesis and osmoregulation in plants. Plant J. 4, 215–223.

Deuschle, K., Funck, D., Forlani, G., Stransky, H., Biehl, A., Leister, D., et al., 2004. The role of delta1-pyrroline-5-carboxylate dehydrogenase in proline degradation. Plant Cell 16, 3413–3425.

Dietz, K.J., Mittler, R., Noctor, G., 2016. Recent progress in understanding the role of reactive oxygen species in plant cell signaling. Plant Physiol. 171 (3), 1535–1539.

Dobrá, J., Vanková, R., Havlová, M., Burman, A.J., Libus, J., Štorchová, H., 2011. Tobacco leaves and roots differ in the expression of proline metabolism-related genes in the course of drought stress and subsequent recovery. J. Plant Physiol. 168 (13), 1588–1597.

Elthon, T.E., Stewart, C.R., 1981. Submitochondrial location and electron transport characteristics of enzymes involved in proline oxidation. Plant Physiol. 67, 780–784.

Emamverdian, A., Ding, Y., Mokhberdoran, F., Xie, Y., 2015. Heavy metal stress and some mechanisms of plant defense response. Sci. World J.:756120. https://doi.org/10.1155/2015/756120.

Fabro, G., Kovacs, I., Pavet, V., Szabados, L., Alvarez, M.E., 2004. Proline accumulation and *AtP5CS2* gene activation are induced by plant-pathogen incompatible interactions in *Arabidopsis*. Mol. Plant Microbe Interact. 17, 343–350.

Fabro, G., Rizzi, Y.S., Alvarez, M.E., 2016. *Arabidopsis* proline dehydrogenase contributes to flagellin-mediated PAMP-triggered immunity by affecting RBOHD. Mol. Plant Microbe Interact. 29 (8), 620–628.

Faes, P., Deleu, C., Ainouche, A., Le Cahérec, F., Montes, E., Clouet, V., Gouraud, A.M., Albert, B., Orsel, M., Lassalle, G., Leport, L., Bouchereau, A., Niogret, M.F., 2015. Molecular evolution and transcriptional regulation of the oilseed rape proline dehydrogenase genes suggest distinct roles of proline catabolism during development. Planta 241 (2), 403–409.

Fariduddin, Q., Khalil, R.R., Mir, B.A., Yusuf, M., Ahmad, A., 2013. 24-Epibrassinolide regulates photosynthesis, antioxidant enzyme activities and proline content of *Cucumis sativus* under salt and/or copper stress. Environ. Monit. Assess. 185 (9), 7845–7856.

Farooq, Z.R., Iqbal, M.Z., Kabir, M., Shafiq, M., 2009. Toxic effects of lead and cadmium on germination and seedling growth of *Albizia lebbeck* (L.) Benth. Pak. J. Bot. 41 (1), 27–33.

Feigl, G., Kumar, D., Lehotai, N., Tugyi, N., Molnár, Á., Ördög, A., Szepesi, Á., Gémes, K., Laskay, G., Erdei, L., Kolbert, Z., 2013. Physiological and morphological responses of the root system of Indian mustard (*Brassica juncea* L. Czern.) and rapeseed (*Brassica napus* L.) to copper stress. Ecotoxicol. Environ. Saf. 94, 179–189.

Fiala, R., Kenderešová, L., Syshchykov, D.V., Martinka, M., Repka, V., Pavlovkin, J., Čiamporová, M., 2013. Comparison of root growth and morphological responses to cadmium and nickel in two maize cultivars. Mod. Phytomorphol. 3, 131–137.

Fichman, Y., Gerdes, S., Kovacs, H., Szabados, L., Zilberstein, A., 2014. Evolution of proline biosynthesis: enzymology, bioinformatics, genetics, and transcriptional regulation. Biol. Rev. Camb. Philos. Soc. 90, 1065–1099.

Fichman, Y., Gerdes, S.Y., Kovács, H., Szabados, L., Zilberstein, A., Csonka, L.N., 2015. Evolution of proline biosynthesis: enzymology, bioinformatics, genetics, and transcriptional regulation. Biol. Rev. 90 (4), 1065–1099.

Filippou, P., Antoniou, C., Obata, T., Van Der Kelen, K., Harokopos, V., Kanetis, L., Aidinis, V., Van Breusegem, F., Fernie, A.R., Fotopoulos, V., 2016. Kresoxim-methyl primes *Medicago truncatula* plants against abiotic stress factors via altered reactive oxygen and nitrogen species signalling leading to downstream transcriptional and metabolic readjustment. J. Exp. Bot. 67 (5), 1259–1274.

Funck, D., Winter, G., Baumgarten, L., Forlani, G., 2012. Requirement of proline synthesis during *Arabidopsis* reproductive development. BMC Plant Biol. 12, 191.

Gajewska, E., Skłodowska, M., 2005. Antioxidative responses and proline level in leaves and roots of pea plants subjected to nickel stress. Acta Physiol. Plant. 27, 329–340.

Gallego, S.M., Benavides, M.P., Tomaro, M.L., 1999. Effect of cadmium ions on antioxidant defense system in sunflower cotyledons. Biol. Plant. 42, 49–55.

Ghars, M.A., Richard, L., Lefebvre-De Vos, D., Leprince, A.S., Parre, E., Bordenave, M., Abdelly, C., Savouré, A., 2012. Phospholipases C and D modulate proline accumulation in *Thelungiella halophile/salsuginea* differently according to the severity of salt or hyperosmotic stress. Plant Cell Physiol. 53 (1), 183–192.

Giberti, S., Funck, D., Forlani, G., 2014. Delta1-Pyrroline-5-carboxylate reductase from *Arabidopsis thaliana*: stimulation or inhibition by chloride ions and feedback regulation by proline depend on whether NADPH or NADH acts as co-substrate. New Phytol. 202, 911–919.

Halušková, L.U., Valentovičová, K., Huttová, J., Mistrík, I., Tamás, L., 2010. Effect of heavy metals on root growth and peroxidase activity in barley root tip. Acta Physiol. Plant. 32 (1), 59–65.

Hare, P.D., Cress, W.A., 1997. Metabolic implications of stress-induced proline accumulation in plants. Plant Growth Regul. 21, 79–102.

Harsh, A., Sharma, Y.K., Joshi, U., Rampuria, S., Singh, G., Kumar, S., Sharma, R., 2016. Effect of short-term heat stress on total sugars, proline and some antioxidant enzymes in moth bean (*Vigna aconitifolia*). Ann. Agric. Sci. 61 (1), 57–64.

Hayashi, F., Ichino, T., Osanai, R., Wada, K., 2000. Oscillation and regulation of proline content by *P5CS* and *ProDH* gene expressions in the light/dark cycles in *Arabidopsis thaliana* L. Plant Cell Physiol 41, 1096–1101.

Hirt, H., Shinozaki, K., 2003. Plant responses to abiotic stress. In: Topics in Current Genetics, vol. 4. Springer-Verlag, Berlin-Heidelberg, pp. 188–189.

Hmida-Sayari, A., Gargouri-Bouzid, R., Bidani, A., Jaoua, L., Savouré, A., Jaoua, S., 2005. Overexpression of Δ1-pyrroline-5-carboxylate synthetase increases proline production and confers salt tolerance in transgenic potato plants. Plant Sci. 169 (4), 746–752.

Hossain, M.A., Hoque, M.A., Burritt, D.J., Fujita, M., 2014. Proline Protects Plants against Abiotic Oxidative Stress: Biochemical and Molecular Mechanisms. Oxidative Damage to Plants. Elsevier, San Diego, USA, pp. 477–522.

Imtiyaz, S., Agnihotri, R.K., Ganie, S.A., Sharma, R., 2014. Biochemical response of *Glycine max* (L.) Merr. to cobalt and lead stress. J. Stress Physiol. Biochem. 10 (3), 259–272.

Iqbal, N., Umar, S., Khan, N.A., 2015. Nitrogen availability regulates proline and ethylene production and alleviates salinity stress in mustard (*Brassica juncea*). J. Plant Physiol. 178, 84–91.

Islam, E., Yang, X., Li, T., Liu, D., Jin, X., Meng, F., 2007. Effect of Pb toxicity on root morphology, physiology and ultrastructure in the two ecotypes of *Elsholtzia argyi*. J. Hazard. Mater. 147 (3), 806–816.

Janas, K.M., Zielińska-Tomaszewska, J., Rybaczek, D., Maszewski, J., Posmyk, M.M., Amarowicz, R., Kosińska, A., 2010. The impact of copper ions on growth, lipid peroxidation, and phenolic compound accumulation and localization in lentil (*Lens culinaris* Medic.) seedlings. J. Plant Physiol. 167, 270–276.

Jayakumar, K., Jaleel, C.A., Vijayarengan, P., 2007. Changes in growth, biochemical constituents, and antioxidant potentials in radish (*Raphanus sativus* L.) under cobalt stress. Turk. J. Biol. 31 (3), 127–136.

Kaplan, F., Kopka, J., Sung, D.Y., Zhao, W., Popp, M., Porat, R., Guy, C.L., 2007. Transcript and metabolite profiling during cold acclimation of Arabidopsis reveals an intricate relationship of cold-regulated gene expression with modifications in metabolite content. Plant J. 50, 967–981.

Karataş, İ., Öztürk, L., Demir, Y., Ünlükara, A., Kurunç, A., Düzdemir, O., 2014. Alterations in antioxidant enzyme activities and proline content in pea leaves under long-term drought stress. Toxicol. Ind. Health 30 (8), 693–700.

Kaur, G., Kumar, S., Thakur, P., Malik, J.A., Bhandhari, K., Sharma, K.D., Nayyar, H., 2011. Involvement of proline in response of chickpea (*Cicer arietinum* L.) to chilling stress at reproductive stage. Sci. Hortic. 128 (3), 174–181.

Kaushal, N., Gupta, K., Bhandhari, K., Kumar, S., Thakur, P., Nayyar, H., 2011. Proline induces heat tolerance in chickpea (*Cicer arietinum* L.) plants by protecting vital enzymes of carbon and antioxidative metabolism. Physiol. Mol. Biol. Plants 17 (3), 203–213.

Kavi Kishor, P.B., Hong, Z., Miao, G.H., Hu, C.A.A., Verma, D.P.S., 1995. Overexpression of Δ1-pyrroline-5-carboxylate synthase increases proline production and confers osmotolerance in transgenic plants. Plant Physiol. 108, 1387–1394.

Kavi Kishor, P.B., Sreenivasulu, N., 2014. Is proline accumulation per se correlated with stress tolerance or is proline homeostasis a more critical issue? Plant Cell Environ. 37, 300–311. doi: 10.1111/pce.12157.

Kesari, R., Lasky, J.R., Villamor, J.G., Marais, D.L.D., Chen, Y.J.C., Liu, T.W., et al., 2012. Intron-mediated alternative splicing of *Arabidopsis* P5CS1 and its association with natural variation in proline and climate adaptation. Proc. Natl. Acad. Sci. USA 109, 9197–9202. https://doi.org/10.1073/pnas.1203433109.

Khan, A.A., McNeilly, T., Collins, J.C., 2000. Accumulation of amino acids, proline, and carbohydrates in response to aluminum and manganese stress in maize. J. Plant Nutr. 23 (9), 1303–1314.

Kiyosue, T., Yoshiba, Y., Yamaguchi-Shinozaki, K., Shinozaki, K., 1996. A nuclear gene encoding mitochondrial proline dehydrogenase, an enzyme involved in proline metabolism, is upregulated by proline but downregulated by dehydration in *Arabidopsis*. Plant Cell 8, 1323–1335.

Kotak, S., Larkindale, J., Lee, U., von Koskull-Döring, P., Vierling, E., Scharf, K.D., 2007. Complexity of the heat stress response in plants. Curr. Opin. Plant Biol. 10, 310–316.

Ku, H.M., Hu, C.C., Chang, H.J., Lin, Y.T., Jan, F.J., Chen, C.T., 2011. Analysis by virus induced gene silencing of the expression of two proline biosynthetic pathway genes in *Nicotiana benthamiana* under stress conditions. Plant Physiol. Biochem. 49 (10), 1147–1154.

Kumar, V., Yadav, S.K., 2009. Proline and betaine provide protection to antioxidant and methylglyoxal detoxification systems during cold stress in *Camellia sinensis* (L.) O. Kuntze. Acta Physiol. Plant. 31 (2), 261–269.

Kumar, P., Tewari, R., Sharma, P., 2007. Excess nickel–induced changes in antioxidative processes in maize leaves. J. Plant Nutr. Soil Sci. 170, 796–802.

Li, M., Zhu, Q., Hu, C.W., Chen, L., Liu, Z.L., Kong, Z.M., 2007. Cobalt and manganese stress in the microalga *Pavlova viridis* (Prymnesiophyceae): effects on lipid peroxidation and antioxidant enzymes. J. Environ. Sci. 19, 1330–1335.

Li, Z.G., Ding, X.J., Du, P.F., 2013a. Hydrogen sulfide donor sodium hydrosulfide-improved heat tolerance in maize and involvement of proline. J. Plant Physiol. 170 (8), 741–747.

Li, X., Yang, Y., Jia, L., Chen, H., Wei, X., 2013b. Zinc-induced oxidative damage, antioxidant enzyme response and proline metabolism in roots and leaves of wheat plants. Ecotoxicol. Environ. Saf. 89, 150–157.

Liang, X., Zhang, L., Natarajan, S.K., Becker, D.F., 2013. Proline mechanism of stress survival. Antioxid. Redox Signal. 19, 998–1011.

Liu, Z., Li, L., Luo, Z., Zeng, F., Jiang, L., Tang, K., 2016. Effect of brassinolide on energy status and proline metabolism in postharvest bamboo shoot during chilling stress. Postharvest Biol. Technol. 111, 240–246.

Lovell, J.T., Mullen, J.L., Lowry, D.B., Awole, K., Richards, J.H., Sen, S., et al., 2015. Exploiting differential gene expression and epistasis to discover candidate genes for drought-associated QTLs in *Arabidopsis thaliana*. Plant Cell 27, 969–983. https://doi.org/10.1105/tpc.15.00122.

Lv, W.T., Lin, B., Zhang, M., Hua, X.J., 2011. Proline accumulation is inhibitory to *Arabidopsis* seedlings during heat stress. Plant Physiol. 156 (4), 1921–1933.

Maksymiec, W., 2007. Signaling responses in plants to heavy metal stress. Acta Physiol. Plant. 29, 177–187.

Mansour, M.M.F., Ali, E.F., 2017. Evaluation of proline functions in saline conditions. Phytochemistry 140, 52–68.

Matsumoto, H., Motoda, H., 2012. Aluminum toxicity recovery processes in root apices. Possible association with oxidative stress. Plant Sci. 185, 1–8.

Meloni, D.A., Oliva, M.A., Martinez, C.A., Cambraia, J., 2003. Photosynthesis and activity of superoxide dismutase, peroxidase and glutathione reductase in cotton under salt stress. Environ. Exp. Bot. 49, 69–76.

Miller, G., Honig, A., Stein, H., Suzuki, N., Mittler, R., Zilberstein, A., 2009. Unraveling delta1-pírrolyne-5-carboxylate-proline cycle in plants by uncoupled expression of proline oxidation enzymes. J. Biol. Chem. 284 (39), 26482–26492.

Mishra, D., Shekhar, S., Agrawal, L., Chakraborty, S., Chakraborty, N., 2017. Cultivar-specific high temperature stress responses in bread wheat (*Triticum aestivum* L.) associated with physicochemical traits and defense pathways. Food Chem. 221, 1077–1087.

Mittler, R., 2006. Abiotic stress, the field environment and stress combination. Trends Plant Sci. 11, 15–19.

Nagajyoti, P.C., Lee, K.D., Sreekanth, T.V.M., 2010. Heavy metals, occurrence and toxicity for plants: a review. Environ. Chem. Lett. 8 (3), 199–216.

Nakashima, K., Satoh, R., Kiyosue, T., Yamaguchi-Shinozaki, K., Shinozaki, K., 1998. A gene encoding proline dehydrogenase is not only induced by proline and hypoosmolarity, but is also developmentally regulated in the reproductive organs of *Arabidopsis*. Plant Physiol. 118, 1233–1241.

Oh, S., Shin, H., Arora, R., Kim, K., Kim, D., 2017. Proline accumulation and related gene expression during spring regrowth in three *Rosaceae* species. Hortic. Environ. Biotechnol. 58 (1), 21–26.

Öncel, I., Keleş, Y., Üstün, A.S., 2000. Interactive effects of temperature and heavy metal stress on the growth and some biochemical compounds in wheat seedlings. Environ. Pollut. 107 (3), 315–320.

Parre, E., Ghars, M.A., Leprince, A.S., Thiery, L., Lefebvre, D., Bordenave, M., Luc, R., Mazars, C., Abdelly, C., Savouré, A., 2007. Calcium signaling via phospholipase C is essential for proline accumulation upon ionic but not nonionic hyperosmotic stresses in *Arabidopsis*. Plant Physiol. 144, 503–512.

Pavlovkin, J., Fiala, R., Čiamporová, M., Martinka, M., Repka, V., 2016. Impact of nickel on grapevine (*Vitis vinifera* L.) root plasma membrane, ROS generation, and cell viability. Acta Bot. Croat. 75 (1), 25–30.

Prasad, T.K., Anderson, M.D., Martin, B.A., Stewart, C.R., 1994. Evidence for chilling-induced oxidative stress in maize seedlings and a regulatory role for hydrogen peroxide. Plant Cell 6, 65–74.

Rejeb, I.B., Pastor, V., Mauch-Mani, B., 2014. Plant responses to simultaneous biotic and abiotic stress: molecular mechanisms. Plants (Basel) 3 (4), 458–475.

Rejeb, I.B., Lefebvre-De Vos, D., Le Disquet, I., Leprince, A.S., Bordenave, M., Maldiney, R., Jdey, A., Abdelly, C., Savouré, A., 2015. Hydrogen peroxide produced by NADPH oxidases increases proline accumulation during salt or mannitol stress in *Arabidopsis thaliana*. New Phytol. 208, 1138–1148.

Rivero, R.M., Ruiz, J.M., García, P.C., López-Lefebre, L.R., Sánchez, E., Romero, L., 2001. Resistance to cold and heat stress: accumulation of phenolic compounds in tomato and watermelon plants. Plant Sci. 160, 315–321.

Rizzi, Y.S., Monteoliva, M.I., Fabro, G., Grosso, C.L., Laróvere, L.E., Alvarez, M.E., 2015. P5CDH affects the pathways contributing to Pro synthesis after ProDH activation by biotic and abiotic stress conditions. Front. Plant Sci. https://doi.org/10.3389/fpls.2015.00572.

Rizzi, Y.S., Cecchini, N.M., Fabro, G., Alvarez, M.E., 2016. Differential control and function of *Arabidopsis* ProDH1 and ProDH2 genes on infection with biotrophic and necrotrophic pathogens. Mol. Plant Pathol. https://doi.org/10.1111/mpp.12470.

Rout, G.R., Das, P., 2003. Effect of metal toxicity on plant growth and metabolism: I. Zinc. Agronomie 23, 3–11.

Satoh, R., Nakashima, K., Seki, M., Shinozaki, K., Yamaguchi-Shinozaki, K., 2002. ACTCAT, a novel cis-acting element for proline- and hypoosmolarity-responsive expression of the *ProDH* gene encoding proline dehydrogenase in Arabidopsis. Plant Physiol. 130, 709–719.

Savouré, A., Jaoua, S., Hua, X.J., Ardiles, W., Van Montagu, M., Verbruggen, N., 1995. Isolation, characterization, and chromosomal location of a gene encoding the delta1-pyrroline-5-carboxylate synthetase in *Arabidopsis thaliana*. FEBS Lett. 372, 13–19.

Savoure, A., Hua, X.J., Bertauche, N., VanMontagu, M., Verbruggen, N., 1997. Abscisic acid-independent and abscisic acid-dependent regulation of proline biosynthesis following cold and osmotic stresses in *Arabidopsis thaliana*. Mol. General Genet. 254, 104–109.

Schertl, P., Cabassa, C., Saadallah, K., Bordenave, M., Savoure, A., Braun, H.P., 2014. Biochemical characterization of proline dehydrogenase in *Arabidopsis mitochondria*. FEBS J. 281, 2794–2804.

Schützendübel, A., Polle, A., 2002. Plant responses to abiotic stresses: heavy metal-induced oxidative stress and protection by mycorrhization. J. Exp. Bot. 53, 1351–1365.

Senthil-Kumar, M., Mysore, K.S., 2012. Ornithine-delta-aminotransferase and proline dehydrogenase genes play a role in non-host disease resistance by regulating pyrroline-5-carboxylate metabolism-induced hypersensitive response. Plant Cell Environ. 35, 1329–1343.

Seregin, I.V., Ivanov, V.B., 2001. Physiological aspects of cadmium and lead toxic effects on higher plants. Russ. J. Plant Physiol. 48, 523–544.

Servet, C., Ghelis, T., Richard, L., Zilberstein, A., Savoure, A., 2012. Proline dehydrogenase: a key enzyme in controlling cellular homeostasis. Front. Biosci. 17, 3947. https://doi.org/10.2741/3947.

Shabala, S., 2013. Learning from halophytes: physiological basis and strategies to improve abiotic stress tolerance in crops. Ann. Bot. 112, 1209–1221.

Sharma, S., Verslues, P.E., 2010. Mechanisms independent of abscisic acid (ABA) or proline feedback have a predominant role in transcriptional regulation of proline metabolism during low water potential and stress recovery. Plant Cell Environ. 33, 1838–1851.

Sharma, S., Villamor, J.G., Verslues, P.E., 2011. Essential role of tissue-specific proline synthesis and catabolism in growth and redox balance at low water potential. Plant Physiol. 157, 292–304.

Shevyakova, N.I., Bakulina, E.A., Kuznetsov, V.V., 2009. Proline antioxidant role in the common ice plant subjected to salinity and paraquat treatment inducing oxidative stress. Russ. J. Plant Physiol. 56 (5), 663–669.

Shinde, S., Villamor, J.G., Lin, W., Sharma, S., Verslues, P.E., 2016. Proline coordination with fatty acid synthesis and redox metabolism of chloroplast and mitochondria. Plant Physiol. 172, 1074–1088.

Signorelli, S., 2016. Thermentation analogy: a point of view for understanding the intriguing role of proline accumulation in stressed plants. Front. Plant Sci. 7, 1339. https://doi.org/10.3389/fpls.2016.01339.

Singh, M., Singh, V.P., Prasad, S.M., 2016. Nitrogen modifies NaCl toxicity in eggplant seedlings: assessment of chlorophyll a fluorescence, antioxidative response and proline metabolism. Biocatal. Agric. Biotechnol. 7, 76–86.

Stoláriková-Vaculíková, M., Romeo, S., Minnocci, A., Luxová, M., Vaculík, M., Lux, A., Sebastiani, L., 2015. Anatomical, biochemical and morphological responses of poplar *Populus deltoides* clone Lux to Zn excess. Environ. Exp. Bot. 109, 235–243.

Strizhov, N., Abraham, E., Okresz, L., Blickling, S., Zilberstein, A., Schell, J., et al., 1997. Differential expression of two P5CS genes controlling proline accumulation during salt-stress requires ABA and is regulated by ABA1, ABI1 and AXR2 in *Arabidopsis*. Plant J. 12, 557–569. https://doi.org/10.1111/j.09607412.1997.00557.x.

Surapu, V., Ediga, A., Meriga, B., 2014. Salicylic acid alleviates aluminum toxicity in tomato seedlings (*Lycopersicum esculentum* Mill.) through activation of antioxidant defense system and proline biosynthesis. Adv. Biosci. Biotechnol. 5 (9). https://doi.org/10.4236/abb.2014.59091.

Sytar, O., Kumar, A., Latowski, D., Kuczynska, P., Strzałka, K., Prasad, M.N.V., 2013. Heavy metal-induced oxidative damage, defense reactions, and detoxification mechanisms in plants. Acta Physiol. Plant. 35 (4), 985–999.

Szabados, L., Savouré, A., 2010. Proline: a multifunctional amino acid. Trends Plant Sci. 15 (2), 89–97.

Szekely, G., Abraham, E., Cseplo, A., Rigo, G., Zsigmond, L., Csiszar, J., et al., 2008. Duplicated P5CS gene of *Arabidopsis* play distinct roles in stress regulation and developmental control of proline biosynthesis. Plant J. 53, 11–28.

Szepesi, Á., Poór, P., Gémes, K., Horváth, E., Tari, I., 2008. Influence of exogenous salicylic acid on antioxidant enzyme activities in the roots of salt stressed tomato plants. Acta Biol. Szeged. 52, 199–200.

Szőllősi, R., Varga, I.S., Erdei, L., Mihalik, E., 2009. Cadmium-induced oxidative stress and antioxidative mechanisms in germinating Indian mustard (*Brassica juncea* L.) seeds. Ecotoxicol. Environ. Saf. 72, 1337–1342.

Takahashi, H., Matsumura, H., Kawai-Yamada, M., Uchimiya, H., 2008. The cell death factor, cell wall elicitor of rice blast fungus (*Magnaporthe grisea*) causes metabolic alterations including GABA shunt in rice cultured cells. Plant Signal. Behav. 3, 945–953.

Tewari, R.K., Kumar, P., Sharma, P.N., Bisht, S.S., 2002. Modulation of oxidative stress responsive enzymes by excess cobalt. Plant Sci. 162 (3), 381–388.

Thiery, L., Leprince, A.S., Lefebvre, D., Ghars, M.A., Debarbieux, Savouré, A., 2004. Phospholipase D is a negative regulator of proline biosynthesis in *Arabidopsis thaliana*. J. Biol. Chem. 279, 14812–14818.

Verslues, P.E., Kim, Y.S., Zhu, J.K., 2007. Altered ABA, proline and hydrogen peroxide in an Arabidopsis glutamate: glyoxylate aminotransferase mutant. Plant Mol. Biol. 64, 205–217.

Verslues, P.E., Sharma, S., 2010. Proline metabolism and its implications for plant-environment interaction. Arab. Book 8, e0140.

Verslues, P.E., Lasky, J.R., Juenger, T.E., Liu, T.W., Kumar, M.N., 2014. Genome-wide association mapping combined with reverse genetics identifies new effectors of low water potential-induced proline accumulation in *Arabidopsis*. Plant Physiol. 164, 144–159. https://doi.org/10.1104/pp.113.224014.

Viswanath, K.K., Palaka, B.K., Ampasala, D.R., 2016. Alleviation of nickel toxicity in finger millet (*Eleusine coracana* L.) germinating seedlings by exogenous application of salicylic acid and nitric oxide. Crop J. https://doi.org/10.1016/j.cj.2016.09.002.

Wang, G., Zhang, J., Wang, G., Fan, X., Sun, X., Qin, H., Xu, N., Zhong, M., Qiao, Z., Tang, Y., Song, R., 2014. Proline responding1 plays a critical role in regulating general protein synthesis and the cell cycle in maize. Plant Cell 26, 2582–2600.

Wani, A.S., Ahmad, A., Hayat, S., Tahir, I., 2016. Is foliar spray of proline sufficient for mitigation of salt stress in *Brassica juncea* cultivars? Environ. Sci. Pollut. Res. 23 (13), 13413–13423.

Wen, J.F., Gong, M., Liu, Y., Hu, J.L., Deng, M.H., 2013. Effect of hydrogen peroxide on growth and activity of some enzymes involved in proline metabolism of sweet corn seedlings under copper stress. Sci. Hortic. 164, 366–371.

Wu, X.X., He, J., Zhu, Z.W., Yang, S.J., Zha, D.S., 2014. Protection of photosynthesis and antioxidative system by 24-epibrassinolide in *Solanum melongena* under cold stress. Biol. Plant. 58 (1), 185–188.

Xu, J., Yin, H., Li, X., 2009. Protective effects of proline against cadmium toxicity in micropropagated hyperaccumulator, *Solanum nigrum* L. Plant Cell Rep. 28 (2), 325–333.

Yadav, S.K., 2010. Heavy metals toxicity in plants: an overview on the role of glutathione and phytochelatins in heavy metal stress tolerance of plants. South Afr. J. Bot. 76, 167–179.

Yoshiba, Y., Kiyosue, T., Nakashima, K., Yamaguchi-Shinozaki, K., Shinozaki, K., 1997. Regulation of levels of proline as an osmolyte in plants under water stress. Plant Cell Physiol. 38, 1095–1102.

Zhang, C.Y., Wang, N.N., Zhang, Y.H., Feng, Q.Z., Yang, C.W., Liu, B., 2013. DNA methylation involved in proline accumulation in response to osmotic stress in rice (*Oryza sativa*). Genet. Mol. Res. (12), 1269–1277.

Zhang, L., Becker, D.F., 2015. Connecting proline metabolism and signalling pathways in plant senescence. Front. Plant Sci. https://doi.org/10.3389/fpls.2015.00552.

Zhao, M.G., Chen, L., Zhang, L.L., Zhang, W.H., 2009. Nitric reductase-dependent nitric oxide production is involved in cold acclimation and freezing tolerance in Arabidopsis. Plant Physiol. 151, 755–767.

Zouari, M., Ben Ahmed, C., Elloumi, N., Bellassoued, K., Delmail, D., Labrouuse, P., Ben Abdallah, F., Ben Rouina, B., 2016a. Impact of proline application on cadmium accumulation, mineral nutrition and enzymatic antioxidant defense system of *Olea europea* L. cv Chemlali exposed to cadmium stress. Ecotoxicol. Environ. Saf. 128, 195–205.

Zouari, M., Elloumi, N., Ben Ahmed, C., Delmail, D., Ben Rouina, B., Ben Abdallah, F., Labrousse, P., 2016b. Exogenous proline enhances growth, mineral uptake, antioxidant defense, and reduces cadmium-induced oxidative damage in young date palm (*Phoenix dactylifera* L.). Ecol. Eng. 86, 202–209.

FURTHER READING

Sheokand, S., Kumari, A., Sawhney, V., 2008. Effect of nitric oxide and putrescine on antioxidative responses under NaCl stress in chickpea plants. Physiol. Mol. Biol. Plants 14, 355–362.

Chapter 18

Flooding Stress in Plants and Approaches to Overcome

Sakshi Tewari[1], **Anurag Mishra**[2]

[1]*Department of Biochemical Engineering and Biotechnology, Indian Institute of Technology, Delhi, Hauz Khas, New Delhi, India;* [2]*NEB Division, Department of Science & Technology, New Delhi, India*

1. INTRODUCTION

Changes in climatic conditions impact not only the resident flora/fauna but also entire mankind (Eigenbrod et al., 2014). An increase in the level of greenhouse gases, such as carbon dioxide (CO_2) and methane (CH_4), due to the industrial revolution is one of the key factors for the alteration in the precipitation pattern and inducing global warming (Hao et al., 2010). These varying environmental conditions have not only affected biotic parameters, but have also favorably induced several abiotic stress factors, such as salinity, drought, chilling, flooding, fluctuation in temperature (high/low), excessive or low irradiance, and oxidative and nutrient stress (Beck et al., 2007; Parvaiz and Satyawati, 2008; Manavalan et al., 2009; Bita and Greats, 2013).

Flooding (also called as waterlogged/ponded/saturated/submerged soil) is one of the crucial stress factors that has an appreciable effect on crop growth, ultimately leading to reduced yield and production of different crop plants (Normile, 2008). It is usually induced as a result of climatic discrepancy, especially when there is an unpredictable, erratic, or irregular rainfall (Bailey-Serres and Voesenek, 2008, 2010). This stress is mainly prevalent in the rain-fed ecosystem, where poor drainage occurs. Globally, nearly 13% of the total land and nearly 10% of the agricultural land are affected adversely because of the problem of flooding (Cramer et al., 2011). The total yield loss in different crops suffering from waterlogging range from 15% to 80%, depending on the type of plant, kind of soil, and extent of the stress (Patel et al., 2014).

Different definitions of flooding have been given by researchers, but Perata et al. (2011) explained flooding as a type of complex stress factor, composed of ongoing cooperative changes within plant tissues triggered by accumulated water around the plant. Flooding is also defined as the complete saturation of soil with water, which facilitates anoxic conditions and provide good opportunity for anaerobes to perform anticipated activities more conveniently (Jackson and Colmer, 2005).

Most of the agriculturally essential crops are incompetent of tolerating this stress. Waterlogging severely affects the yield of farmland by restricting root development and thereby reducing the crop's ability to uptake nutrients and water (Setter and Waters, 2003). The Intergovernmental Panel on Climate Change stated that various anthropogenic activities have increased the frequency of tropical cyclone and heavy rain in plains and arable farmland affecting the poorest sections of the society (IPPC, 2001, 2007). The impact of flooding is observed on a variety of crops, such as maize, oats, corn, wheat, rice, lentils, dry beans, soybean, lotus, peas, and peppers, and based on this, plants are either categorized as tolerant or sensitive toward flooding. Intolerant plants are excluded from flood-prone environment, whereas tolerant plants can be divided into plants that utilize an escape strategy and those that use a quiescence strategy (Colmer and Voesenek, 2009). The escape strategy is based on a set of inducible anatomical and morphological characters that allow reaeration of flooded tissues, whereas the quiescence strategy is composed of characters that conserve the use of carbohydrates for prolonged underwater survival (Bailey-Serres and Voesenek, 2008; Colmer and Voesenek, 2009).

Although a number of reviews have been published on different aspects of abiotic stress tolerance mechanisms (Arora et al., 2012; Tewari and Arora, 2013, 2016; Glick, 2014; Tewari et al., 2016), very little information is available on flooding stress, its tolerance by plants, mechanisms involved, and advance technologies to study and explore their mechanisms further.

Plant Metabolites and Regulation Under Environmental Stress. https://doi.org/10.1016/B978-0-12-812689-9.00018-2

2. IMPACT OF FLOODING

Flooded soil inflicts severe pressure on plant life, mainly because surplus water in the environment can deprive plants of their basic needs, such as O_2, CO_2, and sunlight (Jackson et al., 2009). Reduced oxygen, sunlight, and carbon dioxide in turn hampers the process of photosynthesis and aerobic respiration, resulting in scarcity of carbohydrate thereby limiting growth and development of crop (Voesenek and Bailey-Serres, 2013). Several researchers proved that excess of water results in complex changes such as impeded gas exchange, which subsequently affects the soil microbiome, leading to hypoxia and anoxia (Tewari and Arora, 2013, 2016; Tewari et al., 2016). Hypoxia encourages an increase in the redox potential between waterlogged soil and plants, which leads to the production of reactive oxygen species (ROS) (Fig. 18.1). ROS is necessary for intra/intercellular signaling, but as its concentration rises, it disrupts normal plant metabolism through oxidation of membrane lipids, pigments, proteins, and nucleic acids (Sairam et al., 2008; Arbona et al., 2017). A significant increase in the activity of superoxide radical (SOD) production, lipid peroxidation, and membrane injury was observed in mung, maize, and pigeon pea genotypes in response to flooding (Yan et al., 1996; Kumutha et al., 2009; Sairam et al., 2011; Shabala, 2017). An increase in the level of CO_2 in plant roots is accompanied by phytotoxin accumulation in reduced soil, which inhibits respiration (Tewari and Arora, 2016). Furthermore, reoxygenation of organs and tissues once the flood subsides often leads to oxidative stress (Fukao and Bailey-Serres, 2008).

3. MECHANISMS UNDERLYING FLOOD TOLERANCE

Different plants display incredible diversity toward flooding, but an important question is that how these plants sense, acclimatize, and adapt toward flooding stress. Here we discuss a few mechanisms that help plants in mitigating the detrimental effect of flooding stress (Fig. 18.2).

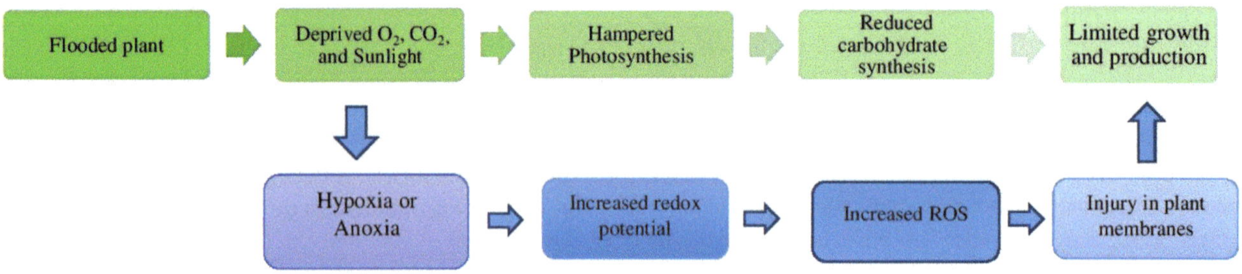

FIGURE 18.1 Deleterious effects of flooding on plants.

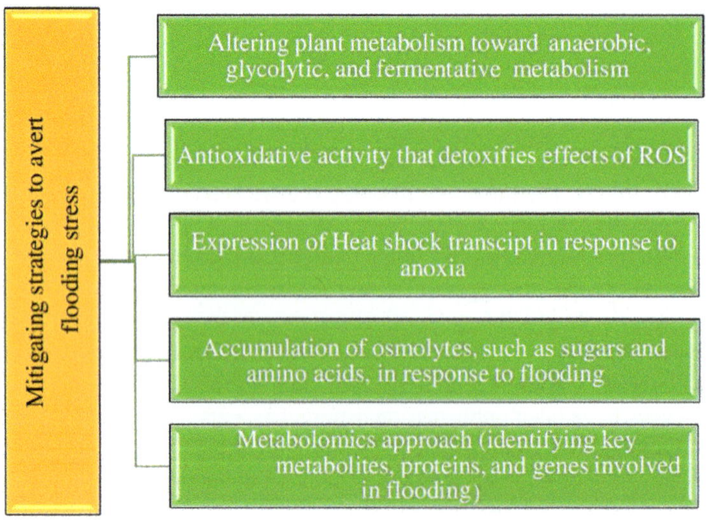

FIGURE 18.2 Mechanism to overcome the effects of flooding stress in plants.

3.1 Shift in Metabolism

During flooding, plants develop a set of morphological, physiological, and anatomical changes to protect themselves from this detrimental stress (Colmer and Voesenek, 2009; Striker and Colmer, 2016). The most common morphological response toward flooding is the development of adventitious roots. These roots have a high porosity that helps in the uptake of water and nutrients under the submerged state (Zhang et al., 2017). Expansion of adventitious roots helps in the oxygenation of submerged tissues through the aerenchyma tissue and thereby increases plant biomass (Colmer, 2003). The generation of aerenchyma in tissues is an anatomical response, which enables the passage of oxygen from shoots to roots (Colmer, 2003). At the physiological level, flooding alters plant carbon fixation and water retention. It also checks closing of stomata, the rate of transpiration, and monitors photosynthesis depending on the tolerance to flooding of each plant species (Mollard et al., 2010). Expansion of adventitious roots in flooded tomato, rice, and *Rumex* is an ethylene-dependent process (Sauter, 2013). Flooding-adapted *Oryza* has constitutively developed adventitious root primordia buried under nodal tissues (Coudert et al., 2010). The interaction of auxin and ethylene leads to the emergence of adventitious roots in rice. Ethylene is an important regulator of several adaptive systemic responses of the waterlogging. 1-Aminocyclopropane-1-Carboxylate (ACC) is the precursor of ethylene; under the flooding state, ACC is transported from the roots to the shoots (Jackson, 2002). This oxygen-mediated conversion of ACC to ethylene triggers adaptive changes typically observed in the shoots of waterlogged plants, such as shoot aerenchyma formation and leaf nastic movements (Sasidharan and Voesenek, 2015). Plants such as *Arabidopsis*, rice, tomato, and *Rumex* exhibit leaf hyponasty during waterlogging to protect themselves from stress by an ethylene-dependent mechanism (Yu and Huang, 2017). Hyponastic growth increases anoxia tolerance by facilitating the leaves to rise above water level and coping with stress (Pierik et al., 2011; Voesenek and Bailey-Serres, 2015).

Plants also tolerate oxidative stress by shifting their metabolism or altering their anatomy. Ethylene responsive factors (ERFs), ERF VII, are a low oxygen sensing group that regulates adaptation toward flooding (Gibbs et al., 2011; Licausi et al., 2011). Developmental adaptation comprises variation in organ and cellular composition for enhancing the accessibility of oxygen (Bailey-Serres and Voesenek, 2008). During flooding when plants are unable to avail oxygen for aerobic respiration, they maintain cell viability by shifting toward anaerobic, glycolytic, and fermentative metabolism. The glycolytic pathway is less efficient and yields two ATP molecules instead of 36 ATP per hexose molecule formed during oxidative phosphorylation. Glycolysis is also limited by the rate of NADH oxidation. Ethanolic fermentation involving alcohol dehydrogenase (ADH) and lactic acid fermentation involving lactic dehydrogenase (LDH) are the major mechanisms of NADH oxidation working under anaerobic conditions (Bailey-Serres and Voesenek, 2008). Energy-demanding progressions, such as replication, transcription, and cell division, are truncated during anoxic conditions, prevalent during flooding. The process of translation involving protein synthesis is finely regulated and encodes proteins involved in anaerobic and ROS metabolism. Different anaerobic proteins (ANPs), such as pyruvate decarboxylase, sucrose synthase, LDH, and ADH, which are synthesized in response to flooding stress, have been characterized by Gibbs and Greenway (2003).

3.2 Antioxidative Activity

Submergence may cause plant senescence, because of the decrease in CO_2 concentration and increase in the concentration of ROS (Puyang et al., 2015). The toxic ROS, including singlet oxygen ($^1O_2$), superoxide (O_2^-), hydroxyl radical ($OH\bullet$), and hydrogen peroxide (H_2O_2), can oxidize proteins, lipids, carbohydrates, and nucleic acids (Smirnoff, 1993). In addition, these radicals can inactivate various Calvin cycle enzymes that are involved in oxidative systems. To ameliorate the oxidative damage and protect tissues and cells, plants have developed a complex antioxidant defensive system (Voesenek and Bailey-Serres, 2015). Enzymatic antioxidants, such as catalase, ascorbate peroxidase, SOD, dehydroascorbate reductase, mono-dehydroascorbate reductase, glutathione reductase, and glutathione S-transferase, and nonenzymatic antioxidants, including α-tocopherol, ascorbic acid, glutathione, and carotenoids, respectively, detoxify the detrimental effects of ROS (Lekshmy et al., 2015). Upregulation of the antioxidant defense system is often correlated with flood tolerance (Dubey, 2014). It was observed that exposing soybean plant to anoxic stress (of 2 h) increased the level of SODs; however, exposing the same plant to anoxia for a longer duration (of 5 h) increased its ability to cope with free oxygen radicals, and it exhibited less injury (Blokhina et al., 2003). Incubating plants with exogenous ascorbic acid alleviated postanoxic injury in soybean roots. Roots and leaves of *Hordeum vulgare* and wheat showed H_2O_2 accumulation under hypoxic conditions (Biemelt et al., 2000). The existence of H_2O_2 in the apoplast and plasma membrane has been visualized under hypoxic conditions by Transmission Electron Microscope (TEM) in different plant species (Blokhina et al., 2003). Indirect evidence of ROS formation (i.e., lipid peroxidation products) under low oxygen has also been reported by the same workers. An increase in the activity of NADPH oxidase and NADPH oxidase mRNA expression under hypoxic condition in pigeon pea genotypes was observed by Sairam et al. (2009) under flooding stress. Monk et al. (1987) reported a 14-fold increment in SOD activity during hypoxia in the *Iris* plant. Similarly, increased activity of glutathione reductase and glutathione content in wheat seedlings mitigated posthypoxia oxidative stress (Ushimaru et al., 1997). Reduced activity of mono-dehydroascorbate reductase, dehydroascorbate reductase, and glutathione reductase enzymes was observed in wheat plants under conditions of anoxia (Biemelt et al., 2000). A study conducted by Thirunavukkarasu et al. (2013) showed transcriptomic changes in flood-tolerant and flood-susceptible maize genotypes by displaying an upregulation of flood-tolerant genes involved in antioxidant defense and fermentation pathway. Flood-tolerant pigeon pea genotypes showed an incessant increase in antioxidant activity after 6 days of flooding; however, in susceptible genotypes the same activity dropped within 2 days of waterlogging (Kumutha et al., 2009). The role of ROS during flooding was reviewed by Steffens (2014); they reported the potential of electron paramagnetic resonance technique to investigate ROS in plants. This study convincingly proved that waterlogging induces an increase in ROS via NADPH oxidase. Mcmanmon and Crawford (1971) proposed the metabolic theory of flood tolerance by explaining the upregulation of enzyme ADH and malic dehydrogenase in plant roots on exposure to waterlogging. Likewise, there are innumerable reports that suggest that the antioxidant potential of crop tremendously increases during flooding (Yiu et al., 2008; Puyang et al., 2015; Wu et al., 2017).

3.3 Heat Shock Transcription Factors

Heat shock transcription factors (HSF) are also known as H_2O_2 sensors in plants. A rise in heat shock protein (HSP) in response to anoxia is conserved in various plant species (Mustroph et al., 2014). Banti et al. (2010) observed the transgenic model plant *Arabidopsis* expressing HSFA2 is more tolerant toward hypoxia compared with the wild type. Researchers also reported the induction of HSFs in response to flooding, which leads to the transcription of genes encoding both high- and low-molecular-weight HSPs in plants. The NADPH oxidase–dependent ROS pathway regulates the ROS-related transcription factor and HSFs under the flooding state (Pucciariello et al., 2012). Transcription factors regulated by HSPs and ROS are not N-end rule pathway–dependent proteolysis; hence ROS signaling is a self-regulating mechanism working under flood tolerance (Lekshmy et al., 2015).

3.4 Osmoprotectants

Flooding is an important stress factor that hampers the phenomena of photosynthesis. The concentration of amino acids, sugars, and sugar alcohols in plant tissue is affected by the impairment in CO_2 assimilation and by complex regulatory mechanisms (Valerio et al., 2011; Krasensky and Jonak, 2012; Iqbal et al., 2016). Plants accumulate different sugars and amino acids to protect themselves from stress. Osmoprotectants, such as glycine betaine and mannitol, play an important role in stress amelioration. Glycine betaine is copiously found in the chloroplast where it provides safety to the thylakoid membranes and preserves the photosynthetic efficiency. Mannitol acts as a free radical scavenger and stabilizes cellular and subcellular structures and plays a significant role in carbon storage during flooding stress (Arora et al., 2012). When rice plant was exposed to flooding, there was an increase in the proline content (under submerged conditions) in comparison with nonsubmerged conditions serving as control (Chanu and Sarangthem, 2015). The accumulation of proline under stress conditions (due to de novo synthesis) is an adaptive mechanism to maintain osmoregulation (Chen and Kao, 1993). Cabellero et al. (2005) reported the role of proline as a compatible solute that adjusts the osmotic potential in the cytoplasm under flooding state (Cabellero et al., 2005). A similar report of proline accumulation in response to flooding in black pepper was reported by Ou et al. (2011). Proline acts as an antidehydrating agent and a cytoplasmic penetrant and can reduce cell osmotic potential under flooded conditions. Proline acts as a compatible solute, and the multifunctional role of proline has been elucidated by several researchers in ROS scavenging, protein stabilization, and DNA stabilization (Arbona et al., 2003, 2008). On the other hand, another group of workers, Wangkhem and Kananbala (2014), stated a significant increase in the accumulation of proline and total sugars after exposing plant to complete submergence for 20 days. Gas chromatography (GC) coupled with electron-impact time-of-flight mass spectrometry (TOF-MS) can be used for the identification of these metabolites (sugars and amino acids). Submerged *Arabidopsis* plants harbor a high amount of soluble sugars and amino acids as a result of improved fatty acid and starch catabolism (Hsu et al., 2011). The function of carbohydrate transporters was upregulated in shoots, following the transfer of carbohydrates from shoots to roots.

Submergence lead to ROS production and the accumulation of sugars, such as sucrose and trehalose, that can ameliorate the effect of oxidative stress (Chen and Hoehenwarter, 2015; Figueroa and Lunn, 2016). During submergence, there is a decline in cytosolic pH and oxygen deprivation, which induces the greatest accumulation of gamma aminobutyric acid proteins. These proteins act as a stress mitigator and protect the plant from flood injuries (Kinnersley and Turano, 2000). Carbohydrate metabolism plays a crucial role in flood tolerance, as it affects photosynthetic activity. Utilization of glucose, fructose, fructan, and starch as energy source was observed during the period of stress (Huynh et al., 2005; Arbona et al., 2013). These sugars also act as osmolytes and maintain cell turgidity and membrane stability, preventing the degradation of proteins (Sharp et al., 2004). In addition, an increase in the concentration of sugar alcohols, such as sorbitol or mannitol, has also been linked to stress tolerance. Accumulation of sugars in the roots of *Inga marginata* was observed when seeds were exposed to flooding stress as reported by Bender et al. (2016). Qin et al. (2013) outlined that reduced consumption and enhancement in the accumulation of soluble sugars are the main mechanisms associated with flood tolerance.

4. ROLE OF METABOLOMICS IN FLOOD TOLERANCE

As climatic conditions are turning hostile day by day, and the incidence of flooding is becoming more common, there is an emergent need for developing new resistant crop varieties for sustainability. This is possible only through genetic engineering and by understanding the detailed mechanisms involved in flooding stress. Combination of contemporary knowledge with traditional breeding approach is essential for improving existing plant varieties with improved technology to nonoptimal environmental conditions. The development of the "omics" technologies, such as genomics, proteomics, transcriptomics, and metabolomics, has transformed plant research and has allowed it holistically to integrate metabolites, proteins, genes, regulatory elements, and fluxes within it (Yuan et al., 2008; Okazaki and Saito, 2016).

Metabolomics is a novel associate to the "omics" family of systems biology (Bino et al., 2004) and is defined as a quantitative and comprehensive evaluation of all molecules in the biological system (Fiehn et al., 2000). In simple terms, metabolomics is a large-scale study of very small molecules (known as metabolites <1500 Da) that are present within cells, tissues, and biofluids of organisms. All together, the cooperative functions of these small molecules and their interactions within the living system are known as the metabolomics. It is the widespread, nonbiased, high-throughput investigation of complex metabolite mixtures that allows preferably the identification and quantification of every single metabolite (Gomez-Casati et al., 2013). Although metabolomics is new compared to proteomics and genomics, it is now firmly recognized as a purposeful genetic approach for understanding the molecular complexity of plant life (Oliver, 1996; Wagner et al., 2003). Metabolomic approaches provide a detailed insight into the metabolic pathways, their links, and connections with each other; it also clarifies the role of genes, proteins, and metabolites that participate in various aspects of functional genomics and systems biology (Mosa et al., 2017).

Metabolomics in an emerging technique that has drawn attention in different areas of plant science, including stress biology, phytochemistry, and environmental physiology. In general, as genomics is rooted in classical molecular biology so is metabolomics deeply related to phytochemistry. The objective of metabolomics study is qualitative and quantitative analysis of metabolites in a plant species at a given developmental stage under particular environmental conditions. Metabolomics study is used for several purposes such as checking the influence of abiotic stress factor on plant metabolism, tracking the effect of certain genes, proteins, and metabolites within the plants that are assisting in stress tolerance (Okazaki and Saito, 2016). The exact composition of metabolites in a given plant is the result of a particular gene expression profile. The activation of metabolic pathways in turn activates intermediates and precursors to produce a bioactive molecule, such as a signaling compound, an antioxidant molecule, a cell structure biosynthesis intermediate, and even a storage compound. The assembly of these complexes can be controlled in turn by other signal molecules that can positively or negatively regulate other metabolic steps. Under a stressed state, metabolites also damage the cell and malondialdehyde, lipid peroxides, and DNA fragments resulting from oxidative damage are produced. Considering the metabolome, the balance between the impaired metabolite, defense, and signaling can be used to evaluate plant tolerance in any stress conditions (Kim et al., 2007; Kerchev et al., 2012).

Several analytical techniques are involved in analyzing different plant metabolites under control or flooded conditions. A separation technique coupled with a detection device, usually MS, is used nowadays (Rodziewicz et al., 2014). However, there are other substitutes to separation techniques, such as flow injection analysis coupled to MS, or analyzers such as nuclear magnetic resonance (NMR) or Fourier transform infrared spectroscopy (FTIR) for fingerprinting purposes. Separation of metabolites involves selective chromatography, such as separation of volatile compounds and primary metabolites (sugars, amino acids, or tricarboxylic acid cycle intermediates) requires GC (Desbrosses et al., 2005). Liquid chromatography (LC) is broadly used to analyze a mixed array of diverse secondary metabolites. Capillary electrophoresis also provides similar features as that of LC; in addition, it can also separate ionic metabolites (Cevallos-Cevallos et al., 2009). These techniques are used in combination since they are mostly complimentary to each independent technique having special coverage of assorted metabolite (Obata and Fernie, 2012). Owing to the high diversity in chemical properties and structures of molecules, so far no sole method has been developed to classify, identify, and enumerate them. In spite of this the most comprehensive procedures distinguish only between 1000 and 2000 molecular features (Giavalisco et al., 2009; Fernie et al., 2004). However, more precise and targeted techniques for the quantitation of plant metabolites are needed to be explored for enhancing specificity and sensitivity in the results (Chiwocha et al., 2005; Segarra et al., 2006).

Several investigations have explained the functions of different key metabolites, genes, proteins, and molecular networks involved in flooding stress (Gomez-Casati et al., 2013; Meena et al., 2017). Successive advancement in dense chromosome mapping allowed researchers to identify corresponding quantitative trait loci (QTLs), which has become a crucial molecular tool for plant breeding programs (Tuberosa and Salvi, 2006). In spite of all these new research, still there are lots of unknown molecular processes occurring in plants that assist in coping stress. Proteomic and metabolomic studies have now become commanding tools to screen the changes in response to numerous environmental stresses. The detailed information regarding the proteins and metabolites involved in growth and development toward flooding conditions provides proper explanation of the data that help in the identification of potential biomarkers related to stress (Salekdeh and Komatsu, 2007). Iqbal et al. (2016) recently reviewed the role of osmolytes in ameliorating flooding stress by utilizing various metabolomics approaches. Although the metabolomic approach has been instrumental in ameliorating flood tolerance, it has not been adopted universally. Some of the plants in which this approach has been successfully implemented are briefly discussed.

In *Arabidopsis* the expression of genes encoding ANPs were involved in sugar metabolism. In model crop *Arabidopsis*, genes encoding ANPs are somewhat involved in sugar metabolism. During hypoxic conditions, signal transduction components that activate calmodium, calmodium-associated proteins, mitochondrial alternative oxidase, and RopGAP4 were upregulated in *Arabidopsis* (Bailey-Serres and Chang, 2005). The energy-sensing protein kinases KIN10 and KIN11 belonging to SnRK1 clade of *Arabidopsis* regulate carbon utilization under hypoxic conditions (Baena-González et al., 2007). KIN10 positively regulates genes encoding enzymes that catalyze carbohydrate and amino acid catabolism in *Arabidopsis* (Baena-González et al., 2007; Cho et al., 2012). One of the KIN10- or KIN11-regulated genes is EXORDIUM-LIKE1, an Hypoxia-responsive unknown protein (HUP) that is essential for carbon management under low-oxygen conditions (Schröder et al., 2011; Tunc-Ozdemir and Jones, 2017). Under low-oxygen stress, ROS have connection with the signal transduction pathway. Screening of gene trap transposon (DsGus) from *Arabidopsis* led to the identification of mutants with increased ADH activity, under hypoxic conditions. The mutant phenotype is formed from the insert of DsGus in the first exon of a gene that encodes Rop (RHO-like small G protein of plants) guanosine triphosphatase (GTPase) activating protein, ROPGAP4. Rop signaling is related to H_2O_2 and Ca^{2+} signaling for flood tolerance. Stimulation of ROP by GTP under low O_2 induces the production of H_2O_2 through a calcium-dependent NADPH oxidase. Accumulation of H_2O_2 is crucial for the induction of ADH and RopGAP4 expression. The level of Rop is regulated by RopGAP4, which participates in negative feedback regulation ROP activity; hence, RopGAP4-1 seedlings fail to resist oxidative stress due to H_2O_2 accumulation (Baxter-Burrell et al., 2002).

Wang et al. (2017) elucidated the impact of flooding on soybean seedling by performing organ-specific analysis through a gel-free/label-free proteomic technique. The activities of ADH and delta-1-pyrroline-5-carboxylate synthase were noticeably increased in roots and leaves of plants exposed to flooding. The tricarboxylic acid cycle was completely suppressed in leaves and roots, whereas proteins related to DNA, RNA signaling, and photosynthesis were badly affected in leaf and roots in response to flooding. Seventeen different proteins, including β-glucosidase 31 and β-amylase 5, were observed in leaves and roots of soybean seeds during flooding. The expression of β-amylase 5 might be involved in carbohydrate mobilization for providing energy to the leaf of soybean plant (Wang et al., 2017).

In the paddy field, flooding is regulated by QTL found on chromosome 9 (Toojinda et al., 2003). The protein *SUB1A* is an ethylene-responsive transcription factor that limits the growth of crop under flooding state (Bailey-Serres and Voesenek, 2008, 2010). Metabolic profiling of the crossbred line M202 (*Sub1*) displayed high tolerance toward flood in comparison with wild strain M202. This suggests that the presence of *SUB1A* in M202 led to the repression of carbohydrate metabolism in shoot tissues (Peng et al., 2004). These findings recommend that in the crossbred line M202 (*Sub1*) with *SUB1A*, carbohydrate metabolism is arranged in such a manner that it suppresses growth if the plant is submerged and further reduces energy loss under unfavorable conditions (Peng et al., 2004). Neeraja et al. (2007) also used molecular markers for *Sub1* gene in a backcross breeding program with recurrent parent Swarna. This *Sub1* provides tolerance to the flood-sensitive mega varieties. This QTL has provided a great opportunity for marker-assisted backcrossing for developing submergence tolerance in mega varieties. Singh et al. (2017) identified unique QTLs and discuss their physiological consequences; the authors also reported that these beneficial QTLs might be taken as targets for molecular breeding and can be further examined to comprehend the underlying molecular mechanisms involved in flood tolerance in rice plant.

The complete transcriptome of maize at three different stages of flooding was analyzed by Thirunavukkarasu et al. (2013). Genes responsible for programmed cell death following aerenchyma formation was selectively upregulated in the HKI 1105 tolerant genotype after submergence. Calcium-binding proteins, such as calmodulin, were overexpressed in HKI 1105 and interact with glutamate decarboxylase, which in turn helps in maintaining the cytosolic pH under anoxia. The gene specific for flooding, XET A, was upregulated in HKI 1105 during severe (16-fold) and moderate (253-fold) stresses, but downregulated in V372 (sensitive). Genes involved in signal transduction and plant hormone biosynthesis were differentially controlled under waterlogged state. There was an increase in the level of ethylene, indole-3-acetic acid (IAA), abscisic acid, and gibberellic acid and a reduction in the production of cytokinin (Zou et al., 2010). ERFs, such as BBM2, protein 1, AIL5-like, and WRI1, were upregulated in HKI 1105. In the tolerant genotype, auxin receptor genes, such as IAA14, IAA16, and IAA3, were unregulated. Signaling pathways between auxin and ethylene increase the formation of lateral and adventitious roots in the flood-tolerant genotype of maize.

Rumex palustris shows hyponastic growth by orientating its leaves in an ethylene-dependent manner (Cox et al., 2006). In *Rumex acetosa*, accumulation of ethylene reduces petiole elongation and inclines the plants toward metabolic rearrangement to minimize carbon consumption (Pierik et al., 2011). Van Veen et al. (2013) employed RNA sequencing technology to investigate the molecular basis of adaptive traits of these two species. There was an enhanced expression of ACC oxidase, enzyme catalyzing ethylene biosynthesis in both the species. In *R. palustris*, the expression of an EIN3 BINDING F-BOX was specifically upregulated. Complete metabolic reprogramming of *R. acetosa* enhanced its survivability in flood-prone area (Van Veen et al., 2013).

Luffa (*Luffa cylindrica*) is a flood-tolerant plant species. Su and Su (1996) monitored the metabolic responses of luffa roots toward flooding stress. It was observed that the respiratory activity of flooded roots reduced to 60% in comparison with the nonflooded ones. The activity of pyruvate decarboxylase and ADH amplified quickly. After exposing the plant to 7 days of flooding, an approximately 50-fold increment in the level of both the enzymes was observed, which indicated strong alcoholic fermentation (Su and Su, 1996). Yet, the level of ethanol did not increase considerably, but declined after 1 day. A high ADH/pyruvate decarboxylase ratio might play an important role in avoiding the accumulation of toxic acetaldehyde. In other fermentative pathways, the activity of phosphoenolpyruvate carboxylase, the key enzyme catalyzing malate synthesis, showed no measurable difference between nonflooded and flooded roots. NADP malic enzyme, catalyzing the decarboxylation of malate, displayed an increase in activity in flooded roots as compared with the nonflooded ones; the amount of malate in flooded roots reduced to 40% of the control ones (Shimamura et al., 2007).

Water lily *Nymphoides peltata* is widely distributed in subtropical and temperate regions of Eurasia (Barrett et al., 1993). Leaves of this plant float on the water surface with connecting petioles and roots are fixed to the bottom mud. As this plant is a nonmodel organism genomic information is very less; so, global gene expression changes of *N. peltata* was monitored both under control and flooded conditions using Illumina RNA-Seq technology (Goodwin et al., 2016). The results gave a comprehensive view of the complex molecular events involved in the response of floating-leaved plants toward flooding. The results of Illumina RNA-Seq technology gave 78,037,588 and 103,266,542 clean reads from the untreated sample (US) and the treated sample (TS), respectively. Association of the reads from TS and US individually generated 95,372 and 87,673 unigenes, respectively. Pooling of all clean reads from US and TS together constituted 124,302 unigenes with an N50 length of 1449 bp after assembly. The number of unigenes from the pooled strategy was observed to be larger than those from separate assemblies, because some unigenes with low expression levels generated with the mutual approach cannot be found when using the strategy of single approach because of fewer reads (Wu et al., 2017). In total, 8883 unigenes were differentially expressed between the normal and flooded states. Among them, most differentially expressed genes were primarily related to photosynthesis, antioxidant process, and carbohydrate metabolism (including glycolysis, tricarboxylic acid cycle, and alanine metabolism). The antioxidative activity and glycolysis are usually induced by a flooded environment, whereas alanine metabolism and photosynthesis are rarely found in flooding-tolerant plants (Wu et al., 2017). These results suggest the significant contributions of both the processes in the active response of *N. peltata* toward flooding stress.

Ferner et al. (2012) observed the metabolite profiling of flood-tolerant oak (*Quercus robur*) and suggested that soluble carbohydrate partitioning constituted the basis of waterlogging. When citrus roots are exposed to flooding, more than 40 flavonoids were induced in the leaves (Munns and Tester, 2008). Metabolite profiling of the samples from control and submerged plants showed variations in the level of flavonoid. The amount of flavonoids was much reduced in the sensitive genotype thereby suggesting an efficient redox equilibrium in the tolerant species (Arbona et al., 2008).

5. METABOLOMIC RESOURCES

Metabolomic study generates a large amount of data, which needs dedicated tools and software for identifying the structure of known and unknown metabolites (Wolf et al., 2010). Metabolomics data could be analyzed with a wide range of statistical and device-learning algorithms, classified as unsupervised and supervised classes, which include hierarchical clustering, component analysis, and self-organizing maps and ANOVA, partial least squares, and discriminant function analysis, respectively. It has been observed that managing, analyzing, and processing of data is the most challenging and time-consuming step in data analysis dealing with metabolomics research (Gomez-Casati et al., 2013). A set of raw chromatograms involves spectrum deconvolution, noise reduction, peak detection, chromatogram alignment, and identification and quantification of compounds from mixtures. Hence, automated data processing software, such as the automated mass spectral deconvolution and identification system (http://chemdata.nist.gov/mass spc/amdis/) software, is utilized for the processing of GC-MS data. Electron spray injection (ESI)-LC-MS data can be analyzed using the component detection algorithm or windowed mass selection method. Different databases for studying plant metabolomics have been developed (http://plantmetabolomics.vrac.iastate.edu), which explain the functions of various genes and metabolites involved in stress tolerance (Moco et al., 2006). A related additional Armec Repository Project was generated to explain flow injection electrospray MS and high-performance liquid chromatography (HPLC)-ESI-MS data (Gomez-Casati et al., 2013).

6. CONCLUSION AND FUTURE PROSPECTS

Flooding is an important abiotic stress factor recurring in plants because of the existing climatic variations caused by global warming. However, this important topic has been little explored by researchers worldwide. As flooding will become a more common phenomenon in the near future, further research in the area of metabolomics can greatly transform the existing traditional knowledge to the contemporary one. Analyzing plant genotype with transcriptomics, metabolomics, and proteomics can draw appreciated conclusions. Some of the well-characterized genes/proteins/metabolites may focus on flood tolerance by either germplasm selection or transgenic manipulation of gene expression. For example, recognition of the *Sub1A* locus from flood-tolerant Indian rice landrace FR13A gave impetus for marker-assisted breeding of this locus into cultured rice varieties. Developed *Swarna-Sub1* can contribute nearly 45% increase in production compared with existing varieties after flood exposure. Therefore, flood-tolerant genes, proteins, and metabolites may be identified with the help of the omics approach and integrated into the host through genetic engineering for constructing flood-tolerant crops. Using metabolomics or integrated metabolomics is a novel thought that improves the complex understanding between the metabolic networks and subsequent biochemical composition of plants under stress conditions. Analytical tools such as LC-MS, GC-MS, UP-LC-MS, HPLC, NMR, and FTIR can be used to profile the impact of flooding on thousands of metabolites simultaneously in huge, complex data sets. Amalgamation of metabolomics with proteomics and genomics has the potential to profile a more complete picture of the crop. Metabolomics in regard to flooding stress is at a nascent stage and much progress is needed on this subject. However, challenges are to deeply understand, interpret, and finally apply the complete knowledge generated from metabolomic research on plant systems for enhancing flood tolerance. The idea of bringing integrated metabolomics will definitely be useful for identifying genes, their functions, and biosynthesis of targeted metabolites.

The future outlook could be the integration of genomics, proteomics, metabolomics, and transcriptomics via integrative pathway analysis or integrative phytochemical array concept that can facilitate the contextualization of probable biomarkers related to flood tolerance. Further research is required to validate the utility of these potential biomarkers participating in flooding stress. Although there are certain limitations, we still believe that many exciting, thrilling, and electrifying developments are to be expected in the coming years, which will subdue the effect of flooding stress on agriculturally important crops.

ACKNOWLEDGMENTS

Authors are thankful to the Director, IIT Delhi, and DST New Delhi for their cooperation and support.

REFERENCES

Arbona, V., Flors, V., Jacas, J., García-Agustín, P., Gómez-Cadenas, A., 2003. Enzymatic and non-enzymatic antioxidant responses of *Carrizo citrange*, a salt-sensitive citrus rootstock, to different levels of salinity. Plant Cell Physiol. 44, 388–394.

Arbona, V., Hossain, Z., López-Climent, M.F., Pérez-Clemente, R.M., Gómez-Cadenas, A., 2008. Antioxidant enzymatic activity is linked to waterlogging stress tolerance in citrus. Physiol. Plant 132, 452–466.

Arbona, V., Manzi, M., Ollas, C.D., Gómez-Cadenas, A., 2013. Metabolomics as a tool to investigate abiotic stress tolerance in plants. Int. J. Mol. Sci. 14, 4885–4911.

Arbona, V., Manzi, M., Zandalinas, S.I., Vives-Peris, V., Pérez-Clemente, R.M., Gómez-Cadenas, A., 2017. Physiological, metabolic, and molecular responses of plants to abiotic stress. In: Stress Signaling in Plants: Genomics and Proteomics Perspective. Springer, pp. 1–35.

Arora, N.K., Tewari, S., Singh, S., Lal, N., Maheshwari, D.K., 2012. PGPR for protection of plant health under saline conditions. In: Maheshwari, D.K. (Ed.), Bacteria in Agrobiology: Stress Management. Springer, Berlin Heidelberg, pp. 239–258.

Baena-González, E., Rolland, F., Thevelein, J.M., Sheen, J., 2007. A central integrator of transcription networks in plant stress and energy signalling. Nature 448, 938–943.

Bailey-Serres, J., Chang, R., 2005. Sensing and signaling in response to oxygen deprivation in plants and other organisms. Ann. Bot. (Lond) 96, 507–518.

Bailey-Serres, J., Voesenek, L.A.C.J., 2008. Flooding stress: acclimations and genetic diversity. Annu. Rev. Plant Biol. 59, 313–339.

Bailey-Serres, J., Voesenek, L.A.C.J., 2010. Life in the balance: a signaling network controlling survival of flooding. Curr. Opin. Plant Biol. 13, 489–494.

Banti, V., Mafessoni, F., Loreti, E., Alpi, A., Perata, P., 2010. The heat-inducible transcription factor HsfA2 enhances anoxia tolerance in *Arabidopsis*. Plant Physiol. 152, 1471–1483.

Barrett, S.C., Eckert, C.G., Husband, B.C., 1993. Evolutionary processes in aquatic plant populations. Aquat. Bot. 44, 105–145.

Baxter-Burrell, A., Yang, Z., Springer, P.S., Bailey-Serres, J., 2002. RopGAP4-dependent Rop-GTPase rheostat control of *Arabidopsis* oxygen deprivation tolerance. Science 296, 2026–2028.

Beck, E.H., Fettig, S., Knake, C., Hartig, K., Bhattarai, T., 2007. Specific and unspecific responses of plants to cold and drought stress. J. Biosci. 32, 501–510.

Bender, B., Capellesso, E.S., Lottici, M.E., Sentkovski, J., Mielniczki-Pereira, A.A., Rosa, L.M.G., Sausen, T.L., 2016. Growth responses and accumulation of soluble sugars in *Inga marginata* Wild. (Fabaceae) subjected to flooding under contrasting light conditions. Braz. J. Biol. https://doi.org/10.1590/1519–6984.11315.

Biemelt, S., Keetman, U., Mock, H.P., Grimm, B., 2000. Expression and activity of isoenzymes of superoxide dismutase in wheat roots in response to hypoxia and anoxia. Plant Cell Environ. 23, 135–144.

Bino, R.J., Hall, R.D., Fiehn, O., Kopka, J., Saito, K., Draper, J., Nikolau, B.J., Mendes, P., Roessner-Tunali, U., Beale, M.H., Trethewey, R.N., Lange, B.M., Wurtele, E.S., Sumner, L.W., 2004. Potential of metabolomics as a functional genomics tool. Trends Plant Sci. 9, 418–425.

Bita, C.E., Greats, T., 2013. Plant tolerance to high temperature in a changing environment: scientific fundamentals and production of heat stress-tolerant crops. Front. Plant Sci. 4, 273.

Blokhina, O., Virolainen, E., Fagerstedt, K.V., 2003. Antioxidants, oxidative damage and oxygen deprivation stress: a review. Ann. Bot. 91, 179–194.

Caballero, J.L., Verderzco, C.V., Galam, J., Jimenez, E.S.D., 2005. Proline accumulation as a symptom of water stress in maize: a tissue differentiation requirement. J. Exp. Bot. 39, 889–897.

Cevallos-Cevallos, J.M., Rouseff, R., Reyes-De-Corcuera, J.I., 2009. Untargeted metabolite analysis of healthy and Huanglongbing-infected orange leaves by CE-DAD. Electrophoresis 30, 1240–1247.

Chanu, W.S., Sarangthem, K., 2015. Changes in proline accumulation, amino acid, sugar and chlorophyll content in leaf and culm of Phourel-amubi, a rice cultivar of Manipur in response to flash flood. Ind. J. Plant Physiol. 20, 10–13.

Chen, C.T., Kao, C.H., 1993. Osmotic stress and water stress have opposite effects on putrescine and proline production in excised rice leaves. Plant Growth Regul. 13, 197–202.

Chen, Y., Hoehenwarter, W., 2015. Changes in the phosphoproteome and metabolome link early signaling events to rearrangement of photosynthesis and central metabolism in salinity and oxidative stress response in *Arabidopsis*. Plant Physiol. 01486.

Chiwocha, S.D.S., Cutler, A.J., Abrams, S.R., Ambrose, S.J., Yang, J., Ross, A.R.S., Kermode, A.R., 2005. The etr1-2 mutation in *Arabidopsis thaliana* affects the abscisic acid, auxin, cytokinin and gibberellin metabolic pathways during maintenance of seed dormancy, moist-chilling and germination. Plant J. 42, 35–48.

Cho, Y.H., Hong, J.W., Kim, E.C., Yoo, S.D., 2012. Regulatory functions of SnRK1 in stress-responsive gene expression and in plant growth and development. Plant Physiol. 158, 1955–1964.

Colmer, T.D., 2003. Long-distance transport of gases in plants: a perspective on internal aeration and radial oxygen loss from roots. Plant Cell Environ. 26, 17–36.

Colmer, T.D., Voesenek, L.A.C.J., 2009. Flooding tolerance: suites of plant traits in variable environments. Funct. Plant Biol. 36, 665–681.

Coudert, Y., Périn, C., Courtois, B., Khong, N.G., Gantet, P., 2010. Genetic control of root development in rice, the model cereal. Trends Plant Sci. 15, 219–226.

Cox, M.C.H., Peeters, A.J.M., Voesenek, L.A.C.J., 2006. The stimulating effects of ethylene and auxin on petiole elongation and on hyponastic curvature are independent processes in submerged *Rumex palustris*. Plant Cell Environ. 29, 282–290.

Cramer, G.R., Urano, K., Delrot, S., Pezzotti, M., Shinozaki, K., 2011. Effects of abiotic stress on plants: a systems biology perspective. BMC Plant Biol. 11, 163.

Desbrosses, G., Steinhauser, D., Kopka, J., 2005. Metabolome analysis using GC-MS. In: Lotus japonicus Handbook. Springer, Netherlands, pp. 165–174.

Dubey, N.K. (Ed.), 2014. Plants as a Source of Natural Antioxidants. CABI.

Eigenbrod, F., Gonzalez, P., Dash, J., Steyl, I., 2014. Vulnerability of ecosystems to climate change moderated by habitat intactness. Glob. Change Biol. 21, 275–286.

Fernie, A.R., Trethewey, R.N., Krotzky, A.J., Willmitzer, L., 2004. Metabolite profiling: from diagnostics to systems biology. Nat. Rev. Mol. Cell Biol. 5, 763–769.

Ferner, E., Rennenberg, H., Kreuzwieser, J., 2012. Effect of flooding on C metabolism of flood-tolerant (*Quercus robur*) and non-tolerant (*Fagus sylvatica*) tree species. Tree Physiol. 32, 135–145.

Fiehn, O., Kopka, J., Dormann, P., Altmann, T., Trethewey, R.N., Willmitzer, L., 2000. Metabolite profiling for plant functional genomics. Nat. Biotechnol. 18, 1157–1161.

Figueroa, C.M., Lunn, J.E., 2016. A tale of two sugars: trehalose 6-phosphate and sucrose. Plant Physiol. 172, 7–27.

Fukao, T., Bailey-Serres, J., 2008. Submergence tolerance conferred by Sub1A is mediated by SLR1 and SLRL1 restriction of gibberellin responses in rice. Proc. Natl. Acad. Sci. USA. 105, 16814–16819.

Giavalisco, P., Köhl, K., Hummel, J., Seiwert, B., Willmitzer, L., 2009. 13C isotope-labeled metabolomes allowing for improved compound annotation and relative quantification in liquid chromatography-mass spectrometry based metabolomic research. Anal. Chem. 81, 6546–6551.

Gibbs, D.J., Lee, S.C., Isa, N.M., Gramuglia, S., Fukao, T., Bassel, G.W., Correia, C.S., Corbineau, F., Theodoulou, F.L., Bailey-Serres, J., Holdsworth, M.J., 2011. Homeostatic response to hypoxia is regulated by the N-end rule pathway in plants. Nature 479, 415–418.

Gibbs, J., Greenway, H., 2003. Mechanisms of anoxia tolerance in plants. I. Growth, survival and anaerobic catabolism. Funct. Plant Biol. 30, 1–47.

Glick, B.R., 2014. Bacteria with ACC deaminase can promote plant growth and help to feed the world. Microbiol. Res. 169, 30–39.

Gomez-Casati, D.F., Zanor, M.I., Busi, M.V., 2013. Metabolomics in plants and humans: applications in the prevention and diagnosis of diseases. Biomed. Res Int. 6, 2013.

Goodwin, S., McPherson, J.D., McCombie, W.R., 2016. Coming of age: ten years of next-generation sequencing technologies. Nat. Rev. Gene 17, 333–351.

Hao, X.Y., Han, X., Ju, H., Lin, E.D., 2010. Impact of climatic change on soybean production: a review. Ying Yong Sheng Tai Xue Bao 21, 2697–2706.

Hsu, F.C., Chou, M.Y., Peng, H.P., Chou, S.J., Shih, M.C., 2011. Insights into hypoxic systemic responses based on analyses of transcriptional regulation in *Arabidopsis*. PLoS One 6 (12), e28888.

Huynh, L.N., VanToai, T., Streeter, J., Banowetz, G., 2005. Regulation of flooding tolerance of SAG12: ipt *Arabidopsis* plants by cytokinin. J. Exp. Botany 56, 1397–1407.

IPCC, 2001. Climate Change 2001 the Scientific Basis. Contribution of Working Group I to the Third Assessment Report of the Intergovernmental Panel on Climate Change. Cambridge University Press, United Kingdom and New York, NY, USA.

IPCC, 2007. In: Solomon, S.D., Qin, D., Manning, M., et al. (Eds.), Climate Change 2007: The Physical Science Basis. Contribution of Working Group I to the Forth Assessment Report of the Intergovernmental Panel on Climate Change. Cambridge University Press, Cambridge, United Kingdom and New York, NY, USA.

Iqbal, N., Nazar, R., Khan, N.A., 2016. In: Osmolytes and Plants Acclimation to Changing Environment: Emerging Omics Technologies. Springer.

Jackson, M.B., 2002. Long-distance signalling from roots to shoots assessed: the flooding story. J. Exp. Bot. 53, 175–181.

Jackson, M.B., Colmer, T.D., 2005. Response and adaptation by plants to flooding stress. Ann. Bot. 96, 501–505.

Jackson, M.B., Ishizawa, K., Ito, O., 2009. Evolution and mechanisms of plant tolerance to flooding stress. Ann. Bot. (Lond) 103, 137–142.

Kerchev, P.I., Fenton, B., Foyer, C.H., Hancock, R.D., 2012. Plant responses to insect herbivory: interactions between photosynthesis, reactive oxygen species and hormonal signalling pathways. Plant Cell Environ. 35, 441–453.

Kim, J.K., Bamba, T., Harada, K., Fukusaki, E., Kobayashi, A., 2007. Time-course metabolic profiling in *Arabidopsis thaliana* cell cultures after salt stress treatment. J. Exp. Bot. 58, 415–424.

Kinnersley, A.M., Turano, F.J., 2000. Gamma aminobutyric acid (GABA) and plant responses to stress. Crit. Rev. Plant Sci. 19, 479–509.

Krasensky, J., Jonak, C., 2012. Drought, salt, and temperature stress-induced metabolic rearrangements and regulatory networks. J. Exp. Bot. 63, 1593–1608.

Kumutha, D., Ezhilmathi, K., Sairam, R.K., Srivastava, G.C., Deshmukh, P.S., Meena, R.C., 2009. Waterlogging induced oxidative stress and antioxidant activity in pigeon pea genotypes. Biol. Plant. 53, 75–84.

Lekshmy, S., JHa, S.K., Sairam, R.K., 2015. Physiological and molecular mechanisms of flooding tolerance in plants. In: Elucidation of Abiotic Stress Signaling in Plants. Springer, New York, pp. 227–242.

Licausi, F., Kosmacz, M., Weits, D.A., Giuntoli, B., Giorgi, F.M., Voesenek, L.A.C.J., Perata, P., van Dongen, J.T., 2011. Oxygen sensing in plants is mediated by an N-end rule pathway for protein destabilisation. Nature 479, 419–422.

Manavalan, L.P., Guttikonda, S.K., Tran, L.S., Nguyen, H.T., 2009. Physiological and molecular approaches to improve drought resistance in soybean. Plant Cell Physiol. 50, 1260–1276.

McManmon, M., Crawford, R.M., 1971. A metabolic theory of flooding tolerance: the significance of enzyme distribution and behaviour. New Phytol. 70, 299–306.

Meena, K.K., Sorty, A.M., Bitla, U.M., Choudhary, K., Gupta, P., Pareek, A., Singh, D.P., Prabha, R., Sahu, P.K., Gupta, V.K., Singh, H.B., 2017. Abiotic stress responses and microbe-mediated mitigation in plants: the omics strategies. Front. Plant Sci. 8, 172.

Moco, S., Bino, R.J., Vorst, O., Verhoeven, H.A., de Groot, J., van Beek, T.A., Vervoort, J., De Vos, C.R., 2006. A liquid chromatography-mass spectrometry-based metabolome database for tomato. Plant Physiol. 141, 1205–1218.

Mollard, F.P., Striker, G.G., Ploschuk, E.L., Insausti, P., 2010. Subtle topographical differences along a flood plain promote different plant strategies among *Paspalum dilatatum* subspecies and populations. Austral Ecol. 35, 189–196.

Monk, L.S., Fagerstedt, K.V., Crawford, R.M., 1987. Superoxide dismutase as an anaerobic polypeptide a key factor in recovery from oxygen deprivation in *Iris pseudacorus*? Plant Physiol. 85, 1016–1020.

Mosa, K.A., Ismail, A., Helmy, M., 2017. Plant Stress Tolerance: An Integrated Omics Approaches. Springer, pp. 1–19.

Munns, R., Tester, M., 2008. Mechanisms of salinity tolerance. Annu. Rev. Plant Biol. 59, 651–681.

Mustroph, A., Barding Jr., G.A., Kaiser, K.A., Larive, C.K., Bailey-Serres, J., 2014. Characterization of distinct root and shoot responses to low-oxygen stress in *Arabidopsis* with a focus on primary C- and N-metabolism. Plant Cell Environ. 37, 2366–2380.

Neeraja, C.N., Maghirang-Rodriguez, R., Pamplona, A., Heuer, S., Collard, B.C., Septiningsih, E.M., Vergara, G., Sanchez, D., Xu, K., Ismail, A.M., Mackill, D.J., 2007. A marker-assisted backcross approach for developing submergence-tolerant rice cultivars. Theor. Appl. Genet. 115, 767–776.

Normile, D., 2008. Reinventing rice to feed the world. Sci 321, 330–333.

Obata, T., Fernie, A.R., 2012. The use of metabolomics to dissect plant responses to abiotic stresses. Cell Mol. Life Sci. 69, 3225–3243.

Okazaki, Y., Saito, K., 2016. Integrated metabolomics and phytochemical genomics approaches for studies on rice. Giga Sci. 5, 1.

Oliver, S.G., 1996. From DNA sequence to biological function. Nature 379, 597–600.

Ou, L.J., Dai, X.Z., Zhang, Z.Q., Zou, X.X., 2011. Responses of pepper to waterlogging stress. Photosynthetic 49, 339–345.

Parvaiz, A., Satyawati, S., 2008. Salt stress and phyto biochemical responses of plants. Plant Soil Environ. 54, 89–99.

Patel, P.K., Singh, A.K., Tripathi, N., Yadav, D., Hemantaranjan, A., 2014. Flooding: abiotic constraint limiting vegetable productivity. Adv. Plants Agric. Res. 1, 00016.

Peng, S., Huang, J., Sheehy, J.E., Laja, R.C., Visperas, R.M., Zhong, X., Centeno, G.S., Khush, G.S., Cassman, 2004. Rice yields decline with higher night temperature from global warming. Proc. Nat. Acad. Sci. USA 101, 9971–9975.

Perata, P., William, A., Laurentius, A.C.J.V., 2011. Plant and flooding stress. New Phytol. https://doi.org/10.1111/j.1469–8137.2011.03702.

Pierik, R., De Wit, M., Voesenek, L.A.C.J., 2011. Growth mediated stress escape: convergence of signal transduction pathways activated upon exposure to two different environmental stresses. New Phytol. 189, 122–134.

Pucciariello, C., Parlanti, S., Banti, V., Novi, G., Perata, P., 2012. Reactive oxygen species-driven transcription in Arabidopsis under oxygen deprivation. Plant Physiol. 159, 184–196.

Puyang, X., An, M., Xu, L., Han, L., Zhang, X., 2015. Antioxidant responses to waterlogging stress and subsequent recovery in two Kentucky bluegrass (*Poa pratensis* L.) cultivars. Acta Physiol. Plant. 37, 197.

Qin, X., Li, F., Chen, X., Xie, Y., 2013. Growth responses and non-structural carbohydrates in three wetland macrophyte species following submergence and de-submergence. Acta Physiol. Plant. 35, 2069–2207.

Rodziewicz, P., Swarcewicz, B., Chmielewska, K., Wojakowska, A., Stobiecki, M., 2014. Influence of abiotic stresses on plant proteome and metabolome changes. Acta Physiol. Plant. 36, 1–19.

Sairam, R.K., Dharmar, K., Chinnusamy, V., Lekshmy, S., Joshi, R., Bhattacharya, P., 2011. NADPH oxidase as the source of ROS produced under waterlogging in roots of mung bean. Biol. Plant. 55, 741–746.

Sairam, R.K., Kumutha, D., Ezhilmathi, K., 2009. Waterlogging tolerance: nonsymbiotic haemoglobin–nitric oxide homeostasis and antioxidants. Curr. Sci. 10, 674–682.

Sairam, R.K., Kumutha, D., Ezhilmathi, K., Deshmukh, P.S., Srivastava, G.C., 2008. Physiology and biochemistry of waterlogging tolerance in plants. Biol. Plant. 52, 401–412.

Salekdeh, G.H., Komatsu, S., 2007. Crop proteomics: aim at sustainable agriculture of tomorrow. Proteomics 7, 2976–2996.

Sasidharan, R., Voesenek, L.A., 2015. Ethylene-mediated acclimations to flooding stress. Plant Physiol. 169, 3–12.

Sauter, M., 2013. Root responses to flooding. Curr. Opin. Plant Biol. 16, 282–286.

Schröder, F., Lisso, J., Müssig, C., 2011. EXORDIUM-LIKE1 promotes growth during low carbon availability in *Arabidopsis*. Plant Physiol. 156, 1620–1630.

Segarra, G., Jáuregui, O., Casanova, E., Trillas, I., 2006. Simultaneous quantitative LC-ESI-MS/MS analyses of salicylic acid and jasmonic acid in crude extracts of I under biotic stress. Phytochemistry 67, 395–401.

Setter, T.L., Waters, I., 2003. Review of prospects for germplasm improvement for waterlogging tolerance in wheat, barley and oats. Plant Soil 253, 1–34.

Shabala, S. (Ed.), 2017. Plant Stress Physiology CABI.

Sharp, R.E., Poroyko, V., Hejlek, L.G., Spollen, W.G., Springer, G.K., Bohnert, H.J., Nguyen, H.T., 2004. Root growth maintenance during water deficits: physiol functional genomics. J. Exp. Bot. 55, 2343–2351.

Shimamura, S., Yoshida, S., Mochizuki, T., 2007. Cortical aerenchyma formation in hypocotyl and adventitious roots of *Luffa cylindrica* subjected to soil flooding. Ann. Bot. 100, 1431–1439.

Singh, A., Carandang, J., Gonzaga, Z.J., Collard, B.C., Ismail, A.M., Septiningsih, E.M., 2017. Identification of QTLs for yield and agronomic traits in rice under stagnant flooding conditions. Rice 10, 15.

Smirnoff, N., 1993. The role of active oxygen in the response of plants to water deficit and desiccation. New Phytol. 125, 27–58.

Steffens, B., 2014. The role of ethylene and ROS in salinity, heavy metal, and flooding responses in rice. Front. Plant Sci. 5.

Striker, G.G., Colmer, T.D., 2016. Flooding tolerance of forage legumes. J. Exp. Bot. 68, 1851–1872.

Su, P.H., Su, C.H., 1996. Metabolic responses of luffa roots to long-term flooding. J. Plant Physiol. 148, 735–740.

Tewari, S., Arora, N.K., 2013. Plant growth promoting rhizobacteria for ameliorating abiotic stresses triggered due to climatic variability. Clim. Change Environ. Sustain 1, 95–103.

Tewari, S., Arora, N.K., 2016. Soybean Production under flooding stress and its mitigation using plant growth promoting microbes. Environ. Stresses Soybean Prod. 14, 2–23.

Tewari, S., Arora, N.K., Miransari, M., 2016. Plant growth promoting rhizobacteria to alleviate soybean growth under abiotic and biotic stresses. In: Miransari, M. (Ed.), Abiotic and Biotic Stresses in Soybean Production. Academic Press, Elsevier, USA, pp. 131–156.

Thirunavukkarasu, N., Hossain, F., Mohan, S., Shiriga, K., Mittal, S., Sharma, R., Singh, R.K., Gupta, H.S., 2013. Genome-wide expression of transcriptomes and their co-expression pattern in subtropical maize (*Zea mays* L.) under waterlogging stress. PLoS One 8, e70433.

Toojinda, T., Siangliw, M., Tragoonrung, S., Vanavichit, A., 2003. Molecular genetics of submergence tolerance in rice: QTL analysis of key traits. Ann. Bot. 91, 243–253.

Tuberosa, R., Salvi, S., 2006. Genomics-based approaches to improve drought tolerance of crops. Trends Plant Sci. 11, 405–412.

Tunc-Ozdemir, M., Jones, A.M., 2017. BRL3 and AtRGS1 cooperate to fine tune growth inhibition and ROS activation. PLoS One 18 (5), 12 e0177400.

Ushimaru, T., Maki, Y., Sano, S., Koshiba, K., Asada, K., Tsuji, H., 1997. Induction of enzymes involved in the ascorbate-dependent antioxidative system, namely, ascorbate peroxidase, monodehydroascorbate reductase and dehydroascorbate reductase, after exposure to air of rice (*Oryza sativa*) seedlings germinated under water. Plant Cell Physiol. 38, 541–549.

Valerio, C., Costa, A., Marri, L., Issakidis-Bourguet, E., Pupillo, P., Trost, P., Sparla, F., 2011. Thioredoxin-regulated beta-amylase (BAM1) triggers diurnal starch degradation in guard cells, and in mesophyll cells under osmotic stress. J. Exp. Bot. 62, 545–555.

van Veen, H., Mustroph, A., Barding, G.A., Vergeer-van Eijk, M., Welschen-Evertman, R.A.M., Pedersen, O., Visser, E.J.W., Larive, C.K., Pierik, R., Bailey-Serres, J., Voesenek, L.A.C.J., Sasidharan, R., 2013. Two *Rumex* species from contrasting hydrological niches regulate flooding tolerance through distinct mechanisms. Plant Cell 25, 4691–4707.

Voesenek, L.A., Bailey-Serres, J., 2015. Flood adaptive traits and processes: an overview. New Phytol. 206, 57–73.

Voesenek, L.A.C.J., Bailey-Serres, J., 2013. Flooding tolerance: O_2 sensing and survival strategies. Curr. Opin. Plant Biol. 16, 647–653.

Wagner, C., Sefkow, M., Kopka, J., 2003. Construction and application of a mass spectral and retention time index database generated from plant GC/EI-TOF-MS metabolite profiles. Phytochemistry 62, 887–900.

Wang, X., Khodadadi, E., Fakheri, B., Komatsu, S., 2017. Organ-specific proteomics of soybean seedlings under flooding and drought stresses. J. Proteom. 162, 62–72.

Wangkhem, S.C., Kananbala, S., 2014. Changes in proline accumulation, amino acids, sugar and chlorophyll content in leaf and culm of phourel amubi, a rice cultivar of Manipur in response to flash flood. Indian J. Plant Physiol. 20, 10–13.

Wolf, S., Schmidt, S., Müller-Hannemann, M., Neumann, S., 2010. In silico fragmentation for computer assisted identification of metabolite mass spectra. BMC Bioinformatics 11, 148.

Wu, J., Zhao, H.B., Yu, D., Xu, X., 2017. Transcriptome profiling of the floating-leaved aquatic plant *Nymphoides peltata* in response to flooding stress. BMC Genomics 18, 119.

Yan, B., Da, Q., Liu, X., Huang, S., Wang, Z., 1996. Flooding induced membrane damage, lipid oxidation and activated oxygen generation in corn leaves. Plant Soil 179, 261–268.

Yiu, J.C., Liu, C.W., Kuo, C.T., Tseng, M.J., Lai, Y.S., Lai, W.J., 2008. Changes in antioxidant properties and their relationship to paclobutrazol-induced flooding tolerance in Welsh onion. J. Sci. Food Agri. 88, 1222–1230.

Yu, Y., Huang, R., 2017. Integration of ethylene and light signaling affects hypocotyl growth in *Arabidopsis*. Front. Plant Sci. https://doi.org/10.3389/fpls.2017.00057.

Yuan, J.S., Galbraith, D.W., Dai, S.Y., Griffin, P., Stewart Jr., C.N., 2008. Plant systems biology comes of age. Trends Plant Sci. 13, 165–171.

Zhang, Q., Huber, H., Beljaars, S.J., Birnbaum, D., de Best, S., de Kroon, H., Visser, E.J., 2017. Benefits of flooding-induced aquatic adventitious roots depend on the duration of submergence: linking plant performance to root functioning. Ann. Bot. 5 mcx049.

Zou, X., Jiang, Y., Liu, L., Zhang, Z., Zheng, Y., 2010. Identification of transcriptome induced in roots of maize seedlings at the late stage of waterlogging. BMC Plant Biol. 10, 189.

Chapter 19

Small Heat Shock Proteins: Structural Assembly and Functional Responses Against Heat Stress in Plants

Aditya Banerjee, Aryadeep Roychoudhury
St. Xavier's College (Autonomous), Kolkata, India

1. INTRODUCTION

Temperature stress directly affects plant metabolism and can lead to worldwide large-scale crop losses (Banerjee et al., 2016). The basic physiology affected by heat stress includes the cell membrane fluidity, photosynthesis, respiration, and water relations. Heat stress also critically hampers the activities of every participating enzyme (Banerjee and Roychoudhury, 2016a,b). The different mechanisms evolved in plants to cope up with heat stress include the maintenance of cell membrane stability, elimination of the reactive oxygen species (ROS), synthesis of antioxidants, accumulation and regulation of osmoticum, induction of kinases that respond to stress, Ca-dependent kinase proteins, and enhancing the transcription and signal transfer of chaperones (Stief et al., 2014; Banerjee and Roychoudhury, 2017).

Protein aggregation is a major threat to cell survival, especially in the crowded environment of the cell. Evolution has designed proteins to be marginally stable and hence there exists a continuous risk of unfolded proteins getting aggregated under such circumstances. The cellular machinery always maintains equilibrium between the already folded and unfolded states of the proteins. Application of heat stress abruptly shifts this equilibrium in favor of the unfolded proteins. This results in protein aggregation, which is toxic for cell survival. To prevent such damages, plants have developed an extensive protein quality control system governed by molecular chaperones (Hartl et al., 2011), which actively maintain the native folded state of the aggregation-prone proteins.

Chaperones can exhibit multiple functions based on their molecular and structural nature. They may aid in folding of newly synthesized proteins or may refold partially unfolded or misfolded proteins usually utilizing molecules of ATP, the cellular energy currency. Other chaperones play an important role in the ubiquitinylation of unfolded proteins, by directing them to the degradation machinery or permitting controlled aggregation (Mayer, 2010). Heat shock proteins (HSPs) are a class of such molecular chaperones. Their nomenclature can be credited to their expression being induced during heat stress. However, modern findings have also reported their upregulation in other stress responses and interestingly, many HSPs are constitutively expressed housekeeping proteins as well (Lambert et al., 2011; Eisenhardt, 2013).

Several classes of HSPs have been reported in plants of which small HSPs (sHSPs) are abundantly found in plants. The sHSPs bind to their target proteins and hold them in a refolding competent state, thus preventing their aggregation. However, sHSPs cannot induce active refolding and so do not require ATP expenditure. The term "small" indicates the monomeric size of these proteins ranging between 12 and 42 kDa, although many sHSPs can form large oligomers with up to 40–50 subunits (Basha et al., 2012). This heterogeneity in macromolecular structure can be attributed to the vast structural diversity of sHSPs in terms of substrate binding, making them difficult to be technically studied (Dong et al., 2015). The sHSPs have also been found in the red alga, *Cyanidioschyzon merolae*, and the green alga, *Chlamydomonas reinhardtii*, where they confer tolerance even against heat shocks at 63°C (Kobayashi et al., 2014). Interestingly, viruses such as cyanophages also express sHSPs, which form a monophyletic clade closer to the bacterial class (Maaroufi and Tanguay, 2013). Several reports have been published regarding the various roles of plant HSPs such as those belonging to the Hsp60 and Hsp70 families. Compared with those, the regulatory functions of

Plant Metabolites and Regulation Under Environmental Stress. http://dx.doi.org/10.1016/B978-0-12-812689-9.00019-4

plant sHSPs in heat stress are still poorly understood. These proteins play crucial roles in developing tolerance against multiple abiotic stresses as well as in generating resistance against pathogen infections such as bacterial leaf streak in rice (Ju et al., 2017). In this chapter, we have reviewed the literary data available regarding these chaperone proteins, which improve plant tolerance against heat stress.

2. SMALL HEAT SHOCK PROTEINS: STRUCTURAL SIGNIFICANCE

The sHSPs comprise an evolutionarily ancient superfamily because of their presence in almost all organisms (Haslbeck and Vierling, 2015). The structurally best known sHSPs are the α-crystallins, αA- and αB-crystallin (also designated as HspB4 and HspB5). These are the most prominent structural proteins in the vertebrate eye lens. Earlier, it was conceptualized that α-crystallins were limited exclusively to lens functioning, until their expression was reported in even other tissues where they acted as molecular chaperones. Overwhelming evidences now show that sHSPs are a ubiquitous class of players in the protein quality control system in cells (Eyles and Gierasch, 2010). The sHSP activities have been speculated to be further more than chaperone activity alone (Basha et al., 2012).

2.1 Structural Evidences of Dimeric Small Heat Shock Proteins: Building Blocks

All members of the sHSP superfamily are structurally characterized by the α-crystallin domain of the human αA- and αB-crystallins. This central domain consists of around 90 amino acids and forms an immunoglobulin domain supported by a β-sandwich (made of seven β-strands). The N-terminal region is highly variable among sHSPs. The C-terminal to the central α-crystallin domain is called the C-terminal extension. This region is also highly diverse except for a conserved I/V/L-X-I/V/L motif, which is obligatory for forming higher order oligomeric assemblies (Rutsdottir et al., 2017).

High-resolution structures have been experimentally prepared using crystallographic techniques, solid-state nuclear magnetic resonance (NMR), small-angle X-ray scattering (SAXS), or a combination. These structures reveal the presence of two α-crystallin domains coming from two monomers in dimeric sHSPs (Stamler et al., 2005). The dimeric sHSPs actually act as building blocks of larger oligomers essential to tackle heat stress–induced protein aggregation. In the Hsp16.9 (a sHSP) crystal structure of *Triticum aestivum*, two α-crystallin domains were bound by a β-strand exchange (Patel et al., 2014). The β-strands of the α-crystallin domain are conventionally named β2–β9, after the first sHsp crystal structure was solved (Wang et al., 2014a). The strand named β6 is located on a loop extending away from the β-sandwich of one monomer, and participates in β-sheet formation by binding strand β7 of the other monomer. The strand β6 of the other monomer interacts again with β7 of the first monomer (Bakthisaran et al., 2015).

2.2 Oligomeric Structures Formed by Small Heat Shock Proteins

Formation of large oligomers is a characteristic feature reported in most of the sHSPs. The dimeric building blocks are held together by several interactive forces. Among these, the most significant is the interaction involving the conserved I/V/L-X-I/V/L motif on the C-terminal extension. The N-terminal region is also believed to play crucial roles in oligomer formation. However, intricate studies such as hydrogen/deuterium exchange and solution-state NMR showed that the N-terminal region is intrinsically disordered (Cheng et al., 2008; Jehle et al., 2011). Thus it is possible that the N-terminal region adopts a molten globule-like structure to support dynamic interactions between dimeric subunits followed by formation of oligomers. Further studies are required to confirm such hypothesis.

The first crystal structure of a sHSP to be decoded was that of Hsp16.5 from *Methanococcus jannaschii*. Since then, every sHSP has been detected with a signature domain called the α-crystallin domain. It forms a β-sandwich, the edge of which is demarcated by the strands β4 and β8. These strands form a hydrophobic groove that is bound by the conserved I/V/L-X-I/V/L motif of a neighboring dimer. This interaction has been a characteristic of all sHSP dimers. However, this interaction has been observed to occur in two opposite orientations among sHSPs. This is possible because of the palindromic nature of the motif (Laganowsky et al., 2010). Actually the large variation of the oligomeric states of sHSPs can be accredited to the flexible C-terminal extension between the α-crystallin domain and the I/V/L-X-I/V/L motif. This allows variable orientations of the dimeric building blocks with respect to each other. It is expected that sHSPs have variable oligomeric states, and a single sHSP can exist as an ensemble of differently sized oligomers. This is difficult to be detected by crystallography, which intrinsically selects for a single or few protein orientations. Alternative techniques such as cryo-electron microscopy (EM), SAXS, and native mass spectrometry (MS) are gaining popularity to structurally characterize the highly polydisperse sHSPs (Mymrikov et al., 2011).

The only two sHSP oligomers that have been successfully crystallized in their native state are the Hsp16.5 from *M. jannaschii* and Hsp16.9 from wheat (Fu et al., 2013). Both of these sHSP oligomers are homogenous in nature. Wheat Hsp16.9 is a dodecamer (12-mer) comprising two stacked hexamers, each made of three dimers. In this structure, the C-terminal extensions are located on the outside of the two hexameric rings. Six of the N-terminal regions are unstructured, but the other six are structured on the inside of the dodecamer and connect dimers from the two different rings. Here, the N-terminal region forms important oligomeric contacts, in addition to the established contacts of the C-terminal extension, binding the hydrophobic groove on the next dimer (Mymrikov et al., 2011; Kumar et al., 2013).

Cryo-EM images were accessed to separate mixed populations of structurally different single particles of Hsp26 from *Saccharomyces cerevisiae*. From these data, the various assembly forms could be distinguished (White et al., 2006). In this study, only the 24-meric form, the most abundant state at room temperature, was reported. However, in reality, yeast HSp26 occupies a range of oligomeric states as has been validated by native MS (Benesch et al., 2010). Similarly, a combination of NMR, native MS, ion mobility, and EM data have provided a real insight into the polydispersity and spatial arrangements of the variable oligomeric structures in human αB-crystallin (Baldwin et al., 2011). Hence, mere crystallographic approaches will not suffice to study the structural significance of dynamic chaperones such as sHSPs. Furthermore, in-solution techniques are required to be adopted to solve the structures of more plant sHSPs. Knowledge in this field is limited only to the wheat Hsp16.9. Popularization of MS methodologies supplemented with nano-electrospray ionization (ESI) and subsequent mass analysis of intact proteins may be promoted to design a useful protocol in the study of sHSP oligomer dynamism (Sharon and Robinson, 2007).

The structure of oligomeric Hsp21 from *Arabidopsis thaliana* was characterized by single particle averaging of negative stain EM images of Hsp21 particles. This yielded a reconstructed image of the dodecamer. The dimeric subunits were reported to be organized in rings of three dimers, two of which were stacked to form the dodecamer. An atomic model based on homology modeling to Hsp16.9 was docked into the density map, which required rotating the two rings with respect to each other. This atomic model exhibited low resolution. The rotated conformation could also indicate some degree of heterogeneity, even though the particles selected for the image reconstruction were of homogeneous size (Lambert et al., 2011).

2.3 The Dynamic Subunit Exchange Model in Small Heat Shock Proteins

The sHSPs in plants exhibit continuous dissociation from and association with larger oligomers. This is a characteristic feature of plant sHSPs and has been referred to as dynamic exchange. This process aids in maintaining the polydispersity among sHSPs by allowing the exchange into various oligomeric transition states. Dynamism has also been noted in the homogenous sHSP oligomers. Several sHSPs have been shown to be capable of heterooligomer formation in vitro. Two sHSPs from different species, Hsp18.1 from *Pisum sativum* (pea) and Hsp16.9 from *T. aestivum* (wheat) were able to form heterooligomers. Native MS was used to measure the rate of subunit exchange among the sHSPs belonging to two different species (Sobott et al., 2002; Wang et al., 2017a).

Separate studies using nano-ESI MS showed real-time analyses of subunit exchange between related sHSPs from *A. thaliana*, Hsp17.6 and Hsp18.1 (Painter et al., 2008). The exchanging subunits in this case were the dimers. However, in the previous example of exchange between pea Hsp18.1 and wheat Hsp16.9, the heterooligomers are composed of odd numbers of sHSPs. This implied participation of even monomeric subunit exchange (Painter et al., 2008; Sobott et al., 2002). These two cases clearly highlight the variation in dynamic exchange that exists among plant sHSPs to yield the huge structural diversity among the proteins.

The dynamic subunit exchange rate for sHSPs (0.04–0.40 min^{-1}) has been reported to be much faster than that of other proteins, which is approximately $4.5 \times 10^{-4} min^{-1}$. This dynamics increases for sHSPs with the increase in temperature, thus suggesting its importance in sHSP function during heat stress (Basha et al., 2012). The exclusive dodecameric structure of *P. sativum* Hsp18.1 undergoes a rapid shift to monomeric, dimeric, and higher oligomeric forms during temperature increases. Such heterogeneous population for the same sHSP must be beneficial for heat shock responses in the plant. The dynamics and polydispersity of sHSPs are integral to chaperone activity. This has been shown in the presence of a substrate protein where a diverse range of differently sized sHSP-substrate complexes with varying stoichiometries have been identified (Stengel et al., 2010). Plant sHSPs structurally have been less investigated. Further information in this regard is required to completely undermine the motifs and domains responsible for such dynamic exchange that acts as the root cause of heterogeneous oligomers in sHSPs (Fig. 19.1). For the human αB-crystallin, NMR and nano-ESI MS experiments revealed that the dissociation events of the C-terminal-extension-to-hydrophobic-groove interaction guide the dynamics of the polydisperse ensemble of αB-crystallin oligomers (Baldwin et al., 2011; Ahrman et al., 2007).

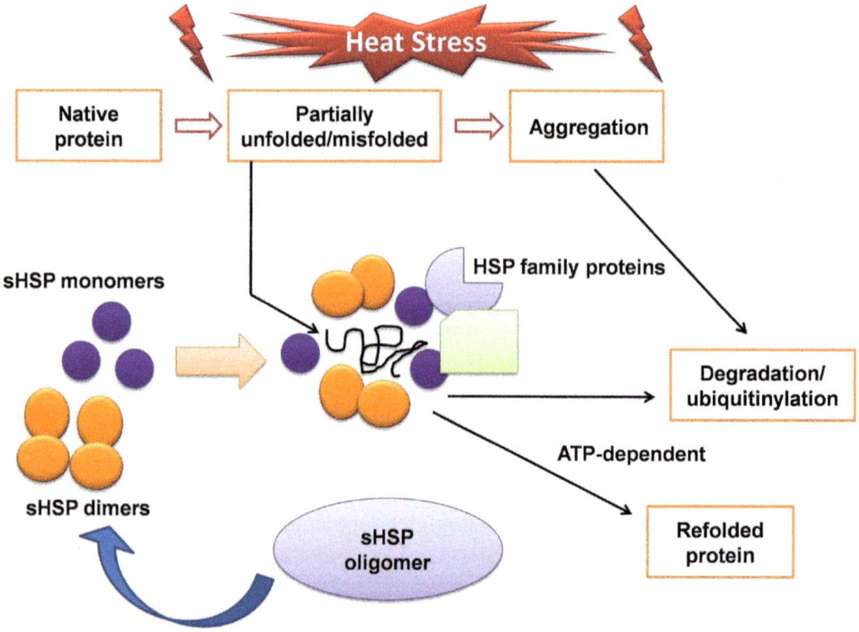

FIGURE 19.1 **The structural and functional significance of small heat shock proteins (sHSPs) during heat stress.** Elevated temperature stress in plants results in partially unfolded/misfolded proteins, which cause aggregation and cytotoxicity. Such unfolded/misfolded proteins are attended by the monomers and dimers of sHSPs, formed from the sHSP oligomers as a heat shock response. The sHSPs stabilize the candidate substrate proteins and promote their refolding with the help of other proteins of the HSP family in an ATP-dependent fashion. Sometimes, the substrate proteins may attain misfolding to such an extent that they cannot be refolded to their native structure. In such cases, these are channelized to the proteasomal machinery where they are ultimately degraded.

3. FUNCTIONAL ASPECTS OF SMALL HEAT SHOCK PROTEINS: CHAPERONE ACTIVITIES

Chaperone activity of a candidate protein is assayed using model substrates (proteins) that usually unfold upon heat stress. Such experiments are performed in vitro. If the candidate sHSP really does have a chaperone activity, it should be able to keep the model substrates soluble even after an increase in temperature. However, in the absence of sHSPs, the substrate protein unfolds and aggregates, which can be detected spectrophotometrically by centrifugation or by gel electrophoresis and subsequent staining (Park et al., 2014). The model substrate proteins, such as citrate synthase (CS), luciferase, and malate dehydrogenase (MDH), that are used are mostly thermosensitive. Fluorescently labeled T4 lysozyme variants with temperature-induced destabilizing mutations have been developed to study the sHSP interactions with a predominantly folded protein (Santhanagopalan et al., 2015). Study of Hsp17 from *Caenorhabditis elegans* showed that the protein existed as a 24-subunit spherical oligomer with tetrahedral symmetry and formed large sheet-like supramolecular assemblies at high temperatures (45–60°C) to exhibit chaperone activities (Zhang et al., 2015).

The intricate molecular mechanisms of sHSP–substrate protein interaction in plants are still largely unknown (Li et al., 2015). The most rational explanation is that sHSPs bind to the large number of hydrophobic surfaces of thermosensitive proteins and stabilize their structures to maintain solubility. For this mechanism, sHSPs should also have hydrophobic surfaces available for interaction with the substrate proteins. Such hydrophobic patches in sHSPs remain buried in the oligomers and become available for substrate binding after the stress-induced dynamic exchanges (McHaourab et al., 2009). The sHSPs prevent nonnative aggregation of the substrate proteins, subsequently facilitating their refolding by ATP-dependent chaperones, such as the DnaK system or ClpB/DnaK. It was reported that Hsp18.1 from *P. sativum* and also Hsp16.6 from *Synechocystis* sp. PCC6803 bind to unfolded and misfolded proteins, further initiating their refolding by the Hsp70/Hsp100 chaperone complexes under in vitro conditions (Wang et al., 2004). ΔClpB1 strain of *Synechocystis* was developed for an in vivo assay of chaperone activity. In this experiment, only the presence of functional Hsp16.6 regulated the thermotolerance mechanism. Some mutations that affected the oligomerization of Hsp16.6 in vitro were identified and introduced in the wild type protein. As a result, the effect of such mutations could be studied in vivo and the link between oligomerization and chaperone activity could be undermined (Giese and Vierling, 2004). Mutants reducing the in vivo thermotolerance of Hsp16.6 were screened and purified to study their oligomeric structure and in vitro chaperone activity. It was inferred that mutations in the α-crystallin domain yielded deformed oligomerization capability and chaperone activity in vitro, whereas mutations in the N-terminal region showed no effect in vitro, thus highlighting the importance of in vivo experiments (Giese et al., 2005).

Substrate recognition by sHSPs may be mediated by the monomeric, dimeric, or oligomeric states and this is highly debated. It has been suggested that dissociation of the oligomers may not be necessary for binding to substrates because of the structural flexibility at both the N- and C-terminal regions. The dynamic exchange model as discussed in the previous section clearly supports the hypothesis that the substrate-binding sites of sHSPs must be continuously exposed. Such continuous exposure should not be supplemented with dissociation into stable suboligomeric sHSP species (Basha et al., 2012). A set of delicate equilibria actually exists between sHSP oligomers, suboligomeric species, the native substrate protein, and the unfolding/misfolding substrate protein (Lambert et al., 2011).

It has been observed that sHSPs recognize substrate proteins at an early stage of misfolding via reversible interactions. Prolonged denaturing conditions promote the formation of large heterogeneous sHSP–substrate complexes, which keep the partially unfolded thermosensitive proteins in a soluble state. The ATP-dependent chaperones then aid the misfolded proteins to fold in the presence of sHSPs (Mogk et al., 2003). Such chaperone activities of sHSPs over thermosensitive proteins remain prevalent even in the postaggregation phase (Ratajczak et al., 2009). The substrate proteins appear to be firmly associated with the dynamic subunits of the sHSPs (Friedrich et al., 2004). Size exclusion chromatography followed by EM was used to study such polydisperse sHSP–substrate complexes, but prevailing heterogeneity has prevented any conclusions from being drawn. Stengel et al. (2010) showed via native MS studies that the pea Hsp18.1 builds up to 300 diverse-sized complexes (varying both in size and stoichiometry) with the substrate luciferase, upon exposure to increasing temperature.

Owing to the immense variability and heterogeneity among sHSPs, no particular sHSP-binding site has been identified for a specific substrate protein. Analogous to the dynamic subunit exchange model, even the sHSP–substrate protein complexes have been regarded to be dynamic in nature. This hypothesis has been supported by the hydrogen/deuterium exchange levels of sHSPs in the absence or presence of the substrate protein, MDH. Thus the flexibility of the N- and C-terminal regions of sHSPs is retained even within the sHSP–substrate complexes (Cheng et al., 2008). In another study with the pea Hsp18.1, the N-terminal region has been considered responsible for substrate binding. In this case, a photoactivable cross-linker was introduced on multiple positions of Hsp18.1. The cross-linking between the N-terminal region and the substrate was more frequently observed than cross-linking with the α-crystallin domain or the C-terminal region (Jaya et al., 2009). Mani et al. (2015) mapped the substrate-binding sites on the intrinsically disordered N-terminal arms of rice sHSPs. Construction of chimeras from pea Hsp18.1 and wheat Hsp16.9 led to the identification of the N-terminal region as an important domain for exhibiting chaperone activity toward substrate proteins, such as CS and luciferase. However, both the N-terminal region and the α-crystallin domain were essential to prohibit misfolding/unfolding of MDH (Basha et al., 2006).

Basha et al. (2010) compared two classes of plant sHSPs and inferred that the chaperone activity and substrate-binding modes for different sHSPs may also vary. The activation cascades for different sHSPs are also different. Phosphorylation affects the oligomerization of vertebrate sHSPs, whereas the yeast Hsp26 has a "middle domain" before the α-crystallin domain that acts as a thermosensor, and induces a switch to a high-affinity oligomeric conformation upon temperature increase (Benesch et al., 2010). This "middle domain" is, however, not universal for all sHSPs. The exact mechanism of activation in the case of plant sHSPs is yet to be undermined.

It is obvious that, owing to their indispensability within the system, sHSPs have evolved successfully into a large and divergent superfamily. Different members of this superfamily adopt a variation of strategies for activation, substrate binding, and protection. However, the ability to sense minute deformities in protein stability under suboptimal conditions has remained a conserved character for all sHSPs known. This property has actually led the system to promote sHSPs as the first line of defense against stress (McHaourab et al., 2009).

4. EVIDENCES OF PLANT SMALL HEAT SHOCK PROTEINS CONFERRING PROTECTION AGAINST HEAT STRESS

The sHSPs are the most abundant among the five conserved families of HSPs (Hsp70, Hsp60, Hsp90, Hsp100, and sHSPs) (Wang et al., 2004). Six nuclear multigene families encode different sHSPs, of which many exhibit high levels of expression. Each gene family encodes respective sHSPs in a particular cellular compartment (cytosol, chloroplast, endoplasmic reticulum, and mitochondrion) (Carra et al., 2017). Almost 13 different sHSPs have been grouped into six classes in *Arabidopsis* based on their intercellular localization and sequence homology. Multidomain proteins encoded by six open reading frames and with one or more regions homologous to the α-crystallin domain have been identified (Carra et al., 2017).

Accumulation of sHSPs has been correlated to plant acclimatization to multiple abiotic stresses, including heat stress. Maize mitochondrial sHSPs (msHSPs) were found to protect the reduced nicotinamide adenine dinucleotide (NADH):

ubiquinone oxidoreductase activity (Complex I) during stress conditions, resulting in improved electron transport and accumulation of energy equivalents. However, these msHSPs could not protect the enzymes associated with Complex II (Hamilton and Heckathorn, 2001). Two sHSPs, ZmHsp17.0 and ZmHsp17.8, were purified and characterized from maize. During heat shock at 43°C, ZmHsp17.8 exhibited higher chaperone activity than ZmHsp17.0 by dissociating into smaller oligomers from its dodecameric form to refold CS (Klevin et al., 2014). Another sHSP from a heat-tolerant variety of *Agrostis stolonifera* was isolated from the chloroplasts as sHsp26.2. A point mutant, sHsp26.2m generated with a premature stop codon isolated from a heat-sensitive variety failed to accumulate and confer protection during heat stress (Wang and Luthe, 2003). Increased thermotolerance was observed in transgenic rice overexpressing rice chloroplast *Oshsp26* gene. Overexpression of this gene in *Escherichia coli* resulted in better scavenging of ROS produced during heat stress. Overexpression of *sHSP17.7* led to improved tolerance of rice plants toward both heat and ultraviolet-B stress (Xu et al., 2011). Another class of stress-associated proteins, viz., SP family proteins, also share some structural features of sHSPs (Wang et al., 2004). This family of proteins seems to have additional characteristics and functions, which are yet to be completely identified and resolved.

Hsp20s are the most abundant sHSPs reported in plants. Yan et al. (2017) identified 63 Hsp20s in switchgrass, *Panicum virgatum*, of which 87% were responsive to heat and other kinds of abiotic stresses. These proteins also conferred protection to the stress-sensitive reproductive organs (Yan et al., 2017). Lopes-Caitar et al. (2013) identified 51 *Hsp20* genes distributed across 15 subfamilies in *Glycine max*. The *Hsp20* genes in bread wheat, *T. aestivum*, were also upregulated during heat stress. About 163 *TaHsp20* genes were recently identified and classified as per their subcellular localization and phylogenetic relation with the genomes of some members (*Oryza sativa*, *Sorghum bicolour*, *Zea mays*, *Brachypodium distachyon*, and *Setaria italica*) of the grass family (Muthusamy et al., 2017). Of the 42 *Hsp20* genes identified in tomato, four genes exhibited housekeeping functions, whereas a majority of the rest of the genes were profusely overexpressed during abiotic stresses, such as heat, drought, and salt (Yu et al., 2016). Tandem duplication events in tomato have caused the expansion of the sHSP gene family. The core set of four homeostasis genes, *Solyc05g014280*, *Solyc03g082420*, *Solyc11g020330*, and *Solyc06g076560*, endowed in the tomato genome confer systemic protection during heat stress (Krsticevic et al., 2016).

Interestingly, heat stress affects two-thirds of chloroplast Hsp21 in plants and induces the protein to interact with the thylakoid membranes. The reason behind such membrane association of sHSPs during heat stress is largely unknown (Bernfur et al., 2017). It has been predicted that Hsp21 interacts with the C4-type zinc finger motif–containing plastid protein, PLASTID TRANSCRIPTIONALLY ACTIVE 5, and forms a complex that maintains the plastid-encoded RNA polymerase complex functions during heat stress in *Arabidopsis* (Zhong et al., 2013). Wang et al. (2017b) used suppression subtractive hybridization technology to characterize Hsp17.7, Hsp18.1, and Hsp21.8 from *Camellia sinensis*. Overexpression of the genes encoding these sHSPs in *Arabidopsis* led to lower malondialdehyde levels, ion leakage, and higher accumulation of proline and antioxidants during heat stress (Wang et al., 2017b). The molecular chaperone encoding gene, *CsHsp17.2*, is upregulated by heat stress in an abscisic acid (ABA)-independent manner in *C. sinensis*. CsHsp17.2 mediates thermotolerance by increasing photochemical efficiencies, scavenging ROS, and inducing the expression of crucial heat-responsive genes (Wang et al., 2017c). Prior exposure to heat stress generates acquired tolerance in *Arabidopsis*. Such "thermomemory" is sustained via high Hsp21 levels regulated by the plastid-localized metalloprotease, FtH6 (Sedaghatmehr et al., 2016). Overexpression of *sHSP1* and *sHSP2* genes in rice resulted in higher germination rates compared with the wild plants only under salt stress and not under heat or drought treatments (Ham et al., 2013). This research indicated the functional specialization of the members of the sHSP family over the evolutionary timeline.

Mu et al. (2013) reported that the heterologous expression of *LimHSP16.45* from David Lily (*Lilium davidii* Raffill var. Willmottiae) in *Arabidopsis* plants resulted in enhanced cell viability under high temperature, salinity, and oxidative stresses in the transgenic lines. Heat treatment led to the formation of heat shock granules (HSGs). These HSGs were found to be aggregates of electron-dense cytoplasmic particles consisting of sHSPs and Hsp70 family of proteins. The assembly of HSGs is well regulated, in which HSP CII promotes stress-induced autoaggregation and recruitment of sHSPs into HSGs (Kirschner et al., 2000; Fu et al., 2016). Mu et al. (2013) found that LimHSP16.45 associates with the membrane as a monomer under normal conditions, but is released into the cytoplasm to form oligomers in the transgenic *Arabidopsis* lines under temperature stress. HSG formation was visualized in the anthers of the transgenic plants even under control conditions. This indicated the natural potential of LimHSP16.45 to protect the pollen mother cells and tapetal cells against heat-induced damages (Mu et al., 2011). Under elevated temperatures, LimHSP16.45 promoted a systemic protection by upregulating the antioxidant genes, such as superoxide dismutase, ascorbate peroxidase (APX), and catalase, in the transgenic *Arabidopsis* lines. The strong expression of *LimHSP16.45* in the guard cells has also been considered responsible for the enhanced tolerance of the *Arabidopsis* plants toward abiotic stresses (Mu et al., 2013).

The primary factor for the decline in summer bentgrass (*A. stolonifera*) is heat stress (Wang et al., 2014b). Lee et al. (2015) overexpressed *sHsp23* from *Medicago truncatula* in creeping bentgrass. The transgenic lines exposed to 42°C for 24h resulted in minimal wilting of the leaves, which retained a healthy green color unlike the control plants. These transgenic lines displayed enhanced activity of the APX isoforms, which helped scavenging the ROS formed during heat stress. Transgenic *Medicago sativa* lines overexpressing *MsHsp17.7* showed reduced accumulation of malondialdehyde and better tolerance to heat and salt stress in comparison with the control plants (Li et al., 2016a). Similarly, overexpression of *PtHsp17.8* from *Populus trichocarpa* conferred heat and salt stress tolerance in *Arabidopsis*. The transgenic lines showed higher seedling survival, root length, relative water content, antioxidant, and compatible solute accumulation under heat and salt stresses (Li et al. 2014, 2016b). Kaur et al. (2015) revealed the identity of OsHsp18.2 in rice via proteomic approaches. It was found that OsHsp18.2 forms homooligomeric complexes and is dodecameric in nature. The chaperone activity of OsHsp18.2 was assayed using CS as the model substrate. The *Arabidopsis* transgenic lines overexpressing *OsHsp18.2* exhibited improved seed vigor, longevity, and seedling establishment. It was reported that OsHsp18.2 acts as an aging responsive protein, functioning as a molecular chaperone and protecting the proteins in seeds by prohibiting ROS accumulation.

Wu et al. (2015) reported a negative correlation between the accumulation of the universal stress hormone, ABA, and sHSPs in maize using a proteomic approach. For this experiment, viviparous5 (vp5) mutants were designed, which had impaired ABA biosynthesis in the seeds. Total protein was extracted from both the mutant and control plants exposed to heat stress and run on two-dimensional gels. The proteins identified by matrix assisted laser desorption/ionization and two analyzers time of flight were sHSPs, late embryogenesis abundant (LEA) proteins, stress-associated proteins, storage proteins, and other enzymes. The LEA proteins and the sHSPs exhibited differential accumulation in the vp5 and wild-type maize embryos. Six of eight identified LEA proteins decreased, whereas nine sHSPs increased in abundance in the vp5 embryos, clearly pointing out that such sHSPs are regulated in an ABA-independent pathway. Further promoter analysis of those *sHSP* genes clarified the absence of ABA response elements (ABREs) unlike the *LEA* genes. The promoter sequences of the *sHSP* genes, however, contained dehydration responsive elements, which might regulate their expression in an ABA-independent fashion. However, Reddy et al. (2014) reported the presence of ABREs in the promoters of heat shock factors (HSFs) acting as the upstream regulatory transcription factors (TFs) of several sHSPs in maize. Ruibal et al. (2013) identified an ABA-regulated gene encoding a 16.4-kDa sHSP in *Physcomitrella patens*. PpHsp16.4 is encoded by two identical genes arranged in tandem. Knock-down of both genes limited the ability of the plants to recover from heat, salt, and osmotic stresses (Ruibal et al., 2013). In another study, transgenic *Arabidopsis* and lettuce plants overexpressing *AtHsp17.8* exhibited ABA hypersensitivity but abiotic stress–tolerant phenotypes (Kim et al., 2013).

5. CONCLUSION

The intricate roles of sHSPs in plants are yet to be elucidated in detail. As evident from the literature and scientific works available in this field, one has a brief conception about the structural significance and diversity of this class of HSPs in plants. It appears that the cooperation between the sHSPs and other members of the multiple HSP families is the pivotal regulator of heat stress response and protein protection in plants. The fate of a protein during the stress phase or the post-stress recovery phase is completely controlled by the regulation of the chaperone machinery.

6. FUTURE PERSPECTIVES

Several unexplained aspects still remain to be answered by the scientific community. It is actually impossible for the sHSPs to randomly find their target substrate protein in such a crowded cellular environment. Especially when the sHSP–substrate interaction is vital for the refolding of the substrate protein, this interaction must be guided by some specific adaptors, which are still unidentified. The sHSPs in association with other members of the HSP family initiate refolding in misfolded/unfolded proteins. However, there must be an extent of misfolding/unfolding beyond which such substrates cannot be refolded by the chaperone machinery and have to be ubiquitinylated. The measure of such extent in biophysical terms is still unknown. The dynamics of protein refolding assisted by the chaperone network also needs to be revisited.

The transcriptional activation of HSPs is still somewhat clear. It occurs through dedicated TFs, the HSFs that trimerize under heat stress, get phosphorylated, and transcribe downstream HSP encoding genes. An imperfect heat shock element has been found to be important for heat induction of *Hsp18.2* in *Arabidopsis* and its homolog, *Hsp16.9C* in rice (Zhang et al., 2016). Gel shift analyses using the specific *sHSP* gene as a probe and the nuclear extract as a source of TFs followed by subsequent mass spectrometric and peptide mass fingerprinting analyses may be adopted to identify potential TFs regulating *sHSP* expression. Personat et al. (2014) overexpressed *HSFA9* from *Helianthus annuus* in *Nicotiana tabacum*. They showed increased seed longevity and enhanced tolerance to dehydration and oxidative stresses. Identification, characterization, and targeted overexpression of such TFs can possibly create heat stress–tolerant crops beneficial for cultivation under suboptimal conditions.

ACKNOWLEDGMENTS

Financial assistance from Council of Scientific and Industrial Research (CSIR), Government of India, through the research grant 38(1387)/14/EMR-II to Dr. Aryadeep Roychoudhury is gratefully acknowledged.

REFERENCES

Ahrman, E., Lambert, W., Aquilina, J.A., Robinson, C.V., Emanuelsson, C., 2007. Chemical cross-linking of the chloroplast localized small heat-shock protein, Hsp21, and the model substrate citrate synthase. Protein Sci. 16, 1464–1478.

Bakthisaran, R., Tangirala, R., Rao, C.M., 2015. Small heat shock proteins: role in cellular functions and pathology. Biochim. Biophys. Acta 1854, 291–319.

Baldwin, A.J., Hilton, G.R., Lioe, H., Bagneris, C., Benesch, J.L., Kay, L.E., 2011. Quaternary dynamics of alphaB-crystallin as a direct consequence of localised tertiary fluctuations in the C-terminus. J. Mol. Biol. 413, 310–320.

Banerjee, A., Roychoudhury, A., 2016a. Group II late embryogenesis abundant (LEA) proteins: structural and functional aspects in plant abiotic stress. Plant Growth Regul. 79, 1–17.

Banerjee, A., Roychoudhury, A., 2016b. Plant responses to light stress: oxidative damages, photoprotection and role of phytohormones. In: Ahammed, G.J., Yu, J.-Q. (Eds.), Plant Hormones under Challenging Environmental Factors. Springer Nature, pp. 181–213.

Banerjee, A., Roychoudhury, A., 2017. Abscisic-acid-dependent basic leucine zipper (bZIP) transcription factors in plant abiotic stress. Protoplasma 254, 3–16.

Banerjee, A., Roychoudhury, A., Krishnamoorthi, S., 2016. Emerging techniques to decipher microRNAs (miRNAs) and their regulatory role in conferring abiotic stress tolerance of plants. Plant Biotechnol. Rep. 10, 185–205.

Basha, E., Friedrich, K.L., Vierling, E., 2006. The N-terminal arm of small heat shock proteins is important for both chaperone activity and substrate specificity. J. Biol. Chem. 281, 39943–39952.

Basha, E., Jones, C., Wysocki, V., Vierling, E., 2010. Mechanistic differences between two conserved classes of small heat shock proteins found in the plant cytosol. J. Biol. Chem. 285, 11489–11497.

Basha, E., O'Neill, H., Vierling, E., 2012. Small heat shock proteins and alpha crystallins: dynamic proteins with flexible functions. Trends Biochem. Sci. 37, 106–117.

Benesch, J.L., Aquilina, J.A., Baldwin, A.J., Rekas, A., et al., 2010. The quaternary organization and dynamics of the molecular chaperone HSP26 are thermally regulated. Chem. Biol. 17, 1008–1017.

Bernfur, K., Rutsdottir, G., Emanuelsson, C., 2017. The chloroplast-localized small heat shock protein Hsp21 associates with the thylakoid membranes in heat-stressed plants. Protein Sci. http://dx.doi.org/10.1002/pro.3213.

Carra, S., Alberti, S., Arrigo, P.A., Benesch, J.L., Benjamin, I.J., et al., 2017. The growing world of small heat shock proteins: from structure to functions. Cell Stress Chaperones 22, 601–611.

Cheng, G., Basha, E., Wysocki, V.H., Vierling, E., 2008. Insights into small heat shock protein and substrate structure during chaperone action derived from hydrogen/deuterium exchange and mass spectrometry. J. Biol. Chem. 283, 26634–26642.

Dong, X., Yi, H., Lee, J., Nou, I.S., Han, C.T., Hur, Y., 2015. Global gene-expression analysis to identify differentially expressed genes critical for the heat stress response in *Brassica rapa*. PLoS One 10, e0130451.

Eisenhardt, B.D., 2013. Small heat shock proteins: recent developments. Biomol. Concepts 4, 583–595.

Eyles, S.J., Gierasch, L.M., 2010. Nature's molecular sponges: small heat shock proteins grow into their chaperone roles. Proc. Natl. Acad. Sci. U.S.A. 107, 2727–2728.

Friedrich, K.L., Giese, K.C., Buan, N.R., Vierling, E., 2004. Interactions between small heat shock protein subunits and substrate in small heat shock protein-substrate complexes. J. Biol. Chem. 279, 1080–1089.

Fu, C., Liu, X.X., Yang, W.W., Zhao, C.M., Liu, J., 2016. Enhanced salt tolerance in tomato plants constitutively expressing heat-shock protein in the endoplasmic reticulum. Genet. Mol. Res. 15.

Fu, X., Shi, X., Yan, L., Zhang, H., Chang, Z., 2013. In vivo substrate diversity and preference of small heat shock protein IbpB as revealed by using a genetically incorporated photo-cross-linker. J. Biol. Chem. 288, 31646–31654.

Giese, K.C., Basha, E., Catague, B.Y., Vierling, E., 2005. Evidence for an essential function of the N terminus of a small heat shock protein in vivo, independent of in vitro chaperone activity. Proc. Natl. Acad. Sci. U.S.A. 102, 18896–18901.

Giese, K.C., Vierling, E., 2004. Mutants in a small heat shock protein that affect the oligomeric state. Analysis and allele-specific suppression. J. Biol. Chem. 279, 32674–32683.

Ham, D.J., Moon, J.C., Hwang, S.G., Jang, C.S., 2013. Molecular characterization of two small heat shock protein genes in rice: their expression patterns, localizations, networks, and heterogeneous overexpressions. Mol. Biol. Rep. 40, 6709–6720.

Hamilton, E.W., Heckathorn, S.A., 2001. Mitochondrial adaptations to NaCl. Complex I is protected by anti-oxidants and small heat shock proteins, whereas complex II is protected by proline and betaine. Plant Physiol. 126, 1266–1274.

Hartl, F.U., Bracher, A., Hayer-Hartl, M., 2011. Molecular chaperones in protein folding and proteostasis. Nature 475, 324–332.

Haslbeck, M., Vierling, E., 2015. A first line of stress defense: small heat shock proteins and their function in protein homeostasis. J. Mol. Biol. 427, 1537–1548.

Jaya, N., Garcia, V., Vierling, E., 2009. Substrate binding site flexibility of the small heat shock protein molecular chaperones. Proc. Natl. Acad. Sci. U.S.A. 106, 15604–15609.

Jehle, S., Vollmar, B.S., Bardiaux, B., Dove, K.K., Rajagopal, P., Gonen, T., Oschkinat, H., Klevit, R.E., 2011. N-terminal domain of alpha B-crystallin provides a conformational switch for multimerization and structural heterogeneity. Proc. Natl. Acad. Sci. U.S.A. 108, 6409–6414.

Ju, Y., Tian, H., Zhang, R., Zuo, L., Jin, G., et al., 2017. Overexpression of *OsHSP18.0-CI* enhances resistance to bacterial leaf streak in rice. Rice 10, 21.

Kaur, H., Petla, B.P., Kamble, N.U., Singh, A., Rao, V., Salvi, P., Ghosh, S., Majee, M., 2015. Differentially expressed seed aging responsive heat shock protein OsHSP18.2 implicates in seed vigor, longevity and improves germination and seedling establishment under abiotic stress. Front. Plant Sci. 6, 713.

Kirschner, M., Winkelhaus, S., Thierfelder, J.M., Nover, L., 2000. Transient expression and heat-stress-induced co-aggregation of endogenous and heterologous small heat-stress proteins in tobacco protoplasts. Plant J. 24, 397–412.

Kim, D.H., Xu, Z.Y., Hwang, I., 2013. *AtHSP17.8* overexpression in transgenic lettuce gives rise to dehydration and salt stress resistance phenotypes through modulation of ABA-mediated signaling. Plant Cell Rep. 32, 1953–1963.

Klevin, R.D., Chidawanyika, T., Tims, H.S., Meulia, T., Bouchard, R.A., Pett, V.B., 2014. Chaperone function of two small heat shock proteins from maize. Plant Sci. 221–222, 48–58.

Kobayashi, Y., Harada, N., Nishimura, Y., Saito, T., Nakamura, M., et al., 2014. Algae sense exact temperatures: small heat shock proteins are expressed at the survival threshold temperature in *Cyanidioschyzon merolae* and *Chlamydomonas reinhardtii*. Genome Biol. Evol. 6, 2731–2740.

Krsticevic, F.J., Arce, D.P., Ezpeleta, J., Tapia, E., 2016. Tandem duplication events in the expansion of the small heat shock protein gene family in *Solanum lycopersicum* (cv. Heinz 1706). G3 (Bethesda) 6, 3027–3034.

Kumar, R.R., Sharma, S.K., Goswami, S., Singh, G.P., Singh, R., et al., 2013. Characterization of differentially expressed stress-associated proteins in starch granule development under heat stress in wheat (*Triticum aestivum* L.). Indian J. Biochem. Biophys. 50, 126–138.

Laganowsky, A., Benesch, J.L., Landau, M., Ding, L., et al., 2010. Crystal structures of truncated alphaA and alphaB crystallins reveal structural mechanisms of polydispersity important for eye lens function. Protein Sci. 19, 1031–1043.

Lambert, W., Koeck, P.J.B., Ahrman, E., Purhonen, P., et al., 2011. Subunit arrangement in the dodecameric chloroplast small heat shock protein Hsp21. Protein Sci. 20, 291–301.

Lee, K.W., Cha, J.Y., Mun, J.Y., Lee, B.H., Kim, Y.G., Lee, S.H., 2015. Heterologous expression of *msHsp23*, a *Medicago sativa* small heat shock protein, enhances heat stress tolerance in creeping bentgrass. J. Anim. Plant Sci. 25, 884–891.

Li, X., Yang, Y., Sun, X., Lin, H., Chen, J., et al., 2014. Comparative physiological and proteomic analyses of poplar (*Populus yunnanensis*) plantlets exposed to high temperature and drought. PLoS One 9, e107605.

Li, J., Xiang, C.Y., Yang, J., Chen, J.P., Zhang, H.M., 2015. Interaction of HSP20 with a viral RdRp changes its sub-cellular localization and distribution pattern in plants. Sci. Rep. 5, 14016.

Li, J., Zhang, J., Jia, H., Li, Y., Xu, X., et al., 2016b. The *Populus trichocarpa PtHSP17.8* involved in heat and salt stress tolerances. Plant Cell Rep. 35, 1587–1599.

Li, Z.Y., Long, R.C., Zhang, T.J., Yang, Q.C., Kang, J.M., 2016a. Molecular cloning and characterization of the *MsHSP17.7* gene from *Medicago sativa* L. Mol. Biol. Rep. 43, 815–826.

Lopes-Caitar, V.S., de Carvalho, M.C., Darben, L.M., Kuwahara, M.K., Nepomuceno, A.L., et al., 2013. Genome-wide analysis of the *Hsp20* gene family in soybean: comprehensive sequence, genomic organization and expression profile analysis under abiotic and biotic stresses. BMC Genomics 14, 577.

Maaroufi, H., Tanguay, R.M., 2013. Analysis and phylogeny of small heat shock proteins from marine viruses and their cyanobacteria host. PLoS One 8, e81207.

Mani, N., Ramakrishna, S., Suguna, K., 2015. Characterization of rice small heat shock proteins targeted to different cellular organelles. Cell Stress Chaperones 20, 451–460.

Mayer, M.P., 2010. Gymnastics of molecular chaperones. Mol. Cell 39, 321–331.

McHaourab, H.S., Godar, J.A., Stewart, P.L., 2009. Structure and mechanism of protein stability sensors: chaperone activity of small heat shock proteins. Biochemistry 48, 3828–3837.

Mogk, A., Schlieker, C., Friedrich, K.L., Schonfeld, H.J., Vierling, E., Bukau, B., 2003. Refolding of substrates bound to small Hsps relies on a disaggregation reaction mediated most efficiently by ClpB/DnaK. J. Biol. Chem. 278, 31033–31042.

Mu, C., Wang, S., Zhang, S., Pan, J., Chen, N., et al., 2011. Small heat shock protein LimHSP16.45 protects pollen mother cells and tapetal cells against extreme temperatures during late zygotene to pachytene stages of meiotic prophase I in David Lily. Plant Cell Rep. 30, 1981–1989.

Mu, C., Zhang, S., Yu, G., Chen, N., Li, X., et al., 2013. Overexpression of small heat shock protein *LimHSP16.45* in Arabidopsis enhances tolerance to abiotic stresses. PLoS One 8, e82264.

Muthusamy, S.K., Dalal, M., Chinnusamy, V., Bansal, K.C., 2017. Genome-wide identification and analysis of biotic and abiotic stress regulation of small heat shock protein (HSP20) family genes in bread wheat. J. Plant Physiol. 211, 100–113.

Mymrikov, E.V., Seit-Nebi, A.S., Gusev, N.B., 2011. Large potentials of small het shock proteins. Physiol. Rev. 91, 1123–1159.

Painter, A.J., Jaya, N., Basha, E., Vierling, E., Robinson, C.V., Benesch, J.L., 2008. Real-time monitoring of protein complexes reveals their quaternary organization and dynamics. Chem. Biol. 15, 246–253.

Park, S.M., Kim, K.P., Joe, M.K., Lee, M.O., Koo, H.J., Hong, C.B., 2014. Tobacco class I cytosolic small heat shock proteins are under transcriptional and translational regulations in expression and heterocomplex prevails under the high-temperature stress condition in vitro. Plant Cell Environ. 38, 767–776.

Patel, S., Vierling, E., Tama, F., 2014. Replica exchange molecular dynamics simulations provide insight into substrate recognition by small heat shock proteins. Biophys. J. 106, 2644–2655.

Personat, J.M., Tejedor-Cano, J., Prieto-Dapena, P., Almoguera, C., Jordano, J., 2014. Co-overexpression of two *Heat Shock Factors* results in enhanced seed longevity and in synergistic effects on seedling tolerance to severe dehydration and oxidative stress. BMC Plant Biol. 14, 56.

Ratajczak, E., Zietkiewicz, S., Liberek, K., 2009. Distinct activities of *Escherichia coli* small heat shock proteins IbpA and IbpB promote efficient protein disaggregation. J. Mol. Biol. 386, 178–189.

Reddy, P.S., Kavi Kishor, P.B., Seiler, C., Kuhlmann, M., Eschen-Lippold, L., et al., 2014. Unraveling regulation of the small heat shock proteins by the heat shock factor HvHsfB2c in barley: its implications in drought stress response and seed development. PLoS One 9, e89125.

Ruibal, C., Castro, A., Carballo, V., Szabados, L., Vidal, S., 2013. Recovery from heat, salt and osmotic stress in *Physcomitrella patens* requires a functional small heat shock protein PpHsp16.4. BMC Plant Biol. 13, 174.

Rutsdottir, G., Harmark, J., Weide, Y., Hebert, H., Rasmussen, M.I., et al., 2017. Structural model of dodecameric heat-shock protein Hsp21: flexible N-terminal arms interact with client proteins while C-terminal tails maintain the dodecamer and chaperone activity. J. Biol. Chem. 292, 8103–8121.

Santhanagopalan, I., Basha, E., Ballard, K.N., Bopp, N.E., Vierling, E., 2015. Model chaperones: small heat shock proteins from plants. In: Tanguay, R.M., Hightower, L.E. (Eds.), The Big Book on Small Heat Shock Proteins, vol. 8. Springer, pp. 119–153.

Sedaghatmehr, M., Mueller-Roeber, B., Balazadeh, S., 2016. The plastid metalloprotease FtsH6 and small heat shock proteinHSP21 jointly regulate thermomemory in *Arabidopsis*. Nat. Commun. 7, 12439.

Sharon, M., Robinson, C.V., 2007. The role of mass spectrometry in structure elucidation of dynamic protein complexes. Annu. Rev. Biochem. 76, 167–193.

Sobott, F., Benesch, J.L., Vierling, E., Robinson, C.V., 2002. Subunit exchange of multimeric protein complexes. Real-time monitoring of subunit exchange between small heat shock proteins by using electrospray mass spectrometry. J. Biol. Chem. 277, 38921–38929.

Stamler, R., Kappe, G., Boelens, W., Slingsby, C., 2005. Wrapping the alpha crystalline domain fold in a chaperone assembly. J. Mol. Biol. 353, 68–79.

Stengel, F., Baldwin, A.J., Painter, A.J., Jaya, N., Basha, E., Kay, L.E., Vierling, E., Robinson, C.V., Benesch, J.L., 2010. Quaternary dynamics and plasticity underlie small heat shock protein chaperone function. Proc. Natl. Acad. Sci. U.S.A. 107, 2007–2012.

Stief, A., Brzezinka, K., Lamke, J., Baurle, I., 2014. Epigenetic responses to heat stress at different time scales and the involvement of small RNAs. Plant Signal Behav. 9, e970430.

Wang, D., Luthe, D.S., 2003. Heat sensitivity in a bentgrass variant. Failure to accumulate a chloroplast heat shock protein isoform implicated in heat tolerance. Plant Physiol. 133, 319–327.

Wang, W., Vinocur, B., Shoseyov, O., Altman, A., 2004. Role of plant heat-shock proteins and molecular chaperones in the abiotic stress response. Trends Plant Sci. 9, 244–252.

Wang, Y., Lin, S., Song, Q., Li, K., Tao, H., et al., 2014a. Genome-wide identification of heat shock proteins (Hsps) and Hsp interactors in rice: Hsp70s as a case study. BMC Genomics 15, 344.

Wang, K., Zhang, X., Goatley, M., Ervin, E., 2014b. Heat shock proteins in relation to heat stress tolerance of creeping bentgrass at different N levels. PLoS One 9, e102914.

Wang, X., Wang, R., Ma, C., Shi, X., Liu, Z., et al., 2017a. Massive expansion and differential evolution of small heat shock proteins with wheat (*Triticum aestivum* L.) polyploidization. Sci. Rep. 7, 2581.

Wang, M., Zou, Z., Li, Q., Xin, H., Zhu, X., et al., 2017b. Heterologous expression of three *Camellia sinensis* small heat shock protein genes confers temperature stress tolerance in yeast and *Arabidopsis thaliana*. Plant Cell Rep. 36, 1125–1135.

Wang, M., Zou, Z., Li, Q., Sun, K., Chen, X., Li, X., 2017c. The CsHSP17.2 molecular chaperone is essential for thermotolerance in *Camellia sinensis*. Sci. Rep. 7, 1237.

White, H.E., Orlova, E.V., Chen, S., Wang, L., et al., 2006. Multiple distinct assemblies reveal conformational flexibility in the small heat shock protein Hsp26. Structure 14, 1197–1204.

Wu, X., Gong, F., Yang, L., Hu, X., et al., 2015. Proteomic analysis reveals differential accumulation of small heat shock proteins and late embryogenesis abundant proteins between ABA-deficient mutant *vp5* seeds and wild-type *Vp5* seeds in maize. Front. Plant Sci. 5, 901.

Xu, Y., Zhan, C., Huang, B., 2011. Heat shock proteins in association with heat tolerance in grasses. Int. J. Proteomics 2011, 529648.

Yan, H., Zhang, A., Chen, J., He, X., Xu, B., et al., 2017. Genome-wide analysis of the PvHsp20 family in switchgrass: motif, genomic organization, and identification of stress or developmental-related Hsp20s. Front. Plant Sci. 8, 1024.

Yu, J., Cheng, Y., Feng, K., Ruan, M., Ye, Q., et al., 2016. Genome-wide identification and expression profiling of tomato *Hsp20* gene family in response to biotic and abiotic stresses. Front. Plant Sci. 7, 1215.

Zhang, K., Ezemaduka, A.N., Wang, Z., Hu, H., Shi, X., et al., 2015. A novel mechanism for small heat shock proteins to function as molecular chaperones. Sci. Rep. 5, 8811.

Zhang, Y., Zou, B., Lu, S., Ding, Y., Liu, H., Hua, J., 2016. Expression and promoter analysis of the *OsHSP16.9C* gene in rice. Biochem. Biophys. Res. Commun. 479, 260–265.

Zhong, L., Zhou, W., Wang, H., Ding, S., Lu, Q., et al., 2013. Chloroplast small heat shock protein HSP21 interacts with plastid nucleoid protein pTAC5 and is essential for chloroplast development in *Arabidopsis* under heat stress. Plant Cell 25, 2925–2943.

Chapter 20

Silicon-Mediated Alleviation of Stresses in Plants

Abbu Zaid[1], Fahima Gul[2], Mohammad A. Ahanger[3], Parvaiz Ahmad[2,4]
[1]*Aligarh Muslim University, Aligarh, India;* [2]*SP College, Srinagar, India;* [3]*PG College, Rajouri, Jammu, India;* [4]*King Saud University, Riyadh, Saudi Arabia*

1. INTRODUCTION

Earth's resources have allowed life to flourish for the past 4.5 billion years. Over this span the world population has advanced at a fast pace, and this has a bearing on food scarcity. The primary food generators, the green plants, have braved rapid industrialization alongside urbanization leading to new stress loci like drought, salinity, heavy metal, and temperature. Depressed plant growth manifests as poor-quality food and has reduced the quantity of food produced. The current agricultural yields appear lacking in feeding the community on a minimum sustainable ground leading to numerous socio-economic problems, reducing the quality life. The need of the hour is to strategize pathways, supporting and enhancing the yield of plants.

Natural plantation or agriculture is dynamic and depends on vital key factors like quality of soil and water, mineral repository, effects of predation and harmful radiation, etc. The fluidity of these key factors at times generates stress. Plant stress is a function of varied factors both biotic and abiotic. Accordingly, there is scope for stress alleviation. The possibility of unitary customized stress alleviation is rare. External and internal determinants may synchronize for stress mitigation. Exogenous stress reducers have an economic bearing. The natural adaptive pathways of plants to stress conditions involving changes at physiological and genetic levels are the best combat. So far silicon has been symbolized for its applicability in stress situations.

Studies on silicon have been carried for more than 150 years; however, the role of silicon in plant stress alleviation has gained much attention only in recent decades (Coskun et al., 2016). Silicon is the second most abundant element on the earth composing about 29% of the earth's crust (Haynes, 2014). Although silicon is indexed as a nonessential element, plant tissue often contains its macronutrient equable. An analogous term, quasiessential element is applied for silicon due to the significant benefits that it confers on plants including added growth, yield and crop quality, rate of photosynthesis, nitrogen fixation, predation protection, heavy metal toxicity tolerance, temperature fluctuations, ultraviolet (UV) radiations, nutrient inconstancy, salinity, and gravity (Epstein, 1994, 1999, 2009; Richmond and Sussman, 2003; Ma, 2004; Liang et al., 2007, 2015; Cooke and Leishman, 2011; Guntzer et al., 2012; Van Bockhaven et al., 2013; Coskun et al., 2016). In higher plants, silicon showed many positive effects on growth and development under different environmental stress conditions (Liang et al., 2003, 2006). Studies on silicon pointing to its beneficial role in unstressed plants has also gained ground in the past decades. Along with stress alleviation, silicon boosts the plants and crops to their maximum yield potential and imparts pathogen and disease resistance. Silicon helps the plants to overcome the harms and fluctuations of temperature, heavy metals, salinity, and water deficiency. Silicon also improves the functioning of enzymes under stress conditions. The delay in elucidating the effects of silicon on plant lifecycle is due to the poor understanding of its possible role and the mechanism by which it alters the physiological and genetic makeup of the plants. This article attempts to investigate these aspects of silicon in relation to plants.

2. SILICON OCCURRENCE, UPTAKE, AND ROLE IN PLANT METABOLISM

Plant growth is determined by a multitude of organic and inorganic factors. About 18% of the total elements found on the earth's crust have been listed as essential elements to all plants. These range from macro- to micronutrients on the basis of their dry biomass content. Elements like silicon, sodium, cobalt, aluminum, and selenium being prerequisites for normal growth and development in some taxa when stressed are labeled as beneficial elements (Marschner, 1995; Willams and Salt, 2009). Of these silicon is the eighth most common element in the universe by mass (Sahebi et al., 2015).

Plant Metabolites and Regulation Under Environmental Stress. https://doi.org/10.1016/B978-0-12-812689-9.00020-0

Silicon is released in soil by chemical as well as biological routes (Gonzalez et al., 2010). The pure form of silicon is very rare; as a compound it comprises 25% of the total earth's crust (Sommer et al., 2006). Chemical combinations of silicon are ubiquitous; after oxygen it is the second most abundant element on the earth's crust (Lindsay, 1979; Epstein, 1994; Gong et al., 2006) with a mean content of 28.8% (weight) and an occurrence that ranges from 0.52 to ~47 wt% (McKeague and Cline, 1963; Wedepohl, 1995). Silicon combines with oxygen as silicon dioxide amounting to 50%–70% of the soil mass. Silicon is also present in rocks ranging from 23% (e.g., basalt) to 46.5% (e.g., orthoquartzite) (Monger and Kelly, 2002). Silicon is an essential element for the diatoms (Ketchum, 1954; Volcani, 1978; Round et al., 1990; Kinrade et al., 2001, 2002). Trace amounts of silicon are also present in carbonaceous rocks such as the limestones and the carbonites (Monger and Kelly, 2002). Silcretes are the component of derived soils that contain significant amounts of silicon (as high as 46%).

Despite the occurrence of silicon in soil solution and mass it is not readily available for plants uptake (Richmond and Sussman, 2003). Silicon (Si) present as silicic acid H_4SiO_4 or in its ionized form $Si(OH)_4$ predominating at alkaline pH greater than 9 is available for plant uptake (Currie and Perry, 2007). The bioavailable form of silicon is silicic acid (monosilicic acid or orthosilicic acid H_4SiO_4, ranging between 0.1 and 0.6 mM in soil) (Epstein, 1994; Weiss and Herzog, 1978; Epstein, 1999; Richmond and Sussman, 2003; Epstein and Bloom, 2005; Coskun et al., 2016), and it is then polymerized into silica gel in shoots ($SiO_2 \cdot nH_2O$). Wiese et al. (2007) reported accumulation if Si at maximum transpiration rate sites to which is it transported passively by transpiration stream. After water transpiration silicic acid converts into its colloidal form, which then forms silica gel or phytolith (amorphous silicic acid). In plants, silica gel polymerizes at high concentrations (>2 mM) into a polymer, thereby imparting protection to the plant (Jones and Handreck, 1967; Gao et al., 2006; Ma and Yamaji, 2006; Van Hoest, 2006). The regular theory, which states that bioavailable Si is chiefly obtained from solvation of primary and secondary minerals and later buffered by the adsorption and desorption of silicate onto sesquioxides, has now been replaced by the view that phytogenic cycling of Si, which involves the uptake by plants, silica formation mainly in leaves, and return to the soil as plant litter, is the main root source of bioavailable form of soil Si in natural ecosystems (Haynes, 2014; see also Pii et al., 2015; Gattullo et al., 2016). In contrast, in agricultural fields the content of silicon is low due to frequent elimination of silicon-rich litter postharvest (Savant et al., 1997; Struyf et al., 2010; Vandevenne et al., 2015). In the natural ecosystem and in agricultural soil the silicon cycle faces biological interventions, which shape its set-up globally. Also, in the absence of proper biosilification reaction in soil the transformation of silicon remains obscure.

3. ROLE IN PLANT METABOLISM

Despite its pervasive presence silicon has been listed out of the essential element class, because silicon does not satisfy one of the criteria of essential element class established by Arnon and Stout (1939), namely, evidence of its possible role in plant metabolism. Redefining the essentiality of elements in plant systems, Epstein and Bloom (2003) set qualification as satisfying at least one of the two basic criteria, viz., (1) the element is a part of a molecule that is an indispensable part of plant's structure or its (2) deficiency causes abnormalities in growth, reproduction, or development, i.e., "performance," when compared with plants that have the element. As per this redefinition silicon is essential for higher plants as its deficiency alters their performance (Ma and Takahashi, 2002). The only rationale behind silicon not being acclaimed as an established essential element appears to be the lack of knowledge about its possible role in plant metabolism (Ma, 2004); however, research on silicon has made great strides.

Silicon has been found to be present in all terrestrial plant tissues ranging from 0.1% to 10% on dry weight basis, depending upon the species (Ma and Takahasi, 2002). Silicon accumulation has been reported to higher but is not limited to monocotyledonous families like Poaceae, Equistaceae, and Cyperaceae (Currie and Perry, 2007). Silicon content can vary within the different parts of the same plant, a good example being the variation ranging from 0.5 g/kg in polished rice, 50 g/kg in rice bran, 130 g/kg in rice straw, 230 g/kg in rice hulls, to 350 g/kg in rice joints (found at the base of the grain) (Van Hoest, 2006).

Silicon is now widely accepted to benefit a range of plants and prominent agricultural crops across the world (Coskun et al., 2016). Ranging from providing pathogenetic resistance to increasing flower size silicon has been listed as one improvement tool for all plants especially under stress conditions. Silicon nutrition was effective in alleviating both biotic and abiotic stresses in a number of plant taxa (Epstein, 2009). Enhanced pathogenic resistance due to silicon was reported in many taxa (Meyer and Keeping, 2001; Shiguro, 2001; Richmond and Sussman, 2003; Reynolds et al., 2009). Drought resistance and heavy metal tolerance due to silicon uptake was reported by Lux et al. (2002), Richmond and Sussman (2003), Neumann and Zur Nieden (2001), Feng et al. (2010), Song et al. (2014), and Tripathi et al. (2015). Administration of silicon improved the quality of the agricultural crops and their yields (Korndorfer and Lepsch, 2001; Richmond and Sussman, 2003).

Many scientists have reported resistance to UV-B radiation and salt due to the presence of silicon in plants under stress (Shen et al., 2010; Moussa, 2006; Ali et al., 2012; Xie et al., 2015).

For its multitude of effects in stress alleviation in plants, research studies on silicon have attracted much attention across the world. Tolerance for abiotic and biotic stresses coupled with growth promotion has been a distinctive effect of silicon application in various plant taxa (Liang et al., 2007). Scientists are exploring the possibility of silicon being a part of structural or functional programs of plants and its role in changing the metabolic or physiological response of plants especially under stress conditions (Liang et al., 2003, 2006). Claims of silicon conferring benefits to stressed plants alone is losing gravity (Fauteux et al., 2006; Chain et al., 2009; Epstein, 2009; Van Bockhaven et al., 2013). In unstressed plants increment in the content of photosynthetic pigments, enhanced gas exchange, and photosystem II activity was reported due to silicon nutrition (Detmann et al., 2012; Pilon et al., 2013; Harizanova et al., 2014). Silicon addition showed little alteration of the transcriptome in unstressed *Arabidopsis* (Fauteux et al., 2006), wheat (Chain et al., 2009) and rice (Watanabe et al., 2004; Brunings et al., 2009; Van Bockhaven et al., 2013), and the proteome of rice (Nwugo and Huerta, 2008). Extensive studies on silicon have shown enhancement in crop yields, even in unstressed rice, caused by shift in major aspects of plant functioning related to curtailed transpiration by spikelets, remodeling of source–sink relationships and increasing photosynthesis, mesophyll conductance, N use competence, and mobilization of photoassimilates and amino acids from vegetative tissue to grains (Tamai and Ma, 2008; Detmann et al., 2012, 2013). The role of silicon as a signaling factor in redirecting the primary metabolism of plants is being recognized, although much work needs to be done in this direction (Coskun et al., 2016).

It has been widely reported that Si imparts drought resistance in plants by activating the antioxidant defense pathway (Agarie et al., 1998; Hattori et al., 2001; Lux et al., 2002; Gong et al., 2005). Gong et al. (2003) concluded that silicon not only improves water availability to the plants during drought but also increases its dry mass content. Silicon follows a two-pronged strategy in mitigating drought stress toxicity in plants by enhancing both the enzymatic (superoxide dismutase [SOD], catalase [CAT], peroxidase [POD], ascorbate peroxidase [APX], and glutathione reductase [GR]) and the nonenzymatic antioxidants (cysteine, glutathione, and ascorbic acid) (Ashraf and Harris, 2004; Gunes et al., 2007; Ashraf, 2009).

Silicon has been reported to enhance resistance against insects, pathogens, drought, and heavy metals and to boost floristic crop quality and yield. Silicon added to vase solutions enhanced the flower quality in many ornamental flowers like *Gerbera* (Kamenidou et al., 2010), sunflower (Kamenidou et al., 2008), roses (Ehret et al., 2005; Hwang et al., 2005), zinnias (Kamenidou et al., 2009), and chrysanthemums (Sivanesan et al., 2013a,b). Silicon-rich vase solutions have been found to increase the fresh mass, promote water uptake, and decrease lipid peroxidation with decline in aminocyclopropanecarboxylate oxidase activity, which postponed senescence in *Gerbera* flowers (Kazemi et al., 2012).

Silicon-mediated resistance against fungal pathogens and insect pests is poorly understood; however, the application of silicon-based fertilizers has shown effective performance (Bowen et al., 1992; Sangster et al., 2001; Ma and Takahasi, 2002). The earliest reports of the use of silicon-containing fertilizers dates back to the 1950s from Japan (Guntzer et al., 2012). Use of silicon fertilizers is widespread these days, with easy extraction from industrial slag (Haynes, 2014) and plant straw, typically from rice (Hossain et al., 2001). Application of silicon-rich fertilizers enhanced the yield and quality of prominent agricultural monocot crops such as rice, wheat, maize, barley, millets, sorghum, and sugarcane as well as dicots like cotton and soybean (Liang et al., 2015; Coskun et al., 2016).

4. SILICON-MEDIATED ALLEVIATION OF STRESSES

Abiotic environmental pressures have adversely affected the growth, morphology, physiology, and overall crop productivity of plants. Stress can also affect the geographical distribution of plants across varying environmental regimes (Zhang et al., 2017). Abiotic stresses along with unpredictable climate change has turned the situation from bad to worse, which in turn has caused enormous damage to agricultural productivity. Different plant-based strategies were thus developed, and one of the significant plant-based adaptation strategies to various abiotic stresses is the upregulation/activation of multiple defense systems that involve various processes at physiological, biochemical, molecular, cellular, and whole-plant levels (Ahmad et al., 2017; Gill and Tuteja, 2010; Atkinson and Urwin, 2012). Silicon (Si) is not listed among the essential elements within the plant kingdom but is considered as a quasiessential element and the second most abundant element after oxygen on the surface of the earth; it makes up as much as 0.1%–10% of the total biomass of the plant body (Manivannan and Ahn, 2017). It plays a well-exemplified role in plant growth and development, but its exact role is poorly understood. The exogenous application of Si has been known to modulate plants' response to an array of environmental stresses as Si showed pronounced effect on various components of plant defense systems. There are increasing evidences regarding the beneficial role of Si in promoting overall plant growth and metabolism and in mitigating biotic and abiotic environmental stresses as Si mediates molecular regulation of genes in plants under both abiotic and biotic stresses (Yin et al., 2016). In the present section we summarize the regulatory interactions of Si in alleviating plant stresses via its coordinated role in vital plant processes.

4.1 Silicon and Antioxidant System

As plants employ sessile lifecycle, stressful environments can induce the expression of various genes associated with stress tolerance, physiobiochemical and metabolic processes, and signal amplification and transduction in plants. Transcription factors are thought to be the primary regulators of signal transduction genes, which are important for plants to resist biotic and abiotic stresses and are influenced by specific cis-acting regulatory elements called "regulons" located in the target gene (Lucas et al., 2011; Qin et al., 2011). Application of Si to various crop plants is known to modulate/upregulate antioxidant defense system and their components, respectively. Si supplementation under normal or environmentally stressed conditions has been shown to influence the antioxidant system or its components. Under stress conditions the production of reactive oxygen species (ROS) is increased, which causes peroxidation of lipids and protein and leakage of ions. Si has been shown to reduce malondialdehyde (MDA) content, which is the final product of lipid peroxidation under stressed environments in plants such as barley (Liang et al., 2003) and wheat (Xu et al., 2017) and helps to keep membrane integrated in its normal form and decrease permeability. Si is known to trigger/increase the activities of key enzymes linked to antioxidant defense systems such as SOD, APX, and CAT, as also of ascorbate (AsA), GR, and total glutathione (GSH) under stressed conditions (Abbas et al., 2015; Shi et al., 2016; Li et al., 2016). Hasanuzzaman et al. (2017) working on rapeseed reported that exogenous Si modulated the antioxidant defense system, including the glyoxalase system. The supplementation of Si in rapeseed plants has a marked effect on diverse antioxidant enzymes like monodehydroascorbate reductase, dehydroascorbate reductase, as well as CAT, APX, and glutathione S-transferase (glyoxalase I and glyoxalase II and redox pools of ascorbate and reduced glutathione). In fact redox homeostasis of antioxidant enzymes by exogenous Si treatment was responsible for reducing the adverse effects of metal stress in rapeseed plants. Different modes and graded levels of Si application to wheat plants exposed to salinity stress enhanced antioxidant mechanisms, which alleviate the saline stress in field conditions. In such plants Si reduced salt stress through the production of antioxidant enzymes such as SOD, CAT, and APX. Similar stimulatory role during abiotic stress condition on antioxidant enzyme activities in maize plants has been reported by Khan et al. (2016). These results compelled us to conclude that Si supplementation can decrease/bring down the adverse impacts of various environmental stresses through the modulation/regulation of antioxidant defense mechanisms in plants.

4.2 Silicon and Osmotic Stress

Plant growth is affected the by presence of ample amounts of Si in some crop plants; it has been reported that deficiency abnormalities occurred in Si-starved plants and that Si application ameliorates plant nutrient deficiency (Rafi and Epstein, 1997; Ma and Yamaji, 2015). Treatment of plants with Si can alleviate decrease in growth, yield, quality, and quantity of crop plants, under both biotic and abiotic environmental stresses, such as herbivory, microbial pathogen stress, UV radiation, gravity, temperature extremities, osmotic stress, heavy metal toxicity, nutrient imbalance and toxicity, and drought stress (Van Bockhaven et al., 2013; Liang et al., 2015 Bityutskii et al., 2017). Osmotic stress is an important abiotic stress factor that posed tremendous limits on normal and sustainable yields and productivity of crop plants worldwide and can affect approximately 800 million hectares (ha) of cultivable arable land (Li et al., 2017; Rios et al., 2017; Khan et al., 2010, 2017). There is an elaborative role of Si in abiotic stress tolerance in general, and with regard to salinity stress in particular (Rengasamy, 2010; Zia et al., 2017; Ashfaque et al., 2017). Detailed reports are available that the supplementation of Si increases tolerance to osmotic stress in diverse crops such as wheat (Bibordy, 2014; Saqib et al., 2008; Tale and Haddad, 2011), barley (Liang, 1999, 2005), rice (Mahdieh et al., 2015; Kim et al., 2014a,b), tomato (Li et al., 2015), soybean (Lee et al., 2010), cucumber (Khoshgoftarmanesh et al., 2014), sugarcane (Ashraf et al., 2010), banana (Henriet et al., 2006.), faba bean (Shahzad et al., 2013; Kardoni et al., 2013), haloxylon (Khan et al., 2000), *Glycyrrhiza* (Li et al., 2016), common bean (Parande et al., 2013), pea (Shahid et al., 2015), sorghum (Sonobe et al., 2011), and alfalfa (Wang and Han, 2007; Wang et al., 2011). Si has a beneficial and ameliorative role in bringing down the ill effects of osmotic stress. In a classical experiment done on barley plants by Hosseini et al. (2017), Si-mediated increase in mineral nutrient status, starch biosynthesis, and modulation of the glycolytic and Krebs cycle pathways were the physiological mechanisms behind osmotic stress tolerance. In cucumber plants subjected to salt stress, Si-mediated reductions in assimilate accumulation in leaves alleviated photosynthetic capacity inhibition and Si-mediated ample assimilate transport was responsible for salt stress tolerance. Application of Si to salt-stressed cucumber plants improved plant growth and photosynthetic inhibition, simultaneously alleviating salt-stress-induced oxidative damage. Supplied Si also decreased the total soluble sugar content in leaves by showing regulatory interactions on the enzyme activities of carbohydrate metabolism (Zhu et al., 2016). In sorghum plants, Liu et al. (2014) found that application of Si alleviates water deficit stress in sand culture. Osmotic stress decreased dramatically the plant dry weight, net photosynthetic and transpiration rate, leaf water content, and stomatal conductance, but silicon application reversed these stress-induced repressions. Their study concluded with the observation that silicon plays a modulating role in improving plant resistance to osmotic stress despite being a physical barrier. Last, seed Si priming for particular duration is also an adaptive strategy for countering osmotic stress whereby seed exposure to Si for only a few hours fortifies plants against future abiotic and biotic pressures. In wheat plants, 6–8 h seeds priming with Si showed significant increases in rate of germination, vegetative and reproductive morphology, overall growth, and crop yield under osmotic stress (Ahmed et al., 2016). Similar results were obtained by Latef and Tran (2016) in maize plants. They observed that alkaline-stressed maize seedlings undergo decrease in growth attributes, photosynthetic pigments, leaf relative water content, soluble sugars, and total phenol contents and increase in proline and MDA levels, together with the increase in activities of CAT, SOD, and POD observed in salt-stressed plants. However, seed-priming application of Si reversed salt stress repressions. Their results indicate that Si priming is useful in overcoming osmotic stress in maize plants.

4.3 Silicon and Phytohormones

Under stressful environments, plant growth regulators (PGRs) including salicylic acid (SA), abscisic acid (ABA), methyl jasmonate (MeJ), indole acetic acid, and gibberellin (GA) enhance abiotic stress plant tolerance (Fahad et al., 2015; Wani et al., 2016). Si has a well-coordinated regulatory interaction with PGRs in lowering the severity of environmental stresses. In barley plants, Hosseini et al. (2017) reported that Si exerts its beneficial effect by mediating homeostasis of phytohormones and concluded that Si may effectively improve stress tolerance under K-deficient condition, in particular when additional stress like osmotic stress interferes as Si regulates metabolite and hormone homeostasis. Si exerts impacts on the levels of endogenous phytohormones, and the responses are commonly monitored during plant responses to various stresses. In *Oryza sativa* plants exposed to heavy metal stress, Si reduced the endogenous biosynthesis of MeJ and SA, whereas ABA showed a varying trend confirming the inverse relationship between ABA and JA/SA biosynthesis (Kim et al., 2014a,b). Kim et al. (2011) also found that Si has negative interaction with JA concentration during wounding stress, whereas Si-treated soybean plants triggered the production of GAs in response to salinity stress (Lee et al., 2010). SA is involved in systemic acquired resistance, whereas the production of MeJ and ethylene (ETH) involve plant resistance against necrotrophic pathogens (Khan and Khan, 2013; Khan et al., 2015; Iqbal et al., 2017). In *Arabidopsis* plants, Fauteaux et al. (2006) observed that Si application caused a threshold increase in biosynthesis of MeJ, SA, and ETH in leaves when seedlings were subjected to biotic stress. In the same manner, on subjecting tomato plants to *Ralstonia solanacearum* stress, prior Si priming results in activation of MeJ and ETH genes for increasing resistance against pathogen stress (Ghareeb et al., 2011). In rice plants, Brunings et al. (2009) also observed that exposure of pathogenic blast fungi–stressed rice plants to Si treatment activates the genes responsible for ETH signaling pathways. Markovich et al. (2017) confirmed the hypothesis that Si specifically promotes cytokinin biosynthesis in *Arabidopsis* and *Sorghum*. In conclusion, regulatory interactions of phytohormones with Si are responsible for modulating vital and complex physiobiochemical processes in diverse groups of crop plants.

4.4 Silicon and Enzyme Activity

During abiotic environmental stress conditions, Si application shows an array of responses to various ROS-scavenging enzymes. In this regard, the activities of various antioxidant enzymes such as CAT, SOD, APX, and guaiacol peroxidase show an oscillating trend depending on the type of plant and the severity of incoming stress. In fact, Si supplementation showed Si-induced abiotic stress tolerance by bringing down the excess ROS biosynthesis or by modulating the activities of principal ROS-scavenging enzymes such as CAT and APX as these catalyze the conversion of hydrogen peroxide (H_2O_2) into H_2O and APX decreases the ROS-stimulated lipid peroxidation. In addition to enzymes associated with oxidative stress signaling, Si also exerts its beneficial activity on other enzymes controlling vital plant activities including enzymes involved in sucrose and starch metabolism. The activities of enzymes like sucrose phosphate synthase, sucrose synthase, and invertases are differentially regulated by Si in both the roots and leaves of salt-stressed *Cucumis sativus* L. seedlings as silicon in the growth medium decreased the soluble sugar levels in leaves by regulating the activities of carbohydrate metabolism enzymes (Zhu et al., 2016). In *Brassica napus* L. plants Hasanuzzaman et al. (2017) discovered that the attenuating effect of exogenous Si during cadmium-induced oxidative stress is achieved through modulation of the enzymes of the glyoxalase system and the Asada–Halliwell pathway (AsA-GSH). They concluded that Si-mediated coordinated actions of enzymes of the AsA-GSH pathway and glyoxalase systems maintained the redox state of AsA and GSH, which neutralizes the damages of Cd-induced oxidative stress.

4.5 Silicon and Cell Wall Synthesis/Protection

Plant being sessile are frequently exposed to dynamic environmental stresses, and for coping with these pressures they possess a range of ubiquitous cellular surveillance systems that enable them to efficiently tolerate these stresses. In addition to the beneficial role of Si in crop abiotic stress tolerance, one of the effects of Si in protecting plants against various biotic and abiotic stresses is the formation of strong barriers in above- and belowground plant parts (Cai et al., 2009). Plant cell wall is regarded as the first target barrier site for reducing the cascade of impacts of stresses on plants (Janicka-Russak et al., 2012). The cell wall components are thought to slow down the inflow of excess freely available ions from the soil solution into the cortex (Faiyue et al., 2010). The development of cell wall components is facilitated by Si supply in diverse crop plants such as *Guizotia abyssinica*, *O. sativa*, *Zea mays*, *Allium cepa*, and *Tradescantia virginiana* (Hinrichs et al., 2017; Fleck et al., 2015). The role of Si is well recognized in strengthening cell walls and providing mechanical support for some plant divisions as Si enhances suberization, lignification, and silicification (Coskun et al., 2016). Plants differ in their capacity to concentrate Si differently; some are known to be strong accumulators, containing Si in concentrations even than those of other nutrients (Epstein, 1994). A hemicellulose bond form of Si has been attributed to provide improved structural rigidity and stability (Ma et al., 2015), which is protective as well as beneficial during osmotic and other environmental stresses. Si also forms an amorphous silica barrier by the polymerization of silicic acid with cell wall, a process aptly called biosilicification (Exley, 2015), which help plants to survive under biotic and abiotic capricious conditions like pathogen entry metal/metalloid penetration (Fauteux et al., 2005, 2006; Ma et al., 2015; Fleck et al., 2015). In roots of wheat plants, Si has been shown to increase cell wall binding of sodium ions in the roots (Saqib et al., 2008; Krishnamurthy et al., 2011; Gong et al., 2006). Hinrichs et al. (2017) concluded their study with the elaborative findings on the role of Si in cell wall synthesis in rice plants. Si-induced formation of Casparian strips helps to reduce radial oxygen loss in rice plants. Furthermore, they confirmed their findings by using several transcriptomic techniques, which revealed potent candidate genes involved in the Si-induced formation of cell wall components such as Class III POXs, transferases and ligases, and ATP-binding cassette (ABC) transporter. Nevertheless, by using overexpression and knockout mutants of these genes they further elaborated the contribution of the ABC transporter (*OsABCG25*) to exodermis cell wall formation, which was also evident in the expression of other transporters of *OsABCG25* in the Si-induced formation of cell wall genes, which corresponds to phenylalanine-ammonia lyase, 4-coumarate-CoA ligase, and diacylglycerol O-acyltransferase of phenylpropanoid pathway, while differential development of different components of cell wall in mutants and supply of Si also regulated the function of cell wall exodermis.

5. CONCLUSION AND FUTURE PROSPECTS

Environmental stresses hamper growth by altering nutrient uptake and assimilation. Improving the tolerance mechanisms using conventional breeding techniques like proper nutritional management can prove a success if the transport proteins involved in their uptake are screened and introgressed in crop plants. Silicon protects plants by improving the uptake and assimilation of key macroelements, thereby optimizing the growth and productivity. Besides this the involvement in providing the mechanical strength to plants determines the significance of silicon as a promising stress-enhancing mineral nutrient. Protecting processes like photosynthesis and respiration, optimizing phytohormone and polyamine synthesis and membrane lipids, and delaying senescence make silicon a promising mineral for stress mitigation in plants. Therefore identifying and developing a set of markers for enhancing the uptake silicon so as to exploit its potential to ameliorate the stress-induced negative effects on growth can be an important step forward in further consolidation of its role in stress tolerance in plants. In addition, the changes in the plant metabolome as induced by deficiency and sufficiency of silicon should be analyzed to find any correlation.

REFERENCES

Abbas, T., Balal, R.M., Shahid, M.A., Pervez, M.A., Ayyub, C.M., Aqueel, M.A., Javaid, M.M., 2015. Silicon-induced alleviation of NaCl toxicity in okra (*Abelmoschus esculentus*) is associated with enhanced photosynthesis, osmoprotectants and antioxidant metabolism. Acta Physiol. Plant 37, 6.

Abdel Latef, A.A., Tran, L.S.P., 2016. Impacts of priming with silicon on the growth and tolerance of maize plants to alkaline stress. Front. Plant Sci. 7, 243. https://doi.org/10.3389/fpls.2016.00243.

Agarie, S., Hanaoka, N., Ueno, O., Miyazaki, A., Kubota, F., Agata, W., Kaufman, P.B., 1998. Effects of silicon on tolerance to water deficit and heat stress in rice plants (*Oryza sativa* L.), monitored by electrolyte leakage. Plant Prod. Sci. 1, 96–103.

Ahmed, M., Qadeer, U., Ahmed, Z.I., Fayyaz-Ul, H., 2016. Improvement of wheat (*Triticum aestivum*) drought tolerance by seed priming with silicon. Arch. Acker Pflanzenbau Bodenkd 62, 299–315.

Ahmad, P., Ahanger, M.A., Alyemeni, M.N., Wijaya, L., Alam, P., 2017. Exogenous application of nitric oxide modulates osmolyte metabolism, antioxidants, enzymes of ascorbate-glutathione cycle and promotes growth under cadmium stress in tomato. Protoplasma. https://doi.org/10.1007/s00709-017-1132-x. Article (in press).

Ali, A., Basra, S.M.A., Iqbal, J., Hussain, S., Subhani, M.N., Sarwar, M., Haji, A., 2012. Silicon mediated biochemical changes in wheat under salinized and non-salinized solution cultures. Afr. J. Biotechnol. 11, 606–615.

Arnon, D.I., Stout, P.R., 1939. The essentiality of certain elements in minute quantity for plants with special reference to copper. Plant Physiol. 14 (2), 371–375.

Ashfaque, F., Inam, A., Iqbal, S., Sahay, S., 2017. Response of silicon on metal accumulation, photosynthetic inhibition and oxidative stress in chromium-induced mustard (*Brassica juncea* L.). South Afr. J. Bot. 111, 153–160.

Ashraf, M., 2009. Biotechnological approach of improving plant salt tolerance using antioxidants as markers. Biotech. Adv. 27, 84–93.

Ashraf, M., Harris, P.J.C., 2004. Potential biochemical indicators of salinity tolerance in plants. Plant Sci. 166, 3–16.

Ashraf, M., Rahmatullah, K., Afzal, M., Ahmed, R., Mujeeb, F., Sarwar, A., Ali, L., 2010. Alleviation of detrimental effects of NaCl by silicon nutrition in salt-sensitive and salt-tolerance genotypes of sugarcane (*Saccharum officinarum* L.). Plant Soil 326, 381–391.

Atkinson, N.J., Urwin, P.E., 2012. The interaction of plant biotic and abiotic stresses: from genes to the field. J. Exp. Bot. 63, 3523–3543.

Bibordy, A., 2014. Interactive effects of silicon and potassium nitrate in improving salt tolerance of wheat. Int. J. Agric. 13, 1889–1899.

Bityutskii, N.P., Yakkonen, K.L., Petrova, A.I., Shavarda, A.L., 2017. Interactions between aluminium, iron and silicon in *Cucumber sativus* L. grown under acidic conditions. J. Plant Physiol. 218, 100–108.

Bowen, P., Menzies, J., Ehret, D., Samuels, L., Glass, A.D.M., 1992. Soluble silicon sprays inhibit powdery mildew development on grape leaves. J. Am. Soc. Hortic. Sci. 117, 906–912.

Brunings, A.M., Datnoff, L.E., Ma, J.F., Mitani, N., Nagamura, Y., Rathinasabapathi, B., Kirst, M., 2009. Differential gene expression of rice in response to silicon and rice blast fungus *Magnaporthe oryzae*. Ann. Appl. Biol. 155, 161–170.

Cai, K., Gao, D., Chen, J., Luo, S., 2009. Probing the mechanisms of silicon-mediated pathogen resistance. Plant Signal Behav. 4, 1–3.

Chain, F., Côté-Beaulieu, C., Belzile, F., Menzies, J.G., Bélanger, R.R., 2009. A comprehensive transcriptomic analysis of the effect of silicon on wheat plants under control and pathogen stress conditions. Mol. Plant Microbe Interact. 22, 1323–1330.

Cooke, J., Leishman, M.R., 2011. Is plant ecology more siliceous than we realise? Trends Plant Sci. 16, 61–68.

Coskun, D., Britto, D.T., Huynh, W.Q., Kronzucker, H.J., 2016. The role of silicon in higher plants under salinity and drought stress. Front. Plant Sci. 7, 1072. https://doi.org/10.3389/fpls.2016.01072.

Currie, H.A., Perry, C.C., 2007. Silica in plant biological, biochemical and chemical studies. Ann. Bot. 100, 1383–1389.

Detmann, K.C., Araújo, W.L., Martins, S.C., Sanglard, L.M., Reis, J.V., 2012. Silicon nutrition increases grain yield, which, in turn, exerts a feed-forward stimulation of photosynthetic rates via enhanced mesophyll conductance and alters primary metabolism in rice. New Phytol. 196, 752–762.

Detmann, K.C., Araújo, W., Martins, S., Fernie, A.R., DaMatta, F.M., 2013. Metabolic alterations triggered by silicon nutrition: is there a signalling role for silicon? Plant Signal Behav. 8, e22523. https://doi.org/10.4161/psb.22523.

Ehret, D.L., Menzies, J.G., Helmer, T., 2005. Production and quality of greenhouse roses in re-circulating nutrient systems. Sci. Hort. 106, 103–113.

Epstein, E., 1994. The anomaly of silicon in plant biology. Proc. Natl. Acad. Sci. USA 91, 11–17.

Epstein, E., 1999. Silicon. Ann. Rev. Plant Physiol. Plant Mol. Biol. 50, 641–664.

Epstein, E., 2009. Silicon: its manifold roles in plants. Ann. Appl. Biol. 155, 155–160.

Epstein, E., Bloom, A.J., 2003. Mineral Nutrition of Plants: Principles and Perspectives, second ed. John Wiley & Sons, New York.

Epstein, E., Bloom, A.J., 2005. Mineral Nutrition of Plants: Principles and Perspectives, second ed. Sinauer Associates, Sunderland, MA.

Exley, C., 2015. A possible mechanism of biological silicification in plants. Front. Plant Sci. 6, 853.

Fahad, S., Hussain, S., Matloob, A., Khan, F.A., Khaliq, A., Saud, S., Faiq, M., 2015. Phytohormones and plant responses to salinity stress: a review. Plant Growth Reg. 75, 391–404.

Faiyue, B., Al-Azzawi, M.J., Flowers, T.J., 2010. The role of lateral roots in bypass flow in rice (*Oryza sativa* L.). Plant Cell Environ. 33, 702–716.

Fauteux, F., Rémus-Borelm, W., Menzies, J.G., Bélanger, R.R., 2005. Silicon and plant disease resistance against pathogenic fungi. FEMS Microbiol. Lett. 249, 1–6.

Fauteux, F., Chain, F., Belzile, F., Menzies, J.G., Bélanger, R.R., 2006. The protective role of silicon in the *Arabidopsis*-powdery mildew pathosystem. Proc. Natl. Acad. Sci. USA 103, 17554–17559. https://doi.org/10.1073/pnas.0606330103.

Feng, J., Shi, Q., Wang, X., Wei, M., Yang, F., Xu, H., 2010. Silicon supplementation ameliorated the inhibition of photosynthesis and nitrate metabolism by cadmium (Cd) toxicity in *Cucumis sativus* L. Sci. Hort. 123, 521–530.

Fleck, A.T., Schulze, S., Hinrichs, M., Specht, A., Wassmann, F., Schreiber, L., Schenk, M.K., 2015. Silicon promotes exodermal Casparian band formation in Si-accumulating and Si-excluding species by forming phenol complexes. PLoS One 10, e0138555.

Gao, X., Zou, C., Wang, L., Zhang, F., 2006. Silicon decreases transpiration rate and conductance from stomata of maize plants. J. Plant Nutr. 29, 1637–1647.

Gattullo, C.E., Allegretta, I., Medici, L., Fijan, R., Pii, Y., Cesco, S., 2016. Silicon dynamics in the rhizosphere: connections with iron mobilization. J. Plant Nutr. Soil Sci. 179, 409–417. https://doi.org/10.1002/jpln.201500535.

Ghareeb, H., Bozsó, Z., Ott, P.G., Repenning, C., Stahl, F., Wydra, K., 2011. Transcriptome of silicon-induced resistance against *Ralstonia solanacearum* in the silicon non-accumulator tomato implicates priming effect. Physiol. Mol. Plant. Pathol. 75, 83–89.

Gill, S.S., Tuteja, N., 2010. Reactive oxygen species and antioxidant machinery in abiotic stress tolerance in crop plants. Plant Physiol. Biochem. 48, 909–930.

Gong, H., Chen, K., Chen, G., Wang, S., Zhang, C., 2003. Effects of silicon on growth of wheat under drought. J. Plant Nutr. 26, 1055–1063.

Gong, H., Zhu, X., Chen, K., Wang, S., Zhang, C., 2005. Silicon alleviates oxidative damage of wheat plants in pots under drought. Plant Sci. 169, 313–321.

Gong, H.J., Randall, D.P., Flowers, T.J., 2006. Silicon deposition in the root reduces sodium uptake in rice (*Oryza sativa* L.) seedlings by reducing by pass flow. Plant Cell Environ. 29, 1970–1979.

Espindola Gonzalez, A., Martínez-Hernández, A.L., Angeles-Chavez, C., Castan, O., Velasco-Santos, C., 2010. Novel crystalline SiO_2 nanoparticles via annelids bioprocessing of agro-industrial wastes. Nanoscale Res. Lett. 5, 1408–1417.

Gunes, A., Inal, A., Bagci, E.G., Coban, S., Sahin, O., 2007. Silicon increases boron tolerance and reduces oxidative damage of wheat grown in soil with excess boron. Biol. Plant 51, 571–574.

Guntzer, F., Keller, C., Meunier, J.D., 2012. Benefits of plant silicon for crops: a review. Agron. Sustain. Dev. 32, 201–213. https://doi.org/10.1007/s13593-011-0039-8.

Harizanova, A., Zlatev, Z., Koleva, L., 2014. Effect of silicon on activity of antioxidant enzymes and photosynthesis in leaves of cucumber plants (*Cucumis sativus* L.). Turk J. Agric. Nat. Sci. 2, 1812–1817.

Hasanuzzaman, M., Nahar, K., Anee, T.I., Fujita, M., 2017. Exogenous silicon attenuates cadmium-induced oxidative stress in *Brassica napus* L. by modulating AsA-GSH pathway and glyoxalase system. Front. Plant Sci. 8, 1061. https://doi.org/10.3389/fpls.2017.01061.

Hattori, T., Lux, A., Tanimoto, E., Luxova, M., Sugimoto, Y., Inanaga, S., 2001. The effect of silicon on the growth of sorghum under drought. In: Presented at Proceedings of the 6th Symposium of the International Society of Root Research, Nagoya, Japan.

Haynes, R.J., 2014. A contemporary over view of silicon availability in agricultural soils. J. Plant Nutr. Soil Sci. 177, 831–844.

Henriet, C., Draye, X., Oppitz, I., Swennen, R., Delvaux, B., 2006. Effects, distribution and uptake of silicon in banana (Musa spp.) under controlled conditions. Plant Soil 287, 359–374.

Hinrichs, M., Fleck, A.T., Biedermann, E., Ngo, N.S., Schreiber, L., Schenk, M.K., 2017. An ABC transporter is involved in the Silicon-induced formation of Casparian bands in the exodermis of rice. Front. Plant Sci. 8, 671. https://doi.org/10.3389/fpls.2017.00671.

Hossain, K.A., Horiuchi, T., Miyagawa, S., 2001. Effects of silicate materials on growth and grain yield of rice plants grown in clay loam and sandy loam soils. J. Plant Nutr. 24 (1), 13.

Hosseini, S.A., Maillard, A., Hajirezaei, M.R., Ali, N., Schwarzenberg, A., Jamois, F., Yvin, J.C., 2017. Induction of barley silicon transporter HvLsi1 and HvLsi2, increased silicon concentration in the shoot and regulated starch and aba homeostasis under osmotic stress and concomitant potassium deficiency. Front. Plant Sci. 8, 1359. https://doi.org/10.3389/fpls.2017.01359.

Hwang, S.J., Park, H.M., Jeong, B.R., 2005. Effects of potassium silicate on the growth of miniature rose 'Pinocchio' grown on rockwool and its cut flower quality. J. Jpn. Soc. Hortic. Sci. 74, 242–247.

Iqbal, N., Khan, N.A., Ferrante, A., Trivellini, A., Francini, A., Khan, M.I.R., 2017. Ethylene role in plant growth, development and senescence: interaction with other phytohormones. Front. Plant Sci. 8, 475. https://doi.org/10.3389/fpls.2017.00475.

Janicka-Russak, M., Kabała, K., Burzyński, M., 2012. Different effect of cadmium and copper on H^+-ATPase activity in plasma membrane vesicles from *Cucumis sativus* roots. J. Exp. Bot. 63 (11), 4133–4142.

Jones, L.H.P., Handreck, K.A., 1967. Silica in soils, plants and animals. Adv. Agron. 19, 107–149.

Kamenidou, S., Cavins, J.C., Marek, S., 2008. Silicon supplements affect horticultural traits of greenhouse-produced ornamental sunflowers. HortScience 43, 236–239.

Kamenidou, S., Cavins, T.J., Marek, S., 2009. Evaluation of silicon as a nutritional supplement for greenhouse zinnia production. Sci. Hortic. 119, 297–301.

Kamenidou, S., Cavins, T.J., Marek, S., 2010. Silicon supplements affect floricultural quality traits and elemental nutrient concentrations of greenhouse produced gerbera. Sci. Hort. 123, 390–394.

Kardoni, F., Mosavi, J.S., Parande, S., Torbaghan, M.E., 2013. Effect of salinity stress and silicon application on yield and component yield of faba bean (*Vicia faba*). Int. J. Agric. Crop Sci. 6, 814–818.

Kazemi, M., Gholami, M., Bahmanipour, F., 2012. Effect of silicon and acetylsalicylic acid on antioxidant activity, membrane stability and ACC-oxidase activity in relation to vase life of carnation cut flowers. Biotechnology 11, 87–90.

Ketchum, B.H., 1954. Mineral nutrition of phytoplankton. Ann. Rev. Plant Physiol. 5, 55–64.

Khan, M.I.R., Khan, N.A., 2013. Salicylic acid and jasmonates: approaches in abiotic stress tolerance. J. Plant Biochem. Physiol. 1 (4), e113.

Khan, M.A., Ungar, I.A., Showalter, A.M., 2000. Effects of sodium chloride treatments on growth and ion accumulation of the halophyte *Haloxylon recurvum*. Commun. Soil Sci. Plant Anal. 31, 2763–2774.

Khan, M.N., Siddiqui, M.H., Mohammad, F., Naeem, M., Khan, M.M.A., 2010. Calcium chloride and gibberellic acid protect linseed (*Linum usitatissimum* L.) from NaCl stress by inducing antioxidative defence system and osmoprotectant accumulation. Acta Physiol. Plant 32 (1), 121.

Khan, M.I.R., Fatma, M., Per, T.S., Anjum, N.A., Khan, N.A., 2015. Salicylic acid-induced abiotic stress tolerance and underlying mechanisms in plants. Front. Plant Sci. 6, 462.

Khan, W.U.D., Aziz, T., Hussain, I., Ramzani, P.M.A., Reichenauer, T.G., 2016. Silicon: beneficial nutrient for maize crop to enhance photochemical efficiency of photosystem II under salt stress. Arch. Agron. Soil Sci. https://doi.org/10.1080/03650340.2016.1233322.

Khan, M.N., Mobin, M., Abbas, Z.K., Siddiqui, M.H., 2017. Nitric oxide-induced synthesis of hydrogen sulfide alleviates osmotic stress in wheat seedlings through sustaining antioxidant enzymes, osmolyte accumulation and cysteine homeostasis. Nitric Oxide. https://doi.org/10.1016/j.niox.2017.01.001.

Khoshgoftarmanesh, A.M., Khodarahmi, S., Haghighi, M., 2014. Effect of silicon nutrition on lipid peroxidation and antioxidant response of cucumber plants exposed to salinity stress. Arch. Agron. Soil Sci. 60, 639–653.

Kim, Y.H., Khan, A.L., Hamayun, M., Kang, S.M., Beom, Y.J., Lee, I.J., 2011. Influence of short-term silicon application on endogenous physiohormonal levels of *Oryza sativa* L. under wounding stress. Biol. Trace Elem. Res. 144, 1175–1185.

Kim, Y.H., Khan, A.L., Waqas, M., Shim, J.K., Kim, D.H., Lee, K.Y., Lee, I.J., 2014a. Silicon application to rice root zone influenced the phytohormonal and antioxidant responses under salinity stress. J. Plant Growth Reg. 33, 137–149.

Kim, Y.H., Khan, A.L., Kim, D.H., Lee, S.Y., Kim, K.M., Waqas, M., Jung, H.Y., Shin, J.H., Kim, J.G., Lee, I.J., 2014b. Silicon mitigates heavy metal stress by regulating P-type heavy metal ATPases, *Oryza sativa* low silicon genes, and endogenous phytohormones. BMC Plant Biol. 14, 1. https://doi.org/10.1186/1471-2229-14-13.

Kinrade, S.D., Hamilton, R.J., Schach, K.C.T.G., 2001. Aqueous hypervalent silicon complexes with aliphatic sugar acids. J. Chem. Soc. Dalton Trans. 7, 961–963.

Kinrade, S.D., Gillson, A.M.E., Knight, C.T.G., 2002. Silicon-29 NMR evidence of a transient hexavalent silicon complex in the diatom *Navicula pelliculosa*. J. Chem. Soc. Dalton Trans. 3, 307–309.

Korndorfer, G.H., Lepsch, I., 2001. Effect of silicon on plant growth and crop yield. In: Datnoff, L.E., Snyder, G.H., Korndorfer, G.H. (Eds.), Silicon in Agriculture. Studies in Plant Science, vol. 8. Elsevier, Amsterdam, pp. 133–147.

Krishnamurthy, P., Ranathunge, K., Nayak, S., Schreiber, L., Mathew, M.K., 2011. Root apoplastic barriers block NaCl transport to shoots in rice (*Oryza sativa* L.). J. Exp. Bot. 62, 4215–4228.

Lee, S.K., Sohn, E.Y., Hamayun, M., Yoon, J.Y., Lee, I.J., 2010. Effect of silicon on growth and salinity stress of soybean plant grown under hydroponic system. Agrofor. Syst. 80, 333–340.

Li, H., Zhu, Y., Hu, Y., Han, W., Gong, H., 2015. Beneficial effects of silicon in alleviating salinity stress of tomato seedlings grown under sand culture. Acta Physiol. Plant 37, 71–80.

Li, Y.T., Zhang, W.J., Cui, J.J., Lang, D.Y., Li, M., Zhao, Q.P., Zhang, X.H., 2016. Silicon nutrition alleviates the lipid peroxidation and ion imbalance of *Glycyrrhiza uralensis* seedlings under salt stress. Acta Physiol. Plant 38, 1–9.

Li, M., Zhang, K., Long, R., Sun, Y., Kang, J., Tiejun, Z., Shihao, C., 2017. iTRAQ-based comparative proteomic analysis reveals tissue-specific and novel early-stage molecular mechanisms of salt stress response in *Carex rigescens*. Environ. Exp. Bot. 143, 99–114.

Liang, Y.C., 1999. Effects of silicon on enzyme activity and sodium, potassium and calcium concentration in barley under salt stress. Plant Soil 29, 217–224.

Liang, Y.C., Chen, Q., Liu, Q., Zhang, W.H., Ding, R.X., 2003. Exogenous silicon (Si) increases antioxidant enzyme activity and reduces lipid peroxidation in roots of salt-stressed barley (*Hordeum vulgare* L.). J. Plant Physiol. 160, 1157–1164.

Liang, Y., Si, J., Römheld, V., 2005. Silicon uptake and transport is an active process in *Cucumis sativus*. New Phytol. 167, 797–804.

Liang, Y., Zhang, W., Chen, Q., Liu, Y., Din, R., 2006. Effect of exogenous silicon (Si) on HC-ATPase activity, phospholipids and fluidity of plasma membrane in leaves of salt-stressed barley (*Hordeum vulgare* L.). Environ. Exp. Bot. 57, 212–219.

Liang, Y., Sun, W., Zhu, Y.G., Christie, P., 2007. Mechanisms of silicon-mediated alleviation of abiotic stresses in higher plants: a review. Environ. Pollut. 147, 422–428.

Liang, Y., Nikolic, M., Bélanger, R., Gong, H., Song, A., 2015. Silicon in Agriculture. Springer, Dordrecht. https://doi.org/10.1007/978-94-017-9978-2.

Lindsay, W.L., 1979. Chemical Equilibria in Soils. Wiley, New York.

Liu, P., Yin, L., Deng, X., Wang, S., Tanaka, K., Zhang, S., 2014. Aquaporin-mediated increase in root hydraulic conductance is involved in silicon-induced improved root water uptake under osmotic stress in *Sorghum bicolor* L. J. Exp. Bot. 65 (17), 4747–4756.

Lucas, S., Durmaz, E., Akpınar, B.A., Budak, H., 2011. The drought response displayed by a DRE-binding protein from *Triticum dicoccoides*. Plant Physiol. Biochem. 49, 346–351.

Lux, A., Luxová, M., Hattori, T., Inanaga, S., Sugimoto, Y., 2002. Silicification in sorghum (*Sorghum bicolor*) cultivars with different drought tolerance. Physiol. Plant 115, 87–92.

Ma, J.F., 2004. Role of silicon in enhancing the resistance of plants to biotic and abiotic stresses. Soil Sci. Plant Nutr. 50, 11–18. https://doi.org/10.1080/00380768.2004.10408447.

Ma, J.F., Takahasi, E., 2002. Soil, Fertilizer, and Plant Silicon Research in Japan. Elsevier, Amsterdam.

Ma, J.F., Yamaji, N., 2006. Silicon uptake and accumulation in higher plants. Trends Plant Sci. 11, 392–397.

Ma, J.F., Yamaji, N., 2015. A cooperative system of silicon transport in plants. Trend Plant Sci. 20, 435–441.

Ma, J., Cai, H., He, C., Zhang, W., Wang, L., 2015. A hemicellulose-bound form of silicon inhibits cadmium ion uptake in rice (*Oryza sativa*) cells. New Phytol. 206, 1063–1074.

Mahdieh, M., Habibollahi, N., Amirjani, M.R., Abnosi, M.H., Ghorbanpour, M., 2015. Exogenous silicon nutrition ameliorates salt induced stress by improving growth and efficiency of PSII in *Oryza sativa* L. cultivars. J. Soil Sci. Plant Nutr. 15, 1050–1060.

Manivannan, A., Ahn, Y.K., 2017. Silicon regulates potential genes involved in major physiological processes in plants to combat stress. Front. Plant Sci. 8, 1346. https://doi.org/10.3389/fpls.2017.01346.

Markovich, O., Steiner, E., Kouřil, S., Tarkowski, P., Aharoni, A., Elbaum, R., 2017. Silicon promotes biosynthesis and delays senescence in *Arabidopsis* and *Sorghum*. Plant Cell Environ. https://doi.org/10.1111/pce.12913.

Marschner, H., 1995. Mineral Nutrition of Higher Plants, second ed. Academic Press, London.

McKeague, J.A., Cline, M.G., 1963. Silica in soil solutions. I. The form and concentration of dissolved silica in aqueous extracts of some soils. Can. J. Soil Sci. 43, 70–82.

Meyer, J.H., Keeping, M.G., 2001. Past, present and future research of the role of silicon for sugarcane in Southern Africa. In: Datnoff, L.E., Snyder, G.H., Korndorfer, G.H. (Eds.), Silicon in Agriculture. Studies in Plant Science, vol. 8. Elsevier, Amsterdam, pp. 257–275.

Monger, H.C., Kelly, E.F., 2002. Silica minerals. In: Soil Mineralogy with Environmental Applications. Soil Science Society of America, Madison, pp. 611–636.

Moussa, H.R., 2006. Influence of exogenous application of silicon on physiological response of salt stressed maize (*Zea mays* L.). Int. J. Agric. Biol. 8, 293–297.

Neuman, D., zur Nieden, U., 2001. Silicon and heavy metal tolerance of higher plants. Phytochemistry 56, 685–692.

Nwugo, C.C., Huerta, A.J., 2008. Effects of silicon nutrition on cadmium uptake: growth and photosynthesis of rice plants exposed to low level cadmium. Plant Soil 311, 73–86.

Parande, S., Zamani, G.R., Syyari Zahan, M.H., Ghaderi, M.G., 2013. Effects of silicon application on the yield and component of yield in the common bean (*Phaseolus vulgaris*) under salinity stress. Intern. J. Agron. Plant Produc. 4 (7), 1574–1579.

Pii, Y., Cesco, S., Mimmo, T., 2015. Shoot ionome to predict the synergism and antagonism between nutrients as affected by substrate and physiological status. Plant Physiol. Biochem. 94, 48–56.

Pilon, C., Soratto, R.P., Moreno, L.A., 2013. Effects of soil and foliar application of soluble silicon on mineral nutrition, gas exchange, and growth of potato plants. Crop Sci. 53, 1605–1614.

Qin, F., Shinozaki, K., Yamaguchi-Shinozaki, K., 2011. Achievements and challenges in understanding plant abiotic stress responses and tolerance. Plant Cell Physiol. 52, 1569–1582.

Rafi, M.M., Epstein, E., 1997. Silicon deprivation causes physical abnormalities in wheat (*Triticum aestivum* L.). J. Plant Physiol. 151, 497–501.

Rengasamy, P., 2010. Soil processes affecting crop production in salt affected soils. Funct. Plant Biol. 37, 613–620.

Reynolds, O.L., Keeping, M.G., Meyer, J.H., 2009. Silicon-augmented resistance of plants to herbivorous insects: a review. Ann. Appl. Biol. 155 (2), 171–186.

Richmond, K.E., Sussman, M., 2003. Got silicon? The non-essential beneficial plant nutrient. Curr. Opin. Plant Biol. 6, 268–272.

Rios, J.J., Martínez-Ballesta, M.C., Ruiz, J.M., Blasco, B., Carvajal, M., 2017. Silicon-mediated improvement in plant salinity tolerance: the role of aquaporins. Front. Plant Sci. 8, 948. https://doi.org/10.3389/fpls.2017.00948.

Round, F.E., Crawford, R.M., Mann, D.G., 1990. The Diatoms: Biology and Morphology of the Genera. Cambridge University Press, Cambridge.

Sahebi, M., Hanafi, M.M., Abdullah, S.N.A., Rafii, M.Y., Azizi, P., Tengoua, F.F., Azwa, J.N.M., Shabanimofrad, M., 2015. Importance of silicon and mechanisms of biosilica formation in plants. Biol. Med. Res. 2015, 396010.

Sangster, A.G., Hodson, M.J., Tubb, H.J., 2001. Silicon deposition in higher plants. In: Datnoff, L.E., Snyder, G.H., Korndorfer, G.H. (Eds.), Silicon in Agriculture. Studies in Plant Science, vol. 8. Elsevier, Amsterdam, pp. 85–113.

Saqib, M., Zörb, C., Schubert, S., 2008. Silicon-mediated improvement in the salt resistance of wheat (*Triticum aestivum*) results from increased sodium exclusion and resistance to oxidative stress. Funct. Plant Biol. 35, 633–639.

Savant, N.K., Snyder, G.H., Datnoff, L.E., 1997. Silicon management and sustainable rice production. Adv. Agron. 58, 151–199.

Shahid, M.A., Balal, R.M., Pervez, M.A., Abbas, T., Aqeel, M.A., Javaid, M.M., Garciasanchez, F., 2015. Foliar spray of phyto-extracts supplemented with silicon: an efficacious strategy to extenuate the salinity-induced effects in pea (*Pisum sativum* L.). Am. Soc. Hort. Sci.

Shahzad, M., Zörb, C., Geilfus, C.M., Mühling, K.H., 2013. Apoplastic Na^+ in *Vicia faba* leaves rises after short-term salt stress and is remedied by Si. J. Agron. Crop Sci. 199, 161–170.

Shen, X., Li, X., Li, Z., Li, J., Duan, L., Eneji, A.E., 2010. Growth, physiological attributes and antioxidant enzyme activities in soybean seedlings treated with or without silicon under UV-B radiation stress. J. Agron. Crop Sci. 196, 431–439.

Shi, Y., Zhang, Y., Han, W., Feng, R., Hu, Y., Guo, J., Gong, H., 2016. Silicon enhances water stress tolerance by improving root hydraulic conductance in *Solanum lycopersicum* L. Front. Plant Sci. 7, 196. https://doi.org/10.3389/fpls.2016.00196.

Shiguro, K., 2001. Review of research in Japan on the roles of silicon in conferring resistance against rice blast. In: Datonoff, L., Korndorfer, G., Synder, G. (Eds.), Silicon in Agriculture. Studies in Plant Science, vol. 8. Elsevier, Amsterdam, p. 8. 277–291.

Sivanesan, I., Son, M.S., Song, J.Y., Jeong, B.R., 2013a. Silicon supply through the sub irrigation system affects growth of three chrysanthemum cultivars. Hortic. Environ. Biotechnol. 54, 14–19.

Sivanesan, I., Son, M.S., Soundararajan, P., Jeong, B.R., 2013b. Growth of chrysanthemum cultivars as affected by silicon source and application method. Kor. J. Hortic. Sci. Technol. 31, 544–551.

Sommer, M., Kaczorek, D., Kuzyakov, Y., Breuer, J., 2006. Silicon pools and fluxes in soils and landscapes – a review. J. Plant Nutr. Soil Sci. 169, 310–329.

Song, A., Li, P., Fan, F., Li, Z., Liang, Y., 2014. The effect of silicon on photosynthesis and expression of its relevant genes in rice (*Oryza sativa* L.) under high-zinc stress. PLoS One 9 (11), e113782.

Sonobe, K., Hattori, T., An, P., Tsuji, W., Eneji, A.E., Kobayashi, S., Kawamura, Y., Tanaka, K., Inanaga, S., 2011. Effect of silicon application on sorghum root responses to water stress. J. Plant Nutr. 34, 71–82.

Struyf, E., Smis, A., VanDamme, S., Garnier, J., Govers, G., VanWesemael, B., 2010. Historical land use change has lowered terrestrial silica mobilization. Nat. Commun. 1, 129. https://doi.org/10.1038/ncomms1128.

Tale, A., Haddad, R., 2011. Study of silicon effects on antioxidant enzyme activities and osmotic adjustment of wheat under drought stress. Czech J. Genet. Plant Breed. 47, 17–27.

Tamai, K., Ma, J.F., 2008. Reexamination of silicon effects on rice growth and production under field conditions using a low silicon mutant. Plant Soil 307, 21–27.

Tripathi, D.K., Singh, V.P., Prasad, S.M., Chauhan, D.K., Dubey, N.K., Rai, A.K., 2015. Silicon-mediated alleviation of Cr(VI) in wheat seedlings as evidenced by chlorophyll fluorescence, laser induced breakdown spectroscopy and anatomical changes. Ecotoxicol. Environ. Saf. 113, 133–144.

Van Bockhaven, J., De Vleesschauwer, D., Hofte, M., 2013. Towards establishing broad-spectrum disease resistance in plants: silicon leads the way. J. Exp. Bot. 64, 1281–1293. https://doi.org/10.1093/jxb/ers329.

Van Hoest, P.J., 2006. Rice straw, the role of silica and treatments to improve quality. Anim. Feed Sci. Technol. 130, 137–171.

Vandevenne, F.I., Barao, L., Ronchi, B., Govers, G., Meire, P., Kelly, E.F., Struyf, E., 2015. Silicon pools in human impacted soils of temperate zones. Glob. Biogeochem. Cycles 29, 1439–1450.

Volcani, B.E., 1978. Role of silicon in diatom metabolism and silicification. In: Bendz, G., Lindquist, J. (Eds.), Biochemistry of Silicon and Related Problems. Plenum, New York, pp. 177–206.

Wang, X.S., Han, J.G., 2007. Effects of NaCl and Si on ion distribution in the roots, shoots and leaves of two alfalfa cultivars with different salt tolerance. Soil Sci. Plant Nutr. 53, 278–285.

Wang, X., Wei, Z., Liu, D., Zhao, G., 2011. Effects of NaCl and silicon on activities of antioxidative enzymes in roots, shoots and leaves of alfalfa. Afr. J. Biotechnol. 10, 545–549.

Wani, S.H., Kumar, V., Shriram, V., Sah, S.K., 2016. Phytohormones and their metabolic engineering for abiotic stress tolerance in crop plants. Crop J. 4 (3), 162–176.

Watanabe, S., Shimoi, E., Ohkama, N., Hayashi, H., Yoneyama, T., Yazaki, J., et al., 2004. Identification of several rice genes regulated by Si nutrition. Soil Sci. Plant Nutr. 50, 1273–1276.

Wedepohl, K.H., 1995. The composition of the continental crust. Geochim. Cosmochim. Acta 59, 1217–1232.

Weiss, A., Herzog, A., 1978. Isolation and characterization of a silicon organic complex from plants. In: Bendz, G., Lindquist, J. (Eds.), Biochemistry of Silicon and Related Problems. Plenum, New York, pp. 109–127.

Wiese, H., Nikolic, M., Roemheld, V., 2007. Silicon in plant nutirion. Effect of zinc, manganese and boron leaf concentrations and compartmentation. In: Sattelmacher, B., Horst, W.J. (Eds.), The Apoplast of Higher Plants: Compartment of Storage, Transport and Reactions. Springer, Dordrecht, pp. 33–47.

Williams, L., Salt, D.E., 2009. The plant ionome coming into focus. Curr. Opin. Plant Biol. 12, 247–249.

Xie, Z., Song, R., Shao, H., Song, F., Xu, H., Lu, Y., 2015. Silicon improves maize photosynthesis in saline-alkaline soils. Sci. World J. 2015, 245072.

Xu, L., Islam, F., Ali, B., Pei, Z., Li, J., Ghani, M.A., Zhou, W., 2017. Silicon and water-deficit stress differentially modulate physiology and ultrastructure in wheat (*Triticum aestivum* L.). Biotech 7 (4), 273.

Yin, L., Wang, S., Tanaka, K., Fujihara, S., Itai, A., Den, X., et al., 2016. Silicon mediated changes in polyamines participate in silicon-induced salt tolerance in *Sorghum bicolor* L. Plant Cell Environ. 39, 245–258. https://doi.org/10.1111/pce.12521.

Zhang, W., Xie, Z., Lang, D., Cui, J., Zhang, X., 2017. Beneficial effects of silicon on abiotic stress tolerance in legumes. J. Plant Nutr. https://doi.org/10.1080/01904167.2017.1346127.

Zhu, Y., Guo, J., Feng, R., Jia, J., Han, W., Gong, H., 2016. The regulatory role of silicon on carbohydrate metabolism in *Cucumis sativus* L. under salt stress. Plant Soil 406 (1–2), 231–249.

Zia, Z., Bakhat, H.F., Saqib, Z.A., Shah, G.M., Fahad, S., Ashraf, M.R., Shahid, M., 2017. Effect of water management and silicon on germination, growth, phosphorus and arsenic uptake in rice. Ecotoxicol. Environ. Saf. 144, 11–18.

Chapter 21

Involvement of Mitogen-Activated Protein Kinases in Abiotic Stress Responses in Plants

Rinukshi Wimalasekera[1], Günther F.E. Scherer[2]

[1]University of Sri Jayewardenapura, Nugegoda, Sri Lanka; [2]Leibniz University of Hannover, Institute of Horticultural Production Systems, Hannover, Germany

1. INTRODUCTION

Mitogen-activated protein kinases (MAPKs) are an evolutionarily highly conserved family of proteins present in all the eukaryotes. MAPK cascades play key roles in signal transduction in a variety of developmental processes, such as proliferation, differentiation, motility, cell cycle regulation, response to hormones, and in stress responses (Zwerger and Hirt, 2001; Jonak et al., 2002; Takahashi et al., 2007; Sinha et al., 2011; Çakır and Kılıçkaya, 2015; Mohanta et al., 2015a; Singh et al., 2015; Tang et al., 2016; Chardin et al., 2017; Singh and Sinha, 2017). Plant MAPKs play significant roles in a wide variety of biotic and abiotic stresses, such as pathogen infection, wounding, temperature, drought, and salinity (Sinha et al., 2011; Pecher et al., 2014; Lin et al., 2015; Xu and Zhang, 2015; Li et al., 2015; Sidonskaya et al., 2016; Aguilar et al., 2017).

MAPK cascades are minimally composed of three protein kinase modules, namely, (1) MAPKs, (2) MAPK kinases (MAPKKs), and (3) MAPK kinase kinases (MAPKKKs) (Zwerger and Hirt, 2001; Sinha et al., 2011; Kong et al., 2013). Addition of phosphate groups to serine and threonine/tyrosine residues of these proteins takes place at the transcriptional, translational, and posttranslational levels and modifications in turn lead to the alterations in catalytic activity, affinity, and interactions of target proteins. Usually, MAPKs function via a cascade of networks, where MAPK is phosphorylated and activated by MAPKK, which itself is activated by MAPKKK. The phosphorylation of proteins is reversible as a result of phosphatase-driven dephosphorylation events taking place in the cells (Zwerger and Hirt, 2001; Mohanta et al., 2015b; Pitzschke, 2015).

MAPK, MAPKK, and MAPKKK are networked in specific signal transduction pathways that affect diverse upstream receptors and downstream target components (Mishra et al., 2006; Rodriguez et al., 2010; Sinha et al., 2011; Raghuram et al., 2014; Kim et al., 2016; Zhang et al., 2016). Recent advances in technological knowledge and identification of loss-of-function and gain-of-function mutants of the MAPK cascade paved the way to identify and characterize the complex gene networks and cross talk of MAPKs in plants (Zhang et al., 2006; Abass and Morris, 2013; Neupane et al., 2013; Persak and Pitzschke, 2013; Sun et al., 2014; Su and Krysan, 2016). The three iconic MAPKs in plants are MPK3, MPK4, and MPK6. Several studies show that the well-characterized MPK3, MPK4, and MPK6 of *Arabidopsis* are involved in diverse signaling cascades in abiotic and biotic stress responses (Colcombet et al., 2016).

2. PLANT MAPK CASCADES

In plants, a distinct array of MAPK genes has been identified, suggesting the complex nature of the plant MAPK cascade. Depending on the amino acid sequences plant MAPKs are grouped into at least four families (Ichimura et al., 2002). MAPKs in family I are mostly involved in hormonal and environmental responses, family II members are involved in cell cycle regulation and environmental stress responses, and the function of family III and IV are not well known. In *Arabidopsis* genome, four different groups of MAPKKs have been identified, with a total of 10 MAPKKs that are likely to activate multiple MAPKs (Ichimura et al., 2002). MAPKKKs comprise a diverse family of kinases. About 80 MAPKKKs, which can be categorized into two subfamilies, have been identified in *Arabidopsis* (Ichimura et al., 2002; Zhang et al., 2006; Moustafa, 2014).

Plant Metabolites and Regulation Under Environmental Stress. http://dx.doi.org/10.1016/B978-0-12-812689-9.00021-2

In *Arabidopsis* genome, 20 MAPK genes have been identified (Zhang et al., 2006 and references therein). MAPK gene families from 40 different plant species have been identified, collectively giving rise to 589 different MAPKs (Mohanta et al., 2015b). Based on the sequences and phylogenetic analysis, all the sequences are being clustered into six different groups (Mohanta et al., 2015b). A higher number of MAPK genes in the genomes is found in species such as *Glycine max* (31 genes), *Brassica campestris* (30 genes), and *Gossypium raimondii* (28 genes) (Mohanta et al., 2015b). Variation in the number of MAPK genes within a family is dependent upon the complexity and ploidy level of the species of plants (Yin et al., 2013; Zhang et al., 2013, 2014; Nitta et al., 2014; Mohanta et al., 2015b).

3. MAPK CASCADES IN ABIOTIC STRESS SIGNALING

Throughout the life cycle, plants encounter an array of biotic and abiotic stresses. Plants are capable of sensing stresses, such as drought, chilling, high temperature, salinity, and a wide variety of pathogen/microbial elicitors. Sensing (perception) of diverse stresses is followed by signal transduction leading to a variety of defense reactions to adapt to metabolic and physiological alterations they encounter during stress conditions. One of the ubiquitous evolutionarily conserved signal transduction mechanisms that lead to stress tolerance is through the MAPK cascade (Pathak et al., 2013; Danquah et al., 2014; Mithoe et al., 2016). The MAPK cascade, which is minimally composed of MAPKKK, MAPKK, and MAPK, is sequentially activated via phosphorylation (Ichimura et al., 2002; Colcombet and Hirt, 2008; Danquah et al., 2014) (Fig. 21.1). MAPKKKs act as adaptors that link upstream signaling steps to the MAPK cascades (Colcombet and Hirt, 2008).

Several MAPK cascade components are known to be activated by a single abiotic and biotic stress. One MAPK module can be used for the transmission of multiple signals in plants. MAPK signaling cascades are activated by abiotic stresses, such as drought, salt, heat/cold, ultraviolet, wounding, and heavy metal (Tena et al., 2001; Zhang et al., 2006; Colcombet and Hirt, 2008; Liu, 2012; Sheikh et al., 2013; Danquah et al., 2014). In *Arabidopsis*, the expression of MPK3, MPK4, and MPK6 is activated by low temperature, osmotic stress, salinity, touch, and wounding but they are involved in specific signaling pathways responding to the environmental stresses. Extensive cross talks among different pathways at the level of MAP kinases are present in plants (Jonak et al., 2002). Several studies support key functions of MAPK cascades in pathogen defenses in an array of plant species, including *Arabidopsis*, parsley, tobacco, rice, and tomato (Rodriguez et al., 2010; Tsuda et al., 2013; Danquah et al., 2014; López-Bucio et al., 2014; Nie and Xu, 2016).

3.1 Drought Tolerance

Over the evolution, plants have evolved multiple mechanisms to tolerate drought stress. Accumulation of metabolites, synthesis of osmoprotectants and antioxidants, and regulation of water uptake and evaporation are important mechanisms

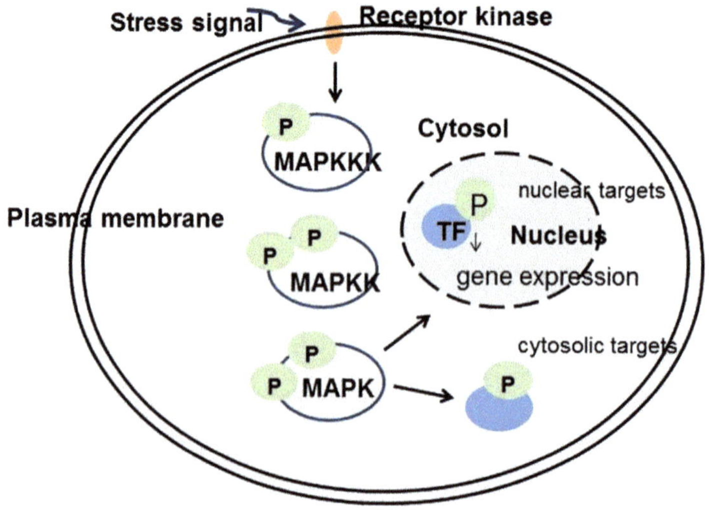

FIGURE 21.1 Schematic representation of a typical mitogen-activated protein kinase (MAPK) cascade. Stimulation of receptor kinase activates the upstream MAPKKKs, which in turn leads to the progressive phosphorylation and subsequent activation of downstream MAPKKs and MAPKs. MAPKs cascade through their kinase activity and transduce environmental signals that drive specific stress response by gene activation/repression and ultimately stress adaptation.

in adapting to drought conditions (Krasensky and Jonak, 2012; Liu et al., 2015). As a component of the signal transduction pathway, MAPK cascades play significant roles in responses to drought stresses (Zhang et al., 2012). MAPK cascades play roles in several abscisic acid (ABA) responses, including stomatal regulation, antioxidant defense, and transcriptional regulation of drought stress-responsive genes (Xing et al., 2008; Jammes et al., 2009; Danquah et al., 2015; Mitula et al., 2015; Li et al., 2017a,b).

Several experimental evidences show that MAPKs regulate the downstream signaling pathways related to drought stress responses. Overexpression of MKK1 and MPK6 in *Arabidopsis* showed an increased tolerance to drought stress (Xing et al., 2008). Wang et al. (2016) reported that overexpressing an MAPKK gene, GhMKK3, from cotton in *Nicotiana benthamiana* leads to enhanced drought tolerance. The drought tolerance resulted mainly by the regulation of water loss by ABA-induced stomatal closure and a decrease in stomatal density. Moreover, silencing of *GhMKK3* of cotton resulted in increased sensitivity to water stress (Wang et al., 2016). The MKK4/MKK5-MPK3/MPK6 cascade in *Arabidopsis* is shown to be a key regulator of stomatal development (Wang et al., 2007). In rice, overexpression of *OsMAPK5* showed enhanced tolerance to drought, whereas suppression of *OsMAPK5* reduced the stress tolerance, and ABA is shown to induce *OsMAPK5* in rice leaves (Xiong and Yang, 2003).

Certain members of the MAPK cascade in a variety of plant species have been reported to be activated by ABA to perform diverse functions leading to water stress tolerance (Xing et al., 2008; Jammes et al., 2009; Li et al., 2013, 2017a,b; Danquah et al., 2015). A number of experimental data indicate that treatment with ABA induced the transcriptional regulation of MPK3, MPK5, MPK7, MPK18, MPK20, MKK9, MAPKKK1 (ANP1), MAPKKK10 (MEKK3), MAPKKK14, MAPKKK15, MAPKKK16, MAPKKK17, MAPKKK18, MAPKKK19, Raf6, Raf12, and Raf35 in *Arabidopsis* (Ortiz-Masia et al., 2007; Menges et al., 2008; Wang et al., 2011; Danquah et al., 2015). Mutants of MKK3 in *Arabidopsis* showed impaired control of water loss in long-term drought conditions indicating the role of MKK3 in ABA signaling (Danquah et al., 2015). In *Arabidopsis*, many of the Ca^{2+}-mediated ABA responses are shown to be regulated by Ca^{2+}-dependent protein kinases, such as AtCPK3, AtCPK4, and AtCPK6 (Brandt et al., 2012; Ronzier et al., 2014). ABA-insensitive protein kinase 1 (AIK1), a novel MAPKKK, is shown to be involved in ABA signaling and has functions in regulating stomatal responses and root growth (Li et al., 2017). Recent experimental evidences indicate ABA-stimulated regulation of several members of the MAPK family of barley, moss, maize, pea, and rice (Danquah et al., 2015 and references therein). Taken together, MAPKs are involved in ABA-mediated opening/closing machinery of stomata during water stress. Although the emerging experimental evidences support the link between ABA and MAPK pathways, the details of the complex cellular signaling are poorly understood.

3.2 Salt Tolerance

Crop production in many parts of the world is affected by salinity stress. Increasing evidences indicate that MAPKs are involved in signal transduction in response to salt stress by transmitting stress signal to at least two MAPK cascades (Zhang et al., 2006; Sinha et al., 2011 and references therein; Zhang et al., 2012; Zaidi et al., 2016).

Several studies have shown that in *Arabidopsis*, MAPKKKs are important signaling elements in salt stress signaling cascades. For example, *AtMPK4* and *AtMPK6* of *Arabidopsis* are induced in response to a variety of environmental stresses and salt stress is reported to be one of the stresses where these two genes are rapidly and transiently induced to regulate the downstream reactions (Mishra et al., 2006 and references therein). In *Arabidopsis*, the overexpression of MKK2 (MAPKKs) resulted in constitutive activity of MPK4 and MPK6 (MAPKs) leading to salt tolerance by upregulation of downstream stress-induced genes (Teige et al., 2004). Reduced tolerance to salinity of mkk2 mutants indicates the contribution of MKK2 in salt stress tolerance (Teige et al., 2004) Overexpression of *AtMKK9* increased the sensitivity to salt stress (Xu et al., 2008).

In other plant species, several components of the MAPK pathway are shown to be regulated in response to salt stress. In alfalfa, SIMK, an MAPK is upregulated by salt stress (Munnik et al., 1999). The components of the MAPK signaling cascade, *GhMPK7* of cotton (Shi et al., 2010); *ZmMPK3*, *ZmMAPK5*, and *ZmSIMK1* of maize (Gu et al., 2010); MAPKK1 and OsMAPK5 (Xiong and Yang, 2003) of rice; and SIPK of tobacco (Mikołajczyk et al., 2000) are known to be activated to alleviate the salt stress conditions.

3.3 Cold and Heat Tolerance

Plants experience high- and low-temperature stresses throughout their life cycles. They have developed various mechanisms to regulate the extreme temperatures. Several studies have shown the role of MAPKs/MAPPKs in both the high- and low-temperature stress tolerance (Mishra et al., 2006; references therein; Danquah et al., 2015 and references therein).

Heat shock–activated regulation of MAPK signaling components has been reported in certain plant species. HAMK from alfalfa is reported to be the first identified heat shock–activated MAPK in plants (Sangwan et al., 2002). Since then, several studies have reported the heat stress–mediated regulation of MAPKs. For example, transcription levels of StMPK1 of potato tubers have been increased by heat treatment (Blanco et al., 2006); in rice, OsMSRMK2 is rapidly induced by high temperatures (Agrawal et al., 2002).

MAPK signaling is a key signal transduction pathway that leads to cold tolerance in plants. Rapid transient activation of AtMPK4 and AtMPK6 enzymes of *Arabidopsis* has been reported in response to low temperature (Ichimura et al., 2000). Further studies showed that the constitutive activity of MPK4 and MPK6 are mediated by an overexpression of MKK2 and results in an increase of cold tolerance in *Arabidopsis* (Teige et al., 2004; Qiu et al., 2008). In maize, ZmMPK3 and ZmMAPK5 transcript levels were increased markedly within 30 min in response to cold temperatures (Ding et al., 2009; Wang et al., 2010). In alfalfa, p44MKK4, an MAPKK, was activated by cold stress (Jonak et al., 1996).

3.4 Reactive Oxygen Species and Ozone

Several signaling pathways regulated the reactive oxygen species (ROS) homeostasis. Xing et al. (2008) showed that MKK1–MPK6 regulate the H_2O_2 metabolism through catalase CAT1. MPK3, MPK6, and multiple MAPKKKs in *Arabidopsis* are known to be involved in regulating ROS metabolism (Kovtun et al., 2000; Li et al., 2015; Liu and He, 2017). MKK3-based MAPK has been upregulated upstream of ROS production in response to wounding and mutants of MPK8 showed reduced ROS production in wounded leaves (Colcombet et al., 2016 and references therein). H_2O_2-induced activation of MAPKs has been reported in other plant species, including maize, pea, and alfalfa (Sinha et al., 2011 and references therein).

Exposure of plants to ozone leads to the activation of MAP kinase signaling pathway in plants (Samuel et al., 2000). Ozone triggers the oxidative burst in cells. For example, increased transcription levels of OsMSRMK2 gene (MAP kinase) in rice OsMSRMK2 (Samuel et al., 2000; Samuel and Ellis, 2002) in response to the exposure to ozone has been observed. In *Arabidopsis*, MPK3 and MPK6 (Ahlfors et al., 2004) are activated by ozone treatments and plants lacking these kinase genes are hypersensitive to ozone (Miles et al., 2005).

3.5 Wounding

In most instances, the activation of defense-related genes in response to physical injuries is known to be induced through the signaling pathways that include MAPKs. After being wounded, an increase in MAPK levels is reported (Sinha et al., 2011 and references therein).

MKK3 and MPK8 in *Arabidopsis* are upregulated in response to wounding (Colcombet et al., 2016; references therein; Genenncher et al., 2016). Several wound-activated MAPKs, including MMK4 (MAPK) of alfalfa, NtMPK4 of tobacco, and StMPK1 of potato, have been identified in various plant species.

4. PERSPECTIVES

MAPKK cascades are highly conserved and several components of the MAPK cascade have been characterized from different plant species. They have emerged as universal signal transduction elements that link different receptors with cellular and nuclear responses. MAPKs, MAPKKs, and MAPKKKs are known to play key roles in cell division, growth, hormone responses, and biotic and abiotic stress responses.

Although there are increasing evidences that several MAPK cascade components play central roles in development and stress responses, the regulatory mechanisms are still unclear and need deeper understanding. With the light of advanced knowledge and techniques, novel approaches can be used to identify the functions of MAPK signaling cascades in plants.

REFERENCES

Abass, M., Morris, P.C., 2013. The *Hordeum vulgare* signalling protein MAP kinase 4 is a regulator of biotic and abiotic stress responses. J. Plant Physiol. 170, 1353–1359.

Agrawal, G.K., Rakwal, R., Iwahashi, H., 2002. Isolation of novel rice (*Oryza sativa* L.) multiple stress responsive MAP kinase gene, OsMSRMK2, whose mRNA accumulates rapidly in response to environmental cues. Biochem. Biophys. Res. Commun. 294, 1009–1016.

Aguilar, E., Del Toro, F.J., Canto, T., Tenllado, F., 2017. Identification of MAPKs as signal transduction components required for the cell death response during compatible infection by the synergistic pair Potato virus X-Potato virus Y. Virology 509, 178–184.

Ahlfors, R., Macioszek, V., Rudd, J., Brosché, M., Schlichting, R., Scheel, D., Kangasjärvi, J., 2004. Stress hormone independent activation and nuclear translocation of mitogen-activated protein kinases in *Arabidopsis thaliana* during ozone exposure. Plant J. 40, 512–522.

Blanco, F.A., Zanetti, M.E., Casalongué, C.A., Daleo, G.R., 2006. Molecular characterization of a potato MAP kinase transcriptionally regulated by multiple environmental stresses. Plant Physiol. Biochem. 44, 315–322.

Brandt, B., Brodsky, D.E., Xue, S., Negi, J., Iba, K., Kangasjärvi, J., Ghassemian, M., Stephan, A.B., Hu, H., Schroeder, J.I., 2012. Reconstitution of abscisic acid activation of SLAC1 anion channel by CPK6 and OST1 kinases and branched ABI1 PP2C phosphatase action. Proc. Natl. Acad. Sci. U.S.A. 109, 10593–10598.

Çakır, B., Kılıçkaya, O., 2015. Mitogen-activated protein kinase cascades in *Vitis vinifera*. Front. Plant Sci. 6, 556.

Chardin, C., Schenk, S.T., Hirt, H., Colcombet, J., Krapp, A., 2017. Review: mitogen-activated protein kinases in nutritional signaling in *Arabidopsis*. Plant Sci. 260, 101–108.

Colcombet, J., Hirt, H., 2008. *Arabidopsis* MAPKs: a complex signalling network involved in multiple biological processes. Biochem. J. 413, 217–226.

Colcombet, J., Sözen, C., Hirt, H., 2016. Convergence of multiple MAP3Ks on MKK3 identifies a set of novel stress MAPK modules. Front. Plant Sci. 7, 1941.

Danquah, A., de Zelicourt, A., Colcombet, J., Hirt, H., 2014. The role of ABA and MAPK signaling pathways in plant abiotic stress responses. Biotechnol. Adv. 32, 40–52.

Danquah, A., de Zélicourt, A., Boudsocq, M., Neubauer, J., FreiDitFrey, N., Leonhardt, N., et al., 2015. Identification and characterization of an ABA-activated MAP kinase cascade in *Arabidopsis thaliana*. Plant J. 82, 232–244.

Ding, H., Zhang, A., Wang, J., Lu, R., Zhang, H., Zhang, J., Jiang, M., 2009. Identity of an ABA activated 46 kDa mitogen activated protein kinase from *Zea mays* leaves: partial purification, identification and characterization. Planta 230, 239–251.

Genenncher, B., Wirthmueller, L., Roth, C., Klenke, M., Ma, L., Sharon, A., Wiermer, M., 2016. Nucleoporin-regulated MAP kinase signaling in immunity to a necrotrophic fungal pathogen. Plant Physiol. 172, 1293–1305.

Gu, L., Liu, Y., Zong, X., Liu, L., Li, D.P., Li, D.Q., 2010. Overexpression of maize mitogen-activated protein kinase gene, ZmSIMK1 in *Arabidopsis* increases tolerance to salt stress. Mol. Biol. Rep. 37, 4067–4073.

Ichimura, K., Mizoguchi, T., Yoshida, R., Yuasa, T., Shinozaki, K., 2000. Various abiotic stresses rapidly activate *Arabidopsis* MAP kinases ATMPK4 and ATMPK6. Plant J. 24, 655–665.

Ichimura, K., MAPK group, et al., 2002. Mitogen-activated protein kinase cascades in plants: a new nomenclature. Trends Plant Sci. 7, 301–308.

Jammes, F., Song, C., Shin, D., Munemasa, S., Takeda, K., Gu, D., et al., 2009. MAP kinases MPK9 and MPK12 are preferentially expressed in guard cells and positively regulate ROS-mediated ABA signaling. Proc. Natl. Acad. Sci. U.S.A. 106, 20520–20525.

Jonak, C., Kiegerl, S., Ligterink, W., Barker, P.J., Huskisson, N.S., Hirt, H., 1996. Stress signaling in plants: a mitogen-activated protein kinase pathway is activated by cold and drought. Proc. Natl. Acad. Sci. U.S.A. 93, 11274–11279.

Jonak, C., Okresz, L., Bogre, L., Hirt, H., 2002. Complexity, cross talk and integration of plant MAP kinase signaling. Curr. Opin. Plant Biol. 5, 415–424.

Kim, H.S., Park, S.C., Ji, C.Y., Park, S., Jeong, J.C., Lee, H.S., Kwak, S.S., 2016. Molecular characterization of biotic and abiotic stress-responsive MAP kinase genes, IbMPK3 and IbMPK6, in sweetpotato. Plant Physiol. Biochem. 108, 37–48.

Kong, X., Pan, J., Zhang, D., Jiang, S., Cai, G., Wang, L., Li, D., 2013. Identification of mitogen-activated protein kinase kinase gene family and MKK-MAPK interaction network in maize. Biochem. Biophys. Res. Commun. 441, 964–969.

Kovtun, Y., Chiu, W.L., Tena, G., Sheen, J., 2000. Functional analysis of oxidative stress-activated mitogen-activated protein kinase cascade in plants. Proc. Natl. Acad. Sci. U.S.A. 97, 2940–2945.

Krasensky, J., Jonak, C., 2012. Drought, salt, and temperature stress induced metabolic rearrangements and regulatory networks. J. Exp. Bot. 63, 1593–1608.

Li, C., Yan, J.-M., Li, Y.-Z., Zhang, Z.-C., Wang, Q.-L., Liang, Y., 2013. Silencing the SpMPK1, SpMPK2, and SpMPK3 genes in tomato reduces abscisic acid—mediated drought tolerance. Int. J. Mol. Sci. 14, 21983–21996.

Li, L., Ye, C., Zhao, R., Li, X., Liu, W.Z., Wu, F., Yan, J., Jiang, Y.Q., Yang, B., 2015. Mitogen-activated protein kinase kinase kinase (MAPKKK) 4 from rapeseed (*Brassica napus* L.) is a novel member inducing ROS accumulation and cell death. Biochem. Biophys. Res. Commun. 467, 792–797.

Li, K., Yang, F., Zhang, G., Song, S., Li, Y., Ren, D., Miao, Y., Song, C.-P., 2017a. AIK1, a mitogen-activated protein kinase, modulates abscisic acid responses through the MKK5-MPK6 kinase cascade. Plant Physiol. 173, 1391–1408.

Li, Y., Cai, H., Liu, P., Wang, C., Gao, H., Wu, C., Yan, K., Zhang, S., Huang, J., Zheng, C., 2017b. *Arabidopsis* MAPKKK18 positively regulates drought stress resistance via downstream MAPKK3. Biochem. Biophys. Res. Commun. 484, 292–297.

Lin, Z.-J.D., Liebrand, T.W.H., Yadeta, K.A., Coaker, G., 2015. PBL13 is a serine/threonine protein kinase that negatively regulates *Arabidopsis* immune responses. Plant Physiol. 169, 2950–2962.

Liu, Y., 2012. Roles of mitogen-activated protein kinase cascades in ABA signaling. Plant Cell Rep. 31, 1–12.

Liu, Y., He, C., 2017. A review of redox signaling and the control of MAP kinase pathway in plants. Redox Biol. 11, 192–204.

Liu, Y., Zhou, M., Gao, Z., Ren, W., Yang, F., He, H., Zhao, J., 2015. RNA-seq analysis reveals MAPKKK family members related to drought tolerance in maize. PLoS One 10, e0143128.

López-Bucio, J.S., Dubrovsky, J.G., Raya-González, J., et al., 2014. *Arabidopsis thaliana* mitogen-activated protein kinase 6 is involved in seed formation and modulation of primary and lateral root development. J. Exp. Bot. 65, 169–183.

Menges, M., Doczi, R., Okresz, L., Morandini, P., Mizzi, L., SolovievM, et al., 2008. Comprehensive gene expression atlas for the *Arabidopsis* MAP kinase signalling pathways. New Phytol. 179, 643–662.

Mikołajczyk, M., Awotunde, O.S., Muszynska, G., Klessig, D.F., Dobrowolska, G., 2000. Osmotic stress induces rapid activation of a salicylic acid-induced protein kinase and a homolog of protein kinase ASK1 in tobacco cells. Plant Cell 12, 165–178.

Miles, G.P., Samuel, M.A., Zhang, Y., Ellis, B.E., 2005. RNA interference-based (RNAi) suppression of AtMPK6, an *Arabidopsis* mitogen-activated protein kinase, results in hypersensitivity to ozone and misregulation of AtMPK3. Environ. Pollut. 138, 230–237.

Mishra, N.S., Tuteja, R., Tuteja, N., 2006. Signaling through MAP kinase networks in plants. Arch. Biochem. Biophys. 452, 55–68.

Mithoe, S.C., Ludwig, C., Pel, M.J., et al., 2016. Attenuation of pattern recognition receptor signaling is mediated by a MAP kinase kinase kinase. EMBO Rep. 17, 441–454.

Mitula, F., Tajdel, M., Cieśla, A., et al., 2015. *Arabidopsis* ABA-activated kinase MAPKKK18 is regulated by protein phosphatase 2C ABI1 and the ubiquitin–proteasome pathway. Plant Cell Physiol. 56, 2351–2367.

Mohanta, T.K., Arora, P.K., Mohanta, N., Parida, P., Bae, H., 2015a. Identification of new members of the MAPK gene family in plants shows diverse conserved domains and novel activation loop variants. BMC Genom. 16, 58.

Mohanta, T.K., Mohanta, N., Parida, P., Bae, H., 2015b. Mitogen activated protein kinase (MPK) interacts with auxin influx carrier (OsAux/LAX1) involved in auxin signaling in plant. Biol. Proced. Online 17, 13.

Moustafa, K., 2014. MAPK transgenic circuit to improve plant stress-tolerance? Plant Signal. Behav. 11, e970101.

Munnik, T., Ligterink, W., Meskiene, I., Calderini, O., Beyerly, J., Musgrave, A., Hirt, H., 1999. Distinct osmo-sensing protein kinase pathways are involved in signaling moderate and severe hyper-osmotic stress. Plant J. 20, 381–388.

Neupane, A., Nepal, M.P., Benson, B.V., MacArthur, K.J., Piya, S., 2013. Evolutionary history of mitogen-activated protein kinase (MAPK) genes in *Lotus*, *Medicago*, and *Phaseolus*. Plant Signal. Behav. 8, e27189.

Nie, S., Xu, H., 2016. Riboflavin-induced disease resistance requires the mitogen-activated protein kinases 3 and 6 in *Arabidopsis thaliana*. PLoS One 11, e0153175.

Nitta, Y., Ding, P., Zhang, Y., 2014. Identification of additional MAP kinases activated upon PAMP treatment. Plant Signal. Behav. 9, e976155.

Ortiz-Masia, D., Perez-Amador, M.A., Carbonell, J., Marcote, M.J., 2007. Diverse stress signals activate the C1 subgroup MAP kinases of *Arabidopsis*. FEBS Lett. 581, 1834–1840.

Pathak, R.K., Taj, G., Pandey, D., Arora, S., Kumar, A., 2013. Modeling of the MAPK machinery activation in response to various abiotic and biotic stresses in plants by a system biology approach. Bioinformation 9, 443–449.

Pecher, P., Eschen-Lippold, L., Herklotz, S., Kuhle, K., Naumann, K., Bethke, G., Uhrig, J., Weyhe, M., Scheel, D., Lee, J., 2014. The *Arabidopsis thaliana* mitogen-activated protein kinases MPK3 and MPK6 target a subclass of 'VQ-motif'-containing proteins to regulate immune responses. New Phytol. 203, 592–606.

Persak, H., Pitzschke, A., 2013. Tight interconnection and multi-level control of *Arabidopsis* MYB44 in MAPK cascade signaling. PLoS One 8, e57547.

Pitzschke, A., 2015. Modes of MAPK substrate recognition and control. Trends Plant Sci. 20, 49–55.

Qiu, J.L., Fiil, B.K., Petersen, K., Nielsen, H.B., Botanga, C.J., et al., 2008. *Arabidopsis* MAP kinase 4 regulates gene expression through transcription factor release in the nucleus. EMBO J. 27, 2214–2221.

Raghuram, B., Sheikh, A.H., Sinha, A.K., 2014. Regulation of MAP kinase signaling cascade by microRNAs in *Oryza sativa*. Plant Signal. Behav. 9, e972130.

Rodriguez, M.C., Petersen, M., Mundy, J., 2010. Mitogen-activated protein kinase signaling in plants. Annu. Rev. Plant Biol. 61, 621–649.

Ronzier, E., Corratgé-Faillie, C., Sanchez, F., Prado, K., Brière, C., Leonhardt, N., Thibaud, J.B., Xiong, T.C., 2014. CPK13, a non canonical Ca^{2+}-dependent protein kinase, specifically inhibits KAT2 and KAT1 shaker K^+ channels and reduces stomatal opening. Plant Physiol. 166, 314–326.

Samuel, M.A., Ellis, B.E., 2002. Double jeopardy: both overexpression and suppression of a redox-activated plant mitogen-activated protein kinase render tobacco plants ozone sensitive. Plant Cell 14, 2059–2069.

Samuel, M.A., Miles, G.P., Ellis, B.E., 2000. Ozone treatment rapidly activates MAP kinase signaling in plants. Plant J. 22, 367–376.

Sangwan, V., Orvar, B.L., Beyerly, J., Hirt, H., Dhindsa, R.S., 2002. Opposite changes in membrane fluidity mimic cold and heat stress activation of distinct plant MAP kinase pathways. Plant J. 31, 629–638.

Sheikh, A.H., Raghuram, B., Jalmi, S.K., Wankhede, D.P., Singh, P., Sinha, A.K., 2013. Interaction between two rice mitogen activated protein kinases and its possible role in plant defense. BMC Plant Biol. 13, 121.

Shi, J., An, H.L., Zhang, L., Gao, Z., Guo, X.Q., 2010. GhMPK7, a novel multiple stress-responsive cotton group C MAPK gene, has a role in broad spectrum disease resistance and plant development. Plant Mol. Biol. 74, 1–17.

Sidonskaya, E., Schweighofer, A., Shubchynskyy, V., Kammerhofer, N., Hofmann, J., Wieczorek, K., Meskiene, I., 2016. Plant resistance against the parasitic nematode *Heterodera schachtii* is mediated by MPK3 and MPK6 kinases, which are controlled by the MAPK phosphatase AP2C1 in *Arabidopsis*. J. Exp. Bot. 67, 107–118.

Singh, P., Sinha, A.K., 2017. Interplay between auxin and cytokinin and its impact on mitogen activated protein kinase (MAPK). Methods Mol. Biol. 1569, 93–100.

Singh, S., Chaturvedi, A., Mani, A., 2015. Cross-family comparative proteomic study and molecular phylogeny of MAP kinases in plants. Interdiscip. Sci. 7, 357–363.

Sinha, A.K., Jaggi, M., Raghuram, B., Tutejya, N., 2011. Mitogen-activated protein kinase signaling in plants under abiotic stress. Plant Signal. Behav. 6, 196–203.

Su, S.H., Krysan, P.J., 2016. A double-mutant collection targeting MAP kinase related genes in *Arabidopsis* for studying genetic interactions. Plant J. 88, 867–878.

Sun, Y., Wang, C., Yang, B., Wu, F., Hao, X., Liang, W., Niu, F., Yan, J., Zhang, H., Wang, B., Deyholos, M.K., Jiang, Y.Q., 2014. Identification and functional analysis of mitogen-activated protein kinase kinase kinase (MAPKKK) genes in canola (*Brassica napus* L.). J. Exp. Bot. 65, 2171–2188.

Takahashi, F., Yoshida, R., Ichimura, K., Mizoguchi, T., Seo, S., Yonezawa, M., et al., 2007. The mitogen-activated protein kinase cascade MKK3-MPK6 is an important part of the jasmonate signal transduction pathway in *Arabidopsis*. Plant Cell 19, 805–818.

Tang, Q., Guittard-Crilat, E., Maldiney, R., Habricot, Y., Miginiac, E., Bouly, J.P., Lebreton, S., 2016. The mitogen-activated protein kinase phosphatase PHS1 regulates flowering in *Arabidopsis thaliana*. Planta 243, 909–923.

Teige, M., Scheikl, E., Eulgem, T., Dóczi, R., Ichimura, K., Shinozaki, K., 2004. The MKK2 pathway mediates cold and salt stress signaling in *Arabidopsis*. Mol. Cell 15, 141–152.

Tena, G., Asai, T., Chiu, W.L., Sheen, J., 2001. Plant mitogen-activated protein kinase signaling cascades. Curr. Opin. Plant Biol. 4, 392–400.

Tsuda, K., Mine, A., Bethke, G., et al., 2013. Dual regulation of gene expression mediated by extended MAPK activation and salicylic acid contributes to robust innate immunity in *Arabidopsis thaliana*. PLoS Genet. 9, e1004015.

Wang, H., Ngwenyama, N., Liu, Y., Walker, J.C., Zhang, S., 2007. Stomatal development and patterning are regulated by environmentally responsive mitogen-activated protein kinases in *Arabidopsis*. Plant Cell 19, 63–73.

Wang, J., Ding, H., Zhang, A., Ma, F., Cao, J., Jiang, M., 2010. A novel mitogen-activated protein kinase gene in maize (*Zea mays*), ZmMPK3, is involved in response to diverse environmental cues. J. Integr. Plant Biol. 52, 442–452.

Wang, R.S., Pandey, S., Li, S., Gookin, T.E., Zhao, Z., Albert, R., et al., 2011. Common and unique elements of the ABA-regulated transcriptome of *Arabidopsis* guard cells. BMC Genom. 12, 216.

Wang, C., Lu, W., He, X., Wang, F., Zhou, Y., Guo, X., Guo, X., 2016. The cotton mitogen-activated protein kinase kinase 3 functions in drought tolerance by regulating stomatal responses and root growth. Plant Cell Physiol. 57, 1629–1642.

Xing, Y., Jia, W., Zhang, J., 2008. AtMKK1 mediates ABA-induced CAT1 expression and H_2O_2 production via AtMPK6-coupled signaling in *Arabidopsis*. Plant J. 54, 440–451.

Xiong, L., Yang, Y., 2003. Disease resistance and abiotic stress tolerance in rice are inversely modulated by an abscisic acid–inducible mitogen activated protein kinase. Plant Cell 15, 745–759.

Xu, J., Zhang, S., 2015. Mitogen-activated protein kinase cascades in signaling plant growth and development. Trends Plant Sci. 20, 56–64.

Xu, J., Li, Y., Wang, Y., Liu, H., Lei, L., Yang, H., et al., 2008. Activation of MAPK kinase 9 induces ethylene and camalexin biosynthesis and enhances sensitivity to salt stress in *Arabidopsis*. J. Biol. Chem. 283, 26996–27006.

Yin, Z., Wang, J., Wang, D., Fan, W., Wang, S., Ye, W., 2013. The MAPKKK gene family in *Gossypium raimondii*: genome-wide identification, classification and expression analysis. Int. J. Mol. Sci. 14, 18740–18757.

Zaidi, I., Ebel, C., Belgaroui, N., Ghorbel, M., Amara, I., Hanin, M., 2016. The wheat MAP kinase phosphatase 1 alleviates salt stress and increases antioxidant activities in *Arabidopsis*. J. Plant Physiol. 193, 12–21.

Zhang, T., Liu, Y., Yang, T., Zhang, L., Xu, S., Xue, L., An, L., 2006. Diverse signals converge at MAPK cascades in plant. Plant Physiol. Biochem. 44, 274–283.

Zhang, L., Li, Y., Lu, W., Meng, F., Wu, C.A., Guo, X., 2012. Cotton GhMKK5 affects disease resistance, induces HR-like cell death, and reduces the tolerance to salt and drought stress in transgenic *Nicotiana benthamiana*. J. Exp. Bot. 63, 3935–3951.

Zhang, S., Xu, R., Luo, X., Jiang, Z., Shu, H., 2013. Genome-wide identification and expression analysis of MAPK and MAPKK gene family in *Malus domestica*. Gene 531, 377–387.

Zhang, X., Wang, L., Xu, X., Cai, C., Guo, W., 2014. Genome-wide identification of mitogen-activated protein kinase gene family in *Gossypium raimondii* and the function of their corresponding orthologs in tetraploid cultivated cotton. BMC Plant Biol. 14, 345.

Zhang, X., Xu, X., Yu, Y., Chen, C., Wang, J., Cai, C., Guo, W., 2016. Integration analysis of MKK and MAPK family members highlights potential MAPK signaling modules in cotton. Sci. Rep. 6, 29781.

Zwerger, K., Hirt, H., 2001. Recent advances in plant MAP kinase signaling. Biol. Chem. 382, 1123–1131.

Chapter 22

Sugar Signaling Under Abiotic Stress in Plants

Giselle M.A. Martínez-Noël[1,2], Jorge A. Tognetti[3,4]
[1]INBIOTEC and FIBA, Mar del Plata, Argentina; [2]Consejo Nacional de Investigaciones Científicas y Técnicas (CONICET), Buenos Aires, Argentina; [3]Universidad Nacional de Mar del Plata (UNMdP), Mar del Plata, Argentina; [4]Comisión de Investigaciones Científicas (CIC), Provincia de Buenos Aires, Argentina

1. INTRODUCTION

When plants encounter adverse (or potentially adverse) environments, primary metabolic processes, such as growth and/or photosynthesis, are, in general, rapidly affected, and the effect depends on the type and magnitude of stress (Arbona et al., 2017). For example, water deficit and low temperature directly inhibit growth, whereas low irradiance directly reduces the photosynthetic rate. The imbalance between carbon fixation and consumption leads to altered sugar levels in cells.

It is thus not surprising that plants have evolved mechanisms that enable them to get information from hazardous environments through the concentration of certain sugars. Not every sugar has been attributed a signaling role in plant stress, but there is compelling evidence that at least glucose (Glc), fructose (Fru), sucrose (Suc), and trehalose-6-P (T6P) may fulfill this role (Van den Ende and El-Esawe, 2014; Li and Sheen, 2016; Sami et al., 2016; Ceusters et al., 2017).

Sugar levels are perceived through sugar sensors that initiate a signaling cascade ultimately resulting in altered gene expression and protein modification. Upon these changes, plants may respond to a specific stressful condition and through this response may improve their chance of success. Many of the sugar-mediated responses do not occur in an independent manner, but rather are orchestrated with other endogenous or environmental stimuli. Two large signaling networks that essentially correspond to metabolically opposite situations have been identified. One of them corresponds to low carbon availability, this is, when C fixation is more affected than consumption or growth: the SnRK family-signaling network (Broeckx et al., 2016). The other corresponds to high C availability, when growth is more affected than photosynthesis: target of rapamycin (TOR) signaling network (Baena-González and Hanson, 2017).

Variation in sugar levels is associated not only with stressful environments, but also with normal functioning (i.e., day/night cycles). Accordingly, a large array of *Arabidopsis* genes (at least 10%) is sugar responsive (Cramer et al., 2011). Therefore, at first glance sugar signaling might appear as a rather unspecific way of responding to external stimuli. To solve this query, plants have evolved complex interplays with stress-related hormones, such as abscisic acid (ABA) and ethylene, as well as with direct environmental stimuli, such as light or mineral nutrients. Furthermore, not all abiotic stresses may be associated with C imbalance. Mechanical stress resulting in plant wounding may also elicit specific sugar-mediated plant responses to ameliorate plant status and to prevent pathogen infection.

In this chapter, we summarize the present status of knowledge about sugar roles in plant responses to abiotic stresses, indicating main research areas and gaps still existing in this knowledge. Furthermore, potential uses of the information gathered toward obtaining crops that may be more productive under unfavorable environmental conditions are briefly discussed.

2. VARIATION IN SUGAR LEVELS IN PLANTS UNDER ABIOTIC STRESS

It has been known for decades that sugar concentrations are severely altered in plants subjected to abiotic stress. Main examples of stresses under which sugars accumulate include water, salt, cold, and nutrient deficiency stresses. Accumulated sugars (which shortly after exposure to stressful condition often include especially Suc and also Glc and Fru) may serve as direct protective agents acting upon membranes or macromolecules (Sami et al., 2016 and references therein) besides initiating a signaling cascade.

Plant Metabolites and Regulation Under Environmental Stress. http://dx.doi.org/10.1016/B978-0-12-812689-9.00022-4

In the case of water stress, a decline in turgor drastically impairs cell elongation, but photosynthesis may still proceed, since stomatal closure due to ABA signaling from roots (which limits CO_2 uptake), occurs usually later (Hsiao, 1973). This imbalance between sugar utilization and synthesis leads to sugar accumulation in different plant organs (Shinohara et al., 1995; Sami et al., 2016). Salt stress, which partially consists of a water stress, also commonly leads to an accumulation of sugars. In some cases, particularly high Suc concentrations can be found, as reported by Saito and Matsukura (2015) for tomato fruits from plants under salt stress.

A similar imbalance occurs in plants subjected to low temperature (Pollock, 1990). Again, growth may be much more affected by cold (mainly because of the kinetic effect on enzymes involved in cell cycle) than photosynthesis, and this phenomenon is especially remarkable in C_3 plants, such as grasses, which may accumulate sugars in very large amounts, in an inverse relationship with temperature (Lorenzo et al., 2015).

Mineral nutrient deficiency is another example of a stressful condition promoting sugar accumulation. Nitrogen deficiency impairs plant growth, whereas photosynthesis may still occur at a relatively high rate. For example, sugar accumulation was reported for barley leaves (Wang and Tillberg, 1996), barley seedlings (Comadira et al., 2015), and radish hypocotyls (Su et al., 2016) under N deficiency. Also phosphorus, potassium, and magnesium deficiencies were found to result in Suc accumulation in leaves, in parallel with a decrease of this sugar in phloem, which is indicative of growth impairment (Marschner et al., 1996). Sugars also accumulate under oxygen deprivation conditions. This has been explained as a consequence of either a decreased growth rate, or a reduction in respiratory rate that is greater than the reduction in C supply via phloem (de Sousa and Sodek, 2002).

On the other hand, environmental conditions that deplete Suc concentration are associated with those that directly limit photosynthesis (i.e., low irradiance or low concentration of carbon dioxide, although the latter may be observed only in enclosures such as greenhouses) or those that promote elongation growth under photosynthetic restriction, such as in the phenomenon called shade avoidance syndrome, which involves phytochromes, and occurs especially in the basal parts of dense canopies. Changes in sugar concentration due to variation in light intensity are generally less marked than those observed, for example, under contrasting temperatures. The reason may be attributed to the promotive effect of light on overall plant growth, which prevents sugars from accumulating. For example, Gent (1986) reported that in tomato plants grown at 370 μmol photons m^2/s, the total nonstructural carbohydrate concentration was just 41% greater than in plants at 110 μmol photons m^2/s. Besides, synthesis or mobilization of stored starch may buffer cell sugar concentration under changes in irradiance, such as those occurring in a day/night cycle.

In the case of phytochrome-mediated plant responses, a change in the relative amounts of different sugars can be usually found. For example, Lercari (1982) observed an increase in the proportion of monosaccharides (Glc and Fru) in both leaf blades and bulbs of onion, which depended on photoperiod. Similarly, Damayanthi Ranwala et al. (2002) reported an increase in the proportion of monosaccharides in the petioles of watermelon plants subjected to a far red treatment at the end of the day, which paralleled an increase in petiole length of these plants, compared with those treated with red light. From these results it is important to note that, although in most cases sugar accumulation in plants is associated with a photosynthetic surplus over C utilization, this is not always true, since in some cases (such as shade avoidance) the levels of specific sugars can increase even under photosynthetic C limitation, because of degradation of stored reserves, mainly starch.

3. SUGARS ACTING AS SIGNALS, AND THEIR SENSORS

Besides their evident role in metabolism, and in the protection of macromolecules, sugars may act as signaling molecules that regulate many important processes throughout the life of the plant (Ruan, 2014). The first described and best characterized sugar intracellular sensor is a hexokinase (HXK1), which is involved in Glc sensing (Jang and Sheen, 1994). It is important to notice that sugar signaling effects may differ among types of cells, tissues, and developmental stages (Tiessen and Padilla-Chacon, 2013).

3.1 Monosaccharides: Glucose and Fructose Sensing

Hexose sugars, Glc and Fru, have been recognized for long as signals that regulate gene expression to support plant growth and development in relation to internal and external cues. A direct Glc signaling pathway is mediated by HXK1, which may act independently of its enzymatic activity, forming a complex with two partners (VHA-B1 and RPT5B) (Cho et al., 2006). Thus a dual role should be assigned to hexokinases, one in Glc metabolism and the other in sugar signaling. Glucose sensing through HXK1 has been reported to affect photosynthesis, chlorophyll production, cotyledon expansion, and early seedling growth (Baena-Gonzalez and Sheen, 2008; Cho et al., 2010; Granot et al., 2013). HXK1 sensing under stress generally involves interactions with hormones, as shown later. Besides, different Glc signaling pathways that are independent from HXK1 have also been described (Hausler et al., 2014 and references therein).

Fructose also induces a specific sugar signaling pathway. In this case, putative sensors are fructose 1,6 biphosphatase and fructokinase (FRK) (Cho and Yoo, 2011; Granot et al., 2013). Similarly, as with HXK1, the sensing role of FRK is independent from its enzymatic activity. The fact that HXK and FRK sensors have isozymes with enzymatic activity complicates the understanding of their involvement in sugar signaling (Tiessen and Padilla-Chacón, 2013).

3.2 Disaccharides: Sucrose and Trehalose Sensing

Sucrose is the main sugar transported in plants and sucrose-specific signaling was demonstrated with numerous evidences (Tognetti et al., 2013). For example, sucrose induces anthocyanin and fructan biosynthesis and these inductions have been broadly characterized. Regarding anthocyanin synthesis, high sucrose levels specifically activate the transcription factor MYB75 through the stabilization of DELLA proteins (Li et al., 2014). In the case of fructan metabolism, sucrose induces the transcription of fructan synthesis enzymes (FSS) in wheat and barley leaves (Martinez-Noël et al., 2001; Nagaraj et al., 2001). In addition to inducing FSS, sucrose has also been found to strongly inhibit 6-fructan exohydrolase, a main enzyme of fructan degradation, as shown by Lothier et al. (2014) in *Lolium perenne* and by Suárez-González et al. (2016) in *Agave tequilana*.

Another disaccharide that is essential for normal growth and development in plants, although it is often found in very low levels, is trehalose. This disaccharide is synthesized by a two-step pathway, T6P being the phosphorylated intermediate. T6P levels are directly related to sucrose levels and for that reason T6P was proposed as a signal of the C status. But it was not until recent years that T6P was recognized as a crucial signal in plants (Figueroa and Lunn, 2016; John et al., 2017). Yadav et al. (2014) have proposed a model in which T6P is a signal and also a negative feedback regulator of Suc levels, conforming Suc and T6P to a system similar to that of insulin and glucagon in animals. One of the best described effects of trehalose is the regulation of starch synthesis (Kolbe et al., 2005).

Regarding sucrose sensing, despite all the research carried out, a sucrose sensor has not yet been identified. However, H^+-sucrose cotransporters (SUT) have been proposed as specific sucrose sensors in sieve elements (Barker et al., 2000). Besides, sucrose synthases associated to membrane and calcium channels have been recognized as other putative sucrose sensors (Li and Sheen, 2016). Undoubtedly, more studies and new techniques are needed to clarify this issue.

3.3 Other Sugars

Substantial evidence indicates that, besides Glc, Suc, and trehalose/T6P, there are other sugar signals and sensors that may regulate plant metabolism and development (Li and Sheen, 2016 and references therein). One very important signaling role has been attributed to a specific type of oligosaccharides, the oligogalacturonides (OGs), which consist of α-1,4-linked galacturonosyl residues. OGs do not occur freely in the plant cells, but rather they are integral components of the cell wall, which may be released by hydrolytic enzymes under mechanical stress or wounding (Ferrari et al., 2013; Savatin et al., 2014). OGs induce plant responses that are similar to those elicited by pathogen infection. Because of their relevance regarding plant defense toward pathogens, they are mainly the subject of biotic stress studies, and will not be further discussed here.

4. SUGAR SIGNALING NETWORKS

After changes in sugar concentration are perceived by specific sensors, sugar signal transduction may directly result in altered gene expression, such as in many responses mediated by the HXK1 signaling pathway. However, sugar signaling may also involve networks that integrate changes in sugar concentration with other endogenous and environmental stimuli, to result in multiple developmental and/or metabolic plant responses. Two main signaling networks are known in plants, namely, the SnRK family and TOR kinases, which control metabolism and growth in a manner opposite to response to sugars (Li and Sheen, 2016). The precise manner in which sugar availability affects SnRK1 and TOR activity is, however, still largely unknown (Baena Gonzalez and Hanson, 2017).

4.1 Sucrose Nonfermenting Related Kinases

In plants, SnRKs are evolutionarily conserved proteins; sucrose nonfermenting 1 (SNF1) and AMP-activated protein kinase (AMPK) are their analogues in yeast and mammals, respectively. SnRKs are energy sensors that are activated under unfavorable conditions and regulate gene expression to cope with stress. These kinases activate catabolism and inhibit anabolism to maintain energy homeostasis in response to energy deprivation (Rodrigues et al., 2013; Nukarinen et al., 2016). Through the analysis of in vivo phosphoproteome, proteome, and metabolome it was found that activation of SnRK1 is crucial for the repression of processes that demand high energy, such as protein synthesis (Nukarinen et al., 2016). Important enzymes of plant metabolism, such as sucrose phosphate synthase and nitrate reductase, and transcription factors are substrates of SnRK1 (Nietzsche et al., 2016). It has also been reported that sugars, such as Glc, G1P, G6P, and Suc, inhibit SnRK activity. Some studies, however, showed that Suc may, in some cases, activate SnRK1, and this discrepancy could be due to the complex regulation of SnRK1 depending on the tissue and the sucrose concentration (Crozet et al., 2014).

SnRK1 is directly involved in response to flooding. Cho et al. (2012) reported that in *Arabidopsis* and rice under hypoxic conditions, SnRK1 activity induced stress-responsive expression of genes coding for alcohol dehydrogenase and pyruvate decarboxylase, which are associated with tolerance to this type of stress. Furthermore, Rodrigues et al. (2013) found that two PP2C protein phosphatases, which are well-established repressors of the ABA hormonal pathway, interact with the SnRK1 catalytic subunit causing its dephosphorylation and inactivation, whereas conversely, PP2C inhibition by ABA resulted in the activation of SnRK1. This result, together with a large overlapping of transcriptional responses induced by SnRK1 and ABA, led the authors to conclude that these pathways are complementary and cooperate to result in a stronger response to stress (Rodrigues et al., 2013). Besides its role in plant stress responses, SnRK1 is involved in cell function and developmental processes (Crozet et al., 2014).

The SnRK family also includes SnRK2 and SnRK3, whose roles in plants under stress are being actively investigated. SnRK2 kinases (especially those belonging to group 3) are considered an important regulator of ABA signaling, promoting plant responses to water deficit (Kulik et al., 2011). These authors pointed out that SnRK2 takes part in phosphorylation of both anion and cation channels, which is necessary for the ABA-dependent stomatal closure in response to water deficit. Furthermore, McLoughlin et al. (2012) reported that two SnRK2 kinases are rapidly and transiently activated in roots exposed to salt stress (150 mM NaCl), and that under these conditions, knockout mutants had either a reduced primary root length (SnRK2.4) or a lower number of lateral roots (SnRK2.10). Regarding SnRK3, it has also been involved in plant responses to salt stress, as reviewed by Chandna et al. (2013).

4.2 Target of Rapamycin

TOR is a protein kinase highly conserved among eukaryotes that is considered as a central regulator of growth in relation to nutritional status (Verma and Chatterjee, 2009; Xiong and Sheen, 2013; Dobrenel et al., 2016). TOR, which forms two complexes (TORC1 and TORC2, but only the first was found in plants) with other proteins (Wullschleger et al., 2006; Cornu et al., 2013), is known to promote a huge number of biological processes, including mRNA translation and anabolic metabolism, while repressing nutrient remobilization and storage (Wullschleger et al., 2006). Most evidence comes from studies on mammals and yeasts, but work from recent years shows that it also plays a pivotal role in plant signaling in relation to energy availability (Xiong and Sheen, 2014). In plants, TOR inhibition results in slower growth of roots, leaves, and shoots, and in a delayed development, ultimately leading to low nutrient uptake and light energy utilization (Ren et al., 2012).

TOR also seems to be involved in plant sensitivity to salt and osmotic stresses (Mahfouz et al., 2006; Deprost et al., 2007; Robaglia et al., 2012). It has also been observed that inhibiting the TORC1 activity results in starch accumulation together with a decrease in biomass production (Dobrenel et al., 2013; Caldana et al., 2013). The accumulation of starch observed in TORC1-deficient *Arabidopsis* plants was found to be accompanied by a decrease in raffinose production. Since both starch and raffinose syntheses depend on the supply of G6P, it appears that TOR controls metabolic partitioning between these carbohydrates (Dobrenel et al., 2013). Moreover, raffinose usually accumulates in response to stress, but stressed *Arabidopsis* plants affected in the activity of the TORC1 complex failed to accumulate raffinose (Ren et al., 2012; Dobrenel et al., 2013).

Most evidence on sugar control of TOR activity in plants involves Glc signaling. Direct evidence for a Glc signaling role of TOR has been gathered by Sheen and coworkers on plant developmental control (Xiong et al., 2013; Xiong and Sheen, 2013). These authors showed that TOR transduces Glc signals by phosphorylating the N terminus of E2Fa, a transcription factor that controls transcription of S-phase genes to activate the root meristem, and thus promotes the proliferation of stem progenitor cells. Another plant developmental feature controlled by Glc-TOR signaling is root hair production (Xiong and Sheen, 2012). Because this Glc regulation was found to be unaffected in a hexokinase 1 Glc sensor mutant, the authors concluded that Glc-TOR signaling controls root hair production through a hexokinase 1–independent pathway.

Glucose-TOR signaling also controls many aspects of plant metabolism. A transcriptome analysis of *Arabidopsis* revealed that Glc-activated TOR signaling modulates genes encoding root growth factor peptides and glutathione synthesis, among other genes (Xiong et al., 2013; Xiong and Sheen, 2013). Since glutathione (GSH; γ-Glu-Cys-Gly) is well known as an essential small molecule maintaining cellular redox status via hydrogen peroxide detoxification, a link between TOR and reactive oxygen species (ROS) signaling has been envisaged (Yokawa and Baluska, 2016).

There is evidence indicating that Suc signaling may also control TOR activity. Zhang et al. (2016) showed that Suc signaling controls the accumulation of the brassinosteroid (BR) signaling transcription factor BZR1, which is essential for growth promotion in relation to hormonal and environmental signals. The authors showed that starvation due to darkness induces BZR1 protein degradation, but Suc together with TOR promotes the stabilization of BZR1 in the dark.

On the other hand, no direct evidence of a link between trehalose or T6P and TOR signaling has yet been described in plants (Figueroa and Lunn, 2016).

4.3 SnRK—Target of Rapamycin Interplay

Master switches SnRK and TOR appear not to operate independently, but rather in an interactive manner (Nietzsche et al., 2016). Through analysis of the in vivo phosphoproteome, proteome, and metabolome of *Arabidopsis* mutants with inhibited SnRK1 activity and subjecting to stress due to prolonged darkness, Nukarinen et al. (2016) found not only that the activation of SnRK1 is only essential for the repression of high-energy-demanding cell processes, such as protein synthesis, but also that these mutants had a constitutively high phosphorylation of ribosomal protein S6 (RPS6), which is in turn a major target of TOR signaling, and its phosphorylation correlates with translation. Further evidence for an antagonistic SnRK1 and TOR cross talk, comparable with that existing in animals, was demonstrated by the in vivo interaction of SnRK1α1 and RAPTOR1B (a member of the TOR complex) in the cytosol, and by phosphorylation of RAPTOR1B by SnRK1α1 in kinase assays (Nukarinen et al., 2016). To further complicate the relationship between SnRK and TOR signaling, it has been observed that, contrary to other processes, plant senescence is affected by TOR and SnRK1 in a similar manner, i.e., inactivating both SnRK and TOR kinase activities induce early senescence (Baena-Gonzalez and Hanson, 2017), although the precise mechanisms that explain this result are unknown. Clearly, much research is still needed to unveil the complexities of SnRK-TOR interplay.

5. INTERACTIONS BETWEEN SUGAR SIGNALS AND HORMONES OR OTHER STIMULI IN PLANTS UNDER STRESS

5.1 Abscisic Acid

ABA is the hormone that is most widely implicated in plant stress signaling. Abundant evidence has been gathered indicating an interaction between ABA and sugars regarding several plant responses to stress (Van den Ende and El-Esawe, 2014). The expression of genes involved in both ABA synthesis and signaling is positively modulated by Glc (Cheng et al., 2002; Arroyo et al., 2003). This is consistent with the fact that plant stresses that drastically impair growth and induce sugar accumulation, such as drought and cold, also result in an increased ABA concentration. Sucrose–ABA interactions are, however, very complex, and this is highlighted by the fact that diverse signal transduction pathways are involved, including HXK1 (Cho et al., 2010), SnRK1 (Rodrigues et al., 2013), SnRK2 (Kulik et al., 2011), and TOR (Dong et al., 2015).

Several studies on the interplay between sugars and ABA have been performed with grapevine plants. In this species, sugar and ABA signaling pathways appear to interact to control sugar transport into fruits. Stress-induced protein kinases that promote the expression of hexose transporters have been shown to be upregulated synergistically by ABA and sugars (Cramer et al., 2011). Also in grapevine, it has been shown that both trehalose and T6P induce stomatal closure, but this response ceases to occur under ABA-induced stomatal closure (Gamm et al., 2015). These authors reported that trehalose applied at low concentrations to grapevine epidermal peels induced a significant closure of stomata in a dose-dependent manner. However, the addition of trehalose to ABA-treated epidermis resulted in a significant reduction of the ABA-induced stomatal closure. In addition, T6P induced a more pronounced stomatal closure than trehalose, but it had a lesser effect on ABA-induced stomatal closure. To add complexity to the trehalose–ABA interplay, it has been found that the expression of several trehalose biosynthetic genes, which are particularly highly expressed in *Arabidopsis* guard cells, is upregulated by ABA (Figueroa and Lunn, 2016).

5.2 Ethylene

There are many reports of interactions between ethylene and sugar signals. Ethylene represses the expression of apoplastic invertase, an enzyme associated with C utilization in growth (Linden et al., 1996). Besides, ethylene inhibits anthocyanin accumulation induced by Suc (Jeong et al., 2010). These authors reported that the transcript level of sugar transporter gene SUC1 was enhanced in *Arabidopsis* ecotype Columbia treated with an ethylene-binding inhibitor as well as in ethylene-insensitive mutants. They proposed that SUC1 represents an integrator for signals provided by sugar, light, and ethylene, and that the suppression of SUC1 expression by ethylene inhibits Suc-induced anthocyanin accumulation.

There are also reports about interactions between ethylene, sugars, and ABA signaling pathways (Ramon et al., 2008). It has been known for many years now that ethylene may antagonize Glc–ABA signaling (Gazzarrini and McCourt, 2001). Glucose-mediated plant responses, such as the inhibition of cotyledon growth under high Glc concentration, may be prevented by exogenous application of the ethylene precursor 1-aminocyclopropane-1-carboxylic acid (ACC) (Ramon et al., 2008). The ethylene–ABA antagonism, however, may solely be observed under relatively high Glc concentrations (Cho et al., 2010). To further complicate this picture, ABA may even promote ethylene biosynthesis in ripening fruits (Zhang et al., 2009).

5.3 Auxin

Sugars directly promote auxin biosynthesis, and a tight relationship between sugar and auxin daily fluctuations has been reported (Sairanen et al., 2012). Genes that encode key auxin biosynthetic enzymes (i.e., YUCCA 8 and YUCCA9) are promoted by Glc or Suc, but differential effects between root and shoot have been reported (Ljung et al., 2015). It has been proposed that the growth-promoting effect of Suc on hypocotyl elongation may occur through the effect of the sugar on auxin, since it could be partially mimicked by directly adding auxin or blocked by adding polar auxin inhibitors (Lilley et al., 2012). Ljung et al. (2015) pointed out that this mechanism is reminiscent of the shade avoidance syndrome, in which shade detected primarily in cotyledons results in increased auxin biosynthesis. However, these relationships are unlikely to be extended to any growth-promoting effect of Suc, since in some cases both signaling molecules (Suc and auxin) may be antagonistic, such as in the case of lateral bud release (Assuero and Tognetti, 2010).

The sugar-mediated activation of auxin biosynthesis and promotion of cell expansion involve HXK1 (Li and Sheen, 2016). Nevertheless, SnRK1 and TOR signaling networks are also modulated by auxin, and TOR, in turn, regulates the expression of auxin signaling components (Baena Gonzalez and Hanson, 2017).

5.4 Cytokinin

This hormone promotes photosynthesis through an increased leaf chlorophyll content and nitrogen partitioning to leaves, and accordingly, its levels correlate with sugar availability in the plant. In transgenic rice, overexpressing a gene (IPT) that encodes a key cytokinin biosynthetic enzyme led to a higher Suc content together with an increased drought stress tolerance (Reguera et al., 2013). Sucrose in turn is known to promote lateral root development; this effect is modulated by light (Kircher and Schopfer, 2012), and root apices are the main source of cytokinins in the plant. The parallelism between sugar- and cytokinin-mediated responses was further investigated by Kushwah and Laxmi (2014) who analyzed gene expression of *Arabidopsis* under Glc and BAP (a cytokinin) exogenous supply, and found that of 941 BAP-regulated genes, 713 genes were affected by Glc treatment alone, and from these, 89% were agonistically regulated by either BAP or Glc. Interestingly, however, the rest of the genes (11%) that were antagonistically regulated by BAP or Glc were predominantly involved in stress responses. These authors also investigated the involvement of the HXK1-dependent and HXK1-independent pathways in Glc signaling in controlling hypocotyl growth, and concluded that at least regarding this response, cytokinin interacts with Glc via the HXK1-dependent pathway (Kushwah and Laxmi, 2014).

5.5 Gibberellins

Contrary to cytokinins, gibberellins are associated with low Suc concentration in plants, and this hormone is associated with remobilization of plant reserves. It has been shown that gibberellic acid inhibits the Suc-mediated induction of anthocyanin synthesis through repression of several genes of this biosynthetic pathway (such as CHS, encoding chalcone synthase) that are induced by Suc (Solfanelli et al., 2006; Loreti et al., 2008). It has been proposed that DELLA proteins, which act as important negative regulators in GA signaling, are also positive regulators in the Suc signaling pathway controlling anthocyanin biosynthesis (Li et al., 2014). These authors also reported an interaction between gibberellin signaling and Suc regarding the inhibiting effect of this sugar upon hypocotyl growth inhibition in the dark. Furthermore, other hormones, namely, jasmonates and BRs, may interact with Suc–gibberellin signaling (Ljung et al., 2015).

5.6 Brassinosteroids

Dark grown *Arabidopsis* hypocotyl elongation is promoted by sugars, likely by Glc, and this promotion is dependent on BRs. It has been shown that sugars positively regulate the transcription of *BZR1*, the gene that encodes a BR-activated transcription factor (BRZ1), and also stabilizes the BZR1 protein (Zhang and He, 2015). These authors proposed that the function of HXK1 is dependent on the presence of BR, which may act downstream of HXK1 to positively regulate Glc-induced hypocotyl elongation in *Arabidopsis* in darkness. More recent evidence, however, points to a pivotal role of TOR in this response, whereas HXK1 has been suggested to play a minor role, since when TOR was inactivated, the hypocotyl elongation response to sugar and darkness was greatly diminished (by 61%) compared with TOR-uninhibited controls (Zhang et al., 2016). Glucose and BR have also been reported to interact on lateral root formation in *Arabidopsis* seedlings (Gupta et al., 2015; Singh et al., 2017).

5.7 Reactive Oxygen Species

ROS are harmful radicals that are generated especially under stress conditions, and they also have a systemic signaling role through the so-called ROS waves (Baxter et al., 2014). The possibility of an interaction between sugar and ROS signaling under stress has been put forward during recent years by Van den Ende et al., since several soluble sugars, including fructans, have ROS-scavenging properties (Van den Ende and El-Esawe, 2014). This novel perspective opens interesting possibilities to be investigated.

5.8 Environmental Stimuli

Several environmental factors have been shown to interact with sugar signaling, but the most deeply studied interaction is that involving light environment. As mentioned earlier, light and Suc interact on root development, and this interaction varies along plant ontogeny (Kircher and Schopfer, 2012). An important role in the sugar–light interplay is played by PIF (phytochrome-interacting factor). PIF is a family of transcription factors that were originally identified through their direct interaction with phytochromes but that are now known to integrate light and temperature cues to regulate photosynthetic genes through an antagonistic regulatory circuit with HY5 (LONG HYPOCOTYL 5) (Ljung et al., 2015). Glucose induction of auxin biosynthesis is enhanced in PIF mutants and repressed in plants overexpressing PIF. In contrast, the higher levels of auxin promoted by Suc supplementation are lost in PIF mutants, even when the auxin biosynthetic enzyme gene *YUCCA8* expression may be enhanced (Ljung et al., 2015).

Although knowledge about interactions between sugar and light signaling pathways is rapidly growing, a large effort is still needed to unveil interactions between sugars and most other environmental factors, such as mineral nutrient availability (Moore et al., 2003) or temperature (Tarkowski and Van den Ende, 2015), about which evidence is still scarce, and scattered.

6. CONCLUDING REMARKS AND FUTURE PROSPECTS

A large progress has been made since the first evidence of sugars playing a signaling role in plants was obtained. Many links between cell C status and molecular responses in plants under abiotic stresses have been established, including sugar roles in hormone synthesis, transport, and signaling. The manipulation of sugar signaling pathways is known to result in alterations in plant tolerance to abiotic stresses. However, a large effort is still needed to elucidate many steps of sugar signaling pathways, including targets for enzymes and transcription factors, which may help understanding how changes in sugar concentrations may affect plant responses at the cellular or whole-plant levels (Sami et al., 2016; Baena Gonzalez and Hanson, 2017).

There is an increasing interest in sugar signaling networks, which is accompanied by the development of new technologies, which in turn provide deeper insights of the mechanisms involved. Because of the multiplicative nature of signaling molecules on plant responses, it appears feasible to manipulate signaling pathways so that great effects on crop yield and resistance under stressful environments are obtained by modifying the concentration of some key molecules. An example of this is given by the chemical intervention strategy that directly modulates T6P levels *in planta*, recently reported by Griffiths et al. (2016). In their work, plant-permeable analogues of T6P were designed, which were exogenously supplied to wheat, which released T6P *in planta* upon sunlight illumination. This treatment led to improved recovery from drought, and to increased yield. This strategy, together with more conventional genetic modification, is an example of the many possibilities that knowledge of sugar signaling pathways offers for crop improvement toward increased production under unfavorable environmental conditions.

In sum, an in-depth knowledge of sugar-mediated plant signaling pathways is essential to identify major targets to be addressed for improving plant tolerance to stresses, such as drought, salt, and cold, that currently limit crop productivity and stability worldwide, in the context of a search for a more sustainable agriculture.

REFERENCES

Arbona, V., Manzi, M., Zandalinas, S.I., Vives-Peris, V., Pérez-Clemente, R.M., Gómez-Cadenas, A., 2017. Physiological, metabolic, and molecular responses of plants to abiotic stress. In: Sarwat, M., Ahmad, A., Abdin, M.Z., Ibrahim, M.M. (Eds.), Stress Signaling in Plants: Genomics and Proteomics Perspective, vol. 2. Springer, pp. 1–35.

Arroyo, A., Bossi, F., Finkelstein, R.R., Patricia, L., 2003. Three genes that affect sugar sensing (abscisic acid insensitive 4, abscisic acid insensitive 5, and constitutive triple response 1) are differentially regulated by glucose in *Arabidopsis*. Plant Physiol. 133, 231–242.

Assuero, S.G., Tognetti, J.A., 2010. Tillering regulation by endogenous and environmental factors and its agricultural management. Am. J. Plant Sci. Biotechnol. 4, 35–48.

Baena-González, E., Hanson, J., 2017. Shaping plant development through the SnRK1-TOR metabolic regulators. Curr. Opin. Plant Biol. 35, 152–157.

Baena-González, E., Sheen, J., 2008. Convergent energy and stress signaling. Trends Plant Sci. 13, 474–482.

Barker, L., Kühn, C., Weise, A., Schulz, A., Gebhardt, C., Hirner, B., Hellmann, H., Schulze, W., Ward, J.M., Frommer, W.B., 2000. SUT2, a putative sucrose sensor in sieve elements. Plant Cell 12, 1153–1164.

Baxter, A., Mittler, R., Suzuki, N., 2014. ROS as key players in plant stress signaling. J. Exp. Bot. 65, 1229–1240.

Broeckx, T., Hulsmans, S., Rolland, F., 2016. The plant energy sensor: evolutionary conservation and divergence of SnRK1 structure, regulation, and function. J. Exp. Bot. 67, 6215–6252.

Caldana, C., Li, Y., Leisse, A., Zhang, Y., Bartholomaeus, L., Fernie, A.R., Willmitzer, L., Giavalisco, P., 2013. Systemic analysis of inducible target of rapamycin mutants reveal a general metabolic switch controlling growth in *Arabidopsis thaliana*. Plant J. 73, 897–909.

Ceusters, N., Van den Ende, W., Ceusters, J., 2017. Exploration of sweet immunity to enhance abiotic stress tolerance in plants: lessons from CAM. Prog. Bot. 78, 145–166.

Chandna, R., Azooz, M.M., Ahmad, P., 2013. Recent advances of metabolomics to reveal plant response during salt stress. In: Ahmad, P., Azooz, M.M., Prasad, M.N.V. (Eds.), Salt Stress in Plants. Signalling, Omics and Adaptations. Springer, pp. 1–14.

Cheng, W.H., Endo, A., Zhou, L., Penney, J., Chen, H.-C., Arroyo, A., Leon, P., Nambara, E., Asami, T., Seo, S., Koshiba, T., Sheen, J., 2002. A unique short-chain dehydrogenase/reductase in *Arabidopsis* abscisic acid biosynthesis and glucose signaling. Plant Cell 14, 2723–2743.

Cho, Y.H., Yoo, S.D., Sheen, J., 2006. Regulatory functions of nuclear hexokinase1 complex in glucose signaling. Cell 127, 579–589.

Cho, Y.H., Sheen, J., Yoo, S.D., 2010. Low glucose uncouples hexokinase1-dependent sugar signaling from stress and defense hormone abscisic acid and C_2H_4 responses in *Arabidopsis*. Plant Physiol. 152, 1180–1182.

Cho, Y.H., Yoo, S.D., 2011. Signaling role of fructose mediated by FINS1/FBP in *Arabidopsis thaliana*. PLoS Genet. 7, e1001263.

Cho, Y.H., Hong, J.W., Kim, E.C., Yoo, S.D., 2012. Regulatory functions of SnRK1 in stress-responsive gene expression and in plant growth and development. Plant Physiol. 158, 1955–1964.

Comadira, G., Rasool, B., Karpinska, B., Morris, J., Verrall, S.R., Hedley, P.E., Foyer, C.H., Hancock, R.D., 2015. Nitrogen deficiency in barley (*Hordeum vulgare*) seedlings induces molecular and metabolic adjustments that trigger aphid resistance. J. Exp. Bot. 66, 3639–3655.

Cornu, M., Albert, V., Hall, M.N., 2013. mTOR in aging, metabolism, and cancer. Curr. Opin. Genet. Dev. 23, 53–62.

Cramer, G.R., Urano, K., Delrot, S., Pezzotti, M., Shinozaki, K., 2011. Effects of abiotic stress on plants: a systems biology perspective. BMC Plant Biol. 11, 163.

Crozet, P., Margalha, L., Confraria, A., Rodrigues, A., Martinho, C., Adamo, M., Elias, C.A., Baena-González, E., 2014. Mechanisms of regulation of SNF1/AMPK/SnRK1 protein kinases. Front. Plant Sci. 5, 190.

Damayanthi Ranwala, N., Decoteau, D.R., Ranwala, A.P., Miller, W.B., 2002. Changes in soluble carbohydrates during phytochrome-regulated petiole elongation in watermelon seedlings. Plant Growth Regul. 38, 157–163.

Deprost, D., Yao, L., Sormani, R., Moreau, M., Leterreux, G., Nicolaï, M., Bedu, M., Robaglia, C., Meyer, C., 2007. The *Arabidopsis* TOR kinase links plant growth, yield, stress resistance and mRNA translation. EMBO Rep. 8, 864–870.

de Sousa, C.A.F., Sodek, L., 2002. The metabolic response of plants to oxygen deficiency. Braz. J. Plant Physiol. 14, 83–94.

Dobrenel, T., Marchive, C., Azzopardi, M., Clément, G., Moreau, M., Sormani, R., Robaglia, C., Meyer, C., 2013. Sugar metabolism and the plant target of rapamycin kinase: a sweet operaTOR? Front. Plant Sci. 4, 93.

Dobrenel, T., Caldana, C., Hanson, J., Robaglia, C., Vincentz, M., Veit, B., Meyer, C., 2016. TOR signaling and nutrient sensing. Annu. Rev. Plant Biol. 67, 261–285.

Dong, P., Xiong, F., Que, Y., Wang, K., Yu, L., Li, Z., Ren, M., 2015. Expression profiling and functional analysis reveals that TOR is a key player in regulating photosynthesis and phytohormone signaling pathways in *Arabidopsis*. Front. Plant Sci. 6, 677.

Ferrari, S., Savatin, D.V., Sicilia, F., Gramegna, G., Cervone, F., Lorenzo, G.D., 2013. Oligogalacturonides: plant damage-associated molecular patterns and regulators of growth and development. Front. Plant Sci. 4, 49.

Figueroa, C.M., Lunn, J.E., 2016. A tale of two sugars: trehalose 6-phosphate and sucrose. Plant Physiol. 172, 7–27.

Gamm, M., Héloir, M.C., Adrian, M., 2015. Trehalose and trehalose-6-phosphate induce stomatal movements and interfere with ABA-induced stomatal closure in grapevine. J. Int. Sci. Vigne Vin 49, 165–171.

Gazzarrini, S., McCourt, P., 2001. Genetic interactions between ABA, ethylene and sugar signaling pathways. Curr. Opin. Plant Biol. 4, 387–391.

Gent, M.P., 1986. Carbohydrate level and growth of tomato plants II. The effect of irradiance and temperature. Plant Physiol. 81, 1075–1079.

Granot, D., David-Schwartz, R., Kelly, G., 2013. Hexose kinases and their role in sugar-sensing and plant development. Front. Plant Sci. 4, 44.

Griffiths, C.A., Sagar, R., Geng, Y., Primavesi, L.F., Patel, M.K., Passarelli, M.K., Gilmore, I.S., Steven, R.T., Bunch, J., Paul, M.J., Davis, B.G., 2016. Chemical intervention in plant sugar signalling increases yield and resilience. Nature 540, 574–578.

Gupta, A., Singh, M., Laxmi, A., 2015. Interaction between glucose and brassinosteroid during regulation of lateral root development in *Arabidopsis thaliana*. Plant Physiol. 168, 307–320.

Häusler, R.E., Heinrichs, L., Schmitz, J., Flügge, U.I., 2014. How sugars might coordinate chloroplast and nuclear gene expression during acclimation to high light intensities. Mol. Plant 7, 1121–1137.

Hsiao, T.C., 1973. Plant responses to water stress. Ann. Rev. Plant Physiol. 24, 519–570.

Jang, J.C., Sheen, J., 1994. Sugar sensing in higher plants. Plant Cell 6, 1665–1679.

Jeong, S.W., Das, P.K., Jeoung, S.C., Song, J.Y., Lee, H.K., Kim, Y.K., Kim, W.J., Park, Y.I., Yoo, S.D., Choi, S.B., Choi, G., Park, Y.-I., 2010. Ethylene s of sugar-induced anthocyanin pigmentation in *Arabidopsis*. Plant Physiol. 154, 1514–1531.

John, R., Raja, V., Ahmad, M., Jan, N., Majeed, U., Ahmad, S., Yaqoob, U., Kaul, T., 2017. Trehalose: metabolism and role in stress signaling in plants. In: Sarwat, M., Ahmad, A., Abdin, M.Z., Ibrahim, M.M. (Eds.), Stress Signaling in Plants: Genomics and Proteomics Perspective. Springer, pp. 261–275.

Kircher, S., Schopfer, P., 2012. Photosynthetic sucrose acts as cotyledon-derived long-distance signal to control root growth during early seedling development in *Arabidopsis*. Proc. Natl. Acad. Sci. U.S.A. 109, 11217–11221.

Kolbe, A., Tiessen, A., Schluepmann, H., Paul, M., Ulrich, S., Geigenberger, P., 2005. Trehalose 6-phosphate regulates starch synthesis via posttranslational redox activation of ADP-glucose pyrophosphorylase. Proc. Natl. Acad. Sci. U.S.A. 102, 11118–11123.

Kulik, A., Wawer, I., Krzywińska, E., Bucholc, M., Dobrowolska, G., 2011. SnRK2 protein kinases—key regulators of plant response to abiotic stresses. Omics J. Integrat. Biol. 15, 859–872.

Kushwah, S., Laxmi, A., 2014. The interaction between glucose and cytokinin signal transduction pathway in *Arabidopsis thaliana*. Plant Cell Environ. 37, 235–253.

Lercari, B., 1982. The effect of far-red light on the photoperiodic regulation of carbohydrate accumulation in *Allium cepa* L. Physiol. Plant 54, 475–479.

Li, L., Sheen, J., 2016. Dynamic and diverse sugar signaling. Curr. Opin. Plant Biol. 33, 116–125.

Li, Y., Van den Ende, W., Rolland, F., 2014. Sucrose induction of anthocyanin biosynthesis is mediated by DELLA. Mol. Plant 7, 570–572.

Lilley, J.L., Gee, C.W., Sairanen, I., Ljung, K., Nemhauser, J.L., 2012. An endogenous carbon-sensing pathway triggers increased auxin flux and hypocotyl elongation. Plant Physiol. 160, 2261–2270.

Linden, J.C., Ehness, R., Roitsch, T., 1996. Ethylene regulation of apoplastic invertase expression in autotrophic cells of *Chenopodium rubrum*. Plant Growth Regul. 19, 219–222.

Ljung, K., Nemhauser, J.L., Perata, P., 2015. New mechanistic links between sugar and hormone signalling networks. Curr. Opin. Plant Biol. 25, 130–137.

Lorenzo, M., Assuero, S.G., Tognetti, J.A., 2015. Low temperature differentially affects tillering in spring and winter wheat in association with changes in plant carbon status. Ann. Appl. Biol. 166, 236–248.

Loreti, E., Povero, G., Novi, G., Solfanelli, C., Alpi, A., Perata, P., 2008. Gibberellins, jasmonate and abscisic acid modulate the sucrose-induced expression of anthocyanin biosynthetic genes in *Arabidopsis*. New Phytol. 179, 1004–1016.

Lothier, J., Van Laere, A., Prud'homme, M.P., Van den Ende, W., Morvan-Bertrand, A., 2014. Cloning and characterization of a novel fructan 6-exohydrolase strongly inhibited by sucrose in *Lolium perenne*. Planta 240, 629–643.

McLoughlin, F., Galvan-Ampudia, C.S., Julkowska, M.M., Caarls, L., van der Does, D., Lauriere, C., Munnik, T., Haring, M.A., Testerink, C., 2012. The Snf1-related protein kinases SnRK2. 4 and SnRK2. 10 are involved in maintenance of root system architecture during salt stress. Plant J. 72, 436–449.

Mahfouz, M.M., Kim, S., Delauney, A.J., Verma, D.P.S., 2006. *Arabidopsis* TARGET OF RAPAMYCIN interacts with RAPTOR, which regulates the activity of S6 kinase in response to osmotic stress signals. Plant Cell 18, 477–490.

Marschner, H., Kirkby, E.A., Cakmak, I., 1996. Effect of mineral nutritional status on shoot-root partitioning of photoassimilates and cycling of mineral nutrients. J. Exp. Bot. 47, 1255–1263.

Martinez-Noël, G., Tognetti, J.A., Pontis, H.G., 2001. Protein kinase and phosphatase activities are involved in fructan synthesis initiation mediated by sugars. Planta 213, 640–646.

Moore, B., Zhou, L., Rolland, F., Hall, Q., Cheng, W.-H., Liu, Y.-X., Hwang, I., Jones, T., Sheen, J., 2003. Role of the *Arabidopsis* glucose sensor HXK1 in nutrient, light, and hormonal signaling. Science 300, 332–336.

Nagaraj, V.J., Riedl, R., Boller, T., Wiemken, A., Meyer, A.D., 2001. Light and sugar regulation of the barley sucrose: fructan 6-fructosyltransferase promoter. J. Plant Physiol. 158, 1601–1607.

Nietzsche, M., Landgraf, R., Tohge, T., Bornke, F., 2016. A protein-protein interaction network linking the energy-sensor kinase SnRK1 to multiple signaling pathways in *Arabidopsis thaliana*. Curr. Plant Biol. 5, 36–44.

Nukarinen, E., Nägele, T., Pedrotti, L., Wurzinger, B., Mair, A., Landgraf, R., Börnke, F., Hanson, J., Teige, M., Baena-Gonzalez, E., Dröge-Laser, W., 2016. Quantitative phosphoproteomics reveals the role of the AMPK plant ortholog SnRK1 as a metabolic master regulator under energy deprivation. Sci. Rep. 6, 31697.

Pollock, C., 1990. The response of plants to temperature change. J. Agric. Sci. 115, 1–5.

Ramon, M., Rolland, F., Sheen, J., 2008. Sugar sensing and signaling. In: The *Arabidopsis* Book. The American Society of Plant Biologists.

Reguera, M., Peleg, Z., Abdel-Tawab, Y.M., Tumimbang, E.B., Delatorre, C.A., Blumwald, E., 2013. Stress-induced cytokinin synthesis increases drought tolerance through the coordinated regulation of carbon and nitrogen assimilation in rice. Plant Physiol. 163, 1609–1622.

Ren, M., Venglat, P., Qiu, S., Feng, L., Cao, Y., Wang, E., Xiang, D., Wang, J., Alexander, D., Chalivendra, S., Logan, D., Mattoo, A., Selvaraj, G., Datla, R., 2012. Target of rapamycin signaling regulates metabolism, growth, and life span in *Arabidopsis*. Plant Cell 24, 4850–4874.

Robaglia, C., Thomas, M., Meyer, C., 2012. Sensing nutrient and energy status by SnRK1 and TOR kinases. Curr. Opin. Plant Biol. 15, 301–307.

Rodrigues, A., Adamo, M., Crozet, P., Margalha, L., Confraria, A., Martinho, C., Elias, A., Rabissi, A., Lumbreras, V., González-Guzmán, M., Antoni, R., 2013. ABI1 and PP2CA phosphatases are negative regulators of snf1-related protein kinase1 signaling in *Arabidopsis*. Plant Cell 25, 3871–3884.

Ruan, Y.L., 2014. Sucrose metabolism: gateway to diverse carbon use and sugar signaling. Annu. Rev. Plant Biol. 65, 33–67.

Saito, T., Matsukura, C., 2015. Effect of salt stress on the growth and fruit quality of tomato plants. In: Kanayama, Y., Kocheto, A. (Eds.), Abiotic Stress Biology in Horticultural Plants. Springer, Japan, pp. 3–19 (Chapter 1).

Sairanen, I., Novak, O., Pencik, A., Ikeda, Y., Jones, B., Sandberg, G., Ljung, K., 2012. Soluble carbohydrates regulate auxin biosynthesis via PIF proteins in *Arabidopsis*. Plant Cell 24, 4907–4916.

Sami, F., Yusuf, M., Faizan, M., Faraz, A., Hayat, S., 2016. Role of sugars under abiotic stress. Plant Physiol. Biochem. 109, 54–61.

Savatin, D.V., Gramegna, G., Modesti, V., Cervone, F., 2014. Wounding in the plant tissue: the defense of a dangerous passage. Front. Plant Sci. 5, 470.

Shinohara, Y., Akiba, K., Maruo, T., Ito, T., 1995. Effect of water stress on the fruit yield, quality and physiological condition of tomato plants using the gravel culture. Acta Hort. 396, 211–218.

Singh, M., Gupta, A., Laxmi, A., 2017. Glucose and brassinosteroid signaling network in controlling plant growth and development under different environmental conditions. In: Pandey, G.K. (Ed.), Mechanism of Plant Hormone Signaling under Stress. Wiley, pp. 443–469.

Solfanelli, C., Poggi, A., Loreti, E., Alpi, A., Perata, P., 2006. Sucrose-specific induction of the anthocyanin biosynthetic pathway in *Arabidopsis*. Plant Physiol. 140, 637–646.

Su, N., Wu, Q., Cui, J., 2016. Increased sucrose in the hypocotyls of radish sprouts contributes to nitrogen deficiency-induced anthocyanin accumulation. Front. Plant Sci. 7.

Suárez-González, E.M., Suárez, P.A., Cruz-Rubio, J.M., Martínez-Gallardo, N.A., Hernández, I.C., Délano-Frier, J.P., Gómez-Leyva, J.F., 2016. Differential fructan accumulation and expression of fructan biosynthesis, invertase and defense genes is induced in *Agave tequilana* plantlets by sucrose or stress-related elicitors. Agri Gene 31, 17–28.

Tarkowski, L.O.P., Van den Ende, W., 2015. Cold tolerance triggered by soluble sugars: a multifaceted countermeasure. Front. Plant Sci. 6, 203.

Tiessen, A., Padilla-Chacon, D., 2013. Subcellular compartmentation of sugar signalling: links among carbon cellular status, route of sucrolysis, sink-source allocation, and metabolic partitioning. Front. Plant Sci. 3, 306.

Tognetti, J.A., Pontis, H., Martínez-Noël, G., 2013. Sucrose signaling in plants: a world yet to be explored. Plant Signal Behav. 8 (3), e23316.

Van den Ende, W., El-Esawe, S.K., 2014. Sucrose signaling pathways leading to fructan and anthocyanin accumulation: a dual function in abiotic and biotic stress responses? Environ. Exp. Bot. 108, 4–13.

Verma, D.P.S., Chatterjee, J., 2009. In: Yoshioka, K., Shinozaki, K. (Eds.), TORing with Cell Cycle, Nutrients, Stress, and Growth, in Signal Crosstalk in Plant Stress Responses. Wiley-Blackwell, Oxford, UK.

Wang, C., Tillberg, J.E., 1996. Effects of nitrogen deficiency on accumulation of fructan and fructan metabolizing enzyme activities in sink and source leaves of barley (*Hordeum vulgare*). Physiol. Plant 97, 339–345.

Wullschleger, S., Loewith, R., Hall, M.N., 2006. TOR signaling in growth and metabolism. Cell 124, 471–484.

Xiong, Y., Sheen, J., 2012. Rapamycin and glucose-target of rapamycin (TOR) protein signaling in plants. J. Biol. Chem. 287, 2836–2842.

Xiong, Y., McCormack, M., Li, L., Hall, Q., Xiang, C., Sheen, J., 2013. Glucose-TOR signalling reprograms the transcriptome and activates meristems. Nature 496, 181–186.

Xiong, Y., Sheen, J., 2013. Moving beyond translation. Glucose-TOR signaling in the transcriptional control of cell cycle. Cell Cycle 12, 1989–1990.

Xiong, Y., Sheen, J., 2014. The role of target of rapamycin signaling networks in plant growth and metabolism. Plant Physiol. 164, 499–512.

Yokawa, K., Baluška, F., 2016. The TOR complex: an emergency switch for root behavior. Plant Cell Physiol. 57, 14–18.

Yadav, U.P., Ivakov, A., Feil, R., Duan, G.Y., Walther, D., Giavalisco, P., Piques, M., Carillo, P., Hubberten, H.-M., Stitt, M., Lunn, J.E., 2014. The sucrose-Tre6P nexus: specificity and mechanisms of sucrose signalling by trehalose-6-phosphate. J. Exp. Bot. 65, 1051–1068.

Zhang, M., Yuan, B., Leng, P., 2009. The role of ABA in triggering ethylene biosynthesis and ripening of tomato fruit. J. Exp. Bot. 60, 1579–1588.

Zhang, Y., He, J., 2015. Sugar-induced plant growth is dependent on brassinosteroids. Plant Signal Behav. 10 (12), e1082700.

Zhang, Z., Zhu, J.Y., Roh, J., Marchive, C., Kim, S.K., Meyer, C., Sun, Y., Wang, W., Wang, Z.Y., 2016. TOR signaling promotes accumulation of BZR1 to balance growth with carbon availability in *Arabidopsis*. Curr. Biol. 26, 1–7.

Chapter 23

Gene Expression Analysis in Medicinal Plants Under Abiotic Stress Conditions

Sanchita, Ashok Sharma
CSIR-Central Institute of Medicinal and Aromatic Plants, Lucknow, India

1. INTRODUCTION

The development of new cultivars of medicinal plants with better yields under adverse environmental conditions is desired if the ever-increasing demand of pharmaceutical constituents is to be matched. Understanding the mechanisms involved in the response of plants to unfavorable environmental conditions is the first step in the generation of plants having tolerance to stresses. The development in the field of functional genomics in the last few years has led to a deeper understanding in areas such as gene expression and its relationship to stress tolerance (Perez-Torres et al., 2009). Abiotic stress signaling in plants involves receptor-coupled phosphorelay, phosphoinositol-induced Ca^{2+} changes, and mitogen-activated protein kinase cascades followed by transcriptional activation of stress-responsive genes (Xiong and Zhu, 2001). The transcriptional control of the expression of stress-responsive genes is a crucial part of a plant's response to abiotic stress (Singh et al., 2002). Nowadays, one of the important topics in plant science is the understanding of responses against abiotic stresses. In the postgenomics era, research on plant abiotic stress tolerance has increased a lot (Hirayama and Shinozaki, 2010). Plants being immobile organisms have to cope with changes in environmental conditions. Stresses cause the accumulation of many intracellular substances, such as nucleic acids, proteins, carbohydrates, and amino acids. These accumulated substances may act as signaling stimuli to the cells. The signaling molecules pass from one cell to another through plasmodesmeta (Zambryski and Crawford, 2000). Different signaling pathways are involved in the transduction of external stimuli to the cells. These signaling pathways may operate independently of each other or share their components to pass signals for the induction of genes in response. This interdependence of various signaling pathways acts as a cascade of events (Kaur and Gupta, 2005). Stress-inducible genes are classified into two major groups. The first includes genes coding for proteins involved in direct cell protection against the stress. These proteins are chaperons, late embryogenesis abundant, antifreeze, and enzymes involved in osmolyte (proline, sugars) biosynthesis and detoxification. The other group of genes code for proteins involved in signal transduction and regulation of gene expression such as protein kinases, phosphatases, transcription factors (TFs), and enzymes of phospholipid metabolism (Seki et al., 2004; Yamaguchi-Shinozaki and Shinozaki, 2005). The genes of the aforementioned categories are used as tools to develop stress-tolerant plants. There may be several factors that lead to differential gene expression. Transcription requires access of DNA to transcriptional complex (polymerase, etc.), and therefore packing of chromatin also has an effect on the expression of a gene. Chromatin acetylation controls DNA binding and thus changes the expression of genes. DNA methylation is also a cause of differential gene expression. The degree of methylation correlates with the degree of transcription (Maunakea et al., 2010). Another important reason for differential gene expression may be RNA polymerase II that binds to the promoter region at the TATA box of the gene of interest. However, RNA polymerase II cannot initiate transcription alone and requires various proteins that bind to the regulatory region. Together these proteins play an important role in the differential expression of genes. TFs are regulatory proteins responsible for the regulation of genes involved in the stress responses in plants (Shen et al., 2017; Phukan et al., 2017). Being capable of activating or repressing the process of transcription, the TFs affect metabolism, physiological balance, progression in cells, and responses of cells to the environment (Mochida et al., 2009). One to three TFs form complex regulatory networks through protein–protein interactions among themselves or with proteins of other classes at the transcriptional region of a gene. Protein–protein interactions may also be formed with other transcriptional regulators such as chromatin remodeling/modifying proteins to recruit or block access of RNA polymerases to the DNA template (Mochida et al., 2009). Sequencing of plant genomes has reported many TF-coding genes. The TFs have

Plant Metabolites and Regulation Under Environmental Stress. https://doi.org/10.1016/B978-0-12-812689-9.00023-6

been found in every organism, and their numbers increase with the genome size. Larger genomes tend to have more TFs per gene (van Nimwegen, 2003). It has been reported that 10% and 5% of the genes code for TFs in *Arabidopsis thaliana* (Riechmann et al., 2000) and soybean (Schmutz et al., 2010), respectively. Most of the TFs are reported to be involved in stress responses. DREB/CBF, AREB/ABF, WRKY, MYC, MYB, bZIP, and NAC are important TFs majorly involved in the stress regulation in plants. The present chapter will mainly focus on understanding the effect of abiotic stresses on medicinally important plants through the utilization of gene expression analysis methods.

2. GENE EXPRESSION ANALYSIS AND ITS IMPACT

Functional genomics is the science of employing genes to determine their functions. Gene expression is an approach to analyze the functional changes in the genes. The expression levels of a gene across different experimental conditions are cumulatively called the gene expression profile, and the expression levels of all the genes under an experimental condition are cumulatively called the sample expression profile. Gene expression profiling globally represents the expression of thousands of genes simultaneously by measuring the expression of messenger RNA (mRNA). Different techniques for gene expression analyses have been used so far. Complementary DNA Amplified Fragment Length Polymorphism (AFLP) (Bachem et al., 1996), serial analysis of gene expression (SAGE) (Lima Neto et al., 2012), and massive parallel signature sequencing (Brenner et al., 2000) are the techniques for expression analysis wherein the genomic or transcriptomic sequence of the organism is not needed. Some advanced and largely used techniques are microarrays (Schena et al., 1995) and real-time polymerase chain reaction (Higuchi et al., 1993) wherein previous knowledge of the sequence is required. These two techniques are those most frequently used for gene expression analyses (Perez-Torres et al., 2009). DNA microarray technology has made it possible to simultaneously monitor the expression levels of thousands of genes during important biological processes and across the collections of related samples (Jiang et al., 2004). The use of miniaturized microarrays for gene expression profiling was first reported by Schena et al. (1995). The use of microarray technique was first reported for the study of complete eukaryotic genome of *Saccharomyces cerevisiae* (Lashkari et al., 1997). The main objective of a microarray experiment is usually to identify differentially expressed gene profiles (Dudoit et al., 2002; Sreekumar and Jose, 2008; Yang et al., 2005). This is mainly done by comparing the expression of a set of genes in a particular condition with reference to genes maintained under normal (control) conditions (Babu, 2004). Differential expression is the identification of genes expressed at different levels at the time of transcription under adverse condition. The expression of genes changes at the level of its transcription from the DNA template to mRNA synthesis (Cramer, 2011). The change in expression of genes at which mRNA are synthesized from a DNA template is regulated by different mechanisms. Microarray data are usually presented in an expression matrix in which each column represents all the gene expression levels from a single experiment and each row represents the expression of a gene across all experiments. Each element in the matrix is a log ratio, which is defined as the logarithmic value of ratio of the gene expression level in the testing sample to the gene expression level in the reference sample. The first objective of microarray data analysis is to identify genes for which the expression level is significantly changed under different experimental conditions. One of the goals of microarray data analysis is to cluster the expressed genes or samples having similar expression profiles to make a meaningful biological inference from a set of genes or samples (Sanchita and Sharma, 2014). The similarity in expression or coexpression implies some kind of regulatory or functional similarity of the genes, or it can be said that functionally similar genes are found to be coexpressed (Sanchita and Sharma, 2016). Thus the finding of genes that might be involved in the same biological process requires clustering of genes into groups based on their similarity in expression profiles. Thus the elucidation of the patterns hidden in gene expression data enhances the understanding of functional genomics. However, the large number of genes and the complexity of biological networks make it difficult to interpret the resulting volume of data, which often consist of millions of measurements. The advent of high-throughput next-generation sequencing has revolutionized the gene expression profiling methods (Jain, 2012).

3. ABIOTIC STRESSES AND ITS EFFECTS

A stress is an adverse effect on an organism that may change its growth, physiology, biochemistry, and metabolism. The genome of a plant species has evolved to cope with environmental changes. Abiotic stresses affect plant growth and potential. Being sessile organisms, plants are constantly exposed to various adverse environmental conditions such as water deficiency (drought stress), extreme temperatures (heat or cold stress), excess of soil water (water logging), and excess of salt (salt stress). Abiotic stress is responsible for reduced production and yield of major crops as well as plants of medicinal importance. Hence it has become necessary to develop stress-tolerant plants that can withstand the fluctuations in environmental conditions. Climate changes generate complex environmental stresses with a wide range of unpredictable impacts on the ecosystem. Up to some extent, plants have evolved effective mechanisms to avoid or reduce potential damages. Plants may respond against these stresses in elastic and plastic manners, i.e., reversible and irreversible manners, respectively (Cramer et al., 2011). High temperature downregulates the responding genes and affects the growth and development of plants. Heat shock proteins (HSPs) or molecular chaperons assist in folding, intracellular distribution, assembly, and degradation of proteins under heat stress conditions (Baniwal et al., 2004). Cold stress also adversely affects the growth and development of plants. Several important crops such as sorghum, maize, soybean, cotton, and tomato are reported to be sensitive to chilling and incapable of cold acclimation (Chinnusamy et al., 2007). The stress responses against the subzero, i.e., freezing, temperature have also been reported in many plants. It is reported that plants under extreme cold conditions survive either by avoiding supercooling of tissue water or by freezing tolerance (Meena et al., 2017). When the tolerance range of abiotic stress is exceeded the plant suffers, and if the stress level is harsh enough the plants may die (Jackson and Ram, 2003). Therefore understanding the responses of these abiotic factors for plants is important with regard to plant stress tolerance development. The stress in the form of signals is first perceived at the membrane of cells by the receptors and then is transduced into the cells. Inside the cells, these signals are responsible for switching on the stress-responsive genes for mediating stress tolerance. At the advent of stress condition, the cells produce reactive oxygen species (ROS) such as singlet oxygen ($^1O_2$), superoxide anion radicals ($O_2{\cdot}^-$), hydroxyl ions ($HO{\cdot}$), and hydrogen peroxide (H_2O_2). The accumulation of these ROS cause extensive cellular damage and inhibition of photosynthesis (Cruz de Carvalho, 2008; Pandey et al., 2017). These are natural by-products of cellular oxidative metabolism (Abdal Dayem et al., 2017). The increased formation of ROS interferes with the redox status of the cell and causes oxidative damage of proteins, lipids, and other biomolecules (Pindi et al., 2017). To neutralize these free radicals many antioxidant enzymes, viz., superoxide dismutase (SOD), catalase, glutathione peroxidase, ascorbate peroxidase, glutathione S-transferase (GST), and glutathione reductase, are produced (Gill and Tuteja, 2010). The recognition of these ROS activates different signal transduction pathways that transmit the information from cell to cell. Various pathways are specialized in signaling of abiotic stresses in plants. Different signaling pathways are involved in the transduction of extracellular signals to intracellular targets in eukaryotes. These extracellular stimuli initiate signal transduction pathways in a cell. Thus responses to stress are not linear pathways but complex, integrated circuits involving multiple pathways. Proteins of these signaling pathways are activated and translocate the secondary messengers. These secondary messengers are involved in the activation of TFs. The activated TFs are responsible for the change in expression of genes (Fig. 23.1).

In response to abiotic stresses, plants have shown a coordinated change at the transcriptional level. Thus transcriptional regulation plays an essential role in the adaptation of cells against environmental challenges? The TFs are proteins with a DNA-binding domain that binds to the *cis*-acting elements present in the promoter regions of a target gene. Most of the TFs belonging to different families have been reported to act as negative as well as positive regulators of stress-responsive

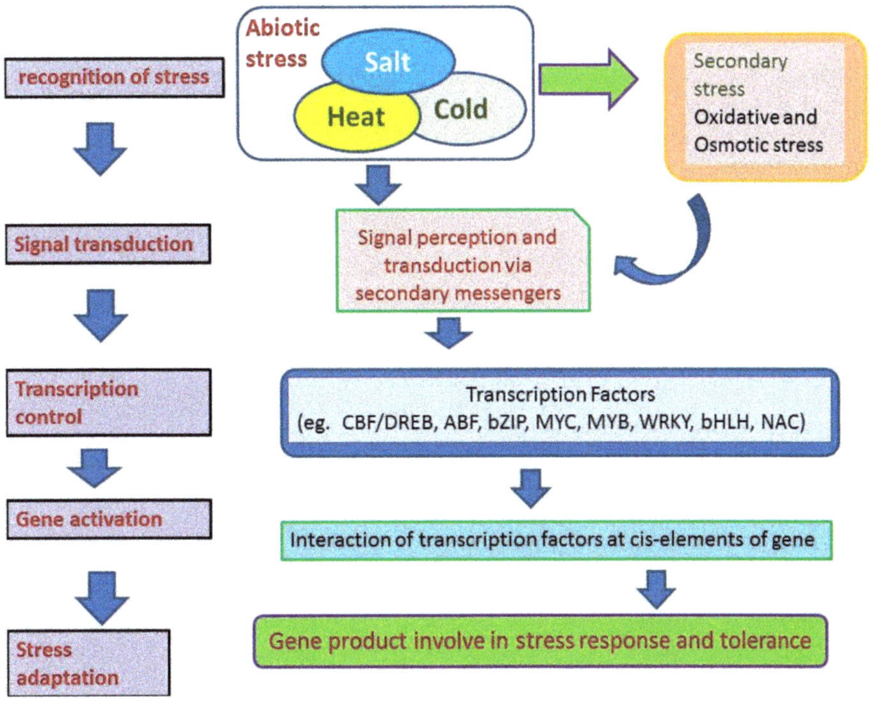

FIGURE 23.1 Schematic representation of abiotic stress signaling.

genes (Santos et al., 2011). The families of TFs are categorized according to their DNA-binding domains (Riechmann et al., 2000). The APETALA2 (AP2) (Akhtar et al., 2012), WRKY (Chen et al., 2012), NAC (Nuruzzaman et al., 2013), and MYB (Roy, 2015) are some of the important TF families that bind to *cis*-element and regulate the expression of stress-responsive genes. Generally, the transcriptional regulation of genes is directly controlled by a network of TFs and TF-binding sites (Chaves and Oliveira, 2004). The effect of abiotic stresses on plants has also been studied from the system biology perspective. The systems biology approach could be utilized for analyzing the responses of different biomolecules (transcripts, proteins, and metabolites) under environmental fluctuations (Cramer et al., 2011).

4. MEDICINAL PLANTS AND SECONDARY METABOLISM

Plants have been a source of traditional medicines for thousands of years in different parts of the world. Plants are considered medicinal because they produce secondary metabolites that are useful for human health. Secondary metabolites produced from these plants have pharmaceutical value besides playing a major role in the adaptation of plants to their environmental fluctuations. Different parts of medicinal plants may possess various medicinal properties. The demand of medicinally important plant products is increasing worldwide (Lipp, 1996). Many secondary metabolites such as alkaloids, terpenoids, and phenylpropanoids are being considered for drug development. These compounds are synthesized through a series of enzyme-catalyzed reactions in the cells. The stepwise pathways, genes, and enzymes involved in the production of these metabolites have been well represented and documented in the KEGG pathway database (http://www.genome.jp/kegg/pathway.html). These secondary metabolites include pharmacologically active natural products with antimalarial, antidiabetic, hepatoprotective, antiulcer, antiinflammatory, and antimicrobial activities. Secondary metabolites are chemical compounds that are not related to normal plant growth and development. They are synthesized in different plant parts in higher proportion in the presence of adverse environmental conditions. Many secondary metabolic pathways and their regulation are highly susceptible to environmental variations. The secondary metabolite content increases under stress conditions in medicinal plants because the genes involved in secondary metabolic pathways are affected in the presence of different stresses. Several environmental stresses causes drastic changes in the growth, physiology, and metabolism of plants leading to the increased accumulation of secondary metabolites (Debnath et al., 2011). The accumulation of these metabolites varies in plant species according to the expression profiles of responsible genes (Sanchita et al., 2014). The transcriptional activity of genes involved in the production of secondary metabolites may be modulated under stress conditions.

5. STUDIES ON CHANGE IN GENE EXPRESSION PATTERNS IN MEDICINAL PLANTS DUE TO ABIOTIC STRESSES

The quality and quantity of secondary metabolites of medicinal plants strongly depend on environmental conditions. Studies on the effects of abiotic stress on medicinal plants through gene expression analyses are gaining importance, since this knowledge will allow to enhance the production of secondary metabolites. The differential expression, i.e., up- and downregulation of genes responsible for increasing the metabolites, has also been analyzed and the influence of abiotic stresses as signals of secondary metabolism has been reviewed (Ramakrishna and Ravishankar, 2011). Several reports are available on the expression analysis of genes responsible for the development of abiotic stress tolerance in medicinal plants. *Cistus creticus* is a perennial, dimorphic, Mediterranean plant, found mainly in arid and warm areas like maquis and garrigue ecosystems. Biosynthesis of terpenoids in *C. creticus* has been reported (Pateraki and Kanellis, 2010). The trichomes of leaves of *C. creticus* excrete a resin rich in several labdanetype diterpenes with verified in vitro and in vivo cytotoxic and cytostatic activities against human cancer cell lines. It has been demonstrated that the expression pattern of genes related to terpenoid biosynthesis depends on tissues as well as the developmental stages. One of the important medicinal plants, *Withania somnifera* has much therapeutic potential, including sedative, narcotic, thyroid stimulation, antiinflammatory, hypnotic, antistress, general tonic, diuretic, antimicrobial, and antitumor activities (Aslam et al., 2017). In *W. somnifera*, the expression analysis of different categories of genes such as osmoregulation (Δ1P5CS), detoxification (GST and SOD), signal transduction (STK and PSP), metabolism (AD and LD), and transcription regulation (HSP, MYB, and WRKY) were considered to show their enhanced responses under drought stress condition (Sanchita et al., 2015). The tuberous roots of *Rehmannia glutinosa* are used to treat fever, nervous conditions, diabetes, and hypertension as well as to strengthen liver function, and have hematopoietic, immunoenhancing, and tonic properties. In this plant the expression response of RghBNG gene has been studied in the presence of different abiotic stresses. The results suggest its role in growth, development, and response to plant growth regulators and abiotic stresses (Zhou et al., 2015). In *Artemisia annua*, upon exposure to chilling, heat shock, or ultraviolet (UV) light, the transcription levels of amorpha-4,11-diene synthase (ADS) and cytochrome P450 monooxygenase (CYP71AV1) genes were upregulated when compared with control plants (Yin et al., 2008). The genes related to chilling? and freezing tolerance in *Hippophae rhamnoides* L., a Himalayan medicinal plant, have been analyzed by deep SAGE technique (Chaudhary and Sharma, 2015). It is reported that this study may provide important gene resources to be exploited for the development of stress-tolerant crop plants in the future. The exposure of hairy root cultures of *Catharanthus roseus* to UV-B light has been studied by Binder et al. (2009). The authors reported that the production of terpenoid indole alkaloids (TIAs) including precursors to vinblastine and vincristine increased in the presence of UV-B light (Binder et al., 2009). The effect of elevated temperature on ginsenoside production in *Panax quinquefolius* has been studied by Jochum et al. (2007). The authors found that due to an increase of 5°C of temperature over normal—optimal values?, the photosynthesis and biomass production decreased, whereas the ginsenoside concentration in the roots increased (Jochum et al., 2007). The sterol C-4 methyl oxidase gene (SMO1) has been identified in *Artemisia annua*, a medicinal plant that produces an antimalarial drug, artemisinin. The SMO1 gene is responsible for the enhanced production of sterols (secondary metabolite) and tolerance against dehydration/drought stress (Singh et al., 2002). The TFs also play an important role in the regulation of stress responses. In *C. roseus*, the expression analysis of WRKY TF has been studied. It has been identified that these WRKY TFs are involved in signal pathways in response to various stress conditions, and are also responsible for the modulation of TIA biosynthesis (Yang et al., 2013). It has also been reported that the AP2/ERF family of TFs contains conserved DNA-binding domains, binding with different *cis*-elements like DRE/CRT, GCC, and JERE (Fujimoto et al., 2000). Due to the ability of binding to multiple motifs, these TFs are effective against various stress responses (Wu et al., 2008). The AP2/ERF TF, isolated from *Papaver somniferum* producing benzylisoquinoline alkaloids, showed overexpression in response to wounding, ethylene, methyl jasmonate, and ABA treatments (Mishra et al., 2015). Another study on *P. somniferum* was carried out to analyze the WRKY family of TFs, which is responsible for the increased accumulation of narcotine and papavarine in response to wounding in the walls of green pods prior to exudation of latex (Mishra et al., 2013). The study of expression profiling of the same gene against multiple abiotic stresses is important. These multiple stresses affect the gene simultaneously for different time periods of stress. Studies on the effects of multiple stresses are generally limited to crop plants only, because of the interest in the development of stress-tolerant varieties. For example, in *Solanum tuberosum*, analysis of genes responsible for more than one abiotic stress condition was carried out (Sanchita et al., 2014). In this study, using microarray data, conserved genes showing differential expression in salt, heat, cold, and drought stresses could be analyzed among several species? of the Solanaceae family (Sanchita et al., 2014). In another study, the transcriptional profiles and protein expression analyses were considered against salinity, dehydration, and cold stresses in salt-tolerant and salt-sensitive varieties of rice (Basu and Roychoudhury, 2014). The gene expression profile study of multiple stress-inducible genes revealed that the expression was higher in the salt-tolerant rice variety. These studies may help in developing plants that are tolerant to multiple stresses. Hence, there is a need for the development of broad-spectrum resistance (i.e., resistance to multiple stresses) in medicinal plants. The cross talk among phytohormone signaling under abiotic stress has been extensively studied (Kohli et al., 2013). Different phytohormones have been reported to understand their role in signal transduction under abiotic stresses (Wani et al., 2016; Fahad et al., 2015). Phytohormones are endogenous messengers for controlling water balance in plants under stress conditions. One of these hormones, abscisic acid (ABA) is critical for plant growth and development under different types of stress. ABA induces leaf stomata closure, thus reducing transpirational water loss and photosynthetic rate in a plant (Tuteja, 2007).

6. CONCLUSION AND FUTURE PROSPECTS

A wide range of environmental stresses such as high and low temperatures, drought, salinity, UV stress, and pathogen infection are potentially harmful to plants. Over the last few years, there has been a surge in gene expression analyses in a variety of signaling processes/abiotic stresses in plants. A study of changes in the expression pattern of genes especially responsible for the regulation of secondary metabolism in medicinal plants under the influence of abiotic stresses is necessary to understand the stress tolerance. Stress-tolerant medicinal plants could survive in adverse condition as well as produce secondary metabolites in large amount. The evolution of knowledge of stress tolerance in plants through functional genomics approaches provides valuable support in this regard. The approaches of functional genomics and assisted breeding of abiotic stress tolerance could produce valuable information for developing stress-tolerant medicinal plants for their ultimate use in pharmaceutical industries. The discovery of novel genes, determination of their expression patterns in response to abiotic stress, and an improved understanding of their roles in stress adaptation and medicinal importance will provide the basis of effective engineering strategies leading to greater stress tolerance.

ACKNOWLEDGMENTS

The financial assistance under the BTISnet programme of DBT, New Delhi, is gratefully acknowledged.

REFERENCES

Abdal Dayem, A., Hossain, M.K., Lee, S.B., Kim, K., Saha, S.K., Yang, G.M., Choi, H.Y., Cho, S.G., 2017. The role of reactive oxygen species (ROS) in the biological activities of metallic nanoparticles. Int. J. Mol. Sci. 18 (1), 120.

Akhtar, M., Jaiswal, A., Taj, G., Jaiswal, J.P., Qureshi, M.I., Singh, N.K., 2012. DREB1/CBF transcription factors: their structure, function and role in abiotic stress tolerance in plants. J. Genet. 91, 385–395.

Aslam, S., Raja1, N.I., Hussain, M., Iqbal, M., Ejaz, M., Ashfaq, M., Fatima, H., Shah, M.A., Rehman, A.U., Ehsan, M., 2017. Current atatus of *Withania somnifera* (L.) Dunal: An endangered medicinal plant from himalaya, Am. J. Plant Sci. 8, 1159–1169. https://doi.org/10.4236/ajps.2017.85076.

Babu, M.M., 2004. An introduction to microarray data analysis. In: Grant, R. (Ed.), Computational Genomics. Horizon Press, UK.

Bachem, C.W., van der Hoeven, R.S., de Bruijn, S.M., Vreugdenhil, D., Zabeau, M., Visser, R.G., 1996. Visualization of differential gene expression using a novel method of RNA fingerprinting based on AFLP: analysis of gene expression during potato tuber development. Plant J. 9, 745–753.

Baniwal, S.K., Bharti, K., Chan, K.Y., Fauth, M., Ganguli, A., Kotak, S., Mishra, S.K., Nover, L., Port, M., Scharf, K.D., Tripp, J., Weber, C., Zielinski, D., von Koskull-Döring, P., 2004. Heat stress response in plants: a complex game with chaperones and more than twenty heat stress transcription factors. J. Biosci. 29, 471–487.

Basu, S., Roychoudhury, A., 2014. Expression profiling of abiotic stress-inducible genes in response to multiple stresses in rice (*Oryza sativa* L.) varieties with contrasting level of stress tolerance. Biomed. Res. Int. 2014, 706890. https://doi.org/10.1155/2014/706890.

Binder, B.Y., Peebles, C.A., Shanks, J.V., San, K.Y., 2009. The effects of UV-B stress on the production of terpenoid indole alkaloids in *Catharanthus roseus* hairy roots. Biotechnol. Prog. 25, 861–865. https://doi.org/10.1002/btpr.97.

Brenner, S., Johnson, M., Bridgham, J., Golda, G., Lloyd, D.H., Johnson, D., Luo, S., McCurdy, S., Foy, M., Ewan, M., Roth, R., George, D., Eletr, S., Albrecht, G., Vermaas, E., Williams, S.R., Moon, K., Burcham, T., Pallas, M., DuBridge, R.B., Kirchner, J., Fearon, K., Mao, J., Corcoran, K., 2000. Gene expression analysis by massively parallel signature sequencing (MPSS) on microbead arrays. Nat. Biotechnol. 18, 630–634.

Chaudhary, S., Sharma, P.C., 2015. DeepSAGE based differential gene expression analysis under cold and freeze stress in seabuckthorn (*Hippophae rhamnoides* L.). PLoS One 10, e0121982. https://doi.org/10.1371/journal.pone.0121982.

Chaves, M.M., Oliveira, M.M., 2004. Mechanisms underlying plant resilience to water deficits: prospects for water-saving agriculture. J. Exp. Bot. 55, 2365–2384. https://doi.org/10.1093/jxb/erh269.

Chen, L., Song, Y., Li, S., Zhang, L., Zou, C., Yu, D., 2012. The role of WRKY transcription factors in plant abiotic stresses. Biochim. Biophys. Acta 1819, 120–128. https://doi.org/10.1016/j.bbagrm.2011.09.002.

Chinnusamy, V., Zhu, J., Zhu, J.K., 2007. Cold stress regulation of gene expression in plants. Trends Plant Sci. 12, 444–451. https://doi.org/10.1016/j.tplants.2007.07.002.

Cramer, G.R., Urano, K., Delrot, S., Pezzotti, M., Shinozaki, K., 2011. Effects of abiotic stress on plants: a systems biology perspective. BMC Plant Biol. 11, 163. https://doi.org/10.1186/1471-2229-11-163.

Cruz de Carvalho, M.H., 2008. Drought stress and reactive oxygen species: production, scavenging and signaling. Plant Signal Behav. 3, 156–165.

Dudoit, S., Yang, Y.H., Callow, M.J., Speed, T.P., 2002. Statistical methods for identifying differentially expressed genes in replicated cDNA microarray experiments. Stat. Sin. 12, 111–139.

Debnath, M., Pandey, M., Bisen, P.S., 2011. An omics approach to understand the plant abiotic stress, *OMICS*. 15, 739–762. https://doi.org/10.1089/omi.2010.0146.

Fahad, S., Hussain, S., Bano, A., Saud, S., Hassan, S., Shan, D., Khan, F.A., Khan, F., Chen, Y., Wu, C., Tabassum, M.A., 2015. Potential role of phytohormones and plant growth-promoting rhizobacteria in abiotic stresses: consequences for changing environment. Environ. Sci. Pollut. R 22, 4907–4921.

Fujimoto, S.Y., Ohta, M., Usui, A., Shinshi, H., Ohme-Takagi, M., 2000. Arabidopsis ethylene-responsive element binding factors act as transcriptional activators or repressors of GCC box-mediated gene expression. Plant Cell 3, 393–404.

Gill, S.S., Tuteja, N., 2010. Reactive oxygen species and antioxidant machinery in abiotic stress tolerance in crop plants. Plant Physiol. Biochem. 48, 909–930. https://doi.org/10.1016/j.plaphy.2010.08.016.

Higuchi, R., Fockler, C., Dollinger, G., Watson, R., 1993. Kinetic PCR analysis: real-time monitoring of DNA amplification reactions. Biotechnol 11, 1026–1030. https://doi.org/10.1038/nbt0993-1026.

Hirayama, T., Shinozaki, K., 2010. Research on plant abiotic stress responses in the post-genome era: past, present and future. Plant J. 61, 1041–1052. https://doi.org/10.1111/j.1365-313X.2010.04124.x.

Jackson, M.B., Ram, P.C., 2003. Physiological and molecular basis of susceptibility and tolerance of rice plants to complete submergence. Ann. Bot. 91, 227–291.

Jain, M., 2012. Next-generation sequencing technologies for gene expression profiling in plants. Brief. Funct. Genomics 11, 63–70. https://doi.org/10.1093/bfgp/elr038.

Jiang, D.X., Tang, C., Zhang, A.D., 2004. Cluster analysis for gene expression data: a survey. Knowl. Data Eng. 16, 1370–1386.

Jochum, G.M., Mudge, K.W., Thomas, R.B., 2007. Elevated temperatures increase leaf senescence and root secondary metabolite concentrations in the understory herb *Panax quinquefolius* (Araliaceae). Am. J. Bot. 94, 819–826. https://doi.org/10.3732/ajb.94.5.819.

Kaur, N., Gupta, A.K., 2005. Signal transduction pathways under abiotic stresses in plants. Curr. Sci. 88, 1771–1780.

Kohli, A., Sreenivasulu, N., Lakshmanan, P., Kumar, P.P., 2013. The phytohormone crosstalk paradigm takes center stage in understanding how plants respond to abiotic stresses. Plant Cell Rep. 32, 945–957. https://doi.org/10.1007/s00299-013-1461-y.

Lashkari, D.A., DeRisi, J.L., McCusker, J.H., Namath, A.F., Gentile, C., Hwang, S.Y., Brown, P.O., Davis, R.W., 1997. Yeast microarrays for genome wide parallel genetic and gene expression analysis. Proc. Natl. Acad. Sci. U. S. A. 94, 13057–13062.

Lima Neto, A.S., de Melo Neto, O.P., Costa, C.H.N., 2012. Application of serial analysis of gene expression to the study of the gene expression profile of *Leishmania infantum* chagasi promastigote. J. Biomed. Biotechnol. 2012, 673458. https://doi.org/10.1155/2012/673458.

Lipp, F.J., 1996. The efficacy, history, and politics of medicinal plants. Altern. Ther. Health Med. 2, 36–41.

Maunakea, A.K., Nagarajan, R.P., Bilenky, M., Ballinger, T.J., D'Souza, C., Fouse, S.D., Johnson, B.E., Hong, C., Nielsen, C., Zhao, Y., Turecki, G., Delaney, A., Varhol, R., Thiessen, N., Shchors, K., Heine, V.M., Rowitch, D.H., Xing, X., Fiore, C., Schillebeeckx, M., Jones, S.J.M., Haussler, D., Marra, M.A., Hirst, M., Wang, T., Costello, J.F., 2010. Conserved role of intragenic DNA methylation in regulating alternative promoters. Nature 466, 253–257. https://doi.org/10.1038/nature09165.

Mishra, S., Phukan, U.J., Tripathi, V., Singh, D.K., Luqman, S., Shukla, R.K., 2015. PsAP2 an AP2/ERF family transcription factor from *Papaver somniferum* enhances abiotic and biotic stress tolerance in transgenic tobacco. Plant Mol. Biol. 89, 173–186. https://doi.org/10.1007/s11103-015-0361-710.1007/s11103-015-0361-7.

Mishra, S., Triptahi, V., Singh, S., Phukan, U.J., Gupta, M.M., Shanker, K., Shukla, R.K., 2013. Wound induced transcriptional regulation of benzylisoquinoline pathway and characterization of wound inducible PsWRKY transcription factor from *Papaver somniferum*. PLoS One 8, e52784. https://doi.org/10.1371/journal.pone.0052784PONE-D-12-15918.

Mochida, K., Yoshida, T., Sakurai, T., Yamaguchi-Shinozaki, K., Shinozaki, K., Tran, L.S., 2009. *In silico* analysis of transcription factor repertoire and prediction of stress responsive transcription factors in soybean. DNA Res. 16, 353–369. https://doi.org/10.1093/dnares/dsp023.

Meena, K.M., Sorty, A.M., Bitla, U.M., Choudhary, K., Gupta, P., Pareek, A., Singh, D.P., Prabha, R., Sahu, P.K., Gupta, V.K., Singh, H.B., Krishanani, K.K., Minhas, P.S., 2017. Abiotic Stress Responses and Microbe-Mediated Mitigation in Plants: The Omics Strategies, Front Plant Sci. https://doi.org/10.3389/fpls.2017.00172.

Nuruzzaman, M., Sharoni, A.M., Kikuchi, S., 2013. Roles of NAC transcription factors in the regulation of biotic and abiotic stress responses in plants. Front. Microbiol. 4, 248. https://doi.org/10.3389/fmicb.2013.00248.

Pandey, S., Fartyal, D., Agarwal, A., Shukla, T., James, D., Kaul, T., Negi, Y.K., Arora, S., Reddy, M.K., 2017. Abiotic stress tolerance in plants: myriad roles of ascorbate peroxidase. Front. Plant Sci. 8.

Pateraki, I., Kanellis, A.K., 2010. Stress and developmental responses of terpenoid biosynthetic genes in *Cistus creticus* subsp creticus. Plant Cell Rep. 29, 629–641. https://doi.org/10.1007/s00299-010-0849-1.

Perez-Torres, E., Paredes, M., Polanco, V., Becerra, V., 2009. Gene expression analysis: a way to study tolerance to abiotic stresses in crop species. Chil. J. Agric. Res. 69, 260–269.

Phukan, U.J., Jeena, G.S., Tripathi, V., Shukla, R.K., 2017. Regulation of apetala2/ethylene response factors in plants. Front. Plant Sci. 8, 150. https://doi.org/10.3389/fpls.2017.00150.

Pindi, W., Mah, H.W., Munsu, E., Wahab, N.A., 2017. Effects of addition of *Kappaphycus alvarezii* on physicochemical properties and lipid oxidation of mechanically deboned chicken meat (MDCM) sausages, *Brit Food J.* 119, 2229–2239. https://doi.org/10.1108/BFJ-10-2016-0501.

Ramakrishna, A., Ravishankar, G.A., 2011. Influence of abiotic stress signals on secondary metabolites in plants. Plant Signal Behav. 6, 1720–1731. https://doi.org/10.4161/psb.6.11.17613.

Riechmann, J.L., Heard, J., Martin, G., Reuber, L., Jiang, C., Keddie, J., Adam, L., Pineda, O., Ratcliffe, O.J., Samaha, R.R., Creelman, R., Pilgrim, M., Broun, P., Zhang, J.Z., Ghandehari, D., Sherman, B.K., Yu, G., 2000. Arabidopsis transcription factors: genome-wide comparative analysis among eukaryotes. Science 290, 2105–2110.

Roy, S., 2015. Function of MYB domain transcription factors in abiotic stress and epigenetic control of stress response in plant genome. Plant Signal Behav. 11 (1), e1117723. https://doi.org/10.1080/15592324.2015.1117723.

Sanchita, Singh, R., Mishra, A., Dhawan, S.S., Shirke, P.A., Gupta, M.M., Sharma, A., 2015. Physiological performance, secondary metabolite and expression profiling of genes associated with drought tolerance in *Withania somnifera*. Protoplasma 252, 1439–1450. https://doi.org/10.1007/s00709-015-0771-z.

Sanchita, Sharma, A., 2014. Future challenges in application of algorithms and tools for clustering of gene expression data. In: Roy, A.K. (Ed.), Emerging Technologies of the 21st Century. New India Publishing Agency, 101, Vikash Surya Plaza, New Delhi-110034, India, pp. 515–531.

Sanchita, Dhawan, S.S., Sharma, A., 2014. Analysis of differentially expressed genes in abiotic stress response and their role in signal transduction pathways. Protoplasma 251, 81–91. https://doi.org/10.1007/s00709-013-0528-5.

Sanchita, Sharma, A., 2016. Computational gene expression profiling under salt stress reveals patterns of co-expression. Genomics Data 7, 214–221. https://doi.org/10.1016/j.gdata.2016.01.009.

Santos, A.P., Serra, T., Figueiredo, D.D., Barros, P., Lourenço, T., Chander, S., Oliveira, M.M., Saibo, N.J., 2011. Transcription regulation of abiotic stress responses in rice: a combined action of transcription factors and epigenetic mechanisms. OMICS 15, 839–857. https://doi.org/10.1089/omi.2011.0095.

Schena, M., Shalon, D., Davis, R.W., Brown, P.O., 1995. Quantitative monitoring of gene expression patterns with a complementary DNA microarray. Science 270, 467–470.

Schmutz, J., Cannon, S.B., Schlueter, J., Ma, J., Mitros, T., Nelson, W., Hyten, D.L., Song, Q., Thelen, J.J., Cheng, J., Xu, D., Hellsten, U., May, G.D., Yu, Y., Sakurai, T., Umezawa, T., Bhattacharyya, M.K., Sandhu, D., Valliyodan, B., Lindquist, E., Peto, M., Grant, D., Shu, S., Goodstein, D., Barry, K., Futrell-Griggs, M., Abernathy, B., Du, J., Tian, Z., Zhu, L., Gill, N., Joshi, T., Libault, M., Sethuraman, A., Zhang, X.-C., Shinozaki, K., Nguyen, H.T., Wing, R.A., Cregan, P., Specht, J., Grimwood, J., Rokhsar, D., Stacey, G., Shoemaker, R.C., Jackson, S.A., 2010. Genome sequence of the palaeopolyploid soybean. Nature 463, 178–183. https://doi.org/10.1038/nature08670.

Seki, M., Satou, M., Sakurai, T., Akiyama, K., Iida, K., Ishida, J., Nakajima, M., Enju, A., Narusaka, M., Fujita, M., Oono, Y., Kamei, A., Yamaguchi-Shinozaki, K., Shinozaki, K., 2004. RIKEN Arabidopsis full-length (RAFL) cDNA and its applications for expression profiling under abiotic stress conditions. J. Exp. Bot. 55, 213–223.

Shen, J., Lv, B., Luo, L., He, J., Mao, C., Xi, D., Ming, F., 2017. The NAC-type transcription factor OsNAC2 regulates ABA-dependent genes and abiotic stress tolerance in rice. Sci. Rep. 7, 40641.

Singh, K.B., Foley, R.C., Onate-Sanchez, L., 2002. Transcription factors in plant defense and stress responses. Curr. Opin. Plant Biol. 5, 430–436.

Sreekumar, J., Jose, K.K., 2008. Statistical tests for identification of differentially expressed genes in cDNA microarray experiments. Indian J. Biotechnol. 7, 423–436.

Tuteja, N., 2007. Abscisic acid and abiotic stress signaling. Plant Signal Behav. 2, 135–138.

van Nimwegen, E., 2003. Scaling laws in the functional content of genomes. Trends Genet. 19, 479–484.

Wani, S.H., Kumar, V., Shriram, V., Sah, S.K., 2016. Phytohormones and their metabolic engineering for abiotic stress tolerance in crop plants. Crop J. 4, 162–176.

Wu, L., Zhang, Z., Zhang, H., Wang, X.C., Huang, R., 2008. Transcriptional modulation of ethylene response factor protein JERF3 in the oxidative stress response enhances tolerance of tobacco seedlings to salt, drought, and freezing. Plant Physiol. 148, 1953–1963. https://doi.org/10.1104/pp.108.126813.

Xiong, L.M., Zhu, J.K., 2001. Abiotic stress signal transduction in plants: molecular and genetic perspectives. Physiol. Plant 112, 152–166.

Yamaguchi-Shinozaki, K., Shinozaki, K., 2005. Organization of cis-acting regulatory elements in osmotic- and cold-stress-responsive promoters. Trends Plant Sci. 10, 88–94.

Yang, Y.H., Xiao, Y., Segal, M.R., 2005. Identifying differentially expressed genes from microarray experiments via statistic synthesis. Bioinformatics 21, 1084–1093.

Yang, Z., Wang, X., Xue, J., Meng, L., Li, R., 2013. Identification and expression analysis of WRKY transcription factors in medicinal plant *Catharanthus roseus*. Sheng Wu Gong Cheng Xue Bao 29, 785–802. https://doi.org/10.1371/journal.pone.0052784.

Yin, L., Zhao, C., Huang, Y., Yang, R.Y., Zeng, Q.P., 2008. Abiotic stress-induced expression of artemisinin biosynthesis genes in *Artemisia annua* L. Chin. J. Appl. Environ. Biol. 14, 1–5.

Zambryski, P., Crawford, K., 2000. Plasmodesmata: gatekeepers for cell-to-cell transport of developmental signals in plants. Ann. Rev. Cell Dev. Biol. 16, 393–421. https://doi.org/10.1146/annurev.cellbio.16.1.393.

Zhou, Y., Zhang, Y., Wei, J., Li, J., Wang, W., Duan, H., Chen, J., 2015. Cloning and analysis of expression patterns and transcriptional regulation of RghBNG in response to plant growth regulators and abiotic stresses in *Rehmannia glutinosa*. Springerplus 4, 60. https://doi.org/10.1186/s40064-015-0830-0.

Chapter 24

Plant Physiological Responses to Nutrient Solution: An Overview

Daniel Baron[1], Amanda C.E. Amaro[2], Felipe G. Campos[2], Carmen S.F. Boaro[2], Gisela Ferreira[2]
[1]*Universidade Federal de São Carlos, Brazil;* [2]*Universidade Estadual Paulista, Brazil*

1. INTRODUCTION

Being sessile, plants have developed adaptive and flexible strategies for the acquisition of nutrients, and these mechanisms are similar for all the essential mineral elements (Marschner, 2012; Nkebiwe et al., 2016). The uptake mechanisms at the root–soil limit are typically multiphasic, with varied affinities for uptake of the mineral elements for vegetative and reproductive development (Miller, 2009).

Plants, generally, need essential "nutrients," with three nonmineral elements, such as carbon (C), hydrogen (H), and oxygen (O), coming from air and/or water and nitrogen (N), phosphorus (P), potassium (K), calcium (Ca), magnesium (Mg), and sulfur (S) being required in large quantities (>0.1% of dry mass) (Maathuis, 2009; Cole et al., 2016). Usually, these minerals are absorbed by the roots in their ionic forms: Ca^{2+}, Mg^{2+}, and K^+ present as free cations; P and S as oxy-anions phosphate (PO_4^{3-}) and sulfate (SO_4^{2-}), respectively; and N as nitrate anion (NO_3^-) or ammonium cation (NH_4^+) (Marschner, 2012; Bacha et al., 2015; Del Amor et al., 2015; Imtiaz et al., 2017). The elements boron (B), copper (Cu), iron (Fe), nickel (Ni), manganese (Mn), molybdenum (Mo), and zinc (Zn) are classified as essential heavy metals (EHMs); they are required in smaller amounts for vital physiological functions of plant (Hansch and Mendel, 2009; Ali et al., 2013; O'carrigan et al., 2014; Moreno-Jiménez et al., 2016).

Mengel and Kirkby (2001) postulate that the essential mineral elements are classified according to their biochemical role and physiological plant function and not by the quantity demanded by the vegetal tissues as reported previously. According to these authors, the first group of essential elements, such as N and S, is responsible for the formation of organic compounds and plants assimilate these nutrients through biochemical reactions involving oxidations and reductions. The second group, such as P, B, and Si, is responsible for energy storage or maintenance of plant structural integrity, and the elements belonging to this group are commonly present in plant tissues in the form of phosphate, borate, and silicate esters. The third group is present in the plant tissue as free ions or is substance-bound, such as pectic acids present in the plant cell wall, and as enzymatic cofactors in the regulation of osmotic potentials, for instance, K, Ca, Mg, Cl, and Mn. The fourth group of mineral elements play important roles in reactions involving electron transport, for example, Fe, Ni, Cu, Mo, and Zn.

Thus localized deficiency or excessive supply for many mineral elements induces morphological adaptations such as lateral root proliferation and changes in the interionic balance of plant cells (Maathuis, 2009). When these elements are in excess, it has been suggested that there is a change in the vacuole and apoplast pools (Gusman et al., 2013); however, when nutritional deficiency occurs, these elements are readily used generally for the maintenance of cytoplasmic requirements, leading their depletion from the vacuole (Taiz et al., 2015). The literature reports that plant growth analysis is an essential tool for physiological studies (Radford, 1967), because environmental factors, such as light, temperature, dioxide carbon (CO_2) concentration, and the availability of essential minerals, specific to each site, affect several physiological indexes, such as leaf area ratio (LAR), net assimilation rate (NAR), relative growth rate (RGR), and leaf weight ratio (LWR), among others. Based on the study of the interactions of these parameters with each environmental factor, in particular the water status, it is possible to know the growth efficiency and plant ability to adapt to the environmental conditions of the species or variety, under the influence of nutritional availability variations (Larcher, 2003; Ben Amor et al., 2005; Peixoto et al., 2006; Masood et al., 2012; Nedjimi, 2014).

In the last decade, the analyses high-throughput metabolomics (omics), specially metabolomics, has enabled an advanced understanding of the biochemical and gene–protein product connections in adverse crop conditions (dehydration, high salinity, and low temperature) in the cultivation of crop plants (Urano et al., 2010; Matsuda et al., 2012; Jorge et al., 2016). The plant metabolome involves a wide variety of compounds with diverse physiological functions, such as

Plant Metabolites and Regulation Under Environmental Stress. https://doi.org/10.1016/B978-0-12-812689-9.00024-8

carbohydrates, amino acids, and fatty acids and, in the literature are reported studies that use nutrient solutions in hydroponics as a cultivation methodology (Ibañez et al., 2016; Llanes et al., 2016; Montes et al., 2017).

This chapter discusses how recent advances of some physiological methods (leaf gas exchange, plant metabolomics, ionic accumulation, and growth analysis) may be used to understand the role of plant nutrients, besides the management of nutrient solution composition (electrical conductivity [EC]) to accomplish these tasks.

2. RELATIONSHIP BETWEEN NUTRIENT SOLUTION AND HYDROPONIC SYSTEM

It is important to note that in plant nutrition study, the soil constitutes a highly complex and interactive medium for analyzing the effects of a given essential mineral element in plant physiology (Epstein and Bloom, 2005; Franco and Prado, 2006). To analyze the effects of a given mineral element, without soil interference, aerated nutrient solutions are employed, and the pioneer among them was proposed by researchers Dennis R. Hoagland and Daniel I. Arnon in 1950 at the Experimental Station of Agriculture, located in Berkeley. However, Resh (2001) reports that the first scientists to cultivate plants in aqueous medium with crop production cycle from "seed to seed" were the Germans Julius von Sachs (1860) and Johann August Ludwig Wilhelm Knop (1861), wherein they used aqueous solutions containing the "macronutrients" and "micronutrients" discovered later, since they were present in nutrient solution as salt contaminants.

For the preparation and maintenance of nutrient solutions used in hydroponic cultivation, the control of EC is of great importance, since this factor determines how much of the mineral elements (ions) are available in the nutrient solution. The higher the ionic concentration, the higher the level of EC, hence EC can be used as an indicator of the ionic strength. There are three variables that determine the availability of nutrient solution ions: (1) ionic strength of the solution, which acts on the ionic activity of the mineral elements; (2) pH, which favors the presence of several ionic species in solution; and (3) the presence of chelating agents responsible for ion sequestration. Summarizing, the higher the EC, higher is the ionic strength (Morgan, 2005; Xu et al., 2009; Baron et al., 2015).

In the literature there are a large number of studies that report the use of the nutrient solution proposed by "Hoagland and Arnon" that investigate physiological responses in the most diverse vegetal models, including fruit plants of commercial interest, such as peach (Shi et al., 1993), quince (Viti and Cinelli, 1993), melon (Asao et al., 2013), fig (Kilinc et al., 2007), kiwi (Sotiropoulos et al., 2003), strawberry (Caruso et al., 2011), and *Annona emarginata* var. *terra-fria* (Baron et al., 2013, 2015); aromatic plants, such as peppermint (David et al., 2009; Búfalo et al., 2016); ornamental plants, such as chrysanthemum and *Ficus* sp. (Soffer et al., 1991); grains and cereals, such as wheat (Saqib et al., 2006) and maize (Doncheva et al., 2006); and vegetables, such as tomato (López-Millán et al., 2009), pepper (Martínez-Ballesta et al., 2004), and cucumber (Csog et al., 2011).

The cultivation technique that employs aerated nutrient solutions is known as hydroponics, and it is rapidly developing as a plant production medium and a study tool (Son et al., 2016). It is the use of nutrient solution in the absence of soil or in the presence of natural and/or artificial substrates ("semihydroponic method"), which was developed from laboratory experiments with considerable degree of environmental control around the root. Hydroponic cultivation has been so successful that the National Aeronautics and Space Administration intends to use it to obtain food, on the occasion of future space missions to Mars (Resh, 2001).

Hydroponics is also used as a tool to solve a wide range of problems, including reducing soil and groundwater contamination, and manipulating nutrient levels in the final product (Rouphael et al., 2008; Colla et al., 2012; Adrover et al., 2013; Santos et al., 2013; Rodrigues et al., 2014; Lima Neto et al., 2014), and Rius-Ruiz et al. (2014) complement that results obtained in hydroponic systems are more precise than those obtained in soil culture.

According to Zhang et al. (2013), the study of plant metabolomics provides valuable insights into the changes that occur in the biochemical mechanisms, in response to biotic and abiotic factors that the cultivated plants are exposed to. Such studies are explored, for example, using multivariate analyzes (Jonsson et al., 2005) under hydroponic cultivation, and this technique is consolidated as a suitable methodological tool to investigate the effects of toxic mineral elements on the productive potential of cultivated plants (Wu et al., 2017).

Metabolites-metabolites correlations can be studied using data matrix generated by metabolomic techniques, which allows a wide variety of studies on regulatory mechanisms. Furthermore, improvements in plant responses and adaptations to a wide range of stresses, such as mineral elements deficiency, mineral toxicity, osmotic potential, and temperature, are investigated using hydroponic cultivation in several crops, such as tomato (Urbanczyk-Wochniak and Fernie, 2005; Sung et al., 2015), bean (Hernández et al., 2007), barley (Roessner et al., 2006), and *Arabidopsis* (Hirai et al., 2004). Other motivations in relation to the technology of hydroponic cultivation are higher yield per area, lower incidence of pests and diseases, easier implementation of cultural dealings, better production scheduling, and shorter cycles, as a result of better environmental control (Savvas and Gizas, 2002; López-Morales et al., 2008; Elmer and Datnoff, 2014).

According to Hoagland and Arnon (1950), there is no ideal solution for all plant species, therefore the choice of nutrient solution for hydroponic cultivation is of fundamental importance, since each plant species has a different nutritional requirement. Thus the development of plants in hydroponics involves the choice of appropriate nutrient solution considering plant age, period of year, and environmental factors; it is also necessary to perform continuous control with addition of salts, temperature monitoring, and controlling the EC of the nutrient solution (Fan et al., 2012). pH control is relevant for the maintenance of membrane integrity and to avoid precipitation of micronutrients, such as Fe, B, and Mn (Roosta and Hamidpour, 2011) and allow replacement of the entire solution when necessary.

The growth-induced variation causes a decrease in the amount of salts available for roots, in addition to leading to a qualitative change, since not all elements are absorbed in the same proportion (Epstein and Bloom, 2005). Even with the existence of required nutritional compositions, there is little probability that any particular combination of concentrations and proportions of salts will prove to be decidedly superior to any other over the growing time of a particular plant, although this is a long-term goal (Rijck and Schrevens, 1999). In addition, the nutrient solution should provide the nutrients at appropriate levels, as the plants do not require high nutrient concentrations, as found in usual nutrient solution formulations; also, the pH should be within an appropriate range, between 5.5 and 6.5 (Epstein and Bloom, 2005).

The reason for the high concentration of nutrients in the standard formulations lies in the experimental difficulty of supplying nutrients from a deposit; as the plants exhaust them, it is easier for researchers to replace a nutrient solution weekly, than hourly (Ingestad, 1982). In the study of plant mineral nutrition it is necessary to consider the uptake of each nutrient, because, in the uptake process, one element can exert an influence on the other, given the possible interactions that can occur, thus altering the mineral composition of the plants (Taiz et al., 2015). However, the diagnosis can be more complex because deficiency of more than one nutrient can occur simultaneously and excess of one nutrient can affect the uptake of another nutrient. Adequate supply of mineral nutrients to crops is one of the most important factors in increasing their productivity (Fageria, 2001).

The first step in the formulation of adequate nutrient solutions is the understanding of the physiological aspects aiming at more favorable conditions for the physiological mechanisms to occur with maximum efficiency. In this context, scientific investigations show a clear relationship between the parameters of gas exchange and plant growth and the variations of the solution ionic strength (Baron et al., 2013, 2015).

Knowledge about the growth dynamics of plant species cultivated in hydroponic systems favors their management, guaranteeing good development, reflecting productivity (Maia et al., 2012). In addition, studies that aim to evaluate the influence of nutrients on plant development are of fundamental importance, since this may influence the density of stomata and their opening as well as the thickness and shape of certain tissues compromising its functionality and interfering in nutrient translocation and, consequently, in the development of the plant (Marschner, 2012).

3. RELATIONSHIP BETWEEN NUTRIENT SOLUTION AND GAS EXCHANGE

Foliar gas exchanges are the most studied physiological processes and are better elucidated in plants, and they are considered essential to plant survival (Taiz et al., 2015). Scientific investigations that report the use of metabolomics to identify the carbon partition and the vegetal water balance have become a powerful tool for understanding the physiological mechanisms that regulate the complex relationship between acclimatization vegetation and abiotic stress, for example, water use efficiency (WUE), stomatal opening, osmotic adjustment, and mineral element action (Marschner, 2012; Flexas et al., 2013; Gago et al., 2014). The biochemical effects on photosynthesis occur because the mineral elements are integral components of enzymes and pigments or, still, direct activators of the photosynthetic process (Larcher, 2003; Hansch and Mendel, 2009; Maathuis, 2009).

In this way, through the use of hydroponic solutions it is possible to measure the interference of certain mineral elements in foliar gas exchange. K, for example, is a mineral element of great importance for the process of photosynthesis, since it interferes in the stomatal conductance of CO_2 and transport of photoassimilates to the phloem, besides participating in the activation of enzymes involved in the photosynthetic process (Zhao et al., 2001). Hafsi et al. (2016) found that K deficiency in *S. carnosa* plants caused the reduction of photosynthesis, reducing the net CO_2 assimilation rate (A_{net}) to 28%; this reduction was mainly due to the 47% decrease in stomatal conductance (g_s), which leads to a substantial decrease in the CO_2 diffusion to the carboxylation sites. In relation to the photosynthetic responses, it is pointed out that the atmospheric CO_2 access to the photosynthetic cells of the mesophyll occurs through the stomatal opening and therefore, changes in g_s directly affect A_{net} and the transpiration rate (E) (Farquhar and Sharkey, 1982); besides that g_s suffers interference from mineral nutrition (Von Caemmerer and Farquhar, 1981; Marschner, 2012; Epstein and Bloom, 2005; Taiz et al., 2015). The reduction of the A_{net} was accompanied by the yellowing of leaves caused by the decrease in the content of chlorophyll, which possibly contributed to the decrease of the photosynthetic rates (Hafsi et al., 2016).

Zn is also a mineral element that can interfere with foliar gas exchange, since it is a cofactor of ribulose biphosphate carboxylase/oxygenase (Rubisco) enzyme (Brown et al., 1993); however, its effects are beneficial only in low concentrations, since it is an EHM, which can cause soil contamination. Rooted cuttings of Eridano and I-214 *Populus* clones were subjected to hydroponic cultivation containing different concentrations of Zn, to verify the capacity of these plants in phytoremediation. During this study, the authors found that increasing concentrations of Zn caused a decline in leaf gas exchange rates in both clones, with marked effects on younger leaves, probably because growing tissues with metabolically active cells are more sensitive to metal excess (Fernàndez-Martínez et al., 2014).

The different ionic strength (*I*) of the nutrient solution may also interfere with leaf gas exchange. Young plants of *A. emarginata* var. *terra-fria* cultivated in 50%*I* presented higher A_{net}, which can be attributed to the better utilization of the mineral elements supplied, resulting in an increase in the photosynthetic rates of the mature leaves or even a greater photosynthetic capacity of the leaves (Baron et al., 2015). On the other hand, the supply of nutrient solution containing 100%*I* to the plants resulted in a decrease in gas exchange and in the accumulation of mineral elements, thus demonstrating that this species does not require large amounts of mineral elements to express its potential vegetative development. In this context, the use of 50%*I* of the ionic force is sufficient for the effective performance of the mineral elements in the different physiological processes, resulting in greater photosynthetic efficiency and consequently higher growth (Baron et al., 2015).

Thus A_{net} and the apparent carboxylation efficiency (A_{net}/C_i) obtained by plants grown in 50%*I* of the "Hoagland and Arnon" nutrient solution were superior compared with those obtained by plants grown in nutrient solutions with 25%*I* and 100%*I*. This finding is in line with the proposal to use lower concentrations than the standard concentrations, as proposed by Zheng et al. (2005) and Rouphael et al. (2008), with 50%*I* reduction of macro- and micronutrients in the nutrient solution without losses in the development of species such as *Geranium* and *Gerbera*, respectively.

In addition, *A. emarginata* var. *terra-fria* cultivated in 50%*I* of "Hoagland and Arnon" nutrient solution presented high values of *E*, which is related to high rates of CO_2 uptake and, probably, large water consumption, promoting an increase in the development of the species. What should be observed, therefore, is that CO_2 diffusion and water loss by transpiration are independent diffusions (Larcher, 2003); it can be regarded as a "necessary cost" associated with the stomatal opening for air diffusion of CO_2 for photosynthesis and to allow the mass flow of nutrients taken up by the roots to the shoot. Plants maintained in nutrient solutions at 25%*I* presented the lowest values for WUE (Baron et al., 2015), which may have reflected directly on the stomatal conductance of the young plants and later on the enzymatic activities (Von Caemmerer and Farquhar, 1981) with consequences on the use of mineral nutrients. In addition, the higher values of intercellular CO_2 concentration (C_i) are a result of the low efficiency of Rubisco in the carboxylation. Thus even with the rate of transpiration ascending in the period and the tendency to increase the stomatal conductance, there is no flow of CO_2 into the interior of the leaf, therefore, failure occurs in the photosynthetic apparatus due to the low ionic availability in the nutrient solution, preventing the reduction of the chemical potential of the internal CO_2 and, consequently, the entry of more CO_2 into the leaf. With the decrease in Rubisco activity, the organic synthesis will be lower, and as the *E* is high, there is a reduction in the WUE.

Hydroponics can also be used to study the effects of salinity on plants, since high levels of saline ions in the apoplast alter the aqueous and ionic thermodynamic equilibrium, which can also influence the gas exchange, as well as the growth and development. In this context, pepper plants were grown in Hoagland and Arnon nutrient solution plus NaCl or KCl for 10 days to verify the effects of salinity on gas exchange and water relation (Martínez-Ballesta et al., 2004). The authors observed that both saline solutions affected foliar gas exchange, and this reduction was attributed to the effects of salt on the stomata, verified by the decrease in the stomatal opening and, consequently, in the g_s, which led to the reduction in A_{net}. However, salinity did not damage the photosynthetic apparatus (Martínez-Ballesta et al., 2004). Schwarz et al. (2002) also observed that different ECs influence the vegetative development of tomato, since the increase in the EC of the nutritive solution can lead to the reduction of the vegetative development, be it of the whole plant or of its individual parts due to the ionic accumulation, directly affecting photosynthesis (Gomez et al., 1992).

4. RELATIONSHIP BETWEEN NUTRIENT SOLUTION AND IONIC ACCUMULATION

Finding an ideal solution for plant cultivation is desirable, but it is important to consider a number of abiotic factors, since, generally, not only the average concentration of nutrients in the solution but also the nutritional concentration on the root surface affects the nutritional uptake (Gorbe and Calatayud, 2010). This fact is closely involved with transpiration because all the dissolved ions in the nutrient solution are transported to the root through the mass flow, which is driven by the transpiration rate (Mengel and Kirkby, 2001). In addition, other issues should be noted, as it is well known that nutrient uptake varies with the climatic seasons throughout the year (Le Bot et al., 1998) and with the vegetative or reproductive stage of the species (Kim et al., 2008).

According to Abdelrahman et al. (2017), the study of plant metabolome can provide an instant overview of the physiological mechanisms under normal and stressful conditions. The studies on the tools known as "omics" are currently quite accessible for global expression analyses, as a predictive marker for validating numerous interdisciplinary investigations, for example, genomics, physiology, and plant biochemistry, in the most varied technological applications (De la Rosa et al., 2017). These same authors report that scientific advances in the area of metabolomics used in nanomaterial engineering have a direct impact on agriculture, since they help in understanding the physiological effects of plant species and their role in the accumulation of mineral elements and growth (Libralato et al., 2016; López-Luna et al., 2016; Wang et al., 2016).

According to Larcher (2003), the availability of nutrients to plants is affected by several factors, including the pH of the medium in which the minerals are found (whether in nutrient solution or soil solution); climatic factors, such as air and temperature (Pardossi et al., 2005); and the nutrient solution temperature (Bassirirad, 2000; Bougoul et al., 2000). These factors, therefore, affect the relationships between the mineral elements present in the nutrient solution, influencing the uptake and transport of a particular nutrient, and indirectly the uptake and translocation of the others (Resh, 2001; Marschner, 2012; Taiz et al., 2015).

In the literature, the importance of the direct effect of the mineral elements, supplied in nutrient solution, directly to the root has been verified, such as N, which is an essential biochemical component of many proteins and nonprotein molecules, such as proteins, coenzymes, specialized metabolites, polyamines, and photosynthetic pigments (Marschner, 2012); besides this the concentration of N plays a fundamental role in the activity of the enzyme Rubisco or in the amount of this enzyme, essential for the photosynthesis. Nitrate (NO_3^-) and ammonium (NH_4^+), forms used to provide nitrogen in nutrient solution, directly influence the uptake of all other ions, that is, they are involved in the maintenance of electroneutrality within the plant (Maathuis, 2009). Also, N changes may cause altered allocation of assimilated carbon used in the synthesis of carbohydrates and/or amino acids, which indicates competition (Nunes-Nesi et al., 2010; Peng et al., 2014).

Another important element in nutrient solution is P, considered crucial for the maintenance of phosphorylation reactions during carbon assimilation, besides positively assisting in other photosynthetic parameters such as g_s and WUE. Amount of P below ideal plant levels results in decreases in carbon assimilation rate, because many intermediate stages of carbon fixation involve phosphate sugars (Taiz et al., 2015). The element K is closely related to stomatal regulation, transpiration, and osmoregulation (Mahouachi et al., 2006); as mentioned in Section 3, and according to Baron et al. (2013), A_{net} decreases with the lower availability of K present in the nutrient solution. This pattern of response was also observed in hydroponics by Terry and Ulrich (1973b) and Basile et al. (2003) in sweet potato and cotton and almond plants, respectively.

The element Ca plays a structural role in cells and signals some responses of the plant in the medium, such as osmotic regulation (Maathuis, 2009; Liu et al., 2013). It is of fundamental importance in the photochemical phase, since it acts on the functioning of photosystem II (PSII), where it promotes the stability and aggregation of the chlorophyll molecules in the antenna complex (Ramalho et al., 1995) and regulation of electron transport and the synthesis of NADP and ATP (Hochmal et al., 2015), is a constituent and stabilizer of the oxygen evolution complex, and is responsible for the photooxidation of water and electron donation to the P680 photosystem, to keep the transport chain of these electrons active (Najafpour et al., 2012). Different from the other elements, P (Terry and Ulrich, 1973a) and K (Terry and Ulrich, 1973b) are required in the biochemical phase of photosynthesis. According to Baron et al. (2015), the accumulated Mg in plants maintained in 50%*I* nutrient solution plays a fundamental role in photosynthesis, particularly in the ability to promote light reactions in the stroma (Liu et al., 2008), besides being required in the stacking of the thylakoid, in the enzymes involved in the assimilation of carbon, and in the partition of assimilates between starch and sugars.

Sulfur also accumulated initially in plants grown in 50%*I* nutrient solution (Baron et al., 2013) is responsible for the formation of sulfolipids, which are usually found in small proportions in the chloroplast thylakoids (Epstein and Bloom, 2005; Maathuis, 2009). However, the reason why photosynthetic membranes require such lipids is unclear, but it is suggested that they are essential for the stabilization of photosynthetic components (Hammond et al., 2003). Some plant proteins contain S, such as ferrodoxins, implicated in photosynthesis, as in the formation of chlorophyll and electron transfer reactions (Epstein and Bloom, 2005).

According to Baron et al. (2017), plants grown in 75%*I* accumulated Zn and showed an increase in A_{net}, A_{net}/C_i, and dry matter. In chloroplasts, zinc-dependent enzymes provide several major functions that repair processes of PSII, by changing photodamaged D1 protein (Ibrahim and Ramadan, 2015; Fu et al., 2015). On the other hand, Baron et al. (2017) using nutrient solution report that low accumulation of Zn in plants grown in 25%*I* could explain the reduction of the A_{net}. Hence Zn might play an important role in protecting carbonic anhydrase from reactive oxygen species–mediated oxidation (Aravind and Prasad, 2005; Hansch and Mendel, 2009). Net CO_2 assimilation rate (A_{net}), as much as apparent A_{net}/C_i, may be a consequence not only of the reduction in Zn but also of the reduction of macronutrients when supplied in 25%*I*, as observed by Baron et al. (2015). Baron et al. (2017) cultivated plants using nutrient solution; they found that there was an inverse relationship between the uptake of Fe and Zn, comparing plants grown in the solutions 75%*I* and 25%*I*. Fe deficiency decreases the uptake or immobilization in the roots and is caused by increased levels of Zn because they are antagonists (Hansch and Mendel, 2009).

5. RELATIONSHIP BETWEEN NUTRIENT SOLUTION AND GROWTH ANALYSIS

The literature justifies that adequate supply of mineral elements in hydroponic systems using nutrient solution promotes good functioning of the vegetal photosynthetic process and implies a greater development of the culture, as can be observed from the calculations of the physiological index of growth analysis in several studies, which report their results using hydroponics as a culture medium.

According to Baron et al. (2015), young plants of *A. emarginata* var. *terra-fria* cultivated in nutrient solution respond to the different ionic strength (*I*) in a clear way, as previously mentioned for gas exchange, with higher growth values for the plants grown in 50%*I* and 75%*I*, lower values for the plants grown in 25%*I*, and intermediate values for those grown in 100%*I*. The same authors affirm that drastic reductions in the supply of mineral elements reduce plant development, in the same way as observed by Batista et al. (2003) in plant species belonging to the Annonaceae family. In this way, there was a greater accumulation of dry matter in relation to the preexisting material, referring to the RGR, with differences between the average values obtained in 50%*I* in relation to the culture in 25%*I* (Baron et al., 2013). Thus the RGR shows similar behavior to that described in the literature, with high values in the first intervals, and subsequent decreases in the final intervals (David et al., 2009).

Caruso et al. (2011) also verified the effect of different ECs of the nutrient solution in alpine strawberry (*Fragaria vesca* L.), to evaluate the most effective nutrient solution strength and its influence on the growth of the plants, evaluating 1.3, 1.6, 1.9, and 2.2 mS/cm ECs. The increase in the nutrient solution ionic strength decreased the growth and productivity of these plants; in agreement with the reports of Baron et al. (2015), the highest EC (2.2 mS/cm) caused the greatest decrease of biomass (23%) and leaf area (31%), compared with the lowest EC (1.3 mS/cm), and decreased the berry mean weight. The authors justify that although the higher ionic forces provide more nutrients, there is an increase in the water deficit.

According to Meziane and Shipley (1999), differences in the RGR among the species, even when cultivated in the same environmental conditions, influence the dynamics and the structure between the species and may be associated with the NAR, in the sense of relating the photosynthetic rate to the amount of leaf area available for light interception and leaf biomass allocation. NAR is the result of the interaction of variations in environmental factors (light, CO_2, temperature, water and nutrient availabilities, leaf age, etc.) with the specific differences of each species according to their genotype (Lambers and Poorter, 1992).

NAR reflects the efficiency of the assimilator system that is involved in dry matter production, estimating liquid photosynthesis (Radford, 1967). It is worth mentioning that higher the photosynthetic rate, greater the plant growth, through the accumulation of dry mass. As explained in Section 3 through the use of hydroponic solutions it is possible to measure the interferences of certain mineral elements in foliar gas exchange and, consequently, in plant growth. As mentioned earlier, in the experiments conducted by Hafsi et al. (2016) the deficiency of K in *Sulla carnosa* plants caused a decrease of photosynthesis. The same authors verified that this decrease in A_{net} reflected in a 50% reduction of the vegetative growth of the organs, with a significant decrease in the RGR of leaves, stems, and roots being observed. The stem was the most affected organ.

According to Marschner (2012), plant species adapted to soils with nutritional scarcity may present a growth strategy in which they are more efficient in acquiring nutrients than "cultivated" species. The most important characteristic of efficient species in nutrient acquisition is their low maximum potential growth rate and less efficient use of nutrients, which is extremely important for their survival and reproduction (Chapin, 1980) even in conditions of high nutritional supply (Fageria, 2001). It is verified that the trend of lower values of NAR, reported by Baron et al. (2013), in plants grown in nutrient solution can be explained by the greater use of the mineral elements provided, since it reflected as increases in the primary production of the species and corroborates other studies developed in hydroponic cultivation with variations of elements in the nutrient solution; among them Alves et al. (2008) and David et al. (2009) carried out studies with beet and peppermint plants, respectively.

The physiological effects of plants grown under the influence of cultivation conditions on hydroponic systems have been investigated in metabolomics (Sinkovič et al., 2015, 2016). Metabolomic analyses have evidenced an increase in the accumulation of amino acids and soluble sugars with direct effect on the growth or biomass of plants upon application of allelopathic substances in *Arabidopsis* (Araniti et al., 2017) and corn (Lupini et al., 2010). In addition, the metabolomic profile of tea leaves [*Camellia sinensis* (L.)] in hydroponic cultivation reveals comprehensive information on the accumulation of essential mineral elements and growth under conditions of nutritional deficiency of Zn (Zhang et al., 2017), K in *Arabidopsis* (Quan et al., 2016), and P or N in barley plants (Armengaud et al., 2009; Huang et al., 2008; Comadira et al., 2015), corn (Schlüter et al., 2012), and tomato (Urbanczyk-Wochniak and Fernie, 2005).

As mentioned earlier, the use of nutrient solution can also be employed to assess the toxic effects of heavy metals. Rooted cuttings of Eridano and I-214 *Populus* clones subjected to hydroponic cultivation containing high concentrations of

Zn presented a decrease in A_{net}, which negatively influenced growth, representing approximately 50% less biomass accumulation in I-214 clones, which are more sensitive to the excess of this heavy metal than the Eridano clones (Fernàndez-Martínez et al., 2014). Tomato plants cultivated in nutrient solution containing cadmium (Cd), also considered a heavy metal, were studied with the objective of verifying its effects on the physiology of this plant. The application of Cd decreased the growth of leaves and stem, showing chlorosis in the leaves, with the appearance of necrotic spots and browning in the roots (López-Millán et al., 2009).

The LWR, which reflects the dry matter fraction exported from the leaf to the other organs of the plant, and therefore the relation of the photosynthetic and total plant biomass, tends to decrease over time, as the cycle of the plants develops (Radford, 1967); however, Baron et al. (2013) reported that the plants kept in nutrient solution at 25%*I* were the ones that tended to show greater export, which may have been reflected in the reduction of leaf thickness due to its reduced development. In *A. emarginata* var. *terra-fria* plants cultivated with nitrogen variation, it was confirmed that the supply of 157.5 mg of N caused an increase in the indexes RGR, NAR, and LAR, indicating higher investment in leaf production, whereas plants grown with intermediate concentrations of 78.75 and 118.12 mg/L of N showed a decrease in RGR and NAR, resulting in a greater accumulation of stem and root dry mass and higher stem diameter. Plants cultivated with 39.37 mg/L N showed a decrease in RGR and NAR, which may indicate faster development (Campos, 2014).

6. CONCLUSION

The increasing advancement of studies that investigate the mineral nutrition of plants in hydroponics and the understanding of the physiological responses, such as leaf gas exchanges, ionic accumulation, and physiological indexes of growth analysis, using aerated nutrient solutions as a study model, with different electrical conductivities ("ionic strength") is notorious. Throughout this review, it is evident that the study of the essential mineral elements availability, through the use of adequate nutrient solutions, provides greater control of salinity levels to the plants, food security, reduction of soil and groundwater contamination, consumption of mineral elements from nonrenewable sources, as well as the understanding of the physiological responses of plants to these solutions and their use in soil phytoremediation, besides ionic balance and storage of ions at the cellular level in the plant.

7. FUTURE PROSPECTS

The authors of this review encourage new studies on the physiological effects of different essential mineral elements availability using other plant species not considered "model plants," because we believe that this type of progress is crucial for the advances in the understanding of plant physiology. The urgent need to diversify the plant species can help understand and exploit more accurately the physiological mechanism underlying the availability of different mineral elements, thus seeking to change the "reductionist view" that scientific researchers have on the physiological effects presented by plants.

The use of nutrient solution is the main technology to test numerous scientific hypotheses about the metabolism of plants, since it allows the researcher to offer to the plant all favorable conditions for its satisfactory development (root temperature, availability of mineral elements, and oxygenation of root). However, there is no ideal solution for plant cultivation that allows the plant to express its full genetic and physiological potential, for example, photosynthesis and biomass production. In this sense, new studies are urgently needed to investigate the exact nutritional needs of the plants, so that the plants can express their full physiological potential and all experimental results come exclusively from what is intended to be tested.

The study of genomics, proteomics, and melabolomics, called "omics," can be an important tool to elucidate and identify environmental stresses associated biochemistry and molecular networks. These approaches are means for the deduction of different physiological pathways that are altered in response to environmental stresses.

ACKNOWLEDGMENTS

The authors of this chapter would like to thank the *Universidade Federal de São Carlos* (*UFSCar*) and *Universidade Estadual Paulista* (*Unesp*) for the opportunity to make this chapter possible.

REFERENCES

Abdelrahman, M., Burritt, D.J., Tran, L.-S.P., 2017. The use of metabolomic quantitative trait locus mapping and osmotic adjustment traits for the improvement of crop yields under environmental stresses. Semin. Cell Dev. Biol. (in press).

Adrover, M., Moyà, G., Vadell, J., 2013. Use of hydroponics culture to assess nutrient supply by treated wastewater. J. Environ. Manag. 127, 162–165.

Ali, H., Khan, E., Sajad, M.A., 2013. Phytoremediation of heavy metals—concepts and applications. Chemosphere 91 (7), 869–881.

Alves, A.U., Prado, R.M., Gondim, A.R.O., Fonseca, I.M., Cecílio Filho, A.B., 2008. Effect of macronutrient omission on growth and nutritional status on table beet. Hortic. Bras. 26 (2), 292–295.

Araniti, F., Scognamiglio, M., Chambery, A., Russo, R., Esposito, A., D'Abrosca, B., Fiorentino, A., Lupini, A., Sunseri, F., Abenavoli, M.R., 2017. Highlighting the effects of coumarin on adult plants of *Arabidopsis thaliana* (L.) Heynh. by an integrated -omic approach. J. Plant Physiol. 213, 30–41.

Aravind, P., Prasad, M.N.V., 2005. Zinc mediated protection to the conformation of carbonic anhydrase in cadmium exposed *Ceratophyllum demersum* L. Plant Sci. 169 (1), 245–254.

Armengaud, P., Sulpice, R., Miller, A.J., Stitt, M., Amtmann, A., Gibon, Y., 2009. Multilevel analysis of primary metabolism provides new insights into the role of potassium nutrition for glycolysis and nitrogen assimilation in *Arabidopsis* roots. Plant Physiol. 150, 772–785.

Asao, T., Asaduzzaman, M., Mondal, M.F., Tokura, M., Adachi, F., Ueno, M., Kawaguchi, M., Yano, S., Ban, T., 2013. Impact of reduced potassium concentrations in nutrient solution on the growth, yield and fruit quality of melon in hydroponics. Sci. Hortic. 164, 221–231.

Bacha, H., Ródenas, R., López-Gómez, E., García-Legaz, M.F., Nieves-Cordones, M., Rivero, R.M., 2015. High Ca^{2+}reverts the repression of high-affinity K^+ uptake produced by Na^+ in *Solanum lycopersicum* L. (var. microtom) plants. J. Plant Physiol. 180, 72–79.

Baron, D., Amaro, A.C.E., Macedo, A.C., Boaro, C.S.F., Ferreira, G., 2017. Is *Annona emarginata* capable of accumulate essential heavy metals without damages in the metabolism? Rev. Bras. Frutic. 39 (4), e-646.

Baron, D., Ferreira, G., Boaro, C.S.F., Rodrigues, J.D., Amaro, A.C.E., Mischan, M.M., 2015. The effect of the ionic strength of nutrient solution on gas exchange, ionic concentration and leaf biomass of *Annona emarginata* (Schltdl.) H. Rainer variety 'Terra-Fria' seedlings. J. Plant Nutr. 38 (12), 1944–1960.

Baron, D., Ferreira, G., Rodrigues, J.D., Boaro, C.S.F., Macedo, A.C., 2013. Gas exchange, physiological indexes and ionic accumulation in *Annona emarginata* (Schltdl.) H. Rainer seedlings in nutrients solution. Rev. Bras. Frutic. 35, 361–376.

Basile, B., Reidelb, E.J., Weinbaumb, S.A., Dejong, T., 2003. Leaf potassium concentration, CO_2 exchange and light interception in almond trees (*Prunus dulcis* (Mill.) D.A. Webb). Sci. Hortic. 98, 185–194.

Bassirirad, H., 2000. Kinetics of nutrient uptake by roots: responses to global change. New Phytol. 147, 155–169.

Batista, M.M.F., Viegás, I.J.M., Frazão, D.A.C., Thomaz, M.A.A., Silva, R.C.L., 2003. Effect of macronutrient omission in growth, symptoms of nutritional deficiency and mineral composition in soursop plants (*Annona muricata*). Rev. Bras. Frutic. 25 (2), 315–318.

Ben Amor, N., Ben Hamed, K., Debez, A., Grignon, C., Abdelly, C., 2005. Physiological and antioxidant responses of the perennial halophyte Crithmum maritimum to salinity. Plant Sci. 168 (4), 889–899.

Bougoul, S., Brun, R., Jaffrin, A., 2000. Nitrate absorption–concentration of Rosa hybrida cv. Sweet Promise grown in soilless culture. Agronomie 20 (2), 165–174.

Brown, P.H., Cakmak, I., Zhang, Q., 1993. Form and function of zinc in plants. In: Robson, A.D. (Ed.), Zinc in Soils and Plants. Kluwer Academic Publishers, Dordrecht, pp. 90–106.

Búfalo, J., Rodrigues, T.M., De Almeida, L.F.R., Tozin, L.R.S., Marques, M.O.M., Boaro, C.S.F., 2016. PEG-induced osmotic stress in *Mentha* x *piperita* L.: structural features and metabolic responses. Plant Physiol. Biochem. 105, 174–184.

Campos, F.G., 2014. *Annona emarginata* (Schltdl.) H. Rainer 'araticum-de-terra-fria' cultivada com variação de nitrogênio: produtividade e perfil químico dos voláteis foliares (MSc. thesis). Faculdade de Ciências Agronômicas, Universidade Estadual Paulista, Botucatu.

Caruso, G., Villari, G., Melchionna, G., Conti, S., 2011. Effects of cultural and nutrient solutions on plant growth, yield and fruit quality of alpine strawberry (*Fragaria vesca* L.) grown in hydroponics. Sci. Hortic. 129, 479–485.

Chapin, F.S., 1980. The mineral nutrition of wild plants. Annu. Rev. Ecol. Syst. 11 (1), 233–260.

Cole, J.C., Smith, M.W., Penn, C.J., Cheary, B.S., Conaghan, K.J., 2016. Nitrogen, phosphorus, calcium, and magnesium applied individually or as a slow release or controlled release fertilizer increase growth and yield and affect macronutrient and micronutrient concentration and content of field-grown tomato plants. Sci. Hortic. 211, 420–430.

Comadira, G., Rasool, B., Karpinska, B., Morris, J., Verrall, S.R., Hedley, P.E., Foyer, C.H., Hancock, R.D., 2015. Nitrogen deficiency in barley (*Hordeum vulgare*) seedlings induces molecular and metabolic adjustments that trigger aphid resistance. J. Exp. Bot. 66, 3639–3655.

Colla, G., Rouphael, Y., Rea, E., Cardarelli, M., 2012. Grafting cucumber plants enhance tolerance to sodium chloride and sulfate salinization. Sci. Hortic. 135, 177–185.

Csog, A., Mihucz, V.G., Tatár, E., Fodor, F., Virág, I., Majdik, C., Záray, G., 2011. Accumulation and distribution of iron, cadmium, lead and nickel in cucumber plants grown in hydroponics containing two different chelated iron supplies. J. Plant Physiol. 168, 1038–1044.

David, E.F.S., Mischan, M.M., Boaro, C.S.F., 2009. Development and essential oil yield of mint (*Mentha* × *piperita* L.) cultivated in nutritive solution under different levels of phosphorus. Biotemas 20 (2), 15–26.

Del Amor, F.M., Piñero, M.C., Otálora-Alcón, G., Pérez-Jimenez, M., Marín-Miñano, M., 2015. Effect of different nitrogen forms and CO_2 enrichment on the nutrient uptake and water relations of pepper plants (*Capsicum annuum* L.). Proc. Environ. Sci. 29, 203–204.

De la Rosa, G., García-Castañeda, C., Vázquez-Núñez, E., Alonso-Castro, Á.J., Basurto-Islas, G., Mendoza, Á., Cruz-Jiménez, G., Molina, C., 2017. Physiological and biochemical response of plants to engineered NMs: implications on future design. Plant Physiol. Biochem. 110, 226–235.

Doncheva, S., Stoyanova, Z., Georgieva, K., Nedeva, D., Dikova, R., Zehirov, G., Nikolova, A., 2006. Exogenous succinate increases resistance of maize plants to copper stress. J. Plant Nutr. Soil Sci. 169 (2), 247–254.

Elmer, W.H., Datnoff, L.E., 2014. Mineral nutrition and suppression of plant disease. In: Alfen, N.K.V. (Ed.), Encyclopedia of Agriculture and Food Systems. Academic Press, Oxford, pp. 231–244.

Epstein, E., Bloom, A.J., 2005. Mineral Nutrition of Plants: Principles and Perspectives, second ed. Sinauer, Massachusetts.

Fageria, V.D., 2001. Nutrient interactions in crop plants. J. Plant Nutr. 24 (8), 1269–1290.

Fan, R., Yang, X., Xie, H., Reeb, M.-A., 2012. Determination of nutrients in hydroponic solutions using mid-infrared spectroscopy. Sci. Hortic. 144, 48–54.

Farquhar, G.D., Sharkey, T.D., 1982. Stomatal conductance and photosynthesis. Ann. Rev. Plant Physiol. 33, 317–345.

Fernàndez-Martínez, J., Zacchini, M., Fernández-Marín, B., García-Plazaola, J.I., Fleck, I., 2014. Gas-exchange, photo- and antioxidant protection, and metal accumulation in I-214 and Eridano *Populus* sp. clones subjected to elevated zinc concentrations. Environ. Exp. Bot. 107, 144–153.

Flexas, J., Niinemets, Ü., Gallé, A., Barbour, M.M., Centritto, M., Diaz-Espejo, A., Douthe, C., Galmés, J., Ribas-Carbo, M., Rodriguez, P.L., Rosselló, F., Soolanayakanahally, R., Tomas, M., Wright, I.J., Farquhar, G.D., Medrano, H., 2013. Diffusional conductances to CO_2 as a target for increasing photosynthesis and photosynthetic water-use efficiency. Photosynth. Res. 117, 45–59.

Franco, C.F., Prado, R.M., 2006. Nutrition solutions in the culture of guava: effect in the development and nutritional state. Acta Sci. Agron. 28 (2), 199–205.

Fu, C., Li, M., Zhang, Y., Zhang, Y., Yan, Y., Wang, Y.A., 2015. Morphology, photosynthesis, and internal structure alterations in field apple leaves under hidden and acute zinc deficiency. Sci. Hortic. 193, 47–54.

Gago, J., Douthe, C., Florez-Sarasa, I., Escalona, J.M., Galmes, J., Fernie, A.R., Flexas, J., Medrano, H., 2014. Opportunities for improving leaf water use efficiency under climate change conditions. Plant Sci. 226, 108–119.

Gomez, I., Navarro-Pedreno, J., Mataix, J., 1992. The influence of saline irrigation on organic waste fertilisation on the mineral content (N, P, K, Na, Ca, and Mg) of tomatoes. J. Sci. Food Agric. 59, 483–487.

Gorbe, E., Calatayud, A., 2010. Optimization of nutrition in soilless systems: a review. Adv. Bot. Res. 53, 1–230.

Gusman, G.S., Oliveira, J.A., Farnese, F.S., Cambraia, J., 2013. Mineral nutrition and enzymatic adaptation induced by arsenate and arsenite exposure in lettuce plants. Plant Physiol. Biochem. 71, 307–314.

Hafsi, C., Falleh, H., Saada, M., Rabhi, M., Mkadmini, K., Ksouri, R., Abdelly, C., Smaoui, A., 2016. Effects of potassium supply on growth, gas exchange, phenolic composition, and related antioxidant properties in the forage legume *Sulla carnosa*. Flora 223, 38–45.

Hammond, J.P., Bennett, M.J., Bowen, H.C., Broadley, M.R., Eastwood, D.C., May, S.T., Rahn, C., Swarup, R., Woolaway, K.E., White, P.J., 2003. Changes in gene expression in *Arabidopsis* shoots during phosphate starvation and the potential for developing smart plants. Plant Physiol. 132, 578–596.

Hansch, R., Mendel, R.R., 2009. Physiological functions of mineral micronutrients (Cu, Zn, Mn, Fe, Ni, Mo, B, Cl). Curr. Opin. Plant Biol. 12 (3), 259–266.

Hernández, G., Ramírez, M., Valdés-López, O., Tesfaye, M., Graham, M.A., Czechowski, T., Schlereth, A., Wandrey, M., Erban, A., Cheung, F., Wu, H.C., Lara, M., Town, C.D., Kopka, J., Udvardi, M.K., Vance, C.P., 2007. Phosphorus stress in common bean: root transcript and metabolic responses. Plant Physiol. 144, 752–767.

Hirai, M.Y., Yano, M., Goodenowe, D.B., Kanaya, S., Kimura, T., Awazuhara, M., Arita, M., Fujiwara, T., Saito, K., 2004. Integration of transcriptomics and metabolomics for understanding of global responses to nutritional stresses in *Arabidopsis thaliana*. Proc. Natl. Acad. Sci. USA 101, 10205–10210.

Hoagland, D.R., Arnon, D.I., 1950. The Water: Culture Method for Growing Plants Without Soil. California Agricultural Experiment Station, Berkeley.

Hochmal, A.K.A., Schulze, S., Trompelt, K., Hippler, M., 2015. Calcium-dependent regulation of photosynthesis. Biochim. Biophys. Acta 1847, 993–1003.

Huang, C.Y., Roessner, U., Eickmeier, I., Genc, Y., Callahan, D.L., Shirley, N., Langridge, P., Bacic, A., 2008. Metabolite profiling reveals distinct changes in carbon and nitrogen metabolism in phosphate-deficient barley plants (*Hordeum vulgare* L.). Plant Cell. Physiol. 49, 691–703.

Ibañez, S., Talano, M., Ontañon, O., Suman, J., Medina, M.I., Macek, T., Agostini, E., 2016. Transgenic plants and hairy roots: exploiting the potential of plant species to remediate contaminants. N. Biotechnol. 33, 625–635.

Ibrahim, E.A., Ramadan, W.A., 2015. Effect of zinc foliar spray alone and combined with humic acid or/and chitosan on growth, nutrient elements content and yield of dry bean (*Phaseolus vulgaris* L.) plants sown at different dates. Sci. Hortic. 184, 101–105.

Imtiaz, M., Rizwan, M.S., Mushtaq, M.A., Yousaf, B., Ashraf, M., Ali, M., 2017. Interactive effects of vanadium and phosphorus on their uptake, growth and heat shock proteins in chickpea genotypes under hydroponic conditions. Environ. Exp. Bot. 134, 72–81.

Ingestad, T., 1982. Relative addition rate and external concentration: driving variables used in plant nutrition research. Plant Cell Environ. 5 (6), 443–453.

Jonsson, P., Johansson, A.I., Gullberg, J., Trygg, J., Jiye, A., Grung, B., Marklund, S., Sjöström, M., Antti, H., Moritz, T., 2005. High-throughput data analysis for detecting and identifying differences between samples in GC/MS-based metabolomic analyses. Anal. Chem. 77, 5635–5642.

Jorge, T.F., Rodrigues, J.A., Caldana, C., Schmidt, R., van Dongen, J.T., Thomas-Oates, J., António, C., 2016. Mass spectrometry-based plant metabolomics: metabolite responses to abiotic stress. Mass Spectrom. Rev. 35, 620–649.

Kilinc, S.S., Ertan, E., Seferoglu, S., 2007. Effects of different nutrient solution formulations on morphological and biochemical characteristics of nursery fig trees grown in substrate culture. Sci. Hortic. 113 (1), 20–27.

Kim, W.S., Roh, M.Y., Lieth, J.H., 2008. Nutrient uptake potential of cut roses (*Rosa hybrida* L.) in soilless culture. Acta Hortic. 7, 45–52.

Lambers, H., Poorter, H., 1992. Inherent variations in growth rate between higher plants: a search for physiological causes and ecological consequences. Adv. Ecol. Res. 23, 188–261.

Larcher, W., 2003. Physiological Plant Ecology. Springer-Verlag, Berlin.

Le Bot, J., Adamowicz, S., Robin, P., Andriolo, J.L., Gary, C., 1998. Modelling nitrate uptake by greenhouse tomato crops at the short and long time scales. Acta Hortic. 456, 237–246.

Libralato, G., Costa Devoti, A., Zanella, M., Sabbioni, E., Mičetić, I., Manodori, L., Pigozzo, A., Manenti, S., Groppi, F., Volpi Ghirardini, A., 2016. Phytotoxicity of ionic, micro- and nano-sized iron in three plant species. Ecotoxicol. Environ. Saf. 123, 81–88.

Lima Neto, M.C., Lobo, A.K.M., Martins, M.O., Fontenele, A.V., Silveira, J.A.G., 2014. Dissipation of excess photosynthetic energy contributes to salinity tolerance: a comparative study of salt-tolerant *Ricinus communis* and salt-sensitive *Jatropha curcas*. J. Plant Physiol. 171, 23–30.

Liu, H., Chen, X., Chen, R., Song, S., Sun, G., 2008. Effects of magnesium deficiency on growth and photosynthesis of flowering Chinese cabbage. Acta Hortic. 767, 169–174.

Liu, Y., Han, X., Zhan, X., Yang, J., Wang, Y., Song, Q., Chen, X., 2013. Regulation of Calcium on peanut photosynthesis under low night temperature stress. JIA 12 (12), 2172–2178.

Llanes, A., Arbona, V., Gómez-Cadenas, A., Luna, V., 2016. Metabolomic profiling of the halophyte *Prosopis strombulifera* shows sodium salt-specific response. Plant Physiol. Biochem. 108, 145–157.

López-Millán, A.F., Sagardoy, R., Solanas, M., Abadía, A., Abadía, J., 2009. Cadmium toxicity in tomato (*Lycopersicon esculentum*) plants grown in hydroponics. Environ. Exp. Bot. 65, 376–385.

López-Morales, V., López-Ortega, O., Ramos-Fernández, J., Muñoz, L.B., 2008. JAPIEST: an integral intelligent system for the diagnosis and control of tomatoes diseases and pests in hydroponic greenhouses. Expert. Syst. Appl. 35 (4), 1506–1512.

López-Luna, J., Silva-Silva, M.J., Martinez-Vargas, S., Mijangos-Ricardez, O.F., González-Chávez, M.C., Solís-Domínguez, F.A., Cuevas-Díaz, M.C., 2016. Magnetite nanoparticle (NP) uptake by wheat plants and its effect on cadmium and chromium toxicological behavior. Sci. Total Environ. 565, 941–950.

Lupini, A., Sorgonà, A., Miller, A.J., Abenavoli, M.R., 2010. Short-term effects of coumarin along the maize primary root axis. Plant Signal. Behav. 5, 1395–1400.

Maathuis, F.J., 2009. Physiological functions of mineral macronutrients. Curr. Opin. Plant Biol. 12 (3), 250–258.

Mahouachi, J., Socorro, A.R., Talon, M., 2006. Responses of papaya seedlings (*Carica papaya* L.) to water stress and re-hydration: growth, photosynthesis and mineral nutrient imbalance. Plant Soil 281 (1–2), 137–146.

Maia, J.M., Ferreira-Silva, S.L., Voigt, E.L., Macedom, C.E.C., Ponte, L.F.A., Silveira, J.A.G., 2012. Activities of antioxidant enzymes and root growth inhibition in cowpea seedlings exposed to different salt levels. Acta Bot. Bras. 26 (2), 342–349.

Marschner, P., 2012. Marschner's, Mineral Nutrition of Higher Plants, third ed. Academic Press, San Diego.

Martínez-Ballesta, M.C., Martínez, V., Carvajal, M., 2004. Osmotic adjustment, water relations and gas exchange in pepper plants grown under NaCl or KCl. Environ. Exp. Bot. 52, 161–174.

Masood, S., Saleh, L., Witzel, K., Plieth, C., Mühling, K.H., 2012. Determination of oxidative stress in wheat leaves as influenced by boron toxicity and NaCl stress. Plant Physiol. Biochem. 56, 56–61.

Matsuda, F., Okazaki, Y., Oikawa, A., Kusano, M., Nakabayashi, R., Kikuchi, J., Yonemaru, J.-I., Ebana, K., Yano, M., Saito, K., 2012. Dissection of genotype–phenotype associations in rice grains using metabolome quantitative trait loci analysis. Plant J. 70, 624–636.

Mengel, K., Kirkby, E., 2001. Principles of Plant Nutrition, fifth ed. Kluwer Academic Publishers, London.

Meziane, D., Shipley, B., 1999. Interacting components of interspecific relative growth rate: constancy and change under differing conditions of light and nutrient supply. Funct. Ecol. 13 (5), 611–622.

Miller, A.J., 2009. Freeways in the plant: transporters for N, P and S and their regulation. Curr. Opin. Plant Biol. 12 (3), 284–290.

Montes, A., Bisson, M.A., Gardella, J.A., Aga, D.S., 2017. Uptake and transformations of engineered nanomaterials: critical responses observed in terrestrial plants and the model plant *Arabidopsis thaliana*. Sci. Total Environ. 607, 1497–1516.

Moreno-Jiménez, E., Fernández, J.M., Puschenreiter, M., Williams, P.N., Plaza, C., 2016. Availability and transfer to grain of As, Cd, Cu, Ni, Pb and Zn in a barley agri-system: impact of biochar, organic and mineral fertilizers. Agric. Ecosyst. Environ. 219, 171–178.

Morgan, L., 2005. Hydroponic classroom experiments. In: Weller, T. (Ed.), The Best of the Growing Edge International 2000–2005. New Moon Publishing, Corvallis, pp. 308–311.

Najafpour, M.M., Moghaddam, A.N., Allakhverdiev, S.I., Govindjee, 2012. Biological water oxidation: lessons from nature. Biochim. Biophys. Acta 1817 (8), 1110–1121.

Nedjimi, B., 2014. Effects of salinity on growth, membrane permeability and root hydraulic conductivity in three saltbush species. Biochem. Syst. Ecol. 52, 4–13.

Nkebiwe, P.M., Weinmann, M., Bar-Tal, A., Müller, T., 2016. Fertilizer placement to improve crop nutrient acquisition and yield: a review and meta-analysis. Field Crop. Res. 196, 389–401.

Nunes-Nesi, A., Fernie, A.R., Stitt, M., 2010. Metabolic and signaling aspects underpinning the regulation of plant carbon nitrogen interactions. Mol. Plant 3 (6), 973–996.

O'carrigan, A., Babla, M., Wang, F., Liu, X., Mak, M., Thomas, R., Bellotti, B., Chen, Z.-H., 2014. Analysis of gas exchange, stomatal behaviour and micronutrients uncovers dynamic response and adaptation of tomato plants to monochromatic light treatments. Plant Physiol. Biochem. 82, 105–115.

Pardossi, A., Falossi, F., Malorgio, F., Incrocci, L., Bellocchi, G., 2005. Empirical models of macronutrient uptake in melon plants grown in recirculating nutrient solution culture. J. Plant Nutr. 27, 1261–1280.

Peixoto, C.P., Cerqueira, E.C., Soares Filho, W.S., Castro Neto, M.T., Ledo, C.A.S., Matos, F.S., Oliveira, J.G., 2006. Growth analysis of different citrus genotypes cultivated under water deficit. Rev. Bras. Frutic. 28, 439–443.

Peng, Y., Li, C., Fritschi, F.B., 2014. Diurnal dynamics of maize leaf photosynthesis and carbohydrate concentrations in response to differential N availability. Environ. Exp. Bot. 99, 18–27.

Quan, X., Qian, Q., Ye, Z., Zeng, J., Han, Z., Zhang, G., 2016. Metabolic analysis of two contrasting wild barley genotypes grown hydroponically reveals adaptive strategies in response to low nitrogen stress. J. Plant Physiol. 206, 59–67.

Radford, P.S., 1967. Growth analysis formulae: their use and abuse. Crop Sci. 7, 171–175.

Ramalho, J.C., Rebelo, M.C., Santos, M.E., Antunes, M.L., Nunes, M.A., 1995. Effects of calcium deficiency on *Coffea arabica*. Nutrient changes and correlation of calcium levels with some photosynthetic parameters. Plant Soil 172 (1), 87–96.

Resh, H.M., 2001. Hydroponic Food Production, sixth ed. Woodbridge Press, Santa Barbara.

Rijck, G., Schrevens, E., 1999. Chemical feasibility region for nutrient solutions in hydroponic plant nutrition. J. Plant Nutr. 22 (2), 259–268.
Rius-Ruiz, F.X., Andrade, F.J., Riu, J., Rius, F.X., 2014. Computer-operated analytical platform for the determination of nutrients in hydroponic systems. Food Chem. 147, 92–97.
Rodrigues, C.R.F., Silva, E.N., da Mata Moura, R., dos Anjos, D.C., Hernandez, F.F.F., Viégas, R.A., 2014. Physiological adjustment to salt stress in *R. communis* seedlings is associated with a probable mechanism of osmotic adjustment and a reduction in water lost by transpiration. Ind. Crops Prod. 54, 233–239.
Roessner, U., Patterson, J.H., Forbes, M.G., Fincher, G.B., Langridge, P., Bacic, A., 2006. An investigation of boron toxicity in barley using metabolomics. Plant Physiol. 142, 1087–1101.
Roosta, H.R., Hamidpour, M., 2011. Effects of foliar application of some macro- and micro-nutrients on tomato plants in aquaponic and hydroponic systems. Sci. Hortic. 129 (3), 396–402.
Rouphael, Y., Cardarelli, M., Rea, E., Colla, G., 2008. Grafting of cucumber as a means to minimize copper toxicity. Environ. Exp. Bot. 63 (1–3), 49–58.
Santos, J.D., Silva, A.L.L., Costa, J.L., Scheidt, G.N., Novak, A.C., Sydney, E.B., Soccol, C.R., 2013. Development of a vinasse nutritive solution for hydroponics. J. Environ. Manag. 114, 8–12.
Saqib, M., Zörb, C., Schubert, S., 2006. Salt-resistant and salt-sensitive wheat genotypes show similar biochemical reaction at protein level in the first phase of salt stress. J. Plant Nutr. Soil Sci. 169 (4), 542–548.
Savvas, D., Gizas, G., 2002. Response of hydroponically grown gerbera to nutrient solution recycling and different nutrient cation ratios. Sci. Hortic. 96 (1–4), 267–280.
Schlüter, U., Mascher, M., Colmsee, C., Scholz, U., Bräutigam, A., Fahnenstich, H., Sonnewald, U., 2012. Maize source leaf adaptation to nitrogen deficiency affects not only nitrogen and carbon metabolism but also control of phosphate homeostasis. Plant Physiol. 160, 1384–1406.
Schwarz, D., Klaring, H.P., Van Iersel, M.W., Ingran, K.T., 2002. Growth and photosynthetic response of tomato to nutrient solution concentration at two light levels. J. Am. Soc. Hortic. Sci. 127, 984–990.
Shi, Y., Byrne, D.H., Reed, D.W., Loeppert, R.H., 1993. Influence of bicarbonate level on iron-chlorosis development and nutrient uptake of the peach rootstock Montclar. J. Plant Nutr. 16 (9), 1675–1689.
Sinkovič, L., Demšar, L., Žnidarčič, D., Vidrih, R., Hribar, J., Treutter, D., 2015. Phenolic profiles in leaves of chicory cultivars (*Cichorium intybus* L.) as influenced by organic and mineral fertilizers. Food Chem. 166, 507–513.
Sinkovič, L., Vidrih, R., Abram, V., Žnidarčič, D., Grdiša, M., Treutter, D., 2016. Leaf phenolic fingerprints of chicory cultivars (*Cichorium intybus* L.) hydroponically forced in different nutrient solutions. Food Sci. Technol. 74, 346–352.
Soffer, H., Burger, D.W., Lieth, J.H., 1991. Plant growth and development of *Chrysanthemum* and *Ficus* in aero-hydroponics: response to low dissolved oxygen concentrations. Sci. Hortic. 45, 287–294.
Son, J.E., Kim, H.J., Ahn, T.I., 2016. Hydroponic systems. In: Niu, G., Takagaki, M. (Eds.), Plant Factory. Academic Press, San Diego, pp. 213–221.
Sotiropoulos, T.E., Therios, I.N., Dimaxi, K.N., 2003. Boron toxicity in kiwifruit plants (*Actinidia deliciosa*), treated with nitrate, ammonium, and a mixture of both. J. Plant Nutr. Soil Sci. 166 (4), 529–532.
Sung, J., Lee, S., Lee, Y., Ha, S., Song, B., Kim, T., Waters, B.M., Krishnan, H.B., 2015. Metabolomic profiling from leaves and roots of tomato (*Solanum lycopersicum* L.) plants grown under nitrogen, phosphorus or potassium-deficient condition. Plant Sci. 241, 55–64.
Taiz, L., Zeiger, E., Moller, I.M., Murphy, A., 2015. Plant Physiology and Development, sixth ed. Sinauer, Massachusetts.
Terry, N., Ulrich, A., 1973a. Effects of phosphorus deficiency on the photosynthesis and respiration of leaves of sugar beet. Plant Physiol. 51, 43–47.
Terry, N., Ulrich, A., 1973b. Effects of potassium deficiency on the photosynthesis and respiration of leaves of sugar beet. Plant Physiol. 51, 783–786.
Urano, K., Kurihara, Y., Seki, M., Shinozaki, K., 2010. 'Omics' analyses of regulatory networks in plant abiotic stress responses. Curr. Opin. Plant Biol. 13, 132–138.
Urbanczyk-Wochniak, E., Fernie, A.R., 2005. Metabolic profiling reveals altered nitrogen nutrient regimes have diverse effects on the metabolism of hydroponically-grown tomato (*Solanum lycopersicum*) plants. J. Exp. Bot. 56, 309–321.
Viti, R., Cinelli, F., 1993. Lime-induced chlorosis in quince rootstocks: methodological and physiological aspects. J. Plant Nutr. 14 (4), 631–641.
Von Caemmerer, S., Farquhar, G.D., 1981. Some relationships between the biochemistry of photosynthesis and the gas exchange of leaves. Planta 153 (4), 376–387.
Wang, J., Fang, Z., Cheng, W., Yan, X., Tsang, P.E., Zhao, D., 2016. Higher concentrations of nanoscale zero-valent iron (nZVI) in soil induced rice chlorosis due to inhibited active iron transportation. Environ. Pollut. 210, 338–345.
Wu, B., Zhu, L., Le, X.C., 2017. Metabolomics analysis of TiO_2 nanoparticles induced toxicological effects on rice (*Oryza sativa* L.). Environ. Pollut. 230, 302–310.
Xu, H., Qin, F., Gauthier, L., Gosselin, A., 2009. Greenhouse tomato growth and physiological responses to high nutrient solution electrical conductivity and low substrate water content. FVCSB 3 (1), 54–61.
Zhang, L., Wang, X., Guo, J., Xia, Q., Zhao, G., Zhou, H., Xie, F., 2013. Metabolic profiling of Chinese tobacco leaf of different geographical origins by GC-MS. J. Agric. Food Chem. 61, 2597–2605.
Zhao, D., Oosterhuis, D.M., Bedmarz, C.W., 2001. Influence of potassium deficiency on photosynthesis, chlorophyll content, and chloroplast ultrastructure of cotton plants. Photosynthetica 39, 103–109.
Zhang, Y., Wang, Y., Ding, Z., Wang, H., Song, L., Jia, S., Ma, D., 2017. Zinc stress affects ionome and metabolome in tea plants. Plant Physiol. Biochem. 111, 318–328.
Zheng, Y., Graham, T.H., Richard, S., Dixon, M., 2005. Can low nutrient strategies be used for pot gerbera production in closed-loop subirrigation? Acta Hortic. 691, 365–372.

Index

'*Note*: Page numbers followed by "f" indicate figures and "t" indicate tables.'

C

D

E

F

G

N

O

P

Q

R

T